BIOCHEMISTRY
A Case-Oriented Approach

REX MONTGOMERY, Ph.D., D.Sc.
THOMAS W. CONWAY, Ph.D.
ARTHUR A. SPECTOR, M.D.

Department of Biochemistry
The University of Iowa College of Medicine
Iowa City, Iowa

With contributions by
BARRY H. GINSBERG, M.D., Ph.D.

Fifth Edition

*with **550** illustrations*

The C. V. Mosby Company

St. Louis • Baltimore • Philadelphia • Toronto   1990

Editor Stephanie Manning
Assistant editor Anne Gunter
Project manager John A. Rogers
Production editors Jolynn Gower, Catherine M. Vale
Book design Candace Conner
Cover design Elise Stimac

Fifth Edition

The C.V. Mosby Company
11830 Westline Industrial Drive, St. Louis, Missouri 63146

Library of Congress Cataloging-in-Publication Data

Montgomery, Rex.
    Biochemistry: a case-oriented approach.

    Rev. ed. of: Biochemistry/Rex Montgomery...
[et al.]. 4th ed. 1983.
    Includes bibliographical references.
    1. Biochemistry. 2. Clinical biochemistry.
I. Conway, Thomas W. II. Spector, Author A.
III. Biochemistry. IV. Title. [DNLM: 1. Biochemistry.
QU 4 M788b]
QP514.2.M66 1990   612'.015   90-5551
ISBN 0-8016-3549-7

Preface to the Fifth Edition

Twenty years ago, the course in biochemistry for medical students at the University of Iowa was deliberately oriented toward the use of biochemical principles and facts for the analysis of clinical cases. After three years of refinement, the course material was collected into the first edition of the present text, the preface of which is reprinted in this edition. Since 1974 the text has evolved through frequent editions to keep abreast of the advances in biochemistry, always with the same objective of developing an understanding in the reader of why biochemistry is important in the health sciences and how it is involved in everyday practice.

The present edition has collected biochemistry in 19 chapters, rather than 13 chapters as in the fourth edition. In part this reflects the recent advances in membrane biochemistry, DNA, RNA and protein metabolism, endocrinology, and lipoprotein structure and function. The chapters are reasonably self-contained, which allows for considerable flexibility in the order in which an instructor may choose to present them. As in previous editions, nutrition is selected as the topic for the first chapter to introduce from the beginning the case-oriented method of study, because most students have already been exposed to the subject in their everyday lives and in previous schooling. A consideration of proteins as biocatalysts, blood and hemoglobin as the vehicles of extracellular transport, and the bioenergetics of intermediary metabolism are considered in the next four chapters. The order in which the metabolism of carbohydrates, lipids, proteins, and nucleic acids are presented is a matter of personal choice.

Believing that the world is becoming increasingly smaller in terms of health care delivery and that communication with colleagues abroad through scholarly journals and international meetings must continue, the International System of Units (SI), used in all countries except the United States of America, are continued in this edition. Other traditional units are given in parentheses. We continue to hope that students can learn and be prepared to communicate with their future colleagues using either system.

A significant new feature of the present text is the clinical comments sections in the body of each chapter to illustrate simple applications of biochemical principles to health science problems. In addition, the last few pages of each chapter contain a series of multiple choice questions using the format of the National Board of Medical Examiners; these questions are often centered around a clinical problem. It is recommended that examinations be based on clinical situations, real or constructed, designed to gauge the student's capacity to deal with applications of basic biochemistry. We believe that this is the principal purpose of including biochemistry in the curriculum of the future health science professional.

We note here since the publication of the last edition the passing of R. L. Dryer, an enthusiastic supporter of the teaching of biochemistry to health science students by the case-oriented approach. His contributions to this edition were missed. B. H. Ginsberg was an invited contributor to the chapter on endocrinology and so added significantly to the clinical applications in this area.

Rex Montgomery
Thomas W. Conway
Arthur A. Spector

Preface to the First Edition

This book is addressed to students of the health sciences and differs significantly from other biochemistry texts. It was written to serve a dual teaching function: first, to make students of the health sciences aware of certain basic biochemical principles and, second, to use these principles to analyze commonly occurring health-related problems in terms of their biochemical components. We feel that the second goal is the ultimate objective of biochemical education for the health science student, an approach that makes this book unique.

The text, although self-standing, is not meant to cover biochemistry in its entirety. The references direct the inquisitive reader to more detailed information. The book has been divided into thirteen subject areas that are presented as separate chapters. The first part of each chapter is a "language" statement, setting forth the biochemical principles needed to understand the chemical and molecular aspects of health science problems. Knowledge of these elementary biochemical principles can be built on and expanded at later stages in the student's training. In addition, these facts and concepts will be of particular value in understanding future discoveries in biochemistry. The second part of each chapter is a presentation and analysis of clinical cases selected to demonstrate the applicability of the material contained in the language statement.

The case studies are an important part of the text because they allow the student to apply the newly learned principles. By means of the case analyses, the student not only learns the principles themselves but also comes to understand *why* biochemistry is important in the health sciences and *how* biochemical principles are involved in day-to-day professional practice. The following are examples of the application of some biochemical principles. The Krebs cycle is a biochemical concept; how it works in the alcoholic is a health-related application. Cholesterol absorption and biosynthesis are principles; how they work in diseases of cholesterol metabolism in the human is an important application. Glycolysis is a principle; its role in the diabetic is the problem. Using this approach, biochemical facts can be extracted from the cases and resynthesized into rational guidelines for explaining physiologic events and disease in chemical terms. To ask the student for a complete analysis of every clinical problem would be more than could be managed in a reasonable time. Therefore students also need the opportunity to work with problems on a smaller dimension. In each chapter several of these shorter cases are presented and followed by questions directed to the particular biochemical subject being covered, with appropriate references to the literature.

The original literature represents the source of most knowledge, and it is clearly the data bank for continuing education. It is imperative that students become accustomed to the method and the value of bibliographic research. For these reasons a number of additional questions and problems have been appended to each chapter.

The last chapter is concerned with broad metabolic interrelationships and illustrates the analysis of cases using all of the biochemical data presented in the book. The cases presented for student analysis also assume mastery of all of the general biochemical principles discussed in earlier chapters, although some library search for particular facts may be necessary. It is hoped that this approach will provide a solid foundation of biochemical information on which the student will continue to build during the remainder of his or her professional life and most certainly during the remainder of his or her professional education.

It should be emphasized that the intent of this book is not to describe clinical practice and management. Rather, it is to demonstrate how biochemistry can and does provide valuable insight for such practice and management.

This method of teaching biochemistry to freshmen medical students has been used for the last three years at The University of Iowa. No less biochemistry has been taught by adopting this approach, and the important dimension of how to use the information has been added. It may be helpful to give some general conclusions from our three-year experience.

In order to introduce the case-oriented method of teaching from the beginning of the course, it was desirable to present case materials during the first week. Nutrition was selected as the first topic, since many students have been exposed to this subject in their everyday living. Regrettably, nutrition is often presented late, passed by quickly, or even omitted in many biochemical courses. Yet the subject is not conceptually difficult, and the involvement of nutritional effects in many diseases demands an early consideration. As soon as possible, acid-base, fluid, and electrolyte control are introduced (Chapter 4). Problems of this sort are also common in many diseases, so the earliest introduction to these subjects allows the broadest selection of later cases. These topics are more difficult conceptually, and a knowledge of the properties of proteins and enzymes is required as background information.

Other subject material can be treated more flexibly insofar as order of presentation is concerned. Each chapter has been kept reasonably self-contained, with references to other portions of the text whenever necessary. The total text is presented in a sixteen-week period, for the most part one chapter per week. Weekly contact has involved four lecture hours and three hours of discussion in small groups.

Material for a book of this kind is not selected without the help of many colleagues. To them, we express our deep appreciation, particularly to Drs. G. N. Bedell, R. L. Blakley, A. R. Boutros, J. D. Brown, G. F. DiBona, J. L. Filer, S. J. Fomon, E. M. Gal, N. Halmi, K. Hubel, R. E. Hodges, H. P. C. Hogenkamp, V. Ionasescu, E. Mason, V. A. Pedrini, R. Roskoski, R. F. Sheets, L. D. Stegink, S. Terman, and H. Zellweger. We are also grateful to our colleague Kenneth C. Moore for Fig. 3-1.

Rex Montgomery
Robert L. Dryer
Thomas W. Conway
Arthur A. Spector

Contents

Chapter 3   **Enzymes and biologic catalysis** 95

Chapter 6   **Carbohydrate metabolism**   247

Chapter 9   **Amino acid and neurotransmitter synthesis** 379

Chapter 10   **Lipid metabolism** 409

Chapter 12

Membranes   497

Chapter 13   **Nucleotide metabolism** 551

Chapter 14   **Structure and synthesis of DNA** 599

Chapter 15   **RNA and protein biosynthesis** 641

Chapter 18

Molecular endocrinology: hormones active at the cell surface 749
Barry H. Ginsberg

Chapter 19

Molecular endocrinology: hormones active inside the cell 797
Barry H. Ginsberg

Abbreviations 837

Chapter 1

Nutrition

Objectives

1 To analyze the biochemical role of a proper diet in maintaining homeostasis
2 To interpret the different dietary demands that result from alterations in workload, age, and normal physiologic conditions
3 To interpret the metabolic basis of some nutritional diseases

As with all other living systems, humans survive only by means of a continual energy flux and the provision of essential nutrients in the diet to maintain good health. These nutrients must be obtained in proper amounts and proportions; any excess or deficiency can lead to disease. In the broadest sense, nutrition provides the body with needed energy and essential constituents that cannot be synthesized de novo. Sound nutrition depends on a proper dietary regimen, or food intake, to sustain normal growth, development, and health. This must include the six major components of the diet, *carbohydrates, proteins, fats, vitamins, minerals,* and *water.* Foods often contain nonnutritive components that, together with intestinal bacteria and waste materials manufactured by our cells, comprise excreta in the form of urine, feces, and sweat.

An average adult maintains a fairly constant body weight despite consuming six to seven times that weight of food each year. For this state of equilibrium to be maintained, energy must be supplied to satisfy the demands of the total body requirements, which include tissue maintenance. In other cases food must also provide for growth, as in children and during pregnancy. These demands vary, depending on the workload and environment, and will change with age and physiologic state. Thus the needs of a hospitalized person will likely change when he or she returns to health, and those of an athlete in training will be different. The nutritional needs of a person nearing retirement also differ from those of an adolescent.

Imposed on these variations are those of biologic individuality. The so-called average 70 kg man is not represented by any one person. Everyone is different, and nutritionally this is expressed at all biochemical and physiologic levels. For example, differences exist in the digestion and absorption of food, the supplementation of essential nutrients by the gastrointestinal flora, the transport of food to the cells, the uptake of the nutrients across the plasma membrane of the cells, and the rate of waste elimination. However, the physiologic and biochemical regulatory mechanisms, responding to all these individual factors, arrive at an equilibrium for each person that is recognized as health. In some diseases, either genetic or acquired, the resulting nutritional deficiencies of the cells cannot be overcome without external assistance.

Homeostasis

An organism as complex as the human body is an ordered aggregation of cells. Each cell obtains the nutrients essential for its well-being from the circulating extracellular interstitial fluid in which it is bathed. This same fluid also serves to remove waste products excreted from the cell. The composition of living cells is remarkably constant as long as the interstitial fluid is normal. The internal cellular composition must be kept constant within narrow limits, or the animal dies. The maintenance of this stable internal environment is termed *homeostasis*.

The biochemical mechanisms by which the compositions of the extracellular and intracellular fluids are controlled represent a major segment of biochemical knowledge and research. The relationship of disease to the breakdown in these controls is a subject of continuing study.

Body water

The average adult is composed of 55% water, 19% protein, 19% fat, less than 1% carbohydrate, and 7% inorganic material. The body water is distributed between two main components, that within the cells *(intracellular fluid)* and that outside the cells *(extracellular fluid)*. The extracellular fluid is subdivided into *interstitial fluid,* which bathes the cells, and the *blood plasma.* On the average, body fluid makes up 65% to 70% of the *lean* body mass—that is, the weight of a person after removal of excess body fat. With such an extensive distribution of water, one might think that transport and metabolism are slow and perhaps inefficient. However, this is not the case. It has been shown in animals injected with a solution of isotonic saline containing either radioactively labeled [^{14}C] bicarbonate or [^{14}C] glucose that the label is distributed and incorporated into lipid and carbohydrate polymer (glycogen) molecules within 30 sec. As seen later, many cellular and subcellular particle membranes have to be crossed and many chemical reactions must occur to achieve such results. That all this takes place in a very short time demonstrates exquisite organization in and between the cells of the multicellular animal.

Principal food components

Our normal food is a mixture of complex plant and animal materials composed largely of protein, lipid, and carbohydrate. These must be reduced to simpler components before they can be used by our tissues. The processes by which food is broken into simpler components are known collectively as *digestion* and involve biologic catalysts called enzymes. The products of digestion are then taken into the bloodstream by the selective processes of *absorption*. A simple introductory description of these food components is necessary to understand how they function in nutrition.

Proteins

Proteins are biopolymers composed of monomeric units called α-amino acids, 20 or so of which are biologically important (Table 2.1). The α-amino acids have the following general formula:

$$
\begin{array}{c}
COO^- \\
| \\
H_3\overset{+}{N}-C-H \\
| \\
R
\end{array}
$$

where R may be an alkyl, aryl, or a heterocyclic group. The polymeric bond is formed through the interaction of the amino group of one amino acid with the carboxyl group of another amino acid to form a *peptide bond*. Thus two amino acids condense to form a dipeptide.

Amino acid   Amino acid   Dipeptide

The further sequential addition of amino acid residues creates a polypeptide and finally a protein. Further details are presented in Chapters 2 and 15.

The sequence of the amino acids in each protein is unique and in some way determines the overall shape of the molecule. The particular shape, or *conformation,* in the living cell is called the *native state.* Proteins in the native state are not easily digested. For this reason the dietary proteins are *denatured* by such means as cooking (heat denaturation), beating (surface denaturation), or chemical transformation (e.g., by acid in the stomach). The native structures are disordered in these processes so that the enzymes of the digestive system can catalyze the hydrolysis of the peptide bonds more rapidly to release the component amino acids. This reaction is initiated in the stomach by the enzymes pepsin and rennin at an acid condition of pH $\sim$ 1. It is completed in the small intestine, where the acidic stomach contents are made alkaline by the pancreatic secretions. The pancreatic juice also contains *proteolytic* (e.g., trypsin and chymotrypsin) and *peptidolytic* (e.g., carboxypeptidase) enzymes. *Aminopeptidases* and *dipeptidases* in the intestinal mucosal cells complete the hydrolysis of the resulting peptides to give amino acids, which are actively transported into the portal blood and then to the cells of the body to be used as needed for homeostasis. The amino acids are quickly removed from the circulation; the liver removes most, but the kidneys and muscle are also significantly involved. Other tissues take up less of this high postprandial amino acid flux in the bloodstream. After the high amino acid concentration subsides, 2 to 6 mmol/L of amino acids are continuously circulating in plasma through all tissues. The digestion of proteins is considered in greater detail in Chapter 8.

Lipids

Many different types of fat are present in our food. However, the bulk of the dietary fat that has nutritional value is *triglyceride,* also called *triacylglycerol,* or neutral fat. Triglyceride is composed of glycerol and fatty acids, the fatty acids being esterified to the three hydroxyl groups of the glycerol. The ester bond is the most common linkage in the lipids.

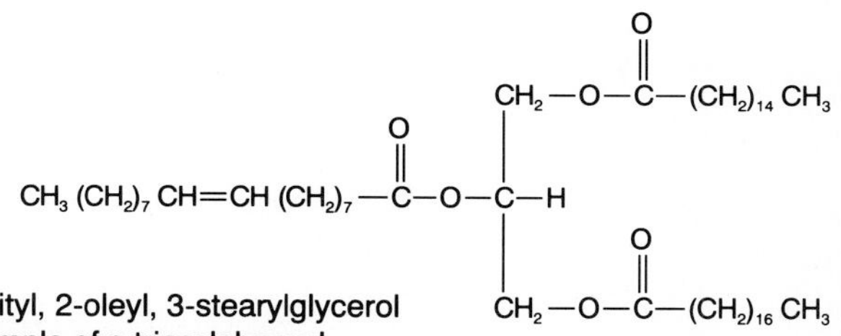

$$CH_3(CH_2)_7\,CH{=}CH\,(CH_2)_7{-}\overset{\displaystyle O}{\overset{\|}{C}}{-}O{-}\underset{|}{\overset{|}{C}}{-}H$$

1-Palmityl, 2-oleyl, 3-stearylglycerol
An example of a triacylglycerol

$$CH_2{-}O{-}\overset{\displaystyle O}{\overset{\|}{C}}{-}(CH_2)_{14}\,CH_3$$
$$CH_2{-}O{-}\overset{\displaystyle O}{\overset{\|}{C}}{-}(CH_2)_{16}\,CH_3$$

A triacylglycerol containing fatty acid residues of palmitic, oleic, and stearic acids, each in ester linkage to a hydroxyl group of glycerol, is illustrated above. One should note that the fatty acids may have a saturated or unsaturated alkyl chain. The more unsaturated the fatty acid, the lower is the melting point for a given carbon chain length. Triacylglycerols that contain a high percentage of unsaturated fatty acid residues are more likely to be oils than solid fat. The unsaturated double bonds in the polyunsaturated fatty acids (those unsaturated fatty acids with two or more double bonds) are three carbons apart. For example, linoleic acid, a biologically important fatty acid, contains two double bonds and has the following structure:

$$CH_3{-}(CH_2)_4{-}CH{=}CH{-}CH_2{-}CH{=}CH{-}(CH_2)_6{-}CH_2{-}COOH$$

Linoleic acid (18:2ω—6)

As discussed more fully in Chapter 10, one nomenclature of the lipids identifies the number of carbon atoms in the fatty acid, the number of double bonds, and the numbered carbon, counting from the CH_3— end, that carries the first double bond. Thus linoleic acid is 18:2ω—6, meaning 18 carbons and two double bonds, with the first at the sixth carbon from the CH_3— end. There are four classes of unsaturated fatty acids, each characterized by the number of carbons from the CH_3— end at which the first double bond occurs. The fatty acids in the diet are modified by metabolism to those fatty acids characteristic of the organism, but modifications of the unsaturated fatty acids never change the group classification of the dietary component. Thus linoleic acid, ω—6 group, cannot be metabolized to oleic acid, ω—9 group, but it can be the source of arachidonic acid, 20:4ω—6.

$$CH_3—(CH_2)_4CH{=}CH—CH_2—CH{=}CH—CH_2—CH{=}CH—CH_2—CH{=}CH—(CH_2)_3—COOH$$

Arachidonic acid (20:4ω—6)

Most lipids are poorly soluble in water and must therefore be emulsified to be digested and absorbed. The dietary triacylglycerol must be emulsified in the duodenum by bile, a fluid that is secreted by the liver and stored for use as demanded in the gallbladder. Digestion occurs through hydrolysis catalyzed by a pancreatic enzyme, *lipase,* which hydrolyzes the triacylglycerols to a mixture of monoacylglycerols and fatty acids. The released fatty acids and monoacylglycerols diffuse into the intestinal mucosa, where resynthesis to triacylglycerols occurs. This process is followed by the release into the lymph of particles called chylomicrons containing lipid and protein, which enter the venous blood and are finally deposited in the tissues. The shorter, more water-soluble fatty acids are absorbed directly into the portal blood and delivered to the liver as fatty acid. (See Chapters 10 and 11 for further discussion of the lipids.)

Carbohydrates

The dietary carbohydrates are principally starch, lactose (milk sugar), sucrose (cane or beet sugar), and glucose (dextrose or blood sugar).

α-D-Glucose

α-D-Glucose is one of the simple building blocks of the dietary carbohydrates. The structure is shown as a six-membered ring with substituents above and below the plane of the ring. The various simple sugars, called monosaccharides, differ in the arrangement of these substituents and sometimes in the size of the ring. For example, α-D-galactose, a constituent of the disaccharide lactose, differs from α-D-glucose at only one position of the ring, whereas β-D-fructose, present in sucrose, has a five-membered ring structure. The structures of these sugars are given next. One should note that the substituent hydrogen atoms on the ring have been deleted, as is the common practice, but one must remember that they are in fact present. The significance of α-D- or β-D- is a subject for discussion in Chapter 6.

α-D-Galactose   β-D-Fructose

The simple monosaccharides are linked one to the other by *glycosidic bonds;* two monosaccharides joined by such a bond form a *disaccharide,* for example, lactose:

Lactose

The addition of a third monosaccharide forms a *trisaccharide* and so on until the high molecular weight polymers, called *polysaccharides,* are formed. The monosaccharides are usually abbreviated by their first three letters; for example, galactose is represented as Gal. The disaccharide sucrose might therefore be written as Glc-Fru, indicating that sucrose is a molecule of glucose linked glycosidically to fructose. The polysaccharide cellulose, in which many glucose units are linked glycosidically to form a polymer, might be represented as Glc-(Glc)$_n$-Glc.

The higher molecular weight carbohydrates can be hydrolyzed with acid to give their component simple sugars. The monosaccharides are characterized by having a reducing group, $>C = O$. They reduce Cu^{++} in Fehling's solution to cuprous oxide (Cu_2O) and ammoniacal silver nitrate to metallic silver. They are therefore identified as *reducing sugars.*

Digestion of the carbohydrates involves splitting (hydrolyzing) all the glycosidic bonds with specific glycosidase enzymes down to the simple monosaccharide units. This action starts in the mouth with the enzyme salivary amylase, which is inactivated by the stomach acidity. The other glycosidases occur in pancreatic secretions (amylases) and in the intestinal mucosal cells so that only monosaccharides are absorbed into the portal system. These are commonly D-glucose, D-fructose, and D-galactose. The latter two sugars are converted in the liver to D-glucose, which is the sugar used by the cells to maintain normal metabolic function.

Overall view of metabolism of principal dietary components

Following the digestion of the proteins and the carbohydrates, their simple building blocks, the amino acids and monosaccharides, are absorbed into the portal blood system. They are delivered mainly to the liver, where some portions are transformed to other molecules and the remainder are supplied to other tissues. At times between meals the glucose and amino acid content of the blood are maintained by their release from liver and other tissues, such as muscle. In this way homeostasis is maintained. The metabolic reactions concerned with the biosynthesis of large molecules from small molecules are usually referred to as *anabolism.* Anabolism is involved, for example, in the maintenance of tissue structures. The reverse metabolic process, the breakdown of molecules, is called *catabolism.*

Lipids

The triacylglycerols are either stored in the adipose tissue or are hydrolyzed to glycerol and fatty acids. These products are transported to the liver, kidneys, and muscles, where they are catabolized as a source of energy. The glycerol is converted principally to D-glucose. By a series of oxidative reactions known as β-oxidation, the fatty acids are catabolized to acetyl coenzyme A (acetyl CoA) (see Chapter 10), which is a thioester of acetic acid and the complex molecule coenzyme A (CoASH).

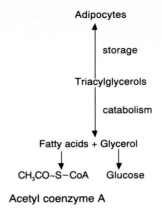

Acetyl coenzyme A

Finally, the acetyl residue passes into the oxidative reactions of the Krebs cycle (also known as the citric acid or tricarboxylic acid cycle), where the final products are carbon dioxide and water. These reactions are central to an understanding of metabolism and are discussed in detail in Chapter 5.

Carbohydrates and amino acids

D-Glucose is either anabolized to other carbohydrate or carbohydrate-containing substances, including stores of glycogen, or it is catabolized to pyruvate. The pyruvate is converted to acetyl CoA and finally to carbon dioxide and water, with the storage of released energy as adenosine triphosphate (ATP). The α-amino acids are either synthesized into tissue proteins or other body components or catabolized by way of their corresponding α-keto acids. These keto acids are either converted to other amino acids or metabolites or are oxidized by way of the Krebs cycle to carbon dioxide and water. In general, the metabolism of the α-amino acids may be summarized as follows:

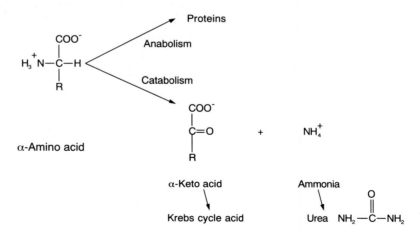

For example, L-alanine is converted to pyruvate, which is an intermediate in the catabolic pathway of D-glucose (p. 270).

Figure 1.1 Overall view of catabolic pathways.

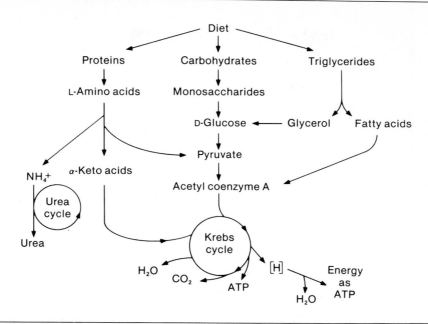

The deamination of L-glutamate, an amino acid present in large amounts in many proteins, produces α-ketoglutarate, which is a component of the Krebs cycle (see Chapter 8).

Krebs cycle and adenosine triphosphate

The overall reactions of catabolism are summarized in Figure 1.1. The Krebs cycle is a focal point of metabolism here, as it is in the interconversions of many metabolites. The Krebs cycle is also greatly involved in the oxidation of substrates, mainly by the abstraction of hydrogen from the molecules. The hydrogen is then oxidized to water in a stepwise manner, with the storage of the released energy as ATP. The energy released when ATP is hydrolyzed to adenosine diphosphate (ADP) and inorganic phosphate is used to drive many energy-demanding reactions and processes in cells.

Adenosine-5'-triphosphate (ATP)

Modifications in response to stress

Many refinements of the simplified presentation in Figure 1.1 are developed as different areas of metabolism are considered in later chapters. Modifications or extensions of these pathways occur in response to physiologic changes. For example, the acute need for energy in muscle contractions during intensive work outstrips the tissues' supply of oxygen for normal oxidative reactions. At such times pyruvate serves as an alternate oxidizing agent, permitting some continuation of glucose metabolism and ATP generation. The pyruvate is converted to lactate, which is released from the muscle cells into the blood for later conversion back to pyruvate in the liver when the supply of oxygen is adequate, as follows:

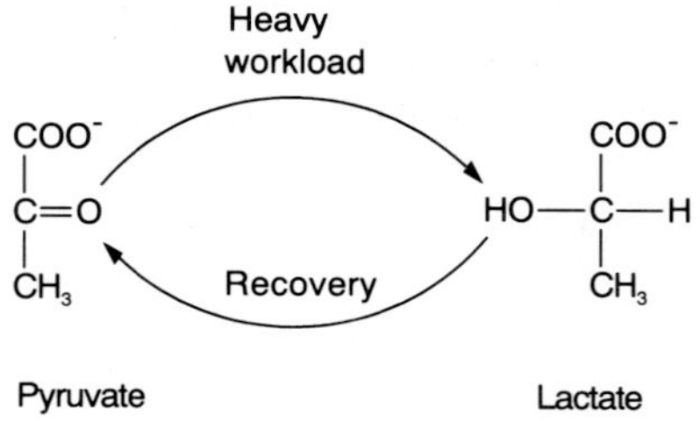

Thus, even during periods of oxygen deficiency, called *oxygen debt,* some energy can still be derived from the metabolism of glucose.

Clinical comment

Metabolic diseases and homeostasis Disease can often be related to the changes in the dynamic equilibrium of body constituents. For example, keto acid is overproduced from the metabolism of lipid when an inadequate supply of carbohydrates is caused by improper eating, dieting, or fasting, as well as by diseases such as uncontrolled diabetes mellitus. The result is elevated blood levels of acetoacetate, β-hydroxybutyrate, and acetone; these compounds are the *ketone bodies.* This condition leads to *ketosis* and *acidosis,* metabolic imbalances that will produce serious illness if they are not corrected quickly. At the least, ketosis represents a loss of calories, since these compounds are lost in the urine. The formation of ketone bodies from the catabolism of fatty acids may be summarized as follows, with the two arrows indicating that several metabolic steps are involved:

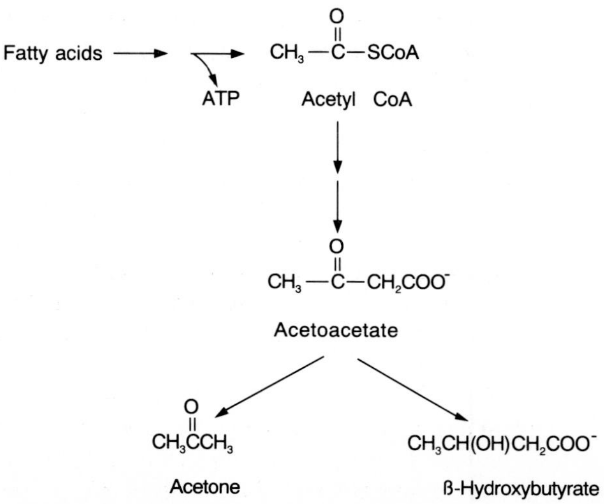

General nutritional
requirements

Control of body functions is exercised in the whole animal at organ, cellular, and sub-cellular levels. The only voluntary step for humans is that of selecting their diet, from which the six major food components provide both energy for the body and basic materials for maintaining the body components. The molecules that make up the body structure or maintain homeostasis result largely from the metabolism of molecules in the food. In some instances, however, the synthesis of a metabolite might occur at a rate too slow to maintain health, or it may not occur at all. In such cases the diet must include these molecules. They comprise the *essential amino acids, essential fatty acids,* and the necessary *vitamins* and *minerals.* From a nutritional standpoint the vitamins differ from the essential fatty acids and essential amino acids only in that much smaller amounts are required. The essentiality of these substances in the diet does not imply that the other metabolites are unimportant, but rather that the human body can synthesize sufficient amounts of the nonessential metabolites from other components of the diet.

**Practical daily
food plan**

A practical and convenient food plan is based on four basic food groups, each of which is chosen to contribute nutrients essential for a complete and balanced diet:
1. Milk group—provides high-quality protein, calcium, phosphorus, riboflavin, and vitamin D
2. Meat group—supplies protein of high biologic value, niacin, thiamine, vitamin B_{12}, heme iron, and minerals
3. Vegetable-fruit group—supplies ascorbic acid, carotene (vitamin A precursor), other water-soluble vitamins, minerals, and fiber ("roughage"); no one vegetable protein has a high biologic value, but when properly mixed it is possible for one protein to complement another
4. Bread-cereal group—high in carbohydrate for energy needs, also includes vitamins, fiber, and iron if the cereals and wheat are not refined; the proteins of this group do not have as high biologic value as animal protein

A variety of selection within these groups is desirable to provide all the factors known to be essential for humans and to exclude excessive components, such as cholesterol, that are injurious to some people.

**Recommended daily
dietary allowances
(RDA)**

Any statement of a minimum daily nutritional requirement will be subject to some debate because, at present, the points of reference are poorly defined. For example, it is possible to omit vitamins from the diet for a few weeks without any apparent ill effects. However, prolonged omission of vitamins results in serious disease. The U.S. Recommended Daily Allowance (U.S. RDA) is derived from the highest value for each nutrient given by the recommended daily dietary allowances, which serve as a guide for decisions affecting food intake for large groups. Thus the optimum nutritional needs of healthy individuals will be met by diets containing the correct RDA. The allowances are not determined for individuals, each of whom will have particular dietary requirements, with the statistical probability that these will be met by the RDA. The newer recommendations would reduce the fat in the diet and replace the calories with carbohydrate (Table 1.1).

It is clear from the table of RDAs (Appendix A) that the requirements for food differ with age, body size, sex, workload, and physiologic condition. For example, during pregnancy the mother provides all the nutrients for the growing fetus, and her dietary intake must necessarily reflect this extra demand.

Clinical comment

RDA changes during illness The RDA in no way reflects the special needs of diseased states or other traumatic situations, such as the severely burned patient or the premature infant. It makes no attempt to describe the source of the nutrients, so that carbohydrate intake may be derived from either refined foods or from those containing more cellulosic

Table 1.1   Mean daily intake of food energy for adults

| | Percent total energy | |
Food	1985*	New recommendations
Fat	36	30
Protein	16	15
Carbohydrates	45	55

From Surgeon General's Report on Nutrition and Health, Public Health Service DHHS Pub No 88-50210, Washington, DC, 1988, US Government Printing Office.
*Data based on 1-day dietary recalls for 2117 adults by U.S. Department of Agriculture in 1985.

fiber, which will be excreted in the feces. Aside from the other nutrients that may be lost in such refining processes (e.g., vitamins and minerals), it is a matter of concern and research to evaluate the need for roughage in the diet. Food should contain some fiber, which plays a part in the peristaltic stimulation of the bowel. Excessive intake of the nonutilizable, ingestible fiber, however, can lead to irritation of the intestinal mucosa. Even normal amounts of roughage may be undesirable in inflammatory diseases of the colon. Some studies suggest that the absence of cellulose roughage may be related to colonic carcinoma and other diseases.

Energy requirements   Almost all the dietary components that can be digested and absorbed are eventually catabolized to provide energy. In these processes the food is oxidized, as summarized in Figure 1.1, and the energy of the diet is transformed to the intermediate chemical form— ATP. ATP is the common currency of energy in the body and is involved in driving various metabolic and physiologic functions such as chemical reactions, electrical work, osmotic work, and mechanical work.

The maximum available energy contained in a food can be measured by burning it in an atmosphere of oxygen. This is done in a calorimeter. The resulting increase in temperature in the calorimeter is determined, and the heat energy produced can be calculated as the thermochemical calorie. In the field of nutrition this thermochemical calorie usually was referred to simply as the "calorie," "kilocalorie," or "large calorie," with the abbreviations of "cal," "kcal," and "Cal," respectively. However, different values and definitions can be found for this calorie, and the joule (J), kilojoule (kJ), or megajoule (MJ) has been used since 1935 to define the thermochemical calorie. The thermochemical calorie is equal to 4.185 J, and both terms are used; for example, 5.02 MJ (1200 kcal), 16.7 kJ (4.0 kcal), and 37.7 kJ (9 kcal).

Living systems oxidize nutrients in more nearly reversible steps, as opposed to the irreversible burning that occurs in the calorimeter. Consequently the energy available (metabolic energy) is less by a few percent. For example, D-glucose has a maximum metabolic energy, which is referred to later as *free* energy for performing useful work. The maximum free energy is about 2% less than the calorimetric, or *enthalpic,* energy. Since the calorimetric measurements are relatively simple and the energy contents of foods are used for purposes that do not demand the greatest precision, the values listed in the dietary tables are those derived by calorimetry, with some adjustments for digestibility. At present most dietary tables in the United States give values in kilocalories per gram, kilocalories per 100 g, or kilocalories per portion of food typically used, such as kilocalories per slice of bread. The nutritional literature, however, is also using joules, and these units are gradually replacing the calorie, particularly outside the United States.

The metabolic energy of food is converted principally to heat and ATP; the overall efficiency for the conversion to ATP is approximately 40%. In practice the proper diet for an adult is the one that maintains health and a constant body weight. This requires approximately 129 to 184 kJ/kg (31 to 44 kcal/kg) of desired body weight, or about 9.03 to 12.9 MJ (2200 to 3000 kcal) for an average 70 kg adult man. These calories provide for (1) the basal metabolic rate, (2) the specific dynamic action of food, and (3) the extra energy expenditure for activity.

Clinical comment

Scientific units All the countries in the world except for the United States have adopted an international system of units (SI units) for scientific and medical purposes. SI units have also been recommended by most scientific journals and the American Medical Association (AMA). Therefore, in preparation for the twenty-first century and international communication, these units have been adopted in the textbook with the older units given in parentheses.

Basal metabolic rate (BMR) The BMR is the energy required by an awake individual during physical, digestive, and emotional rest. It is measured directly by the heat evolved or indirectly by the volumes of oxygen consumed and carbon dioxide evolved per unit time. It is not the minimum metabolism necessary for maintaining life, since the metabolic rate during sleep may be less than the BMR.

The BMR is influenced by many factors. It is changed in some hormone diseases; for example, hyperthyroidism increases and hypothyroidism lowers the BMR. The BMR changes with growth, reaching a maximum at about 5 yr of age; in the adult it varies with body size. Exposure to cold or a regular exercise routine causes increases in the BMR; starvation produces a lowering—some small attempt at natural compensation for a reduced caloric intake. Increased cellular activity produced by disease is reflected in an increased BMR. With fever, an approximate 12% increase occurs in BMR for each degree centigrade rise in temperature.

Many measurements of BMR have been made on humans, but the variations noted have made it difficult to predict a value for any individual. For the purposes of this discussion the energy demand for basal metabolism in an adult will be taken as 100 kJ/day · kg of body weight (24 kcal/day · kg of body weight). These numbers are rounded to the nearest whole number:

$$BMR = \text{Weight (kg)} \times 100 \text{ kJ/day}$$
$$(\text{or } BMR = \text{Weight (kg)} \times 24 \text{ kcal/day})$$

Specific dynamic action (SDA) Although the metabolic processes continuously produce heat, heat production increases during digestion and metabolism of food. This stimulatory effect of food on heat production is called specific dynamic action (SDA). The value is different for proteins, carbohydrates, and fat, being greatest for protein. On the average, this wasted energy amounts to about 10% of the total food intake.

Energy value of nutrients The energy content of ingested nutrients as measured by calorimetry is not totally available to the body for metabolism. Some of the food is not absorbed and is lost in the feces. In addition, losses result from SDA and through the excretion of urea, creatinine, and other organic materials in the urine. Since the concern here is with the amount of energy in food that is available for metabolism, the calorimetric values for each major foodstuff can be corrected for these losses and the metabolic energy contents obtained. The specific values are actually different for each nutrient in each food. The average values, which include the corrections for SDA, are summarized in Table 1.2. Considering the variability of the individual and the differences in composition

Table 1.2   Average metabolic energy of food components

Food components	Average energy value: kJ (Kcal)
Protein	17 (4)
Carbohydrates	17 (4)
Fat	37 (9)

of diet, the precise predictability of the loss or gain of body weight clearly is not possible from tables of average values. They do, however, permit initiation of treatment, which can be periodically corrected for individual responses.

Activity The effect of various types of muscular work on the energy requirements of the body can be measured (Table 1.3). For example, a 90 kg man engaged in heavy activity requires approximately the following dietary energy content:

$$BMR = 90 \times 100 \text{ kJ/day} = 9000 \text{ kJ/day} (2160 \text{ kcal/day})$$
$$\text{Heavy activity} = 50\% \text{ BMR} = 4500 \text{ kJ/day} (1080 \text{ kcal/day})$$
$$\text{Metabolic energy requirement} = (9000 + 4500) \text{ kJ/day} = 13,500 \text{ kJ/day} (3240 \text{ kcal/day})$$

Table 1.3   Activity/energy demands

Activity	Additional Energy Demand
Sedentary	30% BMR
Moderate	40% BMR
Heavy	50% BMR

During the weekend this same man, doing little work, would have a sedentary level of activity, and his daily need would fall to the following:

$$(9000 + 30\% \times 9000) \text{ kJ/day, or } 11,700 \text{ kJ/day} (2708 \text{ kcal/day})$$
$$[(2160 + 30\% \times 2160) \text{ kcal/day, or } 2808 \text{ kcal/day}]$$

All calculations of the energy requirement of a person are inexact, and the daily requirements are usually rounded off to the nearest 100 kJ or 50 kcal. The 90 kg man would theoretically need 13.5 MJ/day (3250 kcal/day) when at work and 11.7 MJ/day (2800 kcal/day) during sedentary periods.

These calculations can be made more reliable if the individual activities are analyzed. Particular examples of additional energy expenditure over and above the BMR for various activities are listed in Table 1.4 for a 70 kg adult. The corresponding energy needs of other adults are proportional to their body weight. Thus a 50 kg woman would expend 1653 kJ (395 kcal) in addition to her BMR requirements when swimming for 1 hr at 2 mph.

Clinical comment   **Energy requirements** A 70 kg man driving a car for 8 hr/day, eating 2 hr/day, writing 6 hr/day, and sleeping 8 hr/day requires approximately the energy content shown in the following equation (no calories except for BMR are needed for sleeping):

Table 1.4   Additional energy expenditure for a 70 kg adult during various activities

	Extra energy	
Activity	**Kilojoules per hour**	**Kilocalories per hour**
Bicycling ∼ 8 mph	732	175
Dishwashing	293	70
Driving car	264	63
Eating	117	28
Lying still, awake	29	7
Running	2051	490
Swimming 2 mph	2314	553
Typing (at high speed, electric typewriter)	293	70
Walking 3 mph	586	140
Walking 4 mph	996	238
Writing	117	28

$$\text{BMR} = (70 \times 100) \text{ kJ/day} = 7000 \text{ kJ/day}$$
$$[\text{or } (70 \times 24) \text{ kcal/day} = 1680 \text{ kcal/day}]$$
$$\text{Activity} = (8 \times 264 + 2 \times 117 + 6 \times 117) \text{ kJ/day} = 3048 \text{ kJ/day}$$
$$[\text{or } (8 \times 63 + 2 \times 28 + 6 \times 28) \text{ kcal/day} = 728 \text{ kcal/day}]$$
$$\text{Metabolic energy requirement} = \text{BMR} + \text{activity} = (7000 + 3048) \text{ kJ/day} = 10.05 \text{ MJ/day}$$
$$[\text{or } (1680 + 728) \text{ kcal/day} = 2408 \text{ kcal/day}]$$

Such a daily activity would be considered moderate, and on this basis the energy need could be estimated as 7000 + 40% of 7000, or 9.80 MJ/day (1680 + 40% of 1680, or 2352 kcal/day). This method of estimation provides an answer that is close to the 10.05 MJ/day (2408 kcal/day) given by the preferred approach of individual analysis of the various activities. This range of 9.03 to 12.9 MJ/day (2200 to 3000 kcal/day) represents a figure that was initially suggested, based on 129 to 184 kJ/day · kg of body weight (31 to 44 kcal/day · kg of body weight).

Lipid requirements   Lipids, found primarily in the adipose tissue, represent a concentrated energy supply in the diet as well as an efficient form of energy storage in the body. There seems to be no limit to the amount of fat that can be stored in adipose tissue from excessive food intake.

     Cholesterol The individual response to increased dietary consumption of foods high in cholesterol, such as whole eggs (200 to 300 mg per egg, contained entirely in the yolk), is variable. Many people have feedback mechanisms that control the amount of cholesterol in the blood by balancing the amount synthesized in the liver and the amount absorbed from the intestine. Some individuals have a less efficient control, with resulting elevation of blood cholesterol, unless dietary restrictions are observed or other measures, such as increased intake of lipid with high contents of polyunsaturated fatty acids, are followed.

     Since the 1950s it has become increasingly clear that for many people large proportions of vegetable fats in the diet produce a sustained reduction in serum cholesterol (see p. 483), whereas the same proportions of animal fats produce opposite effects. As noted from Table 1.5, the vegetable oils have higher contents of polyunsaturated fatty acids in

Table 1.5   Lipid composition of foods

Food group	Percent linoleic acid	P/S ratio*
Milk		
Butterfat	2	1/34
Meat		
Lard	10	1/34
Beef fat	2	1/24
Vegetables and fruits		
Corn oil margarine	32	1/0.45
Corn oil	57	1/0.2
Safflower oil	77	1/0.1
Soybean oil	52	1/0.3
Peanut oil	28	1/0.5
Bread and cereals†		
Wheat flour, white		1/0.3
Rolled oats		1/0.5

*Total polyunsaturated fatty acids/total saturated fatty acids.
†Composition of extracted lipid.

their composition than the animal fats or the hydrogenated oils, as seen from the ratio of total polyunsaturated fatty acid to saturated fatty acid. The relationship between elevated serum cholesterol concentrations and a higher risk of cardiovascular disease thus indicates the advisability of reducing the total dietary fat and replacing saturated animal fat with appropriate vegetable oils, which have high polyunsaturates and no cholesterol (see Table 1.6), or partially hydrogenated vegetable oils (margarines), which have no cholesterol and moderate amounts of polyunsaturated fatty acids.

Essential fatty acids Two groups of polyunsaturated fatty acids must be derived from the diet. All others can be synthesized from other fatty acids or non–fatty acid precursors, mainly carbohydrates. Linoleic acid (18:2ω—6) is synthesized by plants and is the primary

Table 1.6   Cholesterol content of foods

Food group	Cholesterol (mg/100 g)
Cow's milk	
Skimmed	1
Whole	11
Butter	250
Cheese	100
Meat	
Beef, pork, chicken, bacon, fish	60 to 70
Brain	2000
Kidney	375
Liver	300
Sweetbreads	250
Egg yolk	1500
Egg (whole)	550
Oysters	200
Vegetables fruits (all)	0
Bread, cereal (all)	Trace (from additives)

source of one of the groups needed by the body. Linoleic acid must be provided in the diet because it is essential to health; it is an essential fatty acid. It is a component of cell membranes and the precursor of arachidonic acid, the substrate for the synthesis of most of the eicosanoids, a group of 20-carbon fatty acid derivatives with hormonal properties. The eicosanoids include the prostaglandins, thromboxanes, leukotrienes, hydroperoxy-polyunsaturated fatty acids, hydroxypolyunsaturated fatty acids, and lipoxins. The poly-unsaturated fatty acid arachidonic acid, which can be synthesized from linoleic acid, is therefore not required in the diet but is still considered to be a member of the essential fatty acid class.

The second class of polyunsaturated fatty acids that cannot be synthesized de novo is the ω—3 group. Examples are eicosapentaenoic acid (EPA; 20:5ω—3) and docosahex-aenoic acid (DHA; 22:6ω—3). They are synthesized by cold water plants and algae, which are ingested by deep water fish and are therefore abundant in certain fish oils. No precise function has been delineated for EPA or DHA, nor is it known if they are required for optimum health; thus they are not yet identified as essential fatty acids. Both are found in sizable amounts in some tissues, such as brain, and have been prescribed by some for the reduction of coronary heart disease. More research is required, however, to establish their role in the diet.

In general, the dietary supply of essential fatty acids should be approximately equivalent to 1% of the energy intake. Thus a diet of 10.04 MJ/day (2400 kcal/day) should include at least 100 kJ (24 kcal)—that is, (100/37) g or (24/9) g or 2.7 g—of linoleic acid.

Carbohydrate requirements

In theory no need exists for carbohydrate in the diet, since all the carbohydrate components of the body can be synthesized from proteins. However, ketosis and acidosis develop when carbohydrate is unavailable to cells. This occurs in the disease diabetes mellitus; although plenty of glucose is available in the extracellular fluid, the tissues cannot properly absorb it. To avoid ketosis, it is necessary to provide 12 g of carbohydrate for each 1000 kJ, or 5 g for each 100 kcal. This means that a minimum of 20% of the total energy should be provided in the form of carbohydrates.

Protein requirements

Amino acids derived from proteins in the diet serve as the source of the nitrogen for the biosynthesis of nonprotein compounds, such as the porphyrins of hemoglobin and the nucleic acids. Amino acids also provide for the biosynthesis of tissue proteins. As described later, all 20 amino acids must be available together for protein synthesis to occur.

Dietary protein as an energy source If the diet is deficient in carbohydrates, protein will be degraded. The resulting amino acids will be deaminated and the carbon skeletons used for the maintenance of blood glucose concentrations, the formation of Krebs cycle intermediates, and the generation of ATP (see Figure 1.1). For this reason, carbohydrates in the diet have a sparing effect on amino acids. Thus a high-protein diet is a costly way of providing energy. Furthermore, this presents a load on the kidney to eliminate the urea produced from the deamination of the amino acids.

Essential amino acids Although half of the 20 amino acids can be synthesized by the cells, the other 10, the essential amino acids, cannot always be synthesized at a sufficient rate and must be supplied in the diet. Of the 10 essential amino acids, eight are essential at all times during life. The other two are required in the diet during periods of rapid growth, as in childhood. Following is a list of these essential amino acids,* and their chemical formulas are given in Table 2.1.

---

*"*Any help in learning these little molecules proves truly valuable*" is a mnemonic device that may prove helpful.

Essential for adults	Additional amino acids essential for infants
Isoleucine	Arginine
Leucine	Histidine
Lysine	
Methionine	
Phenylalanine	
Threonine	
Tryptophan	
Valine	

One should note that the infant requires all the amino acids that are basic, branched, or aromatic except tyrosine. Tyrosine is essential only when phenylalanine is absent from the diet. A diet that is very low in phenylalanine is prescribed for children suffering from phenylketonuria, a genetic disease in which the normal metabolism of phenylalanine is disturbed. A premature infant may also require tyrosine and cysteine, depending on whether the appropriate enzyme systems for the synthesis of these amino acids have developed at the time of birth.

The inclusion of adequate quantities of nonessential amino acids in the diet will spare the essential amino acids from being catabolized and the fragments used in nonessential ways. Carbohydrates also have a sparing effect on amino acids. Consequently the diet must supply adequate amounts of carbohydrate and essential amino acids. This is particularly important in a vegetarian diet, where the supply of protein comes from various sources. This takes into account, for example, that sweet corn is deficient in tryptophan and lysine, which can be made up by including beans with the corn in the food.

Nitrogen balance An adult is said to be in nitrogen balance or equilibrium when the amount of nitrogen consumed equals the amount of nitrogen excreted in the urine, sweat, and feces. *Positive* nitrogen balance occurs when the nitrogen intake exceeds the nitrogen excreted. A positive nitrogen balance is associated with growth, as in children, during pregnancy, or during convalescence when injured tissues are being repaired. *Negative* nitrogen balance occurs when the intake of nitrogen is less than the nitrogen excreted. This happens during starvation and in malnutrition. It also occurs in a variety of illnesses. Burns, trauma, or surgery will produce a period of negative nitrogen balance.

Biologic value of proteins Generally the most nutritious diet with respect to amino acid content is one that provides a good balance of all the amino acids, both essential and nonessential. In this regard some protein foods are better than others. For example, the proteins of human milk would score higher than beef steak, which is better than rice.

The biologic value of a dietary protein is a measure of the extent to which it satisfies the amino acid requirement for growth or the maintenance of total body function. For the dietary proteins to have biologic value for tissue growth and replacement, they must provide all the essential amino acids. As seen later when considering protein biosynthesis, the process terminates when any one amino acid is not available. In general, animal proteins have a high biologic value. A major exception is gelatin, which lacks the essential amino acid tryptophan and therefore has no biologic value. Vegetable proteins have a low biologic value for humans because each one has a low level of one or more essential amino acids. This is the main reason why plant geneticists try to develop plants with higher levels of essential amino acids in their protein, such as the corn strains high in lysine. The vegetarian clearly demonstrates, however, that a proper mixture of plant foods can provide all the necessary protein requirements in the diet.

Biologic values have been placed on a relative scale, with whole-egg protein or egg albumin scored at 100. The RDA for protein is calculated for a biologic value of 70, which is the biologic value for beef and the average biologic value of protein in an

American diet. Corrections are necessary for dietary proteins of different average values. Thus an adult has an RDA of 56 g of protein. If the dietary protein were of plant origin, with a biologic value of 40 instead of the 70 assumed in the RDA, then the daily intake should be increased to (56 × 70/40) g, or 98 g.

Clinical comment

Adult celiac disease Some foods contain proteins that are toxic under special circumstances or may produce allergic reactions. Patients with celiac disease are sensitive to a certain fraction of wheat protein, called gliadin, a small protein of 88 amino acid residues. In such sensitive persons the intestinal wall becomes inflamed, and the normal absorption of nutrients is impaired. Diarrhea results, but the symptoms disappear with the elimination of wheat from the diet.

Vitamin requirements

Vitamins are small organic molecules in the diet that either cannot be synthesized by humans or are synthesized at a rate less than that consistent with health. The vitamins are usually divided into those that are fat soluble and those that are water soluble. The division may sound artificial, but it does relate to nutrition in the sense that a low-fat diet may be deficient only in the fat-soluble class of vitamins. The role of vitamins in metabolism is considered at a molecular level in later sections. However, in some cases the functions are known only in a descriptive way.

Fat-soluble vitamins

Vitamin A Vitamin A is an alcohol, retinol, that is present as the 11-*cis* or all-*trans* isomers. Vitamin A is found primarily in liver, notably in the fish liver oils.

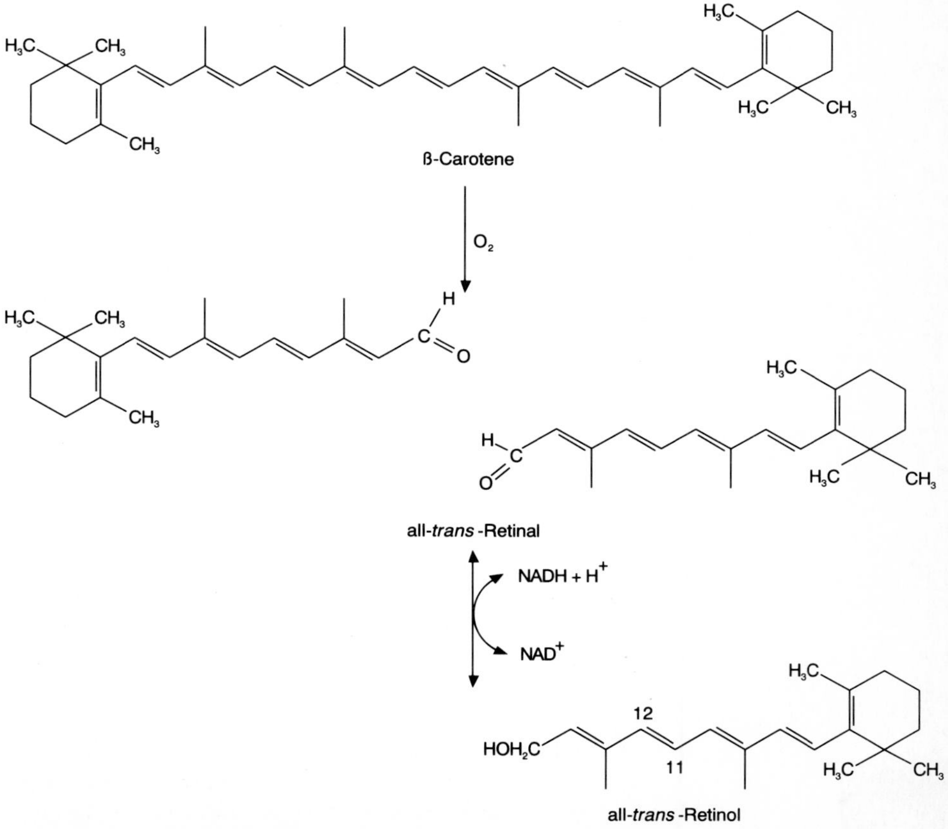

ß-Carotene

all-*trans*-Retinal

all-*trans*-Retinol

Figure 1.2   Visual cycle and storage of vitamin A.

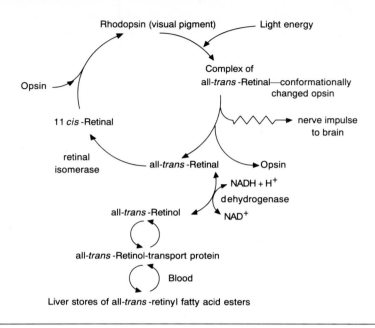

Carotene β-Carotene is an orange pigment found in carrots and other plant tissues. It has the biologic potency of vitamin A. The β-carotene is oxidatively cleaved by the intestinal mucosa into two retinal molecules, each of which is then reduced to all-*trans*-retinol. Free retinol is absorbed into the intestinal mucosa, where it is esterified with a fatty acid such as palmitate and then transported in the lymph to the liver. Retinol is transported in the blood to the tissues by a retinol-binding plasma protein. When this protein contains retinol, it binds to prealbumin, another protein present in small amounts in blood plasma. This forms a protein-protein complex that is thought to protect the retinol in some way. The retinol is removed from the plasma-binding protein by the cells of the retina, and the all-*trans*-retinol is oxidized to all-*trans*-retinal, as noted in the visual cycle (Figure 1.2).

Except for the visual cycle, the mode of action of vitamin A is largely unknown. Vitamin A deficiency leads to disturbances in the growth and remodeling of bone, skin lesions, keratinizing of many epithelial cells, and abnormal cellular function of the adrenal cortex.

Clinical comment   How is night blindness related to vitamin A deficiency?

Visual cycle The biochemistry of vitamin A in the visual process involves both types of photoreceptor cells in the retina. The rods provide black and white vision and respond to dim light. The cones provide color vision and respond to bright light. Both these cells require vitamin A for the formation of the visual pigments. The molecular cycle is summarized in Figure 1.2.

The protein opsin, when complexed with 11-*cis*-retinal by the formation of a Schiff base with a free amino group, produces the visual pigment rhodopsin. Under the influence of light, the retinal moiety is converted to all-*trans*-retinal. This isomerization is associated with a conformational change in the opsin. Such a change is transferred as a nerve impulse by alteration of the permeability to Na^+ or K^+ ions, from the photoreceptors to the optic nerve endings. The rod cell can be stimulated by a single photon. Dissociation of the all-*trans*-retinal is followed by reduction of the aldehyde group to the primary alcohol to give all-*trans*-retinol. Retinol returns to the circulating blood pool of vitamin A that is stored in the liver. The retinol is then converted back to all-*trans*-retinal in the retina, where it is also isomerized back to 11-*cis*-retinal. The cycle is completed when 11-*cis*-retinal reacts with opsin to produce rhodopsin.

Vitamin A replacement For some reason, loss of vitamin A occurs during the photochemical event, so that a continuous dietary requirement of the vitamin is necessary. Deficiency results in lowered blood levels of vitamin A, and the time required for the reformation of rhodopsin after bleaching by light is lengthened. Dark adaptation is thus slow, and treatment with vitamin A is needed to build up both the circulating and the storage levels.

Vitamin A is stored in the liver in such relatively large amounts (0.2 to 2.0 μmol/g of liver) that deficiencies severe enough to cause clinical problems are not likely to develop unless deprivation is extreme and is continued for relatively long periods, at least many months.

Color vision Similar photochemical events occur in both the rod and the cone cells. Each cone cell has one of three different color-sensitive pigments, blue (430 nm), green (540 nm), or red (575 nm). The three pigments contain the same 11-*cis*-retinal moiety, but their opsins are different. Each pigment has a different wavelength of maximum absorbance, just noted in parentheses, and its photochemical reaction responds to one of the three primary color components of the incident light. In each case the visual cycle is the same as that noted for rhodopsin.

Vitamin D Vitamin D, the antirachitic member of the vitamin family, is involved in the metabolism of calcium and phosphate and in the calcification of bone. In deficiency states the growing portions of the bones are affected, producing bowlegs, knock-knees, and enlarged joints. The beneficial effects of sunlight in the prevention of rickets have long been recognized and are known to be caused by the photo-oxidation in the skin of a derivative of cholesterol to a form of vitamin D, cholecalciferol, as follows:

7 - Dehydrocholesterol   Cholecalciferol

This compound is also found in animal products, particularly fish liver oils. Another similar molecule with vitamin D activity is calciferol, which is found in plants. Both are converted to physiologically active forms, as discussed later (see Chapter 19).

Clinical comment

Vitamin toxicity Vitamins A and D are toxic if taken in excess.

Vitamins A, D, E, and K are fat-soluble substances. Thus in nature they are associated with lipids and are found in tissues rich in fat. These vitamins can be stored in very large quantities within the adipose tissue as well as within the lipid component of many other cells. Therefore, if excessive amounts are taken in, the excess is stored within the body and builds up over a time. Eventually this produces toxicity. By contrast, vitamin C is a water-soluble substance. It cannot be stored in large amounts within the body. If an excess is taken in, it is excreted and does not accumulate. Therefore very large doses of vitamin C over a prolonged period will not build up within the body to the point where it will cause toxicity.

Vitamin D is involved primarily in the regulation of calcium and phosphorus metabolism. Therefore these minerals would be most affected by an abnormality in vitamin D concentrations. Many tissues within the body would be damaged in vitamin D toxicity. The kidneys might be affected, since the high level of vitamin D causes an excessive concentration of calcium in the body fluids. This might precipitate the formation of a calcium stone within the kidney's tubules or ureter. Other tissues might also be damaged by precipitation of calcium within the tissue. Finally, the bones, which are storehouses of calcium, probably would become deprived because of excessive calcium mobilization, and the elevated plasma concentration of Ca^{++} would affect blood pressure.

Vitamin A toxicity produces lethargy, abdominal pain, headache, excessive sweating, and brittle nails.

Vitamin E As with vitamins A and D, vitamin E is found in animal or plant fats and oils. The most active vitamin E is α-tocopherol.

α-Tocopherol

β- and γ-Tocopherols, which differ from the α-form in their substituents on the aromatic ring, have 25% and 19% of the activity of α-tocopherol, respectively.

The biologic function of vitamin E in humans is not clear. It probably functions partly as an inhibitor of the peroxidation of the lipid in cell membranes, and it may function to maintain the integrity of cellular membranes. Vitamin E is required for normal reproduction, muscle integrity, and resistance of erythrocytes to hemolysis. No firm evidence indicates that it acts in humans as an antisterility vitamin, as it does in rats. Fortunately, tocopherols are relatively nontoxic in humans, since there is a popular misconception that vitamin E improves sexual performance.

Vitamin K Vitamin K_1 and K_2 are required for the blood-clotting process. Vitamin K is required for the carboxylation of the side chains of some glutamyl amino acid residues in prothrombin. The carboxyl groups so formed serve as tight binding sites for Ca^{++}, which are integral in processes involving Ca^{++} transport, Ca^{++} regulation of metabolism, and for the activation of prothrombin. The latter is discussed in greater detail in Chapter 4. These molecules are derivatives of menadione, a 1,4-napthoquinone that acts more rapidly when injected than vitamin K_1 and is thus often used as an antihemorrhagic pharmaceutic agent.

Menadione, R = H

Vitamin K_1 is present in green leaves and other plant tissues eaten in the diet. Vitamin K_2 is a bacterial product and is produced by the bacterial flora of the intestine. Therefore, under normal circumstances, and human is well provided with this vitamin. The situation may change when prolonged antibiotic therapy is instituted and some of the bacteria in the intestinal lumen are killed. A favorable symbiosis is thus lost.

Since vitamin K is absorbed with the other fat-soluble vitamins in the upper part of the small intestine, the process depends on the presence of bile salts. As a result, any absence of bile formation or any bilary obstruction will be reflected in a deficiency of the fat-soluble vitamins. Thus in obstructive jaundice, when bile cannot pass from the liver to the intestine, prothrombin or clotting time may be prolonged because of inability to absorb adequate amounts of vitamin K.

Water-soluble vitamins

Thiamine (vitamin B_1) Thiamine is present in meat, particularly pork, and in yeast, the outer layers of whole grains, and nuts. A deficiency of thiamine produces beriberi, a disease characterized by extensive damage to the nervous and circulatory systems, muscle wasting, and edema.

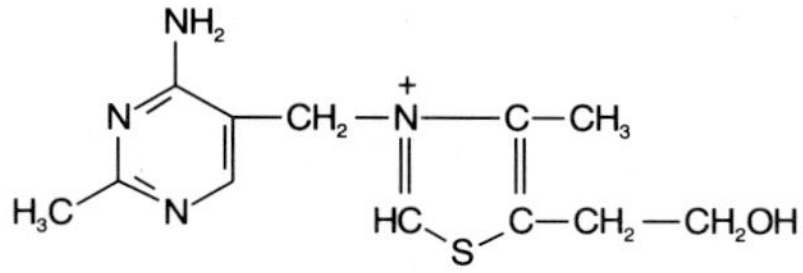

Thiamine

The vitamin is converted in the body to a pyrophosphate ester, which is the metabolically active form. It functions in the conversion of pyruvate to acetyl CoA and in many similar reactions that involve carbon dioxide removal coupled with oxidation, that is, oxidative decarboxylation.

Riboflavin (vitamin B_2) and niacin Riboflavin is found in meat, milk, and plant products along with thiamine and niacin. Riboflavin is also produced by the bacteria that are always present in the intestine. As a result, little is known about the effects of riboflavin deficiency in humans. Niacin, however, is known as the protective dietary component against pellagra. The symptoms of pellagra include a swollen tongue, dermatitis, and nervous and gastrointestinal disturbances.

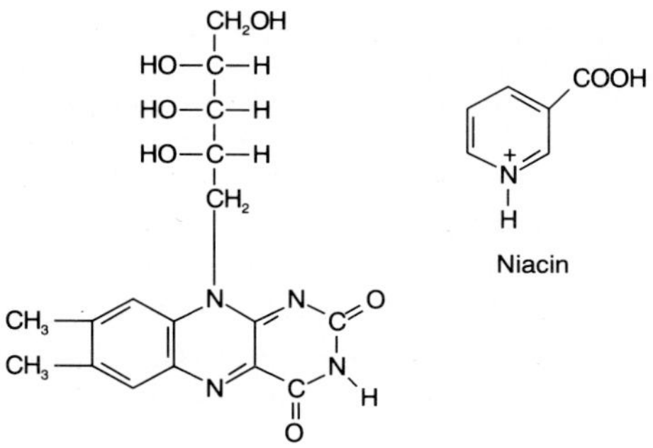

Niacin

Riboflavin

Niacin can be synthesized by humans from the amino acid tryptophan but at a rate that is inadequate to maintain good health. Both vitamins are building blocks for the biosynthesis of compounds that act in oxidation-reduction reactions, namely nicotinamide adenine dinucleotide (NAD) and flavin adenine dinucleotide (FAD). These vitally important compounds are discussed in later chapters.

Pyridoxine (vitamin B₆) Pyridoxine is the common form of vitamin B_6 in most commercial preparations. The other forms are pyridoxal and pyridoxamine.

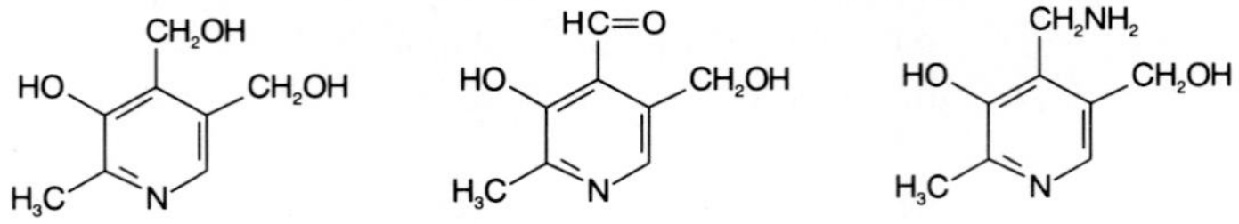

Pyridoxine   Pyridoxal   Pyridoxamine

The phosphate ester derivatives of these forms are involved in many reactions in metabolism, including the conversion of amino acids to the corresponding keto acids.

Pantothenic acid Pantothenic acid is present in many tissues and was first isolated from liver and yeast. This vitamin is necessary for the biosynthesis of coenzyme A, which, as noted earlier in the overall view of biochemical processes, is a most important molecule in the metabolism of many nutrients.

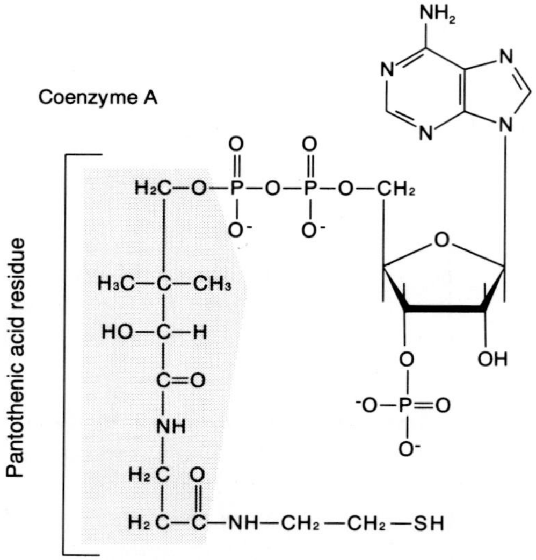

Coenzyme A

The pantothenic acid moiety is shaded in the above structure of coenzyme A, which is central to the anabolic and catabolic processes. Therefore it is understandable that pantothenic acid deficiency would have profound consequences. Fortunately, little evidence exists of pantotheic acid deficiency in humans.

Biotin As with pantothenic acid, biotin is widely distributed in foods so a deficiency state rarely exists. Egg white contain a protein, *avidin,* that binds tightly with biotin. Therefore biotin deficiency can occur in people who eat large quantities of raw egg whites. The relatively simple, bicyclic molecule is involved in biochemical reactions where carbon dioxide is added to a molecule to produce a carboxyl (—COOH) group. This is in contrast to decarboxylations, reactions in which carbon dioxide is removed and thiamine is usually involved.

Biotin

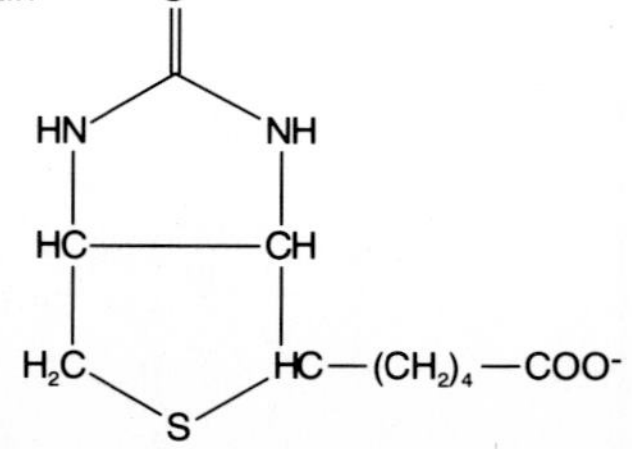

Folic acid Folic acid is composed of residues of glutamic acid, *p*-aminobenzoic acid, and a heterocyclic ring system, a pteridine. It is found in meats and vegetables, particularly green leaves. In most food sources folic acid is combined with one to five additional glutamic acid residues, which are linked together as a peptide.

Folic acid, (R=H)

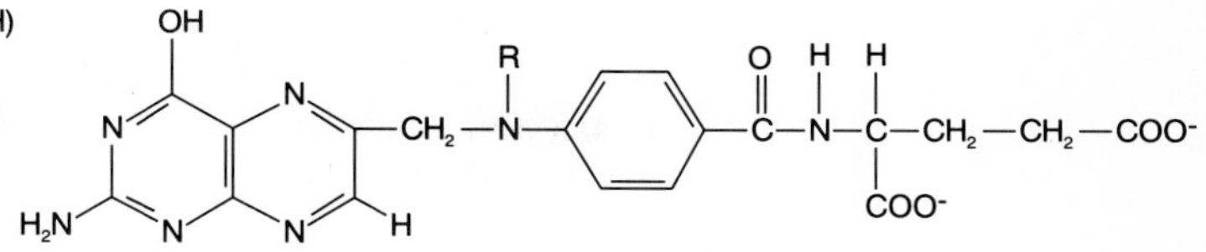

Additional variations in the molecule at substituent R also exist. Some of these derivatives are destroyed by food preparation such as boiling. Folate deficiencies are more widespread than is generally appreciated; this may be a result of low dietary intake, but the situation is accentuated by the destruction of folate in food preparation, in the prolonged heating and rewarming of food, and by poor intestinal absorption. For the elderly living alone who consume food that is highly processed and that may be reheated more than once, the risk of folate deficiency is particularly high.

Folate is involved in many biochemical reactions, particularly those related to the biosynthesis of deoxyribonucleic acid (DNA). Folate deficiency anemia is preceded by a reduction in serum folate levels, which is followed in a few months by a reduced number of erythrocytes. Demands for folate are increased during periods of growth and in hemolytic diseases or parasitic invasions. Growth demands the laying down of more tissue (protein biosynthesis), and hemolytic diseases require the increased rate of replacement of the erythrocytes. Parasites may limit the intestinal absorption of folate.

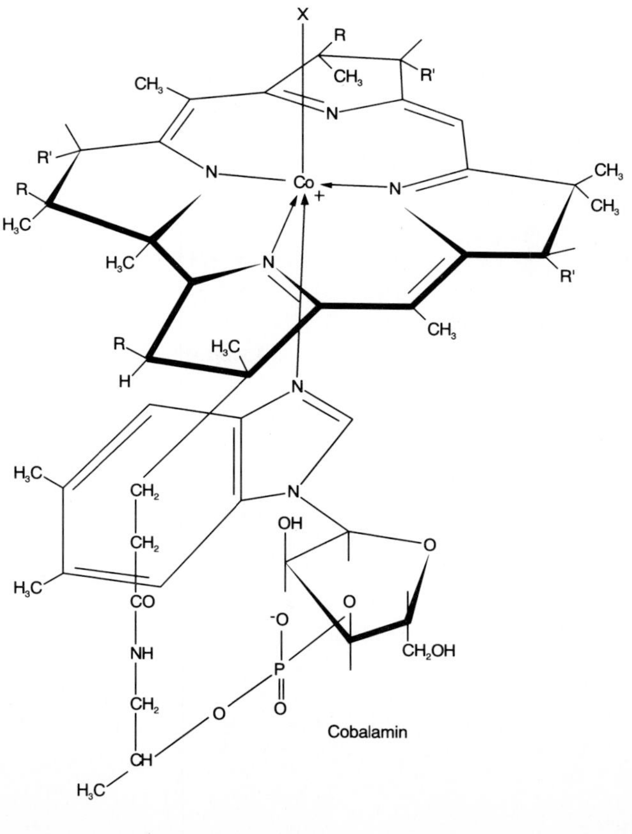

Cobalamin

Cobalamin (vitamin B_{12}) Vitamin B_{12} is found in foods of animal origin such as meat, especially liver and kidney; little, if any, is present in plants. In humans negligible amounts are provided by intestinal flora. It is particularly important, therefore, to supplement the vitamin B_{12} supply in a strictly vegetarian diet containing only plant protein. Vitamin B_{12} is a complex molecule that is isolated as a cyanocobalamin (X = —CN).[23] (See p. 23.) The active forms in vivo are the methyl (X = —CH$_3$) or 5'-deoxyadenosyl derivatives. These substituents are attached directly to the cobalt, as shown in the partial structural formula. The total body content of vitamin B_{12} is very low, about 2 μmol. In humans only two reactions have been shown to be dependent on it (pp. 421 and 563).

Clinical comment

Anemias Deficiency in vitamin B_{12} results in anemic symptoms similar to those seen in folate deficiency. In addition, vitamin B_{12} deficiency causes irreversible degeneration of the spinal cord by demyelinization. Because folate supplementation may mask B_{12} deficiency, it is customary to delete folate from vitamin preparations.

    Pernicious anemia is a disease state produced by a deficiency of vitamin B_{12}. It is caused by the absence of a protein called *intrinsic factor* that is normally synthesized in the gastric mucosa and is required for B_{12} absorption in the intestine. It is therefore necessary to inject B_{12} parenterally to treat pernicious anemia and its associated nervous system defects.

    Ascorbic acid (vitamin C) Ascorbic acid is found in fresh fruits and vegetables, and its deficiency produces scurvy. The condition is corrected by consuming ascorbic acid, more correctly called L-*xylo*ascorbic acid. Ascorbic acid is easily oxidized to dehydroascorbic acid; the ratio of these two forms in the tissues varies somewhat, but always with the reduced form in large excess. Ascorbic acid is involved in the biosynthesis of collagen, the major structural protein of connective tissue (see Chapter 2.)

Ascorbic acid   Dehydroascorbic acid

Clinical comment

Scurvy The symptoms seen in scurvy are pathologic changes in teeth and gums, the tendency to hemorrhage resulting from weakened blood vessels and capillary beds, poor wound healing, and ulceration. The amount of ascorbic acid required to avoid scurvy is usually less than 284 μmol/day (50 mg/day). The exact amount depends on many factors, such as age, pregnancy, and individual variables, but a daily intake of 284 to 341 μmol (50 to 60 mg) is the presently recommended RDA (see Appendix A). Contrary to these recommendations, an intake of several grams per day has been recommended by some researchers to maintain optimum health and maximum defense against infection. Although much of the high doses of ascorbic acid is either excreted or metabolized when the body is saturated with this vitamin, reports are appearing of toxicity that results from megadoses, such as dehydration, increased oxalate synthesis in renal calculi, and increased uptake of heavy metals.

Mineral requirements   The inorganic components of the human body are principally sodium, potassium, calcium, magnesium, iron, phosphorus, chloride, and sulfur. To a large extent they represent the minerals of the skeleton and the buffer ions of the body fluids. They are essential parts of the diet. Other elements present in much smaller amounts, called the trace elements, are also essential dietary components. These include copper, molybdenum, cobalt, manganese, zinc, chromium, selenium, iodine, and fluoride.

Although the specific role of minerals in human health still requires further study, some information is being obtained from investigations with animals and from the prolonged nutritional support of humans through intravenous feeding with defined liquid diets. There are also indications for the inclusion of nickel, silicon, tin, and vanadium in this list, but their roles in human nutrition are unknown. Also found in the ash of the body are other elements, such as barium, strontium, lead, and mercury. These are not known to be essential for metabolism and represent accumulations without controlled absorption or excretion—a body pollution. The presence of aluminum in water used in hemodialysis in time causes an anemia not caused by an iron deficiency. In some cases the process is slowly reversed by using deionized water to prepare dialyzing solutions with low aluminum concentrations.

The following comments are not intended to be extensive, but they do give some indications of the principal physiologic roles of the inorganic ions (see also Table 1.7).

Sodium, potassium, and chloride Sodium, potassium, and chloride represent the major ions of the body fluids. Sodium and chloride are concentrated mainly in the extracellular fluids, whereas potassium is found largely within the cells and is essential for many enzymic processes, for the transmission of nerve impulses, and for the functioning of muscle. The interplay of these ions, the control of body fluid volume and ionic concentration, and the energy involved in maintaining these ions in their respective fluid compartments are critical problems of homeostasis.

Table 1.7   Essential trace elements

Element	RDA	Function	Signs of deficiency
Chromium	1-4 μmol (50-200 μg)		Impaired glucose metabolism
Cobalt		Vitamin B$_{12}$	Anemia, growth retardation
Copper	30-50 μmol (2-3 mg)	Oxidative enzymes	Anemia, skeletal defects, demyelinization and degeneration of nervous system, cardiovascular lesions
Fluoride	50 μM (1 mg/L in water)	Compound of bones and tooth enamel	Caries, abnormal bone structure
Iodide	0.4-0.6 μmol (50-75 μg)	Thyroid hormones	Goiter, depressed thyroid function
Manganese	45-90 μmol (2.5-5 mg)	Mucopolysaccharide metabolism	Growth retardation, impaired glucose tolerance
Molybdenum	5.2-11.6 μmol (0.15-0.5 mg)	Purine metabolism, aldehyde oxidation	Gout-like symptoms, elevation of methionine in blood
Selenium	0.6-2.5 μmol (50-200 μg)	Glutathione peroxidase	Muscular pain, cardiomyopathy
Zinc	190 μmol (12.5 mg)	Nucleic acid metabolism	Loss of appetite, failure to grow, skin changes, impaired regeneration of wounds, decreased taste acuity

Sodium and potassium are present in most diets, and dietary allowances have not been established for these. Dietary deficiencies do not occur with a normal diet.

Calcium and phosphorus A great variety of foods contain calcium and phosphorus. Calcium phosphate in the form of hydroxyapatite is the principal component of the hard structures of bones and teeth. A continuous exchange of these ions occurs between the circulating fluids and the solid skeletal tissues and cell membrane structures. The maintenance of serum calcium concentrations over a narrow range is vital and involves delicate homeostatic control systems. Calcium is involved in nerve and muscle excitability, blood coagulation, mediation of hormonal responses, and some enzyme activities. Phosphate is an essential part, as organic esters, of the reactive form of most intermediary metabolites. It also plays an important role in the storage of chemical energy in the form of ATP.

Magnesium Most foods, especially those of plant origin such as potatoes, whole grain cereals, and fruits, contain magnesium. RDAs for magnesium are about 12 to 19 mmol (300 to 450 mg) for adults (see Appendix A). Magnesium is largely bound with phosphates in the skeleton. It is also essential in metabolism, particularly in those reactions involving ATP. Mg^{++} is a depressant of the central nervous system and also elicits hypotension. High concentrations of magnesium reduce the heart rate and ultimately produce cardiac arrest. Some antagonism is noted between Ca^{++} and Mg^{++}. High concentrations of Ca^{++} tend to aggravate the effects of low Mg^{++}, probably because of interference with Mg^{++}-requiring enzymes.

Sulfur Most of the required dietary sulfur is provided by the dietary protein. Sulfur is a component of the amino acids cysteine and methionine, the coenzyme CoASH, and the vitamins thiamine and biotin. It is present as an organic sulfate ester in some biopolymer components of connective tissue, in complex lipids known as sulfatide glycolipids, and in conjugated bile acids referred to earlier as emulsifying agents for dietary fat in the intestine.

Iron Iron is required for the synthesis of the heme portion of hemoglobin and myoglobin. It is also needed in nonheme iron proteins and other intracellular molecules called the cytochromes, which are involved in the oxidation of metabolites. Approximately 25% of the body iron is stored in the liver, spleen, and bone in the form of ferritin, an iron-globulin compound, and hemosiderin. It is also present in the plasma protein transferrin. Iron is present in a variety of foods of animal and plant origin, particularly liver and legumes; milk is a poor source. Dietary iron is considered in two categories—iron that is complexed in the heme molecule, derived principally from the hemoglobin in meats, and the nonheme iron. The availability for absorption of the latter form of iron depends on the other food components such as ascorbic acid, which increases the absorption, and phosphate, which decreases the availability. A deficiency of dietary iron gives rise to iron-deficiency anemia. Iron is conserved efficiently by the body, but significant losses occur with bleeding. Iron in plasma is bound to a transport protein, transferrin, and in normal adults only 27% to 44% of the transferrin is saturated. The remaining potential for transporting iron is referred to as the latent iron-binding capacity. When the body iron stores are low, most of the iron in tissue is in the form of ferritin. When iron stores are high, it is in the form of hemosiderin. If the control in the absorption of iron from the intestine is impaired, more iron is taken into the tissues than is required. The accumulation builds up, with damage to the organ functions, particularly liver, pancreas, heart, and spleen. The buildup in such diseases as idiopathic hemochromatosis is relieved by periodic phlebotomy so that the loss of iron in the blood will eventually compensate for the improperly high absorption of dietary iron (see Chapter 4).

Iodine A high proportion, 70% to 80%, of the iodine in an adult is concentrated in the thyroid gland, where it is involved in the biosynthesis of the hormones thyroxine and triiodothyronine (see Chapter 19). The natural sources of dietary iodine are vegetables, which vary in their iodine content according to the soil in which they are grown and the mineral content of their water supply. Deficiencies of iodine in the diet result in hyperplasia of the thyroid gland, a condition known as goiter. The natural occurrence of iodine in the water and soil can be correlated with the incidence of endemic goiter. When the natural iodine supply is low, food additives such as iodized salt must be employed. Endemic goiter persists as a serious health problem in some parts of the world despite these simple prophylactic measures that have been widely known for many years.

Fluoride The classification of fluoride as an essential element depends to some extent on the criterion of essentiality. Ample evidence shows that the consumption of fluoride confers increased resistance to dental caries, and proper fluoride intake apparently is related to normal skeletal maintenance. Thus it is reasonable to classify fluoride as an important ingredient of the diet but, as with carbohydrate, not absolutely essential.

Clinical comment

Fluoride and caries Few foods contain more than 1 to 2 parts per million (ppm) of fluoride, and the natural drinking water of many areas contains less than 0.1 ppm. In local areas where the fluoride content of water is very high, greater than 8 ppm, chalky white patches are evident on the surface of the permanent teeth because the enamel is weak and easily pitted. These areas may also become stained. This mottled effect is found only in permanent teeth and develops during their formation. Severe mottling can also be accompanied by abnormal bone mineralization. The incidence of mottling is reduced to near zero as the fluoride level of the drinking water falls to around 1 ppm.

Researchers noted that mottling of the enamel accompanied increased resistance to dental caries. This resistance can be achieved without the mottling effect. The optimum protection against caries without the development of mottling occurs at about 1 ppm in the drinking water. Such concentrations are established in many communities by fluoridation of water with sodium fluoride or fluorosilicate. The fluoride seems to be involved in a modification of the hydroxyapatite crystal in the enamel of the tooth, possibly by substituting for the hydroxyl group.

$$(Ca_3P_2O_8)_3 \cdot Ca(OH)_2 \quad + \quad 2F^- \longrightarrow (Ca_3P_2O_8)_3 \cdot CaF_2 \quad + \quad 2OH^-$$

Hydroxyapatite   Fluorapatite

Whether this is the mechanism by which fluoride imparts resistance to dental caries is still a matter of debate.

Skeletal conditions of demineralization such as osteoporosis, which result in decreased radiologic bone density, have been related in some cases to low fluoride levels in drinking water. Reports of improvements in calcium balance and bone density as a result of fluoride therapy indicate a promising direction for research into such diseases.

Other trace elements Copper, manganese, molybdenum, and zinc have each been shown to function as part of different enzyme systems or to be present in serum, where they are transported as part of different enzyme systems or to be present in serum, where they are transported as metalloproteins. Cobalt must be provided in the form of vitamin

B_{12}. There is demonstrable interplay between some pairs of metals; for example, high levels of copper reduce the intestinal absorption of zinc. Low serum levels of copper have been accompanied by anemia, but the role of copper in hemoglobin synthesis is not known. Zinc is present in several enzymes; its deficiency is sometimes seen in patients fed preprocessed foods in total parenteral nutrition. In the case of pasteurized cow's milk fed to infants, the deficiency may be caused by the destruction of a heat-sensitive factor that promotes intestinal absorption of Zn^{++}.

Since most foods or water contain these trace metals, it is difficult to evaluate daily requirements. Excesses of most minerals result in symptoms of toxicity. In many instances the abnormalities resulting from intoxication are reversible; even heavy metal poisons such as lead tend to be excreted to some extent. More rapid elimination requires the administration of chelating agents such as the use of 2,3-dimercaptopropanol in the chelation of mercury and arsenic.

Special problems in nutritional maintenance
Parenteral nutrition

When a patient is unconscious or unable to take food by mouth, it is necessary to provide nutrients by special feeding. Parenteral (intravenous) feeding has routinely provided calories in the form of glucose, dissolved in isotonic sodium chloride (153 mmol/L), for periods of a few days. A total of 2 to 3 L/day may be infused. This volume may be increased if other fluid losses are to be corrected. Such a fluid diet places the patient in (1) negative nitrogen balance, since some nitrogen is always being excreted in the urine, (2) caloric insufficiency—5% aqueous glucose is eqivalent to 837 kJ/L (200 kcal/L), and (3) deficiency in the bulk minerals such as potassium, magnesium, and calcium.

Total parenteral nutrition (hyperalimentation)

Prolonged parenteral feeding procedures must provide a complete diet that is calorically adequate and nutritionally balanced. This allows the patient to maintain body weight or, if an infant, to grow. Provision of sufficient energy for an adult requires the infusion of large volumes, and this presents the danger of pulmonary edema or cardiac failure. This problem may be solved by providing more concentrated solutions containing glucose and the amino acids contained in a hydrolysate of a nutritionally complete protein such as casein. Clearly, no complex food can be included, such as sucrose or protein, because they cannot be hydrolyzed to the simple components in the blood. Other additives, such as electrolytes, essential minerals (except iron), lipids, and vitamins are included according to need. This solution could be infused by way of the subclavian vein directly into the superior vena cava, where the blood flow is great enough to dilute the hypertonic intravenous solution. The procedure is known as *total parenteral nutrition* or *intravenous hyperalimentation*. A parenteral form of fat emulsion with adequate essential fatty acid is prepared from soybean oil and is infused together with vitamins through the total parenteral nutrition line. Care must be exercised to avoid mineral and trace element deficiency.

Gavage feeding

A more natural feeding method is accomplished by stomach tube when the gastrointestinal system is functioning properly. A typical gastrostomy tube has an internal diameter of approximately 3 to 4 mm, a size that permits homogenates of most foods to be fed. Nutrition can be provided for weeks by this route without danger of venous thrombosis, pulmonary edema, or cardiac failure. The stomach contents can be emptied at any time should this become necessary. Occasionally, however, gavage feeding requires precautions, for example, with unconscious patients who might aspirate if they vomit.

Weight changes

Weight loss The body obeys all the classic laws of thermodynamics and requires a specific amount of energy for homeostasis. Weight loss occurs when there is a deficit of dietary calories. Balance studies indicate that at times of dietary restriction, a rapid loss of stored carbohydrate and an initial loss of tissue protein occur. However, adaptation quickly takes place to conserve protein, and the fat of adipose tissue becomes the predominant source of the missing dietary calories. Adipose tissue contains 85% lipid and 15% water, and therefore 1 kg can be metabolized to produce the following:

$$(1000 \times 85/100 \times 37) \text{ kJ/kg, or } 31,500 \text{ kJ/kg of body fat}$$
$$[(454 \times 85/100 \times 9) \text{ kcal/lb, or } 3500 \text{ kcal/lb of body fat}]$$

From this figure one can predict the rate of weight loss within about 10% over time. Daily and weekly fluctuations are associated largely with water losses.

Weight gain More than 31.5 MJ (7500 kcal) of energy from the diet are required to add 1 kg of body fat. However, it is even more difficult, for several reasons, to predict the rate of weight gain in an adult. The new tissue weight may be fat or protein, depending on the nutritional state of the individual. A well-nourished person will convert the additional food energy into fat, but it is found on the average and with wide individual variability that an excess of 31.5 MJ adds only 0.5 kg of body weight or 3500 kcal adds only about 0.5 lb of weight. An emaciated individual on a balanced diet will gain weight by the development of both muscle protein and adipose tissue. Body protein, however, is approximately 80% water by weight and is equivalent to the following:

$$(1000 \times 20/100 \times 17) \text{ kJ/kg, or } 3400 \text{ kJ/kg of muscle tissue}$$
$$(454 \times 20/100 \times 4) \text{ kcal/lb, or } 365 \text{ kcal/lb of muscle tissue}$$

This is significantly less than adipose tissue.

In general, for initial estimates one may assume that twice as much food energy is required to produce a weight gain as would be expected from the energy content of the tissue. Thus to add 1 kg of adipose tissue requires 63 MJ (15,000 kcal), whereas 1 kg of muscle requires 6.8 MJ (1625 kcal). Just how many calories can be fed will be controlled by the appetite, although forced feeding is possible by gavage if necessary. When a patient should gain weight, the protein content of the diet is usually increased. High-protein diets, however, require good kidney function. In addition, the development of muscle tissue requires exercise.

Alcohol

Ethanol consumption is important because it is abused, it provides energy (29 kJ/g; 7 kcal/g), and its ingestion affects many aspects of nutrition. Chronic alcoholism is often associated with many signs of nutritional deficiency, caused largely by the reduced consumption of other foods.

Alcohol is absorbed by all parts of the gastrointestinal tract, so concentrations in the blood rise rapidly following consumption. It diffuses quickly across all cell membranes and distributes evenly in both the extra and the intracellular fluids. Metabolism occurs in the liver, where ethanol is oxidized to acetaldehyde, then to acetate, and finally to CO_2 and H_2O. The principal enzyme, liver alcohol dehydrogenase, is found in the cytoplasm. A second pathway that uses a microsomal ethanol-oxidizing system becomes more significant at higher alcohol concentrations. The microsomal oxidation system is important in the metabolism of some drugs and explains the interaction of certain drugs and alcohol.

Approximately 95% of the alcohol ingested is oxidized to CO_2 and H_2O, the remainder being excreted or exhaled without metabolism.

Bibliography

Guthrie HA and Braddock KS: Programmed nutrition, ed 2, St Louis, 1978, The CV Mosby Co.

Passmore R and Eastwood MA: Human nutrition and dietetics, ed 8, New York, 1986, Churchill Livingstone, Inc.

Shils ME and Young VR: Modern nutrition in health and disease, ed 7, Philadelphia, 1988, Lea & Febiger.

Williams SR: Nutrition and diet therapy, ed 5, St Louis, 1984, The CV Mosby Co.

The texts listed above present a detailed statement of the problems of human nutrition and its application to dietary planning.

Cahill GF Jr: Starvation in man, N Engl J Med 282:668, 1970.

Dudrick SJ and Rhoads JE: Total intravenous feeding, Sci Am 227:73, 1972. *Describes, with good illustrations, the experiments conducted in one hospital to arrive at a complete parenteral feeding program for prolonged periods.*

Evaluation of protein quality, Pub No 1100, Washington, DC, 1963, National Academy of Sciences, National Research Council. *Deals with methods of determining biologic value.*

Gawthorne JM, Howell JM, and White CL, editors: Trace element metabolism in man and animals, New York, 1982, Springer-Verlag, Inc.

McNamara DJ: Effects of fat-modified diets on cholesterol and lipoprotein metabolism, Annu Rev Nutr 7:273, 1987.

Mertz W: The essential trace elements, Science 213:1332, 1981.

Michel L, Serrano A, and Malt RA: Nutritional support of hospitalized patients, N Engl J Med 304:1147, 1981.

Miller WH: Molecular mechanisms of photoreceptor transduction, New York, 1981, Academic Press, Inc.

Recommended dietary allowances, ed 9, Washington, DC, 1980, Food and Nutrition Board, National Academy of Sciences, National Research Council. *The table of recommended dietary allowances, reproduced in part in Appendix A, is a valuable source of reference and illustrates various nutritional states in an American population. Equivalent tables are published for several other countries.*

Recommended dietary intakes, Am J Clin Nutr 45:661-704, 1987.

Stokes GB: Estimating the energy content of nutrients, Trends in Biochemical Sciences 13:422, 1988.

Stryer L: Cyclic GMP cascade of vision, Annu Rev Neurosci 9:87, 1986.

Surgeon General's Report on Nutrition and Health, Public Health Service DHHS Pub No 88-50210, Washington, DC, 1988, US Government Printing Office.

Clinical examples

A diamond (♦) on a case or question indicates that literature search beyond this text is necessary for full understanding.

Case 1	Obesity

A graduate student, 188 cm (6 ft 3 in) tall and weighing a steady 132 kg (290 lb), damaged his knee while playing volleyball. Toward the end of his recovery, he was convinced that he needed to make an effort to correct his obesity. He was referred to a dietician by the student health service, and his typical daily intake was found to be 140 g of protein, 120 g of fat, and 590 g of carbohydrate. He expressed the desire to reduce to 86 kg, a level close to his ideal weight, without significantly changing his level of activity.

Biochemical questions

1. How is obesity defined?
2. What approaches might be considered to achieve this desired weight loss?
3. What diet would you recommend for weight reduction? Explain.

Case discussion

1. Definition of obesity Obesity is most often defined as an accumulation of excess body fat.

Obesity is not always the reason for a person being overweight; overweight is also seen in the great musculature of an athlete. Obesity is a disease that carries risks, since it is frequently associated with shortened life expectancy, coronary heart disease, and other problems such as backache that arise because the person is carrying too much weight. Obesity often has a familial component. Several studies have shown that if one parent is obese, then approximately 40% to 50% of the children will develop obesity. Everybody, normal or otherwise, loses weight as the result of a restriction of energy intake; such an approach was taken with this patient.

2. Calculation of weight loss Obesity results in an abnormal increase in fat in the adipose tissues.

The patient wishes to lose 46 kg in, for example, 8 months. For each deficit of 31,500 kJ (7527 kcal) of food energy, he will lose 1 kg of body weight. Therefore the caloric deficit must be $46 \times 31,500 = 1,449,000$ kJ (1449 MJ) or $46 \times 7527 = 346,242$ kcal in 8 months. There are approximately 244 days in 8 months. Therefore each day the energy deficit should be $1,449,000/244 = 5939$ kJ (1419 kcal)/day.

Begin by calculating the energy need at the patient's present weight of 132 kg. For the BMR, $132 \times 100 = 13,200$ kJ ($132 \times 24 = 3168$ kcal). Since he is a student, one can assume sedentary activity. Therefore the energy need for activity is 30% of $13,200 = 3960$ kJ (946 kcal)/day. Add the two together, and the total energy need is $13,200 + 3960 = 17,160$ kJ (or $3168 + 946 = 4114$ kcal/day). Therefore, to induce the proper energy deficit, subtract 5939 from 17,160, which gives a value of 11,221 kJ/day ($4114 - 1419 = 2695$ kcal/day). The patient should be started on this diet. In practice he should lose approximately $5939/31,500 = 0.188$ kg/day, as long as the energy deficit of 5939 kJ/day is maintained.

To approach the problem more precisely, one could then assume that the patient on the second day weighs 132 kg minus 0.188 kg, or 131.812 kg. Calculate his energy need at that weight and then subtract 5939 kJ (1419 kcal) from this to obtain the amount that should be fed on the second day. This daily calculation could be continued throughout the 8 months. This would produce a daily weight loss of 0.188kg/day and lead to a linear decrease over 8 months until the weight of 86 kg was reached. Obviously, however, such a scheme is highly impractical. In practice the first diet of 11221 kJ should be maintained for about 2 weeks, at which time the patient will have lost approximately 2.6 kg. He should then be reweighed and his caloric need recalculated based on the lower weight. A new diet should be prepared that would be 5939 kJ less than that calculated for this new weight. This process should be repeated every 2 weeks for the 8 months, and at the end of the time the patient will be close to 86 kg.

An alternative way to achieve this goal is to calculate the amount of food energy that the patient needs when he will be the weight he wishes to achieve, namely 86 kg. This is 8600 kJ (2064 kcal) for the BMR and 2580 kJ (619 kcal) for sedentary activity, for a total of 11,180 kJ (2683 kcal), or rounding off, 11,200 kJ (2700 kcal). The patient then should be fed this 11,200 kJ diet continuously. Eventually he will have reached 86 kg, but it will take him much longer because near the end, when his energy need will be only about 13,000 kJ (3120 kcal), he is being fed 11,200 kJ (2688 kcal); thus the caloric deficit is only 1800 kJ (432 kcal)/day. This will produce much slower weight loss. The advantage however, is that only a single diet needs to be calculated, and throughout the weight reduction period the patient is eating the amount of food energy that will be needed by him after he reaches his ideal weight.

Psychologically, this method is preferable, since it trains the patient in the diet that he will ultimately need to maintain his ideal weight.

3. Recommended diet It is now recommended that diets should provide 15% of the energy from proteins, 55% from carbohydrates, and 30% from fats. If this were the case for a 11,200 kJ diet, then

(1) Protein provides $11200 \times \dfrac{15}{100}$ kJ/day = 1680 kJ/day

And since 1 g of protein is equivalent to 17 kJ of energy, then

$$1680 \text{ kJ from protein} = \frac{1680}{17} \text{ g} = 99 \text{ g/day}$$

Similarly,

(2) Carbohydrate provides $11200 \times \dfrac{55}{100}$ kJ/day = 6160 kJ/day

$$= \frac{6160}{17} \text{ g/day}$$

$$= 362 \text{ g/day}$$

(3) Fat provides $11200 \times \dfrac{30}{100}$ kJ/day = 3360 kJ/day

$$= \frac{3360}{37} \text{ g/day}$$

$$= 91 \text{ g/day}$$

In planning a diet for weight loss, one usually provides a protein intake that is guided by the body weight. An adequate carbohydrate content is then maintained, and the fat intake is minimized. A minimum intake of fat is probably 1 g from each 6 g of meat if this is the source of the protein and can only be reduced further by using such foods as skim milk or egg whites for protein. These points are now considered further.

Nature of protein In a low-energy diet the intake of protein, minerals, and vitamins should not fall below the levels recommended by the National Research Council. For this student such a level requires approximately 56 g of protein with a biologic value of 70 (based on the biologic value of egg protein as 100). The biologic value of a protein measures its nutritional value relative to a standard protein and takes into account its digestibility as well as its amino acid composition. Protein diets that completely lack any one of the essential amino acids will not support protein synthesis; that is, they will have no biologic value. Such a protein (e.g., gelatin), if not supplemented by other proteins to compensate for the essential amino acid deficiency, will be catabolized and used by the body only as a source of energy. On this relative scale, haddock protein is 89, milk protein is 79, vegetable protein is 59, and flour protein is 46. More than 56 g of protein must be consumed to provide the minimum daily requirements if, for example, a vegetarian diet is requested by the patient.

The proposed diet for the student follows:

	Weight (g)			Food energy in
	Protein	Carbohydrate	Fat	kJ (kcal)
Breakfast	26	80	18	2468 (590)
Lunch	32	110	32	3598 (860)
Supper	30	110	40	3860 (922)
Snack in evening	13	60	1	1278 (305)

Total energy: approximately 11.2 MJ (2677 kcal)

	Grams	Approximate percentage of food energy
Total protein	90	15
Total carbohydrate	360	55
Total fat	91	30

Carbohydrate and ketosis Enough carbohydrate should be provided to prevent ketosis and to ensure that the protein can be spared to provide amino acids for tissue maintenance. A minimum amount of carbohydrate to prevent ketosis requires approximately 12 g/1000 kJ in the diet, equivalent to 20% of the caloric intake in the form of carbohydrate. In a 11,200 kJ diet this amounts to 134 g of carbohydrate. However, the reducing diet is planned to allow the patient to lose an average of 46/244 kg day = 188 g/day, which is largely a loss of fat tissue. Since adipose tissue contains about 85% lipids, the body metabolism includes the oxidation of $188 \times 85/100$ (or 160) g of lipids/day, to produce 160×37 (or 5930) kJ/day. This is added to the 11,200 kJ/day in the diet, so that the total metabolic energy actually provided to tissues is (11,200 kJ + 5930 kJ), or approximately 17,000 kJ/day. On the basis of this requirement, the student needs at least 12 g $\times$ 17, or 204 g, of carbohydrate.

Lipids and essential fatty acids A nutritional requirement for essential fatty acids is satisfied, on the average, by 1% of the energy intake (112 kJ), equivalent in this case to about 112/37 = 3 g of linoleic acid.

Vitamins and minerals An adequate supply of vitamins and minerals should be ensured. Since most minerals are derived from vegetables, fruits, cereals, and proteins, most average dietary programs probably satisfy the mineral need. Vitamins also are usually adequately supplied, but these variables should be checked as the diet program progresses.

Water Water is required in the diet for many physiologic functions; an approximate amount is 30 to 45 ml/kg of body weight.

References

Herzog DB and Copeland PM: Eating disorders, N Engl J Med 313:295, 1985.
Stunkard AJ: Conservative treatments for obesity, Am J Clin Nutr 45:1142, 1987.
Wadden TA et al: Very low calorie diets: their efficacy, safety and future, Ann Int Med 99:675, 1983.

Case 2

Protein-calorie undernutrition in anorexia nervosa

A woman, age 36, had suffered from anorexia nervosa for 10 years and was admitted to the Clinical Research Center for study and treatment. The patient was 165 cm (5 ft 5 in) in height. Her weight was 32.9 kg (72.6 lb), which is about 60% of the ideal for this height. She was fed by gavage (nasogastric feeding) with 3 L/day of a nutrient solution that provided 4200 kJ/L (1000 kcal/L) and had the following composition:

Component	Amount/L
Hydrolysate of proteins	0.37 mol N
Fat	42 g
Carbohydrate	125 g
Na$^+$	21.7 mmol
K$^+$	32.0 mmol
Cl$^-$	28.0 mmol
Mg^{++}	8.3 mmol
Phosphate	16.1 mmol

Continued.

Case 2　　　　Protein-calorie undernutrition in anorexia nervosa—cont'd

Ca^{++}	15.0 mmol
Zn^{++}	157 μmol
Mn^{++}	45 μmol
Cu^{++}	15.7 μmol
I$^-$	0.6 μmol
Fe^{++}	161 μmol
Complete vitamin mixture	

No complications resulted from the treatment, and on the basis of the positive nitrogen balance and the positive balances of several of the other nutrient minerals (phosphate, K$^+$, Na$^+$, Cl$^-$, Ca^{++}, and Mg^{++}), it was concluded that the nutrients were being used for the synthesis of normal body mass. After 21 days the patient had increased 9.7 kg in body weight.

Biochemical questions　　1. What is the protein equivalent of the nutrient fluid? What could be concluded concerning its biologic value?
2. What differences might be noted in the nature of the nutrients for gavage feeding as compared to total parenteral feeding (hyperalimentation)?
3. In the light of recent knowledge, what additional components might be added to the diet to ensure optimum nutrition?
4. If the patient is sedentary during the 3 wk of gavage feeding, will this diet provide any excess of dietary food energy?
5. Assuming that there is synthesis of normal body mass, what weight gain might be expected after 21 days?

Case discussion　　1. **Biologic value** On the average the nitrogen content of dietary proteins is 16% by weight; thus

$$\frac{100}{16} \times \text{Nitrogen content (g)} = \text{Protein (g)}$$

In other words, 6.25 × N (g) = Protein (g). We are given the nitrogen intake in moles, not grams. Remember that mol = g/MW. To convert the nitrogen intake into grams, we rewrite the equation as g = mol × MW. Therefore N (g) = 0.37 × 14 = 5.18 g. We now know that 1 L of the solution contains about 5.2 g of nitrogen. This value can be converted into protein by substituting into the second equation, 6.25 × 5.2 g N = 32 g protein. Thus, in 1 L of the nutrient solution, there is the equivalent of 32 g of protein.

　　Note that the patient is in positive nitrogen balance while receiving gavage feeding. Thus nitrogen in the form of amino acids is being biosynthesized into protein. More of the nitrogen fed is retained by the body than is catabolized and excreted as urea. Since all amino acids must be available to cells at the same time for protein synthesis to occur, the protein equivalent of the nutrient given to this patient must have a high biologic value.

　　2. **Gavage feeding** The gavage feeding can contain foods that require digestion in the gastrointestinal tract, such as proteins and fat. On the other hand, the constituents in intravenous feeding must be in a condition similar to that after digestion and absorption, since the material is being delivered directly into the blood. The protein nitrogen is therefore provided in intravenous feeding solutions by a balanced mixture of

amino acids, carbohydrate as glucose, and lipid in the form of a triglyceride emulsion. In the studies from which the present case was taken, the patient was fed for 3 to 4 weeks by hyperalimentation but received no lipid or iron by this route.

3. Trace elements The nutrient solution might be deficient in some trace metals if it were intended that the feeding be continued for a prolonged period. These types of deficiencies are being reported with increasing frequency in patients who have been maintained by total parenteral nutrition for an extended time. The minimum daily requirements for other trace elements suspected of being essential, such as chromium, nickel, and vanadium, probably will become apparent as long-term parenteral nutrition is employed more in clinical treatment regimens.

4. Energy need for increase in body weight

$$
\begin{aligned}
\text{BMR} &= 32.9 \times 100 \text{ kJ/day} \\
&= 3{,}290 \text{ kJ} \\
\text{Activity} &= 30\% \text{ BMR} \\
&= 987 \text{ kJ} \\
\text{Metabolic energy requirement} &= (3290 + 987) \text{ kJ/day} \\
&= 4277 \text{ kJ/day} \\
&\text{or } 4300 \text{ kJ/day}
\end{aligned}
$$

The patient is receiving a total of 12,600 kJ/day in the 3 L of nutrient solution. Since her metabolic requirement is 4300 kJ/day, she is taking in an excess of 8300 kJ/day. This excess stored within the body in the form of muscle mass and adipose and other tissue, produces weight gain.

5. Weight gain Assuming that the additional tissue contains an equal amount of lipid and protein and recalling that adipose tissue is 15% water and muscle is 80% water, then the energy content of 1 kg of muscle tissue (which contains 200 g protein) is 3400 kJ. On this basis 100 g protein is equivalent to 0.5 kg muscle and 1700 kJ. Similarly, 1 kg adipose tissue contains 850 g lipid and 31,500 kJ, or 100 g lipid is equivalent to 0.118 kg adipose tissue and 3700 kJ. Thus, since the average normal body mass contains an equal proportion of protein and lipid, then 1700 + 3700 kJ of energy corresponds to 0.5 + 0.118 kg of tissue, or 5400 kJ corresponds to 0.618 kg.

It has been estimated that twice as much energy is required to add body tissues than corresponds to the energy content released when the tissue is catabolized. Thus 0.618 of normal body mass requires 10,800 kJ for synthesis. The excess dietary energy is 8300 kJ/day, which is equivalent to:

$$
\frac{8300}{10{,}800} \times 0.618 \text{ kg lean body mass per day} = 0.48 \text{ kg/day}
$$

In 21 days the increase in body weight would be 9.9 kg. This compares well with the actual weight gain of 9.7 kg, considering the assumptions involved in making these calculations. In particular, the values used for the average BMR for such a patient, the energy demand for tissue acquisition, and the body composition at the end of the treatment period can vary considerably and are only approximations for any given individual.

References

Heymsfield SB et al: Enteral hyperalimentation: an alternative to central venous hyperalimentation, Ann Intern Med 90:63, 1979.

Hopkins BS et al: Protein-calorie management in the hospitalized patient. In Schneider H, Anderson C, and Coursin D, editors: Nutritional support of medical practice, ed 2, New York, 1983, Harper & Row, Publishers, Inc.

Rudman D et al: Elemental balances during intravenous hyperalimentation of underweight adult subjects, J Clin Invest 55:94, 1975.

Case 3   Ulcerative colitis

A 52-year old man, 183 cm in height, had inflammation and ulceration of the colon. Careful examination showed a diffuse mucosal involvement, which was worse in the rectum. He had increasing episodes of diarrhea over the previous several months and was now having an average of 10 small watery stools a day, often mixed with blood. The cramping colonic pain was relieved by bowel activity. Before onset of the disease he had weighed 65 kg, but his weight had fallen to 56 kg during the last year and his appetite was poor (anorexia). Remission was induced by administration of glucocorticosteroids.

Biochemical questions

1. What type of diet would be recommended for this patient?
2. With the possible associated problems in the intestinal digestion and absorption of nutrients, is any vitamin deficiency possible, and, if so, which would be most acute?
3. How would the absence of some digestive enzymes result in changed composition of the stools?
◆ 4. If a higher than normal concentration of low molecular weight solutes were present in the stools, how might this contribute to the diarrhea?

Case discussion

Ulcerative colitis is a chronic disease characterized by ulceration of the colon and rectum. It frequently causes anemia, hypoproteinemia, and electrolyte imbalance.

1. Recommended diet This patient would be fed a hypercaloric diet to regain his former weight (65 kg). One should note that 63 MJ/kg are required for weight gain; that is, twice the number of calories are needed compared with the amount for 1 kg of weight loss. Thus one should calculate the diet the patient needs to maintain a weight of 65 kg. For the BMR this is 65 × 100, or 6500 kJ. One may assume that the patient is sedentary; therefore, 30% of the BMR for activity, or an additional 1950 kJ, makes a total of 8450 kJ. Although no fixed rule exists, one might add an additional 2000 kJ/day to induce weight gain, and rounding off, arriving at 10,500 kJ/day for the patient.

Several other factors must be considered. The patient's appetite is poor, and thus he may have difficulty in eating this increased amount. In addition, one would want to avoid roughage in foods, which may predispose to more diarrhea. Finally, because accumulation of tissues is desired, giving the patient a slightly higher than usual protein diet might be the best course, for example, 20% rather than the usual 15% of the food energy. One must ensure that the protein is of high biologic quality so that it is rich in essential amino acids. Also, because of the malabsorption of some nutrients, such as fat, it is often recommended that the diet be low in fat. This requires periodic examination for possible essential fatty acid deficiency and fat-soluble vitamin signs.

2. Digestion and absorption of food Deficiencies of many different vitamins may occur. One would suspect that the water-soluble vitamins "might be more acutely deficient" than the fat-soluble vitamins. This is based on the fat-soluble vitamins being stored in much larger quantities within the body; the body can call on these reserves when the intake is low. Deficiencies in fat-soluble vitamins take a relatively long time to become manifest. By contrast, little reserve of the water-soluble vitamins such as the B group and ascorbic acid exists, and these must be constantly absorbed into the body to prevent a deficiency. Therefore one would expect that a deficiency in the B vitamins, folic acid, or ascorbic acid would appear long before a deficiency in the fat-soluble vitamins.

3. Digestive enzymes Since this is a disease of the colon and rectum, no problem should occur with the enzymes in the pancreatic secretion. The problem might develop in the enzymes contained in the intestinal mucosa, such as the dipeptidases and disaccharidases. This would lead to reduced absorption of disaccharides and the breakdown products of proteins. In addition to producing malnutrition and weight loss, one would probably find an elevated amount of disaccharides and dipeptides in the stools. Also, because of the inflammation of the intestinal wall, the absorption of other products might be compromised, even though no specific enzymatic defect existed. This could lead to the accumulation of substances such as fatty acids in the stools.

4. Diarrhea The high concentration of low molecular weight materials, such as disaccharides and dipeptides in the colon, would draw in water through an osmotic effect. This is one of the main causes of the watery diarrhea.

References

Bury KD et al: Nutritional management of colitis with neuration, Can J Surg 15:108, 1972.
Cello JP: Ulcerative colitis. In Sleisinger MH and Fordtran JS, editors: Gastrointestinal disease, ed 4, Philadelphia, 1988, WB Saunders Co.
Vogel CM et al: Intraveneous hyperalimentation in the treatment of inflammatory disease of the bowel, Arch Surg 108:460, 1974.

Case 4	Ascorbic acid deficiency

A housewife and factory worker, 37 yr of age, complained for 3 days of progressive pain and discoloration of both legs. She claimed to eat normally, including fresh fruits and vegetables, but on closer examination it appeared that she had been depressed for several months and in fact had eaten very little. She was unable to walk or stand unaided because of the pain. There were extensive areas of hemorrhage over the backs of the calves. A few follicular keratoses were present below the knees. No other hemorrhagic areas were found on the body, nor were there any changes in the gums. Laboratory analysis revealed 12.5 μmol of ascorbic acid/100 mg of leukocytes. Treatment with ascorbic acid (3.97 mmol/day) was started immediately. After 2 wk the level of urinary excretion of ascorbic acid was 2.27 mmol/24 hr.

Biochemical questions

1. What are the biochemical functions of ascorbic acid?
2. What structural similarities are there between ascorbic acid and monosaccharides?
♦ 3. Since the catabolism of ascorbic acid varies greatly among species, yet recommendations for humans are often extrapolated from other animals, how are excessive amounts of dietary ascorbic acid dealt with in humans compared to animals?
4. How does ascorbic acid function to prevent hemorrhage into various tissues?
♦ 5. What is the argument for high daily intake (grams per day) of ascorbic acid?

References

Baker EM et al: Metabolism of ^{14}C- and ^{3}H-labeled L-ascorbic acid in human scurvy, Am J Clin Nutr 24:444, 1971.

Levine M: New concepts in the biology and biochemistry of ascorbic acid, N Engl J Med
    314:892, 1986.
Recommended dietary intakes, Am J Clin Nutr 45: 693,1987.
Walker, A: Chronic scurvy, Br J Dermatol 80:625, 1968.

Case 5	Obesity*

A young typist, O.B., has been plagued by obesity for a long time. She was heavy
throughout childhood but gained more rapidly during her high school and college
years. She had a family history of both obesity and diabetes. Her father and several
aunts and uncles on her mother's side were diabetic. Her maternal grandmother had
had a coronary occlusion. She herself had had glycosuria at one time.

At the age of 22 she weighed 76 kg (168 lb) and was 175 cm (5 ft 9 in) tall. Her
weight had been quite variable, ranging from a low of 68 kg (150 lb) to a high of 95
kg (210 lb). On her first clinical visit at age 25 she weighed 80 kg (176 lb) and
wished to reduce to 59 kg (130 lb), a good weight for her height. Her activity could
be described as sedentary to moderate. Her dietary habits were poor in that she usually
skipped breakfast, ate a light lunch, and then had a large evening meal followed by
numerous snacks. Although prepared to diet, she did not wish to reduce her protein
intake below the present levels for reasons of palatability. Her dietary history follows:

Meal	Weight (g)			Food energy in kJ (kcal)
	Protein	Carbohydrate	Fat	
Breakfast None				
Lunch	16.6	52.5	19.8	1971 (458)
Supper	46.4	119.2	72.8	5524 (1320)
Snacks in evening	26.7	156.5	46.7	4742 (1133)

Total energy: approximately 12.14 MJ (2900 kcal)

	Grams	% (w/w)	Approximate percentage of food energy
Total protein	89.7	16	12
Total carbohydrate	328.2	59	45
Total fat	139.3	25	43
	557.2		

Biochemical questions	1. Does the history of the patient provide any clues as to the cause of her obesity?

2. What is the main chemical substance that accumulates in the body in obesity, and
   what interconversions of nutrients are responsible for this accumulation?
3. How should the present diet be changed in terms of caloric intake?
4. What rate of weight loss should be recommended, and how should the diet be modified
   periodically to achieve this?
5. What proportions of protein, carbohydrate, and fat should be incorporated into the
   new diet?
6. What kind of protein should be used in the new diet?
7. Should there be a change in physical activity? Explain.

References

Bortz WM: Predictability of weight loss. JAMA 204:101, 1968.
Gordon ES: Metabolic aspects of obesity, Adv Metab Disord 4:229, 1970.

*Case courtesy Dr. R.E. Hodges, Department of Family Medicine, University of California, Irvine.

Hafen BQ, editor: Overweight and obesity, Provo, Utah, 1975, Brigham Young University Press.

Mann GV: The influence of obesity on health, N Engl J Med 291:178, 1974.

Case 6   **Vitamin D toxicity**

J.B., a 71-year old man, had a 3 wk history of weakness, polyuria, intense thirst, difficulty in speaking and understanding commands, staggering walk, confusion, and a 13.5 kg (30 lb) weight loss. For 1 mo he had been taking 200,000 units of vitamin D/day for the self-treatment of severe osteoarthritis. His blood was low in potassium and more alkaline than normal. Serum calcium was 3.37 mmol/L (13.5 mg/dl), and serum phosphorus was 1.13 mmol/L (3.5 mg/dl). Vitamin D intoxication was diagnosed.

Biochemical questions
1. Why did J.B. develop vitamin D intoxication while similar excesses of vitamin C have no apparent toxicity?
2. The metabolism of which minerals might be most affected by vitamin D intoxication?
3. Should any vitamin D therapy be needed with average exposure to the sun?
4. What tissue would be most affected by vitamin D toxicity?
5. Why are vitamins A, D, E, and K found in foods rich in fat rather than fruits and vegetables containing little fat?
◆ 6. How does this condition compare with vitamin A toxicity?

References

DeLuca HF and Schnoes HK: Vitamin D: recent advances, Annu Rev Biochem 52:411, 1983.

Frame B et al: Hypercalcemia and skeletal effects in chronic hypervitaminosis A, Ann Intern Med 80:44, 1974.

Wieland RG et al: Hypervitaminosis A with hypercalcemia, Lancet 1:698, 1971.

Case 7   **Vitamin B_{12} deficiency**

A male infant, 6 mo of age, was admitted to the hospital in a coma. He had been born at term to a woman who had knowingly eaten no animal products for 8 yr and taken no vitamin supplements. The child was breast fed and appeared at first to develop normally, but at 4 mo his development regressed. He lost head control, was lethargic, and became increasingly irritable. Physical examination revealed a pale, flaccid infant who was completely nonresponsive, even to painful stimuli. The electroencephalogram was extremely abnormal. His length (65 cm) was in the fiftieth percentile, and his weight (5.6 kg) was below the thirtieth percentile. Blood chemistry showed hemoglobin (54 g/L) and vitamin B_{12} (10 pg/ml; normal 150 to 1000 pg/ml) below normal, but folate (10 ng/ml; normal 3 to 15 ng/ml), glucose, and electrolyte concentrations within the normal range. A urine sample contained severe increases in methylmalonate, glycine, methylcitrate, and homocystine, substances involved in methionine metabolism.

The mother's milk was low in vitamin B_{12} (75 pg/ml; normal 1000 to 3000 pg/ml), and her urine contained methylmalonate.

Vitamin B_{12} was administered to the infant intramuscularly, 1 mg/day for 4 days. After 4 days he was alert, smiling, and responding to stimuli, and his electroencephalogram was normal.

Biochemical questions
1. What are the normal sources of vitamin B_{12} in the diet?
2. Why is folate customarily deleted from vitamin preparations?
3. How does the absorbance from the diet of vitamin B_{12} differ from that given intramuscularly?

◆4. Why are the metabolites of methionine found in the urine in a vitamin B_{12} deficiency?

References

Scriver CR et al, editors: The metabolic basis of inherited disease, ed 6, New York, 1989, McGraw-Hill Information Services Co.

Vitamin B_{12} in the breast infant of a strict vegetarian, Nutr Rev 37:142, 1979.

Additional questions and problems

1. Why do various food proteins have different nutritional value? Do they have the same biologic energy content?
2. Why are carbohydrates said to have a "protein sparing" effect in the diet?
3. Compare the relative merits of total nutrition when given by intravenous or gavage feeding.
4. Food faddism: "Eggs, bacon, and grapefruit—all you can eat—but nothing else and you will lose weight." What are the possible biochemical consequences of this diet?
5. Bile: Discuss the effects of obstruction of the bile duct on the digestion and absorption of food.
6. Low-fat diets: What are the nutritional consequences of a low-fat diet?
7. Pancreatic secretions: Impairment of pancreatic function that reduces the flow of pancreatic secretions lowers the absorption of some nutrients from the diet. Discuss the nutritional consequences of reduced pancreatic secretion.
8. Why do fats yield more metabolic energy than either carbohydrates or proteins?
9. Discuss the food groups in terms of a balanced diet.
10. Modify a diet to replace 40 g of fat with 10 g of protein and enough carbohydrate to provide an isocaloric substitution.

Multiple choice problems

1. What would be the consequences of a carbohydrate-free diet?
   1. Lack of fiber would affect the stools.
   2. Protein would be metabolized for energy.
   3. Urea formation would increase.
   4. Fatty acid metabolism would be incomplete.
      A. 1, 2, and 3 only are correct
      B. 1 and 3 only are correct
      C. 2 and 4 only are correct
      D. 4 only is correct
      E. All are correct
2. In the absence of salivary α-amylase, the carbohydrate in the stools after consuming an average diet would be higher than normal in
   A. Starch
   B. Lactose
   C. Sucrose
   D. Fructose
   E. None of the above
3. A. 16 kJ (4 kcal)/g
   B. 37 kJ (9 kcal)/g
   C. 100 kJ/day (24 kcal/day)
   D. 30% BMR
   For each numbered item, select the lettered item most closely associated (a lettered item may be used more than once).
   1. Sedentary activity
   2. Metabolic energy that can be obtained from fat
   3. BMR of average person
   4. Metabolic energy for protein

4. An obese person wishes to lose 1.4 kg/wk. The reduction in dietary intake should correspond to:
   A. 3500 kJ/day
   B. 6300 kJ/day
   C. 3150 kJ/day
   D. 6300 kcal/day
   E. 2150 kcal/day

5. Which of the following must be provided in the diet?
   1. Tyrosine
   2. Linoleic acid
   3. Aspartic acid
   4. Ascorbic acid
      A. 1, 2, and 3 only are correct
      B. 1 and 3 only are correct
      C. 2 and 4 only are correct
      D. 4 only is correct
      E. All are correct

A worker reported to a physician with many of the symptoms of pellagra: swollen tongue, dermatitis, and nervous disturbances. The man's diet consisted principally of sweet corn with a small amount of other sources of protein. The identical twin of the man had no similar complaints, and although the twin's diet was high in sweet corn, it was mixed with significant amounts of beans. Questions 6 to 10 refer to this case:

6. Pellagra is caused by a deficiency in
   A. Pyridoxal phosphate
   B. Ascorbic acid
   C. Niacin
   D. Vitamin B_{12}
   E. Riboflavin

7. Sweet corn protein is deficient in
   1. Tryptophan
   2. Glutamic acid
   3. Lysine
   4. Arginine
      A. 1, 2, and 3 only are correct
      B. 1 and 3 only are correct
      C. 2 and 4 only are correct
      D. 4 only is correct
      E. All are correct

8. Match the numbers with the following letters as appropriate:
   A. Sweet corn protein
   B. Bean proteins
   C. Both
   D. Neither
      1. High biologic value
      2. Deficient in tryptophan
      3. Calorie value
      4. Essential fatty acids

9. A vitamin is defined as a compound that is
1. An essential component of the body
2. Only synthesized in plants
3. Not synthesized in adequate amounts in the body
4. Only synthesized in animals
   A. 1, 2, and 3 only are correct
   B. 1 and 3 only are correct
   C. 2 and 4 only are correct
   D. 4 only is correct
   E. All are correct

10. The identical twin, who had no complaints
1. Had sufficient tryptophan to biosynthesize niacin
2. Derived enough niacin from beans
3. Did not cook his food in water
4. Added vitamin C to his diet
   A. 1, 2, and 3 only are correct
   B. 1 and 3 only are correct
   C. 2 and 4 only are correct
   D. 4 only is correct
   E. All are correct

Chapter 2

Protein structure

Objectives

1 To understand the structural elements of protein conformation
2 To analyze the interactions of proteins with large and small molecules
3 To relate the structural properties of particular proteins such as collagen, immunoglobulins, and albumin to their biologic functions in health and disease

An understanding of the structural aspects of proteins is essential to an analysis of the biochemical components of living processes and to an appreciation of the biologic events that take place in the body at a molecular level. Few biochemical reactions occur without catalysis, and almost all biocatalysts, called enzymes, are proteins; one exception to date is a specific ribonucleic acid (RNA). These catalytic events occur by the stereochemical fitting of the reactants to the highly ordered, three-dimensional structures of the enzyme molecules. Many of the nonenzymatic proteins also exhibit remarkable specificity in their interaction with other molecules. For example, antibodies are proteins that react with the antigens that originally stimulated their biosynthesis; in the visual cycle the protein opsin reacts specifically with retinal.

Life, with the cell as its unit, is continued by biochemical processes that occur within the confines of a cell membrane. Selective absorption and transport of all materials must occur across this lipid-proteinaceous envelope, which probably evolved to take part in such highly differentiated cellular functions as muscular contraction, electric impulse transmission, and intestinal absorption. In all such cases a relationship exists between the structure of the protein and its function, details of which are slowing coming to light.

General properties of amino acids and proteins
Amino acids

The α-carbon atoms of the amino acids, except glycine, are each linked to four different chemical groups, which is the characteristic of an asymmetric carbon atom and a chiral center. Considering the general formula for the α-amino acids (see p. 44) and its relation in space to the tetrahedrally arranged valencies of the asymmetric carbon atom, isomers of the molecule can be represented by no more than two three-dimensional models (Figure 2.1). If the R group is identical in each model and does not itself contain other asymmetric centers, the two models are mirror images of each other, and each isomer is optically active. The two isomers differ in the direction in which they rotate the plane of polarized light.

Such pairs of isomers are called enantiomorphs. A compound that rotates the plane of polarized light in a clockwise direction is said to be dextrorotatory ($+$), as opposed to a compound that is levorotatory ($-$). Unfortunately, this property does not relate in any simple way to the spatial arrangements of groups around the chiral center. Therefore one cannot easily predict the optical rotation from the structural configuration, and, conversely, the configuration cannot be predicted from the optic rotation.

The Fischer convention The representation of stereochemical relationships of organic molecules with several chiral centers can become complicated. Emil Fischer first proposed a convention whereby these molecules can be represented in two dimensions. His directions are brief and simple. Literally translated, they are as follows: "One has to construct a model of the molecule (using balls for atoms and springs for the valency bonds) and put it into the plane of the paper in such a way that the carbon atoms are situated in a straight line with the groups of interest lying above the plane of the paper, then the formulae are obtained by projection."* When this convention is used, the straight carbon chain can be at any direction in the paper. The stereomodels in Figure 2.1 can be represented in many ways, one of which follows:

$$
\begin{array}{ccc}
& \text{COOH} & \\
& | & \\
\text{H}_2\text{N} - & \text{C} & - \text{H} \\
& | & \\
& \text{R} &
\end{array}
\qquad\qquad
\begin{array}{ccc}
& \text{COOH} & \\
& | & \\
\text{H} - & \text{C} & - \text{NH}_2 \\
& | & \\
& \text{R} &
\end{array}
$$

α-L-Amino acid   α-D-Amino acid

The "groups of interest" in the α-amino acids are the hydrogen atom and amino group attached to the α-carbon. Because of the way in which the carbon atoms are brought into the plane of the paper with the substituents on the asymmetric carbons above the plane, *rearranging the projected formula by taking it out of the plane of the paper is not permitted*. It cannot be flipped or partially turned out of the plane of the paper but must be rotated in the original plane. Additions have been made to the original convention of Fischer: one common convention implies that the vertical valency lines in the projected formula must refer to groups that lie behind the plane in the stereomodel and that the horizontal lines are restricted to those valencies that lie above the plane. This restriction was not used by Fischer but is now in fairly common usage. It is not appropriate to pursue the implications of this added convention here; it is mentioned only to emphasize the existence of more than one set of criteria. The projection formulas used in this text are true for any set of conventions used to date. In another common representation the chiral carbon atoms in the chain may be indicated by the intersection of two lines, and an equation may be written as shown on p. 45.

Figure 2.1   General models of α-amino acid enantiomorphs.

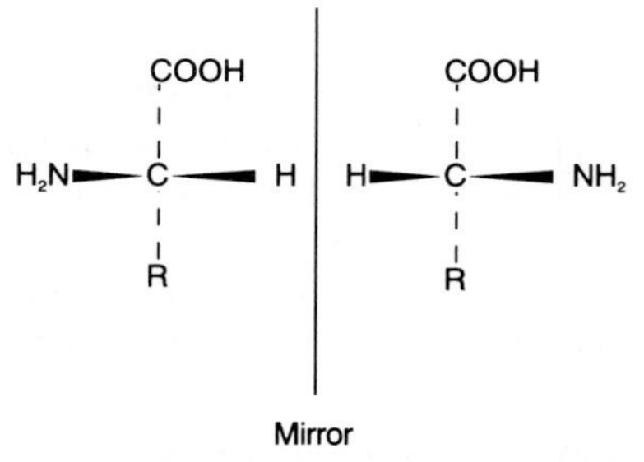

Mirror

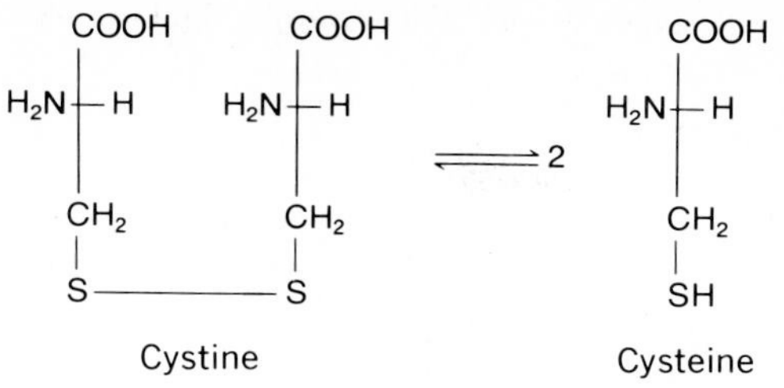

As noted, two optical isomers exist for those α-amino acids with one chiral center. The most commonly occurring amino acids in nature as represented by the Fischer formula are shown with the identification as L-; the mirror image is the enantiomorph designated as D-.

As with the amino acids, simple sugars are chiral molecules, and the Fischer convention has also been applied to these. The absolute configuration of D- and L-glyceraldehyde was established by x-ray crystallography, and the crystallographic findings showed that the dextrorotatory glyceraldehyde in fact had the D configuration. It is not possible to use these data to predict the L or D configuration of other molecules.

For discussion of the R, S designation of chiral centers, the reader is referred to standard texts of organic chemistry.

Classification of amino acids When simple proteins are hydrolyzed to their constituent building units, 19 α-L-amino acids and one L-imino acid are usually produced. The different structures are summarized in Table 2.1, where the R group is identified together with the conventional abbreviations of the amino acid and data on the ionization of the functional groups. All 20 of these residues, including L-asparagine and L-glutamine, are identified by the genetic code for sequential inclusion into the polypeptide chain during protein biosynthesis. Thus, when one speaks of there being 20 amino acids in protein, it is the genetically coded amino acids that are usually meant. During chemical hydrolysis of the proteins, the L-asparagine and L-glutamine residues are converted to L-aspartic acid and L-glutamic acid, respectively.

Amino acid derivatives Other amino acids in protein result from a modification of one or more of the basic 20 amino acid residues after biosynthesis of the polypeptide chain. Cystine is the result of the oxidation of two residues of L-cysteine, an —S—S— linkage being formed from the two —SH groups shown above; a similar reaction occurs between two properly spaced cysteinyl residues in the polypeptide chain. The hydroxylation of L-lysyl or L-prolyl residues in the proteins collagen and elastin results in the introduction of hydroxyl groups in the corresponding R side-chains.

A carboxylation of glutamate to give γ-carboxyglutamate is critical in the regulation of Ca^{++} metabolism in blood clotting (see Chapter 4). Perhaps the most common and reversible modification of proteins is that of enzymes involving the phosphorylation of serine, threonine, and tyrosine residues (Table 2.1). In many cases the resulting phosphoprotein has a different enzymatic property from the nonphosphorylated protein. By this reversible modification, an increasing number of enzymatic reactions are metabolically regulated, such as the biosynthesis and breakdown of glycogen (see Chapter 7).

Reactions leading to all these modifications of amino acid residues are catalyzed by enzymes. There are diseases, however, that lead to elevations in the blood and in some

*Fischer E: Über die konfiguration des traubenzuckers und seiner isomeren, Ber Dtsch Chem Ges 24:2683, 1891.

Table 2.1   Classification of amino acids found in protein

$$H_3\overset{+}{N}-\overset{\displaystyle COO^-}{\underset{\displaystyle R}{\overset{|}{\underset{|}{C}}}}-H$$

Name	Abbreviation Three letters	Abbreviation One letter	R—	pKa* Values —COO⁻	—NH₃⁺	Other
Aliphatic monoamino monocarboxylic acids						
Glycine	Gly	G	H—	2.34	9.60	
Alanine	Ala	A	CH_3—	2.35	9.69	
Valine	Val	V	CH_3-CH- with CH_3	2.32	9.62	
Leucine	Leu	L	$CH_3-CH-CH_2-$ with CH_3	2.36	9.60	
Isoleucine	Ile	I	CH_3-CH_2-CH- with CH_3	2.36	9.68	
Serine	Ser	S	$HO-CH_2-$	2.21	9.15	
Threonine	Thr	T	$CH_3-\overset{H}{\underset{OH}{C}}-$	2.09	9.10	
Aromatic amino acids						
Phenylalanine	Phe	F	⬡—CH_2—	1.83	9.13	
Tyrosine	Tyr	Y	HO—⬡—CH_2—	2.20	9.11	10.07
Tryptophan	Trp	W	indole—CH_2—	2.38	9.38	
Acidic amino acids and their amides						
Aspartic acid	Asp	D	$^-OOC-CH_2-$	2.01	9.93	3.80
Asparagine	Asn	N	$H_2N-\overset{O}{\overset{\|}{C}}-CH_2-$	2.02	8.80	
Glutamic acid	Glu	E	$^-OOC-CH_2-CH_2-$	2.13	9.76	4.31

*See p. 61 for definition.

Table 2.1 Classification of amino acids found in protein—cont'd

$$H_3^+N-\underset{\underset{R}{|}}{\overset{\overset{COO^-}{|}}{C}}-H$$

Name	Abbreviation Three letters	Abbreviation One letter	R—	pK$_a$* Values —COO$^-$	—NH$_3^+$	Other
Glutamine	Gln	Q	$H_2N-\overset{\overset{O}{\|\|}}{C}-CH_2-CH_2$ ——	2.17	9.13	
Basic amino acids						
Lysine	Lys	K	$CH_2-CH_2-CH_2-CH_2$— with NH_3^+	2.18	8.95	10.53
Arginine	Arg	R	$H_3^+N-\overset{\overset{}{}}{\underset{\underset{NH}{\|\|}}{C}}-NH-CH_2-CH_2-CH_2$—	2.17	9.04	12.48
Histidine	His	H	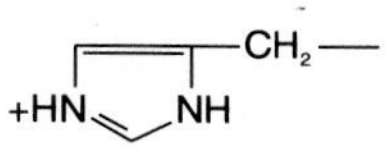 imidazole ring $-CH_2$— $+HN$ NH	1.82	9.17	6.0
Sulfur-containing amino acids						
Cysteine	Cys	C	$HS-CH_2$—	1.91	10.36	8.24
Methionine	Met	M	$CH_3-S-CH_2-CH_2$—	2.28	9.21	
Imino acid						
Proline†	Pro	P	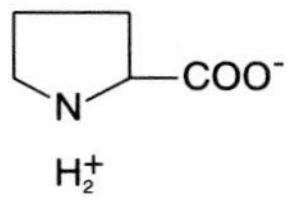 pyrrolidine ring $-COO^-$ N H_2^+	1.95	10.64	
Other amino acids‡						
Hydroxylysine			$CH_2-\overset{\overset{OH}{\|}}{CH}-CH_2-CH_2$— with NH_3^+	2.13	8.62	9.67
Cystine			CH_2— S S CH_2—	1.04 / 2.1	8.02 / 8.71	
Hydroxyproline†			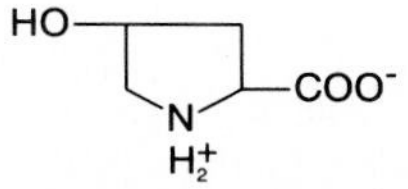 HO pyrrolidine ring $-COO^-$ N H_2^+	1.92	9.73	

†Complete amino acid structure shown; not R group.
‡Amino acids that are formed after the peptide chain has been synthesized.

Continued.

Table 2.1   Classification of amino acids found in protein—cont'd

$$H_3^+N-\underset{\underset{R}{|}}{\overset{\overset{COO^-}{|}}{C}}-H$$

Name	Abbreviation		R—	pKa* Values		
	Three letters	One letter		—COO⁻	—NH₃⁺	Other
γ-Carboxyglutamate						
O-Phosphoserine						
O-Phosphothreonine						
O-Phosphotyrosine						

tissues of small metabolites that react with proteins nonenzymatically. Diabetes mellitus, for example, produces elevations of blood glucose that may be controlled to varying extents with insulin therapy. Over a time, however, the excess glucose reacts more than normally with proteins to produce glycosylated derivatives that may contribute to the long-range pathology of the disease.

The elevation of urea in uremic patients results in identification of the carbamoylation of hemoglobin at amino terminal residues.

The arrangement of amino acids in proteins is unique for each protein. This may be seen in the amino acid composition and, more precisely, in the amino acid sequence. Table 2.2 gives the amino acid composition of the proteins considered in some detail in this chapter. Note the similarity in composition of the subunits that make up hemoglobin

Table 2.2   Amino acid composition of some proteins (residues per 100 residues)

Amino acids	Human hemoglobin		Human immuno-globulin G		Calf skin tropo-collagen	Pig tropo-elastin	Bovine myosin	Human serum albumin
	α	β	L	H				
Essential								
Arg	2.1	2.0	2.3	2.5	5.0	0.7	5.7	4.5
His	7.1	6.2	0.9	2.0	0.5	0	1.4	2.9
Ile	0	0	2.8	2.2	1.2	1.9	3.8	1.7
Leu	12.8	12.3	7.0	6.7	2.7	4.6	10.8	11.6
Lys	7.8	7.5	7.0	6.9	2.5	4.8	9.8	10.8
Met	1.4	0.7	1.4	1.3	0.5	0	2.7	1.1
Phe	5.0	5.5	3.7	3.4	1.3	2.8	3.3	6.1
Thr	6.4	4.8	8.4	7.4	1.5	1.4	4.7	5.4
Trp	0.7	1.4	1.4	1.6	0	0	0.4	0.1
Val	9.2	12.3	7.0	10.1	2.2	12.1	4.5	8.4
Nonessential								
Ala	14.9	10.3	6.1	4.9	10.7	21.8	8.6	—
Asn	2.8	4.1	3.3	3.6	—	—	—	—
Asp	5.7	4.8	4.7	3.4	4.8	0.3	10.4	10.0
Cys	0.7	1.4	2.3	2.5	0	0	1.1	6.7
Gln	0.7	2.0	7.0	4.3	—	—	—	—
Glu	2.8	5.5	4.7	5.1	7.4	1.8	21.3	1.5
Gly	5.0	8.9	6.1	7.4	33.5	33.4	3.9	2.7
Hyl	0	0	0	0	0.7	0	0	0
Hyp	0	0	0	0	9.1	1.1	0	0
Pro	5.0	4.8	4.7	8.3	12.4	10.9	1.8	5.7
Ser	7.8	3.4	14.9	11.9	3.4	0.9	4.3	4.5
Tyr	2.1	2.0	4.2	4.0	0.4	1.6	1.7	3.3
Amide N	—	—	—	—	—	—	—	8.1

Data on hemoglobin and immunoglobin from Dayhoff MO: Atlas of protein sequence and structure, vol 5, Washington DC, 1972, National Biomedical Research Foundation; data on tropocollagen from Balazs EA, editor: Chemistry and molecular biology of the intercellular matrix, New York, 1972, Academic Press, Inc; data on tropoelastin from Stevens FS and Jackson DS: Biochem J 104:534, 1967; data on myosin from Horiaux F, Hamoir G, and Oppenheimer H: Arch Biochem Biophys 120:274, 1967; data on serum albumin from Tristram GR and Smith RH: In Neurath H, editor: The proteins, New York, 1963, Academic Press, Inc.

and human immunoglobulin. Contrast these compositions with those of tropocollagen and tropoelastin, which lack several essential amino acids but contain unusually high amounts of glycine and hydroxyproline.

Analysis of amino acid mixtures A common feature of all amino acids is the presence of basic amino groups and acidic carboxylic groups. The slightly different ionic properties of these groups permit separation of the amino acids. Highly automatic machines, called amino acid analyzers, use ion exchange chromatography to accomplish these separations. The separated amino acids are reacted with ninhydrin to create a purple-blue color.

Ninhydrin   Amino acid   Purple color

The only exceptions are L-proline and L-hydroxyproline, which produce a yellow color. The density of the colors produced can be measured by a spectrophotometer and the corresponding amount of amino acid determined with an accuracy of 2% to 3%.

Analysis of the amino acid composition of proteins or polypeptides is carried out by first hydrolyzing the peptide bonds through which the amino acid residues are linked together; under these conditions the glycosyl and phosphate residues in modified proteins are also hydrolyzed. The amino acid mixture in the hydrolysate is then applied to a column of ion exchange material, which has been previously equilibrated with buffer at a particular pH and temperature. A typical amino acid analysis is shown in Fig. 2.2.

With the development of minicomputers the information in the chromatogram can be calculated automatically, the summary giving the identification and the number of moles of each amino acid in the mixture.

Peptide bond

Nomenclature for peptides A polypeptide is formed as the result of the condensation of amino acids, producing a linear molecule. Usually one end of this string of amino acid residues has a free amino group, and the other end has a free carboxyl group. Polypeptides are named as derivatives of the amino acid with the free carboxyl group; for example, the following would be the tripeptide alanyl-glycyl-serine:

Figure 2.2   Chromatographic analysis of amino acids from a protein hydrolysate.

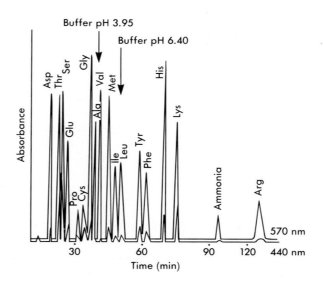

By convention the peptide sequences are written with the NH_3^+-terminus on the left. Each amino acid may be abbreviated by a three-letter symbol, usually the first three letters of its name or by a one-letter symbol (see Table 2.1). The one-letter symbols are often used to abbreviate the sequences of large polypeptides.

The distinction between polypeptide and protein is unclear; the term *protein* usually refers to a biopolymer with a molecular weight greater than several thousand. A polypeptide is of lower molecular weight, and if the number of amino acid residues in the molecule is known, it may be indicated. For example, the hormone glucagon is a non-acosapeptide (29 amino acid residues), and oxytocin and vasopressin are nonapeptides (nine amino acid residues, with cystine being counted as two residues).

The structure of vasopressin, the antidiuretic hormone, is presented next to illustrate several points. This nonapeptide becomes cyclized by the formation of an —S—S— bond between the two cysteine residues. The L-aspartyl and L-glutamyl residues exist in the cyclic peptide as amides and are therefore L-asparaginyl and L-glutaminyl residues. The COOH-terminus of glycine is present as an amide, abbreviated — Gly (NH_2).

Vasopressin

Peptide conformations The condensation of the amino group of one amino acid with the carboxyl group of another forms a peptide bond with a resonance stabilization that results in a planar arrangement of the six atoms involved (Figure 2.3). The partial double-bond character of the C $=$ N bond places a distinct restriction on the shape of the molecule, free rotation being permitted only around the C—R, C—NH_3^+, and C— COO^- bonds.

Figure 2.3   Peptide bond conformation.

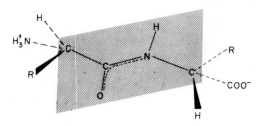

When two or more peptide bonds are joined together, the most flexible movement would be somewhat equivalent to a chain in which each link, representing the peptide $C \doteq N$ bond, can move through a limited angle with respect to the next link. A polypeptide structure with such maximum flexibility is referred to as having a *random coil* conformation, the conformation being the three-dimensional shape of a molecule. Other properties of the polypeptide, however, tend to stabilize the molecule into a more rigid form. Excluding any covalent cross-linkages, such as the —S—S— bonds in a cystinyl residue, these stabilizing elements are hydrogen bonds, ionic interactions, and hydrophobic interactions.

1. *Hydrogen bonds.* Hydrogen bonds may form between —NH— or —OH groups and the $C = O$ groups in the peptide bonds or —COO^- in the R group. For example, two adjacent peptides may form hydrogen bonds, as indicated by the broken lines:

$$
\begin{array}{cc}
\mathrm{C{=}O} - - - \cdot \mathrm{H{-}N} \\
\mathrm{H{-}N} \qquad\quad \mathrm{C{=}O}
\end{array}
$$

2. *Ionic interactions.* Ionic interactions include ionic bonds or repulsions between charged groups such as —NH_3^+, —COO^-, guanidinium groups of L-arginyl residues, —S^- from cysteinyl residues, or phenolic —O^- from tyrosinyl residues. The interactions between oppositely charged groups such as —COO^- and H_3N^+— give rise to salt bridges.
3. *Hydrophobic interactions.* Hydrophobic interactions represent affinities of groups for each other in much the same way that the aliphatic chain of the fatty acid is attracted to the oil droplet in a water-oil emulsion. The hydrophobic groups are driven out of the water. Thus an affinity exists between the R groups of the aliphatic amino acids or the aromatic groups of phenylalanine and tryptophan.

The stabilizing contribution of each ionic interaction to the conformation of a molecule depends greatly on the distance between the groups and could be greater than the stabilization resulting from a single hydrogen bond. In an average protein the number of hydrogen bonds is great, and their total contribution is probably more significant. However, it is likely that the hydrophobic interactions are the most important in determining the conformation. Although the contribution of each interaction may be small, the total effect is a stabilization that is considerable.

It should be noted that these stabilizing factors are noncovalent. Stabilization is also afforded by the covalent disulfide cross-linkages formed between L-cysteinyl residues.

Figure 2.4   Peptide parallel pleated sheet conformation.

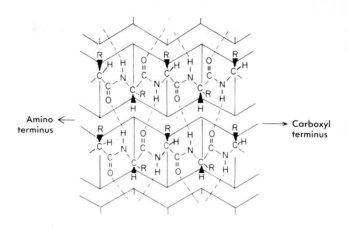

Amino ← terminus   Carboxyl → terminus

Conformational segments of polypeptide chains

The flexible random coil that is formed by a polypeptide can be arranged in either a helix or a pleated sheet to give the maximum number of hydrogen bonds between the peptide linkages. In the absence of any other stabilizing factors, those arrangements would give a more stable conformation than the random coil. A large polypeptide or protein may contain regions of one or more of these conformational segments, identified here as random coil, pleated sheet, or α-helix.

Pleated sheets

1. *Parallel pleated sheets*. Polypeptide chains arranged with their carboxyl terminal ends in the same direction may be represented as shown in Figure 2.4. Each peptide bond is involved in hydrogen bond formation with an adjacent peptide bond from another chain. Such a conformation places the R groups on both sides of the sheet, an important point when considering the layering of one polymolecular sheet on top of another. Hair or wool can be stretched in moist heat to form this type of conformation.
2. *Antiparallel pleated sheets*. A similar conformation is derived if the two adjacent polypeptide chains run in opposite directions. This is less common in human proteins; the best example in nature is silk fibroin.

Alpha helix The polypeptide chain can twist to a limited extent to form a spiral or helix in which the planar elements of the peptide backbone remain but are arranged around a central axis. A right-handed α-helix results when a polypeptide chain is arranged so that the maximum number of intrachain hydrogen bonds is formed between atoms of the peptide groups. Such a structure has an average of 3.6 amino acid residues per turn, with hydrogen bonds between every fourth amino acid. A right-hand α-helix is shown diagrammatically in Figure 2.5; a common wood screw is an example of a right-handed helix. Other helical coils are possible but are less stable.

Up to this point the influence of the R groups on forming these conformational segments has not been considered. These groups have a potential to make their own contribution

Figure 2.5   Peptide α-helix conformation.

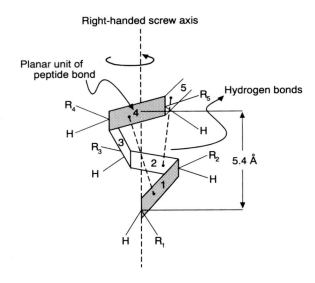

to stabilization and destabilization. This is particularly evident in the effects of different amino acid residues on the α-helix. Those residues that are uncharged permit the formation of stable α-helical units unless the groups are very bulky, as is the leucyl residue. A prolyl residue, having no substituent hydrogen atom on the peptide nitrogen, causes a break in the α-helix and produces a characteristic bend at the junction of the two helical portions of the chain on either side of it. Charged groups of like sign—for example, the carboxylate ions of the aspartyl or glutamyl residues or the —NH$^+_3$ groups of the lysyl and arginyl residues—cause destabilization as a result of charge repulsion. Ionic interactions of oppositely charged groups may modify the helix, as will the hydrogen bonds that may form between R groups or between an R group and the peptide bond. Similarly, all these factors can disrupt the pleated sheet conformations.

Folding of polypeptide chains Now that possible conformational segments and stabilizing factors of a polypeptide chain have been briefly examined, it is useful to consider the possible steps involved in arriving at a more stable structure in aqueous solution. A polypeptide chain might be considered as simply collapsing into a conformation in which hydrophobic interactions dominate. These interactions are numerous and do not depend on forces in any particular direction. Thus the aliphatic R groups would cluster together as much as possible, whereas the hydrophilic residues from glutamate, serine, and the like would be attracted to the solvent water molecules. Such an initial state most probably has many residues in a less than optimum environment for maximum stability. During the constantly occurring natural gyrations of the parts of the polypeptide chain, there are opportunities for a relatively weak hydrophobic interaction to be overcome by the formation of a more stable hydrogen bond or ionic interaction, until eventually the polypeptide chain is either in the most stable conformation possible or in one of several nearly equivalent stable conformations. In such a state every group that can form a hydrogen bond probably has done so in one of the several pairings discussed earlier or with the aqueous solvent molecules. The stable conformation will remain until perturbed by a change in a condition such as temperature, acidity, solvent, or interacting ions that may chelate or form salt bridges. Likewise, a change in the oxidation state may break —S—S— bridges between proximal cysteinyl residues in the stable conformation.

Protein conformation

Each protein or polypeptide has a unique sequence of amino acid residues in its chain, which in large measure determines its biologic and physicochemical properties. Methods of determining the amino acid sequence of proteins are well developed. With this knowledge, it is possible to synthesize chemically the simpler proteins and polypeptides, such as the hormone insulin (51 amino acids). The biologic activity is also related to the way in which these polypeptide chains (polypeptide is used here in an adjectival sense to indicate any biopolymer chain that has a peptide polymer linkage) are arranged in space, that is, the conformation of the molecule.

The sequence of amino acids in a protein chain is referred to as its *primary* structure. The protein molecules assume some three-dimensional shape called the *tertiary* structure, which is stabilized by covalent and noncovalent forces. The component of the tertiary structure that is stabilized by hydrogen bonds is called the *secondary* structure and is recognized in such segments as α-helices or pleated sheets, but the subdivision is somewhat arbitrary. Finally, some proteins are composed of aggregates of simple protein subunits; such a level of organization is referred to as the *quaternary* structure. The preceding levels of organization and structure are sometimes difficult to separate—for example, secondary and tertiary structure—and the student should not be discouraged by the lack of exact definition of these terms.

Some details of important proteins are summarized in Table 2.3. One can see that several proteins containing multiple subunits have quaternary structures. Sometimes all the constituent subunits are the same, but other native proteins are composed of different types of subunits. For example, hemoglobin has two pairs of subunits, designated $\alpha_2\beta_2$, with all the subunits having approximately the same molecular weight (16,100). Insulin, which is composed of two different polypeptide chains covalently linked together through two disulfide bonds, is not considered here to have two subunits. The subunits of the proteins listed in Table 2.3 are not covalently bound together.

The particular three-dimensional shape of a molecule can be determined by various physical methods such as x-ray crystallography. The conformation in the living cell, or that of the isolated protein where it shows maximum biologic activity, is called the *native state*. A protein must be in the native state to best perform its functions, be they structural (e.g., collagen), catalytic (almost all *enzymes* are proteinaceous biocatalysts), or transport (e.g., hemoglobin).

Denaturation Many conformations exist between the native state of a protein molecule and that in which the protein chain is spatially completely random. In the latter conformation the protein is said to be completely denatured. However, complete randomization of the protein chain is not necessary for loss of biologic function to occur. The biologic function has maximum activity in the native state, and any change in structure that leads to the loss of natural function is called denaturation.

Denaturation can be produced by many agents. These include heat and chemicals that destroy hydrogen bonds in the protein, such as urea at a concentration of 6 mol/L, detergents such as sodium dodecyl sulfate, and sulfhydryl reagents such as mercaptoethanol. Denaturation does not involve cleavage of the primary protein structure, so that it is sometimes possible to reverse the denaturation by removing the causative agent.

Properties of proteins in solution

Proteins are polypeptides with an approximate range of molecular weight from 5000 to 1×10^6. Protein molecular weights may be given in daltons or kilodaltons (abbreviated kDa), and the term "molecular weight" is abbreviated as Mr. Some proteins are soluble

Table 2.3   Molecular weights and subunits of selected proteins

Protein	Molecular weight	Sedimentation coefficient (S)	Subunits		
			Number	Molecular weight	S value*
Insulin dimer	11,466	1.95	1†		
Bovine serum albumin	66,300	4.41	1		
Liver alcohol dehydrogenase	82,500	5.39	4	20,000	2.26
Haptoglobin	85,000	4.20	2	40,000	
Hexokinase	102,000		4	27,500	
Fructose 1,6-bisphosphatase	130,000		2	29,000	
			2	37,000	
Lactate dehydrogenase	136,300	7.18	4	36,200	1.55
Aldolase	156,500	7.80	4	42,400	1.35
Ceruloplasmin	151,000	7.50	8	18,000	

*The sedimentation coefficient (S) of the subunits are those for the denatured but not reduced state.
†The two covalently bound peptide chains (molecular weight about 5800) dimerize under the conditions of the sedimentation, and the dimer is considered a single unit for the purposes of calculating the S value.

in water; others require a dilute salt solution for the solvent; and still others, such as the keratin of hair and skin, are insoluble in any aqueous system. Many proteins have been fractionated and purified on the basis of their molecular size and solubility.

Spectroscopy The absorbance of light by proteins is caused principally by the peptide bond, the tyrosyl, tryptophanyl, and phenylalanyl residues, together with whatever spectral properties are contributed by an associated nonprotein groups. Thus the absorbance maxima for human serum albumin are seen in the absorption spectrum to be about 230 nm (peptide) and, in a broad peak, approximately 280 nm because of the summated absorbances of the aromatic residues. A similar protein absorbance may be observed from many simple proteins, whereas in the chromoproteins, such as hemoglobin, the absorbance spectrum is dominated by the absorbance in the visible regions because of the heme group. The absorbance spectrum of an aqueous solution of a protein will vary with pH according to the ionization of the amino acid residues. For example, the ionization of the phenolic hydroxy group of a tyrosyl residue may shift the absorbance maximum from 278 nm to about 290 nm. The change in the oxidation state of the heme group in hemoglobin significantly alters the whole absorption spectrum, as seen in Chapter 4.

Clearly, therefore, the absorbance spectrum of a protein solution is sensitive to several environmental variables. However, under specified conditions the absorbance of a solution at a specified wavelength is directly proportional to the concentration of the protein and is the method frequently used for its assay. Similarly, a mixture of proteins, such as deoxyhemoglobin (Fe II state) and carboxyhemoglobin (Fe II state), can be analyzed from the absorbancies at two wavelenghts, 555 nm and 480 nm, as illustrated in Chapter 4, case 1, on carbon monoxide poisoning. The use of spectroscopy is further illustrated in assays of enzyme activity (see p. 134).

Nuclear magnetic resonance Certain atomic nuclei, such as ^{1}H, ^{13}C, ^{15}N, ^{19}F, and ^{31}P, act as tiny magnets. When placed in a uniform magnetic field, these nuclei align themselves with the field. Irradiation of these oriented nuclei by pulses of radio waves of the proper frequency excites them to reorient. After the pulse, the nuclei then "relax," aligning themselves again with the field of the permanent magnet. In so doing, they transmit radio waves, the wavelength of which yields information on the number of atoms of a particular type that are present, the molecular structure in which the atoms are present, and the chemical environment of the molecules.

When the sample being analyzed is a homogeneous solution of a single molecule, the information can be analyzed as would any other spectrum to reflect structural characteristics of molecules as complex as proteins, by examining the nuclear magnetic resonance (NMR) of the ^{1}H atoms, the ^{13}C or ^{15}N atoms, or in the case of the phosphoproteins, the ^{31}P atoms. With the development of new technology, detailed structures of proteins in solution can be compared with the x-ray data of the crystalline form.

Similar but less detailed spectra can be obtained for molecules in living cells or tissues, such as the key metabolites, sugar phosphates, phosphodiesters, creatine phosphate, adenosine triphosphate (ATP), and inorganic phosphate (P_i) in muscle. Such spectra permit analysis of these types of molecules in vivo, thus reflecting changes in their concentration with time following exercise or drug therapy.

Alternatively the relaxation time for realignment of the nuclei following excitement by radio waves may be analyzed to reflect the chemical environment of the atom, and a map of the internal structure of the organism or tissue can be reconstructed. For example, one can present the different features of the ^{1}H atoms, showing whether these atoms are tightly or loosely bound. These maps are displayed on a computer screen and are superficially similar to an x-ray image of the same tissue. Such maps of the ^{1}H atoms reflect principally the water and lipid in the tissue and so present a different image from x-ray methods.

Clinical comment

Magnetic resonance imaging Using a magnet large enough to surround a human being, with a field strength of 0.5 or greater kilogauss, data can be analyzed from NMR that present unique sectional images. The molecular and chemical differences between one anatomic structure and another frequently differentiate better than x-ray images of the sample body part. For example, magnetic resonance imaging (MRI) of a tumor results in an image with a different intensity from the surrounding tissue; this is caused by the cancer tissue having an increased number of mobile ^{1}H atoms, mostly in water, than normal tissue. It is also possible to image *any* section of the body, such as a sagittal section of the head, some of which cannot be obtained using x-ray techniques.

Besides the ability to obtain more information through MRI than x-ray imaging, this new technology also carries no hazards of ionizing radiation. With the availability of more powerful magnets, MRI will significantly extend the biochemical information on the composition of functional tissues, without the artifacts arising from the isolation of their components.

Gel permeation chromatography Gel permeation chromatography separates fractions according to their molecular weight. A protein mixture in solution is slowly pumped through a long tube packed with synthetic gel particles containing many pores of a known and controlled size. If the molecular weight of a solute protein is such that the molecular size is larger than the pores of the gel, the solute will pass through the column, with the bulk flow of solution relatively unimpeded by the presence of the gel. Solute proteins of a smaller molecular weight and size can penetrate into the gel pores and will spend part of their time within the gel and part outside the gel. The smaller the molecule, the higher is the probability that it will be in the gel. Since the gel is stationary, trapped particles will be relatively retarded or impeded by the presence of the gel, and their bulk flow through the column will be slower. By collecting the fluid emerging from the bottom of the gel column in small fractions, it is possible to separate the solute mixture into individual components. Numerous gels have been prepared and marketed, and this type of chromatography is now an important tool of the biochemist studying proteins.

Since separation by gel permeation chromatography depends on the size and shape of the molecule, and since in similar types of proteins these factors are related to molecular weight, it is possible to estimate rapidly the molecular weight of a protein by the rate at which it moves through a gel column relative to similar proteins of known size.

Ultracentrifugation A more precise method of separating protein molecules according to molecular weight is ultracentrifugation. Under high gravitational forces, molecules sediment at a rate that is proportional to their molecular weight and shape. This principle has been used to measure the molecular weight of many types of molecules, particularly proteins. Ultracentrifuges have been built to operate at force levels as high as 500,000 times the force of gravity. The rotors of these devices are fitted with cuvets through which light of a suitable wavelength may be passed during rotation, so that the light absorbed by a dissolved protein may be measured. At a given level of centrifugal force the dissolved protein molecules will move through the solution until their buoyant force is just equal and opposite to the force developed by the rotational field; thus molecules of different molecular weight can be separated. Although they are extremely valuable, ultracentrifugal techniques are generally slow and require an elaborate and expensive apparatus. The behavior of molecules in a centrifugal field is typically described in terms of their *sedimentation coefficient*.

$$S = \frac{v}{\omega^2 r} \times 10^{13} \text{ Svedberg units}$$

The sedimentation coefficient (S), expressed in Svedberg units ($1S = 10^{-13}$ sec), is equal to the velocity (v) in centimeters per second that the molecule moves at a distance (r) in centimeters from the center of rotation when rotating at ω radians per second. (Recall that there are 2π radians per revolution.) The factor of 10^{13} brings the S values to reasonable numbers (Table 2.3).

Some molecules, which contain bound or associated lipids with densities significantly less than one, will move upward in the centrifugal field. In such cases the rate of movement is expressed in flotation units (S_F).

One can see from Table 2.3 that the molecular weights are not directly proportional to the S values. This is caused by the different overall shapes of each protein in solution, the two extremes being the *globular* proteins, which have an almost spheric shape, and the *fibrous* proteins, which are more rodlike.

Classification of proteins Proteins have been classified in several ways, for example, as globular or fibrous types. One of the earliest classification systems was based on solubility; more recently, classifications have been drawn according to the components in the complex or conjugated protein molecule that are not amino acids.

Although the classification by solubility is not commonly used now, it gave rise to several protein class names, such as albumin and histone, that are still part of protein nomenclature. The solubility properties of the proteins are used on occasion for their preparation and purification. A mixture of proteins, such as might be extracted from a tissue with water or dilute salt solution, can be fractionated by the gradual addition of ammonium sulfate. The globulins will be precipitated first and can be removed by centrifugation or filtration. The albumins are precipitated when the ammonium sulfate is saturating the solution. These salt fractionations, combined with changes in the acidity of the solutions, separate many mixtures of proteins reasonably well. Further purification requires the more delicate chromatographic procedures discussed previously.

The proteins may be divided into two groups, those composed only of amino acids, or *simple* proteins, and those that are *conjugated* proteins. The conjugated proteins have residues other than amino acids that are often linked to the structure by covalent bonds.

They are classified according to the principal nonprotein component, as indicated in Table 2-4. Several of the conjugated proteins are described in greater detail later, as is the more elaborate scheme used for the classification of enzyme proteins.

Ionic properties of amino acids, peptides, and polypeptides

Because of the effect of ionic interactions on the conformation of polypeptides and because biologic systems and their metabolic processes depend critically on the ionization of the acidic and basic groups in the biochemical components, it is important to consider the factors that determine the charges on these molecules. These charges vary according to the H^+ concentration in the solution, which alters the extent to which a carboxyl group is undissociated or carries a negative charge as a $—COO^-$ ion. Similarly, the amino and other basic groups are changed to positively charged ions or neutral groups. Therefore it is necessary to review the properties of weak acids and weak bases and the reversible equilibria involved in such systems.

Reversible reactions The majority of reactions in biochemical systems are reversible. For example:

$$H_2O + CO_2 \rightleftharpoons H_2CO_3$$

Such reversible systems are in fact the summation of two reactions, the forward and the reverse. The hydration of CO_2 yields undissociated carbonic acid, whereas the reverse reaction dissociates carbonic acid into H_2O and CO_2. These reactions are continuous, but at equilibrium the rates are equal and the ratios of the concentrations of H_2O, CO_2, and H_2CO_3 are constant, such that:

$$K_{eq} = \frac{[H_2CO_3]}{[H_2O][CO_2]}$$

where [ ] in each case means "the concentration of . . . ." The previous expression follows from the general reaction:

$$aA + bB + \text{-} \text{-} \text{-} \text{-} \rightleftharpoons xX + yY \text{-} \text{-} \text{-} \text{-}$$

where, at equilibrium:

$$K_{eq} = \frac{[X]^x[Y]^y \text{-} \text{-} \text{-} \text{-}}{[A]^a[B]^b \text{-} \text{-} \text{-} \text{-}}$$

In calculations involving equilibria, it is necessary to have all the concentrations expressed in the same units, which are best converted to moles per liter, since, by convention, tables of equilibrium constants are derived using these units.

Table 2.4   Protein classification by composition

Protein class	Examples	Nonprotein components
Simple proteins	Serum albumin	None
Glycoproteins	Immunoglobulins, mucins, connective tissue proteoglycans	Carbohydrate
Nucleoproteins	Viruses, chromosomes	Nucleic acids
Metalloproteins	Ferritin, carboxypeptidase	Metal ions
Lipoproteins	Chylomicrons	Lipids of various types
Chromoproteins	Hemoglobin, rhodopsin	Colored prosthetic group, for example, heme, riboflavin, and retinal

For example, an important equilibrium reaction in the catabolism of D-glucose involves the isomerization of D-glyceraldehyde 3-phosphate and dihydroxyacetone phosphate. In a solution of 10 ml volume, 1.7 mg of D-glyceraldehyde 3-phosphate is in equilibrium with 42.2 mg of dehydroxyacetone phosphate. This equilibrium may be represented:

$$\text{D-Glyceraldehyde 3-P} \rightleftharpoons \text{Dihydroxyacetone P}$$
$$(1.7 \text{ mg}) \qquad\qquad (42.2 \text{ mg})$$

The molecule weight of each isomer is 169. Therefore the molar concentration of D-glyceraldehyde 3-phosphate is:

$$\frac{1.7 \times 10^{-3}}{169} \times \frac{1000}{10} = 1.0 \times 10^{-3} \text{ mol/L}$$

Similarly, for dihydroxyacetone phosphate:

$$\frac{42.2 \times 10^{-3}}{169} \times \frac{1000}{10} = 25.0 \times 10^{-3} \text{ mol/L}$$

For this equilibrium equation, therefore:

$$K_{eq} = \frac{25.0 \times 10^{-3}}{1.0 \times 10^{-3}} = 25$$

Again by convention, the equilibrium constant is calculated by dividing the product of the components on the right side of the reversible reaction *as written* by the product of the concentrations on the left side. The unqualified statement for an equilibrium constant does not indicate the direction to which the value refers. The K_{eq} for reaction $A \rightleftharpoons B$ is the reciprocal of the K_{eq} for the reaction $B \rightleftharpoons A$. Thus, if the previous equilibrium had been written:

$$\text{Dihydroxyacetone P} \rightleftharpoons \text{D-Glyceraldehyde 3-P}$$

the same concentrations of reactants would have applied to the system described in the example, but:

$$K_{eq} = \frac{1.0 \times 10^{-3}}{25.0 \times 10^{-3}}$$
$$= \frac{1}{25}$$
$$= 4.0 \times 10^{-2}$$

Thus the larger the equilibrium constant, the greater is the tendency of the reaction to proceed as written from left to right.

Since all biochemical reactions to be considered are assumed to occur in dilute aqueous solutions, the concentration of water does not change significantly and is set to unity, rather than to 55.5 mol/L. For the H_2CO_3 system:

$$K_{eq}' = \frac{[H_2CO_3]}{[CO_2]}$$

where:

$$K_{eq}' = K_{eq} \times [H_2O]$$

or:

$$K_{eq}' = 55.5\, K_{eq}$$

This approximation is valid for the dilute solutions of extracellular fluids or cytoplasm. A different situation may exist for those reactions occurring in membranes and similar cellular structures, where the free water may be limited.

Weak acids and weak bases One of the largest groups of reversible reactions is the dissociation of weak acids and bases, which has been defined by Brønsted as follows:

$$HA \rightleftharpoons H^+ + A^-$$
Acid   Conjugate base

$$B + H^+ \rightleftharpoons H^+B$$
Base   Conjugate acid

By this definition, an acid ionizes to produce a proton and conjugate base. A base accepts a proton to form a conjugate acid. Thus the ammonium ion NH_4^+ is the conjugate acid of the base ammonia, NH_3. Similarly, the acetate ion CH_3COO^- is the conjugate base of acetic acid, CH_3COOH. From the Brønsted definitions it is also true that the acetate ion is a base and acetic acid is the conjugate acid, so that the CH_3COOH and CH_3COO^- species are called a conjugate acid–base pair. This relationship is also seen in the dissociation of the α-amino acids or the polyprotic acids, for example, phosphoric acid. Taking glycine, the simplest of α-amino acids, the reactions may be written:

$$NH_3^+\!-\!CH_2\!-\!COOH \rightleftharpoons NH_3^+\!-\!COO^- + H^+ \qquad (2.1)$$
$$(H_2A^+) \qquad\qquad\qquad (HA)$$

$$NH_3^+\!-\!CH_2\!-\!COO^- \rightleftharpoons NH_2\!-\!CH_2\!-\!COO^- + H^+ \quad (2.2)$$
$$(HA) \qquad\qquad\qquad (A^-)$$

HA is the conjugate base of the fully protonated glycine H_2A^+, and the conjugate acid of the glycinate ion A^-. A^- is also the conjugate base of HA.

The equilibria of the glycine species expressed in equations 2.1 and 2.2 are written as stepwise dissociations of the diprotonic acid H_2A^+. The equilibrium constants are 4.47 $\times$ 10^{-3} for equation 2.1 and 1.70×10^{-10} for equation 2.2, as would be expected from the general knowledge that the —COOH groups is a stronger acid (more dissociated) than the —NH_3^+ group. Equilibrium constants for such ionizing molecules are called dissociation constants and are represented as K_a for weak acids.

$$K_a = \frac{[A^-][H^+]}{[HA]} \qquad\qquad (2.3)$$

Weak bases are best considered in the context of dissociation reactions in their conjugate acid form, for which K_a refers to the dissociation constant for the loss of a proton. This is illustrated in equation 2.4. This unifying concept avoids the introduction of K_b, a second type of dissociation constant that refers to the dissociation of the base.

Definition of pX Weak acids are usually considered to have dissociation constants smaller than 10^{-2}; that is, the acid is less than 1% ionized in molar solution. It is advantageous to manipulate very small or very large numbers in their logarithmic form. In general, for a value X:

$$-\log_{10} X = pX$$

so that if X is 10^{-2}, then pX is 2. The convention used for pH is:

$$pH = -\log_{10} [H^+]$$

The convention used for the dissociation constant K_a is:

$$-\log_{10} K_a = pK_a$$

Note that the *weaker* the acid, the *larger* is the pK_a in a *logarithmic* proportion. An acid with pK_a4 is not twice as ionized as one of pK_a8; rather, it is 10^4 times more dissociated, a logarithmic change. Similarly, a change of pH 7.25 to pH 7.55 represents a *halving* of the hydrogen ion concentration.

As examples of the principles introduced to this point, the following situations may be considered.

1. The normal pH of arterial blood is 7.40. That is:

$$-\log_{10}[H^+] = pH$$
$$\text{or}\quad \log_{10}[H^+] = -pH$$
$$\text{in blood}\quad \log_{10}[H^+] = -7.4$$

'Since the mantissa in logarithm tables is always positive, then -7.4 must be written $-8 + 0.6$, the arithmetic equivalent. This is usually written $\overline{8}.6$ to indicate that the characteristic, -8, is negative. Therefore the $[H+]$ for blood is:

$$[H^+] = \text{antilog } -8 \times \text{antilog } 0.6$$
$$= 10^{-8} \times 3.98 \text{ mol/L}$$

Similarly:

$$\text{pH 7.25 corresponds to } 5.62 \times 10^{-8} \text{ mol/L of } H^+$$

and

$$\text{pH 7.55 corresponds to } 2.82 \times 10^{-8} \text{ mol/L of } H^+$$

2. The K_a and pK_a of common weak acids are as follows:

		K_a	pK_a
Acetic acid		1.82×10^{-5}	4.74
Ammonium ion		5.50×10^{-10}	9.26
Glycine	(1)	4.47×10^{-3}	2.35
	(2)	1.70×10^{-10}	9.78

In the case of the ammonium ion:

$$NH_4^+ \rightleftharpoons NH_3 + H^+ \tag{2.4}$$

$$K_a = \frac{[H^+][NH_3]}{[NH_4^+]}$$

The value of K_a is 5.50×10^{-10}. Therefore:

$$pK_a = \log_{10} 5.50 \times 10^{-10}$$
$$= -(\log_{10} 5.50 + \log_{10} 10^{-10})$$

From log tables:

$$\log_{10} = 5.50 = +0.7404$$

and by inspection:

$$\log_{10} 10^{-10} = -10.0000$$

$$pK_a = -(+0.7404 - 10.0000)$$
$$= -(-9.26)$$
$$= 9.26$$

This is the value in the tables.

Henderson-Hasselbalch equation The dissociation of a weak acid is represented as:

$$HA \rightleftharpoons H^+ + A^-$$

$$K_a = \frac{[H^+][A^-]}{[HA]} \tag{2.5}$$

$$= [H^+] \cdot \frac{[A^-]}{[HA]}$$

Taking logarithms:

$$\log_{10}K_a = \log_{10}[H^+] + \log_{10}\frac{[A^-]}{[HA]}$$

Changing the signs throughout, the equation gives:

$$-\log_{10}K_a = -\log_{10}[H^+] - \log_{10}\frac{[A^-]}{[HA]}$$

Since $-\log_{10}K_a = pK_a$ and $-\log_{10}[H^+] = pH$, one may write:

$$pK_a = pH - \log_{10}\frac{[A^-]}{[HA]}$$

or by transposing terms:

$$pH = pK_a + \log_{10}\frac{[A^-]}{[HA]}$$

This form is known as the Henderson-Hasselbalch equation.

One application of the Henderson-Hasselbalch equation is to amphoteric molecules, such as α-L-amino acids, peptides, and proteins, as they relate to pH. The ionizable groups of the peptide and its R groups each have their pK_a values, which as a first approximation are the same as the corresponding amino acid. These are summarized in Table 2.1. Considering the dissociation of the proton from one of these groups, the Henderson-Hasselbalch relationship states that as more of the conjugate acid is converted to the conjugate base, the log of their ratio will increase, the corresponding pH will become larger, and the solution will be more alkaline. At the point at which the concentrations of the conjugate acid and its conjugate base are equal, the pH is that given by pK_a. In terms of a titration curve, when half of a weak acid has been neutralized by a strong base such as NaOH, the pH of the solution is equal to the value of pK_a. This matter is considered again in the discussion of buffers in Chapter 4.

Isoelectric point In polyprotonic and amphoteric molecules it is assumed in the present discussion that the dissociation of each proton is discrete and is not influenced by the other ionizable groups. This is essentially true if the pK_a values are separated by more than two units, which means that in titrating a molecule such as glycine with alkaline, the —COOH group (pK_a 2.34) is completely ionized before the —NH_3^+ group (pK_a 9.60) begins to lose its proton. At the point at which an equal number of negatively and positively charged groups is present, the net charge is zero and the pH of the solution is known as the *isoelectric point* (pI). In simple molecules it is calculated by taking the average of the pK_a values of the two groups that ionize either side of the molecular form with zero net charge. For a diprotic amphoteric molecule such as glycine, it is the average of the two pK_a values. The pI of glycine is $\dfrac{2.34 + 9.60}{2}$ or 5.97. For glutamic acid (see Table 2.1, pK_a values 2.13, 4.31, and 9.76) and similar polyprotonic ampholytes, it is necessary to determine the pair of values to be averaged. The dissociation of glutamic acid may be abbreviated:

$$pI = \frac{2.13 + 4.31}{2}$$
$$= 3.22$$

The ionized form of the molecule that has a zero net charge is associated with the pK_a values of 2.13 and 4.31, which average to 3.22, the pH at which the molecule of glutamic acid is electrically neutral. Approximate pI values can be calculated for proteins and polypeptides from the numbers of each type of ionizable groups, but the assumptions are many. The procedure, however, can be illustrated by the following examples.

The completely protonated form of the antidiuretic hormone vasopressin (see p. 51) is represented as follows, with the corresponding pK_a values taken from Table 2.1:

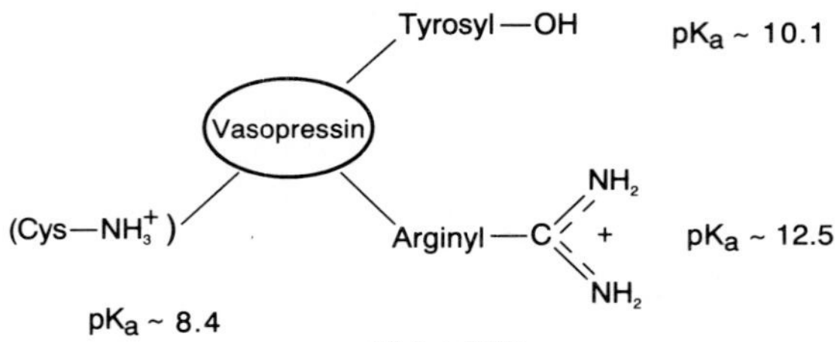

$$pI = \frac{10.1 + 12.5}{2} = 11.3$$

At about pH 10 the $—NH_3^+$ group of the cystinyl residue will be almost completely ionized to an uncharged $—NH_2$; the charge on the vasopressin is then approximately +1. Above pH 12.5 the molecule has a −1 charge. Thus the ionization of the tyrosyl residue, pK_a 10.1, is the one to be used in the pI estimation, together with the final ionization of the arginyl residue, pK_a 12.5, since the molecule will be least charged between pH 10.1 and 12.5 and with zero overall charge at pI 11.3.

Glutathione, γ-L-glutamyl-L-cysteinylglycine, is an important tripeptide found in erythrocytes and other cells. It is an atypical tripeptide in the sense that the γ-carboxyl group, not the α-carboxyl group, of L-glutamic acid is in the peptide bond with the amino group of the cysteinyl residue. Of the five ionic forms of glutathione, the three of importance in calculating the pI may be represented as follows:

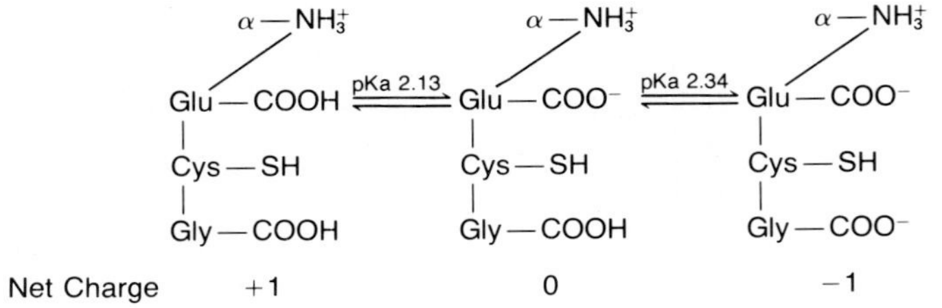

$$pI = \frac{2.13 + 2.34}{2} = 2.23$$

The principle with regard to proteins is the same as that just illustrated, except that the number of ionizable groups is larger and the approximations in the calculation are greater.

Ion exchange chromatography Ion exchange resins are water-insoluble polymers to which charged groups such as $—COO^-$, $—NH_3^+$, $—OSO_3^-$, and $—NH^+R_2$ have been covalently linked. These groups will interact with oppositely charged groups on molecules that are dissolved in an aqueous solution. The reaction is an equilibrium, the molecule (M) being distributed between the solid and the soluble aqueous phases.

$$\underset{\text{Insoluble}}{Resin—X^-} + \underset{\text{Soluble}}{MH^+} \rightleftharpoons \underset{\text{Insoluble complex}}{R—X^- \cdots {}^+HM}$$

The relative distribution depends on the nature of M, the pK_a values of the groups, and the pH of the aqueous system.

These properties are used to separate ionic and amphoteric molecules such as amino acids, peptides, and proteins. The ion exchange resin is placed into a chromatographic column and equilibrated with the solution used for elution; the mixture of materials is then filtered through the column. The molecules are bound to the resin and then move slowly down the column at rates that depend on the equilibrium position of each compound in relation to the resin. Separation is achieved if the relative rates are sufficiently different. Fractionation may require changes in salt concentration or pH, depending on the particular mixture.

Many types of ion exchange resins are available, and many variable conditions can be applied to achieve the needed separations. As a result, ion exchange chromatography is a common technique in the analysis and preparation of biochemicals. An earlier example (see p. 49) described the separation of amino acids in the hydrolysate of a protein.

Electrophoresis If a solution of a mixture of proteins is placed between two electrodes, the charged molecules will migrate to one electrode or the other at a rate that depends on the net charge and, depending on the supporting medium used, on the molecular weight. If the supporting medium is a solid substance such as an inert gel, a strip of paper, or cellulose acetate, sections may be cut out to provide purified or concentrated samples of the individual proteins making up the mixture. Alternatively the entire gel or strip may be stained with reagents that react with proteins, thereby allowing a visual display of the separation. Techniques of this sort, which depend on the movement of charged molecules, are known collectively as electrophoresis. They are sometimes employed as diagnostic aids. Electrophoretic methods are usually simple and rapid and require equipment of modest cost.

More sophisticated equipment permits rapid, free-solution capillary electrophoresis without a supporting medium, using microsamples and the resolution of molecules by differences in charge density.

Structural aspects of specific proteins
Blood plasma and blood serum proteins

When blood is allowed to clot, several plasma proteins contribute in forming the matrix of the clot. The resulting solution, lacking fibrinogen, fibrin, and several clotting factors, is known as *serum*. Blood *plasma* is the supernatant fluid obtained by centrifuging out the blood cells in the absence of clotting. Unlike serum, plasma contains fibrinogen and other clotting factors. Many clinical chemical determinations are made on serum rather than plasma.

Normal plasma contains 60 to 80 g/L of protein. Of this, 30 to 50 g is albumin and 15 to 30 g is composed of a mixture of globulins. The plasma proteins are usually classified on the basis of solubility and electrophoretic separation into the six categories listed in Table 2.5. The plasma proteins are separated and measured chemically by a procedure involving electrophoresis, staining, and densitometric scanning. The electrophoretic mobilities of the main plasma protein fractions are shown in Figure 2.6, which illustrates the manner in which this information is usually presented clinically. The peaks correspond to those similarly identified in Table 2.5; fibrinogen is absent, however, in serum.

Most clinical electrophoretic analyses of serum are carried out with cellulose acetate paper as a support and a barbital buffer of pH 8.6. At this pH all normal serum proteins are negatively charged and migrate toward the anode of the electrophoresis cell; the γ-globulins are the least negatively charged of the serum proteins.

γ-Globulins are synthesized by a class of lymphocytes known as plasma cells, whereas the other major plasma proteins are synthesized in the liver. Therefore one of the hallmarks of severe liver disease is an abnormality, usually a decrease, in one or more of the plasma proteins.

Albumin The most abundant protein in plasma is albumin. This protein has an isoelectric point of 4.8 and thus has a considerable negative charge at a physiologic pH, which explains the relatively high anodal mobility of albumin on the plasma protein electrophoretogram (Figure 2.6). Albumin is a globular protein consisting of a single polypeptide chain, and it has a molecular weight of about 66,300.

The two main functions of albumin are to *transport small molecules* through the plasma and extracellular fluid and to *provide osmotic pressure* within the capillary. Many metabolites, such as free fatty acids and bilirubin, are poorly soluble in water. However, they must be shuttled through the blood from one organ to another so that they can be metabolized or excreted. A carrier is required to enhance the solubility of these substances in plasma so that they can be transported through this aqueous medium. Albumin fulfills this function and serves as a *nonspecific transport protein*. Moreover, albumin binds those poorly soluble drugs such as aspirin, digoxin, the coumarin anticoagulants, and barbiturates so that they also are efficiently carried through the bloodstream.

Albumin-bound cations In addition to carrying these large organic molecules, albumin binds small anions and cations. Indeed, about 50% of the calcium in the plasma exists as a complex with albumin. All the small molecule–albumin interactions occur through physical bonds; that is, covalent linkages are *not* formed. The binding of a small molecule to a protein may be formulated by the following general equation:

$$[P] + [A] \rightleftharpoons [PA]$$

where [P] is the concentration of protein that does not contain any complexed small molecule, [A] is the concentration of the unbound small molecule, and [P] is the concentration of the protein–small molecule complex. In such a system the effective concentration that determines the biologic activity of the small molecule depends on [A], the unbound concentration. For example, the physiologic effectiveness of calcium in the blood depends on the unbound calcium concentration, which in turn is in equilibrium with the calcium that is bound to albumin. This concept is crucial to an understanding of physiologic and pharmacologic function. The following hypothetic examples illustrates the point.

Suppose the clinical data indicate hypercalcemia, but repeated laboratory examinations reveal that the plasma total calcium concentration is only 2.25 mmol/L, the *lower* limit

Table 2.5   Major classes of plasma proteins

Type	Percentage of total plasma protein	Special properties and functions
Albumin	55	Transport of organic anions, osmotic pressure, binds Ca^{++}
α_1-Globulins	5	Glycoproteins, high-density lipoproteins
α_2-Globulins	9	Haptoglobin (hemoglobin transport), ceruloplasmin (Cu^{++} transport), lipoproteins of very low density
β-Globulins	13	Transferrin (Fe^{++} transport), low-density lipoproteins
Fibrinogen	7	Blood clotting
γ-Globulins	11	Immunoglobulins

of normal. One must realize that the so-called normal calcium value is based on the assumption that the plasma albumin concentration is also normal. If the patient had nephrosis, a disease in which the plasma albumin concentration is low, the apparently normal total calcium concentration of the plasma would actually be in the hypercalcemic range as far as the unbound concentration is concerned. Under these conditions the unbound (ionized) calcium concentration would be high because less albumin is available to bind calcium. This is illustrated in Table 2.6 by typical data the might be obtained for such a hypothetic example.

Table 2.6   Effect of albumin binding on "effective" plasma calcium concentration

	Normal patient	Nephrotic patient
Plasma total Ca^{++} (mmol/L)	2.25	2.25
Plasma albumin (g/L)	40	20
Bound Ca^{++} albumin ratio (mmol/g)	0.04	0.04
Bound Ca^{++} (mmol/L)	1.50	0.75
Free Ca^{++} (mmol/L)	0.75	1.50

Figure 2.6   Normal serum protein electrophoretogram and its densitometric tracing.

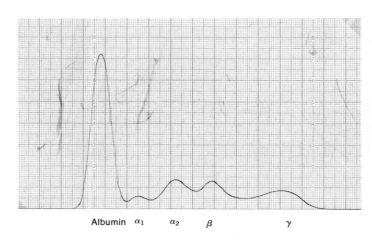

Albumin α_1 α_2 β γ

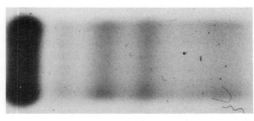

A second vital function of albumin is that it provides 80% of the osmotic pressure effect of the plasma proteins. Osmotic pressure is the main force that draws interstitial fluid back into the capillary at its venous end. Albumin provides much of the osmotic effect for two reasons: it is the most abundant protein in plasma on a weight basis, and it has a low molecular weight relative to the other major plasma proteins. Remember that colligative properties, such as osmotic pressure, depend on the number of particles in a solution. Moreover, the high negative charge that albumin exhibits at pH 7.4 causes water to cluster at its surface, producing a greater osmotic effect than would be predicted simply from the number of solute molecules present in the solution.

Alpha globulins Two classes of α-globulins exist in plasma, known as $\alpha\text{-}_1$ and α_2. The main α_1-globulins are *glycoproteins* and *high-density lipoproteins* (HDLs). The main α_2-globulins are *haptoglobin,* the transport protein for any free hemoglobin that escapes into the plasma; *ceruloplasmin,* the copper transport protein; *prothrombin,* a proenzyme that is involved in blood coagulation; glycoproteins; and *very low–density lipoproteins* (VLDLs).

Clinical comment

Alpha$_1$-antitrypsin Human serum contains a natural trypsin inhibitor called α_1-antitrypsin. The activity of this protein has been shown to be abnormally low in people with a genetic susceptibility to emphysema because of the substitution of one amino acid for another in the α_1-antitrypsin molecule. The lack of this trypsin-inhibiting activity is thought to favor the destruction of lung tissue because α_1-antitrypsin also inhibits the action of elastase and other proteases in the lung. The association of α_1-antitrypsin with emphysema induced by cigarette smoking led to a search for a direct effect of cigarette smoke on the protease inhibitor. Cigarette smoke was found to oxidize a methionine residue near the *N*-terminal of α_1-antitrypsin to produce methionine sulfones and sulfoxides. The oxidation of this methionine leads to loss of activity for inhibiting proteases. Fortunately, enzymes in lung tissue usually are able to reduce the protein-bound sulfones and sulfoxides back to methionine and to restore activity of α_1-antitrypsin.

Beta globulins The major β-globulins are *transferrin* (the iron-transport protein) and *low-density lipoproteins* (LDLs).

Gamma globulins The γ-globulin fraction is made up of *immunoglobulins,* or *antibodies*. There are 10 to 15 g/L of γ-globulins in normal human plasma. The major plasma immunoglobulins that make up the γ-globulin fraction are listed in Table 2.7. The commonly used abbreviation of immunoglobulin is Ig, and this is followed by the letter signifying the particular subclass, for example, IgG and IgA. Other immunoglobulin classes are present in smaller amounts (IgD, IgE, and IgM). These five classes of immunoglobulins are heterogeneous. Indeed, each class is made up of hundreds or thousands of individual immunoglobulins.

The synthesis of a specific immunoglobulin is stimulated by an antigen, a protein or complex carbohydrate that is foreign to the species or individual. The newly synthesized immunoglobulin has the property of recognizing the antigen that stimulated its synthesis and combining very tightly with it. This antigen-antibody interaction is through specific noncovalent bonds. An analogous type of reaction in terms of the bonds involved and their specificity is the reaction of an enzyme with its substrate. Antigen-antibody complexes are marked for elimination from the body. The mechanism for elimination may vary with the complex being considered, but a common mechanism of eliminating these complexes is by phagocytosis.

Immunoglobulin G structure The immunoglobulins have several structural similarities even though at first glance at the molecular weights given in Table 2.7, one might think that each immunoglobulin type is totally different. All types are similar in that they have two small (light) and two large (heavy) polypeptide chains. These are called light and heavy to indicate how the isolated chains sediment in the ultracentrifuge. Most of our knowledge concerning immunoglobulin structure has been obtained from studies with IgG. The IgG fraction is the major immunoglobulin of the plasma, accounting for 80% of the total. Following is the structure of a typical IgG molecule:

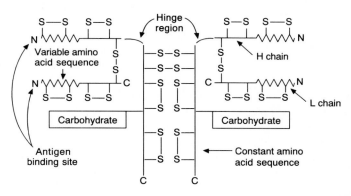

These molecules have a molecular weight of approximately 1.5×10^5 and a sedimentation coefficient of 7S. They are symmetric, the two halves being joined covalently by two disulfide bridges. Each half is composed of a heavy (H) and a light (L) chain. Thus an immunoglobulin can be abbreviated H_2L_2. This configuration of chains makes up the basic four-peptide unit that composes all immunoglobulins. The H chain contains 446 amino acids; the L chain contains 214 amino acids. Each H chain is joined to the corresponding L chain by a single disulfide bond. In addition to the four disulfide bonds that maintain the quaternary structure of IgG, intrachain disulfide bonds help to maintain the tertiary structure of each L and H chain.

It is important to note at this point that the amino acids in a peptide or protein chain are numbered starting at the *N*-terminal residue. The primary sequence of 108 amino acids at the *N*-terminal of the L and H chain of each species of IgG is unique; in other words, the sequence of amino acids of the *N*-terminal end is *different* for each antibody. (Remember that there may be millions of copies of a single antibody or IgG species.) In contrast, the primary structure of the remaining amino acids 109 to 214 is *identical* in the L chains of every IgG molecule in a particular class, that is, κ or λ. Also, the sequence of amino acids 109 to 446 is identical in each class of H chains of every IgG molecule. The antigen-binding site is located at the *N*-terminal end of each pair of H and L chains, that is, at the variable end, which is unique for each antibody. In this way the variable primary structure confers specificity to each antibody for a given antigen or group of closely related antigens, whereas the constant *C*-terminal structure provides the basic framework that all IgG molecules require to carry out their functions.

IgA and IgM structure The gross structural features of IgA and IgM molecules are similar to IgG molecules, but some important differences also exist. There are two basic kinds of light chains, called kappa (κ) and lambda (λ) chains. Two κ- or two λ-chains may pair with two heavy chains to form all the different immunoglobulin classes (Table 2.7). Kappa chains can be distinguished from λ-chains by a difference in the amino acid sequence at position 191. Despite this difference, the molecular weights of κ- and λ-

Table 2.7   Human plasma immunoglobulins

Type	Percentage of total immunoglobulin	Molecular weight × 10⁻³	Sedimentation constant (S)
IgG	80	150	7
IgA	13	160, 320, 480	7, 9, 11
IgM	6	900	19
IgD	1	185	7
IgE	0.002	200	8

Table 2.8   Properties of immunoglobulin types

Type	Heavy chain			Light chain class, molecular weight 22,500
	Class	Subclass	Molecular weight	
IgG	γ	$\gamma_1, \gamma_2, \gamma_3, \gamma_4$	53,000	κ, λ
IgA	α	α_1, α_2	64,000	κ, λ
IgM	μ		70,000	κ, λ
IgD	δ		58,000	κ, λ
IgE	ϵ		75,000	κ, λ

chains are about the same. The heavy chains can also be classified according to their primary structure into broad groups. The different types of heavy chains, unlike the κ- and λ-chains, are characteristic of the immunoglobulin types. For example, IgG has a heavy chain called γ; IgA has a chain called α; and the IgM heavy chain is called μ. The Greek letters δ and ϵ represent the heavy chains in IgD and IgE. Thus a particular IgG molecule might be described as $\gamma_2\kappa_2$ or $\gamma_2\lambda_2$, and IgA as $\alpha_2\kappa_2$ or $\alpha_2\lambda_2$, and so on. The different types of heavy chains also can be distinguished by a characteristic amino acid sequence, primarily in the constant region; however, unlike κ- and λ-chains, the heavy chain types vary considerably in molecular weight from approximately 53,000 for the γ-chain to about 75,000 for the ϵ-chain. Molecular weights of the heavy chains are greatly influenced by their carbohydrate content. For example, α- and γ-chains contain approximately the same number of amino acids, but the α-chain contains more carbohydrate. (See Table 2.8 for a summary of these properties.)

The three major immunoglobulin types, IgG, IgA, and IgM, also differ in the number of basic four-peptide units that make up their structure. The simplest, IgG, has only one such unit (i.e., H_2L_2) whereas IgM has five such units, that is $(H_2L_2)_5$ (see also case 2, Figure 2.12). IgA may have one, two, or three such units. Obviously, the number of basic four-peptide units in an immunoglobulin determines its molecular weight, and the polymeric nature of IgA and IgM is reflected in the molecular weights given in Table 2.7. Polymeric IgA and IgM also contain another glycoprotein called J (molecular weight of 16,000). Only one J chain exists per IgA or IgM polymer.

Monoclonal antibodies Each B lymphocyte that has been stimulated by a complex antigen makes an antibody that recognizes a small part of the structure of the antigen. This portion of the antigen is called an *epitope*. When an animal is immunized with a complex antigen, some B cells will make antibodies against one epitope, whereas other cells will make antibodies against other epitopes in the same antigen. Consequently the antibodies in the bloodstream are actually a mixture of molecules, each specific for a certain epitope and each made by a specific cell. Because these antibodies are derived from several different B cells, they are called *polyclonal* antibodies. In contrast, *monoclonal* antibodies are derived from clones of single cells, where each clone synthesizes one type of antibody specific for just one epitope. Monoclonal antibodies are homogeneous, being composed of exactly the same immunoglobulin. In contrast, polyclonal antibodies are heterogeneous and composed of a mixture of immunoglobulins specific for a variety of epitopes on a single antigen.

Monoclonal antibodies are very useful in both the laboratory and the clinic. In the laboratory they can be used to gather much detailed information about the structure of protein antigens. They can be produced from impure antigens or mixtures of antigens, so it is no longer necessary to use highly purified protein antigens to obtain antibodies that are highly specific for part of a protein of interest. In medicine monoclonal antibodies are being widely used in diagnostic procedures, usually to assay accurately for a specific antigen. It is likely that in the future monoclonal antibodies will be typically used in treatment.

Monoclonal antibodies must be produced in special cells called *hybridomas*. This is because the B lymphocytes that make antibodies have rather short life spans. When these "mortal" cells are mixed and fused with "immortal" myeloma cells in the laboratory and under special conditions, one can isolate hybrid cells, hybridomas, that carry the immunoglobulin gene of the B lymphocyte and the immortal characteristics of the myeloma cells. The resulting hybridoma cells can be cloned, grown in large amounts, and used to produce considerable quantities of monoclonal antibodies.

Fibrous proteins

Most of the enzyme and plasma proteins considered so far have been of the globular type. These have only one or a few polypeptide chains arranged so that the polar groups on the surface make them soluble. In contrast, fibrous proteins, such as fibrinogen, have exceedingly extended structures consisting of several polypeptide chains often tightly associated one to the other. These proteins are relatively insoluble in most physiologic fluids. Examples of fibrous proteins are the keratins, collagens, and elastins. The keratins are the proteins in hair, fingernails, and horny tissues. The collagens are constituents of skin, bones, teeth, blood vessels, tendons, cartilage, and connective tissue. Elastin is also an important structural element of skin, arteries, ligaments, and connective tissue. The primary role of the fibrous proteins is architectural: they provide the strong structure and support that the body requires to perform its many complex functions.

Collagen More is known about the collagen class of fibrous proteins than the others, perhaps because of its wide occurrence and its association with various disease processes. More than 30% of the protein contained in the human body is collagen, making it by far the most abundant protein.

Table 2.9 lists the collagen and elastin content of several tissues. Soft organs such as the liver contain only small amounts of collagen, whereas harder tissues such as skin and tendons are almost entirely collagen. This is consistent with the structural role of collagen (Table 2.10). Although most of the collagen in the adult is metabolically stable, it is far from being inert. Skin, for example, is continuously being broken down and resynthesized. Furthermore, collagen changes appreciably during growth, development, and morphogenesis. It is also important in repair processes such as the formation of scar tissue in wound healing. Its exact role in the regulation of these events is not understood.

Primary structure Early studies showed that the collagen molecule is very large and has a peculiar amino acid composition. More than one third of its amino acid residues are glycine, and about 20% are proline or hydroxyproline (see Table 2.2). Of all proteins studied, only elastin has such a high concentration of glycine. The hydroxylated derivatives of lysine and proline are almost unique to collagen. Elastin contains a smaller amount

Table 2.9   Collagen and elastin content of some tissues (g/100 g of dry weight)

Tissue	Collagen	Elastin
Bone, mineral free	88.0	
Achilles tendon	86.0	4.4
Skin	71.9	0.6
Cornea	68.1	
Cartilage	46 to 63	
Ligament	17.0	74.8
Aorta	12 to 24	28 to 32
Liver	3.9	

Table 2.10   Physical form of collagen in different tissues

Tissue	Arrangement of fibrils and microfibrils*
Tendon	Parallel bundles
Cartilage	Associated with mucopolysaccharides, no distinct microfibril arrangement
Skin	Planar sheets of microfibrils layered at many angles
Cornea	Planar sheets stacked crossways for strength

*Fibers can be seen by the naked eye, fibrils are visible by means of light microscopy, and microfibrils are seen only by means of electron microscopy.

of hydroxyproline but no hydroxylysine. Tyrosine is present in low amounts, and the essential amino acid tryptophan is absent, so collagen is a protein of zero biologic value. Cystine also is absent, and thus no disulfide cross-links are present.

Collagen subunits When collagen fibers are warmed slightly in dilute acid, they dissociate and go into solution. If the solution is cooled, fibers will re-form, but if it is boiled, gelatin results, and the denaturation is irreversible. Ultracentrifugation of the warm acid solution separates the mixture into three fractions designated α, β, and γ. The so-called β-fraction has a molecular weight twice that of the α-fraction, and the γ-fraction is three times as large as the α-fraction. The α-fraction is the basic polypeptide unit that, when covalently cross-linked in a dimeric form, gives the β-fraction and, when cross-linked in a trimeric form gives the γ-fraction. Figure 2.7 illustrates these relationships. Several separate α-chains differ in amino acid composition but are about the same size. They have molecular weights of 97,000 and are abbreviated α1 and α2. The common collagen of skin consists of two α1-chains and one α2-chain twisted together to form a triple helix called *tropocollagen*. The degree of cross-linking in tropocollagen may vary so that differing proportions of α-, β-, and γ-forms will be present.

Tertiary and quaternary structure None of the three protein units of tropocollagen has α-helical conformations because they contain a high proportion of L-prolyl and hydroxy-L-prolyl residues. Also, the large proportion of glycyl residues, which by itself would destabilize the α-helix, are arranged in uniform repeating units so that they contribute to a more extended conformation and make possible a greater number of hydrogen bonds between adjacent chains. These polypeptide chains readily wrap around each other to form a left-handed helix much like a three-stranded rope. Thus the unusual amino acid composition of collagen is directly responsible for creating the triple helix.

Figure 2.7

Sodium dodecyl sulfate (SDS) polyacrylamide gel electrophoresis of guinea pig collagen. The molecular weight of the α1- and α2-chains is approximately 100,000, with the α1-chain being slightly smaller. The proteins move in SDS polyarcylamide gels as an inverse function of their molecular weights. The β1,2-band consists of an α1-chain cross-linked to an α2-chain; the β1,1-band has two α1-chains cross-linked. The β1, 1- and β1,2-bands have molecular weights of about 200,000. The γ-band, or tropocollagen, has a molecular weight of 300,000, and consists of $(\alpha 1)_{2\alpha 2}$, where the chains are cross-linked to one another.

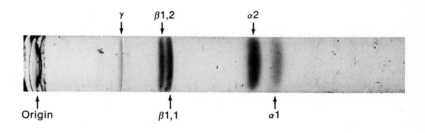

Sequencing data show that, in the most polar regions of the α-chains, the tripeptide Gly-X-Y is repeated in sequence over and over again. The symbols X and Y refer to amino acids other than glycine. The tripeptide Gly-X-Hyp is another common sequence; here Hyp is the abbreviation for hydroxy-L-proline. The small glycine residues are packed tightly by hydrophobic interactions against one another at the center of the triple helix. Each glycine —NH— group hydrogen bonds to a carbonyl group of an amino acid in the X position of an adjacent chain. The charged amino acids and those with bulky groups are on the outer part of the helix.

Tropocollagen is the building material of which the microfibrils of collagen are made. The triple helix is a rod about 3000 × 15 Å. These rods align themselves in the parallel overlapping manner illustrated in Figure 2.8. The tropocollagen molecule has polar regions at either end and at intervals of about 680 Å along its length; these appear as dark vertical lines in the diagram. It is the interaction of these polar regions that contributes to the typical cross-striations seen in electron micrographs of collagen fibers. This appearance is produced by aligning the polar groups of one tropocollagen molecule with those of a parallel neighbor displaced longitudinally by about one fourth of its length (Figure 2.8). The tropocollagen molecules do not quite touch when placed end to end. This creates a so-called hole in the structure next to a region of slight overlap. The overlapped region next to the hole is thought to represent the most electron-dense striations. The hole is believed to be the nidus for the formation of the hydroxyapatite crystals that trigger bone formation.

Cross-linkages When microfibrils are formed in the laboratory by reaggregating tropocollagen and are examined by means of the electron microscope, the resulting structures appear identical to microfibrils isolated from tissue. However, they are lacking both in tensile strength and in interchain covalent cross-linkages. The polypeptide regions close to the native cross-connections are specifically broken by proteolytic enzymes, such as pepsin. The cross-linkages themselves are not cleaved by proteolytic enzymes because they are not peptide bonds. Apparently the peptide bonds of the tightly twined triple helix

Figure 2.8

Diagram of assembly of collagen microfibrils. Each long gray bar represents a tropocollagen molecule, whereas the vertical bars depict evenly spaced polar regions. The polar regions line up to leave spaces or holes.

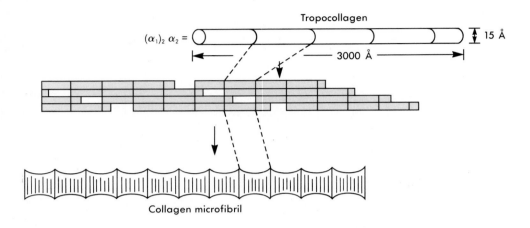

are inaccessible to the proteolytic enzymes. Consequently, it is believed that the cross-links occur in those polar regions of tropocollagen that are less tightly compacted and therefore more accessible to the enzymes. This fits with what is known about the amino acids that participate in making the cross-links. They are lysine or hydroxylysine residues or derivatives of these residues. It is expected that a basic amino acid such as lysine would be found in the polar regions.

Clinical comment

Lathyrism An enzyme called peptidyl lysyl oxidase converts the ε-amino group of certain lysyl residues to aldehyde groups. Peptidyl lysyl oxidase is specific for collagen-like molecules. It contains copper and pyridoxal phosphate, a coenzyme derived from vitamin B_6. The enzyme is drastically inhibited by nitriles, specially β-aminopropionitrile, a compound that causes lathyrism. This disease is associated with deformation of the spine, demineralization of bones, dislocation of joints, and aortic aneurysms. A copper deficiency in pigs produces the same symptoms, undoubtedly because amine oxidase also is affected.

The semialdehydes of lysine form cross-links in a very slow nonenzymatic reaction. Two types of reactions of these aldehydes are possible. For example, the aldehyde reacts with an intact lysyl residue on an adjacent chain to form a Schiff base, a rather unstable compound but one that can be reduced to a very stable secondary amine.

In the other type of reaction, two aldehydes condense in an aldol condensation; this product is quite stable.

$$2 \; R-(CH_2)_2-CH_2-\overset{\overset{\displaystyle O}{\|}}{C}H \longrightarrow R-(CH_2)_2-CH_2-\underset{\underset{\displaystyle OH}{|}}{C}H-\overset{\overset{\displaystyle H{\diagdown}C{\diagup}^{O}}{}}{C}H-(CH_2)_2-R$$

Clinical comment

Cross-linking in aging and disease It is impossible to say when the cross-linking of collagen is completed. Some investigators think that it is never completed but rather continues as one ages, producing increasingly stiffer skin, blood vessels, and other tissues, all of which contribute to the medical problems of aged persons. No evidence, however, suggests that treatment of animals with lathyrogens inhibits the aging process.

Homocystinuria is a genetic disease in which a block in the utilization of homocysteine results in increased plasma and urine levels of both homocysteine and its disulfide homocystine. Skeletal deformities and other symptoms similar to those seen in lathyrism are secondary to this defect. It is believed that the excessive amounts of homocystine accumulated by these patients react chemically with the lysyl semialdehydes to block cross-linking. The similar sulfhydryl-containing compound D-penicillamine, used in the treatment of Wilson's disease and as an antidote for mercury or lead poisoning (see case 3, Chapter 3), produces lathyrism in animals and, after long administration to humans, extravasation of blood into the skin over the elbows and knees.

Marfan's syndrome is another genetic disease that is characterized by lathyritic symptoms. The defect here may be associated with an inability to form the proper substrate for the amine oxidase that makes the aldehydes necessary for cross-linking.

Elastin Some of the similarities between elastin and collagen have been pointed out, but the differences are equally important. Elastin occurs with collagen in connective tissues. It is a yellow, fluorescent protein found mostly in ligaments and blood vessel walls (see Table 2.9), but it also occurs in small amounts in skin, tendon, and loose connective tissue. Unlike collagen, fibers of elastin can be stretched to several times their length, and they snap back almost as would a piece of rubber. Tropoelastin is the basic building unit for the elastic fibers. It contains components having molecular weights ranging from 30,000 to 100,000. It is not clear whether these represent subunits. Tropoelastin was first isolated from the aortas of copper-deficient pigs. Because of the lowered activity of the copper-requiring amine oxidase, this material lacked the stabilizing cross-links between interchain lysyl residues, and it was easily extracted from the insoluble fibers.

The cross-links in elastin are considerably more complex than those in collagen. Condensations between lysyl and lysyl semialdehyde residues can cross-link two to four polypeptide chains. These lysine-derived structures, released by the hydrolysis of elastin, are called desmosine or isodesmosine, depending on the nature of the initial condensation.

Although elastin and collagen contain similarly high amounts of glycine and proline and both lack cysteine and tryptophan, elastin contains less hydroxyproline and no hydroxylysine (see Table 2.2). The glycine content is about the same, but the Gly-X-Y repeating unit of collagen is not present. Consequently, elastin is resistant to hydrolysis by bacterial collagenases, which are proteolytic enzymes highly specific for collagen.

Keratin The keratins are the most visible of all animal proteins. They are the highly insoluble fibrous proteins of hair, feathers, fingernails, hoofs, and horns. They are good examples of proteins with α-helical structures. Other fibrous proteins with high α-helix contents are myosin (the major protein of muscle) and fibrinogen (a protein involved in blood clotting). Keratin is often called α-keratin to emphasize its α-helical structure.

When α-keratin is heated, the strong intrachain hydrogen bonds of the α-helix are broken, and the molecules extend into a parallel pleated sheet (see Figure 2.4). This form is called β-keratin. Although the intrachain hydrogen bonds break, the stable interchain disulfide bonds remain intact. On cooling, β-keratin re-forms its intrachain hydrogen bonds and returns to the α-conformation.

Specific binding of molecules to proteins

The binding of molecules to proteins is of general importance in biomedical studies, particularly in the specific interaction of hormones with cell membranes of the target tissue. This high degree of specificity results from the presence in the cell membranes of specific receptors, which are frequently proteins but may be other membrane constituents such as glycolipids. As a result of this intitial binding of the hormone, the target cell responds in an appropriate biochemical fashion; the "message" from the hormone is transmitted into the cell. For example, insulin is secreted from the pancreas in response to an elevated concentration of glucose. The insulin in the blood is carred in the circulation to the various tissues, where it binds to its specific receptor on the cell membrane. Such a binding on the liver cell membrane generates a stimulus that results in glucose being stored in the cell as glycogen. A more detailed consideration of the biochemical mechanism of action of insulin in such information transfer is presented in Chapter 18.

Hormones are present in blood at concentrations of 10^{-6} to 10^{-13} mol/L. A study of the reaction with their receptors requires very sensitive means of quantitation, which is usually provided by labeling the hormones with radioactive groups. One of the more common procedures for polypeptide hormones is to iodinate the tyrosyl-residues with ^{125}I, which is a powerful emitter of γ-radiation. Such tracers permit the determination of small quantities of the hormone, its location on the receptor, and its disappearance or removal from the receptor.

One of several complications in the study of the binding of a hormone to its specific receptor is the concomitant nonspecific general absorption of the hormone to the cell membrane. It is believed that such nonspecific binding does not result in any transmission of biochemical information to the cell. However, it makes the quantitation of specific receptor binding more difficult. Radioactive hormone *H, when added to a suspension of cells, will bind to its specific receptor, R, and to nonspecific sites (NS). Therefore we could write the following:

$$\text{*H + R + (NS)} \rightarrow \text{*HR + *H(NS)}$$

One of the characteristics defined for the specific receptor is the reversibility of the binding of *H by the addition of a large excess of nonradioactively labeled hormone H. Therefore the cells carrying the *H in the form of *HR and *H(NS) will loose the *H from the receptor on the addition of excess unlabeled H, as follows:

$$\text{*HR + *H(NS) + H}_{excess} \rightarrow \text{HR + *H(NS) + *H}$$

The remaining radioactive label on the cells is then caused by nonspecific binding, the specifically bound *H being displaced into the reaction medium by the excess H. This correction for nonspecific binding must always be applied in quantitative studies.

Quantitative equilibrium studies Because of the difficulty in working with such small concentrations of hormone and the complexity of the structure of the cell membrane, certain assumptions are necessary to simplify the quantitative studies of specific binding to receptors. The following facts are assumed:
1. The receptor sites do not interact with each other.
2. The radioactively labeled hormone and the unlabeled hormone behave the same in their reaction with the cell membrane.
3. Only one hormone molecule reacts with each receptor molecule.

4. Specific binding is a reversible reaction.
5. Complete equilibrium is attained.

The reaction of hormone with the specific receptor can thus be written as follows:

$$H + R \rightleftharpoons HR$$

$$K = \frac{[HR]}{[H][R]}$$

Since the total receptor concentration in the system, R_0, is either unoccupied, R, or bound to H, then:

$$[R_0] = [R] + [HR]$$

and the ratio of bound hormone [HR] to free hormone [H] is:

$$\frac{[HR]}{[H]} = K([R_0] - [HR])$$

This is the equation of a straight line (y = mx + c) and is known as the *Scatchard equation*. The slope of the line is equal to the equilibrium constant, K, for the reaction of the hormone with its receptor. This constant is also called an *affinity* or *association constant*, since its magnitude is proportional to the affinity of the hormone for its receptor. When [HR] is zero, $\frac{[HR]}{[H]}$ is also zero (y = o); then the intersection of the line with the x axis is $[R_0]$, which gives the number of specific receptor binding sites for each cell. Remember that in all this development the practical, experimental correction for non-specific binding is required.

The resulting Scatchard plot for a hormone-receptor system that meets all the assumptions is shown in Figure 2.9 for such systems as the binding of growth hormone to human lymphocytes.

Figure 2.9   Classical Scatchard plot of hormone binding to cells.

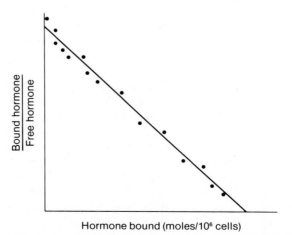

Hormone bound (moles/10^6 cells)

Figure 2.10   Curvilinear Scatchard plot of hormone binding to cells.

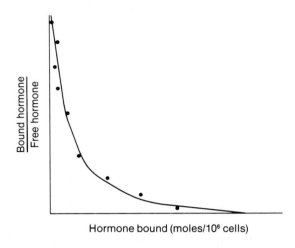

Hormone bound (moles/10^6 cells)

When the same quantitative study is applied to many other hormones, such as insulin, a curvilinear Scatchard plot is obtained (Figure 2.10), one interpretation of which is an interaction of binding sites with each other, a phenomenon called *cooperativity* or *allosterism* (see Chapters 3 and 13). Another analysis would argue for at least two types of receptors, a high-affinity population (steep initial slope) and a low-affinity population (smaller slope).

Even more complicated results have been reported for studies of the binding of other molecules to proteins or receptors. Before credence is given to any one interpretation, it must be stressed that the Scatchard analysis requires complete equilibrium and does not consider the possibility that some of the bound molecules may traverse the cell membrane and be absorbed irreversibly into the cell, or that the molecule may be partially degraded on the cell surface and thus bind differently. These types of reactions are minimized in some studies by conducting the binding at reduced temperatures, 5° or 15° C, where the cell is not metabolically very active and where the binding is not complicated by pinocytosis, phagocytosis, or other membrane changes. Such studies, however, present different problems, such as changes in the fluidity or conformation of the cell membranes. Nevertheless, some approximations are frequently needed in dealing with these complex cellular systems, and valuable data may result.

Clinical comment   **Positron emission tomography (PET)** Short-lived radioactive isotopes, such as ^{11}C, ^{13}N, ^{15}O, and ^{18}F, decay with the emission of two γ-rays at 180-degree directions from the point in space of the isotope. In human applications the compound under study is labeled with the appropriate isotope, injected into the body, and that part of the body (e.g., the head) is positioned in a circular array of γ-detectors. Only when two diametrically opposite detectors are simultaneously stimulated is the decay recorded. Computer analysis of these events can precisely pinpoint the location of the radioactive isotope. In this manner, using labeled opiates for brain studies, it is possible to identify normal or abnormal receptor sites. Similarly, by labeling normal metabolites such as glucose or amino acids with ^{11}C, one can study rates of metabolism under various physiologic or disease conditions. Extensive applications have been made to the study of human brain function and the localization of tumors.

Protein turnover

Why does an adult who is not growing require amino acids for the synthesis of protein? The reason is that protein is continually broken down and resynthesized; it is turning over. The use of radioisotopes makes it possible to measure the "turnover rate" of the proteins of a particular tissue. Most frequently the turnover rates are expressed in terms of the time required for half of the material to be changed. A half-life of 6 days, using a radioactive label as the indicator, means that during this time half of the radioactive protein originally present has been degraded and replaced by newly synthesized protein from the unlabeled amino acids in the body pool.

Most of the proteins of normal nondividing liver cells have half-lives of several days. Structural proteins of the muscle, such as myosin, have longer half-lives of about 180 days, and some of the collagens of connective tissue have very slow turnovers, with half-lives of approximately 1000 days. The average 70 kg man who is not gaining or losing weight synthesizes and degrades about 400 g of protein per day.

Most of the nitrogen metabolized by the body is contained in the proteins. Those proteins that occur in the blood plasma have short half-lives of approximately 10 days, and the erythrocyte has a total life span of only 120 days. Changes in serum nitrogen and protein occur quickly and thus are useful indicators of many disease states. Some common clinical tests measure serum proteins, including albumin and globulins. Blood urea nitrogen (BUN), nonprotein nitrogen (NPN), serum creatinine, serum uric acid, and a variety of other nitrogen-containing substances are also measured.

Other macromolecules besides proteins exhibit turnover. The template or messenger ribonucleic acids (mRNAs) are good examples. The life spans of these substances are important in regulating the amounts of proteins made in response to them. In all cases the metabolic reactions involved in making proteins or nucleic acids are different from those that break them down; therefore the degradative reactions are not simply the reverse of the synthetic ones.

Genetic basis of protein structure

It is common knowledge that the biologic uniqueness of an individual has a genetic basis. The genetic information responsible for determining individual form and substance is stored in the nucleotide sequences of deoxyribonucleic acid (DNA). The information coded in these nucleotide sequences is expressed in a regulated fashion to synthesize proteins whose amino acid sequences are directly determined by the nucleotide sequences in DNA. A few examples of the manner in which genetic variation determines the structure of proteins in humans are considered later, but to put these findings in their proper perspective, it is worthwhile to review the principles of mendelian inheritance.

Mendelism The concept of the gene emerged from the first experiment described in Mendel's classic paper, published in the obscure Czechoslovakian *Journal of the Brno Society of Natural Science* in 1865. In this experiment garden peas that produced round seeds were crossed with another type of peas that had wrinkled seeds. All the seeds that resulted from this first filial generation (F_1) were *round*. The next season Mendel planted the round seeds produced by the previous mating. These plants were allowed to self-fertilize; this second filial generation (F_2) produced 5474 round seeds and 1850 wrinkled seeds, or 2.96 times more round than wrinkled seeds. Several other similar crosses were made, and other characteristics were observed. All these experiments produced the same results. Of the two parental characteristics, only one was seen in the F_1 progeny, but in the second generation the lost characteristic reappeared in one fourth of the progeny. On the basis of this simple experiment, Mendel proposed that each characteristic was produced by a discrete entity that was passed from one generation to the next. We now call this discrete entity a *gene*.

Furthermore, Mendel realized that in the first generation one of the parental characteristics was *dominant;* the other, which failed to appear, was *recessive.* He reasoned that each character was controlled by two genes. For example, the parents with round seeds could be designated as *RR,* whereas those that always gave wrinkled seeds could be called *rr.* In the F_1 generation all the progeny are of the *Rr* type, where *R* is dominant and *r* is recessive. A mating between the hybrids results in a random but *independent segregation* of the genes, so that the progeny of the two *Rr* parents have an equal probability of receiving either an *R* or an *r* gene from either parent. Thus the progeny are evenly distributed between the gene arrangements *RR, rr, Rr,* and *rR*. (*Rr and rR* are actually of identical genetic composition; they are designated in this order to emphasize the four different ways of pairing the two genes.) Only one fourth of the seeds, those with the *genotype* (genetic constitution) *rr,* would be wrinkled. On the other hand, seeds with the different genotypes *RR* and *Rr,* representing three fourths of the whole, would be round; that is, their *phenotype* is the same.

Mendel's interpretation of his data, based on the concept of paired genes, is applicable to the human set of diploid chromosomes. There are 22 sets of paired chromosomes and two sex chromosomes for a total of 46. The sex chromosomes in the female are also paired and are designated as XX, indicating that the somatic cells each contain a pair of X chromosomes. In male somatic cells the sex chromosomes are designated as XY. The other 44 chromosomes are called *autosomal* chromosomes. Many human diseases can be analyzed from family pedigrees in terms of Mendel's rules. Often it can be deducted that a disease is inherited as an autosomal dominant or recessive trait or an X-linked dominant or recessive trait. This information is very useful in genetic counseling.

In one respect Mendel was fortunate. He happened to study characteristics that were clearly either dominant or recessive. Many characteristics are of an intermediate type in which the heterozygote shows a phenotype intermediate between those of the homozygotes. As might be expected, the phenotypes of the human heterozygotes are not always pure dominant or recessive, and in these cases an intermediate severity of the disease may be found.

One gene—one polypeptide concept At first Mendel's ideas about the dominant factors were interpreted in terms of the presence of a dominant gene and the absence of recessive gene in the F_1 generation. However, it subsequently became clear that the recessive gene is not actually absent but is present in either an inactive or a modified form. These changed genes occur as the result of mutations, those rare events caused by chemical changes in the nucleotide bases of the DNA.

The modified gene reflects this change in the phenotype. For many years investigators searched for the most immediate product of the gene's action, little realizing that in 1902 the English physician Garrod, aware of Mendel's ideas, had correctly concluded from his own observations on human genetic diseases that genes were responsible for producing enzymes. He proposed that a defect in a gene would produce an enzyme defect, and that this was the basis of inherited diseases. Some 30 years later the work of Beadle and Tatum conclusively proved this idea, and their names are generally associated with the one gene—one enzyme concept. Because of our more recent knowledge of the subunit structure of proteins and enzymes, we know that the one gene—one enzyme concept should be restated as the one gene—one polypeptide concept, since the amino acid sequence of all proteins is determined by the nucleotide sequence of DNA, the chemical substance composing genes.

Thus we can think of one gene for the α-subunit of hemoglobin and another for the β-subunit. If, as in sickle cell anemia, the gene for the β-subunit is mutated, a person heterozygous for this autosomal disease would have all normal α-chains, but half the β-chains would be normal and half would be abnormal. Consequently, one would predict that half of the hemoglobin in the carriers of this disease would be normal Hb A, and half would be abnormal Hb S. However, the carriers (heterozygotes), who are said to have the sickle cell trait rather than the disease, have somewhat more Hb A than Hb S. It is not completely clear why this is so. For most genetic diseases, however, a carrier will have approximately half the gene product of a normal individual. Because this is often enough of the protein to perform its function, most human genetic diseases are recessive.

Gene dosage and the X chromosome Mutations in the X chromosome are usually not serious in females because they have a pair of X chromosomes; however, X-linked mutations may seriously afflict males, since they have only a single X chromosome. The male is said to be hemizygous in respect to the genes on the X chromosomes. The X chromosome is quite large compared to the Y chromosome and is expected to carry a considerable amount of genetic information. It is clear, however, that females make the same amount of the gene products of the X chromosomes as males, not twice as much. The reason for this is explained by the *Lyon hypothesis,* which states that very early in embryogenesis one of the X chromosomes of female somatic cells becomes metabolically sequestered by forming a Barr body. The Barr bodies are seen in the interphase nuclei of these cells, one for each normal cell. Lyon further suggests that selection of the particular X chromosome to be inactivated is done *randomly,* so that approximately half the cells contain the X chromosome of the female's father and the other half the X chromosome of the mother. Cloning experiments have borne out this prediction. The cells of the female are said to be *mosaic* in respect to the X chromosome, since some cells contain a functioning X chromosome derived from the father, whereas other cells have the functional X chromosome of the mother.

Because the commitment to sequester the X chromosome occurs early in embryogenesis, when only a few cells exist for particular functions, the probability for an even distribution of the X chromosome is lower than it would be if many cells were involved. Consequently, a population of females will show a whole spectrum of activity, from that of the normal to that of the homozygote, for an X-linked product such as glucose 6-phosphate dehydrogenase. However, the number of individuals at each extreme is small; most will show the typical intermediate values of the usual heterozygote.

It is still possible for a female to have certain X-linked diseases. For example, she could have a disease that is inherited as a dominant trait, such as manic depression. She could also inherit an X-linked recessive disease if defective X chromosomes came from both her father and her mother. In this situation the X-linked disease must be mild, since the father, who would also have the disease, must be able to reproduce.

The automosal chromosomes of somatic cells differ from the X chromosomes in that both members of the pair are active. In a few instances individual genes of one member of an autosomal pair of chromosomes may be inactivated without inactivating the rest of the chromosome or the homologous gene on the other member of the pair.

Bibliography

Chothia C: Principles that determine the structure of proteins, Annu Rev Biochem 53:537, 1984.

Chou PY and Fasman GD: Empirical predictions of protein conformation, Annu Rev Biochem 47:251, 1978.

Davies DR and Metzger H: Structural basis of antibody function, Annu Rev Immunol 1:87, 1983.

Dickerson RE and Geis I: The structure and action of proteins, New York, 1969, Harper & Row Publishers, Inc. *Describes in simple terms the relationship of the structure and function of those proteins that have been studied at an atomic level. The illustrations are excellent.*

Karplus M and McCammon JA: The dynamics of proteins, Sci Am 254:42, 1986.

Kreeland JB et al: Magnetic resonance imaging of a fractured temporomandibular disc prosthesis: a case report, J Comput Assist Tomogr 11:199, 1987.

Phelps ME and Mazziotta JC: Positron emission tomography: human brain function and biochemistry, Science 228:799, 1985.

Price BJ: Monoclonal antibodies: the coming revolution in diagnosis and treatment of human disease, Ann Otol Rhinol Laryngol 96:497-504, 1987.

Clinical examples

A diamond (♦) on a case or question indicates that literature search beyond this text is necessary for full understanding.

Case 1   Alpha$_1$-antitrypsin deficiency

A severe form of obstructive lung disease was found in several members of a family. One brother had died earlier of lung disease. Blood plasma from his surviving brother and sister showed abnormally low concentrations of α_1-antitrypsin (3 to 5 μM; normal, 20 to 48 μM). The α_1-plasma fraction also moved abnormally on isoelectric focusing gel electrophoresis.

Biochemical questions   1. Is this obstructive lung disease X linked or autosomal?
2. The destruction of lung tissue seen in emphysema is caused by elastase carried by neutrophils from the bone marrow to the lungs. Does α_1-antitrypsin inhibit neutrophil elastase as well as trypsin? Explain.
3. Where is α_1-antitrypsin made, and what type of molecule is it?
4. How is α_1-antitrypsin activity measured?
5. The unusual electrophoretic pattern suggests that the patients have an abnormal α_1-antitrypsin molecule. The most common defects in such patients result from a substitution of a lysine residue for a glutamate at position 342; this is the Z variant. What is the consequence of this mutation on the structure and function of α_1-antitrypsin?
6. Electrophoresis of α_1-antitrypsin is usually done at pH 4.9. How would the Z variant migrate in respect to the normal molecule, closer to the anode or to the cathode? The S variant of this molecule has a valine substituted for a glutamate at position 264. How would this variant migrate in respect to the normal molecule?
7. Methionine[358] and serine[359] are important residues on α_1-antitrypsin that bind essentially irreversibly to the active site of elastase, thereby inactivating the protease. One effect of cigarette smoke is to oxidize this methionine residue so that it no longer binds tightly to elastase. How is the methionine residue changed chemically?
8. What are the prospects for treatment of this emphysema?

Case discussion   Emphysema is a lung disease characterized by destruction of the alveolar walls. Its causes are complex, but airway infection, cigarette smoking, air pollution, and familial factors may be involved. A deficiency in plasma α_1-antitrypsin leads to the development of emphysema and, less frequently, hepatic disease.
1. **The inheritance of α_1-antitrypsin deficiency** The two patients described in this case are male and female, suggesting that the deficiency is inherited as an autosomal trait. The genetics has been studied in large numbers of patients and family members. Analysis suggests that two independent alleles exist on an autosomal chromosome. The deficiency is inherited as a codominant trait. The two genes that each person carries for α_1-antitrypsin may be diverse; more than 75 variants are known. These have been classified into four broad groups, those that have (a) normal antitrypsin levels; (b) deficient, or less than 35% of normal;

(c) "null," or no activity; and (d) "dysfunctional," meaning the α_1-antitrypsin is present but functions abnormally. The serious deficiency disease of the homozygote is found in 1 of approximately 1700 people of northern European ancestry. Approximately 5% of these people carry the gene as heterozygotes. It occurs less frequently in other racial groups.

2. **The specificity of α_1-antitrypsin** α_1-Antitrypsin was named because it is a good inhibitor in vitro of the pancreatic protease trypsin. The inhibitor should more properly be called α_1-protease inhibitor or α_1-antiprotease to reflect its activity against many serine proteases. Serine proteases have an essential serine residue at their active centers and are composed of enzymes such as plasmin, thrombin, plasminogen, cathepsin G, chymotrypsin, trypsin, and neutrophil elastase. In vivo α_1-antitrypsin primarily inhibits neutrophil elastase; however, it can inhibit many of the proteases released from dying cells and thus serves to protect normal tissues during periods of stress, such as inflammation. In the absence of the inhibitor, elastase will degrade lung tissue. It is one of the few enzymes that can attack insoluble, cross-linked elastin, the structural protein responsible for elastic recoil in the lung.

3. **Origin and composition of α_1-antitrypsin** α_1-Antitrypsin is made primarily in liver. It is secreted into the bloodstream and probably functions as an important protease inhibitor for several tissues. The inhibitor makes up most of the proteins in the α_1-globulin band following the electrophoresis of plasma. A decreased concentration leads to diseases of the lung, although the liver is sometimes affected.

    α_1-Antitrypsin is a single-chain protein of 394 amino acids, Mr 52,000. Three of its asparagine residues have complex carbohydrate side-chains attached. It has an elongated shape; 30% of the amino acids are in α-helices and approximately 40% in β-sheets. A methionyl-seryl group (positions 358 and 359) is present at the point where the inhibitor binds to the active site of the serine proteases. This is illustrated by the diagram in Figure 2.11. Binding to the protease is extremely tight, so α_1-antitrypsin is called a "suicide" inhibitor; once bound to the protease, it is not released.

4. **Measurement of α_1-antitrypsin** The deficiency state was first observed by the virtual absence of the α_1-band on electrophoresis of blood plasma. The antiprotease activity can be measured on whole plasma using synthetic trypsin substrates such as N-benzoyl-DL-arginine-p-nitroanilide, which produces a yellow colored product, since 95% of the antitrypsin activity of plasma results from α_1-antitrypsin. Radioimmunodiffusion is also used, but isoelectric focusing on polyacrylamide gels is more useful in known genetic cases because specific genetic variants can be distinguished.

5. **Consequences of a mutation from Glu to Lys** A change from an acidic to a basic amino acid might be very serious. It is known that the mutant antitrypsin is poorly secreted from the liver cells where it is made. The mutant protein accumulates in the endoplasmic reticulum. It is believed that the normal salt bridge between Glu^{342} and Lys^{290} fails to form, which causes the molecule to fold more slowly into its three-dimensional conformation. This allows usually internal hydrophobic amino acids to interact between chains and thus causes aggregation and secretory failure. However, other interpretations have been proposed.

6. **Electrophoresis of genetic variants of α_1-antitrypsin** The Z variant with a Glu residue substituted with a Lys residue would be expected to make the molecule much less anionic, by two charges. Consequently, compared to the normal protease inhibitor, the Z variant would migrate close to the cathode. The S

Figure 2.11   Representation of structural elements on α_1-antitrypsin and neutrophil
                         elastase. (Used with permission of Crystal RG et al: Chest 95:196,
                         1989.)

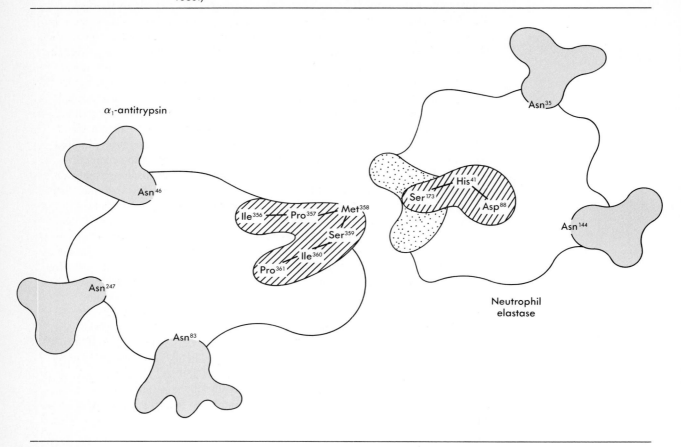

variant, which has a valine in place of a glutamate residue, would also be less
anionic, by one charge. It also would migrate abnormally closer to the cathode,
although not as close as the Z variant.

7. **The oxidation of methionine residues** Methionine residues can be oxidized to
   the corresponding sulfoxide, as shown here:

$$
\begin{array}{ccc}
\text{COO}^- & & \text{COO}^- \\
| & & | \\
\text{H}_3\overset{+}{\text{N}}\text{—CH} & & \text{H}_3\overset{+}{\text{N}}\text{—CH} \\
| & & | \\
\text{CH}_2 & \xrightarrow{[O]} & \text{CH}_2 \\
| & & | \\
\text{CH}_2 & & \text{CH}_2 \\
| & & | \\
\text{S} & & \text{S} = \text{O} \\
| & & | \\
\text{CH}_3 & & \text{CH}_3
\end{array}
$$

Further oxidation to methionine sulfone is possible in vitro, but it is unlikely
that much sulfone is made in vivo. On the other hand, the presence of

methionine sulfoxide leads to the inactivation of several proteins, most notably α_1-antitrypsin. The biologic oxidation of methionine residues in proteins might be accomplished by a hypochlorite ion, hydrogen peroxide, or superoxide and hydroxyl radicals. It is not known how cigarette smoke leads to the oxidation of methionine, but this probably results from an oxidizing agent in the smoke itself. The activity of α_1-antitrypsin from the lung lavage of smokers is about half that from nonsmokers, whereas the total amount of immunologically assayed antitrypsin is about the same. Oxidized methionine residues in α_1-antitrypsin may also contribute to the pathology of rheumatoid arthritis and may play a role in the formation of lens cataracts.

Certain enzymes can reduce methionine sulfoxide residues in vitro back to methionine and restore inhibitory activity to α_1-antitrypsin. Undoubtedly these enzymes are important in vivo, but to what extent is unknown.

8. **Treatment of emphysema** Much recent progress has been made in administering recombinant α_1-antitrypsin as an aerosol to sheep. The 45 kDa protein remains intact and functional after passing through the pulmonary epithelial surface. The recombinant protein has been isolated from yeast cells that were transformed so that they expressed an active gene for human α_1-antitrypsin. These experiments have not yet been performed with humans.

References

Brot N and Weissbach H: The biochemistry of methionine sulfoxide residues in proteins, Trends Biochem Sci 7:137, 1982.

Crystal RG et al: The alpha$_1$-antitrypsin gene and its mutations: clinical consequences and strategies for therapy, Chest 95:196, 1989.

Eriksson S: Alpha$_1$-antitrypsin deficiency: lessons learned from the bedside to the gene and back again, historic perspectives, Chest 95:181, 1989.

Hubbard RC et al: Fate of aerosolized recombinant DNA-produced α_1-antitrypsin: use of the epithelial surface of the lower respiratory tract to administer proteins of therapeutic importance, Proc Natl Acad Sci USA 86:680, 1989.

Resendes M: Association of α_1-antitrypsin deficiency with lung and liver diseases, West J Med 147:48, 1987.

Case 2   Rheumatoid factors in rheumatoid arthritis

A 35-year-old woman was diagnosed as having rheumatoid arthritis. She showed the typical symptoms of morning stiffness and painful, symmetrically swollen joints. Serologic tests for rheumatoid factor were positive. In this test, dilutions of the patient's serum cause the agglutination of sheep red blood cells or polystyrene (Latex) particles that have been previously coated with human IgG.

Biochemical questions

1. Rheumatoid factor can be thought of as an antibody. What is the antigen that stimulates its production?
2. Rheumatoid factor is usually, but not always, IgM. Describe the structure of IgM.
3. IgM can be made to yield monomers by treatment with mercaptoethanol. How are the monomers held together in IgM?
4. Mixtures of IgM and IgG can be separated from one another on a column of Sephadex-G200, a molecular sieve. Which component will emerge first?
5. Treatment of IgG with pepsin, a proteolytic enzyme, produces a fragment called F_{ab} and several small fragments. Such treatment prevents the complexing of IgG with rheumatoid factor. What can you conclude about the location of the antigenic determinants of IgG?

Case discussion

Clinical features Rheumatoid arthritis is a chronic inflammatory disease of the joints that affects mainly middle-aged people. The incidence in females is twice that of males. Its causes are unknown, although infection or trauma is thought to initiate or predispose to an immune response against the patient's own joint tissue.

Rheumatoid factors Rheumatoid factors of the IgM type can be detected in the sera of 60% to 75% of patients with this disease using the Latex agglutination method and in 69% to 98% when assayed by newer enzyme immunoassays. The latex method picks up only IgM rheumatoid factors. More sensitive procedures indicate that all patients have rheumatoid factors of either the IgM or IgG types. False positive reactions are common, however, since the IgM factor can be produced experimentally by repeated injections of dead bacteria into animals. Apparently IgG antibodies formed in response to the microorganisms combine with the bacteria to produce a complex that is itself immunogenic. This complex might cause the production of another set of antibodies (rheumatoid factors) against the IgG bound to the dead bacterial cells. This may explain the high titers of rheumatoid factor in subacute bacterial endocarditis, a chronic infection of the heart valves.

1. Factor stimulating production of rheumatoid factor The initiating stimulus is far from clearly understood. Dead bacteria are not necessarily the cause, but they cannot be ruled out. Mycoplasmas, viruses, and streptococci have all been suggested as initiating factors. The inflammation of the joints may be caused by the presence of rheumatoid factor complexes and IgG within the leukocytes that cause them to rupture and spill out their lysosomal enzymes—a collection of destructive hydrolytic enzymes that break down tissue, produce inflammation, and perpetuate the autoimmune cycle. The release of lymphokines from patients' cells can also be triggered by type II (cartilage) or type III (blood vessel) collagen. Lymphokines are nonantibody mediator substances with a diversity of biologic effects. In this case they activate inflammatory cells.

2. Structure of IgM Some investigators are interested in trying to chemically modify the structure of the rheumatoid factors to inhibit the inflammatory process. To understand the problems involved, it is necessary to describe the structure of the immunoglobulins in more detail in the following paragraphs. When polymeric IgM is gently treated with the reducing agent 2-mercaptoethanol, the 19S molecule is broken into five equal parts that sediment at 7S, much as with IgG.

3. Dissociation of IgM Mercaptoethanol dissociates the molecule by promoting disulfide exchange reactions, which result in breaking disulfide bonds between cysteine residues in adjacent IgM monomers. These bonds are shown as S—S bridges in Figure 2.12.

4. Separation of IgM and IgG Unlike IgG, the IgM monomers are composed of $\mu_2\lambda_2$- or $\mu_2\kappa_2$-chains. Molecular sieves such as gels of Sephadex-G200 will separate any unreacted IgM from its IgG-like monomers. The smaller 7S proteins penetrate the gels, whereas the larger 19S molecules are unable to do so and rapidly flow around the gel particles and emerge from the column first. Because reduction is required to disaggregate IgM, it is thought that the polymeric molecule is held together by disulfide bridges. (One model is given in Figure 2.12.) Rheumatoid factor that has been reduced to yield 7S monomers loses its activity to agglutinate IgG-coated Latex beads. The monomers, however, are inhibitors of IgM-promoted agglutination.

The IgM immunoglobulins can be broken into several small identical fragments, F_{ab}, and one large fragment, F_c, by treatment with trypsin at 60° C. The cleavage points are indicated by the dashed lines in Figure 2.12. F_{ab} fragments still contain enough structure to bind their homologous antigen; thus only the variable portion of the

Figure 2.12

Model of pentameric IgM molecule. Zigzag lines represent variable *N*-terminal regions of H and L chains. Dashed lines indicate sites cleaved by papain to form F_{ab} and F_c fragments.

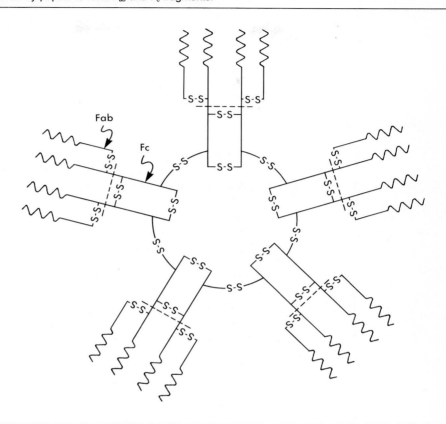

molecule is required for antigen recognition. The subscript "ab" indicates that this F fragment retains the specificity of the intact antibody.

IgG molecules also yield F_{ab} and F_c pieces when treated with trypsin or papain, a proteolytic enzyme from papaya. Because there are many common amino acid sequences contained in the F_c fragments of IgG, F_c molecules can be crystallized, a property that is indicated by the subscript "c."

5. Epitopes of IgG Pepsin treatment of IgG produces a fragment similar to F_{ab} that contains the variable region of the molecule, and it breaks the F_c portion into several peptides. This treatment also prevents complexing of IgG fragments with rheumatoid factor. These experiments suggest that the antigenic portion of IgG, considering rheumatoid factor as the antibody, is located in the nonvariable F_c region of the molecule. The sequences of several IgG species are known, and since the F_c region is similar in most of them, pharmacologic intervention using chemically synthesized portions of this region may become useful in the future for certain therapeutic situations. Remember, however, that the biochemical basis of rheumatoid arthritis is not understood, and this hypothesis should be considered in that light.

References

Dorner RW, Alexander RL, and Moore TL: Rheumatoid factors, Clin Chim Acta 167:1, 1987.
Levinson AI and Martin J: Rheumatoid factor: Dr. Jekyll or Mr. Hyde? Br J Rheumatol 27:83, 1988.

Case 3   Familial amyloid polyneuropathy

The patient was a 35-year-old female living in Lisbon, Portugal. Her first symptoms were episodes of constipation and abdominal colic, which appeared 4 yr earlier. Eventually the week-long attacks became more severe, with alternating periods of constipation and diarrhea. She failed to feel a severe burn on her foot 3 yr earlier, but presently she felt burning sensations in her legs. Analysis of a skin biopsy revealed an amyloid substance in the erector muscles of the hairs and walls of the small vessels. (Adapted from Andrade C: Brain 75:408, 1952.)

Biochemical
questions:

1. What is amyloid? How is it detected?
2. What type of amyloid is present in this case?
3. What are the consequences of amyloid deposition?
4. Why does amyloid form in some people and not in others?
5. Can amyloidosis be successfully treated? Explain.
6. Can the disease be prevented?

Case discussion

Amyloidosis is a syndrome characterized by the extracellular deposition and progressive accumulation of a proteinaceous fibrillar material called amyloid. Amyloid was first observed more than 300 yr ago, but not until 1840 was the substance named by the distinguished pathologist Rudolph Virchow. Virchow called the material amyloid (starchlike) because it turned purple when treated with acidic iodine.

1. Nature of amyloid The name has persisted, but now we know that the substance is mostly protein rather than carbohydrate. The diagnosis of amyloidosis is based on the reaction of these extracellular deposits with the dye Congo red. This forms a product that shows a green birefringence when viewed under a polarizing microscope.

This green color is also displayed by fibrous proteins that are arranged in antiparallel β-pleated sheets. Most, but not all, amyloid has the antiparallel β-sheet structure. This structure is typical of silk fibroin; these structures are not usually found in vertebrates. If vertebrates lack the cellular enzymes necessary to degrade antiparallel β-sheet proteins, this might account for the progressive accumulation of amyloid.

2. Type of amyloid The amyloid in this case is composed of a mutated form of a serum protein called prealbumin. This is the protein that accumulates in type I and type II familial polyneuropathies. Prealbumin was named because it preceded the migration of albumin during the electrophoresis of serum. Actually prealbumin is not related to albumin at all. The more recent literature calls prealbumin *transthyretin,* because it is a *trans*port protein for *thy*roxine and *retin*ol (vitamin A). Even normal transthyretin has a tendency to form fibers; thus the slightest change, such as a mutation of a single amino acid, is enough to accelerate polymerization. The mutant transthyretin forms fibrous deposits in the most common form of amyloid neuropathy. The fibers contain normal as well as mutant transthyretin, a finding not unexpected in this autosomal dominant disease. The gene for transthyretin is carried on chromosome 18.

The amino acid substitutions are known for six of seven variants of the transthyretin gene. All but one of these variants affect the nervous system. The exception is a leucine-to-methionine mutation at position 111 that affects primarily the heart.

Not all amyloid is composed of transthyretin, not even for the inherited amyloidoses. For example, in one type of hereditary amyloidosis, the deposited protein is a mutant fragment of cystatin C. The accumulated fragment of 110 amino acids is derived from cystatin C, a γ-trace protein of 120 amino acids that carries a glutamine residue in place of leucine at position 58. γ-Trace proteins are basic serum proteins that migrate on electrophoresis near γ-immunoglobulins. The name cystatin C has been given to one of the γ-trace proteins that is an inhibitor of cysteine proteases. In other hereditary amyloidoses the accumulated proteins are derived from either apolipoprotein A_1 (type III, Iowa), calcitonin (a hormone), or β-protein. β-Protein (Mr 4200) accumulates in the brain in patients with Down's syndrome and with Alzheimer's disease. The gene for β-protein is located on chromosome 21 near, but not part of, a locus associated with familial Alzheimer's disease. The β-protein is part of a larger 695-residue transmembrane protein.

The amyloid deposits of the presumably acquired amyloidoses consist of a variety of proteins as well. Recent interest has focused on the β_2-microglobulin-containing amyloid that forms in patients with kidney problems whose blood has been dialyzed for several years. β_2-Microglobulin is part of the major histocompatibility complex. It is continuously shed from its location on cell membranes and is normally catabolized in the kidney. In patients with renal insufficiency the protein accumulates in the plasma. Dialysis patients have very high plasma levels. The carpal tunnel syndrome that often develops in these patients after several years of dialysis is probably caused by β_2-microglobulin-containing amyloid.

3. Consequences of amyloid deposition Little is known about the pathogenesis of amyloid formation. Usually the subunit protein is first modified genetically or proteolytically. Fibril formation is often associated with the polymerization of a less-soluble but smaller protein fragment. Deposited amyloid is not removed but continues to accumulate until it interferes with the normal structure and function of particular tissues. The same mutation may affect different tissues in different kindreds or ethnic types; consequently, other genes determine which tissue will be affected. Often the nervous system is compromised, but kidneys, heart, bowel, eyes, skin, and thyroid may be involved as well.

4. Predilection for amyloid deposition Great biologic variation undoubtedly exists in amyloid fibril formation. As mentioned, kidney function is related to the deposition of β_2-microglobulin amyloid. Endocrine amyloidosis (precalcitonin fibrils), cutaneous amyloid (keratin-derived fibrils), and ocular amyloid may represent diseases with a strong hereditary basis.

In those patients who develop transthyretin fibrils, such as the present one, the diseases are autosomal dominant; thus all genetic carriers will develop the disease eventually.

5. Treatment of amyloidosis No effective treatments reduce amyloid deposits; also, it is extremely difficult to analyze the efficacy of a therapy, since it is so difficult to sample the affected tissue short of autopsy. No promise is held for the use of γ-camera imaging to detect ^{123}I-labeled amyloid P component in whole animals. Amyloid P is a small serum protein that makes up 5% to 15% of virtually all amyloid fibrils. Thus γ-radio imaging might be useful to detect the extent of amyloid deposition in vivo, both for diagnosis and for charting the effectiveness of possible therapies.

6. Prevention of the inherited diseases It may be possible to prevent some of the cases of familial amyloid polyneuropathy with genetic counseling. DNA probes are available to detect the transthyretin gene before the clinical disease occurs and before the reproductive age. These tests would be limited to individuals who have known affected relatives. Fortunately the inherited amyloidoses are relatively uncommon.

References

Benson MD: Hereditary amyloidosis—disease entity and clinical model, Hosp Pract, March 15, 1988.

Gorevic PD and Munoz P: Dialysis amyloidosis: beta-2-microglobulin in the context of other amyloidogenic proteins, Blood Purif 6:132, 1988.

Pepys MB: Amyloidosis: some recent developments, Q J Med 67:283, 1988.

Varga J and Wohlgethan JR: The clinical and biochemical spectrum of hereditary amyloidosis, Semin Arthritis Rheum 18:14, 1988.

Case 4	Monoclonal antibodies for tumor targeting

A ruptured adenocarcinoma of the appendix and a tumor coating the liver, diaphragm, and peritoneum was discovered in a 33-year-old woman undergoing surgery for another reason. Three years later most of a large tumor was removed. The patient then became part of a study using monoclonal antibody treatment. Iodine 131–labeled mouse monoclonal antibodies were administered intraperitoneally. Much of the labeled antibody was cleared, but after 8 days, when surgery was done, there was a clear localization of the heavily labeled antibody to a region adjacent to the liver and overlying the spleen.

The monoclonal antibody used was made with an antigen consisting of a membrane-rich fraction of human liver tissue metastasized by a mammary carcinoma. The antibody reacts with *tumor-associated glycoprotein-72,* a protein associated with several breast carcinomas and adenocarcinomas of the colon, ovaries, stomach, endometrium, and pancreas. It is rarely found in other tumors or in normal tissues. (Adapted from Schlom J: JAMA 261:744, 1989.)

Biochemical questions ◆ 1. What functional groups on proteins are modified by iodination? What potential problems might be associated with the iodination of immunoglobulins? Of tropocollagen?
2. Explain how radioactively labeled monoclonal antibodies are useful for both the detection and the treatment of tumors.
3. Explain why a monoclonal antibody may be too specific to be useful.
◆ 4. What other types of molecules might be attached to monoclonal antibodies so that they can be delivered to specific tumor cells?

References

Carrasquillo JA et al: Peritoneal carcinomatosis: imaging with intraperitoneal injection of I-131-labeled B72.3 monoclonal antibody, Radiology 167:35, 1988.

Johnston WW et al: Applications of immunocytochemistry to clinical cytology, Cancer Invest 5:593, 1987.

Schlom J: Innovations in monoclonal antibody tumor targeting: diagnostic and therapeutic implications, JAMA 261:744, 1989.

Case 5	Aspirin-induced alteration of human serum albumin

A 56-year-old woman with rheumatoid arthritis had been taking 4 g of sodium salicylate daily for 4 mo when a serum sample was obtained. This drug was stopped, and she was given 4 g of aspirin daily. Another serum sample was obtained 5 wk later.

The serum albumin was separated from the other serum proteins by sodium sulfate fractionation and was further purified by electrophoresis. The purified albumins, including a normal control sample, were reduced and carboxymethylated. The samples were dialyzed against water and lyophilized. Hydrolysis with trypsin was done in a 2 mol/L urea solution containing 0.1 mol/L ammonium bicarbonate, pH 9.0. The hydrolysate was diluted and applied to a Dowex-50 column. The column was washed with water and the peptides eluted with 4 mol/L NH$_4$OH and dried. About 5 mg of the peptides was applied to paper and chromatographed in one dimension, followed by electrophoresis at pH 3.55 in the other direction. The peptide maps of the patient's albumin showed a unique peptide, called "A," which was not present in tryptic digests from control albumin. As "A" was increased, two other peptides, "L" and "C," were reduced as compared to the control sample.

Biochemical questions
1. Why were the albumins reduced and carboxymethylated? Why was the trypsin treatment done in 2 mol/L urea?
2. Considering the specificity of trypsin, what can be said about the primary structure of the peptides produced?
3. Explain how the peptides are bound to the Dowex-50 column and eluted with 4 mol/L NH₄OH.
4. Peptide "C" was found to contain equimolar amounts of lysine and leucine. Is the structure of the dipeptide Lys-Leu or Leu-Lys? Explain.
5. About 2% of asthmatic patients are hypersensitive to aspirin but not to sodium salicylate. Try to explain this from the results already presented.
6. Aspirin inhibits prostaglandin synthesis by acetylating the enzyme cyclo-oxygenase. What functional groups on the enzyme would you predict are acetylated?

References
Aarons L et al: Aspirin binding and the effects of albumin on spontaneous enzyme-catalyzed hydrolysis, J Pharm Pharmacol 32:537, 1980.
Hawkins D et al: Structural changes in human serum albumin induced by ingestion of acetylsalicylic acid, J Clin Invest 48:536, 1969.
Roth GJ et al: Acetylation of prostaglandin synthase by aspirin, Proc Natl Acad Sci USA 72:3073, 1975.

◆ **Case 6**   **Insulin resistance**

A 13-year-old girl and a 38-year-old woman had acanthosis nigricans. Both had hyperinsulinemia and were intolerant to exogenous insulin. The younger patient showed a marked reduction of the binding of [¹²⁵I] insulin to circulating monocytes (type A). This was caused by a reduction in the number of receptor sites per cell. The older woman had circulating monocytes with a change in the affinity of insulin for cellular receptors (type B). This lower affinity was caused by circulating autoantibodies to insulin receptors.

Biochemical questions
1. Describe experiments that would detect a change from normal in the number of insulin receptors per cell in the 13-year-old patient.
2. How would one test for a change in the insulin affinity of a normal number of receptors per cell?
3. It has been found that F_{ab} and $F_{(ab)2}$ prepared from autoantibodies against insulin receptors have different effects on adipocyte insulin receptors. Do they inhibit insulin action or do they mimic insulin action?
4. How are these antibody fragments prepared?
5. Some patients (not this case) develop a syndrome characterized by hypoglycemia and spontaneous development of insulin autoantibodies. What might account for this?

References
Flier JS: Receptors, antireceptor antibodies and mechanisms of insulin resistance, N Engl Med 300:413, 1979.
Flier JS: Insulin receptors and insulin resistance, Annu Rev Med 34:145, 1983.
Flier JS: Metabolic importance of acanthosis nigricans, Arch Dermatol 121:193, 1985.

Case 7   Multiple myeloma

F.R., a 66-year-old man, had fallen out of a tree 4 yr previously without any apparent fractures. Since that time he complained of pain over the ribs. The pain worsened by coughing or straining and became more severe in the 2 mo before he sought medical help. This patient had had a firm, marble-sized mass in his neck for 10 to 15 yr; the size had remained constant until the previous 6 mo. At that point the mass approximately doubled in size. The patient denied pain associated with the mass, hoarseness, dysphagia, or respiratory obstruction as well as easy bruising or bleeding, anemia, shortness of breath on exertion, palpitation, or tachycardia.

An x-ray film obtained 1 yr previously by the patient's local physican revealed multiple rib fractures, and he was referred to a university hospital. A complete workup was done, the significant finding being that serum and urine immunoelectrophoresis reveal the presence of monoclonal L (light) chains.

Biochemical questions ◆ 1. What does the term *monoclonal* signify with respect to immunoglobulins?
2. How do L and H (heavy) immunoglobulin chains differ?
3. In what way are L and H chains similar?
4. How are L and H chains held together in a typical IgG molecule?
◆ 5. Explain the occurrence of L chains in the urine of this patient.

References

Cohen HJ: Monoclonal gammopathies and aging, Hosp Pract 23:75, 1988.
Durie BGM: Staging and kinetics of multiple myeloma, Semin Oncol 13:300, 1986.
Tonegawa S: The molecules of the immune system, Sci Am 253(8):122, 1985.

Additional questions and problems

1. Would serum albumin levels be elevated or decreased in the following disorders: dehydration, malnutrition, nephrosis, chronic hepatic insufficiency? Explain your answers.
2. A child with congenital agammaglobulinemia, the inability to synthesize appreciable amounts of γ-globulin, was constantly ill with bacterial diseases. Explain the underlying mechanism of this child's illness.
3. A patient with the nephrotic syndrome was losing large quantities of albumin in his urine. His plasma albumin concentration fell to 10 g/L. He subsequently developed edema (swelling) caused by the collection of fluid in extracellular spaces. Explain.
4. Bradykinin is a plasma peptide and a potent vasodilator. It is derived from a larger precursor called kininogen. Bradykinin has the following structure: Arg-Pro-Pro-Gly-Phe-Ser-Pro-Phe-Arg. What is its isoelectric point? What kind of enzyme would produce bradykinin from kininogen? Which amino acids in bradykinin have reactive functional groups in their side-chains? What are these groups?
5. Erythrocytes from patients with anorexia nervosa show increased binding of [^{125}I] insulin (N Engl J Med 300:882, 1979). Explain how a Scatchard plot would be useful in analyzing how binding is increased.
6. (a) Describe the peptide bond. (b) What treatments might lead to the denaturation of proteins in a sample of human plasma? (c) Explain why albumin migrates faster on electrophoresis at pH 8.3 than β-globin. (d) Distinguish simple proteins from complex proteins. (e) What is meant by the primary, secondary, and quaternary structures of proteins?
7. An enzyme, which is a monomer in its natural state, is reversibly denatured. Describe the processes whereby the native conformation returns on removal of the denaturing agent.
8. What is the isoelectric point of asparagine, aspartic acid, and glycyl aspartic dipeptide?
9. Discuss the relative merits of imaging by roentgenography, MRI, and PET.

Multiple choice problems

Informed consent was obtained from a patient undergoing hand surgery for the testing of the effect of a lathyrogen on scar formation. β-Aminopropionitrile fumarate was administered the day before surgery. Problems 1 to 5 refer to this case.

1. The collagen in scar tissue is hard and stiff because the polypeptide chains are:
   1. Long chains of repeating Gly-Pro-X units.
   2. Arranged in long, stable α-helices.
   3. Arranged in antiparallel pleated sheets.
   4. Covalently cross-linked.

   The best answer is:
   A. 1, 2, and 3.
   B. 1 and 3.
   C. 2 and 4.
   D. 4 only.
   E. All are correct.

2. The semialdehyde that reacts with the ε-amino groups of lysyl residues in collagen to form a Schiff base is derived from the amino acid residue:
   A. Lysine.
   B. Hydroxyproline.
   C. Aspartate.
   D. Histidine.
   E. Glycine.

3. These semialdehydes are formed under the action of enzymes characterized as:
   A. Aldolases.
   B. Amine oxidases.
   C. Collagenases.
   D. Elastases.
   E. Transaminases.

4. Almost one third of all the amino acid residues in collagen are:
   A. Lysine.
   B. Hydroxyproline.
   C. Aspartate.
   D. Leucine.
   E. Glycine.

5. The protein subunits that make up the smallest linear chains of the collagen molecule are all approximately the same molecular weight, that is:
   A. 20,000.
   B. 50,000.
   C. 100,000.
   D. 200,000.
   E. 300,000.

The Lesch-Nyhan syndrome is an inherited disease that exclusively affects young boys. The enzyme defect prevents the reutilization of purine bases, which causes a gouty arthritis and the bizarre behavior of self-mutilation. Problems 6 to 10 refer to this syndrome.

6. The gene(s) for the defective enzyme is (are):
   1. Located on the X chromosome.
   2. Located on several chromosomes.

3. Likely inherited from the mother.
4. Likely inherited from the father.
The best answer is:
   A. 1, 2, and 3.
   B. 1 and 3.
   C. 2 and 4.
   D. 4 only.
   E. All are correct.

7. Skin fibroblasts from the mother of a Lesch-Nyhan patient were diluted so that when grown in the laboratory, colonies arose from individual isolated cells, that is, clones. Extracts of cells from many different clones were analyzed for the enzyme affected in this disease. The pattern of *active* enzyme found was:

	Percentage of clones with complete activity	Percentage of clones with no active enzyme
A.	25	75
B.	50	50
C.	75	25
D.	100	0
E.	All clones had about 50% of the normal enzyme activity.	

8. Skin fibroblasts from the patient and the patient's father were cloned and the enzyme from each clone assayed, as was done for the mother. The following pattern of activity was found:

	Percentage of father's clones with complete activity	Percentage of patient's clones with complete activity
A.	50	0
B.	100	50
C.	75	25
D.	All had 50% normal activity.	0
E.	100	0

9. Neither the mother nor the father had symptoms; consequently, this disease is:
   A. Autosomal recessive.
   B. X-linked recessive.
   C. Autosomal dominant.
   D. X-linked dominant.
   E. Inherited, but the presumed father is not actually the patient's father.

10. In one form of the Lesch-Nyhan disease, an arginine residue replaces the normal glycine in the affected enzyme. Arginine differs from glycine in that the arginine residue:
1. Is a much larger residue that could affect enzyme conformation.
2. Is a basic molecule that might form inappropriate salt bridges in the enzyme.
3. Has a charged side-chain that might disrupt hydrophobic interactions.
4. Could change the isoelectric point of the enzyme protein.
The best answer is:
   A. 1, 2, and 3.
   B. 1 and 3.
   C. 2 and 4.
   D. 4 only.
   E. All are correct.

Chapter 3

Enzymes and biologic catalysis

Objectives

1 To describe the nature of enzymes and the process of enzyme catalysis as the basis of biochemical transformations of cellular substances
2 To describe the locations of enzymes within cells and their organelles
3 To explain the relationship among enzymes, their physiologic or metabolic function, and how function is controlled
4 To demonstrate how enzymes frequently operate in sequences or interrelated systems that form the basis of so-called metabolic pathways, which in turn are subject to various types of controls
5 To explain how quantitative assay of selected enzyme activities in blood and other body fluids can assist in the diagnosis and treatment of disease

Enzymes catalyze virtually all biologically important reactions. Important areas of medicine have substantially benefited from the application of enzyme analysis; such diseases as myocardial infarcts, hepatitis, cancer of the prostate, obstructive liver disease, and the muscular dystrophies may be cited as common clinical examples. Enzyme activity may be high in some diseases and low or lacking in others.

Tissue enzymes clearly are distributed in a highly organized fashion; that is, cells are not "loose sacks" of enzymes. However, the products of an enzyme reaction in one tissue component may have significant effects on a separate enzyme process in another component of the given tissue or even in an entirely different tissue. Some detailed knowledge of enzyme distribution within cells and of the chemistry and function of enzymes is therefore essential for a detailed understanding of disease mechanisms and therapies.

What are enzymes?

Enzymes are *biocatalysts* produced by living tissue that increase the *rate* of reactions that may occur in the tissue. For example, CO_2 reacts with H_2O to form carbonic acid (H_2CO_3), part of which immediately ionizes at a physiologic pH to form bicarbonate ions (HCO_3^-). The dissociation, of course, does not depend on enzyme catalysis but is a property of the acid structure.

$$H_2O + CO_2 \rightleftharpoons H_2CO_3 \rightleftharpoons H^+ + HCO_3^-$$

The reaction rate of the uncatalyzed decomposition of carbonic acid in H_2O and CO_2 is slow, and true equilibrium may not be reached for 1 hr or more. If one adds the enzyme carbonic anhydrase to a sample of carbonated water, equilibrium is reached in minutes or less. Red blood cells are especially rich in carbonic anhydrase, which promotes the rapid interconversion of carbon dioxide and bicarbonate through the intermediate form of undissociated carbonic acid.

It should be noted that although enzymes increase reaction rates, they do not change the relative concentrations of reactants and products when equilibrium is attained.

Enzyme structure

Almost all known enzymes are proteins. The molecular weights of enzymes cover a wide range. For example, the enzyme ribonuclease, which hydrolyzes ribose-containing nucleic acids, is relatively small, having a molecular weight of approximately 13,700. In contrast, aldolase, an enzyme involved in glucose metabolism, has a molecular weight of approximately 156,500, composed of four subunits, each with a molecular weight of about 40,000. Pyruvate dehydrogenase, which catalyzes the conversion of pyruvate to acetyl CoA, is a multienzyme complex in which the components are so tightly organized that the entire system can be isolated as a discrete, particulate entity from many tissues. The complex from pig heart has a molecular weight of about 10×10^6; each complex contains no fewer than 42 individual molecules, including several important and essential cofactors. The entire structure of the pyruvate dehydrogenase complex is required for catalysis.

Enzyme cofactors

In addition to the protein component, many enzymes require nonprotein constituents for their function as catalysts. These accessory moieties are variously termed prosthetic group, cofactor, and coenzyme. The term *prosthetic group* applies to any non–amino acid portion of a protein that confers on that protein some particular property. Thus heme is the prosthetic group of hemoglobin, and it confers the property of a high capacity for oxygen binding. In hemoglobin the prosthetic group is held by noncovalent forces; in other heme proteins such as the cytochromes the hemes are held by covalent bonds. The term *cofactor* is also broadly defined. Small organic molecules such as phospholipids are essential to maintain certain enzyme proteins in a conformation suitable for catalysis, as is true of β-hydroxybutyrate dehydrogenase. Accordingly, these lipids are regarded as cofactors even though they do not directly participate in the catalytic event. Some enzymes require

Table 3.1   Common Coenzyme Structures

Vitamin	Coenzyme

Thiamine (B₁)

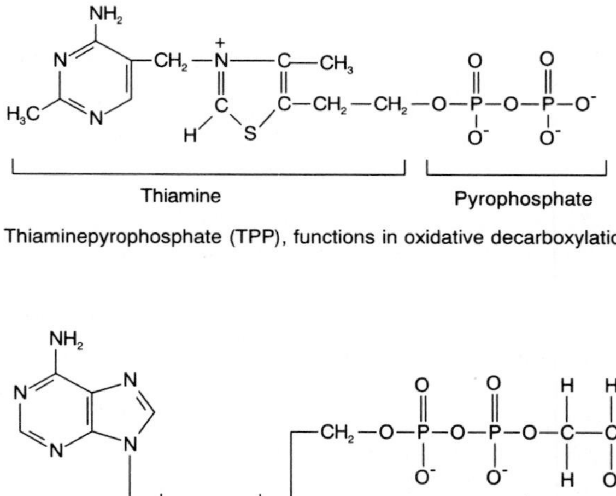

Thiaminepyrophosphate (TPP), functions in oxidative decarboxylations and transketolase reactions

Riboflavin (B₂)

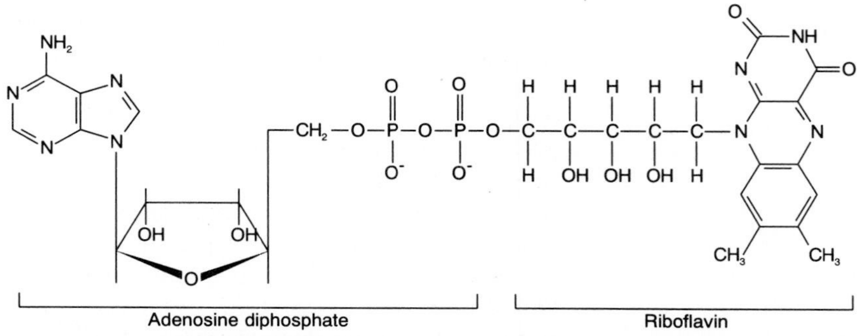

Flavin adenine dinucleotide (FAD), functions in dehydrogenations

*Residue also occurs in nicotinamide adenine dinucleotide phosphate (NADP⁺) where the 2'-hydroxyl of the adenosyl moiety is also phosphorylated.

†Not a vitamin; included for reference.

Table 3.1   Common Coenzyme Structures Cont'd

Vitamin	Coenzyme

Pyridoxine (B$_6$)

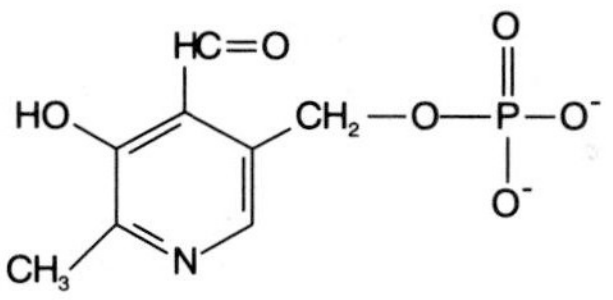

Pyridoxal phosphate, functions in transamination, deamination, decarboxylation, and racemization reactions of amino acids

Niacin*

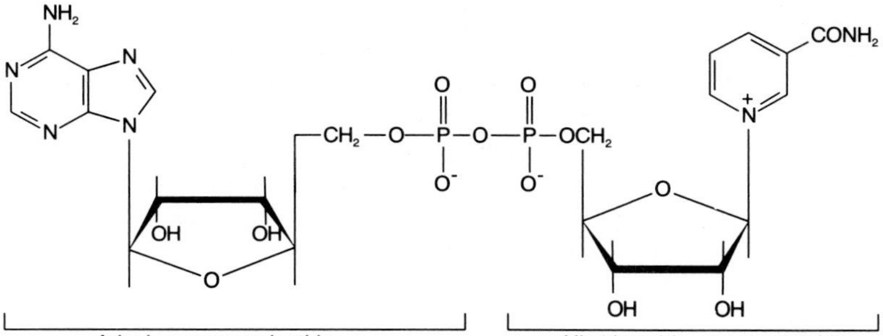

Adenine mononucleotide   Nicotinamide mononucleotide

Nicotinamide adenine dinucleotide (NAD$^+$), functions in dehydrogenations

Lipoic acid†

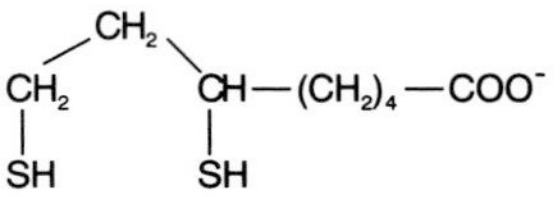

Dihydrolipoate   Dihydrolipoate-protein complex

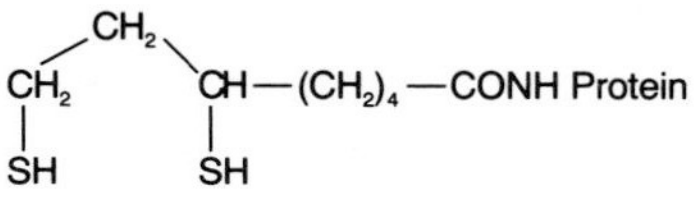

anion or cation cofactors such as the chloride ion. The term *coenzyme* applies to organic molecules, often but not always derived from a vitamin, which are essential for activity of numerous enzymes. Some coenzymes are tightly bound to the protein portion of a given enzyme; indeed, the enzyme may be denatured when attempts are made to remove the coenzyme. In other instances the coenzyme is bound so loosely that simple dialysis will separate it from its protein partner. Coenzymes participate in the catalytic reaction (Table 3.1). The complete functional complex of protein plus all required accessory factors of any kind are known as the *holoenzyme;* the protein part, free of cofactors, is termed an *apoenzyme.* Sometimes, remixing of the separated components is all that is required to restore complete activity. Carbonic anhydrase, mentioned earlier, is one of a class known as *metalloenzymes,* since at least 1 gram atom of zinc per mol of protein is an absolute requirement for its activity. Removal of the zinc will completely inactivate carbonic anhydrase. Humans require numerous trace metals in their diet, and this requirement is a direct expression of the need for certain specific metals in the structures of various enzymes.

Enzyme and cofactor turnover

As with all biologic materials, enzymes are subject to turnover and replacement. Therefore a diet must include sufficient essential amino acids, metals, and vitamins to provide for enzyme replacement, among other needs. Humans have only a limited ability to store most essential metal ions, and the biologic lability of most vitamins requires that the nutritional needs for all the components of enzymes be met on a continuing, everyday basis.

Enzyme classification

To assist in the study of enzymes, an international classification has been established that defines six major classes of enzyme function, each with several subclasses. Within each subclass formal names and individual classification numbers have been assigned to the known enzymes to describe the reactions they catalyze, in much the same way as the names of the Geneva IUPAC conventions describe the structure of organic compounds. Since the trivial names for many enzymes, for example, pepsin, trypsin, and urease, have been deeply embedded in the literature, and since many other named enzymes are still easier to recognize by their trivial names, the newer nomenclature has been less widely adopted than the classification scheme itself. A summary of the international classification of enzymes follows; a more complete form is given in Appendix B.

1. *Oxidoreductases* catalyze a variety of oxidation-reduction reactions and frequently employ coenzymes such as nicotinamide adenine dinucleotide (NAD^+), its phosphate derivative ($NADP^+$), flavin adenine dinucleotide (FAD), or lipoate. Common trivial names include dehydrogenase, oxidase, peroxidase, and reductase.

2. *Transferases* catalyze transfers of groups such as amino, carboxyl, carbonyl, methyl, acyl, glycosyl, or phosphoryl. Kinases catalyze the transfer of phosphoryl groups from adenosine triphosphate (ATP) or other nucleotide triphosphates. Common trivial names include aminotransferase (transaminase), carnitine acyl transferase, and transcarboxylase.

3. *Hydrolases* catalyze cleavage of bonds between a carbon and some other atom by addition of water. Common trivial names include esterase, peptidase, amylase, phosphatase, urease, pepsin, trypsin, and chymotrypsin.

4. *Lyases* catalyze breakage of carbon-carbon, carbon-sulfur, and certain carbon-nitrogen (excluding peptide) bonds. Common trivial names include decarboxylase, aldolase, synthase, citrate lyase, and dehydratase.

5. *Isomerases* catalyze racemization of optical or geometric isomers and certain intramolecular oxidation-reduction reactions. Trivial names include epimerase, racemase, and mutase.

6. *Ligases* catalyze the formation of bonds between carbon and oxygen, sulfur, nitrogen, and other atoms. The energy required for bond formation is frequently derived from the hydrolysis of ATP; the term *synthase* is reserved for this group. Trivial names include thiokinase and carboxylase.

This scheme indicates how the enzymes discussed later in this book are classified in terms of reaction type and substrates employed.

Intracellular location of enzymes

The exquisite organization of living cells first became apparent with the development of the light microscope, and our knowledge has proceded to much greater sophistication with the development of the electron microscope. Figure 3.1 is a reproduction of an electron micrograph of a typical mammalian liver cell, and Figure 3.2 is a diagrammatic representation of a typical mammalian cell. The organized subcellular elements, or organelles, appear to have highly individualized complements of enzymes that are related to the functions that the organelles must perform.

Figure 3.1   Electron micrograph of a rat liver cell. Structures identified include
nucleus *(N)*, nucleolus *(NU)*, nuclear pore *(NP)*, mitochondrion *(M)*,
rough endoplasmic reticulum *(RER)*, lysosome *(L)*, plasma or cell
membrane *(PM)*, bile canaliculus *(C)*, and peroxisome *(P)*, typical of
liver cells. (× 9500.)

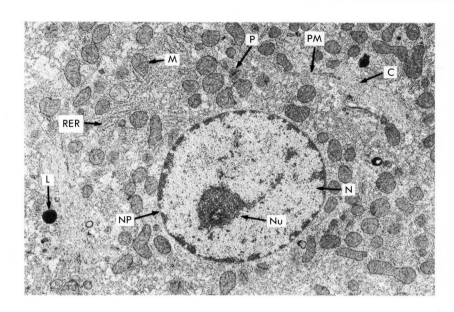

Distribution of enzymes

The subcellular location of enzyme systems may be summarized as follows. Many of the enzymes associated with the nucleus are involved with the maintenance, renewal, and utilization of the genetic apparatus. Most of the enzymes of mitochondria deal with reactions best described as energy-yielding or oxidative reactions that provide the driving force for many types of cellular work.

Enzymes associated with ribosomes promote the biosynthesis of proteins. Microsomal enzymes are responsible for a variety of hydroxylation reactions involved in steroid hormone biosynthesis and drug metabolism or inactivation. Treatment with certain drugs, such as barbiturates, leads to induction of microsomal enzymes if the drug use is continued for a time. One must remember that microsomes are not subcellular organelles but rather fragments of the endoplasmic reticulum.

Lysosomes contain enzymes that catalyze the hydrolytic destruction of materials no longer needed by the cell, which is an intracellular digestive process. Typically, lysosomal enzymes function at a more acid pH than elsewhere in the cell. They are released more rapidly in the acidotic condition associated with cellular injury or death.

The Golgi complex is a system of tubules and vesicles that apparently functions in secretory cells to collect and extrude protein or other cellular products from the cell. In cells that specialize in absorption, for example, in the epithelial cells of the intestinal tract, the Golgi complex appears to work in the opposite manner; it serves to collect material taken up by the cells.

In addition, cytoplasmic enzymes catalyze the type of carbohydrate metabolism known as glycolysis (see Chapter 6). In fractionated extracts these enzymes appear to be simply dissolved, although in the intact cell they may be more highly organized. The enzymes responsible for fatty acid biosynthesis are also cytoplasmic, but unlike those involved in glycolysis, this system of enzymes is highly organized as a single, very long peptide containing all seven of the required enzymes.

Figure 3.2   Generalized cell with principal organelles, as might be seen with an
electron microscope. Each of the major organelles is shown enlarged.
Membranes of organelles are believed to be continuous with, or
derived from, plasma membrane by an infolding process. Structures
of other membranes (of nucleus, endoplasmic reticulum,
mitochondria, etc.) are probably similar to that of plasma membrane,
shown enlarged at lower left. Solid circles carrying short lines
represent a lipid bilayer, and unfilled circles represent surface layers
of proteins or peptides that constitute inner and outer surfaces of
membrane. (From Hickman C and Hickman C Jr: Biology of animals
St Louis, 1972, The CV Mosby Co.)

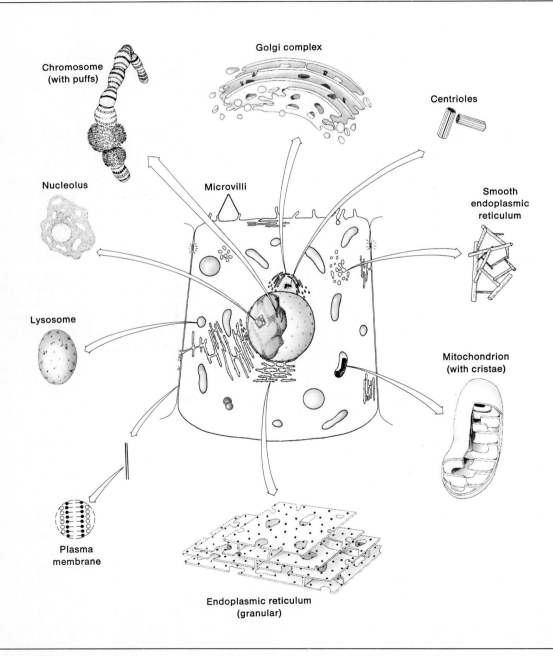

Isolated organelles often exhibit a further detailed architecture of their own. For example, mitochondrial particles have a double-membrane boundary. The outer membrane, the inner membrane, and the space between them constitute regions, each possessing specific groups of enzymes that are localized to perform their functions most efficiently (see Chapter 5). The same is true of the limiting membrane of the cell, or plasma membrane, which is not merely an inert barrier; it contains its own characteristic array of enzymes, most of which facilitate or control the entrance of materials into the cell or the egress of substances from it. Some hormones, which circulate in the blood, exert their earliest metabolic effects on enzymes or receptor proteins that are part of the surface membranes of their target cells (see Chapter 17).

General enzyme properties

Enzymes isolated from their natural sources can be used in vitro to study in detail the reactions they catalyze. Reaction rates may be altered by varying such parameters as pH or temperature, by changing the ionic composition of the medium, or by changing ligands other than the substrate or coenzymes.

Effect of temperature on enzymes

Since protein structure determines enzyme activity, anything that disturbs this structure may lead to a change in activity. The process of protein denaturation described in Chapter 2 also applies to enzymatic proteins, and the denaturants that cause it are the same. For example, enzymes frequently show great thermal fragility. When heated to temperatures greater than approximately 50° C, most but not all enzymes are denatured. High-temperature denaturation is usually irreversible. Even under conditions where denaturation does not occur, most enzymes show an optimum temperature at which activity is maximal. Figure 3.3 shows that changes in activity above or below this temperature are not always symmetric.

pH and ionic dependence of enzymes

Enzyme activity is also related to the ionic state of the molecule and especially of the protein part, since the polypeptide chains contain groups that can ionize to a degree that depends on the prevailing pH. As is true of proteins generally, enzymes have an *isoelectric point* at which their net free charge is zero. The pH of the isoelectric point as a rule is not the same as the pH at which maximal activity is demonstrated. The pH optima shown

Figure 3.3   Effect of temperature on enzyme activity.

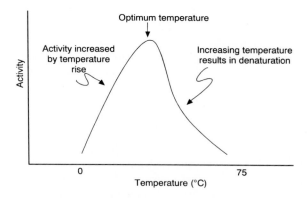

Figure 3.4   pH activity profile of a typical enzyme.

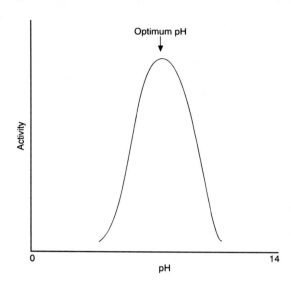

by enzymes vary widely; pepsin, which exists in the acid environment of the stomach, has a pH optimum at about 1.5, whereas arginase, an enzyme that cleaves the amino acid arginine, has its optimum at 9.7. However, most enzymes have optima that fall between pH 4 and 8 (Figure 3.4). Some enzymes show a wide tolerance for pH changes, but others work well only in a narrow range. If any enzyme is exposed to extreme values of pH, it is denatured. The sensitivity of enzymes to altered pH is one reason why regulation of body pH is so closely controlled and why departures form normal may involve serious consequences.

Fructose bisphosphatase and phosphofructokinase are enzymes with reciprocal effects on the interconversion of fructose 1,6-bisphosphate and fructose-6-phosphate. Other factors being equal, phosphofructokinase is more severely inhibited when the pH drops below 7.5, so that mild acidosis favors formation of glucose and diminishes formation of pyruvate and lactate.

The pH is not the only factor that affects the ionic state of enzymes. For example, the concentration of water available to an enzyme bound in a subcellular membrane may be different from the amount of water available to an enzyme in the cytoplasm. The precise requirements for the hydration of enzymes in their physiologic environment may be different from those observed in the isolated and purified state.

Active catalytic site   Enzymes differ from other proteins in that they possess what has been termed an active catalytic site. The active site can be regarded as being composed of a relatively small number of amino acid residues, not necessarily in immediate sequence in terms of primary structure; however, these amino acids interact in a manner that allows catalysis to occur. Because of the peculiar and highly individualized ways in which peptide chains may be folded, amino acids some distance apart in the primary sequence may contribute to the active site. At the same time, if some molecular change occurs, the necessary interaction of the amino acids composing the active site probably will be weakened or lost. This accounts for relatively mild treatment possibly causing denaturation. Nature has established certain means of enforcing, at least to some degree, the folding patterns required for enzyme activity. One important means involves the presence of amino acids that contain —SH groups. These groups can be oxidized to form disulfide (—S—S—) bonds that enforce propinquity between given residues, since the —SH groups from two cysteine residues form cross-links. At times enzyme activity depends on the oxidation of some specific —SH groups to enforce proper folding patterns and on the reduction of others that participate in the active site. Thus one may conclude that many enzymes are sensitive to the oxidation-reduction state of their environment.

Many frequently propose that the substrate fits into the pocket of the active site in the enzyme. Such a model is suggested for chymotrypsin, a protein hydrolase, in which the pocket contains residues of aspartic acid, histidine, and serine that are connected by hydrogen bonds. With the nitrogen in histidine acting first as a general base and then as a general acid, a charge relay system is established. The specific portion of the peptide chain of the protein substrate fits into the active site, which results in hydrolysis of the peptide bond.

Proenzymes, or natural enzyme precursors

Significant portions of some enzymes can be removed without loss of activity. Sometimes activity can even be increased or an inactive protein can be converted to an active enzyme by cutting off a portion of the peptide chain. Typical examples of the latter are the digestive proteinases, pepsin, trypsin, and chymotrypsin, each of which is produced and stored as an inactive *proenzyme,* or a *zymogen.* When the zymogen pepsinogen is released into the gastric juice, it loses a peptide fragment in the acid gastric environment and is converted to active pepsin.

Activation of proenzymes Some of the proteolytic zymogens, notably pepsinogen and trypsinogen, have a detectable proteolytic activity under specific and appropriate conditions of pH.

As discussed in Chapter 8, the digestive proteinase zymogens are activated by trypsin. Trypsin is activated from the trypsinogen through the action of enterokinase, which is under hormonal control. At times a sequential activation of enzymes occurs; one enzyme is activated, which activates the second, then the third, so that a cascade of activations results, such as that found in blood clotting (see p. 167).

Activation of zymogens by scission of a peptide fragment is not limited to proteolytic enzymes. The conversions of the prohormone proinsulin to insulin and of the proenzyme prophospholipase A to the active enzyme are similar cases.

Isomeric enzymes, or isozymes

In many species, including humans, different molecular forms of certain enzymes may be isolated from the same or different tissues. The different molecular forms have been termed isoenzymes, or isozymes. Isozymes were first detected and reported as a result of electrophoretic or column chromatographic attempts to purify certain enzymes, and the early reports were at first discounted as technical or experimental artifacts. However, subsequent studies have clearly demonstrated that the measurable differences in physical properties are accompanied by differences in chemical properties, and thus the concept of isozymes is now firmly established.

Multiple subunit isozymes Lactate dehydrogenase (LDH) and malate dehydrogenase have been thoroughly studied as examples of isozymes. LDH is composed of four subunits. The two subunits, differing in amino acid content and sequence, can be combined into tetramers in five ways. The possible combinations can be separated by electrophoresis, as shown in Figure 3.5. If one subunit type is identified as "M" (the major form found

Figure 3.5

Electropherogram of serum lactate dehydrogenase *(LDH)* isozymes. Sample was applied to the gel at the point marked *O* (origin), and the direction of the applied electric field was shown. The gel was then stained by a method that is sensitive to the dehydrogenase activity but not to any other protein. The arabic numerals directly above the gel represent one system of describing the various isozymes as LDH_1, etc. The subunit composition of the isozymes is given by the symbols H_4, etc., where H represents the subunit most typical of heart muscle and M the subunit typical of skeletal muscle. At the top of the figure is a densitometric scan of the gel. The area under each peak of the scan is proportional to the amount of enzyme in the corresponding band of the gel.

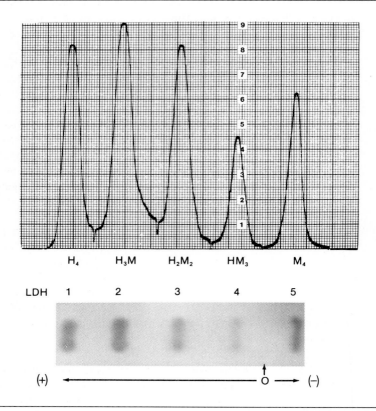

in muscle or liver) and the second as "H" (the major form found in heart), the tetramers could have the compositions M_4, M_3H, M_2H_2, MH_3, or H_4. These can be separated by electrophoresis. Note in Figure 3.5 that the net charge on the isozyme differs; those with an increasing content of the H subunit have an increasingly larger negative charge, whereas the M_4 isozyme has a slightly positive net charge.

Clinical comment

Differential analysis of lactate dehydrogenase LDH catalyzes the reversible interconversion of pyruvate and lactate. In humans the content of several isozymes differs in heart and liver, and this difference is used in diagnostic differentiation of diseases of the liver and myocardium. In both disease states LDH leaks out of the damaged cells, and its concentration in blood serum increases. LDH in the heart can be differentiated from LDH in the liver because of this separation by electrophoresis and because the isozymes from the myocardium are more resistant to heat denaturation than the corresponding hepatic isozymes. In the simpler differential test situation a serum sample may be analyzed twice, once before and once after carefully controlled heat denaturation.

Samples of the serum are assayed as received at 30° C, giving a value, T. An aliquot of the sample is heated to 57° C for 30 min and then assayed at 30° C to give a value, L. A third aliquot is assayed after being heated to 65° C for 30 min, giving a value, H. The difference, $T - L$, represents the heat-labile fraction, which in normal serum amounts to 10% to 25% of the total. This fraction increases to 33% to 80% in patients with liver disease. The heat-stable fraction, designated by H, normally accounts for 20% to 40% of the total. In patients with myocardial infarcts H is 45% to 65% of the total value. The second modification of the LDH assay is based on the reactivity of the enzyme toward α-ketobutyrate instead of α-ketopropionate (pyruvate). Again, a difference exists in the rate at which the various isozymes can attack α-ketobutyrate. By measuring the activity toward both substrates, one can calculate a ratio, LDH/HBDH, in which HBDH refers to dehydrogenase activity using hydroxybutyrate as the substrate. In normal serum this ratio varies from 1.2 to 1.6; in patients with liver disease it increases to 1.6 to 2.5, whereas in patients with myocardial infarcts it decreases to 0.8 to 1.2.

Enzyme specificity

Many inorganic catalysts, for example, charcoal or finely divided platinum, show little specificity toward the substances on which they exert a catalytic effect. Thus a mere handful of selected catalysts is sufficient for much synthetic organic chemistry work practiced in industry or the laboratory. Enzymes, on the other hand, are more specific. Urease, an enzyme that hydrolyzes urea to carbon dioxide and ammonia, has almost an absolute specificity toward urea, and only one other compound that can be split by urease is known. Similarly, catalase is almost completely specific toward hydrogen peroxide, which it converts into water and oxygen. Chymotrypsin, a gastric proteinase, shows a somewhat lesser specificity toward its substrate. It prefers to cleave peptide bonds in which one participant amino acid has an aromatic ring; however, it is indifferent to the other amino acid in the peptide bond. It also preferentially attacks peptide bonds in the interior of a peptide chain, but even this requirement is relative. Other proteolytic enzymes show various types of specificity. Trypsin hydrolyzes only those peptide bonds to which arginine or lysine contribute the carboxyl group.

Clinical comment

Proteolysis and blood clotting A few other proteolytic enzymes have specificity requirements similar to those of trypsin. Several of the factors involved in the process of blood clotting normally exist as proenzymes in the blood. Their activation by tissue injury promotes the formation of thrombin and the production of a clot. To guard against triggering the clotting process by inappropriate activation of the proenzymes, human plasma contains an antitryptic α-globulin that prevents blood from clotting within the vessels under normal circumstances.

Thus the degree of enzyme specificity is variable, ranging from the virtually complete requirement for a single substrate to other enzymes that may function with many different individual molecules, provided they contain some common structural feature. Even when the specificity requirements are high, however, they are not always absolute, since it is possible to synthesize substrate analogues that can block or inhibit many normal enzyme functions.

Mechanism of enzyme catalysis
Catalytic mechanism of enzyme action

Activated or transition states To initiate a conversion of substrate to product, some energy must be expended in most cases; one exception is the decay of radioactive isotopes. In any population of molecules a distribution of energies exists. The conversion of reactant substrate molecules S to product P is proportional to the number of molecules in an

Figure 3.6   Energy diagram for catalyzed and uncatalyzed reactions.

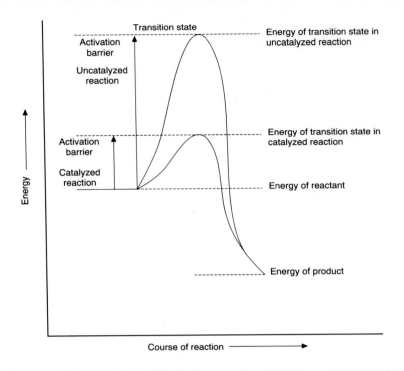

activated state S*, which is greater than the energy barrier that exists between the reactant and product molecules.

$$S \rightleftharpoons S^* \longrightarrow P$$

Thus the addition of energy, such as heat, to a reaction increases the proportion of molecules of reactants in the activated or transition state, and therefore the conversion of reactants to products proceeds more quickly.

An alternate way to increase the rate of the reaction is to reduce the energy barrier of activation. In so doing, the proportion of molecules that can pass the barrier is greater. *Enzymes increase reaction rates by decreasing the activation barrier,* as shown in Figure 3.6.

Enzyme-substrate complex The decrease in the activation barrier in enzyme-catalyzed reactions is achieved by the formation of a complex between the enzyme, E, and the reactants. In the simplest case of single-substrate reactions, a reversible reaction may be represented:

$$E + S \rightleftharpoons ES^* \rightleftharpoons E + P$$

where ES* represents a transition state in the complex with the active site of the enzyme. As noted earlier, the fitting of the substrate to the active site causes the substrate's adoption into the transition state by a charge relay system, as proposed for the hydrolysis of proteins by chymotrypsin; by the conformational distortion of substrate molecules, as seen in the

Figure 3.7   Reaction velocity as function of substrate concentration. See text for details and abbreviations.

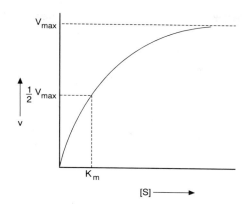

hydrolysis of bacterial cell wall molecules by lysozyme; or by other processes whereby the substrate in the enzyme-substrate complex is brought into the transitional conformation required for conversion to product.

Quantitative analysis of single-substrate enzyme kinetics

The fundamental theory of enzyme catalysis is based on the classic studies of Michaelis and Menten and of Haldane. The quantitative analysis of enzyme action that has been developed depends largely on measured reaction times.

If the *initial* reaction rate, defined as the rate observed for a given amount of enzyme when the concentration of the product formed is nearly zero, is plotted as a function of the substrate concentration, the results appear similar to those shown in Figure 3.7; the curve connecting the observed points would be hyperboloid and would asymptotically approach a maximum value, as shown by V_{max}. This is the maximum initial velocity that can be obtained without increasing the amount of enzyme.

The hyperbola described by a plot of reaction velocities as a function of substrate concentrations is difficult to use. If the reciprocals of the velocities are plotted as a function of reciprocal substrate concentrations, the hyperbola is converted to a straight line. Such linear double-reciprocal plots are far easier to construct and interpret. A transformation of the typical curve shown in Figure 3.7 is presented in Figure 3.8. The double-reciprocal plots are frequently called *Lineweaver-Burk* plots.

If the usual convention is followed, representing concentrations by means of brackets, that is, by letting [S] stand for the molar concentration of the substrate, and if a few assumptions are made regarding the experimental situation, one can obtain a useful mathematical equation that describes the enzyme kinetics. Assume for the present that:

1. The system involves only a single substrate.
2. The system is at a steady state, that is, [ES*] is a constant and the free enzyme E is in equilibrium with ES*.
3. The system is established so that [E] < [S] on a molar basis. Considering the high molecular weight of most enzymes, this is not a stringent limitation.
4. Since the analysis deals with initial reaction rates, [S] ≫ [P] and [P] is negligble under these conditions.

Figure 3.8   Lineweaver-Burk transformation of Michaelis-Menten curves. See text
for details and abbreviations.

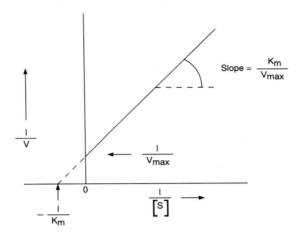

In this case the reaction mechanism may be formulated as follows:

$$E + S \underset{k_2}{\overset{k_1}{\rightleftharpoons}} ES^* \underset{k_4}{\overset{k_3}{\rightleftharpoons}} E + P \tag{3.1}$$

where k_1, k_2, k_3, and k_4 are the respective rate constants. At the steady state the concentration of ES is constant; that is, the rate at which it is being formed is the same as the rate at which it is being broken down. Under these conditions, equation 3.1 can be used to derive the rate equation 3.2.

$$k_1[E][S] + k_4[E][P] = k_2[ES^*] + k_3[ES^*] \tag{3.2}$$

Since the analysis is restricted to initial reaction rates, [P] is negligible and [S] is virtually constant. Thus the term involving [P] can be dropped and those involving [ES*] can be collected to give the following:

$$k_1[E][S] = (k_2 + k_3)\,[ES^*] \tag{3.3}$$

or:

$$\frac{[E][S]}{[ES^*]} = \frac{k_2 + k_3}{k_1} = K_m \tag{3.4}$$

The ratio of rate constants, $(k_2 + k_3)/k_1$, can be replaced by a single constant, K_m.

**Analysis from
concentrations and
reaction velocities**

The *maximum initial* velocity is achieved only when *all* the enzyme (E_t) is in the form of the active complex (ES*), from which it follows that:

$$V_{max} = k_3[E_t] \tag{3.5}$$

Under any other conditions, the observed initial velocity will be the following:

$$v = k_3[ES^*] \tag{3.6}$$

The free enzyme, which does not take part in catalysis, is related to the total quantity of enzyme and to the active complex by the following equation:

$$[E] = [E_t] - [ES^*] \tag{3.7}$$

The information in equations 3.5 and 3.6, when applied to equation 3.7, gives the following:

$$[E] = \frac{V_{max}}{k_3} - \frac{v}{k_3} = \frac{V_{max} - v}{k_3} \tag{3.8}$$

This expression for [E] can now be substituted into equation 3.4:

$$K_m = \frac{(V_{max} - v)[S]}{k_3[ES^*]} \tag{3.9}$$

Using the information in equation 3.6, equation 3.9 is transformed as follows:

$$K_m = \frac{[S]}{v}(V_{max} - v) \tag{3.10}$$

and the relation between K_m, substrate concentration, and reaction velocities is established. Since it is customary to deal with enzymes in terms of reaction velocities, equation 3.10 is usually solved for v and written as follows:

$$v = \frac{[S]V_{max}}{[S] + K_m} \tag{3.11}$$

The constant K_m is generally known as the *Michaelis*, or *Michaelis-Menten*, *constant*. One can see from equation 3.10 that when the initial velocity is equal to half the maximum initial velocity (that is, $v = \frac{1}{2} V_{max}$), then [S] is equal to K_m. Both k_m and [S] are expressed in the same units, moles per liter.

Significance of the Michaelis-Menten constant One can see from equation 3.4 that when $k_2 \ll k_3$, K_m approximates the dissociation constant for the enzyme-substrate complex ES*. The reciprocal of K_m would therefore refer to the binding constant of E for S. This is useful when considering a given enzyme in the presence of two alternate substrates; the substrate with the smaller K_m will be more effectively bound and will successfully compete for the binding site on the enzyme. Both K_m and V_{max} are constants of specific values for any enzyme under the conditions of their measurement. They are the appropriate parameters for comparison of enzyme behavior.

In designing an assay to measure the amount of an enzyme in blood or other material, it is important to ensure that sufficient substrate is present to saturate the enzyme completely, that is, to convert it completely to the enzyme-substrate complex.

The significance of K_m in metabolism centers around its operational definition as the concentration of substrate at which the initial velocity is half its maximum. From this, three important points can be made:

1. Each K_m value is necessarily determined in vitro using purified enzyme. The concentration of substrate in vivo will play a role in the rate of the conversion to product only if its concentration approximates the K_m. If the concentration in vivo is, for example, 10 times that of the K_m, the reaction rate will not be proportional to the substrate concentration; rather, it will be at V_{max}.

2. As an extension of point 1, if an enzyme has two or more substrates that can be converted to their respective products, each substrate having its K_m and V_{max}, the rates of conversion of each substrate can be calculated from the Michaelis-Menten equation (equation 3.11). This calculation requires that the in vivo concentration of each substrate is known. If the in vivo concentration is much less than the K_m, that substrate will not be significantly converted to product.

3. If a substrate can be converted to a product by either of two enzymes, the enzyme

that is physiologically important can be selected, using the K_m, V_{max}, and in vivo concentration.

Turnover number

In any experimental situation, V_{max} depends on the amount of enzyme present. If an experiment could be performed containing *1 mol* of enzyme, the resulting activity would be termed the turnover number, which is expressed as moles of substrate converted to product per minute (or per second) per mole of enzyme. In those cases where the enzyme contains more than one active site per molecule, the turnover number is corrected accordingly. One would then use the turnover per mole of active site of the enzyme. Turnover numbers are of value in comparing the same enzyme from different tissues and in comparing different isozymes.

Significance of maximum initial velocity Even though the V_{max} of a metabolic reaction depends partly on the concentration of the enzyme, the V_{max} is useful in comparing the activity of one enzyme to another. For a series of enzymes in a metabolic pathway, V_{max} indicates the maximum rate of metabolism and where this occurs in the reaction sequences.

Enzymatic activity

Another way of expressing the catalytic activity of an enzyme relates the micromoles of substrate reacted or product formed per minute (or per second) to the weight of total protein in the sampled solution or body fluid. This is called the *specific activity,* which is at its maximum value when all the protein in the sample is enzyme protein. Because the amount of enzyme protein is not readily measurable, the amount of enzyme is frequently expressed in terms of the specific activity in the tissue or fluid. These and similar problems are discussed further in the section on clinical applications later in this chapter.

Kinetic analysis of enzyme inhibition

Lineweaver-Burk plots can be used to assess the nature of enzyme inhibition. If catalysis is to occur, a certain structural correlation must exist between the substrate on the one hand and the active site of the enzyme and its surroundings on the other. Anything that alters or interferes with this "fit" will inhibit or prevent catalysis. Metabolites, drugs, or toxic substances may inhibit enzymes so that the normal catalyzed reaction occurs at a lower rate, if at all. The inhibitors may be classified according to how they react with the enzyme:

1. *Competitive inhibitors* bind reversibly with the enzyme in competition with the substrate. When the inhibitor, I, is bound to the enzyme, the normal substrate cannot form the ES active complex, and thus less enzyme is available for catalysis. Since the competition for the enzyme is proportionate to the concentrations of the substrate and the inhibitor, a sufficient concentration of the substrate will overwhelm the inhibition, and the V_{max} will be the same as with no inhibitor present. At concentrations in which substrate and inhibitor are more comparable, the K_m for the substrate will be reduced. This can be seen in kinetic analysis, expressed as a Lineweaver-Burk plot (Figure 3.9), in which the slopes change but the intercepts ($\frac{1}{V_{max}}$ with the $\frac{1}{v}$ axis) remain the same.

2. *Noncompetitive inhibitors* bind either to the enzyme or the enzyme-substrate complex. In this case the V_{max} is decreased without a change in the K_m for the substrate (Figure 3.10). A high concentration of substrate will not displace the inhibitor.

3. *Uncompetitive inhibitors* bind only to the free enzyme and not to the enzyme-substrate complex. In this case both V_{max} and K_m are changed. The Lineweaver-Burk plots show parallel lines at the different inhibitor concentrations.

Figure 3.9   Competitive inhibition depicted by Lineweaver-Burk plots. *A*, Normal uninhibited reaction. *B* and *C*, Two different inhibitor concentrations *(B < C)*.

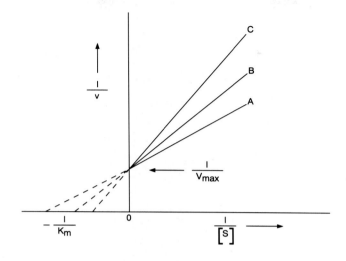

Figure 3.10   Noncompetitive inhibition depicted by Lineweaver-Burk plots. *A*, Normal uninhibited reaction. *B* and *C*, Two different inhibitor concentrations *B < C)*.

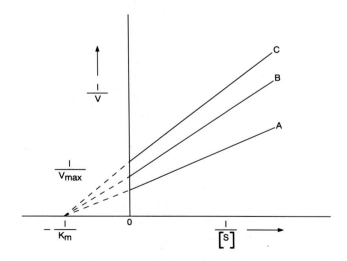

Table 3.2   Comparative Effects of Inhibitors on Enzyme Kinetics

Type of inhibition	Lineweaver-Burk plot	V_{max}	K_m
Competitive	Slope varies; $1/v$ intercept constant	Unchanged	Increased
Uncompetitive	Slope constant; $1/v$ intercept varies	Decreased	Decreased
Noncompetitive	Slope varies; $1/v$ intercept varies	Decreased	Unchanged

Thus, for the single-substrate reactions, the three types of inhibition may be summarized in Table 3.2.

Several examples may be cited to further illustrate these ideas.

Competitive inhibition The mitochondrial enzyme succinate dehydrogenase catalyzes the following reaction:

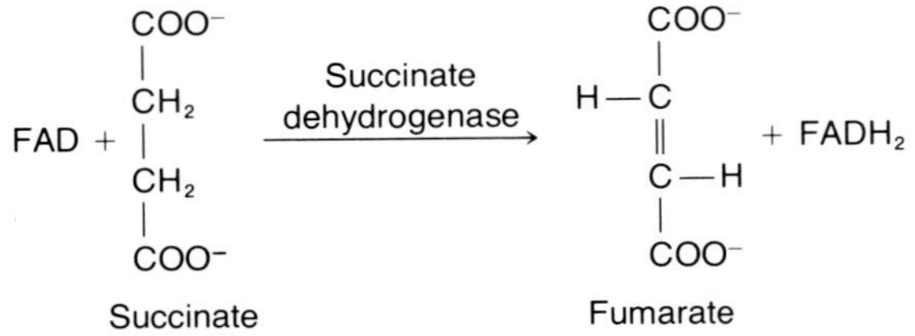

$$\text{FAD} + \begin{array}{c} COO^- \\ | \\ CH_2 \\ | \\ CH_2 \\ | \\ COO^- \end{array} \xrightarrow[\text{dehydrogenase}]{\text{Succinate}} \begin{array}{c} COO^- \\ | \\ H-C \\ \| \\ C-H \\ | \\ COO^- \end{array} + \text{FADH}_2$$

Succinate   Fumarate

Malonate, which has the structure $^-OOC \cdot CH_2 \cdot CCO^-$, can be bound to the enzyme but cannot be dehydrogenated. Malonate meets the stipulations for a competitive inhibitor because its effects can be offset or reversed by sufficiently increasing the concentration of succinate in the presence of the inhibitor.

Mixed competitive, noncompetitive inhibition Another important mitochondrial enzyme is malate dehydrogenase, which reversibly catalyzes the oxidation of L-malate to oxaloacetate. Evidence suggests that this enzyme first forms a complex with the coenzyme and that the complex then reacts with the substrate. An inhibitor known as hydroxymalonate, $^-OOC \cdot CHOH \cdot COO^-$, displays properties that vary according to the direction of the reaction:

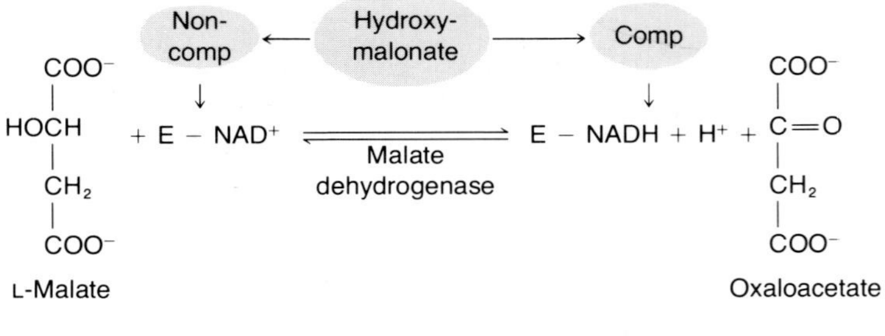

$$\begin{array}{c} COO^- \\ | \\ HOCH \\ | \\ CH_2 \\ | \\ COO^- \end{array} + \text{E}-\text{NAD}^+ \underset{\substack{\text{Malate} \\ \text{dehydrogenase}}}{\rightleftharpoons} \text{E}-\text{NADH} + \text{H}^+ + \begin{array}{c} COO^- \\ | \\ C{=}O \\ | \\ CH_2 \\ | \\ COO^- \end{array}$$

Non-comp ← Hydroxy-malonate → Comp

L-Malate   Oxaloacetate

In other words, hydroxymalonate apparently reacts noncompetitively with the complex of the enzyme and oxidized coenzyme when malate is oxidized to form oxaloacetate. Conversely, it reacts competitively with the complex of enzyme and reduced coenzyme when oxaloacetate is reduced to form malate. This somewhat unusual case emphasizes that the discrimination between natural substrate and inhibitors may result in a different mechanism of action in each direction.

Two-substrate kinetics

The examples just cited are important for another reason. Michaelis-Menten kinetics strictly apply only to those enzymes that employ single substrates. A much larger number of enzymes use more than one substrate at a time. For such multireactant enzymes, illustrated by malate dehydrogenase, the coenzyme NAD^+ is a cosubstrate. The kinetic analysis of multireactant enzyme systems requires an extension of the fundamental principles established by Michaelis and Menten. Since the enzyme and both substrates are simultaneously required for catalysis to occur, it is appropriate to describe the ternary catalytic complex by a symbol ($E \cdot S \cdot NAD^+$). In forming this, two binary complexes are possible, represented by ($E \cdot S$) and ($E \cdot NAD^+$). Following catalytic transformation, ($E \cdot P \cdot NADH$) could equally well produce ($E \cdot P$) and ($E \cdot NADH$); these, in turn, could decompose to liberate free enzyme and the second product.

Kinetic analysis sometimes permits an independent distinction between the two possible sequences of such a reaction mechanism. In the case of 3-phosphoglyceraldehyde dehydrogenase, the mechanism is known as *ordered* because the addition of NAD^+ as the first, or leading, substrate is compulsory (Figure 3.11). Other examples are known in which the sequence of addition is not obligatory, but *random*.

Another pattern for multireactant enzyme systems is known as the "Ping-Pong" mechanism. In this case one substrate must be bound and one product released before the second substrate is bound or the second product released (Figure 3.12). Typical examples of this pattern are found in the aminotransferases.

Clinical comment

Enzyme inhibitors There are significant, practical applications of differential enzyme inhibition. Serum contains a group of acid phosphatases with an optimum activity under slightly acidic (pH 6.2) conditions. The members of this group are produced by many different tissues. The most important one clinically is produced by the prostate gland, since increases in its activity are frequently associated with prostatic carcinoma. With *o*-nitrophenyl phosphate as the substrate, L-tartrate competitively inhibits about 95% of the prostatic phosphatase activity. Tartrate has a much lower inhibitory effect on acid phosphatases from erythrocytes, liver, kidney, or spleen. L-Phenylalanine exerts similar but not identical inhibition. Samples from suspected carcinoma patients can be assayed in the presence and absence of L-tartrate or L-phenylalanine. Formaldehyde, on the other hand, inhibits erythrocyte acid phosphatase but not the prostatic enzyme. Calcium ions are noncompetitive inhibitors of many acid phosphatases, so one usually adds a chelating

Figure 3.11

A simplified scheme of the ordered kinetic mechanism of 3-phosphoglyceraldehyde dehydrogenase. NAD^+ must first bind to the enzyme. Then 3-phosphoglyceraldehyde *(3-PG)* binds to form a ternary complex. Oxidation of the aldehyde is followed by addition of inorganic phosphate (P_i). $NADH + H^+$ are released, and in the last step the product, 1,3-bisphosphoglycerate (1,3-BPglycerate), is released from the enzyme.

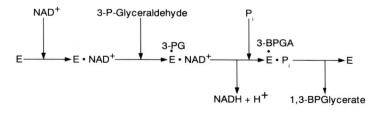

Figure 3.12   "Ping-Pong" kinetic mechanism of aspartate aminotransferase, showing that addition of the first substrate is followed by release of the first product and addition of the second substrate is followed by release of the second product.

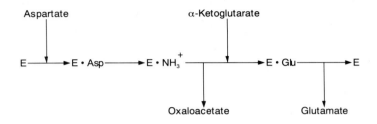

agent to blood samples collected for this assay or to the sera derived form the blood. Ethanol and certain narcotic drugs are also noncompetitive inhibitors of acid phosphatase; one should remember this when interpreting the results of such tests. These tests, although still employed in smaller laboratories, are now being supplanted by physical separation of individual isozymes or by immunochemical methods.

Enzymes may also be inactivated by various chemicals that have clinical relevance. For example, many enzymes depend on essential sulfhydryl groups, which form tight covalent bonds with various heavy metals. For this reason mercury, lead, silver, and other metals are extremely toxic. Even iron and copper, although they are classed as essential minerals, can produce intoxication when ingested in excess. A treatment of syphilis, popular long ago but no longer used, depended on the extreme sensitivity of *Treponema pallidum* to mercury or bismuth ions. Organic mercurials are still employed as disinfectants today.

Another class of inhibitors affects enzymes by introducing foreign alkyl groups into the structure. Most of these substances were initially developed as chemical warfare agents and have high toxicity. The best known is diisopropyl fluorophosphate (DFP). DFP reacts readily with the OH group of serine residues, converting them to diisopropyl phosphate esters. The organic phosphates, of which DFP is only one example, are particularly good inhibitors of acetylcholinesterase, which breaks down the neurotransmitter substance known as acetylcholine. Inactivation of acetylcholinesterase produces violent spasms of the pulmonary system and interferes with normal neuromuscular and cardiac function. Similar agents are employed as insecticides in agriculture and sometimes may be severely or fatally toxic to humans.

Clinical comment   **Coenzyme analogues as drugs** Enzymes can also be inhibited by affecting associated coenzymes or prosthetic groups. This is of considerable importance in designing chemotherapeutic agents. One of the earliest antibiotic drugs was *p*-toluenesulfonamide, an analogue of *p*-aminobenzoic acid, structures of which follow:

$$CH_3 \text{—}\!\!\bigcirc\!\!\text{—} \underset{\underset{O}{\|}}{\overset{\overset{O}{\|}}{S}} \text{—} NH_2 \qquad\qquad H_2N \text{—}\!\!\bigcirc\!\!\text{—} \underset{}{\overset{\overset{O}{\|}}{C}} \text{—} OH$$

p-Toluenesulfonamide   *p*-Aminobenzoic acid

Certain microorganisms produce folic acid, which contains a *p*-aminobenzoyl residue in its structure (see p. 23). Toluenesulfonamide interferes with microbial synthesis of folic acid. Since folic acid is an essential coenzyme involved in the biosynthesis of purines and thymine, sulfonamides inhibit growth of the pathogenic organisms. On the basis of these observations a large series of substituted sulfonamides was produced, some of which are still used in clinical practice.

Similarly, biosynthesis of pyridine nucleotide coenzymes requires incorporation of a nicotinamide moiety. An analogue of nicotinamide is the drug known as isoniazid, shown below. It interferes with the biosynthesis of nicotinamide coenzymes and is particularly useful in slowing growth of the organisms that cause human tuberculosis.

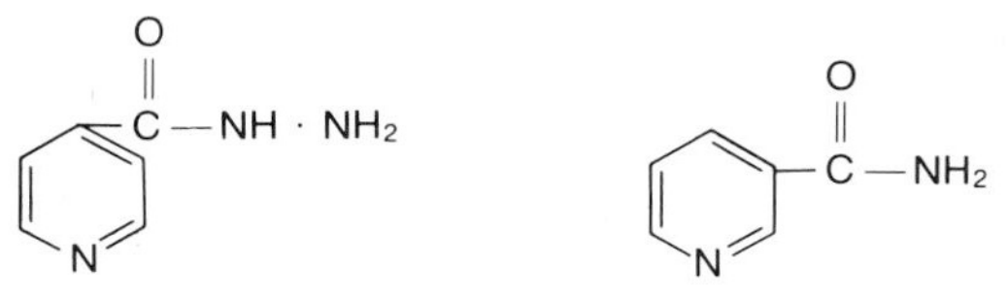

Isoniazid   Nicotinamide

Unfortunately, many pathogenic organisms have become resistant to one or more of these drugs, so the search for new antibiotic agents continues.

Enzyme regulation and control

In a system as complex as a living cell there must be some regulation or control of the multitude of reactions that occur. In the simplest cases one must consider enzyme *activators*. Many enzymes are inactivated or inhibited if cysteinyl sulfhydryl groups (—SH) are oxidized. Glutathione, γ-glutamylcysteinylglycine is a natural activator for enzymes that contain sensitive and essential —SH groups. Note that glutathione is not a specific coenzyme but rather a natural antioxidant that can be synthesized in the cell from readily available precursors. Glutathione *is* a specific coenzyme for a small number of enzymes, including maleyl acetoacetate isomerase, an enzyme involved in the oxidative degradation of phenylalanine and tyrosine. Thus, by controlling the intracellular concentration of glutathione, the activity of several enzymes is affected.

Allosteric enzymes

Some enzymes do not follow the kinetics of the Michaelis-Menten model. A significant group of such enzymes are subject to control by molecules that bind to sites on the enzyme other than the catalytic sites. Such molecules, called *effectors,* influence the binding of the substrate to the catalytic site. These enzymes are known as *allosteric enzymes*.

Some allosteric enzymes are composed of subunits of identical or closely related peptide chains. The quaternary conformation is modified by the appropriate allosteric effectors. One or more of the functional sites on these enzymes may be *catalytic* (C), whereas one or more other sites may be *regulatory* (R) and not identical with the catalytic or active sites. In some instances R and C sites are on different subunits; in other instances the allosteric and catalytic sites are located on the same subunit. When the reaction velocity of an allosteric enzyme is plotted as a function of substrate concentration, a sigmoid rather than a hyperboloid curve is obtained, as shown in Figure 3.13, which provides an example of Michaelis-Menten kinetics for comparison. One can see in Figure 3.13 that the shapes of the allosteric curves are changed considerably by altering the concentration of either positive or negative effectors, as indicated by the dashed arrows. In effect, decreasing the amount of negative effector or increasing the amount of positive effector produces a response equivalent to lowering the K_m of the substrate. In the most general

Figure 3.13   Difference between Michaelis-Menten and allosteric enzymes kinetics.
(*N*, reaction kinetic curve without any allosteric effector present).

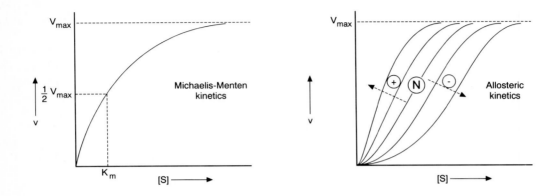

case allosteric kinetics can be represented by the following equation:

$$v = \frac{[S]^n V_{max}}{[S]^n + K}$$

where n is a coefficient that represents the interaction of the binding sites, K represents a measure of the affinity of substrate for the enzyme, and the other symbols have their previously stated meanings. The effect of the binding of the substrate or effector molecule on the binding of the next substrate molecule is expressed as n, the Hill coefficient. In the presence of sufficiently high concentrations of positive allosteric effectors, the binding of one substrate molecule no longer affects the binding of a second or of a third, and n has a value of 1. Under these conditions the plot of initial reaction velocity versus substrate concentration reduces to the form indicated by Michaelis-Menten kinetics, and K is equal to K_m. The reverse is true for negative effectors, which make the plots become increasingly sigmoidal, since they lower the affinity of the enzyme for its substrate (Figure 3.13).

Isocitrate dehydrogenase is a tetramer, and all four of its peptide chains are required for activity, which catalyzes the following reaction:

$$
\begin{array}{c}
COO^- \\
| \\
H-C-OH \\
| \\
^-OOC-C-H \quad + \quad NAD^+ \rightleftharpoons \\
| \\
CH_2 \\
| \\
COO^-
\end{array}
\qquad
\begin{array}{c}
COO^- \\
| \\
O=C \\
| \\
CH_2 \quad + \quad CO_2 \quad + \quad NADH \quad + \quad H^+ \\
| \\
CH_2 \\
| \\
COO^-
\end{array}
$$

Isocitrate   α-Ketoglutarate

NAD$^+$ and ADP are positive allosteric effoctors, whereas NADH and ATP are negative allosteric effectors. NAD$^+$ is a required coenzyme, but it also greatly increases the effect of adenosine diphosphate (ADP) in raising the reaction velocity. ADP increases the affinity of the enzyme for NAD$^+$, and vice versa. The corresponding but opposite effects are

Figure 3.14   Allosteric effectors of isocitrate dehydrogenase.

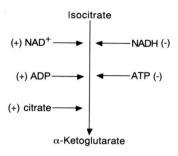

noted with the negative allosteric agents ATP and NADH. Citrate is also a positive effector, and, as seen later, a close relation exists between citrate and isocitrate in the Krebs cycle. Thus this is an example of an enzyme that is controlled by at least three distinct allosteric effectors, one of which is related closely to isocitrate in the Krebs cycle, and one of which happens to be a required coenzyme. A schematic representation of these effectors and the ways in which they alter the activity of isocitrate dehydrogenase is shown in Figure 3.14. Note particularly the opposite effect of NAD^+ and NADH, positive and negative effectors, respectively, and the opposite effects of ADP and ATP.

The major features of allosteric control are summarized as follows. Allosteric enzymes:
1.   Are usually composed of more than one polypeptide chain.
2.   May contain two separate functional centers, one of which is catalytic and the other regulatory.
3.   May be subject to either positive or negative control by one or more factors.
4.   Have effectors that may be coenzymes, substrates, or products.
5.   Undergo changes either in conformation or in the cooperativity of component polypeptides when influenced by appropriate effectors.

Product inhibition   The rate of the reaction slows down as the product(s) accumulate, as would be expected by the law of mass action. In some enzymatic reactions the slowdown is greater than predicted because the product inhibits the enzyme. The hallmark of product inhibition is the reduction of the inhibitory effects of the product(s) when the concentration of substrate is increased. Once again, this follows from the consideration of competitive inhibition.

A given enzyme may be subject to several types of control. The enzyme hexokinase catalyzes the conversion of D-glucose to D-glucose 6-phosphate as follows:

Animal hexokinases are generally subject to product inhibition, as shown in Figure 3.15. Indeed, in the case of animal hexokinases, the product inhibition demonstrated by D-

Figure 3.15   Regulation of glucose phosphorylation by hexokinase. Effect of
glucose 6-phosphate is most frequently described as noncompetitive
inhibition of some mammalian hexokinases; some claim that glucose
6-phosphate is an allosteric effector.

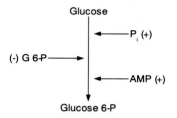

glucose 6-phosphate is a type of mixed control, involving both product inhibition and
allosteric control, since the inhibitory effects are not offset by proportionate increases in
the concentration of free glucose. The inhibitory effect of D-glucose 6-phosphate is
countered by inorganic phosphate and adenosine monophosphate (AMP).

Feedback control   Metabolites may be produced through lengthy sequences of reactions, which together
constitute metabolic pathways. In some of these a product exerts a negative effect either
on the first reaction or on a very early reaction in the pathway. Following is a general
representation of this type of control:

$$A \rightleftharpoons B \;\not\rightleftharpoons\; C - \genfrac{}{}{0pt}{}{\text{Multiple}}{\text{steps}} - \blacktriangleright P$$

$$D$$

As seen in the diagram, the product (P) acts to inhibit some early step in the pathway,
not always the first. Frequently, substance B can be converted to more than one product;
in this case the intermediate products are C and D. By feedback control it is possible not
only to inhibit the production of P, but also to divert the flow of B from one pathway to
another.

In bacteria, feedback control is common in pathways related to amino acid and pyrimidine synthesis (see Chapter 13). In mammalian tissues, feedback control is less clearly
demonstrable. A typical scheme involves pyruvate kinase, which converts phospho-
*enol*pyruvate (PEP) to pyruvate. Major isozymes of this enzyme are found in muscle (M)
and liver (L). The L isozyme is subject to negative feedback as follows:

$$\text{Oxaloacetate} \rightleftharpoons \text{PEP} \;\not\rightarrow\; \text{Pyruvate} \rightleftharpoons \text{Alanine}$$

$$\text{Glucose}$$

Alanine is a feedback inhibitor of pyruvate kinase and diverts the flow of pyruvate from alanine to glucose. Although this pathway is a short one, it fits the requirements set forth for feedback control.

Constitutive and inducible enzymes

Cellular enzymes are either constitutive or induced. *Constitutive enzymes* are present at virtually a constant concentration during the life of a cell. This is caused by a more or less constant relationship between the processes of enzyme synthesis and those of enzyme degradation. *Inducible enzymes* present in a cell at a given moment are variable, and as the need for the enzyme increases, the rate of its synthesis is increased or induced. Thus some of the enzymes responsible for glucose metabolism may be induced by increasing the load of glucose that an animal is required to metabolize. Similarly, some of the enzymes involved in amino acid catabolism are inducible either by loading doses of the amino acid itself or by certain hormones. These subjects are discussed in later chapters. For now it is sufficient to say that induction involves de novo synthesis of enzyme protein.

Regulation by covalent modification

An important means of enzyme regulation involves covalent modification not related to proteolysis. Most frequently the modification involves phosphorylation or dephosphorylation. These transformations are brought about by enzymes known as protein kinases and protein phosphatases, respectively. Protein kinases and phosphatases have other proteins, frequently enzymes, as their substrates. The interconversion of an enzyme from the phosphorylated to the dephosphorylated form usually is associated with a marked change in the activity of the substrate enzyme, sometimes in one direction and sometimes in the other. Serine, threonine, and tyrosine are most frequently the sites to which phosphate groups are attached. Thus, in glucose metabolism, phosphorylation reciprocally activates glycogen breakdown and inhibits glycogen synthesis, so that the control mechanism of covalent modification acts to regulate energy flux through two otherwise competing systems. A growing list of enzymes in the areas of carbohydrate, lipid, amino acid, and protein metabolism emphasizes the importance of regulation by covalent modification.

Figure 3.16

Diagram of a theoretic cascade. *A, B,* and *C* are active forms of different enzymes. *Pro-B* and *pro-C* represent inactive precursors of *B* and *C. Z* is a fragment (or a regulatory subunit) removed from pro-*C* during activation of the cascade.

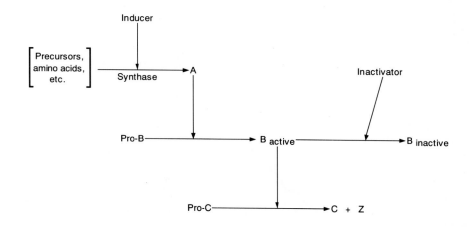

There are some unique advantages to enzyme regulation by covalent modification. It can modulate the extent or direction of energy flux by changing the proportion of an enzyme in the physiologically active form without the waste of removing or replacing the total peptide structure. Thus covalent modification allows for reversible control.

Enzyme cascades

Enzyme cascades are systems in which one enzyme acts on another, usually increasing catalytic effectiveness, in a chain of individual reactions. A generalized example of a cascade is shown in Figure 3.16. Imagine that synthesis of enzyme A is induced by some stimulus and that enzyme A catalyzes conversion of proenzyme B to active B. In turn, B can convert proenzyme C to active C. Assume that A, B, and C have identical turnover numbers; then if 10 molecules of A were newly synthesized, the result of the cascade would be to produce 1000 molecules of active enzyme C. The extraordinary amplification factor of an enzyme cascade is thus a powerful regulatory device, since it can quickly bring about large changes in enzyme activities. Cascades may involve the serial linkage of just a few steps or of many. Initial steps may involve de novo synthesis, as shown in Figure 3.16, or may involve activation of an existing precursor. Negative effectors can act, usually at some intermediate point, to turn the system off.

It is important to note that enzyme induction is a slow process and may not be required for the cascade. The fundamental requirement for a cascade is that one enzyme acts on an enzyme precursor as its substrate, which is a fast process.

Adenylate cyclase and cyclic nucleotide–initiated cascades

In his classic studies of glycogen metabolism, Earl Sutherland proposed that adenylate cyclase, a membrane-bound enzyme, was of central importance. He observed that when certain hormones bound to specific receptors on the external cell surface, the effect was to activate adenylate cyclase, leading to an increased intracellular concentration of 3',5'-cyclic AMP (cAMP). The intracellular cAMP, in turn, causes activation of protein kinases, which eventually leads to the mobilization of glycogen. The cascade that begins with adenylate cyclase activates, in several steps, a wide variety of enzymes by a phosphorylating step (see p. 122). The cyclic nucleotides have become known as "second messengers," a means of transmembrane signaling.

The reaction catalyzed by adenylate cyclase is shown in Figure 3.17; a similar reaction exists whereby guanosine triphosphate (GTP) is converted to cyclic guanosine monophosphate (cGMP).

Figure 3.17   Conversion of ATP to cyclic AMP and pyrophosphate by adenylate cyclase.

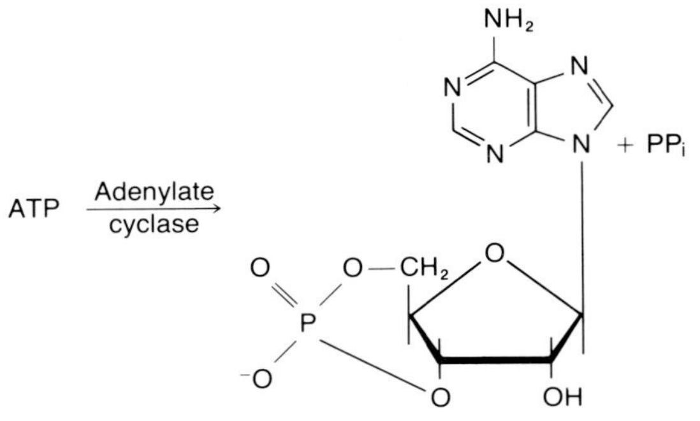

3',5'-Cyclic AMP

Figure 3.18   Structure of a typical protein kinase. Two catalytic *(C)* and two
                             regulatory *(R)* subunits constitute the inhibited (inactive) form of the
                             enzyme. Binding of cyclic AMP to the regulatory subunits releases the
                             two catalytic subunits in active form.

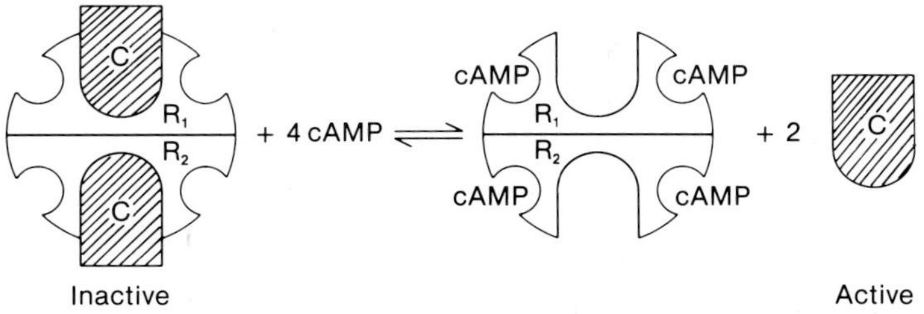

A major protein kinase from muscle exists in an inactive form composed of two regulatory subunits plus two catalytic subunits, as shown in Figure 3.18. Activation involves the binding of 4 moles of cAMP per mole of kinase. Binding of cAMP to the regulatory subunits causes release of the two catalytic subunits, which are then free to catalyze phosphorylation of the dephosphorylated enzyme protein that is the substrate for the particular kinase. This is shown in Figure 3.19. The kinase catalyzes transfer of a terminal phosphate group from ATP to hydroxyl groups of specific serine or threonine residues of the substrate protein. The regulatory subunits allow binding of ATP and Mg^{++} but do not allow binding of the substrate protein. Phosphodiesterase breakdown of cAMP to 5′-AMP is inhibited by some methylxanthines, including caffeine and theophylline. This may account for the stimulatory "lift" of drinking coffee or other caffeine-containing beverages. The generalized organization of cyclic nucleotide cascades is shown in Figure 3.20.

Summary   One can see that control or regulation of enzyme activity takes many forms. At the
                           simplest level it is product inhibition; the effect on an enzyme is exerted largely by its

Figure 3.19   Phosphorylation of a seryl residue of a protein by the active, catalytic
                           form of a protein kinase.

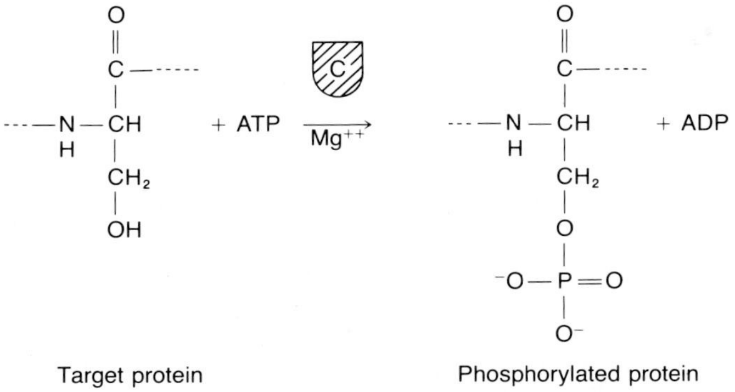

Figure 3.20

A theoretic adenylate cyclase–mediated cascade. Binding of E_x or E_y, external metabolic modulators, activates the cyclase. Increased cyclic AMP concentration in the cytoplasm then activates a protein kinase (R_2C_2), releasing the catalytic subunits that promote protein phosphorylation. Phosphodiesterase is inhibited by methylxanthines *(MX)* such as caffeine or theophylline.

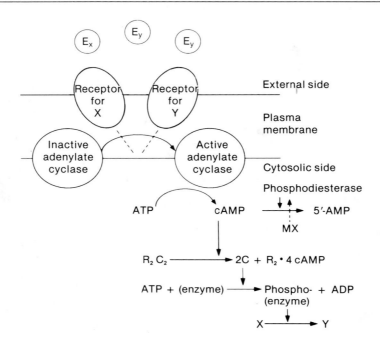

own product. Allosteric control is characteristic of more complex enzymes, and the regulation may be exerted by a substance or substances quite removed from the reaction catalyzed. The mechanism of control is largely a matter of shifts in equilibrium between two quarternary structures of a polymeric enzyme. Another level of control is feedback inhibition, in which an enzyme is affected by the product of an enzymatic reaction that is several steps removed in a multienzyme pathway. Feedback inhibition not only controls a single pathway but also diverts intermediates from one pathway to another. A similar situation exists in control by covalent modification. A greater dynamic range of control is provided by enzyme cascades, wherein enzymes act on other enzymes to multiply effects.

Integration of enzymes into metabolic pathways

In most cases enzymatic reactions are arranged in sequences whereby a building block is converted into a complex product, or a complex material is degraded into building blocks or excretion products. Such integrated series of reactions are known as metabolic pathways. Two examples are used here to illustrate how enzymatic reactions are integrated into such pathways. One, the biosynthesis of the heme portion of hemoglobin, is an anabolic sequence of enzymatically catalyzed reactions that occurs in a single cell, the reticulocyte. The catabolic pathway, hemoglobin degradation following breakdown of the erythrocyte, takes place in several different organs. In this pathway, metabolic intermediates are passed from one organ to the next through the blood plasma and the bile.

Heme biosynthesis Heme proteins are those that contain one or more moles of a metalloporphyrin per mole of protein. Free porphyrins and metalloporphyrins contain extended systems of alternating double and single bonds so that virtually all the metalloporphyrins, including most of the heme proteins, are colored substances. Heme proteins usually serve some function related to cellular respiration, either as electron-transferring proteins or as oxygen carriers. Typical heme proteins found in humans include the cytochromes, myoglobin and hemoglobin. Of these, hemoglobin is quantitatively the most important.

Usually a fairly high turnover of heme proteins occurs. One can calculate from typical hemoglobin concentrations, the average lifetime of circulating erythrocytes, body weight, and blood volume that the daily production of hemoglobin is 5 to 6 g. Given that the molecular weight of hemoglobin is about 67,000, this represents a considerable investment of precursors. The synthesis of hemoglobin occurs in the reticuloendothelial system, largely in the bone marrow. Proerythroblasts or erythroblasts account for most of the synthesis; by the time these cells mature to reticuloyctes, hemoglobin synthesis is much reduced. Mature human erythrocytes have lost all ability to synthesize hemoglobin.

The first step: synthesis of δ-aminolevulinate Synthesis of the porphyrin ring begins with the condensation of succinyl CoA with glycine, shown in Figure 3.21. The product is δ-aminolevulinic acid (ALA), and the reaction is catalyzed by the mitochondrial enzyme ALA synthase. This first step is rate limiting in porphyrin biosynthesis. The amount of ALA produced per day is somewhat greater than the amount of porphyrin produced, since ALA is a minor but significant urinary constituent (2 to 4 mg/day). Situations leading to increased or excessive porphyrin production are usually associated with increased urinary excretion of ALA (the porphyrinurias, porphyrias, and even very severe infections). ALA synthase is an inducible enzyme that requires pyridoxal phosphate. Induction of enzyme synthesis is promoted by hypoxia and by a protein known as erythropoietin, produced in the kidneys. ALA synthase production is subject to repression by free heme.

Formation of porphobilinogen Two moles of ALA condense to form 1 mole of porphobilinogen (PBG); this reaction is catalyzed by the cytosolic enzyme ALA dehydratase (Figure 3.22). Ring closure gives rise to uroporphobilinogen III, a substituted pyrrole, the first cyclic precursor of the porphyrin structure (Figure 3.23). The carboxymethyl and carboxyethyl side-chains on carbons 3 and 4 of the pyrrole ring are typical of some porphyrins. In others such as heme, the side-chains are modifed after completion

Figure 3.21 Formation of δ-aminolevulinate (ALA) by condensation of succinyl coenzyme A and glycine, a mitochondrial reaction.

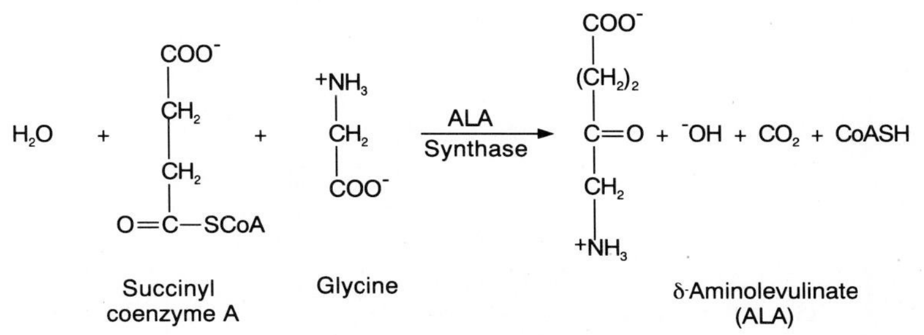

Figure 3.22   Condensation of two ALA, catalyzed by the cytosolic enzyme, ALA dehydratase, to form porphobilinogen. This simple pyrrolic precurosor carries the asymmetric side-chains characteristic of many porphyrin systems. Note that they are both derived from carbon atoms originating in succinyl coenzyme A.

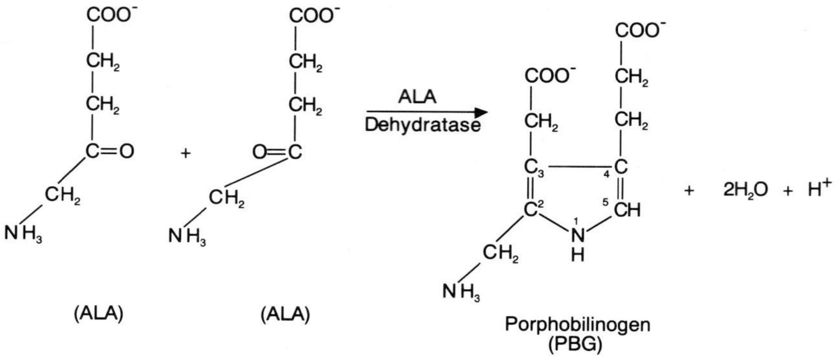

of the porphyrin biosynthesis. Small amounts of free porphobilinogen may also be found in the urine under certain conditions.

Synthesis of isomeric porphyrins The porphyrin structure is generated by joining four molecules of PBG through single-carbon bridges. This gives rise to some interesting possibilities for isomerism, as shown in Figure 3.23. To explain this, we can adopt a convenient shorthand notation for the porphyrin ring, which emphasizes the location of the side-chains as follows. If the side-chains are represented by A and B, respectively, the ordering of these around the periphery of the porphyrin ring could be asymmetric or symmetric, depending on how the pyrroles were arranged at the time of condensation. Based on the pioneering work of Emil Fischer, we now know that both possibilities exist in nature. Type III porphyrins, which have asymmetric distribution of the side-chains, are by far the most abundant forms. Type I porphyrins, which have a symmetric side-chain distribution, constitute only a very small fraction of the total porphyrins in humans under normal conditions.

Figure 3.23   Ordering of side-chains in naturally occurring porphyrins. Type 1 porphyrins have a symmetric ordering of the side-chains, represented by *A* and *B*, whereas the ordering of the side-chains in type III porphyrins is asymmetric.

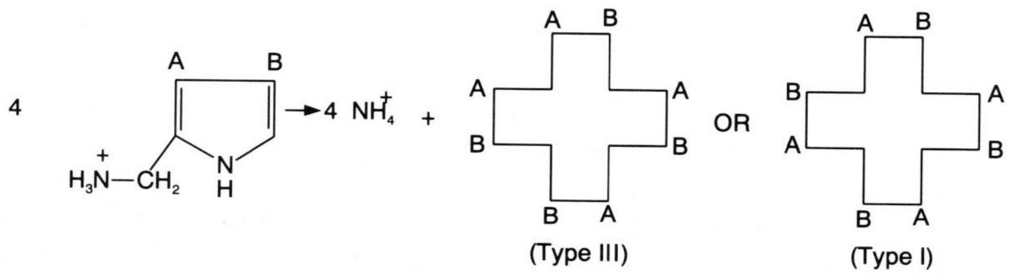

Figure 3.24 Formation of uroporphyriniogen III. Note the asymmetric ordering of the side-chains around the circumference of the ring system, indicating that the molecule is a member of the porphyrin class arbitrarily defined by Emil Fisher as the III series.

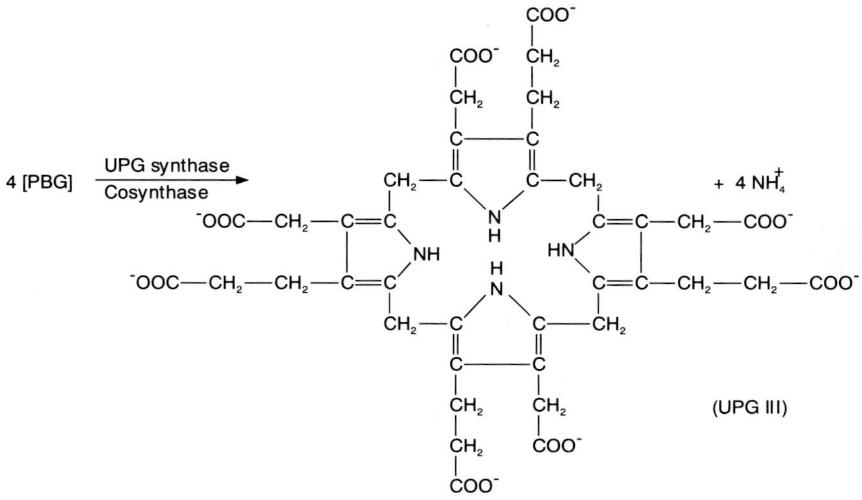

The four molecules of PBG are tightly bound to a cytosolic enzyme, uroporphyrinogen synthase, which is, by itself, relatively inactive. This enzyme requires a second protein, called a cosynthase, to catalyze the formation of uroporphyrinogen III (UPG-III), as shown in Figure 3.24. In this system the cosynthase or coenzyme is itself a protein, which is unusual; few such protein coenzymes are known. The effect of the cosynthase is to "flip" one of the pyrrole rings to produce type III porphyrins. In the *absence* of cosynthase something quite different happens. Now the obligatory flip does not occur, the condensation appears to take place spontaneously, and the result is a type I porphyrin.

Clinical comment

Type I porphyrin A type I prophyrin cannot be converted to protoheme IX, the proper porphyrin component needed to produce hemoglobin or myoglobin. Type I porphyrins, when produced in excess, are quite toxic. They cause severe pain, central nervous system damage, and photosensitivity that leads to extensive skin destruction. These are the overt symptoms of congenital porphyria, the true cause of which is lack of the cosynthase just described. It is claimed that George III of England suffered from congenital porphyria. Some historians argue that his unwillingness to heed even his closest advisors regarding the status of the American colonies was the outgrowth of this disease, which thus led to the events of 1776. Acute congenital porphyria is greatly aggravated by ingestion of barbiturates and certain other sedative drugs, which induce the enzymes involved in the porphyrin biosynthetic pathway. This results from protoporphyrin I being a less effective feedback inhibitor than protoporphyrin III.

Conversion of uroporphyrin to coproporphyrin Early studies, by insufficiently sensitive and specific methods, gave rise to the hypothesis that urine contained only uroporphyrins and that feces contained only coproporphyrins. Today we know that both

Figure 3.25 Decarboxylation of the octacarboxylic uroporphyrinogen III to the tetracarboxylic coproporphyrinogen III, followed by oxidations and further decarboxylations to yield protoporphyrin IX. On insertion of the Fe^{++} ligand, protoporphyrin IX becomes protoheme IX.

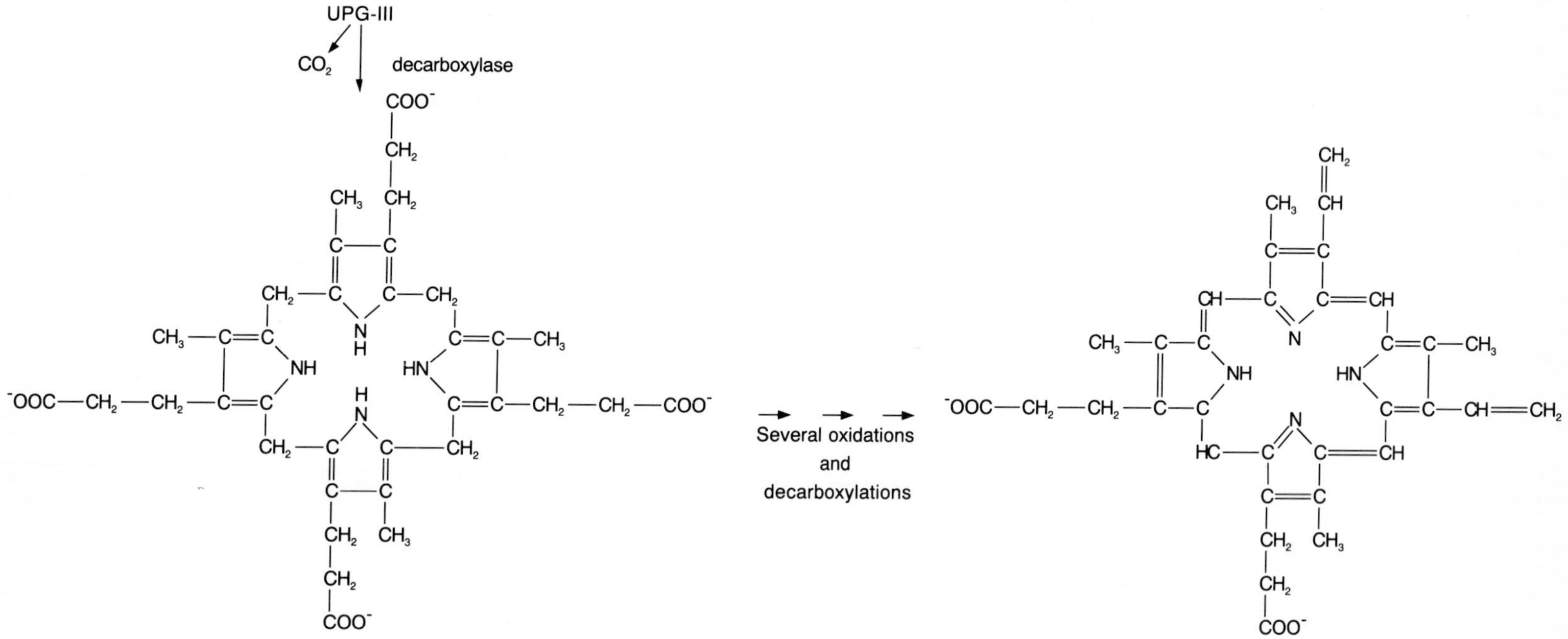

CPG-III

Protoporphyrin IX

Figure 3.26

Conversion of protoporphyrin IX to protoheme IX. Protoheme IX is the characteristic heme moiety of hemoglobin and can be detected in normal human erythrocytes by sensitive methods. Even higher concentrations can be found in normal human reticulocytes.

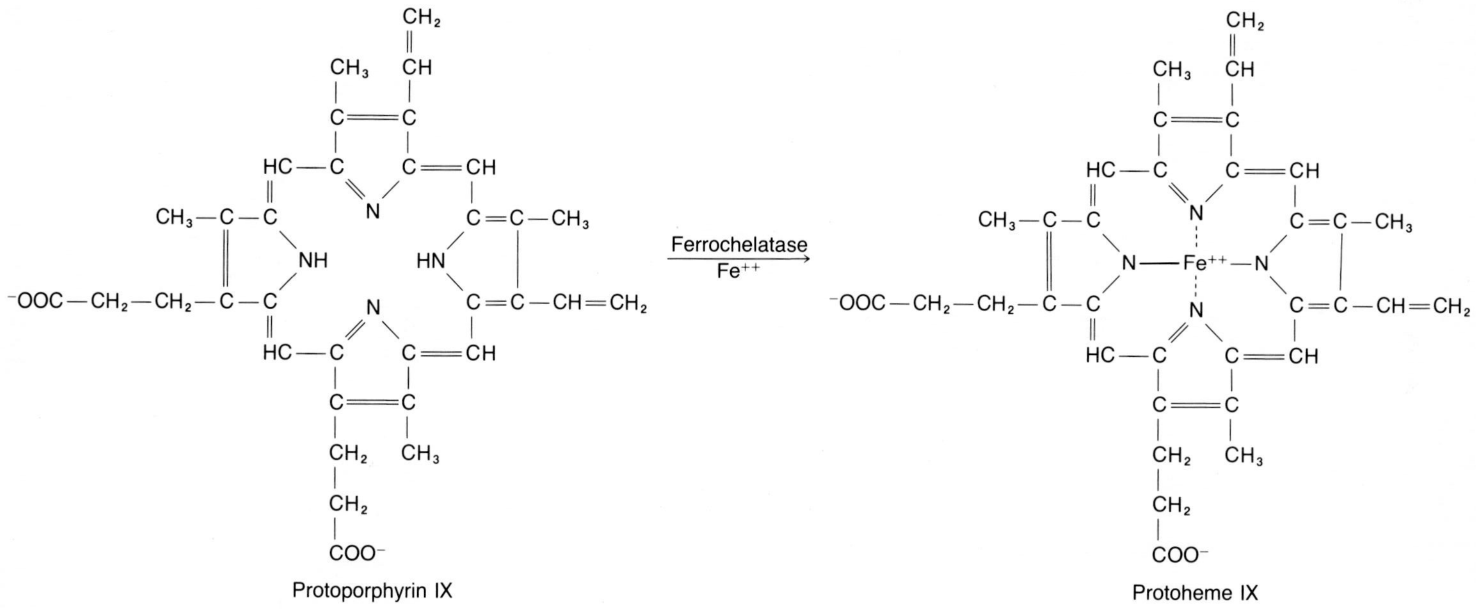

forms are found in either excretory material, but the original designations are too deeply embedded in the literature to discard. The only remaining significance of the terms is that *uro*porphyrins are *octa*carboxylic porphyrins, whereas *copro*porphyrins are *tetra*-carboxylic porphyrins. A sequence of decarboxylating reactions converts uroporphyrinogen III to coproporphyrinogen III, as shown in Figure 3.25. A series of additional decarboxylations and oxidations follows that converts coproporphyrinogen III to protoporphyrin IX.

The final step The conversion of protoporphyrin IX to protoheme IX is catalyzed by the mitochondrial enzyme ferrochelatase, which brings about the insertion of an iron atom into the center cavity of the porphyrin ring. This produces the moiety that can combine with globin to produce hemoglobin, as shown in Figure 3.26. Interestingly, the first and the last steps of this complicated metabolic pathway occur in the mitochondria, whereas the intervening steps are cytosolic. Little or nothing is known about factors that regulate the movement of these intermediates from one cell compartment to another.

Summary of heme biosynthesis The salient features of this pathway can be summarized as follows:
1. Heme biosynthesis begins with simple molecules, succinyl CoA and glycine.
2. The first and last steps are mitochondrial; the other steps are cytosolic.
3. The first step, catalyzed by ALA synthase, is rate limiting.
4. The condensation of 2 moles of ALA produces a substituted pyrrole, which carries a two-carbon and a three-carbon side-chain.
5. As four pyrroles condense to form a porphyrin ring, the ordering of the side-chains is regulated by a protein cofactor of uroporphyrinogen synthase, called cosynthase, producing porphyrins of the III series under normal conditions.
6. Subsequent modification of the side-chains by decarboxylations and oxidations gives rise to uroporphyrins first and then coproporphyrins. The distinction between these forms relates to the number of carboxyl groups on the periphery of the porphyrin ring.
7. The last step is the insertion of an Fe^{++} atom into othe center of the porphyrin ring.
8. The average, healthy adult produces from 5 to 6 g/day of hemoglobin, the most abundant prophyrin-containing protein.

Hemoglobin breakdown and bile pigment metabolism

Mature human erythrocytes lack the genetic information and the organelles for protein biosynthesis. Nevertheless, they have an average lifetime in the circulation of about 120 days. Damaged and devitalized erythrocytes are removed from the circulation by the reticuloendothelial system, especially the spleen. Disposal of erythrocyte debris involves many enzymic reactions. Metabolites produced in the reticuloendothelial system must be transported to the liver for further processing and delivery to the bile, then to the intestine. The last stages of processing are catalyzed by enzymes of the intestinal microbial flora; by virtue of the enterohepatic circulation, some of the bacterial products find their way into the blood and other body fluids.

In discussing the details of hemoglobin breakdown, we must consider the disposition of the iron, of the protein component, and of the porphyrin moiety.

First stages of hemoglobin catabolism The first stages of hemoglobin catabolism occur by oxidation of a methene bridge in the porphyrin ring, as shown in Figure 3.27, where the porphyrin-iron complex is still associated with the protein globin. Apparently, the presence of the central Fe atom makes the porphyrin ring more susceptible to oxidation,

Figure 3.27
First stages in hemoglobin catabolism. Opening of the porphyrin ring by oxidation of a methene bridge, followed by formation of biliverdin. The methene carbon is converted to carbon monoxide and ultimately exhaled.

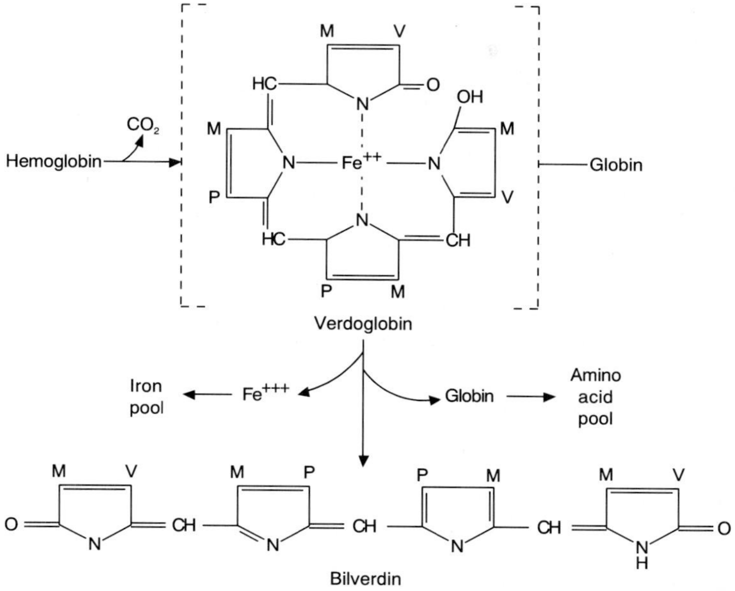

Verdoglobin

Bilverdin

since ring opening of free porphyrins occurs only to a very small extent in mammals. The primary product of ring opening, *verdoglobin,* has a dark, greenish color typical of a bruise or a black eye. The tetrapyrrole is then set free as *biliverdin,* and the globin is degraded by cellular proteases. The liberated amino acids are recycled through the amino acid pools of the body and can be either used again for protein synthesis or catabolized. Since the iron atom cannot form a tight complex with the open-chain tetrapyrrole, it is also returned to the iron pool (for further details of iron metabolism (see Chapter 4, case 4). Reduction of the central methene bridge of biliverdin gives rise to *bilirubin,* which has a yellow-orange color. Almost all the bilirubin formed in the peripheral tissues is transported through the plasma to the liver as a physical complex with albumin, since free bilirubin is not freely soluble in aqueous media (much the same as is true of biliverdin).

Hepatic processing of free bilirubin Reticuloendothelial cells in the liver convert free bilirubin to bilirubin mono- and diglucuronides, the diglucuronide being the predominant form (Figure 3.28). The enzymic reaction by which free bilirubin is converted to the more soluble diglucuronide is sometimes termed a "detoxification" or "conjugation" reaction. The conjugation of slightly soluble substances with sugar acids or with amino acids is a mechanism designed to make them more water soluble. This is certainly the case with bilirubin, as it is with the bile acids (see Chapter 11).

It has long been known that bilirubin forms a colored complex when treated with a solution of diazotized sulfanilic acid; this is the basis of a procedure employed for the determination of bilirubin concentration in serum. The terms *direct* and *indirect* bilirubin relate to the speed with which such a reaction occurs. Direct bilirubin is, in fact, the

Figure 3.28   Structures of biliverdin and two forms of bilirubin. Bilirubin is a reduction product of biliverdin. "Direct" bilirubin is actually bilirubin diglucuronide. "Indirect" bilirubin is not a glucuronide but exists as a physical complex with serum albumin.

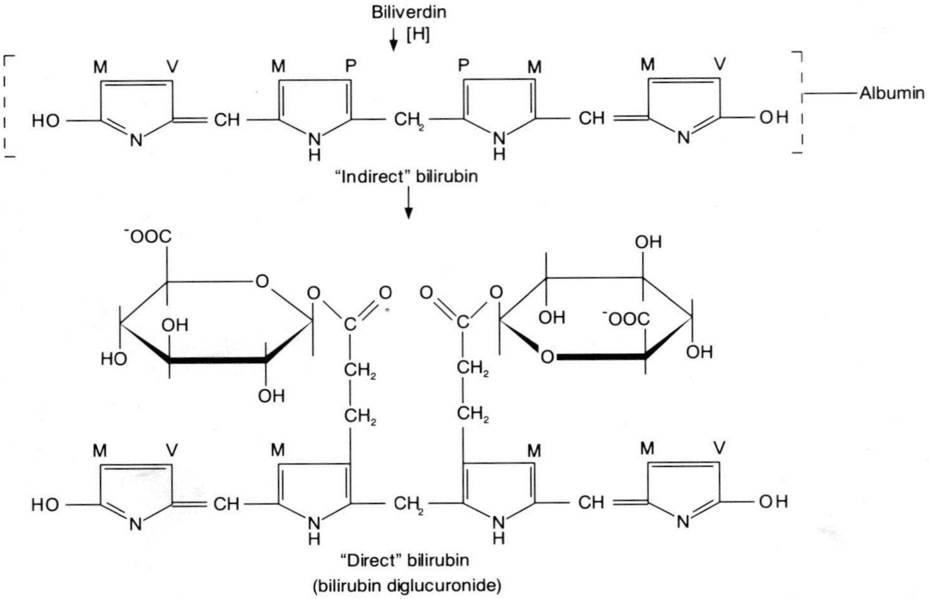

diglucuronide, and indirect bilirubin is free bilirubin; here again the reaction rate is a direct function of solubility, which is a function of the efficiency with which the liver performs the conjugation reaction. Measurement of these two forms of bilirubin, as well as measures of total serum bilirubin, are of considerable importance in the assessment of certain kinds of liver disease.

Late stages of bilirubin metabolism Intestinal microorganisms convert most of the bilirubin delivered via the bile to *urobilinogen*. Some urobilinogen is reabsorbed into the portal circulation along with other fatty products of digestion. Urobilinogen that enters the blood in this way serves no metabolic purpose and is excreted in the urine, to which it imparts the typical amber color. The remainder of the intestinal urobilinogen that was not reabsorbed passes to the lower portions of the bowel, where continued action of microorganisms converts it to stercobilin. This is a deeply pigmented material that gives to the feces their characteristic brown-orange color. When delivery of the bile to the intestine is blocked, urobilinogen is not formed and stercobilin is not produced. In this condition the stools have a chalky, gray color *(acholic stool)* because they lack their natural pigment, stercobilin; this is also a sign of biliary obstruction.

Clinical comment   **Bilirubin metabolism and jaundice** When the blood contains excessive amounts of bilirubin, the sclerae (whites of the eyes) and the skin appear yellowish in color because bilirubin is deposited in the tissues. This is the condition known as jaundice. Although not a disease in itself, jaundice is an important symptom of underlying disease. Measurements of total serum bilirubin and of its major components (free bilirubin and bilirubin diglucuronide) are therefore valuable diagnostic aids. The clinical meaning of these measurements is summarized as follows under the headings of prehepatic, hepatic, and posthepatic jaundice, depending on the cause of the jaundice.

Prehepatic jaundice Diseases or intoxications that result in the breakdown of a greater than normal quantity of erythrocytes cause excessive release of hemoglobin. The heme portion is rapidly converted to indirect (free) bilirubin and transported to the liver. Even if the liver is perfectly healthy, the increased flux of pigment cannot be metabolized rapidly enough to keep the plasma concentration within normal limits. The resultant increase in the total plasma bilirubin concentration is mostly indirect-reacting bilirubin. This condition frequently occurs in infants born with an Rh incompatibility. The total bilirubin may be 10 to 20 times normal, and most of it is indirect. These infants are frequently premature and, in addition to their hemolytic problem, often lack the enzymes required to form the diglucuronide.

Hepatic jaundice In diffuse damage of the liver, as in hepatitis or cirrhosis, the hepatic cells lose some of their ability to extract bilirubin from circulation and may also lose the ability to form the glucuronide. For these reasons the total bilirubin level is frequently elevated, and the indirect fraction is increased. Because the damaged cells allow some of the glucuronide to leak into the circulation, the direct fraction may also be increased.

Posthepatic jaundice Posthepatic jaundice refers to disease that interferes with the delivery of bilirubin to the intestinal tract. The primary consequence is a greatly reduced formation of urobilinogen so that less is found in the urine. The accompanying decrease in stercobilin formation causes the feces to appear chalky gray in color. Continued formation of bilirubin results in increased concentrations of total bilirubin in the serum, primarily resulting from increased direct bilirubin early in the disease process. The bilirubin glucuronide that accumulates in the serum is filtered into the urine, leading to the appearance of a dark brown pigment in the urine.

Clinical applications of enzymes

Measurement of enzyme activity is a useful monitor of overt disease, of genetic tendencies toward a disease state, and of a patient's response to a particular type of therapy. In some instances it is not necessary to measure the enzyme activity itself; determination of a product of an enzyme reaction being present in normal amounts is clinically useful, as is the detection of abnormal isozymes.

Use of enzymes as reagents

A growing number of purified enzymes are becoming commercially available. These can be employed as reagents for the accurate determination of small amounts of such blood constituents as glucose, urea, uric acid, and triglycerides. Frequently these methods are more specific and faster than the chemical determinations used previously. Since many enzymes derived from plant or microbiologic sources can be used to assay components in human blood or tissues, and since the enzymatic methods frequently lend themselves to automation, they have proved invaluable to the health sciences and to the practice of medicine.

Use of enzymes as labeling reagents

Enzymes may be employed as diagnostic aids in yet another way, which takes advantage of their turnover numbers. There is currently great interest in the rapid determination of drugs of abuse, such as morphine, in the urine of suspected addicts. The determination of morphine is a time-consuming and difficult task by classic chemical methods. One can chemically couple morphine to some indifferent protein such as rabbit serum albumin.

Table 3.3   A brief list of enzyme assays known or presumed useful in diagnosis
or treatment of disease

Enzyme	Organ or disease of interest
Commonly assayed	
Acid phosphatase	Prostatic carcinoma
Alkaline phosphatase	Liver, bone disease
Amylase	Pancreatic disease
Glutamate aminotransferase	Liver, heart disease
Aspartate aminotransferase	Liver, heart disease
Alanine aminotransferase	Liver, heart disease
Lactate dehydrogenase	Liver, heart, red blood cells
Creatine kinase	Heart, muscle, brain
Less commonly assayed	
Ceruloplasmin	Wilson's disease (liver)
Aldolase	Muscle, heart
Trypsin	Pancreas, intestine
Glucose 6-phosphate dehydrogenase	Red blood cells (genetic defect)
γ-Glutamyl transpeptidase	Liver disease
Ornithine transcarbamylase	Liver disease
Pseudocholinesterase	Liver (poisonings, insecticides)
Pepsin	Stomach
Hexose 1-phosphate-uridyl transferase	Galactosemia (genetic defect)
Glutathione reductase	Anemia, cyanosis
Lipoprotein lipase	Hyperlipoproteinemia
Elastase	Collagen diseases
Plasmin	Blood-clotting disease

If this coupled preparation is then injected into another species, such as the goat, antibodies to the modified rabbit preparation are produced and can be recovered from its circulation. This antibody recognizes morphine as well as rabbit albumin; when mixed with either, it will form a precipitate of antigen-antibody complex.

Before use, the generated antibody is also modified by coupling it to some indicating enzyme with a fairly high turnover number, such as a phosphatase. The coupling process must not destroy the catalytic capacity of the indicating enzyme. If a sample containing morphine is added to the modified antibody, a precipitate is formed that can be collected, washed, and dissociated. If a suitable phosphatase substrate is added and the mixture properly incubated, a color is produced proportional to the amount of enzyme. That, in turn, is indicative of the presence of morphine. Various commercial immunoassay kits are now available for this purpose.

Physicians and clinical chemists alike have been quick to take advantage of enzyme assays. Table 3.3 lists some of the enzymes that have already been employed on a clinical basis, together with the types of disease states for which the assays are used. Considerable overlap occurs in the use of enzyme assays for specific diseases. For the most part, common clinical practice includes a much smaller list of routine tests; the remainder are employed only in special circumstances. In addition to measurement of the total enzyme activity, the isozyme pattern of certain enzymes is also used. The detection and quantitation of isozymes are of growing importance in differential diagnosis.

What is a valid enzyme assay?

This question can be answered by using the example of lactate dehydrogenase (LDH), and at the same time the mechanism of oxidation-reduction of the nicotinamide adenine dinucleotide (NAD^+) cooenzymes can be explained. LDH reversibly catalyzes the conversion of pyruvate to L-lactate, according to the following equation:

$$\begin{array}{c}CH_3 \\ | \\ C{=}O \\ | \\ COO^-\end{array} + NADH + H^+ \quad \underset{\text{pH 8.8 - 9.8}}{\overset{\text{pH 7.4 - 7.8}}{\rightleftharpoons}} \quad NAD^+ + \begin{array}{c}COO^- \\ | \\ HO{-}C{-}H \\ | \\ CH_3\end{array}$$

LDH

Note that although the reaction is reversible, the optimum pH value is different in each direction. Furthermore, the optimum pH varies somewhat with the temperature and with the substrate and buffer concentrations. LDH is rather sensitive to temperature changes, and if the temperature varies by as little as 3° C from the conventionally accepted assay temperature of 32° C, the observed activity will vary as much as 20%. Thus the assay for plasma LDH must be performed under carefully controlled conditions.

The equilibrium strongly favors the reduction of pyruvate to lactate. Most procedures are based on the reaction just illustrated, even though the required substrate, pyruvate, is less stable than lactate. As mentioned earlier, LDH protein does not undergo oxidation or reduction; that is the province of the coenzyme. The nicotinamide ring of NAD^+ is the point at which reduction takes place, as shown in the following equation:

NAD$^+$
(Oxidized coenzyme)

NADH
(Reduced coenzyme)

Note that the reduced form of the coenzyme accepts only one atom of hydrogen from the substrate and that the second atom remains free in solution as a proton. (The complete structure of the coenzyme is given in Table 3.1.)

Several ways of measuring lactate oxidation and pyruvate reduction exist, but the most widely employed method depends on changes in the coenzyme. This is most practical because a change occurs in the optical properties of the coenzyme as it goes from one oxidation state to the other. NADH readily absorbs light of wavelength 340 nanometers (nm), whereas NAD^+ has a much lower absorbance at this wavelength.

Quantitative measurements are made with a spectrophotometer, a block diagram of which is shown in Figure 3.29. The basic elements consist of a light source *(LS)* that emits "white" light, or light containing many wavelengths. This light passes through a monochromator *(MC)* that selects a narrow band of wavelengths, usually 2 to 20 nm in width, which it transmits through a sample cuvet *(S)*. The amount of radiant energy transmitted through S depends on the concentration of absorbing material and on the length of the path through S. The radiant energy that passes through S then falls on a photodetector *(PD)*, in which the signal is electrically amplified to affect a meter *(M)*, drive a strip chart recorder, or both. In the strip chart recorder the paper chart moves at a fixed rate while the recording pen traces out an instantaneous record of the amplifier output.

The Beer-Lambert law describes the effect of an absorbing substance in a light path as follows:

$$A = \log I_0/I = \epsilon cL$$

where A is the absorbance, or extinction; I_0 is the intensity of light incident on the solution; I is the intensity of light transmitted by the solution; c is the concentration of the absorbing substance; L is the length of the light path, usually expressed in centimeters;

Figure 3.29   Block diagram of a recording spectrophotometer.

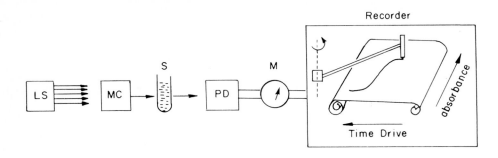

and ϵ is a characteristic of the absorbing substance, termed the extinction, or absorption, coefficient. When c is expressed in moles per liter and L has a value of 1 cm, ϵ is known as the molar absorption of extinction coefficient. The molar absorption coefficient of NADH is 6.22×10^3 L/cm · mole. Consequently, with modern equipment it is possible to measure concentrations in the micromolar range.

Spectrophotometric assay makes it possible to show that for LDH the K_m for pyruvate is 9×10^{-5} mol/L, whereas for lactate the K_m is 5×10^{-6} mol/L; thus it is essential in any assay that the initial respective concentrations of the substrates be fixed well above these values. In the method of Wroblewski and LaDue, at least a 100-fold excess of pyruvate exists in the final assay mixture.

A solution of NADH is added to a suitably buffered and diluted aliquot of serum. Usually a short period occurs in which the absorbance decreases as a result of the consumption of NADH by small amounts of pyruvate in the serum and by other NADH-linked enzymes as well, but the absorbance observed soon becomes constant. When this constant state is reached, the final reaction is initiated by the addition of sodium pyruvate solution, and the time is noted. The decrease in absorbance is then monitored for the next few minutes; if the rate of change is essentially constant, the data can be used to calculate enzyme activity, since each mole of NADH oxidized is exactly equivalent to a mole of pyruvate reduced.

The preceding discussion of an enzyme assay is presented as a typical case and is designed to show how many different properties of an enzyme can be employed to make good clinical use of laboratory determinations. Obviously each enzyme has peculiar properties of its own, and the nature of the reaction may require that different analytic procedures be employed. Correlating chemical findings with clinical or pathologic findings is an ongoing process that occupies a significant portion of the current literature.

Clinical comment

Problems of enzyme assays Enzyme assays in clinical chemistry have been unnecessarily complicated by confusion resulting from arbitrary definitions of enzyme units. Investigators who develop an enzyme assay customarily define their own units. The exact pH, time of incubation, and other variables contribute to the disparate conditions under which assays by different methods are performed. Some idea of the confusion engendered by proprietary units can be gained from a consideration of the data in Table 3.4, which deals solely with phosphatase assays. All the assays mentioned measure the same enzyme activity, but it is clear that the substrates differ, as do the actual substances measured. The pH, the nature of the buffer system, and the time during which the reaction proceeds

Table 3.4 Commonly used phosphatase procedures and associated normal values

Authors	Substrate	Authors' units*	Equivalent international units (IU)†	SI Units (nkat/L)
Bodansky	β-Glycerophosphate	$\dfrac{1.5\text{-}4.0}{dl}$	$\dfrac{8\text{-}22 \text{ mIU}}{ml}$	135-370
Shinowara et al.	β-Glycerophosphate	$\dfrac{2.2\text{-}6.5}{dl}$	$\dfrac{15\text{-}35 \text{ mIU}}{ml}$	250-585
King and Armstrong	Phenyl phosphate	$\dfrac{3.7\text{-}13.0}{dl}$	$\dfrac{25\text{-}92 \text{ mIU}}{ml}$	420-1540
Bessey et al.	p-Nitrophenyl phosphate	$\dfrac{0.8\text{-}2.9}{1000 \text{ ml}}$	$\dfrac{13\text{-}38 \text{ mIU}}{ml}$	220-635
Babson	Phenolphthalein phosphate	$\dfrac{9\text{-}35 \text{ mU}}{ml}$	$\dfrac{9\text{-}35 \text{ mIU}}{ml}$	150-585

*The differences in proprietary units are caused in part by the different substrates, differences in time of incubation, and the pH and composition of the buffers.

†The equivalent international units are not directly calculable from the proprietary units. To be correct over the useful range, each set of equivalent data *must* be determined from the raw data of the analysis.

are not comparable. Similar confusion exists with regard to other enzyme assay methods.

In 1966 the International Union of Pure and Applied Chemistry and the International Federation of Clinical Chemistry joined to establish international unit definitions for all enzyme assays. The fundamental unit proposed is the number of micromoles of substrate transformed per milligram of enzyme per minute. For blood serum or other liquid samples an alternate proposal expresses units in micromoles of substrate transformed per milliliter of sample per minute (or per hour in some cases). As more and more laboratories turn to computer control and machine printout of test results, it becomes imperative to accept a system that is uniform and compatible with machines of various types. The most recent proposal is the International System (SI) of units, proposed by the International Bureau of Weights and Measures. For the standardization of enzyme assays, the SI system proposed a new unit, the *katal* (kat), defined as the moles of substrate transformed per second.

Bibliography

Carrol BJ: Radioimmunoassay of prostatic acid phosphatase in carcinoma of the prostate, N Engl J Med 298:912, 1978.

Clark LC Jr et al: One-minute electrochemical enzymic assay for cholesterol in biological materials, Clin Chem 27:1978, 1981.

Conn RB: New units of measurement in laboratory medicine: the international system of units, J Urol 118:503, 1977.

Everse J and Kaplan NO: Immobilized enzymes in biochemical analysis, Methods Biochem Anal 25:1135, 1979.

Krebs EG and Beavo JA: Phosphorylation-dephosphorylation of enzymes, Annu Rev Biochem 48:923, 1979.

Neurath H: The versatility of proteolytic enzymes, J Cell Biochem 32:35, 1986.

Reid KBM and Porter RR: Proteolytic activation systems of complement, Annu Rev Biochem 50:433, 1981.

Rossman MG and Argos P: Protein folding, Annu Rev Biochem 50:497, 1981.

Young DS: Normal laboratory values (case records of the Massachusetts General Hospital) in SI units, N Engl J Med 292:795, 1975.

Clinical examples

A diamond (♦) on a case or question indicates that literature search beyond this text is necessary for full understanding.

Case 1	Creatine kinase and myocardial infarction

An obese, middle-aged man was brought to the emergency room following an automobile accident. The patient stated that he had been short of breath and very dizzy just before the crash. Examination suggested either a cerebrovascular accident or a myocardial infarction. The patient was admitted for observation, and blood samples for creatine kinase (CK) and other enzyme assays were periodically collected.

Biochemical questions

1. Using procedures described in this chapter, how can CK be assayed?
2. What is the purpose of assaying for CK over a period of time?
3. How could the isozymes of CK be separated?
4. What is the relation of CK activity in the blood to tissue damage?

Case discussion

1. **CK assay** CK catalyzes the following reaction:

$$\text{Creatinine} + \text{ATP}^{4-} \longrightarrow \text{Creatine phosphate}^{2-} + \text{ADP}^{3-} + \text{H}^+$$

At pH 8 the reaction can be followed by titrating the H^+ liberated. The procedure involves adding an aliquot of blood to an aqueous solution of creatine and ATP that had been adjusted to pH with dilute sodium hydroxide. The liberated H^+ is continuously titrated with alkali to maintain pH 8 so that the rate of the reaction can be measured over time, recognizing that for each H^+ produced, an equivalent amount of creatine is phosphorylated. Activity can be expressed as micromoles of creatine transformed per milliliter of blood per minute.

2. **Diagnostic value of CK assays** CK is an enzyme that reversibly catalyzes formation of ATP from ADP, at the expense of creatine phosphate (see Chapter 5). It is probably present in all tissues but is released into the bloodstream, following injury, from only a few tissues. The reason for this is not clear. It has been established that CK is readily released from injured brain and muscle, but little is released from liver. Consequently, CK assays have been of particular use in assessment of myocardial damage in the presence of liver damage, where determinations of aspartate aminotransferase or LDH might be equivocal. As with all laboratory tests, pitfalls are associated with the CK assay. Any appreciable damage to major muscles will produce some increase in *total* serum CK activity. Even repeated intramuscular injections may cause a transient rise. Hypothyroidism is associated with increased serum CK; the reverse is true of hyperthyroidism. Chronic alcoholism, in the stages involving myopathy and neuropathy, may also increase the total CK activity.

3. **Separation of CK isozymes** Three dimeric isozymes of CK are known. Each is composed of SH-dependent monomers, represented by the letters M and B. These letters indicate that one monomer is highly characteristic of muscle, the other of brain. Thus the dimers may have the forms MM, MB, and BB. The unique advantage of CK assays is that, in humans, the singular source of MB isozyme found in the blood is the myocardium. So striking is this finding that the MB isozyme of CK has been termed by some the *myocardial enzyme*. It is therefore not surprising that considerable effort has recently been expended in

devising methods for rapidly and reliably estimating not only the total serum CK but also the MB isozyme.

Electrophoretic separations have been available for some time, but these are discouragingly slow and are not extremely sensitive. Sensitivity is important because the MB isozyme, in normal human heart, comprises only about 15% of the total CK activity, the remainder being MM.

More recently, a radioimmunoassay has been developed, based on [^{125}I]-labeled antibody to the B subunit. This antibody reacts with the MB isozyme, but not with the MM form, even when the latter is present in a 20,000 molar excess. This immunoassay is sufficiently sensitive to detect the MB form in normal blood and requires very small samples. Unlike the earlier procedures, it measures the actual quantity of the isozyme (by virtue of scintillation counting of the [^{125}I] content of the antigen-antibody complex) and does not depend on enzyme activity.

4. **CK activity and tissue damage** The MB-CK isozyme is fairly unambiguous for the detection of myocardial infarction. Its concentration changes rapidly in the serum following myocardial injury. It has been shown that the MB isozyme shows a 100% increase within 4 hr after onset of symptoms, even though the *total* CK activity may still be within normal limits. The peak increase can be found in 8 to 24 hr, after which the values begin to decline. These facts are in sharp contrast to data obtained by measurement of the aminotransferases, where the rise does not begin until 6 to 8 hr following injury, and peak values do not occur until 48 to 60 hr after injury.

A myocardial infarct poses a serious risk to the patient. To maximize the chances for recovery, prompt recognition of the problem and early institution of appropriate therapy are essential. Even during the early stages of treatment, a major concern is whether the disease is spreading or recurring. These newer sensitive and specific tests, which can be employed serially on an hourly basis, have provided cardiologists with excellent tools for management of a major health problem.

The concentration of CK in normal myocardium is quite constant, as is the proportion of the MB isozyme as a fraction of the total. When the myocardium is injured, the release of the MB isozyme causes an increase in the serum activity, which is approximately proportional to the size of the infarct.

References

Bayer PM et al: Immunoinhibition and automated column chromatographs for assay of creatine kinase isozyme MB in serum, Clin Chem 28:166, 1982.

Wicks R et al: Immunochemical determination of CK-MB isozyme in human serum, Clin Chem 28:54, 1982.

Case 2   Antibiotics as enzyme inhibitors

A young girl was brought to the pediatric clinic with a badly infected wound on her knee. The mother was instructed to give the child penicillin G orally and told to return with the child in several days. On the subsequent visit the child had not improved. The mother was given a prescription for oxacillin and again asked to return 5 days later. By the third visit the infection had apparently subsided.

Biochemical questions
1. How is penicillin inactivated?
2. Bacteria sensitive to penicillin do not die if they are grown in an isotonic medium. What does this tell you about a possible mechanism of action of this antibiotic?
3. Significant quantities of some penicillins may be excreted in the urine of individuals given the drug. What does this signify in terms of human metabolism?

Case discussion

The pencillins constitute a valuable class of antibiotics, long used in the treatment of some infections. Typically, penicillins contain a thiazolidine ring fused to a β-lactam ring, with one of a series of organic acids attached to the α-amino group of the lactam ring through an amide linkage, as follows:

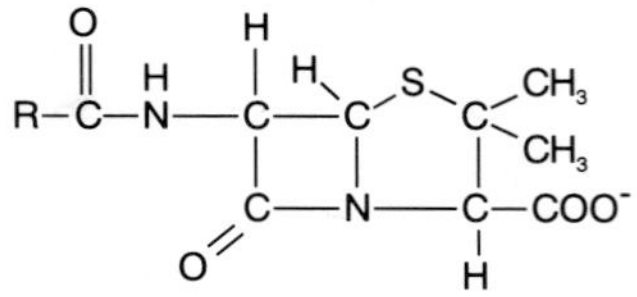

Following are three typical forms of the substituent group, R, identified with the penicillin that contains them:

Penicillin G

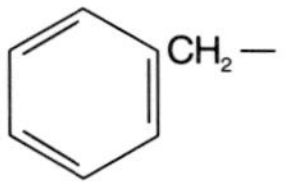

Benzyl group

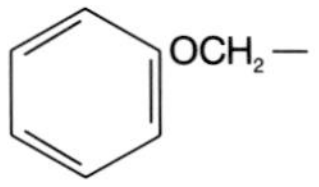

Phenoxymethyl group

Penicillin V

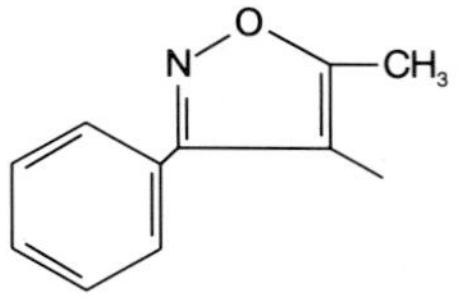

5-methyl-3-phenyl- 4-isooxazolyl group

Oxacillin

   1. **Inactivation of penicillin** Penicillin G is not used orally because it is readily hydrolyzed by the acid gastric juice; thus the first attempt at treatment was unsuccessful. Penicillin V and oxacillin can be given orally, since they resist exposure to the acid gastric environment. Both penicillin G and V are easily destroyed by the enzyme penicillinase, produced by some strains of microorganisms. Penicillinase hydrolyzes the β-lactam ring. The semisynthetic oxacillin is more resistant to penicillinase, and so it can be used against organisms that produce the enzyme. Penicillinase production can be induced in some bacteria by treatment with concentrations of the drug insufficient to kill the invading organisms quickly or by previous exposure to penicillin in other hosts. In other words, penicillinase is an inducible enzyme.

   2. **Excretion in urine** The action of penicillinase on penicillin produces penicillamine, which is without antibiotic action. Penicillamine is useful in clinical medicine because it forms chelates especially well with heavy metal ions. It is frequently employed in the treatments of metal intoxications (see following case 3). Human tissues do not contain large amounts of penicillinase-like activity. A significant part of the administered drug passes into the urine in undegraded form because we lack the means to metabolize it.

   3. **Mechanism of action** Penicillin acts by inhibiting a transpeptidase that cross-links peptidoglycan polymers into a larger molecule, *murein,* which is an essential component of the cell wall in penicillin-sensitive bacteria. By blocking cell wall synthesis, the drug makes the infecting organisms fragile, and they consequently burst and die.

   Murein contains may cross-linkages formed between the terminal D-alanyl-D-alanyl residues of one peptidoglycan chain (R) and the pentaglycyl residue of another peptidoglycan chain (R'). The transpeptidase forms an acyl-enzyme complex intermediate with the elimination of the terminal D-alanyl residue. The active intermediate then reacts with the terminal glycyl residue attached to another chain to form a cross-link between R and R'. The process is repeated many times, with one event represented as follows:

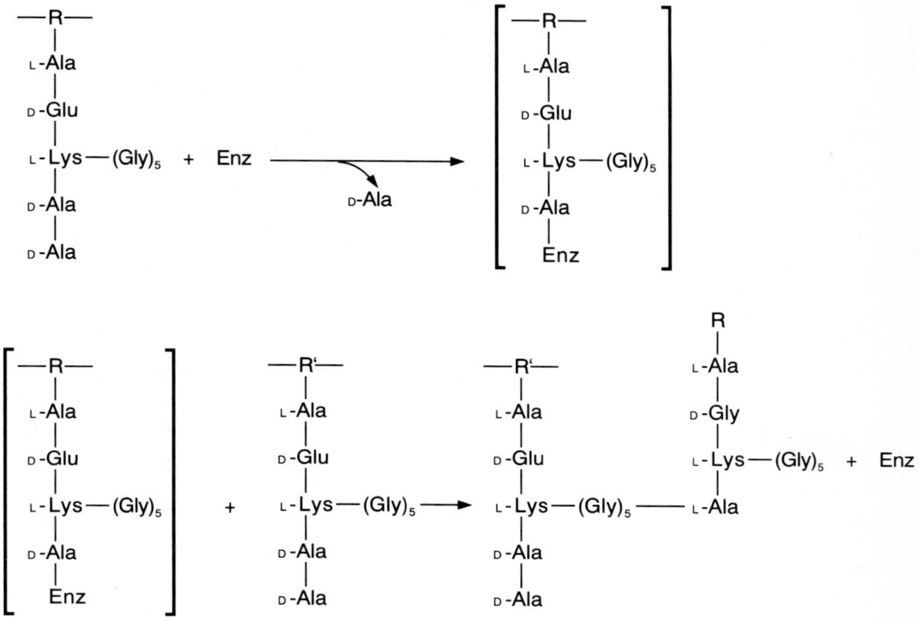

The conformation of penicillin closely resembles that of the D-alanyl-D-alanyl residue with which it reacts, thus blocking the cross-linking and the synthesis of murein.

References

Demain AL: Industrial microbiology, Science 214:987, 1981.
Waxman DJ and Strominger JL: Penicillin-binding proteins and the mechanism of action of β-lactam antibiotics, Annu Rev Biochem 52:825, 1983.

Case 3 — Lead poisoning

An 18-month-old child of migrant farm workers was hospitalized because of weight loss, vomiting, and acute abdominal pain. It was noted that the child had mild muscular incoordination and weakness of the muscles of the feet.

A blood smear showed a moderate but distinct increase in the reticulocyte count. The red blood cell count was 4×10^6 cells/mm^3, and the hematocrit was 37%. A 24 hr urine sample contained 6.4 μmol (840 μg) of aminolevulinic acid and 1.8 μmol (1.2 mg) of coproporphyrin III. Lead poisoning was suspected, and penicillamine therapy was started immediately. A quantitative lead analysis on the urine showed 1.1 μmol (0.24 mg) of lead in the 24 hr sample. X-ray examination of the patient's long bones showed electron-dense deposits at the epiphyses.

Biochemical questions

1. How does lead exert toxic effects on metabolic pathways?
2. Is the enzymatic inhibition produced by lead competitive or noncompetitive?
3. What is the mechanism of the increased urinary excretion of aminolevulinic acid and coproporphyrin III?

4.   Why is lead deposited in the bones?

5.   What is the rationale for treatment of lead poisoning by penicillamine?

Case discussion

1. Effects of lead on metabolism A small amount of lead is always present in the body; normally the tissues contain lead in amounts of 4 to 10 ppm. Individuals with occupational exposures to lead may exhibit distinctly higher levels than normal. Lead is mainly absorbed in the gastrointestinal tract, but it is also absorbed through the lungs, as in chronic exposure to high concentration of automobile exhaust fumes of leaded gasoline. Ingested lead is taken up most avidly by the sternum and long bones, with lesser amounts being deposited in the brain, kidneys, liver, and lungs. The remainder of the absorbed lead is excreted in the urine, which normally contains 19 to 70 nmol/24 hr (4 to 15 µg/24 hr). Accidental lead poisoning results from ingestion of excessive amounts of lead, usually over a considerable time. In this case it is likely that the child was eating or chewing on something in the home that contained lead, such as flakes of old paint from woodwork, furniture, toys, or radiators.

Toxic effects of lead are not limited to the synthesis of hemoglobin. For example, the neuropathy and muscle weakness in this case resulted from the inhibition of essential enzymes and proteins of the nervous system and muscles. The affected proteins clearly contain sulfhydryl groups. Moreover, the heavy metal salts of fatty acids and amino acids are relatively insoluble. This implies that many metabolic reactions could be affected by elevated lead concentrations in the tissues.

The effects of lead on the kidney are also reflected in a decreased ability to excrete uric acid. The uric acid concentration in the blood and other tissues increases and gives rise to the condition known as saturnine gout. In the past, when a high consumption of port wines was popular, saturnine gout was more common, since the wines were frequently contaminated with lead during manufacture and shipping. More recently, illegal stills fabricated with soldered joints have again raised the incidence of saturnine gout among heavy drinkers exposed to illegal alcoholic beverages.

2. Noncompetitive inhibition by lead Lead forms covalent bonds with sulfhydryl groups. Many proteins in the body, including several enzymes, contain cysteine residues with free sulfhydryl groups. The proteins are almost always denatured when lead combines with their sulfhydryl groups. The covalent complex between lead and the protein sulfhydryl groups is very tight, and the changes produced in the proteins are virtually irreversible. A kinetic analysis of the particular enzyme would reveal noncompetitive inhibition. This is the type of inhibition that cannot be overcome simply by raising the substrate concentration. Those enzyme molecules that have formed a complex with lead no longer possess enzymatic activity. The remaining enzyme molecules, which are free of lead, function normally. The decrease in *functional* enzyme concentration results in a lowered V_{max}, but the K_m of the reaction would be unchanged.

Enzymes, being catalysts, are usually present in small amounts. Thus the presence of even a small excess of lead can exert pronounced metabolic effects. All that is required for toxic effects to be manifest is the inactivation of some fraction of one or more key enzymes.

3. Lead and heme biosynthesis A metabolic pathway that is extremely sensitive to the presence of excess lead is heme biosynthesis; in particular, lead affects the enzyme ferrochelatase. Ferrochelatase catalyzes the insertion of ferrous iron into protoporphyrin IX, forming heme. In lead poisoning, aminolevulinate is excreted in increased amounts because when ferrochelatase activity is inhibited, heme production is not rapid enough to take up all the aminolevulinate synthesized. Likewise, the protoporphyrin IX that is formed cannot be adequately converted to heme because of ferrochelatase inhibition.

The porphyrin builds up in the tissues, and a portion of the excess is excreted as coproporphyrin III. The mild anemia noted in this patient resulted in part from inhibition of heme biosynthesis by toxic amounts of lead.

4. Lead and the skeletal system Lead (group IV in the periodic table) has many properties similar to those of calcium (group II). The body does not clearly distinguish between them; thus lead is deposited in the mineral substance of bone, especially in growing children. As with calcium, the greatest amount of lead is laid down at the sites of active bone metabolism. Lead is more electron dense than calcium and appears on roentgenograms as areas of greater density. The ability to deposit lead in the skeletal system may be a protective device, since the deposited lead is not available to react with protein sulfhydryl groups. However, mineral in bone exists in a dynamic equilibrium with other body minerals, so that any need to withdraw calcium from the bony substance will bring with it deposited lead. Therefore treatment should be designed to solubilize deposited lead in a nonionic form that can be excreted without harm to proteins. Diets should contain adequate amounts of calcium and vitamins to minimize the need for calcium resportion from bone.

5. Chelation of lead The child was treated with penicillamine, a nonantibiotic derivative of penicillin that forms stable complexes with lead ions as well as with those of other heavy metals. Penicillamine is one example of a class of compounds known as chelating agents. These compounds have the capacity to combine with certain ligands (in this instance with lead) to produce products that are only slightly dissociated. Other agents in this group include EDTA and 2,3-dimercaptopropanol (BAL). The structures of their lead complexes are shown in Figure 3.30.

Lead in a stable nonionic complex is not toxic; therefore chelating agents serve primarily to lower the toxicity of circulating lead. Second, the chelated form of lead is more soluble than the nonchelated form; lead deposited in the bones will slowly be dissolved if it is bound as a chelate and will then be excreted by the kidneys. Unfortunately, the reaction between the chelating agents and metal ions is not absolutely specific, and various essential ions as well as those that are toxic may be excreted as well. Therapy with chelating agents requires considerable clinical skill and

Figure 3.30 Structures of several lead chelates.

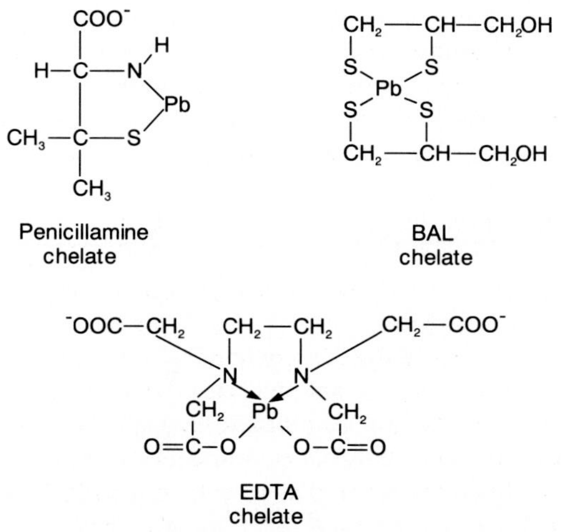

Penicillamine chelate

BAL chelate

EDTA chelate

constant monitoring of the patient so as not to induce deficiencies of essential divalent cations such as Ca^{++}.

Vigorous public health measures are being pursued to minimize exposure to lead, especially among young children. Lead-free gasoline will probably soon become mandatory; older homes and buildings are being scraped and sandblasted. It is ironic that these efforts themselves have produced a rise in lead intoxication in those engaged in urban renewal. Since early signs of lead intoxication can be detected by determination of urinary porphyrins, such tests are indicated for individuals with known exposure to lead.

References

Cavalleri A et al: Biological response of children to low levels of inorganic lead, Environ Res 25:445, 1981.
Feldman RG: Urban lead mining, lead intoxication among deleaders, N Engl J Med 298:1143, 1978.
Lin-Fu JS: Lead exposure and toxicity in children, N Engl J Med 289:1229, 1289, 1973.

Case 4   Serum hepatitis

A 37-year-old man was admitted to a hospital with glomerulonephritis, which was treated conservatively. The disease progressed, however, and over the next 6 months the patient developed an enlarged liver that was tender to palpation. Liver function tests were abnormal; the aspartate aminotransferase (AST) concentration was 1200 IU/L (normal, 7 to 20 IU/L), and total plasma bilirubin concentration was 77 μmol/L (normal, 2 to 19 μmol/L). A diagnosis of serum hepatitis was made.

Biochemical questions

1. What precautions should be taken in collecting blood for assays of enzymatic activity?
2. Why did serum aminotransferase activity increase?
3. What vitamin is part of the coenzymes of AST?
4. What data would you need to know to establish the optimum conditions for the assay of AST?
5. Would the determination of "direct bilirubin" give a better indication of liver function? Explain.

References

Alter HJ: Hepatitis, Semin Liver Dis 6:1, 1986.
Jacobson IM and Dienstag JL: The delta hepatitis agent: viral hepatitis, type D, Gastroenterology 86:1614, 1985.

Case 5   Muscle injury

A 29-year-old laborer developed chest pain while operating a pneumatic hammer on a construction project. The pain was of moderate intensity, but he felt well enough to continue work. As the day continued, he experienced sharp pain on respiration and a tight feeling across the anterior chest wall. He was rushed to the emergency room of a local hospital and was admitted following a brief examination. Electrocardiographic and chest roentgenographic examinations were negative, except for an elevated plasma lactate dehydrogenase (LDH) level of 6.9 μkat/L (400 IU/L). The high plasma LDH persisted during the next 4 days; no other laboratory or physical abnormalities appeared, and the chest pain gradually improved with bed rest.

Biochemical questions

1. What reaction does LDH catalyze?
2. Is an elevation in plasma LDH specific for damage to a given organ or tissue in the body?
3. Explain what isozymes are. How might a LDH isozyme assay help to determine the cause of this patient's illness?
4. In setting up the assay for LDH, the technician added lactate and NAD^+ to the plasma specimen. Why were these substances added? Should lactate and NAD^+ have been added in very small or in excessive amounts? Could NADH or FAD have been substituted for NAD^+ in this assay system?
5. Can you construct an LDH assay that depends on some measurement other than the spectral properties of NAD^+ or NADH? Describe it.
6. How many isozymes of LDH normally occur in serum?
7. Depending on the specific assay method employed, a suitable substrate in an LDH assay could be pyruvate or lactate. Why?
8. How would assay conditions differ with use of pyruvate or lactate as the substrate?

Reference

Bergmeyer HV, editor: Methods of enzymatic analysis, ed 2, New York, 1975, Academic Press, Inc.

◆ Case 6

Unexplained increase of serum creatine kinase MB isozyme in lung cancer

Much value is attached to measurement of isozymes reputed to be derived from specific tissues. However, implicit reliance on isozyme measurements can give rise to misleading conclusions.

A case has been reported, involving a patient with lung cancer, in which persistent increases occurred in the serum activity of the creatine kinase isozymes BB and MB.

Biochemical questions

1. With respect to creatine kinase, how might cancer cells differ from normal cells?
2. Could you expect to obtain reliable results in an enzyme assay performed on tissue samples removed at autopsy? What types of errors might result in such postmortem studies?
3. How might you proceed to more clearly rule out, in biochemical terms, the possibility that this patient had suffered a myocardial infarction?

Reference

Goffman T, Cantrell J, and Schein P: Unexplained increase in serum creatine kinase isozyme MB activity in a lung cancer patient, Clin Chem 27:2068, 1981.

Additional questions and problems

1. What simple experiment could be designed to determine whether an enzyme exhibited classic (Michaelis-Menten) or allosteric kinetics?
2. Describe a simple experiment that would help decide whether an enzyme inhibitor was competitive or noncompetitive.
3. Discuss the role of enzymes in achieving equilibrium of reactions.
4. Why must valid enzyme assays be performed under initial reaction conditions? What kinds of errors might result if this is not done?
5. How might one employ the enzyme urease to measure blood urea nitrogen concentrations in a biologic sample? Could this be made a quantitative measurement?
6. Some dehydrogenases have a strict requirement for NAD^+; others have a strict requirement for $NADP^+$; still others can operate with either coenzyme. How would you rationalize these observations?

7. What argument can be used to explain the apparent need for *acid* as well as *alkaline* phosphatases? Why could one not do as well with only an alkaline phosphatase?

8. Give examples of metabolic pathways mentioned in this chapter that are controlled in part by (a) product inhibition, (b) allosteric enzymes, and (c) protein enzyme modification.

9. What are some metabolic advantages of enzymes in a metabolic pathway being localized in different intracellular locations?

10. How might one explain a mutant enzyme with a greater V_{max} but with an identical K_m compared to the normal enzyme?

Multiple choice problems

An elderly patient, who died after 5 days of hospitalization for pneumonia, showed an unusual electrophoretogram of serum LDH. More than the usual isozymes appeared in the electrophoresis at pH 8.6. A similar result occurred in extracts of several tissues taken at postmortem. Serum from a surviving daughter revealed a similar abnormal pattern.

Reference

Buchholz DH and Donabedian RK: Unusual variant of lactate dehydrogenase isozymes, Clin Chem 21:162, 1975.

The following questions are related to this case.

1. How many different subunits may be present in normal LDH?
   A. One
   B. Two
   C. Three
   D. Four
   E. Five

2. The isozymes of serum and the different tissues:
   1. Have different subunit compositions.
   2. Have different subunits structures.
   3. Are all separable by electrophoresis.
   4. Have the same stability to denaturation by heat.
      A. 1, 2, and 3 only are correct.
      B. 1 and 3 only are correct.
      C. 2 an 4 only are correct.
      D. 4 only is correct.
      E. All are correct.

3. How many isozymes of normal LDH can be identified by electrophoresis at pH 8.6?
   A. One
   B. Two
   C. Three
   D. Four
   E. Five

4. All the isozymes function with the coenzymes:
   1. $NADP^+$.
   2. NADH.
   3. NADPH.
   4. NAD^+.

A. 1, 2, and 3 only are correct.
B. 1 and 3 only are correct.
C. 2 and 4 only are correct.
D. 4 only is correct.
E. All are correct.

5. LDH assays are useful in diagnosing diseases of the:
   1. Heart.
   2. Prostate.
   3. Liver.
   4. Pancreas.
      A. 1, 2, and 3 only are correct.
      B. 1 and 3 only are correct.
      C. 2 and 4 only are correct.
      D. 4 only is correct.
      E. All are correct.

6. Studying the enzyme kinetics of the isolated abnormal isozyme showed that the K_m and V_{max} values for each substrate were different from the normal isoenzymes. This indicates differences in the:
   1. Binding of lactate.
   2. Binding of pyruvate
   3. Turnover number.
   4. Equilibrium constant.
      A. 1, 2, and 3 only are correct.
      B. 1 and 3 only are correct.
      C. 2 and 4 only are correct.
      D. 4 only is correct.
      E. All are correct.

7. The abnormal isozyme need not:
   A. Be an oxidoreductase.
   B. Have any coenzyme.
   C. Require ATP.
   D. Be localized intracellularly.
   E. Be a catalyst.

8. The identification of the abnormal isoenzyme in the patient's daughter suggested that the isozyme was:
   1. Not an isolation artifact.
   2. Autosomally transmitted.
   3. Not disease related.
   4. Was X linked.
      A. 1, 2, and 3 only are correct.
      B. 1 and 3 only are correct.
      C. 2 and 4 only are correct.
      D. 4 only is correct.
      E. All are correct.

9. Assume that this unusual isozyme cannot form hybrid oligomeric enzymes with the normal LDH subunits, but that it can associate with itself to form an oligomer that is more anionic than any normal LDH oligomer. How many bands of enzyme

activity would appear following electrophoresis at pH 8.6?

    A. Three

    B. Four

    C. Five

    D. Six

    E. Eight

10. The chemical forces that bind most coenzymes and substrates to enzymes such as LDH are:

    1. Hydrogen bonds.

    2. Ionic forces.

    3. Van der Waals' forces.

    4. Covalent bonds.

      The *best* answer is:

      A. 1, 2, and 3 only are correct.

      B. 1 and 3 only are correct.

      C. 2 and 4 only are correct.

      D. 4 only is correct.

      E. All are correct.

Chapter 4

Blood, hemoglobin, and acid-base control

Objectives

1. To relate the properties of hemoglobin to its biologic function
2. To interpret the significance of variations from normal in the pH and electrolyte composition of blood
3. To relate information on pH and electrolyte concentrations to possible metabolic or respiratory imbalance
4. To describe the biochemical reactions that control blood loss through blood coagulation

Homeostasis implies a close control of the circulation and the composition of fluids that contain both solids and gases. By means of such circulation, each cell in the body is bathed in a nutrient medium that is optimum for its function. The circulatory fluids also remove metabolic secretions and excretions from the cell. Only a very limited variation in circulating acid and base is consistent with life; therefore even slight changes from the normal acid-base balance require proper and prompt clinical action to correct the cause. The control of the total concentration and volume of the body fluids and the control of specific ion concentrations occur in the normal individual principally through proper function of the lungs and kidneys. These organs interact with each other and all other parts of the body by way of the blood.

Bicarbonate buffer

The bicarbonate (HCO_3^-) and phosphate systems are the most important inorganic buffers in human physiology. The bicarbonate buffer is considered in detail because of its unique property of having a volatile acid component.

Carbonic acid (H_2CO_3), a dibasic acid with pK_a's of 3.88 and 10.22, would have little buffer capacity in humans if it did not dehydrate to CO_2, a gas that can be expired from the body. This reaction is relatively slow in the absence of the enzyme carbonic anhydrase.

$$CO_2 + H_2O \rightleftharpoons H_2CO_3$$
$$\frac{H_2CO_3 \rightleftharpoons H^+ + HCO_3^-}{CO_2 + H_2O \rightleftharpoons H^+ + HCO_3^-}$$

In this coupled system all the CO_2, whether it is physically dissolved as the gaseous CO_2 or in the hydrated form as H_2CO_3, is considered as the acid form. The apparent dissociation constant may be written:

$$K_a' = \frac{[H^+][HCO_3^-]}{[CO_2, \text{ both dissolved as } CO_2 \text{ and in the form of } H_2CO_3]}$$

The coupled system has a pK_a' of 6.1, and the equation relating pH to the concentration of bicarbonate $[HCO_3^-]$, dissolved CO_2, and H_2CO_3 is:

$$pH = 6.1 + \log_{10} \frac{[HCO_3^-]}{[\text{Dissolved } CO_2 + H_2CO_3]} \qquad (4.1)$$

The denominator is in fact the *total* CO_2 in plasma in all forms *less* that present as bicarbonate. The following can be written:

$$pH = 6.1 + \log_{10} \frac{[HCO_3^-]}{[\text{Total } CO_2] - [HCO_3^-]} \qquad (4.2)$$

The total CO_2 content of a sample of plasma is determined by measuring the volume of CO_2 liberated on acidification with strong acid. Throughout such calculations it is necessary to express all concentrations in the same unit, usually millimoles per liter. The calculations and derived conversion factors are developed in millimeters of mercury (mm Hg) pressures instead of the SI unit of kilopascals (kPa). The latter is given in parentheses for comparison but is not used in any calculations. Since CO_2 is a nonideal gas, 1 mmol occupies 22.26 ml at STP, that is, standard temperature (0° C) and pressure (760 mm Hg [101.33 kPa]), so that:

$$\frac{\text{Volume of } CO_2 \text{ in milliliters (STP)}}{2.26 \text{ ml/mmol}} = CO_2 \text{ in millimoles}$$

For example, when a 5 ml sample of plasma was acidified, 0.11 ml of CO_2 was collected. The temperature in the clinical laboratory was 22° C, and the atmospheric pressure was 750 mm Hg (100.00 kPa). Correcting the gas volume to STP:

$$0.11 \times \frac{750}{760} \times \frac{273}{295} \text{ ml of } CO_2 \text{ (STP)}$$

that is:

$$0.10 \text{ ml of } CO_2 \text{ (STP)}$$

or:

$$\frac{0.10}{22.26} \text{ mmol of } CO_2$$

that is:

$$0.00449 \text{ mmol of } CO_2$$

or:

$$4.49 \text{ } \mu\text{mol of } CO_2$$

From the total CO_2 and the pH it is possible to calculate the $[HCO_3^-]$ from equation 4.2. By the mass balance of CO_2 given in the following equation:

$$[\text{Total } CO_2] = [HCO_3^-] + [\text{Dissolved } CO_2 + H_2CO_3]$$

A value for [dissolved $CO_2 + H_2CO_3$] is then obtained. This is now abbreviated [dissolved CO_2].

The concentration of a gas in a liquid is related to the partial pressure of the gas and the Bunsen (or solubility) coefficient. Considering now the concentration of CO_2 in plasma, the solubility coefficient is 0.51 ml of CO_2 (corrected to STP) in 1 ml of plasma at 38° C and 760 mm Hg (101.33 kPa) of CO_2 pressure. Therefore, for each millimeter of mercury partial pressure of CO_2, or pCO_2, there is 0.51/760 ml of physically dissolved CO_2/ml of plasma at 38° C. Now:

$$\frac{0.51}{760} \text{ ml of } CO_2 = 6.71 \times 10^{-4} \text{ ml of } CO_2/\text{mm Hg}$$

To convert this volume to millimoles, it is divided by 22.26:

$$\frac{6.71 \times 10^{-4}}{22.26} \text{ mmol of } CO_2/\text{ml of plasma}$$

Since this is the amount of CO_2 dissolved in 1 ml of plasma/1 mm Hg pCO_2, then the amount dissolved in 1000 ml of plasma is:

$$\frac{6.71 \times 10^{-4}}{22.26} \times 1000 = 0.0301 \text{ mmol/mm Hg } pCO_2$$

As an illustration of the use of these units, consider the equilibrium of a 1 ml sample of plasma at 38° C with alveolar air, which normally has a partial pressure of 40 mm Hg of CO_2; that is, pCO_2 is 40 mm Hg. (NOTE: The abbreviation of the partial pressure of a gas X as pX antedates the pX convention of pH or pK_a and bears no relation to it.)

$$\text{Volume of } CO_2 \text{ dissolved} = 40 \times 6.71 \times 10^{-4} \text{ (measured as STP)/ml of plasma}$$
$$= 2.68 \times 10^{-2} \text{ ml/ml of plasma}$$

Also:

$$\text{Concentration of } CO_2 = 40 \times 0.0301 \text{ mmol/L}$$
$$= 1.204 \text{ mmol/L}$$

Equation 4.1 can be restated so that the acid term is expressed in terms of pCO_2:

$$pH = 6.1 + \log_{10} \frac{[HCO_3^-]}{0.0301 \times pCO_2} \tag{4.3}$$

Alternatively, from equation 4.3 the bicarbonate concentration can be restated in terms of total CO_2 and pCO_2, as in equation 4.4:

$$pH = 6.1 + \log_{10} \frac{[\text{Total } CO_2] - 0.0301 \times pCO_2}{0.0301 \times pCO_2} \tag{4.4}$$

A normal, rested individual will have arterial blood plasma values of pH 7.4; total CO_2, 25 to 28 mmol/L; pCO_2, 40 to 43 mm Hg; and $[HCO_3^-]$, 24 to 27 mmol/L. These values are consistent with each other as calculated from the Henderson-Hasselbalch equation. For example, at pH 7.4 the ratio of the bicarbonate concentration to the acid concentration can be calculated from the pH and pK_a of this system, as in equation 4.5:

$$7.4 = 6.1 + \log_{10} \frac{[HCO_3^-]}{0.0301\, pCO_2} \tag{4.5}$$

or:

$$1.3 = \log_{10} \frac{[HCO_3^-]}{0.0301\, pCO_2}$$

or:

$$\frac{[HCO_3^-]}{0.0301\, pCO_2} = 20$$

Likewise, the average values of the salt and acid components just listed, when substituted in equation 4.5, give the same value, 20:

$$\frac{25}{0.0301 \times 4.1} = \frac{25}{1.23}$$
$$= 20$$

Alternatively, the same value of 20 will be calculated from the total CO_2 and pCO_2 values, using the expression for the ratio of [salt]/[acid] given in equation 4.4.

$$\frac{\text{Total CO}_2 - 0.0301 \, p\text{CO}_2}{0.0301 \, p\text{CO}_2} = \frac{26 - 1.23}{1.23}$$

$$= 20$$

It is the *ratio* of $[\text{HCO}_3^-]/[\text{dissolved CO}_2]$ that is 20 if the pH of the blood plasma is at the normal average value of 7.4. Any disproportionate change of either the $[\text{HCO}_3^-]$ or the $[\text{dissolved CO}_2]$ will change this ratio from 20, and the blood will be either acidic or alkaline with reference to the normal. In such situations the compensatory mechanisms of the body will come into play in an attempt to correct the pH back to 7.4.

Venous blood plasma usually shows about 1.1 mmol/L more HCO_3^- and 0.14 mmol/L more $[\text{dissolved CO}_2]$ than the arterial blood plasma because venous blood transports CO_2 from the peripheral tissues to the lungs. Will the pH of venous blood be more acid or alkaline than arterial blood? The average values for HCO_3^- and dissolved CO_2 in arterial blood are 25 and 1.23 mmol/L, respectively. Therefore in venous blood:

$$\text{pH} = 6.1 + \log_{10} \frac{(25 + 1.1)}{(1.23 + 0.14)}$$

$$= 6.1 + \log_{10} \frac{26.1}{1.37}$$

$$= 6.1 + \log_{10} 19.05$$

$$= 6.1 + 1.28$$

$$= 7.38$$

The ratio of 19.05 in venous blood indicates that it will be slightly more acid, a conclusion borne out by the calculated value of pH 7.38.

Control of pH in the body

The control of the pH of body fluids centers largely around the functions of the lungs and the kidneys, whereby excess H^+ is eliminated. The lungs function to reduce the $p\text{CO}_2$ in the blood, thus increasing the $[\text{HCO}_3^-]/[\text{H}_2\text{CO}_3]$ ratio. The kidneys serve to retain as much HCO_3^- from the blood as necessary and to generate more by converting CO_2 to HCO_3^- and H^+. The H^+ is eliminated by the $\text{HPO}_4^=/\text{H}_2\text{PO}_4^-$ buffer system or as NH_4^+.

Since the H^+ is common to all buffers in the body fluids and is freely exchangeable with the intracellular constituents, all buffering reactions are coupled together.

$$\text{H}^+ \left\{ \begin{array}{l} +\,\text{NH}_3 \rightarrow \text{NH}_4^+ \\ +\,\text{HCO}_3^- \rightarrow \text{H}_2\text{O} + \text{CO}_2 \\ +\,\text{HPO}_4^= \rightarrow \text{H}_2\text{PO}_4^- \end{array} \right.$$

It is therefore not surprising to find that all the above processes are interdependent and closely related to the property of hemoglobin to carry O_2 and CO_2.

Myoglobin and hemoglobin

Hemoglobin is a tetramer, $\alpha_2\beta_2$, in which the β-subunit protein is similar to a simpler oxygen-carrying protein called myoglobin.

Myoglobin

Myoglobin is a chromoprotein found in the muscle cells of mammals. Although there are appreciable quantities in human cardiac muscle, myoglobin is present in large quantities in skeletal muscle, especially the muscles of those mammals that dive deeply in the sea. Human myoglobin is a small globular protein with a molecular weight of 16,700 that contains 152 amino acid residues. It is a single polypeptide chain with L-valine at the

Figure 4.1   Porphyrins.

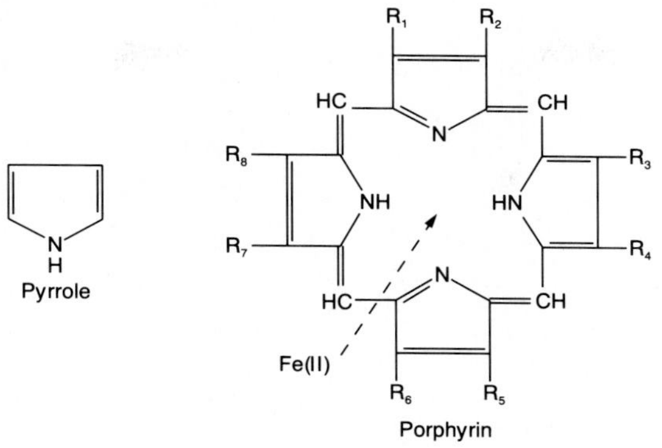

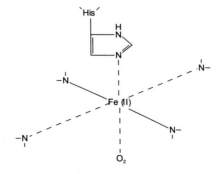

Pyrrole

Porphyrin

NH$_2$ terminus. Coordinated to the protein is a heme residue that, for purposes of the present discussion, may be represented by Figure 4.1. Heme is an iron-containing porphyrin composed of Fe(II), four pyrrole rings linked by methene bridges, and eight sidechains attached to the pyrrole rings (Figure 4.1). The iron is inserted in the center, coordinately linked to the four nitrogen atoms of the pyrrole rings (see Chapter 3). In myoglobin the Fe(II) is also complexed with an imidazole nitrogen atom of a histidine residue in the protein chain, as shown in Figure 4.2.

Myoglobin can react with O$_2$ to form oxymyoglobin (MbO$_2$), which is in equilibrium with deoxymyoglobin (Mb).

$$Mb + O_2 \rightleftharpoons MbO_2$$

The equilibrium position is dependent on the concentration of O$_2$ in the system. Therefore myoglobin may be considered a storage reserve for O$_2$. It is largely in the oxymyoglobin form when the O$_2$ concentration in the cellular fluid is high, but when the O$_2$ supply is reduced, oxymyoglobin releases the bound O$_2$ for cellular use. Its function in deep-diving mammals is to provide a store of O$_2$, and its role in human cardiac muscle during periods of oxygen debt is similar.

Figure 4.2   Representation of oxymyoglobin structure around heme residue. *His,* *Histidine.*

Figure 4.3   Myoglobin molecule. *His*, Histidine; *Val*, valine.

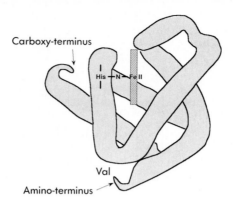

Myoglobin is one of the first proteins whose three-dimensional structure was described through the application of x-ray analysis. It consists of eight sections of relatively straight α-helical segments bent on each other to give a compact globular molecule. Comparing this conformation with the amino acid sequence of the primary structure, one finds that the hydrophobic interactions are maximum, since very few hydrophobic side-chain residues are exposed to the aqueous solvent medium. The peptide chain is further stabilized by the significant formation of α-helical segments. Breaks in the α-helices are found when L-prolyl or other destabilizing amino acid residues occur in the chain. No stability is introduced in myoglobin by — S — S — bonds; cysteine is absent from the molecule. The structure is compact, but space exists for the heme residue in the crevice noted in Figure 4.3; otherwise space exists for only four water molecules in the interior portion of the molecule. The two negatively charged propionate side-chains, R_6 and R_7 (see Figure 4.1), of the heme residue form ionic bonds with two L-lysyl ϵ-NH_3^+ groups in the peptide chain, and this, together with the coordination of the Fe(II) to the L-histidyl nitrogen, binds the heme firmly into the globin.

Hemoglobin

The red color of blood is caused by the hemoglobin content of the erythrocytes. Blood contains 7.8 to 11.2 mmol of hemoglobin monomer/L (12.6 to 18.4 g/dl), depending on the age and sex of the individual. Under normal conditions all the hemoglobin of the blood is present inside the erythrocytes.

Centrifugation of a blood sample causes the cells to pack at the bottom of the centrifuge tube. These cells normally occupy about 35% to 54% of the blood volume, depending on the age and sex of the individual; the percentage value of packed cells is called the *hematocrit*. Since most of these cells are erythrocytes, the hematocrit is a measure of the amount of erythrocyte hemoglobin to the first approximation.

Normal adult hemoglobin, Hb A, contains two globin subunits of one type identified as α-chains and two globin subunits of another type called β-chains. Hb A is therefore represented as $\alpha_2\beta_2$. The molar concentration of hemoglobin is calculated for clinical purposes in terms of quarter-molar units—one subunit per atom of iron. In many ways these α- and β-chains are similar to myoglobin. They have L-valyl amino terminal residues and conformationally have a high proportion of α-helical segments. The α-chain has 141 amino acid residues with an L-arginine carboxyl-end terminus, and the β-chain has 146 residues with an L-histidine carboxyl-end terminus. The four subunits, with their identical heme residues, pack into a tetramer, molecular weight 65,000, which can be dissociated first into two dimeric αβ-units and finally into the mixture of α- and β-monomers. On removal of the dissociating agent, which may be an acid or an alkali, the monomers reassociate to the $\alpha_2\beta_2$-tetramer.

Hemoglobin variants

Normal In the course of a lifetime most humans synthesize five different globin chains. All normal hemoglobins contain two α-chains, which are paired with either β-, γ-, δ-, or ε-chains. The variants are fetal hemoglobin, Hb F($\alpha_2\gamma_2$); Hb A_2 ($\alpha_2\delta_2$), a variant that comprises about 2.5% of the hemoglobin in the adult; and embryonic hemoglobins (e.g., $\alpha_2\epsilon_2$). The primary structures of these normally occurring globins are similar, and each contains a protected pocket into which the heme residue fits.

Abnormal Mutations in the genes for hemoglobin may result in a substitution of a *single* amino acid residue in one the globin chains. These mutations may be innocuous or fatal, depending on the nature of the substitution and on the point at which they occur in the chain. For example, sickle cell anema, a serious and often fatal inherited disease, is caused by the presence of hemoglobin S (Hb S). In Hb S the sixth amino acid residue from the NH_2 terminal end of each β-chain is valine; in the normal β-chain it is glutamic acid. Thus in Hb S an amino acid containing an anionic side-chain is replaced by one with an uncharged hydrocarbon chain. Consequently, the electrophoretic migration rate of Hb S is different from that of Hb A, and the hydrophobic nature of the valine side-chain leads to an aggregation of hemoglobin molecules that causes the erythrocytes to take on abnormal (sicklelike) shapes.

In more than 200 other abnormal hemoblogin variants that have been described, only one amino acid residue is different in the α- or β-chains. Frequently an amino acid residue with an ionic side-chain is substituted so that the charge on the chain is changed and the variant is easily detected by electrophoresis or ion exchange chromatography.

Hemoglobin derivatives Chromatography of the hemolysates of erythrocytes resolves four minor hemoglobin components from the main Hb A fraction. These minor fractions are collectively referred to as Hb A components and constitute approximately 7% of total hemoglobin in normal persons. They result from posttranslational nonenzymatic modification of hemoglobin. The most abundant minor component, Hb A_{1c}, derives from glycosylation of the Hb A and is particularly elevated in patients with diabetes mellitus or galactosemia. In patients with uremia an elevation in Hb A_{1a+2b} results from carbamoylation of hemoglobin. Measurements of the elevations of these abnormal hemoglobins are being used in clinical diagnosis.

Oxygen binding

Hemoglobin reacts reversibly with O_2 in the manner similar to that described earlier for myoglobin. The reaction equilibrium may be expressed in terms of the percentage of the oxygenated form, which will vary with the concentration of O_2. The relationship is represented graphically in Figure 4.4, where the O_2 concentration is expressed as the partial pressure (pO_2) in millimeters of mercury (mm Hg) or kilopascals (kPa).

Oxygen concentration and partial oxygen pressure

As stated in Dalton's law, the total pressure of a gas mixture is the sum of the partial pressures of each of the components. Also, the partial pressures of the gases are proportional to the corresponding mole fractions of each gas in the mixture. Finally, the concentration of the gas in the liquid phase of a gas-liquid system is proportional to its

Figure 4.4   Oxygen saturation curves for myoglobin and hemoglobin.

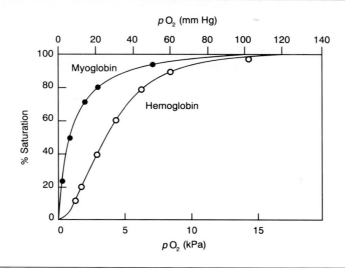

partial pressure in the gas phase, the other factor being the solubility of the gas in the liquid.

Since we are frequently concerned with air at 760 mm Hg (101.3 kPa), which contains O_2, N_2, CO_2, and water vapor, Dalton's law can be stated for air and body gases as follows:

$$P_{total} = pO_2 + pN_2 + pCO_2 + pH_2O$$

Then the mole fraction of any gas, such as O_2, in the mixture is the following:

$$\frac{pO_2}{P_{total}} \text{ or } \frac{pO_2}{760}$$

The amount of O_2 dissolved in liquid at a given temperature is expressed in terms of a *Bunsen coefficient*, the volume (milliliters) of a gas at the specified temperature and 760 mm Hg pressure that will dissolve in 1 ml of liquid. The Bunsen coefficient, although corresponding to a particular temperature, is always corrected to the standard conditions of 760 mm Hg pressure and 273° K. Under such conditions the volume of a gas is easily converted to a molar concentration, with 22.40 ml of an ideal gas being equivalent to 1 mmol.

The use of partial pressures of gases is so common and important in problems of respiration, anesthesiology, and any quantitative consideration involving gases that it is useful to illustrate the interrelationships of gas concentrations. Consider the concentration of O_2 in blood plasma. The O_2 is derived from the inspired air in which the pO_2 is about 158 mm Hg (21.1 kPa). As a result of some mixing with air already present in the lung, as well as a result of its saturation with water vapor, the air in the alveolae of the lungs has a pO_2 of about 100 mm Hg (13.3 kPa). Diffusion of the O_2 into the capillaries of the lungs results in a further lowering of the O_2 tension (pO_2) to about 90 mm Hg (12.0 kPa). Therefore the arterial blood plasma can be considered in contact with a gas that has a pO_2 of 90 mm Hg (12.0 kPa). The temperature of the body is close to 38° C, at which temperature the Bunsen coefficient for O_2 in blood plasma is 0.024. This means that 0.024 ml of O_2, at 760 mm Hg (101.3 kPa) and 273 K (0° C), will dissolve in 1 ml of plasma at 38° C when the pO_2 is 760 mm Hg (101.3 kPa). Since the amount of O_2 dissolved is proportional to its partial pressure and the partial pressure in alveolar air is 90 mm Hg (12.0 kPa), the actual volume of O_2 in arterial plasma at the lungs is the following:

$$0.024 \times \frac{90}{760} \text{ ml } O_2/\text{ml plasma} \quad \left(0.024 \times \frac{12.0}{101.3} \text{ ml } O_2/\text{ml plasma}\right)$$

or:

$$0.00284 \text{ ml } O_2/\text{ml plasma}$$

Since this volume of O_2, derived from the Bunsen coefficient, gives the volume of the dissolved gas corrected to standard conditions, the volume is converted to millimoles by dividing by 22.40.

$$0.00284 \text{ ml } O_2$$

$$= \frac{0.00284}{22.4} \text{ mmol } O_2$$

$$= 1.27 \times 10^{-4} \text{ mmol } O_2$$

Since this is the amount of O_2 in 1 ml of plasma, the concentration of dissolved O_2 is 0.127 μmol/ml, or 0.127 mmol/L.

Similar calculations can be made for any of the other gases in plasma, given their Bunsen coefficients. If, as with CO_2, they are not ideal gases, the volume of 1 mmol of gas at 760 mm Hg (101.3 kPa) and 273° K also must be known.

Oxygen saturation curves of hemoglobin and myoglobin

Considering again the oxygenation curves of myoglobin and hemoglobin, the saturation of myoglobin follows a rectangular hyperbola (Figure 4.4), whereas hemoglobin describes a sigmoidal curve. The reaction of O_2 with myoglobin follows a simple equilibrium where the pK_{eq} is about 6:

$$K_{eq} = \frac{[MbO_2]}{[Mb][O_2]}$$

Throughout the range of pO_2 values, the O_2 is bound more readily to myoglobin than hemoglobin. Most important, the binding of O_2 to each molecule of myoglobin is independent of another molecule because there is only one O_2 binding site on each. Each monomeric subunit of hemoglobin also binds O_2, but in the tetrameric quaternary structure, the binding affinity of O_2 depends on whether an O_2 molecule is bound to one of the other subunits. The binding of O_2 is said to be *cooperative*. The binding of the first O_2 affects the binding of the second O_2, these two O_2 affect that of the third, and so on. This situation reflects a cooperativity in the binding processes. In effect, there are four different K_{eq} values for each of the binding sites, with K_1 being less than K_2 and K_3, and the final equilibrium constant demonstrates that the last O_2 molecule is bound more tightly than the first.

$$Hb_4 + O_2 \overset{k_1}{\rightleftharpoons} Hb_4(O_2)$$

$$Hb_4(O_2) + O_2 \overset{k_2}{\rightleftharpoons} Hb_4(O_2)_2$$

$$Hb_4(O_2)_2 + O_2 \overset{k_3}{\rightleftharpoons} Hb_4(O_2)_3$$

$$Hb_4(O_2)_3 + O_2 \overset{k_4}{\rightleftharpoons} Hb_4(O_2)_4$$

$$K_1 < K_2 \approx K_3 < K_4$$

Stereochemistry of hemoglobin oxygenation

Explanations for the cooperative effects in hemoglobin have been based on evidence obtained by x-ray crystallography of oxyhemoglobins and deoxyhemoglobins. It is known that the conformations of the α- and β-chains are altered when their respective heme residues combine with O_2. The stepwise process may be described using several models, none of which completely explains the facts.

The $(\alpha\beta)_2$ tetramer of hemoglobin has two types of contact interfaces: (1) between the $\alpha_1\beta_1$ subunit and the identical $\alpha_2\beta_2$ subunit and (2) between $\alpha_1\beta_2$, which is identical to $\alpha_2\beta_1$. This can be expressed as:

When hemoglobin becomes oxygenated, a large structural change occurs at the $\alpha_1\beta_2$ and $\alpha_2\beta_1$ interfaces, and much less of a change occurs at the $\alpha_1\beta_1$ and $\alpha_2\beta_2$ interfaces.

The binding of the first O_2 molecule can occur at an α- or a β-subunit; that is, two different binding states are possible. The addition of the second O_2 molecule can occur in four different binding states. Hemoglobin with three O_2 molecules can have two binding states and the fully oxygenated molecule has one state in which all subunits carry an O_2. Thus a total of nine binding states exist. The free energy levels of these nine states fall into three groups, as represented as in Figure 4.5.

These three energy states suggest that in the course of sequential binding, the hemoglobin molecule switches into three molecular forms as opposed to only two, an oxy- and a deoxyhemoglobin conformation. Several conformations probably form the transitions between the two extremes of oxy- and deoxyhemoglobin, with some too fine to measure. Evidence that the switches in conformation are not simple are found in dissociation studies of carboxyhemoglobin and similarly bonded molecules, proton nuclear magnetic resonance spectroscopy, and x-ray studies of partially oxygenated hemoglobin.

These conformation switches occur in response to oxygen binding, and thus the affinity

Figure 4.5

Free energy levels for tetramers of hemoglobin in various binding states. The three geometric symbols represent three conformational intermediate states of the hemoglobin.

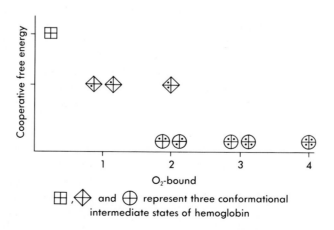

of the O_2 for the hemoglobin states changes, as was noted in the four equilibrium constants. As a result, a sigmoidal saturation curve of Hb_4 for O_2 exists. The conformational switches occurring in the protein also are partly related to the slight change in position of the Fe(II) in the heme when it binds O_2. The conformational changes produce displacement of an L-tyrosyl residue from a pocket in the globular polypeptide. Along with this residue, the NH_2-terminal residue that had been in a salt linkage with nearby negatively charged groups is also displaced. The number of salt linkages broken in the Hb_4 depends on the number of O_2 molecule bound, which again is related to the different affinities for O_2 in the intermediate states. This change is called the Bohr effect and is described more fully in the following discussion.

Oxygen transport   **Binding of 2,3-bisphospho-D-glycerate to hemoglobin** 2,3-Bisphospho-D-glycerate represents approximately 15% of the anionic content of the erythrocyte. It is produced by the catabolism of D-glucose, and its intracellular concentration is about 5 mmol/L, almost equimolar with hemoglobin.

$$\begin{array}{c} \text{COO}- \quad\quad \text{O} \\ | \quad\quad\quad || \\ \text{H}-\text{C}-\text{O}-\text{P}-\text{O}- \\ | \quad\quad\quad | \\ \quad\quad\quad \text{O}- \\ \\ \text{O} \\ || \\ \text{O}- -\text{P}-\text{O}-\text{CH}_2 \\ | \\ \text{O}- \end{array}$$

Its function at such high concentrations was not appreciated until its effect on the affinity of hemoglobin for O_2 was demonstrated.

One mole of 2,3-bisphospho-D-glycerate is bound to 1 mol of hemoglobin. The 2,3-bisphospho-D-glycerate is bound between the two β-subunits by ionic salt bridges. As a result, the equilibrium of the intermediate forms is shifted to favor the deoxyquaternary structure. This reduces the tendency of the partially oxygenated intermediates to convert to the oxyquaternary structure. This results in a saturation curve for hemoglobin that shows an overall lowering of the affinity for O_2 in the presence of 2,3-bisphospho-D-glycerate (Figure 4.6). However, it should be noted that at the pO_2 in the alveoli, about 100 mm Hg (13.3 kPa), hemoglobin is nearly 100% saturated. In contrast, at the pO_2 present in the capillary beds of peripheral tissues (40 mm Hg, or 5.3 kPa), less O_2 is available to bind to hemoglobin. In other words, O_2 can be unloaded more easily in the presence of 2,3-bisphospho-D-glycerate and thus ensure an adequate supply to the tissues. Interestingly, Hb F is not affected in this way by 2,3-bisphospho-D-glycerate.

Carbon monoxide poisoning CO binds competitively at the same site on hemoglobin as does O_2 and shows similar cooperativity and a Bohr effect (see next discussion). CO binding is more than 200 times stronger than O_2 binding. The amount of O_2 that can be carried by the remaining unreacted Fe(II) of the hemoglobin molecule is reduced if CO is present at one of the heme residues. Moreover, the O_2 that *is* associated with hemoglobin molecules that contain some CO is bound more tightly, making it more difficult for the O_2 to be transferred from these hemoglobin molecules to the tissues. Therefore the toxic effect of CO is more severe than might be predicted simply from its concentration in the blood. Subacute CO poisoning gives rise to poorly defined symptoms that mimic those of other illnesses; it has been termed the "great imitator" (see case 1 in the clinical examples).

Figure 4.6   Effect of 2,3-biphospho-D-glycerate on oxygen binding of hemoglobin.

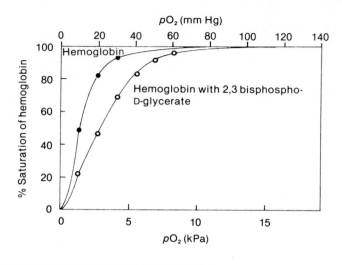

Bohr effect

For each O_2 bound to Hb_4, there is movement of the penultimate L-tyrosyl residues and the breaking of salt bonds. The proton that was held in the salt linkage is now free to dissociate, which it does to a degree that varies according to the pH. In the β-chain the COOH-terminal histidine residue releases a proton. This is not the histidine residue coordinated to the heme (Fe(II). In at least one of the α-subunits the proton arises from the NH_2-terminal L-valine, which, in the deoxy form, was forming a salt link with the COOH-terminal arginine in the other α-unit. Since these protons are arising from — NH_3^+ or imidazole nitrogen atoms, these groups will have a greater tendency to remain protonated as the pH of the system becomes more acidic. Salt bridges are therefore favored, and these tend to stabilize the deoxy form. This is reflected in the O_2 binding curves for Hb_4 (Figure 4.7). The result is a tendency to release O_2 at lower pH values, which is known as the Bohr effect. This effect depends on the quaternary structure of hemoglobin. The Bohr effect does not occur with myoglobin or in the α- or β-monomer subunits of hemoglobin. This and other changes of ionic character cause a change in the overall strength of hemoglobin acidity, such that the apparent pK_a of Hb_4 changes from 7.71 to 7.16 in $Hb_4(O_2)_4$. The oxygenation of hemoglobin is therefore associated with a release of protons. The reaction can be represented as follows:

$$HbH_x + O_2 \rightleftharpoons HbO_2 + xH^+$$

where x is approximately 0.7 mol/mol of hemoglobin. This property is extremely important in the physiology of O_2 transport.

Respiratory control of blood pH

The pCO_2 of blood can be changed quite rapidly by the rate and/or depth of breathing (pulmonary exchange). Slow, shallow breathing *(hypoventilation)* results in a buildup of alveolar pCO_2 and a reduction in the diffusion of CO_2 from the blood into the pulmonary alveolar gas phase. This increases the pCO_2 in the plasma and lowers the blood pH. *Hyperventilation* has the opposite effect. These two effects can be illustrated by the simple example of a normal, healthy person who develops hiccups. Knowing that such an attack

Figure 4.7   Oxygen saturation curves at different pH values.

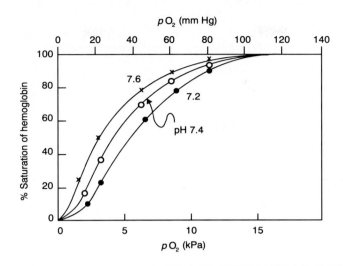

can usually be stopped by increasing the level of CO_2 in the blood, the individual holds his or her breath or breathes with a paper bag over the nose and mouth (i.e., inspires air with high pCO_2) for a few minutes. The CO_2 content of the blood begins to increase. After the hiccups stop, the blood of this individual has a pCO_2 of 60 mm Hg and a total CO_2 content of 30 mmol/L. At this time the blood pH can be calculated:

$$pH = 6.1 + \log_{10} \frac{30 - 0.0301 \times 60}{0.0301 \times 60}$$

$$= 6.1 + \log_{10} \frac{30 - 1.806}{1.806}$$

$$= 6.1 + \log_{10} 15.61$$

$$= 6.1 + 1.19$$

$$= 7.29$$

To eliminate the excess CO_2, the individual breathes rapidly and deeply for a few minutes (hyperventilation). The blood plasma then contains 27.5 mmol/L [HCO_3^-], and the pCO_2 is 30 mm Hg. By how much has the blood pH changed?

$$pH = 6.1 + \log_{10} \frac{27.5}{0.0301 \times 30}$$

$$= 6.1 + \log_{10} \frac{27.5}{0.903}$$

$$= 6.1 + 1.48$$

$$= 7.58$$

Thus the pH has moved from the acid side of normal to the alkaline side, changing 0.29 pH units by going from a state of hypoventilation to one of hyperventilation.

Transport of oxygen and carbon dioxide in blood

Consider the circulation of blood through the lungs where the pO_2 is approximately 100 mm Hg (13.2 kPa) and the pCO_2 is about 36 mm Hg (4.80 kPa). One can see from Figure 4.8 that in this condition the hemoglobin is almost 100% saturated. As this blood

Figure 4.8   Oxygen-hemoglobin dissociation curves, pH 7.4.

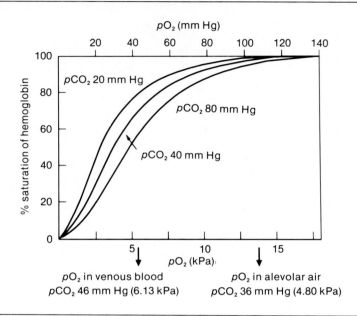

moves to the peripheral capillary bed, the pO_2 is reduced and the pCO_2 is increased because the cells take up and utilize the O_2 to oxidize nutrients to CO_2, and they release the CO_2 that is formed back into the blood. This increases the blood pCO_2, which in turn produces a lowered blood pH. The amount of O_2 combined with hemoglobin at a given pO_2 is further reduced because of the decrease in pH, permitting more O_2 to be taken up by the tissues. Additional release of O_2 in the capillary bed is caused by the effect of 2,3-bisphospho-D-glycerate in the erythrocyte (see Figure 4.6). The overall function of the erythrocyte is therefore to bind the maximum amount of O_2 in the lungs and to release a portion of it to the tissues. Release of O_2 to the tissues is enhanced by the Bohr effect and the presence of 2,3-bisphospho-D-glycerate.

Carbamino hemoglobin The CO_2 that is produced by catabolism must be transported to the lungs, where it is expired. About 10% is transported in the red blood cell as carbamino hemoglobin. In this form the CO_2 is linked covalently to the NH_2-terminal valine residues of the hemoglobin subunits.

$$CO_2 + HbNH_2 \rightleftharpoons HbNH \cdot COO^- + H^+$$

The reaction is rapid, readily reversible, and probably not catalyzed by an enzyme.

Isohydric transport of CO_2 Most of the CO_2 is transported in the plasma as bicarbonate, which is produced in the erythrocyte from CO_2 (Figure 4.9).

Several reactions occur in this transport process.

1. Since oxyhemoglobin is a stronger acid than deoxyhemoglobin, the following must be written:

$$HbH_x^+ + O_2 \rightleftharpoons HbO_2 + xH^+$$

$$pK_a\ 7.71 \qquad\qquad pK_a\ 7.16$$

Figure 4.9 Reactions of gases with erythrocytes.

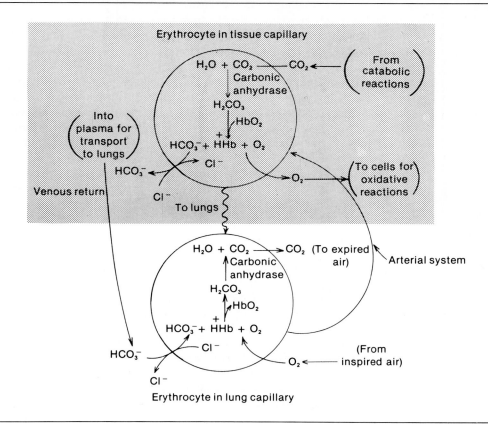

The pK_a values of these two weak acids are such that x = 0.7, approximately. This property of the hemoglobins demonstrates their buffering effect. It also explains the transport of an appreciable quantity of the CO_2 released from the tissues without change in pH—the so-called isohydric transport of CO_2. The oxyhemoglobin, arriving in the red blood cells at the capillary bed, dissociates to release O_2, a process that is facilitated by the state of the tissues relative to incoming arterial blood: low pO_2, lower pH, and higher pCO_2.

$$0.7H^+ + HbO_2 \rightleftharpoons HbH^+_{0.7} + O_2$$

The H^+ in this reaction is derived from the dissociation of H_2CO_3, which is formed from the CO_2 diffusing from the tissue into the plasma and finally into the erythrocyte.

$$H_2CO_3 \rightleftharpoons H^+ + HCO_3^-$$

The uptake of H^+ by hemoglobin buffers the effects of the carbonic acid. The HCO_3^- formed in the erythrocyte diffuses into the plasma and is carried back by the venous blood to the lungs, where reduced hemoglobin is oxygenated. This results in the release of H^+, which reacts with HCO_3^- to give H_2CO_3. The concentration of HCO_3^- in the erythrocyte is therefore being reduced so that HCO_3^- from the plasma returns to the cell. This buffering effect reduces the pH change as a result of the oxygenation of HbH^+. Also, the H_2CO_3 dehydrates (catalyzed by carbonic anhydrase) to form CO_2 for expiration by the lung. The reactions of hemoglobin with CO_2 and O_2 occur in support of each other and of the proper handling of the H^+ load.

2. Added to this H^+ load is the H^+ from the formation of carbamino hemoglobin.

3. The isohydric transport of CO_2 requires the counterdiffusion of Cl^- and HCO_3^- in the erythrocytes to maintain electroneutrality. Thus in the lung the movement of HCO_3^- into the erythrocyte for conversion to H_2CO_3 requires that a negatively charged ion move from the cell into the plasma to take its place. This part is played by Cl^-, so that the $[Cl^-]$ is higher and the $[HCO_3^-]$ lower in arterial than in venous blood. Conversely, in the capillary bed the CO_2 diffuses into the plasma and then into the erythrocyte; HCO_3^- is formed, as explained in the first reaction of the transport process, and moves to the plasma. Its place in the erythrocyte is taken by Cl^- from the plasma, so that venous blood has a lower $[Cl^-]$ and higher $[HCO_3^-]$ in the plasma than does arterial blood. This point can easily be demonstrated in vitro by blowing O_2 over a sample of venous blood.

Not all the H^+ produced by the first two reactions is accommodated by the protonation of deoxyhemoglobin. Therefore the venous blood plasma is more acidic than arterial blood, a state reflected by a higher pCO_2.

Kidney function

The kidney contains approximately 1.3×10^6 *nephrons* that operate in a parallel fashion. Each nephron consists of a *glomerulus,* which is supplied with blood in an arteriolar capillary system such that a high enough filtration pressure exists to effect ultrafiltration of the lower molecular weight materials in the plasma. The *glomerular filtrate* collects in *Bowman's capsule,* which leads successively to the *proximal convoluted tubule,* the *loop of Henle,* the *distal convoluted tubule,* and a branched *collecting duct* that is common to and drains various nephrons. The blood that leaves the glomerulus perfuses the tubules, collecting materials into this peritubular fluid that have been reabsorbed by the tubular cells from the glomerular filtrate (Figure 4.10).

Each part of the renal tubule is long with a narrow lumen, so that at no point is diffusion

Figure 4.10   Ion transport in proximal tubule. Passive diffusion of ions is indicated by broken arrows, ion exchange by a circle in the reversible arrows, and ion pumps involving ATP by ⊝.

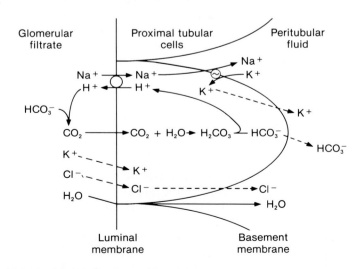

from the filtrate to the tubular cells rate limiting. Each day approximately 160 L of fluid are filtered, but the urinary volume is normally only between 500 to 2500 ml/day, depending on diet and activity. Thus approximately 99% of the filtered water is reabsorbed. Urine has a specific gravity of 1.003 to 1.030, depending on the homeostatic control of total body fluid and solids.

Bicarbonate is reabsorbed in both the proximal and distal tubules, with about 90% of the filtered HCO_3^- being reabsorbed in the proximal segment. By means of a Na^+,H^+ exchange that is not driven by adenosine triphosphate (ATP), H^+ is secreted into the lumen and associates with HCO_3^- in the luminal filtrate to form carbonic acid. In the presence of carbonic anhydrase at the luminal brush border of the tubular cell, the H_2CO_3 is decomposed to CO_2, which, together with that already in the filtrate, freely diffuses into the tubular cell. This CO_2 is partially hydrated to H_2CO_3, catalyzed by intracellular carbonic anhydrase. The newly formed H_2CO_3 then dissociates to HCO_3^- and H^+. These processes are illustrated in Figure 4.11.

As noted in Figure 4.10, Na^+ is pumped out to the peritubular fluid by a Na^+,K^+ pump in the basement membrane so that a low Na^+ concentration is maintained in the tubular cell. The exchange of Na^+ for H^+ (Figure 4.11) occurs with a $1:1$ stoichiometric relationship.

The mechanism for HCO_3^- reabsorption in the distal tubules and collecting ducts is qualitatively the same as in the proximal tubule. The distal segments, however, account for only about 10% of the total HCO_3^- reabsorption.

Partial CO_2 pressure and bicarbonate concentration in blood

An increase in pCO_2 in the blood, and therefore in the glomerular filtrate, results in an increase in H_2CO_3 in the tubular cells (Figure 4.11). Since some of the H_2CO_3 dissociates, this results in a corresponding increase in $[H^+]$ in the tubular cells. Any increase in the $[H^+]$ in the cell permits secretion of H^+ from the intracellular fluid into the luminal fluid so that more HCO_3^- can be reabsorbed. Alternatively the H^+ is excreted in the urine, and the HCO_3^- generated within the cell adds to that already reabsorbed. Thus in hypoventilation (high pCO_2, acidosis) a compensatory increase occurs in $[HCO_3^-]$, as re-

Figure 4.11　　Absorption of bicarbonate from glomerular filtration.

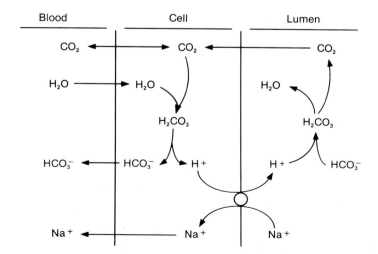

quired by the Henderson-Hasselbalch equation for readjustment of the blood pH. Hyperventilation (alkalosis) results in reduced HCO_3^- reabsorption.

Mechanisms of H^+ excretion

The H^+ secreted by the tubular cells is handled principally in three ways:

1. Reabsorption of HCO_3^-, which has been discussed earlier. The secreted H^+ is neutralized by HCO_3^- in the filtrate to form H_2CO_3 and then H_2O and CO_2.
2. Reaction with the $HPO_4^=/H_2PO_4^-$ buffer, which in effect exchanges one of the Na^+ in Na_2HPO_4 to give NaH_2PO_4. This results in a conservation of Na^+.
3. Formation of ammonia in the tubular cells, which results partly from the catalytic hydrolysis of glutamine with glutaminase and partly from the oxidative deamination of α-amino acids:

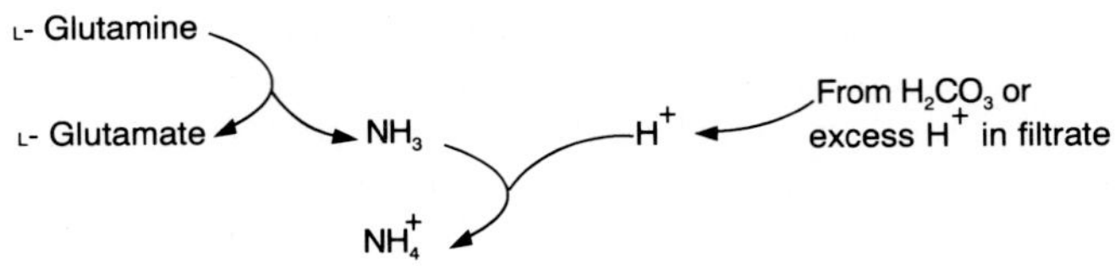

The NH_3 diffuses from the cells to the lumen, where it combines with the H^+ to form ammonium ion. Since the pK_a for NH_4^+ is 9.6 and the pH of urine when there are adequate amounts of acid is 6.0 or less, essentially all the ammonia in the urine is in the form of NH_4^+. Therefore the H^+ is neutralized and excreted as an NH_4^+ anion. The control of deamination in the tubular cells plays an important role in buffering the excess H^+ produced by body metabolism.

Renal threshold

Many solutes in the glomerular filtrate are reabsorbed by mediated transport processes. These transport systems can become saturated. When this happens, the reabsorption capacities of the nephrons are exceeded, and the excess substances are excreted in the urine. The concentration of the substance in the blood plasma above which the transport system is saturated is known as the *threshold value*. For example, the following are threshold values: glucose, 7.77 to 9.44 mmol/L (140 to 170 mg/dl); total CO_2, 27 to 30 mmol/L; and K^+, about 2.8 to 3.1 mmol/L (1.1 to 1.2 mg/dl). The threshold value for urea is very low so that only a small portion is reabsorbed in the tubules. The threshold values vary with the physiologic condition, such as glomerular filtration rate, acid-base balance, and hormone concentrations (parathyroid hormone for phosphate).

Mediated transport in the reabsorption of solutes from the glomerular filtrate also exists for other sugars, vitamin C, phosphate, sulfate, some amino acids, some Krebs cycle organic acids, and uric acid. The transport of sugars and of some amino acids have characteristics in common.

Glucose transport is coupled to Na^+ transport in both the kidney and the intestinal mucosa. Glucose enters the tubular cell by binding to a specific carrier molecule that also binds Na^+. The electrochemical gradient for Na^+ moves the complex across the membrane. The glucose can be transported against a glucose gradient as long as Na^+ is cotransported. The Na^+ that enters the cell is "pumped" out by the Na^+,K^+-ATPase system. Glucose, fructose, galactose, and xylose have the same carrier, with glucose having by far the greatest affinity (see Chapter 12).

Specific transport systems in the intestine and the kidney have been described for about five different groups of amino acids. These are the small neutral, large neutral, acidic, and basic amino acid groups and proline. Each group has its carrier and, as with glucose, Na^+ is required for the transport of the amino acids into the cell. Other mechanisms of transport exist and are presented for the amino acids (see Chapter 8).

Creatinine clearance A fairly constant daily production of creatinine from creatine occurs, the amount of which is determined chiefly by the muscle mass of the person. All the creatinine that enters the glomerular filtrate is completely excreted. The determination of its daily excretion, together with urine volume, is primarily a measure of the glomerular filtration. Therefore creatinine clearance is one measure of kidney function.

Renal control of acid-base balance

The kidneys regulate the concentration of HCO_3^- in blood by adjusting the amount of this anion that is reabsorbed. Under normal conditions little HCO_3^- is excreted because the renal threshold is 26 to 28 mmol/L; normal plasma contains 25 to 26 mmol/L of HCO_3^-.

Metabolic and respiratory disturbances of acid-base balance

The normal pH of blood plasma is approximately 7.4 varying from 7.35 to 7.45. A lower pH identifies the condition of *acidosis,* whereas a state of *alkalosis* exists at higher pH. At pH 7.40 the ratio of $[HCO_3^-]/[$dissolved $CO_2]$ is 20, and the [total CO_2] is 25 to 28 mmol/L. Changes in these values may result from pulmonary or metabolic dysfunction or from both. The ratio can be changed, producing either acidosis or alkalosis as a result of the increase or decrease in either $[HCO_3^-]$ or pCO_2. The [total CO_2] may also be increased or decreased. However, the imbalance may not be completely compensated. Uncompensated acidosis refers to the condition in which the $[HCO_3^-]/[$dissolved $CO_2]$ ratio is less than 20, and uncompensated alkalosis refers to a ratio greater than 20 (Table 4.1).

If the imbalance is caused by an alteration of pCO_2, it is respiratory in origin. Conversely, an alteration HCO_3^- is considered to be of metabolic origin. These four conditions are therefore called *uncompensated respiratory* acidosis or alkalosis and *uncompensated metabolic* acidosis or alkalosis. The changes in the plasma buffer constituents are summarized in Table 4.1, where constituents are shown as either increased or decreased relative to the normal.

The body *attempts* to compensate for abnormal pH conditions by changing the bicarbonate buffer component that was normal. Thus in uncompensated metabolic acidosis the H_2CO_3 concentration is initially normal, but it is reduced by increasing the respiratory rate, so the pH becomes as close to normal as possible. This produces an even more abnormal [total CO_2], and the condition is identified as *compensated metabolic* acidosis. The pH is compensated as much as possible, but the distribution of anions still resembles that in an uncompensated metabolic acidosis. One can note from Table 4.1 that in certain cases the analysis of one constituent cannot identify the nature of the imbalance. For

Table 4.1   Simple disturbances of acid-base balance

	Average normal value	Acidosis				Alkalosis			
		Metabolic		Respiratory		Metabolic		Respiratory	
		U*	C*	U	C	U	C	U	C
pH	7.4†	↓	7.4†	↓	7.4†	↑	7.4†	↑	7.4†
$[HCO_3^-]/[$dissolved $CO_2]$	20	↓	20	↓	20	↑	20	↑	20
$[HCO_3^-]$ (mmol/L)	25-26	↓	↓	25-26	↑	↑	↑	25-26	↓
pCO_2 (mm Hg)	40	40	↓	↑	↑	40	↑	↓	↓
[Total CO_2] (mmol/L)	25-28	↓	↓	26-28	↑	↑	↑	↓	↓

*U, Uncompensated (or primary); C, compensated.
†Approximate values.

example, in both primary metabolic acidosis and primary respiratory alkalosis the [total CO_2] is reduced. More definitive interpretations therefore require that at least two of the parameters be measured.

Respiratory acidosis

Chronic lung disease or depression of the respiratory rate by a disturbance of the nervous system may increase the pCO_2, which produces a decrease in the [HCO_3^-]/[dissolved CO_2] ratio. An uncompensated respiratory acidosis results. The kidney responds by an increased reabsorption of HCO_3^- and generation of HCO_3^- from CO_2, and an increased secretion of H^+ by the tubular cells occurs in response to the high pCO_2. The patient may compensate to a pH that is close to normal, but as long as gas exchange is impaired, the HCO_3^- will stabilize at an elevated concentration and the [total CO_2] will be high. If the onset of uncompensated respiratory acidosis is acute and kidney function is normal, the compensation may take 3 to 5 days. The urine would probably be more acidic, a reflection of the increased excretion of H^+. More Na^+ is reabsorbed in exchange for H^+ and K^+, with the increased H^+ secretion also reflected in greater formation and excretion of NH_4^+ in the urine. The plasma [Cl^-] will be reduced in proportion to the increase in [HCO_3^-], and thus electroneutrality is maintained.

Respiratory alkalosis

Hyperventilation may be seen in patients who have head injuries or who are under the influence of some drugs, such as salicylate poisoning. Hyperventilation for any reason results in a rapid decrease in pCO_2, which decreases the availability of H^+ for secretion into the lumen of the kidney tubules. This leads to a corresponding reduction in the reabsorption of HCO_3^- and Na^+. More secretion of K^+ occurs because a greater concentration of Na^+ is present in the distal segments of the nephron. The overall effect therefore is an increased excretion of Na^+, K^+, and HCO_3^- in the urine with an increased reabsorption of Cl^- to replace the HCO_3^- and thus maintain the anion concentration. When the hyperventilation is controlled, the increase in pCO_2 will reverse these compensatory responses of the kidney. HCO_3^- will be reabsorbed from the filtrate, and more will be generated by the tubular cells to return the [HCO_3^-]/[H_2CO_3] ratio and [total CO_2] to normal.

Metabolic acidosis

A reduction in the concentration of plasma HCO_3^- leading to a metabolic acidosis may be caused by several factors:

1. Increased biosynthesis of metabolic acids such as ketone bodies or ingestion of acids such as salicylic acid or NH_4Cl (NH_4Cl is equivalent metabolically to $HCl + NH_3$, the latter being converted to neutral urea, leaving HCl to be excreted.)
2. Excessive loss of HCO_3^- caused by diarrhea or other conditions, resulting in a loss of pancreatic secretions, which are alkaline and contain higher [HCO_3^-] than the blood plasma
3. Decreased excretion of H^+ by the kidneys resulting from acute kidney failure or an impaired ability to generate NH_3 for excretion of H^+ as NH_4^+

The biochemical mechanisms that result in these imbalances are discussed in the following chapters. The resulting loss of HCO_3^- is compensated by increased pulmonary ventilation, which produces a decrease in pCO_2.

In some types of metabolic acidosis the gap between the [$Na^+ + K^+$] and [$HCO_3^- + Cl^-$] is greater than normal because other nonvolatile anions (fixed acids) are increased. These accumulated nonvolatile anions must be excreted. If the kidney is functioning properly, this excretion is accompanied by an increased loss of water and cations. This may produce dehydration and electrolyte imbalance, two problems that become interrelated.

Table 4.2   Examples of disturbances in acid-base balance

Arterial pCO_2 in		pH	Bicarbonate (mmol/L)	Remarks
mm Hg	(kPa)			
40	(5.33)	7.40	24.5	Normal
25	(3.33)	7.60	24.5	Severe respiratory alkalosis (e.g., patient being ventilated artificially)
31	(4.13)	7.51	24.5	Moderate respiratory alkalosis (e.g., mild hyperventilation)
60	(8.00)	7.22	24.5	Uncompensated respiratory acidosis (e.g., hypoventilation resulting from narcotic overdose)
60	(8.00)	7.37	35.0	Respiratory acidosis partially compensated by metabolic alkalosis; renal in origin (e.g., patient with longstanding chronic obstructive lung disease)
32	(4.27)	7.65	35.0	Combined respiratory and metabolic alkalosis (e.g., patient receiving prolonged mechanical ventilation)
22	(2.93)	7.35	11.0	Metabolic acidosis with secondary respiratory alkalosis (e.g., patient with severe diabetic acidosis)
50	(6.67)	7.07	15.0	Combined metabolic and respiratory acidosis (e.g., patient whose ventilation has been severely depressed by heavy sedation)

Metabolic alkalosis

An increase in plasma $[HCO_3^-]$ occurs when abnormal amounts of alkali are retained. This can develop when salts of metabolic acids (sodium lactate or $NaHCO_3$) are administered or when ethacrynic acid is used to produce diuresis. This state also results when acid is lost, as through vomiting of gastric HCl. The pulmonary compensation for metabolic alkalosis is hypoventilation, which increases the pCO_2.

Mixed disturbances of acid-base balance

In most of these simple situations of acid-base imbalance, compensation by the normal functioning of either the lungs or the kidneys has been assumed. Also, the simultaneous occurrence of two primary disturbances has not been considered; however, such situations often occur. In these cases the simple scheme presented in Table 4.1 does not apply, although the compensatory mechanisms operate to their greatest capacity. The evaluation of the nature and extent of such conditions requires a thorough history and physical examination in conjunction with an analysis of the laboratory data. Examples of such mixed disturbances are given in Table 4.2.

Blood clotting

The most elaborate cascade of enzymatic reactions is found in the process of blood coagulation. Many of the blood plasma proteins that circulate at low concentrations are in fact proenzymes. When blood coagulation is initiated by accident or by disease, activation of these proenzymes occurs. If the response is to an accident, the result is called a clot, which staunches the loss of blood following trauma. If the response is to intrinsic disease of the vessels, the result is called a thrombus. Both clots and thrombi are largely composed of fibrin, an insoluble protein. Intravascular coagulation is normally counteracted by yet another set of enzymes that are fibrinolytic.

Twelve proteins, calcium ions, and phospholipids are intimately involved in the coagulation mechanism. Involvement of the proteins and Ca^{++} has long been appreciated, and each of the factors has been designated by a Roman numeral as well as by a descriptive name. Involvement of the phospholipids and of thromboxane is much more recent, so

Table 4.3 Identifications of factors involved in blood coagulation

Factor*	Name	Function
I	Fibrinogen	Conversion to fibrin
II	Prothrombin	Conversion to thrombin
III	Tissue thromboplastin	Activation of factor X
IV	Calcium ion	Cofactor for several reactions
V	Proaccelerin	Thrombin formation (not an enzyme)
VII	Proconvertin	Factor X activation
VIII	Antihemophilic globulin	Factor X activation
IX	Plasma thromboplastin, or Christmas factor	Formation of factor VIII
X	Stuart factor	Thrombin formation
XI	Plasma thromboplastin antecedent	Factor IX activation
XII	Hageman factor	Factor XI activation
XIII	Transglutaminase	Cross-linking of fibrin

*Note that factor VI does not exist as such; see text for details.

that Roman numerals have not been assigned to them. Table 4.3 shows the clotting factors along with their common eponyms and functions. Except for factor III, the tissue thromboplastin, all the factors circulate in the blood. Note also that Table 4.3 does not contain a factor VI. What was once reported as factor VI was actually an impure activated form of factor V.

It facilitates discussion to split the coagulation mechanism into three parts and to consider the ultimate events first. This portion of the cascade is shown in Figure 4.12. Prothrombin is converted to thrombin under the influence of activated factors V and X, together with Ca^{++}, thromboxane, and other lipids derived from blood platelet membranes. Thrombin is a proteolytic enzyme that cleaves two small peptides from fibrinogen to form fibrin. Although fibrinogen is a soluble protein, fibrin is not, so it readily forms an aggregate of fibers that constitute a rather loose clot. Transglutaminase then cross-links these fibers by formation of amide bonds and produces a firm and adherent clot. Although only three steps occur in this portion of the cascade, note that at least two other activated factors are required.

Factor X may be activated by either or both of two separable portions of the total cascade. One portion, sketched out in Figure 4.13, is known as the *extrinsic system,* since it involves a factor that originates outside the vascular bed. When solid tissues are damaged, tissue thromboplastin is released and activated; its activated form then serves to activate factor X. Note further that thrombin, formed in the later stages of coagulation, also serves to activate factor VII.

Factor X may also be activated by an *intrinsic system,* so called since all its components are found in the circulation. This is diagrammed in Figure 4.14. This is a more elaborate portion of the total cascade. When the wall of a blood vessel is damaged, contact of the Hageman factor with collagen fibers causes activation of the system, ending with formation of X_a. Efficient coagulation is ensured by the redundant loop in which IX_a not only converts X to X_a, but also activates VIII to VIII_a, which promotes activation of X.

This simplified overview of blood coagulation omits certain phenomena, but it does show clearly that Ca^{++} is essential to the process. Thus chelation of Ca^{++} will prevent blood coagulation. Soluble citrates or oxalates, or mixtures of both, are frequently used

Figure 4.12   Ultimate steps in blood coagulation. Roman numerals refer to
individual factors described in Table 4.3. Lower-case letters indicate
activated forms of the given factors.

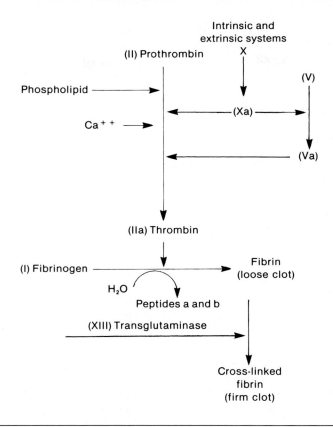

Figure 4.13   Extrinsic system activation of factor X (Stuart factor) in blood
coagulation. Lower-case letters indicate activated forms of the given
factors.

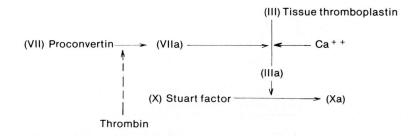

Figure 4.14

Intrinsic system activation of factor X (Stuart factor) in blood coagulation. Lower-case letters indicate activated forms of the given factors.

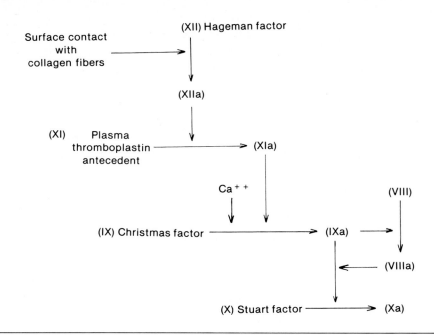

for this purpose. Banked blood is collected in citrate solution. Other anticoagulants related to coumarin interfere with prothrombin biosynthesis. Heparin does not act in this way but interferes with the activation of factors IX, X, XI, and XII. Its primary effect is probably inhibition of thrombin formation from prothrombin.

Bibliography

Friedman JM: Structure, dynamics and reactivity in hemoglobin, Science 228:1273, 1985.
Kilmartin JV: Interaction of hemoglobin with protons, CO_2 and 2,3-bisphosphoglycerate, Br Med Bull 32:209, 1976.
Perutz MF: Myoglobin and hemoglobin: role of distal residues in reactions with heme ligands, Trends in Biochemical Sciences 14:42, 1989.
Weatherall DJ et al The hemoglobinopathies. In Scriver CR et al, editors: The metabolic basis of inherited disease, ed 6, New York, 1989, McGraw-Hill Information Services Co.

Clinical examples

A diamond (◆) on a case or question indicates that literature search beyond this text is necessary for full understanding.

Case 1

Subacute carbon monoxide poisoning

A 67-year-old supervisor at a meat packing plant sought medical attention after 3 days of persistent vertigo, crushing chest pains, nausea, a dry cough, chills, and a headache. His wife had experienced similar malaise and dizziness during the previous week.

On admission to the hospital the patient's blood pressure was 110/90 mm Hg, the pulse was 112 beats/min, and the respirations were 28/min. The patient was admitted to the coronary care unit. Arterial blood gas measurements were made while he was receiving oxygen at 4 L/min: oxygen saturation, 83% (calculated 98%); pO_2, 101 mm Hg; pCO_2, 32 mm Hg (0.992 mmol/L); pH, 7.45; HCO_3^-, 22.1 mmol/L; and carboxyhemoglobin, 15.6%. At the same time a blood sample taken from his wife showed 18.1% carboxyhemoglobin.

The gas furnace in their home was found later to have a rusted-out flue.

Biochemical questions

1. What is the structure of carboxyhemoglobin?
2. How is carboxyhemoglobin determined in blood?
3. What is the effect of CO on the oxygen-carrying capacity of the blood?

Case discussion

1. Structure Carbon monoxide combines with the iron in ferrohemoglobin and ferromyoglobin to form a complex very similar in structure to oxyhemoglobin and oxymyoglobin (see Figure 4.2). Stereochemically the O_2 and CO bind so that the Fe and the C and O atoms of CO are not linear.

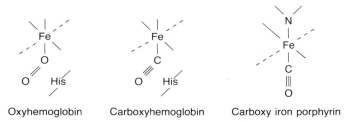

Oxyhemoglobin   Carboxyhemoglobin   Carboxy iron porphyrin

Whereas this structure for the O_2 complex is normal and produces the optimum binding for O_2, the nonlinear structure for the CO complex is forced on it by a histidine residue that is close to the heme group in the globin chain. CO binds to free iron porphyrin so that the Fe, C, and O atoms are linear.

2. Determination of carboxyhemoglobin Carboxyhemoglobin (HbCO) is determined spectrophotometrically, taking advantage of the different spectra for the various hemoglobin derivatives. Oxyhemoglobin (HbO_2) (absorption maxima 577 nm and 541 nm) has a similar spectrum in alkaline solution to HbCO (absorption maxima 570 nm and 539 nm). Hemoglobin has a single broad absorbance band at 555 nm. If weakly alkaline blood is reduced with sodium dithionite ($Na_2S_2O_4$), the HbO_2 is converted to hemoglobin, but HbCO is unaffected. Determination therefore involves diluting a blood sample with ammonium hydroxide and centrifuging it if the blood does not clear. One adds dithionite to an aliquot of this clear solution and measures absorbances at 555 nm and 480 nm against a reference standard containing dithionite in dilute ammonium hydroxide. The ratio of the two absorbancies then is read on a standard curve constructed from known mixtures of hemoglobin and HbCO.

3. Effects of carbon monoxide CO binds 25,000 times as strongly to *free* iron porphyrin as O_2. Because the CO in hemoglobin and myoglobin is forced into a less stable nonlinear structure, the binding to hemoglobin is only 200 times as strong as O_2. Otherwise the toxicity of CO would probably be fatal to smokers, who have 4% to 8% saturation of hemoglobin with CO. Aside from the competitive binding of the CO, which reduces the amount of O_2 that can be carried to the tissues, the loss of the cooperativity of O_2 binding and the consequent tighter binding of O_2 to the hemoglobin means that O_2 is not as readily liberated to the tissues. These combined effects produce significant tissue anoxia, resulting in shortness of breath on exertion (at 10% to 20% CO saturation). At 30% to 50% saturation, headache, confusion, and fainting on exertion occur; concentrations of 80% and greater are rapidly fatal.

The half-life of hemoglobin-bound CO when a person is breathing air is about 5 to 8 hr. CO is released fairly rapidly from hemoglobin if the patient is receiving O_2, as was done in this case. One must also remember that a patient receiving O_2 when transported to a hospital will have a much reduced HbCO concentration, which may confuse the diagnosis.

References

Grace TW and Platt FW: Subacute carbon monoxide poisoning: another great imitator, JAMA 246:1698, 1981.

Myers RA et al: Value of hyperbaric oxygen in suspected carbon monoxide poisoning, JAMA 246:2478, 1981.

Case 2   Thrombophlebitis, anticoagulants, and hemorrhage

A 52-year-old woman consulted her physician because of a painful, swollen left lower limb. The physician correctly diagnosed the condition as acute thrombophlebitis, and the patient was hospitalized immediately. Treatment was begun with heparin, an anticoagulant drug administered by injection. After several days, oral administration of a coumarin type of anticoagulant was begun. The patient's prothrombin time was 12 sec, the control value, before the coumarin was started. This gradually increased to 36 sec as the dosage was raised, and a maintenance dose of courmarin that kept the prothrombin time between 30 and 40 sec was given daily. The patient gradually improved, and the heparin was discontinued. She was discharged after 3 wk of hospitalization with a supply of coumarin, and she was advised to report weekly for laboratory determinations of her prothrombin time. She followed this advice for several months. Because of the inconvenience and expense, she discontinued both her visits to the physician and her weekly blood tests. However, she continued to take the coumarin anticoagulant. Six weeks later she again consulted her physician because she was passing large quantities of bright-red urine. She was hospitalized immediately, and a water-soluble vitamin K analogue, menadione, was administered parenterally. Before the administration of this drug, the patient's prothrombin time was 73 sec. The coumarin anticoagulant was discontinued, and additional menadione and vitamin K were given. Hematuria ceased within 24 hr, and a final prothrombin time was 13 sec.

Biochemical questions
1. What biochemical events are involved in the dissolution of a thrombus?
2. How do the coumarin anticoagulants decrease the tendency of blood to coagulate?
3. Why is vitamin K effective in overcoming the actions of the coumarin anticoagulants?
4. Why was menadione given initially rather than natural vitamin K_2?

Case discussion

This patient's initial problem, thrombophlebitis, was caused by the formation of a blood clot in one of the deep veins of the left leg. The exact reason for the formation of these clots, or thrombi, is not presently understood. Blood will clot if it is withdrawn from the circulatory system. This can be prevented by the addition of an anticoagulant to the blood that is drawn. Most anticoagulants function through the chelation of calcium ions; calcium is essential for several reactions involved in clot formation. Anticoagulants, such as EDTA, cannot be used therapeutically, since ionic calcium is vital to many other physiologic processes. Clots will also form within the blood vessels when, for example, injury occurs to the vessel wall. Unfortunately, the process may also cause illness, such as myocardial infarction secondary to the formation of a thrombus in a coronary artery damaged by atherosclerosis or, as in this patient, obstruction to venous drainage from a limb. A serious complication that may develop from thrombosis of the leg veins is *pulmonary embolism*. This can occur if a portion of the clot dislodges and is carried through the circulation into the pulmonary vessels, where it blocks blood flow through one or more segments of the lung.

Mechanism of blood coagulation Although the exact mechanism causing thrombus formation in thrombophlebitis is unknown, the way in which vascular injury triggers clot formation has been elucidated to some extent. When vascular injury occurs, platelets adhere to collagen and subendothelial fibers in the damaged area of the vessel wall and aggregate with one another. In certain vessels this is associated with vasoconstriction in the injured area. A platelet plug containing trapped neutrophilic leukocytes forms at the point of damage. Platelet aggregation is associated with or causes activation of the plasma coagulation mechanism, leading to the formation of a fibrin clot that seals off the damaged area. The platelets provide phospholipids, probably in the form of a membrane, that activate the clotting mechanism. Activation occurs because the soluble enzymes are provided a surface on which to act. This has the effect of concentrating the enzymes in a localized area, leading to faster reaction rates. The phospholipid material is called *platelet factor III*.

The basic principle is that most of the necessary ingredients required to make fibrin are available in inactive form in the plasma. These circulating clotting factor precursors are proteins and, except for fibrinogen and factor V, are also proenzymes.

Diseases involving clotting factors Diseases that are characterized by abnormal or excessive bleeding result from an inherited deficiency or absence of a clotting factor. *Hemophilia A* results from the lack of factor VIII. *Hemophilia B*, which is caused by lack of factor IX, is also known as *Christmas disease*. Factor XI deficiency produces *hemophilia C*, whereas factor V deficiency results in *congenital parahemophilia*. Deficiencies of the other clotting factors have also been observed clinically. Of the cases resulting from clotting factor deficiencies, approximately 80% involve factor VIII and 10% to 20% involve factor IX. All the other factors combined account for only 1%.

If a bleeding problem is suspected of involving the clotting factors, two tests often are done on the patient's plasma in an attempt to localize the abnormality. The *prothrombin time* measures factors V, VII, and X and prothrombin (see Figures 4.12 and 4.13). If any of these factors is missing or inhibited, the clotting time of the plasma will be prolonged. The second test, the partial thromboplastin time, measures factors VIII, IX, XI, and XII (see Figure 4.14). Once the defect is localized to one of these two segments of the clotting mechanism, more complex tests can be applied to the patient's plasma to pinpoint the precise clotting factor that is deficient or inhibited.

The clotting factors can be thought of as triggering the conversion of the zymogen, prothrombin, to thrombin, the active enzyme. Prothrombin consists of a single polypeptide chain with a molecular weight of 7.2×10^4; thrombin has a molecular weight of 3.7×10^4 and consists of two polypeptide chains joined by disulfide bonds. The β-chain of thrombin is similar to trypsin, and thrombin has trypsinlike activity. In other words, thrombin is a proteolytic enzyme that cleaves specifically at arginyl and lysyl bonds. As with chymotrypsin, its active center contains a free serine hydroxyl group and is inactivated by diisopropylfluorophosphate. One should remember that, as with thrombin, trypsin and chymotrypsin also are biosynthesized and released in zymogen forms.

Fibrinogen is composed of three pairs of nonidentical subunits, Aα, Bβ, an Cγ. Thrombin catalyzes the cleavage of one peptide bond in each of four chains of fibrinogen. Four fibrinopeptides are derived, coming from four of the six N-terminal ends of fibrinogen. The fibrin monomer units so produced polymerize as a result of physical interactions between groups that are unmasked when the fibrinopeptides are removed. The fibrin polymer that is formed by this polymerization process then is transformed into a firm clot through cross-linking of γ-carboxyl groups from glutamine with ε-amino groups of lysine. This *transpeptidation* reaction is catalyzed by factor XIII and Ca^{++}. Thus the bonds that stabilize the fibrin polymer in the form of a firm clot are covalent.

Several types of clinical problems involve fibrinogen. *Afibrinogenemia,* a condition in which no fibrinogen is present in the plasma, is a rare, serious disease with a high risk of hemorrhage. *Dysfibrinogenemia* is caused by production of an abnormal fibrinogen molecule. Several such mutated fibrinogens have been characterized in terms of their primary sequences. Hypofibrinogenemia may occur in certain complications of pregnancy and is associated with premature separation of the placenta and retention of a dead fetus. The cause is not insufficient fibrinogen production, but rather increased clot production and breakdown by fibrinolytic enzymes. The lowered circulating fibrinogen may involve some risk to the mother through hemorrhage and may become serious when the products of conception are expelled.

1. Dissolution of a thrombus (fibrinolysis) Fibrin removal is mediated by the circulating proenzyme, plasminogen, produced in the kidneys. When activated by a protease, *urokinase,* it is converted to the active form, *plasmin.* The latter catalyzes dissolution of the fibrin clot, producing several peptides. Clinical trials of streptokinase, a bacterial activator of plasminogen that acts similar to urokinase, have been used in treatment of thrombotic lesions.

Urokinase is interesting because it can be recovered from urine in the active form, even though its molecular weight has been reported to be at least 31,500. A possible second isozyme can be isolated that has a molecular weight of 54,700. Why these large molecular species pass the renal epithelium is not clear.

2. Coumarin anticoagulants For the patient with thrombophlebitis discussed here, it was useful to intervene in the normal clotting mechanisms. Administration of a coumarin derivative (e.g., dicumarol) interferes with the biosynthesis of prothrombin as well as of factors VII, IX, and X in the liver. One usually estimates the action of coumarin compounds by measurement of the prothrombin time; a diluted plasma sample is fortified with exogenous thromboplastin (factor III) and an optimum concentration of Ca^{++}. The conversion of fibrinogen to fibrin is measured, and the results are expressed as the time required for the earliest appearance of fibrin threads in the test tube. No absolute standards exist for this test; consequently the time of fibrin formation is compared with that of a "normal" plasma run concomitantly as a control. More sophisticated tests exist but are not routinely employed.

Dicumarol

3. Vitamin K and the coagulation process It has long been known that deficiency of vitamin K leads to hemorrhage. Vitamin K increases Ca^{++} binding sites on prothrombin. To be activated and to participate in the coagulation process, prothrombin must be bound to phospholipids, and Ca^{++} is required for this. The Ca^{++} binding sites of prothrombin are formed by the introduction of a second carboxyl group into the glutamyl side-chains located in the amino-terminal region of the protein. When dicumarol is administered, the prothrombin that is produced has a very low Ca^{++}-binding capacity. In this inactive prothrombin, glutamate residues exist in place of γ-carboxyglutamate. Therefore vitamin K facilitates the carboxylation of glutamate residues in prothrombin and perhaps in factors VII, IX, and X. When the action of vitamin K is blocked by dicumarol, Ca^{++} cannot bind to prothrombin because the protein lacks the added carboxyl groups. If it is artificially converted to thrombin; however, the defective molecule is active catalytically. In fact, the segment of prothrombin that is γ-carboxylated and binds calcium actually is removed during the activation process.

4. Menadione The fat-soluble vitamin K is required for normal function of the blood coagulation system. Naturally occurring substances with vitamin K activity are analogues of 2-methyl-1,4-naphthoquinone, and they contain a long aliphatic chain attached to position C_3 (see Chapter 1). However, synthetic 2-methyl-1,4-naphthoquinone, or menadione, can act as vitamin K even though it lacks an alkyl substitution at position C_3. The natural vitamins K are lipid substances. Therefore, to be absorbed from the intestine, natural vitamin K requires adequate emulsification by bile as well as normal function of the fat-absorption is defective (e.g., celiac disease). In these clinical situations vitamin K must be administered parenterally. Humans normally obtain part of their vitamin K requirements from lipid-containing foods in the diet, and the remainder is synthesized by the intestinal bacterial flora. Vitamin K deficiency can lead to internal or external bleeding.

References

Chiu D: Sickled erythrocytes accelerate clotting in vitro: an effect of abnormal membrane lipid asymmetry, Blood 58:398, 1981.

Davie EW and Fujikawa K: Basic mechanisms in blood coagulation, Annu Rev Biochem 44:799, 1975.

Moncada S et al: Prostacyclin and blood coagulation, Drugs 21:430, 1981.

Stenflo J and Suttie JW: Vitamin K–dependent formation of γ-carboxyglutamic acid, Annu Rev Biochem 46:157, 1977.

Case 3   Acquired (toxic) methemoglobinemia

A 26-week-old baby was brought to the pediatric clinic because of increasing lethargy and cyanosis. The infant had been in good health at birth, and the mother had attempted breastfeeding. Because the child has not shown a normal pattern of weight gain, the family physician had recommended a supplementary formula feeding. By admission the mother's lacteal flow had virtually ceased. The mother estimated that most of the child's fluid intake was formula derived.

A blood sample was collected during the initial examination. The resident physician shook an aliquot of the sample with room air for several minutes but noted no change in the sample color. He forwarded the remainder of the sample to the clinical laboratory for further examination and a positive test for methemoglobinemia was obtained. The baby then was treated with intravenous ascorbate and methylene blue. Within 2 days the child was normally alert, and the cyanosis had disappeared. The child was discharged with instructions to the mother concerning formula preparation with distilled water.

Biochemical questions

1. What is the chemical difference between hemoglobin and methemoglobin, and how do their oxygen-carrying capacities compare?
2. What mechanisms normally prevent accumulation of methemoglobin in the blood?
3. What is the cause of the cyanosis associated with toxic methemoglobinemia?
4. How does toxic methemoglobinemia compare with congenital methemoglobinemia?
5. What biochemical tests might the resident physician have ordered from the clinical laboratory to confirm suspicions as to the cause of the cyanosis?
6. What is the biochemical basis for treatment of toxic methemoglobinemia with intravenous ascorbate and methylene blue?

Case discussion

1.   Methemoglobin The primary physiologic function of hemoglobin is the reversible transport of oxygen from the lungs to the peripheral tissues in the form of oxyhemoglobin. During oxygen transport the valence state of the heme iron remains unchanged. If, for any reason, the valence of the heme iron changes from $2+$ to $3+$, the oxidized product is known as methemoglobin, which does not serve as an oxygen carrier. To the extent that a patient's hemoglobin has been oxidized to methemoglobin, that patient will suffer from an oxygen deprivation to the same extent. Under normal circumstances the methemoglobin concentration of human blood is rarely more than 1% of the total hemoglobin, except for a brief period immediately after birth. In a normal neonate the presence of methemoglobin cannot ordinarily be detected by visual observation but requires spectrophotometric analysis.

2.   Factors maintaining the ferrous state of heme iron The iron ligand of hemoglobin is well protected against oxidation. The protective mechanisms can be divided into two classes: nonenzymatic and enzymatic.

The nonenzymatic mechanism depends on the structure of hemoglobin and how the heme fits into pockets formed by the particular folding patterns of the α-and β-chains. Normally the Fe atom is held in place by four bonds to N atoms of the pyrrole rings. Nearby functional groups of globin amino acid residues do not come close enough to the Fe atom to cause oxidation. However, in the mutant hemoglobin M, the iron associated with the mutant peptides is in the Fe^{+++} form, so these mutants do not carry oxygen. The genetic defect is caused by an amino acid substitution that brings an electronegative group close to the Fe atom. In hemoglobin M (Saskatoon) the β63-His is replaced by Tyr, and the OH group of the Tyr can bond to the heme iron; in hemoglobin M (Boston), the α58-His is replaced by Tyr, an equivalent situation. In hemoglobin M (Milwaukee) the β67-Val is replaced by Glu, the free COO^- of which can readily bond to Fe. Hemoglobins M may have defects in either α- or β-subunits, leading to oxidation of the heme iron associated with the affected subunits. Accordingly, their ability to transport oxygen is greatly diminished.

Human erythrocytes contain an enzyme, the flavoprotein methemoglobin reductase, that maintains the $2+$ valence state of heme iron. Earlier reports proposed the existence of both NADH- and NADPH-dependent reductases, but it now appears that the pyridine nucleotide dependence is a function of the concentration of inorganic phosphate. When this is too low, NADH is the preferred nucleotide, but when inorganic phosphate concentration is high, NADPH is preferred. No adequate explanation of this phenomenon is yet available. Regardless of which nucleotide is employed, it serves as an electron donor to a soluble, nonenzymatic cytochrome b_5 (also a heme protein) that in turn reduces methemoglobin to hemoglobin.

Figure 4.15   Coupled enzyme system for reduction of methemoglobin to
                                   hemoglobin. E_1 represents dehydroascorbate reductase and E_2
                                   represents glutathione reductase. The reaction of methemoglobin and
                                   ascorbate is nonenzymatic.

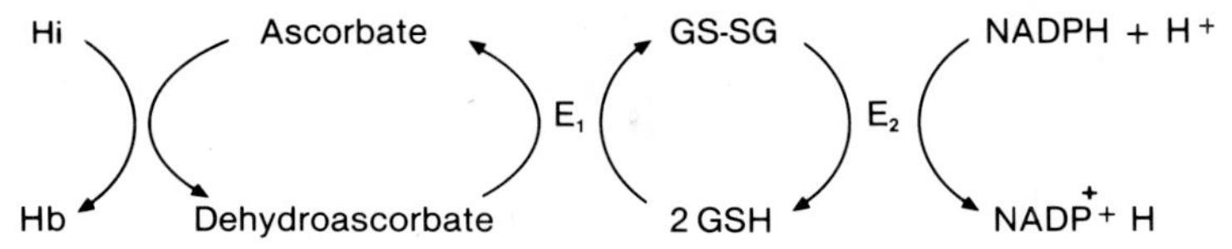

A second enzyme involved in the reduction of methemoglobin to hemoglobin is dehydroascorbate reductase. NADPH can serve as the reducing agent by a series of reactions illustrated in Figure 4.15. Human erythrocytes contain the tripeptide gluthathione, the concentration of which is about 1 mmol/L in the red blood cells. Because it contains an SH group, gluthathione is a naturally effective reducing agent. The ultimate electron source, NADPH, is supplied by any of several metabolic reactions (see Chapter 6).

It is worth noting that the activity of these enzymes may not reach normal values for some time after birth; thus slight cyanosis is sometimes observed in the immediate neonatal period. As neonates develop the ability to produce these enzymes, the cyanosis promptly subsides.

3. Cyanosis and spectral properties of heme proteins Optical spectra of the heme proteins at the wavelengths of visible light are characteristic for hemoglobin and its derivatives. In the porphyrin ring, electron delocalizing effects of systems of alternating double and single bonds occur. Coordination of either Fe^{++} or Fe^{+++} in these rings introduces additional, subtle spectral shifts. Furthermore, binding of other ligands, including O_2, CO, or HCN, induces spectral changes. These are portrayed in Figure 4.16.

Blood is red, but when seen through the skin, it appears blue. This apparent difference in color is a function of the spectra of blood itself and of the tissue through which it is viewed. When blood containing an excess of the methemoglobin is viewed through the skin, it appears to have the slaty, gray-blue color typical of cyanosis; the same blood viewed in a glass tube is more brownish red than normal. These in vivo and in vitro color changes strongly suggest methemoglobinemia, and one should request a precise spectrophotometric examination of the blood. The wavelengths at which absorbance maxima occur are the basis of the definitive qualitative test, the magnitudes of the absorbances provide a quantitative measure of concentrations.

4. Distinction between congenital and acquired methemoglobinemias From the previous discussion one can see that congenital methemoglobinemias may have several causes, including the presence of hemoglobin M or a lack of gluthathione, ascorbate, or a methemoglobin reductase. Hemoglobin M can be detected by sophisticated forms of electrophoresis or by ion-exchange chromatography, but these tests are not generally available in routine clinical laboratories. Test for deficiencies or defects in the protective reductases also are not routinely available. One therefore proceeds, following positive spectrophotometric findings, on the assumption that the methemoglobinemia is acquired. If elimination of common toxic agents and treatment with chemical reducing agents does not result in full disappearance of oxidized hemes, one then should seek a more definitive enzyme study. Since the patient in the present case showed a complete

Figure 4.16

Absorbance spectra of various hemoglobin derivatives. Absorbance values are expressed in terms of quarter-millimolar concentrations (i.e., per millimolar concentration of monomeric units), even though the spectra result from tetrameric compounds. This is done to bring the values observed into agreement with values obtained for free porphyrins, which are the chromophoric groups of the heme species. The left-hand panel shows the spectra over most of the visible region. The right-hand panel shows finer detail in the so-called Soret region, from about 400 to 440 nm, where most porphyrins and porphyrin-containing compounds absorb very strongly. Although it is not shown in the figure, each unit on the right-hand absorbance scale is actually 10 times the value of the left-hand absorbance scale. The individual compounds are represented by the following symbols: oxyhemoglobin —————, deoxyhemoglobin — · — · —, methemoglobin - - -, cyanmethemoglobin — — —, carboxyhemoglobin · · · · · ·. (Modified from van Kampen EJ and Zijlstra WG: Adv Clin Chem 8:142, 1965.)

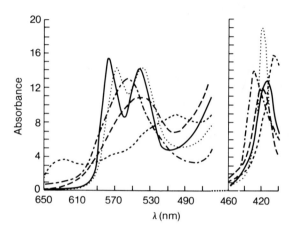

response to treatment, further expensive testing was not considered necessary, and the facts warranted the diagnosis of a toxic methemoglobinemia.

5. Substances known to elicit toxic methemoglobinemia Agents known to induce toxic methemoglobinemia can be divided into three major groups. The first includes numerous common industrial chemicals such as aniline, nitrobenzene, and phenol. All these can be absorbed through the skin as well as through the lungs. Phenol was once employed to sterilize rectal thermometers, even in pediatric nurseries. The practice was abandoned when an unusual incidence of toxic methemoglobinemia was noted in neonates.

Some well-known analgesic and antipyretic drugs, common ingredients of over-the-counter medications, are known to induce toxic methemoglobinemia in certain individuals. These include acetaminophen and acetanilid. Numerous topical anesthetics and several employed in dental anesthesia have been reported as causing methemoglobinemia as well.

Nitrites, contaminants of water supplies, are of particular concern in the pediatric population because bottle-fed infants consume a large volume of water. This is a special concern in rural areas where well water supplies are easily contaminated by runoff from cattle feeding lots. Such well water might not harm an adult but can seriously compromise the health of a very young child. The resulting cyanosis soon becomes evident to other family members. The best solution is to provide a purer water supply, even distilled water.

6.   Biochemistry of remedial measures Intoxication is best treated by removal of the toxic substance. The child's natural defense mechanisms probably would have reduced the methemoglobin to hemoglobin without further measures. The attending physician reasoned, correctly, that although the positive spectrophotometric findings did not rule out the presence of hemoglobin M or a reductase deficiency, a therapeutic trial against toxic methemoglobinemia initially should be attempted. Therefore the physician elected to treat the child with ascorbate and methylene blue. This dye serves as a synthetic electron donor in place of NAD(P)H in the coupled system shown in Figure 4.15. The rapid disappearance of cyanosis as a result of this treatment rules out the existence of hemoglobin M disease. It does not completely settle the question of enzyme deficiencies; this could only be done by further biochemical studies. If the use of clean water in the baby's formula would not have eliminated the problem, such further studies would have been indicated.

References

Das Gupta A et al: Associated red cell enzyme deficiencies and their significance in a case of congenital enzymopenic methemoglobinemia, Acta Haematol (Basel) 64:285, 1980.

Geffner ME et al: Acquired methemoglobinemia, West J Med 134:7, 1981.

Kuma F: Properties of methemoglobin reductase and kinetic study of methemoglobin reduction, J Biol Chem 256:5518, 1981.

O'Donohue WJ et al: Acute methemoglobinemia induced by topical benzocaine, Arch Intern Med 140:1508, 1980.

Shesser R et al: Acute toxic methemoglobinemia following dental analgesia, Ann Emerg Med 10:265, 1981.

Case 4   **Hemochromatosis**

A 54-year-old man had complaints of weakness, lassitude, and moderate weight loss (20 kg in the previous 7 mo). Pulse and respiratory rates were normal, as were the electrocardiographic and chest roentgenographic studies. Blood pressure was 145/75 mm Hg. The liver was firm and moderately enlarged, and the spleen was palpable. The patient remarked that his skin had become darker during the past few years, a change he attributed to time spent outdoors. The patient denied exposure to toxic chemicals or metal fumes. Laboratory results were as follows: fasting serum glucose, 5.6 mmol/L; a random urine glucose, normal; hemoglobin, 2.55 mmol/L; hematocrit, 52%; no methemoglobin detectable; serum iron content was 43 μmol/L, and total serum iron-binding capacity was 56 μmol/L, giving a saturation of approximately 77%; total serum bilirubin content, 20 μmol/L.

On the basis of these observations, a biopsy of the liver was performed. Microscopic examination revealed fatty vacuolization of the hepatocytes and moderate deposits of hemosiderin in the cytoplasm of the cells. A diagnosis of hemochromatosis was made; the iron overload was later confirmed by a differential deferoxamine test.

Biochemical questions

1.   What foodstuffs are particularly rich sources of iron? Should this patient be advised to avoid these?
2.   In what form is iron absorbed from the gastrointestinal tract? How is this absorption regulated?
3.   How is iron transported in the blood? What proteins are involved in iron transport and storage?

4. What is the normal disposition of excess iron?
5. Explain why repeated phlebotomy is used to reduce the body burden of iron.

Case discussion

Primary hemochromatosis is a genetic disorder of iron metabolism characterized by an increased accumulation of dietary iron. The disease generally affects males in their fifties. Iron accumulates in several tissues, including the liver, pancreas, and skin. These accumulations account for diabetes and darkened skin, which are emphasized in the previous name for this disorder, bronze diabetes.

Daily iron needs depend on age and sex. Rapid growth, pregnancy, and lactation pose increased demands for iron. Excessive menstrual flow, hemorrhage, or other significant blood loss and exposure to hypoxic conditions (high altitudes) cause transient increases in iron requirements. For an adult man, iron balance may be maintained by an uptake of 16 μmol (1 mg)/day from an intake of 200 to 250 μmol. The counterbalancing excretion includes urinary loss (1.6 μmol/day), sweating and desquamation of the skin (1.6 μmol/day), desquamation of the intestinal tract and in the bile (4.8 μmol/day), and normal gastrointestinal bleeding (8 μmol/day).

1. Dietary sources of iron Foods rich in iron include red meats (muscle) and fish, liver, raisins, mushrooms, and certain green leafy vegetables. In addition, many foods are fortified with added iron. These include cereal products such as rice, flours, bread products, and various pasta products. However, the availability of iron depends not only on the intrinsic iron content of the diet but also on the presence of other components. For example, phosphates, tannins, and oxalates form insoluble complexes with Fe^{+++}, inhibiting uptake. Conversely, fructose, ascorbate, and citrates tend to form soluble complexes that promote uptake. Alcoholic beverages may contain significant amounts of iron. Thus the iron available for absorption may be only a small fraction of what is ingested. As a consequence, efforts to lower iron uptake by dietary management are not practical. Because iron is such a ubiquitous element, it would be difficult to balance nutritionally a low-iron diet.

2. Mechanism of iron uptake Most of the iron humans consume enters the gastrointestinal tract in the form of Fe^{+++}. Gastric juice contains a glycoprotein, *gastroferrin*, which forms complexes with Fe^{+++}, promoting its uptake in the duodenum and jejunum; absorption from the stomach itself is minimal.

Iron is taken across the brush border of the intestinal mucosal cells as Fe^{++}, but precisely how the reduction occurs is not entirely clear. The mucosal cells contain a *carrier protein* that combines with Fe^{++}; this carrier protein serves to distribute the iron among several possible pathways. The amount of mucosal cell carrier protein is inversely proportional to the quantity of iron entering the cells; it is thus the most important single means of regulating uptake of available iron from within the intestinal lumen. How the relation between iron and the carrier protein is regulated is not known.

3. Iron storage and transport A portion of the iron entering the mucosal cells is taken up by mucosal cell mitochondria, where it is used for formation of iron-sulfur proteins and other iron-containing components of respiratory chains. A second, larger portion of the carrier-bound iron exchanges with apoferritin and is deposited within the mucosal cells as *ferritin*.

Apoferritin is a complex protein composed of 24 polypeptide chains, each with a molecular weight of 18,500. These form a hollow sphere with an outer diameter of 120 Å and an inner diameter of 70 Å. Iron, in the form of mixed micelles of ferric hydroxide and ferric phosphate, can penetrate this shell-like structure. Each ferritin molecule may contain more than 4000 iron atoms, corresponding to an iron content of up to 35%. Ferritin is the primary storage form of iron in the body, not only because of its high capacity to hold iron, but also because of the ready equilibrium that exists between it and the mucosal carrier protein in the intestine or corresponding carriers in other tissues. Under normal conditions the ferritin molecules are finely dispersed and are not visible in the light microscope. In the case of an iron overload and greatly increased ferritin concentration, however, the molecules coalesce into larger siderotic granules that can be seen in the electron microscope even before they become visible with the light microscope. In a proper state of iron balance the ferritin content of the intestinal mucosa is small, since much of the iron is transported by the carrier protein to vessels of the lamina propria, where it is taken up by the major plasma iron carrier, *transferrin*. When iron must be drawn from ferritin stores for transfer to the blood, it must first be reduced to the Fe^{++} state. This is accomplished by ferritin reductase, which requires two coenzymes, NAD^+ and flavinadenine dinucleotide (FAD) or flavin mononucleotide (FMN), as shown here:

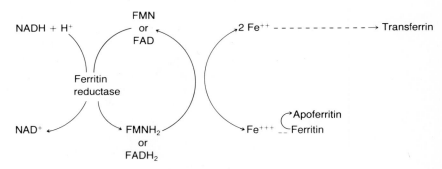

As its name implies, transferrin serves to transport iron from storage sites to sites of usage. It is a glycoprotein made in the liver and comprises from 0.3% to 0.5% of the total plasma proteins, and it has the electrophoretic mobility of a β-globulin. Transferrin has a molecular weight of 76,000 and is composed of a single polypeptide chain with a branched structure. Each transferrin molecule can bind up to two iron atoms, although under physiologic conditions of pH and CO_2 concentration, it is usually only about 30% saturated. Transferrin can bind copper or zinc; however, its major function is the transport of iron to the reticuloendothelial system and then into immature erythrocytes, where hemoglobin is made, or to other tissues for incorporation into iron-containing molecular species of various sorts. Fe-transferrin or Fe_2-transferrin rapidly and specifically attaches to the surface of immature erythrocytes, but not mature erythrocytes. The iron is taken up by an internal reticulocyte iron-carrier protein, and the apotransferrin is then released. It is estimated that as many as 50,000 transferrin molecules may attach to the surface of a single reticulocyte.

The iron carried by transferrin is in the Fe^{+++} state. Before binding by transferrin, Fe^{++} released from ferritin must be oxidized by a ferroxidase. Ceruloplasmin, a protein that contains copper, is a ferroxidase that functions as follows:

Ferroxidase

Reduced
ferroxidase

Fe^{++}

Fe^{+++}

Fe^{+++}-transferrin

Apotransferrin

Ceruloplasmin thus links the metabolism of iron and copper in normal erythropoiesis, and possibly in some other aspects of iron metabolism as well.

Two additional plasma proteins are involved in iron metabolism. *Hemopexin,* a β-globulin, has a high specific affinity for free heme. This complex efficiently scavenges free heme, resulting from erythrocyte breakdown or other tissue injury, and thus spares the need for additional iron uptake. *Haptoglobin* is an α-globulin that forms similar complexes with free plasma hemoglobin. It also spares serum iron-binding proteins and, in addition, protects the kidneys against damage by the free hemoglobin.

4.   Primary hemochromatosis Denaturation of ferritin gives rise to *hemosiderin,* a rather insoluble compound with a large content of ferric hydroxide and with a lower apoferritin content. The reason for denaturation is not entirely clear, but it occurs to a greater extent in iron overload and in the absence of sufficient tissue ascorbate concentrations. Hemosiderin granules frequently become sufficiently large to be visible with the light microscope, especially after staining to form Prussian blue. Tissues that have a high hemosiderin content usually have a high ferritin content as well. Iron can be mobilized from hemosiderin, but somewhat more slowly than from ferritin. The normal ratio of hemosiderin iron to ferritin iron is approximately 0.9; in patients with primary hemochromatosis, this ratio may be increased by a factor of 10. This demonstrates that excessive iron uptake, even though it may amount to only a few milligrams per day, will lead over the years to iron accumulation of up to 50 g or more. This is approximately 12 times the normal iron content of a 70 kg male. Because of the time required to accumulate this much iron, primary hemochromatosis is not a disease of younger persons.

The metabolic problem is brought about by extensive deposition of hemosiderin in the parenchyma of the liver, the pancreas, the pituitary gland, and the adrenal glands. In the most severe form, cardiac damage may be the cause of death. The liver is usually enlarged, with a "hob-nailed" appearance and a rusty color. Initial symptoms include cirrhosis, diabetes mellitus, and skin pigmentation.

Primary hemochromatosis is characterized by an increased serum iron concentration, from 30 to 47 μmol/L, and a latent iron-binding capacity of less than 8.6 μmol/L, to give a transferrin saturation of greater than 75%. The serum transferrin concentration ranges from 15 to 110 nmol/L, whereas the normal range is 0.19 to 3.3 nmol/L. The increased transferrin concentration and its greatly increased saturation emphasize the burden of heightened iron uptake and transport problems.

5.   Treatment of primary hemochromatosis The biochemical basis of treating this problem is twofold. First, patients must be advised against knowingly augmenting their diet with iron-containing medicinals, including vitamin preparations. Second, efforts must be made to reduce the body stores of iron. This can best be accomplished by repeated phlebotomies, withdrawing up to 5 dl of blood/wk. This process must be repeated until the serum transferrin, serum iron, and serum ferritin concentrations are all reduced to approximately normal values, a process that may require several years. Saturation, or near saturation, of the serum iron-binding capacity will follow a course of returning toward normal values that parallels the normal parameters mentioned; thus measurement of iron-binding capacity is a valuable monitor. In some patients managed entirely by phlebotomy, more than 110 g of iron have been removed over years. The hemoglobin content of normal human blood is approximately 2 mmol/L, which is equivalent to 8 mmol/L of iron. Therefore each phlebotomy of 5 dl amounts to the removal of 4 mmol of iron. Even so, if iron had been deposited in excess of need at a rate of only 0.05 mmol/day, up to several years may be required to clear a significant iron burden.

A second, much less valuable, means of lowering iron stores is based on the chelating agent known as deferoxamine. This forms a fairly strong Fe^{+++} chelate known as ferrioxamine, which can be excreted in the urine. Unfortunately, not all the complex is cleared by the kidneys, and some of the chelated iron may even be used in synthesis of iron-containing compounds. More importantly, its therapeutic use would require daily injections of large quantities of the chelator, to which some individuals develop severe reactions. For these reasons, no sound biochemical basis exists for use of this substance in therapy. Deferoxamine is of some value in diagnostic testing. It is also interesting to note that ferrioxamine-like compounds have been isolated and characterized as essential iron-containing growth factors for certain streptomycetes, which has given rise to the supposition that this reagent may be similar to the natural, low molecular weight chelators of mammalian tissues.

References

Aisen P and Listowsky I: Iron transport and storage proteins, Annu Rev Biochem 49:357, 1980.

Axelrod A: Case records of the Massachusetts General Hospital; Case 17-1979, N Engl J Med 300:969, 1979.

Brown EG: Recognition and treatment of iron overload, Adv Intern Med 26:159, 1980.

Fielding J: Differential ferrioxamine test for measuring chelatable body iron, J Clin Pathol 18:88, 1965.

Finch CA and Huebers H: Perspectives in iron metabolism, N Engl J Med 306:1520, 1982.

Case 5   Narcotic overdose

A patient was rushed into the emergency service in a coma and experiencing respiratory depression. Other parts of the history suggested a narcotic overdose. A blood sample (arterial) showed pH 7.22, with a total CO_2 concentration of 26.3 mmol/L.

Biochemical questions

1. What is the acid-base status of the patient?
2. What might be the immediate cause of the acid-base imbalance? Could any improvement be expected if the patient were placed on ventilation with a respirator?

Case discussion

1. Acid-base status of patient It is immediately clear that the blood pH is less than the normal value of 7.4 and that an uncompensated acidosis exists. The history suggests that this is probably an uncontrolled respiratory acidosis, since the narcotic overdose depressed respiration, causing CO_2 to be retained. Support for this thesis is found in the elevated pCO_2, which is calculated after determining the [total CO_2]:

$$[\text{Total } CO_2] = [HCO_3] + [CO_2 + H_2CO_3] \tag{A}$$

From the Henderson-Hasselbalch equation:

$$pH = 6.10 + \log \frac{[\text{Total } CO_2] - 0.0301\, pCO_2}{0.0301\, pCO_2} \tag{B}$$

From B:

$$7.22 = 6.10 + \log \frac{26.3 - 0.0301\, pCO_2}{0.0301\, pCO_2}$$

$$1.12 = \log \frac{26.3 - 0.0301\, pCO_2}{0.0301\, pCO_2}$$

or:

$$\text{antilog } 1.12 = \frac{26.3 - 0.0301\, pCO_2}{0.0301\, pCO_2}$$

$$13.2 = \frac{26.3 - 0.0301\, pCO_2}{0.0301\, pCO_2}$$

Rearrange:

$$13.2 \times 0.0301 \, pCO_2 = 26.3 - 0.0301 \, pCO_2$$

$$0.40 \, pCO_2 + 0.0301 \, pCO_2 = 26.3$$

$$pCO_2 = \frac{26.3}{0.43} \text{ mm Hg}$$

$$= 61 \text{ mm Hg (8.2kPa)}$$

This higher than normal value for pCO_2 is equal to 61×0.0301 mmol/L of CO_2, and from equation A:

$$[HCO_3^-] = 26.3 - (61 \times 0.0301) \text{ mmol/L}$$

$$= 26.3 - 1.8$$

$$= 24.5 \text{ mmol/L}$$

2. Cause of acid-base imbalance One can therefore see that the $[HCO_3^-]$ value is normal and the pCO_2 high because of the hypoventilation induced by the narcotic. Since the kidneys are slower to respond to the increase of the $[HCO_3^-]$ necessary to correct the [salt]/[acid] ratio to about 20, thereby correcting the blood pH, a more rapid reponse would be obtained by assisted ventilation on a respirator, which would reduce the pCO_2 elevation.

References

Davenport HW: The ABC of acid base chemistry, ed 6, Chicago, 1974, University of Chicago Press.

Hudson LD: Diagnosis and management of acute respiratory distress in patients on mechanical respirators. In Moser KM and Spragg RG: Respiratory emergencies, St Louis, 1982, The CV Mosby Co.

Masoro EJ and Siegel PD: Acid-base regulation, Philadelphia, 1978, WB Saunders Co.

Case 6   Poliomyelitis

A 14-year-old boy who had never been immunized against poliomyelitis contracted the disease late in the summer. He was hospitalized and required the use of a respirator during the acute phase of his illness. When he appeared to be recovering, he was taken off the respirator with no apparent ill effects. Several days later, analysis of his blood revealed the following:

Na^+	136 mmol/L
K^+	4.5 mmol/L
Total CO_2	36 mmol/L
Cl^-	92 mmol/L
pCO_2	70 mm Hg (9.33 kPa)
Blood pH	7.32

Biochemical questions

1. What was the acid-base balance of the patient?
2. What anions probably were elevated in this patient's blood?
3. What was the probable cause of this condition?
4. The patient was put back on a respirator and the pCO_2 was reduced to 40 mm Hg. If this rate of ventilation was too rapid, what would be the result?

Case discussion

1. Acid-base balance of patient Comparison of the composition of the patient's blood with normal values shows the following:

a. The concentrations of Na^+ and K^+ are within the normal limits of 136 to 145 and 3.5 to 5.0 mmol/L, respectively.
b. The concentration of Cl^- is slightly below the normal range (100 to 106 mmol/L).
c. Both the [Total CO_2] and the pCO_2 are high. From these values and the blood pH of 7.32, which is on the acid side of the normal range but approaching normal, the [HCO_3^-] can be calculated.

$$CO_2 + H_2CO_2 = pCO_2 \times 0.0301 \text{ mmol/L}$$
$$= 70 \times 0.0301 \text{ mmol/L}$$
$$= 2.11 \text{ mmol/L}$$
$$[HCO_3^-] = [\text{Total } CO_2] - [CO_2 + H_2CO_3]$$
$$= (36.0 - 2.1) \text{ mmol/L}$$
$$= 33.9 \text{ mmol/L}$$

A quick calculation from equation 4.1 shows that the values of [HCO_3^-] and [$CO_2 + H_2CO_3$] are consistent with the experimentally determined blood pH of 7.32.

$$pH = 6.1 + \log_{10} \frac{33.9}{2.11}$$
$$= 6.1 + \log_{10} 16.06$$
$$= 6.1 + 1.21$$
$$= 7.31$$

Where possible, such checks for consistency in laboratory values are useful to exclude analytic errors.

2. Anions in blood The analysis of blood normally shows that the sum of the concentrations of the principal cations K^+ and Na^+ exceeds the sum of the concentrations of HCO_3^- and Cl^- by about 5 to 15 mmol/L. This difference represents the concentrations of the nonvolatile anions: $SO_4^=$, phosphate, lactate, and other organic acids of metabolism. In cases of metabolic imbalances such as uncontrolled diabetes mellitus, the serum concentrations of these organic acids may increase. This rise is associated with relative reductions in the HCO_3^- and Cl^-. In this patient the difference is:

$$([Na^+] + [K^+]) - ([HCO_3^-] + [CL^+])$$
$$= (136 + 4.5) - (33.9 + 92) \text{ mmol/L}$$
$$= 14.6 \text{ mmol/L}$$

This suggests that no severe metabolic imbalance exists.

3. Probable cause of condition Since both total CO_2 and pCO_2 are high, and since an almost compensated pH and no excess of metabolic organic acids exist, the patient had a compensated respiratory acidosis. A patient who required respirator support while recovering from poliomyelitis would have weakness of the chest muscles but no impairment of the efficiency of gaseous exchange across the alveolar membranes. However, there is impairment of pulmonary ventilation, and the pCO_2 builds up. The patient cannot correct this until his chest muscle strength returns. Furthermore, high pCO_2 acts as a narcotic and depresses the respiratory control center. In response to the acidosis and high pCO_2, the proximal and distal tubular reabsorption of HCO_3^- is maximal. Bicarbonate is also generated in the tubular cells by the interplay of the bicarbonate-phosphate buffers.

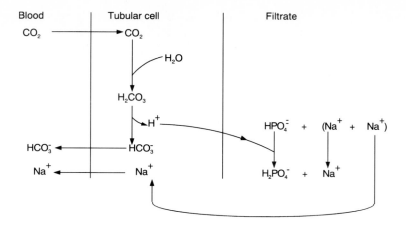

Thus a net loss of H^+ and a conversion of CO_2 to HCO_3^- occur. This raises the $[HCO_3^-]$ so that the $[HCO_3^-]/[H_2CO_3]$ ratio approaches 20, returning the pH toward 7.4. Although this is the correct direction for compensation of the respiratory acidosis, the result is an abnormally high [total CO_2]. As the patient's condition improves, this elevation in [total CO_2] is gradually reduced to normal.

4. Mechanical reventilation Compensation by renal mechanisms is slower by many hours than the changes in pCO_2 that can be effected by the lungs. If the lungs of a patient with a stable pCO_2 elevation are ventilated to reduce pCO_2 more rapidly than the renal compensation can follow, alkalosis would develop. The pCO_2 would approach normal, but the $[HCO_3^-]$ and [total CO_2] would be high (as a residue from the compensated respiratory acidosis). Assuming that there has been no renal response to a reduction in pCO_2 in this patient, the blood pH would be:

$$pH = 6.1 + \log_{10} \frac{33.9}{(0.0301 \times 40)}$$

$$= 6.1 + \log_{10} \frac{33.9}{1.20}$$

$$= 6.1 + 1.45$$

$$= 7.55$$

which is quite alkaline. This could be avoided by slowing the rate of reduction in pCO_2.

References   See references for case 5.

Case 7   Red blood cells and surgery involving a heart-lung bypass

A patient underwent open heart surgery during which an oxygenator was used. The oxygenator's extracorporeal circuit was primed with 1- to 3-day-old acid citrate dextrose blood, diluted with bicarbonate-buffered saline solution to an 8% hemoglobin content. The volume of the priming blood was 40% to 50% of the patient's blood volume, and during the bypass the mixed blood averaged 10% to 11% hemoglobin. The determinations summarized here were performed before, during, and after the bypass. The 2,3-bisphospho-D-glycerate level is expressed as millimoles of phosphorus per liter of packed red blood cells. The $pO_{2_{50}}$ is the oxygen tension in kilopascals at which the hemoglobin was 50% saturated with oxygen; during the determination of the

pO$_2$, it was shown that the 2,3-bisphospho-D-glycerate and pH of the red blood cells (pH$_{RBC}$) did not change. pH$_{RBC}$ was measured in packed, hemolyzed cells. The pH of the whole blood did not change from 7.4 during the entire procedure.

	2,3-Bisphospho-D-glycerate	pO$_{2_{50}}$	pH$_{RBC}$
Priming blood	3.8	1.86	
Patient control	4.8	3.46	7.2
Duration of bypass (min)			
10		2.4	7
20	2.7		
40	2.5		
60	2.88	2.66	6.9
After bypass (hr)			
1	2.4	3.1	7.2
24	4.45	3.6	7.2

On the average, the priming blood contained 3% carbon monoxide, compared to 1% in the patient's blood.

Biochemical questions

1. Can the change in 2,3-bisphospho-D-glycerate content of the red blood cells be explained by admixture of the priming blood with the patient's blood?
2. What is the effect of 2,3-bisphospho-D-glycerate and pH on the oxygen saturation curve of hemoglobin? Explain.
3. As a result of the combined decrease of 2,3-bisphospho-D-glycerate and pH$_{RBC}$ in the patient, what will be the effect on the oxygen transport to tissues? Explain.
4. Would it be reasonable to assume that the donor of the priming blood was a smoker and that the patient smoked less? What is the effect of carbon monoxide on oxygen transport?
♦5. It was noted that the exposure of red blood cells in vitro to elevated levels of oxygen accelerated the decline of 2,3-bisphospho-D-glycerate in the cells compared to pumping the blood without the oxygenator in the circuit. How might this be explained?

References

Bordink JM et al: Alterations in 2,3-diphospho-glycerate and O$_2$ hemoglobin affinity in patients undergoing open-heart surgery, Circulation 43 and 44(suppl. I):1, 1971.

Brewer JG and Eaton JW: Erythrocyte metabolism: interaction with oxygen transport, Science 171:1205, 1971.

Finch CA and Lenfant C: Oxygen transport in man, N Engl J Med 286:407, 1972.

Proctor HJ et al: Alterations in erythrocyte 2,3-diphospho-glycerate in postoperative patients, Ann Surg 173:357, 1971.

Case 8   **Emphysema**

A 57-year-old widow had a medical history of shortness of breath that had progressively worsened over the last 10 yr. On examination, her blood chemistry findings were as follows:

Arterial blood		**Venous blood**	
pO$_2$	45 mm Hg (6.00 kPa)	Blood urea nitrogen (BUN)	7.9 mmol/L
pCO$_2$	65 mm Hg (8.66 kPa)	Total CO$_2$	42.7 mmol/L
pH	7.42	Cl$^-$	87.0 mmol/L
		K$^+$	3.6 mmol/L
		Na$^+$	135.0 mmol/L

Biochemical questions
1. What is the nature of the acid-base balance?
2. What type of disease might be associated with this case?
3. Would unusual findings from the urine analysis be expected?
4. What conclusions can be drawn from the pCO_2 value?

References   See references for case 5.

Case 9   Pulmonary embolism

A 32-year-old man underwent surgery for a radical lymph node dissection on July 1. On July 3 a pulmonary embolism occurred. Shortness of breath developed rapidly, and the patient experienced sweating and a tachycardia of 180/min. Blood chemistry analysis showed the following:

Arterial blood		Venous blood	
pO_2	65 mm Hg (8.66 kPa)	BUN	3.6 mmol/L
pCO_2	30 mm Hg (4.00 kPa)	Total CO_2	23.0 mmol/L
pH	7.50	Cl^-	95.0 mmol/L
		K^+	4.1 mmol/L
		Na^+	143.0 mmol/L

Biochemical questions
1. What is the acid-base condition of this patient?
2. How can the pO_2 and pCO_2 values be reconciled?
3. Is there evidence as to the effectiveness of renal function?
4. What might be concluded from the $[Na^+ + K^+] - [Cl^- + HCO_3^-]$ gap?

References   See references for case 5.

Additional questions and problems
1. How does the kidney conserve Na^+ by removing phosphate from the blood (glomerular filtration) and excreting it in urine at pH 5 to 6?
2. How does the buffer capacity of a solution of bicarbonate, 25 mmol/L at pH 7.4, compare with that of 15 mmol/L at pH 7.2?
3. Predict the direction of electrophoresis at pH 8.6 compared to Hb A of the following abnormal hemoglobins:

Hemoglobin	Amino acid substitution
α-Chains	
M Boston	58 His → Tyr
Bibba	136 Leu → Pro
J Capetown	92 Arg → Gin
β-Chains	
Zurich	63 His → Arg
M Saskatoon	63 His → Tyr
Sydney	67 Val → Ala
M Milwaukee	67 Val → Glu
Kempsey	99 Asp → Asn
Yakima	99 Asp → His

4. In view of the isohydric transport of carbon dioxide in the erythrocyte, is the concentration of Cl^- higher in arterial or venous blood? Explain.
5. Outline the factors that will determine how much compensation has been effected by a patient in response to acid-base imbalance. How is the degree of compensation indicated?

6. Summarize the body's defense mechanisms to an introduction of a strong acid into the blood.
7. Explain the allosterism of oxygen biding to hemoglobin, as well as its absence in myoglobin.
8. Discuss how oxygen binding to hemoglobin is influenced by pH, pCO_2, and 2,3-bisphospho-D-glycerate.
9. Why is compensation of blood pH more rapidly performed by the lungs than by the kidneys?
10. Discuss the potential effects of covalently cross-linking subunits to each other in the hemoglobin tetramer.

Multiple choice problems

A patient with sickle cell disease was a heavy smoker and developed emphysema, thus severely reducing his pulmonary function. Arterial blood measurements were pH, 7.42; pCO_2, 65 mm Hg; and pO_2, 45 mm Hg. In venous blood, the [total CO_2] was 42.7 mmol/L.

The following questions relate to this case.

1. The molecular structure of sickle cell hemoglobin (Hb S) differs from normal hemoglobin in that:
   A. A basic amino acid is replaced by an acidic residue.
   B. An acid amino acid is replaced by a hydrophilic amino acid.
   C. An acidic amino acid is replaced by a neutral residue.
   D. A modification of the heme moiety is made.
   E. A substitution of amino acid residues occurs on the α-subunit.

2. The Hb S can be detected in the blood by:
   1. Electrophoresis.
   2. Abnormal shape of erythrocytes.
   3. Insolubility of reduced Hb S.
   4. Color of erythrocytes.
      A. 1, 2, and 3 only are correct.
      B. 1 and 3 only are correct.
      C. 2 and 4 only are correct.
      D. 4 only is correct.
      E. All are correct.

3. As a result of the poorer oxygenation of the blood:
   1. The sickling of the Hb S will be greater.
   2. The life of the erythrocytes carrying Hb S will be shorter.
   3. The amount of Hb A in blood will be greater.
   4. The erythrocyte will be minimally affected.
      A. 1, 2, and 3 only are correct.
      B. 1 and 3 only are correct.
      C. 2 and 4 only are correct.
      D. 4 only is correct.
      E. All are correct.

4. From the blood chemistry data the most likely condition is:
   A. Metabolic alkalosis.
   B. Compensated respiratory acidosis.
   C. Compensated metabolic acidosis.
   D. Compensated respiratory alkalosis.
   E. Respiratory alkalosis.

5. The factors used to derive the conclusion in problem 4 are:

A. Normal pH.
B. High [total CO_2].
C. Both.
D. Neither.

6. The kidney is functioning in this case by:
   1. Retaining HCO_3^-.
   2. Retaining H^+.
   3. Absorbing CO_2.
   4. Excreting urine with high pH.
      A. 1, 2, and 3 only are correct.
      B. 1 and 3 only are correct.
      C. 2 and 4 only are correct.
      D. 4 only is correct.
      E. All are correct.

7. The conformational abnormality of Hb S causes the:
   A. Loss of allosterism in O_2 binding.
   B. Absence of a Cl^- shift.
   C. Absence of the Bohr effect.
   D. Increased affinity of β-chain interactions.
   E. Increased O_2-binding affinity.

8. The carbon monoxide in the cigarette smoke would:
   1. Bind to the iron of the heme.
   2. Be determined in blood spectrophotometrically.
   3. Be displaced in carboxyhemoglobin by high pO_2.
   4. Affect the blood-clotting process.
      A. 1, 2, and 3 only are correct.
      B. 1 and 3 only are correct.
      C. 2 and 4 only are correct.
      D. 4 only is correct.
      E. All are correct.

9. Hb S can be found:
   1. In the heterozygous state.
   2. In the homozygous state.
   3. Associated with other Hb variants.
   4. In all blood samples.
      A. 1, 2, and 3 only are correct.
      B. 1 and 3 only are correct.
      C. 2 and 4 only are correct.
      D. 4 only is correct.
      E. All are correct.

10. The blood-clotting system, if impaired in the patient, involves:
    A. Hemoglobin.
    B. Albumin.
    C. Calcium.
    D. Transferrin.
    E. Ceruloplasmin.

Chapter 5

Energetics and mitochondrial functions

Objectives

1 To predict direction of a reaction from changes in free energy
2 To explain how the energy from one reaction promotes a second reaction
3 To demonstrate how oxidative phosphorylation produces high-energy bonds
4 To explain how the oxidation of metabolites of carbohydrates, proteins, and fats yield energy
5 To illustrate that mitochondria control and regulate metabolic pathways

Mitochondria

Human cells contain hundreds of mitochondria. These cytoplasmic subcellular organelles are approximately the size of small bacteria. Their double membrane encloses a convoluted structure of inner membranes and a matrix that contains a complex array of enzymes and proteins. The major function of mitochondria is to convert food energy to the chemical energy of the cell, adenosine triphosphate (ATP). The synthesis of most ATP occurs under aerobic conditions by a series of reactions that are tightly linked to the complete oxidation, to carbon dioxide and water, of certain organic acids. These organic acids are derivatives of a few simple compounds that arise primarily from the metabolism of carbohydrates and fat. Acetate in the form of acetyl coenzyme A (acetyl CoA) is the most important of these acids. Molecular oxygen is needed for the oxidation of acetyl CoA; thus highly oxygenated tissues, such as the heart, contain cells with large numbers of mitochondria.

The biochemical reactions of the mitochondria are complex but are understood in considerable detail. Figure 5.1 highlights some of these reactions. Note that cytosolic glucose and fatty acids fuel the mitochondrial furnace. Glucose is converted to pyruvate by a series of reactions in the cytosol. Pyruvate then diffuses to the inner mitochondrial membrane, where it is converted to acetyl CoA. The acetyl group is oxidized to CO_2 and H_2O by Krebs cycle enzymes of the mitochondrial matrix. Cytosolic fatty acids can also yield acetyl CoA by the so-called β-oxidation scheme after they are transported into the mitochondrion as carnitine derivatives. For each acetyl group oxidized, three molecules of reduced nicotinamide adenine dinucleotide (NADH) and one of reduced flavin adenine dinucleotide ($FADH_2$) are produced. These are reoxidized by an electron transport system that is part of the inner mitochondrial membrane. Cytosolic NADH, from the conversion of glucose to pyruvate, is used to reduce substrates that are called *reducing equivalents*. The reducing equivalents, as reduced substrates, shuttle into the mitochondria, where they are reoxidized to yield intramitochondrial NADH. Figure 5.1 also shows the electron transport chain. A series of reactions transfer electrons ultimately to O_2 and in so doing provide the energy that promotes the expulsion of H^+ from the mitochondria. Hydrogen ions reenter the mitochondria via the F_1 ATPase and provide the driving force for the synthesis of ATP from adenosine diphosphate (ADP) and inorganic phosphate (P_i).

This chapter describes many of the mitochondrial reactions in more detail. Some understanding of the principles of energetics is necessary to see how these reactions are orchestrated.

Figure 5.1   Synopsis of some mitochondrial functions. (See also Figure 5.8.)

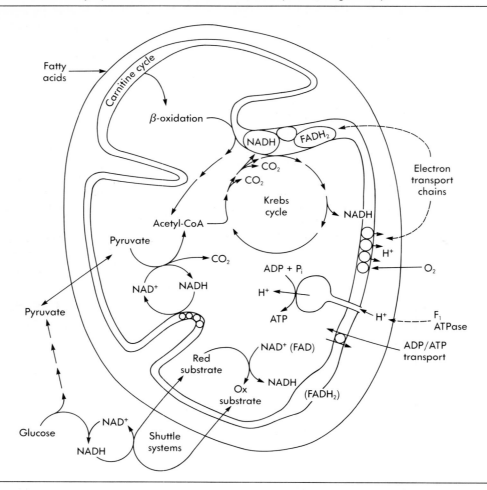

Thermodynamic laws

Experimental thermodynamic systems may be open or closed. *Closed systems* neither lose nor gain matter, but certain forms of energy may enter or leave the isolated system. *Open systems* may gain or lose matter as well as certain forms of energy. Living systems are obviously open systems of great complexity; thus it is difficult to analyze thoroughly the energetics of intact organisms. By dealing with simpler, closed systems, one can gain useful information that is applicable to the more complex situations.

The absolute value of the total energy of a system cannot be measured, but *changes* in the total energy of components that comprise the total energy can be determined. If a suitable reference point exists, comparison of energy changes can be made quantitatively and related to one another.

One of the most useful application of thermodynamics in biochemistry is to determine the direction of enzymatic reactions.

First law

Thermodynamic laws govern the energy in closed systems. The first law states that the total energy of a system is constant. If some amount of heat (Q) is put into the system, it must either do work or increase the total energy of the system. Most biologic systems operate at nearly constant temperature, pressure, and volume; therefore work, as represented by changes in volume and pressure, is not regarded as biologically useful. If the capacity to do biologically useful work is represented by the symbol W, and H stands for *enthalpy,* or heat content, the first law can be expressed as:

$$\Delta H = Q - W \qquad (5.1)$$

Note that the first law applies to an ideal, reversible, closed system; it says nothing about the direction of a chemical reaction.

Second law

The second law states that closed systems, left to themselves, approach a state of equilibrium. Any system, no matter how highly organized, proceeds from an ordered state of low probability to a state of greater disorder, or randomization, which is a state of high probability. Entropy is a measure of this disorder and is designated by the symbol S. Entropy is the part of enthalpy that is *not* available to do useful work. Heat energy cannot be converted completely to other energy forms; some of it is lost as entropy. In chemical terms, entropy is related to the random movements of molecules and is measured by TΔS, where T is the absolute temperature. When a chemical system is at equilibrium, no *net* reaction occurs, the system has no capacity to do useful work, and:

$$Q = T\Delta S \qquad (5.2)$$

This is the condition of maximum entropy. The existence of entropy explains why heat energy cannot be completely converted into other energy forms; some of it is lost as entropy and thus cannot do useful work.

Free energy

Work can be done by systems proceeding toward equilibrium, and a measure of the maximum useful work is given by the following equation:

$$W = -\Delta H + T\Delta S \qquad (5.3)$$

Willard Gibbs introduced the concept of free energy as another measure of the capacity to do useful work. Accordingly, free energy, abbreviated G, is defined as:

$$\Delta G = \Delta H - T\Delta S \qquad (5.4)$$

Note that $\Delta G = -W$, so that when the measure of W is *positive* (meaning that the system is doing useful work), the measure of ΔG is *negative,* and vice versa.

Exergonic and endergonic reactions

We can now qualitatively evaluate energy changes in a chemical reaction and draw some conclusions based on these evaluations. Consider the isomerization reaction:

$$\text{Glucose 1-P} \xrightarrow{\text{Phosphoglucomutase}} \text{Glucose 6-P}$$

Since changes in free energy and enthalpy are related only to the difference between the free energies and enthalpies of the products and reactants, we can characterize the reaction as:

$$\Delta G = G_{\text{Glc 6-P}} - G_{\text{Glc 1-P}}$$

and:

$$\Delta H = H_{\text{Glc 6-P}} - H_{\text{Glc 1-P}}$$

Suitable measurements can be made to reveal the algebraic signs of ΔG and ΔH. The algebraic signs reveal considerable information about reactions. For example, when ΔG is:

1. *Negative,* the reaction is *exergonic.* It will proceed spontaneously from left to right as written.
2. *Positive,* the reaction is *endergonic.* It will not proceed spontaneously as written, but the reverse reaction will be spontaneous.
3. *Zero,* the reaction is at *equilibrium.*

When ΔH is:
1. *Negative,* the reaction is *exothermic.* It will give off heat to its surroundings.
2. *Positive,* the reaction is *endothermic.* It will take up heat from its surroundings.
3. *Zero,* the reaction is *isothermic.* No net exchange of heat occurs with the surroundings.

The predictive value of these expressions is very useful. They also emphasize that change in free energy is the driving force behind all chemical reactions.

Standard state

Since we must deal with *changes* in free energy rather than with absolute values, some reference point must be established. The reference point is defined as the standard free energy and is indicated by ΔG^0. Standard free energies of chemical reactions are calculated at 25° C and at 1 atmosphere (atm) pressure.

The biologic standard free energy change ($\Delta G^{0\prime}$) is more useful in biochemistry. Here the standard conditions are pH 7, 1 M (mol/L) concentration of reactants and products, and 37° C. Tables listing changes in free energy express ΔG^0 or $\Delta G^{0\prime}$ in joules per mole or in calories per mole.

Coupled reaction systems

Two or more reactions are said to be coupled if a product of one reaction is a substrate for another. The concept of coupling applies to any sequence of chemical reactions. Coupling is particularly important in biochemistry because metabolic pathways are made up of many sequential reactions.

Consider an idealized sequence of reactions represented by the following set of equations:

$$
\begin{array}{ll}
A \rightarrow B & (\Delta G_1) \\
B \rightarrow C & (\Delta G_2) \\
\underline{C \rightarrow D} & (\underline{\Delta G_3}) \\
\text{Sum:} \quad A \rightarrow D & (\Delta G_S)
\end{array}
$$

Under conditions where $\Delta G < 0$, material will pass from A to D. This conclusion is valid regardless of the individual values of ΔG_1, ΔG_2, or ΔG_3.

In coupled systems material will flow from left to right if $\Delta G_S < 0$. The number of individual reactions in the coupled sequence is not important, nor is the number of reactants or products in an individual reaction. Only one common intermediate is needed to couple any two consecutive reactions. The common intermediate may be an electron, a proton, or a more complex structure. Many examples of coupled systems in metabolism exist. A notation typically used to indicate coupled reactions is:

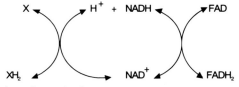

This sequence of reactions is equivalent to:

$$XH_2 + NAD^+ \rightleftharpoons X + NADH + H^+$$
$$NADH + H^+ + FAD \rightleftharpoons FADH_2 + NAD^+$$

Where XH_2 is a substrate from which two hydrogens can be removed, and X is the oxidized product. Note that two common intermediates are present, the free proton and NADH.

ΔG from equilibrium constants

Several ways of determining ΔG exist. One of the most frequently used is from the equilibrium constant of a reaction.

Equilibrium constants For the general reaction,

$$aA + bB \rightarrow cC + dD$$

it can be shown that:

$$\Delta G = \Delta G^0 + RT \ln \frac{[C]^c [D]^d}{[A]^a [B]^b} \tag{5.5}$$

where R is the natural gas constant equal to 8.315 joules/mol/degree (1.987 cal/mol/degree), T is the absolute temperature (°C + 273), ln stands for *natural* logarithm, and ΔG^0 is the standard free energy of the reaction. Note the resemblance of this equation to the Henderson-Hasselbalch equation. ΔG and ΔG^0 are analogous to the relation between pH and pK. Just as pH = pK when the concentrations of the ionized and un-ionized forms of a weak acid are equal, so $\Delta G = \Delta G^0$ when the product $[C]^c[D]^d$ is equal to the product $[A]^a[B]^b$.

At equilibrium, ΔG = 0; consequently, one can write:

$$\Delta G^0 = -RT \ln \frac{[C_{eq}]^c [D_{eq}]^d}{[A_{eq}]^a [B_{eq}]^b} \tag{5.6}$$

Since the equilibrium constant is defined as:

$$K_{eq} = \frac{[C_{eq}]^c [D_{eq}]^d}{[A_{eq}]^a [B_{eq}]^b} \tag{5.7}$$

then:

$$\Delta G^0 = -RT \ln K_{eq} \tag{5.8}$$

It must be stressed that standard thermodynamic conditions probably never apply in living tissues; the typical concentration range of intracellular metabolites is 1 to 5 mmol/L, *not* 1 mol/L concentration that standard conditions require. The energy available from a given reaction depends more on the intracellular concentrations than on the fixed value of ΔG^0.

For example, for the reaction:

$$ATP + H_2O \rightarrow ADP + P_i$$

assume that the intracellular concentration of ATP is 5 mmol/L, ADP is 4 mmol/L, and P_i is 2.1 mmol/L. According to the general equation 5.5, we can write:

$$\Delta G = \Delta G^0 + RT \ln \frac{[ADP] [P_i]}{[ATP] [H_2O]}$$

$$= -7 \times 1.98 \times (273 + 30)2.303 \log_{10} \frac{(4 \times 10^{-3})(2.1 \times 10^{-3})}{5 \times 10^{-3}}$$

$$= -10.85 \text{ kcal/mol, or } -42.7 \text{ kJ/mol}$$

Pure water is in the standard state; so $[H_2O]$ is 1 for dilute solutions. Biologic *standard* state, indicated by $\Delta G^{0\prime}$, is the same except that the pH is 7. Temperature is not stated.

For this example, ΔG^0 (30° C* and pH 7.4) is -29.3 kJ/mol (-7 kcal/mol), whereas ΔG at the more physiologic nucleotide concentrations just given was calculated to be -42.7 kJ/mol (-10.85 kcal/mol). Thus considerably more free energy is available from the hydrolysis of ATP at physiologic concentrations than at standard-state concentrations.

Concentration differences Free energy changes resulting from concentration differences are very important in biochemistry. As mentioned earlier, if molecules at a given concentration (C_1) are *diluted* (and thus allowed to move farther apart) to a new concentration (C_2), the disorder in the system is increased. Consequently, *the free energy decreases*. This can be represented by the equation:

$$\Delta G = RT \ln C_2/C_1$$

The equation indicates that ΔG *will always be greater than zero for concentration processes* and *will always be less than zero for dilution processes*. Many cellular functions involve concentrations, such as the formation of gastric acid from plasma hydrogen ions. Recall that the concentration of hydrogen ions in the plasma is $10^{-7.4}$ M, whereas the [H+] of gastric juice is >0.1 M. Other examples of concentration processes are the pumping of Na^+ from inside cells to the outside, where Na^+ must be pumped from an intracellular solution of 1 to 2 mM (mmol/L) to an extracellular [Na^+] of approximately 140 mM. The formation of tears and urine are two other examples of concentration events.

Oxidation-reduction reactions A whole spectrum of biochemical systems involves oxidation-reduction reactions. Oxidation is defined as the *loss of electrons,* and reduction as a *gain of electrons.* When substrate is oxidized, another substrate is simultaneously reduced. A typical oxidation-reduction system is composed of two half-reactions, as in the following:

1. $NAD^+ + 2H^+ + 2e \rightarrow NADH + H^+$
2. $\underline{\hspace{2cm} XH_2 \rightarrow X + 2H^+ + 2e \hspace{2cm}}$
$$NAD^+ + XH_2 \rightarrow X + NADH + H^+$$

Half-reaction 1 is a reduction, a gain of electrons, whereas half-reaction 2 is an oxidation, a loss of electrons. Describing an oxidation-reduction system in the form of its half-reactions emphasizes that it is a coupled system, since electrons are always products of one half-reaction and reactants of the other. Many biologic oxidations are catalyzed by dehydrogenases, which are enzymes that employ coenzymes that accept or donate hydrogens. NAD^+ is one important oxidation-reduction coenzyme, and FAD is another.

The phenomenon of electron transfer implies that if it were possible to separate the half-reactions and to connect them by a wire, the movement of electrons through the wire would generate a potential difference (voltage) that could be detected by a potentiometer. The sign of the potential difference reflects the direction in which the reaction proceeded, and the magnitude of the potential difference is a measure of the energy driving the reaction. Many enzyme systems have been studied in this manner, by joining one half-reaction with another that is a standard of reference. The reference standard is the *hydrogen electrode,* described by the half-reaction:

$$2H^+ + 2e \rightarrow H_2 \text{ (g)} \qquad E^0 = 0.00 \text{ volts}$$

The standard electrode potential of this half-reaction is arbitrarily defined as zero volts. This is the *primary* international standard. Since virtually no biochemically important reactions occur at the standard [H^+] of 1 mol/L, a *secondary* reference electrode has

---

*Standard values of $\Delta G^{0'}$ are often tabulated at 25° C. This example shows that they can be calculated for any temperature or pH.

been adopted by biologists. The secondary reference is determined at pH 7 ($[H^+]$ of 10^{-7} M) and 30° C. It is given by the half-reaction:

$$2H^+ + 2e \rightarrow H_2 \text{ (g)} \qquad E^{0'} = -0.42 \text{ volts}$$

where the negative potential results from the difference in $[H^+]$ compared to the primary standard. The use of $E^{0'}$ indicates that the pH is 7. By a corresponding convention, $\Delta G^{0'}$ represents free energy values measured at pH 7.

A given half-reaction can serve either as oxidant or reductant, depending on the reaction with which it is coupled. For this reason, standard electrode potentials can be expressed either as oxidation or as reduction potentials, where the algebraic signs in the two cases are opposite. *By convention, biologists list half-reactions as reductions and tabulate these values as reduction potentials.* Table 5.1 lists several standard reduction potentials. The following conventions apply in using the data from the table:

1. Half-reactions with more positive potentials are the oxidizing agents (electron acceptors) for half-reactions with *less positive* potentials; for example, molecular oxygen will oxidize the Fe^{++} of cytochrome *a*.

2. If a half-reaction is written as the reverse of that given in Table 5.1, the algebraic sign of the potential will also be reversed. For example, $E^{0'}$ for the oxidation of malate to oxaloacetate is $+0.17$ volts.

3. The $\Delta G'$ for an oxidation-reduction reaction is related to the electromotive potential by the equation:

$$\Delta G' = -nF (E^{0'}) + RT \ln [\text{oxidant}]/[\text{reductant}]$$

In those cases where the oxidant and reductant concentrations are equal, the equation simplifies to:

$$\Delta G^0 = -nF (E^{0'})$$

where n is the number of electrons transferred (usually n = 2), F is the Faraday constant equal to 96.5 kJ/mol (96,500 coulombs/4.185 joules/cal/mol, or 23,058 cal/volt/mol),

Table 5.1	Standard Reduction Potentials	
	Half-reaction	**$E^{0'}$ (pH 7), volts†**
	$\frac{1}{2}O_2 + 2H^+ + 2e \rightarrow H_2O$	0.82
	Cytochrome *a* (Fe^{+++}) + e $\rightarrow$ Cytochrome *a* (Fe^{++})	0.29
	Cytochrome *c* (Fe^{+++}) + e $\rightarrow$ Cytochrome *c* (Fe^{++})	0.22
	Coenzyme Q(ox) $\rightarrow$ Coenzyme Q(red)	0.10
	Cytochrome *b* (Fe^{+++}) + e $\rightarrow$ Cytochrome *b* (Fe^{++})	0.07
	Fumarate + $2H^+$ + 2e $\rightarrow$ Succinate	0.031
	FAD + $2H^+$ + 2e $\rightarrow$ $FADH_2$	-0.06
	Oxaloacetate + $2H^+$ + 2e $\rightarrow$ Malate	-0.17
	Pyruvate + $2H^+$ + 2e $\rightarrow$ Lactate	-0.19
	Acetaldehyde + $2H^+$ + 2e $\rightarrow$ Ethanol	-0.20
	Acetoacetate + $2H^+$ + 2e $\rightarrow$ L-β-OH-Butyrate	-0.27
	Lipoate + $2H^+$ + 2e $\rightarrow$ Dihydrolipoate	-0.29
	NAD^+ + $2H^+$ + 2e $\rightarrow$ NADH + H^+	-0.32
	$NADP^+$ + $2H^+$ + 2e $\rightarrow$ NADPH + H^+	-0.32
	$2H^+$ + 2e $\rightarrow$ H_2	-0.42

†At 30° C.

and $E^{0\prime}$ is the standard electrode potential at pH 7, that is, -0.42 volt. The other symbols were defined previously.

Summary of energetics

1. Free energy is a measure of the energy available to perform useful work.
2. ΔG can be used to predict the direction of a chemical reaction.
3. Sequential reactions can be joined in coupled systems, if they are linked by a common component.
4. Coupling allows an energetically favored reaction to drive a less favored reaction if the net ΔG for the system is negative.
5. ΔGs measured at physiologic concentrations may differ significantly from those measured at the standard state and when coupled may even have a different sign.

The respiratory chain

As stated earlier, the reoxidation of NADH is coupled to the reduction of FAD to $FADH_2$. This is the first step in a complex system driven by coupled mitochondrial enzymes that function in close physical association. Indeed, the association is so close that the entire assemblage of enzymes can be isolated as submitochondrial particles.

Collectively, the group of enzymes in this assemblage is known as either the *respiratory chain* or the *electron transport chain*. Electron transport chain refers to oxidation and reduction being defined as the loss or gain of electrons, so a system that promotes biologic oxidations involves electron transfer. The system uses a coupled sequence or "chain" of reactions. Respiratory chain emphasizes that the coupled reactions involve the uptake of oxygen, that is, respiration.

The basic components of the respiratory chain are diagrammed in Figure 5.2. Two means of entry to the system exist. These involve the reduced coenzymes NADH or $FADH_2$. One point of entry, that for NADH, is shown at the left and the other, $FADH_2$, toward the top of Figure 5.2. The first entry is the more general, since NADH is produced in many NAD^+-linked reactions, whereas $FADH_2$ is formed by a smaller number of FAD-linked reactions.

The arrangement of components in the respiratory chain are shown (Figure 5.2) in the order of their oxidation-reduction potentials, which have been experimentally determined using isolated mitochondrial fractions.

Major components of the respiratory chain

Three classes of molecules make up the major components of the respiratory chain. Two classes serve as carriers of hydrogens and electrons, whereas the other carries only electrons. Each class is described here.

Flavoproteins and iron-sulfur components Flavoproteins are enzymes of the respiratory chain that contain tightly bound flavin mononucleotide (FMN) or FAD. The flavin coenzymes do not dissociate during their oxidation and reduction, so they behave more as prosthetic groups than coenzymes. The riboflavin portion of the flavin prosthetic group on the flavoprotein Fp_D accepts hydrogens from NAD^+-linked dehydrogenases. Fp_D is part of the enzyme *NADH–cytochrome b reductase* that makes up complex I (Figure 5.2). Another flavoprotein, Fp_S, is an integral part of the mitochondrial inner membrane–associated enzyme *succinate dehydrogenase* that makes up complex II (Figure 5.2). The flavoprotein prosthetic groups that are reduced and accept hydrogens are shown schematically in Figure 5.3. The reactivity of the isoalloxazine ring (the riboflavin ring) is heightened by the presence in the flavoprotein of atoms of nonheme iron and sulfur. A variety of iron-sulfur proteins are known. In mammals these usually contain four atoms each of iron and sulfur arranged as follows (see p. 200):

Figure 5.2

Anatomy of the respiratory chain. The three sites at which $\Delta E^{o'}$ is sufficient to drive ATP formation are shown by the circled numbered arrows. F_p, F_{pD} and F_{ps} are specific flavoproteins, FeS represents iron-sulfur proteins that contain nonheme iron, $DHAP$ means dihydroxyacetone phosphate, and Fe^{++} and Fe^{+++} represent the valence state of the heme iron in the individual cytochromes. See also Figure 5.10.

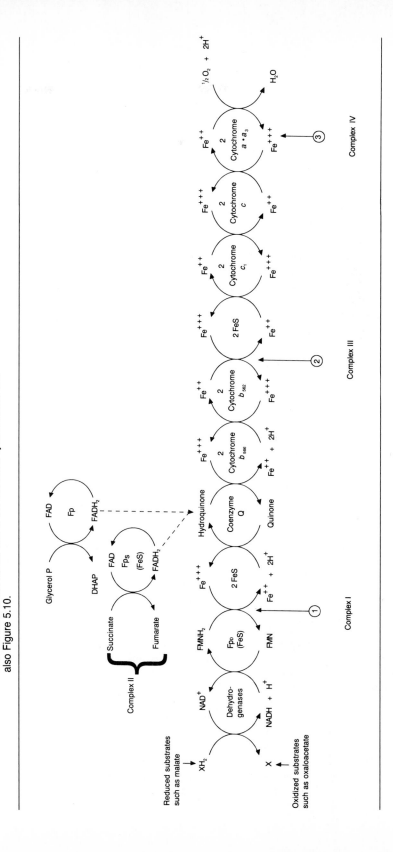

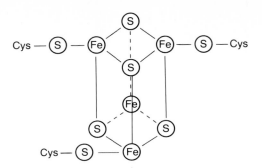

The iron atoms of these substances can undergo oxidation to Fe^{+++} or reduction to Fe^{++}, so they can serve as accessory electron-transferring agents in both mitochondrial and microsomal respiratory chains. The microsomal respiratory chain functions primarily to hydroxylate various substrates, including drugs (see p. 203).

Ferredoxins, such as adrenodoxin, are found in microsomal respiratory chains. They contain fewer atoms of iron and sulfur than the corresponding mitochondrial components. The general nature of the ferredoxin iron-sulfur center is:

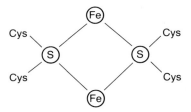

Coenzyme Q, or ubiquinone Coenzyme Q (CoQ), or ubiquinone, is a nonprotein isoprenoid quinone (Figure 5.4). CoQ connects the flavoproteins of complex I and complex II to the cytochromes (see Figure 5.2). This coenzyme exists in various forms, but the one found in mammalian tissues is often CoQ_{10}. The subscript refers to the number of isoprenoid units in the side-chain (Figure 5.4). Bacterial and plant forms of CoQ have fewer isoprenoid units.

CoQ is a small, mobile molecule weakly associated with the inner mitochondrial membrane. It acts as a collector of all the hydrogens that enter the respiratory chain. When CoQ is reoxidized, electrons, rather than hydrogen atoms, are passed to the next component of the chain. The protons are liberated to the mitochondrial matrix. The antibiotic piericidin (Figure 5.5) is a structural analogue of CoQ that blocks the reoxidation of NADH.

Figure 5.3   Oxidized and reduced forms of flavin nucleotides. The numbering
                            system of the isoalloxazine ring is shown at left.

Figure 5.4   Structure of coenzyme Q *(left)* and its reduction product *(right)*.

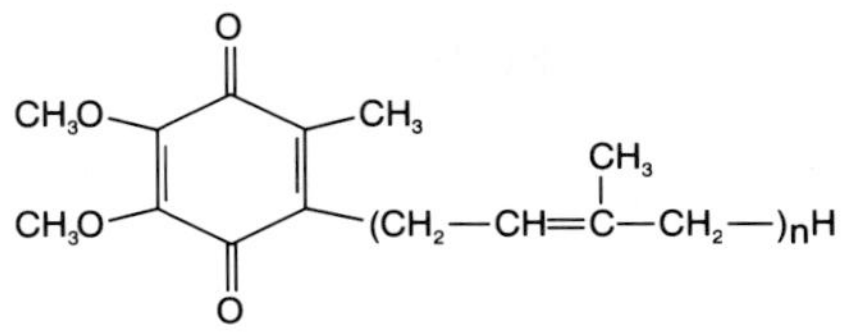

CoQ_6 (n=6)

CoQ_8 (n=8)

CoQ_{10} (n=10)

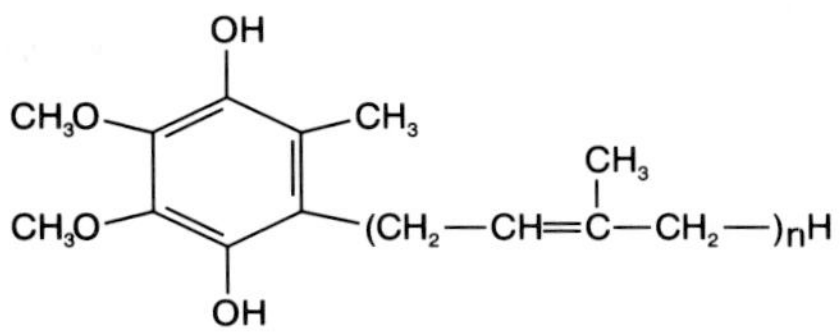

(A dihydroquinone)

Figure 5.5   Structure of piericidin.

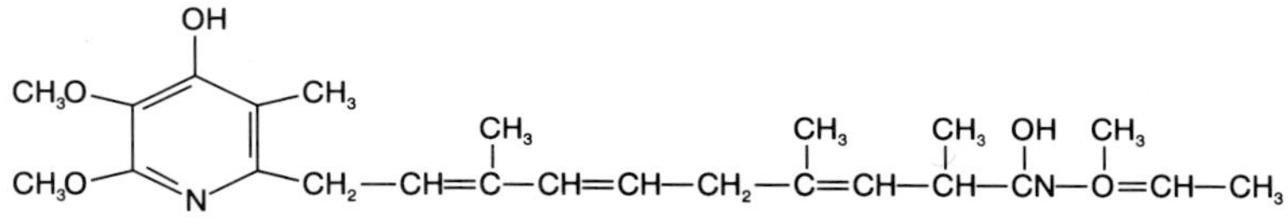

The cytochromes At least four distinct types of cytochromes exist in the mitochondrial respiratory chain. The cytochromes are electron-transferring agents that contain iron bound in a prophyrin ring. This ring resembles the heme ring of hemoglobin, except the side-chains differ. The iron of the cytochromes can undergo reversible change from Fe^{+++} to Fe^{++} and thus are one-electron carriers. Cytochrome c is a small heme protein that is readily extracted from mitochondria; the other cytochromes are tightly bound, intrinsic membrane proteins and are difficult to remove. Cytochromes a, a_3, c_1, and b are spatially oriented across the thickness of the mitochondrial inner membrane.

Other types of cytochromes include cytochrome P_{450}, a component of microsomal, nonphosphorylating respiratory chains (see p. 203). Another, cytochrome b_5, is found in the outer mitochondrial membrane as well as in the microsomes. Both cytochromes are associated with rather specific reductases. These enzyme complexes promote the desaturation or ω-hydroxylation of certain fatty acids.

Some components of the mitochondrial respiratory chain are more abundant than others. For example, the molar ratios of the cytochromes are not identical (Table 5.2), and the relative excess of CoQ is notable. The high ubiquinone concentration may relate to its being less firmly membrane bound than the other components.

Clinical comment

Carbon monoxide and cyanide poisoning As with hemoglobin, the cytochromes can react with either carbon monoxide or cyanide ions. This interferes with the flow of electrons to oxygen and can lead rapidly to death. The best treatment for carbon monoxide poisoning is the administration of high concentrations of oxygen. Cyanide poisoning is treated by administering thiosulfate to convert cyanide to thiocyanate ions. In either situation methylene blue is also given. This substance has an oxidation-reduction potential close to that of the cytochromes, so it forms an artificial bypass for electrons around the poisoned respiratory chain.

Table 5.2　　　Relative Composition of Mitochondria and Submitochondrial Particles

	Submitochondrial particles*		Intact mitochondria
	μmol/g protein	Molar ratio†	μmol/g protein
Cytochrome aa_3	6.08	4	1.30
Cytochrome a_3	3.10	2	—
Cytochrome a	2.98	2	—
Cytochrome c	2.68	1.7	0.45
Cytochrome c_1	1.48	1	0.21
Fp_b	—	—	0.20
Fp_s	—	—	0.50
Nonheme iron	—	—	3.30
Cu^{++}	—	—	1.5
CoQ	—	10-12	3.4-4.0

*Submitochondrial particles are prepared by removing the outer mitochondrial membranes and all the soluble dehydrogeneses.
†The molar value of cytochrome aa_3 is arbitrarily taken equal to 4 to avoid fractions in the remaining values.

Reduction of oxygen Oxidative phosphorylation consumes most of the O_2 of cellular respiration. Through the reduction of O_2, energy is trapped as ATP and water is formed. The complete reduction of one O_2 molecule requires four electrons.

$$4e^- + O_2 + 4H_2O \longrightarrow 2H_2O$$

This might be accomplished by passing two pairs of electrons through the electron transport chain; however, the direct insertion of a pair of electrons into the O_2 molecule is very slow because the unpaired electrons of the O_2 molecule are in separate orbitals and have parallel spin states. The O_2 molecule more likely is reduced by the successive addition of four electrons, one at a time. The exact mechanism of the four electron reduction of O_2 is not known, however, since the partially reduced O_2 intermediates usually stay tightly bound to cytochrome oxidase.

Oxygen toxicity: peroxisomes and the superoxide radical If the reduction of O_2 proceeds by single electron additions, one intermediate would be the superoxide radical, O_2^-. Other reduced products would include H_2O_2 (the addition of two electrons per molecule) and the hydroxyl radical $OH\cdot$. Hydrogen peroxide is a toxic substance, but its concentration is kept very low by the action of catalase, the enzyme that decomposes it to form water and free molecular O_2. This is shown as follows:

$$2H_2O_2 \xrightarrow{\text{Catalase}} 2H_2O + O_2$$

Catalase may constitute as much as 40% of the total protein contained in the organelles known as *peroxisomes*. Peroxisomes are abundant in liver and many other tissues. They contain D–amino acid oxidase and α-hydroxyacid oxidase. Animal, but not human, peroxisomes also contain urate oxidase. Each of these enzymes produces hydrogen peroxide; however, catalase, in the same organelle, promotes the efficient destruction of hydrogen peroxide.

Superoxide dismutase The superoxide radical is more toxic than previously believed. This radical is produced as a by-product of many oxidative reactions, but most of it probably arises as an aberration of the mitochondrial electron transfer chain. It has been estimated that as much as 5% of the O_2 consumed in respiration might be converted to O_2^- in young rat heart mitochondria. This increases with the animal's age. Fortunately, all aerobic cells contain enzymes, *superoxide dismutases,* that scavenge and detoxify O_2^- by catalyzing the dismutation reaction:

$$2O_2^- + 2H^+ \longrightarrow H_2O_2 + O_2$$

These enzymes are found in species ranging from bacteria to mammals. They are found in mitochondria, the cytosol of liver, in erythrocytes and in other tissues. Superoxide dismutases are metalloenzymes, but the metal requirement depends on the enzyme source.

Clinical comment

Oxygen toxicity in premature infants Breathing high concentrations of O_2 over prolonged periods can be dangerous. For example, pure O_2 is lethal to rats and can cause blindness in premature infants. O_2^- is an important cause of O_2 toxicity. Premature infants do not have a fully developed capacity to produce superoxide dismutases and, as a result, may develop retrolental fibroplasia if they are administered high concentrations of O_2 over time.

Clinical comment

Chronic granulomatous disease Active phagocytosis by macrophages involves a burst of O_2 uptake that increases O_2^-. Chronic granulomatous disease results from a rare genetic defect in which cells of the immune system (neutrophils, eosinophils, monocytes, macrophages) do not take up O_2 rapidly during phagocytosis. It is thought that these diseased cells have difficulty killing the organisms they engulf because they have a diminished capacity to make O_2^- and from it, H_2O_2. (See the reference by Curnutte and Babior for more about the biochemistry and genetics of this disease.)

Several explanations may account for the toxicity of O_2^-. Toxicity might cause single-stranded breaks in deoxyribonucleic acid (DNA), depolymerize acid mucopolysaccharides, or initiate the peroxidation of membrane lipids. The antibiotic *streptonigrin* is lethal to sensitive microorganisms such as *Escherichia coli* because the drug greatly increases production of O_2^- by neutrophils. It has also been suggested that one function of the large quantity of glutathione found in erythrocytes is to reduce O_2^-. Clearly the role of O_2^- and its dismutases in human diseases is still poorly understood.

Lipid matrix Mention should be made of the mitochondrial matrix in which the respiratory enzymes are located. The matrix of mammalian mitochondria contains up to 26% by weight of phospholipid. One acidic phospholipid, cardiolipin (see Chapter 10), accounts for almost one sixth of the total phospholipid. This substance is virtually specific for mitochondria, especially those of the heart, where it probably binds the enzymes to the respiratory chain and facilitates the movement of protons through the matrix.

Respiratory chains of the endoplasmic reticulum

Mechanically disrupted endoplasmic reticulum consists of fragments called *microsomes.* Microsomes contain respiratory chains, but their function and structure are unlike those of mitochondria. *Whereas the major function of mitochondria is to couple substrate oxidation to the generation of ATP, microsomal systems hydroxylate several substrates.* For example, steroid nuclei are hydroxylated to produce bile acids or steroid hormones (see Chapter 18), and fatty acids are ω-hydroxylated. Also, foreign substances and many drugs are inactivated by hydroxylation.

Cytochrome P_{450} Microsomal respiratory chains contain cytochrome P_{450}. The 450 subscript indicates that the cytochrome–carbon monoxide complex absorbs light at 450 nm. At least eight forms of cytochrome P_{450} exist, each with limited specificity for the class of compounds to be hydroxylated. As with other cytochromes, cytochrome P_{450} is a one-electron carrier.

Cytochrome P_{450} is a terminal component in the microsomal electron-transport chain. This electron-transfer chain is illustrated in Figure 5.6. In the diagram AH represents a steroid, drug, or other substrate that can be hydroxylated by this enzyme complex. Electrons from NADH are transferred to a flavoprotein (Fp), which is reduced. The reduced Fp is oxidized in a reaction that forms two equivalents of a reduced ferridoxin, adrenodoxin. The reoxidation of the reduced adrenodoxin is complex in that one electron reduces Fe^{+++} to Fe^{++} on the cytochrome P_{450} associated with the substrate, and the other electron reduces O_2 to enzyme-bound O_2^-. One O_2 atom of O_2^- is used to form water, and the other O_2 atom becomes part of the hydroxy group on the steroid product. The overall reaction can be summarized as:

$$NADPH + H^+ + Steroid + O_2 = NADP^+ + Steroid\text{-}OH + H_2O$$

Until recently it was believed these hydroxylating systems were unique to the endoplasmic reticulum. Now mitochondrial ferredoxins and cytochromes P_{450} strongly suggest that mitochondria have both types of respiratory chains. However, the major mitochondrial respiratory function is the generation of ATP, not the hydroxylation of metabolites.

High-energy compounds

The exchange of energy between energy-yielding and energy-using reactions usually requires ATP. That is, an energy-yielding reaction might lead to the synthesis of ATP, whereas an energy-requiring reaction will use ATP. The oxidation of carbohydrates, proteins, or fats yield ATP; muscular exercise, movement, ion transport, or the beating of a heart consume ATP. Consequently, we think of ATP as a "high-energy" compound.

Figure 5.6

Hydroxylation of a steroid or drug by the microsomal respiratory chain. AH represents the steroid or drug and AOH the hydroxylated product; P_{450} indicates cytochrome P_{450}, F_p stands for ferredoxin reductase and $F_R(FeS)_2$ represents ferredoxin. Similar mechanisms use adrenodoxin in place of ferrodoxin.

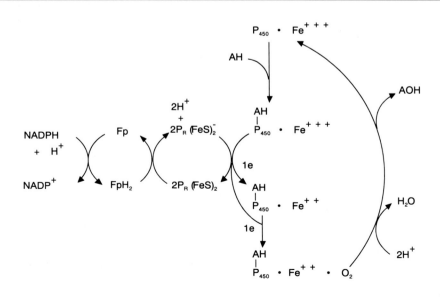

High-energy compounds in biologic systems have high negative free energies of hydrolysis; for example, the hydrolysis of ATP to ADP and P_i has a $\Delta G^{0'}$ of approximately -29.3 kJ/mol (-7 kcal/mol). The bond broken in the hydrolytic reaction is often called a high-energy bond, which is sometimes indicated by a squiggle bond, $\sim$.

Hydrolysis of a low-energy or ordinary bond yields a $\Delta G^{0'}$ of -5 to -20 kJ/mol (-1 to -5 kcal/mol). Hydrolysis of a high-energy bond yields a $\Delta G^{0'}$ of -20 to -60 kJ/mol (-5 to -15 kcal/mol). Although this division is arbitrary, experience indicates that it is useful for classifying the bonds of most compounds as low- or high-energy bonds. Typical data for some high- and low-energy bond types are given in Table 5.3.

All but two of the high-energy compounds listed in Table 5.3 contain phosphate groups. However, not all phosphate-containing compounds are high-energy compounds. Phosphorus-containing high-energy compounds include *enol phosphates, acyl phosphates,* or

Table 5.3 Typical High- and Low-energy Compounds

General formula	Designation	Biochemical example	$\Delta G^{0'}$ kJ/mol	kcal/mol
High-energy compounds				
$\begin{array}{c} NH \\ \| \\ R-C-N\sim PO_3^= \\ \| \\ H \end{array}$	Guanidinium phosphate	Creatine phosphate	-43.9	-10.5
$\begin{array}{c} CH_2 \\ \| \| \\ R-C-O\sim PO_3^= \end{array}$	*Enol* phosphate	Phospho*enol* pyruvate	-61.9	-14.8
$\begin{array}{c} O \\ \| \| \\ R-C-O\sim PO_3^= \end{array}$	Acyl phosphate	Acetyl phosphate	-41.8	-10.1
$\begin{array}{c} O \quad\quad O \\ \| \| \quad\quad \| \| \\ R-P\sim O-P-O^- \\ \| \quad\quad\quad \| \\ O^- \quad\quad O^- \end{array}$	Pyrophosphate	Adenosine triphosphate Uridine diphosphoglucose	-29.3 -30.5	-7.0 -7.3
$\begin{array}{c} O \\ \| \| \\ R-C\sim S-R \end{array}$	Acyl thioester	Acetyl CoA	-31.4	-7.5
$\begin{array}{c} R-S^+-R \\ \| \\ R \end{array}$		*S*-Adenosyl methionine	-29.3	-7.0
Low-energy compounds				
$\begin{array}{c} O \\ \| \| \\ -C-N-R \\ \| \\ H \end{array}$	Peptide bond	Glutathione	-2.1	-0.5
$\begin{array}{c} O \\ \| \| \\ R-C-O-R \end{array}$	Ester	Triacylglycerol	-7.5	-1.8
$\begin{array}{c} O^- \\ \| \\ R-CH_2-O-P-O^- \\ \| \| \\ O \end{array}$	Sugar phosphate	Glucose 6-phosphate	-12.1	-2.9

pyrophosphates. These are often simple or mixed acid anhydrides. A simple acid anhydride is produced when the elements of water are abstracted from two molecules of the same acid, such as acetic anhydride. A mixed acid anhydride results when the elements of water are abstracted from two different acid molecules, such as acetyl phosphate.

A high-energy bond is the result of several factors, including increased ionization of charged groups generated by hydrolysis and an increase in the resonance stabilization of the products compared to the substrate.

Forty percent of food energy can be conserved as high-energy compounds. Conservation is accomplished by coupling the mitochondrial respiratory chain to the phosphorylation of ADP, yielding ATP. In coupled reactions, as in others, the laws of conservation of energy must be obeyed. Table 5.3 indicates that the free energy of hydrolysis of ATP to ADP is approximately -29.3 kJ/mol (-7.0 kcal/mol). The reverse of this reaction requires at least 29.3 kJ/mol for the formation of ATP.

Oxidative phosphorylation

The terminal stages of biologic oxidation catalyzed by the respiratory chain provide three separate sites where the energy change is sufficient to drive the synthesis of ATP from ADP and P_i. ATP synthesis requires coupling to an oxidation step having a change in standard reduction potential of at least 0.15 volt. This is a fairly large energy change, and not many sites in the respiratory chain have potential differences of this magnitude. The three sites are indicated in Figure 5.2 by the vertical arrows. Thus, for each mole of NADH that enters the respiratory chain, 3 moles of ATP are produced. Since this process also involves the consumption of 1 atom of oxygen, the ratio of phosphorylated P atoms to oxygen atoms reduced is 3; that is, the P/O ratio is 3. Figure 5.2 shows that entry to the respiratory chain by way of succinate entails a distinct disadvantage, since this entry misses the first site of phosphorylation. Therefore the oxidation of succinate provides a P/O ratio of only 2.

Control of oxidative phosphorylation

Several points exist at which high-energy bond synthesis might be controlled. For example, the concentrations of substrates, including O_2 and ADP, and the concentration of the product ATP all influence oxidative phosphorylation.

Changes in O_2 uptake by suspensions of mitochondria can be monitored after the addition of various substrates by using the so-called O_2 electrode. Such an experiment is shown in Figure 5.7. At the upper left, mitochondria (Mito) are added to the solution. A decrease in the partial pressure of O_2 occurs with time; this is indicated by the plotted decrease in O_2 concentration. The slow rate of O_2 uptake is caused by the lack of both an oxidizable substrate and ADP. When a suitable substrate is added, O_2 uptake increases only slightly because ADP is still limiting. When ADP is then added, O_2 is rapidly consumed for a brief period, and then its concentration levels off when all of the added ADP is converted to ATP. A second addition of ADP stimulates, but a third addition produces only a slight response. By this time all of the substrate has been oxidized. When more substrate is added, respiration again increases until the O_2 dissolved in the suspending solution is exhausted. Although isolated mitochondria, free of cells, were used in this experiment, similar reactions occur in vivo.

Concept of energy charge

Energy charge is a measure of the relative concentration of high-energy adenylate phosphate compared to the total concentration of adenine mononucleotides. Numerically it is defined as half the average number of pyrophosphate bonds per mole of adenine nucleotide divided by the sum of the molar concentration of all adenine nucleotides. The energy charge is expressed as:

$$\text{Energy charge} = \frac{1}{2} \frac{([ADP] + 2[ATP])}{[AMP] + [ADP] + [ATP]}$$

Figure 5.7

Respiration of isolated mitochondria. As added ADP is converted to ATP, O₂ uptake is greatly slowed.

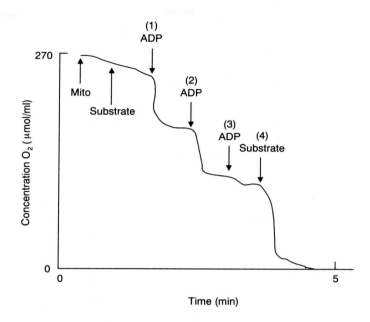

The denominator is the sum of the molar concentration of all the adenine nucleotides in the system, whereas the numerator is that fraction of the total that could, on hydrolysis, release energy equivalent to one or more high-energy bonds. The numeric factor is introduced arbitrarily to make the values of the energy charge range from 0 to 1. For example, if a system contains only ATP, the energy charge would be:

$$\text{Energy charge} = \tfrac{1}{2}\,\frac{2[\text{ATP}]}{[\text{ATP}]} = 1$$

The ADP concentration can control the mitochondrial respiratory rate. When reactions that generate ATP predominate, the concentration of ADP will drop and respiration will diminish; however, if ATP-consuming reactions predominate, the concentration of ADP will rise, as will the rate of respiration. Under conditions approximating the steady state in cells, the energy charge is approximately 0.8. At an energy charge of 1, only ATP is present and oxidative phosphorylation would cease for lack of acceptor ADP. When only ADP is present, the energy charge is 0.5. A value of 0 represents the point where all oxidation ceases, with a consequent cessation of phosphorylation. In fact, the physiologic range is narrow, approximately 0.6 to 0.9. Even in a tissue as active as rat myocardium, the total adenine nucleotide concentration is only 16 μmol/g of wet tissue. This very small amount of adenine nucleotide acts both as the major vehicle for capturing useful energy and as an important agent for controlling O₂ consumption.

Inhibition and uncoupling of oxidative phosphorylation

There are three classes of inhibitors of oxidative phosphorylation.

1. Site-specific inhibitors Although three sites of oxidative phosphorylation exist in the respiratory chain, the three sites are not equivalent because certain compounds inhibit at specific sites in isolated mitochondria. Some properties of a few such inhibitors are given in Table 5.4. Sometimes these are employed as drugs, but the pharmacologic effects are not necessarily related to their action as inhibitors of oxidative phosphorylation.

2. Non–site-specific inhibitors Another set of inhibitors act indiscriminately at any of the three sites. The latter compounds block phosphorylation directly, but in a tightly coupled system oxidation is blocked because of inhibited phosphorylation. Several, but not all, of the non–site-specific inhibitors are antibiotics, such as oligomycin and aureovertin; others, such as atractylate, are toxic plant glycosides.

3. Uncouplers of oxidative phosphorylation Uncoupling agents either loosen or completely destroy the normal tight coupling of oxidation and phosphorylation. Thus the P/O ratio is greatly lowered, sometimes to zero. When uncoupled, oxidation rates increase while phosphorylation decreases. The result is the production of extra heat. Extra heat production might be manifested as *fever*. The diminished formation of ATP resulting from uncoupling can indirectly affect many cellular processes, such as ion transport and membrane permeability.

Uncoupling agents are a diverse group of compounds that are broadly described as lipophilic proton donors or proton acceptors. These agents render the inner mitochondrial membranes abnormally permeable to protons. Common uncouplers include 2,4-dinitrophenol, dicumarol, certain substituted salicylanilides, and even free salicylate, a metabolite of aspirin. The salicylanilides are the most potent. Natural uncouplers include bilirubin, free fatty acids, and perhaps thyroxine; however, these compounds normally are not present in mitochondria at high enough concentrations where they would act as uncouplers.

People are exposed to other substances in the environment that might act as uncouplers. For example, soluble toxins produced by some pathogenic microorganisms can act as uncouplers and contribute to fever. Insecticides used around the preparation of food could be a potential hazard, but since the toxicity of these agents is about the same for humans as it is for the housefly, our larger body mass requires correspondingly larger amounts for uncoupling. Thus human intoxication in this situation is rare.

Clinical comment   **Is bilirubin an uncoupler in erythroblastosis fetalis?** Infants suffering from erythroblastosis fetalis, an Rh incompatibility, may have plasma bilirubin concentrations of 0.7

Table 5.4		Site-specific Inhibitors of Oxidative Phosphorylation	
	Site	**Nature of inhibitors**	**Remarks**
	1	Alkyl guanidines	Guanethidine, a hypotensive agent
		Rotenone	Insecticide
		Chlorpromazine	Tranquilizing drug
		Barbiturates	Seconal, Amytal, etc., used as sleep-inducing drugs
		Progesterone	Hormone, inhibits oxidation but to a lesser degree than it inhibits phosphorylation
	2	Antimycin A	
		BAL	British Anti-Lewisite, an antidote to an old war gas, actually 1,2-dithioglycerol
		Phenethylbiguanides	Phenformin, used as an oral insulin substitute
		Naphthoquinone	
		Hydroxyquinoline-8-oxide	
	3	Carbon monoxide	
		Cyanide ion	
		Azide ion	

mmol/L or more. At these concentrations bilirubin can penetrate the underdeveloped blood-brain barrier and cause permanent brain damage. Although the mechanism of the effect of bilirubin on brain cells is very controversial (see Ostrow JD, editor: Bile pigments and jaundice, New York, 1986, Marcel Dekker, Inc.), some evidence suggests that bilirubin might produce damaging effects by acting as an uncoupler of oxidative phosphorylation.

Structure of the mitochondrial inner membrane

For many years scientists searched diligently, but unsuccessfully, to find soluble high-energy intermediates that would couple oxidation to phosphorylation. To understand our present ideas about how these processes are coupled requires that we examine more closely the physical nature of the mitochondrial inner membrane.

Combined biochemical and electron microscopic data have been useful in understanding the structure and function of the mitochondrion. This complex association of structural and functional entities is illustrated by Figure 5.8, *A*, which represents a cross section of a mitochondrion.

The outer membrane contains an assemblage of enzymes and a pool of coenzymes, including NAD^+ and coenzyme A. These coenzymes do not mix with a similar pool of coenzymes found in the matrix. The matrix is that part of the mitochondrion that is enclosed by the inner membrane. The matrix contains the enzymes of the Krebs cycle, except for succinate dehydrogenase, an enzyme tightly bound to the inner membrane itself. Sometimes the matrix also contains electron-dense granules of calcium phosphate. These granules sequester calcium ions and probably serve as a source of calcium when needed. The inner membrane contains the respiratory chains and numerous translocating systems that regulate the flux of inorganic ions and substrates into and out of the matrix. The inner membrane also contains on the matrix side the ATP-synthesizing structure, often referred to as the F_1 ATPase. This system appears in electron micrographs as knoblike protrusions into the matrix (Table 5.5).

Figure 5.9 illustrates the functional components of the inner membrane. A diagram of

Figure 5.8

Schematic view of mitochondrial membranes. *A*, Relationship of the outer and inner membranes to the matrix. *B*, Location of F_1 ATPase in respect to the inner membranes. *OSCP*, Oligomycin sensitivity-conferring protein.

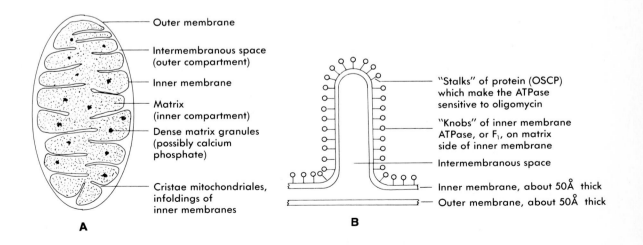

Outer membrane

Intermembranous space (outer compartment)

Inner membrane

Matrix (inner compartment)

Dense matrix granules (possibly calcium phosphate)

Cristae mitochondriales, infoldings of inner membranes

"Stalks" of protein (OSCP) which make the ATPase sensitive to oligomycin

"Knobs" of inner membrane ATPase, or F_1, on matrix side of inner membrane

Intermembranous space

Inner membrane, about 50Å thick

Outer membrane, about 50Å thick

A   **B**

Figure 5.9   Architecture of the mitochondrial inner membrane. The view is of a small portion of crista. A respiratory chain is shown to the right and in the vectorial arrangements believed to exist. The F_1 ATPase is shown at the lower left. (Courtesy Dr. Peter Hinkle, Cornell University; see also Hinkle PC and McCarty RE: Sci Am, March 1978, p 104.)

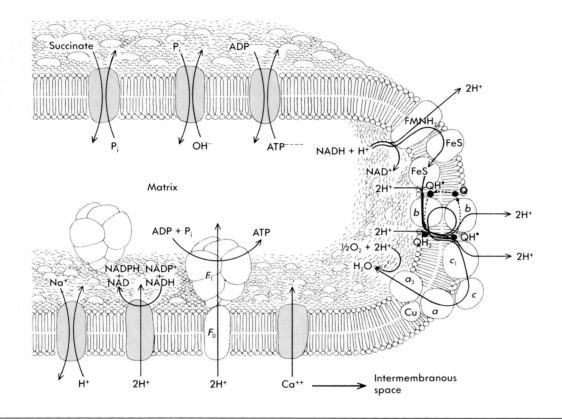

Table 5.5   Location of Some Mitochondrial Enzymes and Coenzymes

Region	Enzyme, coenzymes
Outer membrane	Coenzyme pool
Inner membrane	Pyruvate dehydrogenase complex
	Respiratory chain
	F_1 ATPase
	Succinate dehydrogenase
Matrix	Coenzyme pool
	Most Krebs cycle enzymes
	Calcium phosphate granules

a respiratory chain is depicted at the right. The proteins that make up its beginning (NADH-CoQ reductase) and end (cytochrome a, a_3, or cytochrome oxidase) face the matrix and are tightly bound intrinsic proteins. Cytochrome c faces out toward the intermembranous space; it is an extrinsic protein, readily removed by salt solutions. The flavoproteins, the iron-sulfur proteins, and cytochrome b are intrinsic proteins. CoQ is soluble and mobile in the lipid components of the membrane.

Figure 5.10 Intermediates of the Krebs cycle. Enzymes are omitted for clarity.

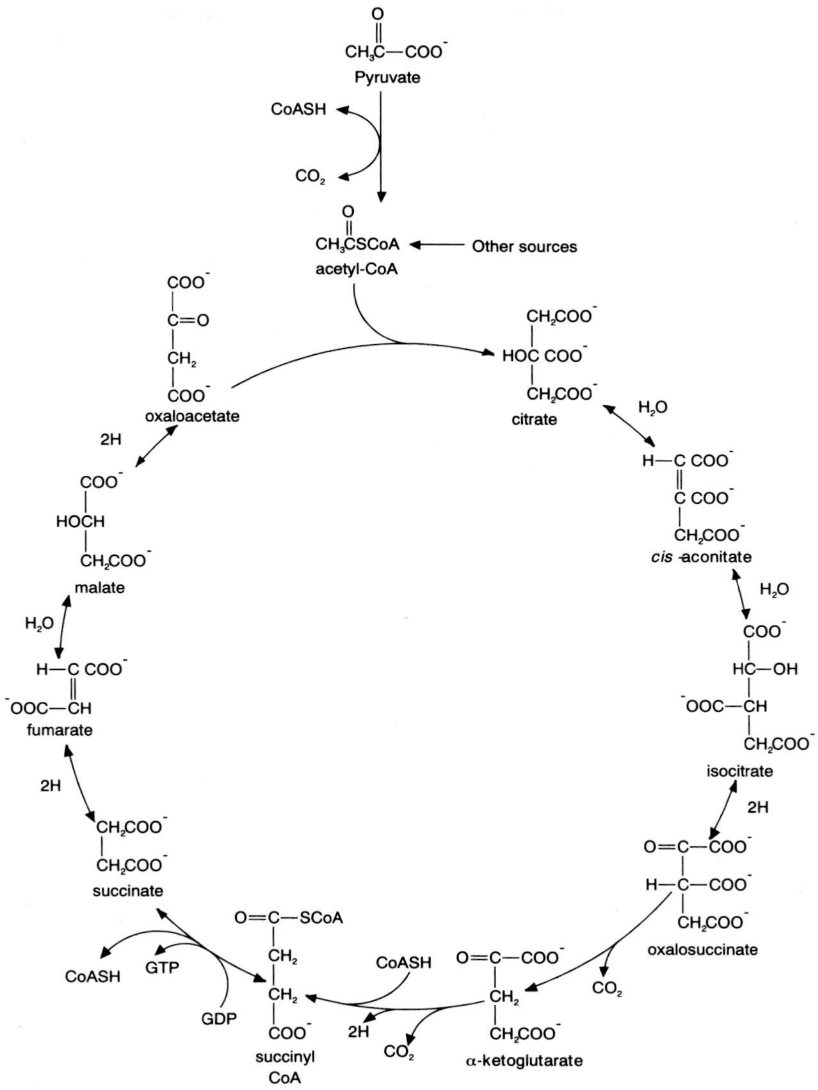

The ATP synthetase The ATP synthetase, sometimes called F_1 ATPase for the reverse reaction, resides in the knoblike structures illustrated both in Figures 5.9 and 5.10. These knobs are labeled F_1. Each knob is joined to the inner membrane by a set of hydrophobic proteins, labeled F_0 in the drawing (Figure 5.10). Translocating systems that promote the flux of substance through the membrane are shown on either side of the F_1 knob. Certain fractions of the ATPase synthetase can be separated from each other and recombined to reconstitute oxidative phosphorylation; these are F_0 and F_1. The F_0 fraction is a transmembrane structure located at the base of the knoblike structure. F_0 is connected to the knob by a stalk. The stalk contains several proteins, one of which (OSCP in Figure 5.8) makes the system sensitive to oligomycin. Oligomycin is an antibiotic that blocks phosphorylation, presumably by preventing reentry of protons into the matrix through the proton channel in F_0. F_1 ATPase catalyzes the synthesis of ATP in the reverse direction. F_1 contains pairs of the following subunits: α, β, γ, δ, and ϵ. F_1 also contains tightly bound ATP. Solubilized F_1 can hydrolyze ATP but cannot operate in the reverse direction, since the isolation process strips away F_0 proteins, which, in the membrane-bound state, promote the enzymic synthesis of ATP.

Mechanism of oxidative phosphorylation

The chemiosmotic theory of Peter Mitchell is a widely accepted mechanism of oxidative phosphorylation. Its basic tenet is that the energy-linked phosphorylation of ADP is driven by a proton gradient across the inner mitochondrial membrane. The oxidation of substrates causes the expulsion of protons from the matrix to the intermembranous space, raising the proton concentration in the latter and causing it to become more acid. We previously mentioned that concentration gradients are a means of storing energy. The negative free energy associated with the dilution of protons returning to the matrix space through F_0 and F_1 drives the synthesis of ATP.

In short, the driving force that couples oxidation to phosphorylation is the proton gradient. It is exactly for this reason that the process *must* be membrane bound; absence of the membrane would allow diffusion of protons, and no concentration gradient could exist. It is also clear that prior efforts to identify soluble high-energy chemical intermediates failed because soluble systems, which lack membranes, cannot support a concentration gradient.

The basic postulates of the chemiosmotic theory of oxidative phosphorylation are:
1. The system must be membrane bound.
2. Ions are transported by energy-linked systems. Free diffusion of ions does not occur.
3. The proton-translocating oxidation-reduction system, that is, the respiratory chain, pumps protons to the exterior of the membrane during oxidation. This concentration phase requires an input of energy.
4. F_1 synthesizes ATP as a result of a reverse flow of protons.
5. Exchange diffusion systems link proton translocation to the exchange of selected ions. (See also the discussion of membrane function in Chapter 13.)

Considering these postulates, the right side of Figure 5.9 shows that NADH passes two electrons and a proton to the flavoprotein. A second proton from the matrix fluid is added, and both protons are expelled across the inner membrane. The two electrons, which cannot cross the membrane, are passed through the iron-sulfur proteins to CoQ, together with another pair of protons from the matrix. These convert the quinone to the semiquinone form, $CoQH\cdot$, which accepts another pair of electrons from cytochrome b and an additional proton from the matrix.

Because CoQ is a small molecule with a high lipid solubility, many believe that it freely passes from one surface of the membrane to the other, thus allowing formation of the fully reduced form of the coenzyme, $CoQH_2$. With oxidation to CoQ, two protons are expelled, and the electrons pass through the series of cytochromes from cytochrome c_1 through cytochrome a, a_3, or cytochrome oxidase. They are taken up by oxygen and, together with a pair of protons from the matrix space, form a molecule of water. Thus, for each molecule of NADH oxidized via the respiratory chain, one pair of matrix protons is used to make one water molecule and two pairs of protons are expelled from the matrix. This forms a proton gradient.

As protons return to the matrix to establish ionic and pH balance, ATP is formed by the coupled F_1 system. The direct coupling of the protons released by electron transport from NADH does not provide sufficient H^+ to drive ATP synthesis, since approximately three or four protons need to pass back into the mitochondria to provide energy for the synthesis of one ATP. The additional protons are probably expelled from the matrix following conformational changes in the highly organized electron transport system. These conformational changes can be visualized as energy-conserving transformations that are tightly linked to the release of protons into the intermembranous space.

The Krebs cycle

Humans eat a large variety of complex plant and animal foods. Digestion reduces this food to a smaller number of simpler substances. These products often are sugars or organic acids. The terminal stages in the oxidation of these substances occurs in a cyclic series of enzymic reactions. This cyclic series of reactions is called the *tricarboxylic acid (TCA) cycle,* the *citric acid cycle,* or the *Krebs cycle.* These names are used because tricarboxylic acids, including citric acid, are prominent intermediates of the cycle, and because Hans Krebs made important experimental and conceptual contributions to the understanding of this cycle.

The tricarboxylic acid cycle, or Krebs cycle, serves five major functions:

1. It produces most of the carbon dioxide made in human tissues.
2. It is the source of much of the reduced coenzymes that drive the respiratory chain to produce ATP.
3. It converts excess energy and intermediates to the synthesis of fatty acids.
4. It provides some of the precursors used in the synthesis of proteins and nucleic acids.
5. Its components control directly (product-precursor) or indirectly (allosteric) other enzyme systems.

The Krebs cycle deals with those features that are common to the metabolic breakdown of carbohydrates, fatty acids, and amino acids. The cycle can be described as a veritable biochemical traffic circle; material coming to it from carbohydrate sources might leave it to form fat, whereas material coming to it from amino acids might leave it to form carbohydrate. The only "road" closed is that leading from fat to carbohydrate.

Cellular location of the Krebs cycle

In mammalian cells all the enzymes of the Krebs cycle are located in mitochondria (Table 5.6). Passage of materials to the inside of the mitochondria, or from the inside to the

Table 5.6   Energetics of Pyruvate Oxidation

Step	$\Delta G^{0'} < 0$		$\Delta G^{0'} > 0$	
	kJ/mol	kcal/mol	kJ/mol	kcal/mol
1. Pyruvate → Acetyl CoA + CO_2	33.5	8.0		
2. Acetyl CoA + Oxaloacetate → Citrate	38.0	9.1		
3. Citrate ⇌ cis-Aconitate			8.5	2.0
4. cis-Aconitate ⇌ Isocitrate	1.9	0.5		
5. Isocitrate → α-Ketoglutarate	7.1	1.7		
6. α-Ketoglutarate ⇌ Succinyl CoA	36.9	8.8		
7. Succinyl CoA ⇌ Succinate	8.9	2.1		
8. Succinate ⇌ Fumarate			0	0
9. Fumarate ⇌ Malate	3.7	0.9		
10. Malate ⇌ Oxaloacetate			28.0	6.7
	−130.0	−31.1	+36.5	+8.7

Net total for all steps $\Delta G^{0'} = -93.5$ kJ/mol (-22.3 kcal/mol)

Net total for Krebs cycle steps $\Delta G^{0'} = -60.0$ kJ/mol (-14.3 kcal/mol)

outside, is a controlled, facilitated process. Mitochondria from different tissues will vary somewhat in their concentration of enzymes or substrates and in the rate at which substances cross the membranes; in general, however, all mammalian mitochondria have similar properties. For example, the enzymes of the cycle are located close to the respiratory chain, on the matrix side of the inner mitochondrial membrane or within the matrix space, loci that facilitate the oxidative aspect of the cycle. Some of the enzymes are bound tightly into the membrane structure, but some are rather easily removed.

Nature of cycle components

Structures of the acid intermediates of the Krebs cycle are illustrated in Figure 5.10. Names of the enzymes that direct the synthesis of these intermediates are omitted for clarity. The overall reaction of the cycle is given by the equation:

$$CH_3COOH + 2O_2 \longrightarrow 2CO_2 + 2H_2O$$

The *net effect* of the cycle is the oxidation of acetic acid to CO_2 and H_2O. The metabolic form of acetic acid used by the cycle is acetyl CoA, which can arise from several sources:

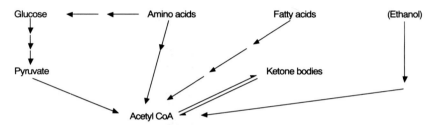

Figure 5.10 shows the system beginnng with pyruvate. Strictly speaking, pyruvate is not a part of the Krebs cycle, but good reasons exist for its inclusion here. First, energy derived from carbohydrate enters the cycle via pyruvate, which is a major source of acetyl CoA. Second, the enzyme complex that decarboxylates pyruvate to acetyl CoA is very similar in subcellular location, composition, and mechanism of action to the α-ketoglutarate dehydrogenase complex of the Krebs cycle, so they are discussed together. Figure 5.11 lists the enzymes that make up the cycle. This figure, together with Figure 5.10, details all the components.

Pyruvate decarboxylation

Pyruvate is the anion of an α-keto acid. Under certain conditions α-keto acids decarboxylate even in the absence of enzymes. The decarboxylation rate is slow in vivo, however, unless catalyzed by the *pyruvate dehydrogenase (PDH) complex*. This enzyme system is called a complex because it contains several enzymes bound together so that they can be isolated as a whole. These particles are found within the mitochondrial inner membrane and have molecular weights of approximately 7 to 8 $\times$ 10^6, large enough to be visible under the electron microscope. Physical and chemical analysis of the particles from *Escherichia coli* reveals that the complex contains a total of 48 individual protein chains:

24 molecules of pyruvate dehydrogenase (E_1)
12 molecules of dihydrolipoyl transacetylase (E_2)
12 molecules of dihydrolipoyl dehydrogenase (E_3)

The E_1 chains each contain a thiamine pyrophosphate, TPP, prosthetic group; the E_2 chains contain lipoic acid covalently bound to the ϵ-amino group of a lysl group; and the E_3 subunits contain a tightly bound FAD. The E_1, E_2, and E_3 subunits derived from *E. coli* can be reassembled in vitro. The mammalian enzyme complex is composed of similar enzymes, but they have not yet been dissociated and reassembled.

A scheme illustrating how the complex operates is shown in Figure 5.12. The drawing

Figure 5.11   Enzymes and coenzymes of the Krebs cycle. Cofactors are shown in parentheses.

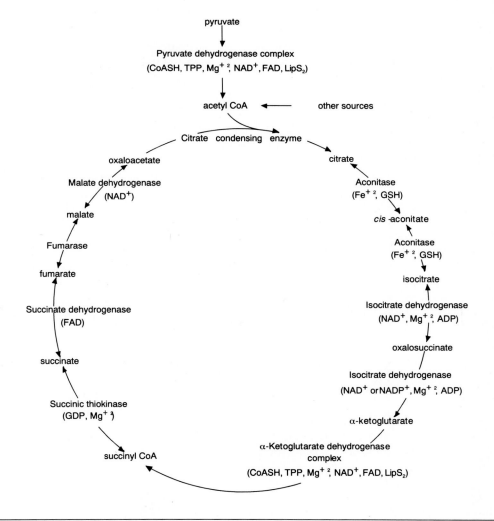

Figure 5.12   Diagram of the action of pyruvate dehydrogenase complex. The three
component enzymes are indicated by E_1, E_2, and E_3. *TPP*,
Thiamine pyrophosphate. Acetylated lipoate and lipoate are shown
attached to E_2.

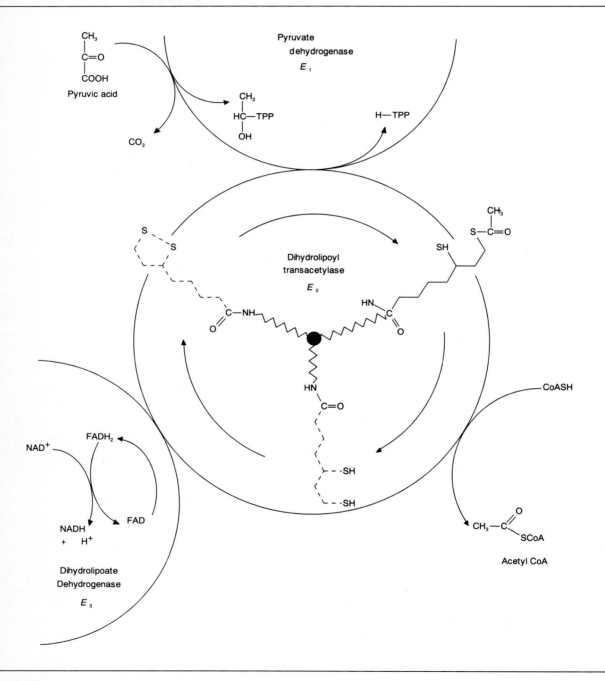

represents only a portion of the entire unit. E_1 signifies the PDH which has an attached α-hydroxyethyl group, a product of the decarboxylation of pyruvate. Next to it, the oxidized lipoate on E_2 swings close to one side of E_1 and to the other side of E_3. As the lipoyl group approaches E_1, the sulfur atom on carbon (C_8) accepts a proton, and the sulfur attached to C_6 accepts the acetyle group, forming an acetyl thioester. The chemical events associated with these two-carbon transfers are illustrated in the following diagram.

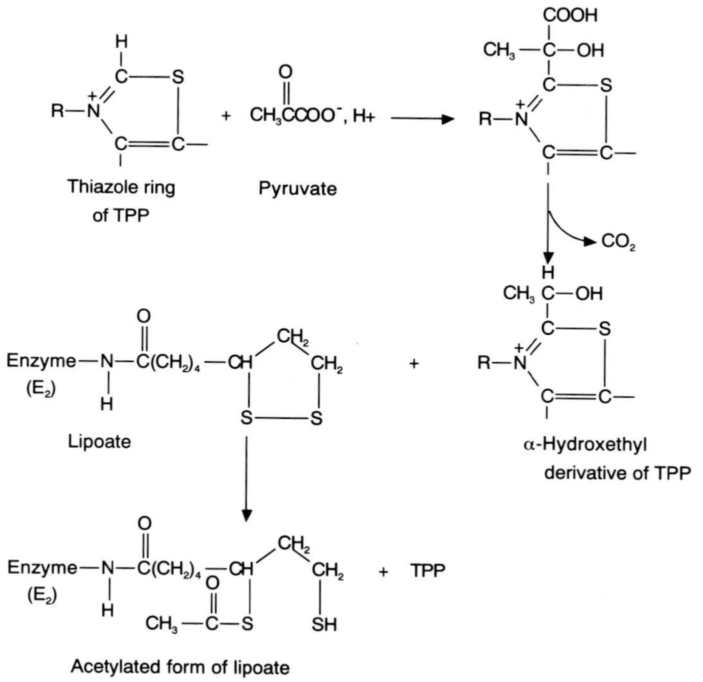

The acetyl thioester is a high-energy compound. In Figure 5.12, as the acetylated lipoate swings away from E_1, E_2 catalyzes the exchange of the acetyl group, with free CoA forming acetyl CoA and dihydrolipoyl-E_2. Acetyl CoA also is a high-energy compound, so the energy released in the decarboxylation reaction is conserved. The lipoate, now in the reduced form, must be oxidized to repeat another round of transacetylation. This oxidation is accomplished by E_3, the dihydrolipoyl dehydrogenase that contains FAD. In the final step, $FADH_2$ is reoxidized by NAD^+, producing NADH, H^+, and FAD. The NADH is ultimately oxidized in the respiratory chain to produce 3 moles of ATP, H_2O, and NAD^+.

Regulation of the pyruvate dehydrogenase complex

Covalent modification of pyruvate dehydrogenase Pyruvate metabolism is regulated in several ways. The mammalian PDH complex is inactivated by phosphorylation and activated by dephosphorylation, as diagrammed in Figure 5.13. These covalent modifications are catalyzed by a *PDH kinase* and a *PDH phosphatase*. The regulating enzymes are integral parts of the dehydrogenase complex. High ratios of $NADH/NAD^+$ or acetyl CoA/CoA are positive effectors of the PDH kinase, but the most crucial stimulating substance is probably high concentrations of ATP. The positive effectors activate the kinase, which phosphorylates the complex, thus shutting it off.

The PDH phosphatase restores the complex to its active state. The phosphatase requires Ca^{++} and Mg^{++} for its activity, and NADH inhibits it. Approximately 50% of the PDH isolated from human tissues is phosphorylated. Simple incubation with Mg^{++} promotes dephosphorylation and reactivation.

Figure 5.13   Covalent modification of the pyruvate dehydrogenase complex. Rapid
regulation is achieved by low molecular weight effectors. These are
shown by broken lines. The phosphorylation-dephosphorylation
mechanism is slower. The PDH kinase is activated by high ratios of
the coenzymes shown and by ATP; this yields a *less* active
(phosphorylated) PDH complex. Factors that stimulate or inhibit the
enzymes are indicated by $+$ and $-$ signs, respectively.

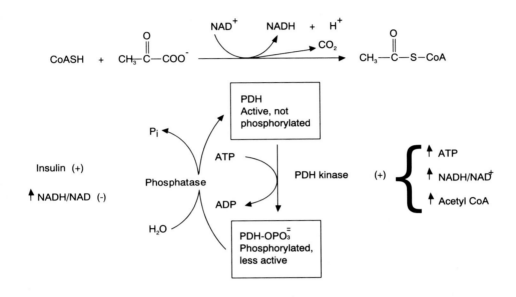

Product inhibition Rapid control of the PDH complex in vivo is by direct product
inhibition. Thus the products of PDH activity, acetyl CoA and NADH, are inhibitory.
Regulation by covalent modification is somewhat slower.

Overview of PDH regulation Two states of the cell are important in the regulation
of PDH by covalent modification. *First, the PDH complex responds to the energy charge
of the cell.* When the ATP concentration is high, glycolysis slows, and the activity of
the PDH complex is lowered. *Second, the PDH complex is sensitive to the oxidation-
reduction state of the cell.* The various intracellular pools of NAD^+, NADH, $NADP^+$,
and NADPH are in equilibrium, to a degree. If these equilibria are altered, as in the
biosynthesis and deposition of fat, the PDH complex will be affected. Strictly speaking,
the PDH complex is not a part of the Krebs cycle, but its regulation enhances the sensitivity
with which the Krebs cycle is controlled.

**Condensing reaction:
start of the Krebs
cycle**

The first reaction involves the condensation of oxaloacetate and acetyl CoA to form
citrate:

Citrate synthase, or the citrate-condensing enzyme, is exclusively a mitochondrial enzyme
in mammals. It is essentially a "one-way" enzyme, since the equilibrium strongly favors
citrate formation. No coenzyme is required.

Figure 5.14　　Three-point attachment of citrate to aconitase, illustrating the biochemical asymmetry of the citrate ion.

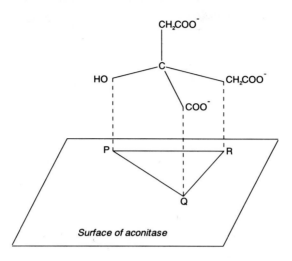

Surface of aconitase

Clinical comment

Fluoroacetate poisoning Citrate synthase is remarkably specific for its substances. Fluoroacetyl CoA is one of a few compounds that can substitute for acetyl CoA. Fluoracetate is sometimes used as a rat poison. Its ingestion by humans can be fatal, since the product of its reaction is fluorocitrate, which cannot be further metabolized by the enzymes of the Krebs cycle. The lethal result emphasizes the importance of the Krebs cycle in metabolism.

The essential nature of the Krebs cycle is borne out by the existence of very few genetic diseases affecting cycle enzymes. Major genetic disturbance of the Krebs cycle often causes severe lactic acidosis and cerebellar ataxia.

Isomerization of citrate

The conversion of citrate to isocitrate takes place in two steps. Both are catalyzed by the enzyme *aconitase*. Aconitase is also highly specific. When aconitase acts on citrate, the elements of water are lost, forming *cis*-aconitate as an enzyme-bound intermediate. Subsequently, *cis*-aconitate is rehydrated to yield isocitrate.

Although citrate has a plane of symmetry such that the terminal carboxylate groups appear to be equivalent, this is not the case from the enzymic point of view. Using labeled citrate, it has been shown that aconitase always operates on that part of the citrate structure derived from oxaloacetate. This is explained by the substrate attaching to the enzyme *at three different binding sites*. This is depicted schematically in Figure 5.14, where citrate is represented in tetrahedral form, with the central carbon atom lying in the center. The idealized surface of aconitase, below the tetrahedron, has the three binding sites identified as P, Q, and R. The dashed lines represent specific affinities between the enzyme and the indicated functional groups of the substrate. Citrate attaches to aconitase *only* if the citrate ion has the orientation shown. This is easily verified with models. Therefore the carbon atoms lost during the first turn of the cycle originated in oxaloacetate, not acetyl CoA.

First decarboxylation

In the next step of the Krebs cycle, isocitrate is oxidized to oxalosuccinate in a reaction mediated by isocitrate dehydrogenase (ICDH). Oxalosuccinate is a transient, enzyme-bound intermediate, which is decarboxylated to α-ketoglutarate.

Isocitrate dehydrogenase occurs in several forms. In most mammalian tissues the extramitochondrial form uses $NADP^+$, whereas the intramitochondrial form uses NAD^+; however, heart mitochondria contain both forms. The NAD^+-linked enzyme exhibits complex allosteric behavior, but stated simply, ADP is a positive effector and ATP a negative effector. ADP causes the enzyme to associate as an active tetramer, whereas ATP dissociates the complex to an inactive dimer.

Another difference between the NAD^+ and the $NADP^+$-linked forms of ICDH is that oxalosuccinate is not a true intermediate of the NAD^+-linked enzyme. Usually oxalosuccinate is shown as a Krebs cycle intermediate because heart mitochondria contain both enzymes. In the laboratory, however, *either* enzyme can decarboxylate added oxalosuccinate; consequently, ICDH is said to catalyze the two separate reactions shown here:

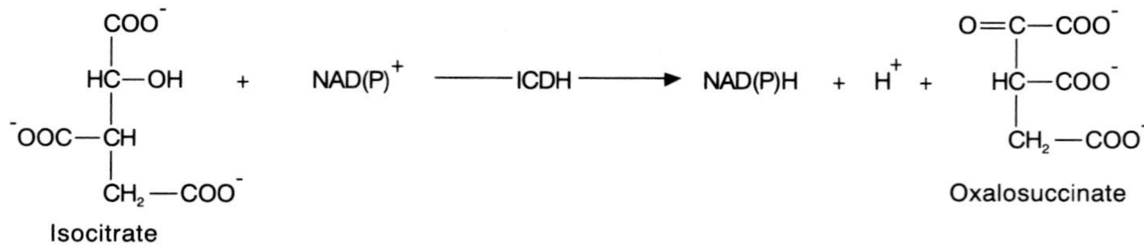

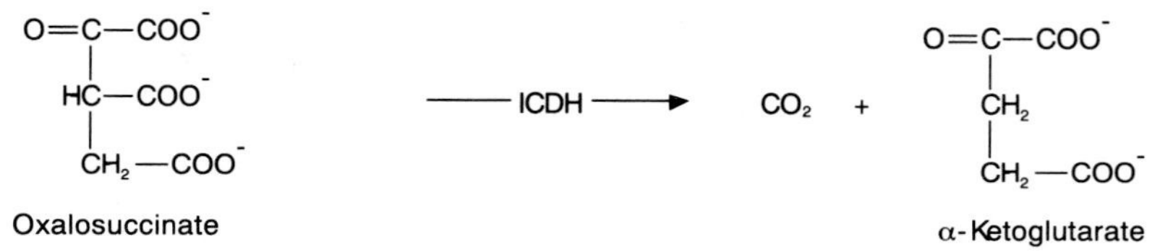

Isocitrate + NAD(P)+ ⟶ ICDH ⟶ NAD(P)H + H+ + Oxalosuccinate

Oxalosuccinate ⟶ ICDH ⟶ CO2 + α-Ketoglutarate

Second decarboxylation

The decarboxylation of α-ketoglutarate differs from that of oxalosuccinate, but it is analogous to the decarboxylation of pyruvate discussed earlier. The activity of this large, multisubunit complex is specific for ketoglutarate instead of pyruvate, but the coenzymatic components and the mechanism of action are the same. The E_3 subunit, the dihydrolipoate dehydrogenase, is exactly the same, although the E_1 and E_2 subunits of the two α-keto acid dehydrogenase complexes have different specificity. The product of the action of the α-ketoglutarate dehydrogenase (α-KGDH) complex is succinyl CoA, a substrate analogous to the acetyl CoA produced by PDH action. The overall reaction is:

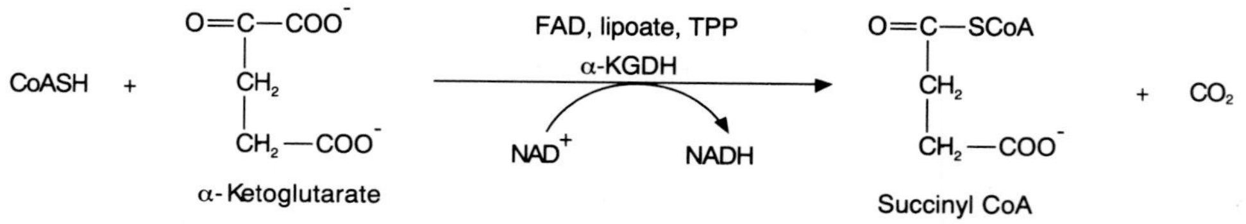

CoASH + α-Ketoglutarate ⟶ (FAD, lipoate, TPP, α-KGDH; NAD+ → NADH) ⟶ Succinyl CoA + CO2

Substrate-level phosphorylation

Succinyl CoA generated by the decarboxylation of α-ketoglutarate is converted to free succinate and CoA by *succinate thiokinase*. This reaction is tightly coupled to the phosphorylation of guanosine diphosphate (GDP) to guanosine triphosphate (GTP):

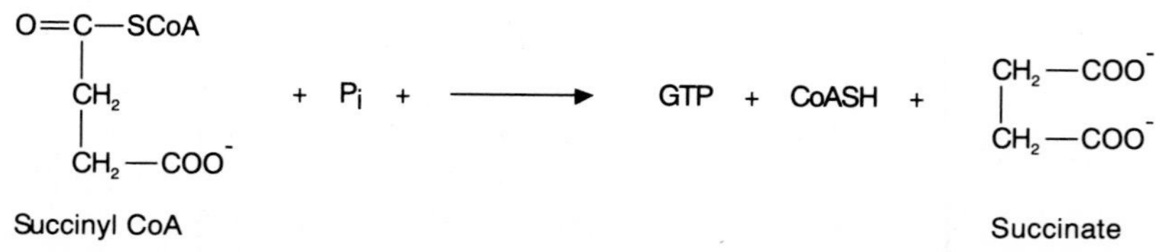

Succinyl CoA + Pi + ⟶ GTP + CoASH + Succinate

This type of phosphorylation, where the respiratory chain is not directly involved, is called *substrate-level phosphorylation*. This is one of the few instances in which GDP is preferred over ADP as the high-energy acceptor. GTP can be visualized as the energy equivalent of ATP, since a nucleoside diphosphokinase can catalyze the reaction:

$$GTP + ADP \rightleftharpoons GDP + ATP$$

Final stages   In the remaining steps of the Krebs cycle, succinate is converted to oxaloacetate by a series of reactions that are mechanistically similar to the conversion of citrate to isocitrate. That is, dehydrogenation results in the formation of a double bond, water is added across the double bond to form an alcohol, and the alcohol is oxidized to the keto compound.

Succinate dehydrogenase   SDH catalyzes the conversion of succinate to fumarate. SDH is a flavoprotein that is tightly bound to the inner mitochondrial membrane, unlike the other Krebs cycle enzymes, which are part of the matrix. The reaction catalyzed is:

Succinate   Fumarate

The tightly bound FAD can be reoxidized only by linking the SDH holoenzyme to the respiratory chain enzymes, which are also part of the inner membrane.

This tight linking was predicted many years ago by Krebs, who specifically inhibited SDH with malonate, a three-carbon dicarboxylic acid that is a competitive inhibitor of the enzyme. Succinate accumulated and respiration was blocked.

Fumarate   Malate

Fumarase   Fumarate is converted to L-malate by the enzyme fumarase, which catalyzes the following dehydration reaction:

The reaction is freely reversible but strongly stereospecific; L-malate is always the product. It requires no coenzyme.

Malate dehydrogenase   The last step in the Krebs cycle is the conversion of malate to oxaloacetate by the enzyme malate dehydrogenase (MDH). The reaction is:

Malate   Oxaloacetate

The reaction as written is highly endergonic ($\Delta^{0\prime} = +6.7$ kcal/mol; see Table 5.6), but it is driven in the direction indicated by the rapid utilization of oxaloacetate by citrate synthase, the enzyme that catalyzes the first reaction in the Krebs cycle.

Six isozymes of MDH have been found in human tissues. Of the six, isozyme IV constitutes more than half of the total activity. Plasma concentrations of MDH, as with those of lactate dehydrogenase (LDH), are sometimes increased because of tissue damage in certain diseases. Since the isozymes of LDH are easier to assay than those of MDH, the clinical assay of the latter is not routine.

**Recapitulation of
Krebs cycle
energetics**

When the free energy changes of the individual reactions of the Krebs cycle are summed (see Table 5.6), a large negative net change occurs, a clear indication that the cycle operates spontaneously in the clockwise direction (see Figures 5.10 and 5.11). Even when the decarboxylation of pyruvate is omitted (see the last line of Table 5.6), the conclusion is the same. Strictly speaking, this calculation assumes that components are at 1 mol/L concentrations, which is clearly not the case. Nevertheless, even at the prevailing tissue concentrations of 1 to 5 mmol/L, the clockwise operation of the cycle is favored.

The following equation gives a chemical summary of the Krebs cycle.

$$\text{Acetyl CoA} + 3\text{NAD}^+ + \text{FAD} + \text{GDP} + \text{P}_i + 2\text{H}_2\text{O} \rightarrow$$
$$2\text{CO}_2 + \text{CoASH} + 3\text{NADH} + \text{H}^+ + \text{FADH}_2 + \text{GTP}$$

For each mole of acetyl CoA burned in the cycle, 12 moles of ATP can be generated.

$$\begin{array}{rl}
3\ \text{NADH}— & 9\ \text{ATP} \\
\text{FADH}_2— & 2\ \text{ATP} \\
\text{GTP}— & \underline{1\ \text{ATP}} \\
Sum: & 12\ \text{ATP}
\end{array}$$

This is equivalent to a $\Delta G^{0'}$ for hydrolysis of -351 kJ (12×-29.3 kJ), or -84 kcal (12×-7 kcal) per mole. Three moles of NADH produce 9 moles of ATP, and 1 mole of FADH$_2$ produces 2 moles of ATP; both of these processes involve the respiratory chain. Finally, 1 mole of ATP is produced by the enzyme-catalyzed exchange of GTP and ADP.

If the same mole of acetyl radical were burned in a bomb calorimeter, the heat liberated would be approximately 875 kJ (209 kcal); thus the biochemical efficiency of trapping energy as ATP is approximately 40%. By itself, the Krebs cycle is the largest single generator of ATP.

**Entry of amino acids
into the Krebs cycle**

Transamination of amino acids Two of the structures of the keto acids of the Krebs cycle, oxaloacetate and α-ketoglutarate, have carbon chains that are homologous to the amino acids aspartate and glutamate (Figure 5.15). Pyruvate is also homologous to alanine. When the supply of these amino acids exceeds the requirements for protein biosynthesis, the excess can readily be converted to Krebs cycle intermediates, and the oxidation of their carbon skeletons can produce energy. On the other hand, when these amino acids are needed, such as for biosynthesis, they can be produced from their Krebs cycle keto acid analogues. Thus the Krebs cycle, which is usually thought of as a catabolic pathway, has anabolic functions under certain conditions.

The reversible interconversions of these α-amino acids and α-keto acids are catalyzed by transaminases or, more properly, *aminotransferases*. These enzymes mediate the exchange of carbonyl and amino groups between oxaloacetate-glutamate and pyruvate-glutamate (Figure 5.16). A special significance of these two aminotransferases is the ease with which they can be measured in the blood serum. For example, these aminotransferases are elevated in the serum of individuals with several different diseases. Other aminotransferases, usually involving glutamate but specific for different amino acids, are also present in the tissues (see Chapter 8).

Anaplerotic reactions

Krebs cycle intermediates are sometimes depleted, for example, by being drawn off for the biosynthesis of aspartate or glutamate. A need for more intermediates can also occur if a large influx of pyruvate or acetyl CoA depletes the supply of the oxaloacetate acceptor required for citrate synthesis. Two anaplerotic, "filling up," reactions provide for the replenishment of cycle intermediates.

1. Pyruvate carboxylase Figure 5.17 shows an abbreviated version of the Krebs cycle where a large influx of substrate is indicated by the heavy arrows. Pyruvate, from glucose,

Figure 5.15 Keto acids related to the Krebs cycle and their homologous amino acids. These pairs are identical except for the groups attached to the α-carbon atom.

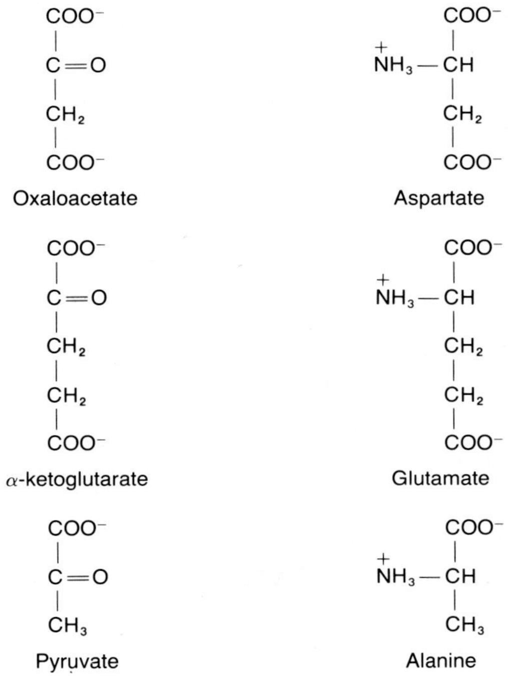

$$
\begin{array}{cc}
\text{COO}^- & \text{COO}^- \\
| & | \\
\text{C}=\text{O} & \overset{+}{\text{NH}_3}-\text{CH} \\
| & | \\
\text{CH}_2 & \text{CH}_2 \\
| & | \\
\text{COO}^- & \text{COO}^- \\
\text{Oxaloacetate} & \text{Aspartate}
\end{array}
$$

$$
\begin{array}{cc}
\text{COO}^- & \text{COO}^- \\
| & | \\
\text{C}=\text{O} & \overset{+}{\text{NH}_3}-\text{CH} \\
| & | \\
\text{CH}_2 & \text{CH}_2 \\
| & | \\
\text{CH}_2 & \text{CH}_2 \\
| & | \\
\text{COO}^- & \text{COO}^- \\
\alpha\text{-ketoglutarate} & \text{Glutamate}
\end{array}
$$

$$
\begin{array}{cc}
\text{COO}^- & \text{COO}^- \\
| & | \\
\text{C}=\text{O} & \overset{+}{\text{NH}_3}-\text{CH} \\
| & | \\
\text{CH}_3 & \text{CH}_3 \\
\text{Pyruvate} & \text{Alanine}
\end{array}
$$

Figure 5.16 The reversible reaction catalyzed by asparate aminotransferase *(AST)* and alanine aminotransferase *(ALT)*. These two enzymes are sometimes called by their previous names: glutamate-oxaloacetate transaminase (GOT) and glutamate-pyruvate transaminase (GPT), respectively. Much of the clinical literature still uses SGOT and SGPT to abbreviate the serum forms of these enzymes.

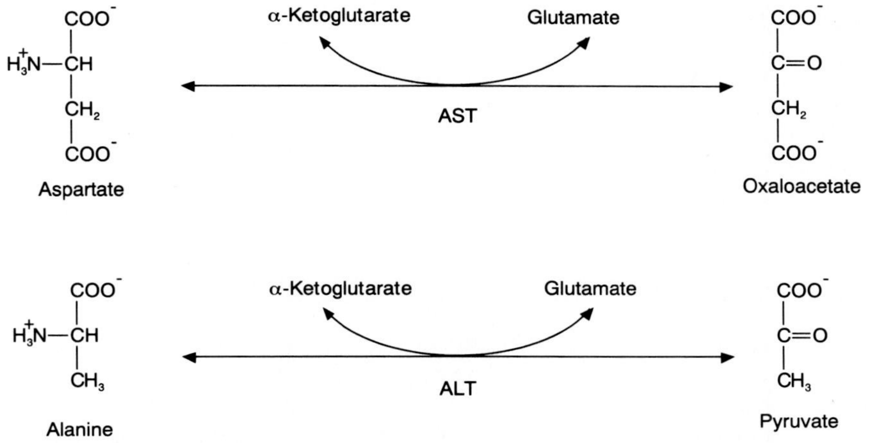

Figure 5.17   Anaplerotic reactions of the Krebs cycle. Reaction **1** is catalyzed by pyruvate carboxylase, for which acetyl CoA is a required positive effector. Reaction **2** is catalyzed by the malate enzyme. These reactions produce the oxaloacetate required to condense with a large influx of acetyl CoA.

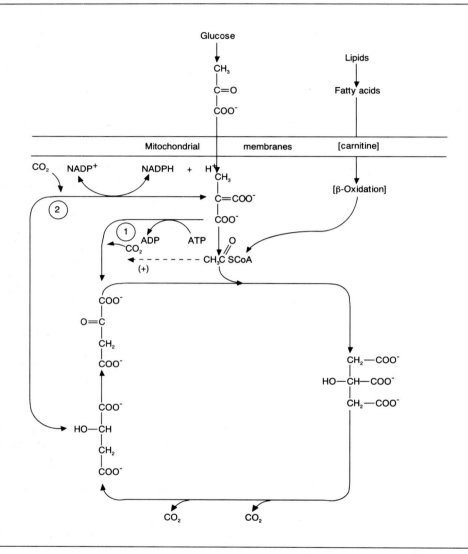

and acetyl CoA, from fatty acids, can be made in great quantity in response to vigorous exercise or epinephrine release by emotional stress. Under these conditions, excessive pyruvate is converted to oxaloacetate by the mitochondrial enzyme pyruvate carboxylase, shown in Figure 5.17 by the circled number 1. Pyruvate carboxylase is an allosteric enzyme for which acetyl CoA is an essential positive effector. The high concentration of acetyl CoA activates pyruvate carboxylase for synthesis of oxaloacetate. In the next step of the Krebs cycle, oxaloacetate accepts an acetyl group from acetyl CoA to yield citrate, which now forms in higher than usual amounts.

2. Malate enzyme A second anaplerotic reaction, shown by a circled number 2 in Figure 5.17, is catalyzed by the so-called malate enzyme, or malic enzyme (not to be confused with MDH). This reaction converts part of the pyruvate to malate by a reductive carboxylation reaction. The extra malate is easily converted to oxaloacetate. Of the two anaplerotic pathways, the pyruvate carboxylase path is usually the more dominant. The malate enzyme, however, is reversible, and in the reverse direction, the cytosolic form of the enzyme provides much of the NADPH needed for fatty acid synthesis.

Mitochondrial compartmentalization

Normal mitochondrial functions require both a sufficient concentration of enzyme intermediates and an osmotic and ionic balance between the mitochondrion and the cytosol. Mitochondrial membranes permit the entrance or exit of several substances under controlled conditions. Not all cytosolic substances can penetrate the outer mitochondrial membrane; for example, cytosolic enzymes cannot because of their large size. Cytosolic coenzymes, such as NAD^+, can penetrate the outer membrane because they are small, but they are unable to penetrate the inner mitochondrial membrane. The outer mitochondrial membrane is permeable to almost all small molecules, and the space enclosed by this membrane is called the *outer mitochondrial compartment*. The *inner mitochondrial compartment* is enclosed by the inner mitochondrial membrane. The term *inside* the mitochondrion refers to the space within the inner membrane, whereas the term *outside* refers to the outer mitochondrial space.

A summary of mitochondrial membrane permeability follows.
1. NAD^+, $NADP^+$, NADH, and NADPH do not cross the inner mitochondrial membrane. There are intra- and extramitochondrial pools that do not mix. The same is probably true for other coenzymes, including CoA, FAD, and FMN.
2. Krebs cycle intermediates, with few exceptions, can move from the outside to the inside of the mitochondria, generally by means of specific carriers or translocases.
3. Amino acids that can give rise to Krebs cycle intermediates or pyruvate by transamination can also penetrate the inner mitochondrial compartment.
4. ATP and ADP move across the mitochondrial membranes by means of a specific translocase.

Nature of translocases

Translocases, or transport systems, have properties similar to enzymes that operate in solution. Since they do not catalyze reactions that result in a covalent change to a substrate, however, they are not usually classified as enzymes. Properties of translocases include the following:
1. *Specificity.* The ATP translocase will not accept uridine, cytidine, or inosine triphosphate (UTP, CTP, or ITP).
2. *Saturability.* A translocase can be saturated with the compound it transports; that is, it has the equivalent of a Michaelis-Menten constant (K_m) or a maximum initial velocity (V_{max}).
3. *Inhibition properties.* Specific inhibitors block the activity of most translocases.
4. *Vectorial nature.* Translocases operate in a spatially directed or vectorial sense. Thus ATP only moves out of the mitochondria, and ADP must move in. This directional specificity has no counterpart in enzymology.

Representation of translocases

Translocases are represented in Figure 5.18 by circles enclosed by parallel horizontal lines that represent the membrane. Arrows show the vectorial movement of the substrate and its counterion.

The more important mitochondrial translocases are listed in Table 5.7 with their substrates, counterions, and inhibitors. Individual translocase systems shown in Figures 5.18, 5.19, and 5.22 are identified by the numbers assigned to them in Table 5.7.

Figure 5.18   Concerted action of four translocases. This system depends on
                                    intramitochondrial and extramitochondrial forms of malate
                                    dehydrogenase (MDH) and aspartate aminotransferase (AST).
                                    Numbers identify systems in Table 5.7. αKG, α-Ketoglutarate.

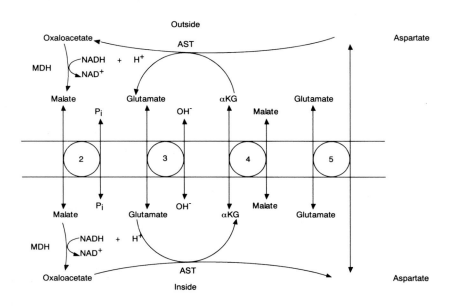

Table 5.7   Major Mitochondrial Translocases

Substrates	Counterions	Inhibitors
1. ADP	ATP	Atractylate
2. Succinate, malate, or malonate	P_i	Chlorosuccinate or 2-butylmalonate
3. Glutamate	OH^-	4-Hydroxyglutamate or 2-amino-adipate
4. α-Ketoglutarate	Malate or malonate	2-Butylmalonate
5. Aspartate	Glutamate or 2-aminoadipate	
6. α-Glycerophosphate	Dihydroxyacetone phosphate	
7. Phosphate, arsenate, or acetate	OH^-	p-Chloromercuribenzoate
8. Citrate, isocitrate, or cis-aconitate	Malate	2-Butylmalonate or benzene-1,2,3-tricarboxylate

    1. The *adenine nucleotide* translocase (Table 5.7; Figure 5.19) transports ATP produced
in the mitochondria to other cellular sites. Inhibition of this system, for example by
atractylate, can be life threatening. An ATP-binding protein of the inner mitochondrial
membrane has been found to strongly bind atractylate. This protein, probably the ATP
translocase itself, is a dimer that composes 10% of the inner membrane mass.

    2. The *dicarboxylic acid* translocase has a divalent phosphate group as a counterion.

    3. and 4. The transport systems for *glutamate* and α-*ketoglutarate,* each having dif-

Figure 5.19   Adenine nucleotide translocase. This system is inhibited specifically
                                         by atractylate. The number refers to the system described in Table
                                         5.7.

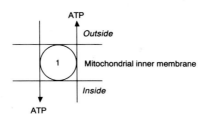

ferent counterions, emphasize the amphoteric nature of the amino acid compared to the
keto acid.

   5. and 6. Specific inhibitors for the transport systems for *aspartate* and α-*glycerophosphate* have not been found.

   7. The system that transports *phosphate* (Table 5.7, number 7) also transports arsenate,
which is an uncoupler of oxidative phosphorylation.

   8. A *tricarboxylic acid* translocase (Table 5.7, number 8) has malate as its counterion.

   These translocating systems underline the concept that movement of materials into or
out of the mitochondria is highly organized and controlled. The translocases are independent systems that work together. Figure 5.18 shows a concerted system that employs
four individual translocation systems. The net effect of these is equivalent to moving
oxaloacetate across the mitochondrial membrane barrier, even though oxaloacetate molecules themselves do not cross the barrier. Clearly, aspartate aminotransferase (AST) is
required, and at least one function of intra- and extramitochondrial MDH is explained
by this system.

Mitochondrial   Acetyl CoA is the primary precursor for the synthesis of long-chain fatty acids. The
function in   enzymes that synthesize fatty acids are found in the cytosol; therefore acetyl CoA made
lipogenesis   in the mitochondria from carbohydrates, amino acids, or other noncarbohydrate precursors
                           must find its way to the cytosol.

   Figure 5.20 diagrams a state of nutritional repletion where glucose and amino acids
are in excess of metabolic needs. The surplus energy is converted to fat and stored in
adipose tissue. In Figure 5.20 direct transamination of amino acids to Krebs cycle intermediates leads to an increased citrate concentration within the mitochondria. The citrate
is exported from the mitochondria to the cytosol, where the enzyme *citrate lyase* cleaves
it to oxaloacetate and acetyl CoA. The oxaloacetate is converted to malate by cytosolic
MDH, and malate is easily transported back into the mitochondrion. The acetyl CoA
derived from the citrate lyase reaction is now available in the cytosol for the biosynthesis
of fatty acids.

   Citrate is not only the major vehicle for the transport of acetyl groups from the mitochondria to the cytosol; it also acts as a positive allosteric effector in the first step of
fatty acid biosynthesis (see Chapter 10).

   Most amino acids cannot enter the Krebs cycle by direct transamination; however,
many of the carbon atoms derived from these amino acids can ultimately enter the Krebs
cycle after several metabolic transformations. The catabolism of these amino acids is
discussed in Chapter 8.

Figure 5.20   Mitochondrial participation in lipogenesis. *AA* represents amino acids
in excess of the requirement for biosynthesis.

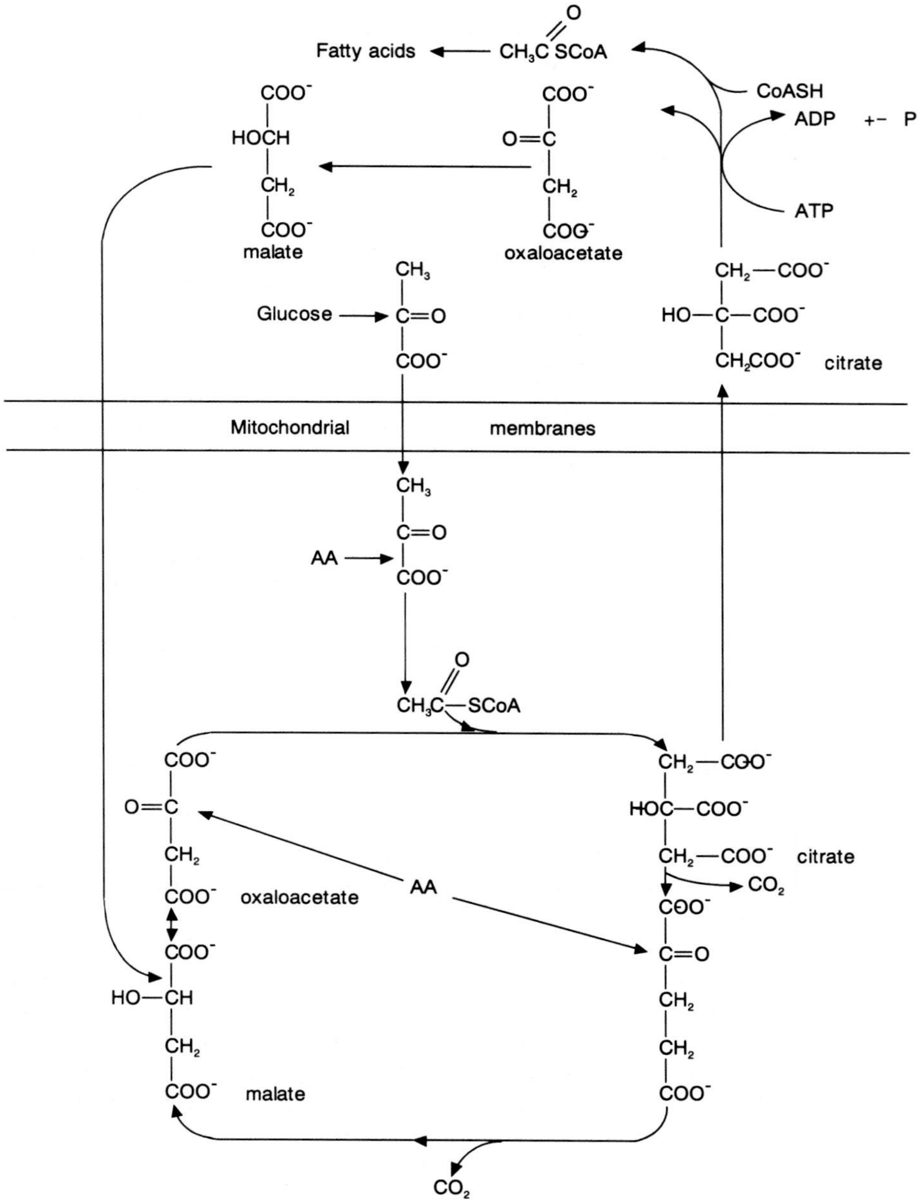

Fatty acid biosynthesis also requires NADPH. When the mitochondria are abundantly
supplied with nutrients, as shown in Figure 5.20, the energy charge of the cells will also
be high. Much ATP will be transported to the cytosol, which shifts the pattern of glucose
oxidation to the pentose phosphate pathway so that NADPH production is substantially
increased (see Chapter 6).

The mitochondrial function in lipogenesis can be summarized as follows:

1. Mitochondria collect compounds containing two or four carbon atoms from many different sources.
2. Citrate at high intramitochondrial concentrations is readily exported to the cytosol.
3. Citrate is the major source of cytosolic acetyl CoA, the primary precursor of fatty acid biosynthesis.
4. Citrate is required as an allosteric effector for the first step in fatty acid biosynthesis.
5. High ATP concentrations shift the pattern of glucose oxidation to the production of the NADPH needed for fatty acid biosynthesis.

Mitochondrial function in gluconeogenesis

Just as the mitochondria play an important role in lipogenesis, they also participate in the production of glucose from noncarbohydrate sources, a process known as *gluconeogenesis* (see Chapter 6). The complete set of gluconeogenic enzymes are found in only a few tissues, most prominently liver and kidney. Glucose produced by this pathway can enter the circulation to nourish those tissues, such as the brain, that require large amounts of glucose. Alternatively, with slight modification the pathway can be used to store glucose as glycogen in liver and skeletal muscle.

Figure 5.21 is a schematic representation of gluconeogenesis as it involves mitochondria. The conversion of phospho*enol*pyruvate (PEP) to glucose phosphate intermediates is a readily reversible process, but the conversion of pyruvate to PEP is not. PEP is a high-energy compound with a $\Delta G^{0'}$ of approximately double (-61.9 kJ/mol; -14.8 kcal/mol) that of ATP. Furthermore, Figure 5.21 shows that the enzymes that convert glucose to pyruvate are cytosolic. This raises a problem similar to that noted with fatty acid biosynthesis—key intermediates must be transported out of the mitochondria. This problem is far from settled, and various theories have been advanced, but the concept presented here is consistent with most experimental results.

Three enzymes play key roles in gluconeogenesis: pyruvate carboxylase, PEP carboxykinase, and pyruvate kinase. These are numbered 1, 2, and 3, respectively, in Figure 5.21. PEP carboxykinase is located both in the mitochondria and in the cytosol. However, this dual location does not hold for all tissues and can differ between species. Also, pyruvate carboxylase is absent from tissues that do not support gluconeogenesis. Unlike the other two enzymes, pyruvate kinase concentrations are low in gluconeogenic tissues that exhibit significant pyruvate carboxylase activity. This enzyme distribution prevents wasteful cycling between pyruvate and PEP, which would consume large amounts of ATP.

Pyruvate carboxylase catalyzes an important anaplerotic reaction (see earlier discussion) that yields oxaloacetate. Its absolute requirement for acetyl CoA is necessary to maintain a tetrameric complex. The enzyme is also subject to end-product inhibition by ADP. Thus this key mitochondrial enzyme is activated by a high-energy charge and by acetyl CoA, a product of fatty acid oxidation. The allosteric requirement is indicated by the dashed arrow in Figure 5.21. Pyruvate carboxylase (reaction 1) converts intramitochondrial pyruvate to oxaloacetate. The oxaloacetate is reduced to malate and exported to the cytosol. Once there, malate can be freely reconverted to oxaloacetate.

PEP carboxykinase is a monomeric enzyme that converts cytosolic oxaloacetate to PEP in a decarboxylation reaction (reaction 2) that requires GTP. Conditions favoring gluconeogenesis cause an increased synthesis of PEP carboxykinase. Fasting, diabetes, or glucocorticoid treatment induces the synthesis of this enzyme.

In summary, mitochondria have several functions in gluconeogenesis:

1. Amino acids enter the mitochondria, where Krebs cycle enzymes convert their keto derivatives to citrate and oxaloacetate.
2. Oxaloacetate gives rise to malate or aspartate for export to the cytosol, where they are reconverted to oxaloacetate.

Figure 5.21   Mitochondrial participation in gluconeogenesis. Note that glucose can be made in several steps from phospho*enol*pyruvate (PEP). Conversion of PEP to pyruvate is irreversible. Reaction **1** is catalyzed by pyruvate carboxylase, reaction 2 by PEP carboxykinase, reaction **3** by pyruvate kinase, and reaction 4 by citrate lyase. Note: reaction **2** can occur inside or outside the mitochondrion. *AA* represents amino acids in excess of that required for biosynthesis; carbon atoms from these are used to make glucose in gluconeogenesis.

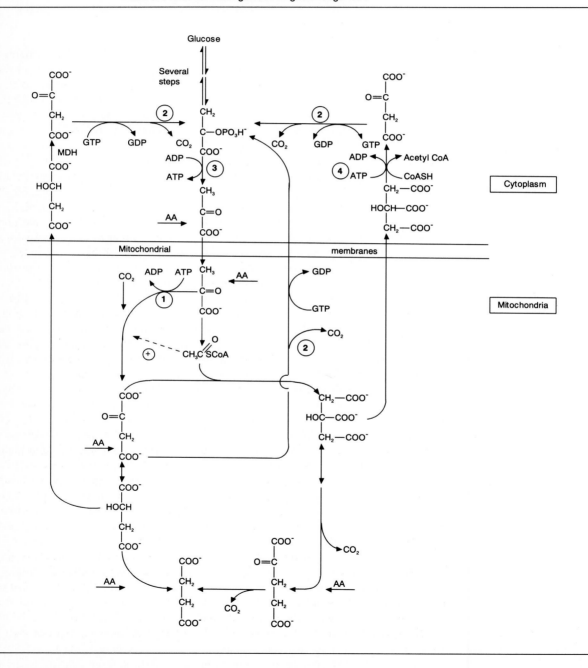

Figure 5.22 Glycerophosphate dihydrogenase–dihydroxyacetone phosphate shuttle. Number 6 refers to the translocase listed in Table 5.7.

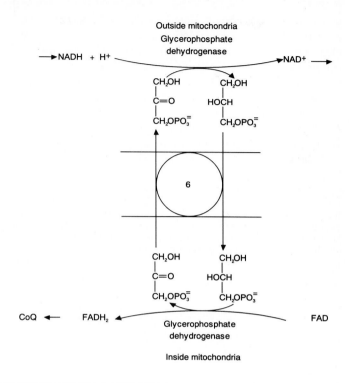

3. Mitochondrial pyruvate is carboxylated to oxaloacetate in a reaction using acetyl CoA as an allosteric activator.
4. Oxaloacetate is decarboxylated to PEP, which is converted to glucose or glycogen.

Transfer of reducing equivalents

Shuttle systems Reduced nicotinamide adenine dinucleotides are oxidized by the respiratory chain. Recall that these nucleotides cannot cross the inner mitochondrial membrane. This raises the question of how reduced cytosolic nucleotides are reoxidized. This is accomplished by the concerted action of the *malate-oxaloacetate* translocase, or shuttle system, shown in Figure 5.18. By employing the oxidation-reduction pairs such as malate-oxaloacetate, the net effect is to move 2H from one side of the membrane to the other. It is the reduced substrate molecules that are transported, not NADH. Figure 5.18 shows transport in only one direction to emphasize how cytosolic reducing equivalents can be reoxidized in the mitochondria; however, this shuttle system is completely reversible.

Another shuttle system depends on the oxidation-reduction pair of dihydroxyacetone phosphate and α-glycerophosphate (Figure 5.22). The extramitochondrial form of the enzyme is different and employs FAD as the coenzyme. Reduced FAD is coupled directly to the respiratory chain through coenzyme Q. This system, the *glycerophosphate–dihydroxyacetone phosphate* shuttle, results in the production of only two ATPs from one cytosolic NADH, whereas three ATPs can be produced when malate-aspartate shuttle is used.

The closed road: why fat is not converted to glucose

Two features of the Krebs cycle are worth emphasis. First, with two carbon atoms entering the cycle as acetyl CoA and two carbons leaving as CO_2, *no net gain* of carbon atoms occurs. Second, the carbon atoms that leave as CO_2 are *not* the same ones as those taken up as acetyl CoA. This is shown in Figure 5.23, where the carbon atoms of acetyl CoA are labeled. The decarboxylation steps involve carbon atoms initially derived from oxaloacetate. Thus, at the end of the first turn of the Krebs cycle, neither of the labeled carbon atoms is lost. Because succinate is free in solution and symmetric with respect to succinate dehydrogenase, all the carbon atoms of oxaloacetate will be labeled at the end of one complete cycle. Therefore, because labeled carbon atoms appear in malate and oxaloacetate, and because these substances contribute to the gluconeogenic pathway, it is possible to find the label in glucose. However, this does not alter the fact that no *net surplus* of carbon atoms from acetyl CoA is transferred to glucose.

A fatty acid with an odd number of carbon atoms is catabolized to yield several

Figure 5.23

Failure of fatty acids to provide *net* synthesis of glucose. The carbon atoms of acetyl CoA are not lost in one turn of the cycle. Note: no label is lost as CO_2. Succinate is symmetric, so oxaloacetate after one turn is labeled in all four carbon atoms.

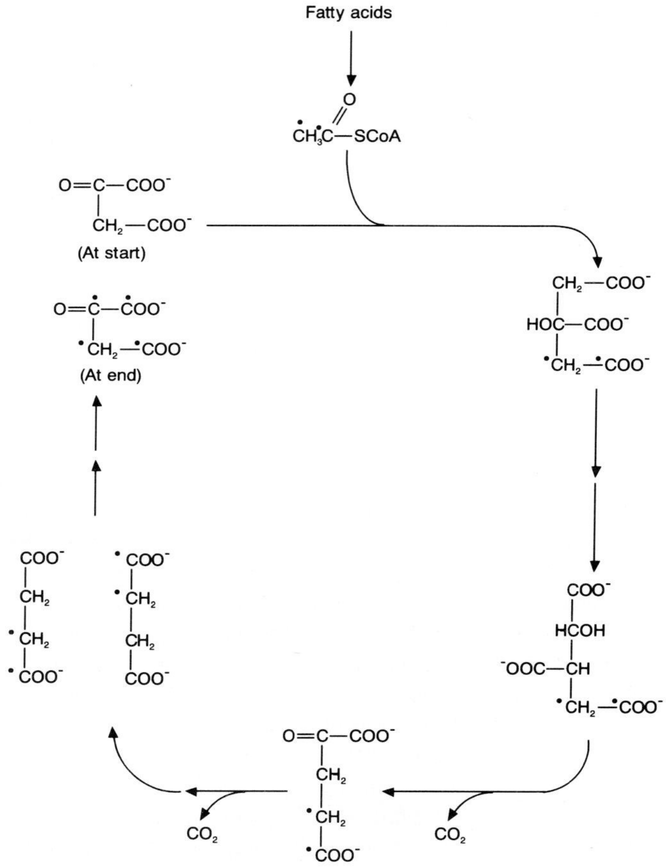

molecules of acetyl CoA and one molecule of propionyl CoA, as shown in Figure 5.24. Propionyl CoA can be carboxylated to methylmalonyl CoA, which in turn is isomerized to succinyl CoA. This succinyl CoA is a precursor of oxaloacetate. Therefore, unlike the acetyl groups, propionyl groups provide a net surplus of carbon atoms for the synthesis of glucose.

Fatty acids with odd numbers of carbon atoms and branched-chain fatty acids make up only a small portion of our total fatty acid intake. Therefore the generalization that fatty acids contribute little to the net synthesis of glucose is still valid.

Figure 5.24   Propionate, derived from odd carbon fatty acids, can provide for net synthesis of glucose, since the three-carbon atoms of propionate enter the Krebs cycle *after* the two steps involving decarboxylations.

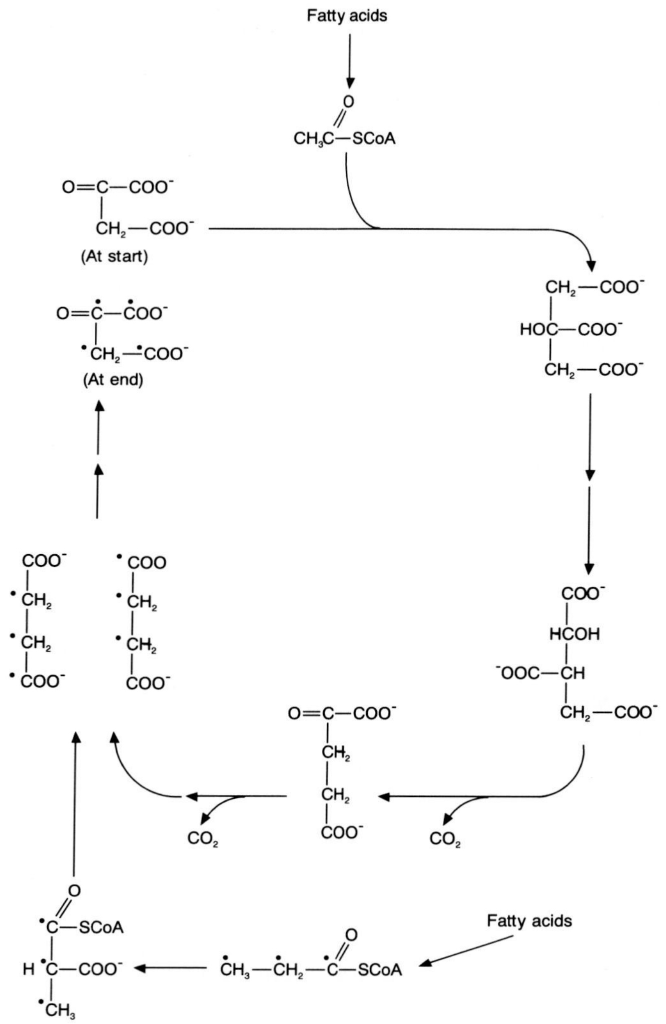

Summary

The Krebs cycle is a means of transforming materials of diverse metabolic origin into a small number of intermediates common to a variety of catabolic and anabolic pathways. Furthermore, a few auxiliary enzymes catalyze anaplerotic reactions that maintain or replenish Krebs cycle intermediates. Finally, the cycle is tightly coupled to the respiratory chain providing the ATP needed for movement, transport, and biosynthesis.

Bibliography

Curnutte JT and Babior BM: Chronic granulomatous disease, Adv Hum Genet 16:229, 1987.

Gautheron DC: Mitochondrial oxidative phosphorylation and the respiratory chain: review, J Inherited Metab Dis 7(suppl 1):57, 1984.

Hatefi Y: The mitochondrial electron transport and oxidative phosphorylation system, Annu Rev Biochem 54:1015, 1985.

Hue L: Gluconeogenesis and its regulation, Diabetes Metab Rev 3:111, 1987.

Mitchell P: Keilin's respiratory chain concept and its chemiosmotic consequences, Science 206:1148, 1979.

Sapega AA et al: Phosphorus nuclear magnetic resonance: a non-invasive technique for the study of muscle bioenergetics during exercise, Med Sci Sports Exerc 19:410, 1987.

Scholte HR et al: Defects in oxidative phosphorylation: biochemical investigations in skeletal muscle and expression of the lesion in other cells, J Inherited Metab Dis 10(suppl 1):81, 1987.

Slater EC: The mechanism of the conservation of energy of biological oxidations (a review), Eur J Biochem 166:489, 1987.

Wieland OH: The mammalian pyruvate dehydrogenase complex: structure and regulation, Rev Physiol Biochem Pharmacol 96:123, 1983.

Clinical examples

A diamond (♦) on a case or question indicates that literature search beyond this text is necessary for full understanding.

Case 1 Genetic defect in the pyruvate dehydrogenase complex

A full-term male infant failed to gain weight, had episodes of vomiting, and showed metabolic acidosis in the neonatal period. A physical examination at 8 mo showed failure to thrive, hypotonia, small muscle mass, severe head lag, and a persistent acidosis, pH 7 to 7.2. Blood lactate (9 mmol/L), pyruvate (2.4 mmol/L), and alanine (1.36 mmol/L) were greatly elevated. Since these symptoms suggested a genetic defect in pyruvate metabolism (Blass, 1983), treatment with thiamine, biotin, bicarbonate, protein restriction, and a ketogenic diet were all tried, but none of the treatments alleviated the lactic acidosis. After informed consent, oral lipoic acid (25 to 50 mg/kg of body weight) greatly improved the lactic and pyruvic acidemia. The patient was doing well 2 years later. (Case description from Matalon R et al: J Pediatrics 104:65, 1984.)

Biochemical questions

1. Why were the plasma concentrations of pyruvate, lactate, and alanine abnormally high?
2. Enzyme activity of the PDH complex, α-ketoglutarate dehydrogenase complex, and dihydrolipoyl dehydrogenase from sonicated skin fibroblasts grown in culture were all low when compared with enzymes from normal fibroblasts. Dihydrolipoyl dehydrogenase was especially low. Explain how a defect in the gene for a single enzyme would lead to these findings in vivo.

3. Lipoic acid added to the fibroblast homogenates did *not* stimulate dihydrolipoyl dehydrogenase activity, yet lipoic acid had a dramatic effect in vivo. Can you explain this?

4. What was the rationale for attempting therapy with thiamine, biotin, bicarbonate, and protein restriction?

5. Restriction of dietary carbohydrate to 40% of the food calories has been fairly effective in alleviating the symptoms of lactic acidosis in patients in whom the genetic defect in the PDH complex is presumably in the E_1 subcomplex (see p. 216). In this patient it caused a worsening of the acidosis. Suggest an explanation.

Case discussion

Genetic defects in pyruvate metabolism may affect either the PDH complex or pyruvate carboxylase, both mitochondrial enzymes. The patient in this case had an inborn error in the gene for one of the enzymes of the PDH complex. Because this complex is so large, containing three enzymatic proteins and five coenzymes, the assay of its component parts in clinical material is subject to various technical difficulties (Stansbie D et al, 1986). It is especially difficult to determine with certainty that the E_1 complex is defective, despite several independent reports of patients who presumably had defective genes for E_1, the PDH complex (Blass, 1983). It may be easier to identify defects in E_2 or E_3. Virtually all the patients with disorders of the PDH complex show a lactic acidosis and abnormalities of the nervous sytem. This is not surprising, since the brain depends more on carbohydrate utilization for energy and biosynthesis than other organs and tissues.

Skin fibroblasts in the diagnosis of genetic diseases It is difficult to determine which tissues are most affected by a genetic defect, and even more difficult to obtain antemortem tissue samples suitable for biochemical study.

The nuclei of somatic cells contain most of the genetic information of a particular organism, but in highly differentiated cells the expression of much genetic information is repressed. Skin biopsies are safe and almost painless. Explants from these can be grown in tissue culture, where they yield large numbers of fibroblasts. These cells proliferate rapidly and are sufficiently undifferentiated to be useful in examining many genetic diseases. Whole fibroblasts or broken cells, as homogenates, are elegant tools for the study of metabolic processes, particularly when used with labeled substrates.

Because of permeability barriers, one usually cannot assay enzymatic activities in intact cells; thus the cultured fibroblasts in this case were disrupted by sonication. This treatment ruptures membranes and exposes enzymes that ordinarily are enclosed by impermeable membranes.

Fibroblasts from this patient oxidized 1-[^{14}C]-pyruvate to $^{14}CO_2$ at a rate approximately one-third that of normal fibroblasts.

1. High plasma concentrations The primary defect in this patient is in the PDH complex, which was much less active than normal. The accumulation of pyruvate in the plasma is consistent with this defect. Much of the accumulated pyruvate was either reduced to lactate or transaminated to alanine, accounting for the accumulation of these substances.

$$CH_3-C-COO^- + NADH + H^+ \rightleftharpoons CH_3-CH-COO^- + NAD^+$$

Pyruvate   Lactate   *Continued on next page.*

$$CH_3-\underset{\underset{O}{\|}}{C}-COO^- \;+\; \text{Glutamate} \;\rightleftharpoons\; CH_3-\underset{\underset{\underset{H_3^+}{|}}{N}}{CH}-COO^- \;+\; \alpha\text{-Ketoglutarate}$$

Pyruvate   Alanine

2. Gene defect A defect in a gene for an enzyme common to both the PDH complex and the α-ketoglutarate dehydrogenase complex would produce these results. The only protein component in common with both of these α-keto acid dehydrogenases is dihydrolipoyl dehydrogenase, E_3, which had exceptionally low activity. Thus it is likely that the gene for E_3, or a gene that affects the activity of E_3, is defective in this patient. Assuming this to be true, one might predict that α-ketoglutarate and its metabolites would also accumulate in the patient; this proved to be true.

3. Effect of lipoic acid Lipoic acid in the PDH complex is covalently attached to E_2, not to E_3. E_2 is dihydrolipoyl transacetylase. Possibly the lipoate-activating enzyme, the enzyme that links lipoic acid to E_2, is partially defective in its ability to use lipoic acid; for example, the K_m for lipoic acid of the mutant enzyme might be higher than normal. Also one must assume that in the absence of a complete E_2, E_3 is unstable and irreversibly inactivated. High concentrations of lipoic acid might compensate for the increased K_m of the mutant lipoate-activating enzyme and allow transfer of lipoate to E_2, which would then stablize E_3. Lipoate would fail to do this in vitro if an unstable E_3 had already been inactivated in vivo.

4. Treatment rationale Therapy with thiamine was attempted because if the patient were defective in E_1 such that the enzyme affinity for thiamine was reduced, excess thiamine might allow the mutant enzyme to function. Biotin was tried for a similar reason. Defects in pyruvate metabolism usually affect either the PDH complex or pyruvate carboxylase. Pyruvate carboxylase, as with many other carboxylases, contains covalently attached biotin. Thus the rationale for biotin therapy was similar to that for lipoic acid therapy. Bicarbonate was used in an attempt to lower the lactic acidosis. This might be reasonable therapy for an acute attack of lactic acidosis, but it is not a very useful therapy for the persistent acidosis experienced by this patient. Protein restriction was tried because the branched-chain amino acids (valine, leucine, isoleucine) are catabolized by α-ketoacid dehydrogenase complexes that are similar to pyruvate and α-ketoglutarate dehydrogenase (see Chapter 8) and, therefore, might be defective in this disease.

5. Dietary carbohydrate and acidosis A genetic defect in E_1, leading to a defect in the conversion of pyruvate to acetyl CoA, might be tolerated with a diet low in carbohydrate, which is catabolized primarily to pyruvate and lactate. The acetyl CoA needed for operation of the Krebs cycle could be derived from the fatty acids in the lipid-rich diet. In this patient, however, the defect is in E_3, or in a gene that affects E_3. Since E_3 is a common component of several α-keto acid dehydrogenases that function in the oxidative decarboxylation of keto acids derived from lipids or from ketogenic amino acids, it is not possible to restore activity to the Krebs cycle by simply lowering dietary carbohydrate and increasing dietary lipids; the α-ketoglutarate dehydrogenase complex remains blocked.

References

Matalon R et al: Lipoamide dehydrogenase deficiency with primary lactic acidosis: favourable response to treatment with oral lipoic acid, J Pediatr 104:65, 1984.
Mituda S et al: Pyruvate dehydrogenase subcomplex with lipoamide dehydrogenase deficiency in a patient with lactic acidosis and branched chain ketoaciduria, Clin Chim Acta 140:59, 1984.

Robinson BH: The lacticacidemias. In Lloyd JK and Scriver CR: Genetic and metabolic disease in pediatrics, London, 1985, Butterworth & Co.

Robinson BH: Lactic acidemia. In Scriver CR et al, editors: The metabolic basis of inherited disease, ed 6, New York, 1989, McGraw-Hill Information Services Co.

Stansbie D et al: Disorders of the pyruvate dehydrogenase complex, J Inherited Metab Dis 9:105, 1986.

Case 2	Defective electron transport in the multiple acyl CoA dehydrogenation disorders

The patient was a 5-year-old girl who had experienced repeated episodes of vomiting, lethargy, and coma with acidosis and hypoglycemia. These symptoms started as early as 7 wk of age, and they suggested ketonic hypoglycemia; however, all tests for that were negative. Analysis of urinary organic acids showed excessive amounts of ethylmalonic acid, adipic acid, and hexanoyl glycine. (Case description from Mantagos S et al: J Clin invest 64:1580, 1979).

Biochemical questions

1. The organic aciduria in this patient suggests a generalized defect in an enzymatic system that oxidizes organic acids for energy. (For details of the oxidation of organic acids, including fatty acids, see Chapter 10). At this point we need only to know that the first step in this oxidative pathway introduces a *trans* double bond into the organic acyl CoA. This enzyme system is similar to the succinate dehydrogenase of the Krebs cycle. Considering this, what metabolic defects might account for a generalized inability to oxidize organic acids?
2. Would mitochondria from this patient be able to reoxidize the NADH produced by reactions of the Krebs cycle? Explain your answer.
3. Would the patient's mitochondria be able to oxidize succiate? Explain.
4. Suggest a dietary treatment for this patient and explain the biochemical rationale for the diet.

Case discussion

The multiple acyl CoA dehydrogenation disorders have at least two subtypes: a mild form, as in this case, called ethylmalonic-adipic aciduria and a more severe variant called glutaric aciduria type II. The severe form can be further subdivided into those patients who show congenital anomalies and those who do not. At first these diseases were baffling to understand because many different organic acids accumulated in the patients' urine. These included glutaric, ethylmalonic, isovaleric, 2-methylbutyric, and isobutyric acids, as well as several other dicarboxylic acids (Vianey-Liaud et al, 1987). Cultured skin fibroblasts from these patients were defective in the oxidation of a variety of organic acyl CoAs, including short-, medium- and long-chain fatty acyl CoAs.

The oxidation of these metabolites requires several different dehydrogenases, each specific for its substrate; however, they share a common complex oxidizing agent, an electron transfer flavoprotein (ETF) that contains tightly bound FAD. The FAD is converted to $FADH_2$ during the dehydrogenase reaction. Human ETF contains two protein subunits, ETF_α (30.5 kDa) and ETF_β (27 kDa) (Frerman and Goodman, 1985). $ETF-FAD_2$ is reoxidized by ETF:ubiquinone oxidoreductase (ETF:QO) EFT:QO is an iron-sulfur flavoprotein (69 kDa) that catalyzes the reoxidation of $ETF-FADH_2$ and the reduction of CoQ to the dihydroquinone, $CoQH_2$. The reduced CoQ connects the flow of electrons to the electron transport chain and eventually to oxygen with the formation of water and the generation of ATP.

1. Organic acids and metabolic defects Succinate dehydrogenase converts succinate to fumarate; thus this part of its action is similar to many other organic acid dehydrogenases. It differs, however, in that succinate dehydrogenase complex (complex II in Figure 5.2) consists of four protein subunits: the succinate dehydrogenase with its two subunits (a 70 kDa iron-sulfur protein containing FAD and a 27 kDa iron-sulfur protein) and two small subunits (15.5 and 13.5 kDa) that are heme-containing cytochromes necessary for CoQ reductase activity. The organic acyl CoA dehydrogenases differ because all their substrates are coenzyme A derivatives, and because the FAD is not part of the dehydrogenase, but rather is linked to an ETF used by all of the acyl CoA dehydrogenases as an oxidizing agent. Another difference is that ETF:QO, an iron-sulfur—containing CoQ oxidoreductase, is required for the metabolism of the acyl CoA derivatives and is not required for succinate oxidation. ETF:QO is highly specific for the FAD bound to ETF. A summary of the relationships of these enzyme systems and the electron transport system follows.

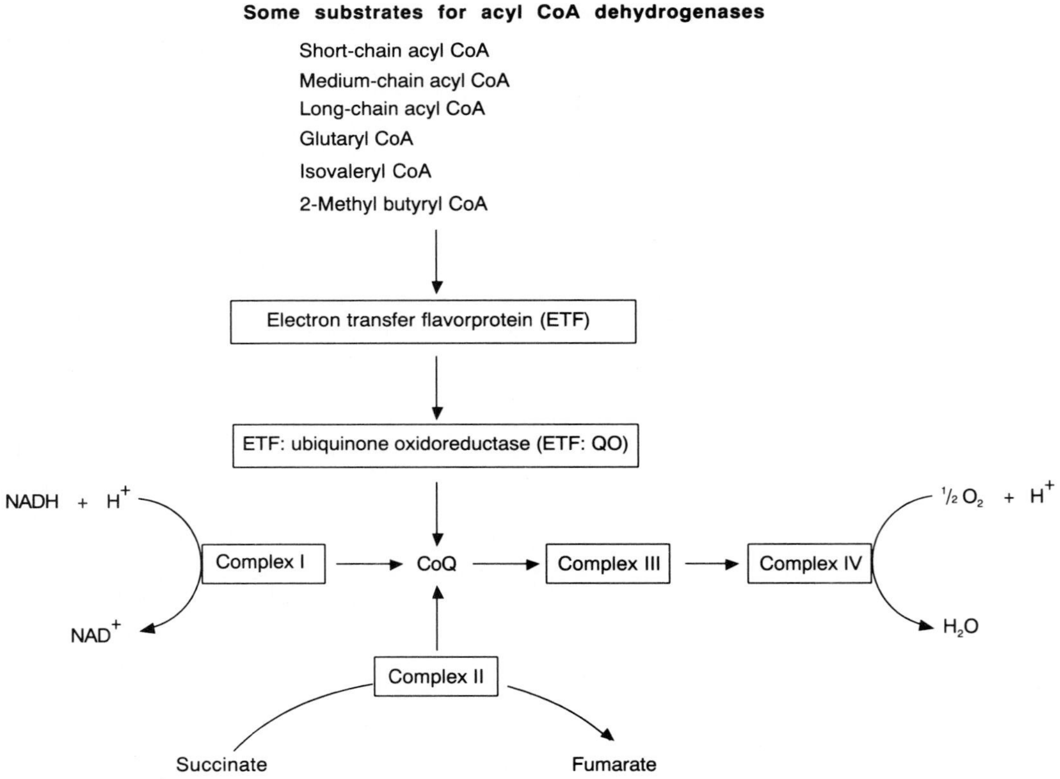

Some substrates for acyl CoA dehydrogenases

Short-chain acyl CoA
Medium-chain acyl CoA
Long-chain acyl CoA
Glutaryl CoA
Isovaleryl CoA
2-Methyl butyryl CoA

From this diagram, one can predict that a multiple acyl CoA dehydrogenase deficiency could result from a defect in any one of the genes that codes for the protein subunits of ETF or EFT:QO. In fact, it has been found that patients with the milder variant have mitochondria that may be deficient in the genes for ETF. Those with the severe variant have a defect in ETF:QO if they have congenital anomalies, but they are defective in ETF if they are without congenital anomalies (Vianey-Liaud et al, 1987).

2. Mitochondria and NADH Mitochondria from the patient should reoxidize NADH normally because the genetic defect is in the genes for proteins involved in the electron flow from reduced FAD to CoQ and not in proteins of complex I (see previous diagram).

3. Mitochondria and succinate The patient's mitochondria should be able to oxidize succinate (Rhead and Amendt, 1984). The enzymes associated with the oxidation of succinate are different from those involved in the oxidation of acyl CoA derivatives; the genetic defect in this patient affects only proteins of the latter system.

4. Dietary treatment Infants with the severe variant, glutaric aciduria type II, usually die within a few days after birth. Infants with the milder form of the disease, such as this patient, would benefit from a diet that is low in fat and relatively high in carbohydrates. Care should also be taken to avoid rapid weight gain as fat, since this fat would eventually have to be oxidized via the defective acyl CoA dehydrogenase system. Energy obtained by the aerobic oxidation of carbohydrates is unaffected.

References

Frerman FE and Goodman SI: Deficiency of electron transfer flavoprotein: ubiquinone oxidoreductase in glutaric acidemia type II fibroblasts, Proc Natl Acad Sci USA 82:4517, 1985.

Goodman SI and Frerman FE: Glutaric acidaemia type II (multiple acyl-CoA dehydrogenation deficiency), J Inherited Metab Dis 7(suppl 1):33, 1984.

Mantagos S et al: Ethylmalonic-acidemia: In vivo and in vitro studies indicating deficiency of activities of multiple acyl CoA dehydrogenases. J Clin Invest 64:1580, 1979.

Przyrembel H: Therapy of mitochondrial disorders. J Inherited Metab Dis 10:129, 1987.

Rhead WJ and Amendt BA: Electron-transferring flavoprotein deficiency in the multiple acyl-CoA dehydrogenation disorders, glutaric aciduria type II and ethylmalonic-adipic aciduria, J Inherited Dis 7(suppl 2):99, 1984.

Vianey-Liaud C et al: The inborn errors of mitochondrial fatty acid oxidation, J Inherited Metab Dis 10(suppl 1):159, 1987.

Case 3

Pyruvate carboxylase deficiency: dysfunction of the Krebs cycle

A 3-month-old female seemed normal until she developed seizures. The infant became progressively worse, showing hypotonia, psychomotor retardation, and poor head control. She had lactic acidosis and an elevated plasma pyruvate level, both more than seven times the normal amount. Plasma alanine concentration was high, and an alanine load failed to induce a normal gluconeogenic response. Pyruvate carboxylase activity was measured using extracts of cultured skin fibroblasts and was found to be less than 1% the normal level. Both the mother and the father had intermediate levels of fibroblastic pyruvate carboxylase. Fibroblasts from the patient accumulated five times greater than normal amounts of lipid.

Biochemical questions

1. What reaction is catalyzed by pyruvate carboxylase? Where is the enzyme located in cells?
2. What is the metabolic function of pyruvate carboxylase?
3. Explain the failure of the alanine load to induce gluconeogenesis in the patient.
4. Why did lipids accumulate excessively in the patient's fibroblasts?
5. Glutamine greatly stimulated the growth of fibroblasts from a patient with pyruvate carboxylase deficiency disease (Oizumi et al, 1986). Why?
6. What treatment would you suggest for a patient with this disease?
7. Most of the patients with pyruvate carboxylase deficiency die at a very early age, and those who survive are mentally retarded. What biochemical findings would you use to do genetic counseling?

Case discussion

As with other diseases of pyruvate metabolism, a deficiency of pyruvate carboxylase has a greater effect on the brain, an organ that is very dependent on carbohydrate metabolism. The greatly reduced biosynthetic function of the Krebs cycle in the brain of patients with this disease suggests a major role for the Krebs cycle in biosynthesis. In tissues such as muscle and heart, the Krebs cycle probably plays a more dominant role in providing the energy needed for motion.

1. Reaction and location of enzyme Pyruvate carboxylase is a tetrameric mitochondrial enzyme of 70 kDa that catalyzes the reaction:

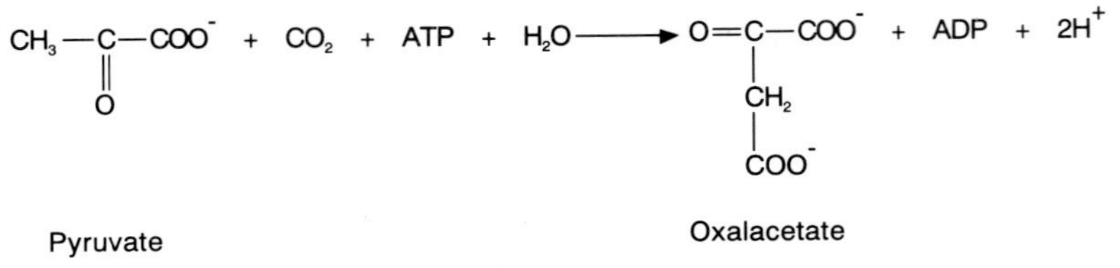

$$CH_3-\underset{\underset{O}{\|}}{C}-COO^- + CO_2 + ATP + H_2O \longrightarrow O=\underset{\underset{\underset{COO^-}{|}}{\underset{CH_2}{|}}}{C}-COO^- + ADP + 2H^+$$

Pyruvate   Oxalacetate

It is completely dependent on activation by acetyl CoA. The enzymatic activity occurs in two steps. The first step involves the activation of bicarbonate ion with ATP to form a carboxybiotinyl enzyme complex. In the second step the carboxy group is transferred to pyruvate to yield oxaloacetate. The structure of carboxybiotin is:

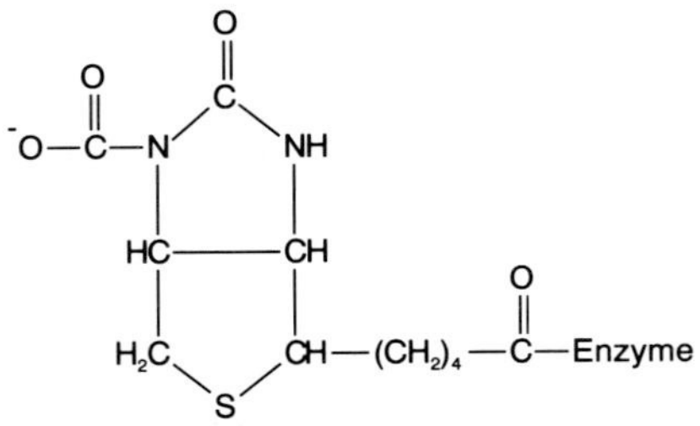

2. Metabolic function The metabolic function of pyruvate carboxylase is to provide the oxaloacetate needed to spark the Krebs cycle. The catalytic functioning of the Krebs cycle requires that oxaloacetate be regenerated with each turn of the cycle. The more oxaloacetate present, the more efficiently the cycle operates. Conversely, without oxaloacetate, as in patients with pyruvate carboxylase deficiency, the cycle ceases to function. The cycle is broken each time oxaloacetate or α-ketoglutrate is transaminated to aspartate or glutamate.

Pyruvate carboxylase is also needed by gluconeogenic organs such as liver and kidney to provide cytosolic oxaloacetate for the synthesis of phospho*enol*pyruvate. As shown in Chapter 6, no enzymatic reaction exists for the direct conversion of pyruvate to phospho*enol*pyruvate.

3. Alanine and gluconeogensis Alanine can be taken up by most cells and transaminated to pyruvate. When the patient was given an alanine load, much of the alanine was taken up by the liver and converted to pyruvate, but the pyruvate could not be used to make the oxaloacetate needed for gluconeogenesis (see answer 2) because of the defective pyruvate carboxylase.

4. Lipid accumulation Lipids accumulated in the patient's cultured fibroblasts because much pyruvate was converted to acetyl CoA by the normal PDH complex. Traces of oxaloacetate, probably from the transamination of aspartic acid in the culture medium, was used to make citrate. The citrate was transported to the cytosol, converted back to oxaloacetate and acetyl CoA, and the cytosolic acetyl CoA was converted to fatty acids (see Figure 5.21). The pathways describing the biosynthesis of fatty acids are analyzed fully in Chapter 10.

5. Glutamine and fibroblast growth Growth of the patient's fibroblasts is undoubtedly limited by a lack of Krebs cycle function. Function of the cycle can be stimulated by providing intermediates in stoichiometric amounts. Glutamine is easily deaminated to glutamate, which can be transaminated to α-ketoglutarate. This provides the oxaloacetate necessary to keep the cycle going and provides intermediates for biosynthetic reactions.

6. Therapy Treatment should be aimed at restoring the products of the reaction catalyzed by pyruvate carboxylase. There has been success with the use of aspartic and glutamic acid therapy. Both amino acids must be amidated in nonneuronal tissues to asparagine and glutamine before they can pass the blood-brain barrier, but amidation is not a problem. Oral administration of glutamine and asparagine would also be effective, but they are more expensive. Similarly, oxaloacetate or malate, which are also dicarboxylic acids, cannot pass the blood-brain barrier.

7. Genetic counseling Genetic counseling can be done if one can accurately determine the carrier (heterozygous) state. Note that the levels of pyruvate carboxylase were intermediate between normal values and those of the patient. Consequently the carrier state probably can be determined quite reliably in other members of the family. The disease is too rare to justify screening large segments of the population, but it would be appropriate to screen the patient's brothers and sisters and their spouses for the carrier state.

References

Oizumi WG et al: Pyruvate carboxylase defect: metabolic studies on cultured skin fibroblasts, J Inherited Metab Dis 9:120, 1986.

Robinson BH: Lactic acidemia. In Scriver CR et al, editors: The metabolic basis of inherited disease, ed 6, New York, 1989, McGraw-Hill Information Services Co.

Case 4 — Fumarase deficiency in mitochondrial encephalomyopathy

A 1-month-old male infant showed signs of hypotonia, lactic and pyruvic acidemia, cerebral atrophy, developmental delay, and failure to thrive. At 3 mo further studies revealed abnormally high amounts of fumarate and succinate in the urine; fumarate was particularly high. Mitochondria from biopsied muscle and liver showed impaired oxidation of glutamate or succinate by skeletal muscle mitochondria, whereas liver mitochondria oxidized these substrates normally; however, fumarase activity was virtually absent in *both* the liver and the muscle mitochondria. Cytosolic fumarase was also missing from homogenates of both tissues. The patient died at 5 mo of age. Both parents and a 2-year-old brother were healthy. There was no family history of unexplained infant death or metabolic disease. The parents were nonconsanguineous.

Mitochondrial encephalomyopathies are mitochondrial myopathies that affect both the skeletal muscle and the brain.

Biochemical questions

1. What reaction is catalyzed by fumarase?
2. This is the first report of a fumarase deficiency disease; consequently, it was recognized too late to initiate therapy. Knowing the biochemical defect in this patient, what therapy would you suggest?
3. How would you establish that this is a genetic disorder? It is known that the fumarase gene is located on the long arm of chromosome 1. Would the disease likely be X linked or autosomal, dominant or recessive?
4. Offer an explanation for the normal oxidative activity of the liver mitochondria in contrast to the impaired muscle mitochondria. Consider that the liver is a highly biosynthetic organ, whereas the muscle is much less so.

References

Williamson JR and Cooper RH: Regulation of the citric acid cycle in mammalian systems, FEBS Lett 117(suppl):K73, 1980.

Zinn AB et al: Fumarase deficiency: a new cause of mitochondrial encephalomyopathy, N Engl J Med 315:469, 1986; 316:345, 1987.

Case 5   Methylcitrate formation in methylmalonic aciduria

An infant was admitted to the hospital with failure-to-thrive syndrome. The attending physician noted signs of mental deterioration and generalized physical disability. The urine from this child was found to contain methylmalonate and methylcitrate.

Biochemical questions
1. Is methylmalonate a normal component of human urine?
2. What enzyme would you suspect to be deficient in this disease, considering the urinary metabolites?
3. What does the presence of methylcitrate in the urine signify?
4. What metabolic pathways other than the Krebs cycle might be affected by the presence of methylcitrate?
5. How might this condition affect the mitochondrial concentration of GTP?
6. What does this case indicate concerning the specificity requirements of citrate synthase and aconitase?

References

Ando T et al: Isolation and identification of methylcitrate, a major metabolic product of propionate in patients with propionic acidemia, J Biol Chem 247:2200, 1971.

Rosenberg LE and Fenton WA: Disorders of propionate and methylmalonate metabolism. In Scriver CR et al, editors: The metabolic basis of inherited disease, ed 6, New York, 1989, McGraw-Hill Information Services Co.

Case 6   Chronic alcoholism: death induced by the acetaldehyde syndrome

A 43-year-old man with a history of chronic alcoholism sought medical assistance. His physician prescribed disulfiram with the caution that severe symptoms would occur if he consumed alcohol while taking the drug. This therapy has been successful for many chronic alcoholics, and in the absence of alcohol the drug is innocuous.

The man prospered for several months, but at an office party he drank a large amount of a punch, not knowing it contained alcohol. He quickly became ill and was rushed to the hospital. He complained of severe chest pain, vertigo, and blurred vision. He died later that evening.

The metabolism of ethanol occurs mostly in the liver and proceeds in several steps; the first involves the NAD^+-linked alcohol dehydrogenase, which oxidizes ethanol to acetaldehyde. Acetaldehyde is moderately toxic, but normally it is rapidly oxidized to acetate by liver aldehyde dehydrogenase. Disulfiram (commonly known by the trade name Antabuse) inhibits the aldehyde dehydrogenase so that the blood concentration of acetaldehyde may rise tenfold above normal. The drug is also a potent inhibitor of dopamine β-hydroxylase, especially in the brain. Since dopamine is a major precursor of norepinephrine (see Chapters 9 and 18), the disulfiram or its metabolites depress synthesis of these hormones.

Biochemical questions
1. Is the conversion of ethanol to acetate an example of a coupled-reaction system?
2. Is the conversion of NAD^+ to NADH and H^+ involved in every coupled system?
3. Assume that the intracellular concentration of NAD^+ is 5 mmol/L, the intracellular pH is 7.4, and that one has consumed enough alcohol to have an intracellular concentration of 1 mmol/L. If the standard oxidation potential for the half-reaction ethanol-acetaldehyde is known to be $+0.197$ volt, what is the free energy made available by the oxidation of 0.5 mole of ethanol?

4. Would the energy yield of this reaction be different if you assumed that the dehydrogenase was linked to NADP+ instead of NAD^+? Explain.
5. The concentration of alcohol in the blood declines linearly with time. When it falls from 30 to 20 mmol/L, the blood acetaldehyde concentration remains virtually constant at approximately 30 mmol/L. In a study of human liver aldehyde dehydrogenase (J Biol Chem 243:6402, 1968) the K_m for acetaldehyde was 7.5×10^{-7} mol/L and 6×10^{-4} for NAD^+. For alcohol dehydrogenase the K_m is 1.8×10^3 mol/L for ethanol and 1.7×10^{-5} mol/L for NAD^+. Using these data, explain the relative constancy of acetaldehyde concentration with a concomitant falling ethanol concentration.

References

Fuller RK et al: Disulfiram treatment of alcoholism: a Veterans Administration cooperative study, JAMA 256:1449, 1986.

Sellers EM et al: Drugs to decrease alcohol consumption, N Engl J Med 305:1255, 1981.

Case 7 — Citrate synthesis by isolated lymphocytes

Production of immunoglobulins by lymphocytes requires a readily available energy source. The Krebs cycle and the respiratory chain are adequate for this purpose in lymphocytes from patients with chronic lymphocytic leukemia.

When 10^8 isolated lymphocytes were incubated with 0.5 μmol of acetate or acetyl CoA and 10 μmol of oxaloacetate, 0.2 μmol/hr of citrate was formed. Increasing the cell population over a fourfold range gave proportionate increases in citrate formation. When the incubation medium contained 20 μmol of monofluoroacetate instead of acetate, an aliquot of 10^8 cells produced 0.8 μmol/hr of citrate.

Biochemical questions

1. Could succinate have been used instead of oxaloacetate in these experiments?
2. How is it possible that acetyl CoA and acetate produced citrate with the same efficiency?
3. What in vivo sources of acetyl CoA are available to lymphocytes?
4. How might the results differ if sonicated mitochondrial extracts were used?
5. Could the experiments have been performed equally well with erythrocytes from leukemic subjects?
6. Why was citrate formation higher when fluoroacetate was added?

Reference

Dixit PK and Cadwell R: Citrate synthesis by lymphocytes, Clin Chem 21:825, 1975.

Additional questions and problems

1. Why is cellular compartmentalization essential for ATP synthesis?
2. What effects might a severe acidosis have on ATP synthesis?
3. ATP synthesis is important not only for energy conservation but also for other metabolic functions. List some of the other uses of ATP.
4. Most biologic oxidations occur by two-electron transfers. Write reactions to show why this is the case.
5. How would deficiencies of water-soluble vitamins affect the operation of the Krebs cycle? Would the effects be the same for all the B vitamins?
6. While conducting an experiment on mitochondrial metabolism, a graduate student accidentally pipetted into his mouth and swallowed an atractylate solution. Explain the mechanism of the convulsions that followed.
7. State two ways in which an increased cellular energy charge would slow the citrate synthase reaction.
8. In what ways does the Krebs cycle contribute to *anabolic*, as opposed to *catabolic*, pathways.

9. Show why there is no net synthesis of oxaloacetate from acetyl CoA, no matter how much of the latter is available.

10. Summarize how an excess of cytoplasmic reducing power is transferred to the mitochondrial respiratory chains.

11. Succinate with methylene groups uniformly radiolabeled with ^{3}H was added to a respiring suspension of isolated adipocytes. Ten minutes later glutamate was isolated from the suspension. Would the glutamate contain any of the label? Where would it be located?

♦ 12. Certain macrocytic ionophores act as antibiotics. What is the basis of this action?

♦ 13. Lipoic acid has been used to treat *Amanita phalloides* mushroom poisoning (Bicker et al: West J Med 125:100, 1976). Can you suggest a biochemical rationale for the function of lipoic acid in this treatment?

Multiple choice problems

Friedrich's ataxia Low activities of the pyruvate and α-ketoglutarate dehydrogenase complexes have been found in disrupted fibroblasts from four patients with Friedreich's ataxia (Blass et al: N Engl J Med 295:62, 1976). Friedreich's ataxia is an inherited neurologic disorder characterized by spinocerebellar degeneration. It is still not clear whether the reduced activity of pyruvate dehydrogenase is directly or only indirectly involved in the pathogenesis of this disease. The following 5 questions apply to this case.

1. Considering these findings, the genetic defect in these patients is likely to be a protein common to both enzyme complexes. Such a protein is:
   A. Pyruvate dehydrogenase.
   B. α-Ketoglutarate dehydrogenase.
   C. Dihydrolipoyl dehydrogenase.
   D. Pyruvate dehydrogenase kinase.
   E. α-Ketoglutarate dehydrogenase phosphatase.

2. When necessary, pyruvate can be converted to oxaloacetate in a reaction catalyzed by pyruvate carboxylase. Pyruvate carboxylase activity is stimulated by an increased concentration of:
   A. Citrate.
   B. Acetyl CoA.
   C. ATP.
   D. cAMP.
   E. Malate.

3. During the reaction catalyzed by normal pyruvate dehydrogenase, an *intermediate* is formed in which a two-carbon derivative at the oxidation level of acetaldehyde becomes covalently bound to a coenzyme. This coenzyme is:
   A. Coenzyme A.
   B. Lipoic acid.
   C. Thiamine pyrophosphate.
   D. Biotin.
   E. FAD.

4. The α-ketoglutarate dehydrogenase reaction of the Krebs cycle results in the formation of a high-energy compound. In the subsequent reaction the energy contained in this compound is conserved as:
   A. Acetyl CoA.
   B. GTP.

   C. NADH.

   D. FADH$_2$.

   E. Succinyl phosphate.

5. It was noted that glycogen had accumulated in some of the tissues examined. To study the metabolic defect, it was necessary to know the ΔG^0 for the following reaction:

UTP + Glucose 1-phosphate → UDP-glucose + Pyrophosphate

		ΔG^0 (kcal/mol)
Glucose + Phosphate	→ Glucose 1-phosphate	+5.0
UDP + Phosphate	→ UPT + Water	+7.3
2 Phosphate	→ Pyrophosphate + Water	+8.0
UDP + Glucose	→ UDP-glucose	+8.0

The ΔG^0 for the reaction in question is approximately (kcal/mol):

   A. -4.3

   B. -3.7

   C. -2.3

   D. $+23.3$

   E. $+3.7$

Hepatitis A junior medical student felt tired, weak, and nauseous following a clinical clerkship. His skin developed a yellow color as the condition worsened. Laboratory studies confirmed that he was jaundiced; hepatitis was suspected. Questions 6 to 10 apply to this case.

6. The concentrations of several enzymes were elevated in the student's blood plasma. These probably leaked out of the inflamed hepatocytes, where it was likely that the enzymes had been:

   A. Attached to the outer mitochondrial membrane.

   B. Attached to the inner mitochondrial membrane.

   C. Bound to the endoplasmic reticulum.

   D. Soluble in the cytoplasm.

   E. Integral proteins of the plasma membrane.

7. To set up a valid assay for plasma lactate dehydrogenase, one must be sure to have the concentration of:

   1. Substrate in excess.

   2. Plasma limiting.

   3. Cofactor in excess.

   4. Both substrate and cofactor limiting.

     The best answer is:

     A. 1, 2, and 3.

     B. 1 and 3.

     C. 2 and 4.

     D. 4 only.

     E. 1, 2, 3, and 4.

8. If mitochondria were prepared from a specimen (biopsy) of the patient's liver, and if the flow of electrons from cytochrome b to cytochrome c were inhibited, one would expect:

   A. Higher than normal oxidative rate.

   B. Lower than normal oxidative rate.

C. No change in oxidative rate.
D. Severe uncoupling of oxidative phosphorylation.
E. Increased oxidation of NADH.

9. If oxidative phosphorylation were uncoupled in these mitochondria, one would expect:
A. A decreased concentration of ADP in the mitochondria.
B. Increased inorganic phosphate in the mitochondria.
C. A decreased oxidative rate.
D. A decreased production of heat.
E. Increased transport of ADP from the cytosol to the mitochondrial matrix.

10. If the mitochondria were blocked at the site of NADH oxidation and were treated with succinate as substrate, the P/O ratio would be:
A. Zero.
B. One less than normally produced by succinate.
C. The same as that normally produced by succinate.
D. One more than that normally produced by succinate.
E. Higher than normal because of the increase in temperature resulting from uncoupling.

Chapter 6

Carbohydrate metabolism

Objectives

1 To interpret the role of carbohydrate metabolism in normal and selected common disease states
2 To relate carbohydrate metabolism to the functioning of the Krebs cycle
3 To review the role of carbohydrates in nutrition and homeostasis

The major source of carbohydrates is found in plants where photosynthetic processes combine carbon dioxide and water to provide the renewable supplies on which animals are dependent. The dependence on plant photosynthesis may be direct in the carbohydrate of the diet, or it may be indirect in the biosynthesis of protein and lipids by the metabolism of other animal, plant, and microbial organisms. This complex food chain, on which humans are totally dependent, usually provides carbohydrate in a polymeric form, such as starch, which enters the human metabolism after hydrolysis to monosaccharides, principally glucose. The glucose concentrations in the body are maintained within limits; this control depends greatly on glycogen, a storage form of carbohydrate. This chapter and Chapter 7 focus on the biochemical reactions involved in the digestion of dietary carbohydrate, the storage of temporary excess carbohydrate as glycogen, the conversion of glucose to other molecules, and the control of these metabolic pathways.

Except for ascorbic acid (vitamin C), carbohydrates are not essential to the diet; through gluconeogenesis the body can synthesize necessary carbohydrates from other materials, principally from certain amino acids. However, low-carbohydrate diets usually result in metabolic imbalances. For example, a high-fat, low-carbohydrate diet results in a metabolic acidosis, whereas a high-protein, low-carbohydrate diet results in a protein imbalance with a high urinary nitrogen output; increasing carbohydrate in the diet prevents this loss of nitrogen, an illustration of the "protein-sparing" effect of carbohydrate.

The term *carbohydrate* originated with the idea that naturally occurring compounds of this class—for example, starch, glycogen, sucrose, and glucose—were hydrates of carbon and could be represented by the formula $C_x(H_2O)_y$. It was soon found that this definition was too rigid, since it did not include the amino sugars such as glucosamine, the deoxy sugars such as 2-deoxyribose, and the sugar acids such as ascorbic acid and glucuronic acid.

Nomenclature

The simple sugars, the monosaccharides, are characterized by the reducing group. $>C=O$. They are also polyhydroxy compounds and in general can be represented by the following structure:

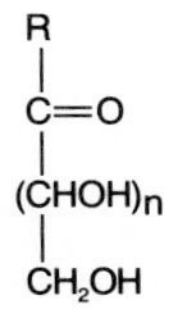

$$
\begin{array}{c}
R \\
| \\
C=O \\
| \\
(CHOH)_n \\
| \\
CH_2OH
\end{array}
$$

A reducing sugar

If R is hydrogen, the monosaccharide has an aldehyde group and is thus one of the *aldoses*. However, if R is —CH_2OH, the monosaccharide is a 2-keto sugar, the most common of the *ketoses*.

$$
\begin{array}{cc}
H & CH_2OH \\
| & | \\
C{=}O & C{=}O \\
| & | \\
(CHOH)n & (CHOH)n \\
| & | \\
CH_2OH & CH_2OH \\
\text{An aldose} & \text{A ketose}
\end{array}
$$

We can further define the monosaccharides by the number of carbon atoms, for example, the *trioses* with three carbons and the *tetroses* with four carbons. These will be either *aldotrioses, aldotetroses, ketotrioses,* or *ketotetroses,* depending on whether they have an aldehyde or a keto group as the reducing function. Glucose is an aldohexose. Fructose is a ketohexose.

$$
\begin{array}{cc}
CHO & CH_2OH \\
| & | \\
(CHOH)_4 & C{=}O \\
| & | \\
CH_2OH & (CHOH)_3 \\
 & | \\
 & CH_2OH \\
\text{Aldohexose} & \text{Ketohexose} \\
\text{(for example, glucose)} & \text{(for example, fructose)}
\end{array}
$$

Galactose, mannose, glucose, and five other sugars are all aldohexoses represented by the same formula just illustrated. They are isomers that differ in the arrangement of the groups around the four chiral carbons in the molecule. When such asymmetry results in a *molecular* asymmetry, it gives rise to an *optically active molecule*.

Every optical isomer has a mirror image, the two isomers differing in the direction in which the plane of polarized light rotates. A compound that rotates the plane of polarized light in a clockwise direction is said to be dextrorotatory ($+$), whereas that which rotates the plane of light in a counterclockwise direction is said to be levorotatory ($-$). Each of the carbohydrates is one or the other of the two possible optical isomers.

By convention, the family of sugars designated D, when represented by the planar Fischer formula (see p. 40), has the hydroxyl group on the chiral carbon farthest from the reducing group on the right-hand side of the carbon chain. The family of L-sugars has this hydroxyl group on the left.

$$
\begin{array}{ccc}
R & & R \\
\backslash & & \backslash \\
C{=}O & & C{=}O \\
| & & | \\
-C- & & -C- \\
| & & | \\
(-C-)x & & (-C-)x \\
| & & | \\
H-C-OH & & HO-C-H \\
| & & | \\
CH_2OH & & CH_2OH \\
D & \text{Sugar} & L
\end{array}
$$

Note, however, that the D- and L-sugars are not necessarily dextrorotatory and levorotatory, respectively. In other words, the configuration around the chiral carbon farthest from the reducing group does not determine the rotation of the plane of polarized light.

The simplest of the aldose sugars is glyceraldehyde; the naturally occurring form is the D-isomer. The corresponding ketotriose is a symmetric molecule, dihydroxyacetone. The first ketose with molecular asymmetry is a ketotetrose, for example, D-erythrulose.

$$\begin{array}{c} ^1CHO \\ | \\ H-^2C-OH \\ | \\ ^3CH_2OH \end{array} \qquad \begin{array}{c} ^1CH_2OH \\ | \\ ^2C=O \\ | \\ H-^3C-OH \\ | \\ ^4CH_2OH \end{array}$$

D-Glyceraldehyde   D-Erythrulose

As just illustrated, each carbon atom is identified by number; the reducing group carbon is C_1 for aldoses and C_2 for the common ketoses.

Other ketoses discussed here in relation to the metabolic paths of the carbohydrates are D-xylulose, D-ribulose, and D-fructose.

$$\begin{array}{c} CH_2OH \\ | \\ C=O \\ H-\!\!\!\!|-OH \\ HO-\!\!\!\!|-H \\ | \\ CH_2OH \end{array} \qquad \begin{array}{c} CH_2OH \\ | \\ C=O \\ HO-\!\!\!\!|-H \\ H-\!\!\!\!|-OH \\ | \\ CH_2OH \end{array} \qquad \begin{array}{c} CH_2OH \\ | \\ C=O \\ H-\!\!\!\!|-OH \\ H-\!\!\!\!|-OH \\ | \\ CH_2OH \end{array} \qquad \begin{array}{c} CH_2OH \\ | \\ C=O \\ HO-\!\!\!\!|-H \\ H-\!\!\!\!|-OH \\ H-\!\!\!\!|-OH \\ | \\ CH_2OH \end{array}$$

L-Xylulose   D-Xylulose   D-Ribulose   D-Fructose

Ring structures

It is important to note that the aldoses and the ketoses, as just written in the open-chain form, are reactive molecules. Although it is this particular form that reacts in some cases, the open-chain form is usually present in small amounts in aqueous solution. In most sugar molecules the $>C=O$ group has reacted with a hydroxyl group in the same molecule to form a ring that may, for purposes of stability, be either five- or six-membered. These two different forms may be interconverted, as shown for glucose in the reaction sequence presented in Figure 6.1. The most immediate consequence of this ring formation is that the previously symmetric $>C=O$ group is now a new chiral carbon; thus there can be two forms of each ring sugar.

$$HO-C-H \qquad\qquad H-C-OH$$

β-D α-D

The new hydroxyl group, as just represented by the planar (Fischer) formula, can be on either side of the carbon chain. In the series of D-sugars the ring form with the hydroxyl group on the right side is designated the α-D-anomer and has the most positive optical rotation of the pair. The other ring form is called the β-D-anomer and has a more negative rotation.

The six-membered ring structures, composed of five carbons and one oxygen, may be related to a similar six-membered ring compound, pyran. Haworth suggested that such sugars be identified as *pyranoses* to differentiate them from the five-membered ring forms, which are called *furanose* sugars. Although aqueous solutions of the reducing sugars may contain five different molecular forms (the open chain, the two pyranose rings, and the two furanose rings), it should be noted that only the pyranose sugars of the pentoses and hexoses are stable enough to exist to any large extent in aqueous solution. However, all forms are in equilibrium with each other; if a reaction occurs to remove one of them, the principle of LeChetalier demands that equilibrium be reestablished by the interconversion of all forms. This equilibrium can be demonstrated by observing the optical rotation when

Figure 6.1   Ring forms of D-glucose.

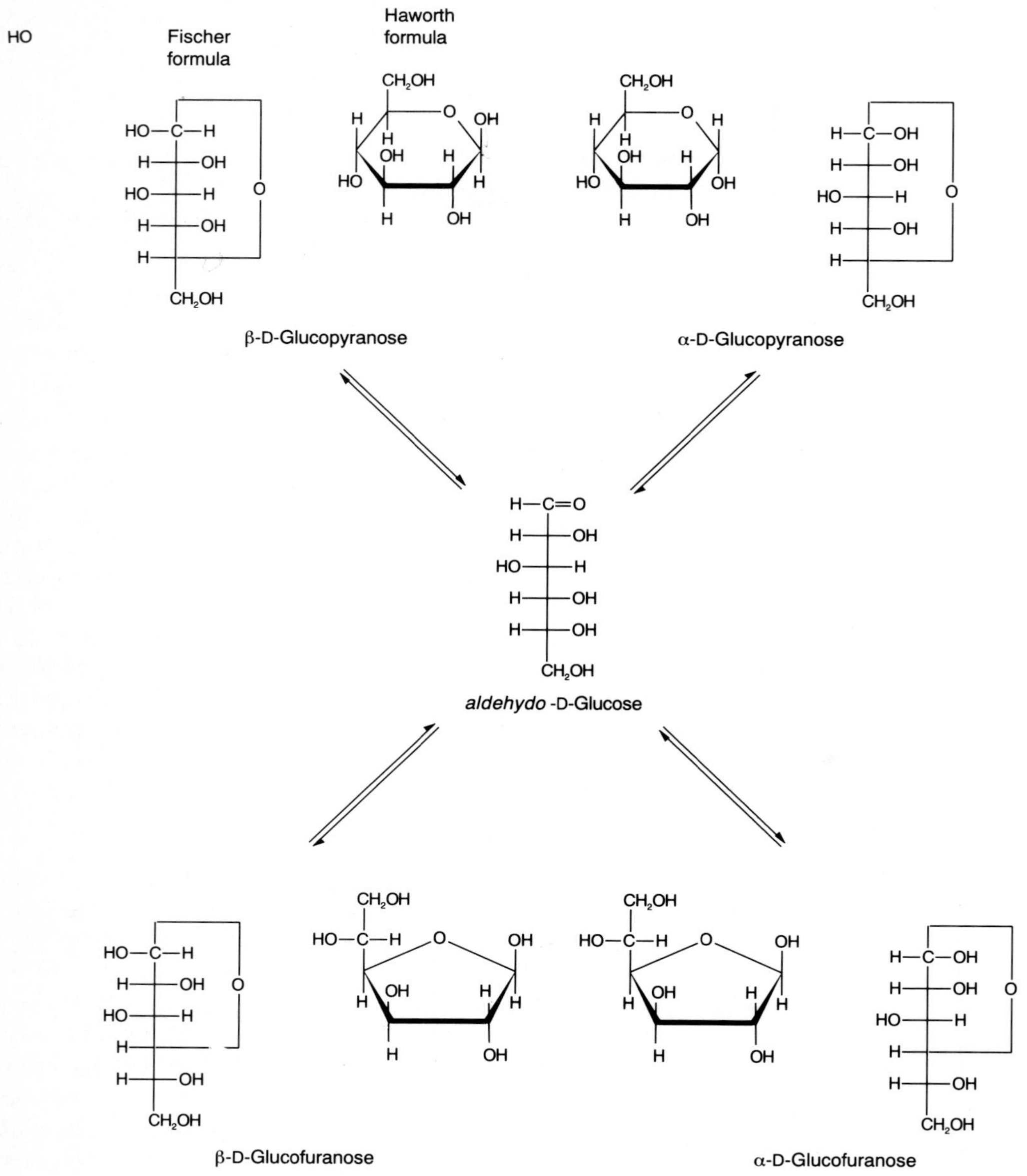

β-D-Glucopyranose

α-D-Glucopyranose

aldehydo -D-Glucose

β-D-Glucofuranose

α-D-Glucofuranose

one dissolves crystalline dextrose (α-D-glucopyranose) in water. Such a change from a single form to an equilibrium mixture that includes its other forms is called *mutarotation*. Biologically, this change is catalyzed by the enzyme *mutarotase*.

The carbon atoms in the ring are identified by the same numbers as were given to them in the planar structures. Thus, for β-D-glucopyranose, α-D-galactopyranose, and β-D-fructofuranose, the formulas are:

β-D-Glucopyranose 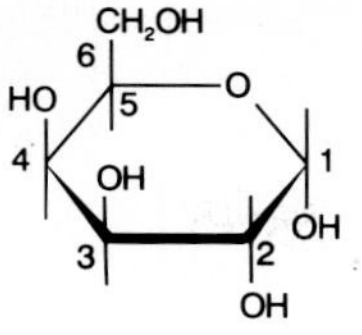   β-D-Galactopyranose 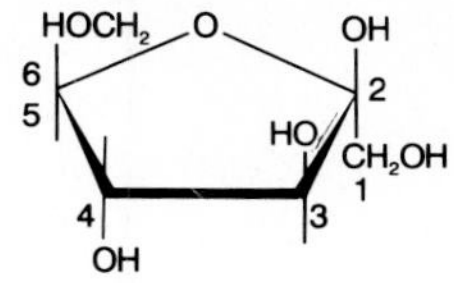   β-D-Fructofuranose

Note, as mentioned in Chapter 1, that the hydrogen atoms on the chiral carbons are usually omitted from the formula.

Clinical comment

Glucose oxidase The substrate for glucose oxidase is β-D-glucopyranose. Blood glucose, which is an equilibrium mixture of the α- and β-anomers of D-glucose, is quantitatively determined by the stoichiometric formation of hydrogen peroxide by the reaction:

$$\text{D-Glucose} + O_2 \xrightarrow{\text{Glucose oxidase}} \text{D-Gluconic acid} + H_2O_2$$

This requires that the α-D-glucopyranose be rapidly isomerized by mutarotation into the β-D-anomer. This reaction is reasonably fast without catalysis, but it is very rapid in the presence of mutarotase. One of the most common clinical chemistry measurements is blood glucose. This test is used to determine whether carbohydrate metabolism is normal and whether diabetic persons are under good therapeutic control. The assay is done in the laboratory by adding an excess of the bacterial enzyme, glucose oxidase, to a fixed amount of the patient's blood. All the glucose present in the blood is converted by the enzyme to gluconic acid and H_2O_2. The latter is measured colorimetrically. For home use the glucose oxidase and reagents are absorbed onto a paper, to which is added a sample of blood. The reaction is complete in a few minutes, and the color of the paper is measured in a simple colorimeter.

Sugar derivatives
Glycosides

Probably the most important derivatives of the reducing sugars are formed by their reaction with alcohols. Being a hemiacetal, the reducing sugar in the ring form will react with an alcohol to produce a *glycoside,* either the α-D- or the β-D-isomer. The hydroxyl group can be from another sugar, a sterol, an alkaloid, a protein, an inositol, or any similar compound, designated ROH. If the hydroxyl group comes from another sugar, the glycosidic bond will join the two monosaccharides to form a *disaccharide*. For example, the reaction of D-galactose with the ROH compound ethanol may be summarized in the manner shown in the following reaction. The D-galactose is in both α-D- and β-D-anomeric

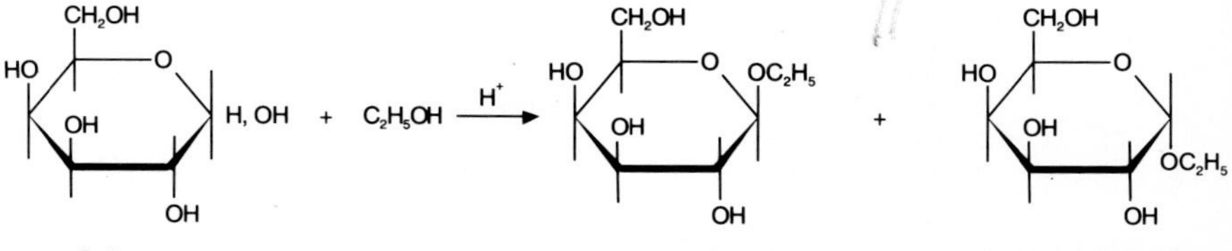

D-Galactose   Ethyl-β-D-galactoside   Ethyl-α-D-galactoside

forms in solution during the reaction (indicated in the formula for the reducing sugar by not specifying the configuration around C_1; the arrangement is left as ⊰ H, OH). When referring to sugar derivatives in general, the generic prefix *glyc-* is used. This is replaced by the identifying name of the sugar when appropriate, for example, *gluc*oside or *gal*actoside. The noncarbohydrate moiety of a glycoside is called the *aglycone*.

Disaccharides

Although there are many possible combinations of pairs of monosaccharides and glycosidic linkages, four disaccharides are particularly important to our discussion; these are maltose, lactose, isomaltose, and sucrose.

Maltose Maltose is a major product of the action of α-amylase on starch or glycogen. It is composed of two D-glucose residues, one of which is joined through its C_1 to the hydroxyl at C_4 of the other D-glucose. The glucosidic linkage is of the α-D-anomeric form. Maltose is a reducing sugar because the one D-glucose has a potentially free aldehyde group.

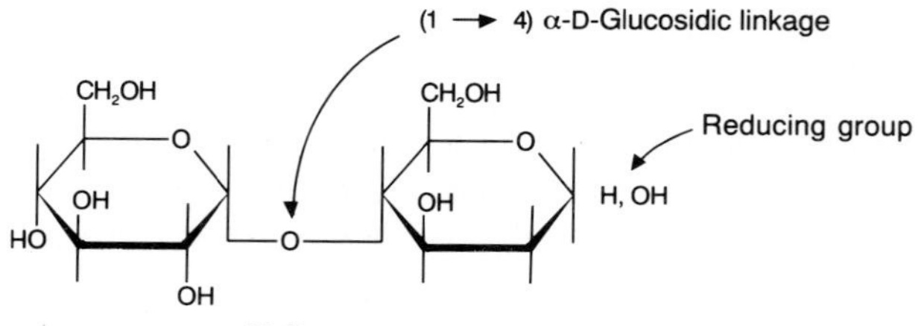

Maltose

Lactose Lactose, the reducing disaccharide of milk, has a $(1 \rightarrow 4)$-β-D-galactosidic linkage to D-glucose.

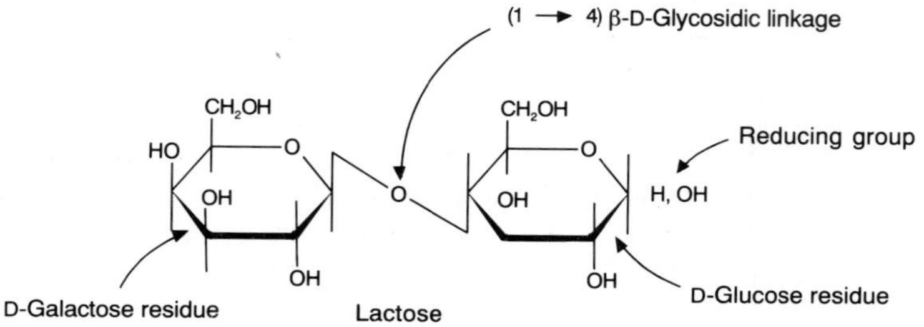

Lactose

Note that the structures of the D-galactose and D-glucose pyranose rings differ only in the configuration around C_4. These sugars are said to be epimeric at C_4, a point that is important in the metabolism of D-galactose and therefore of lactose.

Isomaltose Isomaltose, a disaccharide derived from the branch point of starch, has a $(1 \rightarrow 6)$-α-D-glucosidic linkage to a second D-glucose residue.

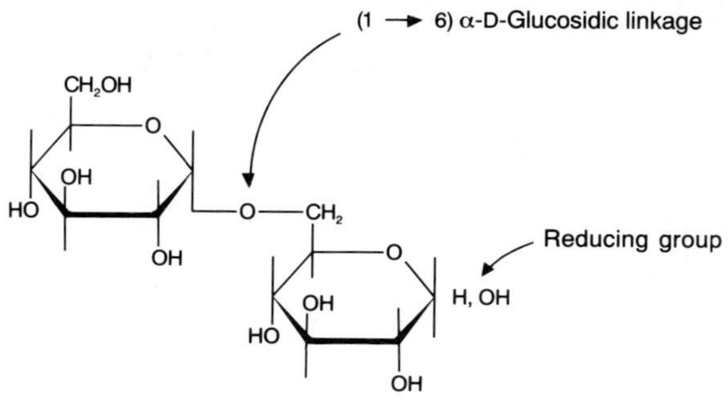

Isomaltose

Sucrose Sucrose is produced commercially from sugar cane or sugar beet. It is a nonreducing sugar, since the reducing groups of α-D-glucopyranose and β-D-fructofuranose are glycosidically linked together.

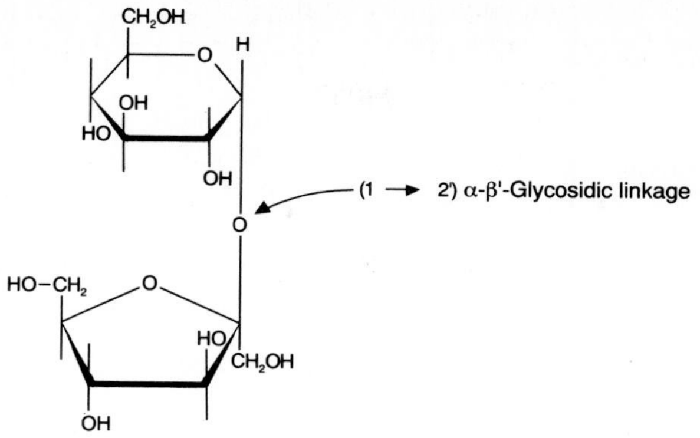

Sucrose

Oligosaccharides and polysaccharides

In further biologic polymerization of monosaccharides the glycosidic linkages continue to be formed by the sequential addition of monosaccharides; thus the trisaccharides, oligosaccharides, and eventually the high molecular weight polysaccharides are synthesized. Hexose monomers are multifunctional compounds with four hydroxyl groups and a hemiacetal group (with α- or β-anomeric configuration) in the ring structures. Many possible structures can be proposed theoretically for oligosaccharides and polysaccharides with the same sugar composition. Thus 11 disaccharides have been synthesized from D-glucose; 176 trisaccharides from D-glucose are possible.

Polysaccharides of great interest are starch and glycogen. Both are composed of D-glucose and are quite similar in general structure, but differences in their finer details show them to be two distinct materials. In both cases the principal glycosidic linkage is between the C_1 of one D-glycopyranosyl residue and the hydroxyl at C_4 of the adjacent residue, the anomeric form being α-D.

$$\alpha\text{-D-Glc}(1 \rightarrow 4)—[\alpha\text{-D-Glc}(1 \rightarrow 4)]_n—\alpha\text{-D-Glc}$$

(Nonreducing end)   (Reducing end)

Short chains of glucose residues linked in this manner are cross-linked through the hydroxyl groups at C_6 of some of the residues.

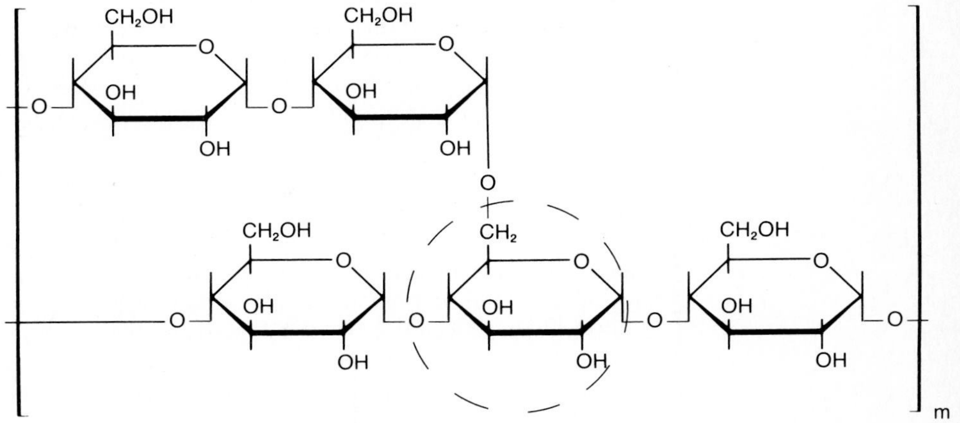

Cross-linking
glucosyl residue
(branch point)

Figure 6.2 Terminal segments of starch or glycogen.

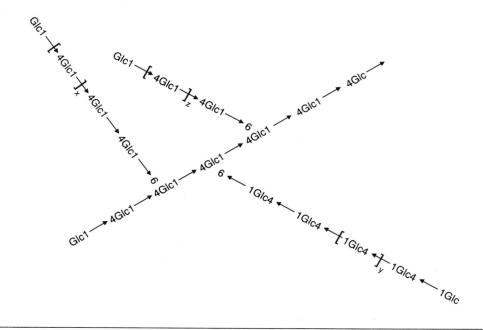

Therefore the structures of starch and glycogen resemble a branching tree. A small portion of the molecule is shown in Figure 6.2.

Polysaccharides share several general structural features. All the polymeric linkages involve the original reducing carbon of the monosaccharide units and a hydroxyl group of an adjacent monosaccharide. Consequently, only one sugar residue in the polysaccharide can have a free reducing group. Branching in the polysaccharide is achieved by additional glycosidic linkages with a particular sugar residue in the existing chain. Since the biosynthesis of polysaccharides usually takes place through addition of monosaccharide units to the nonreducing ends of existing chains, every branch point represents an additional locus for polymer growth by chain elongation. In a highly branched polysaccharide, such as glycogen, the molecule is constructed of short chains whose average length can be calculated. In glycogen the average chain length is about 12 glucose units. Starch is actually a mixture of two polysaccharides. One, amylose, is essentially an unbranched chain of α-D-glucopyranosyl units. The other, a highly branched component called amylopectin, has an average chain length of approximately 25 glucose units.

Unlike proteins or nucleic acids, the polysaccharides are not biosynthesized on a template. This means that the size of the polysaccharide molecule is not fixed; rather a range of molecular weights exists. Thus the molecular weight of normal glycogen may range from approximately 1×10^6 to more than 100×10^6, with most molecules having molecular weights in the vicinity of 5×10^6 to 10×10^6. These values depend on the tissue of origin, the state of nutrition, and the presence of disease.

Other naturally occurring sugar derivatives

Sugar phosphate esters Sugar phosphate esters include α-D-glucopyranose 1-phosphate and D-fructose 1,6-bisphosphate.

α-D-Glucopyranose 1-phosphate 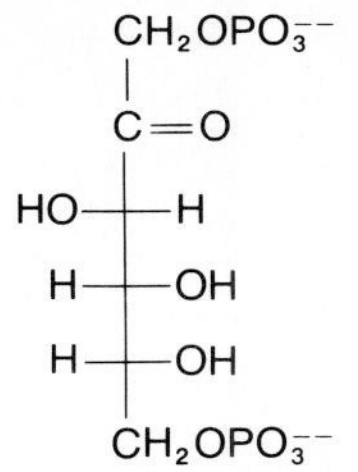   D-Fructose 1,6-bisphosphate

The sugar phosphates are formed by reaction of the sugar with ATP in the presence of the appropriate sugar kinase. When the phosphate residue is attached to the hydroxy group of the reducing carbon, a glycoside is produced and the ring form is fixed; no mutarotation occurs. Esterification of any other hydroxyl groups in the sugar leaves the reducing function free, and the sugar phosphate may react in the open-chain form, as just shown for D-fructose 1,6-bisphosphate.

Glyconic acids The oxidation of the aldehyde group at C_1 in aldoses to the corresponding carboxyl function creates a class of sugar acids called glyconic acids. The name for the glyconic acid of any particular aldose is obtained by replacing -*ose* with -*onic acid,* as in glucose and gluconic acid. Several derivatives of the glyconic acids are important in biochemistry; for example, ascorbic acid is an oxidation product of L-gulonic acid. The metabolism of D-glucose by one pathway involves the 6-phosphate ester of D-gluconic acid.

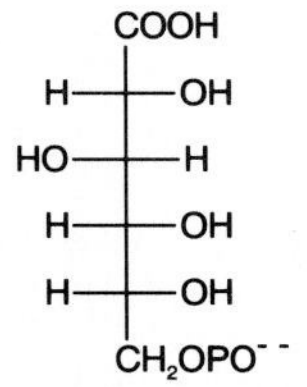

D-Gluconic acid 6-phosphate

Uronic acids Uronic acids are produced through the oxidation of the primary hydroxy groups of sugars. One member of this group, D-glucuronic acid, is shown.

D-Glucuronic acid

The names of these compounds are formed by replacing the -*ose* with -*uronic acid.* These derivatives are particularly important in detoxification mechanisms, as seen earlier in discussions of bilirubin excretion (p. 130), where the bilirubin was linked glycosidically to D-glucuronic acid. The uronic acids D-glucuronic, D-galacturonic, D-mannuronic, and L-iduronic are important residues in polysaccharides. D-Mannuronic acid is the 2-epimer of D-glucuronic acid, and L-iduronic acid is the 5-epimer of D-glucuronic acid.

Polyols Polyhydric alcohols such as mannitol, sorbitol (glucitol), dulcitol (galactitol), and xylitol result from the reduction of the corresponding reducing sugars. Each aldose gives rise to one polyol, whereas a ketose reduces by chemical means to two epimeric polyols. For example, the reduction of D-glucose gives D-glucitol; D-fructose gives a mixture of D-glucitol and D-mannitol, the latter being a symmetric molecule and therefore optically inactive.

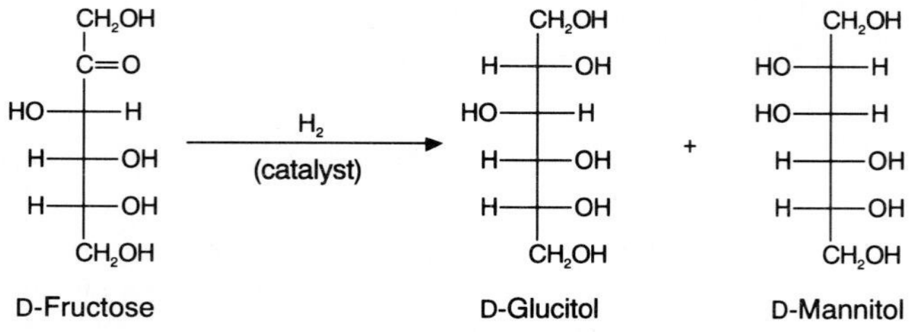

D-Fructose   D-Glucitol   D-Mannitol

The dehydrogenase enzymes that catalyze these reductions are stereospecific so that in the example just given, an equilibrium exists between D-glucitol and D-fructose with the NAD-linked glucitol dehydrogenase; D-mannitol epimer is not a product.

Amino sugars Amino sugars are formed through the substitution of a hydroxyl group by an amino group on the sugar ring. The most common amino sugar is D-glucosamine, which occurs in many polysaccharides.

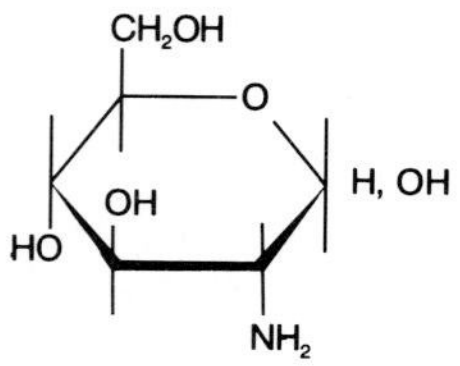

D-Glucosamine

The —NH_2 is usually acetylated (—NH—$COCH_3$) but may also be sulfated (—NH—SO_3^-), as in the anticoagulant polysaccharide heparin.

Many polysaccharides could be formed using only those sugars and sugar derivatives already mentioned. The carbohydrate polymers are also found in covalent linkage with either proteins or lipids. They function in a structural capacity, as in chondroitin sulfate–protein complexes of extracellular ground substance, the hyaluronic acid—protein complex of synovial fluid, and the glycoproteins and glycolipids of cell membranes. All these molecules are complex and play important biologic roles.

Deoxypentoses The most important member of the deoxypentoses is 2-deoxy-D-ribose. 2-Deoxy-D-ribofuranose is the ring form found in the deoxyribonucleosides and deoxyribonucleotides. These pentoses form the backbone of deoxyribonucleic acid (DNA).

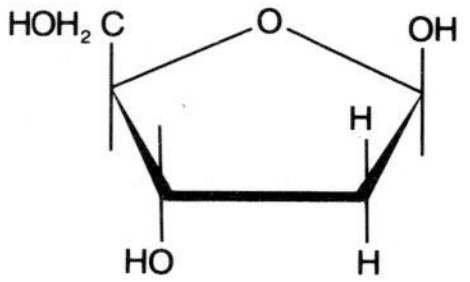

2-Deoxy-β-D-ribfuranose

Nucleosides and nucleotides A sugar that is linked through its reducing carbon to a purine or pyrimidine base is called a nucleoside. Phosphorylation of the sugar residue converts the nucleoside to a nucleotide. Common examples are adenosine and adenosine 5′-phosphate (AMP); 5′ indicates that the phosphate is on carbon atom 5 of the sugar and not on the base, adenine.

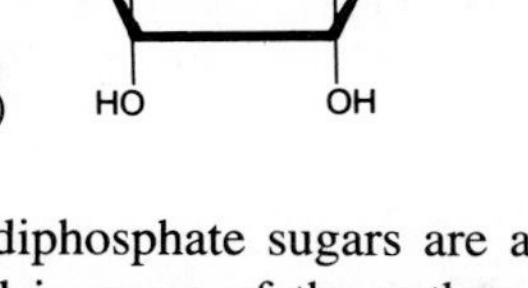

HOH$_2$C   O   Adenine $^-$O$_3$POH$_2$C   O   Adenine

Adenosine
(nucleoside)   HO   OH   HO   OH   5' - AMP
(nucleotide)

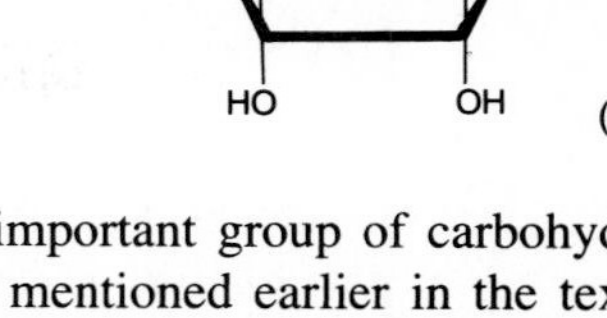

Nucleoside diphosphate sugars and carbohydrate biosyntheses

The nucleoside diphosphate sugars are an important group of carbohydrate derivatives that are involved in some of the pathways mentioned earlier in the text, for example, bilirubin diglucuronide biosynthesis and D-galactose metabolism. Several nucleosides are found in these sugar derivatives, but the most common is uridine (Figure 6.3).

The nucleoside diphosphate sugars are activated forms of sugars, since they contain the necessary energy to form a glycosidic bond. Since the free energy of formation of a glycosidic bond is positive, the reaction must be coupled to a chemical driving force. The coupled reactions involved in disaccharide biosynthesis are:

$$\text{Sugar}_1 + \text{ATP} \rightarrow \text{Sugar}_1\text{-P} + \text{ADP}$$
$$\text{Sugar}_1\text{-P} + \text{UTP} \rightarrow \text{UDP sugar}_1 + \text{PP}_i$$
$$\text{H}_2\text{O} + \text{PP}_i \rightarrow 2\,\text{P}_i$$
$$\underline{\text{UDP sugar}_1 + \text{Sugar}_2 \rightarrow \text{Sugar}_1 - \text{Sugar}_2 + \text{UDP}}$$
$$\text{(Disaccharide)}$$

$$\text{Sugar}_1 + \text{Sugar}_2 + \text{ATP} + \text{UTP} + \text{H}_2\text{O} \rightarrow \text{Disaccharide} + \text{UDP} + \text{ADP} + 2\text{P}_i$$

Most glycosidic bonds are synthesized in this manner. The acceptor of the glycosyl residue from the uridine diphosphate (UDP) sugar may be a substance to be excreted— for example, bilirubin, or a growing polysaccharide chain, or another simpler sugar derivative.

Clinical comment

Lactase synthase Milk production occurs when the mammary glands are activated hormonally. Lactose is a major nutrient contained in the milk, and it is synthesized by the mammary tissue.

Figure 6.3   Uridine diphosphate (UDP) sugar.

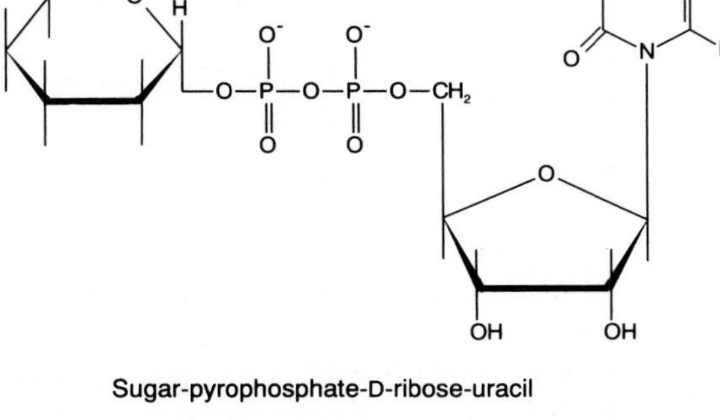

Sugar-pyrophosphate-D-ribose-uracil

One of the more interesting systems for the biosynthesis of disaccharides is that of lactose synthase, which is a β-D-galactosyl transferase. The galactosyl residue is transferred from uridine diphosphate galactose to glucose to form lactose, but only in the presence of the protein α-lactalbumin.

$$\text{UDP-Galactose} + \text{D-Glucose} \xrightarrow[\text{α-Lactalbumin}]{\text{Transferase}} \text{Lactose} + \text{UDP}$$

In the absence of a α-lactalbumin, the β-D-galactosyl residue will transfer to N-acetyl-D-glucosamine groups, either as the free sugar or as nonreducing terminal residues in biopolymers (p. 314) to form the same type of (1→4)-β-D-galactosyl linkage, as in lactose. The catalytic enzyme subunit is found in the mammary gland, liver, and small intestine. However, the modifying regulatory protein α-lactalbumin is synthesized only by the mammary gland under the hormonal control that is in effect during lactation. Prolactin, the hormone that stimulates milk production, is discussed further in Chapter 17.

Digestion of carbohydrates

The hydrolysis of the glycosidic bonds of oligosaccharides and polysaccharides is catalyzed by glycosidases to give the reducing sugar components. The glycosidases are usually specific for the structure of the glycosyl portion of the glycoside, including its anomeric form, but show little specificity for the aglycone. Thus lactase will hydrolyze lactose and many other β-D-galactopyranosides.

For the purpose of our discussion of the digestion of carbohydrates in humans, the fate of starch, lactose, sucrose, and cellulose in the diet will be followed. The principal locations for digestion are the mouth, the lumen of the small intestine, and the brush border of the epithelial cells of the intestinal mucosa. The digestive process is diagrammed in Figure 6.4.

As food is masticated in the mouth into a proper bolus for swallowing, the salivary α-amylase acts on the starch in a random manner. The (1→4)-α-D-glucosidic bonds are split, producing maltose, some D-glucose, and smaller units of the starch molecule called starch dextrins, which contain all the original (1→6)-α-D-glucosidic bonds. At this stage the starch has been reduced in size to an average chain length of approximately eight glucose units, provided the food has been chewed thoroughly.

When the bolus of food meets a high acidity as it enters the stomach, the action of the α-amylase stops. Little hydrolysis of the carbohydrate occurs in the stomach; perhaps sucrose undergoes a slight hydrolysis as a result of the greater sensitivity of its β-D-fructofuranoside linkage to acidity.

The pH of the material entering the small intestine from the stomach is rendered alkaline by secretions from the pancreatic duct. The digestion of the starch dextrins is continued by the action of the pancreatic α-amylase, which is similar to the salivary enzyme except that it has an absolute requirement for chloride ions. When the pancreatic α-amylase completes its hydrolysis of the starch, the intestinal lumen contains principally D-glucose, maltose, isomaltose, α-limit dextrins containing one or more (1→6)-α-D-glucosidic bonds, and the dietary lactose and sucrose. The ingested cellulose is a polysaccharide with (1→4)-β-D-glucosidic linkages for which there is no hydrolytic enzyme in humans; it is nondigestible.

The disaccharides are hydrolyzed at the brush border by specific α-D-glucosidases that are contained in the cell membrane.

$$\text{Glc 1} \xrightarrow{\alpha} \text{4 Glc} \qquad \text{Glc 1} \xrightarrow{\alpha} \text{6 Glc} \qquad \text{Glc 1} \xrightarrow{\alpha} \text{4 Glc 1} \xrightarrow{\alpha} \text{4 Glc}$$

| Maltose | Isomaltose | Maltotriose |

Continued on next page.

Figure 6.4   Pictorial representation of the digestion of carbohydrates.

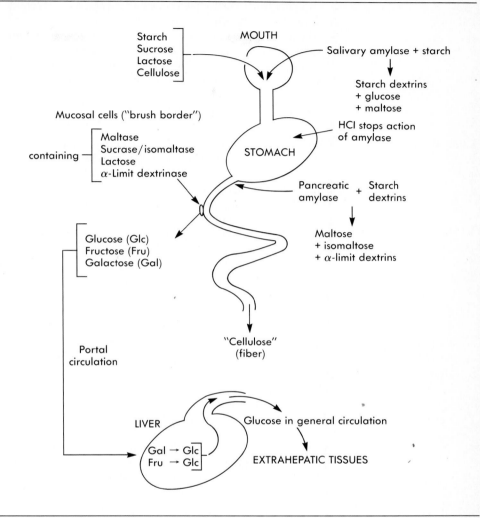

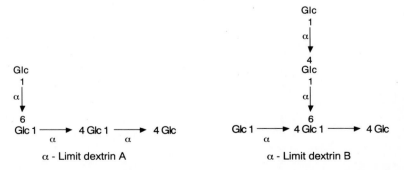

These brush border enzymes are α-limit glucosidase, which will hydrolyze the external (1→4)-α-D-glucosidic linkages to the α-limit dextrin A, in which the (1→6)-α-D-glucosidic linkage is exposed to hydrolysis by isomaltase to give glucose and maltotriose. The latter is hydrolyzed by (1→4)-α-D-glucosidase, which may be maltase, sucrase, or α-limit dextrinase, the last two being active maltases. The sucrase-isomaltase enzymes are in the brush border membrane as a complex, which can be isolated and separated into the two active enzymes.

The monosaccharides so formed and the glucose in the lumen pass into the portal blood system, through which they are transported first to the liver and then to the remainder of the body. Any hydrolysis of the disaccharides that occurs in the lumen of the small intestine is caused by the sloughing of the brush border into this space rather than the secretion of the enzymes into the intestinal fluid. However, depending on the nature of the ingested food, there could be high levels of D-glucose and D-fructose in the lumen.

Clinical comment

Disaccharidase deficiency Mucosal disaccharidase deficiencies, except maltase deficiency, are well-described clinical entities. The most common is lactase deficiency, which varies greatly among different age and ethnic groups. Asians and Africans show a high incidence of lactose intolerance. Primary lactase deficiency is genetically determined. Secondary lactase deficiency can be precipitated by various disturbances in the gastrointestinal tract, including tropical sprue. Removal of lactose (found only in milk or milk products) from the diet is a practical means of alleviating the problem; other carbohydrates, such as sucrose, can be substituted in the diet.

Another rare disaccharidase deficiency is found as an autosomal recessive condition in which sucrase-isomaltase activity is absent in the brush border, leading to sucrose intolerance.

Absorption of carbohydrates from intestine

The mechanism by which sugars are absorbed from the intestine is complex and not completely understood. Some sugars, most commonly the pentoses, pass across the intestinal barrier by simple passive diffusion. The other sugars (notably D-glucose, D-fructose, D-galactose, and possibly D-mannose) can be transported against a concentration gradient; the last traces of these sugars will be absorbed from the intestine in spite of higher concentrations existing in the blood. However, under physiologic conditions the concentration of sugars is greater in the intestine than in the blood during much of the absorption process, and this downhill concentration gradient does not require energy. It can be shown experimentally that the hexoses mentioned previously can be absorbed against a concentration gradient and that some active transport mechanism is available. It is likely that for 90% to 95% of the sugar absorption such a mechanism is not needed. Indeed, it functions mainly to prevent sugars from leaking back into the intestinal lumen.

The sugars absorbed by facilitated diffusion involve carrier proteins.

An Na^+-dependent transport system, specific for D-glucose and D-galactose, causes the active cotransport of these sugars with Na^+. The energy for transport is derived from the concentration gradient of Na^+ across the luminal plasma membrane. This system is inhibited by phlorizin, a plant glycoside. The concentration of the Na^+ in the cells is kept at approximately 40 mmol/L by an active sodium pump that maintains Na^+ transport into the blood, in which the concentration is approximately 140 mmol/L.

Another transport system actively mediates the movement of glucose and galactose from the luminal cell to the blood; this system is inhibited by phlorizin and cytochalasin B. The inhibitors phlorizin and cytochalasin B also affect the active transport of sugars in other tissues.

The digestion of the disaccharides and the absorption of the sugars so produced occur in the brush border, mostly in the upper jejunum. The two processes are largely supportive of each other, since little of the monosaccharide produced by the disaccharidases leaks back into the lumen.

Clinical comment

Oral rehydration therapy Diarrheal diseases kill millions of children before their fifth birthday; those in poor and developing countries are primarily affected. The loss of fluids and electrolytes and the resulting hypovolemia have been particularly well documented for cholera, which is an acute diarrheal disease in which *Vibrio cholerae* are present in large numbers in the liquid stool. Replacement of fluid is essential for survival, and oral rehydration has replaced intravenous administration in many situations. When adequate water and electrolyte replacement is instituted, a cholera patient may have diarrhea for 4 to 6 days, during which time a volume of fluid equal to one to two times the body weight may be lost. Rehydration and any correction of acidosis can be brought about in a few hours.

The basis for oral rehydration is the cotransport of Na^+ with glucose as they travel across the luminal cells of the intestine. The oral fluid cannot exceed the osmolarity of the plasma; or if it does, the diarrhea may increase and plasma Na^+ concentrations may become seriously elevated. These problems are overcome by feeding starch (cooked cereals) and NaCl, a suitable mixture being cooked rice, 50 to 80 g/L; sodium chloride, 3.5 g/L; sodium bicarbonate, 2.5 g/L; and potassium chloride, 1.5 g/L. The last two electrolytes are not essential for successful oral rehydration.

Rate of glucose absorption

After a meal containing sugars the processes of digestion and absorption occur gradually, the quantity of sugars absorbed being approximately 1 g/kg of body weight per hour. It appears that the absorption rate is constant in the small intestine regardless of the amount (within wide limits) of sugar present or the concentration in which it is introduced. In passing through the liver, fructose and galactose are either metabolized to other compounds according to homeostatic need or converted to glucose, the usual circulating blood sugar. Within 30 to 60 min after a meal the blood sugar concentration usually reaches a maximum of approximately 7.2 mmol/L (130 mg/dl) and then decreases in 2 to 2½ hr to approximately 3.9 to 5.0 mmol/L (70 to 90 mg/dl).

Clinical comment

Glucose tolerance test The ability to handle a carbohydrate load may be determined by a glucose tolerance test. The blood glucose concentration is followed over time after the ingestion of 50 or 100 g of glucose (40 g of glucose/m² of body surface). A typical glucose tolerance curve for a normal adult is shown in Figure 6.5. If for any reason the blood glucose concentration exceeds approximately 8.9 to 10.0 mmol/L (160 to 180 mg/dl), the sugar is not completely reabsorbed from the glomerular filtrate. Thus the kidney threshold is exceeded and glycosuria results.

Since the total blood volume in a 70 kg man is approximately 6 L and the dose of 50 to 100 g of glucose is absorbed in about 90 min, it follows that the equilibrium concentration of 3.9 to 5.0 mmol/L (70 to 90 mg/dl) is reached through either the storage or the use of some of the carbohydrate. The process of storage involves the synthesis of glycogen or conversion to fat, whereas the process of use primarily involves glycolysis or an alternative oxidative process, the pentose phosphate pathway. The metabolic pathways are now considered separately.

Interconversion of D-glucose, D-fructose, and D-galactose

A typical meal containing sucrose and lactose places a load of D-galactose and D-fructose on the liver. These sugars must be converted to D-glucose. The postabsorptive blood concentration of these sugars decline rapidly, approaching zero in 1 to 2 hr. The overall isomerization reactions are summarized in Figure 6.6.

There are several enzymes called kinases—for example, glucokinase, fructokinase, and galactokinase—that catalyze the phosphorylation of hexoses. Each kinase (Figure

Figure 6.5   Normal glucose tolerance curve.

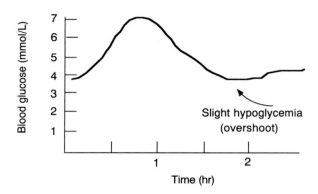

Figure 6.6   Interconversion of hexoses.

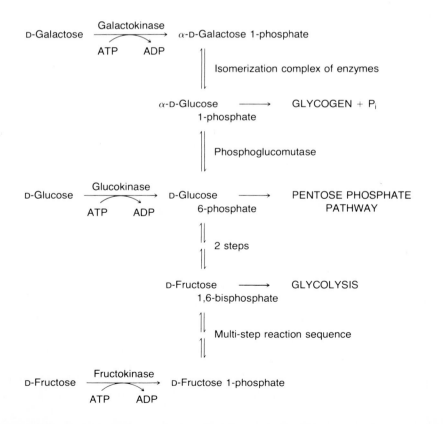

Table 6.1   Brain hexokinase (K_m values)

Sugar	K_m(mol/L)
D-Glucose	8×10^{-6}
D-Fructose	2×10^{-3}
D-Mannose	5×10^{-6}
D-Glucosamine	8×10^{-5}

6.6) is specific for its particular hexose. In addition, there are hexokinase isozymes in D-fructose, D-mannose, and D-glucosamine to give the corresponding 6-phosphate esters. Each hexokinase is different in its Michaelis constant (K_m) for each sugar and for ATP. The data for brain hexokinase are given in Table 6.1. The hexokinases are subject to product inhibition, whereas glucokinase is not. Furthermore, the hexokinases are constitutive enzymes, being present at a fairly constant concentration in the cells, whereas glucokinase is induced in the liver by insulin.

The conversion of both D-galactose and D-fructose to D-glucose 1-phosphate or D-glucose 6-phosphate involves several intermediate steps. In each case homeostatic regulatory mechanisms determine the ultimate fate of the sugars and, as seen in Figure 6.6, the point at which the intermediate enters some other pathway. For example, if the cells of the body are "charged" with glucose and its intermediary metabolites, excess carbohydrate available to the cells will be converted to glycogen or lipid for storage. The D-galactose would therefore be converted to D-glucose 1-phosphate for glycogenesis and not to D-glucose 6-phosphate in the liver cell. If, on the other hand, carbohydrate were needed immediately for energy production, the D-galactose would be converted to D-glucose 6-phosphate and D-fructose 1,6-bisphosphate.

Utilization of D-glucose

Insulin is not required for blood glucose to penetrate liver cell membranes. In contrast, glucose entry into muscle and adipose tissue cells is greatly enhanced by insulin. In liver, insulin induces the synthesis of the enzyme glucokinase, which is present in very low concentrations in the starved state or in diabetes mellitus. Glucokinase is found only in hepatic tissues and is the principal kinase for glucose in the parenchymal cells of liver when the body has received an average carbohydrate diet. Liver, muscle, and adipose hexokinases are unaffected by insulin.

In the preprandial state the blood glucose concentration is approximately 5 mmol/L. The tissues take up glucose from the extracellular fluids as it is available and as insulin is available for muscle, adipose, and other insulin-responsive tissues. The rate of glucose uptake is controlled in part by the feedback inhibition of hexokinase by glucose 6-phosphate. After eating and the concomitant rise in blood glucose concentration to 7 to 10 mmol/L, the glucokinase ($K_m = 1 \times 10^{-2}$ mol/L of glucose) becomes effective. Glucokinase does not exhibit feedback inhibition, and converts blood glucose to D-glucose 6-phosphate even at high glucose 6-phosphate concentrations. Therefore, at times of elevated blood glucose, the catalytic activity of glucokinase is less affected than hexokinase by the glucose load and can better return conditions toward normal. As the glucose concentration in blood falls, the contribution of the glucokinase to the homeostatic mechanisms is reduced. However, as noted in Figure 6.5, a short hypoglycemic period is produced by the residual effects of insulin and glucokinase. During this adjustment period, glucose release from the liver is stimulated by a second hormone, glucagon, and the blood glucose level fluctuates until it becomes stabilized in a dynamic equilibrium. With the return to normal glucose levels, the hexokinases, with their greater affinity (lower K_m) for glucose, take over the phosphorylation of D-glucose as it enters the cell.

Clinical comment

Carbohydrate metabolism in starvation In starvation or diabetes, glucokinase is present in the liver in low concentrations, if at all. Administration of insulin to a diabetic animal results in the biosynthesis of glucokinase, but the maximum concentration of enzyme is not attained for several days. Similarly, a high-carbohydrate diet returns the glucokinase concentration in a starved animal to normal in a few days. Thus, under normal circumstances the liver responds to insulin by maintaining fairly constant levels of glucokinase, levels that do not fluctuate widely at mealtimes. This characteristic is also recognized when a glucose tolerance test is carried out; the patient is primed with a high-carbohydrate diet for more than 3 days before the test. The initial steps in the utilization of blood glucose by the various tissues are summarized in Figure 6.7.

Utilization of D-fructose

The utilization of D-fructose from the diet illustrates the many ways in which an intermediate may be used in metabolism, depending on the homeostatic needs of the tissues. In liver the D-fructose is converted to its 1-phosphate, which is cleaved into two triose fragments by *aldolase,* as shown in Figure 6.8. The reactions catalyzed by aldolase occur with a small change in free energy and have an equilibrium constant of nearly one. The enzyme forms a Schiff base intermediate between an ε-amino group of a lysyl residue and the dihydroxyacetone phosphate. This intermediate may either transfer the dihydroxyacetone phosphate to water, and so drive the reaction in the direction of cleavage, or be transferred to an aldose, for example, D-glyceraldehyde. In the reversible reaction, where there may be alternate aldose acceptors, any of the aldoses may condense with

Figure 6.7 Utilization of blood glucose by tissues.

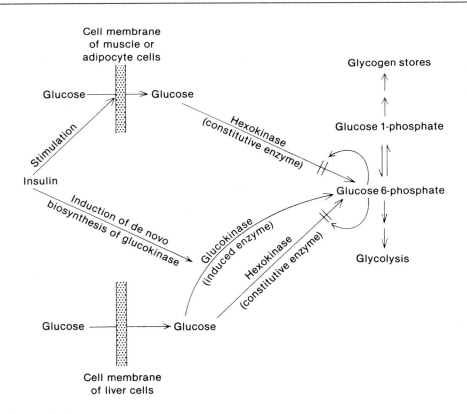

the dihydroxyacetone phosphate residue for which the enzyme is highly specific.

In the liver the liberated D-glyceraldehyde, which is derived from carbon atoms 4, 5, and 6 of the D-fructose 1-phosphate, is phosphorylated by ATP and a triose kinase to D-glyceraldehyde 3-phosphate, an alternate aldose acceptor for the dihydroxyacetone phosphate. Furthermore, the triose phosphates are reversibly interconverted by *triose phosphate isomerase,* an important enzyme also involved in the catabolism of D-glucose.

The condensation of the triose phosphates produces D-fructose 1,6-biphosphate, the 1-phosphate group of which is specifically hydrolyzed by D-fructose 1,6-biphosphate 1-phosphatase. The D-fructose 6-phosphate so formed is isomerized to D-glucose 6-phosphate, in which the phosphate group must be changed to position 1 if the sugar is to be converted to glycogen. The position of the phosphate ester group can be transferred through the catalytic action of a mutase, in this case a *phosphoglucomutase*. This enzyme requires the presence of a coenzyme (D-glucose 1,6-biphosphate) and functions by the

Figure 6.8   Metabolism of D-fructose in liver.

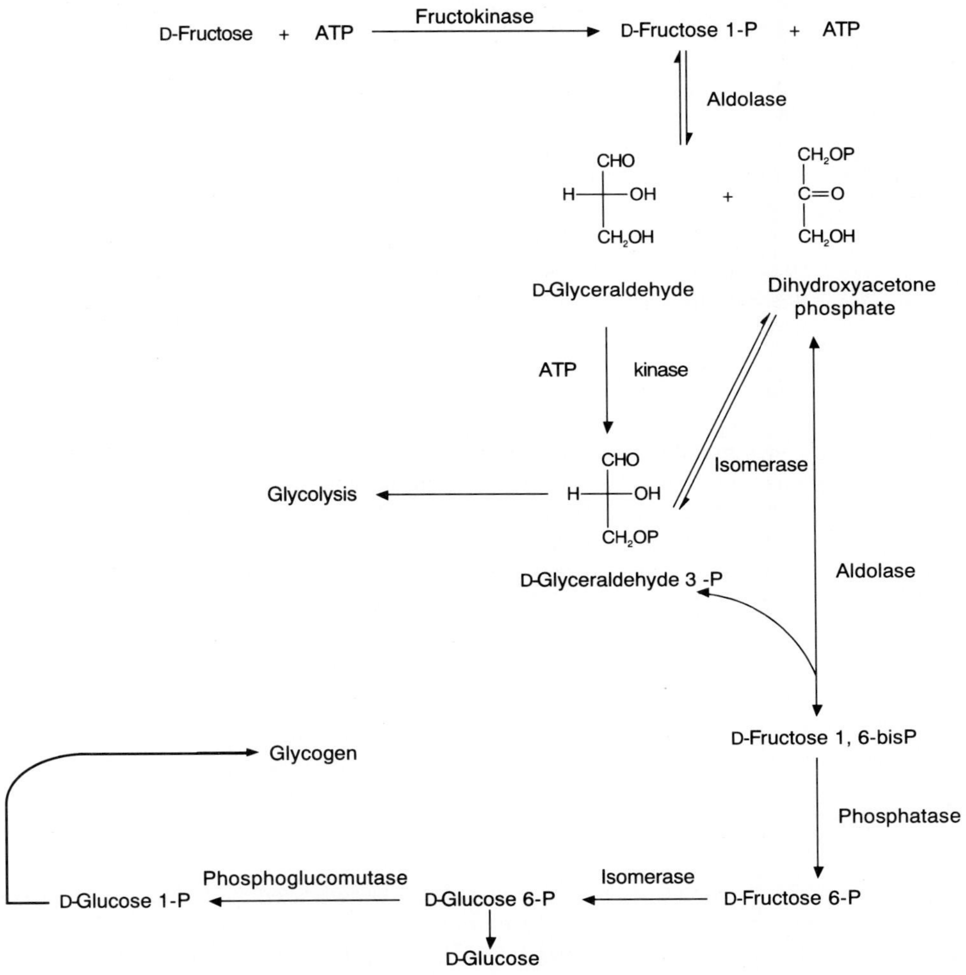

formation of an intermediate phosphoenzyme. This activated enzyme contains an *O*-phosphorylated seryl residue that may donate the phosphate to either position 1 or 6 of a D-glucose phosphate. The 1,6-biphosphate formed in this manner transfers one of its phosphates back to the seryl group. Which of the two phosphate groups is transferred to the enzyme depends on the relative concentrations of the two phosphate esters. Following is the sequence of events, starting with D-glucose 6-phosphate.

$$\text{D-Glucose 6-P} + \text{EnzP} \rightleftharpoons \text{D-Glucose 1,6-bisP} + \text{Enz}$$
$$\Updownarrow$$
$$\text{D-Glucose I-P} + \text{Enz-P}$$

Phosphoglucomutase was one of the first enzymes proved to function by formation of a protein seryl-phosphate intermediate.

Evidence for this apparently circuitous route of fructose metabolism in liver to give D-glucose has been attained through isotopic labeling of D-fructose and determination of the position of the label in the resulting D-glucose. A [^{14}C] label at C_1 of the D-fructose subsequently appears at positions C_1 and C_6 of D-glucose.

In the short time during which D-fructose circulates in the blood before removal by the liver, adipose tissue and muscle are probably the main extrahepatic tissues that make use of it. The pathway for the conversion of D-fructose to D-glucose in extrahepatic tissues is shown in Figure 6.9. Because hexokinase has a relatively high K_m for D-fructose, this pathway is not a major one unless a high level of fructose is maintained in the blood. The hexokinase of muscle directly produces D-fructose 6-phosphate, which is isomerized to D-glucose 6-phosphate. Muscle does not contain D-glucose 6-phosphatase; therefore once a hexose has entered the muscle cell, it cannot be released into the blood as glucose to maintain the circulating sugar levels.

As will be seen later, the metabolism of D-fructose is closely interrelated with glycolysis (p. 268) and the pentose phosphate pathway (p. 277). Since all these pathways occur in the cytoplasm, the common branch points in these schemes (for example, the reactions involving D-glucose 6-phosphate) are subject to metabolic controls.

Utilization of D-galactose

D-Galactose is readily converted to D-glucose in the liver by the metabolic process shown in Figure 6.10. The transformation is rapid. After ingestion of about 40 g of D-galactose, a healthy adult will show a maximum concentration in the blood after 1 hr. All the D-galactose will have disappeared from the blood within 2 hr of ingestion.

The metabolism of D-galactose in the liver is initiated by its phosphorylation to D-galactose 1-phosphate, catalyzed by a specific galactokinase. An exchange reaction by the enzyme D-galactose 1-phosphate uridyl transferase catalyzes the transfer of the glycosyl residues between UDP-glucose and D-galactose 1-phosphate to form D-glucose 1-phosphate and UDP-galactose.

D-Galactose 1-P ⟶ UDP-glucose
D-Galactose 1-phosphate uridyl transferase
UDP-galactose ⟶ D-Glucose I-P

The D-glucose 1-phosphate so formed is converted to D-glucose 6-phosphate and either enters the metabolic pathway of glycolysis or is hydrolyzed to glucose and passes into the circulation. The UDP-galactose undergoes epimerization of the hydroxyl group at C_4 of the D-galactose by formation of an intermediate 4-keto-D-galactose.

Figure 6.9 Alternative metabolism of D-fructose.

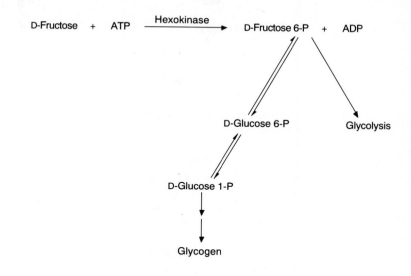

Figure 6.10 Metabolism of D-galactose.

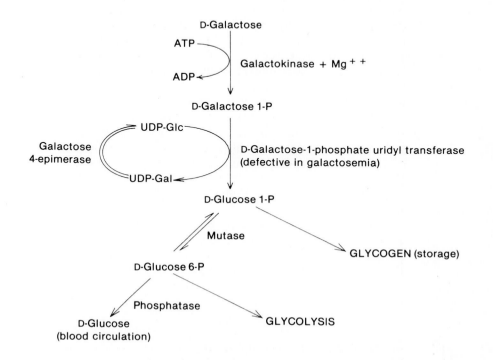

The NAD$^+$ is tightly bound to the enzyme, UDP-galactose-4-epimerase, and accepts the two hydrogens at C_4 from the sugar, only to transfer these same hydrogens back to the 4-keto-intermediate, except that the stereochemistry of the new hydroxyl group may be such that UDP-glucose is formed if UDP-galactose is in excess of that required by the equilibrium. However, the 4-keto intermediate may also be reduced to form UDP-galactose if that is called for. From isotopic studies it is clear that the epimerization of the two sugar residues retains the same hydrogens in the inversion of the hydroxyl groups.

In the metabolism of D-galactose, the UDP-galactose is converted to UDP-glucose that reacts with more D-galactose 1-phosphate, as summarized in Figure 6.10.

Clinical comment

Galactosemia Several pathologic conditions prevent the utilization of D-galactose derived from dietary lactose. Two such diseases involve a congenital deficiency in the metabolism of D-galactose to D-glucose. Von Reuss was the first to report such a disorder when he described an 8-month-old child with what are now recognized as characteristic symptoms of galactosuria and a cirrhotic liver. (The later condition may also have been related to the fact that the infant was fed cognac.) In subsequent cases it was noted that dramatic relief occurred if the patient was given a galactose-free diet.

The most common type of galactosemia is caused by a defect in an autosomal recessive gene that results in the reduction or absence of D-galactose 1-phosphate uridyl transferase. Although the heterozygote will show a reduction in this enzyme, the level usually is sufficient to maintain health unless continuously challenged by high-galactose diets. The homozygote, however, because of the absence of uridyl transferase enzyme, must be recognized at birth so that milk, which contains lactose, can be excluded from the diet. Fortunately, the erythrocytes reflect the enzymatic status of these patients, and cord blood can be used to assay D-galactose 1-phosphate uridyl transferase levels. Such a test is particularly important in families with a history of this disease. Within several weeks or months after birth the consequences of galactosemia are evident and irreversible. A galactose-free diet avoids the difficulties of a high level of galactose 1-phosphate in all tissues. The galactose that is necessary for the biosynthesis of cell membranes, cerebrosides, glycoproteins, and other cell components can be formed from glucose 1-phosphate, as shown in Figure 6.11. The UDP-Gal that is formed is used directly in the biosynthetic processes.

A less common congenital form of galactosemia is caused by a deficiency in galactokinase.

In galactosemia the increased concentration of galactose in blood and the aqueous humor leads to greater entry of the sugar into the lens. There it is converted partly to the corresponding sugar alcohol, galactitol (dulcitol), which eventually causes cataract formation.

Glycolysis

Glycolysis has three principal features:

1. It is the degradative pathway whereby D-glucose is oxidized to pyruvate, which is metabolized further by either of two routes. When the supply of oxygen is inadequate

Figure 6.11   Biosynthesis of D-galatose.

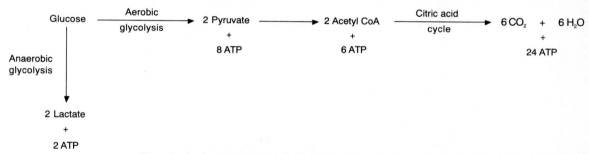

for complete oxidation, the pyruvate is reduced to lactate to maintain the necessary levels of NAD^+ (*anaerobic* glycolysis). Under *aerobic* conditions the pyruvate formed by glycolysis is decarboxylated to acetyl CoA, which enters the citric acid cycle, where it is oxidized to carbon dioxide and water.

2. Glycolysis is integrated into many of the metabolic processes that take place in the cell with intermediates, either six or three carbon atoms; these intermediates are common to other pathways, such as the pentose phosphate pathway. The intermediate compounds also provide sources of starting materials for the biosynthesis of substances such as triacylglycerol from glyceraldehyde 3-phosphate, L-alanine from pyruvate, and glycogen from glucose 1-phosphate.

3. Glycolysis is accompanied by the formation of ATP, although this is only about a quarter of the ATP that can be derived from the complete oxidation of glucose to carbon dioxide and water. The energy conserved in the ATP from glycolysis, however, is critical.

Glycolysis thus can proceed either under aerobic conditions, when the supply of oxygen is adequate to maintain the necessary levels of NAD^+, or under anaerobic (hypoxic) conditions, when the level of NAD^+ cannot be maintained by way of the mitochondrial cytochrome system and depends on the temporary expedient of converting pyruvate to lactate (Figure 6.12). Anaerobic glycolysis, the temporary reliance on pyruvate, is the means by which the body awaits the return of adequate oxygen. Therefore the condition is called *oxygen debt*.

As was noted in Chapters 1 and 5, the maintenance of certain oxygen and carbon dioxide levels in cells is essential to their normal function. Abnormal situations do arise, however, when a stress is placed on the body. Such a stress might be a high-energy demand—for example, extreme exercise or the hyperventilation of encephalitis, when the rate at which oxygen is brought to the cell is not sufficient to keep pace with the oxidative catabolic reactions producing ATP. Since these oxidation reactions are linked

Figure 6.12   Regeneration of NAD⁺ by pyruvate.

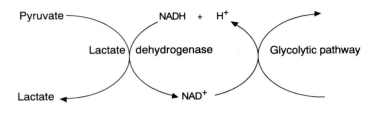

Figure 6.13   Blood lactate levels following exercise, shown here as a 4-minute mile.

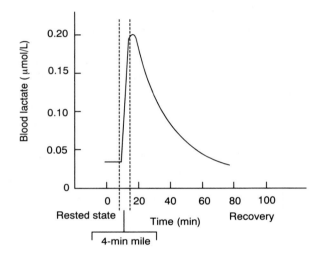

to oxygen through $NAD^+/NADH$ and the cytochrome system and since they cannot proceed unless $NADH + H^+$ is converted to NAD^+, an emergency step involving pyruvate is called into play. This results in the conversion of pyruvate to lactate. When the lactate level in the blood increases, the pH falls, and the expected signs of rapid breathing and exhaustion set in. Variations in blood lactate levels accompanying changes in physical activity are illustrated in Figure 6.13. The lactate that is produced and released into the blood is converted back to pyruvate in the liver when sufficient oxygen is made available.

Pathway of glycolysis   All the reactions leading to the formation of pyruvate and lactate are catalyzed by enzymes present in the cytoplasm. (The various types of enzymatic reactions in glycolysis have been considered previously in this chapter.)

Glycolysis may be considered to take place in three stages. The first stage is concerned with the formation of D-glucose 6-phosphate, which may come about by either the phosphorolysis of glycogen to D-glucose 1-phosphate followed by isomerization to the 6-phosphate or by the isomerization of other hexose phosphates, as described earlier in this chapter, or by the uptake of blood glucose into the cell through its phosphorylation catalyzed by hexokinase or glucokinase. The cell membrane is impermeable to sugar phosphates, and only in the case of liver and kidney is there a D-glucose 6-phosphatase in the endoplasmic reticulum membrane to release the trapped sugar to the blood.

Stage 1 The first series of reactions is represented as follows:

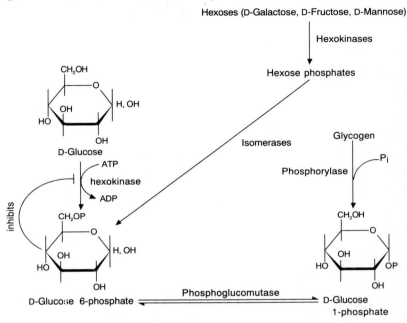

The cellular concentration of the 6-phosphate acts as a product inhibitor of hexokinase and activates glycogen synthase D (see p. 305). Therefore mechanisms are present that limit the entry of unnecessary glucose into glycolysis.

Stage 2 The second stage in glycolysis results in the splitting of the six-carbon chain of D-glucose 6-phosphate into two molecules of D-glyceraldehyde 3-phosphate. Following the isomerization of D-glucose 6-phosphate into D-fructose 6-phosphate, a second phosphorylation occurs to produce D-fructose 1,6-bisphosphate. At this point the sugar is irreversibly committed to glycolysis. This irreversible step is carefully controlled. A large cellular energy charge (see p. 206) and a high concentration of citrate in the cytoplasm, which result from respiratory chain activity and thus ATP production, represent negative allosteric effectors for phosphofructokinase. Phosphofructokinase is also greatly stimulated by D-fructose 2,6-bisphosphate, which increases the affinity of the enzyme for its substrate D-fructose 6-phosphate and relieves the inhibition by ATP. Its stimulation acts synergistically with AMP. D-Fructose 2,6-bisphosphatase is a positive effector of both liver and muscle phosphofructokinase and inhibits D-fructose 2,6-bisphosphatase. Its concentration in liver is greatly increased under conditions of active glycolysis and is decreased by glucagon. It appears therefore to be a major regulator of both glycolysis and gluconeogenesis in the liver. The inactive phosphorylated form of D-fructose 2,6-bisphosphatase is reactivated by hydrolysis of the phosphate group by protein phosphatase 1.

The D-fructose 1,6-bisphosphate is split by aldolase to form the two triose phosphates, which are interconverted by triose phosphate isomerase. At equilibrium this reaction favors dihydroxyacetone phosphate to approximately 95%. However, in glycolysis the D-glyceraldehyde 3-phosphate is continuously oxidatively phosphorylated to 1,3-bisphospho-D-glycerate. The reactions in the second stage of glycolysis are as follows:

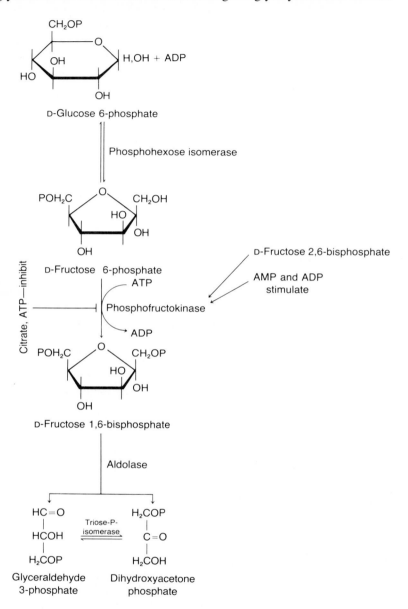

Stage 3 The final stage of glycolysis is notable for the oxidation of 3-phospho-D-glyceraldehyde to pyruvate with the conservation of the energy derived from chemical oxidations by conversion of ADP to ATP (stage 3). The first reaction in this stage is represented in simple terms by Figure 6.14, in which an —SH group of the enzyme is shown to react with the aldehyde group to form a thiohemiacetal. The NAD$^+$ coenzyme that is part of the enzyme complex oxidizes the thiohemiacetal to a reactive thioester, which undergoes phosphorolysis with inorganic phosphate to liberate the enzyme and form a reactive 1,3-bisphospho-D-glycerate ($\Delta G^{0'}$ for hydrolysis is approximately -50.2 kJ/mol; -12 kcal/mol). The phosphate ester can be transferred to ADP to form ATP.

Alternatively, in the erythrocyte 1,3-bisphospho-D-glycerate may be isomerized to the 2,3-bisphospho-D-glycerate, a molecule that plays an important role in the transport of oxygen by hemoglobin (see p. 157).

Glycolysis proceeds as phosphoglycerate mutase catalyzes the formation of 2-phospho-D-glycerate, which by the elimination of water yields 2-phospho*enol*pyruvate. Phospho*enol*pyruvate is another molecule with a high negative free energy of hydrolysis ($\Delta G^{0'}$ is approximately -62.8 kJ/mol; -15 kcal/mol). It reacts with ADP to form ATP and

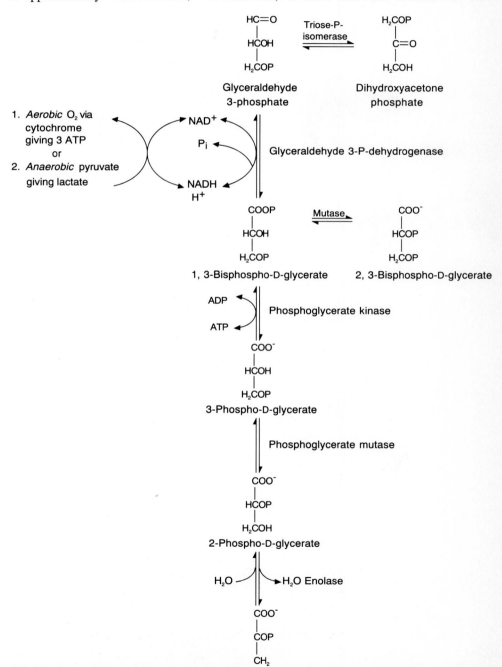

Continued on next page.

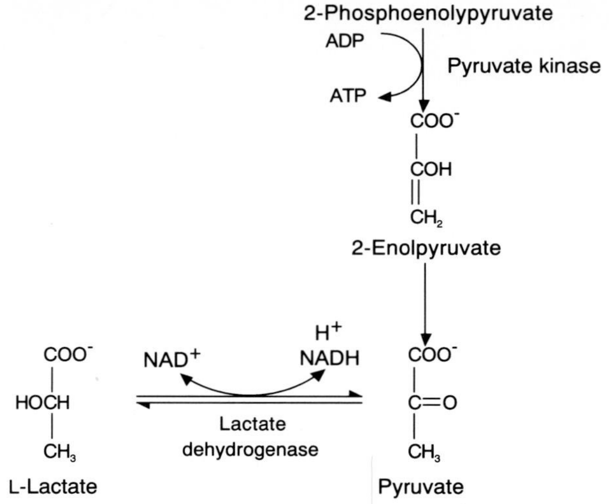

Figure 6.14   The oxidation and phosphorylation of D-glyceraldehyde 3-phosphate.

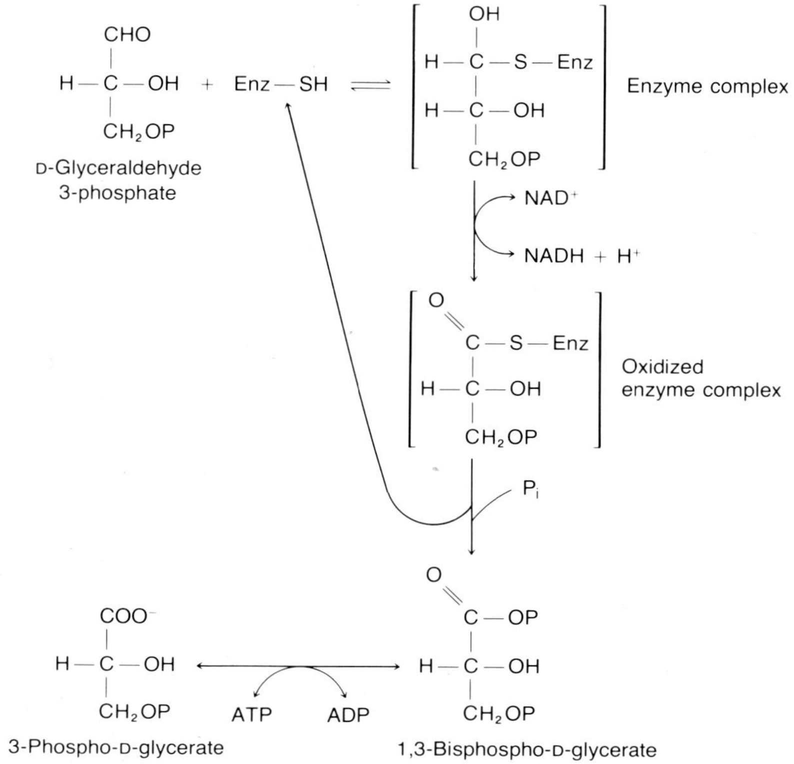

Energetics of glycolysis

As noted in the first reaction of stage 3 in the glycolytic pathway, the NADH may be oxidized under aerobic conditions by coupling the reaction eventually to oxygen through the mitochondrial system (see Chapter 5) or by reduction of pyruvate to lactate anaerobically (see earlier discussion). The overall reaction for glycolysis occurs completely in the cytosol of the cell. The anaerobic pathway yields 2 mol of ATP/mol of glucose.

$$\text{Glucose} + 2ADP + 2P_i \rightarrow 2 \text{ Lactate} + 2ATP + 2H_2O$$

The aerobic glycolysis involves the transfer of hydrogens from NADH, H^+ into the mitochondria. If this occurs by way of the glycerophosphate dehydrogenase–dihydroxyacetone phosphate shuttle (see Chapter 5), the hydrogens appear in the mitochondria as flavin adenine dinucleotide H_2 (FADH$_2$) and there is a loss to the cell of 1 mol of ATP/ NADH, H^+ transferred. Therefore the overall reaction for aerobic glycolysis yields 6 mol of ATP/mol of glucose.

$$\text{Glucose} + 6ADP + 6P_i \rightarrow 2 \text{ Pyruvate} + 6ATP + 6H_2O$$

However, if the transfer goes by way of the concerted action of several translocases, such as was described in Chapter 5 for the oxaloacetate-malate system, no such loss of reducing equivalents is experienced, and mitochondrial oxidative phosphorylation yields 3 mol of ATP per NADH, H^+ transferred. The overall yield of ATP to the cell is then 8 mol/mol of glucose oxidized to pyruvate. The steps involving ATP may be summarized as follows, assuming no loss in ATP from the translocation of the reducing equivalents of NADH, H^+ from cytosol to mitochondria.

Aerobic glycolysis	ATP per mole of D-glucose
Phosphorylation of D-glucose	-1
Phosphorylation of D-fructose 6-P	-1
2(1,3-BisP-Glycerate → 3-P-glycerate)	$+2$
2(Phospho*enol*pyruvate → Pyruvate)	$+2$
2(NADH + H$^+$ → NAD$^+$)	$+6$
TOTAL	$+8$

The disposition of pyruvate can take many paths (see Chapter 5). It cannot, however, be reconverted directly to phospho*enol*pyruvate, as explained in the discussion of gluconeogenesis in Chapter 7.

The $\Delta G^{0'}$ for the oxidation of 1 mol of D-glucose to 2 mol of pyruvate is approximately 502 kJ/mol (-120 kcal/mol). Part of this is conserved as 6 or 8 mol of ATP through the mitochondrial oxidation of NADH and the substrate-level phosphorylation. The $\Delta G^{0'}$ of hydrolysis of 8 mol of ATP is equal to approximately -234 kJ (-56 kcal). Therefore, of the energy available (-502 kJ/mol of D-glucose), slightly less than 50% is conserved as ATP for subsequent use. It should be noted, however, that much of the energy of the D-glucose remains in the 2 mol of pyruvate, since the total oxidation of D-glucose is associated with a $\Delta G^{0'}$ of -2.87 MJ/mol (-686 kcal/mol). As demonstrated in Chapter 5, the metabolism of 1 mol of pyruvate to acetyl CoA yields 3 mol of ATP, and the oxidation of acetyl CoA in the Krebs cycle and respiratory chain adds another 12 mol of ATP. Therefore the 2 mol of pyruvate produced from 1 mol of D-glucose corresponds to 30 mol of ATP, giving a total of 38 mol of ATP per mol of D-glucose metabolized completely to $CO_2 + H_2O$. Thus the overall energy conservation as ATP is about 40%. Your car should be so efficient!

During periods of anaerobic glycolysis, the ATP from mitochondrial oxidative phosphorylation is not available. Consequently, the net gain in ATP per mole of D-glucose converted to 2 mol of L-lactate is reduced to 2 mol. Although this is low in comparison to aerobic glycolysis, in times of stress it is better than nothing and may suffice. If the

Regulation of glycolysis

stress is not too prolonged, the total loss of calories is not serious. However, prolonged anaerobic glycolysis results in loss of lactate in the urine and thus a wasting of calories.

The glycolytic pathway, as with any other metabolic scheme, cannot be considered in isolation from the other biochemical reactions in the cell. Many of the molecules in glycolysis are reactants or products of enzymatic reactions in other schemes, and glycolysis must be studied in this framework of metabolism.

The regulation of glycolysis is related to the following.

1. *Availability of substrates.* Glycolysis requires at least a supply of D-glucose or D-glucose 1- or 6-phosphate, ADP, inorganic phosphate, and NAD^+ for the subsequent utilization of pyruvate. The need for ATP is not included because, if glycolysis proceeds at all, there is a net generation of ATP independent of mitochondrial oxidation and any other pathway.

NAD^+ and inorganic phosphate are required in the oxidation and concomitant phosphorylation of D-glyceraldehyde 3-phosphate. The NADH so formed is regenerated either by the reaction with pyruvate to produce lactate or aerobically through the respiratory chain. Inorganic phosphate (P_i) comes from those many reactions that are energy demanding and coupled to the hydrolysis of ATP, such as ion transport, muscle contraction, and biosynthetic reactions. In the broadest sense, therefore, one may look on glycolysis as being coupled to the hydrolysis of ATP for the provision of P_i and ADP. As a result, some regulation of the pathway occurs.

2. *Oxidation-reduction condition of the cell.* Since glycolysis is an oxidative process, it is controlled in part by the relative concentrations of $NAD^+/NADH$, H^+ and pyruvate/lactate. These in turn are dependent upon the respiratory chain and the availability of oxygen.

3. *Enzyme activity.* With the possible exception of the aldolase reaction, it is improbable that any of the individual reactions in glycolysis are at equilibrium in living cells. The enzymes catalyzing the essentially irreversible steps—hexokinase, 6-phosphofructokinase, and pyruvate kinase—are the important points of control. This is usually achieved by allosteric inhibition of these enzymes by ATP. The inhibition is relieved by those molecules that reflect a low energy charge of the cell, ADP and AMP. The step that commits a hexose to glycolysis is the phosphorylation of 6-phosphofructose.

6-Phosphofructokinase In most animal tissues 6-phosphofructokinase (PFK-1) is a tetramer composed of three different subunits, which are assembled in various ways to produce isoenzymes that are characteristic of the tissue. These isoenzymes have basically the same catalytic properties, but they differ quantitatively in their sensitivity to ATP inhibition and to other effectors. Thus muscle PFK-1 is less sensitive to fructose 2,6-bisphosphate than is liver PFK-1. The inhibitory effect of ATP on PFK-1 is diminished by fructose 2,6-bisphosphate, which also increases the affinity of PFK-1 for its substrate, fructose 6-phosphate.

Fructose 2,6-bisphosphate is not one of the substrates of glycolysis, but it is produced by the phosphorylation of fructose 6-phosphate with 6-phosphofructokinase 2 (PFK-2). Hydrolysis of the 2-phospho group is catalyzed by fructose bisphosphatase 2 (FBP-2).

D-Fructose 6-phosphate   6-Phosphofructose 2-kinase   ATP   ADP   Fructose bisphosphatase 2 P_i D-Fructose 2, 6-bisphosphate

Both enzymatic activities, PFK-2 and FBP-2, are present in the same polypeptide, a bifunctional enzyme. These enzymatic activities are controlled by the phosphorylation of a seryl residue (in the bifunctional enzyme) by a cyclic AMP–dependent protein kinase. In the liver this results in simultaneous and opposite changes in the two enzymatic activities, the phospho-bifunctional enzyme is inactive as a kinase and active as a phosphatase. This accounts for the decrease in fructose 2,6-bisphosphate concentration in liver by glucagon, which stimulates adenylate kinase, to give the following series of reactions:

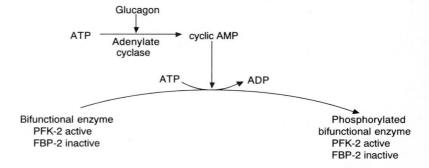

The activities are reversed by hydrolysis of the seryl phosphate by protein phosphatase to give an active PFK-2 and an inactive FBP-2.

The bifunctional enzyme in muscle is different from the liver enzyme in that the PFK-2 kinase activity is stimulated by phosphorylation so that epinephrine, an activator of adenylate kinase in muscle, increases the concentration of fructose 2,6-bisphosphate. However, as noted earlier, the PFK-1 of muscle is not as sensitive to the effects of fructose 2,6-bisphosphate as is the liver enzyme, and thus the control of glycolysis in extrahepatic tissue depends less on the flux of fructose 2,6-bisphosphate concentration in the cytosol.

Other pathways The interplay among the glycolytic and other pathways is extensive. Depending on the overall metabolic state, other pathways may pass intermediates into a segment of the glycolytic pathway for conversion to pyruvate or glucose. The reverse may also occur; for example, although other tissues can phosphorylate glycerol, dihydroxyacetone phosphate from glycolysis is the only source of glycerol 3-phosphate in adipose tissue (see Chapter 10).

Pentose phosphate pathway (hexose monophosphate shunt)

An important branch point in carbohydrate metabolism is the oxidation of D-glucose 6-phosphate by the pentose phosphate pathway. This pathway is the source of D-ribose for nucleotide synthesis and a major source of cytoplasmic NADPH, the biochemical reducing agent for many anabolic reactions. The pathway is composed of two parts.

Part 1

The first part involves the oxidation of the reducing carbon of D-glucose 6-phosphate to CO_2 by two equivalents of $NADP^+$ to give D-ribulose 5-phosphate and two equivalents of $NADPH + H^+$ (Figure 6.15).

D-Glucose 6-phosphate dehydrogenase is widely distributed in tissues and, under physiologic conditions, uses $NADP^+$ as the coenzyme. The oxidation to the 6-phospho-D-gluconolactone is followed by its irreversible hydrolysis to 6-phospho-D-gluconate, which is oxidized and decarboxylated to give D-ribulose 5-phosphate. The 6-phospho-D-gluconate dehydrogenase also requires $NADP^+$, and the reaction proceeds through the intermediate formation of a 3-keto-D-gluconate 6-phosphate. A dephosphorylated isomer, 3-keto-L-gulonate, is important in another carbohydrate pathway to be discussed on p. 284.

The overall reaction for part 1 of the pentose phosphate pathway is:

Figure 6.15   Pentose phosphate pathway.

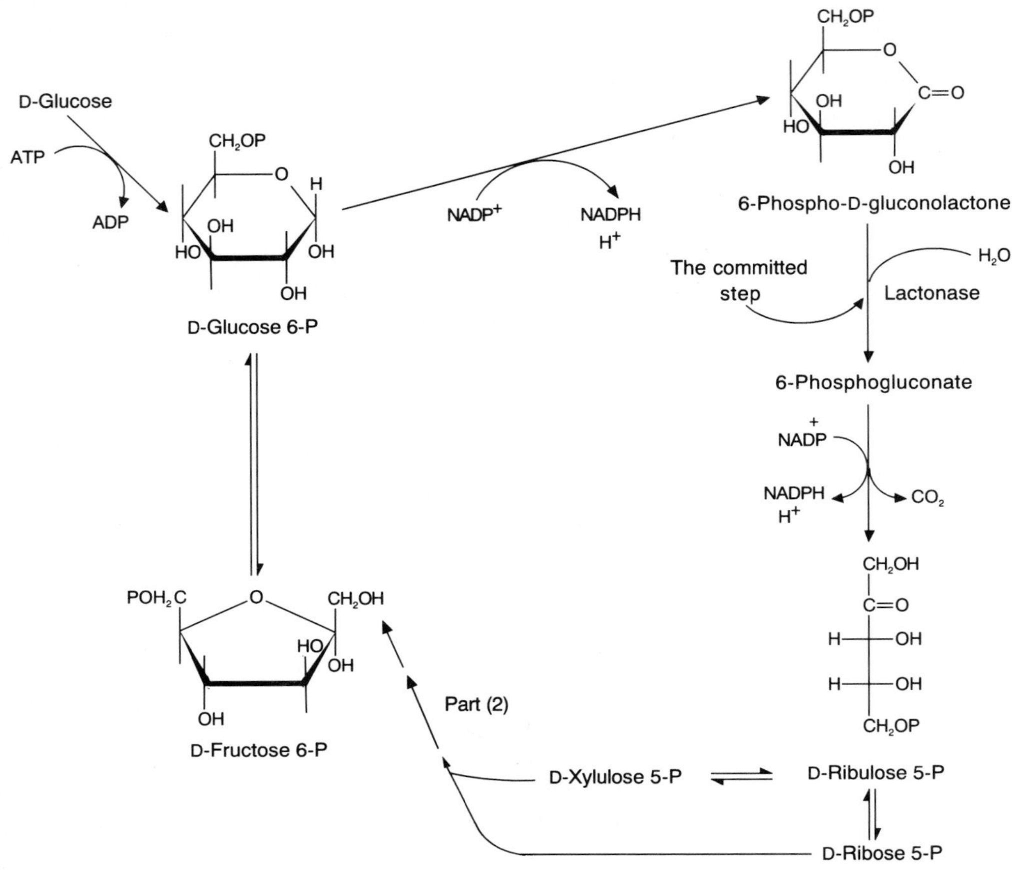

$$6 \text{ Hexose P} + 12\text{NADP}^+ + 6\text{H}_2\text{O} \rightarrow 6 \text{ Pentose P} + 12\text{NADPH} + 12\text{H}^+ + 6\text{CO}_2$$

In each pathway there is an irreversible reaction that, once passed, commits the molecule to its fate. In the pentose phosphate pathway it is the irreversible lactonase-catalyzed step; in glycolysis it is the formation of D-fructose 1,6-bisphosphate.

Part 2   The second part of the pathway entails the reorganization of the pentose phosphate product back to D-glucose 6-phosphate. This step is difficult to demonstrate unless the pentose phosphate pathway is assumed to commence with 6 mol of hexose phosphate. Furthermore, on the basis of experiments with different tissues using radioactively labeled sugars and following the distribution of intermediate sugar phosphates, more than one route for regenerating hexose phosphate is possible.

The most significant difference in the two types of pathway is seen in the properties of the aldolase, which in the F-type pathway behaves in a manner similar to that found in glycolysis. The L-type pathway is characterized by the additional actions of the aldolase and an epimerase that interconverts D-ribose 5-phosphate and D-arabinose 5-phosphate.

1 (a) *F-type pathway; rearrangements in adipose tissue*

$$4 \text{ Pentose P} \xrightleftharpoons{\text{Transketolase}} 2 \text{ Heptulose P} + \text{Triose P}$$

$$2 \text{ Heptulose P} + 2 \text{ Triose P} \xrightleftharpoons{\text{Transaldolase}} 2 \text{ Hexose P} + 2 \text{ Tetrose P}$$

$$2 \text{ Pentose P} + 2 \text{ Tetrose P} \xrightleftharpoons{\text{Transketolase}} 2 \text{ Hexose P} + 2 \text{ Triose P}$$

$$2 \text{ Triose P} \xrightleftharpoons[+ \text{ (Phosphatase)}]{\text{Aldolase}} \text{Hexose P} + P_i$$

Sum: $6 \text{ Pentose P} \longrightarrow 5 \text{ Hexose P} + P_i$

1(b) *L — type pathway: rearrangements in liver:*

By comparison, the rearrangements in liver follow another pathway involving the formation of octulose, 2,8-bisphosphate.

$$4 \text{ Pentose P} \xrightleftharpoons{\text{Transketolase}} 2 \text{ Heptulose P} + 2 \text{ Triose P}$$

$$2 \text{ Pentose P} + 2 \text{ Triose P} \xrightleftharpoons{\text{Aldolase}} 2 \text{ Octulose } P_2$$

$$2 \text{ Octulose } P_2 + 2 \text{ Heptulose P} \xrightleftharpoons{\text{Phosphotransferase}} 2 \text{ Octulose P} + 2 \text{ Heptulose } P_2$$

$$2 \text{ Heptulose } P_2 \xrightleftharpoons{\text{Aldolase}} 2 \text{ Tetrose P} + 2 \text{ Triose P}$$

$$2 \text{ Octulose P} + 2 \text{ Tetrose P} \xrightleftharpoons{\text{Transketolase}} 4 \text{ Hexose P}$$

$$2 \text{ Triose P} \xrightleftharpoons{\text{Aldolase}} \text{Hexose P} + P_i$$

Sum: $6 \text{ Pentose P} \longrightarrow 5 \text{ Hexose P} + P_i$

Part 2 is presented to show that the rearrangement of pentose to hexose is possible and that the triose phosphates are common substrates in the glycolytic and gluconeogenic (see Chapter 7) pathways. The two enzymes that have yet to be discussed, transketolase and transaldolase, function by transferring a two- or three-carbon fragment from a "donor" molecule to an "acceptor." Aldolase has an analogous action.

When either of the pathways in part 2 is coupled with part 1, the overall reaction is:

$$\text{Hexose P} + 12\text{NADP}^+ + 7\text{H}_2\text{O} \rightarrow 6\text{CO}_2 + 12\text{NADPH} + 12\text{H}^+ + P_i$$

The biochemical reactions involved in the two variations of part 2 use four types of enzymes.

Epimerase Epimerase enzymes interchange the groups around one chiral center in a molecule. In the L-type pathway the reaction is:

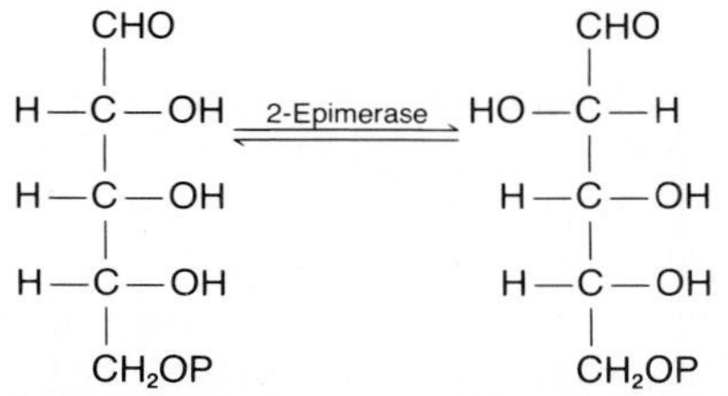

D-Ribose 5-phosphate   D-Arabinose 5-phosphate

Aldolase Aldolase has been discussed previously under glycolysis (p. 271). The reversible aldol condensations demonstrated in the pentose phosphate pathways are similar.

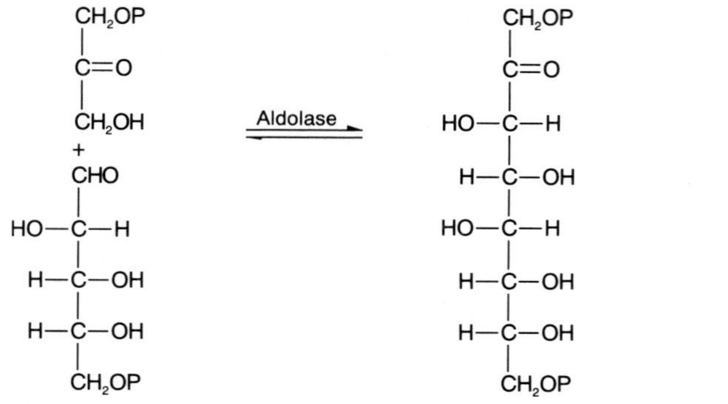

D-Arabinose 5-phosphate   An octulose 1, 8-bisphosphate

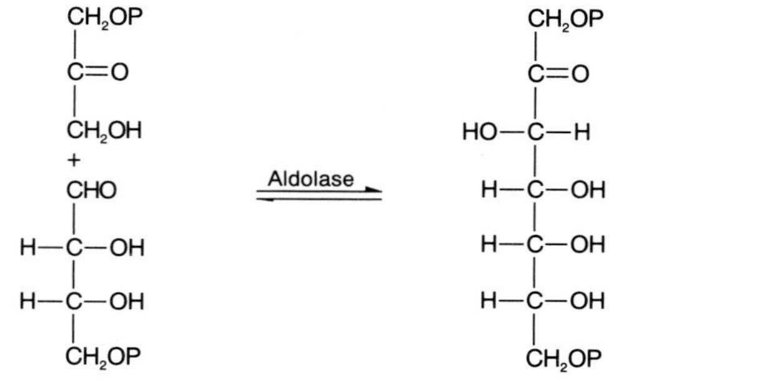

D-Erythrose 4-phosphate   D-Sedoheptulose 1, 7-bisphosphate

Transketolase Transketolase transfers a —CO—CH$_2$OH group according to the reaction:

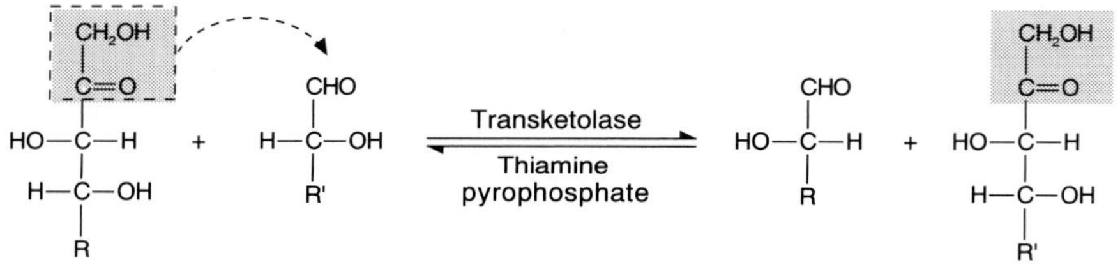

Donor   Acceptor

The thiamine pyrophosphate functions in the transfer in much the same manner as in pyruvate dehydrogenase (see Chapter 5), but in this case an intermediate α,β-dihydroxyethyl thiamine pyrophosphate is formed. The reaction is reversible. The donor molecules have an L configuration at C_3, and the hydroxy groups on positions 3 and 4 are in a D-*threo* configuration; that is, they are *trans*. Thus, as shown next, D-xylulose 5-phosphate donates the two-carbon fragment to D-ribose 5-phosphate to form a seven-carbon sugar (D-sedoheptulose 7-phosphate) as well as D-glyceraldehyde 3-phosphate.

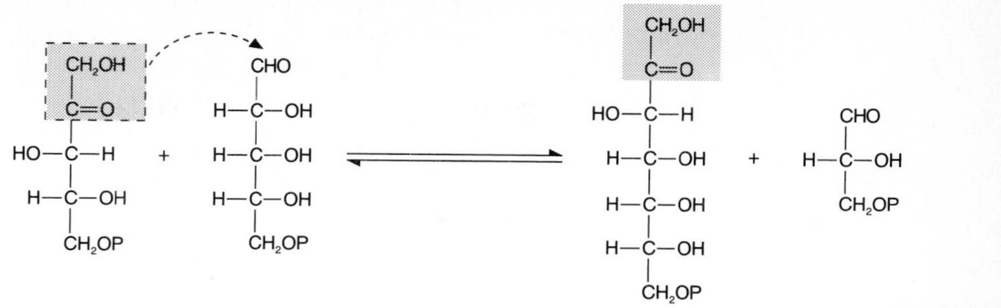

D-Xylulose 5-P   D-Ribose 5-P   D-Sedoheptulose 7-P   Glyceraldehyde 3-P

Clinical comment

Wernicke-Korsakoff syndrome Chronic thiamine deficiency causes serious clinical problems. Such patients may have signs of ataxic stance and gait, weakness or paralysis of eye movement, and deranged mental function. This group of clinical symptoms resulting from thiamine deficiency is called the Wernicke-Korsakoff syndrome. From analysis of cultured fibroblasts, such patients have been shown to have a transketolase with a much reduced affinity for thiamine pyrophosphate (TPP), whereas the other TPP-dependent enzymes are normal. Therefore, in the thiamine-deficient state, the transketolase enzyme has a much reduced activity, leading to serious effects on behavior and motor function.

Transaldolase Transaldolase catalyzes the reaction between D-sedoheptulose 7-phosphate and D-glyceraldehyde 3-phosphate to produce D-fructose 6-phosphate and a tetrose (D-erythrose 4-phosphate). The transfer, shown next, involves a three-carbon fragment but does not have any identified coenzyme or cofactor.

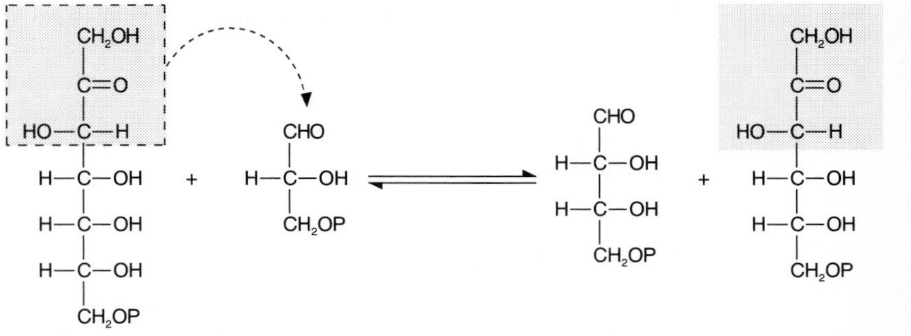

D-Sedoheptulose 7-P   D-Glyceraldehyde 3-P   D-Erythrose 4-P   D-Fructose 6-P

Energetics of the pentose phosphate pathway

The combination of the reactions just illustrated, as shown in Figure 6.15, results in the oxidation of D-glucose to CO_2 and H_2O, with the formation of 12 mol of NADPH + H^+ per mole of hexose. If these reduced nucleotides could be oxidized with NAD^+ to form $NADP^+$ and NADH, they would be equivalent to 12×3, or 36 mol, of ATP by reoxidation of NADH in the respiratory chain. Thus the conservation of energy in the pentose phosphate pathway is similar to that in glycolysis and the Krebs cycle.

Regulation of the pentose phosphate pathway

The pentose phosphate pathway is of minimum importance in muscle but is significant in erythrocytes, liver, adipose tissue, and kidney. Pentose phosphate pathway reactions occur in the cytoplasm. The regulation of the pentose phosphate pathway occurs mainly in the dehydrogenation of D-glucose 6-phosphate. The enzyme concerned, D-glucose 6-

phosphate dehydrogenase, is increased in amount when the diet contains large amounts of carbohydrate. A tenfold increase is noted in liver when the subject passes from a starved state to one in which excess dietary carbohydrate is being converted to fatty acid. This coarse regulation coexists with a finer control of enzyme activity by NADPH acting as a competitive inhibitor, $K_i = 7$ μmol/L. The inhibition is more than 90% when the ratio of the concentrations of NADPH to $NADP^+$ is greater than 10. The inhibition is reversed by oxidized glutathione (GSSG), which acts specifically and not just by oxidizing NADPH to $NADP^+$. These observations go far to explain why the D-glucose 6-phosphate dehydrogenase is active in liver cells, where the concentration of NADPH is often 100 times that of $NADP^+$. The cellular concentration of D-glucose 6-phosphate is approximately 1 mmol/L, which places little constraint on the dehydrogenase activity, for which its K_m is in the micromolar range.

Depending on the needs of the cell, the conversion of pentose phosphate to hexose phosphate (part 2 of the pathway) may not be obligatory. The reactions may favor the formation of glycerol-3-phosphate for biosynthesis of phosphoglycerides or triacylglycerols or the conversion of triose phosphates to 2,3-bisphospho-D-glycerate in the erythrocyte, where the pentose phosphate pathway is very significant.

Clinical comment

Erythrocyte viability The functioning of the pentose phosphate pathway in the human red blood cell is important for the maintenance of its viability. The normal erythrocyte has a life span of approximately 120 to 135 days in the circulation. When mature, the erythrocyte is unable to synthesize protein and is devoid of mitochondria. It requires energy for the synthesis of simple compounds—for example, glutathione, coenzymes, and ATP—and this energy is derived from the metabolism of glucose. The maintenance of ATP levels is achieved by anaerobic glycolysis with the formation of lactic acid, since oxidative phosphorylation through the respiratory chain is not possible. The formation of NADPH by the pentose phosphate pathway provides for the reduction of methemoglobin (the Fe^{+++} form of hemoglobin that does not bind oxygen but is continuously being formed) and for the maintenance of glutathione in its reduced form.

Glutathione functions as a reducing agent to maintain other molecules in the reduced form, for example, enzymes with an essential —SH group at their active site. As a result, the glutathione is oxidized to GS—SG, from which the reduced form, GSH, is regenerated by NADPH. The reaction is catalyzed by *glutathione reductase*.

$$GS—SG + NADPH + H^+ \rightarrow 2GSH + NADP^+$$

The $NADP^+$ is converted back to NADPH in the pentose phosphate pathway.

It is therefore clear that the supply of ATP and the maintenance of the proper oxidation-reduction state of the erythrocyte depend on the catabolism of glucose. The interrelationship of these pathways may be summarized as shown in the illustration that follows.

The normal erythrocyte hexokinase level is low compared to the other enzymes, so that the initial step, which is common to both pathways, is usually rate limiting. The relative rate of the two pathways is influenced by blood pH, [NADP]/[NADPH] ratio, glucose 6-phosphate dehydrogenase activity, and the rate of GSH oxidation. Glycolysis is shown at the left, the pentase phosphate pathway is on the right.

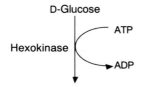

Continued on next page.

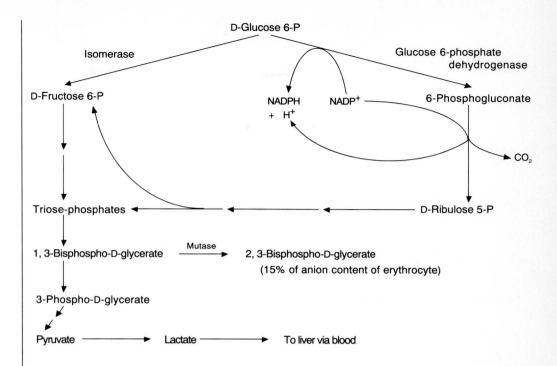

Crossover between these pathways is achieved through the recycling of the pentose phosphates, involving transaldolase and transketolase, to form the fructose phosphates and glyceraldehyde-3-phosphate. Conceivably, therefore, the pentose phosphate and glycolytic pathways run in series to generate NADPH and ATP. The substrate phosphorylations would occur in glycolysis after the formation of triose phosphate. Such a serial sequence also permits the maintenance of 2,3-bisphospho-D-glycerate, which is important in oxygen transport (see Chapter 4).

GSH deficiency in red blood cells results in serious impairment of essential metabolic processes and produces hemolysis. Probably less than 1% of the glutathione is normally present as GS—SG. It therefore follows that the pentose phosphate pathway must function efficiently, requiring an adequate level of enzymes, including glucose 6-phosphate dehydrogenase. The level of this enzyme is reduced in older erythrocytes.

Clinical comment

Glucose 6-phosphate dehydrogenase deficiencies Several types of inherited deficiency of this enzyme have been recognized, all of which result in sensitivity to certain drugs. This induces hemolysis and leads to anemia. The breakthrough in the understanding of this clinical condition came with the in-depth study of the hemolytic effects of primaquine, the 8-aminoquinoline antimalarial agent. Subsequently, many more drugs have been found with actions similar to primaquine. Enzyme-deficient cells have a lower rate of NADPH production, resulting in a deficiency of reduced glutathione (GSH), which is essential to maintain the integrity of the erythrocyte membrane. Such membranes are more liable to destruction by drugs such as primaquine. Since the deficiency of GSH is greatest in older cells, it is these cells that hemolyze. This explains the higher percentage of circulating young cells found in the blood of such patients, the shorter half-life of their erythrocytes, and the larger percentage of reticulocytes (immature erythrocytes) present in their blood.

Patients who have glucose 6-phosphate dehydrogenase deficiencies have a lesser tendency to contract malaria. The malaria parasite, *Plasmodium falciparum*, infects the red blood cell, where it depends on the GSH in the cytoplasm. Understandably, therefore, persons with glucose 6-phosphate dehydrogenase deficiency cannot support growth of this parasite and thus are less prone to malaria than the normal population.

D-Glucuronate and polyol pathways

In studying intermediary metabolism, one must think in terms of the total organism and homeostasis. Even though a very small segment of metabolism in a particular tissue and cell may be under discussion, one must remember that there is a dynamic equilibrium in each cell that is constantly affected by many other aspects of metabolism. Thus, after a segment of intermediary metabolism is understood in detail, it should be considered in relation to the other parts.

It is difficult to identify many of the metabolites of D-glucose catabolism with any one pathway; this problem is illustrated by a series of reactions that are particularly important in the understanding of pentosurias and in the metabolism of ascorbic acid.

D-Glucuronate, formed from D-glucose by way of UDP-glucose, is metabolized by reduction of the aldehyde group to form L-gulonate. It is this hexonic acid that undergoes lactonization and oxidation to form ascorbic acid in many mammals. Humans, however, are deficient in the L-gulonolactone oxidase and thus are absolutely dependent on diet to supply this vitamin. The metabolic fate of ascorbic acid depends greatly on the species, and it is not possible to extrapolate animal data to humans. Humans do not have the lactonase enzyme, unique to the guinea pig, that permits ascorbic acid and dehydroascorbic acid to enter the catabolic process by way of 2,3-diketo-L-gulonate. In humans, little if any ascorbic acid is catabolized to CO_2, but evidence suggests that some (less than 10%) of the ascorbic acid is cleaved to oxalate and a four-carbon intermediate. When an excess of ascorbic acid is ingested, certain mechanisms prevent its accumulation in the body. First, limited intestinal absorption of the vitamin occurs, and second, there is an efficient urinary excretion of any excess.

The alternate pathway of L-gulonate is oxidation to 3-keto-L-gulonate, which undergoes decarboxylation to yield L-xylulose in much the same manner as was seen in the pentose phosphate pathway for D-ribulose, except that the coenzyme is NAD^+. The reactions to this point are shown next, with the original C_1 of the D-glucuronate identified by an asterisk to clarify the interconversions of the sugars.

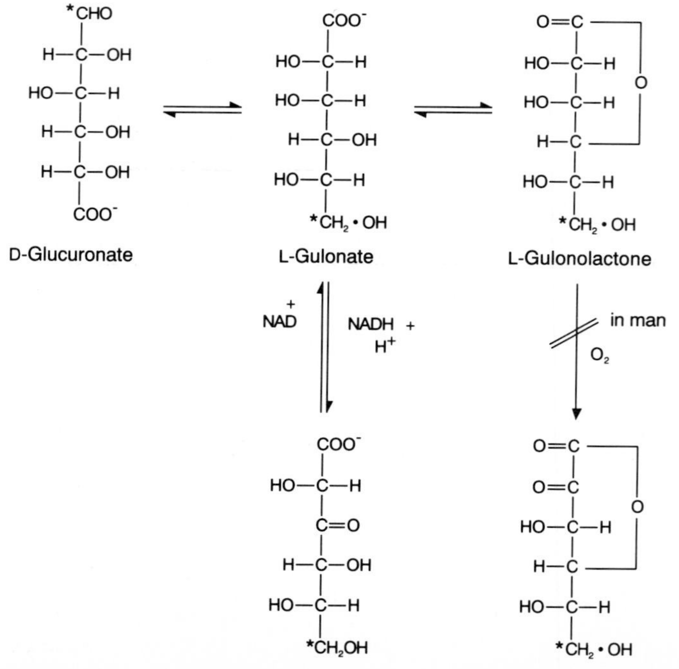

3-Keto-L-gulonate   2-Keto-L-gulonolactone

CO₂

CH₂OH   O=C
| |
C=O   HO—C
| ‖   O
H—C—OH   HO—C
| |
HO—C—H   H—C
| |
*CH₂OH   HO—C—H
|
*CH₂OH

L-Xylulose   L-Ascorbic acid

Through the action of two stereospecific xylitol dehydrogenases, L-xylulose is converted to D-xylulose through the intermediate pentitol xylitol, which is a symmetric molecule.

CH₂OH   CH₂OH   CH₂OH
| | |
C=O   HO—C—H   C=O
| NADPH | NAD⁺ |
H—C—OH   ⇌   H—C—OH   ⇌   HO—C—H
| NADP⁺ | NADH |
HO—C—H   H—C—OH   H—C—OH
| | |
CH₂OH   CH₂OH   CH₂OH

L-Xylulose   Xylitol   D-Xylulose

Clinical comment

Pentosuria Each dehydrogenase has a different coenzyme. D-Xylulose can be phosphorylated with ATP and a kinase to D-xylulose 5-phosphate, which enters the pentose phosphate pathway. The enzyme converting L-xylulose to xylitol is absent or deficient in pentosuria. As a result, L-xylulose is excreted in the urine. This genetic defect does not appear to involve any serious physiologic consequences. However, the test for reducing sugars in the urine should be checked to differentiate the excretory material from D-glucose.

The penittol pathway has become more important recently, since xylitol, which is found naturally in foods such as plums (6.2 mmol/100 g), spinach (0.7 mmol/100 g), and carrots (0.6 mmol/100 g) is being used as a sweetener in some specialty items such as "sugarless" chewing gum. Xylitol is absorbed more slowly from the intestinal tract than other sugars, and its metabolism, which does not depend on insulin, occurs principally in the liver, where it is converted to D-glucose.

A reaction analogous to the for the xyluloses is the interconversion of D-glucose and D-fructose by way of D-glucitol (sorbitol).

*CHO   *CH₂OH   *CH₂OH
| | |
H—C—OH   H—C—OH   C=O
| | |
HO—C—H   NADPH   HO—C—H   NAD⁺   HO—C—H
| ⇌ | ⇌ |
H—C—OH   NADP⁺   H—C—OH   NADH   H—C—OH
| | |
H—C—OH   H—C—OH   H—C—OH
| | |
CH₂OH   CH₂OH   CH₂OH

D-Glucose   D-Glucitol   D-Fructose

This polyol pathway may be the source of D-fructose in cerebrospinal and seminal fluid. The pathway has been described in many tissues, including the peripheral nerves, and it may be important in diabetes, in which hyperglycemia is common. A theory presently being evaluated is that elevated levels of sorbitol cause some of the complications in diabetic persons, particularly the malfunction of the peripheral nerves (neuropathy).

Ethanol

Ethanol is not a carbohydrate, nor is it a precursor for the biosynthesis of carbohydrates. However, ethanol can replace sizable amounts of carbohydrates as an energy source when large amounts are ingested. It is present in the blood of most humans, being produced by the intestinal flora. People ingest ethanol in variable amounts in beverages and fermented fruits. Ethanol is metabolized in the liver to acetate and adds to the caloric content of the diet. Ethanol has an energy equivalent of 29.3 kJ (7 kcal)/g, so that 100 ml of table wine has ethanol corresponding to about 300 kJ (72 kcal) and a "jigger" of whiskey furnishes approximately 500 kJ (120 kcal).

When ethanol is metabolized in the liver, a cytosolic alcohol dehydrogenase oxidizes it first to acetaldehyde.

$$CH_3CH_2OH + NAD^+ \rightarrow CH_3CHO + NADH + 2H^+$$

The acetaldehyde is oxidized further to acetate.

$$CH_3CHO + NAD^+ + H_2O \rightarrow CH_3COO^- + NADH + 2H^+$$

This further oxidation is accomplished by acetaldehyde dehydrogenases, principally the mitochondrial enzyme, which has a much lower K_m (10 μM) than the cytosolic enzyme (1.0 mM).

A small fraction of the alcohol may be oxidized by other systems, including a specific microsomal cytochrome P_{450} oxidase, which is induced by ethanol and is also involved in the detoxification of many drugs.

$$CH_3CH_2OH + NADPH + H^+ + 2O_2 \rightarrow CH_3CHO + 2H_2O_2 + NADP^+$$

The other ethanol-oxidizing system is the catalase (K_m 10 mM) present in peroxisomes.

$$CH_3CH_2OH + H_2O_2 \rightarrow CH_3CHO + 2H_2O$$

The acetate produced from ethanol largely escapes from the liver and is converted in extrahepatic tissues to acetyl CoA and then to carbon dioxide by way of the Krebs cycle. The acetyl CoA that stays in the liver may act as a precursor for lipid biosynthesis.

A significant consequence of the hepatic metabolism of ethanol is the twofold to threefold increase in the NADH/NAD$^+$ ratio, with the resultant shift in the oxidation-reduction potential of the cytoplasm. With higher concentrations of blood alcohol, the shuttle systems involved in the transport of reducing equivalents into the mitochondria, the malate-aspartate and the glycerol phosphate shuttles, are saturated. As a result, the concentration of NADH remains high in the cytoplasm, and the availability of NAD$^+$ in the cytosol drops and limits both the further oxidation of ethanol and the normal functioning of other metabolic pathways, such as gluconeogenesis.

Activity of the liver alcohol dehydrogenase probably is the major rate-limiting step in the metabolism of ethanol, although this has been difficult to establish. The rate of reoxidation of NADH, shuttle transport of reducing equivalents into the mitochondria, and the activity of the mitochondrial acetaldehyde dehydrogenase are partially rate-limiting.

Clinical comment

Alcohol and drug toxicity Most people metabolize ethanol at approximately 100 mg/kg body weight/hr; alcoholics may have a slightly higher rate, about 120 mg/kg body weight/hr. Persons past the "social stage" of alcohol consumption have blood alcohol concentrations greater than 10 mM, and since the ethanol is equally distributed throughout the body water, the concentration is about 10 times greater than the K_m for the cystolic alcohol dehydrogenase. At this point the other alcohol-oxidizing systems play increasingly important roles. Since one of these systems also is used to detoxify many drugs, ethanol oxidation is competing, and the circulating concentration of the drug in question, such as some of the drugs used for hypertension, may increase to toxic levels. Thus the warning on some prescription lables: "Do not take alcohol while using this drug."

Chronic consumption of significant amounts of alcohol (e.g., 80 to 140 g daily) may lead to a "fatty liver," in which excess triacylglycerol is deposited. This is caused by several contributing factors: reduced triacylglycerol secretion from the liver, reduced rates of fatty acid oxidation, and increased rates of lipid biosynthesis. These processes are concomitant with the increased acetyl CoA and NADH/NAD$^+$ ratio in the liver that results from ethanol oxidation.

Bibliography

Carpenter CCJ et al: Oral-rehydration—the role of polymeric substrates, N Engl J Med 319:1346, 1988.

El-Maghrabi MR et al: Regulation of 6-phosphofructo-2-kinase activity by cyclic AMP–dependent phosphorylation, Proc Natl Acad Sci USA 79:315, 1982.

Felig P: Disorders of carbohydrate metabolism. In Bondy PK and Rosenberg LE, editors: Metabolic control and disease, ed 8, Philadelphia, 1980, WB Saunders Co.

Li T-K: Enzymes of human alcohol metabolism, Adv Enzymol 45:427, 1977.

Lieber CS: Biochemical and molecular basis of alcohol-induced injury to liver and other tissues, N Engl J Med 319:1640, 1988.

Lloyd ML and Olsen WA: A study of the molecular pathology of sucrose-isomaltase deficiency, N Engl J Med 316:438, 1987.

Newsholme EA, Chaliss RAJ, and Crabtree B: Substrate cycles: their role in improving metabolic control, Trends Biochem Sci 9:277, 1984.

Randle PJ, Steiner DF, and Whelan WJ, editors: Carbohydrate metabolism and its disorders, vol 3, New York, 1981, Academic Press, Inc.

Robinson S: Physiology of muscular exercise. In Mountcastle VB, editor: Medical physiology, ed 14, vol 2, St Louis, 1980, The CV Mosby Co. *A good discussion of the relationship of carbohydrates to work demands.*

Rodriguez IR, Taravel FR, and Whelan WJ: Identification of the enzymes responsible for the conversion of starch into glucose in the mammalian digestive tract, CRC Crit Rev Biotechnol 5:243, 1987.

VanSchaftingen E: Fructose 2,6-bisphosphate, Adv Enzymol 59:315, 1987.

Wood J: The pentose phosphate pathway, New York, 1985, Academic Press, Inc.

Clinical examples

A diamond (♦) on a case or question indicates that literature search beyond this text is necessary for full understanding.

Case 1   Pediatric gastroenteritis

R.D., a 6-week-old girl, was born after a normal pregnancy and weighed 3.2 kg at birth. Her parents and two older siblings were in good health. She was breast fed for 4 wk, and her weight gain had been normal. The composition of human milk is as follows:

Fat	3.5% to 4.0%
Lactose	7.0%
Total protein	1.35%
Lactalbumin	0.70%
Casein	0.5%
Total minerals	0.2%
Water	85% to 88%
Energy content	2760 kJ/L

At 4 wk of age, breastfeeding was discontinued, and a common baby formula was substituted. As a result of poor initial formula preparation, the child developed a viral gastroenteritis and after several days exhibited fussiness, watery diarrhea, and vomiting. At the age of 6 wk she was admitted to the hospital.

Physical examination at the time of admission revealed a moderate degree of dehydration. Body weight was 3.4 kg. Urinalysis yielded a 1+ reaction for reducing substance and no reaction for glucose.

The infant was hydrated with intravenous fluids over 24 hr and then with fluids orally for an additional 24 hr. During this time the diarrhea subsided, and her weight increased to 3.7 kg. She was then fed formula X (2760 kJ/L). Within 24 hr her stools became watery and were passed at the time of each feeding.

On the fourth day of hospitalization, formula Y was substituted for formula X. The number of stools decreased; they contained no reducing substance and became semiformed in character. The infant gained 40 g/day during the next 4 days. However, because of the incovenience and expense of formula Y, a single feeding of formula X was once again given as a trial. The infant demonstrated some fussiness but did not pass a stool. A second feeding of formula X 4 hr later was followed by several explosive, watery stools. The stool pH was 5.0, and it gave a positive reaction for reducing substance. Data on formulas X and Y are given in the following discussion. Both formulas contain 2760 kJ/L.

Biochemical questions

1. Was there any significant difference between the breast milk and the baby formulas?
2. How did the gastroenteritis affect the digestion of carbohydrates?
3. Was the gastroenteritis related to the diarrhea? What might have caused the explosive, acid, watery stools containing reducing substances?

Case discussion

1. **Comparison of breast milk and baby formulas.** At 6 wk of age, R.D. suffered from a disorder that was not present during the first month of life, when her development was normal. This period coincided with breastfeeding, which indicated that the infant had been able to handle all the nutrients in her mother's milk. The deficiency may have developed gradually, but this is unlikely unless for some reason the mother was producing abnormal milk. Again, the nutritional adequacy of the milk is indicated by the infant's normal weight gain.

It is reasonable to conclude that this problem was not inherited but rather caused by the effects of viral gastroenteritis. The evidence of dehydration was to be expected in light of the excessive water losses through vomiting and diarrhea. Thus the principal question concerned the infant's intolerance of formula X. The following aspects were noted:

a. Intravenous feeding followed by oral feeding of simple solutions reduced the diarrhea and produced a weight gain through rehydration.
b. Each feeding of formula X caused diarrhea.
c. Substitution of formula Y relieved the problem and produced an acceptable weight gain.
d. Feeding of formula X once again produced diarrhea. This points rather directly

to a relation between some components of formula X and the diarrhea.

Several differences existed between formulas X and Y, as indicated by the summary of their compositions. However, further study of the case history helped limit the possibilities. When the infant was admitted to the hospital, urinalysis indicated the presence of a reducing substance or substances that were not glucose. Thus a diabetic type of overload could be ruled out. Reducing substances were also present in the watery stools induced by formula X, but these disappeared from the stools when formula Y was substituted. The most common reducing substances are the carbohydrates, which were present in formula X as lactose and in formula Y as sucrose. Lactose is a reducing substance; sucrose is not. It would seem therefore that the gastroenteritis had caused at least a temporary inability to digest lactose.

Normally, the digestion of carbohydrates is initiated in the mouth, where the salivary α-amylase enzyme acts on starch to initiate the partial hydrolysis to glucose, maltose, and simpler materials called starch dextrins (oligosaccharides). This process is shown in the illustration. Digestion of carbohydrates ceases in the stomach as a result of the extreme gastric acidity that is produced by hydrochloric acid secretion. When

Starch   [X = approx. 5-10 x 10^3]

Salivary amylase

α-D-Glucose   α-D-Maltose   +   Starch dextrins with x ~ 8-10

the gastric contents are emptied into the small intestine, bile and pancreatic juice neutralize the hydrochloric acid. Digestion of the starch dextrins then continues as a result of the secretion of pancreatic α-amylase. At this stage all the digestible carbohydrate, not including cellulose and similar roughage, has been cleaved into smaller fragments. These fragments include monosaccharides, principally glucose and perhaps some fructose; disaccharides, principally maltose, sucrose, and lactose; and a small amount of the α-limit dextrins of starch. Final hydrolysis of the disaccharides occurs in the brush border of the intestinal mucosa. Several disaccharidases (maltase, sucrase, lactase, isomaltase) are present in the membranes of these brush border cells. The resulting monosaccharides are then actively transported by carrier systems, probably located in the brush border membranes, into the cell for eventual passage into the portal system and then to the liver. Some portion of the monosaccharides resulting from the disaccharidase activity are not immediately picked up by the carrier system; these leak back into the lumen, from which they are completely reabsorbed before reaching the colon.

2. Gastroenteritis and digestion Viral gastroenteritis damages the mucosa, and some of the brush border cells have a significant proportion of their disaccharidases destroyed. Since there are at least four different maltases, however, serious deficiencies of maltose hydrolysis are rare. It is generally accepted that only one type of sucrase exists. However, mucosal damage does not usually affect sucrose hydrolysis, probably because a high level of sucrase is normally present. In contrast, lactase activity is sensitive to mucosal damage, and lactose hydrolysis is significantly reduced. Because of the damage to the microvilli in the intestinal mucosa, some lactose is absorbed without hydrolysis, whereas a large proportion passes on to the colon.

3. Diarrhea The previous facts explain many of the symptoms exhibited by R.D. when she was fed formula X. The lactose that was absorbed intact could not be hydrolyzed anywhere in the body other than the mucosa and was thus excreted intact in her urine. This produced the 1+ reaction for reducing substances.

The infant was fed 500 kJ/kg of body weight each day. Thus, based on a weight of 3.5 kg and 2760 kJ/L of formula X, her daily intake was:

$$\frac{500 \times 3.5}{2760} \times 1000 \text{ ml/day} = 627 \text{ ml/day}$$

With five equal feedings per day, this would be equivalent to about 125 ml per feeding. Formula X contains 66 g of lactose/L, so that $66 \times \frac{125}{1000}$ g, or about 8 g, of lactose is fed at each meal. A large proportion of this lactose would pass into the colon, where the bacterial flora would ferment part of it to CO_2 and lactic acid. The presence of these extra low molecular weight materials in the colon would draw water in from the blood, producing an osmotic diarrhea and discomfort ("fussiness"). This explains explosive (CO_2), watery (osmotic), acidic (lactic acid) stools containing reducing substances (lactose); therefore the advisability of maintaining the infant for some time on a formula that is lactose free is apparent.

Mucosal disaccharidase deficiencies, except maltase deficiency, are well-described clinical entities. The most common is lactase deficiency, which varies greatly among different age and racial groups. Primary lactase deficiency is genetically determined. Secondary lactase deficiency, such as that described in this case, can be precipitated by various disturbances of the gastrointestinal tract, including tropical sprue.

References

Appleton H and Higgins PG: Viruses and gastroenteritis in infants, Lancet 1:1297, 1975.
Semenza G and Aunicchio S: Small intestinal disaccharidases. In Scriver CR, et al, editors: The metabolic basis of inherited disease, ed 6, New York, 1989, McGraw-Hill Information Services Co.

Case 2   Galactosemia

The patient, a boy, was the first child of healthy parents without known consanguinity. Delivery was normal and birth weight was 3.78 kg. From the third day of life the child developed an increasing degree of jaundice and at the same time became indolent and difficult to feed. No blood group incompatibility could be demonstrated. At 6 days of age he had a serum bilirubin of 504 μmol/L, and his weight was 15% below normal weight. He was readmitted to the hospital on the seventh day after birth. Muscular tonus was increased, and the patient later began to convulse. Between the seventh and ninth days, exchange blood transfusion was performed three times, but the serum bilirubin concentration still remained high. On the ninth day of life the boy began vomiting, liver enlargement was noted, and the cerebral symptoms became accentuated.

A postive test for reducing sugars in urine had already been obtained on the sixth day after birth. A repeated test was performed on the seventh day, and this was positive, whereas at the same time a Clinistix test, specific for D-glucose, was negative. Hereditary galactosemia was then suspected, and special tests that were performed on the eighth day of life confirmed the diagnosis.

Hemoglobin	12.4 mmol/L (200 g/L)
AST	299 Karmen-Ordell units
ALT	202 units (normal after seventh day)
Bilirubin (max)	549 μmol/L (at seventh day)
Galactose 1-P uridyl transferase (in erythrocytes)	0 (normal 2 to 31 units/g of hemoglobin)

Milk feeding was stopped on the ninth day, and it was replaced by intravenous glucose administration. From the tenth day of life, a galactose-free diet was introduced. With this treatment the patient improved dramatically.

Biochemical questions

1. What are the biochemical effects of galactosemia?
2. Is there an alternate source of tissue galactose for patients on a galactose-free diet?
3. Would a mother who is homozygous for galactosemia be able to produce lactose in her milk?
4. How would a D-galactose tolerance curve have appeared if it had (unfortunately) been given to the patient?
5. What are the interpretations of the laboratory results?
6. Studies also indicate galactitol in the urine. By what mechanisms do you think galactitol was synthesized?
7. What differences would have been noted if the deficient enzyme had been galactokinase?

Case discussion

Galactosemia reflects a reduced ability, or inability, to convert galactose to glucose.

1. **Biochemical effects of galactosemia** The biochemical defect usually found in galactosemia is a deficiency of the enzyme, galactose 1-phosphate uridyl transferase. Initially, galactose 1-phosphate accumulates in the tissues. Because of this, less galactose is converted to galactose 1-phosphate, and free galactose also accumulates in the tissues and blood. As more and more galactose and galactose 1-phosphate accumulate within cells, damage to the cells results. In this case the major organ damaged by galactose accumulation was the liver.

2. **Alternate source of galactose** An alternate source of galactose exists even if the patient is fed a galactose-free diet. The galactose can be formed from glucose. First, glucose is converted to glucose 1-phosphate and then to UDP-glucose. Next, an epimerase converts the UDP-glucose to UDP-galactose.

3. **Milk production** A galactosemic mother can produce milk containing galactose. The galactose needed for lactose synthesis can be formed from glucose by the pathway described in number 2.

4. **Galactose tolerance test** A galactose tolerance test would have been very abnormal in this patient. There would have been a much higher than normal rise in blood galactose after the galactose was administered, and the elevation would have remained for a much longer period than normal. The galactose tolerance test in a normal person shows the level in the blood rising to a maximum at about ½ hour after galactose administration. It then returns to the baseline value within 1 hr.

5. **Interpretation of laboratory data** The laboratory results for this patient indicate a malfunctioning of the liver. Both the serum transaminases are elevated, suggesting the possibility of liver cell damage. More importantly, bilirubin is

elevated. Since the hemoglobin value is not low, it is unlikely that the elevated bilirubin results from excessive hemolysis. A more likely explanation is that the liver cannot handle normal amounts of bilirubin because of the galactose and galactose 1-phosphate accumulation. One might predict that a high proportion of the bilirubin is unconjugated; the bilirubin probably has difficulty in penetrating the hepatocyte and, in addition, the function of the glucuronyl transferase is probably affected adversely by the liver cell damage. Finally, the diagnosis can be made by the demonstration that the galactose 1-phosphate uridyl transferase is absent in erythrocytes.

6. **Biosynthesis of galactitol** Galactitol is an alcohol that is made from galactose. The pathway is similar to that for the conversion of glucose to glucitol (sorbitol; see p. 285). Because of the excessive quantity of galactose in this patient, some of it is converted to galactitol, which builds up in the body and is excreted in the urine. Since it is an alcohol and not a sugar, galactitol is not a reducing substance. Therefore the positive test for reducing substances in the urine is caused by galactose and not galactitol. Galactitol appears to be toxic at high concentration and produces tissue injury, especially cataract formation.

7. **Galactokinase deficiency** The defect would have been much more severe if it was in the enzyme galactokinase, instead of in the uridyl transferase. People with a transferase deficiency eventually have a less severe disease because an alternate enzyme for galactose utilization develops during childhood. This enzyme enables the galactose 1-phosphate to bypass the uridyl transferase step. Instead of reacting with UDP-glucose, the galactose 1-phosphate reacts with UTP to form UDP-galactose. This alternate pathway can handle only a fraction of the galactose that the uridyl transferase reaction can. It handles enough, however, to enable the person to have some galactose in the diet. Therefore the patient does not have to maintain as severely a restrictive diet as he grows older. If the galactokinase were deficient, the alternate pathway could not function later in life because the patient could not produce galactose 1-phosphate. Therefore he would have to remain on a much more restrictive diet throughout life. Fortunately, very few cases of galactosemia are caused by galactokinase deficiency.

References

Dahlquist A, Jagenburg R, and Mark A: A patient with hereditary galactosemia, Acta Pediatr Scand 58:237, 1969.

Segal S: Disorders of galactose metabolism. In Scriver CR, et al, editors: The metabolic basis of inherited disease, ed 6, New York, 1989, McGraw-Hill Information Services Co.

Case 3

Hereditary fructose intolerance

A child had nausea, vomiting, and symptoms of hypoglycemia: sweating, dizziness, and trembling. It was reported that these attacks occurred shortly after eating fruit or cane sugar. This was resulting in a strong aversion to fruits, and the mother was therefore providing large supplementations of multivitamin preparations. The child was below normal weight and was an only child who had been breast fed, during which time none of these symptoms was evident. The clinical findings included some cirrhosis of the liver, a normal glucose tolerance test, and reducing substances in the urine that did not react positively with glucose test papers, in which glucose oxidase was used as the basis for test. A fructose tolerance test was ordered, using 3 g fructose/m^2 of surface, given intravenously in a single, rapid push. Within 30 min the child displayed the symptoms of hypoglycemia. Blood glucose analysis confirmed this and revealed that the hypoglycemia was greatest after 60 to 90 min. Fructose concentrations reached a maximum (3.3 mmol/L) after 15 min and gradually decreased to zero in 2½ hr. P_i concentrations fell by 50%, and AST and ALT elevations were noted after 1½ hr. The urine was positive for fructose.

Biochemical questions

1. Explain why the fructose concentration in the blood remained elevated for an extended period.
2. What evidence exists to suggest that the aldolase for D-fructose 1-phosphate and D-fructose 1,6-bisphosphate is the same protein?
3. Explain the elevations noted for AST and ALT in the fructose tolerance test.
4. Why are the symptoms of hypoglycemia not found in essential fructosuria (type 1)?
5. What would be the consequences of a deficiency of phosphofructokinase in the patient?

Case discussion

Hereditary fructose intolerance is caused by a deficiency of fructose 1-phosphate aldolase.

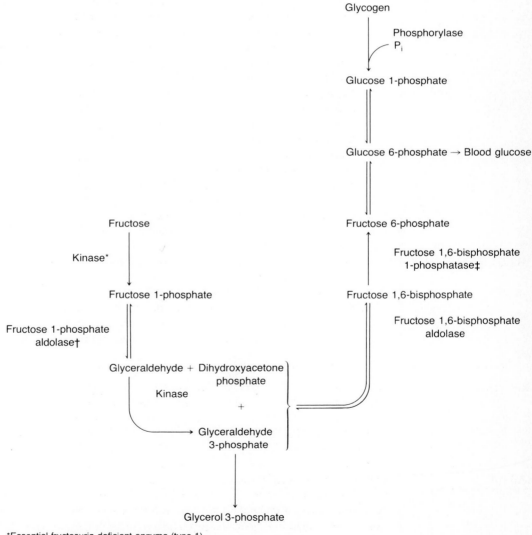

*Essential fructosuria deficient enzyme (type 1).
†Hereditary fructose intolerance deficient enzyme (type 2).
‡Fructose intolerance for patient X.Y. deficient enzyme (type 3).

1. **Blood fructose** The metabolism of fructose is initiated by its phosphorylation in liver to D-fructose 1-phosphate, as catalyzed by fructokinase (see p. 265). With a deficiency of aldolase, D-fructose 1-phosphate accumulates and inhibits the kinase reaction, resulting in a slower removal of fructose from the blood. With high fructose concentration in the blood, some may be phosphorylated in muscle and adipose tissue by hexokinase. The D-fructose 6-phosphate so produced is converted to D-glucose 6-phosphate, which may enter the catabolic pathways or be converted to glycogen. It cannot be released to the blood, however, to overcome the hypoglycemia that develops because muscle and adipocytes do not contain glucose 6-phosphatase.

2. **Aldolase** Crystalline human liver aldolase functions with both D-fructose 1-phosphate and D-fructose 1,6-bisphosphate substrates. However, liver tissue from patients with hereditary fructose intolerance shows significant fructose 1,6-bisphosphate aldolase activity but reduced or no fructose 1-phosphate aldolase activity. Tissue from normal subjects shows approximately equal activities. The aldolase in this type of fructosuria may be a mutant of the normal enzyme in which fructose 1,6-bisphosphate aldolase activity remains. It is known, for example, that antibodies against normal liver aldolase react to an extent of 30% with the liver enzyme of patients with hereditary fructose intolerance.

3. **Enzyme elevations** Fibrosis or cirrhosis of the liver are some of the chronic results of the disease. After ingestion of D-fructose, the liver hepatocytes show histologic changes after 1 to 1½ hr and release increased amounts of AST(GOT) and ALT(GPT).

4. **Essential fructosuria, type 1** Essential fructosuria is caused by the deficiency of fructokinase, without which there would be no accumulation of D-fructose 1-phosphate and no subsequent enzyme inhibition of the glycogen pathways or depletion of P_i or ATP. Without this enzyme deficiency, mechanisms to respond to hypoglycemia would be available, and this condition would not arise.

5. **Phosphofructokinase deficiency** The glycolytic pathway is functioning normally in this patient. Therefore the fructose 1,6-bisphosphate aldolase can catalyze the breakdown of fructose 1,6-bisphosphate, which arises from the phosphorylation of D-fructose 6-phosphate. D-Fructose 6-phosphate is an intermediate in the catabolism of D-galactose, D-glucose, and other hexoses. If phosphorfructokinase were also deficient in this case, the patient would be in a very difficult situation. The catabolism of all the hexoses in the diet could not proceed through glycolysis, so they would be converted to glucose and accumulate as glycogen. As more glycogen is built up, the liver would be damaged further. The absence of phosphofructokinase, which has not been reported clinically, is probably fatal to the fetus.

References

Gitzelmann R et al: Disorders of fructose metabolism. In Scriver CR et al, editors: The metabolic basis of inherited disease, ed 6, New York, 1989, McGraw-Hill Information Services Co.

Salvatore F et al: Aldolase gene and protein families: structure expression and pathophysiology. In Blasi F, editor: Human genes and diseases: Horizons in biochemistry and biophysics, New York, 1986, John Wiley & Sons Inc.

Case 4   Diabetes mellitus and dental care

A 61-year-old man had a periodontal checkup that showed signs of severely ulcerated and inflamed mucosal surfaces of the lips, severe fissuring of the tongue, generalized edema of the gingiva, extensive calculus, and nine new carious lesions of the teeth.

The possibility of uncontrolled diabetes mellitus was recognized and supported by laboratory tests: urine had a specific gravity of 1.030 and contained albumin, glucose, and occasional casts. The blood glucose was 12.9 mmol/L. There was no evidence of diabetes in the family history. The patient was referred to his physician for therapy before dental treatment was pursued. He returned to the dental clinic after a few months, by which time his blood glucose concentration was 9.7 mmol/L. He was placed on a 7.5 MJ liquid diet for 24 hr preoperatively.

Biochemical questions

1. What other biochemical tests would be needed to confirm the diagnosis of diabetes mellitus? Explain.
◆2. From the most recent theories of caries formation, what dietary components should be reduced? Salivary amylase activity is generally increased in the diabetic person.
◆3. What is the specificity of the glucose test strips that depend on glucose oxidase?
◆4. Is the frequency of supragingival calculus higher in the diabetic individual? What might be the possible explanation?

References

Foster DW: Diabetes mellitus. In Scriver CR, et al, editors: The metabolic basis of inherited disease, ed 6, New York, 1989, McGraw-Hill Information Services Co.
Kjellman O et al: Oral conditions in 105 subjects with insulin-treated diabetes mellitus, Swed Dent J 63:99, 1970.
Lavine MH: Diagnosis and management of the diabetic patient, Oral Surg 24:16, 1967.

Case 5 — Diabetic ketoacidosis

A 24-year-old man who has had diabetes for 5 yr had been well until 2 days before admission to the hospital. He had developed fever, nausea, and vomiting. He had not taken his insulin for 24 hr because he was unable to eat. On admission he was semiconscious. Acetone could be smelled on his breath. He had the physical findings of moderate to severe dehydration. A urine specimen was 4+ for glucose and strongly positive for ketone bodies. An arterial blood sample was sent for determination of blood pH and pCO_2. The blood serum was strongly positive for ketones. The patient was immediately given 100 units of insulin. An intravenous drip of hypotonic saline solution with 1 ampule of sodium bicarbonate (50 ml containing 44 mmol) was started. In approximately 45 min the laboratory results were available. The data from the blood serum were as follows: blood sugar, 55 mmol/L; BUN, 6.5 mmol/L; creatinine, 260 μmol/L; sodium, 138 mmol/L; potassium, 5.9 mmol/L; chloride, 94 mmol/L; and total CO_2 concentration, 3 mmol/L. The arterial pCO_2 was 18 mm Hg (2.3 kPa), and the blood pH was 7.05. Serum osmolality was 390 mOsm (normal = 285 to 295 mOsm). Because of the low CO_2 and the high blood sugar, an additional 200 units of insulin was administered, and the rapid intravenous drip of hypotonic saline solution was continued. One ampule of sodium bicarbonate was given by rapid intravenous injection. During the first 2 hr of therapy the patient received a total of 3½ L of hypotonic saline solution and 2 ampules of sodium bicarbonate. After 2 hr, blood was again drawn and sent to the laboratory. Results of this study were blood sugar, 48 mmol/L; total CO_2, 5 mmol/L; and pH, 7.1. Plasma ketones were strongly positive. The patient was still semiconscious but appeared better hydrated and had begun to excrete urine. Because of the continued low CO_2 concentration, the presence of severe ketonemia, and the continued elevation of blood sugar levels, another 300 units of insulin was given, half intravenously and half intramuscularly. Administration of hypotonic saline solution was continued intravenously but at a slower rate. Two hours after the second injection of insulin, additional blood studies were ordered. The results then showed blood sugar, 22 mmol/L and total CO_2, 10 mmol/L. The patient's condition was definitely improved, and he was no longer hyperventilating. He was now conscious. No further insulin was given at this time, and the intravenous infusion was changed to a solution containing hypotonic saline solution, 5% dextrose, and 50 mmol/L of potassium phosphate. This was run at a rate of about 250 ml/hr. Another blood sample was sent to the laboratory 2 hr later. The blood sugar was now 14 mmol/L, and the serum potassium was 4.0 mmol/L. The total CO_2 was 16 mmol/L, and the pH was 7.35. The plasma ketones were only trace positive in undiluted serum.

Biochemical questions

1. Why did severe hyperglycemia develop in this patient?
2. How did ketosis develop in the face of an elevated blood glucose concentration?
3. How did the administered insulin function to help correct the metabolic abnormality? How was the enzyme induction involved in the therapeutic effect of insulin?
4. Describe the mechanism that caused the acidosis to develop, and trace the logic of the therapy with respect to fluid and electrolyte administration.

Reference

Foster DW and McGarry JD: The metabolic derangements and treatment of diabetic ketoacidosis, N Engl J Med 309:159, 1983.

Additional questions and problems

1. How does the supply of inorganic phosphate affect carbohydrate metabolism?
2. Describe, with reactions, how the ingestion of xylitol could provide an energy source for muscular activity.
3. Explain why pyruvate is central to metabolic pathways of carbohydrates, amino acids, and lipids.
4. Discuss the role of the energy charge of the cell in the control of carbohydrate metabolism.
5. Describe how D-ribose may be produced in animals from glucose without the concomitant generation of NADPH.
6. Explain how carbohydrate catabolism in animals responds to changes in (a) the energy charge and (b) the oxidation-reduction condition of the cell.
7. Explain how the increasing concentration of glucose in the blood is converted to glucose 6-phosphate.
8. It has been reported that a human fetus can survive 30 min without oxygen. Discuss the metabolism of glucose under these conditions.
9. Discuss the interplay of glycolysis and the pentose phosphate pathway in the human erythrocyte. Why is the glycolysis only anaerobic?

Multiple choice problems

1. D-Glucose, D-galactose, and D-fructose are all:
   1. Isomers.
   2. Epimers.
   3. Monosaccharides.
   4. Aldohexoses.
      The best answer is:
      A. 1, 2, and 3.
      B. 1 and 3.
      C. 2 and 4.
      D. 4 only.
      E. 1, 2, 3, and 4.

2. The digestion of disaccharides occurs:
   A. In the gastric mucosa.
   B. By the action of salivary α-amylase.
   C. In the intestinal lumen.
   D. At the brush border membrane.
   E. By the involvement of ATP.

3. Pancreatic α-amylase:
   1. Hydrolyzes starch completely to glucose.
   2. Hydrolyzes α-limit dextrins.
   3. Is secreted as a zymogen.
   4. Hydrolyzes 1→4-α-D-glucosidic bonds
      The best answer is:
      A. 1, 2, and 3.
      B. 1 and 3.
      C. 2 and 4.
      D. 4 only.
      E. 1, 2, 3, and 4.

4. In a patient with galactosemia who is on a galactose-free diet, the D-galactose that is required for cell membrane and other biopolymers is formed by:
   A. Epimerization of UDP-D-glucose.
   B. Isomerization of glucose 1-phosphate.
   C. Aldolase condensation of triose phosphates.
   D. Decarboxylation of D-heptonic acid.
   E. Epimerization of D-fructose 6-phosphate.

5. A. 6-Phospho-D-gluconolactone lactonase.
   B. 6-Phospho-D-fructose kinase.
   C. Both.
   D. Neither.
      The enzyme(s) are committed steps in:
      1. Pentose phosphate pathway.
      2. Gluconeogenesis.
      3. Aerobic glycolysis.
      4. Anaerobic glycolysis.

6. Common intermediates in the glycolytic and pentose phosphate pathways are:
   1. Glucose 6-phosphate.
   2. Fructose 6-Phosphate.
   3. Glyceraldehyde 3-phosphate.
   4. Glyceric acid 3-phosphate.
      The best answer is:
      A. 1, 2, and 3.
      B. 1 and 3.
      C. 2 and 4.
      D. 4 only.
      E. 1, 2, 3, and 4.

7. A 3-year old patient with mild mental retardation, was found to have cloudiness of the lens, indicative of cataracts. When an abnormally high blood concentration of a sugar alcohol was found, the child was immediately placed on a milk-free diet. The enzyme most likely defective in the child is:
   A. Hexokinase.
   B. Galactose-1-phosphate uridyl transferase.
   C. Galactokinase.
   D. Lactase.
   E. Liver alcohol dehydrogenase

8. A patient is diagnosed as having a glucokinase deficiency. One of the consequences might be:
   A. Total inability to metabolize glucose.
   B. Requirement of a carbohydrate-free diet.
   C. Advantageous to maintain a high-protein diet.
   D. Advisability to replace carbohydrate by unsaturated fats.
   E. Spillage of all glucose from diet into urine.

9. The potential futile cycle, glucose $\underset{\text{Phosphatase}}{\overset{\text{Kinase}}{\rightleftharpoons}}$ glucose 6-phosphate, is reduced by the following fact:
   A. Glucokinase depends on insulin; glucose 6-phosphatase does not.
   B. Glucose 6-phosphatase depends on glucagon; glucokinase does not.
   C. Glucokinase is in the cytosol; glucose 6-phosphatase is in the endopasmic reticulum.
   D. Glucokinase is inhibited by glucose 6-phosphate; glucose 6-phosphatase is not.
   E. Glucose 6-phosphatase is inhibited by inorganic phosphate; glucokinase is not.

10. Control of enzymatic activities by phosphorylation/dephosphorylation:
    1. Always involves ATP.
    2. Includes Ca^{++} binding.
    3. Is stimulated by cyclic AMP.
    4. Is inhibited by caffeine.
       The best answer is:
       A. 1, 2, and 3.
       B. 1 and 3.
       C. 2 and 4.
       D. 4 only.
       E. 1, 2, 3, and 4.

Chapter 7

Carbohydrate synthesis and biopolymers

Objectives

1 To relate carbohydrate metabolism to gluconeogenesis, glycogen synthesis, and glycogen breakdown
2 To integrate carbohydrate metabolism and regulation
3 To discuss briefly the conjugated carbohydrate-containing biopolymers, such as glycoproteins and proteoglycans

Following the digestion of an average 8.4 MJ (2000 kcal) meal, approximately 250 g, or 1.4 mol, of glucose is loaded into the body. If the blood volume of the person were 7 L and the hematocrit 45%, then the glucose, without other distribution into body fluids, would increase in concentration to approximately 400 mM. Clearly this does not happen, the highest concentration reached is about 6.5 mM and this returns within 1 to 1½ hr to a resting concentration of around 4 mM (see Figure 6.5). Such control is accomplished by the synthesis of glycogen in all tissues up to a maximum, which is 1% to 2% of the tissue weight in muscle and 4% to 6% in liver. When concentrations fall, blood glucose homeostasis is maintained by the release of glucose from the liver and, to a lesser extent, from the kidney. Glucose is also synthesized from noncarbohydrate precursors by gluconeogenesis, which adds to the fine tuning of carbohydrate homeostasis. As a final control, excess carbohydrate leads to the accumulation of triacylglycerol in adipose tissue. There appears to be no upper limit as to how much triacylglycerol can be stored in the adipose tissue. These various metabolic processes are discussed in this chapter so that an integrated concept can be developed.

Glycogen metabolism

The synthesis and breakdown of glycogen occur through different metabolic paths that are delicately controlled in an interrelated manner to ensure the following:

1. Maintenance of blood sugar levels from the liver and, to a smaller extent, kidney glycogen stores. All other tissues either have no D-glucose 6-phosphatase or possess a physiologically insignificant amount and cannot hydrolyze the intracellular D-glucose 6-phosphate to D-glucose. Without such an enzyme, the cells must metabolize all the carbohydrate within the cell, and none can be released to maintain the blood glucose level.

2. Intracellular availability of D-glucose 6-phosphate for glycolysis and adenosine 5′-triphosphate (ATP) production.

3. Relief of a hyperglycemic state by glycogen biosynthesis. There is, however, a limit to the amount of glycogen that can be stored in normal tissues. When this limit is exceeded, the excess glucose is converted to fat.

Metabolic interrelationships

Before proceeding, it should be emphasized that except for the isolation of reactions by subcellular compartmentalization, the pathways of biochemistry are primarily conveniences for discussion rather than isolatable metabolic gears in a cellular machine. For example, the D-glucose 6-phosphate present in the cytoplasm is involved by coupled reactions in all the metabolic pathways of D-glucose. A change in this concentration affects all the pathways. This interrelationship is also supported by *normal* human homeostatic controls, which do not call for the simultaneous stimulation of oppositely directed metabolic pathways. Thus the synthesis of liver glycogen does not occur while the liver is responding to hypoglycemia. Similarly the synthesis of D-glucose from noncarbohydrate sources (gluconeogenesis) is not stimulated during hyperglycemia, when the body is trying to reduce the glucose levels. These concepts are illustrated by the pathways of glycogen breakdown (glycogenolysis) and glycogen synthesis (glycogenesis).

Glycogenesis The breakdown and synthesis of glycogen occur at the nonreducing terminal ends of the branched D-glucose polymer (see Figure 6.2). The chemical changes are shown in Figures 7.1 and 7.2, in which a two-branched segment of the molecule is used for purposes of illustration. During synthesis the outer chains of glycogen are elongated by the transfer of D-glucosyl residues from uridine diphosphate (UDP) glucose. The enzyme responsible for this biosynthesis, *glycogen synthase,* is closely bound to the particulate glycogen stores in the tissue. The reaction is essentially irreversible and results in the synthesis of (1→4)-α-D-glucosidic linkages. Glycogen itself is the most efficient acceptor, but smaller (1→D-glucose oligosaccharides can also serve this purpose. As a result of the sequential addition of glycosyl residues to the nonreducing terminal residues, the chains are longer than those in the original acceptor, which is most often glycogen. In the absence of any other enzyme, an "amylose type" of glycogen would result. However, when any chain has grown more than approximately 11 glucose units from the last branch point, a *branching enzyme* (α-D-glucosyl 4:6 transferase, also called oligo-1,4→1,6-glucantransferase) transfers segments of approximately seven glucose units in length from the outer chains onto position 6 of an α-D-glucosyl residue in another chain of the glycogen molecule. A (1→6)-α-D- linkage is thus formed, and a new branch point is produced. The oligosaccharide segments are transferred in a single step and not by sequential transfer of single α-D-glucosyl units. The reaction is irreversible and produces new chains, all of which can be elongated subsequently by the glycogen synthase reaction.

Glycogenolysis Glycogenolysis occurs by a different pathway. In the presence of the enzyme phosphorylase *a,* for which pyridoxal phosphate is an essential component, inorganic phosphate (P_i) cleaves the nonreducing terminal glucose residues of glycogen one at a time to produce D-glucose 1-phosphate. The reaction is analogous to hydrolysis in that the elements of phosphate (H—OPO_3H), instead of water (H—OH), are added to the cleaved glycosidic bond; the reaction is thus described as *phosphorolysis.* The action of phosphorylase *a* is terminated near the branch point, three to four D-glucosyl residues away from it on the average. The intermediate product is called a phosphorylase limit dextrin. A second enzyme (oligo-1,4→1,4-glucantransferase) transfers a segment of all but one glucosyl unit from the chain stubs that are linked to a glucosyl residue at position 6. The segments are transferred to the nonreducing terminal residues of the other chain stubs. The remaining single glucose units, the original branch points in the glycogen, are now hydrolyzed by α-D-1,6-glucosidase to D-glucose. The branch point that was stopping the further action of the phosphorylase *a* has been removed, and phosphorolysis can proceed to the next branch point. This reaction sequence is represented in Figure 7.2.

The combination of the α-D-1,6-glucosidase and the oligo-1,4 → 1,4-glucantransferase thus removes the blockage to further action of phosphorylase *a.* Together the transferase and glucosidase enzymes have been referred to as the *debranching enzyme.*

In summary, the synthesis and degradation of glycogen occur by different paths. Depending on the process affecting it, the glycogen molecule becomes smaller or larger,

Figure 7.1   Glycogen synthesis pictorialized. Glycogen$_1$ and glycogen$_2$ represent molecules of differing molecular weight; glycogen$_2$ is larger than glycogen; *UDPGlc*, Uridine diphosphate–glucose.

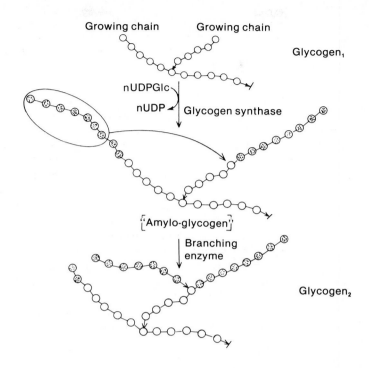

but rarely if ever is it completely degraded: even in starved animals the glycogen stores are never completely depleted. A nucleus of glycogen remains to act eventually as an acceptor for new glycogen to be built up when adequate supplies of carbohydrate are provided. Approximately 90% of the D-glucose produced by glycogen degradation is in the form of the 1-phosphate, whereas 10% is in the form of free sugar.

Control of glycogen metabolism

Almost every cell is capable of metabolizing glycogen. The enzymes of different tissues may vary in terms of cofactor requirements, quaternary structure, and inhibition, but they catalyze similar reactions (Table 7.1). The processes are cyclic, with the glycogen being synthesized or broken down according to the demands of the cell or the body. In its simplest form the intracellular cycle contains D-glucose 1-phosphate being converted to UDP-glucose, which is synthesized into glycogen for eventual phosphorolysis into D-glucose 1-phosphate (Figure 7.3). The two parts of this cycle, each of which is physiologically irreversible, are concerned with glucose utilization or storage. Whether synthesis or breakdown occurs at any particular time depends on the hormonal and substrate control of the various enzymes.

From the simplified scheme of glycogen metabolism represented in Figure 7.3, one can see that the active and inactive forms of glycogen synthase and phosphorylase differ in being either phosphorylated or dephosphorylated. Furthermore, the phosphorylase is active when phosphorylated, whereas the glycogen synthase is deactivated by phosphor-

Figure 7.2   Glycogen degradation pictorialized.

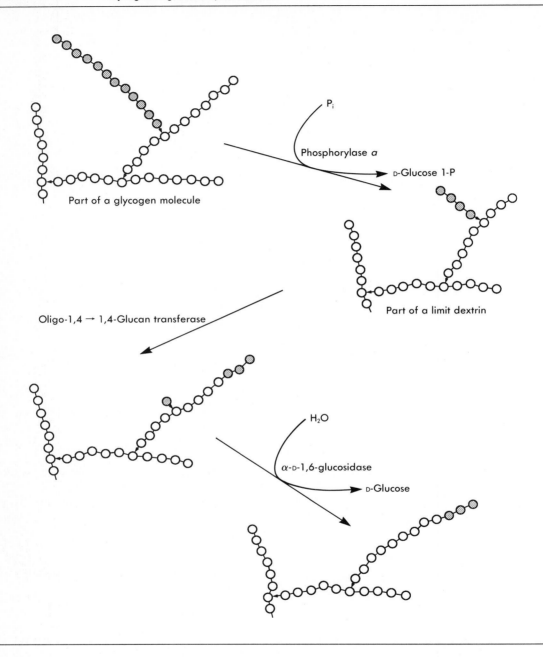

ylation. Thus the metabolic signals for anabolic and catabolic paths act in concert. These signals derive from hormonal and neural sources and act principally by way of protein kinases dependent on cyclic adenosine monophospate (cAMP) and calcium-activated calmodulin. To understand these controls and the action of insulin, one must analyze in greater detail the reactions of protein phosphorylation and dephosphorylation.

Figure 7.3   Glycogen metabolism.

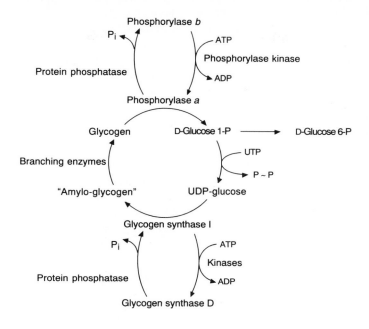

Table 7.1   Enzymes of glycogen metabolism

Enzyme	Role in glycogen metabolism
Phosphorylase *b*	"Inactive" form of phosphorylase *a*
Phosphorylase *a*	Phosphorolysis of glycogen to Glc-1-P
Phosphorylase kinase	Conversion of phosphorylase *b* to *a*
Protein kinase	Activation of phosphorylase kinase or inactivation of glycogen synthase I
Protein phosphatase 1	Hydrolyzes phosphate group from glycogen synthase I, phosphorylase *a,* and phosphorylase kinase
Protein phosphatase inhibitor I	Inhibits protein phosphatase 1
Glycogen synthase I	Addition of glycosyl groups to glycogen
Glycogen synthase D	"Inactive" form of glycogen synthase I, depends on Glc 6-f
Branching enzyme	Introduces branches into glycogen chains
Debranching enzymes	Remove branches that block action of phosphorylase *a*
Oligo 1, 4 → 1, 4 glucantransferase	Transfers small oligosaccharide unit from phosphorylase limit dextrin
α-1, 6-Glucosidase	Removes 1, 6-α-D-glucosyl residues

Cyclic AMP Various hormones bind to specific receptors on the plasma membrane of cells and stimulate adenylate cyclase, which catalyzes the formation of cyclic AMP from ATP. The cyclic AMP binds to the regulatory subunit, R, of protein kinase and causes a dissociation of the free catalytic subunit, C, in which form it is active to catalyze the phosphorylation of a number of proteins (see Chapters 3 and 17). It must be stressed, however, that relatively few proteins are phosphorylated at *significant* rates. Thus the cyclic AMP–dependent protein kinase is quite specific in its effect on metabolic pathways, activating some by phosphorylating inactive forms of enzymes, such as phosphorylase, and deactivating others by similar phosphorylation, such as glycogen synthase I. This is summarized in Table 7.2.

Calmodulin

It has been recognized for many years that free Ca^{++} flux was important in the physiologic function of tissues and cellular regulation. A small, thermostable protein called calmodulin was discovered in the 1970s that was ubiquitous in its distribution and bound four Ca^{++} ions. In the Ca^{++}-complexed state, calmodulin activates several enzymes, some of which phosphorylate proteins. Troponin-C appears to be a specialized form of calmodulin in the muscle, the two proteins having extensive homology in their amino acid sequence.

These hormonal and neural types of metabolic control can therefore be generally represented as:

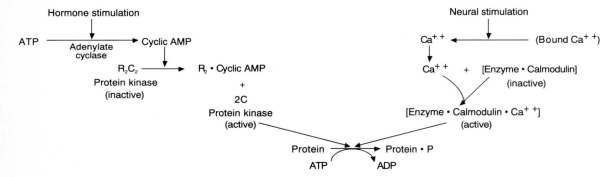

The specificity of the protein that is phosphorylated by one or other of the active kinases can be illustrated by the enzymes of glycogen metabolism.

Activation of phosphorylase kinase Phosphorylase kinase is composed of four protein subunits, α, β, γ, δ, which are themselves combined to give a quaternary structure $(\alpha\beta\gamma\delta)_4$. The α- and β-subunits can be phosphorylated by protein kinase, the γ-subunit is the catalytic subunit, and the δ-subunit is calmodulin. Phosphorylase kinase is phos-

Table 7.2   Phosphorylation-dephosphorylation of the enzymes in glycogen metabolism

	Enzyme	Phosphorylated	Dephosphorylated
	Glycogen synthase	Inactive	Adds glycosyl residues to glycogen
	Phosphorylase	Phosphorolysis of glycogen to Glc 1-P	Inactive
	Phosphorylase kinase	Phosphorylates phosphorylase *b*	Inactive
	Protein phosphatase-1 inhibitor	Inhibits protein phosphatase-1	Inactive

phorylated by a cyclic AMP–dependent protein kinase at two seryl residues, one on the α-subunit and one on the β-subunit, the latter phosphorylation being the more rapid and most closely correlated to the increased activity of the phosphorylase kinase. The activity of phosphorylase kinase also depends on Ca^{++}, which may be the sole controlling factor in glycogenolysis during muscle contraction.

Activation of phosphorylase The activated phosphorylase kinase phosphorylates each subunit of phosphorylase *b* at a single seryl residue, and in muscle, two dimeric phosphorylase *b* molecules combine to form phosphorylase *a,* which is a tetramer. The activation of phosphorylase may therefore be represented as:

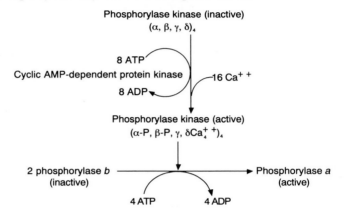

Deactivation of glycogen synthase There are two forms of glycogen synthase. One form, synthase D, is inactive except in the presence of D-glucose 6-phosphate. The activity of the other form, synthase I, is not increased by D-glucose 6-phosphate. Glycogen synthase D is formed from glycogen synthase I by phosphorylation, which is catalyzed by three different kinases. One is a cyclic AMP–dependent protein kinase that phosphorylates two seryl residues, another is phosphorylase kinase that phosphorylates a different seryl residue, and a third, glycogen synthase kinase, which is not cyclic AMP–dependent or activated by Ca^{++}, phosphorylates three different seryl residues. Phosphorylation of all six seryl residues is necessary to completely inactivate the glycogen synthase I in the absence of D-glucose 6-phosphate.

Dephosphorylation of phosphoenzymes Phosphorylase *a,* glycogen synthase D, and the β-subunit of phosphorylase kinase can be dephosphorylated by a single phosphatase, called protein phosphatase-1. The dephosphorylation of the β-subunit of phosphorylase kinase is slow unless the α-subunit is phosphorylated. The α-subunit is dephosphorylated by a different phosphatase, protein phosphatase-2.

Protein phosphatase-1 is inhibited by a protein inhibitor-1 when the latter has been phosphorylated with a cyclic AMP–dependent protein kinase. Interestingly, phosphorylated inhibitor-1 does not inhibit its own dephosphorylation by protein phosphatase.

The interconversions of the enzymes involved in glycogen metabolism are summarized in Figure 7.4.

Increasing evidence suggests that one action of insulin is to control the activity of phosphorylated inhibitor-1, which inhibits the activity of protein phosphatase-1. The latter is the active enzyme in the dephosphorylation of several enzymes controlled by phosphorylation-dephosphorylation. This includes several enzymes in addition to those involved in glycogen metabolism, such as pyruvate dehydrogenase, D-fructose 1,6-biphosphatase, pyruvate kinase, ATP-citrate lyase, and others involved in lipid and amino acid

Figure 7.4   Regulation of enzymes in glycogen metabolism by phosphorylation-
dephosphorylation. The phosphorylated enzymes are shaded.

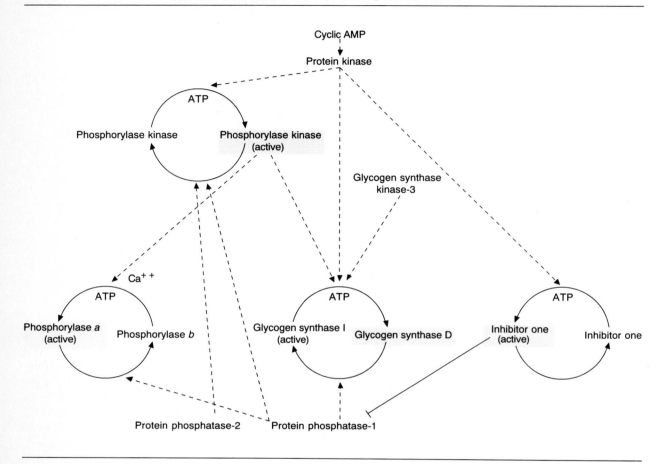

metabolism. If such were the case, the broad hormonal effects of insulin in the human could be understood in a unified way.

The metabolism of glycogen and the control by phosphorylation-dephosphorylation of enzymes and inhibitors has been elucidated most clearly from skeletal muscle. The details are not so well understood for the liver. Liver glycogen synthase appears to be similar to the corresponding muscle enzyme; it exhibits D and I forms, the former being activated by D-glucose 6-phosphate. Similarly the interconversion of liver phosphorylase *a* and *b* occurs, but the inactive phosphorylase *b* is not activated by AMP and does not form a tetrameric *a* form when activated, as occurs with muscle phosphorylase.

Glycogen metabolism cascade As a result of the appropriate hormonal interactions, adenylate cyclase forms cyclic AMP, which activates a protein kinase, which activates another enzyme that activates yet another, so that the resulting *cascade effect* is multi-plicative.

It is necessary to have an effective terminator of such a powerful cascade; this is present in the form of a phosphodiesterase that hydrolyzes the cyclic AMP to 5′-AMP. The

Table 7.3 Factors causing hypoglycemia or hyperglycemia

	Hormone or metabolite	Enzyme activity changed		Blood glucose
	Factors causing hypoglycemia			
	Elevated insulin	Glucose 6-phosphatase	↓	↓
		Glucokinase	↑	↓
	Elevated glucose 6-P	Glycogen synthase D	↑	↓
	Factors causing hyperglycemia			
	Elevated glucagon	Adenylate cyclase	↑	↑
	Elevated epinephrine	Adenylate cyclase	↑	↑
	Elevated cyclic AMP	Protein kinase	↑	↑

synthesis and catabolism of glycogen are regulated exquisitely by intracellular cyclic AMP concentration, which is controlled hormonally.

Added to these hormonal controls are those of enzyme activity, which are exercised either by metabolites—for example, D-glucose 6-phosphate and UDP-glucose—or by cofactors such as AMP. The physiologic consequences of these interacting controls can best be illustrated by a consideration of hypoglycemia and hyperglycemia. The hormone and metabolite effects on blood glucose are summarized in Table 7.3.

Clinical comment

Enzymatic response to hypoglycemia and hyperglycemia

Hypoglycemia In response to hypoglycemia, glucagon is secreted from the pancreas and stimulates adenylate cyclase in the liver to produce cyclic AMP. This in turn stimulates protein kinase and cascades to shut off any synthase I by converting it to the D form, which is relatively inactive in the absence of D-glucose 6-phosphate. Protein kinase affects the phosphorylase cascade to bring the active form, phosphorylase *a*, into action as quickly as possible, producing D-glucose 1-phosphate. This product is converted to the 6-phosphate, which is hydrolyzed to D-glucose by the D-glucose 6-phosphatase present only in liver, kidney, and intestine.

The muscle and other extrahepatic tissues have a reduced glycogen content during hypoglycemia. The levels of ATP may be low and those of AMP may be high. A high level of AMP stimulates phosphorylase *b*, so glycogenolysis is stimulated to provide glucose for glycolysis and ATP production. If concentrations of both glucose 6-phosphate and ATP increase significantly in the cell, they competitively inhibit the AMP stimulation of phosphophosphorylase *b*, and glycogenolysis will be controlled at the cofactor and substrate levels. Epinephrine can provide the additional stimulus for glycogenolysis when required.

Hyperglycemia During hyperglycemia the concentration of glucagon in the blood is low, but the concentration of circulating insulin is elevated. The high insulin level increases the rate of glucose transport across the extrahepatic cell membranes and also stimulates the de novo biosynthesis of glucokinase in the liver. Insulin also reduces the concentrations of phosphorylase *a* and glycogen synthase D. The concentrations of phosphorylase *b* and synthase I are correspondingly increased. All tissues are directed toward glycogen synthesis, which demands the production of UDP-glucose from UTP and D-glucose. In the liver, UDP-glucose inhibits phosphorylase, whereas high concentrations of UDP would inhibit the synthase. The regulation of glycogen synthesis is therefore delicately balanced by the levels of UDP-glucose, UDP, ATP, and even glycogen itself, which inhibits the synthase D phosphatase.

Role of D-glucose 6-phosphate and AMP

Muscle contraction requires the continuous supply of ATP, which is produced by the oxidative catabolism of D-glucose, lipids, and α-L-amino acids or, in extreme activity, may be formed from two molecules of ADP, as shown below:

$$2ADP \xrightarrow{\text{Adenylate kinase}} ATP + AMP$$

This transphosphorylation reaction not only generates ATP for further muscular contraction but also raises the concentration of AMP, which results in the activation of phosphorylase *b*. The continuous degradation of glycogen is thus possible without hormonal intervention as long as ADP is being formed and extreme activity is not demanded.

The use of the D-glucose 1-phosphate for glycolysis involves its conversion to the 6-phosphate. If D-glucose 6-phosphate accumulates, glycogen synthase D will be activated, and the glucose will be laid down again as glycogen. Such an occurrence in the muscle is quite possible, for example, when an excessive release of epinephrine occurs that subsequently is not needed. The false alarm would stimulate phosphorolysis of glycogen in the muscle, and the cell could be flooded with the D-glucose phosphates.

Clinical comment

Enzymatic response to muscle contraction At rest, most of the phosphorylase in muscle is in the *b* form. Creatine phosphate, ATP, and glucose 6-phosphate concentrations are such that glycolysis is not activated, and ADP and AMP concentrations are low, as is P_i. Phosphorylase *b* activity also is low.

When the muscle is called on to contract, several molecular events occur quickly.

1. ATP is hydrolyzed by the myofibrillar ATPase that has been activated by the Ca^{++} released through neural stimulation of the muscle membrane.

2. Creatine phosphate phosphorylates ADP to ATP through the creatine kinase reaction until the reserve of creatine phosphate has disappeared. The creatine phosphate disappearance precedes the reduction of ATP concentrations.

3. Some regeneration of ATP occurs from ADP by adenylate kinase, thus increasing the concentration of AMP.

4. From the elevation of P_i, ADP, and AMP, phosphorylase *b* will become more active. To play much of a role in the phosphorolysis of muscle glycogen, however, large increases in these external regulators of the enzymatic activity must occur.

5. There is a significant activation of the cascade of reactions involving the stimulation of adenylate cyclase, the formation of cyclic AMP, and the activation of protein kinase, phosphorylase kinase, and finally phosphorylase *a*. This occurs more rapidly than the activation of phosphorylase *b*. Activation of the phosphorylase kinase by phosphorylation of seryl residues in the α- and β-subunits reduces the concentration of Ca^{++} required to bind the γ-subunit. Binding of Ca^{++} to the γ-subunit is required to activate the enzyme fully. The Ca^{++} concentration required to activate the myofibrillar ATPase and the phosphorylated phosphorylase kinase is about the same (1 μM), so the hormonal and neural stimuli act in concert.

6. Glycogen is degraded to glucose 1-phosphate. This is converted via the mutase to glucose 6-phosphate and glycolysis proceeds, forming ATP. As the exercise continues, lactate is formed by anaerobic glycolysis, and some level of oxygen debt develops.

7. Recovery after exercise follows removal of the hormonal and neural stimuli, the conversion of lactate to glucose by way of the Cori cycle (see p.270), and the deactivation of phosphorylase *a* to *b*.

Abnormal glycogen metabolism

The confidence with which one can describe the intricate balance of glycogen metabolism in vivo results in large measure from the careful description of inherited diseases of glycogen metabolism by clinicians. Although diseases of glycogen metabolism are rare, examples involving each enzyme in the pathways have been studied. The structure of

Table 7.4   Glycogen storage diseases*

Name	Type	Principal tissue affected	Enzyme deficiency
von Gierke's disease	I	Liver and kidney	Glucose 6-phosphatase
Pompe's disease	II	Liver, heart, and muscle	1,4-α-D-Glucosidase (lysosomal)
Limit dextrinosis	III	Liver and muscle	α-D-1,6-glucosidase or oligo- 1,4 → 1,4-glucantransferase
Amylopectinosis	IV	Liver	"Branching" enzyme
McArdle's disease	V	Muscle	Phosphorylase
Hers' disease	VI	Liver	Phosphorylase
	VII	Muscle	Phosphofructokinase
	VIII	Liver	Phosphorylase kinase
	IX†	Liver	Glycogen synthase

*The disease entities listed here are those for which enzyme deficiencies have been identified.
†A deficiency in glycogen results.

the glycogen has been described in each case, as have the enzyme levels in the affected tissues. The biochemical aspects of these glycogen storage diseases are summarized in Table 7.4, and an example is described in case 2 at the end of the chapter.

Gluconeogenesis

D-Glucose is essential to the proper function of most cells; it is an absolute necessity for the nervous system and the erythrocytes. If adequate amounts of D-glucose are not provided in the diet, the blood glucose concentration falls and the body responds by synthesizing D-glucose from noncarbohydrate precursors. This process is called gluconeogenesis.

The carbon source for gluconeogenesis is several glucogenic precursors that are derived principally from L-amino acids (Figure 7.5). How the precursors are converted to phospho*enol*pyruvate is discussed in Chapter 5 and illustrated in Figure 5.21. The conversion of amino acids to pyruvate and to other Krebs cycle intermediates is discussed further in Chapter 5.

The formation of phospho*enol*pyruvate from pyruvate in the gluconeogenic pathway is not a simple reversal of the corresponding reaction in glycolysis. It takes place in two compartments of the cell, the cytosol and the mitochondria. Pyruvate passes from the cytosol into the mitochondria, where it enters several metabolic paths, one of which is its conversion to oxaloacetate by the enzyme pyruvate carboxylase. This enzyme has covalently attached biotin, and this coenzyme mediates the carboxylation of pyruvate.

$$\text{Biotin−pyruvate carboxylase} + \text{ATP} + \text{HCO}_3 \underset{}{\overset{\text{Acetyl CoA}}{\rightleftharpoons}} \text{ADP} + \text{P}_i + \text{CO}_2 \sim \text{biotin−pyruvate carboxylase}$$

$$\text{CO}_2 \sim \text{biotin−pyruvate carboxylase} + \text{Pyruvate} \rightleftharpoons \text{Oxaloacetate} + \text{Biotin−pyruvate carboxylase}$$

The carboxylation of the biotin only occurs if an acyl CoA, such as acetyl CoA, is bound to the enzyme as an allosteric effector. The oxaloacetate formed is reduced to malate inside the mitochondrion by NADH and malate dehydrogenase. The malate is then translocated from the mitochondrion to the cytosol, using one of the shuttle systems (see Chapter 5), where it is oxidized again to oxaloacetate. The oxaloacetate is converted to phospho*enol*pyruvate by its carboxykinase. This reaction, which involves phosphorylation and loss of CO_2, requires guanosine triphosphate (GTP).

Two other steps are mediated by enzymes that differ from those in glycolysis. One is

Figure 7.5 Carbon sources for gluconeogenesis.

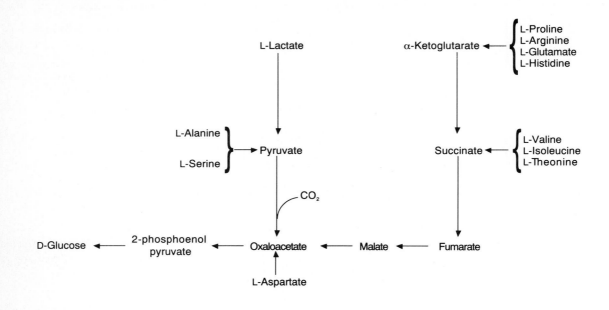

the hydrolysis of D-fructose 1,6-bisphosphate to D-fructose 6-phosphate, catalyzed by fructose-1,6-bisphosphate 1-phosphatase. The other involves the enzyme glucose 6-phosphatase. The pathway is summarized in Figure 7.6, which makes no attempt to show the subcellular compartmentalization of the reactions (see Figure 5.21).

Gluconeogenesis provides glucose during periods when circulating blood glucose levels are low. This pathway occurs principally in the liver and kidney. It does not occur to any physiologic extent in muscle, which has no glucose 6-phosphatase and therefore cannot convert glucose 6-phosphate to glucose for release into the blood.

Regulation of gluconeogenesis

Pyruvate carboxylase is found in the mitochondria of liver and kidney, whereas the pyruvate kinase occurs in the cytoplasm. Without such a compartmentalization of these enzymes, an energy-demanding and futile cycling of pyruvate might occur. As shown in Figures 5.21 and 7.6, mitochondrial pyruvate carboxylase catalyzes the carboxylation of pyruvate to oxaloacetate, which is transported to the cytoplasm via malate and then converted to phospho*enol*pyruvate. These steps involve the use of 1 mol each of ATP and GTP and 2 mol of nucleotide coenzymes. The phospho*enol*pyruvate may then be available for reconversion to pyruvate by pyruvate kinase, with the subsequent regeneration of ATP. However, this cycle is pointless and is prevented by the conversion of the pyruvate kinase into an inactive phospho- form by a glucagon-stimulated cyclic AMP–dependent protein kinase.

Gluconeogenesis is regulated primarily by four key enzymes: pyruvate carboxylase, phospho*enol*pyruvate carboxykinase, D-fructose 1,6-bisphosphate 1-phosphatase, and D-glucose 6-phosphatase. The first enzyme is stimulated by acetyl CoA (an absolute requirement) and inhibited by ADP. The 1-phosphatase is strongly inhibited by AMP and ADP, whereas the D-glucose 6-phosphatase is subject to product inhibition by P_i and

Figure 7.6   Gluconeogenesis. Dashed lines indicate increased glucogenesis; blunted solid lines indicate reduced activity.

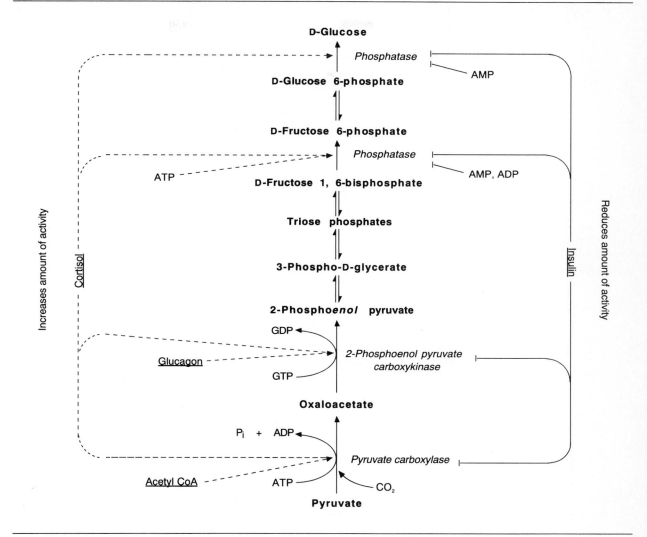

D-glucose. Glucagon stimulates gluconeogenesis by stimulating the activity of the rate-limiting enzymes, particularly phospho*enol*pyruvate carboxykinase.

Glucagon also indirectly facilitates gluconeogenesis by inhibiting glycolysis. This is mediated by a decrease in D-fructose 2,6-bisphosphate, which is a key effector of phosphofructokinase (PFK-1; see Chapter 6).

The biosynthesis of the four key enzymes is also influenced by insulin and the glucocorticoid hormone cortisol. Insulin represses the synthesis of all four enzymes, whereas cortisol induces their de novo synthesis. These effects seem reasonable when the physiologic conditions that necessitate glucogneogenesis are considered. If for some reason there is an inadequate supply of glucose—for example, starvation, high-fat diet, or hormonal imbalance—glucose must be provided by amino acids generated from protein catabolism, which is also stimulated by cortisol. Most of the energy for gluconeogenesis is probably derived from fatty acid oxidation. The fatty acids are produced by lipolysis of triglycerides, which is also potentiated by cortisol. Some evidence suggests that fatty acids have a stimulating effect on gluconeogenesis, but further research is needed to confirm this in vivo. As the glucogenic amino acids enter the gluconeogenesis pathway (see Figure 7.5), the Krebs cycle is involved to a significant extent. Malate, which is produced from oxaloacetate in the mitochondria using $NADH,H^+$ in the Krebs cycle, can be translocated from the mitochondria by the aspartate-malate shuttle. After moving into the cytoplasm, malate may be oxidatively decarboxylated to pyruvate by an $NADP^+$-linked malate enzyme. The pyruvate readily diffuses into the mitochondria. Therefore the ratios of $NAD^+/NADH,H^+$; $NADP^+$; and ATP/ADP, as well as the concentrations of Krebs cycle acids and acetyl CoA, all play a role in the control of gluconeogenesis. Fatty acid oxidation in the mitochondria requires NAD^+ and results in increased $NADH,H^+$ and acetyl CoA. Furthermore, the levels of Krebs cycle acids are increased by amino acid transamination. All these processes favor gluconeogenesis.

Hormonal interaction to control glucose metabolism

The control of glucose concentration in the blood and subsequent glucose metabolism depends on the proper functioning of the islets of Langerhans in the pancreas to produce three hormones. Within the islets are located α-cells that secrete glucagon, β-cells that secrete insulin, and δ-cells that produce somatostatin. These three types of cells are anatomically contiguous so that coordinated secretion of these polypeptide hormones seems to occur, particularly the two antagonists, glucagon and insulin. Thus the maintenance of proper glucose concentrations is achieved through an interactive secretion of insulin and glucagon, both secretions being inhibited by somatostatin; the secretion of somatostatin is stimulated by glucagon. The primary stimulus for these interactions is the concentration of glucose in the blood.

Insulin-induced hypoglycemia stimulates glucagon secretion from the α-cells of the pancreatic islets. Glucagon has no effect on muscle glycogen, but severe hypoglycemia elicits epinephrine secretion from the adrenal medulla. This mobilizes muscle glycogen, but not liver glycogen, with the eventual formation of lactate. By way of the Cori cycle the lactate is converted to glucose in the liver, and blood glucose is elevated. If this results in hyperglycemia, a subsequent increased secretion of insulin occurs. Thus a balance of two or three antagonistic hormones produces glucose homeostasis. This regulation is often lost in conditions that cause insulin insufficiency, such as diabetes mellitus, severe burns, and hemorrhagic shock. Further, the blood concentration of glucagon is reduced by somatostatin, a hormone that also inhibits the secretion of growth hormone. Consequently, in patients with high circulating concentrations of growth hormone, a condition found in acromegaly, treatment with somatostatin concomitantly reduces the levels of glucagon and blood glucose. Combined with the stimulation of gluconeogenesis by cortisol and glucagon, the interplay of hormones in the control of glucose metabolism is of considerable importance in the understanding of metabolism and the treatment of disease.

Biopolymers containing carbohydrate

Aside from glycogen and blood glucose, carbohydrate is present in the body as a part of the glycolipids and glycoproteins. In the latter, carbohydrate may be present as oligosaccharide groups attached to a protein core—for example, fibrinogen, the immunoglobulins, the blood group substances, and the mucous secretions. Some glycoproteins are composed of polysaccharide-protein complexes, such as the mucopolysaccharides, the chondroitin sulfates, dermatan sulfate, and keratan sulfate. The larger carbohydrate groups are present in molecules that have more of a structural function. By contrast, the function of the glycoproteins that contain oligosaccharides is related to molecule or cell

Table 7.5 Distribution and function of some glycoproteins

Presumed function	Name
Structural	Collagen, bacterial cell wall peptidoglycans, proteoglycans (mucopolysaccharides)
Food reserve	Plant pollen allergens, κ-casein
Enzyme	Ribonuclease B, prothrombin, β-D-glucuronidase
Transport	Ceruloplasmin, transferrin
Hormone	Thyroglobulin, erythropoietin, chorionic gonadotropin
Plasma and body fluids	α-Acid glycoprotein, fibrinogen
Immune systems	γ-Globulins, blood-group substances, HLA antigens
Cell membranes	Glycophorin of erythrocyte

Table 7.6 Structure of mucopolysaccharides (See text for abbreviations)

Mucopolysaccharide	Repeating unit	Occurrence
Hyaluronic acid	→4)-β-D-GlcA-(1→3)-β-D-GlcNAc-(1→	Vitreous tissue, joint fluids, skin, umbilical cord
Chondroitin 4-sulfate	→4)-β-D-GlcA-(1→3)-β-D-GalNAc-(1→ | 4-sulfate	Cartilage, skin, bone
Chondroitin 6-sulfate	→4(-β-D-GlcA-(1→3)-β-D-GalNAc-(1→ | 6-sulfate	Cartilage, nucleus pulposus, skin
Heparin	→4)-α-L-IdoA*-(1→4)-α-D-GlcNSO$_4$-(1→ | | | 2-sulfate 3,6-disulfate	Lung, spleen, liver, muscle
Keratan sulfate	→3)-β-D-Gal-(1→4)-β-D-GlcNAc-(1→ | 6-sulfate	Cornea, nucleus pulposus, cartilage
Heparan sulfate	→4)-α-L-IdoA-(1→3)-β-D-GlcNAc-(1→ | | 4-sulfate 4-sulfate	Skin, lung
Dermatan sulfate†	→4)-α-L-IdoA*-(1→3)-β-D-GalNAc-(1→ | | 4-sulfate 4-sulfate	Skin, lung

*D-Glucuronic acid (unsulfated) also present.
†Previous nomenclature; dermatan sulfate is also known as chondroitin sulfate B and β-heparin.

recognition. For instance, the immunochemistry of the blood group substances involves only a few sugar residues at the nonreducing termini of the oligosaccharide groups. Thus blood group B substance is converted to blood group H substance simply by removing its terminal residues of D-galactose. Examples of glycoproteins are given in Table 7.5.

In abbreviated structural representations of complex carbohydrates, the suffix A (for acid) is added to the symbol for the monosaccharide to represent the uronic acids; for example, GlcA is D-glucuronic acid, and L-IdoA is L-iduronic acid. The *N*-acetyl amino sugars have NAc added as a suffix to the monosaccharide symbol; for instance, D-Gal NAc is *N*-acetyl-D-galactosamine. Similarly, S is added for sulfate. In arrays that represent the structures of the polysaccharides, the positional numerals and anomeric prefixes are added at the appropriate places. The numbers are separated by arrows pointing from the glycosyl group to the hydroxyl group where the glycosidic linkage is made. These are illustrated in Table 7.6.

Glycoproteins   The covalent linkage between the carbohydrate groups and the polypeptide chain in glycoproteins may be of two types: (1) through the amide group of L-asparaginyl residues, such as ribonuclease B, and (2) to the hydroxyl group of L-seryl (submaxillary mucin), L-threonyl, hydroxy-L-lysyl (collagen), or hydroxy-L-prolyl (plant cell wall glycoprotein) residues. In the first type the linkage is most frequently made to an N-acetyl-D-glucosamine unit, whereas those of the second type are more varied; N-acetyl-D-galactosamine, D-galactose, and D-xylose are common linkage units. These linkages are shown in Figure 7.7). The other carbohydrate units are glycosidically linked as oligosaccharides or polysaccharides, which are attached to the sugar in the linkage areas as shown in Figure 7.7.

The number of carbohydrate groups attached to each protein varies; for example, only one is attached in ribonuclease B, several are attached in the immunoglobulins, and many are attached in the epithelial mucins and the blood group substances. The proportion of carbohydrate in glycoproteins varies widely. Collagen usually contains less than 1% carbohydrate; ribonuclease B, 8%; fetuin, 20%; serum α-acid glycoprotein, about 38%; submaxillary mucins, about 50%; blood group substances, more than 80%; and the mucopolysaccharides, approaching 100%.

Sialic acid   An important and ubiquitous family of sugars found in glycoproteins and glycolipids are the sialic acids. These are derivatives of neuraminic acid, where R may be an acetyl (CH_3CO—) or glycolyl ($HOCH_2CO$—) group and some of the hydroxyl groups may also be acetylated. The biosynthesis of neuraminic acid is catalyzed by an aldolase that condenses N-acetyl-D-mannosamine 6-phosphate with phospho*enol*pyruvate, producing the 2-keto sugar acid.

Neuraminic acid

$$N\text{-Acetyl-D-mannosamine} + \text{ATP} \xrightarrow{\text{Kinase}} N\text{-Acetyl-D-mannosamine 6-phosphate} + \text{ADP}$$

$$N\text{-Acetyl-D-mannosamine 6-phosphate} + \text{Phospho}enol\text{pyruvate} \xrightarrow{\text{Aldolase}} N\text{-Acetyl neuraminic acid 9-phosphate} + P_i$$

$$N\text{-Acetyl neuraminic acid 9-phosphate} \xrightarrow{\text{Phosphatase}} N\text{-Acetyl neuraminic acid} + P_i$$

Through the formation of the "activated" cytidine monophosphate N-acetyl neuraminic acid (CMP-NANA), the sialic acid residues are transferred to terminal sugar residues, such as galactose in the biopolymers, and may serve as terminators to any further glycosyl-chain growth.

$$N\text{-Acetylneuraminic acid} + \text{CTP} \xrightarrow{\text{Transferase}} \text{CMP-NANA} + P \sim P$$
$$\text{(NANA)}$$

$$\text{Protein} - (\text{Sugar})_x\text{-Gal} + \text{CMP-NANA} \xrightarrow{\text{Sialyltransferase}} \text{Protein} - (\text{Sugar})_x\text{-Gal-NANA} + \text{CMP}$$

The sialic acid residues in carbohydrate groups of biopolymers are hydrolyzed by neuraminidase.

$$\text{Protein} - (\text{Sugar})_x\text{-Gal-NANA} + H_2O \xrightarrow{\text{Neuraminidase}} \text{Protein} - (\text{Sugar})_x\text{-Gal} + \text{NANA}$$

Figure 7.7 Carbohydrate-protein linkages in glycoproteins.

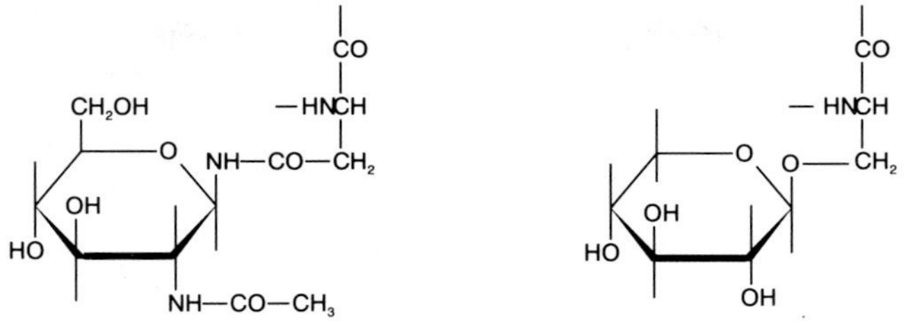

L-Asparaginyl-*N*-acetyl-
β-D-glucosaminyl linkage

L-Seryl-β-D-xylosyl linkage

The turnover of plasma glycoproteins is controlled by desialylation. Thus a serum glycoprotein such as ceruloplasmin is rapidly removed from the circulation by the reticuloendothelial system after the terminal galactosyl residues in the carbohydrate groups have been exposed by hydrolysis of the blocking NANA residues with neuraminidase.

Biosynthesis of glycoproteins The biosynthesis of the carbohydrate portion of glycoproteins is preceded by the biosynthesis of the polypeptide chain on the ribosome. It is presently believed that the oligosaccharides attached to the polypeptide chains through the hydroxyl groups of seryl, threonyl, hydroxyprolyl, or hydroxylysyl residues are synthesized by the sugar residues being added one at a time to the growing oligosaccharide chain.

Except for CMP-NANA, donors of the sugar units are the nucleotide diphosphate sugars. The first transglycosylase must be specific for the amino acid residue to which the sugar will be linked. Each transglycosylase is specific for the sugar residue that is transferred. To varying degrees each enzyme is also specific for the sugar residue that acts as the acceptor. Although no template exists for the synthesis, the specificities of the transglycosylases can properly order the sugar residues. Additional control is provided in the biosyntheses of the nucleotide phosphate sugars. CMP-NANA inhibits the formation of *N*-acetyl-D-mannosamine from UDP-*N*-acetyl D-glucosamine, which inhibits the formation of D-glucosamine 6-phosphate from D-fructose 6-phosphate. This is illustrated in Figure 7.8, where the nucleotide phosphate sugars are shown transferring the glycosyl group to the growing carbohydrate on the protein.

The mechanism of biosynthesis of the asparaginyl-linked glycoproteins involves a carbohydrate-lipid intermediate, a dolichol-diphosphate oligosaccharide. The oligosaccharide is then transferred to a protein acceptor, following which this structure is rapidly degraded by a series of glycosidases to a basic core unit onto which sugar residues are added one at a time to produce the final glycoprotein. The steps in the biosynthesis of the asparaginyl-linked oligosaccharide units are therefore (1) synthesis of a pro-oligosaccharide linked to dolichol diphosphate, (2) transfer of this unit to the polypeptide chain, (3) rapid processing of the pro-oligosaccharide to a core unit, and (4) resynthesis of the oligosaccharide to its final form by transglycosylation from nucleotide diphosphate sugars. These steps are summarized in Figure 7.9, where it is seen that step 1 is initiated by the transfer of two *N*-acetyl β-D-glucosamine residues from UDP-*N*-acetyl D-glucosamine to dolichol P $\sim$ P and then a β-D-mannose residue from GDP-Man to form a core

Figure 7.8   Feedback control in liver glycoprotein biosynthesis. See text for abbreviations.

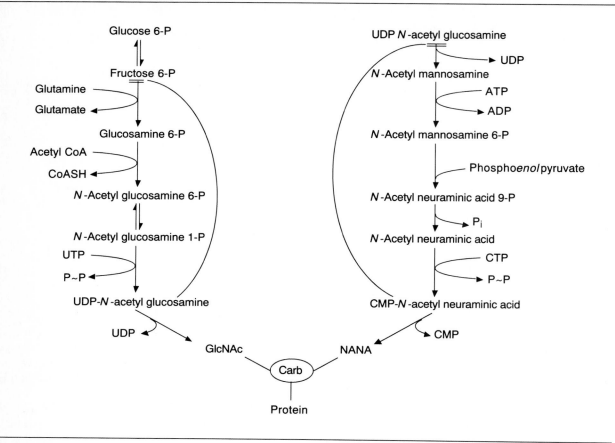

trisaccharide that is common to all glycoproteins of this type. To this core oligosaccharide is added several α-D-mannose residues from dolichol phosphate mannose, followed by one to three α-D-glucose residues from dolichol phosphate glucose. This whole pro-oligosaccharide is transferred to the amide nitrogen of an asparaginyl residue in the acceptor protein in step 2. Several sequential hydrolytic steps catalyzed by specific glycosidases comprise step 3 and result in the degradation of the pro-oligosaccharide unit to a pentasaccharide, which has the following structure:

$$\begin{array}{c} \text{Man I} \\ ^{\alpha\searrow} \\ ^{6}\text{Man I} \xrightarrow{\beta} 4\text{GlcNAc I} \xrightarrow{\beta} 4\text{GlcNAc I} \rightarrow \text{Asn Protein} \\ _{3} \\ _{\alpha\nearrow} \\ \text{Man I} \end{array}$$

The final mature glycoprotein is synthesized in step 4 in the Golgi apparatus by the successive involvement of highly specific glycosyl transferases.

The dolichol pathway in step 1 occurs in the membrane of the rough endoplasmic reticulum, and the pro-oligosaccharide unit is transferred to the nascent peptide while it is still being synthesized on the ribosome. Figure 7.10 illustrates the process of membrane

Figure 7.9   Biosynthesis of asparaginyl-oligosaccharide units in glycoproteins
where GlcNAc, Man, and Glc represent *N*-acetyl-D-glucosaminyl,
D-mannosyl, and D-glucosyl residues, and Asn Protein identifies the
L-asparaginyl residue in the acceptor protein to which the
pro-oligosaccharide is attached.

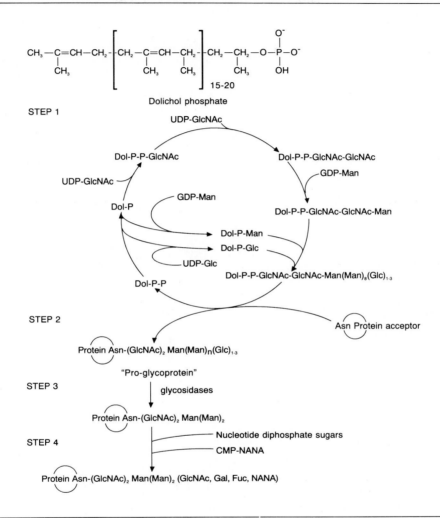

and secreted glycoprotein biosynthesis. The pro-oligosaccharide peptide passes into the
lumen of the rough endoplasmic reticulum, through the smooth endoplasmic reticulum,
and toward the Golgi apparatus. During this time the molecule is processed, and the
mature glycoprotein results. Those glycoproteins that will be secreted from the cell are
free in the lumen, whereas those molecules that become a part of the cell membrane
attach to the membrane of the Golgi apparatus from which secretory vesicles are formed,
move into the cytoplasm and then move to the cell membrane, where fusion occurs. The
inside surface of the vesicle becomes the outside of the cell membrane, the free glycoproteins are released, and the new portion of the cell membrane is properly oriented with
the oligosaccharide units on the outer surface.

Figure 7.10   Biosynthesis of glycoproteins. (After W.J. Lennarz.)

1. Rough ER

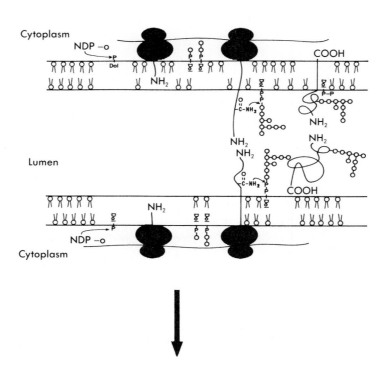

2. Smooth ER and Golgi

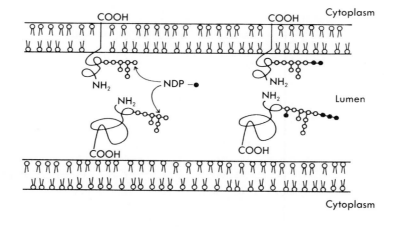

3. Secretory vesicle in cell

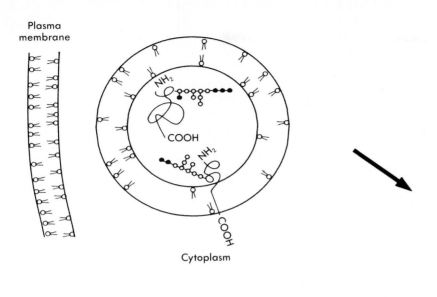

Plasma membrane

NH₂

COOH

Cytoplasm

4. Secretory vesicle fused with cell membrane

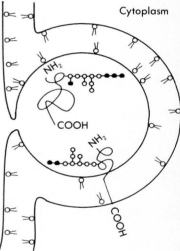

Cytoplasm

NH₂

COOH

NH₂

COOH

5. Secreted glycoprotein

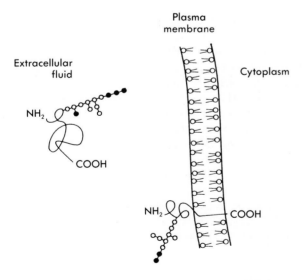

Plasma membrane

Extracellular fluid

Cytoplasm

NH₂

COOH

NH₂

COOH

Glycoproteins in membranes For many years the functions of several glycoproteins and particularly the carbohydrate groups were an enigma. It is now increasingly clear that the carbohydrate alters the physical properties of some glycoproteins, for example, the viscosity of the mucins. It is known that the glycoproteins are important, biologically active components of cell membranes. As discussed in greater detail in Chapter 12, the glycoproteins of cell membranes are embedded in the lipid bilayer. The glycoprotein molecule is exposed on one or both surfaces of the membrane, but the carbohydrate is present almost exclusively on the exterior surface. One of the best examples is that of glycophorin, one of several glycoproteins in the human erythrocyte membrane. It comprises about 5% of the total protein of the membrane and itself consists of about 40% protein and 60% carbohydrate; the latter includes 25% *N*-acetyl neuraminic acid. The molecule extends through the membrane, the *N*-terminal segment in the external environment carrying all the carbohydrate in 16 oligosaccharide groups, 15 of which are linked through the hydroxyl group of seryl or threonyl residues and one of which is in β-asparaginyl linkage. The *C*-terminal segment of glycophorin is exposed to the cytoplasm of the erythrocyte and contains a high proportion of hydrophilic amino acids. The segment of the protein within the membrane lipid bilayer is hydrophobic in composition.

The structures of the oligosaccharides in glycophorin are similar to those that serve as carriers of blood group determinants. Glycophorin is therefore representative of those glycoproteins that have important roles in membrane biology, the carbohydrate groups being similar to those of receptors for some peptide hormones and viruses, blood group antibodies, some other immunoglobulins, and phytohemagglutinins. The accessibility to the external environment of carbohydrate groups in cell membranes enables them to react with specific plant proteins, called lectins, such as concanavalin A. This ability to react is increased in the transformed, malignant cell. Similarly the distribution of the surface glycoproteins on the cell plasma membrane changes in the various stages of cell division.

The glycolipids (see Chapter 12) that have ABH and Le blood groups activities are minor constituents of the erythrocyte membrane. However, as expected, the antigenic determinant groups are the same as those noted in the glycoprotein blood group substances.

Blood group substances Human erythrocyte membranes contain antigenic substances, more than 300 of which can be classified into approximately 18 blood groups. The ABO and Lewis (Le) P_i and I_i systems are known to be oligosaccharides that are attached as glycoproteins, glycolipids, or proteoglycans. Antigenic substances present in saliva, gastric mucin, cystic fluids, and other body secretions can also be characterized by their blood group properties. Thus the sera or secretions of an individual who belongs to group A would agglutinate the erythrocytes of B and AB types. Carbohydrate accounts for 80% to 90% of these molecules, and blood group specificity is determined by the sugar residues close to the nonreducing termini. The existence of precursor structures that have similar antigenic determinants to those in the polysaccharide from the cell wall of a type XIV pneumococcus has been proposed. The pathways of biosynthesis follow the lines indicated in Figure 7.11. Similarly a series of transglycosylations to a Gal $\xrightarrow{\beta 1,4}$ GlcNAc $\xrightarrow{\beta 1,3}$ Gal precursor has been postulated to provide a parallel family of molecules with the same blood group activites. Since every individual maintains the same inherited blood group during life, it is clear that the transglycosylations cannot be random; otherwise, the blood type of each person would not be constant.

Clinical comment

Abnormalities in glycoprotein degradation The oligosaccharides of glycoproteins are degraded in the lysosome by the exoglycohydrolases that remove the specific end sugars in the chains. This is accomplished by an endo-β-D-*N*-acetyl-glucosaminidase that cleaves at the glycosidic linkage of the chitobiose unit linked to the asparaginyl residue and by

Figure 7.11   Possible pathways for biosynthesis of blood group substances.

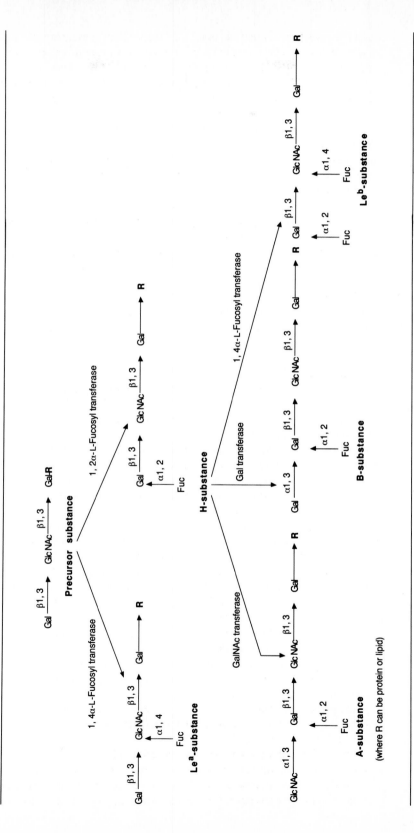

(where R can be protein or lipid)

an aspartylglycosaminidase that hydrolyzes the oligosaccharide from the protein. Lysosomal enzyme defects occur for each of these enzyme groups, giving rise to accumulation of the unhydrolyzed carbohydrate fragments in tissues and in the urine. These disorders are all autosomal recessive genetic defects that give rise to conditions similar to those seen with the mucopolysaccharidoses.

Proteoglycans (mucopoly-saccharides)

Proteoglycans (mucopolysaccharides) occur in many tissues and fluids. The proteoglycans of the ground substances and tissues—for example, cartilage, bone, cornea, and synovial fluid—are polysaccharides that are linked to protein through a xylosyl serine linkage (see Figure 7.7). The polysaccharide portion of the proteoglycans is referred to in general as a *glycosaminoglycan,* which may be classified into one of six types (see Table 7.6). The simplest is *hyaluronic acid,* present in synovial fluid of joints, in vitreous humor of the eye, and on cell membranes. The structure is a repeating unit of a disaccharide, composed of D-glucuronic acid and *N*-acetyl-D-glucosamine.

Hyaluronic acid

$[\longrightarrow$ 4)-β-D-Glc A-(1 $\longrightarrow$ 3)-β-D-GlcNAc-(1 $\longrightarrow]$

Hydrolysis of hyaluronic acid with hyaluronidases of mammalian origin gives rise to a tetrasaccharide consisting of two of the repeating units. A more complex reaction occurs when bacterial hyaluronidase is used, forming an unsaturated disaccharide in which a double bond is formed between positions 4 and 5 of the glucuronyl residue of the repeating unit. A similar unsaturated disaccharide is formed from the mucopolysaccharides chondroitin 4-sulfate, chondroitin 6-sulfate, and dermatan sulfate, except that *N*-acetyl-D-galactosamine replaces the *N*-acetyl-D-glucosamine.

The *chondroitin sulfates* are the most abundant mucopolysaccharides in the body. The repeating sugar unit is the same for each of them (see Table 7.6), but the sulfate is esterified to either the 4-hydroxyl or the 6-hydroxyl of the *N*-acetyl-D-galactosamine residue. Both polysaccharides are hydrolyzed to their corresponding tetrasaccharide units by mammalian hyaluronidase.

Keratan sulfate always accompanies the chondroitin sulfate in adult mammalian cartilage. It is composed of repeating units of → 3)-β-D-Gal-(1 → 4)-β-D-GlcNAc-(1 → with the sulfate ester at position 6 of the hexosamine residue and sulfate esters on some of the galactose residues.

Heparin and *dermatan sulfate* both contain L-iduronic acid residues. D-Glucuronic acid is a minor component of dermatan sulfate but is more significant (about 50%) in heparin. Heparin is not present in connective tissue but can be isolated from lung, where it occurs with dermatan sulfate and a less well-defined mucopolysaccharide called *heparan sulfate*. Heparan sulfate contains more *N*-acetyl groups and thus less *N*-sulfate groups than heparin, and it also has less *O*-sulfate. Unlike heparin, heparan sulfate has little if any blood anticoagulant activity. Heparin and heparan sulfate apparently are at opposite ends of a spectrum of a family of mucopolysaccharides.

These glycosaminoglycans, with the possible exception of hyaluronic acid, are covalently linked to a protein chain through a linkage region composed of —Gal—Gal—Xyl.

The D-xylosyl residue is linked to the protein through a seryl hydroxyl group (see Figure 7.7). In this manner a complex aggregate can be built, one example of which is the proteoglycan aggregate from cartilage containing hyaluronate, keratosulfate, chondroitin sulfate, and two types of protein, one of which can be isolated as part of a proteoglycan unit. The basic repeating unit of the proteoglycan aggregate is represented in Figure 7.12. The aggregate is formed by association of hyaluronic acid, proteoglycan subunits, and link proteins. The link proteins are located at the points where the proteoglycan units bind to the central filament of hyaluronic acid and serve to stabilize the binding. The proteoglycan units are composed of a core of protein, one end of which is the binding region to the hyaluronic acid. Numerous chondroitin sulfate chains are attached to the outer portion of the core protein, and the remaining protein segment is a keratosulfate-rich region. The structure is represented in Figure 7.12. Figure 7.13 shows an electron micrograph of an aggregate from epiphyseal cartilage with many proteoglycan subunits of various lengths spreading laterally from a central filament of hyaluronic acid.

Biosynthesis of proteoglycans The biosynthesis of the mucopolysaccharides may be illustrated using chondroitin 4-sulfate as an example (Figure 7.14). The polysaccharide is built on a D-xylose unit that is attached to the protein through the hydroxyl side-chain of an L-seryl residue. Two D-galactosyl units are then added. Following this, D-glucuronosyl residues and *N*-acetyl-D-galactosaminyl residues are added alternately. Finally, the carbohydrate chain is sulfated in certain positions by 3'-phosphoadenosine 5'-phosphosulfate (PAPS), a molecule performing a function similar to that of ATP in phosphorylation.

Depending on the type of sulfate transferase, the sulfate ester may be located in various positions of the chondroitin sulfate. The most common species are the 4- and 6-chondroitin sulfates, in which the sulfate group is at either position 4 or 6. The generation of the nucleoside diphosphate sugar precursors for the biosynthesis of chondroitin 4-sulfate is shown in Figure 7.14. The dashed lines indicate the points of insertion of the appropriate residues as chain elongation proceeds. In every case except hyaluronic acid, the linkage of the oligosaccharide to the protein is similar to that shown for chondroitin 4-sulfate (Figure 7.14), and the biosyntheses are presumed to be similar.

Figure 7.12   Representation of unit proteoglycan-hyaluronate, according to J.A. Buckwalter, University of Iowa.

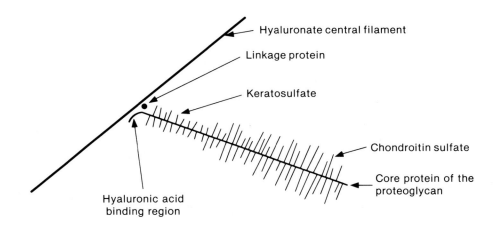

Figure 7.13   Electron micrograph of a proteoglycan aggregate from epiphyseal cartilage. (Courtesy J.A. Buckwalter, University of Iowa.)

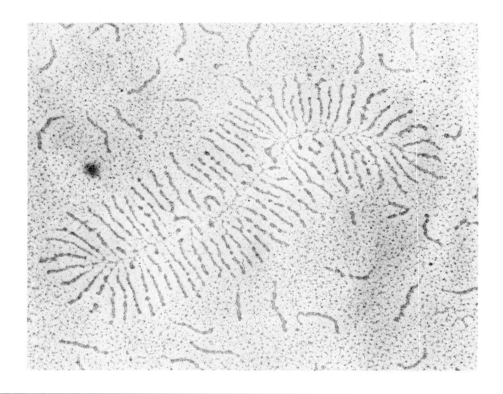

Although some of the proteoglycans have a long biologic half-life, the normal turnover requires the proteolysis of the protein core, which involves the lysosomal cathepsin enzymes. The mucopolysaccharide chains are hydrolyzed to smaller oligosaccharides by enzymes such as hyaluronidase, followed by the sequential removal of terminal sugars by glycosidases; for example, α-L-iduronidase hydrolyzes terminal α-L-iduronic acid residues.

Clinical comment   **Abnormalities of mucopolysaccharide metabolism** The mucopolysaccharidoses represent a group of diseases in which the enzymatic defects are in the degradation of the proteoglycans. Thus the proteoglycans accumulate and the disease results from deposition of these biopolymers in the lysosomes of various tissues. In general the diseases are characterized by mental retardation and the urinary excretion of the particular glycosaminoglycans.

Fortunately, these genetic diseases are relatively rare. Some of the mucopolysaccharidoses are summarized in Table 7.7, which also notes the tissues that are affected by deposits of particular mucopolysaccharides. The deposits are often present in skin fibroblasts, which can be grown in tissue culture. By this method a deficiency of α-L-iduronidase was found in Hunter's and Scheie's syndromes; the addition of this enzyme to the

Figure 7.14   Biosynthesis of chondroitin 4-sulfate.

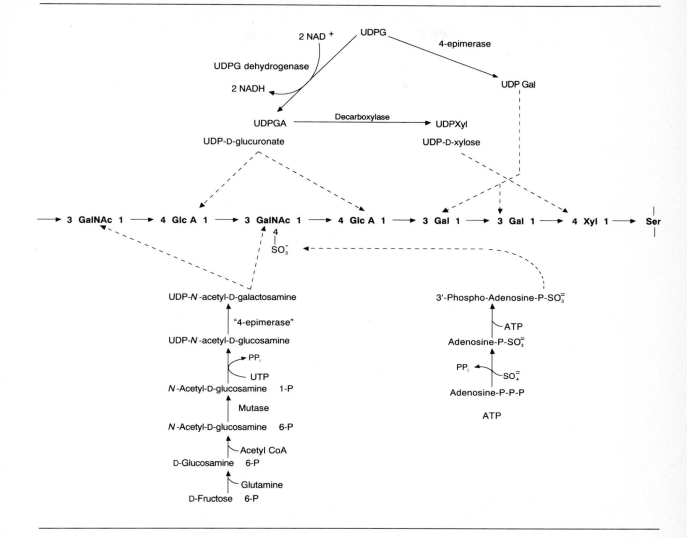

tissue culture media caused the catabolism of the dermatan and heparan sulfate to return to normal, and the fibroblasts grew normally. Similarly, corrective factors in the form of deficient hydrolases were identified for the other mucopolysaccharidoses: Hunter's, iduronate sulfatase; Sanfilippo's, heparan *N*-sulfatase or *N*-acetyl-α-D-glucosaminidase; Maroteaux-Lamy, *N*-acetyl-D-galactosamine 4-sulfatase; and Morquio's, *N*-acetyl-D-galactosamine 6-sulfatase. These enzymes, when added to the culture fluid, are presumably taken up into the lysosomes, which contain the stored, excessive amounts of the corresponding mucopolysaccharides. Normal catabolism can then ensue. Although such findings suggest the potential value of enzyme therapy, many technical problems have precluded its use.

Table 7.7   Some of the mucopolysaccharidoses (MPS)

Syndrome	Type	Clinical characteristics	Biochemical findings	Genetics	Enzyme defect
Hurler's	MPS IH	Severe mental retardation, skeletal deformities, corneal opacity, severe somatic changes	Dermatan sulfate* and heparan sulfate† in urine and tissues, dermatan sulfate in fibroblasts	Autosomal recessive	α-L-Iduronidase
Hunter's	MPS II	Moderate mental retardation, severe skeletal deformities, no corneal clouding, early deafness, extreme somatic changes	Dermatan sulfate and heparan sulfate in urine and tissues, dermatan sulfate in fibroblasts	X-linked recessive	Iduronidate sulfatase
Sanfilippo's (at least two types exist)	MPS III	Severe mental retardation, mild skeletal changes, corneal clouding questionable	Heparan sulfate in urine and tissues, dermatan sulfate in fibroblasts	Autosomal recessive	Heparan N-sulfatase N-acetyl α-D-glucosaminidase; N-adetyl-α-D-glucosaminide 6-sulfate sulfatase
Morquio's (two types exist)	MPS IV	Severe skeletal deformities, severe spondylepiphyseal dysplasia, no mental retardation, corneal opacities may occur	Keratan sulfate and chondroitin sulfates in urine	Autosomal recessive	Galactosamine sulfatase; β-D-galactosidase
Scheie's	MPS IS	Mild skeletal changes, no mental retardation, severe corneal opacity	Dermatan sulfate in urine	Autosomal recessive	α-L-Iduronidase
Maroteaux-Lamy	MPS VI	Severe skeletal deformities, gross corneal opacity, no mental retardation	Dermatan sulfate in urine	Autosomal recessive	N-Acetyl β-D-galactosamine 4-sulfatase
Sly	MPS VII	Mental retardation	Dermatan sulfate and heparan sulfate in urine	Autosomal recessive	β-D-Glucuronidase

*Dermatan sulfate is also known as chondroitin sulfate B and β-heparin.
†Heparan sulfate is also known as heparin sulfate, heparin monosulfate, and N-acetyl heparin sulfate.

Bibliography

Beaudet AL and Thomas GH: Disorders of glycoprotein degradation: mannosidosis, fucosidosis sialidosis, and aspartylglycosaminuria. In Scriver CR et al, editors: The metabolic basis of inherited disease, ed 6, New York, 1989, McGraw-Hill Information Services Co.

Blackshear PJ, Nairn AC, and Kuo JF: Protein kinases, 1988; a current perspective, FASEB J 2:2957, 1988.

Boden G et al: Severe insulin-induced hypoglycemia associated with deficiencies in the release of counterregulatory hormones, N Engl J Med 305:1200, 1981.

Felig P: Disorders of carbohydrate metabolism. In Bondy PK and Rosenberg LE, editors: Metabolic control and disease, ed 8, Philadelphia, 1980, WB Saunders Co.

Fletterick RJ and Madsen NB: The structures and related functions of phosphorylase a, Annu Rev Biochem 49:31, 1980.

Geddes R: Glycogen: a metabolic viewpoint, Biosci Rep 6:415, 1986.

Goodner CJ, Hom FG, and Koerker DJ: Hepatic glucose production oscillates in synchrony with the islet secretory cycle in fasting rhesus monkeys, Science 215:1257, 1982.

Hers HG and Hue L: Gluconeogenesis and related aspects of glycolysis, Annu Rev Biochem 52:617, 1983.

Klee CB, Crouch TH, and Richmond P: Calmodulin Annu Rev Biochem 49:489, 1980.

Kornfeld R and Kornfeld S: Assembly of asparagine-linked oligosaccharides, Annu Rev Biochem 54:631, 1985.

Lis H and Sharon N: Lectins as molecules and as tools, Annu Rev Biochem 55:35, 1986.

Muenzer J and Neufeld EF: Mucopolysaccharidoses. In Scriver CR et al, editors: The metabolic basis of inherited disease, ed 6, New York, 1989, McGraw-Hill Information Services Co.

Pastan IH and Willingham MC: Journey to the center of the cell: role of the receptosome, Science 214:504, 1981.

Randle PJ, Steiner DF, and Whelan WJ, editors: Carbohydrate metabolism and its disorders, vol 3, New York, 1981, Academic Press, Inc.

Sly WS: The mucopolysaccharoidoses. In Bondy PK and Rosenberg LE, editors: Metabolic control and disease, ed 8, Philadelphia, 1980, WB Saunders Co.

Steer CJ and Ashwell G: Hepatic membrane receptors for glycoproteins, Prog Liver Dis 8:99, 1986.

Clinical examples

A diamond (◆) on a case or question indicates that literature search beyond this text is necessary for full understanding.

Case 1	Diabetes mellitus and obesity

D.M., a 24-year-old, 178 cm tall graduate student, was seen in a diabetes outpatient clinic. His chief complaints were fatigue, weight loss, and increases in appetite, thirst, and frequency of urination. He had considered himself well until 6 mo before his visit. At that time he tired easily and tended to fall asleep in class, especially after the noon meal. He had attributed this to a heavy workload, the stress of a difficult research project, and rather high living. The symptoms continued and, during the previous 2 mo, he had lost approximately 6.8 kg from his normal weight of 72.7 kg. His appetite was good; in fact, it was excessive. For the previous 3 to 4 wk he had been excessively thirsty and had to urinate every few hours. He had begun to get up three to four times a night to urinate. With the continued weight loss, muscle weakness, fatigue, and frequent urination, it finally became apparent to D.M. that his life-style was not the cause of these problems. He went to the student health service and was referred to the diabetes outpatient clinic.

His family history was significant in that his maternal grandfather had had diabetes mellitus and had died at the age of 50 yr of a myocardial infarction. The patient had one older sister, 40 yr of age, who was obese and had recently been diagnosed as as having adult-onset diabetes.

Since the age of 15 yr the patient had experienced frequent episodes of weakness, during which he had felt tremulous and faint. He had found that eating a candy bar relieved the symptoms, which always occurred several hours after a meal. A routine physical examination at the age of 18 yr revealed a trace of sugar in his urine. His physician obtained a postprandial blood sugar specimen, which was normal at 5.0 mmol/L (90 mg/dl). Follow-up urinalysis revealed no glycosuria. The earlier analysis was attributed to a contaminated urine jug and dismissed.

The patient's physical examination was essentially within normal limits. Laboratory studies revealed a normal blood count. Urinalysis revealed a specific gravity of 1.040 and was 4+ positive for glucose using a Clinitest tablet; the glucose content in a 24 hr analysis was 0.19 mol (35g). The glycosylated hemoglobin was 14%. The urine also showed a moderate amount of ketone bodies as determined by an Acetest tablet. A glucose tolerance test was performed, and a blood sample was analysed for other substances, as indicated here.

	Patient D.M.	Normal
Glucose tolerance test		
Fasting	8.9 mmol/L	3.3 to 5.6 mmol/L
30 min	13.3 mmol/L	<10 mmol/L
1 hr	18.1 mmol/L	<10 mmol/L
90 min	16.9 mmol/L	<7.8 mmol/L
2 hr	15.8 mmol/L	<6.7 mmol/L
Electrolytes		
Na^+	140 mmol/L	136 to 145 mmol/L
K^+	4.1 mmol/L	3.5 to 5.0 mmol/L
Cl^-	98 mmol/L	100 to 106 mmol/L
Total CO_2	25 mmol/L	24 to 30 mmol/L
BUN	5.7 mmol/L	3.6 to 7.2 mmol/L
Creatinine	80 μmol/L	66 to 133 μmol/L

Biochemical questions

1. What is the basis for the symptoms of the patient?
2. What is glucosylated hemoglobin? What do elevated levels reflect in this patient?
3. How does the response to insulin of the obese diabetic person compare with that of the nonobese diabetic person?
4. Would you deduce from the electrolyte values that the patient was acidotic, dehydrated, or suffering from another type of imbalance?
5. Could the symptoms and laboratory values be the result of inadequate insulin levels?
6. What role does the polyol pathway play in disturbances of carbohydrate metabolism?
7. On his entering the hospital, what would you expect the patient's tissue glycogen levels to be?

Case discussion

1. **Diabetic signs** In his glucose tolerance test, D.M. demonstrated an inability to handle a normal glucose load and a hyperglycemia that exceeded the kidney threshold values. Such findings might have been expected in light of the earlier urinalysis and the familial history of diabetes.

   Evidence is now overwhelming that the potential to develop diabetes mellitus is inherited. Although it had remained latent for years, there had been earlier indications of the disease in this patient at the ages of 15 and 18 yr. At 24 yr of age his physical examination was normal, but he reported an increased appetite and excessive fluid intake and fluid loss. These symptoms suggested that his energy stores were being wasted and that frequent urination was required for the elimination of catabolic end products. In short, his tissues were starving in the presence of excessive amounts of circulating glucose. His muscle mass was producing approximately normal levels of creatinine.

   The major forms of diabetes are identified as *juvenile-onset* diabetes and *adult-onset* diabetes. The juvenile form is characterized by an abrupt presentation and requires insulin to control the hyperglycemia. The adult-onset diabetes produces milder symptoms, develops gradually, and can usually be controlled by diet. These patients are often obese. Either form can be found at any age despite their names.

2. **Glycosylated hemoglobin** D-Glucose reacts nonenzymatically with the free amino residues in proteins. In one of these products the glucose reacts with the free amino groups of the lysine residues or of the amino terminal acid to form a Schiff base (an aldimine), which may undergo an Amadori rearrangement to a more stable

ketoamine. This reaction takes place continuously during the life span of the red blood cell and can be detected by analysis of the hemoglobin. The degree of hyperglycemia is reflected in the amount of glycosylated hemoglobin in the blood sample and therefore presents an important diagnostic tool.

Ketoamine

In a similar way other proteins in the red cell membrane and in the lens of the eye may become modified. Furthermore, elevations of other sugars, such as galactose in untreated galactosemic patients, can give rise to glycosylated proteins, particularly glycosylated hemoglobin.

The glycosylated hemoglobin is determined by lysing a sample of erythrocytes that have been maintained in isotonic saline for a specified time to remove any free glucose in the cell. The lysed cells are centrifuged to remove cellular debris and the clear supernatant is chromatographed on an ion exchange column. The eluate is measured spectrophotometrically at 410 nm, and it is seen that hemoglobin and its derivatives resolve as two peaks, $HbA_{1a + b}$ and HbA_{1c}, separated from the major hemoglobin A (HbA_2) peak (Figure 7.15). The components HbA_{1a}, and HbA_{1b}, and HbA_{1c} result from the nonenzymatic glycosylation of hemoglobin; the most abundant derivative, HbA_{1c}, is formed from the condensation of glucose with amino terminal valine in the β-chains of hemoglobin A. The $HbA_{1a + b}$ peak also is elevated in patients with uremia because of carbamoylation of HbA.

The integrated area under the peaks determines the amounts of the components in these peaks and, in this patient, peak HbA_{1c} represents 14% of the total hemoglobin. This is significantly higher than for the normal person, in whom HbA_{1c} represents 5% to 9% of the total hemoglobin.

The glycosylation reaction is slow and irreversible and, when considered with the 120-day biologic half-life of the normal erythrocyte, any elevation of glycosylated hemoglobin must reflect an integrated elevation in the concentration of blood glucose over approximately the previous 2 to 3 mo. The measurement of HbA_{1c} is therefore a good clinical indicator of the time-averaged control of the blood glucose.

Figure 7.15   Determination of glycosylated hemoglobin by chromatography on Biorex 70. The nonglycosylated hemoglobin is designated HbA$_2$, the glycosylated hemoglobin as HbA$_{1c}$. Other modifications of hemoglobin, which are eluted together as a small peak after that of HbA, are designated HbA$_{1a}$ + A$_{1b}$. (Courtesy B.H. Ginsberg, University of Iowa.)

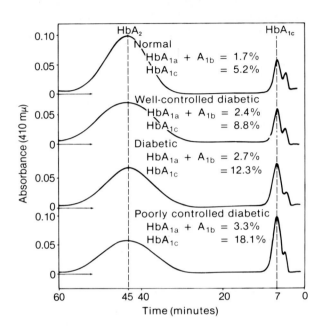

Although the glycosylation of hemoglobin may interfere with the binding of 2,3-bisphospho-D-glycerate, the oxygen-binding capacity of HbA$_{1c}$ is approximately equal to that of HbA, whereas the HbA$_{1a + b}$ capacities are lower.

3. **Glucose and insulin response in blood** The *essential* sign of diabetes mellitus is sustained hyperglycemia. Diabetes involves many other biochemical components. These abnormalities often are attributed to an inadequate secretion of insulin. As noted in the discussion of obesity (see Chapter 1), which is often associated with adult-onset diabetes, the change in the plasma concentrations of insulin is abnormal in response to carbohydrate loads in both obese and diabetic subjects. When plasma insulin concentrations are followed in relation to blood glucose concentrations, it is found that the overt diabetic individual has a slower, reduced secretion of insulin in response to glucose intake. The obese person has a nearly normal glucose tolerance, but it is achieved with high insulin levels. The obese, mild diabetic individual shows a slower but still high rise in plasma insulin, which is associated with a hyperglycemia and a slow recovery. These findings are summarized in Figures 7.16 and 7.17. Thus it would appear that the concentration of circulating insulin, measured by radioimmunoassay is not always proportional to the insulin effect.

It has been clearly demonstrated (see Chapter 17) that cells responding to insulin have receptor molecules in their plasma membranes to which insulin binds specifically. The concentration of these receptors on the cell surface can be determined by the use of radioactively labeled insulin to measure the amount of label that is specifically bound under standard conditions.

Figure 7.16

Glucose tolerance tests—blood sugar response. *MD-NO,* Mild diabetic person, nonobese; *MD-O,* mild diabetic person, obese; *ND-O,* nondiabetic person, obese; *N,* normal person.

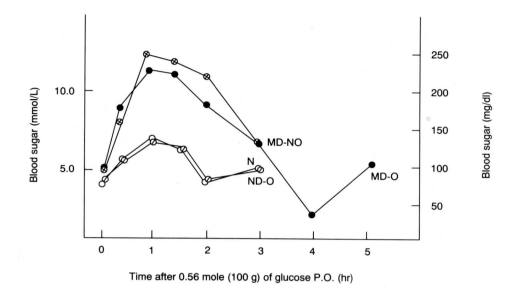

Figure 7.17

Glucose tolerance tests—insulin response. *MD-NO,* Mild diabetic person, nonobese; *MD-O,* mild diabetic person, obese; *ND-O,* nondiabetic person, obese; *N,* normal person.

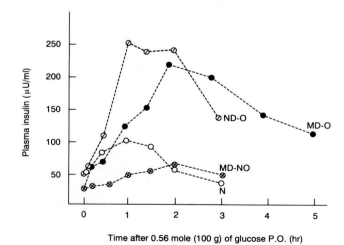

By such quantitative methods it is found that the number of receptor sites on a cell in the obese person with glucose intolerance is reduced. As the obesity is corrected, the concentration of receptors increases. These findings help to explain why obese individuals are refractory to the action of insulin (Figures 7.16 and 7.17.)

4. **Acid-base balance** The hyperglycemia noted after the ingestion of 100 g of glucose by this patient, who is not obese, was more severe than that in Figure 7.16. Therefore one might conclude that his circulating concentrations of insulin were low. Despite his hyperglycemia and glycosuria (urine loss of 0.19 mol of glucose/ 24 hr), his cells were deficient in glucose and were living essentially on a low-carbohydrate, high-fat energy supply, which resulted in a ketosis and compensated metabolic acidosis. No physical signs of acidosis existed; however, no plasma pH was recorded. The total CO_2 was within normal limits, but the serum Cl^- was slightly changed. The Na^+ and K^+ were within normal limits, indicating minimum dehydration. The water concentration mechanisms were functioning properly, as seen by the high urine specific gravity

5. **Insulin concentrations** when the patient entered the hospital, the level of circulating insulin was insufficient to prevent glycosuria but not so low that it precipitated a coma. He felt faint at times, and it seems reasonable to assume that glucose was not being maintained at adequate levels in the extrahepatic tissues, where the rate of transport across the cell membranes is abnormally low in the absence of insulin.

6. **Polyol pathway** Some tissues do not require insulin to assimilate glucose, so the glucose concentrations within these cells approach those in the blood. Some of this glucose may be reduced to sorbitol, which may then be oxidized to D-fructose, both of which may stay within the cell after the blood glucose has returned to normal. These reactions have been noted earlier, and the enzymes, particularly the NADPH-linked sorbitol dehydrogenase, which produces sorbitol from glucose, is found in high concentration in lens epithelium, the Schwann cell in peripheral nerve, the papillae in kidney, and the islets of Langerhans in the pancreas. These localizations identify with areas of pathology in the diabetic person, such as cataract formation and diabetic neuropathy.

7. **Glycogen** The glycogen levels in these tissues were low, inducing a state that resembled starvation. Although glucose entered the liver cells, it could not be metabolized at a sufficient rate because of the low level of glucokinase. Hexokinase cannot compensate for this deficiency because it is competitively inhibited by a high level of D-glucose 6-phosphate. To the extent that hexokinase can function, glycogen stores would be maintained.

Hyperglycemia was controlled in this patient through regular injections of insulin.

References   Brownlee M and Cerami A: Biochemistry of the complications of diabetes mellitus, Annu Rev Biochem 50:385, 1981.

Bunn HF: Evaluation of glycosylated hemoglobin in diabetes patients, Diabetes 30:613, 1981.

Foster DW: Diabetes mellitus. In Scriver CR et al, editors: The metabolic basis of inherited disease, ed 6, New York, 1989, McGraw-Hill Information Services Co.

Case 2   **Von Gierke's disease**

The patient was a 12-year-old girl who had a grossly enlarged abdomen. She had a history of frequent episodes of weakness, sweating, and pallor that were eliminated by eating. Her development had been somewhat slow; she sat at 1 yr of age, walked unassisted at 2 yr, and was doing poorly in school.

Physical examination revealed blood pressure, 110/58 mm Hg (14.7/7.7 kPa); temperature, 38° C; weight, 22.4 kg (low); and height, 128 cm (low). The patient had clear lungs and a normal heart. Slight venous distention was present over the prominent abdomen. The liver was enlarged, firm, and smooth and was descended into the pelvis. The spleen was not palpable, nor were the kidneys. The remainder of the physical examination was within normal limits except for "poor musculature."

Following are laboratory findings for a fasting blood sample.

	Patient	Normal values
Glucose (mmol/L)	2.8	3.9 to 5.6
Lactate (mmol/L)	6.6	0.56 to 2.0
Pyruvate (mmol/L)	0.43	0.05 to 0.10
Free fatty acids (mmol/L)	1.6	0.3 to 0.8
Triglycerides (g/L)	3.15	1.5
Total ketone bodies (mg/L)	400	30
pH	7.25	7.35 to 7.44
Total CO_2 (mmol/L)	12	24 to 30

A liver biopsy specimen was obtained through an abdominal incision. The liver was huge, buff-colored, and firm but not cirrhotic. Histologically the hepatic cells were bulging and dilated. The portal areas were compressed and shrunken. No inflammatory reaction was present. Stain for carbohydrate revealed large amounts of positive material in the parenchymal cells that was removed by digestion with salivary amylase. Glycogen content was 11 g/100 g of liver (normal up to 8%), and lipid content was 20.2 g/100 g of liver (normal less than 5 g). Hepatic glycogen structure was normal.

Following are results of enzyme assays performed on the liver biopsy tissue.

Enzyme	Patient (units per gram of liver N)	Normal range (units per gram of liver N)
Glucose-6-phosphatase	22	214 ± 45
Glucose-6-phosphate dehydrogenase	0.07	0.05 to 0.13
Phosphoglucomutase	27	25 ± 4
Phosphorylase	24	22 ± 3
Fructose-1,6-bisphosphatase	8.4	10 ± 6

Biochemical questions

1. What is a normal structure for hepatic glycogen?
2. What changes in this structure would accompany a deficiency in branching enzyme?
3. Which other tissues might be expected to accumulate excessive amounts of glycogen?
4. Explain the reasons for the fasting hypoglycemic episodes.
5. To what might be ascribed (a) the elevated free fatty acid, (b) the ketonemia, and (c) the metabolic acidosis?
6. What would you predict would be the result of continuous parenteral feeding? Explain.
7. What is the nature of the acidosis?

Von Gierke's disease is caused by a deficiency of glucose 6-phosphatase, resulting in the deposition of glycogen in liver and kidneys.

1. Structure of normal glycogen Glycogen is a high molecular weight polymer, a homopolymer of glucose with a branched structure. The principal glycoside linkage is α-D-1,4 bonds. Branch points are α-D-1,6-glycoside linkages.

2. **Structure of glycogen from patient with von Gierke's disease** The branching enzyme takes a portion of the growing α-D-1,4 chain and moves it in its entirety to form the α-D-1,6 branched linkage. Segments of about six residues are moved at a time. Single glucose units are then added in α-D-1,4 linkage one at a time to the end of this branch chain. If the branching enzyme was deficient, the branched structure of glycogen could not be formed. Therefore glycogen would not have its highly compact structure. In addition, it would not be synthesized as readily because synthesis normally occurs at the ends of each branch. Without the branching enzyme, only one continuous chain would exist. Likewise, removal of glucose units from the glycogen could not occur readily because, normally, the removal occurs simultaneously from all the many ends of the branches. In summary, an abnormality of a branching enzyme would lead to a very serious disease because glycogen of the proper structure could not be deposited in the tissues. It should be stressed that this is not the defect in this patient with von Gierke's disease. In this patient nothing is wrong with a branching enzyme or in the ability of the cell to form glycogen. The problem is a deficiency of glucose 6-phosphatase, the terminal enzyme present primarily in the liver that converts glucose 6-phosphate to free glucose.

3. **Sites of glycogen accumulation** Because of this enzyme deficiency, the liver is the principal organ that accumulates glycogen. Only a few other tissues in the body contain glucose 6-phosphatase. These are kidney and intestine; therefore one might predict that cells within the kidney and intestine also might accumulate glycogen. On the other hand, no evidence in the clinical material of this case indicates any intestinal or kidney malfunction. Poor musculature was reported on physical examination. It should be stressed, however, that the muscle does not normally contain glucose 6-phosphatase, and therefore this clinical finding has to be explained on another basis besides a deficiency of this enzyme.

4. **Hypoglycemia** In normal individuals hypoglycemia is prevented in a fasting state by release of glucose from the liver. This occurs in two ways. First, stored glycogen is broken down, and the ultimate product, glucose 6-phosphate, is converted to glucose through the action of the enzyme glucose 6-phosphatase. Second, glucose 6-phosphate can be synthesized by gluconeogenesis from the amino acid carbon atoms, and subsequently it also is converted to glucose through the action of glucose 6-phosphatase. Because of the deficiency of the glucose 6-phosphatase, the liver is unable to convert the glucose 6-phosphate that is produced to glucose. Because of this, the liver cannot release glucose into the blood, and hypoglycemia develops.

5. **Metabolic imbalances** Hormonal responses are triggered by the low blood glucose, and these stimulate the adipocytes to release free fatty acid. This provides an alternate metabolic fuel to replace the circulating glucose. In addition, the low level of circulating glucose also enables the adipocytes to release increased amounts of free fatty acid. Ketonemia develops because the liver takes a large portion of the circulating free fatty acid and converts it into ketone bodies. This is a protective response, since the ketone bodies can be utilized by the brain in place of glucose as an alternate substrate. The metabolic acidosis results from the excessive release of ketone bodies by the liver.

6. **Dietary management of patients with von Gierke's disease** Continuous intravenous feeding with a glucose-containing solution should be beneficial because it will prevent the episodes of hypoglycemia. Since glucose is continuously added to the circulation, the liver would not have to release glucose into the blood. Therefore some of the problems in this patient would be

improved. The symptoms of hypoglycemia, such as weakness and sweating, would not occur. Likewise, the presence of sufficient glucose in the blood would stop the excessive free fatty acid release, and this would stop the ketonemia. Once the ketosis was controlled, the metabolic acidosis would be corrected. On the other hand, the continuous infusion of glucose probably would lead to greater storage of glycogen within the tissue, and this might lead to even more tissue damage. In addition, the metabolic acidosis in this patient is caused partly by excessive lactate production as a result of glycolysis. This is explained next. In summary, the glucose administration will help some of the metabolic problems, but it will not help or might even worsen others associated with this disease.

7. **Acidosis** The acidosis is a metabolic acidosis which is not well compensated. It is caused by two problems: (a) as just mentioned, the excessive ketone body production results from the high levels of circulating free fatty acid and the low blood glucose concentration, and (b) an excessive amount of glucose is being converted to lactate. All the glucose that is formed within the liver and reaches glucose 6-phosphate cannot be released from the liver as glucose because of the deficiency of the enzyme glucose 6-phosphatase. Therefore more of the glucose 6-phosphate enters the glycolytic pathway, and more of it is converted to lactate. This leads to an excessive circulating lactate concentration, which contributes to the acidosis. Both the ketone bodies and the lactate are relatively strong acids and are buffered with bicarbonate. This lowers the bicarbonate concentration in the blood and leads to the metabolic acidosis. Because of the drop in bicarbonate concentration, the $p\mathrm{CO_2}$ is probably reduced as an attempted compensation for the low bicarbonate. Unfortunately the attempted compensation is not adequate, and the pH remains at 7.25, which is below the level of compensation.

References   Folk CC and Green HL: Dietary management of Type 1 glycogen storage diseases, J Am Diet Assoc 84:293, 1984.

Folkman J et al: Portacaval shunt for glycogen storage disease: value of prolonged intravenous hyperalimentation before surgery, Surgery 72:306, 1972.

Schwenk W and Haymond MW: Optimal rate of enteral glucose administration in children with glycogen storage disease type 1, N Engl J Med 314:682, 1986.

Case 3   Glycoprotein biosynthesis

A 24-year-old man had a lifelong history of diarrhea, abdominal pain, and distention on ingestion of sucrose. Sucrase-isomaltase deficiency was confirmed by enzymatic assay of a biopsy sample of his intestinal mucosa. The small intestinal mucosa was histologically normal. The sucrase-isomaltase was isolated from the biopsy sample and found to be a glycoprotein with a normal protein moiety but abnormal carbohydrate chains. The abnormality in the carbohydrate structure was caused by the lack of enzymatic activity.

Biochemical questions   1. What is the fate of undigested sucrose in the gastrointestinal tract?

2. Would the isomaltase activity of the enzyme be necessarily absent?

3. What would be the structural characteristics of oligosaccharide chains in a glycoprotein that was not completely processed following initial glycosylation of the protein precursor?

4. Where in the cell does preglycoprotein processing occur?

References

Hauri H-P et al: Transport to cell surface of intestinal sucrase-isomaltase is blocked in the Golgi apparatus in a patient with congenital sucrase-isomaltase deficiency, Proc Natl Acad Sci USA 82:4423, 1985.

Lloyd ML and Olsen WA: A study of the molecular pathology of sucrase-isomaltase deficiency, N Engl J Med 316:438, 1987.

Case 4	Hypoglycemia

A 5½-year-old girl, X.Y., had been admitted to her community hospital at 6 mo of age because of "pneumonitis." Fever, dyspnea, severe metabolic acidosis, and hepatomegaly were noted. She responded to therapy with intravenous fluids and antibiotics, but the liver enlargement persisted.

The family history is of interest. A male sibling had died at age 6 mo. He had seemed to be a normal child until 24 hr before death, when he developed unexplained severe metabolic acidosis. Necropsy revealed only the presence of "severe fatty changes" in the liver. Four other siblings are alive and healthy. No evidence of parental consanguinity exists.

On her first hospital admission the patient had been alert, and height and weight were normal. Her abdomen was protuberant, and a soft liver edge was easily palpable 8 cm below the right costal margin. Serum albumin, globulin, and cholesterol levels were normal, as were the AST and ALT. Fasting hypoglycemia was easily elicited, and her blood glucose was 0.5 mmol/L (8 mg/dl) after an 8 hr overnight fast. Glucagon (1 mg intramuscularly) given at that time caused no rise in blood glucose levels; however, when the same dose of glucagon was given 1 hr after a feeding, the blood glucose concentrations rose to 3.3 mmol/L (60 mg/dl) in 30 min. A diagnosis of a limit dextrinosis variety of glycogen storage disease (type III) was considered, and a liver biopsy was performed. The glycogen content was 1.4% of the wet weight, a level that is not excessive. Activities of hepatic glucose 6-phosphatase, phosphorylase, α-D-1,6-glucosidase, acid phosphatase, α-D-glucosidase, and phosphoglucomutase were normal. Histologic examination revealed ballooning of the hepatic cells by fat-containing vacuoles without inflammatory or fibrotic changes. This picture was similar to that found at necropsy in the liver of the deceased sibling.

The child was discharged, without a specific diagnosis, on a regimen of frequent feedings. Over the next 2 yr she was frequently admitted to the hospital for episodes of symptomatic hypoglycemia and severe metabolic acidosis, which were often associated with an intercurrent infection. Blood glucose concentrations less than 0.6 mmol/L (10 mg/dl) and arterial pH less than 7.15 were noted repeatedly.

After several admissions to the hospital, she was finally discharged on a diet that excluded all fructose, sucrose, and sorbitol. On a routine hospital evaluation at 5½ yr of age she was noted to be above the fiftieth percentile for both height and weight. The physical examination was completely normal. The liver was no longer palpable. Neurologic examination was completely within normal limits, and she was doing well in kindergarten.

The effect of fasting was also investigated at this time. Following a 12 hr overnight fast the values noted in Table 7.8 were seen. These changed precipitously over the next several hours of continued fasting. The glucose tolerance test was normal, as were the plasma insulin levels. Both fructose and glycerol tolerance tests produced hypoglycemia.

Hepatic enzyme activities were found to be normal *except* for the absence of fructose 1,6-bisphosphate 1-phosphatase.

Table 7.8   Effect of fasting on patient X.Y.

Value	12 hr	18 hr	21 hr
D-Glucose (mmol/L)	3.1	1.9	0.6
pH	7.4	—	7.17
Total CO_2 (mmol/L)	25.0	—	9.3

Biochemical questions

1. What explanation can be provided for the fasting hypoglycemia in this patient?
2. Why was the diagnosis of type III glycogen storage disease dismissed?
3. Aside from glucose, what other sugars or sugar alcohols would relieve the fasting hypoglycemia? Explain.
4. What are the causes of metabolic acidosis in this patient?

Reference

Gitzelmann R et al: Disorders of fructose metabolism. In Scriver CR et al, editors: The metabolic basis of inherited disease, ed 6, New York, 1989 McGraw-Hill Information Services Co.

Case 5   Mannosidosis

The patient was born of unrelated parents, neither of whom showed abnormalities of the nature seen in the child as he developed. At birth the baby appeared normal, but his development was slow. At 7 mo many vacuolated lymphocytes were seen on his peripheral blood smear, but the urinary mucopolysaccharide excretion was normal. At 3 yr he showed profound mental retardation, enlarged spleen and liver, cataracts, and continued vacuolized lymphocytes. Skin biopsies were taken from the patient and his parents; the fibroblasts from these biopsies were grown in tissue culture and examined for several glycohydrolase enzymes, the activities of which are summarized here.

| | n Moles of substrate hydrolyzed per hour per milligram protein | | |
	α-D-Mannosidase	β-L-Fucosidase	N-Acetyl-β-D-glucosaminidase
Normal controls	94 (range 41 to 162)	32 (12 to 79)	5104 (2909 to 7892)
Patient	3	21	5784
Mother	45	15	4780
Father	34	22	4870

Biochemical questions

1. What is the subcellular location of the glycohydrolase enzyme?
2. What justification is there for analyzing fibroblast cells obtained from the skin in attempting to make the diagnosis in this patient?
3. What type of molecules would be expected to accumulate in tissues and be excreted in excessive amounts in the urine of this patient?
4. What is the type of genetic inheritance for the disease?
5. What are the consequences of a β-D-mannosidase deficiency?

References

Aylsworth AS et al: Mannosidosis, J Pediatr 88:814, 1976.
Neufeld EG and Muenzer J: The mucopolysaccharide storage diseases. In Scriver CR et al, editors: The metabolic basis of inherited disease, ed 6, New York, 1989, McGraw-Hill Information Services Co.
Wenger DA et al: Human β-mannosidase deficiency, N Engl J Med 315:1201, 1986.

Additional questions and problems

1. What is the regulatory role of fructose 2,6-bisphosphate in gluconeogenesis?
2. Explain the role of triose phosphate isomerase in gluconeogenesis.
3. Discuss the effects of a triose phosphate isomerase deficiency on the metabolism of carbohydrates.
4. What is the role of pyridoxal phosphate in phosphorylase *a*?
5. Discuss the consequence of a genetic defect that severely reduced or eliminated the activity of glycogen phosphorylase kinase.
6. What enzymes that mediate carbohydrate metabolism are affected by phosphorylation-dephosphorylation of the protein?
7. Discuss the degradation of serum glycoproteins.
8. Discuss the assembly of the proteoglycan-hyaluronate complex in cartilage.
9. What is the role of calmodulin in glycogen metabolism?
10. How is the glycogen store maintained if a patient is placed on a carbohydrate-free diet?

Multiple choice problems

A girl, J.V., appeared normal at birth but developed signs of liver dysfunction and muscular weakness at the age of 3 mo. Enzyme assays confirmed that the glycogen branching enzyme was absent in leukocytes and liver. The clinical condition of the child was deteriorating rapidly, and six doses of a α-D-glucosidase were given intravenously over 3 days. The liver glycogen content was 6% of net weight of the tissue, and glycogen has accumulated in all the tissues. Two brothers of J.V. had died earlier of the same disease; the parents were second cousins.

The following six questions relate to this case.

1. What is the structure of the abnormal glycogen?
   A. Terminal chains longer than normal.
   B. Terminal chains shorter than normal.
   C. Smaller molecular size.
   D. Contains glucosidic linkages that are not 1,4-α-D- or 1,6-α-D-.
   E. Nonreducing end is phosphorylated.

2. What carbohydrate hydrolase enzymes, such as α-D-glucosidase, are usually present in the lysosomes?
   1. α-L-Iduronidase
   2. Hexosaminidase
   3. β-D-Glucuronidase
   4. α-D-Mannosidase
      The best aswer is:
      A. 1, 2, and 3.
      B. 1 and 3.
      C. 2 and 4.
      D. 4 only.
      E. 1, 2, 3, and 4.

3. α-D-Glucosidase hydrolyzes glycogen to give:
   1. Maltose.
   2. Glucose.
   3. Glucose 1-phosphate.
   4. Glycogen limit dextrin.
      The best answer is:
      A. 1, 2, and 3.

   B. 1 and 3.
   C. 2 and 4.
   D. 4 only.
   E. 1, 2, 3, and 4.

4. The biosynthesis of normal glycogen requires:
   A. Phosphorylase.
   B. Debranching enzyme.
   C. 1,6 α-D-glucosidase (amylo-1,6-glucosidase).
   D. Glycogen synthase.
   E. ATPase.

5. In what tissues would the injected α-D-glucosidase likely concentrate in the body?
   1. Spleen
   2. Muscle
   3. Liver
   4. Heart
   The best answer is:
   A. 1, 2, and 3.
   B. 1 and 3.
   C. 2 and 4.
   D. 4 only.
   E. 1, 2, 3, and 4.

6. The genetics of the disease that:
   A. Either the mother or father is homozygous for the disease.
   B. The disease is sex linked.
   C. Either the mother or father is heterozygous for the disease.
   D. The disease is autosomal recessive.
   E. No genetic association exists.

The fibroblasts from a patient with Hunter's syndrome were grown in tissue culture and noted to accumulate dermatan sulfate. Inclusion of the enzyme α-L-iduronate sulfatase in the culture medium corrected the excess accumulation of the proteoglycan in the cells.

The following questions relate to this experiment.

7. The deficient enzyme in Hunter's syndrome, α-L-iduronate sulfatase, is normally present in:
   A. Nucleus.
   B. Mitochondria.
   C. Cytoplasm.
   D. Lysosomes.
   E. Cell membrane.

8. The biosynthesis of dermatan sulfate uses the following substrates:
   1. UDP-L-iduronate.
   2. UDP-N-acetyl-D-galactosamine.
   3. PAPS, 3'-phosphoadenosine 5'-phosphasulfate.
   4. Dolichol pyrophosphate.
   The best answer is:
   A. 1, 2, and 3.

    B. 1 and 3.
    C. 2 and 4.
    D. 4 only.
    E. 1, 2, 3, and 4.

9. A. D-Glucuronosyl units
   B. L-Iduronosyl units
   C. Both
   D. Neither
     For each numbered item listed next, indicate its association with the items in A to D above.
     1. Chondroitin sulfate
     2. Dermatan sulfate
     3. Heparan sulfate
     4. Hyaluronic acid

10. Sulfate esters of proteoglycans are:
   A. Introduced into the substrate monomers before polymerization.
   B. Linked only to hydroxyl groups.
   C. Positively charged.
   D. Involved in neutralizing the acidic COO^- groups of the associated proteins.
   E. Synthesized from additions from 3'-phosphoadenosine 5'-phosphosulfate.

11. Dephosphorylation of enzymes by phosphatase *always:*
   A. Activates the catalytic effect.
   B. Inhibits the catalytic effect.
   C. Is metabolically reversible.
   D. Is cyclicAMP dependent.
   E. Acts on the tyrosyl phosphate residues.

Chapter 8

Amino acid metabolism

Objectives

1 To describe digestion of dietary protein and the absorption of the constituent amino acids and peptides
2 To explain how amino acids are deaminated and the nitrogen converted to urea
3 To demonstrate how the carbon skeletons of the amino acids are used to obtain energy or to assist in glucose or fatty acid synthesis
4 To analyze deviations in these reactions occurring with certain diseases

The most serious nutritional problem in the world is the deficiency of dietary protein. Although devastating in underdeveloped countries, protein deficiency also occurs in developed countries, especially among pregnant and lactating women, the newborn, poor and aged persons, and those who habitually avoid foods rich in protein. In the strict sense people do not require dietary protein, but rather the amino acids that make up these proteins. Consequently it is important to understand how the dietary amino acids are used for the biosynthesis of other amino acids and how they serve as precursors of other nitrogen-containing metabolites.

Dietary protein requirements

Dietary proteins serve three broad functions: (1) their constituent amino acids are used for the synthesis of the body's proteins; (2) the carbon skeletons of the amino acids can be oxidized to yield energy; and (3) their carbon and nitrogen atoms may be used to synthesize other nitrogen-containing cellular constituents as well as many non–nitrogen-containing metabolites. The use of amino acids as a source of energy is described in this chapter; the biosynthesis of the amino acids and their functions in the synthesis of biogenic amines are considered in Chapter 9; and the use of amino acids in protein synthesis is developed in Chapter 15.

Although biologic energy requirements can be satisfied by the oxidation of lipids and carbohydrates, the amino acids are uniquely required for protein biosynthesis. Therefore one must eat sufficient protein-containing foods of adequate biologic value.

Unlike carbohydrates and lipids, proteins and amino acids are not stored by the body in particular cells, such as the lipids, which are stored in adipocytes. Rather, proteins are present in all cells. Nevertheless, some of the body's proteins can be mobilized during fasting or starvation. The carbon skeletons of the mobilized amino acids are either burned for energy or converted to glycogen or triacylglycerol, which can be stored.

Some of the many metabolic products of dietary amino acids are summarized in Figure 8.1. The amino acid "pool" shown in the diagram is not a storage place for amino acids; rather this conceptual "pool" is a convenient way to indicate that small amounts of amino acids are present in cells or circulate in the blood.

During starvation much of the plasma proteins, especially albumin, are utilized first. The resulting amino acids, particularly their amino groups, are reutilized. Rapidly metabolizing tissues, such as liver, pancreas, and intestinal mucosa, also tend to lose their proteins quickly. Muscle is slower to yield amino acids, but because muscle is approx-

Figure 8.1   Summary of protein metabolism.

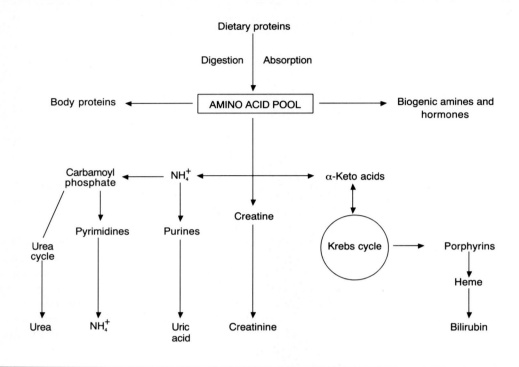

imately 60% protein, it represents the largest protein reservoir. When fat supplies are depleted, the body may lose as much as 6% of its protein mass per day for energy.

Digestion of proteins   Dietary proteins are digested to their constituent amino acids by proteolytic enzymes and peptidases in the gastrointestinal tract. A few small proteins and several peptides are absorbed directly from the intestine, but most of the digestion products circulate as amino acids. Measurement of plasma amino nitrogen is a good index of the content of amino acids in the blood. After a short fast, plasma amino nitrogen is approximately 4 mmol/ L (5 mg/dl), whereas shortly after a meal rich in protein, plasma levels increase to 6 mmol/L (8 mg/dl).

Activation of pepsinogen in the stomach With the exception of the intestinal pepti-dases, all proteolytic enzymes are activated by the conversion of inactive large precursors called *zymogens* to functional enzymes. The first zymogen that comes into play is pep-sinogen. Pepsinogen is produced in the chief cells, which are secretory cells located in the mucosa of the stomach wall. Pepsinogen release is stimulated by the hormone gastrin, which is released from the antrum (the distal end of the stomach) following feeding and gastric distention. The released gastrin is carried in the blood to the body of the stomach, where it triggers the release of pepsinogen from the chief cells. Simultaneously, gastrin acts on the parietal cells in the body of the stomach to cause the release of hydrochloric acid. The secreted pepsinogen is activated autocatalytically by the low pH of the stomach contents so that a small amount of pepsinogen is cleaved into pepsin and inert peptides (Figure 8.2). The pepsin produced is an active proteolytic enzyme that activates the

Figure 8.2   Activation of pepsinogen and action of pepsin.

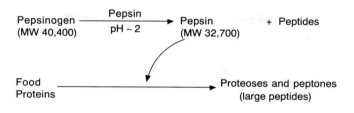

Figure 8.3   Activation of pancreative zymogens.

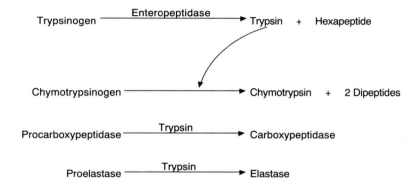

remainder of the pepsinogen molecules by proteolytically converting them to pepsin. As many as seven different pepsinogens may exist, each giving rise to a distinct pepsin enzyme. The pepsin-catalyzed phase of protein digestion is important but not essential. For example, some patients with a total gastrectomy assimilate proteins sufficiently well to maintain a positive nitrogen balance. The hydrolysis products of pepsin action are mostly large polypeptides that cannot be efficiently utilized until they are hydrolyzed by other intestinal proteolytic enzymes.

Activation of proteolytic enzymes in the intestines The pancreas releases its digestive juice when stimulated by secretin, a hormone produced in the duodenum in response to chyme (the complex contents of the stomach). Secretin causes the secretion of an almost protein-free electrolyte solution that has a high concentration of bicarbonate.

Pancreozymin, or cholecystokinin-pancreozymin, is a hormone that causes release of an enzyme-rich juice from the pancreas. Pancreozymin itself is released from the mucosa of the duodenum and proximal jejunum. The acid chyme of the stomach is made alkaline in the duodenum by bile and pancreatic juice. The latter contains the zymogens chymotrypsinogen, trypsinogen, proelastase, and procarboxypeptidase. These zymogens are activated as shown in Figure 8.3.

Table 8.1   Specificity of some proteolytic enzymes

Enzyme	Occurrence	pH optimum	Major site of action
Trypsin	Intestine	7.5 to 8.5	Arginyl, lysyl bonds
Chymotrypsin	Intestine	7.5 to 8.5	Aromatic amino acyl bonds (Phe, Trp, Tyr)
Pepsin	Stomach	1.5 to 2.5	Wide range of specificity
Carboxypeptidase	Intestine	7.5 to 8.5	C-terminal amino acid
Aminopeptidase	Intestinal mucosa		N-terminal amino acid

Trypsin, chymotrypsin, and elastase are *endopeptidases;* that is, they cleave proteins and polypeptides at internal sites, usually at specific amino acid residues. Carboxypeptidases (A or B) are *exopeptidases* that cleave amino acids from the carboxyl ends of polypeptide chains.

These proteolytic enzymes have some remarkable specificities for cleaving protein chains at certain amino acid residues. The specificities are summarized in Table 8.1 The products of these enzymes are free amino acids, dipeptides, and small peptides. The residual peptides are hydrolyzed in the intestinal mucosal cells by *aminopeptidase* and *dipeptidases.*

Activation of enzymes versus enzyme synthesis A clear distinction should be made between the process of enzyme activation and that of enzyme synthesis. An increase in the concentration of a functional enzyme is accomplished in both cases but by different mechanisms. Activation depends on the existence of an inactive precursor protein that can be quickly converted to the active enzyme. This conversion can take various forms, such as phosphorylation or a simple hydrolytic reaction that removes part of the peptide chain. Enzyme synthesis refers to the formation of new enzyme molecules from amino acids, that is, new, or de novo, synthesis. In this case an agent, such as a hormone, stimulates the synthesis from amino acids of new, fully active enzyme molecules. The body often uses enzyme activation when the response must be fast, as in digestion or blood clotting. Enzyme synthesis is a slower process.

Absorption of amino acids and peptides

For a long time it was thought that the digestion of proteins must go to completion, all the way to free amino acids, before the hydrolysis products could be absorbed into the intestinal mucosa. Now it is clear that a substantial amount of small peptides, in addition to amino acids, are absorbed by stereospecific transport systems. The peptides, however, are rapidly hydrolyzed by peptidases located in the absorptive cells so that only amino acids are released into the portal blood. See Appendix F for a table of normal blood amino acid concentrations.

An exception to this generalization is the newborn. The fetal or neonatal intestine is capable of absorbing some milk proteins intact. Thus immunoglobulins from colostrum are absorbed without loss of biologic activity so that they provide passive immunity to the infant. The newborn rapidly loses this ability to transport substantial quantities of protein approximately 48 hr after birth. Normal adult intestinal mucosa can absorb only trace amounts of proteins, but sometimes these are antigenic, such as food allergens, which can cause health problems.

The uptake mechanisms for peptides are separate from those for amino acids. At least five stereospecific transport systems for the amino acids are present in human kidney and intestine (Table 8.2). One system transports small neutral amino acids, another carries

Table 8.2   Stereospecific transport systems for amino acids

Amino acid specificity	Examples of amino acids transported	Human disease
1. Small neutral amino acids	Alanine, serine, threonine	
2. Large neutral and aromatic amino acids	Isoleucine, leucine, valine, tyrosine, tryptophan, phenylalanine	Hartnup disease
3. Basic amino acids	Arginine, lysine, ornithine, cystine	Cystinuria
4. Proline, glycine		Glycinuria
5. Acidic amino acids	Glutamic and aspartic acids	

large neutral and aromatic amino acids, a third transports the basic amino acids and cystine, a fourth is specific for neutral and aromatic amino acids, and the fifth is specific for acidic amino acids. Each system transports amino acids that are structurally similar. For example, in Hartnup disease (see case 2 at the end of the Chapter) the transport system for the large neutral and aromatic amino acids is defective. Despite this, most patients with Hartnup disease have only minor health problems, since essential neutral aromatic amino acids are absorbed as peptides from the intestine by separate transport systems.

γ-Glutamyl cycle

We know more about the transport of amino acids than about the transport of peptides. Amino acid transport involves several carrier-mediated transport systems, usually coupled to sodium ion transport. According to one scheme proposed by Alton Meister, amino acids are transported as dipeptides of glutamic acid. The important feature of this transport system is that the tripeptide glutathione serves as a donor of a γ-glutamyl group that is transferred to the amino group of the amino acid selected for transport. The structure of glutathione, or γ-glutamylcysteinylglycine, is shown next.

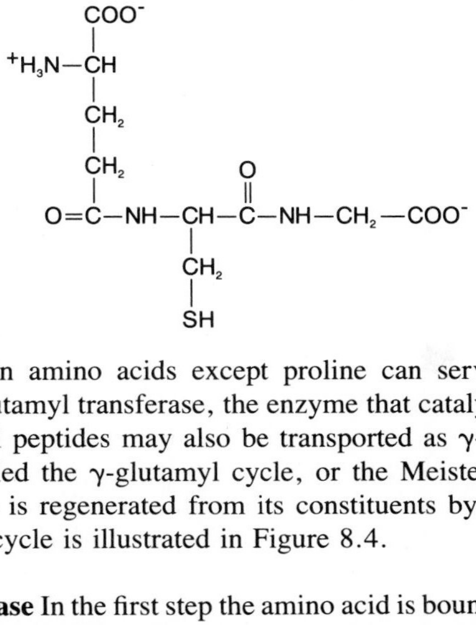

$$
\begin{array}{c}
COO^- \\
| \\
{}^+H_3N-CH \\
| \\
CH_2 \\
| \\
CH_2 \qquad\quad O \\
| \qquad\qquad \parallel \\
O=C-NH-CH-C-NH-CH_2-COO^- \\
| \\
CH_2 \\
| \\
SH
\end{array}
$$

All the common protein amino acids except proline can serve as substrates for the membrane-bound γ-glutamyl transferase, the enzyme that catalyzes the formation of the dipeptide. Several small peptides may also be transported as γ-glutamyl peptides. This transport scheme is called the γ-glutamyl cycle, or the Meister cycle, since after it is hydrolyzed, glutathione is regenerated from its constituents by the action of five cytoplasmic enzymes. The cycle is illustrated in Figure 8.4.

γ-Glutamyl transferase In the first step the amino acid is bound to a specific membrane site, probably under the influence of a membrane-associated protein that recognizes amino acids having common structural features. After the amino acid is bound, the γ-glutamyl transferase of the membrane catalyzes the transpeptidation of a γ-glutamyl

Figure 8.4   The γ-glutamyl cycle. γ-Glutamyl transpeptidase is located in the cell
                        membrane; all other enzymes are cytosolic. These are numbered as
                        1, peptidase; 2, γ-glutamyl cyclotransferase; 3, 5-oxoprolinase; 4, γ-
                        glutamylcysteine synthase; and 5, GSH (γ-glutamylcysteinylglycine,
                        glutathione) synthase.

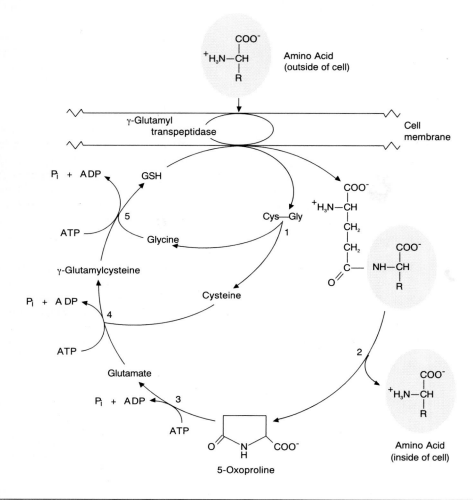

residue from glutathione to the amino acid, forming a γ-glutamyl amino acid and cys-
teinylglycine, the remaining portion of glutathione. This is shown in the top portion of
Figure 8.4. As a result, the γ-glutamyl dipeptide is transported into the cell. Once inside
the cell, the cytoplasmic enzyme γ-glutamyl cyclotransferase (reaction 2, Figure 8.4)
promotes an internal transpeptidation that releases the amino acid and cyclizes the γ-
glutamyl group so that it forms a peptide linkage with itself. The cyclized glutamyl
compound is called 5-oxoproline, or sometimes pyroglutamic acid. It is quite stable; thus
adenosine 5′-triphosphate (ATP) is required to convert it back to glutamate (reaction 3,
Figure 8.4).

Resynthesis of glutathione To complete the γ-glutamyl cycle, the glutathione consumed in the initial reaction must be reformed. However, glutathione is synthesized from γ-glutamylcysteine. No known enzyme uses the cysteinyl glycine dipeptide except a peptidase that cleaves the molecule to cysteine and glycine (reaction 1, Figure 8.4). This reaction and the reaction catalyzed by 5-oxoprolinase yield the component amino acids needed for the synthesis of glutathione. The formation of each of the two peptide bonds in glutathione requires the hydrolysis of ATP. In the first reaction, γ-glutamylcysteine is formed in a reaction catalyzed by γ-glutamylcysteine synthase (reaction 4, Figure 8.4; see also Clinical comment). In the second reaction the glycyl moiety is added under the influence of glutathione synthase (reaction 5, Figure 8.4), regenerating glutathione so that the cycle is completed. The transport of amino acids via the γ-glutamyl cycle is costly in terms of the energy required; three ATPs are hydrolyzed for each amino acid transported.

Specificity of the cycle The stereospecific transport of amino acids is still poorly understood. Multiple transport systems exist, such as those listed in Table 8.2, that are both amino acid and tissue specific. A single amino acid might be transported by several different systems, or it might be transported as part of a small peptide. Thus it should be emphasized that the γ-glutamyl cycle is only one of several different transporting systems and as such cannot account for all the experimental data on amino acid transport. Furthermore, the cycle is at least as important in resynthesizing the glutathione that is consumed in many detoxification reactions. Possibly other membrane proteins function with the γ-glutamyl transferase to give it specificity, and these might be the proteins mutated in the diseases mentioned in Table 8.2. Furthermore, none of the cycle enzymes is activated by sodium ions, which function in the cotransport of many amino acids. According to this view, the amino acids enter along a Na^+ concentration gradient, and the sodium ions are subsequently pumped out of the cell by the Na^+, K^+-ATPase. Probably these cations function at another step in transport or are required by a different transport system.

Clinical comment

Genetic abnormalities of the γ-gultamyl cycle Defects in human genes can affect three of the enzymes of the cycle. A defect in γ-glutamylcysteine synthase (reaction 4, Figure 8.4) leads to symptoms of hemolytic anemia, probably because the defective enzyme prevents the synthesis of glutathione, which is needed to maintain the integrity of erythrocyte membranes. Another genetic abnormality leads to 5-oxoprolinuria. The enzymatic defect in these patients is for glutathione synthase (reaction 5, Figure 8.4). γ-Glutamylcysteine, the substrate for this enzyme, accumulates; as with other γ-glutamyl amino acids, it also is converted to 5-oxoproline. The third genetic abnormality in the γ-glutamyl cycle is associated with the membrane-bound γ-glutamyl transferase. Patients with this inborn error of metabolism excrete excessive amounts of glutathione in their urine (see case 5).

Amino acid catabolism

The degradation of either dietary or biosynthesized amino acids usually starts with the removal of the α-amino group from the rest of the molecule. Because metabolism branches at this early point, it is convenient to consider the fate of the amino groups before examining the metabolism of the carbon skeleton.

Fate of the nitrogen atoms

The major end products of nitrogen metabolism in humans are mostly eliminated in the urine. The relative concentrations of these end products in normal urine are given in Table 8.3. Clearly most of the nitrogen is excreted as urea. However, the most immediate

Table 8.3   Nitrogen-containing components of normal urine

End product	Excreted nitrogen (%)
Urea	86.0
Creatinine	4.5
Ammonium	2.8
Uric acid	1.7
Other compounds	5.0

product of amino acid metabolism is not urea but ammonium ions (NH_4^+). This section describes how amino acids are deaminated to form ammonium ions and then how most of the ammonium ions are converted to urea.

Amino acid deamination Several ways exist in which amino groups may be removed from individual amino acids, but in one general way all amino acids can be deaminated. This mechanism couples the oxidative deamination of glutamic acid to several amino acid–specific amino transferases.

The amino transferases catalyze the general reaction:

$$\text{Amino acid}_1 + \text{Keto acid}_2 \rightleftharpoons \text{Amino acid}_2 + \text{Keto acid}_1$$

Individual amino transferases are specific for a particular amino acid or a structurally similar group of amino acids, but almost all these enzymes use glutamic acid and α-ketoglutarate as one of the amino acid–keto acid pairs. Thus the following reactions apply for most protein amino acids.

(1) α-Ketoglutarate + Pyrdoxamine phosphate (enzyme bound) ⇌ Pyridoxal phosphate (enzyme bound) + Glutamate

(2) Many amino acids + Pyridoxal phosphate (enzyme bound) ⇌ Pyrdoxamine phosphate (enzyme bound) + Many keto acids

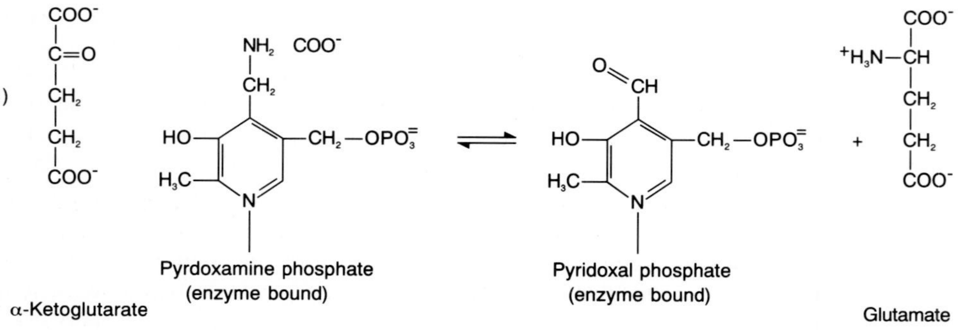

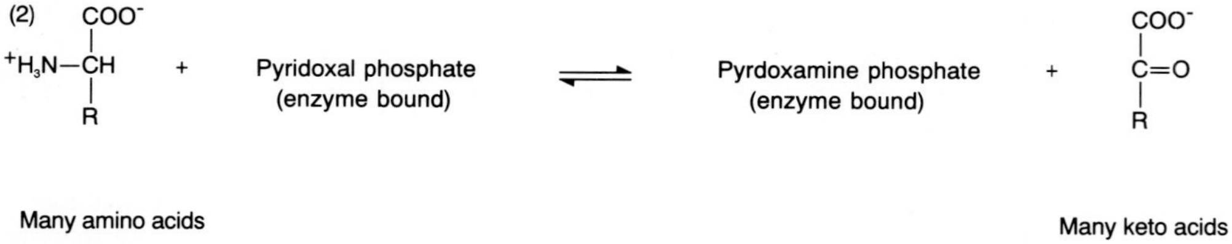

Sum of (1) and (2):

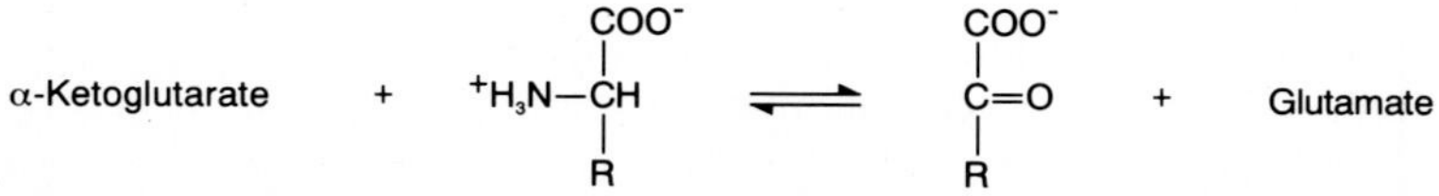

α-Ketoglutarate + ⁺H₃N—CH(COO⁻)(R) ⇌ C=O(COO⁻)(R) + Glutamate

The sum of these reactions shows that almost all amino acids can be transaminated to their keto acid derivatives, with all their amino groups transferred to glutamate.

The oxidative deamination of glutamic acid is catalyzed by the enzyme glutamate dehydrogenase. This reaction regenerates α-ketoglutarate with the production of ammonium ions and NADH:

$$\text{Glutamate} + \text{NAD}^+ \rightleftharpoons \alpha\text{-Ketoglutarate} + \text{NADH} + \text{NH}_4^+$$

By coupling the glutamate dehydrogenase reaction with many different amino transferases, it is possible to deaminate a large variety of amino acids.

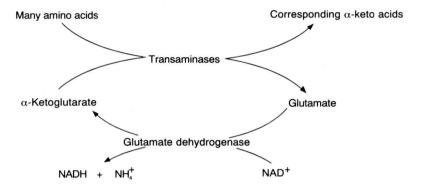

The aspartate amino transferase catalyzes the transfer of amino groups to oxaloacetate. This reaction is of special importance, since, as shown later, one of the nitrogen atoms of urea is derived from ammonium ions, but the other comes from aspartic acid.

Amino acids can also be deaminated by the *L*-amino acid oxidases of kidney and liver. These enzymes, however, are of much lesser importance in the catabolism of amino acids. They have a broad specificity, so many different amino acids can serve as substrates, but their relative rates of oxidation are slow. The tightly bound coenzyme of the *L*-amino acid oxidase from kidney is flavin mononucleotide (FMN). Figure 8.5 illustrates the reaction catalyzed by *L*-amino acid oxidase.

Ammonium ion metabolism Virtually all tissues produce ammonia, which is present predominantly as ammonium ions. These ammonium ions arise primarily from the catabolism of amino acids. Ordinarily the ammonium ion concentration in the peripheral blood is maintained at a very low level. Normal concentrations of ammonium ions in plasma are 25 to 40 μmol/L. Much higher levels are often found in severe hepatic disease. Regardless of their source, ammonium ions (as ammonia) are exceedingly toxic to the central nervous system; consequently, they must be detoxified and eliminated.

Detoxification in the brain The brain detoxifies ammonium ions by converting them to glutamine. The brain is a rich source of glutamine synthase, which catalyzes the reaction:

Figure 8.5   Reaction catalyzed by L-amino acid oxidase.

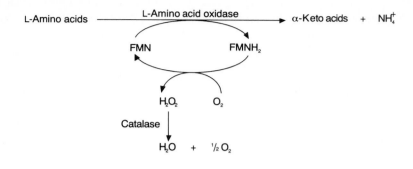

Other tissues also contain glutamine synthase, and the high levels of glutamine found in the blood after ingestion of foods rich in protein may represent a storage and transport form of ammonia. Much of the circulating glutamine is eventually hydrolyzed to glutamate and ammonium ions by a glutaminase in the kidney. Chapter 4 shows how this reaction is important in acid-base balance.

Transport of ammonium ions to the liver as glutamine Alternatively the glutamine of the blood may be hydrolyzed by liver glutaminase to yield ammonium ions, which can be used by that organ for urea synthesis. Most of the ammonia is transported to the liver as glutamine, and the plasma concentration of this amino acid is almost twice that of any other single amino acid (approximately 0.5 to 0.6 mmol/L, or 7.5 mg/dl).

Detoxification in the liver Most ammonium ions in the portal blood are detoxified in the liver; there they are converted to urea, a form of nitrogen much less toxic to the central nervous system. If the urea-synthesizing system should fail as a result of a malfunctioning liver or because of portal obstruction, ammonium ions will pass into the systemic circulation and ammonia intoxication results. Ammonia intoxication produces blurred vision, tremors, slurred speech, and ultimately coma and death. However, this condition, called hepatic coma, is complex and may be precipitated by many other factors.

Urea synthesis A schematic view of urea synthesis by the urea cycle is shown in Figure 8.6. The first step in the synthesis of urea produces *carbamoyl phosphate* from carbon dioxide, ammonium ions, and ATP. Carbamoyl phosphate is not only a key intermediate in the synthesis of urea but also is involved in the synthesis of pyrimidines. However, the carbamoyl phosphate used for pyrimidine synthesis is produced by a different enzyme (see Chapter 13). The carbamoyl phosphate used in urea synthesis is produced in the mitochondria in a reaction catalyzed by the enzyme carbamoyl phosphate synthase. Acetyl glutamate is an essential cofactor.

1. CO_2 + NH_4^+ + 2ATP $\xrightarrow[\substack{\text{N-acetyl glutamate} \\ \text{(mitochondrial)}}]{\substack{\text{Carbamoyl phosphate} \\ \text{synthase}}}$ $\underset{O}{\overset{NH_2}{\diagdown}}\!\!C\!=\!O$ + 2ADP + P_i

Carbamoyl phosphate

Note that the reaction requires 2 mol of ATP for the synthesis of 1 mol of carbamoyl phosphate, which also is a high-energy compound. Because of the ATP requirement, the reaction is irreversible, and the uptake of the very toxic ammonium ions is ensured.

Carbamoyl phosphate synthase, ornithine transcarbamoylase, and the enzyme that synthesizes the *N*-acetyl glutamate cofactor are the only enzymes of the urea cycle found in the mitochondria. As might be expected, no difficulty exists in transporting amino acids and ammonium ions into and out of mitochondria.

In the second step (reaction 2), carbamoyl phosphate, a high-energy mixed anhydride, condenses with ornithine to form citrulline and inorganic phosphate (P_i).

2.

L-Ornithine   L-Citrulline

Citrulline condenses with aspartic acid to form *argininosuccinate* in an ATP-requiring reaction (reaction 3). This is followed by a hydrolytic reaction (4) that cleaves argininosuccinate to *arginine* and *fumarate*. The fumarate produced in the cytoplasm can be converted to malate by a cytoplasmic fumarase, and the malate is transported into the mitochondria by the malate-phosphate translocase system (see Chapter 5).

3.

L-Citrulline   L-Aspartate   Argininosuccinate synthase   L-Argininosuccinate

4.

Argininosuccinase   Fumarate   Glucose

L-Arginine

Finally, the hydrolytic cleavage of arginine by the enzyme arginase (reaction 5) yields ornithine and urea. Ornithine becomes the acceptor for another molecule of carbamoyl phosphate.

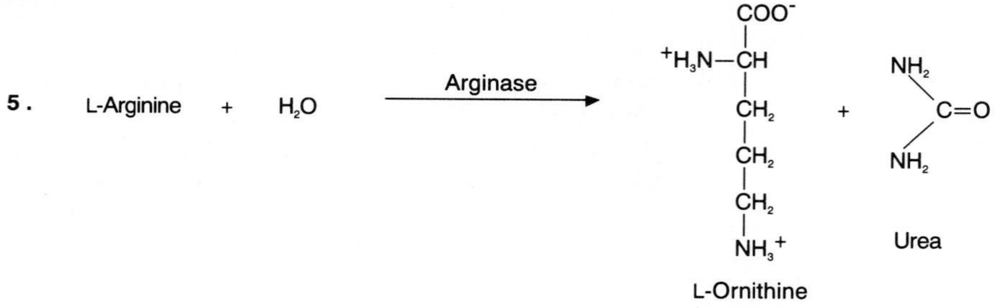

5.   L-Arginine   + H_2O $\xrightarrow{\text{Arginase}}$

$$
\begin{array}{c}
COO^- \\
| \\
{}^+H_3N-CH \\
| \\
CH_2 \\
| \\
CH_2 \\
| \\
CH_2 \\
| \\
NH_3{}^+
\end{array}
\quad + \quad
\begin{array}{c}
NH_2 \\
\diagdown \\
C=O \\
\diagup \\
NH_2
\end{array}
$$

L-Ornithine   Urea

These five reactions can be combined to form a cycle in which urea is produced from ammonium ions and carbon dioxide and in which all the other intermediates are regenerated. Figure 8.6 summarizes the reactions and represents them as a urea cycle.

Clinical comment   **Inherited disease of the urea cycle** *Hyperammonemia* is often associated with inherited abnormalities of urea cycle enzymes. In congenital hyperammonemia type I the defective enzyme is carbamoyl phosphate synthase, and in type II it is ornithine transcarbamoylase. The accompanying symtoms of episodic vomiting, psychomotor retardation, and stupor respond to a restricted protein diet. The glutamine concentration of the blood is often elevated in these patients. In hyperammonemia type II, pyrimidine metabolites appear in the urine. Two other inherited diseases, *citrullinemia* and *argininosuccinic acidemia,* are

Figure 8.6   Urea cycle.

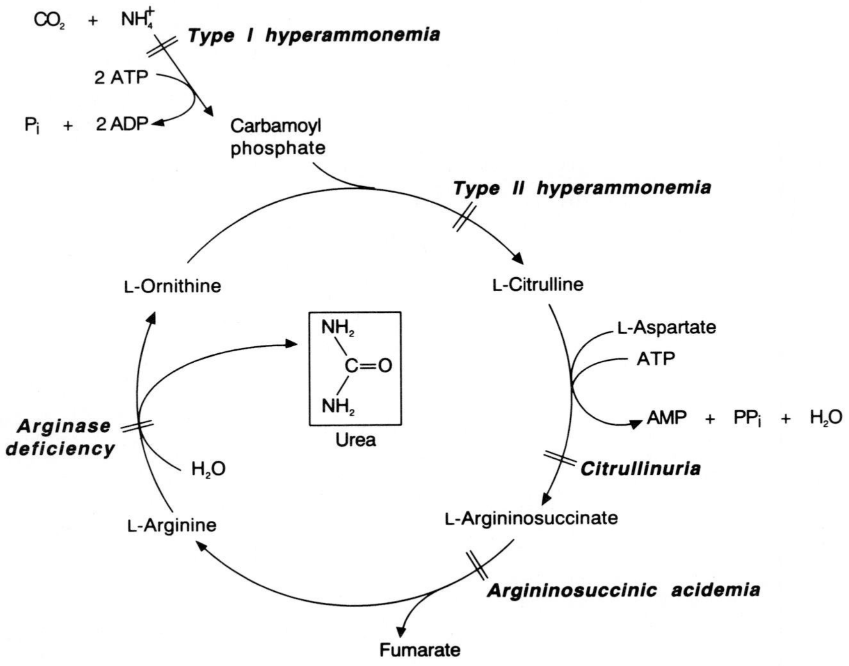

caused by defects in the enzymes that ordinarily act on citrulline and argininosuccinate, respectively. Figure 8.6 shows the location in the urea cycle of the reactions associated with each of these diseases. (Citrullinuria is a manifestation of citrullinemia.) See case 3 for more inherited diseases of the urea cycle.

Fate of the carbon atoms—glycogenic and ketogenic amino acids

The carbon skeletons of the amino acids may be used to produce energy. Some amino acids give rise to ketone bodies and are called ketogenic amino acids. The carbon atoms of other amino acids are converted to glucose or glycogen and are called glycogenic amino acids. Still other amino acids may be both ketogenic and glycogenic.

The detailed metabolism of the amino acids is best studied by enzymatic experimentation; nevertheless, to emphasize their metabolic functions in humans, it is convenient to group the amino acids according to this scheme, recognizing that the conditions of the experiment or of the patient can greatly influence metabolism.

Most amino acids are glycogenic; only *leucine* and *lysine* are completely ketogenic. Four amino acids, *isoleucine, phenylalanine, tyrosine,* and *tryptophan,* are considered to be both ketogenic and glycogenic.

Figure 8.7   Summary of the metabolic fates of carbon skeletons for the amino acids.

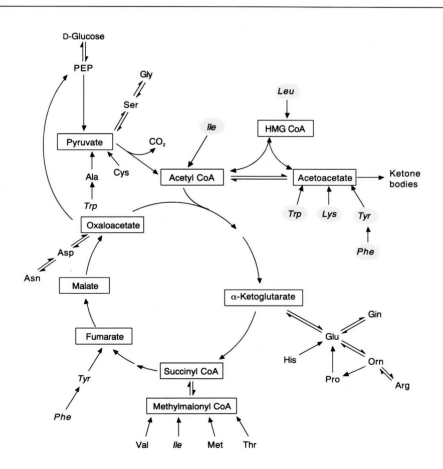

Summary of amino acid catabolism A summary of amino acid catabolism resulting from the breakdown of proteins is shown in Figure 8.7. The Krebs cycle and reactions leading to and from it are the dominant features of this diagram. The names of intermediates produced by the catabolism of the amino acids are in boxes. Those that form Krebs cycle intermediates can be thought of as glycogenic, since oxaloacetate can be converted to phospho*enol*pyruvate (PEP) and then to glucose (see Chapter 6). The names of the ketogenic amino acids are shaded and italicized. Recall that several of these are both glycogenic and ketogenic. Those portions that are ketogenic (shaded) can give rise to either acetoacetate or acetyl CoA. Of greater significance than whether an amino acid is glycogenic or ketogenic is that either of these substances can be mobilized for energy. Thus amino acids are not only important in the synthesis of other nitrogen-containing metabolites and in the synthesis of proteins, but they are also important sources of energy.

Catabolism of the ketogenic amino acids

Leucine Leucine is converted to 3-hydroxy-3-methyl-glutaryl CoA (HMGCoA) by a series of reactions summarized in Figure 8.8. The 2-ketoisocaproate dehydrogenase complex is analogous to the pyruvate and ketoglutarate dehydrogenase complexes (see Chapter 5) in that thiamine pyrophosphate, lipoic acid, CoA, flavine adenine dinucleotide (FAD), and NAD^+ are all involved. This complex is defective in individuals with *maple syrup urine disease*. In this disease the urine takes on a characteristic maple syrup smell as result of accumulated keto acids.

HMGCoA, the last product in the leucine catabolic pathway, is a precursor of cholesterol (see Chapter 11), so one might think that leucine could be used for the synthesis of steroids. Actually, leucine catabolism occurs in the mitochondria, whereas sterol synthesis is cytoplasmic; consequently, leucine produces acetoacetate and acetyl CoA exclusively. Acetoacetate is a ketone body, but any amino acid that can form acetyl CoA may be considered ketogenic, since the acetoacetyl CoA thiolase reaction is somewhat reversible.

Lysine Lysine is one of the few amino acids that is not deaminated in its initial catabolic reaction. Two pathways exist for its metabolism in animals; both share a modification of the ε-amino group before deamination. Some still question the relative importance of the pathways. A few intermediates are omitted in Figure 8.9. Both routes ultimately yield 2-aminoadipate. The oxidative decarboxylation of this substance is catalyzed by an enzyme complex similar to the pyruvate and α-ketoglutarate dehydrogenase complexes. Further decarboxylation and oxidation yield crotonyl CoA, which is also an intermediate in fatty acid oxidation. This intermediate can be split to give acetoacetyl CoA and eventually acetyl CoA or acetoacetate. The pathway through saccharopine is probably important in humans, since patients with hyperlysinemia also accumulate saccharopine.

Catabolism of amino acids that are both ketogenic and glycogenic

Isoleucine Isoleucine is catabolized by a series of reactions similar to those for leucine. As shown in Figure 8.10, the reactions before the formation of the unsaturated acid are entirely analogous to those for leucine (Figure 8.8). The enzymes are similar for each pathway, but only the 2-ketoisocaproate dehydrogenases are identical. In isoleucine catabolism the formation of acetyl CoA results from the action of a thiolase. The propionyl CoA product can be carboxylated to form methylmalonyl CoA (see Chapter 5). Methylmalonyl CoA can be isomerized to succinyl CoA, which enters the Krebs cycle and ultimately yields pyruvate, glucose, or glycogen.

Phenylalanine and tyrosine Phenylalanine is converted to tyrosine by the enzyme phenylalanine hydroxylase:

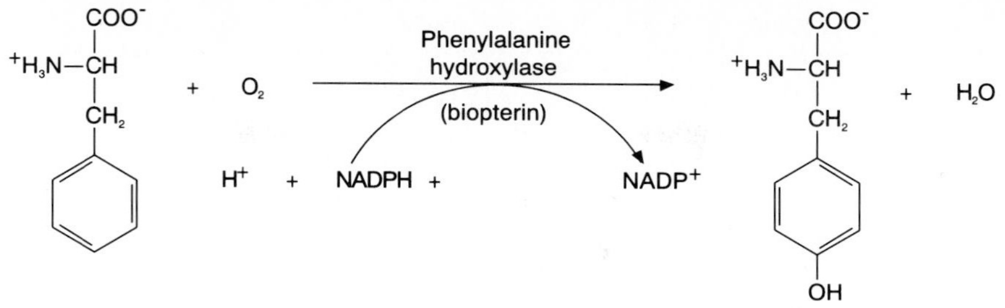

Figure 8.8   Catabolism of leucine.

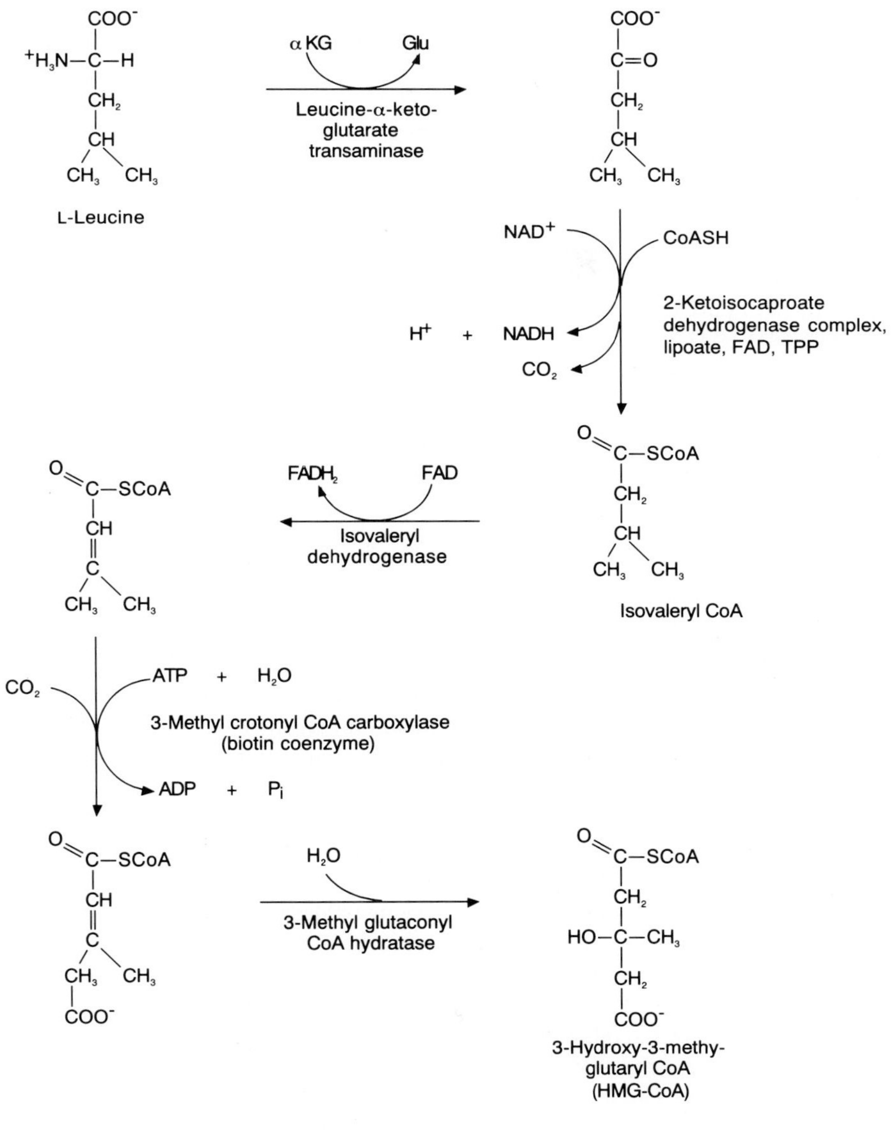

Figure 8.9   Catabolism of lysine.

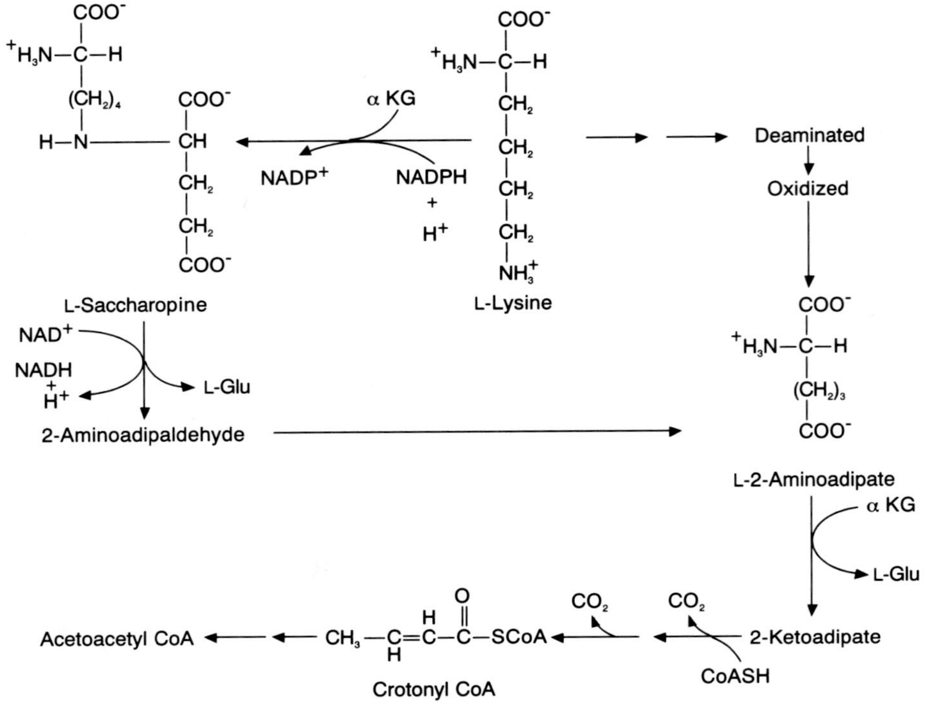

L-Saccharopine

NAD$^+$
NADH + H$^+$ → L-Glu

2-Aminoadipaldehyde

L-Lysine

α KG
NADP$^+$ NADPH + H$^+$

Deaminated
Oxidized

L-2-Aminoadipate

α KG → L-Glu

2-Ketoadipate

CoASH   CO$_2$ CO$_2$

Crotonyl CoA   CH$_3$—C=C—C—SCoA

Acetoacetyl CoA

After this reaction the catabolism of phenylalanine is the same as that of tyrosine.

The enzyme tyrosine aminotransferase (tyrosine transaminase) mediates the deamination of tyrosine; this is shown in the first reaction of Figure 8.11. As with many other aminotransferases, α-ketoglutarate (αKG) is the amino acceptor. The synthesis of tyrosine aminotransferase can be induced in animal liver by glucocorticoids, which are steroid hormones that stimulate protein catabolism and thus increase the concentration of blood glucose. This induction leads to an increase in the messenger ribonucleic acid (mRNA) for the enzyme. The keto acid product of tyrosine transaminase action, *p*-hydroxyphenylpyruvate (Figure 8.11), is subsequently hydroxylated, decarboxylated, and its side-chain rearranged. The product of this reaction is homogentisate, a hydroquinone. Another oxidation, catalyzed by homogentisate oxidase, opens the phenyl ring. This reaction gives rise to maleoylacetoacetate and ultimately acetoacetate and fumarate.

Clinical comment

Alkaptonuria—the first genetic disease A metabolic defect in homogentisate oxidase causes the excretion in the urine of homogentisate in the disease *alkaptonuria*. When the patient's urine is exposed to the air, the homogentisate is dramatically oxidized to black products. The disease itself is rather benign by comparison to other genetic disorders. In middle age, arthritis usually occurs as a result of the accumulation of the black pigment in the connective tissues.

Figure 8.10   Catabolism of isoleucine.

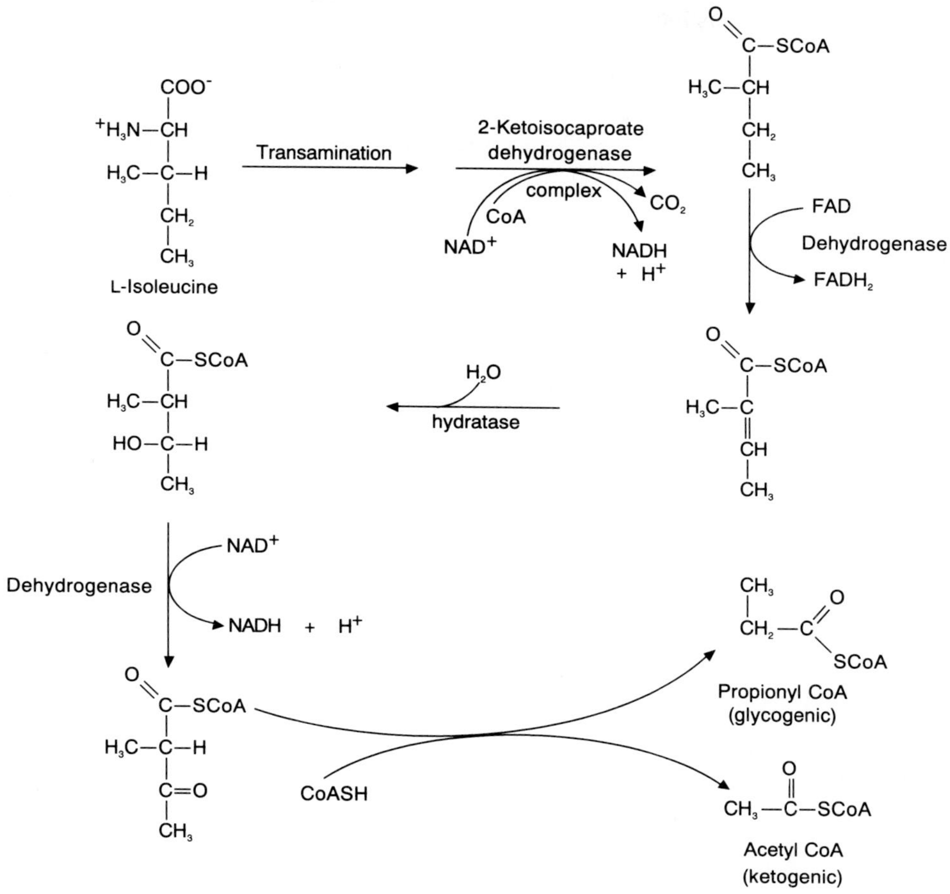

Thirty years before the one gene–one enzyme concept of Beadle and Tatum, the English physician Garrod correctly predicted that alkaptonuria resulted from an inherited enzymatic defect. He first proposed this in his book *Inborn Errors of Metabolism,* published in 1909.

Tryptophan Tryptophan has a complex metabolism that can give rise to both alanine, which is glycogenic, and acetyl CoA, which is ketogenic. The initial reaction (Figure 8.12) is an oxidation catalyzed by the enzyme tryptophan pyrrolase (tryptophan 2,3-dioxygenase). As with tyrosine aminotransferase, the synthesis of this enzyme is induced by the administration of glucocorticoids. Dietary tryptophan also increases the concentration of the enzyme in liver, which is the result of a decrease in the rate of enzyme degradation. In the reactions that follow, the tryptophan side-chain is cleaved to give alanine and 3-hydroxyanthranilate. The latter compound is decarboxylated and reduced to α-ketoadipate. The subsequent reactions that yield acetoacetyl CoA from α-ketoadipate are the same as those shown for lysine (see Figure 8.9).

Figure 8.11   Catabolism of tyrosine.

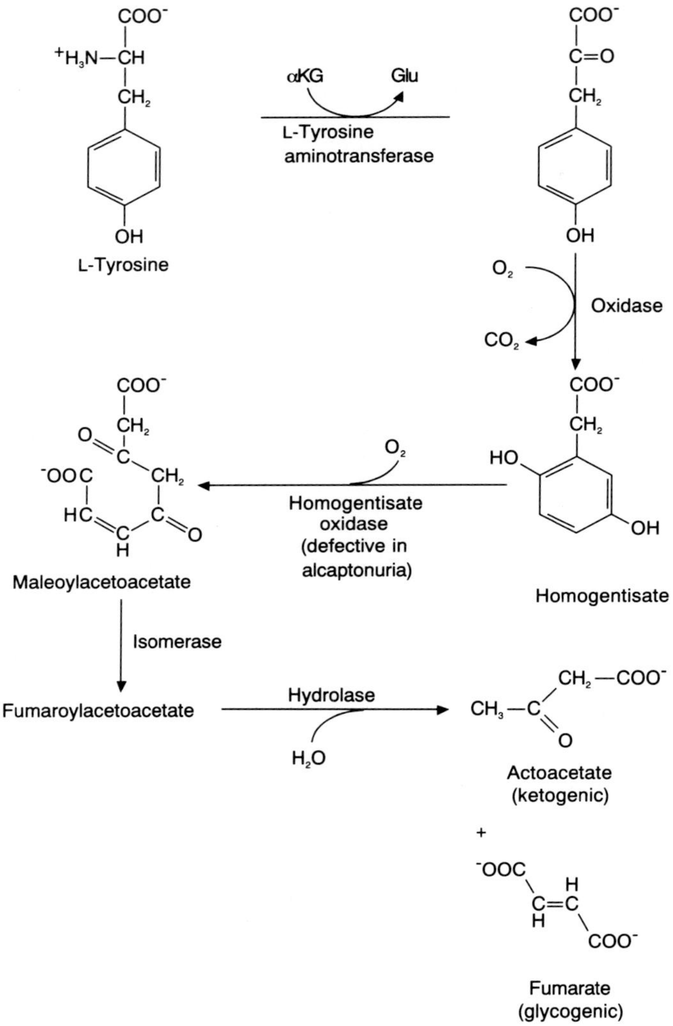

L-Tyrosine

L-Tyrosine
aminotransferase

Oxidase

Homogentisate
oxidase
(defective in
alcaptonuria)

Homogentisate

Maleoylacetoacetate

Isomerase

Fumaroylacetoacetate

Hydrolase

Actoacetate
(ketogenic)

Fumarate
(glycogenic)

Clinical comment   **Tryptophan spares the dietary requirement for niacin** Tryptophan can also be converted to NAD^+ by using a portion of the catabolic pathway shown in Figure 8.12. More
than 50 years ago it was observed that the symptoms of pellagra could be successfully
treated with tryptophan. Pellagra is the disease caused by a deficiency of niacin. The
recommended dietary allowance for niacin depends on the extent to which tryptophan is
converted to NAD^+. In the average normal person, 60 mg of tryptophan is approximately
equivalent to 1 mg of niacin. During pregnancy an increased conversion of tryptophan
to niacin occurs. However, a deficiency in one vitamin often results from eating a poor
diet that may lack other vitamins. For example, an individual with pellagra might eat a

Figure 8.12   Catabolism of tryptophan. Several intermediates after 3-hydroxyanthranilate have been omitted. The brackets indicate that one is a precursor of NAD^+.

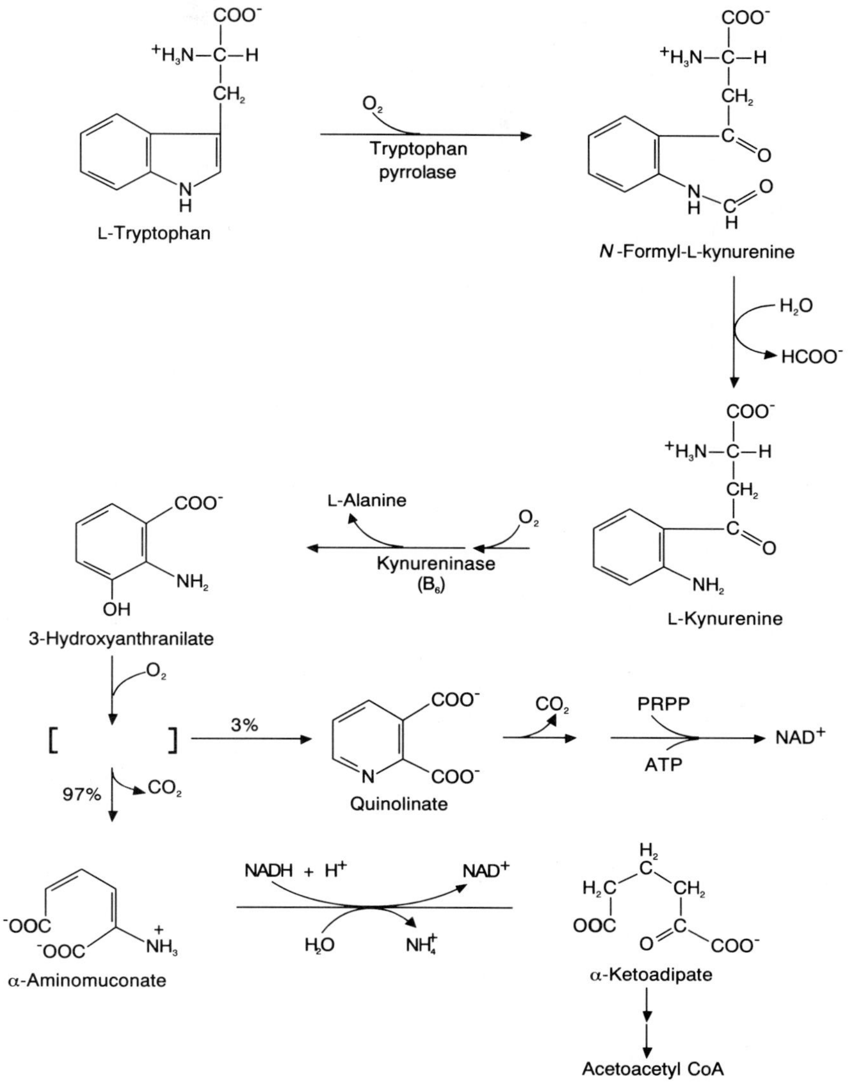

diet low in niacin but not in tryptophan. However, this person might be unable to synthesize niacin from tryptophan if the diet were also deficient in vitamin B_6. This is because the enzyme kynureninase, which catalyzes a reaction leading to NAD^+ synthesis (Figure 8.12), requires a vitamin B_6–derived coenzyme.

People in India who eat diets rich in jowar *(Sorghum vulgare)* develop pellagra, but those who eat diets rich in rice do not; however, the total intake of niacin and tryptophan are similar for both groups of people. The high concentration of leucine in jowar is responsible, since leucine, both in vivo and in vitro, *increases* the activity of tryptophan pyrrolase, which catalyzes the first step in the degradation of tryptophan. Consequently, those who eat jowar are impaired in the synthesis of niacin from tryptophan.

In patients with Hartnup disease the inability to absorb tryptophan can lead to pellagra-like symptoms.

Tryptophan metabolism in disease In carcinoid syndrome, a malignancy of the entero-chromaffin or argentaffin cells that produce serotonin, excessive amounts of the tryptophan metabolites involved in the synthesis of serotonin are excreted, as well as serotonin itself. Chapter 9 shows that the biogenic amine serotonin (5-hydroxytryptamine) is synthesized from tryptophan.

Catabolism of the glycogenic amino acids Chapter 5 discusses how *alanine, aspartate,* and *glutamate* can be metabolized via Krebs cycle enzymes by reversible transamination reactions. The α-keto acids resulting from these transaminations are pyruvate, oxaloac-etate, and α-ketoglutarate, respectively. All these substances are glycogenic.

Glutamine and asparagine Glutamine and asparagine can be converted to glutamate and aspartate by hydrolytic reactions that yield ammonia and the two glycogenic amino acids, glutamate and aspartate. The diagram summarizes the conversions of these amino acids.

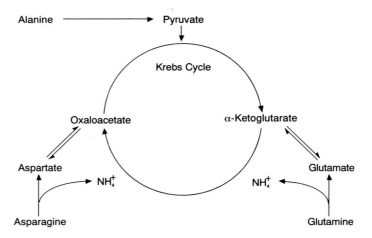

The ammonia produced by the kidney in response to metabolic acidosis originates from glutamine and is released by the enzyme glutaminase. However, the amide nitrogen of glutamine is not always released as an ammonium ion. The nitrogen can be transferred enzymatically to a variety of acceptors; for example, the formation of amino sugars and the synthesis of the purine and pyrimidine rings require transfers of amide nitrogen from glutamine.

Glycine, serine and cysteine Glycine, serine, and cysteine can all be metabolized to pyruvate, as shown in Figure 8.13. The enzyme serine dehydratase causes the direct conversion of serine to pyruvate. Serine can be synthesized from glycine by the transfer of a hydroxymethyl group from 5,10-methylene THF (tetrahydrofolate). Additional trans-formations of the folic acid coenzymes are described in Chapter 13. The sulfur atom of cysteine is first oxidized, the product transaminated, and the sulfite residue hydrolyzed away, leaving pyruvate.

Figure 8.13   Catabolism of glycine, serine, and cysteine.

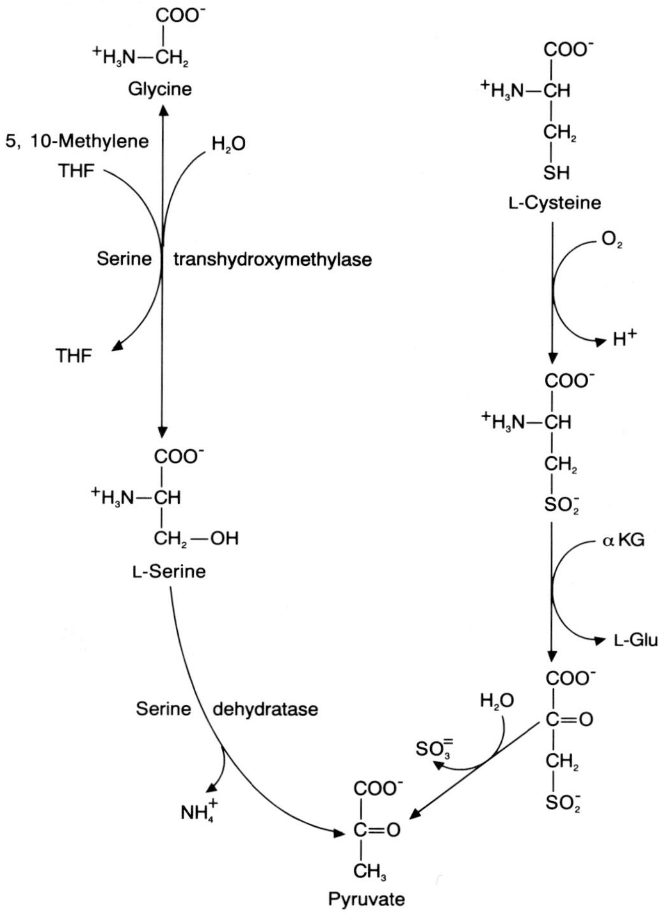

Threonine Threonine dehydratase, an enzyme similar to serine dehydratase, produces α-ketobutyrate, as shown in Figure 8.14. The enzyme is also induced by glucocorticoids. Subsequently, ketobutyrate is oxidatively decarboxylated to yield propionyl CoA by an enzyme complex similar to the pyruvate dehydrogenase complex. The propionyl CoA formed (see Chapter 5) is then carboxylated to methylmalonyl CoA, which in turn is isomerized to succinyl CoA. Succinyl CoA enters the Krebs cycle and gives rise to pyruvate.

Methionine Methionine is another amino acid that produces α-ketobutyrate, and thus it has the same fate as threonine. Methyl groups derived from methionine can be transferred to various acceptor molecules, such as RNA and deoxyribonucleic acid (DNA). The reactions involved are considered in more detail in Chapter 9. More importantly, however, the transfer of the methionine methyl groups leaves homocysteine as the other product of the reaction. Homocysteine has the structure:

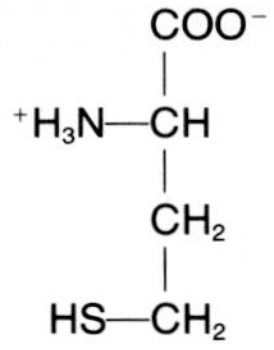

$$
\begin{array}{c}
COO^- \\
| \\
{}^+H_3N-CH \\
| \\
CH_2 \\
| \\
HS-CH_2
\end{array}
$$

Chapter 9 also describes how the sulfur atom of homocysteine is transferred to serine to yield cysteine and homoserine. The carbon atoms of homoserine are the ones derived from methionine, through homocysteine. Homoserine is then deaminated and dehydrated in much the same way as threonine (Figure 8.14) to yield α-ketobutyrate and eventually propionyl CoA and succinyl CoA, substances that are glycogenic.

Arginine Arginine is converted to glutamate-γ-semialdehyde according to the sequence:

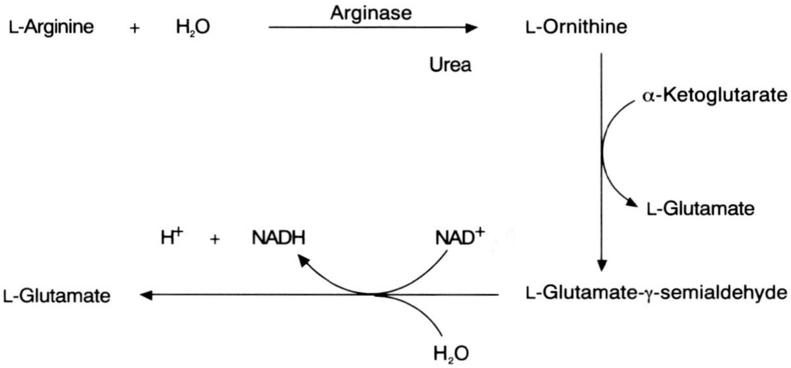

Figure 8.14   Catabolism of threonine.

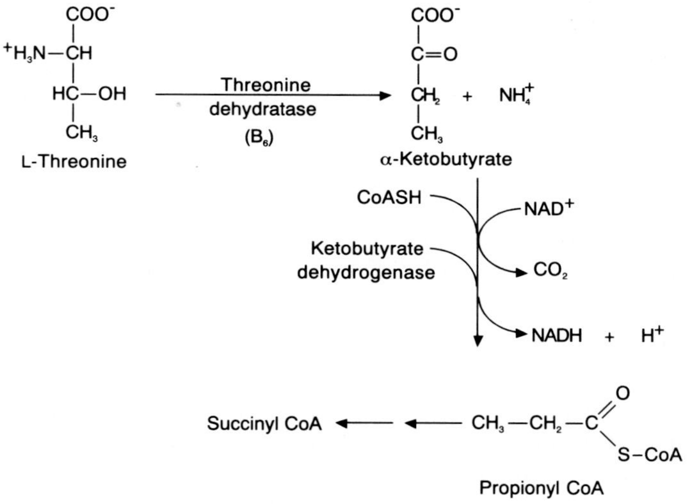

Note that arginase of the urea cycle participates. Glutamate semialdehyde is then oxidized to glutamate, and the glutamate metabolized as described earlier.

Proline Proline catabolism differs from that of most of the amino acids in that the molecule undergoes two oxidations without being deaminated. The first oxidation is mediated by a mitochondrial flavoprotein (FAD) to form an unsaturated intermediate; further oxidation using NAD^+ opens the ring to yield glutamate.

Valine As with other branched-chain amino acids, valine is first transaminated and then oxidatively decarboxylated. In this case the final product is methylmalonyl CoA, which can be isomerized to succinyl CoA (Figure 8.15). Unlike the other branched-chain amino acids, which are at least partly ketogenic, the catabolism of valine is purely glycogenic.

In Figures 8.8 and 8.10, one sees that the branched-chain amino acids leucine and isoleucine undergo similar oxidative decarboxylations after an initial transamination to the corresponding keto acids. The enzyme complexes that catalyze these reactions resemble pyruvate dehydrogenase in that they utilize the same five coenzymes. The α-ketoisocaproate dehydrogenase functions with keto acids derived from either leucine or isoleucine, and a defect in this complex causes maple syrup urine disease. Although the dehydrogenase of the valine pathway is not genetically affected in the disease, keto acids derived from valine are found in the urine; the accumulated keto acids from leucine and isoleucine can inhibit the ketovaline dehydrogenase.

Histidine The breakdown of histidine is shown in Figure 8.16. The enzyme histidase deaminates the amino acid to produce the unsaturated intermediate urocanate. Hydration causes the opening of the imidazole ring to give formiminoglutamate. Perhaps the most distinctive reaction in the pathway is the subsequent transfer of the formimino group to THF, producing glutamate and 5-formimino THF.

Clinical comment

Histidinemia The genetic disease histidinemia is associated with a defect in histidase, the first enzyme of this series. Patients with the disease have difficulty with speech and with the development of language skills. However, the relationship of these symptoms to histidine metabolism is unknown. Treatment by restricting dietary histidine lowers the blood levels of the amino acid but also impairs growth and development.

Figure 8.15   Catabolism of valine.

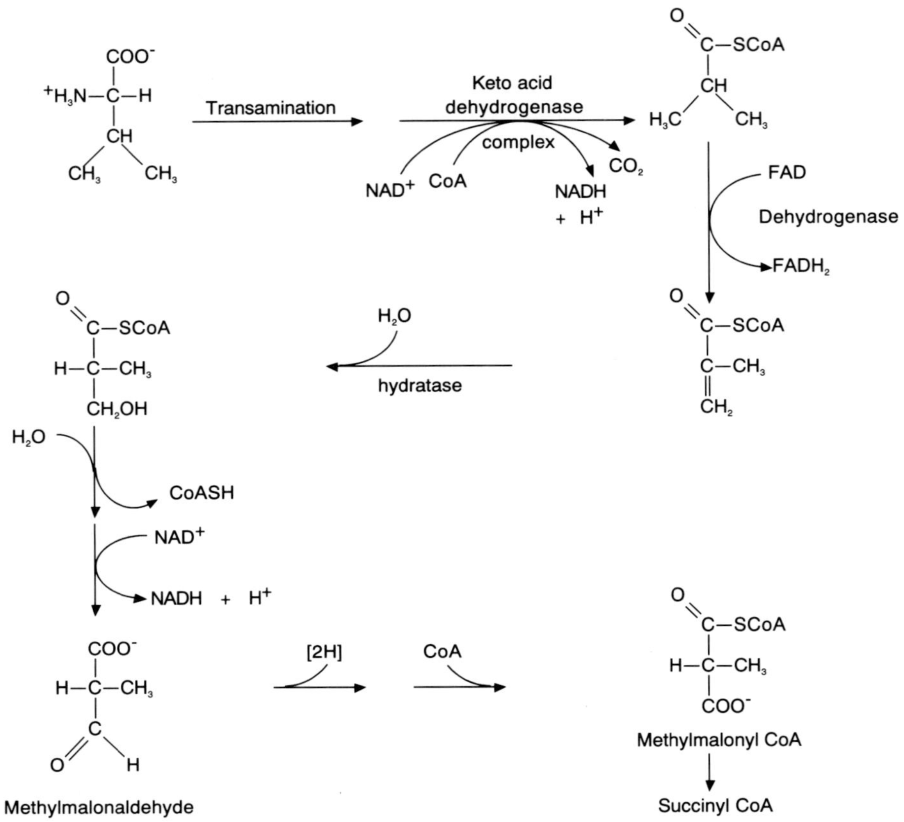

Bibliography

Adibi SA et al, editors: Branched chain amino and keto acids in health and disease, Basel, 1984, S Karger AG.

Batshaw ML et al: Treatment of inborn errors of urea synthesis, N Engl J Med 306:1387, 1982.

Bender DA: Amino acid metabolism, ed 2, New York, 1985, John Wiley & Sons, Inc.

Christensen HN: Interorgan amino acid nutrition, Physiol Rev 62:1193, 1982.

Nyhan WL, editor: Abnormalities in amino acid metabolism in clinical medicine, Norwalk, Conn, 1984, Appleton-Century-Crofts.

Sleisenger MH and Kim YS: Protein digestion and absorption, N Engl J Med 300:659, 1979.

Scriver CR et al, editors: The metabolic basis of inherited disease, ed 6, New York, 1989, McGraw-Hill Information Services Co.

Wellner D and Meister A: A survey of inborn errors of amino acid metabolism and transport in man, Annu Rev Biochem 50:911, 1981.

Figure 8.16   Catabolism of histidine. The formimino group is shown in brackets.

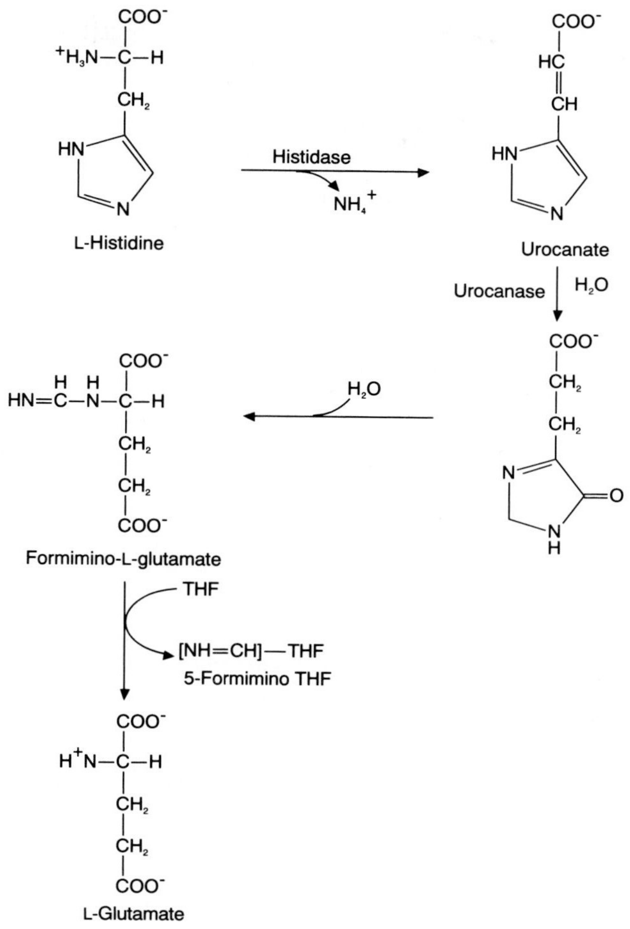

Clinical examples

A diamond (♦) on a case or question indicates that literature search beyond this text is necessary for full understanding.

Case 1   Amino acid metabolism in starvation

An obese patient volunteered to go on a starvation diet as part of a study of amino acid metabolism (Felig, 1975). Blood samples were taken and analyzed for plasma amino acids for as long as 5 to 6 wk following the fast. Valine, leucine, isoleucine, methionine, and α-aminobutyrate concentrations were transiently increased during the first week but dropped below initial levels later. Glycine, threonine, and serine levels increased more slowly, whereas 13 other amino acids eventually decreased. The decrease was largest for alanine, which dropped 70% in the first week. Total plasma amino nitrogen concentration decreased only 12%.

Biochemical questions

1. What changes in carbohydrate and lipid metabolism occur at the beginning of a fast?
2. Explain the ketosis and acidosis observed in starvation.
3. Is the decreased plasma alanine concentration related to gluconeogenesis? What is the alanine-glucose cycle?
4. In animals, circulating branched-chain amino acids are metabolized by muscle to yield energy. What are the catabolic products of leucine, valine, and isoleucine?
5. What might cause an increase in plasma branched-chain amino acids after 5 days of starvation?
6. The branched-chain amino acids stimulate the production of both alanine and glutamate in animal muscle. Explain this observation, recognizing that muscle is rich in aspartate aminotransferase.

Case discussion

Sources of nutrients Humans derive their calories from three main classes of nutrients: carbohydrates, fats, and proteins. These nutrients are metabolically transformed, depending on body requirements. Thus glucose may be either stored as glycogen, oxidized through the pentose phosphate and glucuronate pathways, or converted to pyruvate. Likewise, fatty acids may be oxidized, stored as triacylglycerols, or incorporated into structural lipids. Previous discussions considered the metabolism of each nutrient class separately. However, these processes occur in a concerted manner, especially during starvation or fasting, when all metabolic needs must be satisfied by the breakdown of stored materials.

1. Glucose and fat utilization during fasting The transformation of glucose and fatty acids to acetyl CoA is carefully regulated. When acetyl CoA rises, pyruvate carboxylase is activated. Pyruvate, arising from glycogenic amino acids, is converted to oxaloacetate instead of acetyl CoA. This provides more acceptors for the acetyl CoA coming from the β-oxidation of fatty acids (Figure 8.17), enabling acetyl groups to enter the Krebs cycle, or alternatively the oxaloacetate can be used for gluconeogenesis. In some tissues fatty acids are degraded by cyclic adenosine monophosphate (cAMP)–activated lipases. When the glucose concentration is low, as in starvation, insulin levels drop. Glucagon increases, stimulating a rise in cellular cAMP, and lipolysis is stimulated.

Obese patients can live on starvation diets for as long as 200 days, but this may have serious effects. The primary need of the starving person is fuel for energy. The initial physiologic response to the lack of food is to increase blood glucose concentration. Glucose is especially needed by the brain, which consumes about 65% of the total circulating glucose (approximately 1700 to 2500 kJ/day, or 400 to 600 kcal/day). At this time the liver supplies most of the blood glucose. Later the kidneys become important. Liver glycogen can supply sufficient glucose for several hours, but even after a short fast, gluconeogenesis by the liver requires substrates from other tissues. These substrates are almost always glycogenic amino acids or derivatives of glycogenic amino acids, mostly from the muscle.

After a few days of fasting, most of the energy requirements of the body are met by increased fat catabolism. Weight reduction diets allow the conversion of excessive body fat to energy. Fat cannot be converted to glucose in significant amounts, but it

Figure 8.17   Regulation of pyruvate metabolism by acetyl CoA.

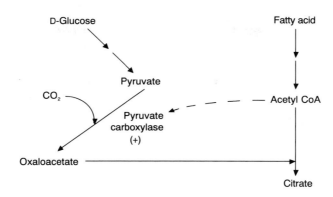

can supply energy that would otherwise come from amino acids released following the degradation of proteins. Most of these amino acids are glycogenic amino acids. Thus fat can spare the glucose requirement by providing approximately 16 g of the 100 to 145 g of glucose that would otherwise be needed each day. Nevertheless, a starving person would use up body protein and die within 3 to 4 wk if the remainder of the glucose requirement continued to come from protein catabolism. Fortunately the brain, the major consumer of glucose, adapts to the starving state by increasing its ability to use fat-derived ketone bodies for energy. The concentration of ketone bodies is normally quite low in healthy and fed individuals but increases significantly during starvation. This increased concentration of ketone bodies, however, can lead to ketosis and acidosis.

2. Ketosis in starvation Ketone bodies made in the liver from acetyl CoA (see Chapter 10) are secreted in large quantities when fat is mobilized and when glucose is low. Normally, when acetyl CoA concentrations rise sufficiently, triacylglycerols are resynthesized, with L-glycerol 3-phosphate serving as the acceptor of the acyl groups. In starvation, insufficient glucose exists for the synthesis of glycerol phosphate. As a result, acetyl CoA exceeds the ability of the Krebs cycle to oxidize it and is shunted into ketone bodies. Usually this is not wasteful, since ketone bodies are taken up by many organs and tissues, where they serve as a good source of calories.

Acidosis in starvation The ketone bodies acetoacetic and β-hydroxybutyric acids lower the pH of the plasma. Bicarbonate (HCO_3^-) neutralizes the acids with the formation of CO_2, which is exhaled. When the bicarbonate buffering capacity is exceeded, the pH decreases, causing metabolic acidosis. Since the pH is related to the $[HCO_3^-]/[CO_2]$ ratio and the bicarbonate concentration $[HCO_3^-]$ has decreased, it is necessary to decrease the dissolved CO_2 accordingly. This is accomplished by hyperventilation.

3. Protein catabolism: gluconeogenesis in starvation The liver maintains glucose output, but the amino acids needed for gluconeogenesis are derived mainly from muscle. The amino acids must be transported from the muscle to the liver. Some of the amino acids produced by protein breakdown are used for energy in the muscle itself. Recall that the first step in the catabolism of amino acids is a deamination or transamination reaction that removes the α-amino group. Because muscle, unlike liver,

Figure 8.18   Alanine-glucose cycle.

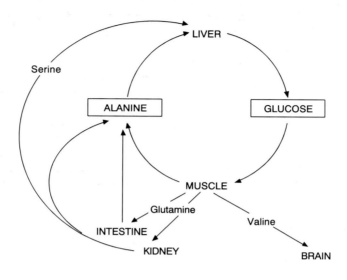

is incapable of synthesizing urea, most of the amino nitrogen is transferred to pyruvate to form alanine. The alanine enters the blood and is taken up by the liver. The amino groups are removed to form urea, and the resulting pyruvate is converted to glucose, which is secreted to the blood, taken up by the muscle, and catabolized to pyruvate. The pyruvate acts again as the acceptor for another amino group. The result of these reactions is a cycle that transports amino groups from a variety of muscle amino acids to the liver. This scheme is called the alanine-glucose cycle, or the Cahill cycle. Figure 8.18 illustrates some features of amino acid metabolism related to this cycle.

Alanine concentration drops late in starvation as release from muscle and other tissues decreases to conserve protein. Glutamine is another important vehicle for transferring amino groups from muscle to the liver. In this case ammonium ions are coupled with glutamate to yield glutamine. The glutamine is released from the muscle to the blood and taken up by the kidneys and the intestine (Figure 8.18). The kidneys release both alanine and serine as transport vehicles for amino groups.

4. Catabolism of branched-chain amino acids Most of the branched-chain amino acids stay in the muscle, where they are used for energy. Leucine breakdown produces HMGCoA and ketone bodies (see Figure 8.8), whereas isoleucine yields propionyl CoA and acetyl CoA (see Figure 8.10). Valine, however, is released to the blood. The brain is capable of directly using valine, a glucogenic amino acid that produces methylmalonyl CoA (see Figure 8.15).

5. Plasma branched-chain amino acids In this case the branched-chain amino acids exhibited a transient increase in serum concentration that peaked at 5 days. As fasting continued, the blood valine level dropped far below the initial level in the fed patient. Increased plasma concentrations of branched-chain amino acids are seen in other conditions besides starvation. For example, branched-chain amino acids accumulate in the plasma of diabetic patients. Alternatively the transient increase in the circulating levels of the branched-chain amino acids may result from their increased release by the liver. This is known to occur in diabetic, fasted, or uremic rats.

6. Alanine and glutamate in muscle Branched-chain amino acids stimulate the formation of glutamate, alanine, and glutamine in experimental animals. Because muscle transaminases specific for the branched-chain amino acids are much more active for α-ketoglutarate than for pyruvate, the amino groups of these amino acids are first transferred to α-ketoglutarate to give glutamate. The glutamate is then transaminated with pyruvate to form alanine by the aspartate aminotransferase, a very active enzyme in muscle.

Further evidence of a coupling of the metabolism of branched-chain amino acids with alanine synthesis is obtained from patients with maple syrup urine disease, who fail to decarboxylate the branched-chain amino acids. This causes a three- to tenfold decrease in plasma alanine concentration. These patients are hypoglycemic even though insulin levels are normal. The low alanine level results from the inability of these patients to metabolize the branched-chain amino acids. If alanine is administered, the hypoglycemia is alleviated. The coupling of the alanine cycle to the oxidation of branched-chain amino acids is important for glucose balance in several physiologic conditions and satisfies as much as 14% of the energy needs of animal muscle.

References

Adibi SA et al, editors: Branched chain amino and keto acids in health and disease, Basel, 1984, S Karger AG.

Felig P: Amino acid metabolism in man, Annu Rev Biochem 44:933, 1975.

Kerndt PR et al: Fasting: the history, pathophysiology and complications, West J Med 137:379, 1982.

Meguid MM et al: Uncomplicated and stressed starvation (a review), Surg Clin North Am 66:529, 1981.

Snell K: Muscle alanine synthesis and hepatic gluconeogenesis, Biochem Soc Trans 8:205, 1980.

Case 2   **Hartnup disease**

Hartnup disease was named for a 12-year-old boy, E. Hartnup, who was admitted to a London hospital with a red, scaly rash and mild cerebellar ataxia. His mother thought that he was suffering from pellagra because the same symptoms in her older daughter had been so diagnosed earlier. The boy did not have the usual *dietary*-deficiency form of pellagra, but large amounts of free amino acids were found in his urine. When an older daughter had a recurrent attack of ataxia, it was found that her urine also contained excessive amounts of amino acids. Two other siblings had the same aminoaciduria; however, four others were normal. The parents were asymptomatic, but a family history revealed that they were first cousins.

Biochemical questions

1. Assuming that this disease is inherited, what abnormality could account for the unusual amounts of aromatic and neutral amino acids in the urine?
2. From the data given, characterize the disease in terms of its dominance and whether it is X linked or autosomal.
3. What is the probability of these parents having normal, heterozygous, or homozygous children?
4. What is the relationship between the high levels of urinary excretion of the aromatic amino acids and the pellagra-like symptoms?
5. Symptoms of pellagra sometimes appear in people who habitually eat diets rich in corn. Explain.

6. Most patients with Hartnup disease outgrow the symptoms. It is surprising that the disease is so mild considering that the absorption of an oral load of an essential amino acid such as phenylalanine is only 25% of normal. Obviously phenylalanine and other amino acids must be absorbed in some other way or in some other form. Considering the specificity of the defect, how might the essential amino acids be absorbed?

Case discussion

1. Aromatic and neutral amino acids in the urine Hartnup disease is an inherited aminoaciduria that results in a reduced ability to transport several large neutral and aromatic amino acids across cell membranes. Transport in both the kidney tubule and intestine is impaired. A defect in the gene for a single protein that is part of a membrane transport system for these amino acids could account for the symptoms. Although the disease is neither serious nor common, it has provided fresh insight into amino acid transport, renal absorption, and protein digestion.

Originally the amino acids in the urine of these patients were determined by two-dimensional paper chromatography (Figure 8.19). Today the amino acids are more often analyzed by high-pressure liquid chromatography of their derivatives. Patients with Hartnup disease excrete 5 to 20 times more alanine, serine, threonine, leucine, isoleucine, valine, tyrosine, tryptophan, and phenylalanine than do control subjects. Several other amino acids are modestly elevated, but proline, methionine, and arginine levels are normal.

2. Genetics Without more information, it cannot be said with certainty that the same allele is defective in every family with a member who has Hartnup disease. However, every case seems to involve an autosomal recessive trait. In the case of E. Hartnup the disease was recessive, as evidenced by neither parent having symptoms. The possibility of the disease being X linked is eliminated because his sister was affected.

3. Inheritance of the disease If Hartnup disease is inherited in a strict Mendelian way, half the couple's children would be carriers, one-fourth would have the disease,

Figure 8.19

Analysis by two-dimensional paper chromatography of urinary amino acids from a patient with Hartnup disease. *Left, BuAc* is butanol:acetic acid:water. *Right, BuPyr* is butanol:pyridine:water. In both diagrams phenol:ammonia *(PhAm)* was the solvent used in the second dimension. (From Jepson JB: Hartnup disease. In Stanbury JB et al, editors: The metabolic basis of inherited disease, ed 4, New York, 1978, McGraw-Hill Book Co.)

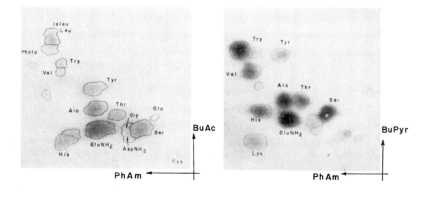

and one-fourth would be normal. Unfortunately, half the children in this family had the disease. Although these predictions are statistically valid for large populations, they are by no means reliable for the distributions in a single family. Since the carrier state cannot be detected, it is impossible to tell if the children who do not have symptoms are normal or carriers.

4. Tryptophan metabolism and pellagra Tryptophan can serve as a substrate for the synthesis of much of the body's niacin requirement (see Figure 8.12). The inability to transport dietary tryptophan and to reabsorb it in the kidney leads to a niacin deficiency that is responsible for the dermatologic and neurologic symptoms of pellagra.

5. Pellagra and diet Pellagra appears sporadically as a result of infection or unusual diets, especially those consisting chiefly of corn. Corn has a very low tryptophan content, and a deficiency can occur unless adequate amounts of tryptophan are supplied by other dietary proteins. The symptoms of pellagra can be treated with niacin and a diet high in protein.

6. Alternate amino acid transport mechanisms At least five separate stereospecific transport systems exist for different groups of amino acids in humans (see Table 8.2). Failure of three of these systems can result in genetic diseases. The systems are not absolutely specific because they show some overlap for different amino acids. The transport system for the monoaminomonocarboxylic acids is defective in Hartnup disease. In *cystinuria* the transport of cystine, lysine, arginine, and ornithine is impaired, whereas in *glycinuria* the transport of glycine, proline, and hydroxyproline is defective.

In view of the excessive urinary loss of several essential amino acids, it is surprising that Hartnup disease is relatively benign. The best explanation is that humans absorb sufficient amounts of these amino acids as small peptides. Transport systems also exist for peptides, and these differ from transport systems for amino acids. Since the amino acid requirements are the highest during growth, symptoms are more apt to be evident during childhood.

Hartnup disease may give rise to a variety of neurologic and psychiatric symptoms in some individuals, which might be the most serious aspect of the disease. It has been speculated that Julius Caesar and the Julian emperors such as Caligula and Nero who succeeded him may have been affected by Hartnup disease. If true, this relatively benign and rare disease may have contributed to the sadistic, bizarre behavior of these men and in turn substantially affected the course of history.

Interestingly, a mother with Hartnup disease had normal children; consequently, it is possible that amino acid transport in the placenta may not be as strongly affected as it is in the kidney or intestine. Furthermore, amino acid transport in skin fibroblasts from patients with either Hartnup disease or cystinuria is normal, indicating that amino acid transport is very complex and probably tissue specific.

References

Dirckx JH: Julius Caesar and the Julian Emperors: a family cluster with Hartnup disease? Am J Dermatopathol 8:351, 1986.

Jepson JB: Hartnup disease. In Stanbury JB et al, editors: The metabolic basis of inherited disease, ed 4, New York, 1978, McGraw-Hill Book Co.

Levy HL: Hartnup disorder. In Scriver CR et al, editors: The metabolic basis of inherited disease, ed 6, New York, 1989, McGraw-Hill Information Services Co.

Mahon BE and Levy HL: Maternal Hartnup disorder, Am J Med Genet 24:513, 1986.

Scriver CR et al: The Hartnup phenotype: Mendelian transport disorder, multifactorial disease, Am J Hum Genet 40:401, 1987.

Case 3   Hereditary hyperammonemia

A 6-month-old infant began to vomit occasionally and ceased to gain weight. At age 8½ mo he was readmitted to the hospital. Routine examination and laboratory tests were normal, but after 1 wk he became habitually drowsy, his temperature rose to 39.4° C, his pulse was elevated, and his liver was enlarged. The electroencephalogram was grossly abnormal. Since the infant could not retain milk given by gavage feeding, intravenous glucose was administered. He improved rapidly and came out of the coma in 24 hr. Analysis of his urine showed abnormally high amounts of glutamine and uracil, which suggested a high blood ammonium concentration. This was confirmed by the laboratory.

Biochemical questions

1. Hereditary hyperammonemia can result from defects in genes for urea cycle enzymes. Which enzymes might be affected?
2. Considering the data, which enzyme may be defective in this patient?
3. Why was the urine glutamine concentration elevated?
4. Offer a genetic explanation for the observation that this disease is usually lethal in males but not in affected females.
5. This patient was treated using procedures available at the time (see Goldstein AS et al: Pediatr Res 8:5, 1974). He was given a daily diet of 1.5 g of protein/kg body weight. After 2 yr on this diet, his height and weight were judged to be normal for his age. What is the effect of diet on a growing child in terms of nitrogen balance?
◆6. How would you treat a similar patient today?

Case discussion

1. Defective enzymes of the urea cycle Hyperammonemia in a newborn or very young infant is the characteristic sign of an inherited defect in a gene for an urea cycle enzyme. Genetic diseases have been found associated with deficiencies in all the urea cycle enzymes, but the most frequent abnormality is in ornithine transcarbamoylase, a mitochondrial enzyme encoded by a nuclear gene on the X chromosome.

2. Enzyme defect Other clinical signs of urea cycle diseases are a low to normal blood urea nitrogen (BUN), lethargy, irritability, hypotonia, and later, convulsions and coma. Death may quickly follow if the patient is left untreated. The enzyme affected in this patient is ornithine transcarbamoylase because of the excretion of uracil. Excessive excretion of uracil, or its precursor orotic acid, results from an accumulation of carbamoyl phosphate in the mitochondria. In the absence of ornithine transcarbamoylase, carbamoyl phosphate accumulates and leaks out into the cytoplasm, where it can be used to make carbamoyl aspartic acid, the first intermediate in the pathway to pyrimidine synthesis.

This case is unusual in that the symptoms took so long to appear. In most patients, extracts made from needle biopsies of the liver reveal a total lack of ornithine transcarbamoylase; however, in a few males (DiMagno et al, 1986), and in many females, enzyme activity can range from 0% to 30% of normal in males and even higher in females. The disease is described as X-linked dominant because most females are somewhat affected. Females usually respond well to dietary treatment.

3. Glutamine excretion Urine glutamine concentration increases because it exceeds the kidney's ability to hydrolyze glutamine to glutamate and ammonia. Glutamine is another carrier for ammonium ion in the blood. It is excreted in compensation for the inoperative urea cycle.

4. Inheritance Because the disease is X linked and males have only one X chromosome and females have two, one would expect that the disease would be much more severe in males than in females. Unfavorable Lyonization (see Chapter 2), however, can produce female patients that are almost as severely affected as males.

5. Nitrogen balance An adult, who is not growing, requires comparatively little dietary protein to maintain good health. A growing child, however, requires considerably more. The load on the urea cycle is directly proportional to the amount of protein in the diet. Consequently, the growing child of ½ to 1 yr of age must have 2 g of high-quality protein/kg/day to maintain good growth, that is, to stay in positive nitrogen balance. In this case the patient was given somewhat less protein to ensure that large amounts of ammonium ions would not accumulate. For this patient the low-protein diet prevented ammonia intoxication but allowed normal growth. This treatment is not sufficient for most patients with a defect in the gene for ornithine transcarbamoylase.

6. Modern treatment Therapy today involves hemodialysis and transfusion as soon as possible to avoid the irreversible brain damage that can result if the plasma ammonium concentration remains high for a long period. This is followed by intravenous treatment with sodium benzoate and phenylacetate. These substances act to trap ammonium ions, as glycine and glutamine, and convert them to forms that are readily excreted by the kidney. The enzymatic reactions involved are:

Benzoate + Glycine → Hippuric acid + H_2O

Phenylacetic acid + Glutamine → Phenylacetylglutamine + H_2O

Both glycine and glutamine are nonessential amino acids that can be synthesized from nonprotein metabolites. Glycine eliminates one nitrogen molecule per mole, but glutamine disposes of two per mole. These two substances have proved to be very useful in the treatment of hyperammonemia. Long-term treatment involves a low-protein diet with arginine or citrulline supplementation. In patients with a defective urea cycle, arginine becomes an essential amino acid required in larger amounts than normal.

References

Batshaw ML et al: Treatment of inborn errors of urea synthesis: activation of alternative pathways of waste nitrogen synthesis and excretion, N Engl J Med 306:1387, 1982.

Brusilow SW: Disorders of the urea cycle, Hosp Pract, Oct 15, 1985, p 65.

Brusilow SW and Horwich AL: Urea cycle enzymes. In Scriver CR et al, editors: The metabolic basis of inherited disease, ed 6, New York, 1989, McGraw-Hill Information Services Co.

DiMagno EP et al: Ornithine transcarbamoylase deficiency—a cause of bizarre behavior in a man, N Engl J Med 315:744, 1986.

Case 4 Methylmalonic acidemia

Earlier this female patient had a child who died at age 3 mo of severe acidosis and dehydration. A diagnosis of methylmalonic acidemia was made posthumously. The patient was now pregnant and concerned that the fetus might have the same disease. At 19 wk gestation, amniotic fluid cells were assayed for propionate and succinate oxidation, vitamin B_{12} (cobalamin) coenzyme biosynthesis, and methylmalonyl CoA mutase activity. Propionate oxidation was much lower than controls, but succinate oxidation was normal. Radioactivity from ^{57}Co-labeled cobalamin was found in methyl cobalamin but not in deoxyadenosyl cobalamin; however, methylmalonyl CoA mutase activity was normal in cell lysates *supplemented in vitro with substrates and cofactors.* This enzyme requires adenosyl cobalamin as a coenzyme. The patient's urine contained large amounts of methylmalonic acid, which it did not before pregnancy.

Biochemical questions

1. What amino acids produce propionyl CoA in humans? What would happen if a person ate a diet lacking these amino acids?
2. How can you explain the high concentration of methylmalonic acid in the urine when amniotic cells showed normal activity of methylmalonyl CoA mutase? Assume that all metabolic defects are present in the fetus and not in the mother.
3. Why was one cobalamin coenzyme labeled after administration of radioactive vitamin B_{12} and not the other coenzyme? (See also Chapter 13, which describes a reaction of methyl cobalamin.)
4. At 32 wk gestation the mother was excreting approximately six times the normal amount of methylmalonate in her urine. She was given oral cyanocobalamin, 10 mg/day. One week after therapy was started, the methylmalonate concentration dropped to threefold above normal and by delivery at 41 wk was almost normal. The infant was born in excellent condition; however, cultured skin fibroblasts from the infant displayed the same abnormalities as the amniotic cells. The infant was maintained on a low-protein diet (1.5 g/kg/day) but was not treated further with cobalamin. Why was the low-protein diet used, considering that the infant would be growing rapidly?
5. During the next 6 wk the infant cleared the vitamin B_{12} absorbed from the mother before birth. At the same time the protein in the diet was increased to 2.5 g/kg/day, and cobalamin was administered intramuscularly for 11 days. Considering the enzymes involved, explain how high levels of vitamin B_{12} result in decreased excretion of methylmalonate.
6. At 19 mo of age the child was in the 75th percentile for weight, length, and head circumference, and development was normal. She was being maintained on a low-protein diet, and continuous cobalamin therapy had not been necessary. Do you think that this child was homozygous or heterozygous for methylmalonic acidemia? Explain.
7. If, as an adult, this girl became obese and desired to lose weight, how and for what reasons would you treat her differently from other obese patients?

References

Fenton WA and Rosenberg LE: Inherited disorders of cobalamin transport and metabolism. In Scriver CR et al, editors: The metabolic basis of inherited disease, ed 6, New York, 1989, McGraw-Hill Information Services Co.

Nyhan WL: Abnormalities in amino acid metabolism in clinical medicine, Norwalk, Conn, 1984, Appleton-Century-Crofts.

Stacey TE: Effect of mutation on maternal-fetal metabolic homeostasis: general concepts. In Lloyd JK and Scriver CR, editors: Genetic and metabolic disease in pediatrics, London, 1985, Butterworth & Co.

Case 5   Glutathionuria

As part of a routine screening of institutionalized individuals for defects in amino acid metabolism, a mildly retarded man was found to excrete abnormally large amounts of glutathione. Serum glutathione concentration was also abnormally high; however, both serum and renal concentrations of individual amino acids were normal. γ-Glutamyl transpeptidase activity was undetectable in cultured skin fibroblasts.

Biochemical questions
1. What is the role of γ-glutamyl transpeptidase in humans?
2. What is the γ-glutamyl cycle? Illustrate your answer with equations.
3. Offer explanations for the lack of serious health problems in this patient. Consider enzyme stability, peptide transport, and alternative functions for this enzyme.

References
Bannai S and Tateishi N: Role of membrane transport in metabolism and function of glutathione in mammals, J Membr Biol 89:1, 1986.

Meister A and Larsson A: Glutathione synthetase deficiency and other disorders of the γ-glutamyl cycle. In Scriver CR et al, editors: The metabolic basis of inherited disease, ed 6, New York, 1989, McGraw-Hill Information Services Co.

Wellner D and Meister A: A survey of inborn errors of amino acid transport in man, Annu Rev Biochem 50:911, 1981.

Case 6   Defective urea cycle

A newborn infant was found to have hyperammonemia with respiratory alkalosis. An inherited defect in the urea cycle or transient hyperammonemia was suspected.

Biochemical questions
1. Plasma citrulline levels were checked and found to be 5 μmol/L; normal is 50 μmol/L. This indicated a defect in the urea cycle rather than transient hyperammonemia. It also ruled out the involvement of certain urea cycle enzymes. Which ones were ruled out?
2. Urinary orotate concentration was measured and found to be normal. This eliminates the possibility that another urea cycle enzyme was defective. Which one is ruled out and why?
3. The patient was treated with sodium benzoate and sodium phenylacetate. What is the rationale for this treatment?
4. Extracts of a liver biopsy were assayed for N-acetylglutamate synthase. No activity was found. Is this enzyme cytosolic or mitochondrial? What relationship does this enzyme have to the urea cycle?
♦ 5. With this information at hand, the patient was maintained on a diet that contained arginine and carbamoylglutamate, 2 g/kg/day. What reasoning is behind the use of each of these substances?

References
Bachmann C et al: N-Acetylglutamate synthetase deficiency: a disorder of ammonia detoxification, N Engl J Med 304:543, 1981.

Brusilow SW: Disorders of the urea cycle, Hosp Pract 10:65, 1985.

Brusilow SW and Horwich AL: Urea cycle enzymes. In Scriver CR et al, editors: The metabolic basis of inherited disease, ed 6, New York, 1989, McGraw-Hill Information Services Co.

Saheki T, Kobayashi K, and Inoue I: Hereditary disorders of the urea cycle in man: biochemical and molecular approaches, Rev Physiol Biochem Pharmacol 108:22, 1987.

Additional questions and problems

1. What are the intermediates of the urea cycle? What is the origin of the nitrogen atoms of urea?
2. Bacitracin is a peptide antibiotic that can be administered orally or intramuscularly. What route of administration would you predict to be more effective? Why?
3. Thienylalanine is a growth inhibitor of some bacteria and a structural analogue of phenylalanine. Phenylalanine will competitively reverse the growth inhibition, whereas glycylphenylalanine will noncompetitively reverse inhibition by thienylalanine. Explain these data in terms of amino acid and peptide transport.
4. The keto forms of valine, isoleucine, and leucine were administered to a patient with hyperammonemia. Explain why the plasma ammonia level fell twofold.
5. Describe the possible effects on the metabolism of valine, isoleucine, threonine, and methionine in an infant who is breast-fed by a strict vegetarian mother. (*Hint:* see Figure 8.7 or Heaton D: N Engl J Med 300:202, 1979.)

Multiple choice problems

Problems 1 to 5 refer to a patient with glomerular nephritis.

1. The patient had a high BUN. This was caused by:
   A. Damage to urea cycle enzymes.
   B. Lowered filtration rate of the kidney.
   C. Excessive synthesis of urea from dietary ammonium ions.
   D. Decrease in the function of kidney glutaminase.
   E. Breakdown of the patient's urease.

2. The patient was maintained on a low-protein diet. This diet must contain adequate amounts of:
   A. All the amino acids.
   B. Glucogenic amino acids, but not ketogenic amino acids.
   C. Ketogenic amino acids, but not glycogenic amino acids.
   D. Essential amino acids.
   E. Nonessential amino acids.

3. Portions of valine, isoleucine, methionine, and threonine derived from the patient's diet can be converted to methylmalonyl CoA. Methylmalonyl CoA is metabolized by being isomerized by a mutase that requires a coenzyme derived from:
   A. Cobalamin (vitamin B_{12}).
   B. Biotin.
   C. Niacin.
   D. Thiamine.
   E. Ascorbic acid (vitamin C).

4. The methylmalonyl CoA derived from these amino acids can enter the Krebs cycle after the mutase catalyzes the formation of:
   A. Acetyl CoA.
   B. Succinyl CoA.
   C. Propionyl CoA.
   D. Acetoacetyl CoA.
   E. β-Hydroxybutyryl CoA.

5. Digestion of dietary proteins by the patient requires several proteolytic enzymes. Many of these are synthesized by the pancreas. However, an important proteolytic enzyme originates in the stomach. This enzyme is:

A. Trypsin.
B. Chymotrypsin.
C. Dipeptidase.
D. Pepsin.
E. Carboxypeptidase.

A 1½-month-old boy was admitted to the hospital because of vomiting and failure to gain weight. During hospitalization he became drowsy, lethargic, and finally comatose with convulsions. He improved after receiving intravenous glucose and came out of the coma. A few days later his plasma ammonium ion concentration was 180 μmol/L (normal, 11 to 50 μmol/L).

Problems 6 to 10 refer to this case.

6. The abnormally high level of ammonium ions is consistent with a genetic defect in:
    A. The synthesis of carbamoyl phosphate.
    B. The formation of N-acetylglutamate.
    C. The synthesis of glutamine from glutamate.
    D. The synthesis of citrulline.
    E. Any of the above.

7. High blood ammonium ion concentrations in some patients can be lowered by treatment with arginine. The reason for this is that:
    A. The defective enzyme prevents arginine biosynthesis.
    B. A defect in the utilization of arginine in protein biosynthesis requires excess arginine.
    C. The enzyme arginase is defective.
    D. Arginine increases the concentration of an acceptor for a carbamoylation reaction.
    E. Arginine is an essential amino acid for infants.

8. With informed consent, the patient was challenged with a high-protein load. After the high-protein diet, one might expect that some ketone bodies would form. Ketone bodies can be obtained from C atoms of the following amino acids:
    1. Arginine
    2. Isoleucine
    3. Valine
    4. Leucine
    The best answer is:
    A. 1, 2, and 3.
    B. 1 and 3.
    C. 2 and 4.
    D. 4 only.
    E. 1, 2, 3, and 4.

9. The patient's protein intake was maintained at a low level. One must be careful, however, that the diet contains sufficient amounts of the essential amino acids:
    1. Isoleucine
    2. Phenylalanine
    3. Threonine
    4. Tyrosine
    The best answer is:

A. 1, 2, and 3.
B. 1 and 3.
C. 2 and 4.
D. 4 only.
E. 1, 2, 3, and 4.

10. The urea cycle:
   A. Occurs in all tissues.
   B. Involves only mitochondrial enzymes.
   C. Results in the formation of 3 mol of ATP per mole of urea synthesized.
   D. Accepts nitrogen from aspartate and releases the aspartate carbon atoms as fumarate.
   E. Accepts nitrogen from N-acetylglutamate and releases the glutamate as α-ketoglutarate.

Chapter 9

Amino acid and neurotransmitter synthesis

Objectives

1 To learn how the nonessential amino acids are synthesized by humans
2 To describe how genetic defects in the biosynthesis of amino acids result in metabolic diseases
3 To demonstrate how pharmacologically active amines are synthesized from amino acids
4 To understand how important metabolites such as polyamines, carnitine, and creatine are synthesized from amino acids

Measurement of nitrogen-containing substances in blood and urine is useful in the diagnosis and treatment of many diseases. Nitrogen metabolism changes significantly during growth, as well as during pregnancy or following recovery from starvation, malnutrition, febrile diseases, burns, trauma, and surgery. The nitrogen-containing substances of the plasma or urine result from the continual biosynthesis and degradation of proteins and amino acids or their metabolites. Even in normal people the concentrations of these substances may change with diet, exercise, or environmental stress. Chapter 8 emphasizes that plasma or urine concentrations of ammonium ions, urea, and amino acids are greatly influenced by the catabolism of protein and amino acids. This chapter considers the effects of amino acid biosynthesis on dietary nitrogen requirements and how some amino acids are converted to amine-containing metabolites. Several of these metabolites can appear in the plasma and urine, where they are useful signals of health or disease.

Most of the biosynthetic amino acids and many of the dietary amino acids are used for protein synthesis. However, amino acids serve as precursors for the synthesis of several other nitrogen-containing compounds. The box on p. 380 lists some of the major biosynthetic functions of the amino acids. Chapter 4 shows that amino acids are required for the synthesis of porphyrins, particularly heme, a substance that must be continually synthesized because it turns over so rapidly. In the same way, amino acids are used for the synthesis of choline and ethanolamine, which are part of the phospholipids (see Chapter 10). The hexosamines of connective tissue (see Chapter 7) also require an amino acid for their synthesis, and substantial amounts of amino acids are used for the synthesis of small nucleotides such as adenosine 5'-triphosphate (ATP) and greater amounts for the synthesis of the nucleic acids, deoxyribonucleic and ribonucleic acids (DNA and RNA; see Chapters 14 and 15).

This chapter emphasizes the functions of the amino acids in the synthesis of various biologically active amines and other small nitrogen-containing substances.

Biosynthesis of nonessential amino acids

In most diets the mixtures of amino acids generated by digestion are not present in the proportions required by the body. Consequently it is necessary to rearrange them metabolically. The human body can do this unless the diet is severely imbalanced. Amino acids released to the portal blood from the gut already show changes in composition. For

---

Major Functions of Amino Acids Derived from Dietary Protein

Oxidation (see Chapters 6 and 8)

Glycogenic amino acids—Blood glucose—Energy

Ketogenic amino acids—Acetyl CoA—Stored fat—Energy

Biosynthesis of nitrogen-containing metabolites

1. Heme (see Chapter 4)

    Succinyl CoA + Glycine—δ-Aminolevulinate—Porphobilinogen

2. Choline (see Chapter 10)

    Phosphatidyl serine—Phosphatidyl ethanolamine—Phosphatidyl choline

3. Glycosamine (see Chapter 7)

    Fructose 6-phosphate + Glutamine—Glutamate + Glucosamine 6-phosphate

4. Nucleotides (see Chapter 13)

    Glycine, glutamine, aspartic acid, amino acid nitrogen—Purine and pyrimidine nucleotides

5. Protein synthesis (see Chapter 15)

    All amino acids—Proteins

6. Biogenic amines (see Chapters 9 and 18)

    Amino acids—Hormones, neurotransmitters, pharmacologic amines

7. Carnitine (see Chapters 9 and 10)

    Lysine + Methionine—Carnitine

8. Creatine phosphate (see Chapter 9)

    Arginine, glycine, methionine—Creatine phosphate

---

example, alanine levels in the portal blood exceed the amount of alanine ingested. On the other hand, glutamate and aspartate concentrations are lower. Thus a large amount of the ingested nitrogen is delivered to the liver as alanine.

Recall from Chapter 1 that the amino acids are grouped according to whether they are essential or nonessential components of the diet. The dietary essential amino acids for adult humans are:

Phenylalanine

Valine

Tryptophan

Threonine

Isoleucine

Methionine

Histidine

Arginine

Leucine

Lysine

Reading down the column of first letters gives *Pvt. Tim Hall,* a useful mnemonic.

Remember that an essential amino acid is one that *cannot be synthesized in sufficient amounts.* Humans totally lack the enzymes necessary for the synthesis of some essential amino acids, such as the aromatic amino acids, but they may be able to synthesize small but insufficient amounts of other essential amino acids. For example, arginine can be synthesized by urea cycle enzymes but not in sufficient amounts to satisfy the needs of the human body.

The biosynthesis of the essential amino acids occurs in plants and microorganisms, so although it is important that we eat proteins that contain these amino acids, it is less important that health scientists know the details of their synthesis. Instead, this chapter considers only the biosynthesis of the nonessential amino acids, those amino acids that can be synthesized by humans.

| Metabolism supporting amino acid biosynthesis | Several of the 10 nonessential amino acids are synthesized from the carbon skeletons of intermediates in metabolic pathways such as the Krebs cycle and the glycolytic pathway. Although these metabolic schemes are very important in producing the energy needed to maintain life, their intermediates are also used in biosynthetic reactions. |

Synthesis of amino acids by transamination

Alanine, aspartate, and glutamate Aspartate and glutamate are synthesized from Krebs cycle intermediates by simple transamination reactions. Chapter 8 points out that aspartate and glutamate can be transaminated to yield oxaloacetate and α-ketoglutarate, both Krebs cycle intermediates. Because these transamination reactions are readily reversible, the Krebs cycle intermediates can also be used for the biosynthesis of these amino acids. Similarly, pyruvate produced from the catabolism of glucose can be transaminated to give alanine. These reactions are shown in Figure 9.1.

Note that the donor of the amino groups is always glutamate; thus sufficient glutamate must be available for the net synthesis of amino acids by transamination. The action of glutamate dehydrogenase, a widely occurring enzyme, can provide the necessary glutamate by catalyzing the reaction:

$$NH_4^+ + \begin{matrix} COO^- \\ | \\ C=O \\ | \\ CH_2 \\ | \\ CH_2 \\ | \\ COO^- \end{matrix} \quad \xrightarrow[\text{dehydrogenase}]{\text{NADH} \quad\quad \text{NAD}^+ \\ \text{Glutamate}} \quad \begin{matrix} COO^- \\ | \\ {^+}H_3N-CH \\ | \\ CH_2 \\ | \\ CH_2 \\ | \\ COO^- \end{matrix} + OH^-$$

The amino transferases shown in Figure 9.1 are specific enzymes that catalyze reversible reactions. Thus amino acids can enter the Krebs cycle and keto acids can be drawn out of the cycle by reactions catalyzed by the same enzymes. The distribution of specific amino transferases can vary widely between different tissues. The significance of these enzymes in clinical diagnosis is discussed in Chapter 3.

Asparagine and glutamine The amides of aspartate and glutamate, asparagine and glutamine, are both synthesized by humans. Despite the similarity of the products, the reactions involved are quite different. The enzyme glutamine synthase catalyzes the formation of a γ-glutamyl phosphate intermediate:

$$\begin{matrix} COO^- \\ | \\ {^+}H_3N-CH \\ | \\ CH_2 \\ | \\ CH_2 \\ | \\ COO^- \end{matrix} + NH_4^+ \quad \xrightarrow[\text{Glutamine synthase}]{\text{ATP} \quad\quad \text{ADP} + P_i} \quad \begin{matrix} COO^- \\ | \\ {^+}H_3N-CH \\ | \\ CH_2 \\ | \\ CH_2 \\ | \\ O=C-NH_2 \end{matrix}$$

L-Glutamate   L-Glutamine

This mixed anhydride provides the driving force for the transfer of the acyl group to ammonium ion (NH_4^+) with the formation of a rather stable amide group. Adenosine diphosphate (ADP) is the other product; P_i is inorganic phosphate.

Asparagine synthase works by a different mechanism. The intermediate is an aspartyl adenylate that is formed from aspartate and ATP. Pyrophosphate is a product of the reaction. The enzyme-bound acyl intermediate reacts with glutamine to yield asparagine, glutamate, and adenosine monophosphate (AMP).

Figure 9.1   Biosynthesis of amino acids from Krebs cycle intermediates. *AST*,
Aspartate aminotransferase; *ALT*, alanine aminotransferase; *OAA*,
oxaloacetate.

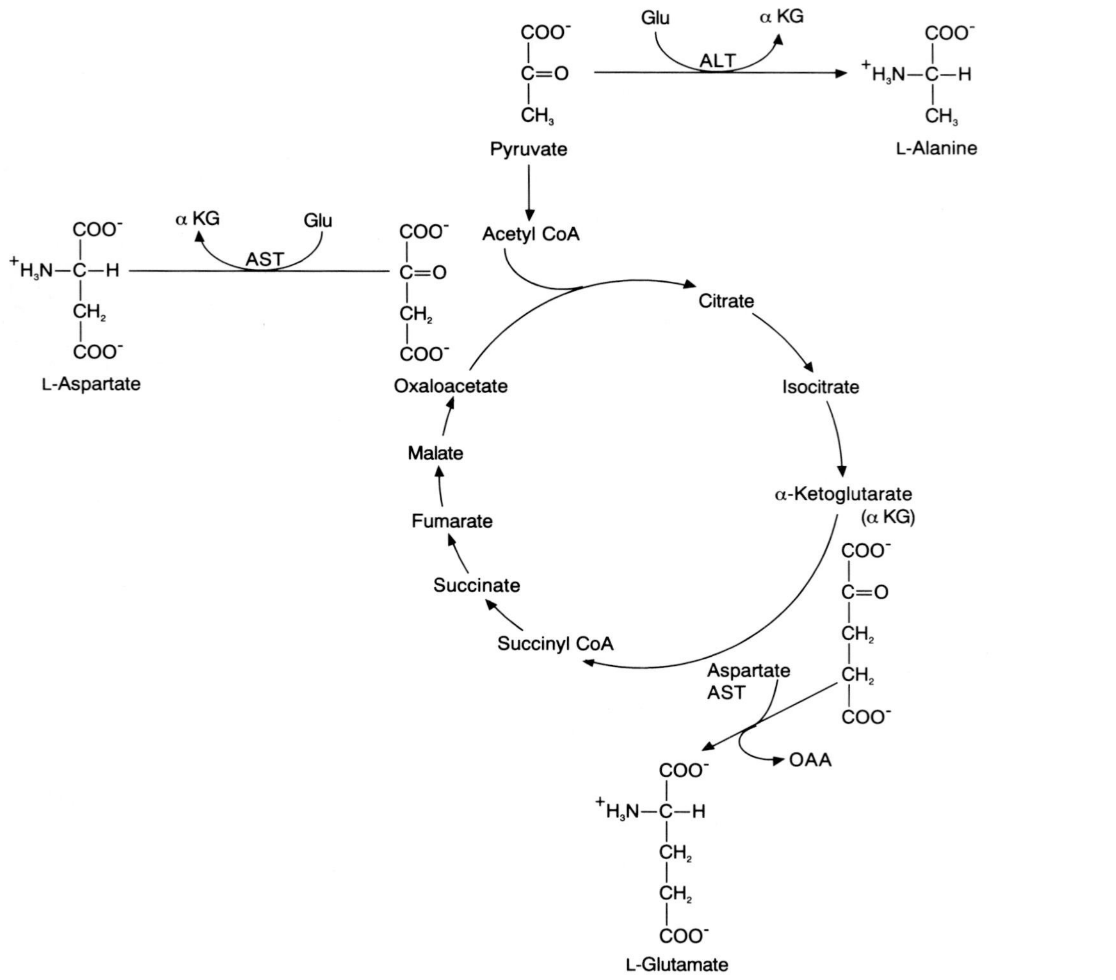

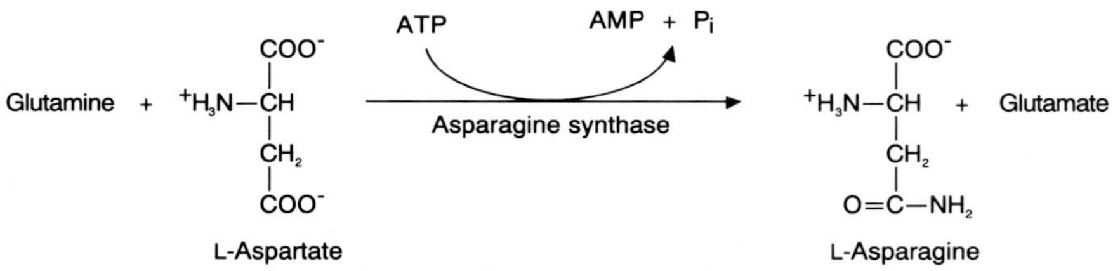

Synthesis of amino acids from intermediates of monosaccharide metabolism

Serine Serine is synthesized from 3-phosphoglycerate, an intermediate of the glycolytic pathway. 3-Phosphoglycerate can also be derived from 3-phosphoglyceraldehyde, an intermediate of the pentose phosphate pathway. Two separate routes exist for the synthesis

Figure 9.2   Synthesis of serine by the glycolytic pathway.

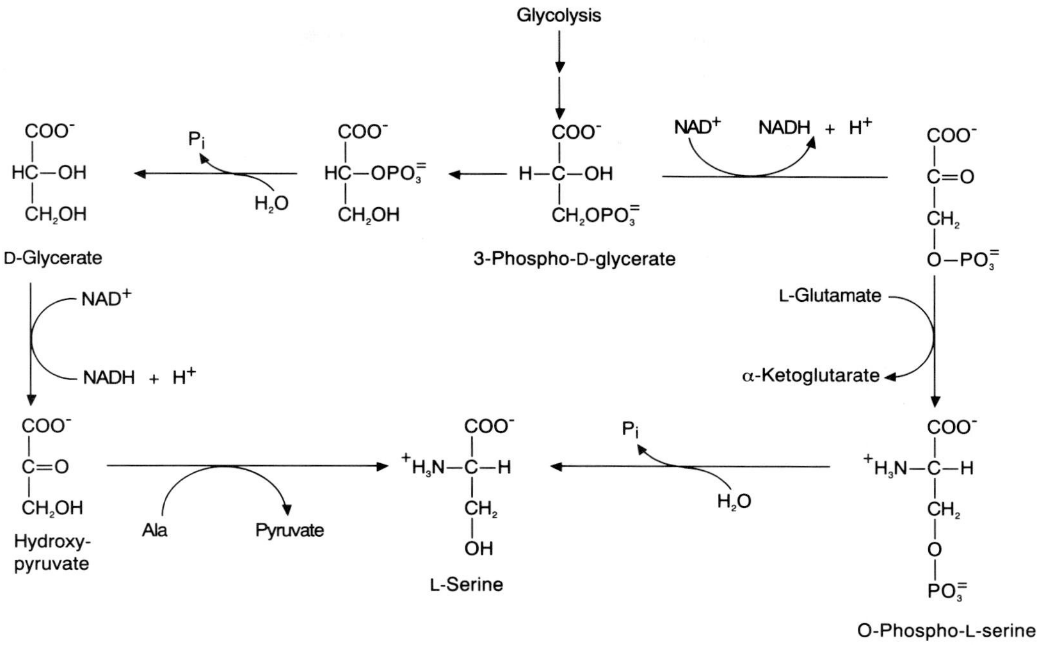

of serine from 3-phosphoglycerate. One pathway is shown at the left in Figure 9.2; the other is shown at the right. Note that the pathway on the right contains phosphorylated intermediates, whereas the one on the left does not. Both pathways are reversible if kinases are available to phosphorylate the appropriate intermediates.

Clinical comment   **L-Glyceric aciduria** A patient with L-glyceric aciduria had no detectable D-glyceric acid dehydrogenase in leukocytes. This enzyme converts D-glycerate to hydroxypyruvate, as in Figure 9.2. However, excessive hydroxypyruvate was not found, and the serum concentration of serine was normal.

Note that L-glyceric acid accumulates, not the D-isomer. Hydroxypyruvate is a good substrate for lactate dehydrogenase (LDH), so it has been suggested that LDH converts hydroxypyruvate to L-glycerate, the substance that accumulates. This means that the pathway shown on the left of Figure 9.2 is primarily a gluconeogenic route from serine to glucose, and the pathway on the right, with its phosphorylated intermediates, is the major biosynthetic pathway. Recall from Chapter 8 that serine is also catabolized via serine dehydratase to pyruvate. Furthermore, serine dehydratase is induced by glucocorticoids.

Most human diets contain substantial amounts of serine, so the biosynthetic pathway may not be essential. However, its existence emphasizes the importance of serine in many biosynthetic reactions, such as in the synthesis of ethanolamine, choline, betaine, glycine, cysteine, sarcosine, pyruvate, and sphingosine.

Glycine Serine is the precursor of glycine. The hydroxymethyl group of serine is transferred to tetrahydrofolate (THF) to form 5,10-methylene THF, water, and glycine.

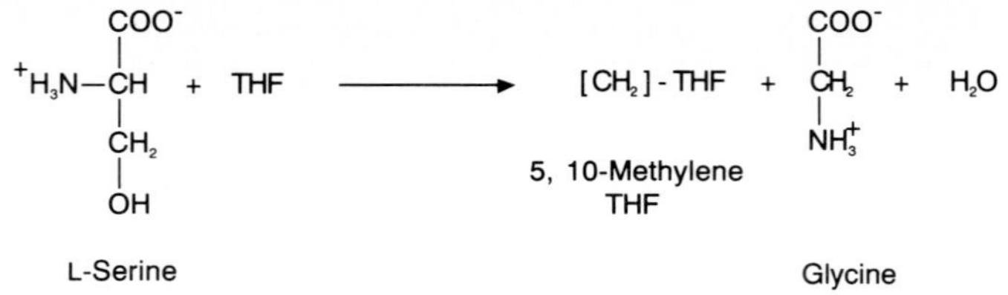

$$^+H_3N-CH \quad + \quad THF \longrightarrow [CH_2]-THF + CH_2 + H_2O$$

L-Serine

5, 10-Methylene THF

Glycine

A more complete description of the metabolism of folic acid derivatives is given in Chapter 13.

Glycine can also be synthesized from glyoxalate. Cytoplasmic enzymes in the liver decarboxylate serine to ethanolamine, which is transaminated to glycoaldehyde. Oxidation of this aldehyde yields glyoxylate. Transaminases then produce glycine from glyoxylate. These transformations are summarized as:

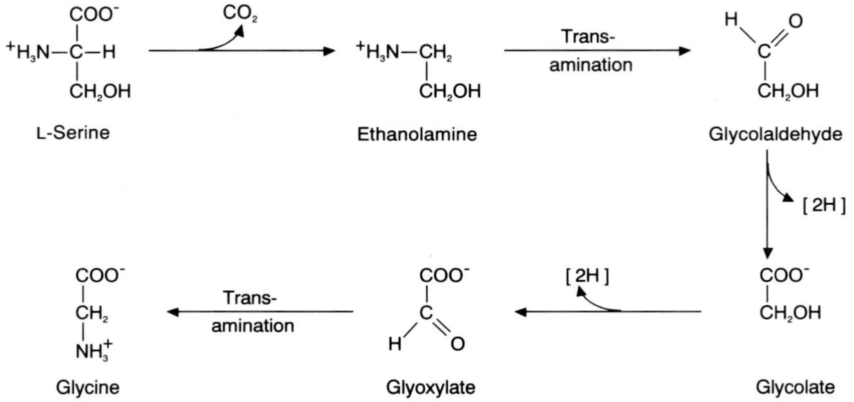

L-Serine → Ethanolamine → Glycolaldehyde → Glycolate → Glyoxylate → Glycine

Clinical comment

Hyperoxaluria, type 1 A block in the metabolism of glyoxalate produces a rare disease called hyperoxaluria, type 1, which is characterized by an excessive excretion of oxalate derived from the oxidation of the accumulated glyoxylate.

Glyoxylate → Oxalate + [2H]

Methyl group transfer

S-Adenosylmethionine: a methylating agent Cysteine biosynthesis cannot be understood without a description of the function of methionine in the transfer of methyl groups. Through these reactions homocysteine, a more immediate precursor of cysteine, is produced.

The transfer of one-carbon units other than carbon dioxide is generally achieved, whether through a derivative of methionine or by a coenzymatic form of the vitamin folic acid. Nucleotide biosynthesis requires several one-carbon transfers that are mediated by folic acid coenzymes (see Chapter 13). The methionine transfer reactions are more involved with amino acid metabolism.

The methionine adenosyl transferase of liver catalyzes the formation of *S*-adenosylmethionine (*S*-AdoMet) from methionine and ATP (Figure 9.3). The reaction is unusual in that all three of the phosphates of ATP remain as an enzyme-bound triphosphate when the adenosyl group is transferred to methionine. The same enzyme then catalyzes the cleavage of the triphosphate to inorganic phosphate and pyrophosphate. This makes the

Figure 9.3   Synthesis and utilization of *S*-adenosylmethionine.

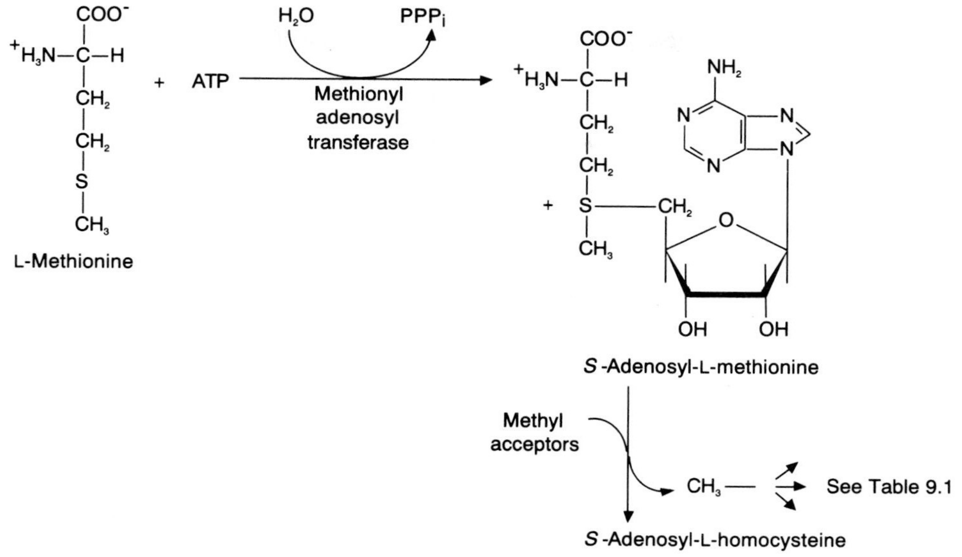

S-Adenosyl-L-methionine

S-Adenosyl-L-homocysteine

Table 9.1   Methyl transfer reactions using *S*-adenosylmethionine as
                                  methyl donor

	Acceptor	Product
	Guanidoacetate	Creatine
	Phosphatidylethanolamine	Phosphatidylcholine
	Ribosomal and transfer RNA	Methylated RNA
	DNA	Methylated DNA
	Norepinephrine	Epinephrine
	Protein-bound lysine	Carnitine

reaction essentially irreversable. The other product, *S*-AdoMet, is the principal methyl donor in the body. The methionine methyl group of *S*-AdoMet can be transferred to amino and hydroxy groups on a variety of acceptor molecules. Some of these acceptors and their products are listed in Table 9.1.

Reactions of homocysteine transmethylation The other product of the methylation reactions that use *S*-AdoMet (Table 9.1) is *S*-adenosylhomocysteine. A hydrolase cleaves the molecule to yield homocysteine and adenosine.

$$H_2O + \text{\textit{S}-Adenosyl-L-homocysteine} \rightarrow \text{L-Homocysteine} + \text{Adenosine}$$

Homocysteine itself can function as a methyl acceptor but in a slightly different way (Figure 9.4). Betaine is the methyl donor and homocysteine the methyl acceptor. Betaine is formed from the oxidation of dietary choline. From the information shown in Figure 9.4, one might predict that the essential amino acid methionine could be synthesized if

Figure 9.4　　Transmethylation.

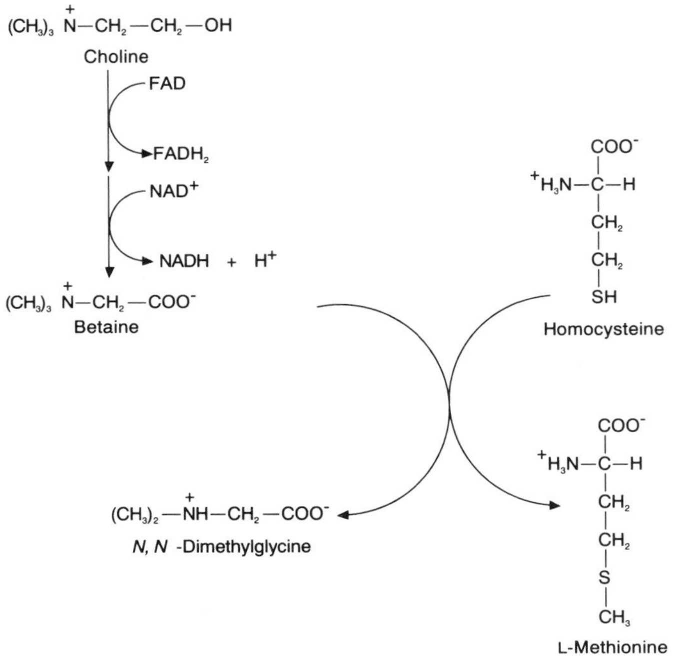

sufficient choline and homocysteine were present in the diet. This proves to be true, since humans are able to synthesize all but the homocysteine portion of methionine.

　　Homocysteine can also be converted to methionine by another pathway. This reaction is important in the metabolism of both folate and vitamin B_{12}, a relationship discussed fully in Chapter 13. For present purposes, the reaction may be illustrated by the equation:

$$\text{L-Homocysteine} + \text{5-Methyl THF} \xrightarrow[\text{coenzyme}]{\text{Corrinoid}} \text{L-Methionine} + \text{THF}$$

The enzyme that catalyzes the reaction is 5-methyl THF:homocysteine transmethylase, sometimes called 5-methyl THF methyltransferase. The corrinoid coenzyme is a derivative of vitamin B_{12}.

Biosynthesis of amino acids from dietary essential amino acids

Cysteine synthesis The sulfhydryl group of cysteine is derived from methionine. More specifically, it is the homocysteine portion of the molecule that furnishes the sulfhydryl group. The carbon skeleton of cysteine comes from serine. The two amino acids condense under the influence of *cystathionine β-synthase,* a pyridoxal-containing enzyme. The β designation indicates that the electrophilic attack is by a carbonium ion at the β position (—CH_2OH) of serine. This distinguishes the enzyme from cystathionine γ-synthase, which is an enzyme found in plants and microorganisms and that synthesizes cystathionine from cysteine.

L-Homocysteine   L-Serine   L-Cystathionine

Cystathionase (cystathionine γ-lyase) is a pyridoxal phosphate–containing enzyme that cleaves cystathionine to give ammonium ion, α-ketobutyrate, and cysteine.

α-Ketobutyrate   L-Cysteine

Cysteine, besides being required for the synthesis of proteins, is also used to make the tripeptide glutathione (γ-glutamylcysteinylglycine) and the aminosulfonic acid, taurine.

Clinical comment

Homocystinuria and cystathioninuria Deficiencies of the enzymes involved in cysteine synthesis occur in two human genetic diseases. In *homocystinuria* large amounts of homocystine are found in the urine. The gene for cystathionine synthase is defective in this disease, so homocysteine accumulates. Homocysteine is readily oxidized to the disulfide compound homocystine. The structure of homocystine is analogous to that of cystine (see case 2 at the end of the chapter).

In *cystathioninuria* the genetic defect involves the step that cleaves cystathionine to produce cysteine; consequently, large amounts of cystathionine are found in the blood and urine. The genetic defect in cystathioninuria is interesting in that an active enzyme protein is produced, but the protein has a much reduced affinity for its essential coenzyme, pyridoxal phosphate.

Tyrosine Given enough of the essential amino acid phenylalanine, humans can synthesize adequate amounts of tyrosine. The enzyme phenylalanine hydroxylase catalyzes this transformation.

L-Phenylalanine   L-Tyrosine

Phenylalanine hydroxylase is an oxygenase that requires tetrahydrobiopterin, a pteridine coenzyme similar to folic acid.

Tyrosine has several important functions in biosynthesis. It serves as a precursor of thyroxine (see Chapter 17) and the melanin pigments. A later section in this chapter considers its role in the synthesis of norepinephrine and epinephrine.

Clinical comment

Phenylketonuria Phenylalanine hydroxylase is defective in most patients with phenyl-ketonuria, a disease caused by an autosomal recessive gene carried by 2% of the population. Phenylketonuria occurs at an incidence of one in 10^4 births and is the most common of the aminoacidurias. Abnormal phenylalanine metabolites, such as phenylpyruvate, phenylacetate, phenyllactate, and phenylacetylglutamine, as well as phenylalanine itself, are found in the patient's urine. Mental retardation associated with the disease is lessened by placing the phenylketonuric infant on a diet low in phenylalanine. The cause of the mental retardation is not clear, but it may be caused by high concentrations of phenyl-alanine inhibiting the enzyme that decarboxylates 5-hydroxytryptophan to form serotonin, a biogenic amine (see later discussion). Serotonin levels are known to be low in these patients. Less frequently, phenylketonuria is caused by a defect in the synthesis of the dihydrobiopterin coenzyme.

L-Proline Proline is synthesized from arginine according to the scheme in Figure 9.5. The first reaction in this series is catalyzed by arginase and results in the conversion of arginine to urea and ornithine. The δ-amino group of ornithine is transaminated to form glutamate-γ-semialdehyde. The formation of a Schiff base closes the ring, and a subsequent reduction yields L-proline.

Arginine and histidine Arginine and histidine are considered by some authorities to be nonessential for the adult human, but their status is not as clear as that of the other nonessential amino acids, and we listed them as essential at the beginning of the chapter. Both amino acids are required in the diet for the rapid growth of infants. As described in Chapter 8, arginine can be made from ornithine, a urea cycle intermediate. Glutamate is the precursor of the carbon skeleton of ornithine. Glutamate is reduced to glutamate-

Figure 9.5   Biosynthesis of proline.

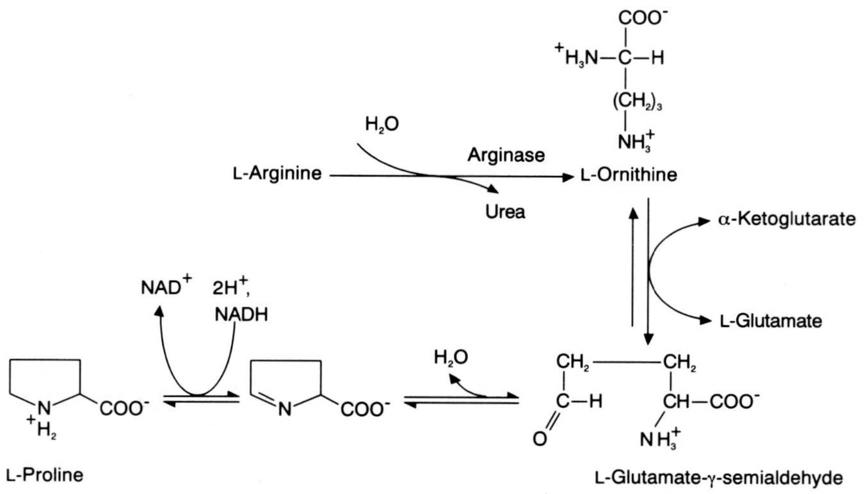

γ-semialdehyde, which is then transaminated to yield ornithine. Infants cannot obtain sufficient arginine by this pathway.

Histidine probably is an essential amino acid for adult humans, as it is for rats. The determination of the essential amino acids for rats was made by measuring large losses of weight that occurred when the animals were denied a single essential amino acid. Obviously humans cannot be used for such experiments. Instead, the nitrogen equilibria of healthy volunteer subjects was measured. When these young men were denied the amino acid histidine, their nitrogen output equaled their nitrogen input, and histidine was scored as nonessential. More than likely these subjects mobilized histidine from carnosine and anserine, substances that are stored in substantial quantities in muscle.

Carnosine · L-Histidine · Anserine

Carnosine is ordinarily synthesized from dietary histidine, and rats denied dietary histidine can use the stored carnosine in place of histidine for a short time. Since the human experiments were conducted for only brief periods, the human subjects probably also used stored carnosine in place of histidine; thus histidine is probably also essential for humans.

Summary of the biosynthesis of amino acids

A listing of the amino acids that can be synthesized by humans is given in Table 9.2, together with the precursors of these amino acids. Important or unusual features of the biosynthetic pathways, as well as the names of some human diseases associated with genetic defects in these pathways, are listed in the right-hand column. Note that two essential amino acids, methionine and arginine, are listed. Both can be synthesized by humans under certain conditions. Methionine is synthesized if the diet contains homocysteine, and arginine can be synthesized, as in the urea cycle, but it cannot be made in sufficient amounts by young animals and is essential for them.

Essential amino acids cannot by synthesized at all or cannot be synthesized at a sufficient rate in humans. The biosynthesis of the essential amino acids involves enzymes that are unique to bacteria and plants; accordingly, their synthesis is not considered here.

Amino acids as precursors of metabolites—amine synthesis

Animals contain various amine compounds derived chiefly from the amino acids. Although the concentrations of these amines are low, their actions are very important. The synthesis of several of these substances, such as serotonin, the catecholamines, and choline, is regulated partly by the type of food one eats. For example, high dietary tryptophan may stimulate serotonin synthesis sufficiently to make one drowsy.

Nervous system amines

Some of the most important amines derived from amino acids are those found in the nervous sytem. Many of these are neurotransmitters. A nerve can be thought of as transmitting a stimulus from one part of the body to another by means of a moving wave of ions. When the wave reaches the end of a neuron, it causes the release of a transmitter that migrates to a receptor cell, where it triggers the propagation of another wave.

Table 9.2   Summary of the biosynthesis of amino acids in animal cells

Amino acid	Precursor	Distinguishing features of pathways
Alanine	Pyruvate	By transamination
Glutamate	α-Ketoglutarate	By reductive amination
	α-Ketoglutarate	By transamination
Glutamine	Glutamate	ADP + P_i are products
Aspartate	Oxaloacetate	By transamination
Asparagine	Aspartate and glutamine	AMP + PP_i are products
Serine	3-Phospho-D-glycerate	Phosphoserine intermediate
	3-Phospho-D-glycerate	Hydroxypyruvate intermediate
	Glycine	Requires 5,10-methylene THF
Glycine	Serine	Requires THF
	Serine	Glyoxalate intermediate (hyperoxaluria, type 1)
Methionine	Homocysteine	Requires 5-methyl THF
	Homocysteine	Requires betaine
Cysteine	Serine and homocysteine	Cystathionine intermediate (cystathioninuria, homocystinuria)
Tyrosine	Phenylalanine	Biopterin coenzyme (phenylketonuria)
Proline	Arginine	Glutamate-γ-semialdehyde intermediate
Arginine	Glutamate	Ornithine intermediate

Neuron → Chemical transmitter → Receptor cell, muscle fiber, or another neuron.

Acetylcholine *Cholinergic* receptors such as the motor end-plates of muscles are sensitive to acetylcholine. Once the desired response has been provoked, the acetylcholine must be removed. The enzyme acetylcholinesterase accomplishes this. Figure 9.6 summarizes the biochemical reactions involved in the synthesis of choline from serine and methionine and shows the reactions involved in the cycling of acetylcholine and choline, which occurs during the functioning of the amine.

Catecholamines Catecholamines are derivatives of 3,4-dihydroxyphenethylamine (dopamine).

$$HO-\bigcirc-CH_2-CH_2-NH_3{}^+$$
$$\quad\quad OH$$

3,4 Dihydroxphenethylamine (dopamine)

Catecholamines include norepinephrine, epinephrine, and L-dopa. They are produced in the brain, sympathetic nerve endings, chromaffin cells of peripheral tissues, and the adrenal medulla. The latter is an endocrine gland that stores the hormones epinephrine and norepinephrine in chromaffin granules (named for their affinity for dichromate). These hormones permit the body to react to stress by increasing cardiac output and increasing blood flow to the muscles, lungs, and brain. They also stimulate cellular metabolism.

Biosynthesis The amino acid precursor of the catecholamines is tyrosine. Figure 9.7 shows the pathway of epinephrine synthesis from tyrosine. The first reaction is catalyzed

Figure 9.6   Synthesis and hydrolysis of acetylcholine. *S-AdoMet, S-*
adenosylmethionine; *S-AdoHcys, S*-adenosylhomocysteine.

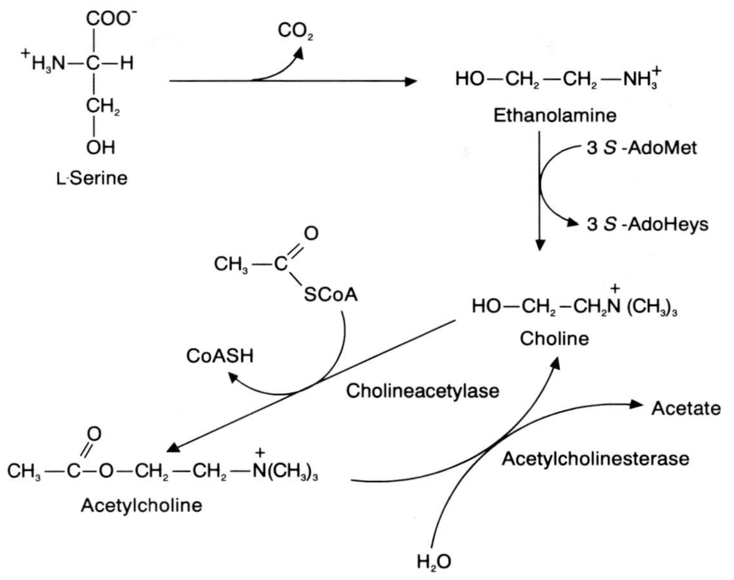

by tyrosine hydroxylase, an enzyme similar to phenylalanine hydroxylase in its mechanism but with a different substrate specificity. The product of the reaction is 3,4-dihydroxyphenethylamine. In previous literature this compound was called *di*oxyphenyl*al*anine and was abbreviated dopa. Although the chemical nomenclature has been improved, the popular abbreviation has been retained. L-Dopa is used with remarkable success to treat the symptoms of Parkinson's disease. L-Dopa is a precursor of the melanin pigments of hair and skin and of dopamine. The adrenal medulla contains an enzyme that hydroxylates the side-chain of dopamine to give norepinephrine, a transmitter that functions at the terminals of sympathetic nerves as well as with the nerves of the central nervous system (adrenergic nerves). The hormone epinephrine is formed through methylation of the amino group of norepinephrine; *S*-AdoMet is the methyl donor (Figure 9.7).

Metabolism Norepinephrine is stored in chromaffin granules, the organelles located near the synapses of adrenergic nerves. The transmitter is discharged by a process of exocytosis, a form of reverse pinocytosis. Implicit in the action of norepinephrine is its inactivation, so that after transmission has occurred, the effect of the chemical signal is rapidly terminated. The immediate termination of action is nonenzymatic; the transmitter is bound to a receptor that facilitates its removal from the interneuronal cleft. After it is taken up by the postsynaptic receptor cell, norepinephrine is oxidized by a mitochondrial enzyme, monoamine oxidase (MAO) (Figure 9.8), a degradative enzyme. The action of MAO occurs long after the termination of the adrenergic impulse. MAO is a flavin enzyme that produces 3,4-dihydroxymandelaldehyde. This aldehyde is transported to the liver, where it is further oxidized to the corresponding acid and where position 3 of the ring is *O*-methylated to form 3-methoxy-4-hydroxymandelate. The trivial name for this substance is vanillylmandelate.

Figure 9.7   Biosynthesis of the catecholamines.

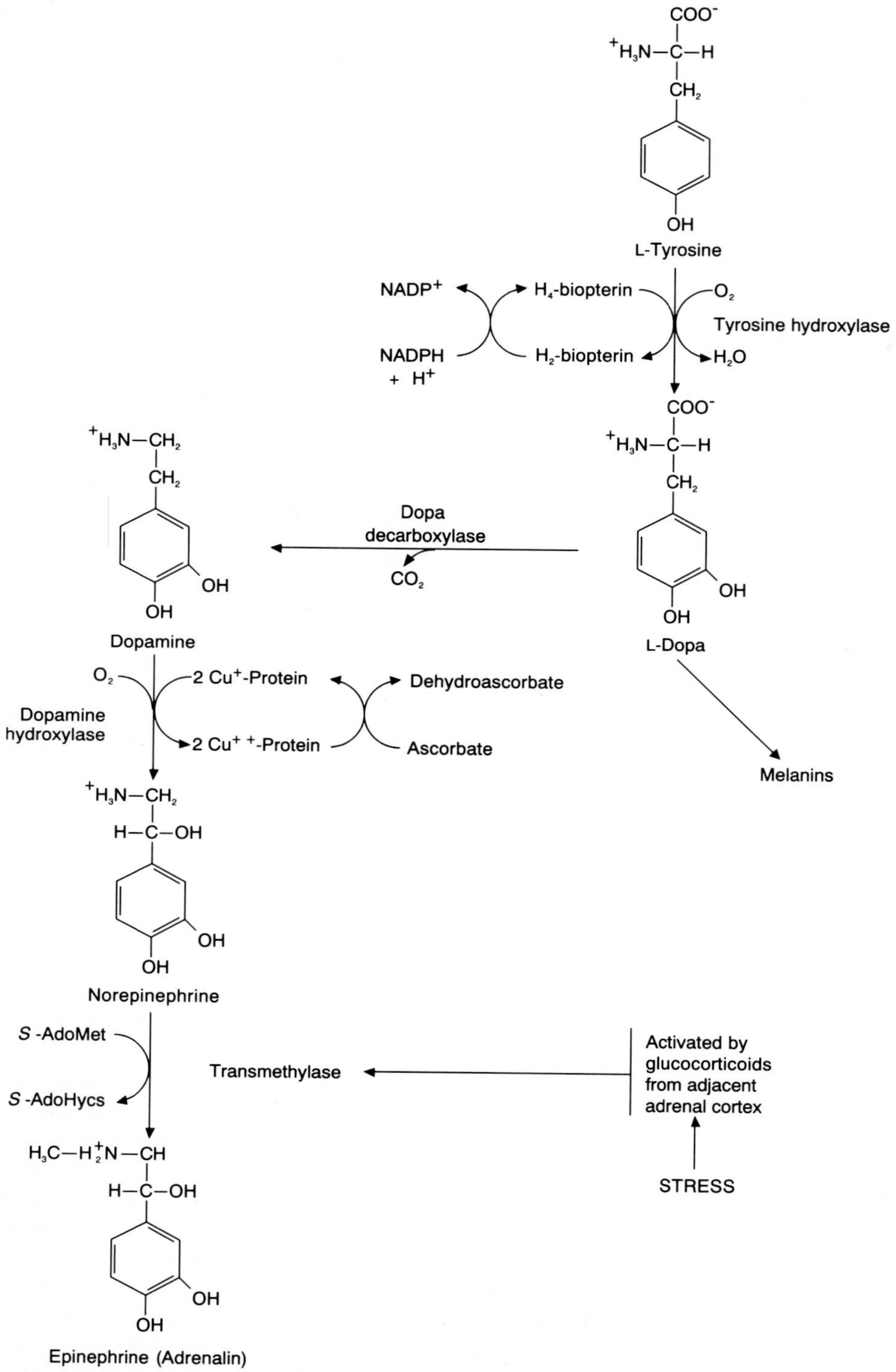

Figure 9.8   Inactivation of norepinephrine. *FAD*, Flavin adenine dinucleotide.

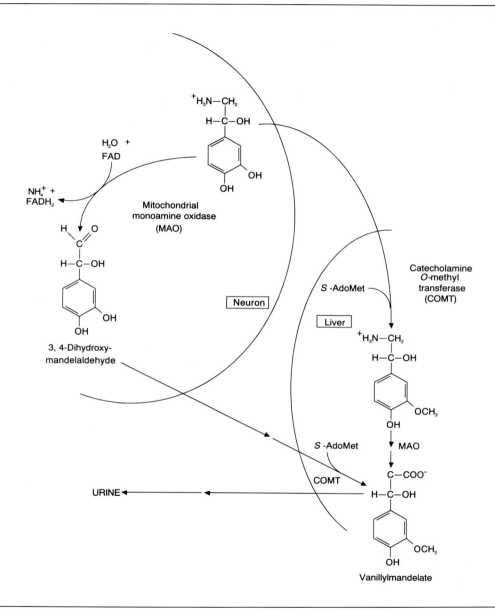

Some of the norepinephrine goes directly into the blood and is taken up by the liver, where it is *O*-methylated by the cytoplasmic enzyme catecholamine *O*-methyl transferase (COMT), which uses *S*-AdoMet as the methyl donor. The methylated epinephrine is then oxidized by liver enzymes to vanillylmandelate, the form ultimately excreted in the urine (Figure 9.8). Both MAO and COMT are widely distributed in the body; the highest concentrations are found in the liver and kidney. Norepinephrine that is rapidly released is oxidized first by MAO. In either case, as shown in Figure 9.8, the ultimate product is vanillylmandelate.

Epinephrine acts as a hormone to stimulate oxygen consumption and increase blood glucose and lactate levels through the mediation of cyclicAMP.

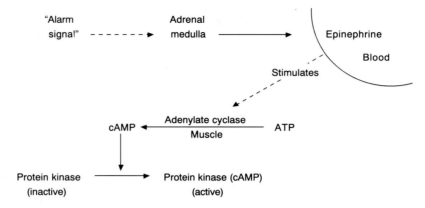

Serotonin (5-hydroxytryptamine) Serotonin is found in cells of the central nervous system, where it functions as a transmitter in relation to sleep. It is also produced in the intestinal mucosa. The drug lysergic acid dielthylamide (LSD) probably competes with serotonin, since LSD toxicity can be treated by serotonin administration. Biosynthesis of the amine, as with that of norepinephrine and epinephrine, begins with a hydroxylation reaction followed by a decarboxylation. Although phenylalanine hydroxylase isolated from liver can synthesize the intermediate 5-hydroxytryptophan from tryptophan, a specific tryptophan hydroxylase functions in the brain. Decarboxylation of 5-hydroxytryptophan yields serotonin.

γ-Aminobutyrate The decarboxylation of glutamate produces γ-aminobutyrate (GABA); it is present in very high concentrations in the brain, where it appears to inhibit synaptic transmission.

It has been suggested that several other amino acids or their derivatives serve as neurotransmitters. Probable excitatory transmitters include aspartate and glutamate, whereas glycine, as with GABA, may be an inhibitor. (For a further discussion of transmitter actions, see Chapter 17.)

Histamine Decarboxylation of histidine produces histamine, which is made and stored in the mast cells as well as in many other cells of the body. It is released in anaphylaxis and as a result of allergies, although drugs and some chemicals can also trigger release.

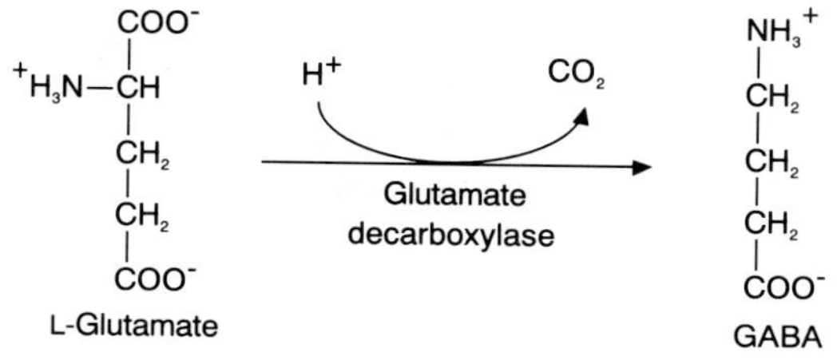

L-Glutamate ⟶ GABA (Glutamate decarboxylase; H^+, CO_2)

Circulating histamine is inactivated in liver by methylation and oxidation reactions.

The amine causes expansion of capillaries, probably by constricting the smaller veins that lead from them. This causes a local edema and an increase in the volume of the vascular bed. The resulting drop in blood pressure, if severe enough, may induce shock. *Antihistamines* are compounds that, because of their structural similarity to histamine, can prevent the physiologic changes produced by histamine released during allergic reactions. The structures of two typical antihistamines, diphenhydramine and chlorpheniramine, are shown next.

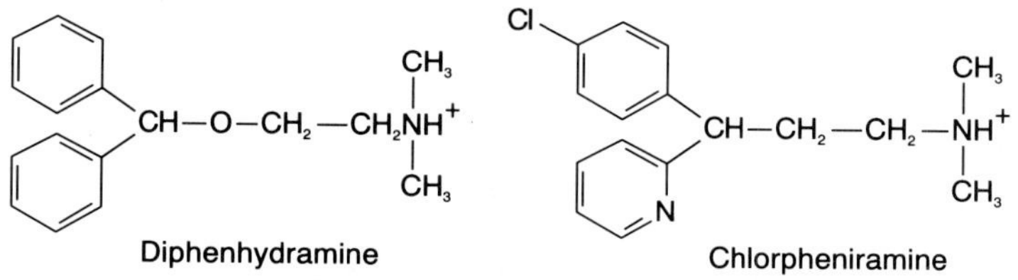

Diphenhydramine   Chlorpheniramine

Polyamines

Putrescine, or decarboxylated ornithine, is a four-carbon diamine that is the precursor of *spermidine* and *spermine,* polyamines found in association with nucleic acid and present in high concentrations in semen. Their functions are still very obscure, but their association with nucleic acids has suggested several hypothetic roles. One such role involves the first enzyme in the biosynthetic pathway, ornithine decarboxylase. This enzyme has been found associated with RNA polymerase, but still no solid evidence exists for how it might function in RNA synthesis.

One can see in Figure 9.9 that methionine (as S-AdoMet) and ornithine are the amino acid precursors of spermine. Notice that S-AdoMet does not donate methyl groups in this reaction; instead, the main carbon chain is decarboxylated and transferred, both reactions being catalyzed by the same enzyme. The other product of the reaction is 5'-methylthioadenosine (for simplicity its structure is not shown in Figure 9.9).

Table 9.3 summarizes the biosynthesis of some biogenic amines.

Creatine synthesis

Glycine and arginine are involved in the synthesis of creatine, an important constituent of muscle. Information of the rate of creatine turnover and its excretion as creatinine is useful in many clinical situations. When the muscles are at rest, almost all the creatine is phosphorylated (see Chapter 3). It is the phosphocreatine that spontaneously and nonenzymatically produces creatinine by cyclizing and splitting out inorganic phosphate (P_i).

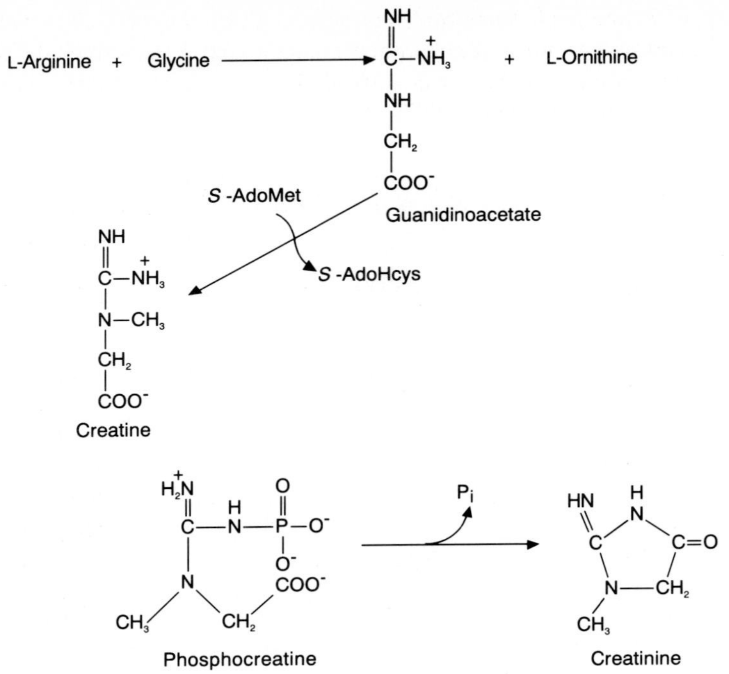

Figure 9.9   Synthesis of spermine.

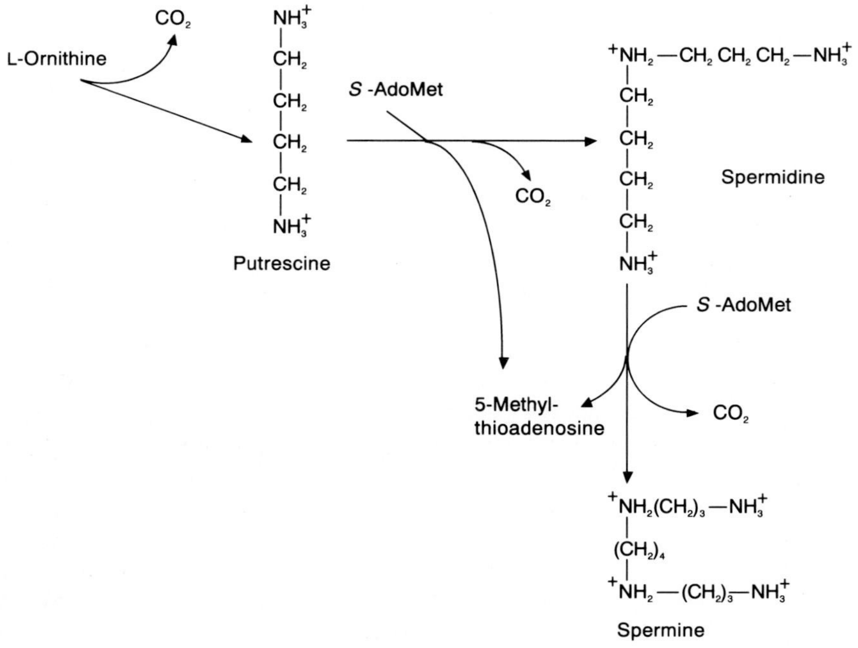

Carnitine synthesis   Carnitine is a nitrogen-containing, low molecular weight compound that serves to shuttle fatty acyl groups across mitochondrial membranes (see Chapter 10). It is synthesized from protein-bound lysine (Figure 9.10).

Table 9.3   Summary of the biosynthesis of some important amines

Amine	Amino acid precursor	Distinguishing features of pathways
Acetylcholine	Serine, methionine	S-AdoMet* is methylating agent
Norepinephrine	Tyrosine	L-Dopa is intermediate and precursor of melanins
Epinephrine	Tyrosine, methionine	S-AdoMet–dependent tyrosine amino-transferase induced by glucocorticoids
Serotonin	Tryptophan	5-Hydroxytryptophan intermediate
γ-Aminobutyrate (GABA)	Glutamate	Decarboxylation reaction
Histamine	Histidine	Decarboxylation reaction
Spermine	Ornithine, methionine	Spermidine is intermediate
Creatine	Arginine, glycine, methionine	Guanidino group transferred to glycine

* S-adenosylmethionine.

Figure 9.10   The biosynthesis of carnitine from protein-bound lysine.

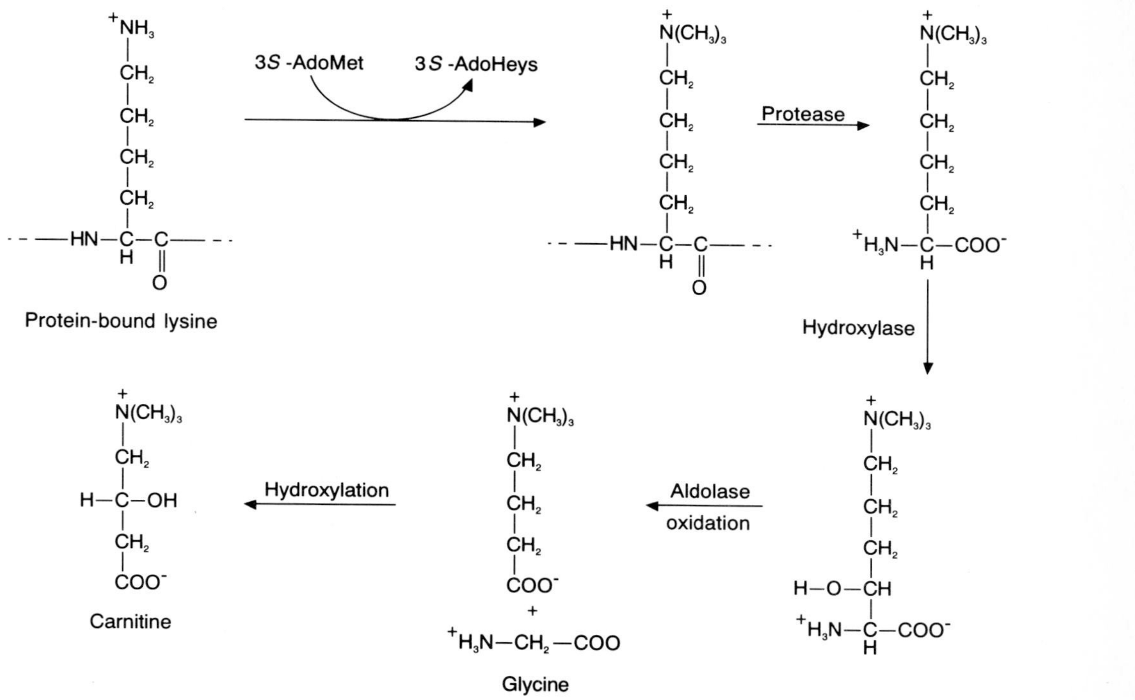

Bibliography   Bender DA: Amino acid metabolism, ed 2, New York, 1985, John Wiley & Sons, Inc.
Nyhan WL: Abnormalities in amino acid metabolism in clinical medicine, Norwalk, Conn,
    1984, Appleton-Century-Crofts.

Clinical examples

*A diamond (♦) on a case or question indicates that literature search beyond this text is necessary
for full understanding.*

Case 1   Phenylketonuria

The patient was a 2-week-old female who responded positively to a test for
phenylketonemia administered on discharge from the hospital following her birth. She
was called back for further testing. Serum phenylalanine concentration was found to be
1.82 mM (30 mg/dl), and tyrosine was 0.11 mM (2 mg/dl). The urine was positive
for phenolic acids, and $FeCl_3$ detected ketones. A diagnosis of classic phenylketonuria
(PKU) was made, and the child was maintained on a diet low in phenylalanine.

Biochemical questions   1. What enzymatic reaction(s) are defective in the patient with PKU?
2. What are the physiologic consequences of PKU, and why should it be detected as
   early as possible?
3. What is the incidence of PKU in the populations studied?
4. What is the treatment for the patient with PKU?
5. Describe the genetics of the disease and its variants.

Case discussion   **Human amino acidopathies** Human diseases that affect the metabolism of the amino
acids are almost always associated with amino acid catabolism rather than amino acid
biosynthesis. Sufficient amounts of all the amino acids are present in the diets of well-
nourished infants, so that deficiency of a nonessential amino acid, caused by a defect
in its biosynthesis, does not usually occur. On the other hand, the failure to catabolize
dietary amino acids may force the accumulation of amino acid derivatives or
intermediates to the point that they become toxic. A defect in the synthesis of tyrosine
might first be regarded as an exception to this generalization; however, PKU is also a
defect in phenylalanine catabolism. This defect reflects the usual pattern of amino
acidopathies that cause damage because of accumulated amino acid derivatives.
Generally, patients with PKU do not have deficiencies of tyrosine, which is an
essential amino acid for the phenylketonuric person. Nevertheless, it is appropriate to
consider PKU in this chapter because the affected biochemical reaction is a
biosynthetic one.

   1. Defective reactions in PKU Phenylketonuria results from an inability to convert
phenylalanine to tyrosine. The reaction affected is shown on p. 387. This
transformation has a direct requirement for two enzymes, phenylalanine hydroxylase
and dihydropteridine reductase, and PKU can result from genetic defects in either of
these enzymes. Most patients with PKU are defective in the gene for phenylalanine
hydroxylase; these have been designated types I, II, and III. Type IV is caused by a
genetic defect for dihydrobiopterin reductase. A rare variant, called type V, results
from a defect in the gene for one of the enzymes involved in the biosynthesis of the
coenzyme biopterin. Since biopterin is required for the hydroxylation of phenylalanine,
tyrosine, and tryptophan, one might predict that type V PKU would have serious
effects on other metabolic pathways. Clinical evidence supports this prediction.

Consequently the treatment for patients with type V PKU includes not only the restriction of dietary phenylalanine but also supplementation with tyrosine, L-dopa, and 5-hydroxytryptophan, which are products of the three biopterin-requiring reactions. Animal studies indicate that administered tetrahydrobiopterin does not cross the blood-brain barrier, so treatment with the biogenic amine precursors is favored.

   2. Abnormalities of the phenylketonuric patient When the conversion of phenylalanine to tyrosine is blocked, several metabolites of phenylalanine accumulate, since phenylalanine is metabolized by alternate pathways. Some of these are illustrated in Figure 9.11. The major excessive metabolite is phenylalanine itself, if the newborn is receiving the usual hospital diet. Plasma phenylalanine concentrations may range up to 0.606 mM (10 to 60 mg/dl); normal is approximately 0.061 mM (1 mg/dl). In classic PKU, plasma phenylalanine concentration is always greater than 1.21 mM (20

Figure 9.11   Catabolic products of phenylalanine that accumulate in phenylketonuria.

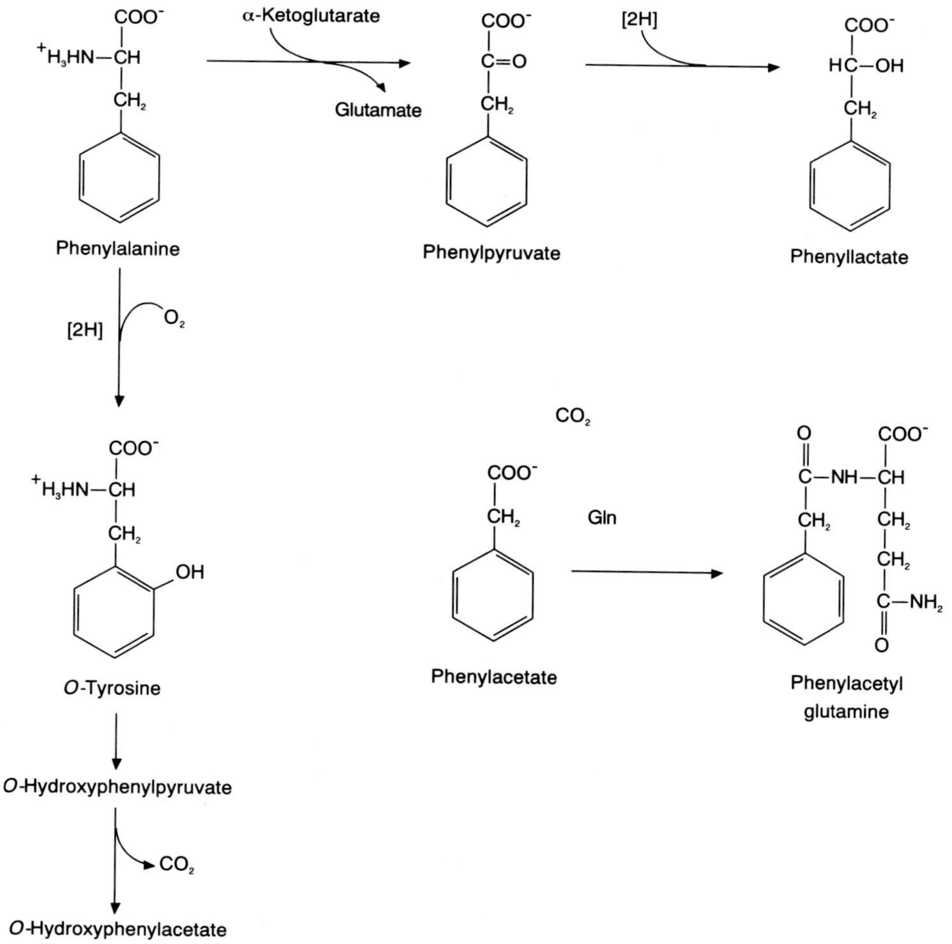

mg/dl). Values lower than 1.21 mM are associated with types IV and V PKU or with variants that have reduced, but not absent, activity of liver phenylalanine hydroxylase.

The most serious consequence of PKU is an inhibition of brain development, presumably because of the accumulated phenylalanine and its metabolites. Other reactions are also inhibited by these substances. For example, tyrosinase is inhibited by phenylalanine to the extent that pigmentation may decrease. Also, serotonin synthesis is lowered by inhibition of 5-hydroxytryptophan decarboxylase. Glutamate decarboxylase is also affected, which lowers the GABA in the brain. Decreases in all the catecholamines are also possible.

Diagnosis and early detection The effects of phenylalanine and its metabolites on brain development occur very rapidly; thus patients with classic PKU must be treated as quickly as possible. It has been estimated that a patient may lose 5 IQ units for each 10 wk that treatment is delayed. If treatment is delayed for 3 yr, mental retardation is usually very severe, and the initiation of treatment at this time has no effect on brain development. Most states and many countries have laws requiring the testing of all newborn infants for PKU.

3. Incidence PKU is the most common of all the aminoacidopathies. The incidence of PKU is approximately 1 per 15,000 births in the United States. In western Europe the incidence varies from approximately 1 per 6500 births to 1 per 16,000. Variation is related to the frequency of consanguinity.

4. Treatment of PKU The accepted treatment for PKU is to start the patient immediately on a diet low in phenylalanine. The diet is adjusted to achieve a serum phenylalanine concentration of 0.18 to 0.91 mM (3 to 15 mg/dl). Commercial dietary preparations are available. Plasma phenylalanine levels must be frequently monitored to ensure that they do not drop too low or rise too high. Phenylalanine is an essential amino acid, and proper growth requires that the diet contains a sufficient amount to achieve normal growth and development. A low-phenylalanine diet may be required up to about the age of 15 yr or until the plasma phenylalanine concentration remains below 0.73 mM (12 mg/dl) in a patient on a normal diet.

5. Genetics PKU is an autosomal recessive disease. As mentioned, the condition can result from defects in the genes for either of three enzymes: phenylalanine hydroxylase, dihydrobiopterin reductase, and dihydrobiopterin synthase, an enzyme required for the biosynthesis of dihydrobiopterin. However, more is known about the genetics of classic PKU and the enzyme phenylalanine hydroxylase than about the genes for the other two enzymes. Patients with defects in the gene for phenylalanine hydroxylase may show partial activity of the enzyme, but most have neither measurable enzymatic activity nor material that cross-reacts immunologically with antisera against the enzyme protein. Clinically, it is very difficult to measure cellular phenylalanine hydroxylase activity, since the enzyme is found only in liver.

Recently the complementary DNA for the phenylalanine hydroxylase has been cloned and expressed in bacteria so that substantial amounts of enzyme protein soon will be available for study. The mutant cloned cDNAs from 200 alleles show no gross structural deletions or rearrangements. Most of the PKU mutations must be point mutations, frame-shifts, or small additions or deletions. The gene for phenylalanine hydroxylase is a single-copy gene with no pseudogenes. The genomic DNA contains 13 exons in approximately 90 kb of DNA. It is located on chromosome 12.

Two different mutant phenylalanine hydroxylase genes have been found. One is defective in the splice donor site 3' to exon 12. This results in an mRNA with exon 12 missing and a protein that should lack 52 amino acids at the carboxyl terminal. The second mutant has a replacement at position 311, which results in a proline residue for a leucine. The molecular biology of PKU should be very useful in carrier detection and in the prenatal diagnosis of the disease. Eventually it may even lead to genetic therapy.

References

Ledley FD, DiLella AG, and Woo SLC: Molecular biology of phenylalanine hydroxylase and phenylketonuria, Trends Genet, November 1985, p 309.
Nyhan WL: Abnormalities in amino acid metabolism in clinical medicine, Norwalk, Conn, 1984, Appleton-Century-Crofts.

Case 2   Homocystinuria

A 6-year-old girl was brought to the hospital with vision problems. She was found to have a downward dislocation of the left lens. Her mother indicated that the girl's birth was normal but that she lagged in development. She was unable to crawl until 1 yr old and did not walk until 2 yr. Speaking was also delayed. She had long, thin bones; on roentgenographic examination the lower femur showed signs of osteoporosis. An older brother had similar symptoms but had been diagnosed as having Marfan's syndrome. A simple cyanide-nitroprusside test of the patient's urine was positive, suggesting homocystinuria, not Marfan's syndrome. This was confirmed by amino acid analysis of the plasma, which revealed homocystine, an abnormally high methionine level, and other sulfur-containing compounds that were derivatives of homocysteine. The patient was treated with a low-methionine diet supplemented with folic acid and pyridoxine.

Biochemical questions

1. What is the origin of the homocystine excreted in this disease?
2. What are some of the metabolic substances formed by the enzymatic reactions that use S-adenosylmethionine as the methylating agent?
3. What are some causes of homocystinuria in humans?
4. How would one test for a deficiency of cystationine β-synthase in this patient?
5. Explain why pyridoxine is useful in the treatment of some patients with homocystinuria.
6. What effect would a diet low in folate have on this patient?
7. What might account for the homocystinuria of an apparently normal infant (not in this case) with severe megaloblastic anemia and who was exclusively breast-fed by a strict vegetarian mother?
8. Describe the genetics of homocystinuria (cystathionine β-synthase deficiency).

Case discussion

Homocystinuria, caused by a deficiency of cystationine β-synthase, results in many of the same clinical symptoms as those in Marfan's syndrome. The metabolic defect in Marfan's syndrome is not known, but it probably involves a defect in collagen metabolism. In this case the biochemical tests for homocystine and homocysteine clearly ruled out Marfan's syndrome and suggested a defective synthesis of cystathionine, the intermediate in the synthesis of cysteine from methionine.

1. Origin of homocystine The origin of the disulfide homocystine is homocysteine. Homocysteine is a product of the many methylating enzymes that use S-adenosylmethionine as their substrates. A summary of the reactions pertinent to this case follows (THF is tetrahydrofolate).

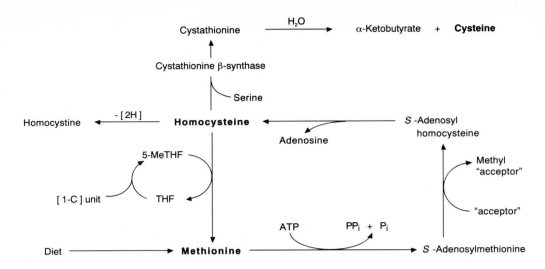

The oxidation of homocysteine to homocystine is probably nonenzymatic. Relatively more homocystine is found in urine than in the blood of individuals defective in the enzyme cystathionine β-synthase.

2. Methylations by *S*-adenosylmethionine Many substances are methylated by specific enzymes that use *S*-adenosylmethionine as the methylating agent. These include certain bases on RNA and DNA, the amino terminal groups of many proteins, and precursors of melatonin, anserine, creatine, epinephrine, phosphatidylcholine, carnitine, and methyl cobalamin.

3. Causes of homocystinuria Much of the methionine in cells is used for methylation reactions, so it must be resynthesized through homocysteine. Anything that compromises the utilization of homocysteine for either the resynthesis of methionine or the synthesis of cystathionine could potentially cause homocystinuria. The symptoms of homocystinuria, or homocystinemia, can result from a deficiency of cystathionine β-synthase, the administration of 6-azauridine, artifactually from bacterial action on cystathionine in the urine, or from an impairment in the synthesis of methionine from homocysteine catalyzed by 5-methyl THF:homocysteine methyltransferase. The anticancer agent 6-azauridine is also a potent inhibitor of pyridoxal phosphate-dependent enzymes, so inhibition of cystathionine β-synthase by this agent can cause a rise in plasma methionine and homocysteine levels. The artifactual synthesis of homocystine in urine can be ruled out by measuring the plasma concentration of homocystine. The 5-methyl THF:homocysteine methyltransferase can be affected by various causes; for example, by genetic defects in the resynthesis of the required folate and cobalamin coenzymes, by defective absorption of either folate or cobalamin, or by a nutritional lack of either B vitamin.

4. Cystathionine β-synthase deficiency A defect in this enzyme is the usual cause of homocystinuria, and it probably accounts for the symptoms observed in this patient. Proof would require that the enzyme activity be measured in extracts of cultured skin fibroblasts. The defective enzyme is genetically heterogeneous. Some patients show partial enzyme activity, but others may have none. Normally this enzyme consists of two identical subunits of 48 kDa each, both containing tightly bound pyridoxal phosphate. Addition of pyridoxal phosphate to the in vitro enzyme assay sometimes stimulates activity. Detection of residual activity strongly correlates with the effectiveness of pyridoxine therapy.

5. Pyridoxine therapy The homocystinuria of many patients who show partial activity of cystathionine β-synthase can be greatly reduced or eliminated by treatment with relatively large amounts of pyridoxine, or vitamin B_6. Improvement in behavior and IQ has been reported even after the appearance of clinical abnormalities. Other dietary measures recommend diets low in methionine. Meals should be smaller and more frequent to prevent a large methionine load. Choline administration has been somewhat successful in sparing the amount of methionine needed for its synthesis, and often additional vitamin B_{12} and folic acid are recommended. All these treatments emphasize the importance of the homocysteine interrelationships shown in the previous reactions.

6. Low-folate diets A diet low in folic acid might seriously affect the reaction catalyzed by 5-methyl THF:homocysteine methyltransferase. This enzyme requires both folate and cobalamin coenzymes and can be impaired by a deficiency of either vitamin. Such deficiencies are serious in normal people but are especially troublesome for those with a defect in cystathionine β-synthase. Some patients respond to pyridoxine therapy only after a folate deficiency has been corrected.

7. Homocystinuria secondary to a vitamin B_{12} deficiency Homocystinuria has been reported in an infant with no genetic abnormality but who had been exclusively breast-fed by a strict vegetarian mother. Vegetarians who eat no eggs or dairy products, and who scrupulously wash fruits and vegetables before eating them, may develop a deficiency of cobalamin that can result in permanent neurologic damage. The homocystinuria in this breast-fed infant was not genetic but was caused by a deficiency in vitamin B_{12}. As a result, the infant could not synthesize the cobalamin coenzyme required for the synthesis of methionine from homocysteine, and the excessive homocysteine was oxidized to homocystine and excreted.

8. Genetics of cystathionine β-synthase deficiency Cystathionine β-synthase deficiency is caused by defects in a gene on chromosome 21. The resulting homocystinuria is autosomal recessive, with an estimated prevalence in newborns of 1:45,000. Genetically the disease is very heterogenous, which probably reflects the existence of several mutable loci and suggests that clinically homozygous patients may carry gene pairs with mutations at different alleles.

References

Kraus JP et al: Cloning and screening with nanogram amounts of immunopurified mRNAs: cDNA cloning and chromosomal mapping of cystathionine β-synthase and the β subunit of propionyl-CoA carboxylase, Proc Natl Acad Sci USA 83:2047, 1986.

Nyhan WL: Abnormalities in amino acid metabolism in clinical medicine, Norwalk, Conn, 1984, Appleton-Century-Crofts.

Scriver CR et al, editors: The metabolic basis of inherited disease, ed 6, New York, 1989, McGraw-Hill Book Co.

Skovby F: Homocystinuria: clinical, biochemical and genetic aspects of cystathionine β-synthase and its deficiency in man, Acta Paediatr Scand (Suppl) 321:1, 1985.

Case 3

Autonomic failure

Evaluation of a 33-year-old woman for orthostatic hypotension (low blood pressure while standing) revealed the following:

Catecholamine	Plasma concentration (pg/ml)	
	Supine	Upright
Norepinephrine		
Patient	12	15
Normal person	206	440
Dopamine (free)		
Patient	248	583
Normal person	53	59

Continued.

Plasma epinephrine concentration was normal. Urinary vanillylmandelic acid level was slightly below normal, but urinary homovanillic acid, a dopamine metabolite, was significantly elevated. Serum dopamine β-hydroxylase was undetectable. The patient's condition was first noted when she was a child.

Biochemical questions

1. What would be the biochemical consequences of a genetically defective dopamine β-hydroxylase?
2. What would be the rationale if this patient were treated with megadoses of vitamin C?
◆ 3. What is homovanillic acid, and how is it derived from dopamine?

References

Biaggioni I et al: Dopamine β-hydroxylase deficiency, N Engl J Med 317:1415, 1987.
Robertson D et al: Isolated failure of autonomic noradrenergic neurotransmission: evidence for impaired β-hydroxylation of dopamine, N Engl J Med 314:1494, 1986.

Case 4   Glomerulonephritis

A 30-year-old man was admitted to hospital with headache, pain in the flanks, anorexia, and the passage of red-colored urine. Examination revealed a mild edema around the eyes. The red-colored urine contained large numbers of hyaline and red cell casts. The electrocardiogram indicated hypertrophy of the left ventricle, which was confirmed by chest roentgenography. Hemorrhages and exudates were observed around constricted retinal arteries; the optic discs were normal. A mild edema was seen around the eyes. On the basis of these observations and the results of laboratory tests, a diagnosis of acute glomerulonephritis was made.

The patient received a protein-free diet with reduced sodium and water (1 L/day). His blood pressure was lowered by administration of antihypertensive drugs, and after 8 days the blood urea nitrogen (BUN) returned to normal.

Biochemical questions

1. Why was the dietary protein lowered? What levels of lipid and carbohydrate should be present in this diet? Would you recommend an all-vegetable diet? Explain.
2. If amino acids were used to replace the protein in the diet, which amino acids might be used? Why? Would this be practical?
3. If this patient's condition were left untreated, would you expect the development of acidosis or alkalosis? Explain.
◆ 4. Renal disease–associated hypertension is often alleviated with a diuretic and α-methyldopa. How does α-methyldopa function to do this?

Reference

Glasscock RJ: Pathophysiology of acute glomerulonephritis, Hosp Pract 23:163, 1988.

Additional questions and problems

1. Extracts of a liver biopsy from a child with PKU showed only a modest decrease in phenylalanine hydroxylase activity but no detectable dihydrobiopterin reductase activity. How does lack of this enzyme cause PKU? What effect would this have on other reactions that use the same coenzyme?
2. Another form of PKU shows a decrease in neither phenylalanine hydroxylase nor dihydrobiopterin reductase. Suggest an explanation for the failure of such a patient to hydroxylate phenylalanine.
3. Why are amino acid analogues of limited usefulness in the treatment of cancer?

4. What neurotransmitters are synthesized from tyrosine?
5. Considering the structural similarity of diphenhydramine (see p. 395) and epinephrine, what type of side effects might be produced by the antihistamine?
6. What is the origin of urinary vanillylmandelate?
◆ 7. What are the mechanisms of action of antihistamines?
◆ 8. How would one show that a compound is a neurotransmitter?
◆ 9. What are the mechanisms for termination of a catecholamine stimulus?

Multiple choice problems

1. PKU type V is a rare variant form of the disease caused by a genetic defect in dihydrobiopterin synthase; however, PKU can also result from primary genetic defects in:
   1. Phenylalanine hydroxylase.
   2. Tyrosine aminotransferase.
   3. Dihydrobiopterin reductase.
   4. Synthesis of melanin pigments.
      The best answer is:
      A. 1, 2, and 3.
      B. 1 and 3.
      C. 2 and 4.
      D. 4 only.
      E. 1, 2, 3, and 4.

2. In type V PKU one would expect that any enzymatic reaction that required the biopterin coenzyme would be compromised. Consequently the patient might synthesize lower amounts of:
   1. Tyrosine.
   2. 5-Hydroxytryptophan.
   3. Dopa.
   4. GABA.
      The best answer is:
      A. 1, 2, and 3.
      B. 1 and 3.
      C. 2 and 4.
      D. 4 only.
      E. 1, 2, 3, and 4.

3. Tetrahydrobiopterin resembles the coenzyme derived from:
   A. Vitamin B_6.
   B. Niacin.
   C. Folic acid.
   D. Vitamin B_{12}.
   E. Pantothenic acid.

4. Examination of the urine from a type V PKU patient shows *lower* amounts of:
   A. Phenylpyruvate.
   B. Urea.
   C. Vanillylmandelic acid.
   D. Phenylacetate.
   E. Creatinine.

5. A low-phenylalanine diet is used to treat patients with classic PKU. The diet, however, must provide sufficient but not excessive amounts of the amino acids that are essential to normal humans, and it also should contain:
   A. Proline.
   B. Glutamic acid.
   C. Glycine.
   D. Cysteine.
   E. Tyrosine.

Ketotic hyperglycinemia is a secondary manifestation of propionic acidemia, but the nonketotic form of the disease is caused by a genetic defect in glycine metabolism. Problems 6 to 10 refer to this statement.

6. Glycine, the simplest of the amino acids, is used by humans for the synthesis of:
   1. Glutathione.
   2. Creatine.
   3. Heme.
   4. Threonine.
      The best answer is:
      A. 1, 2, and 3.
      B. 1 and 3.
      C. 2 and 4.
      D. 4 only.
      E. 1, 2, 3, and 4.

7. Glycine is also metabolized by a cleavage reaction that converts it to CO_2, NH_4^+, and 5,10-methylene THF. By coupling the product of this reaction with another enzymatic reaction, two equivalents of glycine can be converted to one equivalent of CO_2 and one equivalent of:
   A. Alanine.
   B. Serine.
   C. Aspartic acid.
   D. Acetate.
   E. γ-Aminobutyric acid.

8. The glycine cleavage reaction just mentioned is the major pathway for the catabolism of glycine in humans. The enzyme that catalyzes this reaction is deficient in *nonketotic* hyperglycinemia. *Ketotic* hyperglycinemia is a secondary manifestation of propionic acidemia or other disorders of organic acid metabolism (see Chapter 5). To distinguish nonketotic from ketotic hyperglycinemia, one should measure the serum concentration(s) of:
   1. Methylmalonic acid.
   2. Pyruvic acid.
   3. Propionic acid.
   4. THF.
      The best answer is:
      A. 1, 2, and 3.
      B. 1 and 3.
      C. 2 and 4.
      D. 4 only.
      E. 1, 2, 3, and 4.

9. The glycine cleavage reaction is catalyzed by an enzyme complex consisting of four enzymes and four different coenzymes: pyridoxal phosphate, lipoic acid, tetrahydrofolic acid, and NAD$^+$. If vitamin therapy were tried, the supplement should include:
   1. Thiamine.
   2. Pyridoxine.
   3. Riboflavin.
   4. Folic acid.
      The best answer is:
      A. 1, 2, and 3.
      B. 1 and 3.
      C. 2 and 4.
      D. 4 only.
      E. 1, 2, 3, and 4.

10. Patients with ketotic hyperglycinemia have serious damage to the central nervous system and may die within a few days of birth. Since glycine is an inhibitory neurotransmitter and benzodiazepines compete for glycine receptors, treatment with diazepam (Valium), choline, and a B vitamin has successfully controlled some seizures. Choline was used as part of the treatment because it:
    A. Is a neurotransmitter.
    B. Is esterified to glycine and eliminated in the urine.
    C. Lowers the amount of 5,10-methylene THF needed for choline synthesis.
    D. Resembles diazepam and blocks glycine receptors.
    E. Blocks glycine transport across the blood-brain barrier.

Chapter 10

Lipid metabolism

Objectives

1 To describe the structure of certain lipid components of human tissues
2 To discuss the major pathways for lipid biosynthesis, catabolism, and storage
3 To describe the biochemistry of diseases associated with abnormalities in lipid metabolism

Lipids are organic compounds that are poorly soluble in water but readily dissolve in organic solvents such as benzene or chloroform. In the human body, lipids function as a metabolic fuel, as storage and transport forms of energy, and as structural components of cell membranes. The most common disease involving lipids is obesity, the excessive accumulation of adipose tissue in the body. Defects in lipid metabolism also occur in diabetes. Among the less frequent abnormalities are the lipid storage diseases, or sphingolipidoses, which usually appear in infancy or early childhood. This chapter describes the chemistry and metabolism of the five important types of lipids, as listed in Table 10.1. Additional lipid substances that play an important role in normal bodily function and certain diseases, such as fat-soluble vitamins, cholesterol, bile acids, steroid hormones, prostaglandins, and lipoxygenase products, are discussed in other chapters. Furthermore, Chapter 16 is devoted to plasma lipid transport because abnormalities in this process, the hyperlipidemias, are prevalent diseases in humans.

Fatty acids

Fatty acids are compounds represented by the chemical formula R—COOH, where R stands for an alkyl chain composed of carbon and hydrogen atoms. One method of classifying fatty acids is according to their chain length, that is, the number of carbon atoms that they contain. An arbitrary but widely accepted classification is listed in Table 10.2. Most of the fatty acids present in the blood and tissues of humans are of the long-chain variety. The names and structural properties of the most important fatty acids in mammalian systems are given in Table 10.3. Almost all the fatty acids in natural products contain an *even* number of carbon atoms.

Ionization

The pK_a of the fatty acid carboxyl group is about 4.8. Under ordinary conditions the pH of the plasma is 7.4, and the pH of the intracellular fluid is about 7.0. Therefore almost all (99.9%) of the free fatty acid molecules present in body fluids are ionized; that is, the fatty acid is present as an *anion*. This imparts detergent-like properties to the long-chain fatty acids when they exist in unesterified form in biologic fluids, since the ionized carboxyl group interacts with aqueous media, whereas the hydrocarbon tail seeks a nonpolar environment.

$$RCOOH \rightleftharpoons RCOO^- + H^+$$

Saturation

Fatty acids are either saturated or unsaturated. In a saturated fatty acid the alkyl chain does not contain any double bonds.

$$
\begin{array}{ccccc}
 & H & H & H & H \\
-&C&-C&-C&-C- \\
 & H & H & H & H
\end{array}
$$

Table 10.1   Classification and Functions of Lipids*

Lipid	Function
1. Fatty acids	Metabolic fuel, building blocks for other lipids
2. Acylglycerols	Fatty acid storage and transport, metabolic intermediates and regulation
3. Phospholipids	Membrane structure, membrane signal transduction, storage of arachidonic acid
4. Sphingolipids	Membrane structure, surface antigens
5. Ketone bodies	Metabolic fuel

*Additional lipids: Fat-soluble vitamins (Chapter 1); cholesterol and bile acids (Chapter 11); steroid hormones (Chapter 19); prostaglandins and lipoxygenase products (Chapter 12).

Unsaturated fatty acids contain one or more double bonds. Those that contain one unsaturated bond are known as *monounsaturated* fatty acids; those that contain two or more unsaturated bonds are known as *polyunsaturated*.

$$
\begin{array}{cc}
\begin{array}{c}
\text{H H H H} \\
\text{—C–C=C–C—} \\
\text{H} \qquad \text{H}
\end{array}
&
\begin{array}{c}
\text{H H H H H H H} \\
\text{—C–C=C–C–C=C–C—} \\
\text{H} \qquad \text{H} \qquad \text{H}
\end{array}
\\
\text{A monounsaturated acid} & \text{A polyunsaturated acid}
\end{array}
$$

Mammals and plants contain both monounsaturated and polyunsaturated fatty acids, whereas all the fatty acids containing double bonds that are present in bacteria are monounsaturated. Plant and fish fats contain more polyunsaturated fatty acids than animal fats. The double bonds in a polyunsaturated fatty acid are neither adjacent nor conjugated, since this would make the structure too easily oxidizable when exposed to an environment containing oxygen. Rather, the double bonds are three carbons apart; this provides somewhat greater protection against oxidation. The oxidation of unsaturated bonds in fatty acids when they are exposed to oxygen in the environment is referred to as either *autooxidation* or *peroxidation*. Rancid fats are those that contain an appreciable amount of peroxidized fatty acid.

General nomenclature

The carbon atoms of an acid are numbered (or lettered) either from the *carboxyl group* (Δ numbering or Greek lettering system) or from the carbon atom farthest removed from the carboxyl group (n or ω numbering system) as follows:

ω-terminus	CH_3—CH_2—CH_2—CH_2—CH_2—CH_2—CH_2—CH_2—CH_2—COOH	Carboxyl terminus

C numbering	10	9	8	7	6	5	4	3	2	1
n or ω numbering	1	2	3	4	5	6	7	8	9	10
Letter designation	ω	ω—1				δ	γ	β	α	

Greek letters also are used to indicate the various carbon atoms. The α-carbon is adjacent to the carboxyl group, and the ω-carbon atom is the one farthest from the carboxyl group.

Fatty acids are abbreviated as illustrated in Table 10.4. Thus palmitoleic acid is abbreviated as either 9-16:1 or 16:1Δ9. The number 9 in this classification system signifies the position of the double bond relative to the carboxyl end. For example, in 16:1Δ9 the single double bond is nine carbon atoms away from the carboxyl group; that is, it is between carbon atoms 9 and 10, counting the carboxyl carbon atom as carbon atom number 1. In the n or ω numbering system, palmitoleic acid is referred to as 16:1n-7 or

Table 10.2   Fatty Acid Classification According to Chain Length

Type	Number of carbon atoms
Short chain	2 to 4
Medium chain	6 to 10
Long chain	12 to 26

Table 10.3   Important Fatty Acids in Mammalian Tissues

Descriptive name	Systematic name	Carbon atoms	Double bonds	Position of double bonds*	Unsaturated fatty acid class†
Acetic		2	0		
Lauric	Dodecanoic	12	0		
Myristic	Tetradecanoic	14	0		
Palmitic	Hexadecanoic	16	0		
Palmitoleic	Hexadecenoic	16	1	9	ω-7
Stearic	Octadecanoic	18	0		
Oleic	Octadecenoic	18	1	9	ω-9
Linoleic	Octadecadienoic	18	2	9, 12	ω-6
Linolenic	Octadecatrienoic	18	3	9, 12, 15	ω-3
γ-Homolinolenic	Eicosatrienoic	20	3	8, 11, 14	ω-6
Arachidonic	Eicosatetraenoic	20	4	5, 8, 11, 14	ω-6
EPA‡	Eicosapentaenoic	20	5	5, 8, 11, 14, 17	ω-3
DHA‡	Docosahexaenoic	22	6	4, 7, 10, 13, 16, 19	ω-3

*Position of the one or more double bonds listed according to the Δ numbering system. In this numbering system, only the first carbon of the pair is listed; that is, 9 means position 9, 10 starting from the carboxyl end.
†In the ω numbering system, only the first double bond from the methyl end is listed and, as just stated, only the first carbon of the pair is written.
‡The commonly used descriptive name is an abbreviation.

$16:1\omega$-7. This indicates that the acid has 16 carbon atoms and one unsaturated bond that is located seven carbon atoms away from the ω-carbon atom. All of these numbering systems are in current usage, and therefore it is necessary to become familiar with each of them.

The unsaturated fatty acids are divided into four classes.

Class	Parent fatty acid	Structure
ω-7	Palmitoleic acid	9-16:1
ω-9	Oleic acid	9-18:1
ω-6	Linoleic acid	9,12-18:2
ω-3	Linolenic acid	9,12,15-18:3

Each class is made up of a family of fatty acids, and all members of that family can be synthesized biologically form the parent fatty acid. For example, arachidonic acid ($20:4\omega$-6) is synthesized from the parent of the ω-6 class, linoleic acid ($18:2\omega$-6). A fatty acid of one class, however, cannot be converted biologically to another class; that is, no member of the oleic acid class (ω-9) can be converted to either linoleic acid or any other member of the ω-6 class.

Table 10.4   Fatty Acid Nomenclature

Fatty acid structure	Descriptive name	Abbreviation system			
		Numeric	Δ	n	ω
$CH_3 - (CH_2)_{14} - COOH$	Palmitic acid	16:0			
$CH_3 - (CH_2)_5 - CH = CH$ $- (CH_2)_7 - COOH$	Palmitoleic acid	9-16:1	16:1Δ9	16:1n-7	16:1ω-7
$CH_3 - (CH_2)_4 - CH = CH - CH_2 -$ $CH = CH - (CH_2)_7 - COOH$	Linoleic acid	9,12-18:2	18:2Δ9,12	18:2n-6	18:2ω-6

The hydrocarbon chain of a saturated fatty acid usually exists in an extended form, since this linear, flexible conformation is that state which possesses a minimal energy. By contrast, unsaturated fatty acids have rigid bends in their hydrocarbon chains because the double bonds do not rotate, and a 30-degree angulation in the chain is produced by each of the *cis* double bonds that are present. In general, human cells contain at least twice as much unsaturated as saturated fatty acids, but the composition varies considerably among the different tissues and depends to some extent on the type of fat contained in the diet.

Composition of dietary fats

Fat from natural sources is composed of a mixture of fatty acids. Most of these fatty acids are present in triacylglycerols. Table 10.5 lists the fatty acid composition of some common dietary fats and oils. Each one is made up of a combination of saturated and unsaturated acids. In general, the fats derived from animal sources, tallow, butter, and lard, are more saturated than those derived from plants. This is not an absolute rule, however, since coconut oil is made up mostly of 12- and 14-carbon atom saturated fatty acids and contains only 5% monounsaturated and 1% to 2% polyunsaturated fatty acids. Although considerable individual variation exists, the animal fats have about 40% to 60% saturated fatty acids, 30% to 50% monounsaturated fatty acids, and relatively little poly-unsaturated fatty acids. By contrast, the plant oils have approximately 10% to 20% saturated fatty acids and 80% to 90% unsaturated fatty acids. The composition of the unsaturated fatty acid varies, ranging from 79% oleic acid, a monounsaturate, in olive oil to 76% linoleic acid, a polyunsaturate, in safflower oil. Little of the linolenic (ω-3) class of fatty acids is present in the typically used dietary oils, but the n-3 or ω-3 polyunsaturates comprise 25% to 35% of the fatty acid contained in certain fish oils.

Clinical comment

Dietary polyunsaturated fat The ordinary dietary oils that are rich in polyunsaturated fatty acids contain predominantly linoleic acid, the parent fatty acid of the ω-6 class. Diets that are high in polyunsaturated fats are employed therapeutically for the treatment of hypercholesterolemia. Corn or safflower oil often is used in these therapeutic diets and in commercial food preparations that are advertised as being high in polyunsaturates.

The ω-3 class of polyunsaturated fatty acids is produced by vegetation that grows in cold water. Fish that feed on these organisms contain large amounts of ω-3 polyunsaturates in their tissues. Fish oils, such as menhaden and salmon oil, are rich in ω-3 polyunsatrates, especially eicosapentaenoic acid, 20:5ω-3. There is now some interest in marine oils that are rich in ω-3 polyunsaturates instead of the terrestrial plant oils that contain ω-6 polyunsaturates for therapeutic diets. Not only are the ω-3 polyunsaturates effective in lowering the plasma triacylglycerol concentration, but they also appear to protect against thrombosis and certain inflammatory diseases.

Table 10.5   Fatty Acid Composition of Some Edible Fats*

		Composition (%)								
Fatty acid	Structure	Tallow	Butter	Lard	Coconut oil	Olive oil	Cotton-seed oil	Corn oil	Soybean oil	Safflower oil
Lauric	12:0		3		54					
Myristic	14:0	3	11	2	18					
Palmitic	16:0	26	31	25	8	11	20	10	10	5
Palmitoleic	16:1ω-7	3	3	3						
Stearic	18:0	25	14	15	2	2	2	2	4	2
Oleic	18:1ω-9	36	30	45	5	79	18	31	24	17
Linoleic	18:2ω-6	2	2	9	1	7	60	56	54	76
Others		5	6	1	12	1		1	8	

*Where no numeric value is listed, the amount is <0.5% of the total fatty acids.

Isomerism

The presence of double bonds in fatty acids restricts rotation of the alkyl chain. This allows for isomerism around the double bond, recognized as a *cis* or *trans* configuration.

Cis   *Trans*

Naturally occurring unsaturated fatty acids in mammals are all the *cis* configuration.

Trans fatty acids

Even though *trans* fatty acids are not natural products, they are present in the tissues of people who eat Western culture diets. They arise as the result of catalytic hydrogenation of fats, a process used in the manufacture of certain commercially prepared foods such as margarine and peanut butter. Hydrogenation is used to produce some solidification of these fat-containing foods to make them easier to handle. In the hydrogenation process, some of the naturally occurring *cis* double bonds are converted to the *trans* configuration. When ingested by humans, *trans* fatty acids are either oxidized or incorporated into structural lipids. Although their presence in the body in the amounts ordinarily eaten does not appear to produce any ill effects, the question of whether they might be injurious in a subtle way has not yet been resolved.

Other forms of isomerism

Another form of isomerism that can occur in unsaturated fatty acids concerns the *position* of the double bond in the alkyl chain. For example, the double bond can be present in position 9 in one monounsaturated acid; in another monounsaturated acid with the same chemical formula the double bond may be in position 11. The mammalian organism is quite specific in its requirement for certain positional isomers.

Another kind of isomerism that can occur in fatty acids results from the presence of a branch in the alkyl chains.

C—C—C—C—C—C—COOH
Straight chain

C—C—C—C—C—COOH
Branched chain

The fatty acids that are normally present in mammalian tissue are almost all the straight-chain variety. An exception is the sebaceous glands, which normally synthesize considerable amounts of branched-chain fatty acids. These fatty acids are found in the lipid secretion of the skin. Branched-chain fatty acids also may be ingested in certain foods, for example, *phytanic acid* in butter. *Phytol,* a branched-chain alcohol contained in green vegetables, is converted to phytanic acid after ingestion. The body is able to degrade branched-chain acids, and they do not accumulate to any appreciable extent under ordinary conditions. However, in at least one disease, *Refsum's disease,* serious neurologic defects result from the accumulation of phytanic acid in the nervous system.

$$CH_3-\underset{\underset{CH_3}{|}}{CH}-CH_2-CH_2-CH_2-\underset{\underset{CH_3}{|}}{CH}-CH_2-CH_2-CH_2-\underset{\underset{CH_3}{|}}{CH}-CH_2-CH_2-CH_2-\underset{\underset{CH_3}{|}}{CH}-CH_2-COOH$$

Phytanic acid

This 20-carbon atom acid has a 16-carbon atom chain with methyl groups at C_3, C_7, C_{11}, and C_{15}.

Hydroxy and hydroperoxy fatty acids The central nervous system contains some fatty acids that have a hydroxyl group attached to the alkyl chain. As exemplified by *cerebronic acid,* the hydroxyl group in these acids is attached to the α-carbon atom.

$$CH_3-(CH_2)_{21}-\underset{\underset{H}{|}}{\overset{\overset{OH}{|}}{C}}-COOH \qquad CH_3-(CH_2)_5-\underset{\underset{H}{|}}{\overset{\overset{OH}{|}}{C}}-CH_2-CH=CH-(CH_2)_7-COOH$$

Cerebronic acid   Ricinoleic acid

Other hydroxy fatty acids that occur in natural products can have the hydroxyl group attached at other locations. In *ricinoleic acid,* the main fatty acid in castor oil, the hydroxyl group is attached to C_{12}. Castor oil is a cathartic, and this effect is somehow produced by its ricinoleic acid content.

Hydroxy and hydroperoxy derivatives of the 20-carbon-atom polyunsaturated fatty acids are formed by several tissues, including leukocytes, macrophages, and platelets. For the most part, these are formed from arachidonic acid, 20:4ω-6. Some representative examples are 5-hydroxyeicosatetraenoic acid (5-HETE), 5-hydroperoxyeicosatetraenoic acid (5-HPETE), and 12-hydroxyeicosatetraenoic acid (12-HETE). Neutrophils produce 5-HETE and 5-HPETE; macrophages and platelets form 12-HETE. These and related isomeric forms of hydroxy and hydroperoxy derivatives of arachidonic acid are discussed further in Chapter 12.

Essential fatty acids

In addition to obtaining fat from the diet, humans can biosynthesize many fatty acids, including the saturated and monounsaturated varieties. However, mammals cannot synthesize all the necessary types of polyunsaturated fatty acids. Those polyunsaturated fatty acids that cannot by synthesized must be obtained from the diet—they are termed *essential fatty acids*. Two of the four classes of unsaturated fatty acids, the linoleic (ω-6) and linolenic (ω-3) acid classes, cannot be biosynthesized by mammals but are synthesized by plants. One of these, linoleic acid, is essential for health, probably because its longer and more highly unsaturated derivatives are precursors of certain prostaglandins, hydroxy and hydroperoxy derivatives of 20-carbon-atom polyunsaturated fatty acids, and the leukotrienes. As long as adequate amounts of linoleic acid are available, the mammals can synthesize other required members of the ω-6 class.

Essential fatty acid deficiency In animals a disease state known as *essential fatty acid deficiency* occurs if essential fatty acids are exluded from the diet for long periods. The

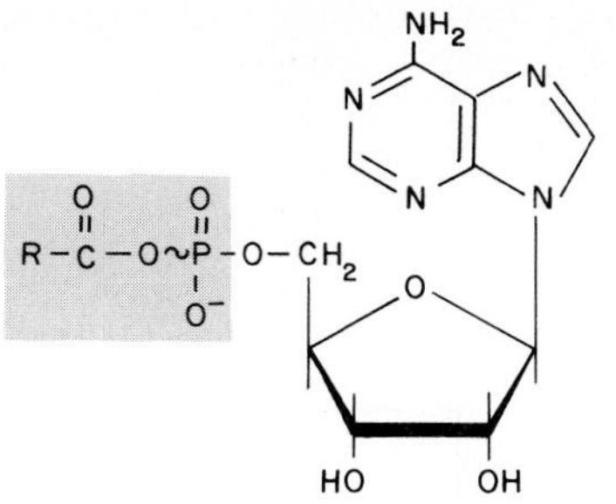

Acyladenylate

A guanosine triphosphate– (GTP) linked acyl CoA synthase is present in the matrix of the mitochondria. Unlike the ATP-linked synthase, the products of the GTP reaction are guanosine diphosphate (GDP) and P_i, not the mononucleotide and pyrophosphate.

$$FA + CoASH + GTP \rightleftharpoons FA{\sim}SCoA + GDP + P_i$$

The GTP-linked synthase activates any free fatty acid that is generated inside the mitochondria.

Acylcarnitine formation

The long-chain acyl CoA formed in the synthase reaction cannot penetrate through the inner mitochondrial membrane to the site of the fatty acid β-oxidation enzyme system. To cross this barrier, the acyl group is transesterified from CoASH to carnitine. This reaction is catalyzed by *carnitine acyltransferase*. The reaction is reversible, indicating that the energy contained in the acyl CoA bond is not dissipated by the formation of acylcarnitine and that acylcarnitine is a high-energy acyl ester. Two forms of the enzyme exist. One, carnitine acyltransferase I (CAT I), is located either on the outside surface of the inner mitochondrial membrane or on the outer mitochondrial membrane. It catalyzes the transfer of the fatty acyl group from CoA to carnitine:

$$FA{\sim}S{-}CoA + \text{L-Carnitine} \rightarrow FA{\sim}O\text{-Carnitine} + CoASH$$

The acyl group is translocated across the inner mitochondrial membrane in the form of the acylcarnitine ester. A second form of the enzyme, carnitine acyltransferase II (CAT II), is located on the matrix surface of the inner mitochondrial membrane. It catalyzes the reverse reaction, the transfer of the fatty acyl group from carnitine to CoASH that is present in the mitochondrial matrix.

$$FA \sim O\text{-Carnitine} + CoASH \rightarrow FA \sim S\text{-CoA} + Carnitine$$

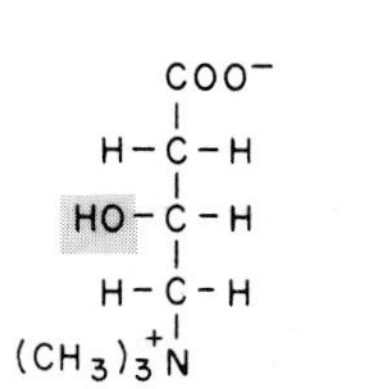

L-Carnitine 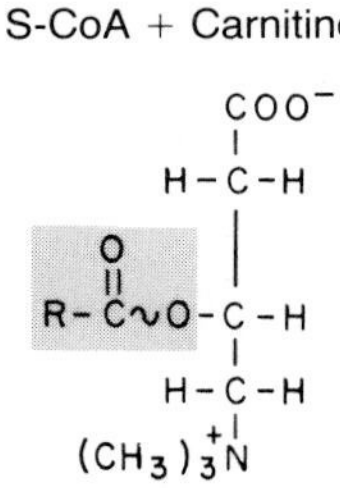   Acylcarnitine

Through these reactions, the fatty acyl group is brought across the inner mitochondrial membrane to the site of the β-oxidation enzymes, the mitochondrial matrix, as shown in Figure 10.2.

β-Oxidation sequence

Once the acyl CoA is inside the mitochondrion, four successive reactions occur. They, together with the enzymes catalyzing them, are shown in Figure 10.3.

Figure 10.2   Role of carnitine in the translocation of long-chain fatty acyl groups
across the mitochondrial membrane. *CAT*, Carnitine acyltransferase.

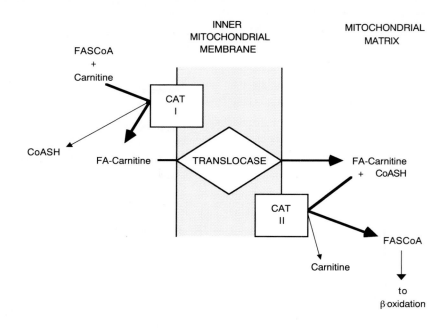

In the first dehydrogenation step, flavin adenine dinucleotide (FAD) is reduced and a
trans-unsaturated acyl CoA intermediate is formed. It in turn transfers a pair of electrons
to the electron-transferring flavoprotein (FP_2), which moves the electrons into the respiratory chain at the level of coenzyme Q. The second step involves hydration of the
unsaturated bond, forming an L-β-hydroxyacyl CoA. A similar β-hydroxyacyl intermediate occurs in fatty acid synthesis, but it is of the D configuration, not the L configuration
as in the oxidation pathway. The dehydrogenase catalyzing the third reaction reduces
NAD^+, and it subsequently transfers the pair of electrons into the initial segment of the
respiratory chain. The final reaction, thiolytic cleavage, forms acetyl CoA and an acyl
CoA that is two carbon atoms shorter than the original acyl CoA that entered this β-
oxidation sequence. This reaction is highly exergonic as written from left to right and
serves to pull the β-oxidation sequence in the proper direction. The acyl CoA that is
generated reenters the β-oxidation cycle, and the process is repeated until the entire chain
is degraded to acetyl CoA. Therefore these reactions can be thought of in terms of a β-
oxidation cycle, as shown in Figure 10.4.

The $FADH_2$ formed in the first step is oxidized in the respiratory chain to yield 2ATP.
The NADH formed in the third step is also oxidized in the electron transport chain,
yielding 3ATP. In other words, if a 16-carbon-atom fatty acid is oxidized by β-oxidation,
eight acetyl CoA units would be generated following seven cleavages, and 35ATP would
be formed in the seven β-oxidation steps.

Figure 10.3   β-oxidation reactions.

First dehydrogenation —acyl CoA dehydrogenase:

$$R-CH_2-CH_2-\underset{\underset{O}{\|}}{C}\sim SCoA \ + \ FAD \ \rightleftharpoons \ R-\underset{\underset{H}{|}}{\overset{\overset{H}{|}}{C}}=C-\underset{\underset{O}{\|}}{C}\sim SCoA \ + \ FADH_2$$

Δ_2 *trans* —Enoylacyl CoA

Hydration —enoylthydratase:

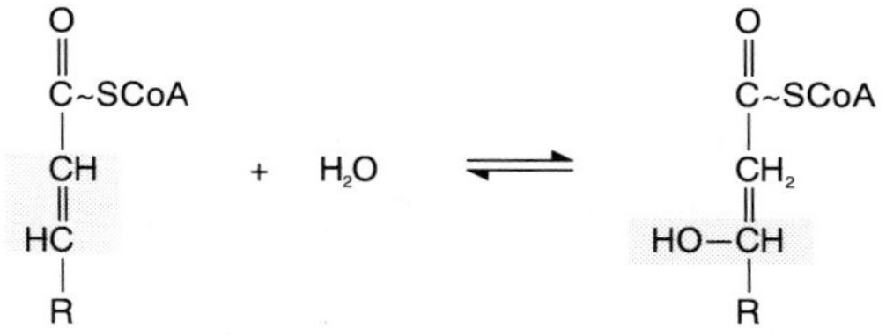

$$\begin{array}{c} O \\ \| \\ C\sim SCoA \\ | \\ CH \\ \| \\ HC \\ | \\ R \end{array} \ + \ H_2O \ \rightleftharpoons \ \begin{array}{c} O \\ \| \\ C\sim SCoA \\ | \\ CH_2 \\ | \\ HO-CH \\ | \\ R \end{array}$$

L-β-Hydroxyacyl CoA

Second dehydrogenation —L-β-hydroxyacyl dehydrogenase

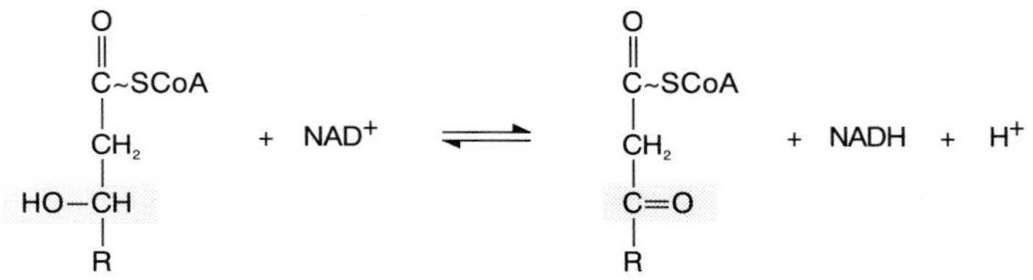

$$\begin{array}{c} O \\ \| \\ C\sim SCoA \\ | \\ CH_2 \\ | \\ HO-CH \\ | \\ R \end{array} \ + \ NAD^+ \ \rightleftharpoons \ \begin{array}{c} O \\ \| \\ C\sim SCoA \\ | \\ CH_2 \\ | \\ C=O \\ | \\ R \end{array} \ + \ NADH \ + \ H^+$$

β-Ketoacyl CoA

Thiolytic cleavage —thiolase:

$$R-\underset{\underset{O}{\|}}{C}-CH_2-\underset{\underset{O}{\|}}{C}\sim SCoA \ + \ CoASH \ \longrightarrow \ R-\underset{\underset{O}{\|}}{C}\sim SCoA \ + \ CH_3-\underset{\underset{O}{\|}}{C}\sim SCoA$$

Acyl CoA   Acetyl CoA

| Third β-oxidation | Second β-oxidation | First β-oxidation |

$$R-CH_2-CH_2-CH_2-CH_2-CH_2-CH_2-\underset{\underset{O}{\|}}{C}\sim SCoA$$

FADH$_2$	FADH$_2$	FADH$_2$
NADH + H$^+$	NADH + H$^+$	NADH + H$^+$
5 ATP	5 ATP	5 ATP

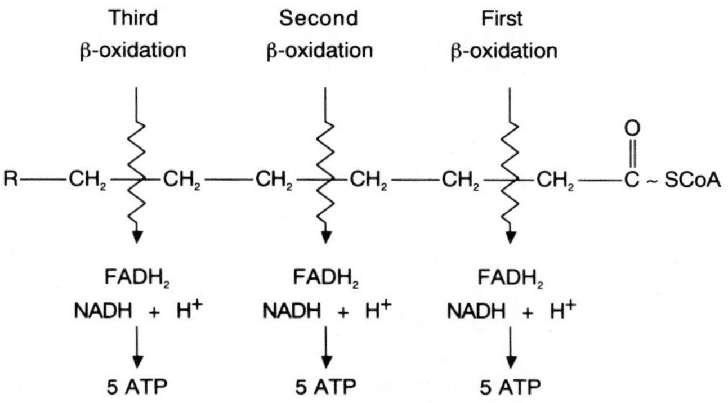

Figure 10.4   The β-oxidation cycle.

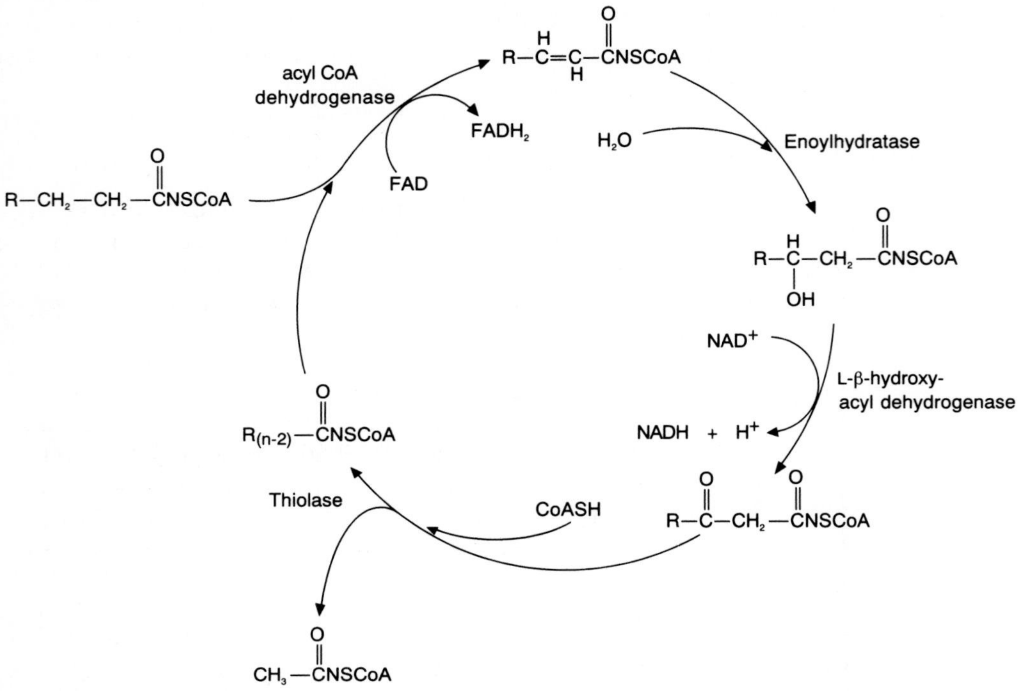

In three situations, auxiliary reactions are required to carry out the β-oxidation sequence. The situations involve the oxidation of monounsaturated acids, polyunsaturated acids, and fatty acids with odd numbers of carbon atoms.

Enoyl CoA isomerase When a naturally occurring monounsaturated fatty acid undergoes β-oxidation, an intermediate acyl CoA eventually is formed that contains a double bond in the incorrect location. Furthermore, this double bond is in the *cis,* not the *trans* configuration that ordinarily is present after the first reaction in β-oxidation, as seen in Figure 10.3. For example, consider the β-oxidation of oleic acid, 9-18:1. Recall that all naturally occurring unsaturated fatty acids have double bonds in the *cis* configuration. After three β-oxidation cycles, the remaining acyl CoA is a Δ3-*cis*-enoyl CoA.

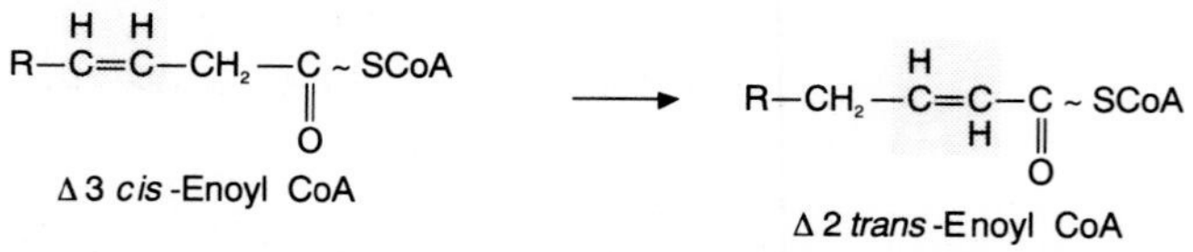

However, the required enoyl intermediate for the next reaction in the β-oxidation sequence must have a *trans* double bond in position Δ2. The potential difficulty is solved by the enzyme Δ3-*cis*-Δ2-*trans*-enoyl CoA isomerase. This enzyme catalyzes the isomerization of the double bond to position Δ2. In shifting positions, the double bond is converted to a *trans* configuration, producing the correct intermediate for β-oxidation (Δ2-*trans*-enoyl CoA), which then proceeds normally through β-oxidation.

2,4-Dienoyl CoA reductase Another problem arises when the common polyunsaturated fatty acids such as linoleic acid (9,12-18:2) undergo β-oxidation. After the first four β-oxidations, the remaining acyl CoA has a 4-*cis* double bond.

R—CH = CH—CH$_2$—CH$_2$—   R—CH = CH—CH = CH—   R—CH$_2$—CH$_2$—CH = CH—
                CO~SCoA   CO~SCoA   CO~SCOA
Δ4-*cis*-Enoyl CoA   Δ2-*trans*-4-*cis*-Dienoyl CoA   Δ2-*trans*-Enoyl CoA

When this undergoes β-oxidation, the first intermediate is Δ2 *trans*-4-*cis*-dienoyl CoA. This conjugated double-bond structure, which cannot proceed through the β-oxidation sequence, is reduced at the Δ4-position by 2,4-dienoyl CoA reductase. NADPH is the reducing agent. The product, Δ2-*trans*-enoyl CoA, then continues through β-oxidation without difficulty.

Energy yield from unsaturated fatty acids Unsaturated fatty acids can be oxidized just as readily as saturated fatty acids. For each unsaturated bond initially present, however, two fewer ATP will be generated because the FAD-linked acyl CoA dehydrogenase step (p. 419) is not required for a β-oxidation sequence when a double bond already is present. Stearic acid, 18:0, will undergo eight β-oxidations to yield 40ATP. Oleic acid, 9-18:1, also undergoes eight β-oxidations but yields only 38ATP from this process.

***Trans* fatty acids** As noted earlier, some *trans* unsaturated fatty acids are taken into the body when commercially prepared hydrogenated fats are eaten, for example, margarine or peanut butter. These *trans* fatty acids are oxidized just as readily as the corresponding *cis* unsaturated fatty acids. The Δ3-*trans*-enoyl CoA, which is generated when elaidic acid (9-*trans*-18:1) is oxidized, apparently is converted to the Δ2-*trans*-enoyl CoA and therefore poses no special problem in terms of β-oxidation.

Odd-carbon fatty acids Acids containing an odd number of carbon atoms also undergo normal β-oxidation. However, the ω-terminal fragment that is generated by the β-oxidation process is *propionyl CoA,* not acetyl CoA. The propionyl CoA is carboxylated to a four-carbon atom intermediate, methylmalonyl CoA, in a reaction involving biotin. In the next step *methylmalonyl CoA* is isomerized to *succinyl CoA,* a reaction that requires vitamin B$_{12}$ in coenzyme form. Succinyl CoA enters the Krebs cycle and is oxidized.

Other types of fatty acid oxidation

There are other pathways for fatty acid oxidation: *α-oxidation,* in which only one carbon atom is removed at a time from the carboxyl end, and ω-*oxidation,* in which the ω-terminus is oxidized to an alcohol and subsequently to a carboxyl group to form a *dicarboxylic acid*. The substrate for both of these pathways is the fatty acid, not the acyl CoA derivative.

α-Oxidation is a peroxisomal process that involves NAD$^+$ and ascorbate. An α-hydroxy fatty acid is an intermediate. *Peroxisomes* also contain β-oxidation enzymes and are able to oxidize palmitoyl CoA to acetyl CoA. However, these enzymes are different from the mitochondrial β-oxidation enzymes, and H$_2$O$_2$ is generated by the peroxisomal β-oxidation process.

ω-Oxidation is a microsomal process that requires NADPH and involves the microsomal mixed-function oxidase that contains cytochrome P_{450}.

Energy yield of fatty acid β-oxidation

Fatty acid β-oxidation is one of the main pathways through which cells derive energy for ATP synthesis. The fatty acid β-oxidation process produces acetyl CoA in the mitochondria, and this is followed by complete oxidation of the acetyl CoA in the Krebs cycle to produce CO_2 and H_2O. Four steps occur in which energy is either utilized or trapped. These are illustrated for the complete oxidation of palmitic acid, the commonly occurring 16-carbon-atom saturated fatty acid:

$$\Delta \sim P$$

$$\text{Palmitate} + \text{CoA} + \text{ATP} \xrightarrow{Mg^{++}} \text{Palmitoyl CoA} + \text{AMP} + PP_i \qquad -1$$

$$PP_i + H_2O \longrightarrow 2P_i \qquad -1$$

$$\text{Palmitoyl CoA} + 7 \text{ CoA} \longrightarrow 8 \text{ Acetyl CoA} \qquad +35$$

$$8 \text{ Acetyl CoA} + 16 \text{ O}_2 \longrightarrow 16 \text{ CO}_2 + 8 \text{ H}_2O + 8 \text{ CoA} \qquad +96$$

Two high-energy phosphate bonds are used to activate the palmitate groups, one in the acyl CoA ligase reaction and the other when the pyrophosphate that is formed in this reaction is hydrolyzed. The initial expenditure by the cell, however, is well rewarded. In the β-oxidation cycle, the palmitoyl CoA is converted to eight acetyl CoA units. This requires seven β-oxidations, and each β-oxidation sequence produces 5ATP. Therefore 35ATP are gained from the β-oxidation reactions. Furthermore, 12 high-energy phosphate bonds are formed when each acetyl CoA unit is oxidized in the Krebs cycle. Actually, 11ATP and 1GTP are formed, but this can be considered as 12ATP, since GTP and ATP are energetically equivalent. Therefore the oxidation of the eight acetyl CoA units produced will yield 96ATP. The total ATP (new high-energy bonds) formed in the complete oxidation of palmitoyl CoA is 131. Since two high-energy bonds were broken in the overall process needed to form palmitoyl CoA, the net yield is 129 high-energy phosphate bonds, considered as 129ATP. This represents about 40% of the energy that is released when palmitic acid is burned to CO_2 and H_2O in a bomb calorimeter. Although 40% efficiency may appear wasteful, it is not so, since the remainder of the energy serves to maintain the body temperature at 37.5° C.

Regulation of substrate utilization

Several regulatory mechanisms exist to prevent the wasteful expenditure of substrates and energy. Two of these, involving fatty acid β-oxidation, are described here to illustrate how several metabolic pathways are interrelated and controlled. These regulatory mechanisms are presented schematically in Figure 10.5.

Consider the situation where substantial amounts of fatty acids are being oxidized. In this metabolic state it would be wasteful also to channel glucose into acetyl CoA, since the Krebs cycle already is receiving enough acetyl CoA from β-oxidation. This condition is shown on the left side of Figure 10.5. To prevent glucose conversion to acetyl CoA, *pyruvate dehydrogenase* is inhibited. The inhibition is mediated partly by fatty acyl carnitine, the form in which the fatty acid crosses the inner mitochondrial membrane to reach the β-oxidation system. Therefore the metabolite feeding the β-oxidation system to form acetyl CoA acts to prevent the simultaneous formation of acetyl CoA from glucose. Acetyl CoA, the product of β-oxidation, also inhibits pyruvate dehydrogenase. The cell is not harmed by this inhibition because it can obtain all the intramitochondrial acetyl CoA that it needs from β-oxidation.

Alternatively, consider the metabolic state where carbohydrate is being used as the substrate for fatty acid synthesis in the liver. This is illustrated on the right side of Figure 10.5. It would be wasteful for the liver to expend energy to synthesize fatty acids and, at the same time, use these fatty acids for oxidation. Therefore when an excess of glucose is available and is being converted to fatty acids, the β-oxidation of fatty acid is inhibited. The inhibition is mediated by *malonyl CoA,* an intermediate in fatty acid biosynthesis. Malonyl CoA inhibits *carnitine acyltransferase I,* the first enzyme involved in the trans-

Figure 10.5 Regulation of substrate utilization. In the metabolic state shown on the left, fatty acid oxidation is occurring and pyruvate dehydrogenase is inhibited by acylcarnitine and acetyl CoA. In the metabolic state shown on the right, carbohydrate is being converted into fatty acid. Under these conditions fatty acid β-oxidation is inhibited through the action of malonyl CoA, an inhibitor of carnitine acyltransferase I.

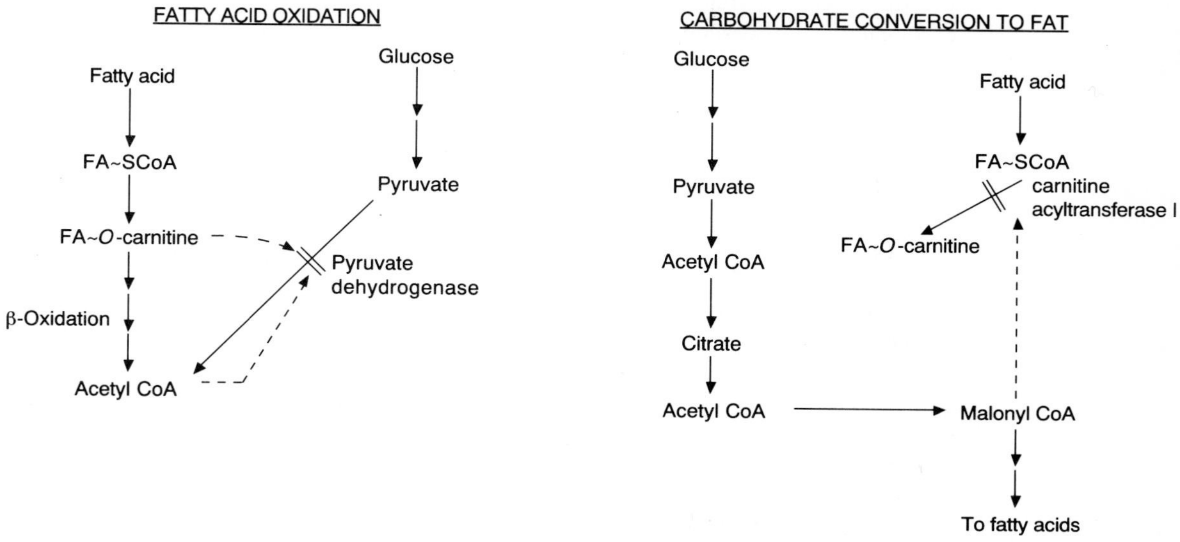

location of the fatty acyl group across the inner mitochondrial membrane. Since fatty acids cannot reach the site of the β-oxidation system when this enzyme is inhibited, they are not oxidized to acetyl CoA. The hepatocyte is not harmed by this, since it can obtain all the intramitochondrial acetyl CoA that it needs from glucose by way of pyruvate.

Fatty acid synthesis

Humans synthesize fatty acids using acetyl CoA derived primarily from carbohydrate. This pathway is known as complete or de novo synthesis and it occurs in the cell cytoplasm. The main sites of fatty acid de novo synthesis are the liver and adipocytes. This process occurs when the cells have an abundance of energy and therefore have a high ATP content. In addition, acetate fragments can be added to an existing fatty acid by a process of *chain elongation*. This reaction occurs in both the mitochondria and microsomes. Finally, *cis* double bonds can be introduced into fatty acids by the process of desaturation, a reaction that occurs in the microsomes.

Source of acetyl CoA

Glucose is the main source of the acetyl CoA used for fatty acid synthesis. Glucose is converted to pyruvate in the cytoplasm. The pyruvate enters the mitochondria, where it is converted to acetyl CoA through the action of the pyruvate dehydrogenase enzyme complex. As already described, acyl CoA derivatives, including acetyl CoA, cannot cross the inner mitochondrial membrane. The acetyl CoA leaves the mitochondria in the form of *citrate*, the initial intermediate of the Krebs cycle, and the transfer across the inner mitochondrial membrane is mediated by the citrate translocase. Remember that when the ATP content of a cell is elevated, the rate-limiting Krebs cycle enzyme isocitrate dehydrogenase is inhibited (see Chapter 5). When this occurs, the equilibrium of the enzyme aconitase is such that citrate accumulates in the mitochondria and becomes available in large quantities for efflux to the cytoplasm. In the cell cytoplasm, citrate is cleaved to acetyl CoA and oxaloacetate in an ATP-requiring reaction catalyzed by *citrate lyase:*

$$\text{Citrate} + \text{ATP} + \text{CoASH} \rightarrow \text{Acetyl CoA} + \text{Oxaloacetate} + \text{ADP} + P_i$$

This process is illustrated in Figure 10.6. One acetyl CoA unit enters the fatty acid synthase pathway directly. This acetyl CoA is the primer or acceptor of two-carbon units derived from *malonyl CoA*. The necessary *malonyl CoA* is synthesized through carboxylation of acetyl CoA.

$$\begin{array}{c} COO^- \\ H_2C \diagdown \\ \quad C \sim S\text{-}CoA \\ \quad \| \\ \quad O \end{array}$$

Carboxylation of acetyl CoA

The rate-limiting reaction in fatty acid synthesis is the carboxylation of acetyl CoA to malonyl CoA.

$$\text{Acetyl} \sim \text{SCoA} + \text{HCO}_3^- + \text{ATP} \rightarrow \text{Malonyl} \sim \text{SCoA} + \text{ADP} + P_i$$

This reaction is catalyzed by an enzyme complex, acetyl CoA carboxylase, that contains *biotin* and utilizes *bicarbonate*. Acetyl CoA carboxylase is an allosteric enzyme that is activated by citrate. Therefore the rate-limiting enzyme in fatty acid synthesis actually is primed to utilize acetyl CoA by the compound in which the acetate moiety is delivered from the mitochondria, citrate. An intermediate in the carboxylation reaction is CO_2 linked covalently to enzyme-bound biotin, as a *carboxybiotin* complex. This is attached to a lysine residue of the enzyme.

Biotin   N-Carboxybiotin enzyme complex

Acetyl CoA carboxylase is activated by conversion of a monomeric enzyme complex with a molecular weight of 4×10^5 to a polymer having a molecular weight of approximately 6 to 8×10^6. Citrate causes polymerization that activates the enzyme; palmitoyl CoA causes the polymer to disaggregate and become inactive. Therefore citrate stimulates and palmitoyl CoA inhibits the reaction. In addition to allosteric control, this reaction is

Figure 10.6   Conversion of glucose to acetyl CoA.

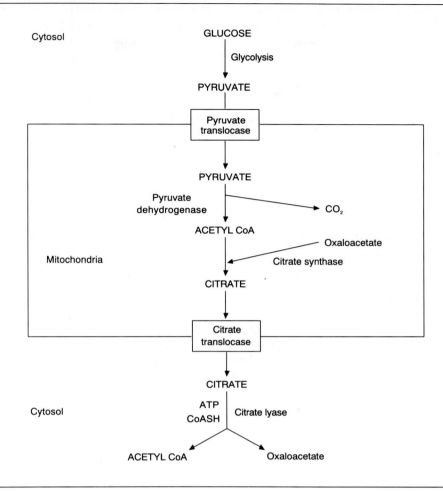

regulated by controlling enzyme production. A high-carbohydrate–low-fat diet stimulates the synthesis of the enzyme.

Acetyl CoA carboxylase consists of two components. One contains two proteins, a biotin carboxyl carrier protein (BCCP) and a biotin carboxylase. The second component is a transcarboxylase that catalyzes the transfer of CO_2 from BCCP to acetyl CoA. The partial reactions are as follows:

$$BCCP + HCO_3^- + ATP \;\rightarrow\; BCCP—CO_2 + ADP + P_i$$
$$BCCP—CO_2 + Acetyl\ CoA \;\rightarrow\; Malonyl\ CoA + BCCP$$

The net result is the carboxylation of acetyl CoA to malonyl CoA.

Fatty acid synthase   Fatty acid synthesis takes place in a complex that is composed of seven separate enzymes and a carrier protein that holds the growing acyl chain. This complex is called the fatty acid synthase. It is present in the cell cytoplasm and has a molecular weight of about 500,000. With affinity chromatography, the animal complex can be separated into two large apparently identical subunits. The two subunits are tightly fitted together in the complex and operate in concert, the monomer being inactive in catalysis. One proposed mechanism pictures the phosphopantetheine group of one subunit being positioned less than 2 Å from the cysteinyl sulfhydryl group of the β-keto synthase, also known as the condensing enzyme, of the other subunit. According to this model, the growing fatty acyl chain actually shuttles back and forth between the pantetheinyl sulfhydryl group of one subunit and the cysteinyl sulfhydryl of the other as it is being produced. The entire subunit is synthesized from a single messenger RNA, which has a molecular weight of about 3.4×10^6 and contains about 1×10^4 nucleotides.

As with acetyl CoA carboxylase, the amount of fatty acid synthase in the cell is regulated according to the need for fatty acid production. This occurs through control of the synthesis of the enzyme. Fasting lowers the amount of synthase present in the liver. Refeeding returns the activity to the normal level by stimulating enzyme production. A fat-free diet increases the enzyme content in the liver far above normal, again by stimulating enzyme production.

4′-Phosphopantetheine group The fatty acid synthase contains a 4′-phosphopantetheine group esterified to a hydroxyl group of serine. This provides the sulfhydryl group to which the growing fatty acid chain is attached as it is synthesized. A 4′-phosphopantetheine group also is the functional group of CoASH.

There is one 4′-phosphopantetheine carrier group for each subunit. These groups remain firmly attached when the complex is isolated. To facilitate communication, the fatty acid synthase 4′-phosphopantetheine group is abbreviated as PPant-SH.

Fatty acid synthase reactions The reactions through which the acetyl and malonyl CoA groups are joined to form a four-carbon product are illustrated in Figure 10.7. Initially the acetyl group of acetyl CoA is transferred to the 4′-phosphopantetheine sulfhydryl group of PPant-SH. The reaction is catalyzed by the *acetyl CoA-transacylase*.

Acetyl ~ SCoA + PPant-SH ⇌ Acetyl ~ S—PPant + CoASH

Next, the acetyl group originally attached to this sulfhydryl group is momentarily transferred to a cysteinyl sulfhydryl group of the β-keto synthase present in the other subunit of the enzyme complex.

Acetyl ~ S—PPant + Cys—SH ⇌ Acetyl ~ S—Cys + PPant—SH

This frees the 4′-phosphopantetheine sulfhydryl group to accept the next incoming group, a malonyl group from malonyl CoA. The malonyl transfer is catalyzed by *malonyl CoA acyltransferase*.

Malonyl ~ SCoA + PPant—SH ⇌ Malonyl ~ S—PPant + CoASH

The acetyl residue then is transferred from its temporary site on the cysteinyl sulfhydryl group of the second subunit to condense with the malonyl residue attached to the PPant-SH of the first subunit. In this reaction the free carboxyl group of malonate is discharged as CO_2 so that the product contains four rather than five carbon atoms.

Acetyl ~ S—Cys + Malonyl ~ S—PPant → Acetoacetyl ~ S—PPant + Cys—SH + CO_2

Figure 10.7

Reactions leading to growth of the fatty acyl chain in fatty acid synthesis. The fatty acid synthase is shown as a dimer, with the 4'-phosphopantetheine sulfhydryl *(PPant-SH)* of one subunit operating in concert with the cysteine sulfhydryl *(Cys-SH)* of the other subunits.

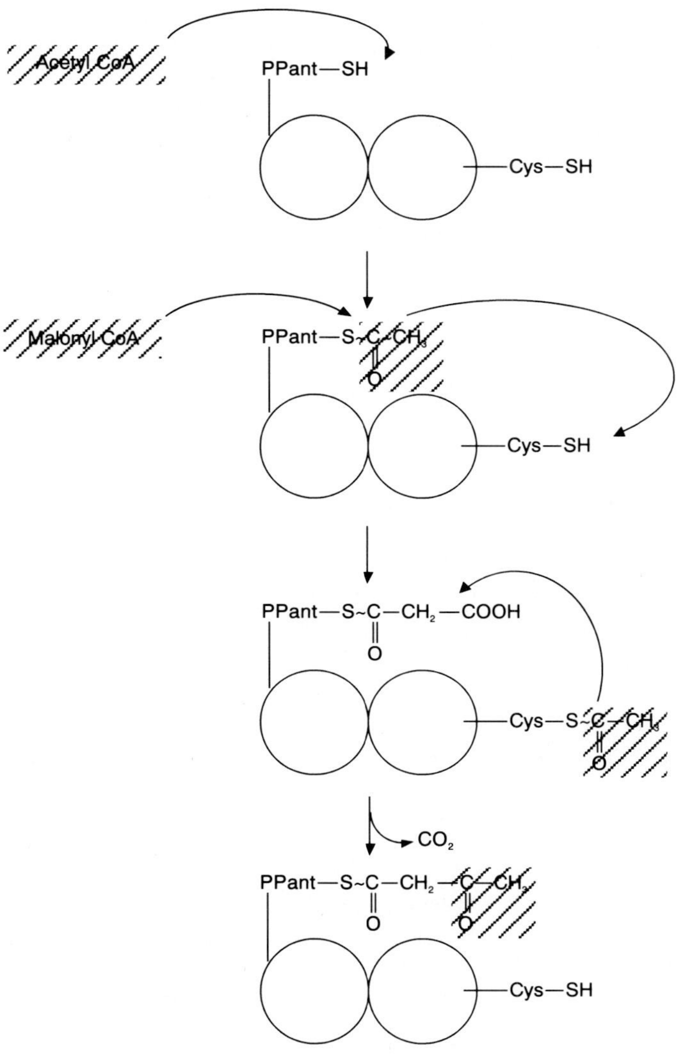

The condensation of the acetyl and malonyl groups is catalyzed by the β-keto synthase. The process that drives the condensation reaction is the decarboxylation of malonate. Therefore the expenditure of ATP in the acetyl CoA carboxylase reaction is not wasteful, since the energy is used subsequently for the condensation reaction.

The product of these reactions, *acetoacetyl~S—PPant*, is reduced to *butyryl* S~PPant by the remaining enzymes in the fatty acid synthase complex, *β-ketoacyl reductase, enoyl dehydratase,* and *crotonyl reductase,* as shown in Figure 10.8.

Figure 10.8   Reductions performed by fatty acid synthase.

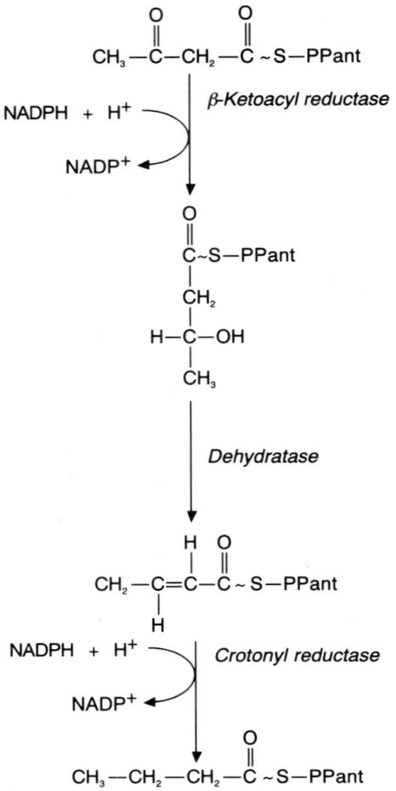

The β-hydroxyl intermediate, unlike that involved in β-oxidation, is of the D-configuration. The Δ2 double bond of the crotonyl~S-PPant intermediate is in the *trans* configuration. NADPH is the reducing agent for both the acetoacetyl~S-PPant and crotonyl~S-PPant intermediates.

Second cycle Following this series of events, the entire sequence is repeated. First, the butyryl group is transferred from the first subunit to the cysteinyl sulfhydryl group of the second subunit, just as was the original acetyl group. Malonyl CoA then adds once again to the free 4'-phosphopantetheine sulfhydryl group of the first subunit, and the butyryl group on the β-keto synthase condenses with the malonyl~S-PPant residue. In this condensation CO_2 is discharged, and the product now has six carbon atoms, with carbon atom 5 being a carbonyl group. The same reduction pathway just described utilizing two NADPH takes place, forming a six-carbon-atom saturated acyl~S-PPant intermediate.

Additional cycles This sequence is repeated, with a two-carbon unit from malonyl CoA being added, followed by reduction of the carbonyl group, until a 16-carbon atom chain is built. Even though the enzyme is a dimeric complex, only one fatty acyl chain is formed at a time. In each case the acyl residue leaves the 4'-phosphopantetheine-sulfhydryl group of one subunit and moves to the cysteinyl sulfhydryl group of the other subunit momentarily so that the malonyl group can add. It then condenses with the 4'-phospho-pantetheine-linked malonate residue, which becomes the carboxyl end of the growing fatty acid. CO_2 is released from the malonyl residue in each condensation. In this way the original acetate residue remains as the ω-end of the growing fatty acid, and each malonate residue that is added successively becomes the carboxyl end.

$$CH_3-CH_2-CH_2-CH_2-CH_2-CH_2-CH_2-\overset{\overset{\displaystyle O}{\|}}{C}\sim S-PPant-\text{synthase complex}$$

From acetyl CoA

From malonyl CoA

De novo fatty acid synthesis always results in a product containing an even number of carbon atoms, since the acetate primer contains two carbon atoms and each malonyl unit that condenses with it donates two carbon atoms to the chain. Fatty acid synthesis usually stops after the chain is 16 carbon atoms long. This occurs partly because the enzyme that removes the acyl group from the 4'-phosphopantetheine group, a *thiolase,* exhibits maximum activity with 16-carbon-atom fatty acids. It also occurs because the cysteinyl sulfhydryl binding site to which the acyl chain is transferred from 4'-phosphopantetheine cannot hold chains that contain more than 14 carbon atoms very well. Much of the NADPH needed for the reduction is obtained from the initial reactions of the pentose phosphate pathway (hexose monophosphate shunt), in which D-glucose 6-phosphate is oxidized (see Chapter 6). The remainder of the NADPH is supplied by the oxidation of malate in the cytoplasm, catalyzed by the *malic enzyme.*

Regulation of de novo biosynthesis

Two types of control mechanisms regulate the complete synthesis of fatty acids by the *de novo* pathway. One is *long-term control.* This involves changes in the content of the fatty acid biosynthetic enzymes within the cell. The *production* of enzymes involved in fatty acid de novo synthesis, including acetyl CoA carboxylase, the fatty acid synthase complex, citrate lyase, glucose 6-phosphate dehydrogenase, and the malic enzyme, is stimulated in the liver when glucose is fed. In an insulin-deficient diabetic animal, the production of these enzymes is stimulated when insulin is administered. By contrast, the content of these enzymes in the liver decreases during starvation or in diabetes associated with insulin deficiency. These enzymes have a relatively short half-life, so that their content in a cell depends on the rate at which they are synthesized. In this way the cell has large quantities of the enzymes needed for conversion of carbohydrate into fatty acid only when carbohydrate and ATP are available. Because it requires enzymes production, long-term control exerts its effect relatively slowly, requiring several hours to manifest itself completely.

A *short-term control* mechanism also acts on acetyl CoA carboxylase, the rate-limiting enzyme in fatty acid de novo biosynthesis. It involves modulation of the activity of acetyl CoA carboxylase. Citrate activates the enzyme by causing aggregation, whereas palmitoyl CoA inactivates it by causing it to disaggregate. The activity of acetyl CoA carboxylase also is controlled by phosphorylation. Glucagon stimulates phosphorylation through a mechanism that involves cyclic AMP. The enzyme is less active when it is phosphorylated, and dephosphorylation increases the activity. Since short-term control involves modulation of the activity of already existing enzyme, its effects manifest themselves very rapidly, that is, within minutes.

Table 10.6   Main differences between fatty acid de novo diosynthesis and fatty acid oxidation

Parameter	Fatty acid oxidation	Fatty acid synthesis
Intracellular location	Mitochondria	Cytoplasm
Intermediates	Acetyl CoA	Acetyl CoA, malonyl CoA
Thioester linkage	CoASH	PPant-SH
Coenzymes for electron transfer	FAD, NAD$^+$	NADPH
Configuration of β-hydroxy intermediate	L	D
Bicarbonate dependence	No	Yes
Energy state favoring the process	High ADP	High ATP
Citrate activation	No	Yes
Acyl CoA inhibition	No	Yes
Highest activity	Fasting, starvation	Carbohydrate fed

Differences between synthesis and oxidation

Although some overall similarities exist between fatty acid synthesis and oxidation, the details of these processes really are very different. To understand each of these very important metabolic pathways more clearly, it is helpful to summarize their main differences. These are listed in Table 10.6 as a review.

Fatty acid chain elongation

Acetate units can be added to the *carboxyl end* of an existing fatty acyl chain. In the mitochondria the acetate group is added in the form of acetyl CoA. In the endoplasmic reticulum, malonyl CoA provides the two-carbon atom fragment. These pathways are called chain elongation. The newly formed β-keto acid is then reduced, and a saturated fatty acid is the final product. Reduction of the β-keto group occurs through the same three steps as in de novo fatty acid biosynthesis, as shown in Figure 10.8, and two NADPH are required. In those mammalian cells that have been studied, the endoplasmic reticulum pathway is much more active than the mitochondrial pathway and accounts for most of the fatty acid that is elongated. The endoplasmic reticulum pathway can be represented as:

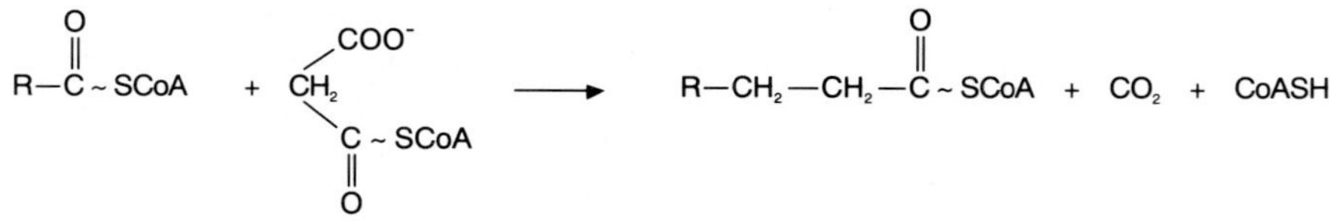

The chain elongation pathway requires the activated form of fatty acid, acyl CoA, as the substrate. It can elongate either saturated or unsaturated fatty acids. More than one elongation can take place so that fatty acids containing up to 26 carbon atoms can be formed. In each elongation step, the fatty acid is lengthened by two carbon atoms. This is true even though malonyl CoA is the elongating agent, for just as in the de novo synthetic pathway, CO_2 is released when the malonyl CoA condenses with the acyl CoA substrate. Therefore the elongation mechanism also leads to the production of fatty acids having an *even* number of carbon atoms.

Fatty acid desaturation

Double bonds can be introduced into fatty acids by desaturating enzymes that require O_2 and NADH. Figure 10.9 is a schematic representation of the fatty acid desaturation mechanism. A *cis* configuration around the double bond always results. This reaction utilizes NADH and O_2, and it occurs in the endoplasmic reticulum. Cytochrome b_5,

Figure 10.9

Fatty acid desaturase system of microsomes. This complex involves three enzymes. Stearoyl CoA desaturase contains nonheme iron, which forms a complex with molecular oxygen. Cytochrome b_5 is a one-electron transferring protein, whereas the flavoprotein transfers two electrons at a time. Unlike cytochrome P_{450}, cytochrome b_5 presumably does not directly interact with molecular oxygen.

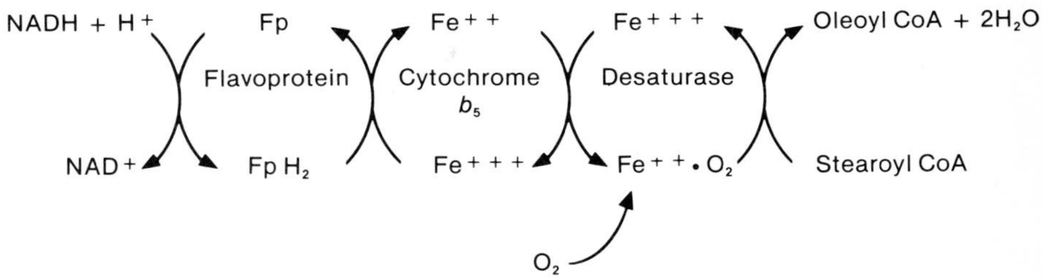

cytochrome b_5, reductase, and a desaturase that is tightly bound to the membrane are required for the desaturation reaction. Both NADH and the fatty acid are oxidized, and the two pairs of electrons are transferred to O_2 to form $2H_2O$. These reactions are a part of what is called the microsomal electron transport chain.

The desaturase enzymes use fatty acyl CoA as substrates. These can be either saturated or unsaturated, depending on the specificity of the desaturase. At least four separate desaturases exist, the $\Delta 9$-, $\Delta 6$-, $\Delta 5$-, and $\Delta 4$-fatty acid desaturases, named according to the position in the acyl CoA chain that they desaturate. Thus the $\Delta 9$-desaturase introduces a double bond between carbon atoms 9 and 10, counting from the fatty acid carboxyl group. The different pathways through which the various unsaturated fatty acids are formed are shown in Figure 10.10.

The $\Delta 9$-fatty acid desaturase, also called *stearoyl CoA desaturase,* can only utilize saturated acyl CoA. The other three desaturases utilize unsaturated acyl CoA substrates. They always insert the double bond between the thioester group and the double bond closest to it, leaving a three-carbon gap. None of these desaturases can operate between the ω-terminus and the double bond closest to it. Therefore $9,12$-18:2 can only be desaturated in the $\Delta 6$ position and is never desaturated between the $\Delta 12$ double bond and the ω-terminus. Because of these specificities, $9,12$-18:2 and $9,12,15$-18:3 cannot be synthesized in the body. They can be lengthened and desaturated, however, as shown in Figure 10.10. In this way linoleic acid ($9,12$-18:2) is converted to arachidonic acid ($5,8,11,14$-20::4).

Regulation of desaturation As with fatty acid synthesis, desaturation is regulated according to the needs of the body. Most of the available information has been obtained for the stearoyl CoA desaturase of liver, the enzyme that acts primarily in the conversion of 18:0 to 9-18:1. This activity decreases in starvation and increases greatly on refeeding carbohydrate. These changes enusre that the fatty acids synthesized after refeeding carbohydrate contain an adequate amount of unsaturation, particularly 9-18:1, to maintain an optimum fluidity state of the membrane lipids. Remember that the de novo fatty acid synthesis pathway produces only saturated fatty acids, almost entirely palmitic acid (16:0). If a large quantity of unsaturated fat is fed, adequate amounts of unsaturation are

Figure 10.10   Pathways for fatty acid elongation and desaturation under normal metabolic conditions. Although not shown in the figure, all the reactions utilize acyl CoA substrates. The enzymes carrying out the various desaturations are shown.

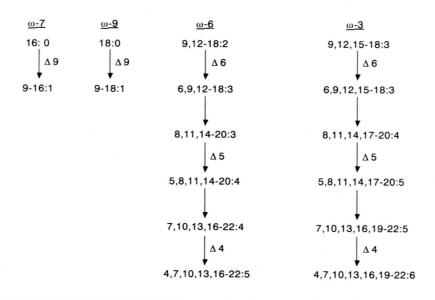

available from the diet, and more of the endogenously produced fatty acid must remain saturated to maintain optimum fluidity. Under these conditions the activity of stearoyl CoA desaturase decreases. This regulation involves neither the NADH-cytochrome b_5 reductase nor cytochrome b_5. Instead, it is exerted on the terminal component of the system, the desaturase enzyme.

Summary of fatty acid pathways The body can synthesize fatty acids from acetyl CoA derived from nonlipid substrates. This pathway produces saturated fatty acids, predominantly palmitic acid (16:0). As shown in Figure 10.10, palmitic acid can be desaturated, forming the ω-7 class of unsaturated fatty acids. Alternatively, the palmitic acid can be lengthened to 18:0 and then desaturated, forming the ω-9 class of unsaturated fatty acids. Under ordinary metabolic conditions these two classes are not further lengthened or desaturated to any appreciable extent. The other two classes of unsaturated fatty acids, the ω-6 and ω-3 classes, are derived from dietary polyunsaturated fats. After incorporation into the body, each of these classes can be further lengthened and desaturated. None of the four classes of unsaturated fatty acids, however, is interconvertible.

Formation of ω-9 eicosatrienoic acid Mammalian tissues require polyunsaturated fatty acids. When the usual diet containing terrestrial plant foods is eaten, most of the polyunsaturated fatty acids are either linoleate (9,12-18:2) or elongation and desaturation products of the linoleate class (ω-6 class). If a fat-deficient diet is eaten, the body will become depleted of these polyunsaturated fatty acids, eventually leading to a condition known as essential fatty acid deficiency (see previous discussion). In such a situation the

Figure 10.11   Pathway for the synthesis of the ω-9 eicosatrienoic acid in essential
fatty acid deficiency.

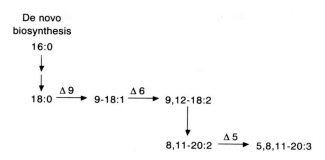

De novo
biosynthesis

16:0

18:0 $\xrightarrow{\Delta 9}$ 9-18:1 $\xrightarrow{\Delta 6}$ 9,12-18:2

8,11-20:2 $\xrightarrow{\Delta 5}$ 5,8,11-20:3

body attempts to compensate for the lack of ω-6 fatty acids by synthesizing polyunsaturated fatty acids de novo. As shown in Figure 10.11, palmitic acid (16:0) made by de novo biosynthesis is elongated to stearate (18:0), which is desaturated to oleate (9-18:1). Oleic acid then is desaturated and elongated, and the major product that accumulates is an eicosatrienoic acid of the ω-9 class. This part of the pathway is mediated by the Δ6- and Δ5-fatty acid desaturases, enzymes that ordinarily do not utilize oleic acid or its derivatives.

When this polyenoic fatty acid, 5,8,11-20:3 or 20:3ω-9, appears in appreciable quantities in tissues, it is a sign of essential fatty acid (ω-6 class) deficiency. The elongation of oleic acid, 9-18:1, is not shown in Figure 10.10 because it is not a pathway that occurs to any appreciable extent under ordinary conditions. It takes place only when an essential fatty acid deficiency exists. This again illustrates the point that the unsaturated fatty acids within a given class, for example, ω-9, are interconverted but that no interconversion occurs between the different classes; that is, an ω-9 fatty acid is not converted to an ω-6 fatty acid.

Retroconversion In addition to inserting double bonds into fatty acids, mammalian tissues can hydrogenate unsaturated bonds already present in a fatty acid. This process, known as retroconversion, has been observed in the liver and testes. It usually is associated with the removal of two carbon atoms from the carboxyl end of the fatty acid. Thus 4,7,10,13,16-22::5 is converted to 5,8,11,14-20::4 by a combination of hydrogenation of the original Δ4 double bond and removal of carbon atoms 1 and 2 from the original 22-carbon-atom fatty acid. With the release of the carboxyl-terminal acetate group, each unsaturated bond is located two carbon atoms closer to the carboxyl end, and the resulting acid is 5,8,11,14-20::4, arachidonic acid.

Acylglycerols
Chemistry

Fatty acid esters of glycerol, the acylglycerols, are typically known as glycerides. The fatty acid moiety in lipid esters is known as an acyl group. The class of glyceride depends on the number of glycerol alcohol groups that are esterified. Three general types of glycerides occur, *monoglycerides, diglycerides,* and *triglycerides*. These compounds also are called *monoacylglycerols, diacylglycerols* and *triacylglycerols,* respectively. The terms are used interchangeably; for example, triacylglycerol is employed to a greater extent in chemical and biochemical usage, whereas triglyceride predominates in the health sciences. This convention is used in this book; for example, triacylglycerol is used in describing biochemical reactions and triglyceride in the clinical discussions.

In monoacylglycerols the single acyl group may be linked to either the primary or the secondary alcohol groups. Therefore two forms of monoacylglycerol can occur, a 1- or 2-monoacylglycerol. The stereospecific numbering (sn) of the glycerol carbon atoms is shown for the monoacylglycerols. This stereospecific numbering scheme applies to all the acylglycerols.

1-Monoglyceride, or 1-monoacylglycerol

2- Monoglyceride, or 2-monoacylglycerol

Monoacylglycerols are important in digestion and as metabolic intermediates.

Two types of diacylglycerols also occur, depending on whether the acyl groups are attached to the 1,2- or 1,3-alcohol groups of the glycerol moiety.

1,2-Diglyceride, or 1,2- diacylglycerol

1,3-Diglyceride, or 1-3-diacylglycerol

Triglyceride, or triacylglycerol

Diacylglycerols are metabolic intermediates, and the 1,2-isomer activates an important enzyme that phosphorylates proteins, protein kinase C.

Triacylglycerols are the most prevalent of the acylglycerols, since they are quantitatively the major storage and transport form of fatty acids. The fatty acid residues in natural acylglycerols occur in many combinations. Thus, although each class of acylglycerols is considered to be "single" species in everyday usage, they actually represent a family of molecules of varying fatty acid composition.

Triacylglycerol synthesis

Triacylglycerol is the end-product of the acylglycerol synthesis pathway. Two separate pathways exist, as shown in Figure 10.12. The intestine utilizes a pathway beginning with 2-monoacylglycerol, one of the products of lipid digestion. Only two acyl CoA are needed to form triacylglycerol in this pathway. Other tissues utilize a pathway beginning with glycerol 3-phosphate. This is derived from glucose, which undergoes glycolysis to dihydroxyacetone phosphate. The dihydroxyacetone phosphate is reduced to L-glycerol 3-phosphate in a reaction requiring NADH.

Figure 10.12   Pathways for triacylglycerol synthesis.

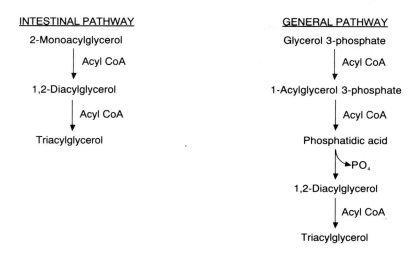

INTESTINAL PATHWAY   GENERAL PATHWAY

2-Monoacylglycerol   Glycerol 3-phosphate

    ↓ Acyl CoA   ↓ Acyl CoA

1,2-Diacylglycerol   1-Acylglycerol 3-phosphate

    ↓ Acyl CoA   ↓ Acyl CoA

Triacylglycerol   Phosphatidic acid

                            ↓ →PO$_4$

                         1,2-Diacylglycerol

                          ↓ Acyl CoA

                         Triacylglycerol

Dihydroxyacetone phosphate + NADH+H$^+$ $\xrightarrow{\text{Glycerol phosphate dehydrogenase}}$ L-Glycerol 3-phosphate + NAD$^+$

Three acyl CoA are needed to form triacylglycerol in this pathway.

Diacylglycerol formation

Diacylglycerol is an intermediate in both of the triacylglycerol synthetic pathways shown in Figure 10.12. Another route for diacylglycerol formation is the hydrolysis of inositol phosphoglycerides, reactions mediated by phospholipase C (see p. 443). Inositol phosphoglyceride hydrolysis is the origin of the diacylglycerol that activates protein kinase C.

Acylglycerol hydrolysis

The acylglycerols are catabolized by hydrolytic enzymes called lipases. These enzymes remove the acyl group from the glycerol backbone by hydrolysis of the ester linkage. Separate lipases exist for mono-, di- and triacylglycerols. The most widely studied are those that hydrolyze triacylglycerols, usually called triacylglycerol lipases. Five separate triacylglycerol lipases play important roles in metabolism. These enzymes, together with their actions and properties, are listed in Table 10.7.

Phosphoglycerides Chemistry

Acylglycerols that contain phosphoric acid esterified at the C$_3$-hydroxyl group are termed phosphoglycerides. They form bilayers when dispersed in an aqueous solution and, in

Table 10.7　　　　　Triacylglycerol lipases

Name	Location	Function	Special properties
Pancreatic lipase	Pancreatic juice	Digestion of dietary triacylglycerols	Hydrolyzes triacylglycerols in mixed micelles; activity enhanced by colipase
Hormone-sensitive lipase	Adipocytes	Fat mobilization	Activated by phosphorylation through the action of a cyclic AMP–dependent protein kinase
Acid lipase	Lysosomes	Intracellular catabolism of lipoproteins	pH optimum of about 5.0
Lipoprotein lipase	Capillaries	Utilization of triacylglycerols in lipoproteins	Released into plasma by heparin, inhibited by protamine
Hepatic lipase	Liver	Lipoprotein catabolism	Release into plasma by heparin, resistant to protamine

this form, are the main structural components of cell membranes, as described in Chapter 12. Phosphoglycerides have the following general structure:

$$X = (CH_3)_3\overset{+}{N}-CH_2-CH_2-\quad (Choline)$$

$$X = NH_2-CH_2-CH_2\quad (Ethanolamine)$$

$$X = \quad (Inositol)$$

General formula for a diacylphosphoglyceride

where X represents a group derived from an alcohol, such as *choline, ethanolamine, serine, inositol,* or *glycerol,* giving rise to *phosphatidylcholine, phosphatidylethanolamine,* and so on. The common name for phosphatidylcholine is *lecithin,* and this term is still in use in the health sciences and the food industry. *Phosphatidic acid* is the name of a class of phosphoglycerides in which the group represented as X in the general structural formula for a diacylphosphoglyceride is a hydrogen. Thus it is the simplest form of diacylphosphoglyceride. As with acylglycerols, the fatty acid residues of a phosphoglyceride usually vary. Therefore a family of lecithins actually exists, for example, palmitoyl-oleyl-phosphatidylcholine and stearoyl-linoleyl-phosphatidylcholine. In general, the acids present in position 1 are more saturated than those contained in position 2, as is implied in these examples. For the sake of simplicity, each phosphoglyceride class is usually considered as a single entity, for example, phosphatidylcholine, without subclassifying as to fatty acid composition. The phosphoglycerides are important structural components of cellular membranes.

Stereospecific numbering

Phosphoglycerides are derivatives of glyceryl phosphoric acid. Carbon atom 2 of this glycerol derivative is asymmetric, and the structure can be considered as derived from either D-glycerol 1-phosphate or L-glycerol 3-phosphate according to the conventions discussed in Chapter 2. The latter configuration has been selected by convention. Therefore, as just illustrated, the phosphate is written as being at positon 3 of the glycerol moiety, and the hydroxyl group attached to carbon atom number 2 is projected to the left. This convention, as noted earlier, is known as the stereospecific numbering (*sn*) system.

Clinical comment

Pulmonary surfactant A lipoprotein material secreted by pulmonary type II epithelial cells, called *surfactant,* reduces the surface tension in the alveoli of the lung. It is needed for the lung to function properly in respiratory gas exchange. Surfactant contains a characteristic phosphoglyceride, *dipalmitoyl phosphatidylcholine*. This form of phosphatidylcholine is unusual because it contains two saturated fatty acids. Most naturally occurring phosphatidylcholines contain an unsaturated fatty acid, usually polyunsaturated, in the *sn*-2 position. It is a common misconception that dipalmitoyl phosphatidylcholine is synonymous with surfactant. The surface active material actually is a lipid-protein complex that contains only 80% to 90% lipid. Moreover, dipalmitoyl phosphatidylcholine makes up only about half the surfactant lipid. Another important lipid component of surfactant is *phosphatidylglycerol,* a structure in which the X in the general formula for diacylphosphoglyceride is a glycerol residue. This and other surfactant lipids have important influences on the surface properties of dipalmitoyl phosphatidylcholine and therefore help to facilitate gas exchange in the lung.

Cardiolipins

Cardiolipins are complex phosphoglycerides, present in the inner mitochondrial membrane and in chloroplast membranes of plants. Cardiolipins are composed of two molecules of phosphatidic acid joined together by a glycerol bridge. As with phosphatidic acid, cardiolipins are more acidic than other diacylphosphoglycerides.

Phosphatidic acid   Cardiolipin

Lysophospho-glycerides

In a lysophosphoglyceride one of the two acyl groups is removed. Several types of phosphoglycerides have lyso derivatives, the most common of which are *lysophospha-*

tidylcholine (lysolecithin), and *lysophosphatidylethanolamine*. There are two forms of lysophosphoglycerides. One contains the acyl group in the *sn*-1 position, and the other contains the acyl group in the *sn*-2 position.

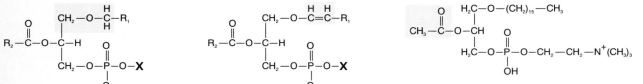

1-Acyl lysophosphoglyceride   2-Acyl lysophosphoglyceride

Alkyl ethers and plasmalogens

In alkyl ethers the hydrocarbon chain present at position 1 is attached to the glycerol moiety through an *ether*, not an ester, linkage. Plasmalogens are special alkyl ether phosphoglycerides in which the ether-linked alkyl chain contains a double bond between C_1 and C_2; it is an *alk-1-enyl ether*, or a vinyl ether, with the double bond in the *cis* configuration. Ethanolamine is the most prevalent base in the plasmalogens. The hydrocarbon chain present in ether linkage usually contains 16 carbon atoms; it is derived from palmitic acid.

Alkyl ether phosphoglyceride   Plasmalogen   Platelet-activating factor (PAF)

Platelet-activating factor

A naturally occurring phosphatidylcholine derivative has been discovered that has several potent physiologic actions. One is to activate blood platelets, causing them to aggregate. Because of this, the compound is called platelet-activating factor (PAF). PAF also reduces blood pressure and mediates inflammation. There are two interesting aspects regarding the structure of PAF. A saturated fatty acyl chain, usually containing 16 carbon atoms, is attached to the *sn*-1 position of the glyceryl phosphorylcholine backbone in ether linkage. Therefore PAF is a 1-*O*-alkyl phosphoglyceride, that is, an alkyl ether. The second striking feature is that the acid attached in ester linkage at the *sn*-2 position is acetate, not a long-chain fatty acid as in other phosphoglycerides. PAF is thus a 1-*O*-alkyl-2-acetyl-glyceryl phosphorylcholine.

l-*O*-hexacetyl-2-acetyl-glyceryl phosphorylcholine (PAF)

Complete (de novo) synthesis

In mammals the main pathway for complete synthesis of phosphoglycerides involves cytidine diphosphate (CDP) and 1,2-diacylglycerol intermediates, as shown in Figure 10.13. The CDP compound is a derivative of either ethanolamine or choline. These are

Figure 10.13   Pathways for de novo biosynthesis of phosphatidylcholine,
phosphatidylethanolamine, and triacylglycerol. This pathway for
triacylglycerol synthesis occurs mainly in the liver and adipose tissue.

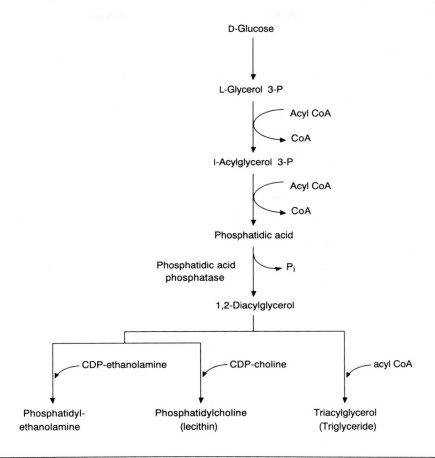

abbreviated as *CDP-ethanolamine* and *CDP-choline,* respectively. The structure of CDP-choline is shown in Figure 10.14. Phosphatidylcholine (lecithin) and phosphatidylethanolamine together make up 75% or more of the phosphoglycerides in most tissues and body fluids. Therefore this pathway involving the diacylglycerol intermediate is the main route for complete phosphoglyceride synthesis in humans. As shown in Figure 10.13, the diacylglycerol usually is derived from L-glycerol 3-phosphate. In a few situations, however, dihydroxyacetone phosphate serves as the first intermediate instead of L-glycerol 3-phosphate. In both cases, however, glucose is the ultimate precursor of the triose phosphate intermediate.

This pathway also produces triacylglycerol. In this case the diacylglycerol interacts with an acyl CoA rather than a CDP compound and the reactions are the same as in the general pathway shown in Figure 10.12. All the steps in this pathway beyond L-glycerol 3-phosphate formation are catalyzed by enzymes associated with the endoplasmic reticulum, both in phosphoglyceride and in triacylglycerol synthesis.

Figure 10.14   Structures of CDP-choline and CDP-diglyceride.

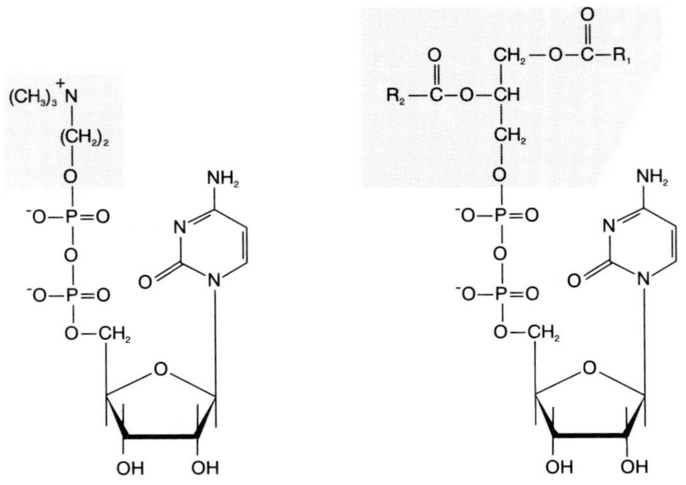

CDP-choline

CDP-diglyceride

A second pathway for the complete synthesis of phosphoglycerides exists, which is illustrated in Figure 10.15. This is called the phosphatidic acid pathway. As shown in Figure 10.13, phosphatidic acid is derived from glucose through L-glycerol 3-phosphate and is the immediate precursor of diacylglycerol. In the phosphatidic acid pathway, the phosphatase does not operate, and phosphatidic acid instead of diacylglycerol is the direct precursor of the phosphoglycerides. This pathway also involves a cytidine intermediate, *cytidine diphosphate diacylglycerol* (CDP-diglyceride). The structure of CDP-diglyceride is shown in Figure 10.14. The phosphatidic acid pathway produces primarily *cardiolipin* and *phosphatidylinositol*. In bacteria, however, it also forms phosphatidylserine, phosphatidylethanolamine by decarboxylation of phosphatidylserine, and phosphatidylcholine through methylation of phosphatidylethanolamine.

Partial synthesis

Acylation Phosphoglycerides also are synthesized by modifying a precursor compound. In the first of these partial synthesis pathways, an acyl residue is added by reaction with acyl CoA in the presence of an acyltransferase. This acylation pathway probably serves an important function in membrane structure, since it is likely that fatty acids are constantly being hydrolyzed from phosphoglycerides and subsequently being replaced by acylation of the lysoderivative that is formed in the course of normal membrane function. In this way the cell has the opportunity to quickly alter the fatty acyl composition of its membrane phosphoglycerides. This may permit rapid regulation of certain enzymes or carriers located within the membrane, since phospholipid fatty acid chains interact with these proteins. The types of fatty acyl groups inserted into the lysoderivative are controlled by the specificity of the acyltransferase. In this pathway it is usually the fatty acid in the *sn*-2 position 2 that is removed and replaced. Acyltransferases that act on the *sn*-2 position have greater specificity for polyunsaturated fatty acids, so that the fatty acyl group inserted into the lysoderivative in this pathway usually is polyunsaturated. Phospholipase A_2

Figure 10.15   Phosphatidic acid pathway for biosynthesis of phosphoglycerides.
                                        CMP, Cytidine monophosphate.

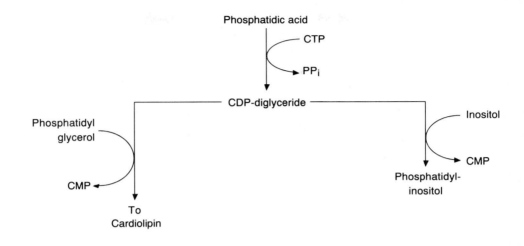

catalyzes the hydrolysis of the fatty acid ester from the *sn*-2 position. This enzyme depends on calcium. Therefore changes in intracellular calcium content may regulate the operation of this pathway. This is probably one of the mechanisms whereby calcium regulates cellular function. In this context it is considered likely that in at least some tissues the phospholipase A_2 reaction, because it releases primarily polyunsaturated fatty acids, provides the arachidonic acid for prostaglandin synthesis. The deacylation-acylation interconversion is illustrated in Figure 10.16.

Condensation The second partial pathway is specific for phosphatidylcholine synthesis and involves the condensation of two lysoderivatives. One of these accepts an acyl residue from the other.

2 Lysophosphatidylcholine   →   Phosphatidylcholine + Glycerylphosphorycholine

Unlike the acylation pathway, this condensation reaction does not play a major role in phosphatidylcholine metabolism.

Methylation In the third partial synthesis pathway, phosphatidylethanolamine is converted to phosphatidylcholine by three methylation reactions. S-*Adenosylmethionine* is the methyl donor for this conversion. Two methyltransferases are involved, and they are located in the microsomal fraction of cellular homogenates. These enzymes appear to be located on opposite sides of the membrane. The first methyltransferase adds one methyl group to phosphatidylethanolamine, and then the second enzyme, which depends on Mg^{++}, adds two more methyl groups to the monomethyl intermediate. This sequential methylation may provide a mechanism for the asymmetric distribution of phosphoglycerides in the phospholipid bilayer of membranes. As described in Chapter 12, phosphatidylcholine and phosphatidylethanolamine are enriched in opposite leaflets of the lipid bilayer. Through this methylation process the phosphatidylcholine that is formed from phosphatidylethanolamine would end up on the opposite side of the lipid bilayer. The

Figure 10.16   The acylation-deacylation cycle.

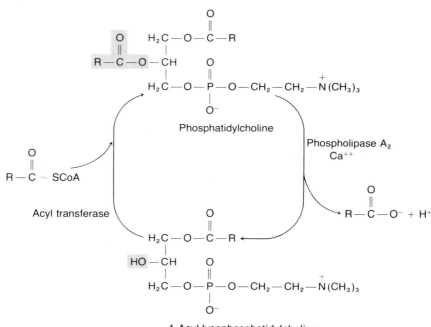

methylation pathway also has been implicated in transmembrane signaling, but the process appears to be too slow to operate effectively in signal transduction.

Base exchange The final partial synthesis pathway is base exchange. This reaction is calcium dependent and involves the replacement of the existing base of a phosphoglyceride by another base; for example, the replacement of ethanolamine in phosphatidylethanolamine by serine, without modification of the remainder of the molecule. Base exchange appears to be the only mechanism for *phosphatidylserine* synthesis in mammalian tissue.

Phosphoglyceride hydrolases

Phosphoglycerides are degraded by phospholipases. As shown in Figure 10.17, four types of phospholipases exist. Each hydrolyzes a different linkage and forms different products, such as fatty acids, lysophosphoglycerides, diacylglycerol, or phosphatidic acid. Phospholipase A_2 requires calcium for activity. Lysophospholipases that hydrolyze lysophospholipids also are present in the tissues.

Phosphatidylinositol and its phosphorylated derivatives

Phosphatidylinositol (PI) can be converted to two phosphorylated derivatives, phosphatidylinositol 4'-phosphate (PIP) and phosphatidylinositol 4', 5'-bisphosphate (PIP$_2$). These structures are shown in Figure 10.18. The phosphatidylinositols are rapidly hydrolyzed and subsequently resynthesized in a cyclic series of reactions, the phosphatidylinositol cycle, which is discussed in Chapter 17.

Figure 10.17   Sites of hydrolysis by phospholipases and products formed.

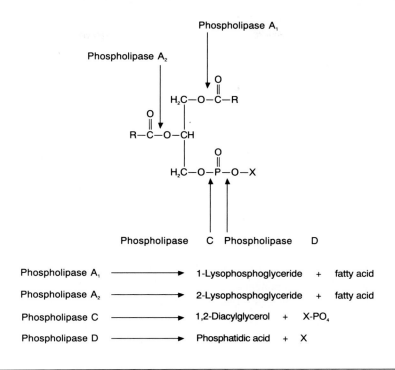

Phospholipase A$_1$	$\longrightarrow$	1-Lysophosphoglyceride	+	fatty acid
Phospholipase A$_2$	$\longrightarrow$	2-Lysophosphoglyceride	+	fatty acid
Phospholipase C	$\longrightarrow$	1,2-Diacylglycerol	+	X-PO$_4$
Phospholipase D	$\longrightarrow$	Phosphatidic acid	+	X

Alkyl ether and plasmalogen metabolism

The alkyl ether linkage discussed earlier is formed by exchange of the acyl group of *1-acyl-dihydroxyacetone phosphate* with a *fatty alcohol.* The fatty alcohol is formed by reduction of the corresponding fatty acid. The resulting *1-alkyl-dihydroxyacetone phosphate* is reduced to a *1-alkyl-glycerol-3-phosphate,* and this intermediate is converted to phosphoglycerides through the pathways illustrated in Figure 10.13. Plasmalogens are synthesized from the 1-alkyl-phosphoglycerides by dehydrogenation of the alkyl chain in position 1,2. The alkyl ether bond is more stable than the corresponding ester bond. Therefore the presence of alkyl ether derivatives may provide additional stability to membrane phospholipids.

The alkyl ether bond is cleaved during the degradation of these compounds by a microsomal enzyme system that requires oxygen and a reduced pteridine. One of the products is a fatty aldehyde.

Sphingolipids Sphingosine derivatives

Another series of lipids that occur in humans are derivatives of sphingosine, an 18-carbon atom, dihydric alcohol, that contains an amino group at C$_{17}$. When written in the following form, sphingosine bears some resemblance to a glycerol moiety containing a *trans*-alk-1-enyl residue:

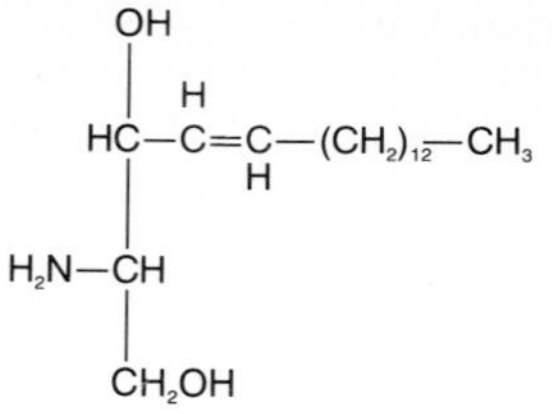

Sphingosine

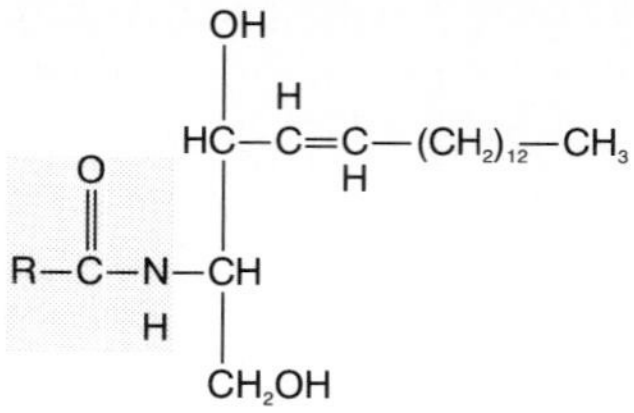

Ceramide

Dihydrosphingosine is a similar long-chain base that does not contain any double bonds. Fatty acids, usually containing 18 or more carbon atoms, can attach to the amino group in amide linkage. Sphingosine containing an amide-linked fatty acid is known as *ceramide*. Three types of ceramide derivatives are present in human tissue: *sphingomyelin*, which contains a phosphocholine group; *galactosylceramides,* which contain a β-D-galactosyl residue; and *glucosylceramides,* which contain a β-D-glucosyl residue.

Sphingomyelin

Because it contains phosphorus, sphingomyelin usually is considered as being a member of the phospholipid class. In sphingomyelin a phosphorylcholine residue is linked to the terminal hydroxyl group of ceramide. The structure is abbreviated as Chln-P-Cer. It is a major lipid component of some biologic membranes; for example, it accounts for 22% to 30% of the phospholipid in the human erythrocyte membrane.

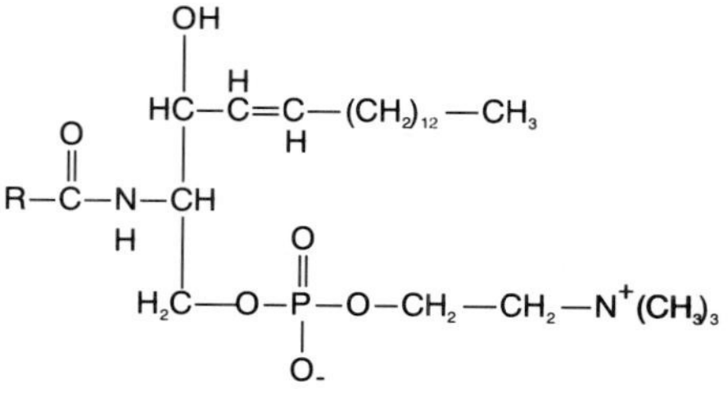

Sphingomyelin

Figure 10.18   Phosphatidylinositol and its phosphorylated derivatives.

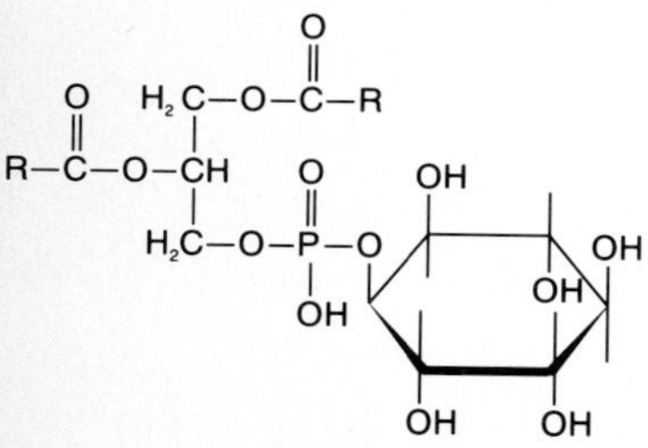

Phosphatidylinositol
(PI)

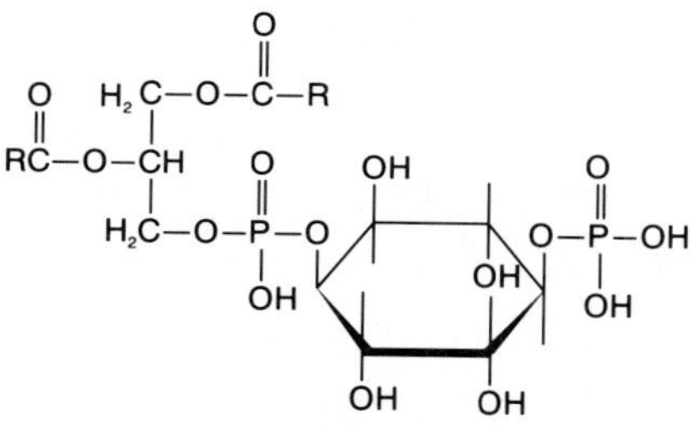

Phosphatidylinositol
4'-phosphate (PIP)

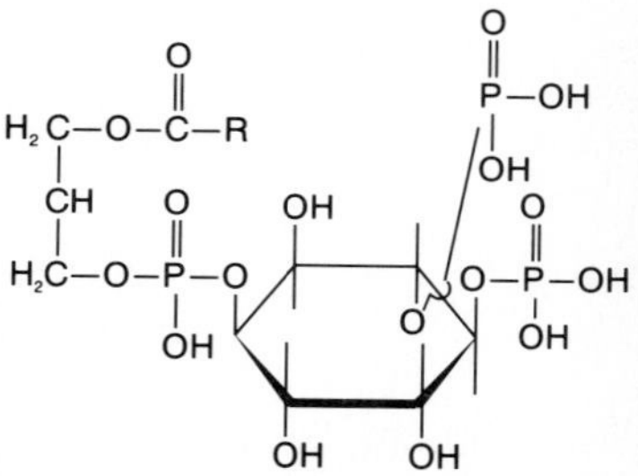

Phosphatiylinositol
4',5'-bisphosphate (PIP₂)

Glycosphingolipids

Glycosphingolipids are ceramides derivatives that contain one or more sugar residues.

Galactosylceramides Ceramides that contain a D-galactose residue at the terminal hydroxyl group are known as galactosylceramides. The galactose is linked to this hydroxyl through a β-D-glycosidic bond, and these compounds are abbreviated as Gal-Cer.

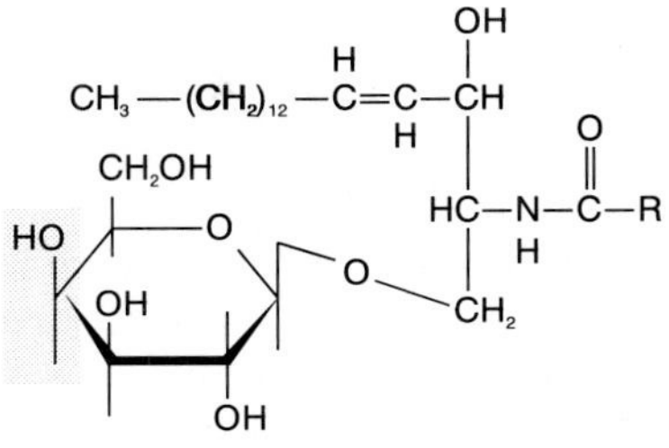

Galactosylceramide

Galactosylceramides do not contain phosphorus and are therefore not phospholipids. Ceramides that contain one monosaccharide residue are known as *cerebrosides*. These glycolipids are present in the brain and peripheral nervous system, particularly in the myelin sheath. A *sulfatide* is formed when a sulfate group is attached to one of the hydroxyl groups of the galactose residue. This form of sulfatide is abbreviated as S-Gal-Cer and is a sulfogalactosyl ceramide.

Glucosylceramides Ceramides that contain a D-glucose residue at the terminal hydroxyl group are known as glucosylceramides. Those that contain only the single glucose residue also are called cerebrosides. There are abbreviated as Glc-Cer. The glucose residue is atached through β-D-glycosidic linkage.

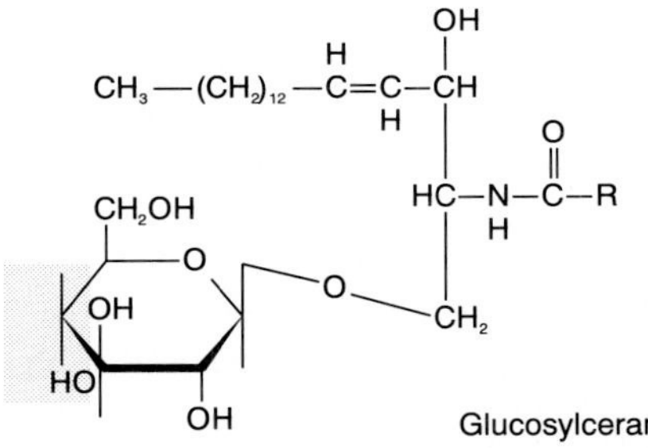

Glucosylceramide

Additional monosaccharide moieties can be attached to the terminal glucose residue, which is shaded, giving rise to the more complex glycosphingolipids, whose basic structures are presented in abbreviated form.

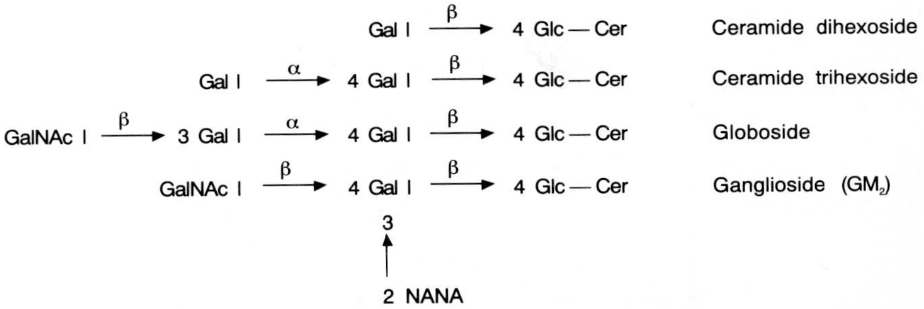

Each of these complex glycosphingolipids has a glucose residue attached directly to the ceramide. The globosides and gangliosides contain *N*-acetylgalactosamine (GalNAc), and the gangliosides also contain *N*-acetylneuraminic acid (NANA). These glycosphingolipids are contained in cell membranes, with their carbohydrate residues projected out into the surrounding extracellular fluid. Gangliosides are particularly prevalent in the gray matter of the brain and at nerve synapses.

Blood group antigens The glycosphingolipids present at the surfaces of cells are involved in cell-cell recognition and are antigenic, accounting for certain of the blood-group substances. The presence of glycosphingolipid and glycoprotein antigens on the surface of cells makes it necessary to match blood or tissue types before carrying out either a blood transfusion or a tissue transplantation.

Plants and bacteria contain glycolipid structures in which a monosaccharide is attached in glycosidic linkage to a 1,2-diacylglycerol. This class of glycolipid, which is known as the glycosylacylglycerols, is not present in mammals.

Sphingolipid metabolism

Sphingolipids, as just discussed, are derived from sphingosine, which in turn is synthesized from palmitoyl CoA and serine. *Pyridoxal phosphate,* a B vitamin derivative, is needed for sphingosine synthesis. NADPH is required in this reaction sequence. Dihydrosphingosine, which does not contain any unsaturated bonds, is an intermediate in this pathway. It is oxidized by a flavoprotein enzyme to sphingosine.

Ceramide is formed by *N*-acylation of sphingosine. A long-chain acyl CoA is the acyl group donor in ceramide synthesis.

$$\text{Sphingosine} + \text{FA}\sim\text{SCoA} \rightarrow \text{Ceramide} + \text{CoASH}$$

As shown in Figure 10.19, ceramide is the key intermediate in sphingolipid synthesis. This is the pathway employed for the synthesis of sphingomyelin (Chln-P-Cer) and the cerebrosides galactosylceramide (Gal-Cer) and glucosylceramide (Glc-Cer). *Gangliosides* are formed from glucosylceramides by the addition of hexose residues to the glucose. These hexose units are added sequentially from nucleoside diphosphate hexose intermediates.

Alternative pathways for sphingolipid synthesis have been described. In one pathway the phosphorylcholine or galactose residue is added to sphingosine instead of to ceramide. The resulting sphingosine derivative is then *N*-acylated by acyl CoA. *Psychosine* is the galactosylsphingosine intermediate that is formed in this pathway.

In another pathway the phosphorylcholine group is transferred from phosphatidylcholine to ceramide. Products of this reaction are sphingomyelin and diacylglycerol.

$$\text{Phosphatidylcholine} + \text{Ceramide} \rightarrow \text{Sphingomyelin} + \text{Diacylglycerol}$$

The formation of sphingomyelin by phosphocholine transfer can occur rapidly in response to stimulation in some cells, suggesting that this pathway may be involved in certain forms of membrane signal transduction.

Sulfatides are synthesized from galactosylceramides or glucosylceramides by sulfation of a hydroxyl group of the hexose. The sulfate donor is 3'-phosphoadenosine-5'-phosphosulfate (PAPS).

Sphingolipid degradation Sphingomyelin is degraded by *sphingomyelinase,* an enzyme that removes the phosphorylcholine residue. Cerebrosides and gangliosides are hydrolyzed by hexosidases, which remove one sugar residue at a time from the nonreducing terminus of the carbohydrate chain. These enzymes are contained in the lysosomes of the cell.

Figure 10.19   Sphingolipid synthesis through the ceramide (Cer) pathway. *Chln,* Choline.

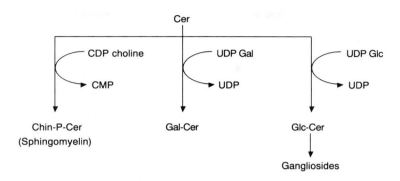

Chin-P-Cer
(Sphingomyelin)   Gal-Cer   Glc-Cer

                                          Gangliosides

Table 10.8   Common forms of lipid storage disease

Disease	Lipid accumulation	Enzyme deficiency	Primary organ involvement
Gaucher's	Glucocerebroside	Glucosylceramide β-D-glucosidase	Liver, spleen, brain
Niemann-Pick	Sphingomyelin	Sphingomyelinase	Brain, liver, spleen
Krabbe's	Galactocerebroside	Galactosylceramide β-D-galactosidase	Brain
Metachromatic leukodystrophy	β-Sulfogalactocerebroside	Sulfatide sulfatase	Brain
Fabry's	Ceramide trihexoside	α-D-Galactosidase	Kidney
Tay-Sachs	Ganglioside GM_2	β-D-Hexosaminidase A	Brain

Lipid storage diseases These diseases, which are also known as the *lipidoses,* result from inherited deficiencies of a sphingolipid catabolic enzyme. For example, sphingomyelinase is deficient in Niemann-Pick disease, and the β-D-glucosidase that hydrolyzes glucosylceramide is deficient in Gaucher's disease. This results in intracellular accumulation of the sphingolipid that ordinarily would be degraded by the enzyme that is deficient, often leading to death in early childhood. Table 10.8 lists the lipid storage diseases, the sphingolipid that accumulates in each case, the enzyme deficiency responsible for each disease, and the organs that are most affected.

Digestion and absorption of dietary fat

The only absolute dietary lipid requirement is that about 1% of the caloric intake be in the form of essential fatty acid. However, in practice, between 20% and 40% of the dietary caloric intake is lipid. Most of the dietary lipid intake is triacylglycerol, but small amounts of phosphoglycerides, cholesteryl esters, and cholesterol also are ingested. These lipids must be emulsified in the intestinal lumen, digested by hydrolytic enzymes, and absorbed into the intestinal mucosal cells. Triacylglycerol and phosphoglyceride digestion and absorption are discussed here; cholesterol is discussed in Chapter 11.

Emulsification of dietary lipids

Emulsification of the lipids present in the aqueous chyme occurs in the duodenum, where lipids interact with bile. The constituents of the bile that produce emulsification are the conjugated bile acids, phosphatidylcholine, and cholesterol. A complete discussion of bile is presented in Chapter 11. Emulsification functions to drive the poorly soluble dietary lipids into mixed micelles.

Micelles are aggregates formed in aqueous solution by a substance composed of both polar and nonpolar groups. Phosphatidylcholine is an example of a compound that forms micelles. It contains nonpolar fatty acid chains as well as the polar phosphorylcholine group and is therefore *amphipathic*. The nonpolar components orient themselves inside the aggregate, whereas the polar groups are on the outside, where they interact with the surrounding water molecules. *Mixed micelles* are micelles made up of more than one compound. For example, the lipid constituents of the bile—phosphatidylcholine, cholesterol, and conjugated bile acids—exist together in aggregates that are mixed micelles. Dietary triacylglycerols are not amphipathic and do not form micelles. However, the mixed micelles composed of the bile lipids are able to take up these very nonpolar materials. In other words, the mixed micelles offer triacylglycerol a suitable nonpolar environment within the interstices of the micellar structure and in this way function to disperse the dietary lipids in the aqueous intestinal chyme. In micellar form these lipids can be acted on by the digestive enzymes. After hydrolysis, the products diffuse from the micelle to the intestinal mucosal cell membrane.

Hydrolysis of dietary triacylglycerol

Pancreatic lipase Pancreatic lipase catalyzes the partial hydrolysis of triacylglycerols containing long-chain fatty acids.

$$\text{Triacylglycerol} + 2H_2O \rightleftharpoons \text{2-Monoacylglycerol} + \text{2 Fatty acid}$$

Almost all the ordinary dietary triacylglycerols are of the long-chain fatty acid variety and contain mostly 16- and 18-carbon-atom saturated and unsaturated fatty acids. Pancreatic lipase is specific for the fatty acid residues at positions 1 and 3 of the glyceryl moiety. Digestion of triacylglycerol largely stops at the 2-monoacylglycerol because the pancreatic triacylglycerol lipase exhibits very low activity toward this substrate. Both the 2-monoacylglycerol and the released fatty acids can pass through cell membranes, and they are absorbed by diffusion into the mucosal cells of the jejunum and ileum.

A protein cofactor that activates the pancreatic lipase, called *colipase,* also is produced by the pancreas. Pancreatic lipase acts on the dietary triacylglycerol after it has been incorporated into the mixed micelles in the intestinal lumen. The lipase acts at interfaces between water and the triacylglycerol molecules, and interfacial adsorption of the enzyme is an important step in the catalytic process. Colipase binds to the mixed micelle containing the ingested triacylglycerol and facilitates adsorption of the lipase to the complex, thereby activating triacylglycerol hydrolysis.

Phospholipase A$_2$ The phosphoglycerides present in the diet are digested by pancreatic phospholipase A$_2$. This enzyme catalyzes the hydrolysis of the fatty acid residue contained at the *sn*-2 position of the phospholipid, forming a 1-acyl lysophosphoglyceride.

$$\text{Phosphoglyceride} + H_2O \rightleftharpoons \text{1-Acyl lysophosphoglyceride} + \text{Fatty acid}$$

It is not known at present whether the lysophosphoglycerides enter the mucosal cells or are degraded further by *lysophospholipases,* enzymes that remove the remaining fatty acid residue from lysophosphatides. In terms of fulfilling caloric needs, phospholipids serve only a minor role relative to triacylglycerol.

Absorption and reesterification

Absorption of the lipid hydrolysis products from the mixed micelles into the mucosal cells is a passive process that occurs through diffusion. The main function of the intestinal mucosa in terms of lipid metabolism is to resynthesize the absorbed fatty acids and 2-monoacylglycerol into triacylglycerol (see Figure 10-12), since the dietary long-chain fatty acid is absorbed into the body only after it is reconverted into triacylglycerol. This is an energy-dependent process requiring 2 mol of ATP per mol of triacylglycerol synthesized. Both high-energy bonds of each ATP are hydrolyzed in this process. Therefore four high-energy bonds actually are expended for the resynthesis of one triacylglycerol molecule. There appear to be two reasons for this seemingly wasteful expenditure of energy; first, the intact triacylglycerol cannot efficiently diffuse to or through the mucosal cell membrane, and, second, the organism is given the opportunity of using endogenous fatty acids to alter the fatty acid composition of the triacylglycerol that is resynthesized and absorbed into the body. Triacylglycerols are synthesized in the intestine through the *2-monoacylglycerol pathway:*

$$\text{2-Monoacylglycerol} + \text{FA}{\sim}\text{SCoA} \rightarrow \text{1,2-Diacylglycerol} + \text{CoASH}$$
$$\text{1,2-Diacylglycerol} + \text{FA}{\sim}\text{SCoA} \rightarrow \text{Triacylglycerol;} + \text{CoASH}$$

Secretion and utilization of dietary triacylglycerols

Lipid droplets made up almost entirely of triacylglycerols accumulate in the mucosal cells. They are released into the lymph in the form of a lipoprotein, the *chylomicron*. This is described in detail in Chapter 16. Chylomicrons are secreted into the lymph and pass from it into the venous blood. Their triacylglycerol content is removed through the action of a hydrolytic enzyme on the surface of the capillary endothelial cells, *lipoprotein lipase* (see Table 10.7), and the released fatty acids are taken up by the tissues. Much of the absorbed dietary fatty acid is deposited in the adipose tissue.

If the plasma triacylglycerol concentrations are measured in a healthy person who has fasted overnight and then eaten a meal containing 50 to 100 g of triacylglycerol; a relationship similar to that shown in Figure 10.20 is observed. In this situation, we use triglyceride instead of triacylglycerol because this is the common clinical usage. Before a meal, the plasma triglyceride concentration in a normal person usually is less than 1.7 mmol/L. After the meal, this gradually rises to a peak within 3 to 4 hr and then returns

Figure 10.20

Normal triglyceride absorption study. *TG* on ordinate refers to triglycerides.

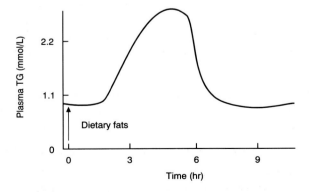

to normal within 6 to 8 hr. The triglyceride increment above fasting is mainly that absorbed in chylomicrons. Development of hypertriglyceridemia is normal after a fatty meal; prolonged and excessive hypertriglyceridemia after eating fat is abnormal.

Clinical comment

Medium-chain triglycerides The triglycerides present in the ordinary foods that we eat contain long-chain fatty acids. Medium-chain triglycerides have been used in certain therapeutic diets. These are triglycerides, prepared synthetically, that contain medium-chain fatty acids having eight- or ten-carbon atoms instead of the usual long-chain acids. Medium- and long-chain triglycerides are digested and metabolized differently. Much of the ingested medium-chain triglyceride is absorbed intact either onto the mucosal cell villi or actually into the cells, and lipolysis is mediated by an intracellular lipase rather than by the pancreatic lipase secreted into the intestinal lumen. The medium-chain triglycerides are completely degraded to fatty acid and glycerol, another difference between them and the ordinary dietary triglycerides. Finally, the medium-chain fatty acids that are produced are *not* reesterified and secreted into the lymph in the form of chylomicrons. Instead, they pass directly into the portal vein as fatty acids, bind physically to plasma albumin, and are delivered directly to the liver as fatty acids by the portal circulation. The differences in digestion and metabolism between the medium-chain and the ordinary dietary triglycerides are shown schematically in Figure 10.21.

Figure 10.21   Differences in utilization of ordinary dietary triglycerides and medium-chain triglycerides.

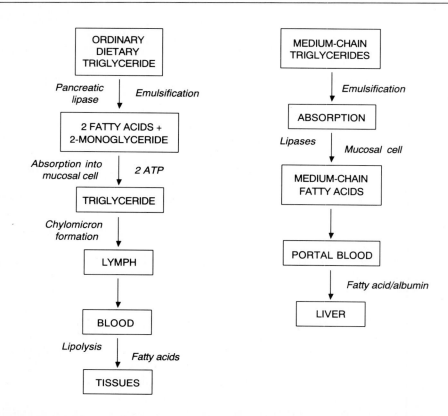

Adipose tissue   Fat in the form of triacylglycerol is the major storage form of energy in humans. Of the three main nutrients, only fat can be stored in large quantities. This occurs because the body contains specialized mesenchymal cells, *adipocytes,* that are devoted solely to the function of storing fat. Obesity, excessive accumulation of fat in the body, is associated with an increase in the number of adipocytes as well as an increase in the size of the individual adipocytes resulting from packing with triacylglycerol. It is therefore both a hypertrophy and a hyperplasia of the adipose tissue. Through mechanisms that as yet are not understood, the presence of excessive numbers of adipocytes signals the body to synthesize more triacylglycerol so that they can be filled, leading ultimately to an excess of total stored fat in the body.

As opposed to fat, no true storage site exists for protein or its constituent amino acids in the body. In times of severe need such as starvation, tissue and plasma proteins are catabolized to supply essential amino acids, but this is clearly a pathologic situation. Also, no specialized cell exists for glycogen storage; it is simply crowded into the cytoplasm of cells, particularly in the liver and muscles. Because of this, the amount of glycogen that can be stored is limited, a total of approximately 0.5 kg. A primary metabolic drive in the body in times of abundant food supply is to convert excess calories from carbohydrate into fat so that they can be kept in reserve in the large and expandable adipose storage depots.

Triacylglycerol accumulation in adipocytes The glyceryl moiety of triacylglycerol is derived from the glucose that is delivered to the adipocyte through the blood. The transport of glucose into the adipocyte is stimulated by insulin. Some fatty acid that is incorporated into the triacylglycerols is synthesized within the adipocyte from glucose. The remainder is delivered through the blood in the form of triacylglycerol contained in plasma lipoproteins, either chylomicrons or very low–density lipoproteins (VLDL). In both cases the lipoprotein triacylglycerols must be hydrolyzed by lipoprotein lipase so that their fatty acid content can enter into the adipose cell. Insulin also facilitates this process by stimulating the production of lipoprotein lipase. Figure 10.22 summarizes the process of triacylglycerol accumulation in adipocytes.

Fatty acid mobilization from adipose tissue To leave the adipocyte, the triacylglycerol must be hydrolyzed to fatty acids and glycerol. The fatty acids are released into the blood and are transported as free fatty acid (FFA) in a physical complex with albumin. Glycerol also is released from the adipose tissue during lipolysis. The glycerol produced by lipolysis cannot be used by adipocytes because they lack the enzyme glycerol kinase. Therefore the glycerol is released and transported to the liver, an organ that contains sufficient amounts of glycerol kinase to efficiently metabolize glycerol to glucose by the process of gluconeogenesis. FFA is mobilized in large amounts from adipose tissue during periods of fasting, anxiety, or physical exertion. In this way the tissues of the body are ensured a constantly available circulating supply of fat, either from the diet after eating or from the adipose tissue in the postabsorptive state or in stressful situations. Conversely, fat is accumulated in the adipose tissue when food is plentiful and the individual is calm and resting.

The general aspects of adipose tissue function and lipid release are summarized in Figure 10.23. Fasting, sympathetic nervous system discharge (norepinephrine), or release of many different hormones into the circulation (epinephrine, adrenocorticotropic hormone [ACTH], growth hormone, or glucagon) stimulates FFA release from the adipose tissue. Norepinephrine, epinephrine, ACTH, and glucagon activate the adipocyte triacylglycerol lipase, also called the *hormone-sensitive lipase.* These hormones combine with receptor sites on the cell membrane and activate *adenylate cyclase.* Cyclic AMP is formed and

Figure 10.22   Triacylglycerol accumulation in adipocytes. *TG,* Triacylglycerol; *DHAP,*
dihydroxyacetone phosphate; *FA,* fatty acid.

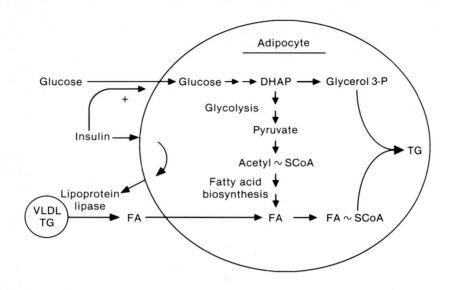

activates a protein kinase that in turn catalyzes an ATP-induced phosphorylation of the
hormone-sensitive lipase and transforms it from an inactive to an active form. The
hydrolysis catalyzed by the hormone-sensitive lipase is the rate-limiting step in triacylglycerol catabolism. Hydrolysis of the resulting diacylglycerol and, subsequently, the
monoacylglycerol is mediated by a single enzyme, *monoacylglycerol lipase.*

The other hormones that stimulate lipolysis, such as thyroxine and growth hormone,
act more slowly. They probably operate by increasing the synthesis of a regulatory protein
rather than by activating existing adenylate cyclase. Glucocorticoids also stimulate lipolysis, but they facilitate the action of other fat-mobilizing hormones and do not exert
a direct lipolytic effect.

Regulation of lipolysis High levels of glucose and insulin in the blood stimulate
triacylglycerol accumulation in adipose tissue. When this occurs, the cAMP content in
the adipocyte is reduced. Conversely, low blood glucose and insulin concentrations enhance the mobilization of fatty acid from the adipocyte. In addition, methylxanthine drugs
such as caffeine and theophylline enhance fatty acid mobilization from adipose tissue.
These substances inhibit phosphodiesterase, the enzyme that inactivates cAMP. As a
result, the cAMP level in the cell increases, helping to maintain the hormone-sensitive
lipase in active form. Prostaglandin E_2, insulin, and nicotinic acid inhibit fatty acid
mobilization by unknown mechanisms. Since the adenylate cyclase reaction and cAMP
are central to the understanding of many hormonal mechanisms, these subjects are dealt
with more comprehensively in Chapter 17.

Clinical comment   **Obesity and weight reduction** When a person is overweight, an excessive amount of
triglyceride is stored in the adipose tissue. The triglyceride accumulates as shown in
Figure 10.22. This occurs when the caloric intake exceeds the amount needed for body
function and the amount of work being done.

Figure 10.23   Fatty acid mobilization from adipocytes.

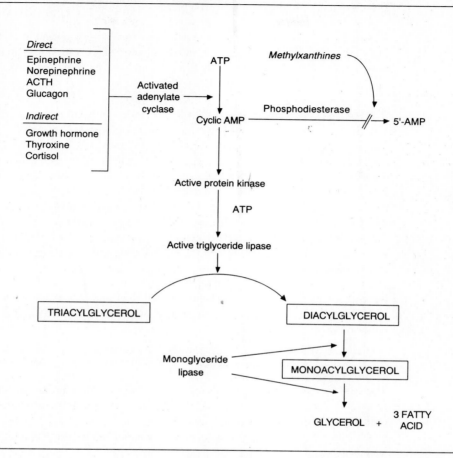

Obesity is treated with a weight-reducing diet. The aim is to feed less calories than the body requires. To meet the deficit, some of the stored triglyceride in the adipose tissue must be broken down. The fatty acid is released into the circulation and utilized throughout the body as a source of energy. This occurs by the mechanism shown in Figure 10.23. Weight loss occurs as the amount of triglyceride in the adipose tissue decreases.

Greater weight loss will occur if energy utilization can be increased to a higher level. This can be achieved through exercise, such as walking, swimming, jogging, or bicycle riding. Extra calories must be expended for the added activity. This makes the deficit between dietary caloric intake and body expenditure greater. To make up the larger deficit, more triglyceride stored in the adipose tissue must be mobilized. This leads to greater, more rapid weight loss.

Ketone bodies

Three metabolic products, *acetoacetate*, D *β-hydroxybutyrate*, and *acetone*, are referred to as ketone bodies (see p. 8). They are produced by the liver mitochondria from acetyl CoA when excessive amounts of fatty acid are being oxidized and glucose availability is limiting, for example, in starvation and diabetic ketoacidosis. Ketone body synthesis is inhibited when an excess of carbohydrate is available.

Ketone body synthesis

Ketone bodies are released by the liver when fatty acid is the main metabolic sulstrate. They are produced in large amounts during periods of starvation or ketoacidosis to supply energy to the heart, skeletal muscles, and brain. The pathway for ketone body formation in the liver is shown in Figure 10.24.

3-Hydroxy-3-methylglutaryl CoA lyase pathway Ketone bodies are synthesized inside liver mitochondria from 3-hydroxy-3-methylglutaryl CoA (HMGCoA) using acetyl CoA generated primarily by the β-oxidation of fatty acids. Initially, two molecules of acetyl CoA condense to form acetoacetyl CoA, which then condenses with another acetyl CoA molecule to form HMGCoA. The HMGCoA is then cleaved through the action of *HMGCoA lyase* to yield acetoacetate and acetyl CoA, the latter being available to reenter the pathway and combine with another molecule of acetoacetyl CoA. Some of the acetoacetate that is formed is released into the plasma, whereas the rest of it is reduced through the action of a NADH-dependent reductase to β-hydroxybutyrate and released from the liver in this form. When the acetoacetate concentration is elevated, some of it is spontaneously decarboxylated to acetone.

Ketone body oxidation

Ketone bodies secreted into the blood by the liver and are taken up by the brain, heart, and skeletal muscles as a source of energy. Acetoacetate is activated in these tissues by either of two reactions:

$$\text{Acetoacetate} + \text{Succinyl CoA} \rightleftharpoons \text{Acetoacetyl CoA} + \text{Succinate}$$
$$\text{Acetoacetate} + \text{CoASH} + \text{ATP} \rightleftharpoons \text{Acetoacetyl CoA} + \text{AMP} + \text{PP}_i$$

The first is a transacylase involving succinyl CoA; the other is an acyl CoA synthase reaction linked to ATP. Acetoacetyl CoA that is formed in these reactions undergoes β-oxidation, producing two acetyl CoA.

β-Hydroxybutyrate is activated through an acyl CoA synthase reaction that also is ATP linked:

Figure 10.24   Ketone body biosynthetic pathway.

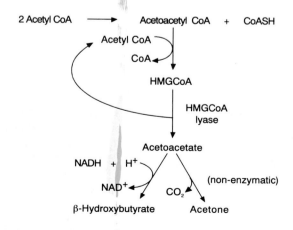

$$\text{β-Hydroxybutyrate} + \text{CoASH} + \text{ATP} \rightleftharpoons \text{β-Hydroxybutyryl CoA} + \text{AMP} + \text{PP}_i$$

The resulting β-hydroxybutyryl CoA is oxidized to acetoacetyl CoA through the action of a NAD-linked dehydrogenase, and the acetoacetyl CoA is utilized as already noted. Acetone, the third ketone body, cannot be further metabolized by the tissues.

Clinical comment

Ketosis The amounts of ketone bodies that are used by the tissues are proportional to their arterial concentration until this exceeds 70 mg/dl. Above this concentration the oxidation process is saturated, the concentration in the glomerular filtrate exceeds the maximum tubular reabsorption rate, and large amounts of ketone bodies are excreted in the urine. This condition is known as *ketonuria*. Acetone also is excreted by the lungs when the arterial ketone body concentration is high, and this odor is easily detected in the breath of patients suffering from *ketosis*. In addition, ketosis produces a metabolic acidosis because the buildup of acetoacetic and β-hydroxybutyric acids exceeds the buffering capacity of the plasma.

Bibliography

Gurr MI and James AT: Lipid biochemistry: an introduction, London, 1980, Chapman & Hall.

Hanahan DJ: Platelet activating factor: a biologically active phosphoglyceride, Annu Rev Biochem 55:483, 1986.

Lands WEM, editor: Proceedings of the American Oil Chemists' Society short course on polyunsaturated fatty acids and eicosanoids, Champaign, Ill, 1987, American Oil Chemists' Society.

Majerus PW et al: The metabolism of phosphoinositide-derived messenger molecules, Science 234:1519, 1986.

Mangold HK and Paltauf F, editors: Ether lipids: biochemical and biomedical aspects, New York, 1983, Academic Press, Inc.

McGarry JD and Foster DW: Regulation of hepatic fatty acid oxidation and ketone body production, Annu Rev Biochem 49:395, 1980.

Vance DE and Vance JE, editors: Biochemistry of lipids and membranes, Menlo Park, Calif, 1985, The Benjamin/Cummings Publishing Co, Inc.

Wakil SJ, Stoops JK, and Joshi VC: Fatty acid synthesis and its regulation, Annu Rev Biochem, 52:537, 1983.

Wion KL: Human lipoprotein-lipase complementary DNA sequence, Science 235:1638, 1987.

Clinical examples

Case 1

Acetyl CoA carboxylase deficiency

An infant girl was hospitalized for severe respiratory problems. Subsequently she was found to have myopathy, severe brain damage, and poor growth. When she was fed a high-carbohydrate diet, the urine contained large quantities of short-chain fatty acids. A liver biopsy revealed that almost no acetyl CoA carboxylase activity was present, whereas propionyl CoA carboxylase activity was observed. Analysis of cultured skin fibroblasts confirmed this finding; the acetyl CoA carboxylase activity was very low, whereas the propionyl CoA carboxylase activity was within the normal range.

Biochemical questions

1. What metabolic processes would be defective in this child as a result of the deficiency of acetyl CoA carboxylase activity?

2. Is it possible that a biotin deficiency could be responsible for a loss of acetyl CoA carboxylase activity?

3. What evidence is presented that the defect is not caused by a biotin deficiency?

4. Why might a high-carbohydrate diet accentuate the excretion of short-chain fatty acid in this case?

5. What lipid abnormality might be related to the neonatal respiratory difficulties?

Case discussion

Acetyl CoA carboxylase is the enzyme that catalyzes the carboxylation of acetyl CoA to malonyl CoA. This is the rate-limiting step in de novo fatty acid synthesis. It is also the step at which many of the regulatory effects are exerted. For example, the enzyme is activated by citrate and inhibited by palmitoyl CoA. These substances regulate the activity by affecting the interconversion of the active polymeric form and the inactive protomeric form. In addition, the nutritional state influences the activity; this is increased when carbohydrate is fed and reduced during fasting or when the diet is high in fat content.

1. Defective metabolic processes An acetyl CoA carboxylase deficiency will affect the capacity of the body to synthesize fatty acids de novo. One must also remember that the main fatty acid elongation pathway in the cell, which occurs in the endoplasmic reticulum, utilizes malonyl CoA as the elongating agent. Therefore fatty acid chain elongation also would be defective if there is a deficiency of acetyl CoA carboxylase.

2. Biotin deficiency Biotin is a component of acetyl CoA carboxylase and is the cofactor involved in the carboxylation mechanism. Because of this, it is reasonable to suspect that a biotin deficiency might be the cause of the observed deficiency in acetyl CoA carboxylase activity.

3. Evidence against biotin deficiency There can be no biotin deficiency in this case because no defect exists in propionyl CoA carboxylase activity. Propionyl CoA carboxylase, the enzyme that converts propionyl CoA to methylmalonyl CoA, also requires biotin and would not have normal activity if a biotin deficiency were present. However, it is possible that the enzyme linking biotin to this specific carboxylase is deficient.

4. High-carbohydrate diet Normally, de novo fatty acid biosynthesis is most active when the diet has a high content of carbohydrate. In this situation, glucose is converted to pyruvate and then to acetyl CoA, the latter being the substrate for fatty acid synthesis. Because of the deficiency of acetyl CoA carboxylase, it is likely that acetyl CoA would accumulate when the carbohydrate input to the liver was elevated. This could lead to excretion of excessive amounts of acetic acid in the urine and possibly the formation of large quantities of ketone bodies. Increased ketone body production would result in the excretion of acetoacetate and β-hydroxybutyrate in the urine.

5. Respiratory difficulties Since the many types of cells in the lung undoubtedly require fatty acid to function properly, defects in fatty acid de novo biosynthesis and chain elongation could compromise respiratory function in numerous ways. Although one cannot pinpoint exactly what caused the respiratory problems, the possibility that it somehow involved surfactant is very high on the list. Surfactant is a lipoprotein material produced and secreted by the alveolar type II cells. It lowers the alveolar surface tension, thereby facilitating gas exchange. Lack of adequate amounts of surfactant is a major cause of the respiratory distress syndrome in the newborn. The major lipid component of surfactant is dipalmitoyl phosphatidylcholine. Palmitic acid is also the main product of the fatty acid de novo synthesis pathway. Whereas palmitic acid can be obtained from dietary fat, it is possible that enough saturated fatty acid was not available in the feedings during the early neonatal period. If this were the case and the lung depended on fatty acid de novo biosynthesis, it is easy to see how an acetyl CoA carboxylase deficiency might lead to respiratory problems.

Reference

Blom W and Keizer SMPF deM: Acetyl-CoA carboxylase deficiency: an inborn error of de novo fatty acid biosynthesis, N Engl J Med 305:465, 1981.

Wolf B and Heard GS: Disorders of biotin metabolism. In Scriver CR et al, editors: The metabolic basis of inherited disease, ed 6, New York, 1989, McGraw-Hill Information Services Co.

Case 2

Glucosylceramide lipidosis (Gaucher's disease)

A 34-year-old woman was admitted to the hospital because of easy bruising and excessive bleeding. Examination of her abdomen revealed a firm, nontender mass in the left upper quadrant that was judged to be an enlarged spleen. A mass also was present in the right upper quadrant of the abdomen, probably an enlargement of the liver. Examination of the blood revealed *pancytopenia,* a decrease of all blood cells. Coagulation tests indicated that a prolonged bleeding time was the only abnormality. Examination of the bone marrow revealed the presence of Gaucher's cells.

Biochemical questions

1. What are the lipidoses?
2. How is the finding of Gaucher's cells in the bone marrow related to the metabolic defect in this disease?
3. What biochemical tests can be performed to identify individuals with a form of lipidosis? Will these tests detect the carrier state as well as the patient with overt disease?

Case discussion

1. Lipidoses The lipidoses are a group of lipid storage diseases that occur in humans. Each is characterized by the accumulation of a particular lipid in the cells of the reticuloendothelial system and central nervous system. All the lipid storage diseases result from an inherited deficiency of the enzyme that catalyzes the degradation of the lipid that accumulates. The lipids that are involved are sphingolipids, including cerebrosides, gangliosides, sulfatides, or sphingomyelin. Table 10.8 lists the common lipid storage diseases, the lipid that accumulates in each, the enzymatic deficiency, and the organs that are most severely involved.

2. Gaucher's cells Gaucher's disease is one of the most common forms of lipid storage disease. The most severe types occur in infancy and lead to early death. However, as is illustrated by this case, milder forms of the disease do occur in the adult. Gaucher's disease is characterized by the accumulation of a β-glucosylceramide. This is a result of a deficiency or absence of β-glucosylceramide glucosidase. The lipid accumulates in the tissues, causing splenic and hepatic enlargement as in this patient. Enlarged cells also appear in the bone marrow resulting from glucosylceramide accumulation, and these are diagnostic for this disease. The enlarged spleen acts as a trap for blood cells, filtering them out of the blood and thereby producing the pancytopenia. Bruising and bleeding resulted from the fact that the blood *platelets* are one of the cells filtered out by the enlarged spleen. Platelets are important in the initiation of blood coagulation, and bleeding often results when a platelet deficiency exists. In severe cases the lipid also accumulates in the central nervous system, causing brain damage.

3. Biochemical tests One can test for the activity of this glucosidase and most of the other enzymes that are involved in the various lipid storage diseases by examining blood *leukocytes.* These cells are readily available from a small sample of venous blood. In contrast, the tissues where the lipid accumulation occurs are often difficult or impossible to biopsy for biochemical study. Leukocytes are isolated from the blood by centrifugation. After disruption, their contents are incubated with the appropriate substrate; that is, glucosylceramide if Gaucher's disease is suspected. The enzyme in question, a β-D-glucosidase, catalyzes the removal of the β-D-linked glucose residue from the ceramide. Therefore the amount of glucose that is released would be analyzed; this is a measure of the cellular β-D-glucosidase activity. Leukocytes obtained from a normal individual are used as a reference. This test is important in making a definitive diagnosis as well as in detecting the *carrier state,* since less than the normal amount of the enzyme is present in the cells of carriers. In this way one can determine whether apparently normal people are carriers of these diseases.

Reference

Brady RO and Baranger JA: Glucosylceramide lipidosis: Gaucher's disease. In Scriver CR et al, editors: The metabolic basis of inherited disease, ed 6, New York, 1989, McGraw-Hill Information Services Co.

Case 3	Carnitine deficiency

A 19-year-old girl was referred to a university medical center because of easy fatigability and poor exercise tolerance. Careful neurologic examination revealed some muscle weakness in her extremities. Several muscle biopsies were performed. Microscopic examination indicated that the muscle was filled with vacuoles containing lipid. Chemical measurements indicated that these muscle specimens contained greatly elevated amounts of triacylglycerol, but only one-sixth as much carnitine as biopsy specimens obtained from other patients who did not have any primary muscle disease.

Biochemical questions

1. What is the main intracellular function of carnitine?
2. Would you expect that fatty acid β-oxidation is impaired in this patient?
3. Would you expect that the oxidation of pyruvate (derived from glucose) might be impaired in this patient?
4. How might the carnitine deficiency account for the triacylglycerol accumulation in the muscles?

Case discussion

1. Function of carnitine Carnitine is involved in the β-oxidation of long-chain fatty acids. Its structure is shown on p. 417. It operates in the mitochondrial translocase system that transports long-chain fatty acyl groups across the inner mitochondrial membrane. In order that β-oxidation of fatty acids can occur, the fatty acyl chain must be translocated into the mitochondrial matrix, the site of the β-oxidation system. Fatty acids are activated to acyl CoA thioesters either in the endoplasmic reticulum or in the outer mitochondrial membrane. Acyl CoA thioesters, however, cannot cross the inner mitochondrial membrane to reach the site where β-oxidation occurs. To accomplish this, the acyl group is transferred across as the acylcarnitine ester, being first transesterified to carnitine and then transesterified back to CoA after crossing the membrane in the form of a carnitine ester.

2. Fatty acid β-oxidation This patient's muscles were deficient in carnitine, and therefore she could not efficiently transport long-chain fatty acyl groups into her mitochondria for β-oxidation. Mitochondrial fatty acid β-oxidation is a major source of energy for many tissues, including the skeletal muscles. The patient's muscular weakness and intolerance toward exercise are explained by her inability to derive sufficient energy from fatty acid β-oxidation for muscular work. Although the β-oxidation system itself is not defective, it is not operating effectively because adequate amounts of long-chain fatty acid substrates cannot gain access to it.

3. Pyruvate As opposed to fatty acid oxidation, one would not expect any defect in glucose or pyruvate oxidation in this patient. As with fatty acids, glucose is converted to acetyl CoA before oxidation in the Krebs cycle. The carbon atoms from glucose enter the mitochondria in the form of pyruvate. Pyruvate dehydrogenase, the enzyme complex that catalyzes the conversion of pyruvate to acetyl CoA, is located on the matrix side of the inner mitochondrial membrane. Therefore the pyruvate is converted to acetyl CoA after it already has crossed the inner mitochondrial membrane. The carnitine-dependent step is not involved in pyruvate transport, and a carnitine deficiency would not be expected to impair either glucose or pyruvate oxidation. In fact, to replace some of the energy ordinarily derived from fatty acid β-oxidation, more than the usual amounts of glucose and pyruvate were oxidized via the Krebs cycle in this patient.

4. Triacylglycerol accumulation It is likely that triacylglycerol accumulated in this patient's muscles because of the defect in fatty acid oxidation produced by the carnitine deficiency. No impairment occurred in the activation of fatty acid in the muscle, so that long-chain acyl CoA could be formed. Not being able to enter the oxidative pathway, more than the usual amounts of fatty acyl groups were diverted into other pathways. Muscle, as well as other tissues, store fatty acids as triacylglycerol. Therefore it is expected that any excess of fatty acyl groups within the muscle might be incorporated into this storage form.

Reference

Engel AC and Angelini C: Carnitine deficiency of human skeletal muscle with associated lipid storage myopathy: a new syndrome, Science 179:899, 1973.

Roe CR and Coates PM: Acyl-CoA dehydrogenase deficiences. In Scriver CR et al, editors: The metabolic basis of inherited disease, ed 6, New York, 1989, McGraw-Hill Information Services Co.

Case 4	Obesity

A 19-year-old woman sought medical help because she was 30 kg overweight. Most of her excess weight was in the form of adipose tissue triacylglycerol. A dietary history revealed that her diet was extremely poor. Much of her caloric intake was carbohydrate—candy, cookies, cake, soft drinks, and beer; her dietary fat intake was actually quite moderate.

Biochemical questions

1. How is it possible to form excess amounts of triacylglycerol in the body if a diet contains predominantly carbohydrate?
2. How does acetyl CoA generated inside the mitochondria reach the cytoplasm for use by the fatty acid de novo biosynthetic pathway?
3. Why is bicarbonate required for fatty acid synthesis?
4. What is the rate-limiting enzyme in fatty acid de novo biosynthesis?
5. How might the carbohydrate ingested by this patient supply the reducing equivalents needed for fatty acid biosynthesis?
6. Devise a test that would indicate whether this patient could mobilize the triacylglycerol that is stored in her adipose tissue.

Reference

Brindley DN: Metabolism of triacylglycerols. In Vance DE and Vance JE, editors: Biochemistry of lipids and membranes, Menlo Park, Calif, 1985, The Benjamin/Cummings Publishing Co, Inc.

Case 5   Lipogranulomatosis (Farber's disease)

A 9-month-old-girl was admitted to the pediatric unit of a hospital because of poor weight gain and psychomotor retardation. Although the child appeared well at birth, at 5 mo of age she developed these problems and became progressively worse. Physical examination confirmed the nutritional failure and psychomotor retardation. Subcutaneous nodules, hepatomegaly, and splenomegaly were observed. Despite vigorous supportive therapy, she deteriorated rapidly and died 3 wk after admission. Tissue specimens were obtained for histologic and chemical analysis during the postmortem examination. Large quantities of lipid-staining material were observed in many tissues, and this was demonstrated chemically to be ceramide.

Biochemical questions
1. In what types of lipids is ceramide found?
2. How is sphingosine related to ceramide?
3. Are fatty acids present in ceramide?
4. Based on our current knowledge concerning the pathologic mechanisms causing lipid-storage diseases, would you expect the cause of this problem to be excessive synthesis of ceramide?
5. How is pyridoxal phosphate related to ceramide?
6. What amino acid is involved in ceramide synthesis?

Reference
Moser HW et al: Ceramidase deficiency: Farber's lipogranulomatosis. In Scriver CR et al, editors: The metabolic basis of inherited disease, ed 6, New York, 1989, McGraw-Hill Information Services Co.

Case 6   Cystic fibrosis

An 8-year-old boy was referred to a pediatrician because he was considerably underweight for his size. A history revealed that he suffered repeated lower respiratory tract infections, many of which required treatment with antibiotics. Although he was eating reasonably well, he had several bowel movements each day and repeated episodes of diarrhea. Analysis of the feces indicated that they contained a large amount of fat. Chloride analysis of the sweat confirmed the diagnosis of cystic fibrosis.

Biochemical questions
1. How is dietary fat digested and absorbed?
2. Cystic fibrosis causes pancreatic damage. How would this lead to malabsorption of fat?
3. Analysis of the plasma lipids demonstrated a low content of linoleic acid (9,12-18:2). What is the cause of this abnormality? Would you also expect a deficiency in the plasma content of palmitic acid (16:0), oleic acid (9-18:1), or arachidonic acid (5,8,11,14-20:4)?
4. An attempt was made to treat this boy with a very low-fat, high-carbohydrate diet. Should such a diet be rich in glucose, sucrose, or starch? Explain.
5. The boy also was given a pancreatic enzyme tablet with each meal in an attempt to enhance his ability to utilize dietary fat. Which enzyme should be prescribed?

Reference
Talamo RC, Rosenstein BJ, and Berninger RW: Cystic fibrosis. In Scriver CR et al, editors: The metabolic basis of inherited disease, ed 6, New York, 1989, McGraw-Hill Information Services Co.

Additional questions and problems

1. Which of the following substances would most likely increase the glycerol concentration of the blood plasma: insulin, glucose, or epinephrine? Explain your answer.
2. During a 24-hour fast, the plasma ketone body concentration rises. How does this occur, and what is the metabolic role of ketone bodies?
3. How would biotin deficiency affect fatty acid synthesis? Would fatty acid chain elongation and desaturation also be affected?
4. Would a severe pantothenic acid deficiency affect fatty acid synthesis? Would it also affect fatty acid β-oxidation, chain elongation, or desaturation?
5. When large amounts of fatty acid are undergoing β-oxidation, what happens to glucose oxidation through the Krebs cycle?

Multiple choice problems

1. A 14-carbon fatty acid undergoes complete β-oxidation. Which of the following is true?

	Number of β-oxidation cycles	Number of acetyl CoAs formed
A.	6	6
B.	7	6
C.	6	7
D.	7	7
E.	7	14

2. How many ATP (high-energy phosphate bonds) are produced when palmitoyl-coenzyme A, a 16-carbon atom saturated acyl CoA, is oxidized completely to CO_2 and H_2O?
   A. 35
   B. 40
   C. 96
   D. 131
   E. 136

3. ATP can be formed as a result of β-oxidation of fatty acids to acetyl CoA because, in the β-oxidation process:
   1. NADH is formed.
   2. NADPH is formed.
   3. $FADH_2$ is formed.
   4. Lipoic acid is reduced to dihydrolipoic acid.
      A. 1, 2, and 3.
      B. 1 and 3.
      C. 2 and 4. only
      D. 4 only.
      E. All are correct.

4. When fatty acids are converted into fatty acyl coenzyme A so that they can undergo β-oxidation:
   A. ADP is converted to ATP.
   B. Bicarbonate is a substrate for the reaction.
   C. Choline is a cofactor for the reaction.
   D. A high-energy acyl thioester bond is formed.
   E. ATP is converted to ADP plus P_i.

5. Which of the following compounds are required to transport fatty acid groups across the inner mitochondrial membrane?
   1. Biotin
   2. Succinyl CoA
   3. Malonyl CoA
   4. Carnitine
      A. 1, 2, and 3.
      B. 1 and 3.
      C. 2 and 4 only
      D. 4 only.
      E. All are correct.

6. The rate limiting step in liver fatty acid biosynthesis is the conversion of acetyl CoA to malonyl CoA. This reaction:
   1. Requires carnitine.
   2. Requires a biotin containing enzyme.
   3. Is stimulated by palmitoyl CoA.
   4. Utilizes ATP.
      A. 1, 2, and 3.
      B. 1 and 3.
      C. 2 and 4.
      D. 4 only.
      E. All are correct.

7. Assume that this 10-carbon-atom fatty acid was synthesized by the de novo fatty acid synthase system. The 2-carbon-atom fragments are numbered as:

$$CH_3-CH_2-CH_2-CH_2-CH_2-CH_2-CH_2-CH_2-CH_2-COOH$$

   $$\underline{\qquad}\quad\underline{\qquad}\quad\underline{\qquad}\quad\underline{\qquad}\quad\underline{\qquad}$$
   $$\quad 1 \qquad\quad 2 \qquad\quad 3 \qquad\quad 4 \qquad\quad 5$$

   Which statement correctly describes the biosynthetic process?
   A. Each of the 5 fragments was derived from malonyl CoA, with fragment 1 entering the chain first.
   B. Fragments 2, 3, 4, and 5 were derived from malonyl CoA; fragment 1 from acetyl CoA, with fragment 1 entering the chain first.
   C. Fragments 2, 3, 4, and 5 were derived from malonyl CoA; fragment 1 from acetyl CoA, with fragment 5 entering the chain first.
   D. Fragments 1, 2, 3, and 4 were derived from malonyl CoA; fragment 5 from acetyl CoA, with fragment 5 entering the chain first.
   E. Fragments 1, 3, and 5 were derived from acetyl CoA; fragments 2 and 4 from malonyl CoA, with fragment 5 entering the chain first.

8. Which of the following statements is most correct regarding fatty acid biosynthesis?
   A. Dietary carbohydrate is the usual source of the carbon atoms.
   B. The process is activated when the cytoplasmic citrate content is elevated.
   C. The process is inhibited when the palmitoyl CoA content of the cell is elevated.
   D. 16:0 is the main product of complete (de novo) biosynthesis.
   E. All of the above are correct.

9. Which of the following facts are correct concerning fatty acid desaturation?
   1. It occurs in the mitochondria.
   2. It requires O_2.
   3. *Trans* double bonds are produced between 5% and 20% of the time.
   4. NADH is required.
      A. 1, 2, and 3.
      B. 1 and 3.
      C. 2 and 4.
      D. 4 only.
      E. All are correct.

10. Which fatty acid can be synthesized in the animal cell from palmitic acid, 16:0?
      A. 8,11-20:2
      B. 9,12-18:2
      C. 9,12,15-18:3
      D. 12-18:1
      E. 11-16:1

Chapter 11

Cholesterol

Objectives

1. To describe the structure of cholesterol and other sterols that are important in normal metabolism and in disease states
2. To discuss the biosynthesis, metabolism, and excretion of cholesterol and the bile acids
3. To discuss the function of the bile and the relationship of cholesterol to gallstone formation
4. To consider the role of cholesterol in the development of atherosclerosis and the relationship of hypercholesterolemia and dietary fat intake in this disease

Cholesterol is the major sterol in the human body. It is a structural component of cell membranes and plasma lipoproteins, and it is also the starting material from which bile acids and steroid hormones are synthesized. An abnormality in either cholesterol metabolism or transport through the plasma appears to be related to the development of atherosclerosis, the form of hardening of the arteries that can lead to myocardial infarction, stroke, aneurysm, or gangrene. In addition, the gallstones that occur most commonly in inhabitants of Western nations are made up predominantly of cholesterol. These are commonly occurring, serious diseases. Therefore health science students should have a basic understanding of cholesterol metabolism.

Steroid chemistry

Steroids are derivatives of the perhydrocyclopentanophenanthrene ring system.

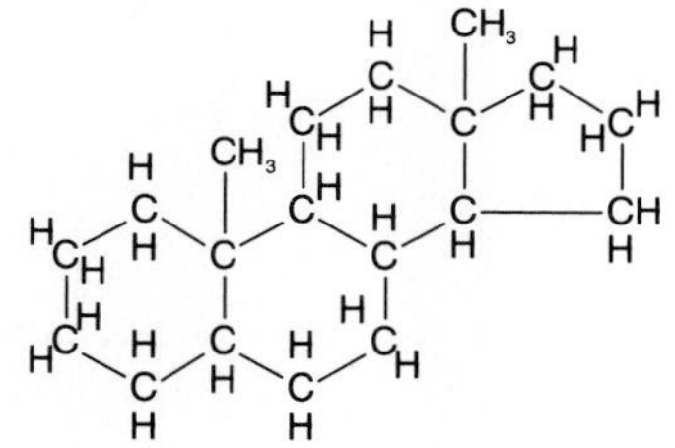

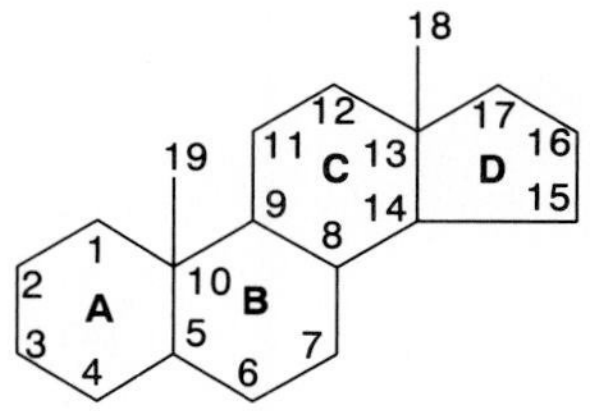

The complete structure, including carbon and hydrogen atoms is shown at the left; the commonly used line drawing in which carbon and hydrogen atoms are omitted is shown for comparison at the right. In the line drawing the four rings are identified by the letters A to D, and the 19 carbon atoms are identified by number. Three of the rings contain six carbon atoms, whereas the D ring contains five. An angular methyl group, C_{19}, is attached at the junction of the A and B rings, and a second angular methyl group, C_{18}, is attached at the junction of the C and D rings.

The steroid nucleus has a conformation that is approximately planar. The side groups such as the angular methyl groups at C_{10} and C_{13} that are above the plane of the rings, as shown on p. 466 are designated as being in a β-orientation. This is indicated by their connection to the ring structure with a solid line. Those side groups that project below the plane of the rings are designated as being in an α-orientation, indicated by their connection to the ring structure with a dashed line. The hydrogen atoms or hydroxyl groups that are attached to the ring carbon atoms also project either above or below the plane of the rings and are also designated as being in either an α- or a β-orientation.

β-Orientation   α-Orientation

The angular methyl groups, C_{18} and C_{19}, are always in the β-configuration.

Sterols

Sterols are a class of steroids that contain a hydroxyl group at C_3 and an aliphatic chain of at least eight carbon atoms attached to C_{17}.

Aliphatic hydrocarbon chain of sterols

Cholesterol is the main sterol in human tissue. It has an eight-carbon-atom hydrocarbon chain that is numbered 20 to 27 as a continuation of the steroid nucleus. Some of the cholesterol that is present in humans is esterified; that is, the hydroxyl group that projects from C_3 is attached to a fatty acid residue in ester linkage. The hydroxyl group in cholesterol is β-oriented, as indicated by the solid line between it an C_3 and the eight-carbon-atom hydrocarbon chain is also in a β-orientation. There is a double bond in the B ring between positions 5 and 6.

Cholesterol   Cholesteryl ester

Dietary cholesterol Foods that are derived from animal products contain cholesterol. Those foods which are particularly rich in cholesterol include eggs, dairy products such as butter, cheese, and cream, and most meats (see Chapter 1). Some of the cholesterol that is contained in these animal products is in the form of cholesteryl esters. Therefore the ordinary diet contains a mixture of cholesterol and cholesteryl esters.

Cholesteryl esters The fatty acid moieties of cholesteryl esters usually contain 16 to 20 carbom atoms and often are unsaturated. Among the most abundant of the esters of cholesterol in humans are cholesteryl oleate and cholesteryl linoleate. These compounds are present in appreciable amounts in the plasma lipoproteins, adrenal cortices, and liver. About 80% of the total cholesterol in low-density lipoproteins (LDL) and 90% of the total cholesterol in high-density lipoproteins (HDL) is in the form of cholesteryl esters (see Chapter 16). They are the most abundant form of lipid that accumulates in the arterial wall in atherosclerotic lesions. Cholesteryl esters are the most nonpolar lipids that occur in human tissues, and they function as a storage form of sterol. Unlike cholesterol itself, cholesteryl esters do not exchange readily between cell membranes and plasma lipoproteins or among the various classes of plasma lipoproteins.

Sterols of nonanimal species Plants do not contain cholesterol; they have other sterols that are known as *phytosterols*. The most abundant of the phytosterols is β-sitosterol. Although the ring structure of β-sitosterol is identical to that of cholesterol, the chain attached to C_{17} contains 10 rather than 8 carbon atoms, for there is an ethyl group attached to C_{24} of the chain. Ergosterol, the sterol of yeast, contains a methyl group at C_{24}, a double bond in the chain, and a second double bond in the B ring. It is the precursor of the form of vitamin D called vitamin D_2. Unlike plants and animals, bacteria do not contain sterols.

β-Sitosterol   Ergosterol

Dietary cholesterol
Absorption

Humans can readily absorb cholesterol contained in the diet. Most people in Western societies eat between 400 and 600 mg (1.1 and 2.1 mmol)/day of cholesterol and absorb from 300 to 400 mg (0.8 to 1.0 mmol)/day. When the dietary intake is relatively small, absorption is efficient. However, when the intake exceeds approximately 500 mg (1.3 mmol)/day, cholesterol absorption becomes somewhat less efficient. If one is reasonably cautious and avoids foods that are rich in cholesterol, it is relatively easy to reduce the dietary cholesterol intake to about 400 to 500 mg (1.0 to 1.3 mmol)/day. However, about 200 to 300 mg (0.5 to 0.8 mmol)/day will nonetheless be absorbed under these conditions. To reduce absorption further, it is necessary to lower the dietary intake of cholesterol to the range of between 100 and 300 mg (0.25 to 0.8 mmol)/day. This requires severe dietary restrictions, because the the average person's daily intake of meat, eggs and dairy products provides at least 500 mg (1.3 mmol).

Plant sterols Unlike cholesterol, plant sterols are absorbed poorly by humans. Indeed, feeding of large quantities of plant sterols such as β-sitosterol actually inhibits cholesterol absorption. This fact has been exploited clinically, for β-sitosterol was administered in the past to patients with *hypercholesterolemia* (excessive blood plasma cholesterol levels) in an attempt to reduce cholesterol absorption. Unfortunately, this treatment was not very successful in practice.

Digestion

All of the dietary cholesterol is incorporated into micelles that are formed from the amphipathic constituents present in the bile. These micelles contain *conjugated bile acids* and phospholipids in addition to cholesterol. Emulsification is necessary because cholesterol is poorly soluble in the chyme, the aqueous medium that is present in the intestinal lumen. It must be brought into a physical state suitable for uptake by the intestinal mucosa. Any esterified cholesterol is hydrolyzed within the intestinal lumen by an enzyme secreted in the pancreatic juice, *cholesteryl esterase*. Hydrolysis of the cholesteryl esters by cholesteryl esterase occurs on or within the micelle. Cholesterol is absorbed by diffusion from the micelles into the intestinal mucosal cells, where much of it is subsequently reconverted into cholesteryl esters. Cholesterol absorption occurs mostly in the jejunum.

Figure 11.1   Digestion and absorption of cholesterol. Mucosal cells constantly degenerate and are discharged into the intestinal lumen. This is noted as *sloughing*. *C,* Cholesterol; *CE,* cholesteryl esters.

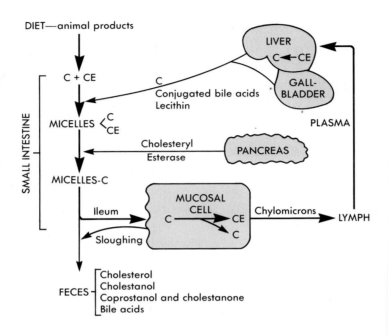

Chylomicrons The cholesteryl esters that are synthesized in the mucosal cells, together with some unesterified cholesterol, are incorporated into large lipid-protein particles that are released into the lymph. These particles are lipoproteins called *chylomicrons,* and they transport cholesterol as well as other dietary lipids into the plasma from the lymph via the thoracic duct (see Chapters 10 and 12). Eventually the cholesterol is deposited in the tissues, mostly in the liver. Cholesterol digestion, absorption, and excretion are illustrated schematically in Figure 11.1.

Excretion

Cholesterol is excreted in the feces. Cholesterol is delivered from the liver into the intestine in the bile, and additional amounts are derived from the sloughed intestinal mucosal cells. Moreover, some dietary cholesterol is excreted without being absorbed. Some of the cholesterol in the bowel is acted on by intestinal bacterial enzymes and converted to other *neutral sterols* before excretion in the feces. In humans the main neutral sterol products in the stool are coprostanol and cholestanone. Another neutral sterol reduction product that is excreted in the feces is cholestanol. It is a metabolic product of cholesterol that is formed in the liver and delivered into the intestine in the bile. Coprostanol and cholestanol are isomers, the only difference being that the hydrogen atom between the A and B rings has a β-orientation in coprostanol and an α-orientation in cholestanol.

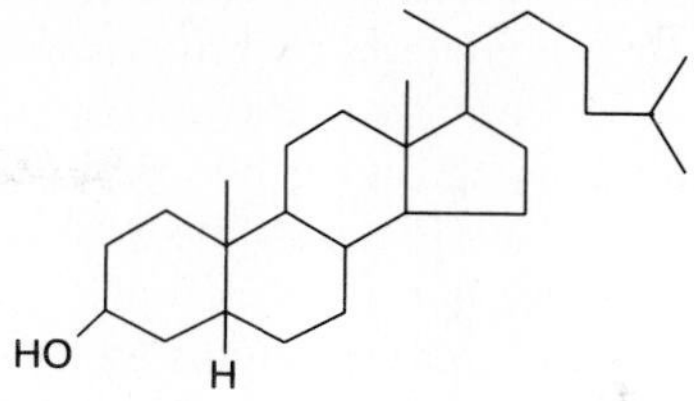

Coprostanol

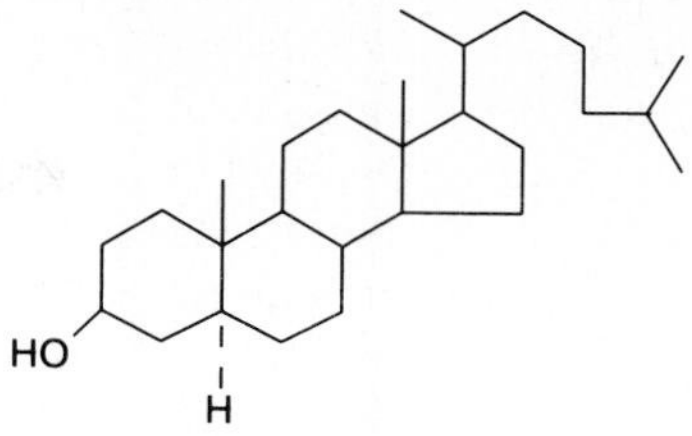

Cholestanol

Bile acids

The main metabolic products of cholesterol in terms of the amounts formed are the bile acids. The major bile acids in humans are cholic, chenodeoxycholic, deoxycholic, and lithocholic acids.

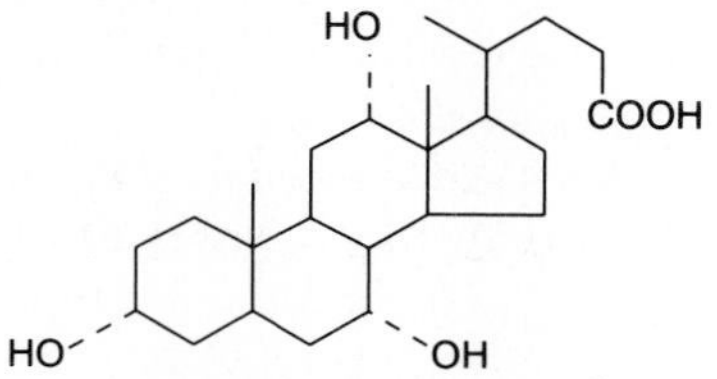

Cholic acid

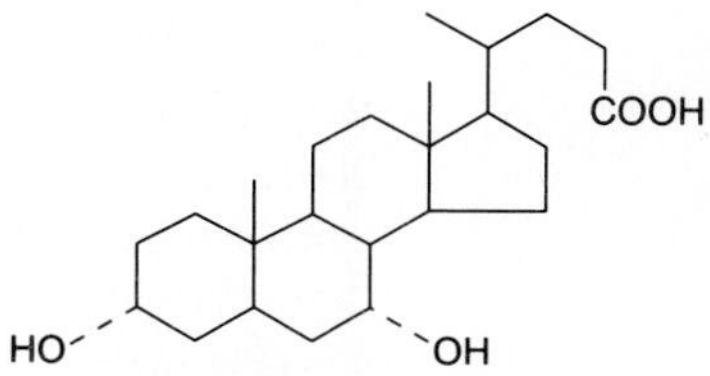

Chenodeoxycholic acid

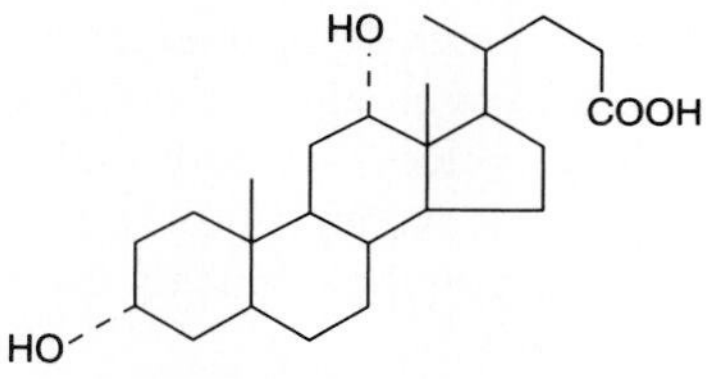

Deoxycholic acid

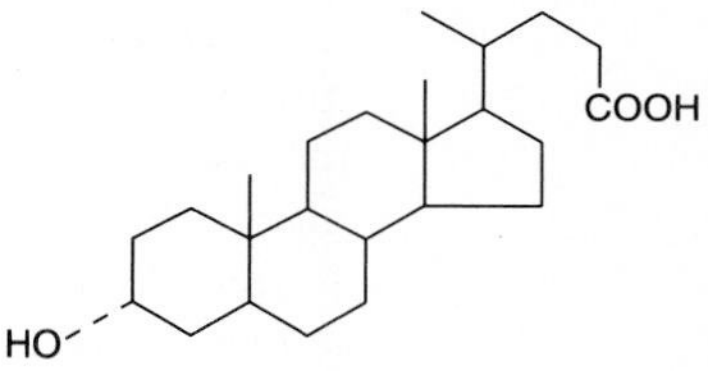

Lithocholic acid

Bile acids contain 24 carbon atoms; the terminal three carbon atoms of the cholesterol side-chain are removed during their synthesis from cholesterol. The double bond that is present in the B ring of cholesterol is also reduced in the synthetic process. The four major bile acids differ only in the number of hydroxyl groups that are attached to the steroid nucleus. Cholic acid has three hydroxyl groups, and they are attached to the carbon atoms at positions 3, 7, and 12. Chenodeoxycholic and deoxycholic acids have two hydroxyl groups attached at positions 3,7 and 3,12, respectively, and lithocholic acid has only one, which is attached at position 3. Notice that all the hydroxyl groups are in the α-orientation.

Conjugation Bile acids are condensed with either glycine or taurine in the liver to form glyco- or tauro-conjugated bile acids. Both glycine and taurine are jointed in amide linkage with the carboxyl group (C_{24}) of the five-carbon atom chain attached to C_{17} of the steroid nucleus.

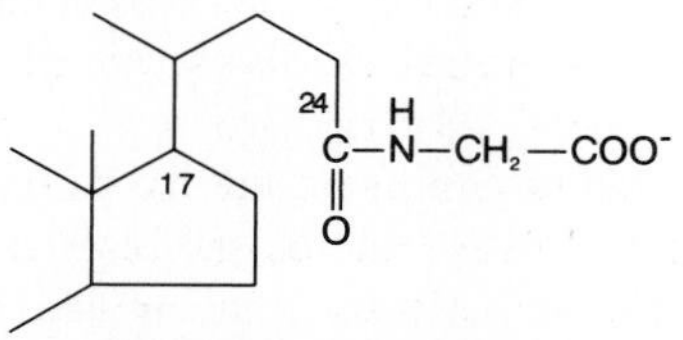

Glycine conjugate

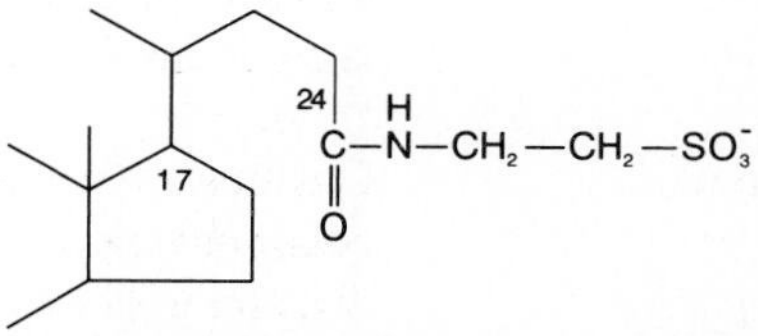

Taurine conjugate

Synthesis of the primary bile acids

The pathways for bile acid synthesis are shown in Figure 11.2. Three types of reactions are involved. First, hydroxyl groups are inserted at specific positions on the steroid nucleus in α-orientation. In this process, the 3-β-hydroxyl group of cholesterol is converted to a 3-α-hydroxyl group. Depending on the bile acid being formed, a total of either two or three hydroxyl groups is added. Second, the double bond present at position 5,6 of the steroid nucleus of cholesterol is reduced. Finally, the hydrocarbon chain of cholesterol is shortened from eight to five carbon atoms by oxidative and thiolytic cleavage, introducing a carboxyl group at the end of the chain. All these reactions occur in the liver.

The rate-limiting reaction in the pathway is the insertion of the first hydroxyl group, which occurs at position C_7 of the cholesterol nucleus. This reaction is catalyzed by the enzyme cholesterol 7-α-hydroxylase, which is located in the endoplasmic reticulum of the hepatocyte.

The 3-β-hydroxyl of cholesterol then is converted into a 3-α-hydroxyl group. A 3-keto group is an intermediate in this isomerization reaction. When cholic acid is formed, a hydroxyl group also is inserted at the C_{12} position. The double bond of the cholesterol B ring is reduced. Next, C_{26} of the cholesterol side chain is hydroxylated, forming 5-β-cholestane-3,7,12,26-tetrol. The C_{26} alcohol group is oxidized to an acid, forming 3,7,12-trihydroxy-5β-cholestanoic acid. The C_{24} group also undergoes oxidation, and the terminal three carbon atoms, C_{25} to C_{27}, are released as propionic acid. In this cleavage reaction, CoASH adds to C_{24}. The product, cholyl CoA, contains three hydroxyl groups.

The other product that is formed in the parallel series of reactions, chenodeoxycholyl CoA, is not hydroxylated at C_{12} and therefore contains only two hydroxyl groups projecting from the steroid nucleus. For both derivatives the CoASH at C_{24} is then replaced by either glycine or taurine, which form an amide linkage with C_{24}, the terminal carbon atom of the bile acid chain. These synthetic reactions take place in the liver, and a mixture of glycocholic, glycochenodeoxycholic, taurocholic, and taurochenodeoxycholic acids is secreted into the bile. Almost all the bile acids that are released from the liver are present in conjugated form.

Regulation Bile acid biosynthesis is regulated by the amount of bile acid that is returned from the intestine to the liver. The main intestinal site of bile acid reabsorption is the ileum. Biosynthesis *decreases* as more bile acid is reabsorbed and as the bile acid return to the liver increases. Bile acid synthesis also is controlled by the amount of cholesterol that is transported from the intestine to the liver; it *increases* as cholesterol absorption increases.

Role in cholesterol excretion Bile acid production is the most important catabolic pathway for cholesterol from the quantitative standpoint. Continuous conversion of cholesterol into bile acid in the liver prevents the body from becoming overloaded with cholesterol. Excessive accumulation of cholesterol in the tissues is harmful. Unlike many other metabolites, cholesterol cannot be destroyed by oxidation to carbon dioxide and water, because mammalian tissues do not have enzymes capable of catabolizing the steroid nucleus. Therefore the only way to rid the body of excess cholesterol is to excrete it either as intact sterols or after conversion to bile acids. In normal humans, roughly equal amounts of cholesterol are excreted in the fecal matter as bile acid and intact sterols, the latter being in the form of neutral sterols (see p. 469 and Figure 11.1).

Metabolism

Conjugated bile acids either pass from the liver directly into the duodenum through the common bile duct or are stored in the gallbladder when not needed immediately for digestion. They form a part of the bile that, in addition, contains phospholipids, cholesterol, salts, water, and excretory metabolites such as bilirubin. Bile that is stored and

Figure 11.2 Biosynthesis of bile acids.

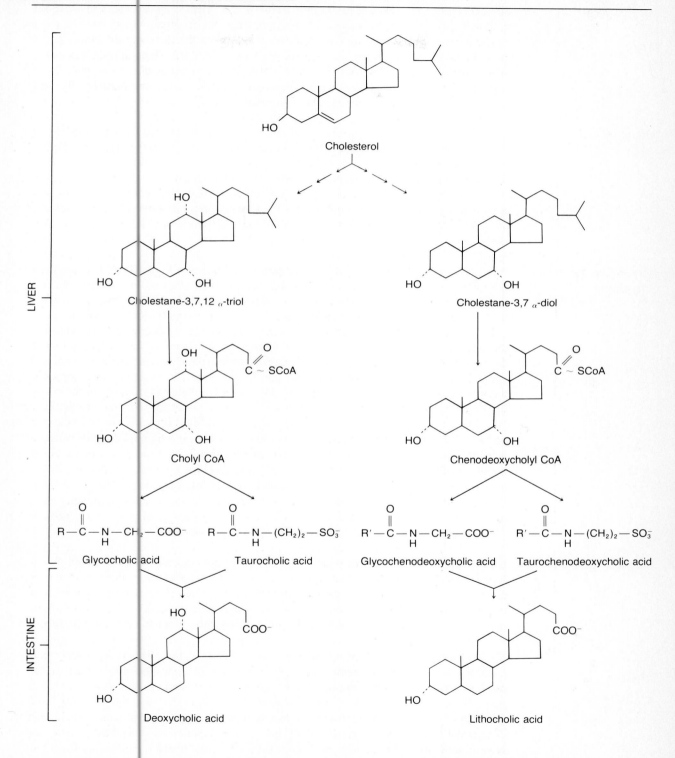

concentrated in the gallbladder is released into the intestine via the cystic and common bile ducts when it is needed to aid in the digestion of dietary lipids. The bile acids aid in the emulsification of the ingested lipids, a process that facilitates enzymatic digestion and absorption of dietary fat. The secretion of bile from the liver and the emptying of the gallbladder are processes that are under hormonal control. *Hepatocrinin* stimulates bile secretion by the liver, and *cholecystokinin* causes the gallbladder to empty. These hormones are synthesized by the intestine and are released when the partially digested food passes from the stomach into the duodenum.

Secondary bile acids

Deoxycholic and lithocholic acids are the secondary bile acids. They are synthesized in the intestine through the action of bacterial enzymes on the primary bile acids (Figure 11.2). Only a portion of the primary bile acids present in the intestine is converted to secondary bile acids. First, the primary acids are deconjugated; that is, the amide linkage at C_{24} is hydrolyzed and glycine or taurine is released. Next, the hydroxyl group present at C_7 is removed. In this way cholic acid (3,7,12-trihydroxy) is converted to deoxycholic acid (3,12-dihydroxy), and chenodeoxycholic acid (3,7-dihydroxy) is converted to lithocholic acid (3-hydroxy).

Enterohepatic circulation

Bile acids recycle between the liver and intestine, as shown in Figure 11.3. Passage from the liver to the intestine takes place through the common bile duct. In humans the bile usually is stored in the gallbladder. Release into the intestine is episodic and occurs when food is eaten. Flow from the intestine back to the liver occurs through the portal vein.

Most of the bile acids present in the intestine (that is, the remaining primary acids and the newly formed secondary acids) are deconjugated and reabsorbed into the portal blood. There are three intestinal sites at which bile acid reabsorption occurs. In the jejunum and colon, absorption occurs by passive diffusion. In the ileum, the major site of absorption, the process is one of active transport. The bile acids are removed from the portal blood by the liver, reconjugated with either glycine or taurine, and then secreted into the bile. Bile that is released from the liver therefore contains all four bile acids, not only the two primary acids that actually are synthesized by the liver. The recycling of bile acids between intestine and liver is called the *enterohepatic circulation* of bile acids. In the portal blood the bile acids are carried as a noncovalent, physical complex by the main plasma protein, *albumin*. From 15 to 30 g/day of bile acids is passed into the intestine from the liver. Only a small fraction, approximately 300 mg/day, is lost in the feces. This quantity must be synthesized daily from cholesterol to maintain the bile acid pool at its optimum size. The bile acids present in the feces are known as *acidic sterols*.

Clinical comment

Cholelithiasis The disease in which the gallbladder contains gallstones is called cholelithiasis. Most of the stones that form in the gallbladders of adults in Western countries are made up predominantly of cholesterol. The sterol is laid down around a central core containing protein and bilirubin. Because of an unknown metabolic defect, bile of abnormal composition is secreted by the liver, causing cholesterol to crystallize and form a stone.

Cholesterol is insoluble in the aqueous salt medium of the bile and is brought into solution by molecular association with the bile salts and phospholipids. Most of the phospholipid in bile is phosphatidylcholine, also known as lecithin. Mixed micelles made up of phosphatidylcholine, bile salts, and cholesterol are formed, and the relatively large amounts of cholesterol ordinarily present in human bile are held in solution. This allows cholesterol to be transported harmlessly through the biliary tract into the intestine, including temporary storage in a solubilized form in the gallbladder. The mixed micelles, however, have a limited capacity to solubilize cholesterol. The amount of cholesterol that

Figure 11.3   Enterohepatic circulation of bile acids.

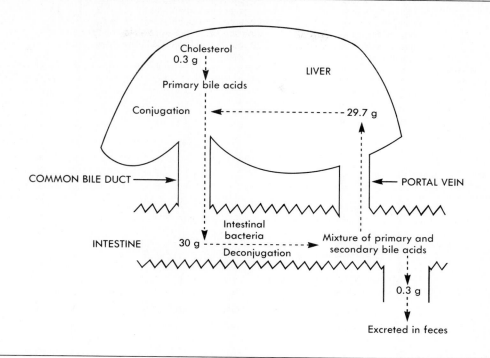

can actually be held in micellar aggregates depends on the relative proportions of phosphatidylcholine, bile salts, and cholesterol as well as on the water content of the bile.

A convenient method for describing the defective bile composition that leads to cholesterol precipitation and gallstone formation is shown in Figure 11.4. In any given sample of bile the molar percentage of biliary cholesterol, phosphatidylcholine, and bile salt may be plotted on triangular coordinates and their point of intersection determined. If this point falls within the small shaded area in the lower left corner, the cholesterol present in the sample of bile will exist in soluble form, that is, in the liquid micellar state. However, if the point of intersection of the three coordinates falls outside of the shaded area, cholesterol will crystallize in the form of discrete aggregates. Cholesterol gallstones will form in bile that is incapable of completely solubilizing all of its cholesterol content. For example, a bile containing 5% cholesterol, 15% phosphatidylcholine, and 80% bile salts will have all the cholesterol in the micellar state, whereas bile containing 20% cholesterol will form gallstones independently of the relative phosphatidylcholine and bile salt concentrations.

Cholesterol metabolism

Cholesterol can undergo a number of metabolic reactions in humans. It can be esterified, and the resulting cholesteryl esters can be hydrolyzed. It can be transformed into bile acids, as described in the previous section. Conversion to bile acids is the main catabolic pathway in terms of the quantity of cholesterol that is metabolized. Cholesterol also is the substrate for steroid hormone synthesis (see Chapter 19). Although the amount of cholesterol converted into steroid hormones is small in terms of total cholesterol utilization, these reactions are of enormous importance in terms of bodily function. Cholesterol is

Figure 11.4   Phase diagram of lipid component composition of bile. The shaded
area represents the region in which cholesterol exists as a micellar
liquid. Cholesterol forms crystals if concentrations of the three
components fall outside the shaded area. The point shown on the
phase diagram represents a composition of 20 mol % cholesterol, 20
mol % phosphatidylcholine, and 60 mol % bile salts. (Modified from
Small DM: N Engl J Med 278:588, 1968.)

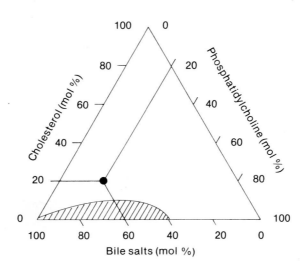

either obtained from the diet or synthesized completely from acetyl CoA. As noted
previously, however, the one metabolic reaction involving cholesterol that cannot be
carried out by human or other mammalian tissues is the catabolism of the sterol nucleus.
The potential problems imposed by this limitation will become apparent when diseases
associated with abnormalities in cholesterol metabolism are discussed.

Ester formation   Cholesteryl ester synthesis occurs through two separate reactions. One is operative in
tissues, the other in plasma. The tissue pathway involves two reactions. First, fatty acid
is activated by incorporation into thioester linkage with CoASH to form an acyl CoA.
Formation of the thioester bond requires the expenditure of one ATP molecule as the
source of energy. This reaction is mediated by an acyl CoA synthase, as described in
Chapter 10. The acyl CoA thioester then reacts with cholesterol through the action of
acyl CoA cholesterol acyltransferase (ACAT), an enzyme localized in the endoplasmic
reticulum.

$$\text{Acyl CoA + Cholesterol} \xrightarrow[\text{acyltransferase}]{\text{Acyl CoA cholesterol}} \text{Cholesteryl ester + CoASH}$$

Major sites for cholesteryl ester synthesis by the ACAT pathway are the liver, intestine,
and adrenal cortex. This reaction also can take place in the arterial wall, and it is thought
to be the main route for the formation of the cholesteryl esters that accumulate in atherosclerosis.

The second pathway of cholesteryl ester synthesis occurs in the plasma and is mediated
by the enzyme *lecithin-cholesterol acyltransferase* (LCAT). In this reaction a fatty acid

contained in the *sn*-2 position of phosphatidylcholine (lecithin) is transferred directly to cholesterol without passing through either a fatty acid or acyl CoA intermediate.

$$\text{Phosphatidylcholine} + \text{Cholesterol} \xrightarrow[\substack{\text{cholesterol} \\ \text{acyltransferase}}]{\text{Lecithin-}} \text{Cholesteryl ester} + \text{Lysophosphatidylcholine}$$

Most of the *plasma* cholesteryl esters are synthesized by this reaction. In other words, the cholesterol that is released by the liver as a constituent of plasma lipoproteins is mostly in nonesterified form. It is esterified subsequently in the plasma through the action of LCAT. This enzyme is produced in the liver and secreted into the plasma, where it functions. The reaction takes place within HDL and uses the cholesterol and phosphatidylcholine contained in the HDL. Esterification serves to trap cholesterol within the lipoprotein, preventing it from moving to the tissues by surface transfer.

The *sn*-2 position of phosphatidylcholine usually contains polyunsaturated fatty acids. Therefore the LCAT pathway produces cholesteryl esters that always have polyunsaturated fatty acyl groups, the most common one being linoleic acid.

Ester hydrolysis

Cholesteryl esters are hydrolyzed to cholesterol and fatty acids through the action of cholesteryl esterase.

The reaction equilibrium is in the direction of hydrolysis. Cholesteryl esterases are present in many tissues. There is a cholesteryl esterase in the pancreatic juice that functions in the digestion of dietary cholesteryl esters (see Figure 11.1). Tissues contain two types of cholesteryl esterase. One is present in the lysosomes and has an acid pH optimum. Cholesteryl esters taken up by the cells as a part of plasma lipoproteins are hydrolyzed by the lysosomal enzyme. The other intracellular cholesteryl esterase is located in the endoplasmic reticulum and has a neutral pH optimum. This reaction is reversible and may catalyze the synthesis of cholesteryl esters under certain conditions. Some of the cholesteryl esters formed in the intestinal mucosa appear to be synthesized through a reversal of the cholesteryl esterase reaction.

Cholesterol biosynthesis

In addition to being taken in from the diet, cholesterol can be synthesized by human tissues. Cholesterol is synthesized from acetyl CoA, which can be derived from carbohydrates, amino acids, or fatty acids. The liver is the main site of cholesterol synthesis, but the intestine also is an important site of synthesis in humans. In addition, cholesterol is synthesized in glands that produce steroid hormones, for example, the adrenal cortex, testes, and ovaries. All the synthetic reactions occur in the cytoplasmic compartment of the cell, but some of the required enzymes are bound to the membranes of the endoplasmic reticulum.

The isoprenoid biosynthetic pathway The isoprenoid biosynthetic pathway, through which cholesterol is produced, has a number of branch points, each of which leads to the synthesis of biologically important compounds. In mammals these include ubiquinone, dolichol, and the isopentenyl group that is added to certain transfer RNAs. Additional compounds are produced by this pathway in plants, including the carotenes and terpenes. All of these substances are derivatives of an intermediate in the pathway called the isoprene unit, which has five carbon atoms and a branched chain. It is illustrated in an abbreviated format in Figure 11.5, which shows only the segments that operate in mammalian tissues.

The isoprenoid synthetic pathway can be thought of as occurring in three stages with respect to cholesterol biosynthesis. In the first stage, acetyl CoA is converted into a six-carbon-atom thioester intermediate, *3-hydroxy-3-methylglutaryl CoA (HMGCoA)*. The second stage involves the conversion of HMGCoA to squalene, an acyclic hydrocarbon containing 30 carbon atoms. An intermediate in this stage is the five-carbon-atom isoprene unit, which occurs in two isomeric phosphorylated forms in this pathway, Δ_3-*isopentenyl pyrophosphate* and *3,3-dimethylallyl pyrophosphate*. In the third stage, squalene is cyclized and converted to the 27-carbon-atom sterol, cholesterol. The series of reactions from squalene to cholesterol occur in the endoplasmic reticulum. Several of the intermediates in the final segment of the pathway are bound physically to cytoplasmic proteins, the *sterol carrier proteins*. A schematic representation of the three stages in the cholesterol biosynthetic pathway is given in Figures 11.6 to 11.8.

HMGCoA formation Two molecules of acetyl CoA condense in the initial step to produce *acetoacetyl CoA*, as shown in Figure 11.6. This reaction occurs in the cytosol.

$$2 \quad CH_3-\overset{\overset{O}{\|}}{C}\sim SCoA \quad \xrightarrow[\text{Synthetase}]{\text{Acetoacetyl CoA}} \quad CH_3-\overset{\overset{O}{\|}}{C}-CH_2-\overset{\overset{O}{\|}}{C}\sim SCoA + CoA$$

Acetyl CoA   Acetoacetyl CoA

Remember that the acetoacetyl group is the four-carbon intermediate involved in fatty acid biosynthesis, a reaction sequence that also occurs in the cytosolic compartment of the cell (see Chapter 10). However, in fatty acid synthesis the intermediates are bound covalently to the phosphopantetheine sulfhydryl group of the fatty acid synthase complex, not to CoASH as in the cholesterol synthesis pathway. In the next reaction, acetoacetyl CoA condenses with another acetyl CoA molecule, forming HMGCoA.

Figure 11.5   The isoprenoid biosynthetic pathway.

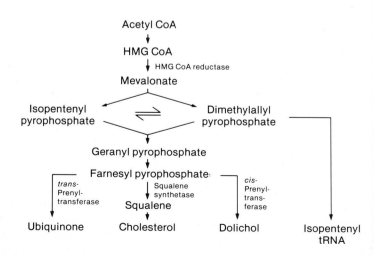

Figure 11.6　First stage of cholesterol biosynthesis. In this sequence of reactions, acetyl CoA is converted to HMGCoA.

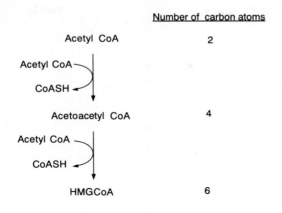

Figure 11.7　Second stage of cholesterol biosynthesis. This sequence of reactions involves conversion of HMGCoA to squalene.

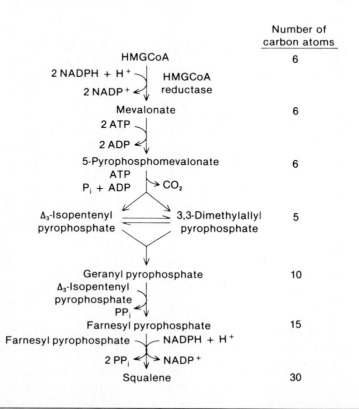

Figure 11.8   Third stage in cholesterol biosynthesis. In this sequence of reactions,
squalene is converted to cyclized intermediates, which eventually are
transformed into cholesterol. These reactions occur while the
intermediates are bound to sterol carrier protein.

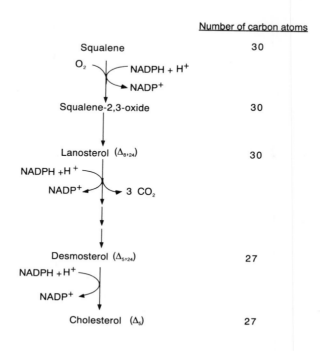

Notice that in this reaction the acetate unit adds to the β-carbon atom of acetoacetate.
The HMGCoA that is formed is present in the cytosol. Thus it differs and is separate
from the HMGCoA that is an intermediate in ketone body synthesis (see p. 454), which
occurs inside the mitochondrion.

Conversion of HMGCoA to squalene Figure 11.7 shows the second stage of the
pathway, the conversion of HMGCoA to squalene. In the first step, HMGCoA is reduced
to mevalonate and the thioester group is converted to an alcohol.

This reaction utilizes two molecules of NADPH as the reducing agent and is catalyzed by HMGCoA reductase, an enzyme attached to the endoplasmic reticulum. It is the rate-limiting step in cholesterol synthesis.

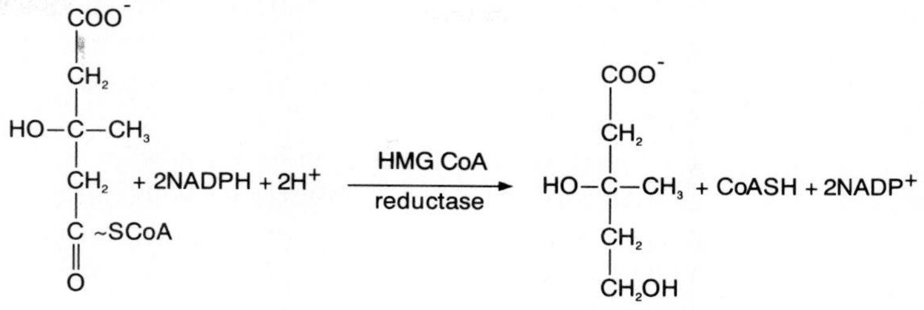

Mevalonate is converted to *5-pyrophosphomevalonate* in two steps, each of which requires the expenditure of one molecule of ATP.

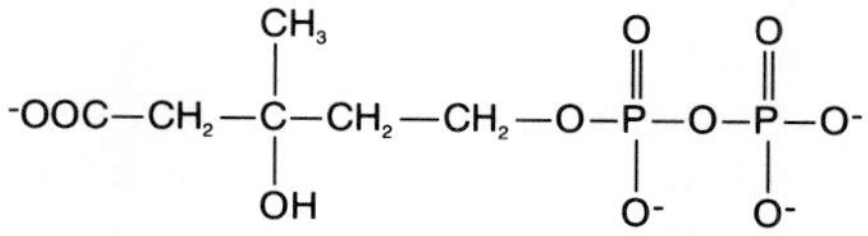

5-Pyrophosphomevalonate

The five-carbon-atom *isoprene unit* is formed from this pyrophosphorylated intermediate. One additional molecule of ATP is required for this decarboxylation step. Actually two isomeric isoprene units are formed, isopentenyl pyrophosphate and dimethylallyl pyrophosphate; they are interconvertible.

Δ_3-Isopentenyl pyrophosphate   3,3-Dimethylally pyrophosphate

These isomers condense, head to tail, forming a 10-carbon-atom intermediate, *geranyl pyrophosphate*. Another molecule of isopentenyl pyrophosphate condenses with the newly formed geranyl pyrophosphate, forming *farnesyl pyrophosphate,* a 15-carbon-atom intermediate.

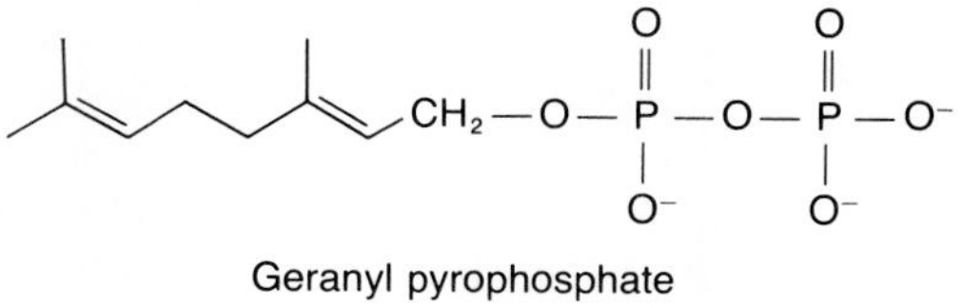

Geranyl pyrophosphate

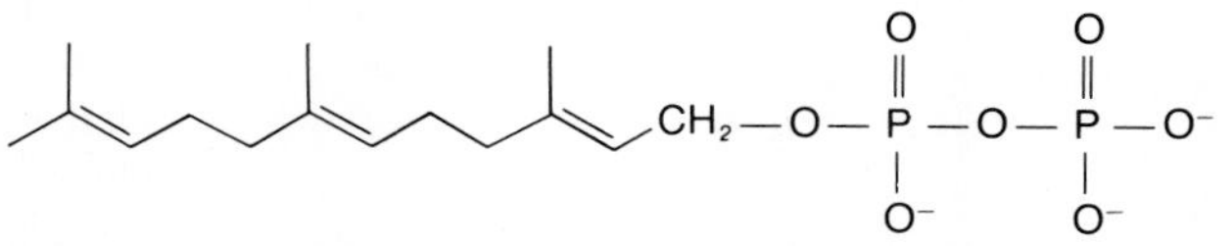

Farnesyl pyrophosphate

Two molecules of farnesyl pyrophosphate then combine head to tail in a series of reactions in which NADPH is utilized and both pyrophosphate groups are eliminated. The product is squalene, a 30-carbon-atom hydrocarbon that exists in a folded conformation but is not cyclized.

Conversion of squalene to cholesterol The last stage in the pathway, the conversion of squalene to cholesterol, is shown in Figure 11.8. Squalene binds to a specific cytoplasmic protein carrier, *sterol carrier protein*$_1$ (SCP$_1$). The conversion of squalene to lanosterol occurs while the intermediates are bound to this carrier. First, an oxygen atom is attached to both positions C_2 and C_3 of squalene, a reaction that requires NADPH as shown in Figure 11.9. The resulting *squalene-2,3-oxide,* an epoxide, then cyclizes in a complicated series of reactions to form *lanosterol*. Cyclization is accomplished by migration of the double bonds in four segments of the squalene oxide that will become the four rings of the sterol. The first cyclized intermediate that is formed has methyl groups projecting from positions C_8 and C_{14} of the ring structure. In the next step, two 1,2-methyl shifts occur. The methyl group originally attached to C_{14} migrates to C_{13}. This becomes the angular methyl group, C_{18}, that projects between the C and D rings of the steroid nucleus. In addition, the methyl group originally attached to C_8 migrates to C_{14}, where it now projects below the plane of the ring structure. The next intermediate that is formed, lanosterol, still contains 30 carbon atoms. The β-hydroxyl group is present at C_3, but there are double bonds between C_8 and C_9 and in the chain between C_{24} and C_{25}. In addition, there are two methyl groups projecting in β-orientation from C_4 as well as the one in α-orientation from C_{14}.

Lanosterol is bound to a second sterol carrier protein, SCP$_2$, and the remaining reactions then take place. In a series of reactions involving oxygen and NADPH, the three methyl groups at positions C_4 and C_{14} are released as carbon dioxide and the double bond in the B ring is moved to position C_5-C_6, forming a 27-carbon-atom intermediate, *desmosterol*.

Desmosterol

HO

Desmosterol still contains the side-chain double bond at C_{24}-C_{25}, but it possesses many of the properties of the final product, cholesterol. In the last step the C_{24}-C_{25} double bond is reduced by NADPH, forming cholesterol.

Stoichiometry The synthesis of a single molecule of cholesterol represents the expenditure of considerable substrate, energy, and reducing equivalents. The overall process requires 18 mol of acetyl CoA, 36 mol of ATP, and 16 mol of NADPH to form 1 mol of cholesterol. Although six molecules of HMGCoA are utilized, only 27 of the 36 carbon atoms are retained in the final product, cholesterol. Of the 36 mol of ATP required, 18 are used for acetyl CoA formation. These are expended in the *citrate lyase* reaction. The acetyl CoA that is used for cholesterol synthesis is derived almost entirely from the mitochondria. It leaves the mitochondria in the form of citrate, which is subsequently cleaved in the cytoplasm to acetyl CoA and oxaloacetate by citrate lyase (see Chapter 10). ATP provides the energy for the cleavage of citrate, so that one ATP is required for each acetyl CoA that is formed in this reaction. The other 18 mol of ATP are utilized for the conversion of mevalonate to the pyrophosphorylated form of the isoprene unit.

Transmethylglutaconate shunt   It used to be thought that once HMGCoA was converted to mevalonic acid, the carbon atoms were committed irreversibly to isoprenoid synthesis. However, it was found that dimethylallyl pyrophosphate, one of the isoprene units formed from mevalonate (see p. 479), can be dephosphorylated. In a series of reactions involving alcohol and aldehyde

Figure 11.9   Cyclization of squalene and its conversion to lanosterol.

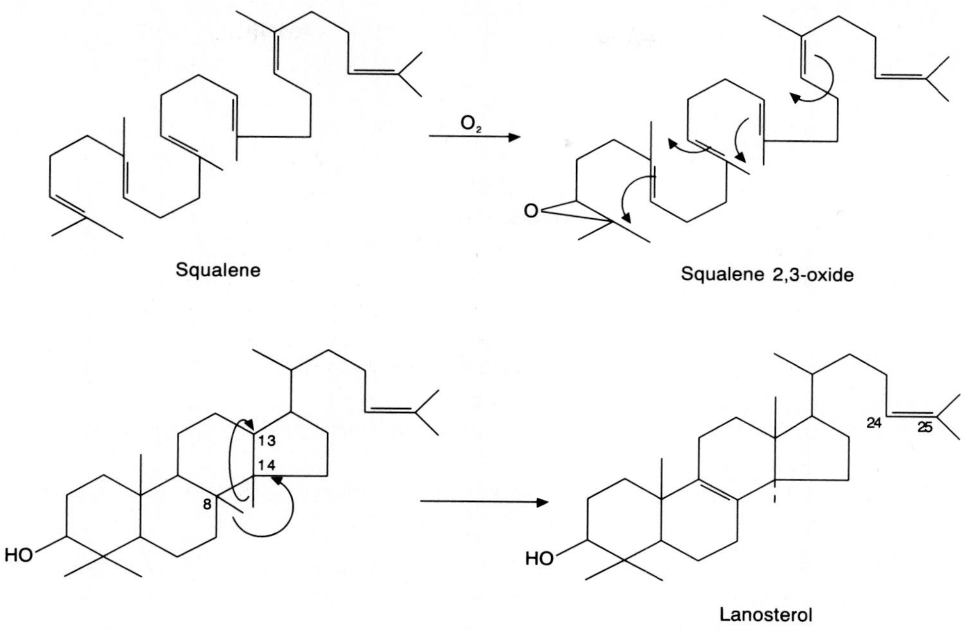

dehydrogenases and a carboxylase, the carbon atoms derived from the dimethylallyl alcohol are converted to acetoacetate and acetyl CoA, intermediates that can be either catabolized to carbon dioxide and water or used for biosynthetic reactions, including fatty acid synthesis. This pathway, called the transmethylglutaconate shunt, offers the potential of reducing cholesterol synthesis by shunting carbon atoms originally destined for cholesterol into other metabolic pathways.

Regulation of cholesterol synthesis

Cholesterol synthesis is regulated by the dietary cholesterol intake, the caloric intake, certain hormones, and bile acids.

    Feedback inhibition Dietary cholesterol itself does not inhibit intestinal cholesterol synthesis, but it does have a strong "feedback inhibitory" effect on cholesterol synthesis in the liver. This occurs through regulation of the enzyme HMGCoA reductase, as is illustrated schematically in Figure 11.10. Cholesterol absorbed from the diet inhibits the liver biosynthetic pathway by decreasing synthesis at the HMGCoA reductase reaction. The dietary cholesterol is delivered to the liver in chylomicron remnants, catabolic products of chylomicrons that are rich in cholesteryl esters (see Chapter 16). Regulation occurs through changes in HMGCoA reductase content. The half-life of HMGCoA reductase is about 4 hr. Therefore, if synthesis of the enzyme is reduced or interrupted, the content within the hepatocyte decreases appreciably after a few hours. Cholesterol biosynthesis slows because of depletion of HMGCoA reductase, the rate-limiting enzyme in the pathway.

    Studies with oxygenated sterols and cultured cells indicate that certain oxygenated

Figure 11.10   Dietary regulation of hepatic cholesterol synthesis.

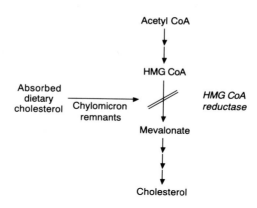

derivatives of cholesterol, including 7α-hydroxycholesterol, 7β-hydroxycholesterol, 7-ketocholesterol, and 25-hydroxycholesterol, are much more potent inhibitors of HMGCoA reductase than cholesterol itself. These studies raise the important question of whether cholesterol must be converted to some oxygenated metabolic product before it exerts its negative feedback effect on HMGCoA reductase production.

Role of plasma lipoproteins in feedback inhibition Feedback inhibition is mediated by cholesterol delivered to the hepatocytes by plasma lipoproteins. Receptors located on the plasma membrane of the hepatocytes recognize the lipoproteins, bind them, and internalize them by a process known as receptor-mediated endocytosis (see Chapter 16). The cholesterol contained within the lipoproteins is released after internalization into the hepatocyte and inhibits HMGCoA reductase formation. There are two sources of this cholesterol. One is dietary cholesterol delivered to the hepatocytes by chylomicron remnants. The other is cholesterol contained in the blood plasma in the form of LDL. Different receptors mediate the uptake of chylomicron remnants and LDL into the hepatocyte.

Covalent modification of HMGCoA reductase HMGCoA reductase also is regulated by covalent modification through phosphorylation and dephosphorylation. The enzyme is inactivated by phosphorylation in a reaction requiring ATP and Mg^{++}, as shown in Fig. 11.11. The inactivating kinase is present in both the cytosol and microsomes of the liver. Removal of the phosphate group by a phosphatase present in the cytosol reactivates the enzyme. Both the activating phosphatase and the inactivating kinase are themselves regulated by phosphorylation and dephosphorylation, as in the case of glycogen synthase. Phosphorylation is thought to be the first step in the process whereby HMGCoA reductase is inactivated and ultimately degraded.

Circadian rhythm Cholesterol synthesis also varies at different times during the day. This effect, known as a circadian rhythm, occurs predominantly in the liver. Synthesis reaches a peak about 6 hr after dark and passes through a minimum about 6 hr after reexposure to light. The activity of hepatic HMGCoA reductase exhibits an identical diurnal variation—highest at midnight and lowest at noon. Therefore the circadian vari-

Figure 11.11   Regulation of HMGCoA reductase by phosphorylation.

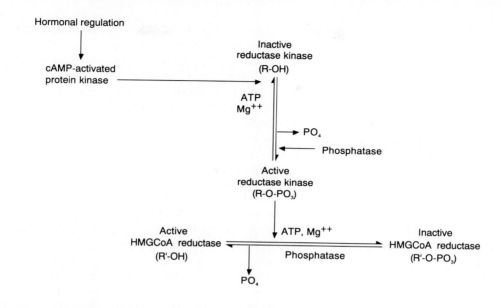

ation in cholesterol synthesis is secondary to changes in the activity of HMGCoA reductase, the rate-limiting enzyme in the cholesterol biosynthetic pathway. Because the half-life of this enzyme is about 4 hr, the circadian rhythm can be explained entirely on the basis of changes in the rate of enzyme synthesis.

Hormonal regulation In addition to the diet, a number of hormones have regulatory actions on cholesterol synthesis in the liver. They, too, operate through effects on HMGCoA reductase activity, probably by controlling either the production of the enzyme or the covalent modifications produced by phosphorylation and dephosphorylation. Insulin or triiodothyronine administration increases HMGCoA reductase activity, whereas glucagon and cortisol decrease the enzyme activity. These findings, together with the dietary and circadian effects, illustrate the enormous complexity with which HMGCoA reductase and therefore cholesterol synthesis is controlled in the liver.

Clinical comment

HMGCoA reductase inhibitors Cholesterol synthesis plays an important role in maintaining the plasma cholesterol concentration. Therefore one approach to reducing plasma cholesterol is to inhibit cholesterol synthesis. The most effective inhibitors are those directed at the rate-limiting step in synthesis, the HMGCoA reductase reaction. Recently, clinically useful HMGCoA reductase inhibitors have been developed, such as *mevinolin,* that can reduce the plasma cholesterol concentration by 30% to 50%. These drugs appear to have only minimum toxicity and are an effective means of reducing plasma cholesterol elevations.

Regulation of cholesterol levels in humans

Most of the clinical information about cholesterol regulation in humans is of necessity based on measurements of the plasma cholesterol concentration. Plasma cholesterol, which is contained in lipoproteins, interacts with tissue cholesterol. However, there are still many uncertainties introduced when attempts are made to extrapolate from plasma cholesterol values to tissue contents or enzymatic regulatory mechanisms.

Clinical comment

Plasma cholesterol concentration The normal cholesterol concentration in plasma obtained from a fasting human is considered by most clinical laboratories to be between 3.1 and 5.7 mmol/L (120 to 220 mg/dl). In young, healthy adults the mean value for plasma cholesterol is about 4.5 mmol/L (175 mg/dl). About 65% of the cholesterol in the plasma of normal fasting subjects is esterified. These measurements usually are made after an overnight fast of 12 to 14 hr. The plasma of a normal patient will have no chylomicrons, which are derived from dietary fat, after this period of fasting. Therefore the main lipoproteins present in normal fasting plasma are LDL, and about 70% of the total cholesterol is present in LDL in normal fasting plasma. Males have relatively more LDL than premenopausal females, and therefore males tend to have higher plasma cholesterol concentrations than premenopausal females.

Origins of body cholesterol Studies with biopsy specimens of human liver clearly demonstrate that feedback inhibition occurs in humans. However, if patients are placed on a cholesterol-free diet for a prolonged period, there is only a 10% to 25% decrease in the plasma cholesterol concentration. These findings lead to several important conclusions. First, although hepatic feedback inhibition occurs, it appears not to be able to compensate completely for dietary cholesterol restriction. Therefore some plasma cholesterol reduction can be achieved by dietary restriction. However, large decreases in circulating cholesterol usually cannot be produced simply by restriction of dietary cholesterol. This can occur only through inhibition of cholesterol biosynthesis.

Clinical comment

Dietary cholesterol restriction A modest reduction of the plasma cholesterol concentration almost always can be achieved by reducing the dietary cholesterol intake. When the diet is rich in cholesterol, the plasma concentration is maintained from 10% to 25% higher than when the dietary intake is restricted. Dietary cholesterol restriction is recommended for patients with hypercholesterolemia. There is a general feeling that the entire population would benefit over the long term by ingesting less cholesterol, and diets low in cholesterol are now being recommended by many physicians for everyone, beginning even in childhood. Epidemiologic studies indicate that persons with plasma cholesterol concentrations above 5.2 mmol/L (200 mg/dl) have a greater tendency to develop atherosclerosis, the arterial disease that leds to myocardial infarction and stroke. It is currently recommended that those patients with elevated plasma cholesterol concentrations go on a low-cholesterol diet to decrease the plasma concentration. Many people with hypercholesterolemia have a relatively mild form of the disease in which the plasma cholesterol concentration is between 6.2 and 7.3 mmol/L (mmol/L (240 to 280 mg/dl). It is in this group that dietary cholesterol restriction probably is most helpful, because the modest reductions in plasma cholesterol concentrations that often can be achieved by dietary modification are sufficient to reduce their values to 6.2 mmol/L or less. If dietary restriction does not produce an adequate reduction in cholesterol, it is necessary to add a hypocholesterolemic drug to the treatment regimen.

Clinical comment

Saturation of dietary fat and plasma cholesterol concentration When humans are fed diets that are rich in saturated fatty acids, the plasma cholesterol concentration is increased. When the saturated fat is replaced by a fat that is rich in polyunsaturated fatty acids, such as linoleic acid, a decrease in the plasma cholesterol concentration occurs. The mechanism through which polyunsaturated fats produce this effect is not known at present. Because the plasma cholesterol concentration is lowered, diets that contain higher percentages of polyunsaturated fatty acids are considered by many to be beneficial to human health.

The present recommendation is to raise the polyunsaturated fat intake and correspondingly lower the saturated fat intake so that the ratio of polyunsaturated to saturated fat in the diet (P/S ratio) becomes 1.0. Recent studies indicate that replacement of saturated fat in the diet by monounsaturated fat also is effective in reducing plasma cholesterol. Olive oil, which is rich in oleic acid, is a good source of monounsaturated fat for this purpose.

Dietary fat saturation is an independent factor regulating the plasma cholesterol concentration. However, in ordinary diets saturated fat is associated with high cholesterol content and, by contrast, polyunsaturated fat is associated with low or absent cholesterol. This occurs because saturated fat is relatively high in animal products that contain cholesterol. Therefore meat and dairy products are, in general, high in both cholesterol and saturated fat. Plant products, which are often high in polyunsaturated fat, do not contain cholesterol. Plants contain sterols such as β-sitosterol (see p. 467), but these phytosterols are poorly absorbed and are not harmful like cholesterol. It is most common that people who eat a diet high in saturated fat are also ingesting a high-cholesterol diet. Conversely, those who eat a diet that is relatively high in polyunsaturated fat tend to have lower cholesterol intakes. Because of this, fat saturation and cholesterol intake tend to be linked together in clinical practice, even though it is clear that each is an independent variable regulating the plasma cholesterol concentration.

Cinical comment

Familial hypercholesterolemia Studies with skin fibroblasts taken from patients who have the homozygous form of familial hypercholesterolemia indicate that feedback inhibition of cholesterol synthesis is defective in this disease. These patients have severe hypercholesterolemia, with plasma cholesterol concentrations as high as 20 to 23 mmol/L (800 to 900 mg/dl) and often develop coronary heart disease in childhood. The fibroblast studies indicate that there is a defect in the ability of LDLs to bind to the surface of the cell, and, as a result of this, they do not cause feedback inhibition of cholesterol synthesis. The reason that feedback inhibition does not occur in the cultured fibroblasts of patients having homozygous familial hypercholesterolemia is that their cells either lack the high-affinity LDL receptor or have defective receptors.

Patients having heterozygous familial hypercholesterolemia have plasma cholesterol concentrations in the range of 350 to 500 mg/dl and are seriously at risk of developing coronary heart disease as young adults. Fortunately, less than 10% of patients with hypercholesterolemeia have this extremely serious genetic form of the disease.

Bibliography

Bloch K: The biological synthesis of cholesterol, Science 150:19, 1965. *A classical discussion of the synthetic pathway.*

Brown MS and Goldstein JL: A receptor–mediated pathway for cholesterol homeostasis, Science 232:34, 1986.

Brown MS and Goldstein JL: Multivalent feedback regulation of HMG CoA reductase, a control mechanism coordinating isoprenoid synthesis and cell growth, J Lipid Res 21:505, 1980.

Goldstein JL and Brown MS: Progress in understanding the LDL receptor and HMG-CoA reductase, two membrane proteins that regulate the plasma cholesterol, J Lipid Res 25:1450, 1984.

Hofmann AF and Roda A: physicochemical properties of bile acids and their relationship to biological properties: an overview of the problem, J Lipid Res 25:1477, 1984.

Schroepfer GJ Jr: Sterol biosynthesis, Annu Rev Biochem 50:585, 1981.

Schroepfer GJ Jr: Sterol biosynthesis, Annu Rev Biochem 51:555, 1982.

Clinical examples

A 36-year-old man was found to have hypercholesterolemia. A diet evaluation indicated that he was consuming about 1.6 mmol (600 mg)/day of cholesterol, and that his blood plasma cholesterol concentration on two separate occasions was approximately 8.5 mmol/L (330 mg/dl). Ultracentrifugal analysis revealed that the cholesterol elevation was contained in plasma LDL. He was treated with a synthetic cholesterol-free formula diet for 3 mo, but the fasting plasma cholesterol level decreased to only 7.7 mmol/L (300 mg/dl). Subsequently he was treated with colestipol hydrochloride, a bile acid–binding resin. This resin is not absorbed and remains in the intestinal lumen, where it binds bile acids, causing increased amounts to be excreted in the feces. The drug treatment was successful in lowering the fasting plasma cholesterol concentration to the range of 5.8 to 6.4 mmol/L (220 to 250 mg/dl), a value that was considered to be acceptable for this patient.

Biochemical questions

1. Can the body use dietary cholesterol?
2. How is it possible for a patient to continue to have high plasma cholesterol after being on a cholesterol-free diet for 3 mo?
3. What connection is there between bile acids and cholesterol?
4. How does colestipol hydrochloride, the bile acid–binding resin, lower the plasma cholesterol concentration?

Answers

   1. Cholesterol absorption Cholesterol is absorbed from the diet. Most of us have a dietary cholesterol intake of 1.3 to 1.6 mmol/day (500 to 600 mg) and absorb 0.8 to 1.0 mmol/day (300 to 400 mg). Any dietary cholesterol that is present in esterified form is hydrolyzed to the free sterol through the action of cholesteryl esterase, a pancreatic digestive enzyme. Hydrolysis occurs within or on the surface of micelles that are formed through mixture with the conjugated bile acids and phosphatidylcholine contained in the bile. Cholesterol is transferred from the micelles to the intestinal mucosal cells by passive diffusion. Most of the absorption occurs in the lower jejunum and ileum. Once inside the mucosal cell, most of the cholesterol is reesterified and released into the lymph in chylomicrons. These lipoproteins enter the bloodstream, and most of their triacylglycerol content is hydrolyzed through the action of lipoprotein lipase (see Chapter 10). The cholesteryl esters remain in the resulting particle called a chylomicron remnant, which is taken up by the liver. The absorptive process may continue for 8 to 10 hr after ingestion of a fatty meal.
   2. Cholesterol synthesis It is perfectly reasonable for this patient to have high cholesterol in his blood plasma and, for that matter, in all his tissues in spite of the fact that he has been on a cholesterol-free diet for a prolonged period. The diet is not the only source of cholesterol for humans. A great deal of the cholesterol present in the body is biosynthesized, not derived from the diet. Acetyl CoA is the substrate for cholesterol synthesis. This metabolic intermediate is produced from the catabolism of carbohydrates, fatty acids, and some of the amino acids. Therefore almost any food that humans ingest will provide the necessary building blocks for cholesterol synthesis.

It is true that less cholesterol is synthesized when the diet contains adequate amounts of this sterol. This occurs because the biosynthetic pathway in the liver is inhibited when the dietary cholesterol supply is adequate. Inhibition occurs because the production of HMGCoA reductase, the rate-limiting enzyme in the cholesterol synthetic pathway, is reduced. However, cholesterol synthesis probably occurs even when the diet contains cholesterol.

3. Cholesterol excretion The main route of cholesterol metabolism from the quantitative standpoint is conversion to bile acids. About 0.8 mmol/day of bile acids are lost in the feces. This is replaced daily, for the body is geared to maintain the size of the bile acid pool at a constant level between 15 and 30 g. Replacement of bile acids is accomplished through synthesis from cholesterol in the liver.

4. Colestipol and cholesterol concentration Administration of a bile acid–binding resin decreases bile acid reabsorption from the intestine and hence increases the loss of bile acids in the feces. To compensate for the greater losses, more cholesterol must be converted into bile acids in the liver. If the dietary intake of cholesterol is restricted to a modest level, about 0.8 mmol/day or less, much of the cholesterol required for bile acid synthesis will have to be supplied from within the body. The capacity of the liver to synthesize cholesterol is limited, and some of the extra cholesterol that is needed for bile acid synthesis will have to come from the cholesterol contained in the plasma lipoproteins, primarily from LDL. Eventually a new steady state develops in which the plasma cholesterol concentration is lower than it was before administration of the bile acid–binding resin. This is thought to be beneficial.

References

Dorr AE et al: Colestipol hydrochloride in hypercholesterolemic patients—effect on serum cholesterol and mortality, J Chronic Dis 31:5, 1978.

Goodman DS et al: Report of the National Cholesterol Education Program expert panel on detection, evaluation and treatment of high blood cholesterol in adults, Arch Intern Med 148:36, 1988.

Case 2

Hypercholesterolemia and atherosclerosis

A 33-year-old man was referred to a cardiologist because of severe intermittent chest pains. He had begun to have these pains in his late twenties, and they gradually became worse. They occurred typically after mild exertion such as brisk walking, carrying a bag of groceries up a flight of stairs, or mowing the lawn. A coronary artery angiogram was performed and revealed severe narrowing and irregularity of all three coronary arteries. A diagnosis of coronary artery atherosclerosis was made. Blood was obtained after an overnight fast on two occasions, and these samples contained 10.0 and 10.6 mmol/L (385 to 410 mg/dl) of cholesterol. Most of the plasma cholesterol elevation was present in LDL. A careful physical examination revealed the presence of firm masses on the extensor tendons of the hands. These were thought to be xanthomas of the tendon sheaths. While in the office, the patient developed severe chest pain. An ECG taken during the episode of pain showed changes that were consistent with myocardial ischemia. The patient was taken immediately to the hospital where coronary artery bypass surgery was performed.

Biochemical questions

1. How is hypercholesterolemia thought to be related to the development of atherosclerosis, myocardial infarction, and tendon xanthomas?
2. Why would lowering of the plasma cholesterol concentration be potentially beneficial for this patient?

Case discussion   **1. Hypercholesterolemia and other conditions** High plasma cholesterol concentrations, particularly in young and middle-aged men, are associated with an increased incidence of *coronary artery disease* and its end result, *myocardial infarction*. In most cases the elevated plasma cholesterol is present in the LDL class of plasma lipoproteins. The underlying disease process is *atherosclerosis*. Figure 11.12 illustrates the prevalent theory concerning the role of hypercholesterolemia in this disease.

Atherosclerosis The general term for hardening of the arteries is *arteriosclerosis*. The most prevalent form of arteriosclerosis is characterized by the accumulation of lipid, particularly cholesteryl esters, in the arterial intima. This form of arteriosclerosis is called *atherosclerosis*. These lipid deposits initially occur inside arterial smooth muscle cells and macrophages. At this stage the lipid deposits are reversible, and no permanent damage to the vessel wall results. This early lesion is called a *fatty streak*. Subsequently, lipids begin to accumulate extracellularly in the arterial intima. When the lesion progresses to this stage, it is no longer completely reversible. Yet recent studies on diet-induced atherosclerosis in monkeys indicate that some lipid can be removed from even advanced lesions so that the atherosclerotic areas actually can be made to shrink. This shrinking process is known as *regression*.

Figure 11.12   Mechanism producing atherosclerosis and tissue infarction.

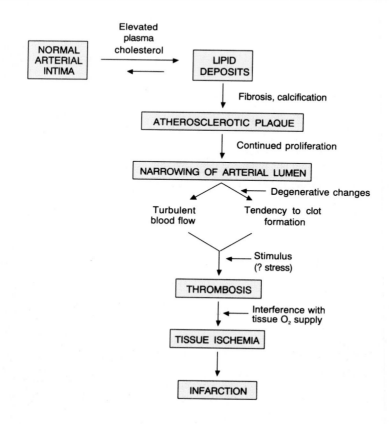

Hypercholesterolemia Elevated plasma cholesterol concentrations, particularly the form resulting from elevations in the LDL class of plasma lipoproteins increase the tendency toward atherosclerosis. The plasma LDLs actually enter the arterial wall and deposit their lipid content there, causing the accumulation of cholesteryl esters. The higher the plasma LDL concentration, the greater the tendency toward lipid deposition. The relation between LDL elevation and arterial wall lipid accumulation has been tied together by the LDL binding defect discovered in skin fibroblasts of patients with familial hypercholesterolemia (Figure 11.13). Studies by Brown and Goldstein show that cultured skin fibroblasts from patients with familial hypercholesterolemia either lack or have defective plasma membrane high-affinity receptors for LDL. Homozygotes for this disease usually exhibit a complete absence of LDL receptors, whereas heterozygotes have a deficiency of these receptors. This is illustrated schematically in Figure 11.13. Because of the deficiency or absence of LDL receptors, LDL uptake by the high-affinity receptor mechanism is defective. LDL accumulates in the plasma because the normal clearance mechanism, which operates through the LDL receptor, is defective.

As the plasma LDL concentration increases, more accumulates in the arterial intima, the space between the endothelial and smooth muscle cells. This attracts macrophages, probably derived from blood monocytes, into the area. Some of the lipids contained in the deposited LDL appear to undergo auto-oxidation, converting the LDL into a form that can be taken up by the macrophages. The lipid-rich macrophages are called foam cells and are a part of the developing atherosclerotic lesion.

Figure 11.13
Schematic model of the role of the LDL receptor in the regulation of intracellular cholesterol metabolism. The normal situation where adequate numbers of LDL receptors are present on the cell surface is shown on the left. In this figure *reductase* refers to HMGCoA reductase. The panel in the center represents heterozygous familial hypercholesterolemia, where there is a reduced number of receptors on the cell surface. The right panel illustrates the defect in the homozygous form of familial hypercholesterolemia, where there is an absence of high-affinity LDL receptors. (From Goldstein JL and Brown MS: Am J Med 58:147, 1975.)

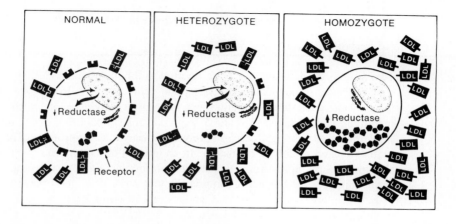

Advanced atherosclerotic lesion Ultimately the lipid deposit triggers a fibrotic reaction in the arterial wall, and collagen is deposited to form a fibrous plaque. At this point the process is no longer completely reversible. Calcium is deposited within the plaque and the lesion grows; the result is a decrease in the diameter of the channel through which the blood flows. Subsequently the plaque can degenerate, producing a roughened intimal surface. This, together with the increased turbulence of blood flow resulting from the narrowed arterial lumen, predisposes to clot formation within the blood channel. If the occlusion of the channel persists, the cells that are normally supplied with oxygen by the vessel will become *anoxic* and ultimately die; the death of tissue is termed an *infarction*.

Atherosclerosis can occur in many different areries, and the clinical disease that results depends on which vessels are involved. The most frequent location in humans is the coronary arteries, leading to myocardial infarction, which is known commonly as "heart attack."

Xanthomatosis Xanthomas are collection of lipid, mostly cholesteryl ester, in the skin or tendon sheaths. Xanthomas are often associated with and are a sign of severe hypercholesterolemia and premature atherosclerosis.

2. Treatment of hypercholesterolemia Lowering the plasma cholesterol concentration in patients who have hypercholesterolemia is considered by many physicians to be helpful in preventing atherosclerosis and in treating existing atherosclerotic lesions. Increased plasma LDL cholesterol concentrations, particularly in middle-aged men, have been correlated with an increased incidence of myocardial infarction, one of the most severe consequences of the atherosclerotic process. Modern medical treatment is therefore aimed at lowering plasma LDL cholesterol concentrations, particularly in individuals such as this 33-year-old man in whom a serious elevation is known to exist.

The rationale is that if the plasma LDL can be reduced, there will be a lesser tendency for LDL to accumulate within the arterial wall and trigger either the formation of new atherosclerotic lesions or the growth of existing lesions. Furthermore, if sufficient LDL reduction can be achieved, evidence suggests that some of the existing lesions may even undergo regression and shrink. Thus effective treatment of hypercholesterolemia may not only prevent atherosclerosis from progressing but also may lead to improvement in already diseased arteries.

References   Armstrong ML: Evidence of regression of atherosclerosis in primates and man, Postgrad Med J 52:456, 1976.

Small DM: Progression and regression of atherosclerotic lesions: insights from lipid physical biochemistry, Atherosclerosis 8:103, 1988.

Steinberg D: Lipoproteins and the pathogenesis of atherosclerosis, Circulation 76:508, 1987.

Case 3   **β-Sitosterolemia**

Two sisters, aged 19 and 17, were referred to the metabolism clinic of a university hospital because they had extensive tendon xanthomatosis. Repeated measurements of their plasma cholesterol concentrations after an overnight fast revealed values that were within the normal range, 5.3 and 5.0 mmol/L (205 and 190 mg/dl), respectively. Therefore it was thought most unlikely that plasma cholesterol could be responsible for such extensive xanthoma formation. In an attempt to explore the cause of this abnormality, the lipids from the patients' plasma were analyzed by gas-liquid chromatography. β-Sitosterol comprised about 10% of the total plasma sterols in each case, whereas it accounts for only about 0.2% in normal people. About 60% of the plasma β-sitosterol was in esterified form. From 75% to 85% of the plasma β-sitosterol was recovered in the LDL fraction. Analysis of specimens from tendon xanthomas removed from these patients demonstrated that β-sitosterol was present in these lesions.

Biochemical questions

1. What is β-sitosterol?
2. What is the most likely origin of the β-sitosterol found in these patients' plasma and tissues?
3. Present a likely mechanism for the formation of β-sitosteryl esters in the blood plasma.
4. Suggest a possible mechanism for xanthoma formation in these patients.

Case discussion

1. Definition of β-sitosterol β-Sitosterol is a plant sterol. Its structure, shown on p. 467, is similar to cholesterol except for the presence of a two-carbon-atom branch at C_{24}. This is located in the hydrocarbon tail that is attached to the sterol nucleus at the apex of the D ring, C_{17}. β-Sitosterol is the predominant sterol in many plants, but it is not the only one. The others include campesterol and stigmasterol, which differ only slightly in structure from β-sitosterol. More detailed analyses of these patients' plasma and tissues revealed that campesterol and stigmasterol also were present, although in smaller amounts than β-sitosterol. For example, in the plasma of one of the patients, the β-sitosterol value was 0.6 mmol/L, campesterol was 0.25 mmol/L, and stigmasterol was 0.01 mmol/L.

2. Origin of β-sitosterol Cholesterol is the only sterol synthesized by animal tissues. Plant products, however, are contained in the usual diet. Therefore the most likely source of the β-sitosterol in these patients was the diet. Studies with isotopically labeled β-sitosterol revealed that these sisters absorbed about 25% of a test dose. Similar tests in normal volunteers revealed absorptions of much less than 5%. On the basis of this information, it was concluded that these two patients with β-sitosterolemia have an inherited metabolic defect that causes them to absorb excessive quantities of plant sterols.

3. β-Sitosteryl esters in blood Most of the cholesteryl esters present in the blood plasma are formed through the action of lecithin-cholesterol acyltransferase (see p. 475). This enzyme catalyzes the transfer of a fatty acid group from position *sn*-2 of phosphatidylcholine to cholesterol, using lipoprotein lipids as substrates. The reaction occurs in the blood plasma. Although the enzyme is produced in the liver, it acts only after it is released into the blood. β-Sitosteryl esters also are formed in the plasma by the action of the lecithin-cholesterol acyltransferase, with the β-sitosterol contained in plasma lipoproteins substituting for cholesterol as the substrate. This actually was demonstrated by incubating radioactive β-sitosterol with plasma from these patients.

4. Xanthoma formation At present, there is no definitive explanation as to why the high level of β-sitosterol in the plasma caused xanthomatosis. Relative to the usual plasma cholesterol concentrations, the amount of β-sitosterol in the plasma of these patients actually is low. Therefore it is unlikely that the xanthoma formed simply through a mass-action effect. Although plant sterols were recovered in the xanthoma, the lesions still contained primarily cholesterol. Why the presence of plant sterols caused cholesterol accumulation in the tendon sheaths when the plasma cholesterol concentration itself was not elevated is unknown at this time.

References

Bhattacharyya AK and Connor WE: β-Sitosterolemia and xanthomatosis: a newly described lipid storage disease in two sisters, J Clin Invest 53:1033, 1974.

Björkhem I and Skrede S: Familial diseases with storage of sterols other than cholesterol: cerebrotendinous xanthomatosis and phytosterolemia. In Scriver CR et al, editors: The metabolic basis of inherited disease, ed 6, New York, 1989, McGraw-Hill Information Services Co.

| Case 4 | Gallstones |

A 39-year-old woman consulted her physician because of intermittent abdominal distress. The discomfort usually followed the ingestion of a large meal, often one that contained greasy or fried foods. The pain was located in the upper abdomen and sometimes radiated to her chest. The patient felt bloated during these episodes and thought that she obtained some relief from belching. Occasionally she became severely nauseated and vomited during one of these acute episodes. She had not experienced any previous episodes of jaundice or gastrointestinal bleeding. Initially a diagnosis of irritable bowel syndrome was entertained, and her physician prescribed antacids and a bland diet. This treatment produced no relief. A cholecystogram demonstrated the presence of numerous gallstones in the gallbladder. A cholecystectomy was performed, and the gallstones were found to be composed predominantly of cholesterol.

Biochemical questions

1. What is the function of bile in digestion?
2. What is the metabolic relationship between cholesterol and bile acids?
3. How is cholesterol kept in the soluble state in normal human bile?
4. What physical-chemical factors cause the formation of cholesterol gallstones?
5. Can gallstones be dissolved by feeding certain bile acids? What other alternatives to surgery are presently available?

Case discussion

   1. Bile and lipid digestion Bile emulsifies the dietary fat after it enters the small intestine. This permits the enzymes that hydrolyze the dietary lipids to act more effectively, including the pancreatic lipase, phospholipase A_2, and cholesteryl esterase. The emulsifying compounds in the bile are the bile acids and phosphatidylcholine.
   2. Cholesterol and bile acids Cholesterol is the substrate for bile acid synthesis. The two primary bile acids, cholic and chenodeoxycholic acid, are synthesized in the liver from cholesterol.
   3. Solubilization of cholesterol Cholesterol is kept soluble in the bile as a result of its incorporation into mixed micelles. In addition to cholesterol, these micelles contain phosphatidylcholine and bile acids.
   4. Gallstone formation Only a relatively small quantity of cholesterol can be kept in the soluble state in bile. The maximum solubility is about 15 mol%. If the amount of cholesterol exceeds this it will precipitate, producing a cholesterol gallstone.
   5. Dissolving gallstones Gallstones usually form in the gallbladder, where the bile is concentrated and stored after it is formed and released by the liver. The usual way of treating gallstones is to surgically remove the gallbladder. Alternatively, gallstones have been treated by feeding one of the bile acids, usually ursodeoxycholic acid or chenodeoxycholic acid. Feeding of the bile acid raises the mole percentage of bile acid in the bile. This increases the capacity of the mixed micelles to incorporate cholesterol, which has the effect of solubilizing some of the precipitated cholesterol. Eventually, the gallstones may dissolve.
   Other nonsurgical approaches for dissolving gallstones are available. One is to insert a catheter into the gallbladder and perfuse the gallstones with an organic solvent such as methyl tertiary butyl ether. Cholesterol is soluble in this solvent, and the stones dissolve. Alternatively, if a large single gallstone is present, it can be pulverized by extracorporeal shock wave lithotripsy. The fragments that are produced are small enough to pass harmlessly out of the gallbladder through the bile duct.

References

Allen MJ et al: Rapid dissolution of gallstones in humans using methyl tert-butyl ether, N Engl J Med 312:217, 1985.

Sackmann M et al: Shock-wave lithotripsy of gallbladder stones: the first 175 patients, N Engl J Med 318:393, 1988.

Van Sonnenberg E and Hofmann A: Horizons in gallstone therapy, AJR 150:43, 1988.

Case 5	Carotid artery atherosclerosis

A 61-year-old woman was referred to a university hospital because of severe headaches and fainting spells. She was noted to have hypertension (blood pressure 200/100 mm Hg), and bruits were heard over the carotid arteries. An angiogram of the right carotid artery demonstrated a narrowed lumen. The fasting plasma cholesterol level was 10 mmol/L (385 mg/dl), and the concentration of LDL was elevated.

Eight days after her admission to the hospital a right carotid endarterectomy was performed. This produced significant improvement in the patient's condition. She was discharged from the hospital with instructions to follow a low-cholesterol diet and take mevinolin, a drug that inhibits HMGCoA reductase.

Biochemical questions

1. How is dietary cholesterol absorbed?
3. How is cholesterol biosynthesis normally regulated?
3. What role does HMGCoA reductase play in cholesterol biosynthesis?
4. What is the purpose of prescribing a drug that inhibits HMGCoA reductase? Is there a physiologic basis for targeting this enzyme in hypocholesterolemic drug therapy?
5. Why should the patient be placed on a low-cholesterol diet if she also is taking a hypocholesterolemic drug such as a HMGCoA reductase inhibitor?
6. Why are reductions in plasma cholesterol concentration beneficial in the treatment of hypercholesterolemia?

Reference

Goldstein JL, and Brown MS: Familial hypercholesterolemia. In Scriver CR, et al, editors: The metabolic basis of inherited disease, ed 6, New York, 1989, McGraw Hill Information Services Co.

Case 6	Lecithin: cholesterol acyltransferase deficiency

A 33-year-old female was admitted to the hospital because of proteinuria. Further examination revealed that she was anemic and had diffuse, grayish corneal opacities. She also had hyperlipidemia. Analysis of her plasma lipid levels revealed an elevated amount of cholesterol and almost no measurable cholesteryl esters; normally, about 65% of the plasma cholesterol is in the form of cholesteryl esters. No lecithin-cholesterol acyltransferase (LCAT) activity was detected in the patient's plasma.

Biochemical questions

1. What are cholesterol esters?
3. Where does the LCAT reaction occur, and what is the role of phosphatidylcholine in this rection?
3. How does the synthesis of cholesteryl esters in the tissues differ from the LCAT reaction?
4. What is the difference between the type of cholesteryl esters formed in the tissues, as compared with those formed in the LCAT reaction?
5. Can severe liver disease produce LCAT deficiency?
6. Does the hydrolysis of cholesteryl esters occur through the reverse of the LCAT reaction?

References

Albers JJ et al: Defective enzyme causes lecithin-cholesterol acyltransferase deficiency in a Japanese kindred, Biochim Biophys Acta 835:253, 1985.

Norum KR, Gjone E, and Glomset JA: Familial lecithin: cholesterol acyltransferase deficiency, including fish eye disease. In Scriver CR et al, editors: The metabolic basis of inherited disease, ed 6, New York, 1989, McGraw-Hill Information Services Co.

Additional questions and problems

1. A patient with severe hypercholesterolemia who did not respond to drug therapy was treated surgically by excision of a part of the intestine. What particular part of the intestinal tract might be selected for excision to produce the maximum decrease in cholesterol levels?

2. During a surgical procedure a sample of bile was obtained for research purposes from the gallbladder of a patient whose biliary and intestinal tracts were normal. Which bile acids would be recovered in this specimen?

3. Would a drug that interfered with biotin-mediated carboxylation reactions be expected to directly block cholesterol biosynthesis?

4. How is the metabolism of glucose by the pentose phosphate pathway related to cholesterol synthesis?

5. If a drug blocked the synthesis of HMGCoA in the cell cytoplasm, would it be expected to interfere directly with ketone body synthesis?

6. Explain the mechanism whereby diets that contain cholesterol suppress cholesterol synthesis in the liver.

7. Mevinolin reduces hepatic HMGCoA reductase activity. Explain why this also might reduce glycoprotein biosynthesis in the liver.

8. What metabolic pathway of cholesterol is associated with hydroxylation at position 7 of the steroid nucleus?

Multiple choice problems

1. The rate-limiting reaction in cholesterol biosynthesis is the conversion of:
   A. Acetyl CoA → acetoacetyl CoA.
   B. Acetyl CoA → malonyl CoA.
   C. Acetoacetyl CoA → hydroxymethylglutaryl CoA (HMGCoA).
   D. Hydroxymethylglutaryl CoA (HMGCoA) → mevalonate.
   E. Mevalonate → mevalonate pyrophosphate.

2. Squalene is
   A. An enzyme involved in cholesterol biosynthesis.
   B. An isomer of the mammalian isoprene unit.
   C. A 30-carbon-atom hydrocarbon intermediate in cholesterol biosynthesis.
   D. One of the sterol intermediates formed after the hydrocarbon chain has cyclized into the four-ring steroid structure.
   E. The carrier protein that binds the steroid intermediates during cholesterol synthesis.

3. Mevalonate contains 6 carbon atoms and cholesterol contains 27. How many mevalonate molecules are required to synthesize one molecule of cholesterol?
   A. Two
   B. Five
   C. Six
   D. Eight
   E. Ten

4. Which of the following are characteristics of the isoprene unit used in the cholesterol biosynthetic pathway?
   1. It contains 5 carbon atoms.
   2. It contains phosphate.
   3. It is a branched-chain compound.
   4. It is present in two isomeric forms.
       A. 1, 2, and 3.
       B. 3.
       C. 4.
       D. 4 only.
       E. All are correct.

5. The conversion of primary to secondary bile acids:
   1. Occurs in the liver.
   2. Involves the removal of three carbon atoms from the hydrocarbon chain.
   3. Involves the addition of glycine to the bile acid.
   4. Involves removal of the hydroxyl group at position 7 of the steroid ring structure.
       A. 1, 2, and 3.
       B. 1 and 3.
       C. 2 and 4.
       D. 4 only.
       E. All are correct.

6. How is the *trans*-methylglutaconate shunt related to cholesterol biosynthesis?
   1. It shunts carbon atoms into the cholesterol synthetic pathway, thereby increasing cholesterol synthesis.
   2. It produces feedback inhibition of cholesterol biosynthesis.
   3. It prevents the conversion of lanosterol to desmosterol and thereby reduces cholesterol formation.
   4. It diverts carbon atoms out of the cholesterol biosynthetic pathway at a step beyond the formation of mevalonate.
       A. 1, 2, and 3.
       B. 1 and 3.
       C. 2 and 4.
       D. 4 only.
       E. All are correct.

7. Which of the following compounds are involved in the synthesis of cholesteryl esters intracellularly?
   1. Phosphatidylcholine
   2. Fatty acyl CoA
       A. 1 only.
       B. 2 only.
       C. 1 and 2.
       D. Neither 1 nor 2.

8. Match the numbered statement with *one* of the letter options.
   A. Coprostanol
   B. Farnesyl pyrophosphate
   C. β-Sitosterol
   D. Taurine
       1. An intermediate in cholesterol biosynthesis

2. Combines with a bile acid through an amide linkage
3. An excretion product of cholesterol found in the feces
4. A plant sterol

Chapter 12

Membranes

Objectives

1. To describe the composition and properties of biologic membranes
2. To discuss membrane function, including transport and prostaglandin formation
3. To explain the biochemical basis of diseases involving membrane function

Cell structure and the cytoskeleton

As noted earlier (see Chapter 3), an animal cell is composed of many subcellular compartments. Each of these compartments, as well as the whole cell, is surrounded by a membrane. The outer cellular membrane, called the plasma membrane, is anchored to the *cytoskeleton,* which is a network of microfilaments and microtubules that interact extensively with each other and with the components of the plasma membrane (Figure 12.1). Among other functions, the cytoskeleton is responsible for the shape of the cell, for its mobility, and for the separation of chromosomes during cell division.

The proteins of the cytoskeleton networks, listed in Table 12.1, fall into three classes: (1) actin filaments, formed by the polymerization of 42,000-dalton G-actin subunits; (2) two types of tubulin, α and β, each 55,000 daltons; as α,β-dimers, they assemble into microtubules; (3) intermediate filament proteins, so-called because the diameter of the filaments is between that for actin filaments (7 nm) and microtubules (11 nm). Although different intermediate filament proteins are found in different cell types, they can be classified into five major groups:

1. *Keratins* are found in epithelial cells.
2. *Neuronal filaments,* composed of three proteins of 200,000, 150,000, and 68,000 daltons, occur in close association with axonal microtubules.
3. *Desmin* filaments are found predominantly in muscle cells.
4. *Glial fibrillary acidic protein* (GFAP) is found exclusively in glial cells.
5. *Vimentin*-containing filaments are associated with mesenchymal cells.

The intermediate filaments, which show a high degree of amino acid sequence homology, associate with the actin and tubulin filament systems in the cell to develop a cytoskeletal network that determines the morphology and mobility of the cell. Other proteins may also be involved, such as myosin in muscle to give it a contractile property and dynein with microtubules to give them mechanical properties. The cytoskeleton is also closely associated with the cell membranes, particularly the plasma membrane.

Clinical comment

Spherocytosis Hereditary spherocytosis results in abnormally shaped erythrocytes, which are caused by a less rigid cytoskeleton. This causes blebbing of the membrane and thus more rapid removal of the erythrocytes by the spleen. The resulting depletion of the erythrocytes from the blood produces anemia.

The erythrocyte membrane and cytoskeleton are highly specialized and are not typical of all cells. The membrane has two principal glycoproteins, glycophorin, which is 64% carbohydrate, and Band 3, the protein of the anion channel. These are the only integral membrane proteins that are exposed to the outer surface of the membrane, and they provide the erythrocyte with a complex carbohydrate surface. The other integral membrane proteins—spectrin, ankyrin, and actin—are exposed on the cytoplasmic side of the membrane and form the cytoskeleton of the cell. The cytoskeleton of the erythrocyte

Figure 12.1   Cytoskeleton, membrane proteins, and lipid bilayer. Peripheral
membrane proteins are illustrated by *A*. They are shown
schematically as sitting on top of the polar head groups of the
membrane phospholipid bilayer, on both surfaces. Electrostatic
interactions are primarily responsible for their binding to the lipid
bilayer. Integral membrane proteins are illustrated schematically by *B,
C, D,* and *E.* They are embedded in the lipid bilayer; the primary
binding forces are nonpolar interactions with the fatty acyl chains of
the phospholipids in the bilayer. Integral proteins either penetrate only
part of the distance through the bilayer, *C* and *E,* or completely
across it, Bands *B* and *D.* Proteins such as *D,* which span the bilayer,
are known as *transmembrane proteins.* The phospholipid polar head
groups in the lipid bilayer are shown as small circles, *L,* and their fatty
acyl tails are illustrated by wavy lines, *G* and *M.* A portion of the
cytoskeleton is indicated by proteins *(N)* that anchor the filamentous
cytosolic proteins *(H)* to the membrane. Actin *(J)* is also indicated and
is attached to the membrane and the skeletal proteins, thus causing
movement of the cell.

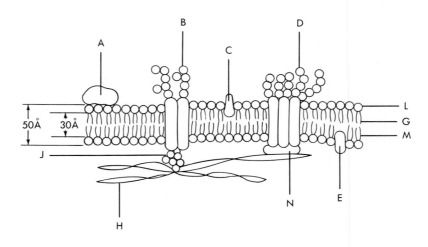

Table 12.1   Cytoskeletal proteins

Protein	Property/function
Actin filaments	Formed by polymerization of G-actin subunits
Tubulin (α and β)	Assembles into α,β-dimers, which form microtubules
Intermediate filament proteins	Differ in different types of cells
Keratins	Epithelial cells
Neuronal filaments	Composed of three proteins; associated with axon microtubules
Desmin	Muscle
Glial fibrillary acidic proteins	Glial cells
Vimentin	Mesenchymal cells

constitutes a network under the plasma membrane, and the plasma membrane binds to it through the cytoplasmic segments of Band 3 protein. The erythrocyte is exposed to significant stresses and shape contortions while circulating through the heart, blood vessels, and capillaries. The loss of mechanical strength and the resulting deformed shape of erythrocytes in hereditary spherocytosis results from a genetic defect in spectrin that affects the structure of the cytoskeleton.

Cell membranes

Biologic membranes separate cells from their external environment and divide the interior of the cell into compartments. They are 75 to 90 Å thick. The chemical composition of cell membranes varies widely, as shown in Table 12.2. As a general estimate, a representative membrane is made up of about 50% protein, 45% lipid, and 5% carbohydrate. Approximately 10% of the membrane proteins are glycoproteins. A major exception to this general chemical composition is the myelin sheath of nerves, one of the most widely studied biologic membranes. Myelin is made up of 20% protein, 75% lipid, and 5% carbohydrate.

The basic organization of biologic membranes is illustrated in Figure 12.1. A bilayer of lipid forms the central structure of the membrane. The bilayer is composed primarily of phospholipids and is held together by physical forces, not covalent bonds. Proteins are interspersed throughout the bilayer. Some of the proteins are attached to the surface, whereas others are embedded within the lipids or penetrate completely through the bilayer and are exposed on both surfaces. Many of the membrane proteins are enzymes. Others are recognition factors, ion channels, transporters, or receptors.

Table 12.3 lists various types of cell membranes and their major functions.

Membrane lipids

The lipid composition of the human erythrocyte membrane, which is representative of the plasma membrane of most human cells, is given in Table 12.4. For comparison, the composition of the myelin sheath of a neuron also is shown. A much higher percentage of glycosphingolipids exists in myelin and correspondingly, much less phospholipid. However, both the erythrocyte membrane and the myelin are *plasma membranes;* that is, they are located at the cell surface and separate the extracellular fluid from the cytoplasm. This emphasizes the diversity in the lipid composition of cell membranes. Some of the intracellular membranes also have specialized lipid composition. For example, inner mitochondrial membranes contain almost no cholesterol and are the only mammalian membranes that contain appreciable amounts of cardiolipin.

Table 12.2   Composition of cell membranes

Membrane	Composition by weight		
	Protein (%)	Lipid (%)	Carbohydrate (%)
Myelin	20	75	5
Erythrocyte	49	43	8
Hepatocyte plasma membrane	54	39	7
Outer mitochondrial membrane	50	46	4
Inner mitochondrial membrane	75	23	2

Table 12.3   Membranes of the mammalian cell

Membrane	Major functions
Plasma membrane	Barrier between extra- and intracellular fluid; transport of ions, nutrients; signal transduction
Mitochondria	
Inner membrane	Energy transduction; proton pump; adenosine triphosphate (ATP) formation; transport of nutrients
Outer membrane	Barrier to maintain mitochondrial intermembrane space
Endoplasmic reticulum	
Rough	Translation; initial protein processing
Smooth	Synthesis of complex lipids; hydroxylation; desaturation
Golgi complex	Processing of proteins for secretion; modification of glycoproteins
Nuclear membrane	Attachment of chromatin; transport of RNA to cytoplasm; entry of nucleotides, DNA-binding proteins
Lysosomes	Proton pump; compartment for hydrolytic enzymes
Peroxisomes	Fatty acid oxidation; compartment for oxidative enzymes
Secretory vesicles	Fusion with plasma membrane to mediate secretion

Table 12.4   Membrane lipid composition

Lipid	Erythrocyte membrane		Myelin	
	Weight (%)	Molar ratio	Weight (%)	Molar ratio
Phospholipids	69	1.0	43	1.0
Cholesterol	26	0.8	27	1.3
Glycosphingolipids	5	0.1	30	0.7

Phospholipids

As shown in Table 12.4, the main lipids in myelin and the erythrocyte membranes are phospholipids. This is true of all biologic membranes; the basic structure of membranes is a bilayer composed of phospholipids (Figure 12.1). In this arrangement the hydrocarbon chains of the phospholipid fatty acyl groups project into the center of the bilayer and are shown schematically in Figure 12.1 as wavy lines. The hydrophilic glyceryl-phosphoryl— base components of the phospholipids are called *head groups* and are located on the outside of the bilayer, where they interact with water or other polar and charged molecules. These polar groups are represented schematically by the circles to which the wavy lines (the fatty acyl chains) are attached. The lipid bilayer is composed of two leaflets, the outer phospholipid leaflet that faces the extracellular fluid and the inner phospholipid leaflet that faces the cytoplasm. Each leaflet is 25 Å thick, with the head group occupying 10 Å and the fatty acyl chains 15 Å. The total thickness of the bilayer is 50 Å, 30 Å of which is comprised of the hydrocarbon core containing the fatty acyl chains of both leaflets.

Phospholipid composition   The membrane lipid bilayer contains a mixture of phospholipids, most of which are glycerol derivatives called phosphoglycerides. Sphingomyelin is the one exception; it is a phospholipid derivative of sphingosine. Table 12.5 lists the phospholipid composition of various membrane fractions obtained from liver. In

Table 12.5   Phospholipid composition of rat liver membranes

Phospholipid	Phospholipid composition (%)						
	Plasma membrane	Rough endoplasmic reticulum	Smooth endoplasmic reticulum	Golgi complex	Nuclear membrane	Outer mitochondrial membrane	Inner mitochondrial membrane
Choline phosphoglycerides	43	59	54	45	52	49	40
Ethanolamine phosphoglycerides	21	20	22	17	25	35	38
Serine phosphoglycerides	4	3	4	4	6	1	1
Inositol phosphoglycerides	7	10	8	9	14	9	2
Sphingomyelin	23	2	6	12	6	2	2
Cardiolipin	—*	1	2	—	—	4	17
Minor components†	2	5	4	13	7	—	—

*Either not detected or less than 1% of the total.
†Minor components include lysophosphoglycerides and phosphatidic acid.

Table 12.6   Fatty acid composition of human erythrocyte phospholipids

Fatty acid*	Composition (%)			
	Choline phosphoglycerides	Ethanolamine phosphoglycerides	Serine phosphoglycerides	Sphingomyelin
16:0	31	13	3	24
18:0	12	12	38	6
18:1	19	18	8	—
18:2	23	7	3	—
20:4	7	24	24	—
22:0	—†	—	—	10
22:4	—	8	4	—
22:5	—	4	3	—
22:6	2	8	10	—
24:0	—	—	—	13
24:1	—	—	—	24

*The fatty acids are abbreviated as number of carbons/number of double bonds.
†Less than 1% of total.

each fraction the choline phosphoglycerides are the most prevalent phospholipid. The ethanolamine phosphoglycerides are the second most abundant phospholipid, and the serine and inositol phosphoglycerides comprise about 15% of the total. Sphingomyelin, which contains a phosphorylcholine head group, is rich in the plasma membrane. Cardiolipin, a *bis*(phosphatidyl)glycerol, is present in substantial amounts only in the inner mitochondrial membrane.

Phospholipid fatty acid composition The fatty acid composition of the different phospholipids varies considerably, as illustrated in Table 12.6. For example, the choline phosphoglycerides are rich in palmitic (16:0) and linoleic (18:2) acids, whereas the ethanolamine and serine phosphoglycerides are rich in arachidonic acid (20:4) and the 22-carbon polyunsaturated fatty acids. By contrast, sphingomyelin is rich in saturated fatty acids and 24-carbon fatty acids. The fatty acids also are not evenly distributed between the *sn*-1 (*sn*, stereospecific numbering) and *sn*-2 positions of the glycerophospholipids. Saturated fatty acids are more prevalent in the *sn*-1 position and polyunsaturated fatty acids in the *sn*-2 position (Figure 12.2). Monounsaturated fatty acids tend to be more evenly distributed among both positions. Ether-linked hydrocarbon groups, when present, always are in the *sn*-1 position. They occur in the alkyl ether phosphoglycerides and plasmalogens (see Chapter 10).

Figure 12.2   Phosphoglycerides of cell membranes. The fatty acid groups project into the interior of the lipid bilayer and form the hydrocarbon core. Conversely, the head group, which is composed of the phosphate and organic group attached to the *sn*-3 carbon of glycerol, are located at the surface of the bilayer, where these polar structures can interact with the surrounding aqueous medium. Saturated fatty acids are more prevalent at the *sn*-1 carbon, whereas polyunsaturated fatty acids are localized at the *sn*-2 carbon. Monounsaturates are more evenly distributed. When ether bonds are present, such as in the alkyl ether or plasmalogen phosphoglycerides, they always are present at the *sn*-1 carbon.

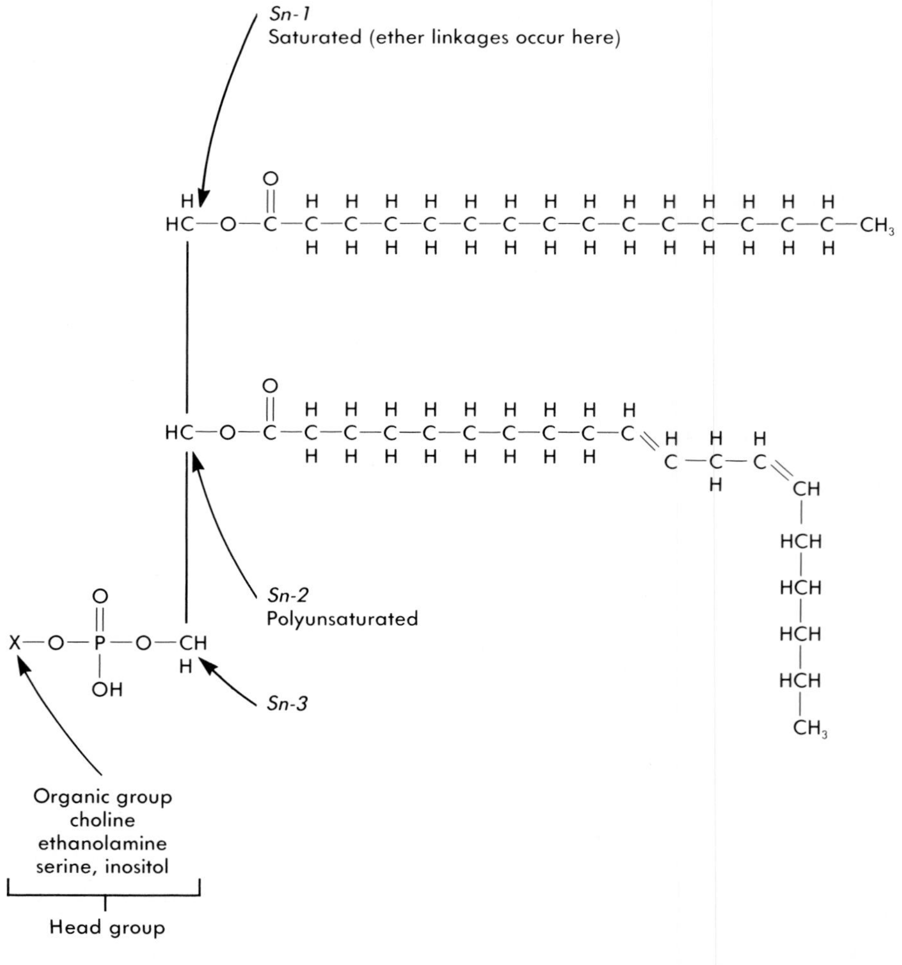

Clinical comment

Saturation of the dietary fat The membrane fatty acid composition is not fixed; it can vary to some degree depending on the type of fat in the diet. If the diet is highly saturated, containing primarily lard, cocoa butter, or coconut oil, the membrane phospholipids will contain somewhat higher percentages of saturated and monounsaturated fatty acids. By contrast, if the diet is rich in plant oils that contain a large amount of linoleic acid, such as corn or safflower oil, the membrane lipid will contain a higher percentage of omega-6 polyunsaturated fatty acids. If the diet is high in fish oils, such as salmon or cod liver oil, the membrane lipids will become enriched in omega-3 polyunsaturated fatty acids. Thus, within limits, the membrane phospholipid fatty acid composition adapts to the composition of the fat available in the environment. In two situations such changes appear to be beneficial. A relative increase in dietary omega-6 polyunsaturated fatty acid intake usually leads to a reduction in the plasma cholesterol concentration. Likewise, an increase in the intake of omega-3 polyunsaturated fatty acids may reduce platelet aggregation and thrombosis. Both these actions reduce the risk of coronary heart disease. It is uncertain, however, whether these apparently beneficial actions are related to the changes that occur in membrane phospholipid fatty acid composition.

Cholesterol

Cholesterol is inserted into the lipid bilayer between phospholipid molecules, in both leaflets of the lipid bilayer. Its hydroxyl group is oriented toward the aqueous environment and interacts with the polar head groups of the phospholipds (Figure 12.3). The nonpolar rings and hydrocarbon tail of cholesterol are positioned so that they interact with the

Figure 12.3

Insertion of cholesterol into the membrane lipid bilayer. The steroid nucleus of cholesterol has a rigid, planar conformation. It fits between the fatty acyl chains of adjacent phospholipids, projecting to the level of about 10 carbons distant from the phospholipid head group. The hydroxyl group of cholesterol faces outward and is located at the level of the phospholipid head groups. The eight-carbon hydrocarbon chain of cholesterol has considerable mobility and projects deep within the hydrocarbon core of the bilayer.

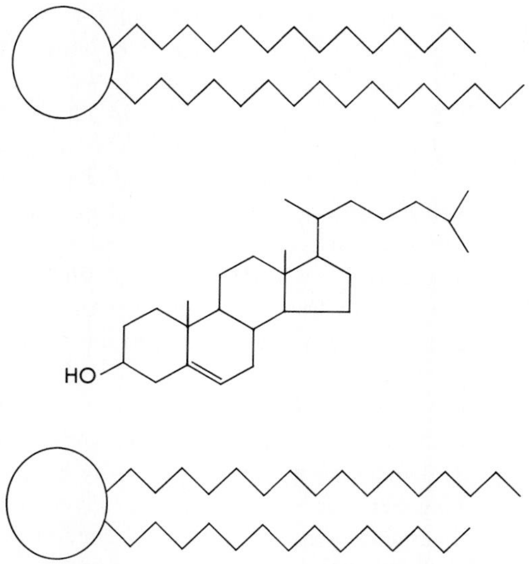

hydrocarbon chains of the phospholipid fatty acyl groups. The planar ring structure of the steroid nucleus penetrates to a depth of about the first 10 carbons of the phospholipid fatty acyl chains. The hydrocarbon chain of cholesterol occupies the region between carbon 11 and the methyl-terminus of the fatty acid. The amount of cholesterol contained in the various cell membranes differs considerably. For example, cholesterol comprises about 25% of the lipids by weight in the plasma membrane, where the molar ratio of cholesterol to phospholipid is about 0.5 to 0.8, but it is not present in the inner mito-chondrial membrane. Likewise, all the cholesterol in the plasma membrane is in the free or nonesterified form, whereas both cholesterol and cholesteryl esters are contained in the endoplasmic reticulum.

Cholesterol exchange and surface transfer Cholesterol is held in the lipid bilayer by physical interactions, primarily between the planar steroid nucleus and the adjacent phos-pholipid fatty acid hydrocarbon chains. Because it is not held in the bilayer through covalent bonds, cholesterol can move in and out of the plasma membrane. In some cases, cholesterol in the membrane exchanges with cholesterol in the surface coats of plasma lipoproteins, and no overall change occurs in membrane cholesterol content. It is possible, however, to transfer cholesterol in or out of a cell by such a process provided that the cholesterol that moves is immediately channeled into another pathway, such as conversion to a cholesteryl ester. If an accumulation or release of cholesterol occurs, the process is called *surface transfer*. The factor that determines whether a net transfer of cholesterol occurs, and, if so, in which direction, is the molar ratio of unesterified cholesterol to phospholipid in the two structures. Net transfer will occur from the structure containing the higher to the one containing the lower cholesterol/phospholipid molar ratio. Surface transfer may be an important mechanism for cholesterol movement, particularly for the efflux of cholesterol from cells to prevent excessive accumulation in tissues such as the walls of arteries. High-density lipoproteins (HDLs) are the main acceptors in the extra-cellular fluid for cholesterol released from cells by the surface transfer mechanism.

Glycosphingolipids

As shown in Table 12.2, carbohydrates comprise 2% to 8% by weight of the membrane constituents. The carbohydrate is not free; it is a component of either glycolipids or glycoproteins. All the glycolipids in animal cells are glycosphingolipids. They are de-rivatives of ceramide and therefore contain sphingosine and a long-chain fatty acid in amide linkage (see chapter 10). This is the same structure contained in sphingomyelin. With the glycosphingolipids, however, the phosphorylcholine group of sphingomyelin is replaced by one or more carbohydrate residues.

Glycosphingolipids are inserted into the membrane lipid bilayer in the same manner as cholesterol. This is illustrated in Figure 12.4, which also indicates the structure of several carbohydrate chains in the commonly occurring glycosphingolipids. The ceramide group is contained within the lipid bilayer, with the sphingosine and fatty acid hydrocarbon chain parallel to and interacting with the fatty acyl chains of the phospholipids. By contrast, the carbohydrate group projects out from the surface of the bilayer, interacting with the phospholipid head groups and the surrounding water.

Clinical comment

Glycosphingolipidoses The lipid storage diseases are described in Chapter 10. In most of these abnormalities, glycosphingolipids accumulate intracellularly. For example, gan-glioside G_{M2} accumulates in Tay-Sachs disease, and glucocerebroside accumulates in Gaucher's disease. Many other glycosphingolipid abnormalities have been characterized. These diseases are caused by genetic deficiencies of lysosomal hydrolases that remove carbohydrate residues from the respective glycosphingolipid. For example, hexosamini-dase A is deficient in Tay-Sachs disease, and glucocerebrosidase is deficient in Gaucher's

Figure 12.4

Insertion and orientation of glycosphingolipids in the membrane lipid bilayer. The ceramide group is contained within the phospholipid bilayer, and the carbohydrate chains project outward into the surrounding aqueous environment. When present, glycosphingolipids are contained almost entirely in the outer leaflet of the lipid bilayer. Only a few representative carbohydrate chains are illustrated. In some cases the carbohydrate chains can be large; for example, the blood group A glycosphingolipid contains 14 monosaccharide residues.

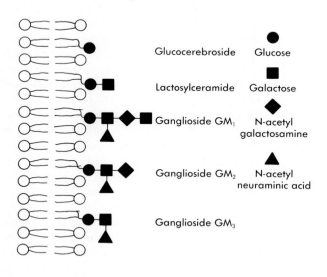

disease. Normally the glycosphingolipids are catabolized in lysosomes to ceramide and then to sphingosine and fatty acid. If a hydrolytic enzyme is deficient, the glycosphingolipid substrate accumulates intracellularly to the point where the cell is injured. The liver and spleen usually enlarge because of the glycosphingolipid accumulation. In addition, the nervous system often is affected, causing mental retardation and motor defects. These genetic abnormalities are evident early in life and the glycosphingolipid storage diseases usually appear in the pediatric age group.

Membrane proteins

As an initial approximation, membrane proteins can be divided into two general types, peripheral and integral. *Peripheral* proteins are bound loosely to the membrane and can be removed by mild treatment with solutions of high ionic strength, chelating agents such as ethylenediamine tetraacetate, or treatment with enzymes such as phospholipase C. They are soluble in aqueous solution and do not contain tightly adherent lipid. Peripheral proteins comprise about 30% of the membrane proteins. Protein A in Figure 12.1 is a peripheral protein. The remainder of the membrane proteins, *integral* proteins, are tightly bound and removed only by such drastic treatments as extraction with detergents. Lipid adheres to the integral proteins when they are removed from the membrane, and these proteins usually are insoluble when they are introduced into aqueous media unless detergent is present. Proteins B, C, D, and E and in Figure 12.1 are integral proteins. Many different individual proteins make up each of these two classes of membrane proteins, and their molecular weights vary widely.

Integral membrane proteins

Two general types of proteins are embedded in the lipid bilayer, as illustrated in Figure 12.5. One type spans the membrane lipid bilayer only once. Such proteins include the low-density lipoprotein (LDL) receptor (see Chapter 16) and glycophorin, the main membrane glycoprotein of the erythrocytes. These proteins contain a single α-helical membrane-spanning segment composed of 18 to 22 nonpolar amino acid residues that interact with the phospholipid fatty acid chains in the lipid bilayer. Most of the structure of these proteins is contained outside the lipid bilayer, both in the extracellular fluid and in the cytoplasm. The second type of integral protein is illustrated by the Band 3 protein of erythrocytes, which is an anion transporter that exchanges bicarbonate for chloride ions. Each of the two subunits of this transporter crosses the lipid bilayer 12 times, and the membrane-spanning segments are connected by hairpin loops. In this case most of the protein structure is contained within the lipid bilayer. Other membrane proteins that span the bilayer several times include rhodopsin, cytochrome P_{450}, Ca^{++}-ATPase, and the β-adrenergic receptor. Many of the integral proteins are glycoproteins that contain a number of carbohydrate chains. The carbohydrate chains are attached to the extracellular domain and project into the surrounding fluid.

Glycoproteins Many biologically active proteins are glycoproteins; some are secreted from the cell, such as antibodies, and others become a part of a membrane. The initial steps in the biosynthesis of the carbohydrate prosthetic groups occur in the lumen of the

Figure 12.5

Structural differences in two intrinsic membrane proteins. Glycophorin, the major membrane glycoprotein of the erythrocyte, has a single membrane-spanning domain containing 21 amino acids in α-helical configuration. Band 3, the bicarbonate-chloride transporter contains two subunits, each of which spans the membrane lipid bilayer 12 times.

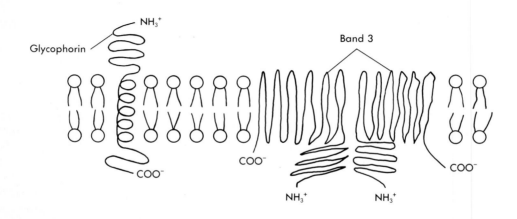

endoplasmic reticulum (ER). The growing peptide chain emerges from the ribosome on the ER and inserts through the membrane of the ER into the lumen under the influence of a signal recognition protein (SRP), which is bound to an SRP receptor in the ER membrane.

The proteins being synthesized are targeted for a particular location in the cell by the first approximately 10 to 40 amino acid residues emerging from the ribosome. The sequence is recognized by the SRP and the ribosome and thus becomes attached to the ER. The sequence continues to translate the messenger ribonucleic acid (mRNA) message into a polypeptide chain, which is inserted through the ER membrane into the lumen.

Much remains to be learned about these processes. For example, many protein chains that are part of a membrane have internal sequences of hydrophobic amino acid residues. These segments, as α-helices, remain embedded in the lipid bilayer of the membrane. Depending on the number of these hydrophobic segments, which can be calculated from the free energy transfer of each successive amino acid from a lipid environment to water, the protein loops back and forth from the cytoplasmic to the luminal side of the subcellular particle or from the cytoplasmic to the extracellular side of the plasma membrane.

Carbohydrate residues are attached to the polypeptide segments in the lumen of the ER, as discussed in Chapter 7. Following transport to the Golgi complex, the carbohydrate groups are further processed before delivery to predestined target sites, which may be membranes or extracellular secretions.

Peripheral proteins

The peripheral proteins are contained entirely in the aqueous environment and are attached to the surface of the lipid bilayer. As illustrated in Figure 12.1, some of the attachment occurs through ionic interactions between charged amino acid residues and the head groups of the phospholipids. Ions such as Ca^{++} often form a bridge between an anionic phospholipid head group such as serine and an anionic amino acid group such as aspartate. In other cases the surface protein is attached to the membrane phospholipid head group through a covalent linkage.

Phosphatidylinositol glycan anchors Many peripheral proteins are attached covalently to the lipid bilayer through phosphatidylinositol. These include alkaline phosphatase, 5′-nucleotidase, acetylcholinesterase, and the Thy-1 antigen. The proteins are linked from their C-terminal amino acid residue to phosphoethanolamine, which is linked to a chain of carbohydrate residues called a glycan. Components of the carbohydrate chain include mannose, glucosamine, galactose, and N-acetylgalactosamine. The glycan chain is attached covalently to the inositol residue of phosphatidylinositol, which is a part of the membrane lipid bilayer (Figure 12.6).

Proteins attached to the cell surface by phosphatidylinositol glycan anchors are synthesized with a leader sequence consisting of nonpolar amino acid residues at the C-terminal end. During processing, this leader sequence is removed, and the resulting C-terminal amino acid group (cysteine in the case of the Thy-1 antigen) becomes attached to the phosphorylethanolamine glycan chain.

Peripheral proteins attached through phosphatidylinositol glycan anchors are released from the cell in response to certain stimuli. These stimuli activate a phosphatidylinositol-specific phospholipase C that hydrolyzes the phosphorylinositol glycan group from the diacylglycerol backbone. In some cases the released diacylglycerol or the glycan structure may act as a second messenger.

Fatty acylation of membrane proteins Many proteins attached to cell membranes contain covalently bound palmitic or myristic acid (Figure 12.6). The palmitic acid is in ester or thioester linkage to an internal amino acid residue, whereas the myristic acid is

Figure 12.6   Attachment of peripheral proteins to the membrane lipid bilayer. The phosphatidylinositol glycan covalent linkage is shown on the left. The C-terminal amino acid residues linked to the ethanolamine phosphate in different proteins are cysteine, glycine, or aspartate. Glycan chains in animal cells contain five to seven monosaccharides consisting of mannose, glucosamine, galactose, and N-acetylgalactosamine (shown as circles). A fatty acylation attachment is shown in the center, in which a palmitic acid residue that penetrates into the lipid bilayer is linked by a thioester linkage to an internal cysteine residue of the protein. On the right a protein is covalently linked through the amide bond of a N-terminal glycine residue to myristic acid, which is embedded in the lipid bilayer.

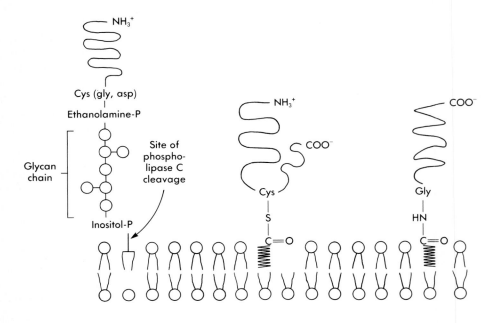

in amide linkage to an N-terminal glycine residue. The fatty acyl group probably facilitates the attachment of the protein to membranes by penetrating into the lipid bilayer. Membrane proteins that contain fatty acyl groups include the transferrin receptor, rhodopsin, the nicotinic acetylcholine receptor, and the Ca^{++}-ATPase of the sarcoplasmic reticulum.

Membrane lipid bilayer

The phospholipid bilayer is composed of two rows of phosphoglycerides that have their fatty acyl groups pointed toward each other and their glyceryl-phosphoryl–base head groups oriented outward to the extracellular and cytoplasmic surfaces. Therefore the inside of the bilayer is composed of nonpolar fatty acyl hydrocarbon chains, whereas the outside surfaces that interact with the aqueous environment contain the polar phospholipid head groups. At body temperature the lipid bilayer is in a fluidlike physical state analogous to an oil droplet. This is called the liquid crystalline state (Figure 12.7). If the membrane

Figure 12.7

Physical states of membrane lipids. Increasing temperature promotes transition of the solid gel state to the liquid crystalline state. Nonbilayer structures at the gel–liquid crystalline interface are hexagonally arranged phospholipid (H_I) or phosphatidylethanolamine derivatives that form inverted hexagonal (H_{II}) domains in the lipid bilayer.

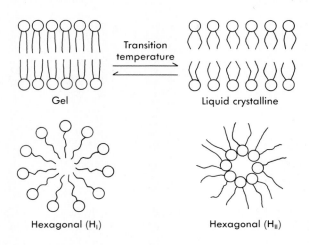

is cooled, the lipids pass into a solid, or gel, state. The temperature at which the lipid bilayer changes from the liquid crystalline to the gel state is known as the *phase-transition temperature*. Under physiologic conditions the membrane is above the phase-transition temperature, and the liquid crystalline state predominates. Some regions of the bilayer are in the gel state, however, and both states coexist in most membranes. In other words, domains of gel structure exist, probably rich in cholesterol, sphingomyelin, and phospholipids with saturated fatty acyl chains, interspersed among the liquid crystalline domains.

Nonbilayer structures

There also may be regions of the membrane that are not in bilayer structure. Areas rich in lysophospholipids can form a hexagonal (H_I) structure, whereas those rich in phosphatidylethanolamine containing highly polyunsaturated chains can form in inverted hexagonal (H_{II}) phase (Figure 12.7). Regions of nonbilayer structure would tend to occur at the interface between gel and liquid crystalline domains. They also are likely to occur at points of membrane fusion.

Lipid motion

As illustrated in Figure 12.8, considerable motion occurs in the lipid bilayer. The fatty acyl chains flex rapidly back and forth. In addition, the phospholipid can rapidly rotate around its long axis. Phospholipids also can move very rapidly laterally within each of the leaflets of the lipid bilayer. Thus a phospholipid molecule can exchange places with the one on either side of it within fractions of a second. In this process the phosphoglyceride remains within the same leaflet of the lipid bilayer, and it does not cross from the extracellular fluid half to the cytoplasmic half of the bilayer structure. The opposite process, movement of a phospholipid molecule between the extracellular and cytoplasmic

Figure 12.8   Mobility of phospholipids in the membrane lipid bilayer.

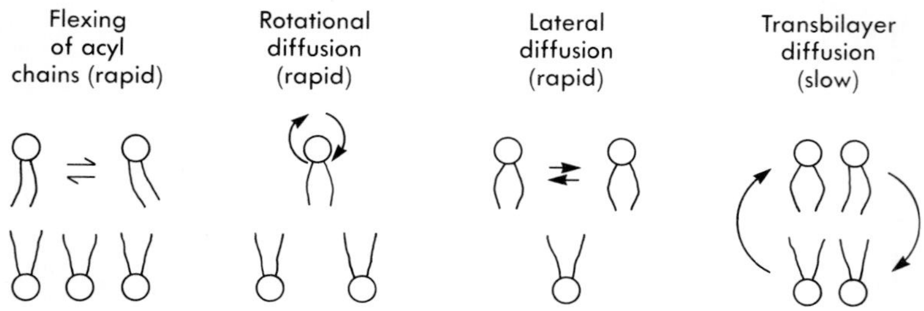

leaflets of the bilayer, known as *flip-flop,* occurs slowly compared to lateral movement within the bilayer leaflet. It is unfavorable energetically to move the polar phospholipid head group through the central nonpolar hydrocarbon region, a process that would be required for the phospholipid to cross to the opposite side of the bilayer. When flip-flop does occur, it may be facilitated by certain membrane proteins that penetrate into the lipid bilayer.

Membrane fluidity

The degree of motion of the hydrocarbon chains within the lipid bilayer is called fluidity. As motion increases, fluidity increases. At body temperature the lipid bilayer exists in a fluid state. The more fluid the bilayer, the more permeable is the membrane. Unsaturated fatty acids present in the membrane phospholipids increase the fluidity of the membrane and make it more permeable. By contrast, saturated fatty acids decrease the fluidity and permeability of the membrane. Integral proteins that penetrate through the membrane modulate the fluidity of the lipid bilayer. Cholesterol also modulates the fluidity, decreasing it in regions of the membrane that contain many unsaturated fatty acids and increasing fluidity in the regions composed primarily of saturated fatty acids. Therefore cholesterol can be considered as a modulator of membrane fluidity and permeability. Cholesterol forms clustered regions within the membrane lipid bilayer; some areas contain 1 mole of cholesterol per mole of phospholipid, whereas others contain almost no cholesterol. This gives the membrane a patchy effect, with solid regions coexisting with adjacent fluid domains. In this way, areas within a membrane can have very different physical and permeability properties.

Clinical comment

Refsum's disease Refsum's disease, heredopathia atactica polyneuritiformis, is a genetic defect in which the patient has serious central nervous system abnormalities. These include cerebellar ataxia, motor weakness, and distal sensory loss. The disease is associated with the accumulation of phytanic acid in tissues, including the nervous system. Phytanic acid is a 16-carbon saturated fatty acid containing methyl branches at carbons 3, 7, 11, and 15. It is derived from phytol, a branched-chain alcohol present in many plant products. To break down phytanic acid by β-oxidation, it is necessary to perform an initial α-oxidation. This process, which occurs in the peroxisomes, is defective in Refsum's disease. As a result, phytanic acid cannot be degraded effectively and accumulates in the

Figure 12.9   Asymmetry of phospholipids in the membrane lipid bilayer.
PC, Choline phosphoglycerides; *PE,* ethanolamine phosphoglycerides;
PS, serine phosphoglycerides; *PI,* inositol phosphoglycerides;
SPM, sphingomyelin.

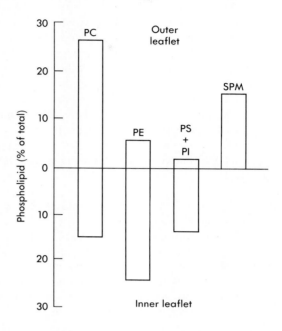

tissues, including to some extent the membrane phospholipids. It is not known how this accumulation causes the neurologic abnormalities. However, insertion of branched-chain fatty acid groups in the lipid bilayer would be expected to perturb the usual tight packing relationships and increase fluidity in certain domains of the neuronal membranes. Based on this, we can speculate that this alteration in membrane bilayer properties may be somewhat responsible for the serious neurologic abnormalities that occur in Refsum's disease.

Membrane asymmetry

Recent studies using cross-linking reagents and enzymes that degrade membrane constituents have indicated that biologic membranes are asymmetric. This is true for the protein, carbohydrate, and lipid components of the membranes. Different peripheral proteins are present on the two surfaces of the lipid bilayer. Likewise, the transmembrane proteins such as the large subunit of the Na^+,K^+-adenosine triphosphatase (ATPase) are asymmetric, with the Na^+ and ATP-binding sites located on the surface exposed on the cytoplasmic side and the K^+- and ouabain- (an inhibitor) binding sites located on the region exposed at the extracellular fluid surface. The carbohydrate chains of the glycoproteins also are asymmetrically distributed; they are oriented so that they project out into the extracellular fluid.

Furthermore, the lipid bilayer itself is asymmetric (Figure 12.9). Phosphatidylcholine and sphingomyelin are concentrated to a greater extent in the leaflet of the bilayer that faces the extracellular fluid. Conversely, phosphatidylethanolamine, phosphatidylserine, and phosphatidylinositol are concentrated in the leaflet that faces the cell cytoplasm.

Phospholipid exchange proteins The cell cytoplasm contains proteins that catalyze the transfer of phospholipids between different membranes. They are called phospholipid exchange proteins. These cytoplasmic proteins have molecular weights between 16,000 and 30,000, and most have isoelectric points about pH 5.0. Each phospholipid exchange protein is fairly specific for a given phospholipid class. Although they are known as exchange proteins, they have been shown to catalyze the net transfer of phospholipid from one membrane to another, for example, from microsomes to mitochondria. Therefore one of their main functions probably is to move phospholipids from the ER, where they are synthesized, to sites where new membrane is being formed. When phospholipid exchange proteins interact with a membrane, they remove or add phospholipids only in the half of the lipid bilayer that they face. Because of this, they probably contribute to the asymmetric distribution of phospholipids across the membrane lipid bilayer.

Fluid mosaic model of membrane structure

The most widely accepted explanation of membrane structure is the fluid mosaic model (Figure 12.10).

According to this model, the phospholipid bilayer forms the matrix of the membrane. Proteins are interspersed in the lipid bilayer of the plasma membrane, producing a mosaic effect. Peripheral proteins are attached through electrostatic interactions and hydrogen bonding with the polar head groups of the phospholipids, or through covalent phosphatidylinositol glycan linkages. The integral proteins are embedded within the lipid bilayer. Integral proteins that penetrate completely through the bilayer, known as *transmembrane proteins,* are the transport carriers for ions as well as water-soluble substrates such as glucose. Transmembrane proteins are exposed on both the extracellular fluid and the cytoplasmic surfaces of the bilayer and in this way can form a channel or pore through the lipid phase. The portions of these proteins exposed to the aqueous environment on both surfaces of the membrane are composed of polar amino acid side-chains such as glutamate and serine, whereas the central segment of the protein that exists within the lipid phase contains primarily nonpolar amino acid side-chains such as valine and leucine. The proteins are not fixed in position within the lipid bilayer. Rather, they have lateral mobility in two dimensions so that they are free to diffuse from place to place within the plane of the bilayer. The proteins also rotate about their longitudinal axes at high speeds. However, the proteins cannot tumble from one side of the lipid bilayer to the other. In other words, the surface of the protein exposed to the extracellular fluid does not roll over and become exposed to the cytoplasm.

Clinical comment

Liposomes and drug delivery Liposomes are artificially formed phospholipid vesicles that are finding increasing usefulness in basic studies of membranes and for introducing substances into cells. When phospholipids are dispersed in a water solution, they form multilamellar, concentric vesicles. The phospholipids in each lamella are in the form of a bilayer, and the lamallae are separated by the aqueous solution. These multilayer vesicles are about 1 μm in diameter. If multilamellar vesicles are exposed to sonic irradiation, they break up and form smaller vesicles that contain only a single lipid bilayer with a diameter of about 250 nm. Unilamellar vesicles also form when ethanolic solutions of phospholipids are injected into water. Both the unilamellar and the multilamellar vesicles are called liposomes.

Most liposomes in use today are made up mostly of phosphatidylcholine. An acidic phosphoglyceride such as phosphatidylserine often is added, and cholesterol also can be

Figure 12.10

Fluid mosaic model of membrane structure. A segment of the lipid bilayer is illustrated schematically in a three-dimensional projection. Various types of proteins are associated with the lipid bilayer: *A* and *B*, peripheral proteins associated with the surfaces of the bilayer; *C*, a transmembrane integral protein viewed in cross section; *D*, the exposed surface of an integral protein viewed from the extracellular fluid surface of the lipid bilayer; *E*, an integral protein exposed on the extracellular fluid surface that does not pass entirely through the bilayer, viewed in cross section; and *F*, an integral protein exposed on the cytoplasmic surface that also does not pass through the entire bilayer, viewed in cross section.

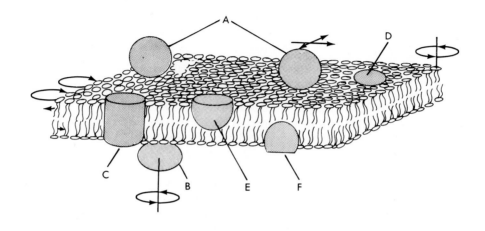

added. Hydrophobic proteins and enzymes can be incorporated into the phospholipid bilayer of the liposomes. In addition, water-soluble drugs or enzymes can be trapped in the aqueous compartment of the vesicles. Because of these properties, liposomes are being tested as carriers for introducing drugs and enzymes into cells. Liposomes are potentially useful for this function because they can fuse with the cell membrane. In this process the contents of the aqueous compartment of the liposome are emptied into the cell interior. This can enable a drug that otherwise could not readily enter a cell to be taken up in large amounts. Furthermore, by incorporating specific antigens or recognition factors into the phospholipid bilayer, the liposome may be targeted to a specific tissue, such as a cancer cell. Although still in the developmental stage, liposomes offer a new approach to the treatment of many different diseases.

Membrane Receptors

Cell-to-cell communication in multicellular organisms is required to maintain stringent, overall control. This is sometimes achieved by extracellular ligands, such as hormones and neurotransmitters, that bind to specific membrane receptors. The bound ligand may cause the receptor to react directly with a response system, as exemplified by some steroids that cross the plasma membrane and interact with nuclear receptors to modify deoxyribonucleic acid (DNA) transcription. Alternatively, an indirect response results with the formation of second intracellular messages, such as cyclic adenosine monophosphate (AMP) or Ca^{++}. The mechanisms of hormone action are detailed in Chapters 17, 18, and 19.

Many plasma membrane receptors are glycoproteins. The assembly of the active receptors involves several posttranslational events that can take from minutes up to 2 to 3 hours. Thus epidermal growth factor (EGF), which stimulates the growth of epidermal and epithelial cells, binds to a plasma membrane receptor that is initially synthesized in the ER. The protein subsequently is processed to a 170,000-dalton glycoprotein before insertion into the membrane as an active receptor. The biologic half-time for these events is around 90 minutes. For the assembly of the α_2, β_2-subunits of an active insulin receptor, the corresponding time is 180 minutes. Part of these processes occur in the ER and Golgi complex and may involve fatty acid acylation and disulfide bond exchange, with consequent conformational changes in the binding domain.

Not all receptors are proteins or glycoproteins. One of the more clearly understood events of receptor action is the binding of cholera toxin to a G_{M1} ganglioside on the plasma membrane (see case 4).

Membrane transport

Lipid-soluble substances diffuse through the membrane lipid bilayer without the need for a carrier mechanism. This process is called *simple diffusion*. As would be expected from the predominantly lipid nature of membranes, however, the movement of water-soluble molecules and ions across them requires specific transport systems that (1) regulate the internal environment of the cell, (2) concentrate substances from the surrounding fluid, (3) excrete toxic substances, and (4) generate ionic gradients. This process is called *facilitated diffusion*. Many of these facilitated diffusion processes are energy demanding and are coupled with ATPases such as the Na^+,K^+-ATPase.

Simple diffusion of such molecules as ethanol, oxygen, or carbon dioxide occurs at a rate that is directly proportional to the concentration gradient across the membrane (Figure 12.11, *A*). Facilitated diffusion is characterized by a rate of transport that is very similar to single-substrate enzyme kinetics (Fig 12.11, *B*). The molecule binds to a transporter protein in the membrane that forms a channel through which it can cross the lipid bilayer. An example of facilitated diffusion is the movement of glucose across the plasma membrane, which is mediated by glucose permease. The rate of transport is exponentially related to the concentration of the solute, and the system eventually becomes saturated, at which point a maximum rate is reached. These diffusion processes are not coupled to the movement of other ions; they are known as *uniport* transport processes (Figure 12.12, *A*).

For certain molecules an obligatory co-transport of other molecules occurs, such as Na^+ co-transported with amino acids or glucose (see Chapters 6 and 8). This has been used clinically in the treatment of cholera patients (see case 4.) Such a co-transport process is called *symport,* since both substances move across the membrane in the same direction (Figure 12.12, *B*). This is opposed to an *antiport* system, in which each substance moves in opposite directions (Figure 12. 12, *C*). An example of antiport is the exchange of chloride for bicarbonate, catalyzed by the Band 3 anion-exchange protein. In other antiport systems an exchange of ions occurs, such as Ca^{++} and Na^+ in cardiac muscle cells or Na^+ and H^+ in fibroblasts. In such antiport systems the pairs of ions, such as Ca^{++} and Na^+, exchange for each other through a single membrane protein.

In all these cases the transport processes occur through specific transporter proteins driven by an electromotive force. This may be a Na^+ motive force for Ca^{++}, sugars, and amino acids, or an electron motive force for proton pumping in the mitochondrial oxidative phosphorylation process.

ATPases and ion pumping

In contrast to passive diffusional processes, *active transport* uses the energy coupled to ATP hydrolysis to move some ions against a concentration gradient. Thus Na^+ that would

Figure 12.11   Rate of transport of molecules across membranes. *A,* Simple
diffusion; **B,** facilitated diffusion. V_{max}, Maximum velocity.

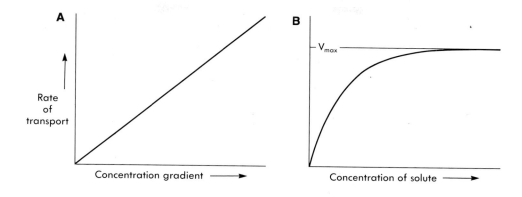

Figure 12.12   **A,** Uniport; **B,** symport; and **C,** antiport forms of membrane transport
processes.

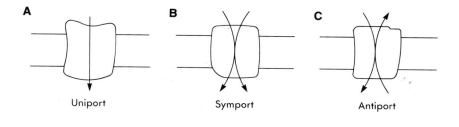

accumulate in the cell following a symport with glucose is pumped out by a Na^+,K^+-ATPase. Similar specific ATPase pumps exist for Ca^{++} and H^+.

General properties of some ion-pumping ATPases All ion-pumping ATPases are firmly bound to one type of membrane or another. They drive the movement of ions across these membranes. Cations must be pumped against a concentration gradient. This thermodynamically unfavorable situation requires energy from the breakdown of ATP to adenosine diphosphate (ADP) and inorganic phosphate (P_i).

A summary of some important cation-pumping ATPases is given in Table 12.7. All require Mg^{++} ions. The first listed in Table 12.7 is the mitochondrial ATPase responsible for generating most of the mitochondrial ATP. It is characterized as an ATPase, but when it is normally embedded in the mitochondrial inner membrane, the enzyme contains a

Table 12.7   Cation specificity and cellular location of some ATPases

Enzyme	Cation pumped	Cellular localization
Mg^{++}-dependent ATPase	H^+	Inner mitochondrial membranes of animal cells; chloroplast membranes of green plants
Mg^{++}-dependent Ca^{++}-ATPase	Ca^{++}	Sarcoplasmic reticulum membranes of muscle
Mg^{++}-dependent Na^+,K^+-ATPase	Na^+,K^+	Plasma membranes of animal cells

subunit that inhibits ATP hydrolysis; thus it functions as an ATP synthase. When isolated from the membrane in soluble form, however, the inhibitory subunit is lost, and the system then acts predominantly as an ATP hydrolase, giving it the name ATPase. Important ATPases also are present in the sarcoplasmic reticulum, which pumps Ca^{++}, and in the plasma membrane, which is an antiport system that pumps Na^+ and K^+.

Na^+, K^+-ATPase

Properties Na^+,K^+-ATPase is composed of two dissimilar, but rather hydrophobic, subunits, α and β. They are closely associated, probably as dimeric pairs $(\alpha, \beta_2.)$ The smaller β-subunit is a glycoprotein, which orients the enzyme in the membrane. The larger catalytic α-subunit penetrates the entire thickness of the membrane. It has been demonstrated that acidic phospholipids such as serine phosphoglycerides are in some way essential for activity. Na^+,K^+-ATPase contains two high-affinity sites for ATP, one on each α-subunit. It has been experimentally determined that these sites face the cytoplasmic surface of the membrane. The Na^+-binding sites also face the cytoplasmic surface. By contrast, the K^+-binding sites face the exterior surface of the membrane. The same is true of the ouabain or digoxin-binding sites.

Catalytic cycles A model for Na^+ and K^+ ion transport is shown in Figure 12.13. The catalytic system binds ATP with a fairly high affinity ($K_d = 0.2 \mu mol/L$). It also has two ion-binding sites. The site at which K^+ ions bind, the potassium site, faces the outside of the plasma membrane and presumably involves a specific conformation identified as E_2. The site where Na^+ ions bind, the sodium site, faces the inside of the plasma membrane and involves a specific conformation identified as E_1.

In the presence of Na^+ ions and Mg · ATP, the α-subunits of the E_1 conformer are phosphorylated at the β-COOH of a specific aspartic acid residue. The product is an acyl phosphate and thus a high-energy compound, $E_1 \sim P$. As ADP is released, $E_1 \sim P$ undergoes a shift to $E_2 - P$. As E_2—P binds K^+ ions, it is dephosphorylated, giving rise to $K^+ \cdot E_2$. This changes to free K^+ ions and E, completing the catalytic cycle. Oligomycin blocks the breakdown of the high-energy phosphate, as it does in the mitochondrial respiratory chain, and thereby inhibits the process.

For each ATP consumed three Na^+ ions move in one direction and two K^+ ions move oppositely. As a result, this ATPase generates a charge across the membrane. Note that this situation is essentially the inverse of that in ATP synthesis.

Clinical comment   **Cardiac glycosides** Ouabain and digoxin are members of a class of compounds known as cardiac glycosides. Digoxin, a component of the digitalis leaf, has been used in treatment of chronic heart failure for some time. This class of compounds binds to the

Figure 12.13   Model of the catalytic cycle of Na$^+$,K$^+$-ATPase. E_1 and E_2 represent different conformational states of the transmembrane catalytic subunits of the enzyme. Sites of action of the inhibitors, ouabain, digoxin, other cardiac glycosides, and oligomycin are noted. Although not shown, the stoichiometry for K$^+$ and Na$^+$ transport is not the same; this gives rise to a transmembrane charge as a result of ion movement.

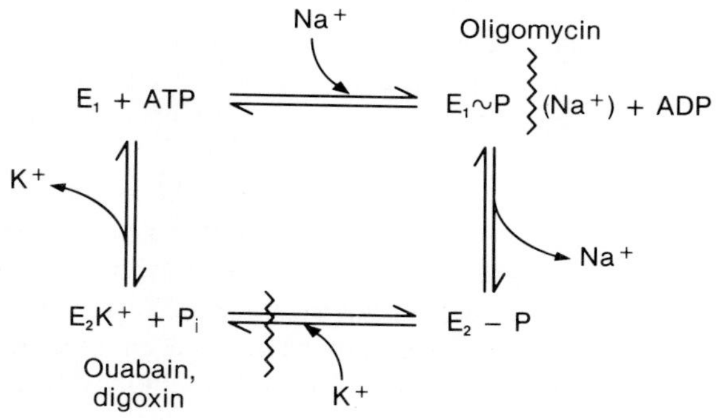

α-subunits of the ATPase, 1 mole of glycoside per mole of K$^+$-binding sites. The glycosides appear to act as competitive inhibitors of K$^+$ ion–binding to the E_2 conformer. The effectiveness of the cardiac glycosides appears to correlate with the magnitude of their binding constants.

The cardiac glycosides are beneficial in heart failure because they strengthen the contractility of the cardiac muscle. This action is mediated in two ways. First, by reducing Na$^+$,K$^+$-ATPase activity, more ATP is available in the heart muscle for the contractile process. Second, by inhibiting the Na$^+$,K$^+$-ATPase, the glycosides cause an increase in intracellular Na$^+$. This leads to Na$^+$/Ca^{++} exchange, raising the cytoplasmic Ca^{++} concentration. Muscle contraction is enhanced by the increase in Ca^{++}.

Mg^{++}-ATPase Mammalian Ca^{++},Mg^{++}-ATPase is most abundant in the sarcoplasmic reticulum and in the sarcolemma of muscle. It is also found in erythrocyte membranes, blood platelets, placenta, salivary glands, nerve cells, and brain microsomes. As with the Na$^+$,K$^+$-ATPase, it is lipid dependent. The large polypeptide obtained from rabbit muscle has a molecular weight of approximately 100,000, whereas the peptide obtained from erythrocyte membranes has a molecular weight of about 150,000. The enzyme has one ATP-binding site, one phosphorylation site, and two Ca^{++}-binding sites. The sarcoplasmic enzyme accumulates Ca^{++}, whereas the erythrocyte enzyme extrudes Ca^{++}. In both instances ATP hydrolysis occurs on the side of the membrane from which Ca^{++} ions are removed. Obviously the orientation of the enzyme is opposite in these two tissues. The enzyme has a high affinity for Ca^{++} ions (K_m = 0.1 to 1 μmol/L), but concentrations greater than 10 μmol/L are inhibitory.

A reaction scheme similar to that of the Na$^+$,K$^+$-ATPase has been proposed for the Ca^{++},Mg^{++}-ATPase. This involves binding of the substrate (which may be Ca^{++} · ATP), phosphorylation of the enzyme, and a conformational change that results in movement of the Ca^{++} to the opposite side of the membrane. A final dephosphorylation step completes the cycle.

Electrically excitable membranes

Virtually all cells in the body have an electrical potential across the plasma membrane. This results from the concentration gradients of the ions, particularly Na^+ and K^+, that are maintained by the Na^+,K^+-ATPase pump and Cl^-. The plasma membrane is more permeable to K^+ than to Na^+ or Cl^-. In neurons and muscle cells, changes in electrical potential can be caused by controlled sequential changes in the permeability of the membrane to Na^+ and K^+. The ion changes in muscle cells, particularly Ca^{++}, are frequently brought about by an electrical impulse from a nerve.

An electrical impulse in a nerve is produced by the rapid influx of Na^+ because of the opening of Na^+ channels in the membrane by a change in membrane voltage. The sequence of events is as follows. The nerve cell, at a resting potential across the membrane of -60 mV, is stimulated at its junction with another nerve cell by a chemical transmitter substance such as acetylcholine. Na^+ ions begin to flow into the cell at this region, resulting in depolarization of the membrane. This results in more Na^+ channels opening until the membrane potential reaches approximately $+30$ mV. At this point the entry of Na^+ stops because the thermodynamic driving force is zero. The peak of the depolarization, the peak action potential, is reached in about 1 msec. The region where the action potential occurs affects the adjacent region, with the opening of its Na^+ channels, so that the electrical impulse passes rapidly down the neuron as a wave.

The Na^+ channel spontaneously closes when the inward flow of Na^+ ceases and the K^+ channels open. This allows K^+ to flow out of the cell until the K^+ equilibrium potential of -75 mV is reached and the cell is hyperpolarized. Following this, the resting state of the membrane (-60 mV) is slowly reestablished. The whole cycle takes only 2 to 3 msec. Myelination of the nerves increases the rate of conduction of the action potential wave. Myelin is composed of plasma membranes that wrap around the nerve 50 to 100 times and thus form very effective electrical insulation.

Na^+ channel protein

The Na^+ channel protein has been isolated and purified. It is a single polypeptide, molecular weight approximately 250,000 to 270,000, with four homologous domains. Each domain folds into clusters of transmembrane helices, one of which is high in lysine or arginine residues and thus positively charged. It is thought that the four clusters of helices come together in the membrane to form a channel that can be opened and closed by shifts in their relative position as a result of changes in the membrane potential.

Various neurotoxins act on the Na^+ channels (Table 12.8). Tetradotoxin (from puffer fish) and saxitoxin (red marine dinoflagellates that create the "red tide" in the ocean) bind to the voltage-dependent Na^+ channel protein in nerve cells and block movement of Na^+ through them. Veratridine (from seeds of Mexican lily sabadilla) and batrachotoxin (from skin of Columbian tree frog) bind to different parts of the protein and keep the channel

Table 12.8   Neurotoxins and their effects on the Na^+ channel and acetylcholine

Neurotoxin	Action
Tetradotoxin	Blocks movement of Na^+ through Na^+ channel
Saxitoxin	Blocks movement of Na^+ through Na^+ channel
Veratridine	Keeps Na^+ channel permanently open
Batrachotoxin	Keeps Na^+ channel permanently open
α-Bungarotoxin	Blocks interaction of acetylcholine with its receptor
β-Bungarotoxin	Alters rate of acetylcholine release
Cobrotoxin	Blocks interaction of acetylcholine with its receptor
Botulinus toxin (Clostridium botulinum)	Inhibits acetylcholine release
Black widow spider venom	Depletes nerve terminal of acetylcholine

permanently open. These toxins do not affect the K^+ channel but clearly inhibit the electrical stimulation of the nerves and conduction of an impulse, sometimes with deadly affects.

Nerve conduction

Neural-neural, neural-muscle, and neural-cell communication in general occurs through synaptic junctions from which neurotransmitters are released from synaptic membrane vesicles. The presynaptic membrane is separated from the postsynaptic membrane at the junction by a gap or synaptic cleft. The neurotransmitter is released into the cleft on arrival of the nerve impulse at the end of the neuron. Several simple chemicals function, or have been said to function, as neurotransmitters: acetylcholine, some amino acids such as glutamate, and amino acid derivatives such as γ-aminobutyric acid. The best characterized is acetylcholine, for which receptors are present on the postsynaptic membranes of the synaptic cleft. When acetylcholine binds to its receptor, depolarization of the postsynaptic neuron is initiated, which leads to propagation of the impulse along the cell. At the neuromuscular junction the nerve impulse leads to the release of Ca^{++} from the sarcoplasmic reticulum, which causes contraction of the muscle. The depolarization signal of acetylcholine is terminated by its hydrolysis to choline and acetate, as catalyzed by acetylcholinesterase. This enzyme is localized in the synaptic cleft, where it is bound to a collagen-glycosaminoglycan fiber network in the cleft.

The acetylcholine receptor is composed of five subunits: α_2, β, γ, and δ. Each subunit transverses the postsynaptic membrane five times. Similar to the Na^+ channel protein, one of the transmembrane helices in each subunit is highly charged. The charged face of each subunit appears to form a channel in the membrane sufficient to permit the movement of both Na^+ and K^+ ions. The channel becomes permeable to Na^+ and K^+ when 2 moles of acetylcholine bind to the α-subunits. The mechanism whereby this occurs is unknown.

Toxins, such as α-bungarotoxin and cobrotoxin from snakes, bind to the acetylcholine receptor, thereby blocking the interaction with acetylcholine and inhibiting neurotransmission. Other toxins act at the presynaptic face of the cleft: botulinus toxin inhibits acetylcholine release; β-bungarotoxin causes an initial increase of acetylcholine release followed by a decreased rate of release; and black widow spider venom causes a massive increase in acetylcholine release, which depletes the terminal of the neurotransmitter and results in no further release of acetylcholine. These toxins (Table 12.8) have highly specific, high-affinity binding sites. They have been used extensively in studying the biochemical action of the acetylcholine receptor and the functioning of the neuromuscular junction. Exposure of humans to these toxins can be fatal.

Eicosanoids

A group of bioactive compounds that modulate cellular function is derived from 20-carbon polyunsaturated fatty acids. These compounds are called eicosanoids. Figure 12.14 illustrates the main classes of eicosanoids. They include the prostaglandins (PG), thromboxanes (TX), fatty acid hydroperoxides (HPETE), fatty acid hydroxides (HETE), fatty acid epoxides (EET), fatty acid dihydroxides (diHETE), leukotrienes (LT), and lipoxins.

Relationship to membranes

The polyunsaturated fatty acid substrates used for eicosanoid formation are stored intracellularly in membrane phospholipids. In many cells the main storage forms appear to be phosphatidylcholine or the 1-O-alkyl ether form of the choline phosphoglycerides.

These are not the only phospholipid storage forms; in some cells the inositol or ethanolamine phosphoglycerides appear to be important sources of the polyunsaturated fatty acids. When cells are exposed to certain stimuli, phospholipases that hydrolyze these membrane phospholipids are activated. The polyunsaturated fatty acid either is released

Figure 12.14 Classes of eicosanoids. Each class is illustrated by a representative compound.

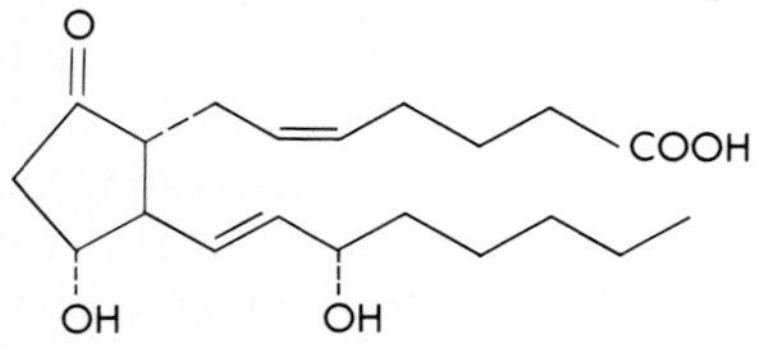

Prostaglandin (PGE$_2$)

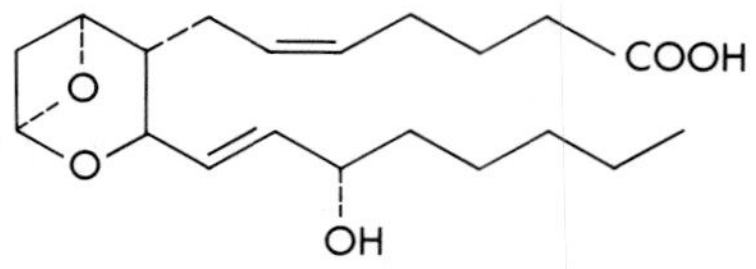

Thromboxane (TXA$_2$)

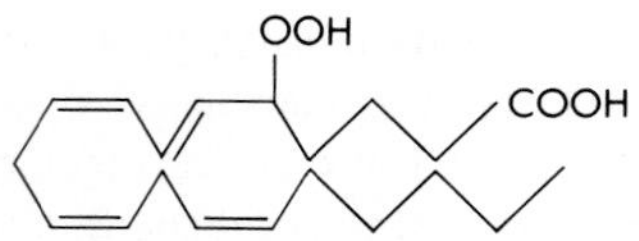

Hyproperoxide (5-HPETE)

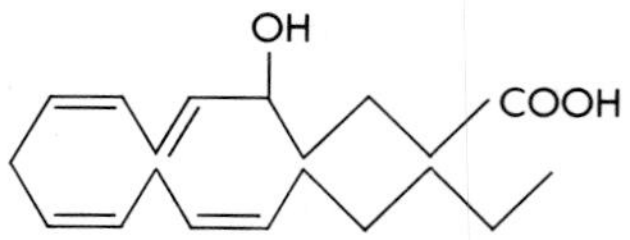

Hydroxide (5-HETE)

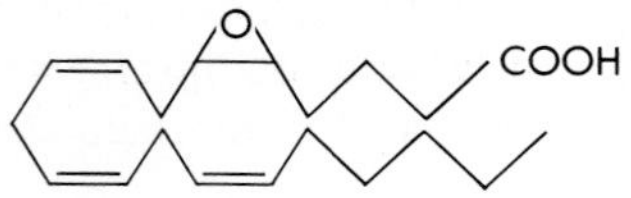

Epoxide (5,6-EET)

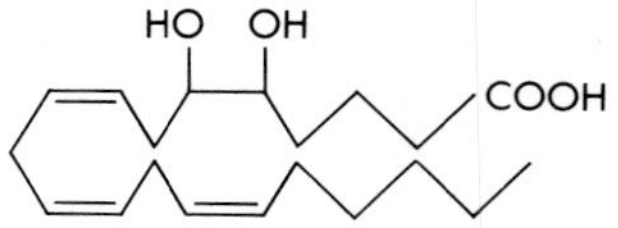

Dihydroxide (5,6-diHETE)

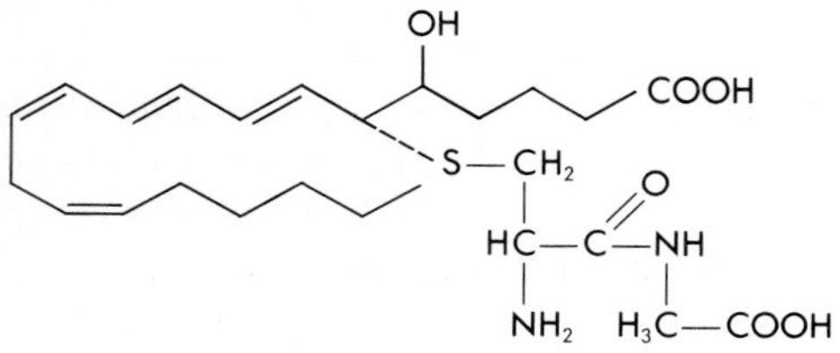

Leukotriene (LTD$_4$)

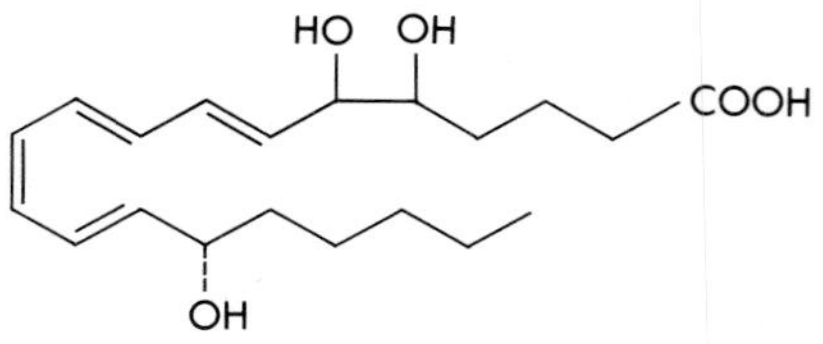

Lipoxin (Lipoxin A)

directly by this hydrolysis or in a subsequent hydrolysis of the product that is formed initially. Most phospholipid substrates are contained in the ER, but phospholipids present in other membranes also may be used in certain cells. In addition, many of the key enzymes that mediate eicosanoid formation are membrane bound, including cyclooxygenase, prostacyclin synthase and cytochrome P_{450}.

Polyunsaturated fatty acid substrates

Eicosanoids can be formed from three 20-carbon atom polyunsaturated fatty acids. Two of these, eicosatrienoic acid (20:3ω-6) and eicosatetraenoic acid (20:4ω-6), are derivatives of linoleic acid (18:2ω-6) and are members of the omega-6 class of essential fatty acids. The common name for 20:4ω-6 is arachidonic acid. Eicosapentaenoic acid (20:5ω-3), a member of the omega-3 class of polyunsaturated fatty acids derived from linolenic acid (18:3ω-3), also can serve as a substrate for the formation of eicosanoids.

Central role of arachidonic acid

Under ordinary conditions the membrane phospholipids contain very little 20:3ω-6 or 20:5ω-3; most of the 20-carbon atom polyunsaturated fatty acid in the tissues of humans and other terrestrial animals is arachidonic acid. Therefore, when the cell phospholipases are activated, the 20-carbon polyunsaturated fatty acid that is released is mainly arachidonic acid. This is a major reason why the bulk of the eicosanoids formed in humans and terrestrial animals is derived from arachidonic acid. In addition, the central enzyme in the prostaglandin synthetic pathway, cyclooxygenase, utilizes arachidonic acid more effectively than eicosapentaenoic acid.

Release of arachidonic acid

The arachidonic acid in membrane phospholipids is contained primarily in the *sn*-2 position. Two pathways hydrolyze the arachidonic acid when cells are exposed to certain external signals, as shown in Figure 12.15.

   Phospholipase A₂ Intracellular phospholipase A_2 is a membrane-bound enzyme activated by Ca^{++}. It is specific for the *sn*-2 position of phosphoglycerides. Although it has activity against several types of phospholipids, it appears to release arachidonic acid primarily from the choline phosphoglycerides, including the 1-*O*-alkyl ether form. Phospholipase A_2 inhibitors such as mepacrine and quinacrine exist, but they are not entirely specific for this enzyme. Phospholipase A_2 activity also is reduced by glucocorticoids, but this is probably a result of formation of the inhibitory peptide lipocortin rather than a direct interaction of the steroid with the enzyme.

   Phospholipase C At least five immunologically distinct forms of phospholipase C exist. The form that may be involved in eicosanoid production is specific for the inositol phosphoglycerides. This enzyme hydrolyzes the phosphorylinositol group, generating a diacylglycerol that still contains the arachidonic acid initially present in the inositol phosphoglyceride. Therefore at least one additional hydrolytic reaction is needed to release the arachidonic acid. This may occur in two ways. One is through direct hydrolysis by diacylglycerol lipase. In the other, the monoacylglycerol product resulting from the action of diacylglycerol lipase retains the arachidonic acid and is subsequently released by a monoacylglycerol lipase. Although a substantial amount of evidence points to a role for phospholipase C in eicosanoid synthesis, some uncertainty remains because of the observation that at least some forms of phospholipase C are not Ca^{++} activated. In almost all systems, eicosanoid formation can be triggered by a calcium ionophore, indicating that Ca^{++} must play a key role in the arachidonic acid release mechanism.

Figure 12.15   Pathways for the hydrolysis of arachidonic acid from membrane
phospholipids.

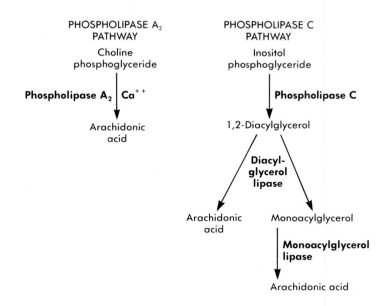

Pathways of arachidonic acid metabolism

Three main pathways for arachidonic acid metabolism exist (Figure 12.16). One is the cyclooxygenase pathway, which produces the prostaglandins and thromboxanes. Prostaglandin G_2 (PGG$_2$), an endoperoxide, is the first intermediate in this series of reactions. Another is the lipoxygenase pathway, which produces leukotrienes, hydroxyeicosatetraenoic acids (HETEs), and lipoxins. Hydroperoxyeicosatetraenoic acids (HPETEs) are the intermediates from which these products are formed. The third pathway, cytochrome P_{450}, forms epoxides, which subsequently are converted to HETEs or diHETEs. Most cells do not contain all these pathways. Usually a cell primarily makes one of these types of products. For example, endothelial cells convert arachidonic acid primarily to prostaglandins, whereas the main compounds formed by neutrophil leukocytes are lipoxygenase products. However, there are exceptions to this; for example, platelets form substantial amounts of both a cyclooxygenase and a lipoxygenase product.

Prostaglandins

The cyclooxygenase pathway produces prostaglandins (PG), 20-carbon atom acids that have hormonelike properties. They are potent smooth muscle agonists and are involved in many important bodily functions. Prostaglandins are synthesized primarily from arachidonic acid. The need for prostaglandins is probably a major reason that the omega-6 polyunsaturated fatty acids are essential to the human diet.

Prostaglandins were discovered in the 1930s when it was noted that extracts prepared from the vesicular glands of sheep or from human semen contained substances that caused contraction of intestinal or uterine strips. The extracts from the sheep seminal glands also

Figure 12.16 Eicosanoids formed from arachidonic acid. *PGG₂*, Prostaglandin G_2; *HPETE*, hydroperoxyeicosatetraenoic acid; *HETE*, hydroxyeicosatetraenoic acid.

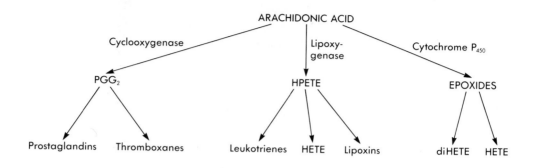

produced a fall in systemic arterial blood pressure. The active substances were characterized as acidic lipids and named prostaglandins. Not until the early 1960s were these compounds identified as cyclic unsaturated fatty acids. Initially they were divided into two classes, E and F, but a more detailed classification system became necessary when the complexity of their structure was revealed.

Prostanoic acid

Nomenclature

The prostaglandins are derived from prostanoic acid, a 20-carbon atom fatty acid that contains an internal five-carbon atom saturated ring. The seven-carbon atom chain attached to the ring at C_8 projects below the plane of the ring, as denoted by the dashed line connecting it to C_8. The eight-carbon atom chain attached at C_{12} projects above the plane of the ring, as denoted by the solid line connecting it to C_{12}.

Prostaglandins are further designated by a capital letter, a number subscript, and in one instance a Greek letter, for example, PGE_1 and $PGF_{2\alpha}$. Except in one case, the capital letter refers to the type of ring substitutions that are present in an individual prostaglandin molecule. Seven types of rings are found in the naturally occurring prostaglandins, giving rise to prostaglandins of the A, B, D, E, F, G, H, and I series. PGA and PGB have keto groups at C_9 and a double bond in the ring. PGD has a hydroxyl group at C_9 and a keto group at C_{11}. PGE has the opposite substitutions; it has a keto group at C_9 and a hydroxyl group at C_{11}. PGF has two hydroxyl groups, one at C_9 and the other at C_{11}. The nomenclature becomes somewhat inconsistent at this point because *both* PGG and PGH have the *same* ring structure, the cyclopentane endoperoxide. They differ only in the group substituted at position 15 of the side-chain. PGG has a hydroperoxide group at C_{15}, whereas PGH has a hydroxyl group. PGI has a double-ring structure, with the oxygen attached to C_9 of the cyclopentane ring being linked to C_6 of the upper hydrocarbon chain to form a second five-membered ring. The ring structures are illustrated next. In these illustrations the seven-carbon atom chain that is attached to the ring at the C_8 position is abbreviated as R_7, and the eight-carbon-atom chain that is attached to the ring at the C_{12} position is abbreviated as R_8.

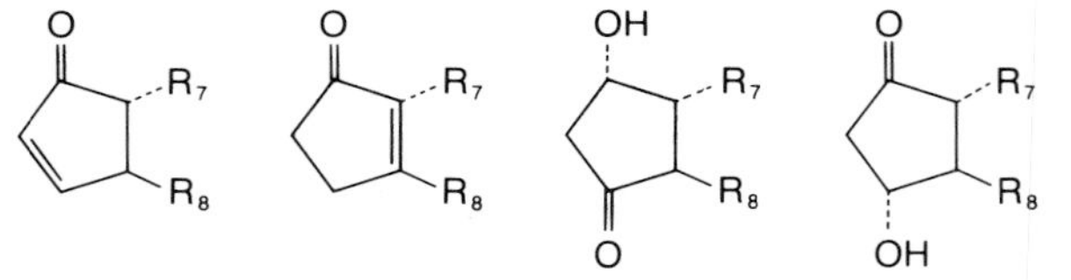

PGA   PGB   PGD   PGE

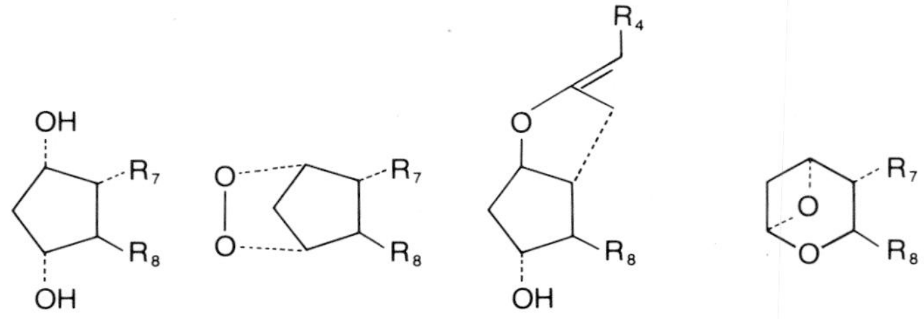

PGF$_\alpha$ PGG or PGH   PGI   TXA

The Greek letter applies only to the F series and refers to the configuration of the hydroxyl group at C_9. In the F_α series the group projects below the plane of the ring in the same orientation as the hydroxyl group at C_{11}. Only the F_α series is naturally occurring.

Thromboxanes The structure of the ring of thromboxane A (TXA), the active form of thromboxane, also is illustrated. It differs from the prostaglandins in that an oxygen atom is incorporated into the ring. Therefore thromboxanes have a six-membered ring consisting of five carbon atoms and one oxygen. In all other respects the structure of the thromboxanes is similar to that of the prostaglandins: they are formed from arachidonic acid, the synthetic pathway also involves cyclooxygenase, and they have similar metabolic properties. For these reasons the thromboxanes are considered here with the prostaglandins and can be thought of as a prostaglandin analogue.

Hydrocarbon chain structure The subscript number after the capital letter denotes the number of unsaturated bonds that a prostaglandin contains, for example, PGE$_1$, PGE$_2$, or PGE$_3$. This refers to double bonds contained in the hydrocarbon chains only, not to those within the ring structure. In the 1-series the double bond is at position 13,14. Double bonds are present at positions 13,14 and 5,6 of the 2-series and at positions 13,14; 5,6; and 17,18 of the 3-series. The 13,14 unsaturation is *trans,* whereas the 5,6- and 17,18- unsaturations are *cis.* In addition to the double bonds present in the hydrocarbon chains, all the naturally occurring active prostaglandins contain a hydroxyl group at C_{15} that projects below the plane of the ring structure. It is connected to C_{15} in structural representations by a dashed line. These hydrocarbon chain structures are illustrated for the E class of prostaglandins.

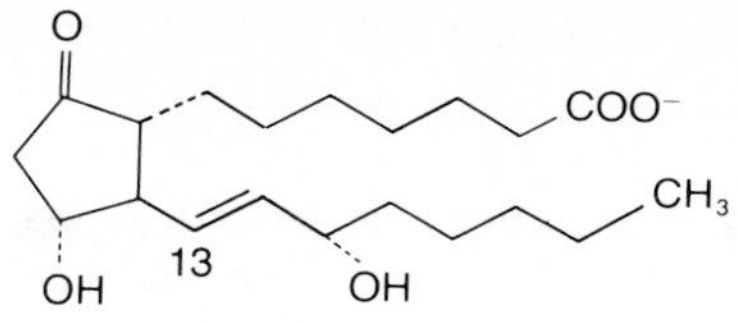

PGE$_1$

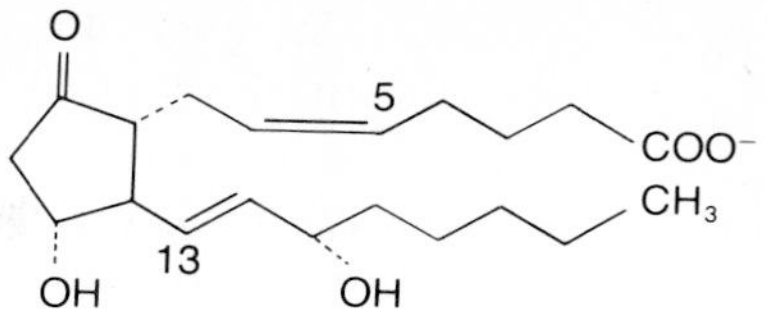

PGE$_2$

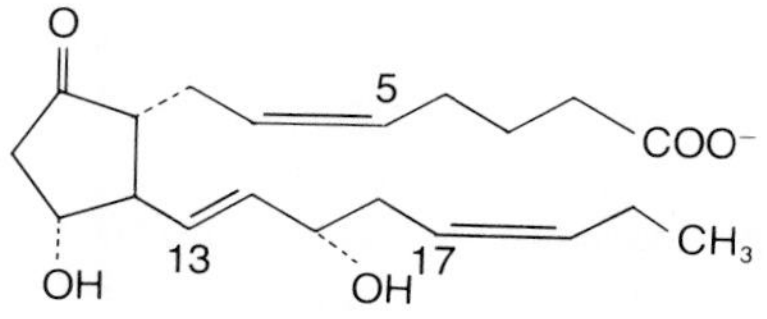

PGE$_3$

Primary prostaglandins

PGG and PGH, the endoperoxide forms, are intermediates in the synthesis of the other prostaglandins. Their endoperoxide structure is cleaved in further biosynthetic reactions, and these intermediates are converted to PGA, PGB, PGE, PGF, PGI, or other products. Only five of these products are widely distributed in the body. These are known as the primary prostaglandins and are all the 2-series: PGD$_2$, PGE$_2$, PGF$_{2\alpha}$, PGI$_2$, and thromboxane A$_2$.

Prostacyclin and thromboxane The prostaglandin PGI$_2$ is called prostacyclin. It is synthesized from arachidonic acid and is the main prostaglandin produced by the vascular endothelium. Prostacyclin is a vasodilator, particularly for the coronary arteries, and also prevents platelets from aggregating and adhering to the endothelial surface. Prostacyclin is very unstable and is rapidly converted to inactive products, primarily 6-keto-prostaglandin F$_{1\alpha}$. All the currently available assays for prostacyclin actually measure 6-keto-prostaglandin F$_{1\alpha}$.

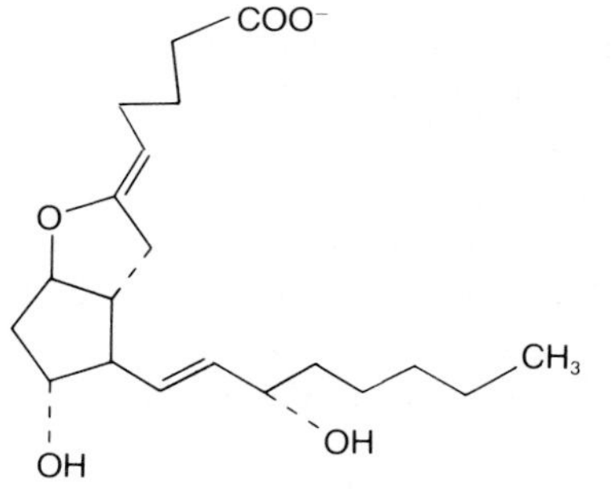

Prostacyclin (PGI$_2$)

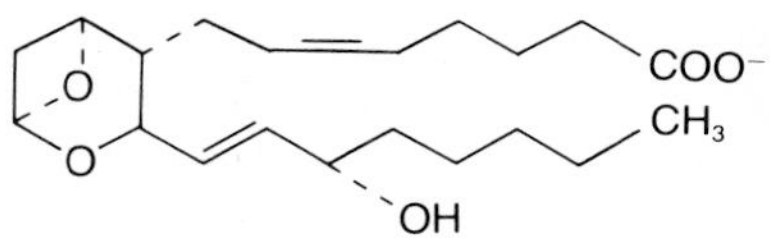

Thromboxane A$_2$ (TXA$_2$)

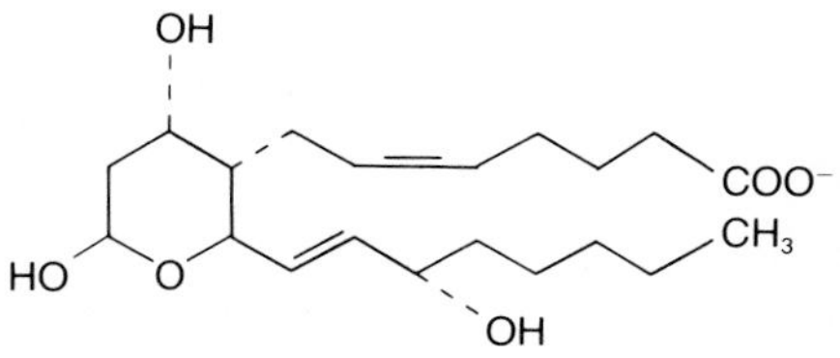

Thromboxane B$_2$ (TXB$_2$)

Thromboxane A$_2$ (TXA$_2$), which also is synthesized from arachidonic acid, is the main prostaglandin produced by the platelets. It is a very short-acting substance, with a half-life of 30 sec. TXA$_2$ has opposite effects from prostacyclin, since it contracts arteries and triggers platelet aggregation. It is rapidly converted to thromboxane B$_2$ (TXB$_2$), an inactive metabolite. Both the thromboxanes contain six-membered rings consisting of five carbon atoms and one oxygen atom. TXA$_2$ contains an additional oxygen atom attached to both C$_9$ and C$_{11}$ of the heterocyclic ring and projecting below the plane of this ring.

This internal oxygen ring is not present in TXB_2, which has in its hydroxyl groups attached to C_9 and C_{11}.

Biosynthesis

Prostaglandins are biosynthesized from three 20-carbon atom, polyunsaturated fatty acids. Eicosatrienoic acid is the precursor of prostaglandins of the 1 series, such as PGE_1, arachidonic acid of the 2-series, and eicosapentaenoic acid of the 3-series. This information is summarized in Table 12.9.

Prostaglandins of the 1- and 3-series occur in certain biologic systems. For example, the ram seminal vesicle produces PGE_1, and certain fish produce PGE_3. A small amount of series-1 and -3 prostaglandins have been found in humans, and human tissues have been reported to produce these classes of prostaglandins if they are enriched with either eicosatrienoic or eicosapentaenoic acid. Under ordinary conditions, however, the tissues contain only very small amounts of these two fatty acids, the predominant 20-carbon-atom polyunsaturate being arachidonic acid. Because arachidonic acid is present in very much larger quantities than either of the other two eicosaenoic acid substrates, almost all the prostaglandins produced by mammalian tissues are derived from arachidonic acid and are series-2-prostaglandins, that is, PGE_2, PGI_2, TXA_2, and so on. Arachidonic acid predominates to such an extent that, except for a few specific instances, we consider here only the conversions of this fatty acid to its prostaglandin products.

The general scheme for stimulation of prostaglandin production is illustrated in Figure 12.17. According to this view, an agonist interacts with its specific receptor located on the plasma membrane of the cell. This interaction mobilizes intracellular Ca^{++}, which activates a specific type of phospholipase. This initiates the reactions that release arachidonic acid, $20:4\omega-6$, from the *sn*-2 position of membrane phospholipids. Exactly which phospholipids and phospholipases are involved in generating the arachidonic acid still is not precisely known. Both the phospholipase A_2 and the C pathways have been implicated in various tissues, and the phospholipids that have been reported to supply the arachidonic acid include phosphatidylcholine, 1-*O*-alkyl choline phosphoglyceride, phosphatidylinositol and its phosphorylated derivatives, phosphatidylethanolamine, and phosphatidic acid.

Cyclooxygenase

In the next step, arachidonic acid is converted to a prostaglandin endoperoxide, PGH_2, by the enzyme cyclooxygenase (Figure 12.16). This is a membrane-bound enzyme that has been found in the ER and nuclear membrane. Whether the phospholipases and phospholipid substrates are located in the same membranes or are released from another membrane or region of a membrane and then pass to the site of the cyclooxygenase (as depicted in Figure 12.17) is not known. Cyclooxygenase has a molecular weight of

Table 12.9   Fatty acid precursors of the prostaglandins

	Prostaglandin series		
	1	2	3
Positions of double bonds	13-*trans*	5-*cis*, 13-*trans*	5-*cis*, 13-*trans*, 17-*cis*
Biosynthetic precursor			
Fatty acid	Eicosatrienoic	Arachidonic	Eicosapentaenoic
Structure	$20:3\omega-6$	$20:4\omega-6$	$20:5\omega-3$
Dietary precursor			
Fatty acid	Linoleic	Linoleic	Linolenic
Structure	$18:2\omega-6$	$18:2\omega-6$	$18:3\omega-3$

approximately 125,000 and is composed of two subunits with a molecular weight of about 72,000 each. Cyclooxygenase binds heme, which serves as a cofactor. The enzyme carries out two reactions: introduction of the endoperoxide at C_9 and C_{11} as well as introduction of the hydroperoxide group at C_{15}. Formation of an internal five-carbon-atom ring occurs when the endoperoxide group is introduced. At this stage the intermediate is PGG_2. Next, the hydroperoxide group is reduced to form an alcohol group at C_{15}, creating the final product, PGH_2, as shown in Figure 12.18. The latter reaction also is mediated by cyclooxygenase. An interesting property of the cyclooxygenase is that it self-inactivates after operating for 15 to 30 sec. Self-inactivation is prevented by hydroquinone, suggesting that the inactivation actually may be caused by the hydroperoxy group that the enzyme introduces. This mechanism prevents overproduction of the prostaglandins. Cyclooxy-

Figure 12.17

Schematic representation of prostaglandin production in response to an extracellular stimulus. The agonist binds to a specific receptor on the cell surface. Binding somehow triggers an increase in the Ca^{++} concentration, perhaps by permitting Ca^{++} flux across the plasma membrane or releasing Ca^{++} from an intracellular storage site. Phospholipase activity is activated by the increase in Ca^{++} in the environment. At present it is questioned whether the phospholipase that is activated initially is phospholipase A_2 or phospholipase C and, furthermore, in which cellular membrane the phospholipase is located. The activated phospholipase then acts on a membrane phosphoglyceride, designated in the figure as the structures with a head group and fatty acyl tails located in the same membrane structure as the phospholipase. Arachidonic acid, *20:4ω-6,* is either released directly or in a series of reactions initiated by the phospholipase. The arachidonic acid is converted by the cyclooxygenase, located in cell membranes to PGH_2, the prostaglandin endoperoxide. PGH_2 is converted to one of the primary prostaglandins by a synthase that also is membrane-bound. The type of prostaglandins produced by a particular cell depends on the synthases it contains. Evidence suggests that the phospholipase, phosphoglyceride substrate, cyclooxygenase, and prostaglandin synthases may be located close together in a single membrane, but this has not been proved conclusively. Therefore the location of the membranes with which these structures are associated is not indicated in the figure.

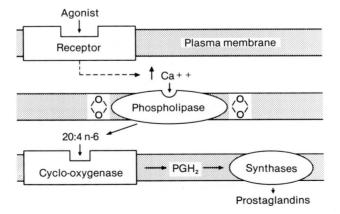

Figure 12.18   Cyclooxygenase reaction.

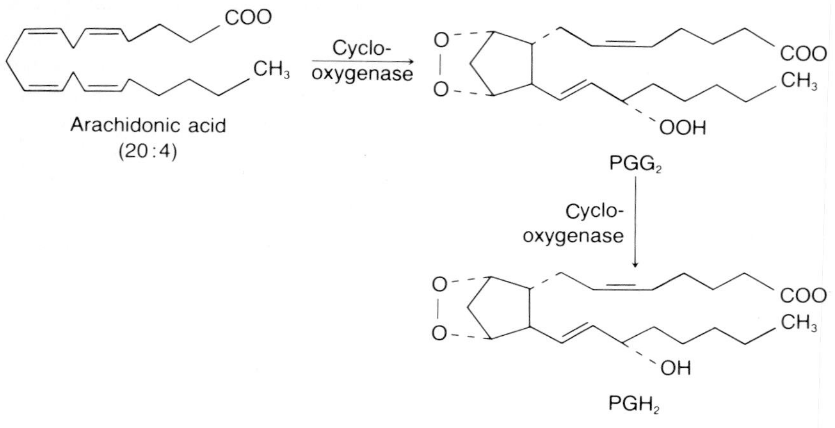

Arachidonic acid
(20:4)

PGG$_2$

PGH$_2$

genase is inhibited by various antiinflammatory drugs. Aspirin is an irreversible inhibitor, whereas indomethacin and ibuprofen are reversible inhibitors.

Prostaglandin synthases

As shown in Figure 12.17, the PGH$_2$ that is generated passes to other membrane-bound enzymes that convert it to the several prostaglandin products manufactured by the particular cell. There are at least five such enzymes, and each has a separate name. For clarity in Figure 12.17, we have called them by a collective name, prostaglandin synthases, remembering that each produces a separate product, for example, PGD$_2$, PGE$_2$, PGF$_{2\alpha}$, PGI$_2$, or TXA$_2$. In addition, PGE$_2$ can be converted, after it is formed to PGF$_{2\alpha}$, by the enzyme PGE$_2$-9-ketoreductase. The mixture of prostaglandins that a cell releases depends on the prostaglandin synthases it contains; for example, PGI$_2$ is the main product released by endothelial cells, whereas TXA$_2$ is the main product of platelets.

Hydroxyheptadecatrienoic acid and malondialdehyde In some cells, particularly the platelet, not all of the PGH$_2$ that is formed is converted into prostaglandins. Some PGH$_2$ is diverted into a 17-carbon atom degradation product, 12-L-hydroxy-5,8,10-heptadeca-trienoic acid (HHT). The remaining three carbon atoms of PGH$_2$ are released as malondialdehyde in this reaction. At present, what controls the conversion of PGH$_2$ to prostaglandins as opposed to degradation to HHT and malondialdehyde is not known. Similarly, the function of HHT and malondialdehyde is not understood.

Inactivation

Two enzymatic reactions are involved in the inactivation of the prostaglandins. One is *15α-hydroxy prostaglandin dehydrogenase,* which catalyzes the conversion of the 15α-hydroxyl group to a keto group. The other is a *Δ13-prostaglandin reductase,* which catalyzes the reduction of the *trans* double bond in position 13,14.

Function

Prostaglandins have numerous functions, some of which are seemingly contradictory. However, certain conflicting actions are understandable in light of there being 19 known naturally occurring prostaglandins, several of which are opposing pairs. Two general physiologic effects have been attributed to the prostaglandins: an effect on the contractile

Table 12.10   Functions regulated by prostaglandins

Function	Tissue	Effect
Muscular contraction	Heart	Contractility
	Arteries	Blood pressure
	Bronchi	Diameter of air passages
	Gastrointestinal tract	Motility
	Uterus	Contractility
Lipolysis	Adipocytes	Inhibition
Secretion	Stomach	HCl production
Permeability	Capillaries	Inflammation, swelling
Electrolytes	Kidney tubule	Na^+, H_2O retention
Blood coagulation	Platelets	Aggregation

state of smooth muscle and a modulating effect on target tissues that respond to adeno-hypophyseal trophic hormones or that have β-adrenergic receptors and respond to catecholamines. These observations imply that the prostaglandins may regulate cellular responses to external stimuli. The entire subject is extremely complex, however, for even the same prostaglandin can have opposite effects on different cells. For example, PGE_1 is an activator of the adenylate cyclase in many tissues, but it is an inhibitor of adenylate cyclase in adipocytes, where it prevents lipolysis.

Prostaglandins are among the most potent biologic substances discovered thus far; as little as 1 ng/ml causes contraction of animal smooth muscle preparations. Some of the main physiologic functions that are modulated by the prostaglandins are listed in Table 12.10. They have potential therapeutic use in such diverse areas as the treatment of hypertension, prevention of conception, relief of bronchial asthma and nasal congestion, and healing of peptic ulcer. The use of a natural prostaglandin often produces unwanted side effects in addition to the desired effect; for example, diarrhea often accompanies the antihypertensive action.

Actions through adenylate cyclase

Many actions of the prostaglandins are mediated through the adenylate cyclase–cyclic AMP–protein kinase A system. Specific membrane receptors are involved in producing these effects. In some cases prostaglandins directly activate the adenylate cyclase system. For example, PGI_2 binding is sufficient to cause a large increase in the cyclicAMP content of human skin fibroblasts. Likewise, when PGI_2 binds to platelet membrane receptors, adenylate cyclase is activated, cyclicAMP concentrations increase, and platelet aggregation is inhibited.

In other systems prostaglandins modulate the response of the adenylate cyclase system to different signals. The extent to which the adenylate cyclase responds to a stimulus is either amplified or reduced, depending on the type of prostaglandin that is bound by a regulatory component. In many cases the component probably is a regulatory subunit of a GTP-binding protein (G protein; see Chapter 17), a membrane integral protein that couples the agonist receptor to adenylate cyclase. Thus, depending on the type of prostaglandin, there can be either up or down regulation of G_s, a stimulatory form of G protein, or G_i, a form that inhibits the coupling of the receptor to adenylate cyclase. Thus many types of modulatory responses can occur, enabling the organism to exert extremely fine control over the adenylate cyclase–cyclic AMP–protein kinase A system.

Autocrine and paracrine regulation

Prostaglandins mediate autocrine and paracrine regulatory processes. Autocrine regulation occurs when a substance released from a cell acts on the same cell to exert a regulatory effect. A regulatory factor that exerts its effect on the same type of cell in which it is formed is called an *autocoid*. An example is TXA_2 formed by platelets, which acts on the same or other platelets to produce aggregation. Following release, the autocoid binds to receptors on the surface of the same cell, or adjacent cells of the same type, and thereby influences cellular function.

Paracrine regulation occurs when a substance released from one cell regulates a process in another type of cell without entering the circulation. Usually the two types of cells are adjacent to one another in the same organ or tissues. Examples of paracrine regulation include PGI_2 release from endothelial cells, preventing platelet aggregation, or TXA_2 release from platelets, causing vascular smooth muscle contraction.

Lipoxygenase products

The arachidonic acid utilized by the lipoxygenase pathway also is stored intracellularly in membrane phospholipids and released by hydrolysis when cellular phospholipases are activated. Unlike cyclooxygenase, however, which is membrane bound, lipoxygenases are cytosolic enzymes. They catalyze the addition of a single oxygen molecule to a double bond, forming a hydroperoxy group. In the process the double bond isomerizes to a position that is one carbon removed from the hydroperoxy group and changes from the *cis* to *trans* configuration. As shown in Figure 12.19, a 1,4-*cis*,*cis*-pentadiene structure is converted to a 1-hydroperoxy-2-*trans*,4-*cis*-pentadiene.

Three main forms of lipoxygenase exist in animal cells; 5-lipoxygenase, 12-lipoxygenase, and 15-lipoxygenase. They operate in the same manner, but they insert the oxygen

Figure 12.19   Mechanism of the lipoxygenase reaction. (From Spector AA et al: Prog Lipid Res 27:271, 1988.)

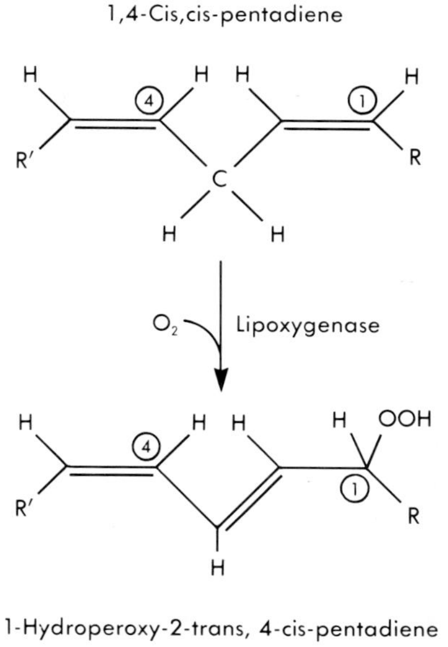

Figure 12.20   Main lipoxygenase pathways present in mammalian tissues. *HPETE*,
                                     Hydroperoxyeicosatetraenoic acid; *HETE*, hydroxyeicosatetraenoic
                                     acid. (From Spector AA et al: Prog Lipid Res 27:271, 1988.)

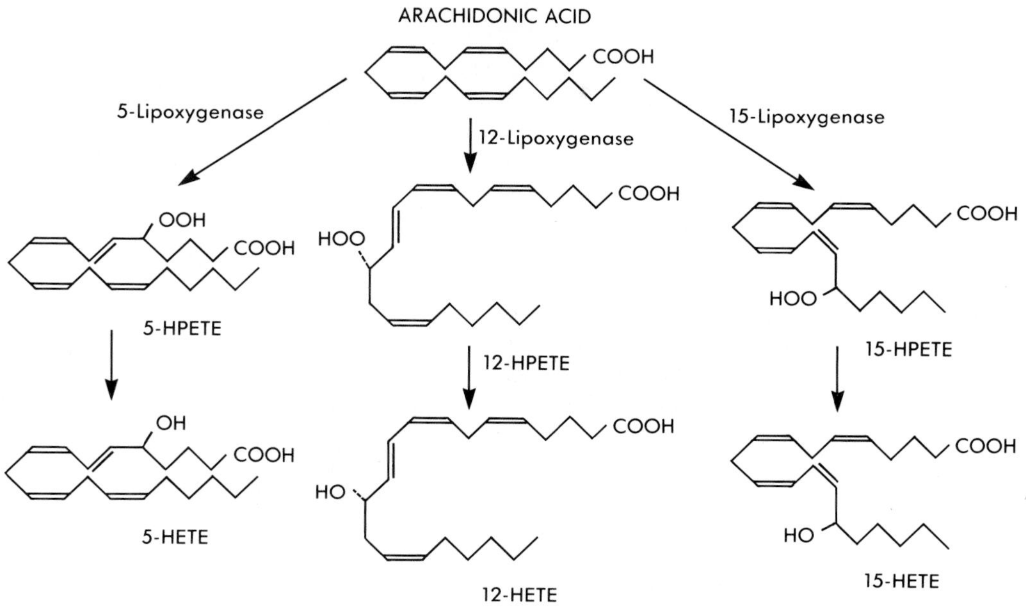

at different places in the arachidonic acid chain, as indicated in Figure 12.20. Different cells contain different lipoxygenases.

For example, neutrophil leukocytes are rich in 5-lipoxygenase, platelets are rich in 12-lipoxygenase, and eosinophil leukocytes are rich in 15-lipoxygenase.

The hydroperoxide derivatives of arachidonic acid that are formed, HPETEs, are very short lived. They are either converted to products such as the leukotrienes or lipoxins, or reduced to the corresponding hydroxy derivatives, HETEs, as shown in Figure 12.20.

Leukotrienes   Leukotrienes are formed from 5-HPETE by cells that contain the 5-lipoxygenase pathway. They were given this name because they are made in leukocytes and contain three *conjugated* double bonds. The first step in leukotriene formation is the addition of a 5-hydroperoxy group to arachidonic acid by the 5-lipoxygenase. As in the case of prostaglandin formation, the arachidonic acid employed for leukotriene synthesis is contained in membrane phospholipids and is released by the action of one or more phospholipases when the cell is exposed to appropriate stimuli. The leukotrienes are abbreviated LT, and all leukotrienes formed from arachidonic acid have the subscript 4 to signify that they contain four double bonds.

Leukotriene synthesis Figure 12.21 illustrates the leukotriene biosynthetic pathway. Leukocytes contain 5-lipoxygenase which converts arachidonic acid to 5-HPETE. This is converted to an epoxide, leukotriene A_4 (LTA$_4$). Two metabolic routes are available to LTA$_4$. One is conversion to LTB$_4$ by the addition of water, forming a 5,12-dihydroxy

Figure 12.21   Leukotriene synthesis. *HPETE,* Hydroperoxyeicosatetraenoic acid; LT, leukotriene.

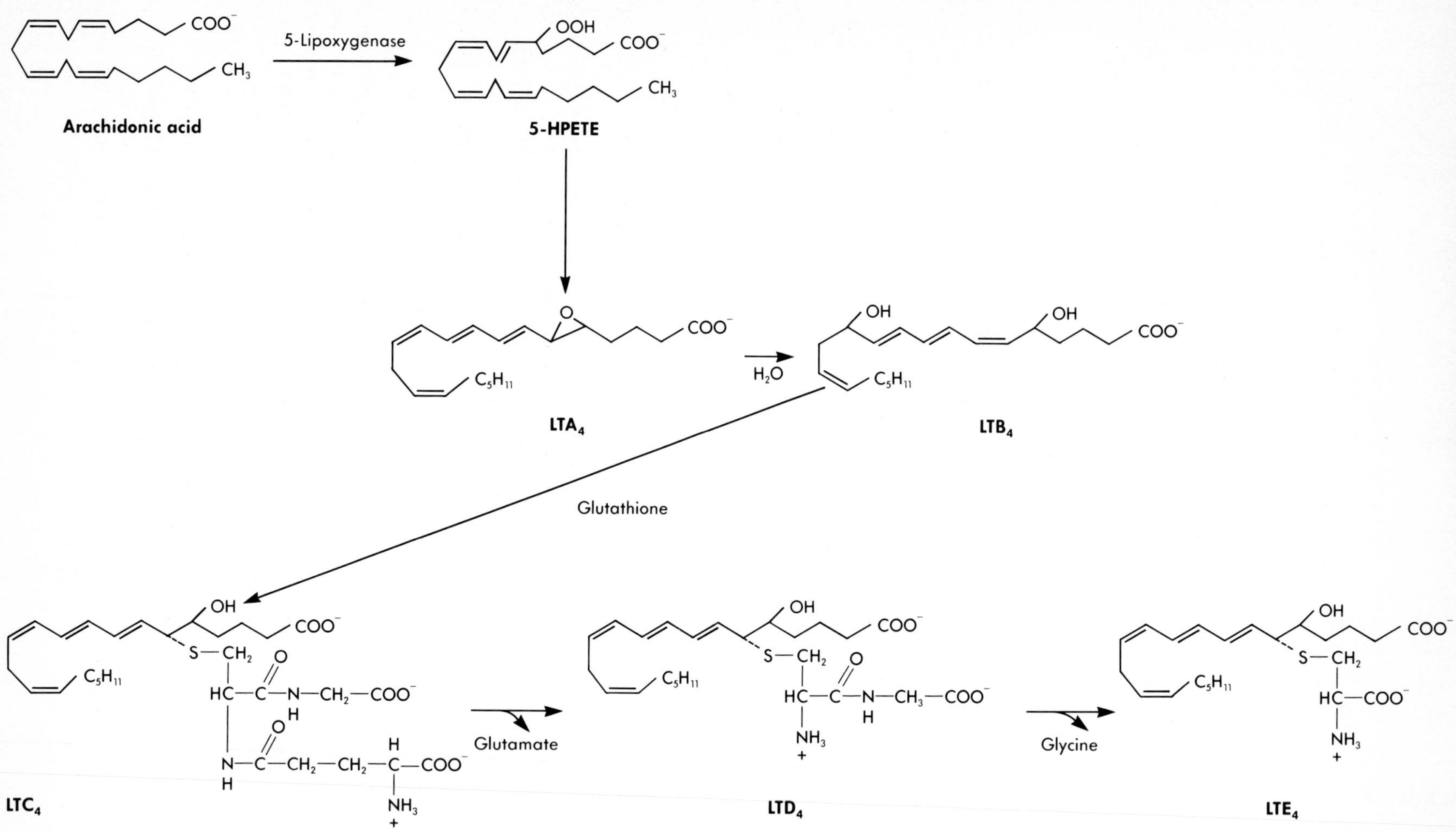

derivative. Alternatively, glutathione can combine with C_6, forming LTC_4. Subsequently the glutamate residue of glutathione can be removed through the action of γ-glutamyl transpeptidase, forming LTD_4. Finally the glycine residue can be removed, creating the cysteinyl derivative LTE_4.

Leukotriene function The leukotrienes facilitate chemotaxis, inflammation, and allergic reactions. LTC_4 and LTD_4 are the active components of the slow-reacting substance of anaphylaxis, SRS-A. They cause smooth muscle contraction and are as much as 1000 times more potent than histamine in constricting the pulmonary airways. They also increase fluid leakage from small blood vessels and constrict the coronary arteries. LTB_4 attracts neutrophils and eosinophils, leukocytes that are present in large amounts at inflammatory sites.

Hydroxyeicosa-tetraenoic acids (HETEs)

HETEs are formed intracellularly by reduction of the corresponding HPETEs, as shown in Figure 12.20, and then released from the cell. It was thought initially that the HETEs are inactivation products and do not have important biologic functions. According to this view, the HPETEs are the bioactive compounds that regulate intracellular processes by carrying out specific oxidations. In the process the HPETEs are reduced to HETEs, the inactivated form released from the cell for catabolism and excretion. The main forms of HETE are 5-, 12-, and 15-HETE (Figure 12.20).

Although this may be correct, increasing evidence suggests that the HETEs themselves may have important biologic functions. They are readily taken up by cells and incorporated into membrane phospholipids. This has led to the theory that HETEs may be paracrine or autocrine regulators that act by perturbing the structure of certain membrane microenvironments in the recipient cell. Furthermore, certain cells can convert HETEs to oxidized products, either by the insertion of additional hydroxyl groups or by chain shortening through fatty acid β-oxidation. It is not yet known whether any of these products may have biologic activity.

Lipoxins

Lipoxins are trihydroxy derivatives of arachidonic acid. They are formed through the sequential oxidation of arachidonic acid by the 15- and 5-lipoxygenase. In the case of lipoxin A (see Figure 12.14), the hydroperoxy group introduced by 15-lipoxygenase is reduced to a hydroxyl group. The hydroperoxy group introduced by 5-lipoxygenase, however, is converted to a 5,6-epoxide, and the latter is hydrated to a 5,6-diol.

Lipoxins appear to be specific intracellular regulators. Lipoxin A stimulates superoxide anion generation in neutrophil leukocytes, causes chemotaxis, produces spasm in the microvasculature, and is an activator of protein kinase C.

Cytochrome P_{450} products

The products formed from arachidonic acid oxygenation by cytochrome P_{450} include epoxides (see Figure 12.14), diol forms of diHETEs, and certain forms of HETEs, including ω-hydroxy and (ω-1)-hydroxy derivatives. The functions of most of these products is not known; however, 5,6-epoxyeicosatrienoic acid (5,6-EET) is reported to be an inhibitor of the Na^+, K^+-ATPase.

Bibliography

Bennett V: The membrane skeleton of human erythrocytes and its implications for more complex cells, Annu Rev Biochem 54:273, 1985.

Bergelson LD and Barsukov LI: Topological asymmetry of phospholipids in membranes, Science 197:224, 1977.

Brisson A and Univin PNT: Quaternary structure of acetylcholine receptor, Nature 315:474, 1985.

Hille B: Ion channels of excitable membranes, Sunderland, Mass, 1984, Sinauer Associates, Inc.

Jay D and Cantley L: Structural aspects of the red cell anion exchange protein, Annu Rev Biochim 55:511, 1986.

Low MG: Glycosyl-phosphatidylinositol: a versatile anchor for cell surface proteins, FASEB J 3:1600, 1989.

Magee AI and Schlesinger MJ: Fatty acid acylation of eucaryotic cell membrane proteins, Biochim Biophys Acta 694:279, 1982.

Marchesi VT: Stabilizing infrastructure of cell membranes, Annu Rev Cell Biol 1:531, 1985.

Needleman P et al; Arachidonic acid metabolism, Annu Rev Biochim 55:69, 1986.

Olson TS and Lane MD: A common mechanism for posttranslational activation of plasma membrane receptors? FASEB J 3:1618, 1989.

Rhee SG et al: Studies of inositol phospholipid–specific phospholipase C, Science 244:546, 1989.

Siegel GF et al, editors: Basic neurochemistry, ed 3, Boston, 1981, Little, Brown & Co.

Singer SJ and Nicolson GL: The fluid mosaic model of the structure of cell membranes, Science 175:720, 1972.

Stroud RM and Finer-Moore J: Acetyl choline receptor structure, function and evolution, Annu Rev Cell Biol 1:317, 1985.

Stubbs CD and Smith AD: The modification of mammalian membrane polyunsaturated fatty acid composition in relation to membrane fluidity and function, Biochim Biophys Acta 779:89, 1984.

Clinical examples

Case 1	Resistance to cancer chemotherapy

A 46-year-old female with metastatic breast carcinoma was treated initially with a chemotherapy regimen. One of the drugs used was methotrexate. The initial response was excellent: some tumor regression was observed, and the overall clinical condition was much improved. During the next year, however, the patient's condition began to deteriorate, and it became clear that the tumor was no longer responding to these chemotherapeutic drugs. She then was switched to a different chemotherapeutic regimen, containing vincristine and daunorubicin, drugs that are totally different from those initially used. However, the tumor also was resistant to these drugs, and the rapid downhill course continued.

Biochemical questions
1. How do chemotherapeutic drugs such as methotrexate enter cells?
2. What is the basis for the development of multidrug resistance in this tumor?

Case discussion

The development of resistance to cytotoxic drugs is the major reason why chemotherapy that starts out successfully subsequently fails. As noted in this case, the tumor often becomes resistant not only to the drugs that are given initially, but also to other structurally and functionally unrelated drugs.

1. Drug transport into cells Many chemotherapeutic drugs are taken up by cells through a carrier-mediated process. This is a facilitated diffusion process involving a transporter located in the plasma membrane. The transporter is an integral membrane protein that penetrates completely through the lipid bilayer. It binds the drug and facilitates its transfer across the membrane into the cell interior. The transporter protein loops through the membrane many times and in this way forms a channel or tunnel along which the drug can pass through the hydrocarbon interior of the lipid bilayer. Since the process is diffusion, the drug can be moved in either direction, either in or out of the cell, depending on the concentration gradient.

2. Drug resistance Multidrug resistance is caused by the induction of a membrane transport protein called the P-glycoprotein, or P170. The protein is encoded by the MDR1 gene. It is an integral membrane protein containing 1280 amino acid residues. The protein is a transporter that passes through the plasma membrane 12 times. Although small loops of the polypeptide chain protrude into the extracellular fluid, the bulb of the protein structure is contained in the cytoplasm below the lipid bilayer and includes two putative ATP-binding sites. The transporter utilizes the energy of ATP to pump a variety of different drugs out of the cell, thereby protecting the tumor against chemotherapeutic drugs. When the tumor is exposed for long periods to these cytotoxic drugs, cancer cells that survive are selected because they express high levels of the MDR1 gene product. Therefore the tumor that remains is resistant to all chemotherapy because the cancer cells contain large numbers of the multidrug transporter, which prevents cytotoxic levels of the drugs from accumulating intracellularly.

Reference	Gottesman MM and Pastan I: The multidrug transporter, a double-edged sword, J Biol Chem 263:12263, 1988.

Case 2   Cystinuria

A 14-year-old boy was admitted to the hospital because of sudden, severe pain in the left flank. Urinalysis revealed the presence of hexagonal crystals in the urinary sediment. Renal stones were detected on roentgenography. A 24 hr urine collection indicated that the patient was excreting large amounts of cystine. Furthermore, the urine contained abnormally high amounts of ornithine, lysine, and arginine.

Biochemical questions

1. What is the molecular basis for cystinuria?
2. How is cystine transported across cell membranes?
3. What are the properties of the cystine transport system?
4. Why is there increased excretion of ornithine, lysine, and arginine?

Case discussion

Cystinuria is an inherited, autosomal recessive disease affecting amino acid transport by epithelial cells of the kidney tubules and the gastrointestinal tract. Cystine and other dibasic amino acids are excreted in excessive amounts in the urine. When cystine is present in high concentrations and the urine is acidic, it tends to precipitate and form stones in the kidney.

1. The molecular basis of cystinuria There is a genetic deficiency for the transporter that moves cystine across epithelial cell membranes. Normally, much of the cystine filtered through the glomerulus is reabsorbed into the blood through the renal tubular epithelium. The failure of transport in the renal tubules is the cause of the increased cystine concentration in the urine.

2. Cystine transport Cystine is transported by carrier-mediated diffusion. The transporter is a transmembrane integral protein that penetrates through the membrane lipid bilayer.

3. The properties of cystine transport Cystine transport properties are those exhibited by facilitated diffusion systems. Transport occurs along a concentration gradient, from solutions of high to low concentration. In the renal tubule the concentration is higher than in the adjacent capillary, and cystine moves from the tubular lumen across the tubular epithelial membrane without the need for any energy input. Since cystine has two carboxyl and two α-amino groups, it is polar and requires a transmembrane protein carrier to cross the epithelial plasma membrane. The transporter has a high affinity for binding cystine; that is, it has a low K_m for cystine. Since cystine must bind to the transporter to cross the membrane, the process is saturable.

4. Excretion of amino acids Not every amino acid has its own, separate transporter. Instead, amino acids of similar structure tend to share the same transporter. The cystine transporter is shared by other basic amino acids, including ornithine, lysine, and arginine. Since this transporter is deficient or defective in cystinuria, these other basic amino acids cannot be adequately reabsorbed and also are excreted in excessive amounts in the urine. Only cystine precipitates and forms stones, so this is the amino acid that causes the clinical difficulties.

Reference   Segal S and Thier SO: Cystinurias. In Scriver CR et al, editors: The metabolic basis of inherited disease, ed 6, New York, 1989, McGraw-Hill Information Services Co.

Case 3   Abetalipoproteinemia

A 4-year-old boy was referred to a pediatric neurologist for consultation because of ataxic neuropathy and retinal degeneration. He was very small, far below the average weight for his age. His mother stated that his stools were very large and foul smelling. Analysis of the plasma revealed a very low cholesterol content, and no low-density lipoproteins (LDLs) were detected on electrophoresis. No apolipoprotein B (apo B) was detected in the plasma with polyclonal antibodies.

Biochemical questions
1. Discuss the relationship of cholesterol to cell membranes.
2. What is the functional role of cholesterol in membranes?
3. How do membranes obtain cholesterol?
4. Explain a hypothetic mechanism for the nervous system disorders affecting this patient.

Case discussion   This patient has familial abetalipoproteinemia. The cause of the disease is a genetic defect in the synthesis of apo B. Apo B is a necessary component for chylomicron synthesis by the intestine (see Chapter 16). Chylomicron formation is needed for the absorption of dietary fat. Poor growth and excessive production of stools is caused by the inadequate absorption of dietary fat. LDLs also contain apo B. The LDLs are formed in the plasma from very low–density lipoproteins (VLDLs), which also require apo B for their synthesis in the liver. The failure to produce VLDL, and consequently LDL, is the cause of the low plasma cholesterol concentration.

1. Role of cholesterol in membranes Cholesterol is a component of some cell membranes, particularly the plasma membrane, where it is often present in an almost 1:1 molar ratio with the phospholipids. It is contained in both leaflets of the plasma membrane lipid bilayer. Almost all the cholesterol in membranes is present in the free (unesterified) form.

2. Function of cholesterol in membranes Cholesterol is inserted between the phospholipds of the lipid bilayer. Its hydroxyl group projects outward, in the direction of the polar head groups of the phospholipids. The ring structure of the steroid nucleus is planar and rigid, and it interacts with the fatty acyl chains of the surrounding phospholipids. Cholesterol is inserted into both gel and liquid crystalline membrane domains. Its main function appears to be the modulation of the motion of the fatty acyl chains in the lipid bilayer. In this way it modulates the fluidity of the membrane. Cholesterol tends to reduce the fluidity of highly liquid domains and, conversely, to decrease the tight packing and thereby raise the fluidity of domains in the gel-like state.

3. Cholesterol sources for membranes Membranes can obtain cholesterol either by synthesis within the cell through the isoprenoid synthetic pathway or by uptake from plasma lipoproteins. In many cells LDL is the main lipoprotein that supplies cholesterol from the plasma.

4. Abetalipoproteinemia and neuronal degeneration The mechanism whereby abetalipoproteinemia causes neuronal degeneration is not known, but several possibilities exist. One is that in certain areas of the nervous system, cholesterol must be obtained from LDL for the membranes to function properly. In these areas cholesterol synthesis may not be adequate to maintain a sufficient supply of cholesterol for normal excitability of the axonal membranes. Another possibility is that essential polyunsaturated fatty acids are not available in adequate amounts because of the inability to synthesize chylomicrons, causing fat malabsorption. Because of a lack of these essential fatty acids, the neural membranes in the affected areas may not have the proper structure and fluidity properties to function normally. A third possibility is that the neural abnormalities result from inadequate absorption of fat-soluble vitamins, particularly vitamin E, which protects against lipid peroxidation in membranes. Fat-soluble vitamins are not well absorbed if chylomicrons cannot be formed, such as occurs in abetalipoproteinemia. Which of these possibilities is the main cause of the neurologic defects remains to be determined.

Reference   Kane JP and Havel RJ: Disorders of the biogenesis and secretion of lipoproteins containing the B-apolipoproteins. In Scriver CR et al, editors: The metabolic basis of inherited disease, ed 6, New York, 1989, McGraw-Hill Information Services Co.

Case 4	Cholera

A 42-year-old man, weighing 80 kg and living in the southern part of the United States, developed malaise, anorexia, abdominal pain, and watery diarrhea. The following day he became weak and severely nauseated, and the vomiting and the diarrhea continued; both the vomit and diarrhea were copious and clear. He was admitted to the hospital showing postural hypotension. The laboratory blood values were as follows:

Na^+	140 mmol/L
K^+	4.5 mmol/L
Cl^-	107 mmol/L
pCO_2	28 mm Hg (3.73 kPa)
pH	7.19
HCO_3^-	9 mmol/L
Glucose	10 mmol/L (180 mg/dl)
Total protein	101 g/L
Plasma specific gravity	1.040
Plasma osmolarity	326 mosmol/L
Hematocrit	47%

The patient rapidly improved with fluid replacement and oral tetracycline, and he was discharged after 14 days. Toxigenic *Vibrio cholerae* was isolated from the stool.

Biochemical questions

1. What membrane processes are affected by *V. cholerae* to produce cholera?
2. Explain why glucose must be present in fluid and electrolyte replacements by the oral route in this case.
3. What were the signs of dehydration?
4. What was the acid-base condition of the patient on admission?

Case discussion

1. **Biochemistry of cholera** Cholera is defined as an acute diarrheal disease during which *V. cholera* are present in large numbers in the liquid stool. The loss of fluid from intestinal cells is cyclic AMP dependent, as are many biochemical reactions. The action of many hormones are mediated through the formation of cyclic AMP by the activation of adenylate cyclase. However, the activation of adenylate cyclase by these hormones is not direct; it depends on two other proteins:

   a. A receptor on the external surface of the cell membrane that binds specifically the hormone in question

   b. Guanosine nucleotide regulatory protein (G_s protein) that (1) binds GTP (guanosine triphosphate) or GDP (guanosine diphosphate), its α-subunit, the GTP displacing bound GDP at a rate that is increased by the hormone-receptor complex but not the receptor alone; (2) is a GTPase, hydrolyzing GTP to GDP and inorganic phosphate; and (3) is in the form of the G_s protein:GTP complex, activating adenylate cyclase to catalyze the conversion of ATP to cyclic AMP

   Thus, if the concentration of hormone is increased in the blood, more receptors are occupied and the GTP rapidly displaces GDP formed by the GTPase action of the G_s protein. More of the G_s protein is in the GTP complexed state, and the formation of cyclic AMP is stimulated; the cyclic AMP reflects the hormone concentration.

   V. cholerae produces a proteinaceous cholera toxin, 87,000 daltons, consisting of an A_1-subunit linked by a disulfide bond to an A_2-subunit, and complexed to five B-subunits, which bind tightly to a specific molecule, called a G_{M1} ganglioside in the cell membrane. While bound to the membrane, the A_1-subunit is reductively detached from the toxin and enters the cell. This A_1 subunit catalyzes the modification of the G_s protein by the ADP-ribosylation of an arginyl residue according to the reaction:

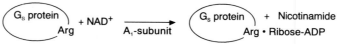

   The ADP-ribosylation of the α-subunit of the G_s protein effectively blocks the breakdown of the GTP complex so that the adenylate cyclase is continuously activated independently of the state of the hormone receptor (Figure 12.22). Excessive production of the cyclic AMP is associated with the loss of isotonic fluid, characteristic of cholera diarrhea and vomit. The latter is secreted from the villous cells of the upper portion of the small bowel.

2. **Fluid therapy for cholera** If adequate water and electrolyte replacement is instituted, the cholera patient may have diarrhea for 4 to 6 days, during which time a volume of fluid equal to one to two times the body weight may be lost. However, fatality occurs in less than 1% of affected adults, principally as a result of proper fluid replacement and control of the acidosis. The rehydration and the correction of the acidosis can be brought about in a few hours.

   From our earlier discussion it is clear that the replacement fluid must contain Na^+, Cl^-, K^+, and a bicarbonate-generating compound—such as lactate or acetate, which are oxidized to bicarbonate—or bicarbonate itself. A suitable fluid for intravenous therapy might be 130 mmol/L Na^+, 4 mmol/L K^+, 109 mmol/L Cl^-, and 28 mmol/L lactate for a fluid concentration of 217 mOsm/L. If the patient had

Figure 12.22 Action of cholera toxin. Inhibition of GTPase activity of ADP-ribosylated G_s protein is indicated by the broken arrow.

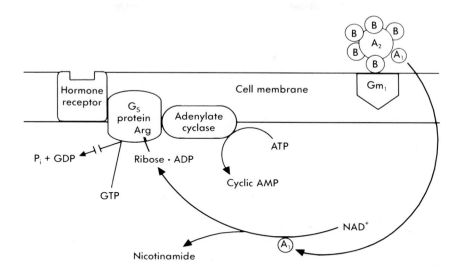

a fluid deficit of 4 L, 2 L could be given in the first hour and 2 L over the next 4 hr.

3. **Dehydration** The loss of fluid by vomiting and diarrhea is manifeset in the hemoconcentration of plasma proteins and glucose, in the high specific gravity of the plasma (normal 1.025 to 1.027), in the plasma osmolarity (normal 280 to 300 mosmol/L), and in the hematocrit. It is not possible to estimate the degree of dehydration from the Na^+ concentration in the plasma because the fluids lost had the same Na^+ concentration as normal plasma. The severity of fluid loss is indicated by the clinical signs, such as postural hypotension of the patient who was not comatose, the latter often occurring after 10% to 12% of the body weight is lost. For this patient, 10% fluid loss would correspond to 8 L, and since he was said to be weak, not comatose, it is likely that his fluid loss was 5% to 8% of body weight, or about 4 L. Since the fluid loss in cholera reaches a maximum in 24 hr and could be 500 ml/hr, it is easy to develop severe dehydration in a relatively short time. In the hospital the best way to follow the fluid loss is by direct measurement; in the past a "cholera cot" was especially designed for this purpose.

4. **Acid-base balance** The acid-base balance of the patient indicates an acidosis in which the bicarbonate concentration is greatly reduced and the pCO_2 is below normal but consistent with the measured blood pH of 7.19.

$$\text{pH} = 6.10 + \log_{10} \frac{[\text{HCO}_3^-]}{0.0301 \times pCO_2}$$

$$= 6.10 + \log_{10} \frac{9}{0.0301 \times 28}$$

$$= 6.10 + \log_{10} \frac{9}{0.843}$$

$$= 7.13$$

The condition would be identified as metabolic acidosis, which is only partially compensated. Although not noted in this patient, the respiration of untreated cholera patients typically is deep and rapid (Kussmaul respiration) in an attempt to reduce the pCO_2.

The concentration in the plasma of Na^+, K^+, and Cl^- are within normal range, suggesting that there is a loss of isotonic electrolyte fluid that is relatively rich in bicarbonate. The kidney has not responded to this loss by regenerating bicarbonate.

References

Holmgren J: Actions of cholera toxin and the prevention and treatment of cholera, Nature 292:413, 1981.

Kahn RA and Gilman AG: ADP-ribosylation of G_s promotes the dissociation of its alpha and beta subunits, J Biol Chem 259:6235, 1984.

Case 5   Myasthenia gravis

Myasthenia gravis (MG) is an acquired autoimmune disorder associated with a deficiency of acetylcholine receptors at the neuromuscular end-plate. The patient in this case had weakness and fatigue of skeletal muscles.

Biochemical questions

1. How is myasthenia gravis related to the acetylcholine receptor?
2. What are the types of muscle cells involved in contraction?
3. Which are the contractile proteins in muscle?
4. What are the biochemical events occurring in muscle contraction and relaxation?
5. Describe the regulation of contraction processes.
6. What are the sources of energy for contraction?

Case discussion

1. Acetylcholine receptor Patients with myasthenia gravis have an antibody against the acetylcholine receptors circulating in the blood (autoantibodies). The numbers of functional receptors in these patients are therefore reduced. As a muscle is repeatedly "fired," less and less acetylcholine is released into the neuromuscular junction. Eventually an insufficient amount of acetylcholine exists to depolarize the membrane to the threshhold for firing the muscle action potential. Thus patients with myasthenia gravis experience fatigue of skeletal muscles during sustained work. Some myasthenic syndromes may appear as a result of an overdose of certain drugs that potentiate neuromuscular blocking agents. For example, succinylcholine, used to produce muscular relaxation during surgery, binds to the acetylcholine receptor. Slow hydrolysis of succinylcholine is catalyzed by acetylcholinesterase and by a less specific plasma cholinesterase so that neurotransmission resumes after the drug infusion is terminated.

Patients with myasthenia gravis may be treated with pyridostigmine bromide or neostigmine methyl sulfate; both inhibit acetylcholinesterase and thus extend the transient time of acetylcholine in the neuromuscular gap junction.

More potent inhibitors of either the acetylcholine receptor (turbocurarine, which competes with acetylcholine binding) or the inhibitors of acetylcholinesterase (fluorophosphates such as the insecticides parathion and sarine) produce an irreversible paralysis.

2. Types of muscle cells

Smooth muscle Smooth muscle consists of relatively small cells. Each cell contains a single nucleus, some mitochondria, and some granules filled with glycogen. The cytoplasm, or *sarcoplasm,* is relatively homogeneous, but in suitably treated preparations, numerous, fine, threadlike structures, or *myofibrils,* can be seen embedded in it. The myofibrils constitute the contractile apparatus.

Striated or skeletal muscle The second major type of muscle is known as striated or skeletal muscle because the cells appear to be filled with distinct and regular transverse markings or striations. When viewed under polarized light, the fibers of striated muscle are seen to have alternating bands that are either optically isotropic (I bands) or anisotropic (A bands). The isotropic bands have uniform properties regardless of the direction in which they are observed, whereas the visible properties of the anisotropic bands depend on the direction of polarization of the light with which they are viewed. The I bands are thinner than A bands, and the I bands have, in their centers, a denser zone known as the Z line. The length between Z lines is known as the *sarcomere,* the fundamental repeating unit of muscle structure. The A bands are bisected by a clearer zone called the H band, and the center of the H band contains another apparent feature known as the M line. These observational features are shown in Figure 12.23. Note that these striations run at right angles to the length of the muscle.

Other faint striations can be seen running parallel to the fiber length, although little detail can be observed with the light microscope. With the electron microscope, more detail is evident. The lengthwise markings can now be seen to define the limits of individual fibrils, each of which is surrounded by a thin membrane known as the *sarcolemma.* The fibrils are composed of many molecules of actin, myosin, and several related regulatory proteins, whose chemistry and physiology are discussed in answer 3.

In contrast to smooth muscle, striated muscle cells are multinucleated and rather large, with diameters of 10 to 100 μm and lengths as great as 40 mm up to several centimeters. Spacing between the I and the A bands is variable; in general, the faster a muscle contracts, the shorter the interval between bands and the greater the number of nuclei and mitochondria per fiber. Furthermore, very active muscles are darker in color than those that are less active; this is because of the larger number of mitochondria and the greater concentration of myoglobin.

Cardiac muscle The third major muscle type is cardiac muscle, which shares some of the properties of smooth and striated muscle. Its fibers do not exist as isolated muscle cells but exhibit numerous anastomoses between segments of fibers. In reality the myocardium is a syncytium, a multinucleate mass of fused cells, through which

Figure 12.23   Typical arrangement of myofilaments in striated muscle. Correlation of optical properties with chemical structure.

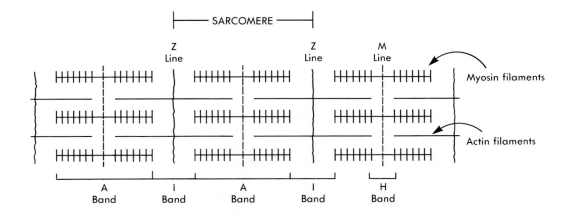

run the *Purkinje fibers;* these are highly specialized fibers that confer the property of rhythmic, independent contractility. Myocardial fibers exhibit transverse striations very similar to those of skeletal muscle.

3. Contractile proteins

Myosin Striated muscle contains from 16.5% to 20.9% of protein by weight, much of which can be extracted by dilute salt solutions. Such extracts contain three major proteins; myosin accounts for approximately 54%, actin for approximately 25%, and tropomyosin for approximately 11%. Myosin is precipitated on dilution of the salt solution and can be selectively redissolved with 0.6 mol/L KCl. Repeated application of this and similar procedures yields a substance, myosin, with a molecular weight of approximately 470,000. It is rather elongated molecule (about 1600 Å) and it is composed of six polypeptide chains, one pair of long chains and two pairs of short chains. Each of the long chains contains about 1800 amino acid residues, making this peptide one of the longest known. Myosin contains several proline residues in its primary sequence. Proline interferes with α-helix formation. It is therefore not surprising to find that although much of the length of the myosin peptide is in the form of an α-helix, a significant portion is of a globular nature. The two pairs of short chains are both associated with the globular portion of the molecule, as is shown in Figure 12.24. The light chains can bind Ca^{++} and can be phosphorylated.

Figure 12.24   Model of interacting myosin and actin systems. The troponin complex is composed of troponins I, C, and T. The complex attaches to tropomyosin, which regulates actin. Actin greatly stimulates the ATPase activity of myosin. This is regulated by Ca^{++} binding and phosphorylation of its light chains, which are located around the globular heads of the myosin structure. (From Adelstein RS and Eisenberg S: Regulation and kinetics of the actin-myosin-ATP interaction. Reproduced, with permission from the Annual Review of Biochemistry, vol 49. © 1980 by Annual Reviews Inc.)

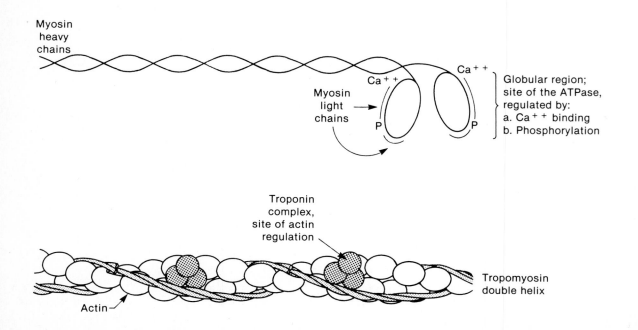

In the presence of Mg^{++}, myosin is an ATPase. Furthermore, it can be shown that when the globular portion of myosin is separated from the fibrous portion by gentle proteolysis, the enzymic activity is totally localized in the globular portion of the molecule. One pair of the associated light chains appears to be essential for ATPase activity, but the other pair does not. Although myosin is, by itself, an effective ATPase, the *rate* at which it can hydrolyze ATP is only about 1% of the rate observed when myosin forms a complex with actin, known as *actomyosin*.

Careful experiments based on protein extraction and fiber reconstitution show that the thick fibers, which run parallel to the length of a striated muscle cell, are largely composed of myosin. These fibers are made up of many molecules, and the globular portions, or heads, of the myosin molecule are randomly located along the length of the thick fibers, somewhat resembling the barbs of barbed wire. Furthermore, the arrangement of these heads is highly ordered in that within the sarcomere, the myosin molecules have one or two possible orientations, so the A band represents bare, fibrous portions of cabled myosin molecules, relatively free of globular heads.

Actin Extraction experiments indicate that the thin filaments are composed of actin and tropomyosin, together with a small group of additional regulatory proteins.

Actin occurs in two forms, G-actin and F-actin. G-actin is a globular protein consisting of a single polypeptide with a molecular weight of approximately 46,000. A molecule of actin tightly binds 1 atom of Ca^{++} and can also bind 1 molecule of either ATP or ADP. Binding of ATP quickly converts G-actin to F-actin, a highly polymerized form. Each time a monomeric G-actin unit is added to the growing fibrous polymer, a molecule of ATP is expended. F-actin is a supercoiled double helix, the diameter of which is very nearly the same as the thin filaments of the striated muscle cell. Actin has some other unusual structural features. It contains a relatively large number of L-proline residues, seven L-cysteine residues, and one residue of ϵ-*N*-methyl-L-lysine, the function of which is not clear.

Actin also occurs in noncontractile cells, where it participates in formation of the cytoskeleton, in assembly of the mitotic spindle, and probably in other essential processes as well.

Tropomyosin Tropomyosin from skeletal, cardiac, and smooth muscle is very much the same. It has a molecular weight of about 66,000 and is composed of two α-helical subunits that wind about each other. This double helical assembly then winds in the groove between the two strands of the actin filaments (see Figure 12.24). Tropomyosin appears to act in concert with troponin as a regulatory device in many muscles. In skeletal and cardiac muscle a Ca^{++}-independent, actin-activated myosin-Mg^{++} ATPase is repressed by the binding of troponin-tropomyosin to actin. This is derepressed by binding of Ca^{++} ions to troponin.

4. Events of the contraction-relaxation cycle Myosin molecules in the thick fibers are organized such that the bipolar arrangement of the myosin heads is toward the ends of the sarcomere. This bipolarity accounts for the M line. Numerous L-proline residues exist in the myosin structure. These cluster in two distinct areas of the molecule, and it is presumed that they may act as two "hinges," one located at the junction between the globular head and the fibrous portion and the second at a point along the fibrous portion of the peptide. These hinges allow the angle of the heads, with respect to the fiber length, to be altered during the events of the contraction-relaxation cycle.

Figure 12.25 diagrams the contraction-relaxation cycle. At the top of the figure is a representation of the resting state. As long as sufficient supply of ATP is present and as long as the concentration of free Ca^{++} in the sarcomere is sufficiently low ($<10^{-7}$ mol/L), the myosin heads do not engage the actin filaments. On excitation, as the

Figure 12.25

Postulated cycle of events in muscle contraction. *M* represents myosin. Calcium is the direct "trigger" of contraction. Note that M·ADP and P_1 are released separately. The relative positions of actin and myosin fibers change as the events of the cycle occur. M·ADP·P_1 represents myosin to which ATP has bound. It is presumed that in skeletal muscle the myosin is not phosphorylated. However, the representation of ATP as ADP·P_1 serves to indicate that the ATP is not readily removed from the myosin moiety at this stage of the cycle.

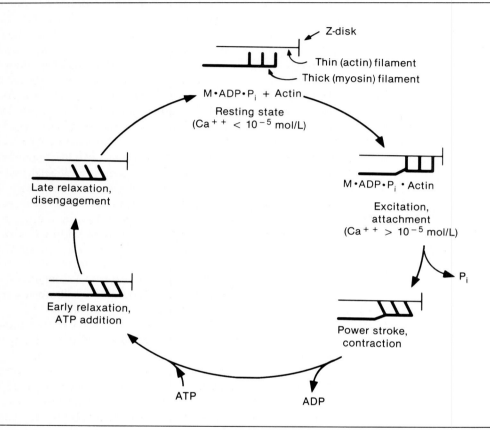

concentration of free Ca^{++} increases (to $>10^{-5}$ mol/L), a pronounced change occurs in the conformation of the myosin molecules. Now the heads engage the actin filaments. During the power stroke, ATP is hydrolyzed and the relative angle of the myosin heads, with respect to the actin filaments, becomes more acute. As a result, the physical length of the myosin fiber decreases. The change in sarcomere length is a result of the actin fibers being pulled toward one another by the changing conformation of the myosin molecules. Because of the bipolar arrangement of the myosin molecules in thick fibers, the overall change is significant in magnitude. Actin greatly increases ADP release, compared to the release rate when ATP is split by myosin alone; thus the actomyosin complex is especially efficient.

Once the ADP has been released, the relaxation phase begins. The acute bend in the myosin molecules diminishes, and subsequently the myosin heads disengage from the actin filaments as the concentration of ATP increases. One of the last events is the release of the inorganic phosphate (P_i) generated by ATP hydrolysis during the power stroke. The system finally returns to the resting state as the concentration of free Ca^{++} ions is reduced by various means.

5. Regulation of contractile tissues Regulation of muscle contraction can be classified as myosin-based or as actin-based, but few contractile tissues depend on either of these alone. In other words, the regulation of muscle usually depends on multiple mechanisms.

Actin-based regulation Skeletal and cardiac muscle contain several previously mentioned accessory proteins. These include tropomyosin and three troponins: troponin I (for inhibitory troponin), troponin C (for Ca^{++}–binding troponin), and troponin T (for tropomyosin-binding troponin). The three troponins form clusters that are found at periodic intervals along the length of the tropomyosin helices (see Figure 12.24). Addition of tropomyosin, troponin I, and troponin T inhibit the action of myosin-Mg^{++} ATPase. This inhibition is independent of the concentration of Ca^{++}. Addition of troponin C, in the presence of Ca^{++} ($>10^{-5}$ mol/L), promptly relieves this inhibition. Therefore the presence of the tropomyosin-troponin system in conjunction with actin apparently serves as an effective regulatory device based on small changes in Ca^{++} concentrations. In this regard it is worth mentioning that troponin C, which has four potential binding sites for Ca^{++}, has a molecular weight of approximately 18,000 and is, in other ways, very similar to calmodulin. Calmodulin, which has a molecular weight of 16,500, is a calcium-binding protein that is ubiquitous throughout the animal kingdom. The similarity in structure and function of these two proteins is striking. Troponin I and troponin T from skeletal and cardiac muscle (but not troponin C) can also be phosphorylated by a cyclic AMP–dependent protein kinase; the exact purpose of this reaction is not clearly understood.

Myosin-based regulation A myosin molecule from practically any source can be phosphorylated on one pair of its light chains. Phosphorylation takes place on a serine residue, usually not more than 20 residues from the amino terminus. The reaction is catalyzed by a myosin kinase that is cyclic AMP–dependent through a typical cascade. The kinase is made up of two peptides; the larger peptide confers specificity to the kinase, which has only the myosin light chain as a suitable substrate, and the smaller of the two peptides is probably identical with calmodulin. The larger peptide is completely inactive in the absence of calmodulin. In smooth muscle, which is relaxed by cyclic AMP, phosphorylation of the myosin kinase results in decreased activity, perhaps because of a decrease in the binding constant for calmodulin of the phosphorylated large subunit of the kinase. Specific phosphatases remove the phosphate group from myosin kinase and from myosin light chains, respectively. The precise role of myosin phosphorylation in skeletal muscle remains puzzling and somewhat obscure. It can be said that myosin phosphorylation is not required for actin-based regulation in skeletal muscle but that skeletal muscle myosin phosphorylation may be a means of modulating force generation.

Summary of regulatory mechanisms Ca^{++} concentration changes are the primary means of initiating muscle contraction. Regulation of these concentrations involves a delicate interplay between the Ca^{++},Mg^{++}-ATPase, which keeps the concentration within the physiologically desirable range; calmodulin, which binds Ca^{++} and serves as a coenzyme for myosin kinase; and the binding of Ca^{++} to troponin C. The Na^+,K^+-ATPase is responsible for conducting the nerve impulse to the motor endplate that signaled the muscle to contract. These observations clearly emphasize the fundamental importance of ion pumps.

6. Sources of muscle energy Life or death may frequently depend on intense bursts of muscle activity, and therefore muscles must have an abundant supply of energy at all times. The ready availability of energy has been ensured by a redundant system.

Muscle is unique in containing significant quantities of creatine and creatine. Most of the creatine exists as phosphocreatine, a high-energy compound. Phosphocreatine has a very large negative change in free energy when it is hydrolyzed; thus in time of need it can be used to replenish depleted stores of ATP, according to the following equation:

$$\text{Creatine} \sim \text{P} + \text{ADP} \rightleftharpoons \text{Creatine} + \text{ATP}, \Delta G^{0\prime} = -12.5 \text{ kJ/mol} (-3 \text{ kcal/mol})$$

The reaction is catalyzed by the enzyme ATP-creatine transphosphorylase, more commonly known as *creatine kinase* (CK). In muscle performing moderate work, the concentration of phosphocreatine drops before any appreciable diminution of ATP occurs, indicating that phosphocreatine is a primary means of maintaining the ATP level. Following exertion and the depletion of energy stores, creatine is once more phosphorylated to phosphocreatine. This is further substantiated by the treatment of intact isolated muscle with fluoro-2,4-dinitrobenzene, which is known to inactivate CK. Muscle poisoned in this manner shows a rapid drop in ATP when it is made to work.

References

Drachman DB: The biology of myasthenia gravis, Annu Rev Neurosci 4:195, 1978.

Massoulie J and Bon S: The molecular forms of cholinesterase and acetylcholinesterase in vertebrates, Annu Rev Neurosci 5:57, 1982.

Sokmann BC et al: Role of acetylcholine receptor units in gating of the channel, Nature 318:538, 1985.

Case 6   Omega-3 fatty acid supplementation

A 42-year-old male with a strong familial history of coronary heart disease was presently asymptomatic. He read several articles in national magazines and in the newspaper indicating that fish oil capsules may help to prevent coronary thrombosis. Such dietary supplements are available over the counter without a prescription. In an attempt to protect himself against coronary heart disease, he purchased 500 mg fish oil capsules containing 50% ω-3 polyunsaturated fatty acids and ingested six capsules daily.

Biochemical questions

1. What are ω-3 polyunsaturated fatty acids?
2. How are fatty acids related to membranes?
3. What changes in membrane composition and physical properties may result from dietary supplementation with large amounts of fish oil capsules?
4. How might the function of membrane proteins be affected by dietary supplementation with fish oil capsules?
5. How are ω-3 fatty acids related to eicosanoid synthesis?

Reference

Gorlin R: The biological actions and potential clinical significance of dietary ω-3 fatty acids, Arch Intern Med 148:2043, 1988.

Case 7   Aspirin treatment following coronary artery angioplasty

A 48-year-old male developed acute chest pain. He was taken to the hospital and diagnosed as having angina pectoris. A coronary artery angiogram revealed coronary artery disease, with a large atherosclerotic plaque obstructing the left anterior descending coronary artery. An angioplasty procedure relieved this obstruction. The patient then received daily aspirin therapy, and following discharge, was told to continue taking a single aspirin a day for the next 6 mo. The rationale was to reduce thromboxane production by platelets, thereby decreasing the tendency for platelet aggregation and thrombosis in the newly opened coronary artery.

Biochemical questions

1. Thromboxane is formed from intracellular arachidonic acid. What are the intracellular storage forms of arachidonic acid?
2. How is the arachidonic acid released from these storage forms?
3. How does aspirin prevent thromboxane synthesis?
4. Is aspirin a specific inhibitor of thromboxane synthesis, or might the formation of other eicosanoids also be reduced?
5. What is the enzymic mechanism through which aspirin reduces the tendency for platelets to aggregate?

Reference

Oates JA et al: Clinical implications of prostaglandin and thromboxane A_2 formation, N Engl J Med 319:689, 761, 1988.

Additional questions and problems

1. Adriamycin, a drug used in cancer chemotherapy, can induce peroxidation in cell membranes. What component of a membrane would be most susceptible to peroxidation?
2. Poisonous snake venoms often contain phospholipase A_2. Hemolysis, the loss of hemoglobin out of erythrocytes, occurs if an individual is bitten by such a snake. Explain the mechanism of hemolysis.
3. Cholesterol esterification does not occur in patients who have a genetic deficiency of the enzyme lecithin:cholesterol acyltransferase (LCAT). As a result, cholesterol accumulates in cell membranes. How would this affect membrane properties and function?
4. Spectrin, a major cytoskeletal protein, has been observed as being deficient in some patients with elliptocytosis and anemia. How might this genetic deficiency account for the anemia?
5. Explain how a drug that keeps the K^+ gate open would affect nerve conduction.
6. What other transport systems for ions and metabolites have been described in this text?
7. List the putative and proved neurotransmitter substances. Is their mechanism of action different from acetylcholine?
8. Is the synthesis of all the eicosanoids decreased in patients who are treated with high doses of glucocorticosteroids?

Multiple choice problems

1. Which of the following phospholipids is localized to a greater extent in the outer leaflet of the membrane phospholipid bilayer?
   A. Choline phosphoglycerides
   B. Ethanolamine phosphoglycerides
   C. Inositol phosphoglycerides
   D. Serine phosphoglycerides
   E. Both choline and ethanolamine phosphoglycerides

2. All the following processes occur rapidly in the membrane lipid bilayer except:
   A. Flexing of fatty acyl chains.
   B. Lateral diffusion of phospholipids.
   C. Transbilayer diffusion of phospholipids.
   D. Rotation of phospholipids around their long axes.
   E. None is correct; all these processes occur rapidly.

3. Which of the following statements are correct about membrane cholesterol?
   1. All subcellular membranes contain about the same molar ratio of cholesterol to phospholipid.

   2. The steroid nucleus forms a rigid, planar structure.
   3. The hydroxyl group is located near the center of the lipid bilayer.
   4. The hydrocarbon chain is flexible and is oriented toward the inside of the lipid bilayer.
      A. 1, 2, and 3.
      B. 1 and 3.
      C. 2 and 4.
      D. 4 only.
      E. All are correct.

4. If a membrane has a high fluidity:
   1. It tends to have a highly saturated fatty acyl group content.
   2. It is rich in cholesterol.
   3. It has a high proportion of domains in the gel state.
   4. The permeability tends to be high.
      A. 1, 2, and 3.
      B. 1 and 3.
      C. 2 and 4.
      D. 4 only.
      E. All are correct.

5. Which of the following are correct about the phosphatidylinositol glycan linkage in membrane proteins?
   1. The protein is connected to the glycan chain through its C-terminal amino acid residue.
   2. Most of the protein structure is located inside the membrane lipid bilayer.
   3. Diacylglycerol is generated in the membrane when the cell is stimulated and the protein is released.
   4. The glycan ordinarily contains between 50 and 100 monosaccharide units.
      A. 1, 2, and 3.
      B. 1 and 3.
      C. 2 and 4.
      D. 4 only.
      E. All are correct.

6. Which of the following cytoskeletal components are *not* contained within the lipid bilayer?
   1. Glycophorin
   2. Spectrin
   3. Band 3, the bicarbonate-chloride transporter
   4. Ankyrin
      A. 1, 2, and 3.
      B. 1 and 3.
      C. 2 and 4.
      D. 4 only.
      E. All are correct.

7. Which of the following are correct concerning glucose transport into a cell?
   1. This occurs by facilitated diffusion.
   2. The process is saturable.
   3. The process is mediated by a transmembrane carrier.
   4. ATP is utilized by the transport process.

    A. 1, 2, and 3.
    B. 1 and 3.
    C. 2 and 4.
    D. 4 only.
    E. All are correct.

8. Alanine is taken up into a cell together with Na^+. This occurs from a high extracellular Na^+ concentration to a low intracellular concentration. The process is:
    A. Active transport.
    B. Passive diffusion.
    C. Uniport facilitated diffusion.
    D. Antiport facilitated diffusion.
    E. Symport facilitated diffusion.

9. What membrane has the highest protein content?
    A. Myelin sheath
    B. Plasma membrane
    C. Golgi membrane
    D. Endoplasmic reticulum
    E. Inner mitochondrial membrane

10. What compound would be formed if the carbohydrate chain of a membrane ganglioside were totally removed?
    A. Diacylglycerol
    B. Sphingosine
    C. Sphingomyelin
    D. Ceramide
    E. Phosphatidylinositol

11. A specific 15-lipoxygenase inhibitor could reduce the synthesis of which of the following eicosanoids?
    1. Leukotriene A_4
    2. Thromboxane A_2
    3. 12-HETE
    4. Lipoxin A
        A. 1, 2, and 3.
        B. 1 and 3.
        C. 2 and 4.
        D. 4 only.
        E. All are correct.

Chapter 13

Nucleotide metabolism

Objectives

1 To explain how the structural features of the nucleic acids allow them to store and express biologic information
2 To explain how folic acid coenzymes function in purine and pyrimidine biosynthesis
3 To explain how folate deficiency leads to megaloblastic anemia
4 To explain how drugs used to treat cancer and gout interfere with nucleotide biosynthesis

Central to a discussion of growth or reproduction at the molecular level is a consideration of the functions of the nucleic acids. To understand these functions, one must study the chemical and physical properties of the nucleic acids. Because most of their constituent parts are synthesized rather than provided by the diet, it will be necessary to consider both the biosynthesis of the nucleotide precursors and the assembly of these precursors into information-laden polymers. This chapter considers only the biosynthesis of the nucleotides, and Chapter 14 deals with the synthesis of nucleic acids from mononucleotides.

The one-carbon metabolism of folic acid derivatives is important for the synthesis of several cellular constituents, but primarily it is needed for the synthesis of the purine and pyrimidine nucleotides. Understanding the metabolic functions of folic acid and its derivatives is important for the treatment of tropical sprue and megaloblastic anemia. Also the synthesis of folic acid compounds is important in explaining the effectiveness of folate analogues and sulfonamides against some bacterial infections.

The usefulness of folic acid analogues in the treatment of leukemia reflects the vitamin's functions in the biosynthesis of purine and pyrimidine nucleotides in rapidly growing cells. Furthermore, some cases of the rather common disease gout and the rare Lesch-Nyhan syndrome share a genetic defect in purine nucleotide metabolism, a knowledge of which is needed to understand these conditions.

Nucleic acids

In 1868 the Swiss physician Miescher described the first isolation of nucleoprotein from the pus cells of bandages discarded by the surgical clinic of a nearby hospital. Later he isolated a similar substance from salmon sperm and showed that it consisted of a basic protein, *protamine*, and a polymeric acidic substance now known to be nucleic acid. Common sources of nucleic acid used by these early workers were thymus glands and yeast. When thymus nucleic acid was hydrolyzed, it yielded purine and pyrimidine bases, deoxy-D-ribose, and phosphate. Yeast nucleic acid also yielded purine and pyrimidine bases and phosphate, but its sugar was primarily D-ribose rather than deoxyribose. It was subsequently found that the thymus nucleic acid was mostly *deoxyribonucleic acid* (DNA), whereas the yeast nucleic acid was mostly *ribonucleic acid* (RNA). All cells contain both DNA and RNA, but some, for example, yeast and thymus, contain more of one type than the other. A particular cell may contain a whole variety of distinct nucleic acids, so that the terms *DNA* and *RNA* are generic in the same sense as the word *protein*, which can refer to either a mixture of several individual substances or to a single protein.

Nomenclature and hydrolysis products

Some of the first experiments performed on nucleic acids by nineteenth-century chemists consisted of hydrolyzing these macromolecules to understand their chemical composition. Complete hydrolysis of a nucleic acid (Figure 13.1) yields purine and pyrimidine bases, deoxyribose or ribose, and phosphoric acid.

DNA and RNA both contain the purine bases adenine (A) and guanine (G). DNA contains the pyrimidines thymine (T) and cytosine (C), whereas RNA contains cytosine and uracil (U). The structure of these purine and pyrimidine bases are shown in Figure 13.2. Although these are the predominant bases of the nucleic acids, other minor bases are found in small amounts. These minor bases are important constituents of transfer RNA (tRNA) where they account for a little less than 5% of the bases. They are often methyl derivatives of the major bases. The structures of a few minor bases found in tRNA are:

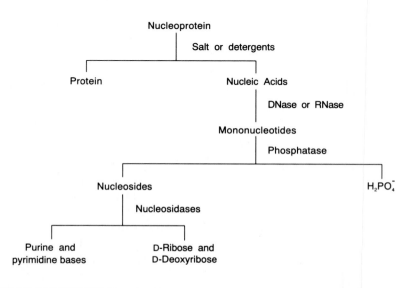

| Dihydrouracil | 2,2- Dimethylguanine | 5-Methylcytosine |

Both purine and pyrimidine bases with keto or hydroxy side-chains can undergo keto-enol tautomerization. The keto form is usually predominant.

The hydrolysis of nucleic acids by enzymes, such as deoxyribonuclease (DNase) and ribonuclease (RNase), can yield *mononucleotides*. The mononucleotides contain equimolar amounts of a nitrogenous base, a sugar, and phosphoric acid. Further hydrolysis by phosphatases yields inorganic phosphate and *nucleosides*. In the nucleosides the bases are linked to the pentose sugars in a β-D-configuration. For example, the structures of deoxyadenosine and thymidine may be represented as:

Figure 13.1   Hydrolysis of nucleic acids.

Nucleoprotein
Salt or detergents

Protein   Nucleic Acids
                                 DNase or RNase
                              Mononucleotides
                                 Phosphatase

Nucleosides $H_2PO_4^-$
                Nucleosidases

Purine and   D-Ribose and
pyrimidine bases   D-Deoxyribose

Deoxyadenosine

Thymidine

Note that all the positions on the purine and pyrimidine rings are numbered and that the positions on the sugars carry a prime symbol to distinguish them from the ring positions of the bases. The other bases are linked to either D-ribose or deoxy-D-ribose in an analogous way. The pentose sugar in RNA is D-ribose and in DNA is 2-deoxy-D-ribose. The hydroxyl group at the 2'-position accounts for the greater ease with which RNA is degraded by alkali. The names of the nucleosides are given in Table 13.1.

The nucleosides may be either deoxyribonucleosides or ribonucleosides. Similarly, the nucleotides may be either deoxyribonucleotides or ribonucleotides. The names of the common mononucleotides are given in Table 13.2. Polynucleotides consist of mononucleotides linked one to another by phosphodiester bonds between the 3'- and 5'-hydroxyl groups of the adjacent D-ribose or deoxy-D-ribose residues, as shown by the dinucleotide represented on p. 554.

Figure 13.2   Structures of purine and pyrimidine bases.

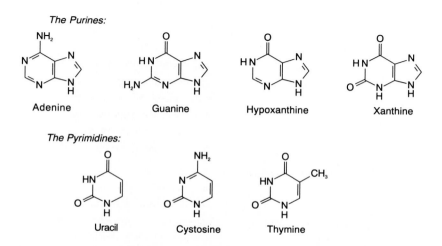

The Purines:

Adenine   Guanine   Hypoxanthine   Xanthine

The Pyrimidines:

Uracil   Cystosine   Thymine

Abbreviated: pUpC

Table 13.1   Names of nucleosides

Base	Ribonucleoside	Deoxyribonucleoside
Adenine	Adenosine	Deoxyadenosine
Guanine	Guanosine	Deoxyguanosine
Cytosine	Cytidine	Deoxycytidine
Uracil	Uridine	Deoxyuridine
Thymine	Thymine riboside	Thymidine

Table 13.2   Alternate names of nucleotides derived from nucleic acids

Acids	Phosphates	Abbreviation*
Ribonucleotides		
2'-Adenylic acid	Adenosine-2'-monophosphate	2'-AMP
3'-Adenylic acid	Adenosine-3'-monophosphate	3'-AMP
5'-Adenylic acid	Adenosine-5'-monophosphate	AMP
3'-Guanylic acid	Guanosine-3'-monophosphate	3'-GMP
3'-Cytidylic acid	Cytidine-3'-monophosphate	3'-CMP
3'-Uridylic acid	Uridine-3'-monophosphate	3'-UMP
Deoxyribonucleotides		
Deoxyadenylic acid	Deoxyadenosine-5'-monophosphate	dAMP
Deoxyguanylic acid	Deoxyguanosine-5'-monophosphate	dGMP
Deoxycytidylic acid	Deoxycytidine-5'-monophosphate	dCMP
Thymidylic acid	Thymidine-5'-monophosphate	dTMP

*When an abbreviation does not have a numerical prefix, it is assumed to be the 5' derivative.

By convention, nucleic acids and oligonucleotides are written with the 5′-terminal at the left and the 3′-terminal at the right. A common form of abbreviation is to use a capital letter such as A, G, U, or C to represent a ribonucleoside (dA, dG, etc., for deoxyribonucleosides) and a lowercase p to represent the phosphate group. Thus the compound shown above would be abbreviated pUpC. In a similar way 5′-AMP might be abbreviated as pA and 5′-ATP as pppA.

Abbreviations of longer nucleotides are simplified by omitting the internucleotide *p*, so that pppApTpGpCpA would be simplified to pppATGCA; the lowercase *d* for deoxyribonucleotides is also dropped, because the presence of the T residues tells the reader that the nucleotide is a deoxyribonucleotide.

Structure of the nucleic acids

Hydrolysis Both DNA and RNA can be hydrolyzed by strong acids, but only RNA is hydrolyzed by alkali. A wide variety of hydrolysis products are possible, but the final hydrolysis products of the nucleic acids are sugars, phosphate, and purine and pyrimidine bases. This process is employed to determine the base composition of a particular nucleic acid.

After hydrolysis the purine and pyrimidine bases can be separated from each other by thin-layer, paper, or high-performance liquid chromatography (HPLC). Estimation of the molar concentration of the separated bases is most conveniently done by spectrophotometry, since the bases all have spectra that show a major peak of absorption near 260 nm.

Nucleases Enzymatic hydrolysis produces mononucleotides and oligonucleotides relatively free of decomposition products. The nucleases are usually classified as either DNases or RNases, but some of these enzymes can hydrolyze both DNA and RNA. Nucleases can be further classified as exonucleases, which cleave mononucleotides one at a time from the ends of a polynucleotide, or as endonucleases, which cleave polynucleotides at internal positions. Both exonucleases and endonucleases are phosphodiesterases in that they cleave phosphodiester bonds. Other nucleases cleave only phosphomonoester bonds and are called phosphomonoesterases. Many of these hydrolytic enzymes function in nucleic acids, especially precursors of RNA.

Nucleases are specific for cleaving phosphoester bonds at either the 3′- or 5′-hydroxyl of the sugar portion of the polynucleotide chain. For example, some nucleases produce 3′-mononucleoside phosphates, whereas others produce 5′-nucleoside phosphates. The specificity for the 3′- or 5′-phosphoester bond applies to both phosphodiesterases and phosphomonoesterases. Deoxyribonucleases may have additional specificity for double- or single-stranded DNA. Table 13.3 lists a few common nucleases with their specificities.

Restriction endonucleases Certain endonucleases from bacteria are useful in the study of DNA structure because of their unusual specificities. These enzymes are called restriction endonucleases. They are used for gene mapping, genetic engineering, and prenatal diagnosis of some genetic diseases. Because of their stringent specificity they are able to hydrolyze foreign DNA but not the DNA of the bacterial cell in which they reside. In this way they are said to *restrict* the development of certain DNA viruses. Of the different kinds of restriction enzymes, type II is the most useful for research and in clinical diagnosis. Type II restriction endonucleases cleave specific residues in DNA; they also require that the DNA have a symmetry in the bases around a given point. For example, the restriction endonuclease from *Escherichia coli,* EcoR1, requires a DNA substrate with the following sequence:

$$\begin{array}{c} \downarrow \quad * \\ 5' \ldots \text{GAA TTC} \ldots 3' \\ \bullet \\ 3' \ldots \text{CTT AAG} \ldots 5' \\ * \quad \uparrow \end{array}$$

Table 13.3   Specificity of some nucleases

Enzyme	Source	Substrate	Specificity
Ribonuclease A	Pancreas	RNA	An endonuclease that splits 5'-hydroxyl O—P phosphodiester bonds connected to pyrimidines
Ribonuclease T$_1$	*Aspergillus* (mold)	RNA	An endonuclease that splits 5'-hydroxyl O—P phosphodiester bonds connected to guanine residues
Deoxyribonuclease I	Pancreas	DNA	An endonuclease that splits some 3'-hydroxyl O—P bonds
Phosphodiesterase	Spleen	RNA or DNA	An exonuclease, starts at 5' end, splits 5'-hydroxyl O—P bonds
Phosphodiesterase	Snake venom	RNA or single-stranded DNA	An exonuclease, starts at 3' end, splits 3'-hydroxyl O—P bonds
Restriction endonuclease, EcoR1	*Escherichia coli*	DNA	Hydrolyzes GAATTC (see text) ↑

Notice that the nucleotide sequence of the two DNA strands is identical, if each strand is written with the 5' end at the left. Also note that the two DNA strands are symmetric around the reference dot. The specific sites cleaved are marked by arrows. A large number of restriction endonucleases exist that have slightly different sequence specificities, but all require a symmetric sequence of from four to six bases. Because these short sequences may occur only a few times in small DNAs, judicious use of the enzymes will yield fragments of sizes that can be easily sequenced. By sequencing the fragments produced by different restriction endonucleases, overlapping fragments can be arranged so as to determine the sequence of larger DNA oligonucleotides.

One might wonder why the DNA of a cell that synthesizes restriction endonucleases is not hydrolyzed. The reason is that the host organism has developed highly specific methylating enzymes that methylate bases in the DNA sequences ordinarily recognized by the restriction enzyme. This can be illustrated by the example of the *E. coli* enzyme EcoR1. A methylating enzyme in *E. coli* attaches methyl groups to the N^6 position of the adenine bases within the DNA sequence (just shown) recognized by the EcoR1 enzyme. The methylated adenines are those represented in the diagram with asterisks on either side of the reference dot. The EcoR1 nuclease is unable to cleave the DNA chain when these adenines are methylated.

Base composition of DNA The base composition of DNA offers some insight into the structure and function of the hereditary material. No matter what cells are used to prepare DNA, the molar concentration of cytosine always equals that of guanine and the concentration of thymine always equals that of adenine. The only known exception to this rule is the base composition of the DNA from bacteriophages such as ΦX174. This DNA is single stranded rather than double stranded.

Even though the bases of DNA occur in ordered pairs (A = T and G = C), the base composition of DNA may vary widely from one species to another. When base compositions are compared, the data are usually given as the percentage of guanine plus cytosine or sometimes the ratio of A + T/G + C. The base composition of animal DNA covers a narrow range from about 39% to 44% G + C, as seen in Table 13.4. In contrast,

Table 13.4   Guanine plus cytosine content of DNA from various species*

Organism or tissue	Guanine + cytosine content (%)
Dictyostelium discoideum	22.0
Staphylococcus aureus	33.0
Bacillus cereus	36.0
Calf thymus	39.0
Human cell line D98	39.5
Mouse liver	40.0
Bull sperm	41.0
Mouse spleen	44.0
Escherichia coli	50.0
Brucella abortus	56.0
Pseudomonas aeruginosa	68.0
Mycobacterium phlei	73.0

*Taken in part from Handbook of biochemistry, Cleveland, 1968, The Chemical Rubber Co.

the composition of DNA from microorganisms covers a broad range. The only exception is the satellite DNA of various animal species. Satellite DNA comprises only about one tenth of the total cellular DNA and contains highly repetitive sequences of up to ten base pairs, each unit repeated from 10^5 to 10^7 times. Satellite DNA is heterochromatic and concentrated in the centromeres of certain chromosomes, for example, the Y chromosome.

Structure of DNA Combining the chemical observations of Chargaff with the x-ray diffraction data of Wilkins and other workers, Watson and Crick proposed a structure for DNA in which the hydrogen bonds bind adenine to thymine and guanine to cytosine (Figure 13.3). Bases in one strand of DNA were paired with bases in the complementary strand in an antiparallel arrangement, and both were twisted into a right-handed double helix. Antiparallel strands were arranged so that the 5′ end of one strand lay next to the 3′ end of the other.

$$5′ \ldots p\, A\, p\, C \ldots 3′$$
$$\| \quad \| \|$$
$$3′ \ldots T\, p\, G\, p \ldots 5′$$

The result of such an arrangement is a molecule in which the planar hydrogen-bonded base pairs are tightly stacked to form a rigid linear molecule. This structure is highly susceptible to breakage by shearing forces and makes for very viscous solutions. The stability of the DNA molecule results more from the strong interactions between the stacked bases than from the hydrogen bonds between paired bases. The bases along one strand of the helix lie on top of one another. The hydrophobic interactions between these stacked flat aromatic bases stabilize the helical structure. However, the base pairs are obviously very important for the remarkable specificity required in the replication and expression of the hereditary material.

Thermal denaturation The forces that hold the double helix together can be disrupted by heating; consequently, DNA is like protein in that it can be denatured. When denaturation occurs, both the stacking forces and the specific hydrogen bonds between the bases are scrambled to the extent that all biologic or template activity is lost. The ordered

Figure 13.3   Base pairs present in DNA.

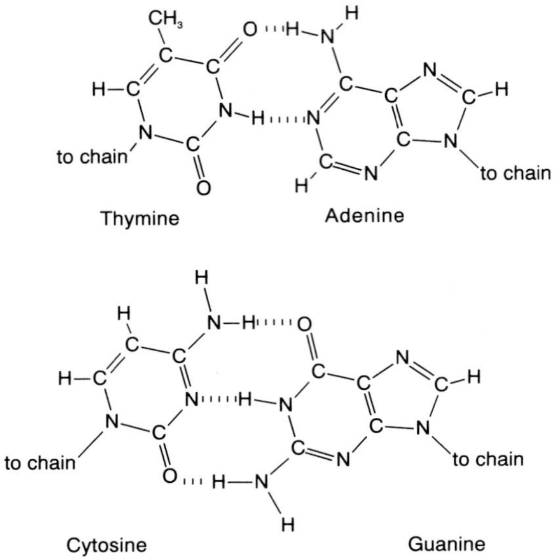

helix is said to take on the conformation of a random coil. Coincident with denaturation is an increase in the absorption of light by the bases at 260 nm, which the powerful stacking forces in the native helix tend to suppress. When a sample of pure or homogeneous DNA is slowly heated and the absorption at 260 nm recorded, one finds that the DNA abruptly denatures when a characteristic temperature, the *melting temperature* (T_m), is reached. The T_m is defined as the temperature at which a given DNA is midway between the helix and coil forms, as judged by the absorption at 260 nm. This is illustrated in Figure 13.4. The T_m is dependent on variables such as the kind and concentration of solvents used, yet it can be used as a rough measure of the base composition of a particular DNA. The T_m is proportional to the percentages of G + C (see Figure 13.4 and Table 13.4). This measurement can be used to estimate the G + C content of a DNA molecule.

Hybridization Denatured DNA is capable of being renatured; the hot denatured DNA will reform an intact helix if cooled slowly. However, if cooling is rapid, the DNA molecule remains denatured for the most part. This property of regaining the duplex structure is useful because it is highly specific for complementary strands and because hybrid helices can form with complementary strands of RNA. This process, called *hybridization,* occurs when a single strand of a denatured DNA molecule is allowed to renature in the presence of single-stranded RNA that has a complementary sequence. It can be used to determine whether a certain DNA molecule is capable of serving as a template for the synthesis of a particular RNA. This will be considered further in Chapter 15 when RNA synthesis is described.

Structure of RNA There are three types of cellular RNA. tRNA molecules are the smallest and have a molecular weight of about 28,000. In rapidly growing cells, ap-

Figure 13.4 Thermal denaturation of DNA.

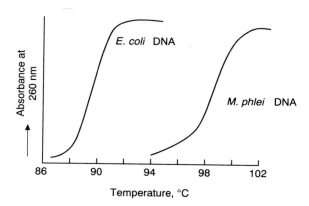

proximately 18% of the RNA is tRNA. Originally tRNA was called soluble RNA (sRNA). However, the newer name better describes the function of these molecules, which is to carry amino acids to the ribosomes where proteins are synthesized. Several tRNAs have been sequenced. This structural information helps elucidate the functions of tRNA and are discussed when protein synthesis is considered in Chapter 15. Messenger RNA (mRNA) represents only 2% of the RNA content of a cell. Yet these single-stranded polymers are the templates for the synthesis of all the cell's proteins. Their sizes vary (400 to 6000 bases), but often they are large, with molecular weights of approximately 10^6. About 80% of the cellular RNA is found in the ribosomes. This material is known as ribosomal RNA (rRNA). The ribosomes are the subcellular particles on which proteins are synthesized. Three kinds of RNA are found in animal ribosomes, and they are designated according to their sedimentation coefficients, which reflect their sizes. Thus the ribosomes contain 28S, 18S, and 5.8S RNAs, (4.7 kilobases (kb), 1.8 kb, and 0.1 kb) with molecular weights of 1.8×10^6, 0.7×10^6, and 36,000, respectively.

Because of their large variation in size and their sensitivity to hydrolysis, RNA molecules are isolated by rather special techniques. Zone centrifugation through a steep density gradient of sucrose will separate the three rRNAs from one another. This separation can also be performed by electrophoresis on polyacrylamide gels or by absorption column chromatography. Messenger RNAs, because of their large number, low concentration, and lability, are exceedingly difficult to isolate in pure form. The mRNA from purified preparations of RNA viruses, however, is easier to obtain and is often the source of the mRNA used in the laboratory.

A summary of the similarities and differences in the physical and chemical properties of DNA and RNA is presented in Table 13.5.

Digestion of dietary nucleic acids

The purine or pyrimidine bases are not essential components of the diet; most mononucleotides are synthesized de novo within the human body. Dietary nucleic acids are digested by pancreatic ribonuclease and deoxyribonuclease to mononucleotides. In the intestine these are converted to nucleosides by mononucleotidases and to free bases by nucleosidases. Figure 13.5 outlines these conversions. The tissue nucleosidases are of two types: one functions with pyrimidine nucleosides to simply hydrolyze the glycosidic

Figure 13.5   Digestion of nucleic acids.

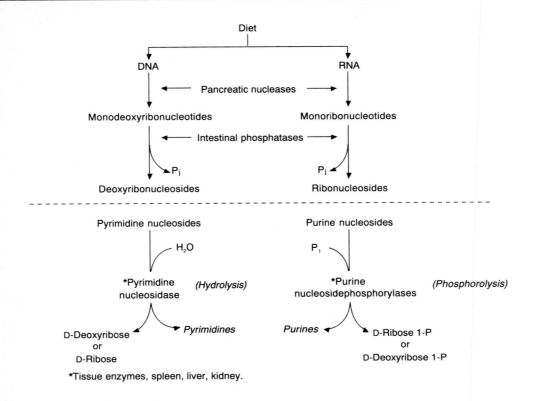

*Tissue enzymes, spleen, liver, kidney.

Table 13.5   Comparisons of DNA and RNA

	DNA	RNA
Bases		
Purines	Adenine Guanine	Adenine Guanine
Pyrimidines	Cytosine Thymine	Cytosine Uracil
Sugar	Deoxy-D-ribose	D-Ribose
Intranucleotide linkage	3'-5'-Phosphodiester	3'-5'-Phosphodiester
Size (molecular weight)	$>2 \times 10^9$	tRNA: ~28,000 mRNA: ~10^6 rRNA: 0.5 to 1×10^6
Shape	Base-paired double-stranded helix	Single-stranded random coils (tRNA is about 75% base paired)
Associated substances	Histones, protamines, spermine	Ribosomal proteins with rRNA
Sensitivity to alkaline hydrolysis	Insensitive	Sensitive
Presence in viruses	Animal or bacterial viruses	Animal, bacterial, or plant viruses
Base distribution	Purines = Pyrimidines; A = T; G = C; contains few unusual bases	No predictable distribution except in double-stranded viral RNA; may contain several unusual bases
Cellular distribution	Mostly in nucleus; some in mitochondria	Highest in cytoplasm but much in nucleus and nucleolus

bond, whereas the other is specific for purine mononucleosides. The latter enzymes catalyze a phosphorylytic cleavage of the glycosidic bond to yield either 2-deoxy-D-ribose 1-phosphate or D-ribose 1-phosphate. Guanosine and inosine are better substrates than adenosine for the purine nucleoside phosphorylases.

The phosphate and sugars produced by the digestion of nucleic acids (Figure 13.5) can be reused, as can the base adenine, but most of the other purine and pyrimidine bases are catabolized and excreted. Purines and their derivatives are efficiently converted to uric acid by an active xanthine oxidase of the intestinal mucosa. Nevertheless, small amounts of dietary nucleosides and nucleotides may be absorbed and used directly for nucleic acid synthesis, but these amounts are so low that the bulk of the purine and pyrimidine nucleotides are synthesized from smaller molecules.

Mononucleotides are important for all living things. They serve as precursors for the synthesis of nucleic acids and other metabolites, and they function as energy sources for metabolic reactions.

Biosynthesis of purine and pyrimidine nucleotides
Folic acid functions

For many years the structures of the intermediates in purine synthesis were unknown. The first intermediate was identified in a culture medium of *E. coli* that had been treated with *sulfanilamide*. This antibiotic, still useful in the treatment of some kidney infections, is highly toxic to bacteria that must synthesize their own folic acid. *p*-Aminobenzoic acid, a precursor of folic acid biosynthesis, will competitively reverse sulfonamide inhibitions.

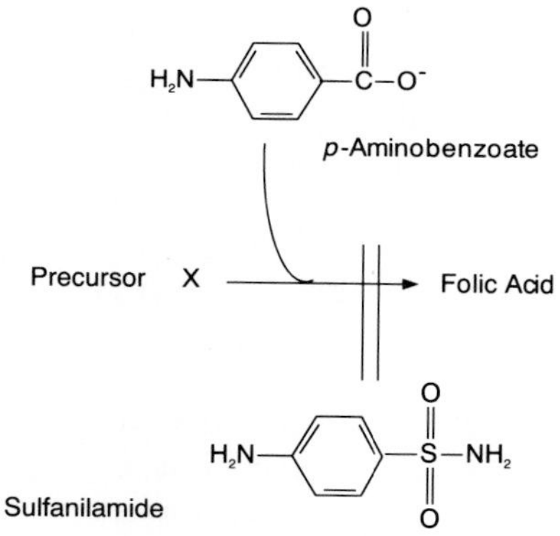

The intermediate in the synthesis of purine nucleotides that accumulates in these cells was 5-aminoimidazole-4-carboxamide ribotide:

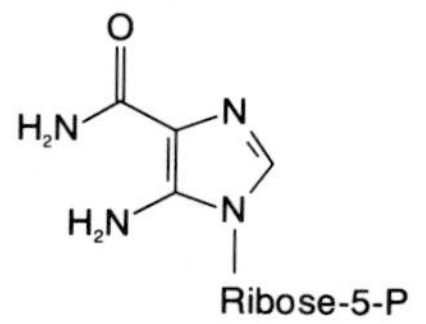

This compound is not an intermediate in folate synthesis but rather accumulates because of decreased amounts of a folate coenzyme required for purine synthesis. Sulfanilamide is toxic to microorganisms that synthesize folate, but because humans cannot synthesize folate, it is relatively innocuous to them. Recall that folate is a vitamin for humans and must be supplied in the diet.

Folate in nutrition Lack of folic acid is the most common vitamin deficiency in the world. People in underdeveloped countries or in low economic classes are most often affected, although deficiencies may be caused by infections, hemorrhage, pregnancy, or certain drugs. In short, a deficiency may occur if the diet is poor or if an individual is unable to absorb folic acid.

Because blood cells turn over rapidly, folate deficiencies are usually seen first as megaloblastic anemias in which hemoglobin levels are low and the bone marrow shows abnormally high numbers of megaloblastic cells (large abnormal immature erythrocytes).

Utilization of folic acid Dietary folic acid is converted to tetrahydrofolate (THF) by dihydrofolate (DHF) reductase, which uses NADPH as the reducing agent. DHF reductase participates in both of the reductions shown in Figure 13.6, and it is the enzyme most sensitive to the folic acid analogues (antimetabolites) used in cancer chemotherapy. Following are the structures of two common folate analogues:

Methotrexate

Aminopterin
R= see Fig.13.6

The portions of the molecules that differ from folate are shaded (compare with Figure 13.6). Because of folate's many functions in synthesizing the precursors of nucleic acids (Table 13.6), its inhibitory analogues will slow down nucleic acid synthesis in all rapidly growing cells, including cancer cells. All the coenzymatic forms of folic acid are derivatives of THF (Figure 13.7). Consequently, inhibition of DHF reductase by folate analogues would be expected to affect every reaction requiring folate coenzymes.

When some cells become resistant to antifolates, they do so by greatly amplifying the genes for dihydrofolate reductase. This results in excessive synthesis of dihydrofolate reductase, which requires massive amounts of folate analogue for inhibition.

Structural relationships between folate and its analogues The THF derivative of the vitamin is the coenzymatic acceptor of one-carbon units. The one-carbon units come from a variety of sources and are linked to THF to form four important coenzymes, one at the oxidation level of methanol (5-methyl THF), one at the level of formaldehyde (5,10-methylene THF), and two at the level of formic acid (5,10-methenyl THF and 10-formyl THF). The structures of these compounds are shown in part in Figure 13.7. A derivative of less importance, 5-formimino THF, results from histidine catabolism (Figure

Table 13.6	Some reactions where folate coenzymes function		
	Coenzyme	**Reaction**	
	1. 5-Methyl THF	Homocysteine	→Methionine
	2. 5, 10-Methylene THF	Glycine	→Serine
	3. 5, 10-Methylene THF	dUMP	→dTMP
	4. 10-Formyl THF	Glycine amide ribotide	→Formylglycine amide ribotide
	5. 10-Formyl THF	Aminoimidazole Carboxamide ribotide	→Formamidoimidazole carboxamide ribotide
	6. 10-Formyl THF	Met-tRNA	→Formyl Met-tRNA

Figure 13.6   Reactions catalyzed by dihydrofolate reductase.

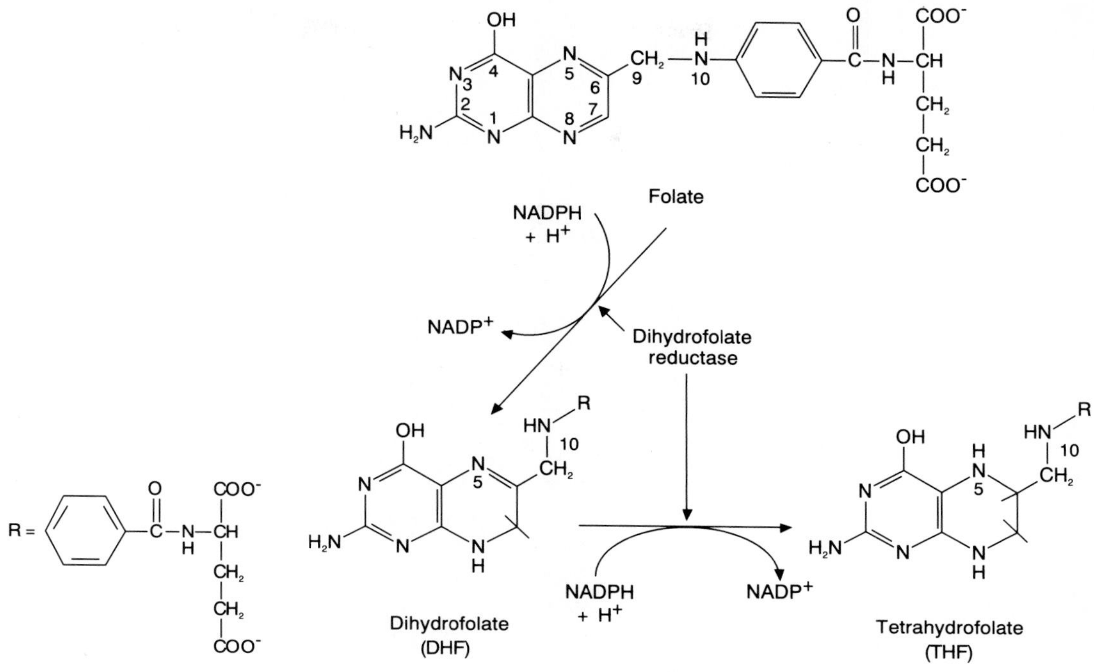

8.16). Some of the more important reactions in which folic acid coenzymes participate are summarized in Table 13.6. Several of these reactions occur during the biosynthesis of the purine (reactions 4 and 5) or pyrimidine (reaction 3) nucleotides.

Methyl-trap hypothesis 5-Methyl THF is involved in the synthesis of methionine by the liver. It is formed and utilized as follows:

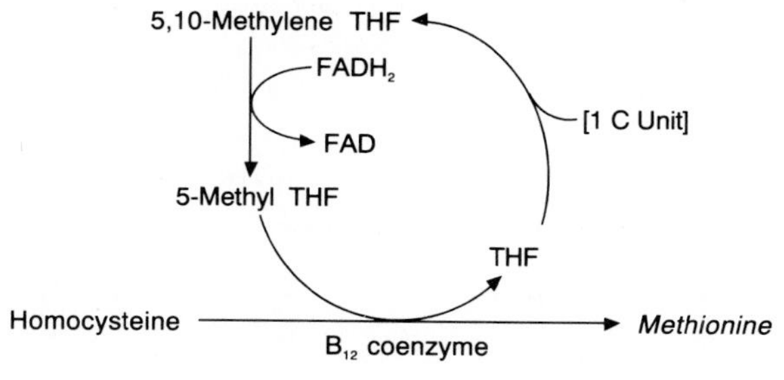

It should be recalled that although methionine is an essential amino acid, only its homocysteine portion cannot be synthesized by humans.

The enzyme 5-methyl THF methyltransferase has a requirement for a methyl B_{12} coenzyme. This enzyme catalyzes the methylation of homocysteine shown in the above reaction, which is the only significant metabolic reaction in humans that uses both folate

Figure 13.7   Folic acid coenzymes.

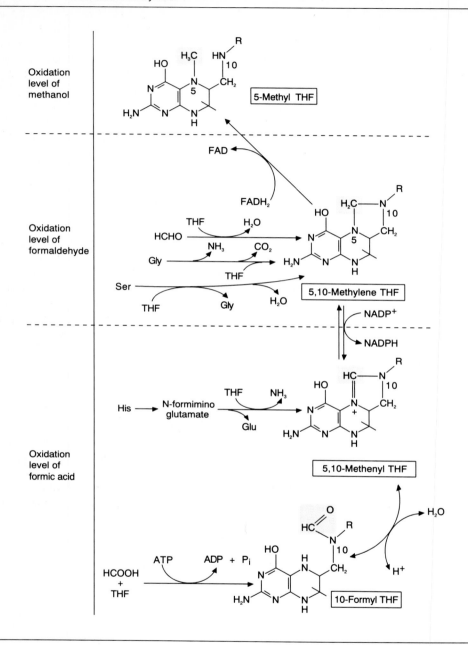

and B_{12} coenzymes. Consequently, a vitamin B_{12} deficiency may result in the conversion of all THF to 5-methyl THF. The 5-methyl THF methyltransferase reaction is the major, and perhaps the only, metabolic reaction that can use 5-methyl THF. If this reaction is blocked or slowed down as a result of a deficiency of vitamin B_{12}, 5-methyl THF accumulates and the THF needed for other folate-requiring reactions is not formed. Eventually, most of the folate in the body ends up as 5-methyl THF. Even in normal individuals, the major blood and tissue form of folate is 5-methyl THF; thus any impairment of its use could lower greatly the concentration of other folate coenzymes.

This hypothesis, the *methyl-trap hypothesis,* has been offered to explain why certain anemic conditions respond to either vitamin B_{12} or folic acid and particularly to a combination of the two. For example, the administration of folate will alleviate the hematologic symptoms of the pernicious anemia caused by vitamin B_{12} deficiency but does not prevent the irreversible neurologic damage caused by a continued lack of vitamin B_{12} (see also case 3). Although the methyl-trap hypothesis offers a convenient explanation for the interrelationships of these two vitamins, the situation may be more complicated. Some evidence indicates that 5-methyl THF may participate in reactions other than methionine synthesis. Consequently, 5-methyl THF would not be expected to accumulate. Other evidence indicates that a B_{12} deficiency somehow hinders the absorption of folate derivatives and affects the rapidly growing hematopoietic cells so that megaloblastic anemia results.

Purine nucleotide synthesis

One of the first applications of isotopes to biologic systems showed that the purine ring was labeled when simple, small radioactive molecules were fed or injected into laboratory animals. Considerably fewer nucleic acids were labeled when the animal was fed radioactive purines. See the diagram below. Three atoms of the ring are derived from glycine, while two carbon atoms are labeled when ^{14}C formic acid is administered. Two of the nitrogen atoms come from the amide nitrogen of glutamine, and the remaining nitrogen originates in aspartate. Finally, the carbon atom at position 6 is provided by carbon dioxide.

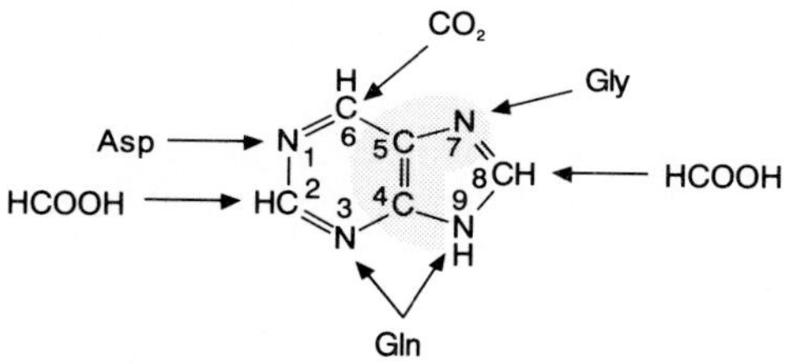

The rate-limiting reaction in purine synthesis In humans, all the enzymes of purine metabolism are found in the cytoplasm. Synthesis starts with the formation of 5-phospho-α-D-ribosyl-1-pyrophosphate (PRPP).

$$\alpha\text{-D-Ribose } 5 - P + ATP \rightarrow PRPP + AMP$$

The PRPP is then aminated in the following reaction:

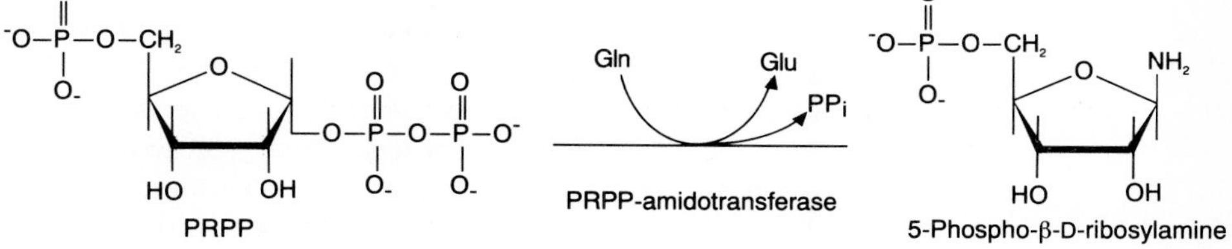

In this reaction, catalyzed by the enzyme PRPP amidotransferase, the pyrophosphate group of PRPP is displaced by the amide group of glutamine with an inversion to yield the β-D configuration found in all nucleotides. PRPP functions in several other ribosylating reactions, including pyrimidine nucleotide synthesis, NAD^+ synthesis, histidine biosynthesis, and the conversion of guanine to GMP.

This initial step in purine nucleotide synthesis is at a branch point in PRPP utilization and thus represents the first committed, essentially irreversible step in purine nucleotide biosynthesis. As is true for many other branch points in metabolic pathways, the reaction is susceptible to feedback inhibition. The feedback inhibitors in this case are the purine nucleotides, ATP, ADP, AMP, GTP, GDP, GMP, or IMP. When the cells overproduce purine nucleotides, these same nucleotides function to limit their own biosynthesis. In some cases excessive synthesis of *uric acid* (see Figure 13.15), the final catabolic metabolite of purines excreted in the urine, results from an overproduction of purine nucleotides caused by a breakdown in the feedback control of the PRPP amidotransferase.

Ribose-phosphate intermediates The other steps leading to purine nucleotide biosynthesis are illustrated in Figures 13.8 and 13.9. Note that the intermediates are all ribose-phosphate derivatives. This is consistent with the poor utilization of dietary purines, which in part is related to a poor conversion of the free purine bases to the ribose-phosphate derivatives.

The first atoms incorporated into the purine ring are derived from glycine. The enzyme glycinamide ribotide synthetase catalyzes the addition of glycine to 5-phosphoribosylamine. ATP provides the driving force, probably by forming a high-energy glycyl phosphate intermediate. The reaction is reversible, however, because the product, glycine amide ribotide (GAR), is a high-energy compound. GAR is formylated at the expense of 10-formyl THF, which can in turn be formed from formic acid.
This accounts for the fact that radioactive formate labels position 8 of the purine ring.

Another amination reaction using glutamine as the nitrogen donor allows the imidazole portion of the purine ring to be closed by an enzymatic reaction driven by ATP. The subsequent carboxylation reaction occurs on an enzyme that does not appear to contain biotin. The amination of this carboxyl group occurs in two steps in which aspartate is the nitrogen donor. The aspartate carbon atoms are released as fumarate (Figure 13.8), and 5-aminoimidazole-4-carboxamide ribotide (AICAR) is formed. Recall that AICAR is the substance that accumulates in sulfanilamide-inhibited bacterial cells, because the subsequent reaction requires the folate coenzyme 10-formyl THF. The formylated intermediate, 5-formylamidoimidazole-4-carboxamide ribotide (FAICAR), undergoes enzymatic ring closure with the formation of IMP, the purine nucleotide that is the common and immediate precursor of both AMP and GMP.

IMP: The precursor of AMP and GMP AMP synthesis requires the amination of position 6 of IMP, which is accomplished in a GTP-requiring reaction in which the amino group is derived from aspartate. This reaction is similar to the amination that formed AICAR (Figure 13.8). GMP synthesis requires an oxidation at the expense of NAD^+ followed by an amination reaction in which glutamine is the nitrogen donor. AMP and GMP can be further phosphorylated by ATP or by direct mitochondrial oxidative phosphorylation to form the purine ribonucleoside triphosphates.

Thus AMP formation occurs in a series of reactions requiring GTP, while GMP is formed at the expense of ATP. It is likely that the control of the synthesis of AMP and GMP from their common intermediate, IMP, depends on the levels of ATP and GTP; that is, the adenine nucleotides control the synthesis of guanine nucleotides, and the guanine nucleotides control the extent to which the adenine nucleotides are synthesized.

It should be emphasized that two of the steps in purine nucleotide biosynthesis are *indirectly* sensitive to aminopterin and methotrexate, both of which are antileukemic agents (see p. 562) that primarily inhibit dihydrofolate reductase.

Pyrimidine nucleotide synthesis

The rate-limiting steps in formation of carbamoyl phosphate Unlike synthesis of the purine nucleotides, the pyrimidine ring is formed before the β-D-ribose 5-phosphate moiety is attached. Yet both pathways are similar in that the initial reaction controls the whole sequence of the following reactions. In pyrimidine synthesis this first reaction is catalyzed by carbamoyl phosphate synthetase II, a cytoplasmic enzyme that uses the amide group of glutamine as the nitrogen donor. Recall that the carbamoyl phosphate synthetase I that functions in the urea cycle is mitochondrial and uses ammonium ion as the nitrogen donor. When UTP is increased above the steady-state concentration, carbamoyl phosphate synthetase II is inhibited.

Figure 13.8   Biosynthesis of purine nucleotides. (See also Figure 13.9.)

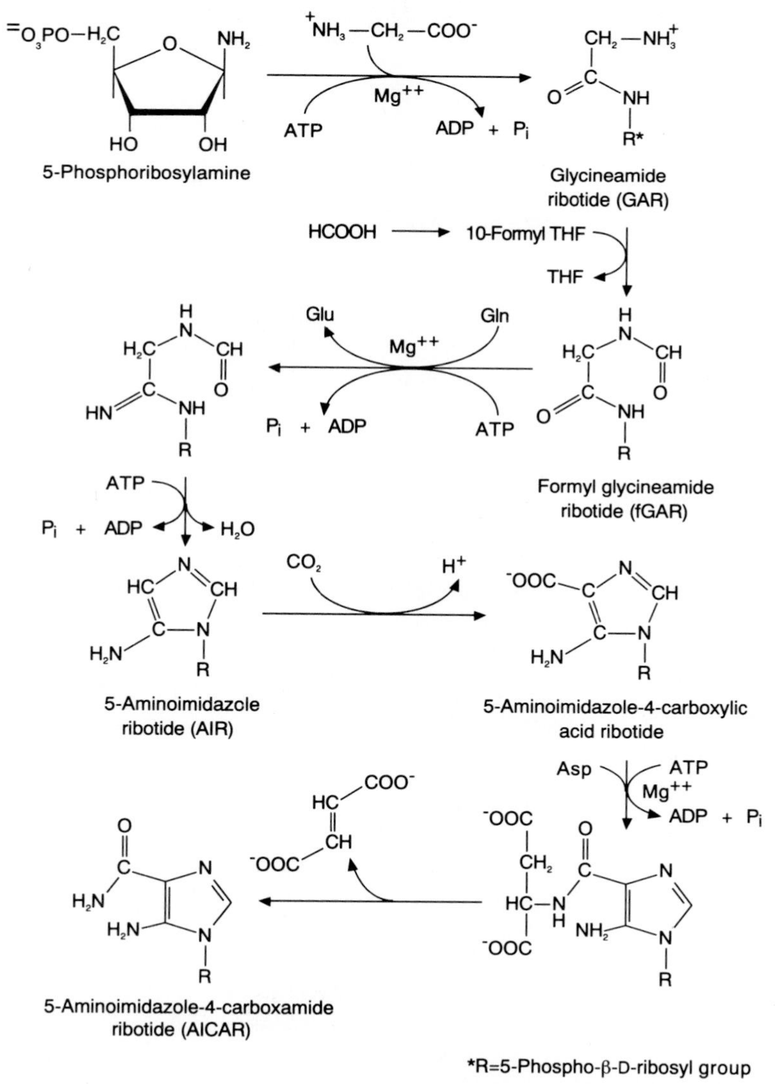

5-Phosphoribosylamine

Glycineamide ribotide (GAR)

Formyl glycineamide ribotide (fGAR)

5-Aminoimidazole ribotide (AIR)

5-Aminoimidazole-4-carboxylic acid ribotide

5-Aminoimidazole-4-carboxamide ribotide (AICAR)

*R=5-Phospho-β-D-ribosyl group

Figure 13.9 Biosynthesis of purine nucleotides.

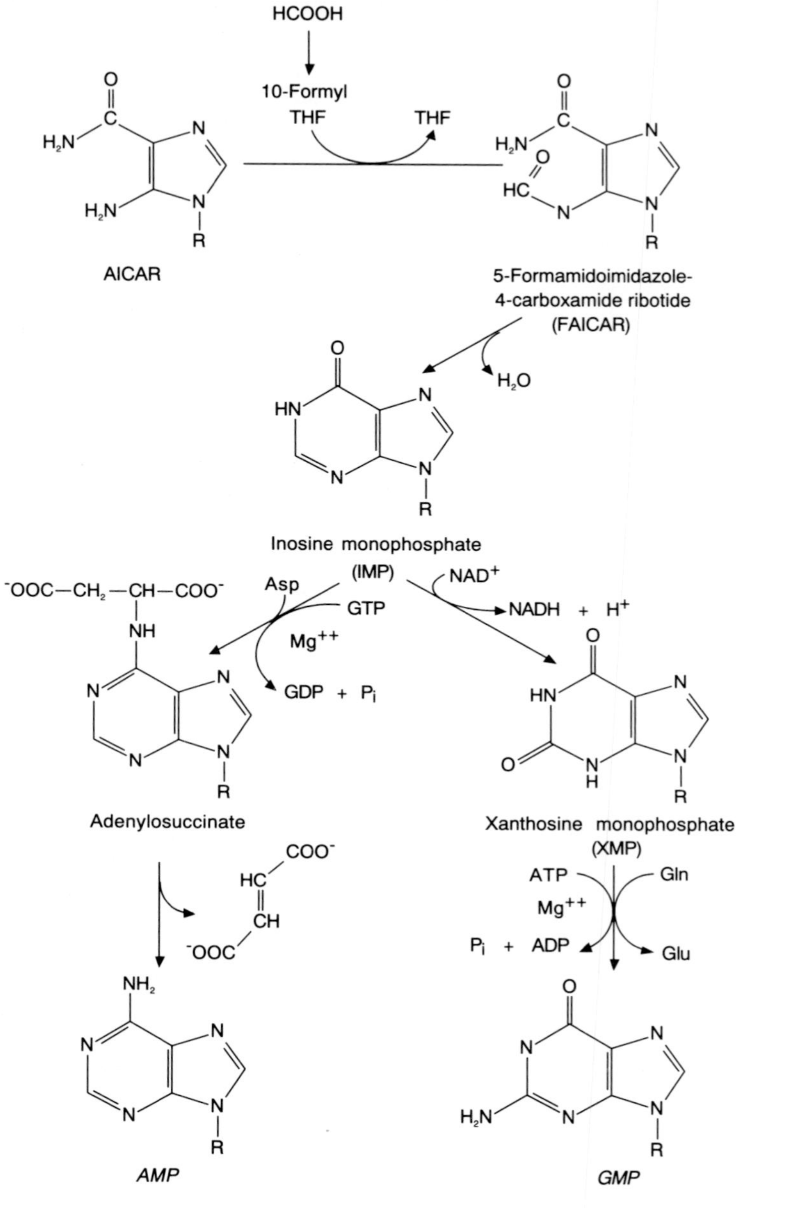

$$CO_2 + 2ATP + H_2O + \text{Glutamine} \xrightarrow[\text{Synthetase}]{\text{Carbamoyl phosphate}} H_2N{-}\overset{\displaystyle O}{\overset{\|}{C}}{-}OPO_3^{=} + 2ADP + Pi + \text{Glutamate}$$

The enzyme activities that catalyze the first three steps in pyrimidine synthesis (see also Figure 13.10) are controlled in an unusual way; they are all part of the same protein molecule. This large (215,000 Mr) molecule contains the activities of carbamoyl phosphate synthetase II, aspartate transcarbamoylase, and dihydro-orotase. This structural arrangement allows for efficiency because only limited diffusion of intermediates is needed. This enzyme, like dihydrofolate reductase, is amplified by gene duplication.

In *E. coli,* pyrimidine biosynthesis is controlled by the second enzyme in the pathway, aspartate transcarbamoylase. This allosteric enzyme is regulated in a novel way. It contains a regulatory protein subunit as well as a catalytic subunit; the catalytic subunit can function in vitro in the absence of the regulatory subunit. When both subunits exist together, as they do in vivo, the regulatory subunit, by binding CTP, can decrease the affinity with which aspartate binds to the catalytic subunit, thus slowing the rate of pyrimidine synthesis.

Clinical comment

Orotic aciduria: genetic and acquired forms of the disease As seen in Figure 13.10, orotic acid is an intermediate in pyrimidine nucleotide synthesis. This intermediate accumulates in the urine of individuals with a rare hereditary metabolic disease, *orotic aciduria*. Individuals who inherit this disease lack orotidylate pyrophosphorylase and orotidylate decarboxylase activities, because both activities are catalyzed by the same polypeptide. An acquired form of orotic aciduria mayoccur in patients receiving the antineoplastic agent 6-azauridine. When converted to 6-azauridylic acid, this drug is a competitive inhibitor of the decarboxylase activity. The inherited disease has been treated by oral administration of uridine. In spite of the poor use of dietary nucleosides, it appears that enough uridine is absorbed to satisfy the patient's requirements for pyrimidine nucleotides.

Salvage synthesis The importance of using pyrimidine derivatives from the diet has not been established. Some dietary uracil can be used for UMP synthesis, but high concentrations are required. The enzyme uracil phosphoribosyl transferase uses uracil and PRPP to produce UMP. Although dietary uridine is better used in its conversion to UMP by a uridine kinase, high concentrations are again required.

The first pyrimidine nucleotide synthesized is UMP, and all the others are derived from this common precursor. The sequence of these transformations can be illustrated as follows:

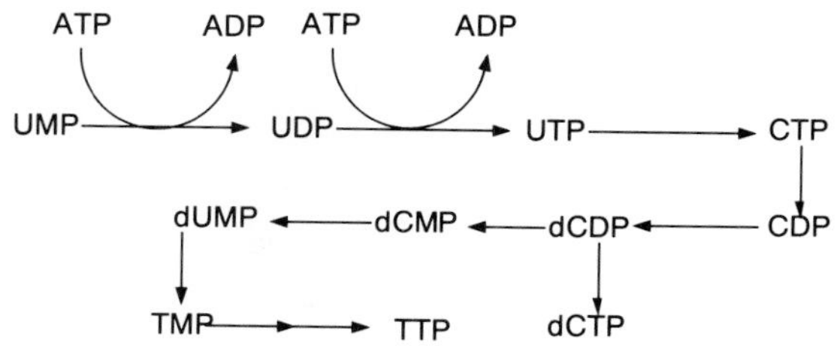

Kinases are required to convert UMP to UTP. The amination of UTP is catalyzed by the enzyme CTP synthetase. In animal cells the amino donor for the CTP synthetase reaction is the amide group of glutamine (Figure 13.11), whereas the bacterial enzyme uses ammonium ion.

Deoxyribonucleotide synthesis

All the deoxyribonucleotides are synthesized from ribonucleotides. The structures of the four deoxyribonucleotides are the same as those of the ribonucleotides except for the different sugar. In addition, thymidylic acid, which is present only in DNA, is formed by the methylation of deoxyuridine monophosphate. The formation of the deoxyribonu-

Figure 13.10   Biosynthesis of uridine 5'-monophosphate.

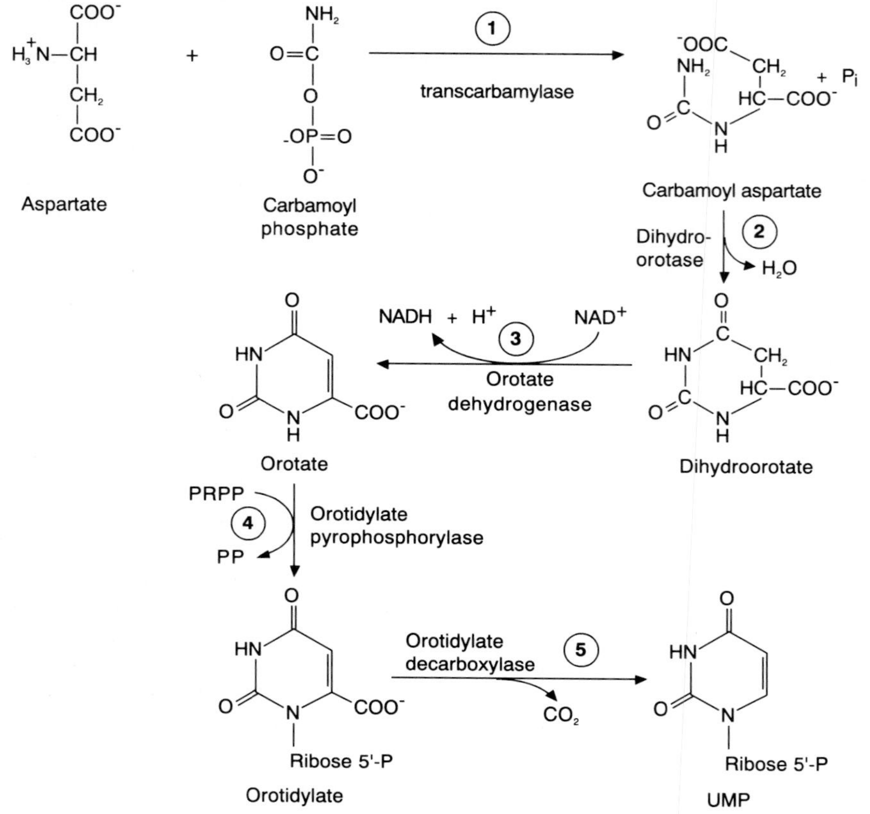

a. **Steps 1, 2, and 3:**
Catalyzed by the
same enzyme
polypeptide.

b. **Steps 4 and 5:**
Catalyzed by the
same enzyme
polypeptide.

Figure 13.11   Synthesis of cytidine triphosphate.

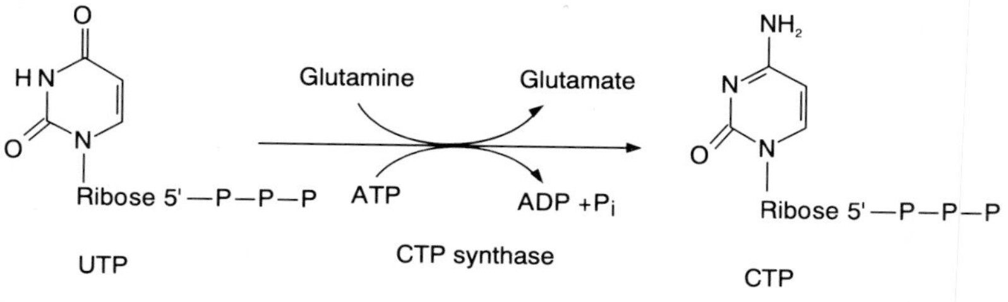

cleotides from ribonucleotides is shown in Figure 13.12. The enzyme ribonucleotide reductase requires ribonucleoside diphosphates as substrates. However, it has loose specificity with respect to the particular base, for ADP, GDP, CDP, and UDP are all reduced by the enzyme. The proximal reducing agent is *thioredoxin,* a small protein that contains two free sulfhydryl groups positioned in such a way that a disulfide bond can be easily formed between them. A separate enzyme, thioredoxin reductase, uses NADPH to regenerate reduced thioredoxin.

A rather unusual ribonucleotide reductase has been found in certain lactobacilli and a few other species. These enzymes contain a cobamide coenzyme and use ribonucleoside triphosphates instead of the diphosphates. Ribonucleotide reductases from animal cells and *E. coli,* however, do not contain B_{12} coenzymes. Interestingly, thioredoxin from *E. coli* can be used by the ribonucleotide reductases isolated from either lactobacilli or animal cells. Ribonucleotide reductase is inhibited by hydroxyurea, a substance sometimes used in cancer chemotherapy.

Thymidylate synthesis

The precursor of thymidylic acid is dUMP, which in *E. coli* is synthesized from dUTP. However, in regenerating rat liver the enzyme dCMP deaminase is induced slightly before dTMP formation, a finding that suggests a role of this enzyme in the synthesis of thymidylic acid in animal cells. This enzyme may be one of the rate-limiting reactions in DNA synthesis because it must be induced. Figure 13.13 illustrates the sequence of reactions from dCMP to dTMP. The methylation of dUMP is catalyzed by thymidylate synthase and requires 5,10-methylene THF as the one-carbon donor. However, this folate derivative is at the oxidation level of formaldehyde (Figure 13.7) and must be further reduced to form the methyl group. The reducing agent in this case is THF; the oxidized product, DHF, is converted to THF by DHF reductase (Figure 13.6). It is primarily at this reaction that antifolates are most effective against cancer, because for each thymidylate molecule synthesized, of which there are many in fast-growing tissue, one molecule of DHF must be reduced to THF.

Figure 13.12   Synthesis of deoxyribonucleotides.

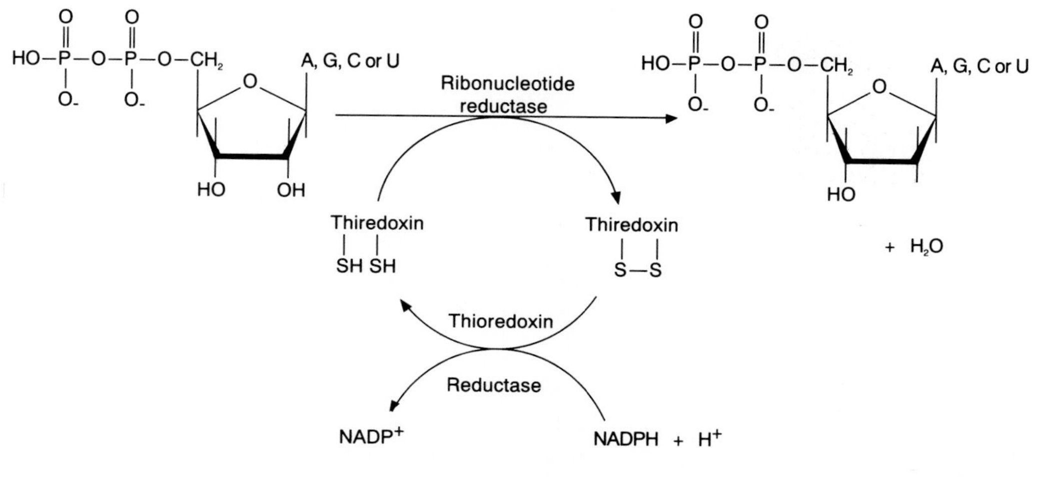

Figure 13.13   Synthesis of thymidine 5′-monophosphate.

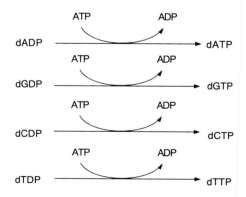

The immediate precursors for DNA synthesis are the deoxyribonucleoside 5′-triphos-phates, which are formed by phosphorylating the products of the ribonucleotide reductase–catalyzed reactions.

Analogues of
pyrimidines and
purines
Pyrimidine analogues

Several analogues of pyrimidines or of their ribonucleosides or deoxyribonucleosides have
been used in cancer chemotherapy. Because the fluorine atom is approximately the size
of a hydrogen atom, 5-fluorouracil is usually thought of as a uracil analogue, whereas
bromodeoxyuridine and trifluorothymidine have larger groups at the 5'-position and are
thymidine analogues:

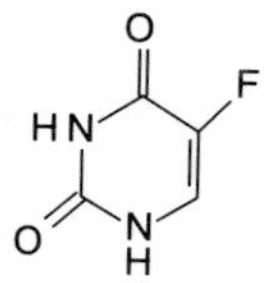

| 5-Fluorouracil | 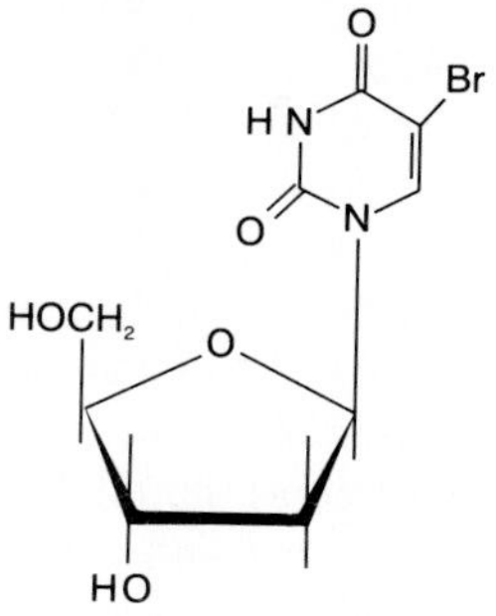 Bromodeoxyuridine | 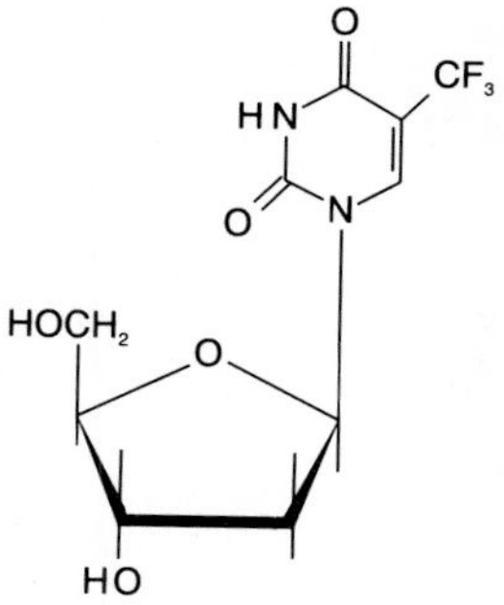 Trifluorothymidine |

Other pyrimidine analogues useful in chemotherapy are β-D-arabinofuranosylcytosine and
azauridine.

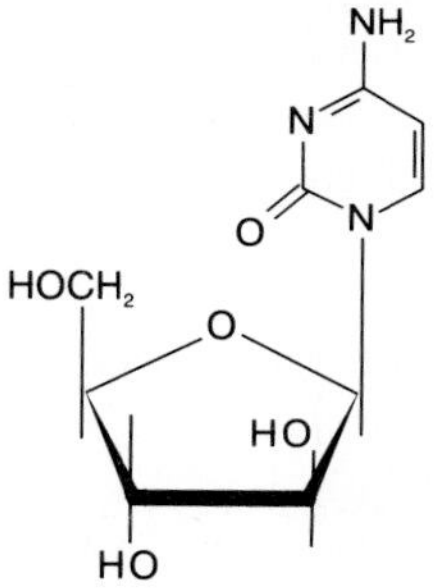

β-D-Arabinofuranosylcytosine
(Ara-C)

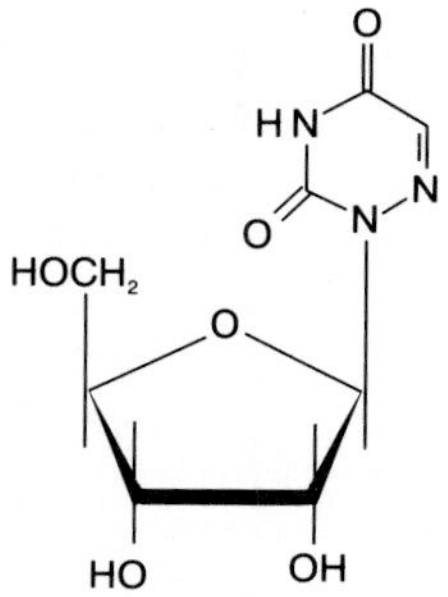

Azauridine

Purine analogues

Examples of purine analogues are 6-thioguanine and 6-mercaptopurine.

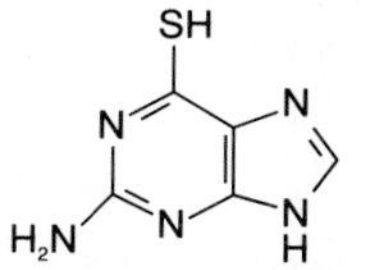

6-Thioguanine

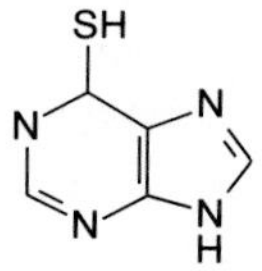

6-Mercaptopurine

**Conversion of
analogues to analogue
nucleotides**

Pyrimidine and purine analogues are best utilized as the free bases or as nucleosides,
because the phosphate derivatives are transported poorly across membranes. Also, the
phosphate group is rapidly hydrolyzed by cell surface enzymes. Consequently, these
antimetabolites must be converted to their nucleotide derivatives if they are to be toxic
to the rapidly growing cancer cells. In the case of the purine analogues, conversion to
the nucleotide derivative is catalyzed by the enzymes hypoxanthine-guanine phosphoribosyltransferase or adenine phosphoribosyltransferase (see case 1 at end of the chapter).
In the case of the pyrimidine analogues, uracil phosphoribosyltransferase serves to attach
the ribose phosphate moiety. In all cases the ribose phosphate donor is PRPP.

The action of 5-fluorouracil The mechanism by which these analogues are effective in cancer chemotherapy is understood only in part. Generally the analogues are effective because nucleic acid synthesis in cancer cells is more rapid or extensive than it is in normal cells. Often DNA synthesis is primarily inhibited, but the analogues can also block RNA synthesis. For example, it was thought for several years that 5-fluorouracil functioned by inhibiting DNA synthesis. The 5-fluorouracil was known to be converted to 5-fluorodeoxyuridylic acid, dFUMP, which inhibits thymidylate synthase by acting as a structural analogue of dUMP. Although these reactions can be easily demonstrated in the laboratory, they do not seem to account for the effectiveness of 5-fluorouracil in cancer chemotherapy, because administration of both thymidine and 5-fluorouracil is more effective than administration of 5-fluorouracil alone. The specific toxicity of the analogue to cancer cells is believed to result from the incorporation of 5-fluorouracil into RNA. The fluorouracil-containing RNA is apparently more detrimental to cancer cells than it is to normal cells, whereas the inhibition of thymidylate synthase by dFUMP shows no selectivity between normal and cancer cells. Thus the simultaneous administration of thymidine protects both normal and cancer cells, but the incorporation of 5-fluorouracil into RNA is more toxic to cancer cells.

Biosynthesis of nucleotide-containing coenzymes
Coenzyme A

Many coenzymes are derived from vitamins provided by the diet. All the B vitamins must be converted to coenzymatic forms before they can function in enzymatic reactions. Many of these coenzymes are produced in reactions involving ATP; for example, the vitamin pantothenic acid undergoes several reactions with ATP to form CoA. Figure 13.14 outlines these reactions.

FAD and FMN

Similarly, ATP is required to phosphorylate riboflavin in order to form flavin mononucleotide (FMN) or riboflavin phosphate:

$$ATP + Riboflavin \rightarrow Riboflavin\ phosphate\ (FMN) + ADP$$

ATP is also required for the synthesis of flavin adenine dinucleotide (FAD), but in this transformation the AMP moiety is donated to the riboflavin phosphate to form the dinucleotide:

$$ATP + FMN \rightarrow FAD + PP_i$$

NAD, NADP, TPP and pyridoxal phosphate

The synthesis of NAD^+ requires that the vitamin niacin be first converted to nicotinic acid nucleotide by adding a ribose phosphate. An adenylate group is then added and the carboxyl of the nicotinic acid moiety is amidated to yield NAD^+:

$$H_2O + Niacin \rightarrow Nicotinic\ acid + NH_3$$
$$Nicotinic\ acid + PRPP \rightarrow PP_i + Nicotinic\ acid\ nucleotide$$
$$Nicotinic\ acid\ nucleotide + ATP \rightarrow Deamido\text{-}NAD$$
$$Deamido\text{-}NAD + Glutamine + ATP \rightarrow P_i + Glutamate + NAD^+$$

As might be expected, $NADP^+$ is synthesized from NAD^+, and the phosphoryl group is derived from ATP; ADP is the other product. In similar reactions, thiamine pyrophosphate and pyridoxal phosphate are formed from their vitamin precursors by phosphorylations using ATP.

Biotin and lipoic acid

As with fatty acids, biotin and lipoic acid must first be activated to acyladenylates. However, instead of being transferred to CoA, the vitamins are attached in covalent linkage to the ε-amino groups of specific lysyl residues contained in their respective apoenzymes.

Figure 13.14   Biosynthesis of coenzyme A.

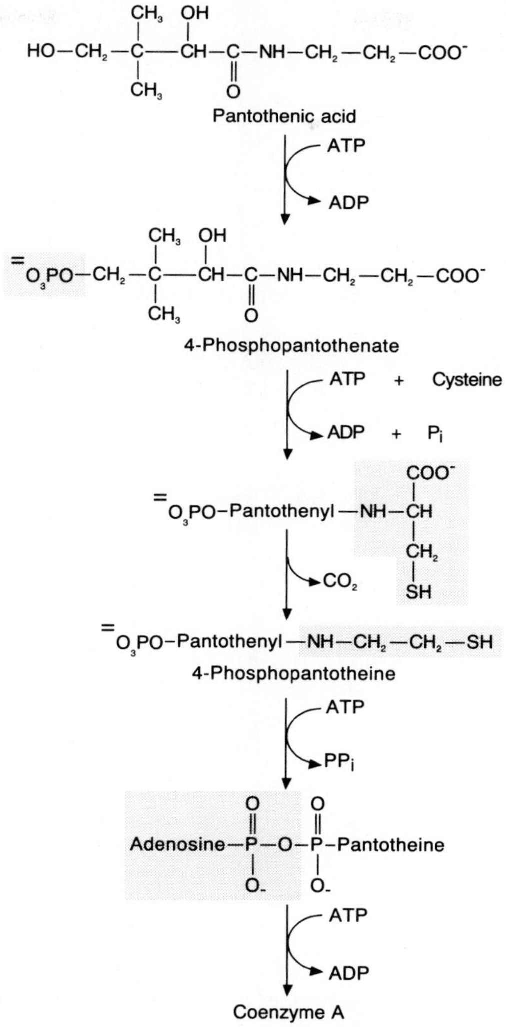

$$\underset{\substack{\displaystyle |\\ CH_3}}{HO-CH_2-\overset{\displaystyle CH_3}{\underset{\displaystyle |}{C}}-\overset{\displaystyle OH}{\underset{\displaystyle |}{CH}}-\overset{\displaystyle |}{\underset{\displaystyle O}{C}}-NH-CH_2-CH_2-COO^-}$$

Pantothenic acid

ATP → ADP

$$\underset{\substack{\displaystyle |\\ CH_3}}{^=O_3PO-CH_2-\overset{\displaystyle CH_3}{\underset{\displaystyle |}{C}}-\overset{\displaystyle OH}{\underset{\displaystyle |}{CH}}-\overset{\displaystyle |}{\underset{\displaystyle O}{C}}-NH-CH_2-CH_2-COO^-}$$

4-Phosphopantothenate

ATP + Cysteine → ADP + P_i

$$^=O_3PO-Pantothenyl-NH-\overset{\displaystyle COO^-}{\underset{\substack{\displaystyle |\\ CH_2\\ |\\ SH}}{CH}}$$

→ CO_2

$$^=O_3PO-Pantothenyl-NH-CH_2-CH_2-SH$$

4-Phosphopantotheine

ATP → PP_i

$$Adenosine-\overset{\displaystyle O}{\underset{\displaystyle O_-}{P}}-O-\overset{\displaystyle O}{\underset{\displaystyle O_-}{P}}-Pantotheine$$

ATP → ADP

Coenzyme A

Biotin + ATP → Biotin-adenylate + PP_i

Biotin-adenylate + Apoenzyme → AMP + Holoenzyme

Lipoate + ATP → Lipoate-adenylate + PP_i

Lipoate-adenylate + Apoenzyme → AMP + Holoenzyme

Catabolism of nucleotides
Purines

The ribonucleotides and deoxyribonucleotides derived from the hydrolysis of nucleic acids are catabolized to form the corresponding sugar, phosphate, and purine and pyrimidine bases. In humans and other primates the purine bases are catabolized to uric acid

Figure 13.15 Catabolism of purines.

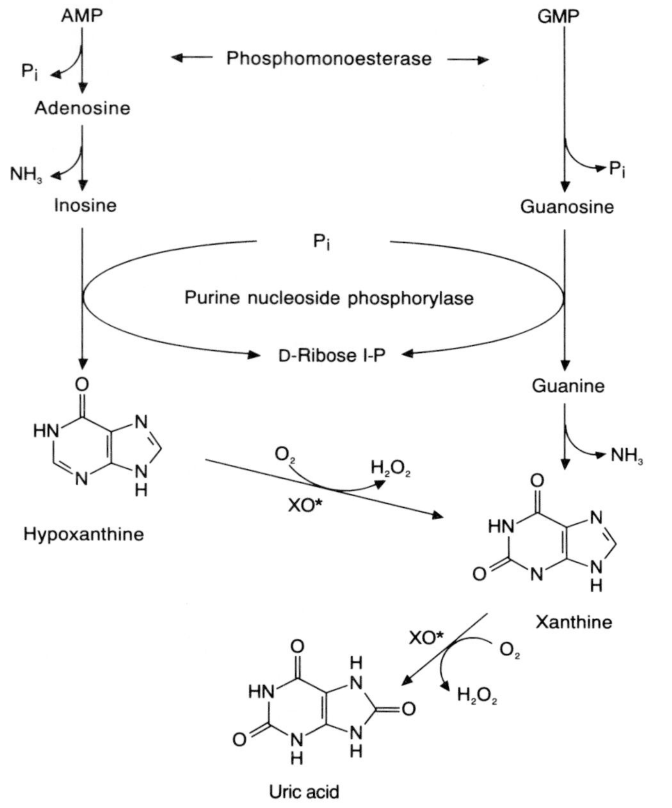

(Figure 13.15). Exactly how the nucleic acids are broken down intracellularly has not yet been fully elucidated. However, the available evidence suggests that the pathways from the mononucleotides to uric acid follow the scheme presented in Figure 13.15.

Phosphomonoesterases convert AMP and GMP, or their analogous derivatives, to the respective nucleosides. Adenosine deaminase catalyzes the synthesis of inosine. The purine ribonucleosides or deoxyribonucleosides are converted to the free bases by the enzyme purine nucleoside phosphorylase. Mechanistically, this enzyme acts like glycogen phosphorylase, as it removes a sugar 1-phosphate derivative by a phosphorolytic reaction that uses inorganic phosphate.

The bases guanine and hypoxanthine have two fates; they may be reconverted to their 5'-ribonucleotides (see case 1) or they may be converted to xanthine (Figure 13.15). The oxidation of xanthine to uric acid and the oxidation of hypoxanthine to xanthine are catalyzed by xanthine oxidase, an enzyme found in liver, and the intestinal mucosa. This enzyme is a metalloflavoprotein that contains FAD, molybdenum, and nonheme iron. Electrons from the substrates are passed to molybdenum, FAD, iron, and finally molecular oxygen, which is reduced to hydrogen peroxide.

Figure 13.16   Catabolism of pyrimidines.

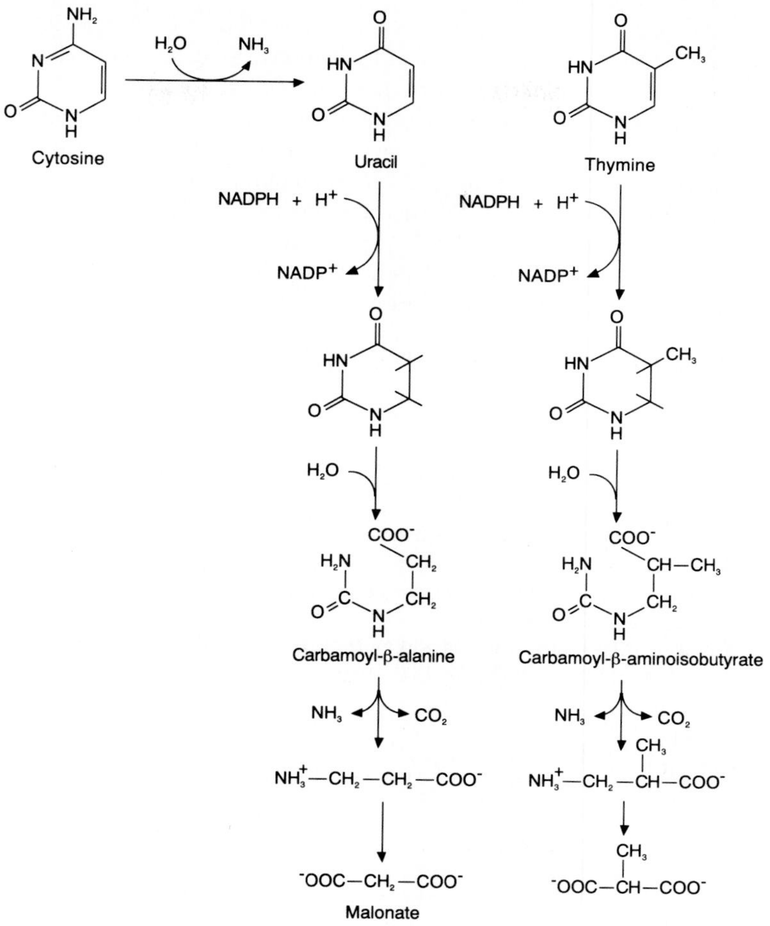

Cytosine → (H₂O, NH₃) → Uracil   Thymine

Carbamoyl-β-alanine   Carbamoyl-β-aminoisobutyrate

Malonate

Clinical comment   **Gout** Increased levels of uric acid in the blood and urine are characteristic of the disease known as gout. However, high levels of uric acid excretion also may occur in various groups of normal individuals, for example, those with a large body surface area, those of certain ancestries, and some persons in high-stress positions. (See case 1 for a more detailed analysis of the biochemical defects in gout.)

Pyrimidines   The pyrimidine bases are catabolized in the liver through reduction by NADPH. As shown in Figure 13.16, the catabolic enzymes operate on pyrimidine bases that do not contain ring amino groups. Although the diagram shows only cytosine as being deaminated, it is possible that deamination reactions also occur with cytidine and cytidylic acid. The reduced rings are opened in a hydrolytic reaction to form carbamoyl derivatives. This reaction is much like the reverse of the reactions involved in pyrimidine synthesis except that the intermediates are not derivatives of orotic acid. The carbamoyl groups are eliminated as ammonia and carbon dioxide, yielding β-alanine in the case of uracil or cytosine and β-aminoisobutyrate in the case of thymine. These amino acids are transaminated to yield the corresponding aldehydes, which in turn are oxidized to malonate and methylmalonate, respectively.

When the diet is high in DNA-containing foods, the intermediate of thymine catabolism, β-aminoisobutyrate, is excreted in the urine. Some individuals are genetically constituted so that even on normal diets they excrete large amounts of β-aminoisobutyrate. They probably have a defect in one of the enzymes that convert this substrate to methylmalonate. As might be expected, excretion of β-aminoisobutyrate is also high in leukemia patients and in patients undergoing radiation therapy; in both conditions there is an abnormally high rate of cell destruction and therefore pyrimidine catabolism.

The excretory products derived from the pyrimidines are more soluble than those produced by purine degradation. This is probably the reason that there is no disease associated with excessive pyrimidine breakdown, which would be analogous to gout.

Bibliography

Adams RLP et al: The biochemistry of the nucleic acids, ed 10, London, 1986, Chapman & Hall.

Blakley RL and Benkovic SJ: Folates and pterins, vols 1 and 2, New York, 1985, John Wiley & Sons, Inc.

Reichard P: Interactions between deoxyribonucleotides and DNA synthesis, Ann Rev Biochem 57:349, 1988.

Schimke RT: Methotrexate resistance and gene amplification: mechanisms and implications, Cancer 57:1912, 1986.

Scriver CR et al, editors: The metabolic basis of inherited disease, ed 6, New York, 1989, McGraw-Hill Book Co.

Wilson JM, Young AB, and Kelley WN: Hypoxanthine-gaunine phosphoribosyltransferase deficiency, N Engl J Med 309:900, 1983.

Clinical examples

A diamond (♦) on a case or question indicates that literature search beyond this text is necessary for full understanding.

Case 1	Gout

A 61-year-old member of the medical faculty was awarded an honorary degree 4 days before admission to the hospital. Following the award ceremony he spent the evening with friends where, as he put it, the "conviviality flowed extensively." The next morning he noticed a dull pain in the upper left flank, which worsened until hospitalization was required.

Physical examination revealed no overt signs of disease. A urine sample obtained on admission had a pH of 4.5 and was positive for protein. Microscopic examination of the centrifugal sediment from the sample revealed fine crystalline material and numerous casts. A 24 hr urine sample contained 115 mg of protein and 1.52 g (9 mmol) of uric acid. The serum uric acid content was 0.70 mmol/L (11.8 mg/dl).

Biochemical questions

1. Could useful information be obtained from a family history of gout? Explain.
2. Are purines and pyrimidines required in the diet? Explain.
3. What foods are high in purines and pyrimidines?
4. Would you expect a diet high in protein to be harmful to this patient? Explain.
5. What is the ionic form of uric acid in this patient's urine? What is its solubility?
6. What is the biochemical basis for the action of drugs used in the treatment of gout?

Case discussion   Gout is associated with either increased formation of uric acid or its decreased renal excretion. Its incidence is relatively high, occurring in about 0.3% of the population.

1. Family history Gout is classified into two broad types: *primary* and *secondary*. Primary gout, of which there are several subtypes, is inherited. The familial incidence of all cases of gout may be as high as 75% to 80%. Secondary gout is brought on by a variety of disorders such as leukemia (increase in leukocytes) and polycythemia (increase in RBC mass) or by antimetabolites used in the treatment of cancer. Primary gout is most often found in men over 30 yr of age. When women are affected, the age of onset is usually postmenopausal. Secondary gout occurs in both sexes and at younger ages.

Most cases of primary gout are idiopathic. There are probably several genetic forms of the disease, some of which may be polygenic. For example, individuals with glucose 6-phosphatase deficiency develop a glycogen storage disease; because they cannot make glucose from phosphorylated sugars, hypoglycemia develops. As a consequence, lactic acid and ketone bodies, such as β-hydroxybutyrate, build up increasing acid concentrations. Tubular urate secretion is inhibited, and hyperuricemia and gout result. Whatever the cause, gout is associated with hyperuricemia (although hyperuricemia is not always associated with gout). The disease produces a painful arthritis, particularly in the joints of the extremities.

Glutathione reductase Some people with an autosomal dominant variant of glutathione reductase develop gout. This variant has *increased* activity as compared to the normal enzyme. Glutathione is a widely distributed tripeptide, γ-glutamylcysteinylglycine. The peptide functions in the disulfide-sulfhydryl exchange reactions that maintain cellular proteins in their reduced forms. It also protects protein sulfhydryl groups by serving as a substrate for glutathione peroxidase, an enzyme that removes the hydrogen peroxide formed in certain oxidase reactions (Figure 13.15). In the absence of glutathione, hydrogen peroxide accumulates. Glutathione is kept in the reduced state by the reaction shown below, which is catalyzed by glutathione reductase. People with the superactive variant of glutathione reductase rely more heavily on the pentose phosphate pathway to provide the NADPH used by the reductase. As a result, this leads to increased production of ribose 5-phosphate, and 5-phosphoribosyl pyrophosphate (PRPP) is produced in excess. PRPP then drives the excessive synthesis of purine nucleotides, which must be catabolized to uric acid.

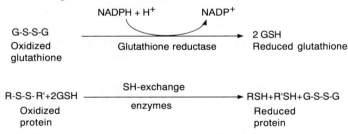

Defects in purine metabolism Most cases of primary gout are caused by excessive purine synthesis rather than increased purine nucleotide breakdown. Consistent with this view is the observation that some patients with gout have PRPP amidotransferases (see p. 565) that are resistant to feedback inhibition by purine nucleotides. The regulatory sites on this enzyme are like those of other allosteric enzymes in that they are separate from the catalytic sites. Consequently, a defect in a regulatory site as a result of a mutation could lead to the overproduction of purines as seen in gout.

Gout and Lesch-Nyhan syndrome There is a tremendous overproduction of purines in the Lesch-Nyhan syndrome. This severe X-linked disease has a very early age of onset and is characterized by extremely aggressive behavior that generally leads to self-mutilation. The defect in the Lesch-Nyhan disease is in the gene for the enzyme hypoxanthine-guanine phosphoribosyltransferase (HPRT). The reaction catalyzed is shown below. Initially, this reaction was considered to function as a minor "salvage" pathway that permitted reutilization of purine bases that otherwise would be oxidized. However, the severity of the Lesch-Nyhan syndrome suggests a more important role for the phosphoribosyltransferase. Very likely the enzyme has an essential role in nonhepatic tissues where de novo synthesis of purines occurs at a very low rate. The nonhepatic tissues contain the phosphoribosyltransferases but depend on circulating purine bases or nucleosides derived from the liver. They take up the circulating purines and, through the action of HPRT, form nucleotides. A similar enzyme, adenine phosphoribosyltransferase (APRT), produces adenine nucleotides (see case 12).

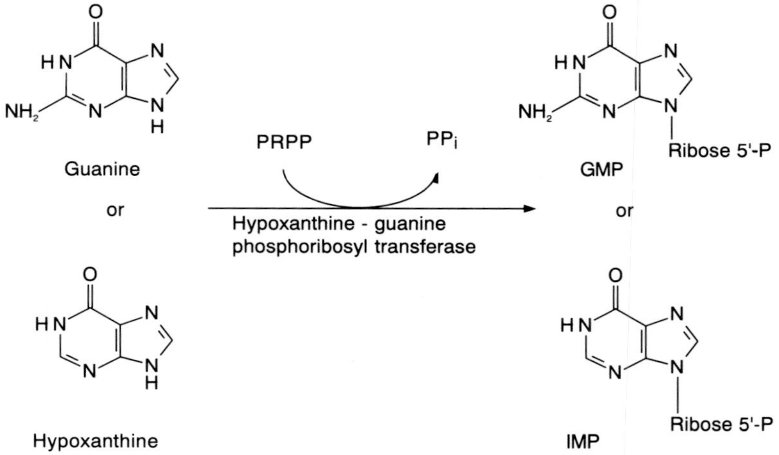

Figure 13.17 summarizes the catabolic interconversions of the purines as well as their ultimate conversion to uric acid. Humans can deaminate adenosine to inosine but not adenine to hypoxanthine. In addition, the purine nucleoside phosphorylase favors guanosine and inosine over adenosine. Consequently, all the catabolized purines are funneled through hypoxanthine and guanine en route to uric acid. Recall that the biosynthesis of the purine nucleotides is very sensitive to the levels of the GMP and IMP pools. This sensitivity occurs at the level of the PRPP amidotransferase reaction, the first enzyme in the purine pathway. In the Lesch-Nyhan syndrome, excessive purine synthesis might result from two effects. The levels of GMP and IMP might drop as a result of the inability of the defective HPRT to "salvage" guanine and hypoxanthine; consequently, the PRPP amidotransferase might respond by producing excessive amounts of 5-phospho-D-ribosylamine. Alternatively, the defective HPRT might cause PRPP to accumulate and become available for the stimulation of the PRPP amidotransferase. In either case, abnormally high levels of 5-phospho-D-ribosylamine would result, and purine production would be excessive. The available evidence seems to favor the second interpretation—that involving PRPP accumulation.

The lack of HPRT activity in patients with the Lesch-Nyhan syndrome is virtually complete. Because some of these patients also develop gouty arthritis, and because this disease is X-linked recessive and thus limited to males, a search was made to find patients with gout but not the severe neurologic symptoms of the Lesch-Nyhan syndrome. It was thought that their levels of HPRT might be between the normal values and those found in patients with the Lesch-Nyhan syndrome. Such patients have been found, but it is unlikely that they represent a significant percentage of the total

Figure 13.17   Summary of purine metabolism.

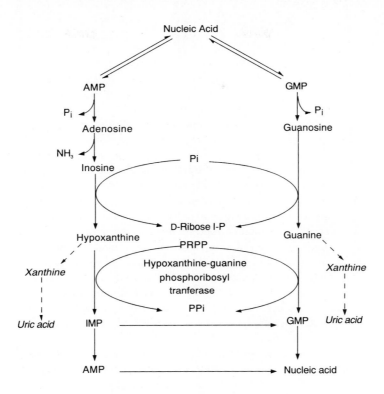

number of patients with primary gout. This emphasizes the fact that primary gout is probably inherited in a number of ways.

2. Diet Gout has been associated with wealthy and talented people. For example, noteworthy sufferers include the Medici, Charles IV, Samuel Johnson, Benjamin Franklin, Martin Luther, John Calvin, Isaac Newton, and Charles Darwin. In the traditional view, gout is associated with rich foods and high living. A rich diet and an excessive life-style probably do contribute to the expression of the disease in genetically susceptible individuals. However, diets high in protein are consumed in such a broad spectrum of today's population that this appears to be a less important causative factor in the twentieth century than it might have been in the past.

Animal experiments have shown that few of the purines derived from dietary nucleic acids are incorporated into the animal's nucleic acids. Xanthine oxidase, the enzyme that converts hypoxanthine to xanthine and xanthine to uric acid, is found in the liver and intestinal mucosa. Most of the dietary purine bases are converted to uric acid by intestinal xanthine oxidase. Thus a diet that includes 4 g of yeast RNA per day will raise blood urate levels to those found in gout.

3. Foods rich in nucleic acids A sustained intake of large amounts of food rich in nucleic acids, for example, sweet breads, liver, yeast, anchovies, kidneys, and sardines, could raise the serum urate levels to over 0.4 mmol/L (7 mg/dl). On the other hand, a diet low in purines would decrease the urate level by only about 60 μmol/L (1 mg/dl), because most of the purines in the body are synthesized de novo.

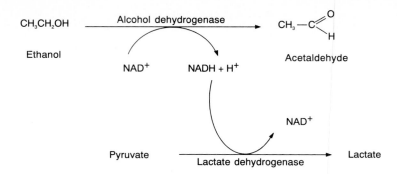

4. High protein diets In treating the disease a diet adequate but not high in protein is advised. High-protein foods might also contribute to an accelerated synthesis of purines, because several amino acids are required for their synthesis. Obesity and dehydration should also be avoided. Obesity contributes to the urate level because of the excessive amount of food required to maintain body weight.

Dehydration should be avoided, because urate crystallization is very dependent on concentration. Alcohol causes diuresis leading to dehydration. A high rate of alcohol metabolism might contribute to a lactic acidosis, thus causing the precipitation of sodium urate crystals in the joints. Alcohol dehydrogenase promotes the production of NADH, which is used to form lactate from pyruvate. The resultant lactic acidemia suppresses the tubular secretion of uric acid, which is highly insoluble in its undissociated form.

5. Ionic forms of uric acid Crystals of sodium urate initiate a series of reactions that produce the inflammation associated with acute gouty arthritis. The first step is probably the activation of the Hageman factor (see Table 3.2), a protease zymogen. In a series of poorly understood reactions the activated protease causes leukocyte infiltration and phagocytosis of the urate crystals. The phagocytosis causes lysosome destruction and the subsequent release of hydrolytic enzymes, which causes destruction of tissue and inflammation. The urate crystals found in the fluid of the joints are at a nearly neutral pH. Accordingly, they are composed of monosodium urate, because the pKa's of uric acid are 5.7 and 10.3. On the other hand, renal stones are more likely to be uric acid crystals because the urine may have a pH of 4.5 to 5.0.

6. Treatment of gout Treatment of gout can differ with the particular case, depending on whether it is acute or chronic. Details of treatment are beyond the scope of this discussion, but it is useful to consider the biochemical action of two drugs.

Colchicine Colchicine has been used in the treatment of gout for many years. It is highly specific for this condition, but its mechanism of action is understood only in part. The drug binds to the subunits of proteins that make up the microtubules of polymorphonuclear leukocytes. The disaggregation of the microtubules inhibits leukocyte locomotion, adhesiveness, and metabolism during phagocytosis. Although colchicine produces clinical improvement, it does not affect serum urate levels.

Allopurinol The drug that most effectively inhibits the formation of urate is allopurinol, a competitive inhibitor of xanthine oxidase. Hypoxanthine and xanthine, which are more soluble than uric acid, are excreted during allopurinol therapy. However, they are not excreted in amounts stoichiometric with the decrease in uric acid seen in those patients with gout who have normal phosphoribosyltransferase levels. In individuals deficient in this enzyme, hypoxanthine and xanthine are excreted instead of urate. Allopurinol, as with guanine and hypoxanthine, can be converted to its ribotide by phosphoribosyltransferase. This reaction would be expected to consume additional PRPP, and the analogue ribotide might even inhibit PRPP amidotransferase. Furthermore, the high levels of hypoxanthine resulting from inhibition of xanthine oxidase may cause phosphoribosyltransferase to reutilize the base and thus further inhibit de novo purine synthesis. Purine nucleotide synthesis is lowered regardless of the mechanism. Reducing the formation of uric acid through administration of allopurinol relieves the symptoms and decreases the possibility that uric acid kidney stones will form. However, allopurinol is a purine antagonist and must be used with caution in treating secondary gout in leukemic patients undergoing therapy with purine analogues. This is especially true when purine analogues are given orally in combination with allopurinol. The intestinal xanthine oxidase normally degrades many of the administered purine analogues, but when it is inhibited by allopurinol, more of the analogues can enter the bloodstream. This could prove dangerous, because most antimetabolites are used at very high levels, just below their toxic concentrations.

Hypoxanthine   Allopurinol

References

Becker MA: Clinical aspects of monosodium urate monohydrate crystal deposition disease (gout), Rheum Dis Clin North A 14(2):377, 1988.

Lo B: Hyperuricemia and gout (topics in primary care medicine), West J Med 142:104, 1985.

Scriver CR et al, editors: The metabolic basis of inherited disease, ed 6, New York, 1989, McGraw-Hill Book Co.

Case 2

Tropical sprue

A 35-year-old man was hospitalized 3 mo after returning from a 1 mo visit to Puerto Rico. He had a recent history of watery diarrhea, anorexia, 13.6 kg (30 lb) weight loss, and progressively severe weakness. He had a megaloblastic anemia with pancytopenia. Serum folate was 7.5 nmol/L (3.3 ng/ml) (normal, 15.9 to 45.3 nmol/L [7 to 20 ng/ml]), and serum vitamin B_{12} was 27.2 pmol/L (37 pg/ml) (normal, 111 to 663 pmol/L [150 to 900 pg/ml]).

The patient was found to have malabsorption of both folic acid and B_{12}. He was treated with tetracycline, which caused cessation of the diarrhea.

Three years later the patient was hospitalized with the same symptoms. Once again he was found to have megaloblastic changes in the bone marrow and a pancytopenia. The pharmacologic doses of folic acid and vitamin B_{12} administered during the course of absorption studies resulted in a hematologic and clinical remission.

Biochemical questions

1. What are folic acid polyglutamates? How are the glutamate residues linked together in folic acid polyglutamates?
2. Can folic acid polyglutamates be absorbed from the intestine?
3. What reaction is catalyzed by the enzyme γ-glutamyl-carboxypeptidase?
4. What would have happened if foods rich in folic acid had been given to the patient?

5. Cholic acid is not inhibitory, but deoxycholate is an inhibitor of conjugase. What does this tell you about the cause of this patient's disease?

6. How might a toxin function in this disease? What would be its source? What might the toxin be?

Case discussion

Types of sprue Sprue is a syndrome characterized initially by diarrhea, anorexia, and weight loss. The intestinal morphology as determined by x-ray analysis or biopsy is abnormal, and the absorption of a variety of nutrients such as glucose, fat, and vitamins is impaired. Megaloblastic anemia usually develops because of the inability to absorb adequate amounts of folate or vitamin B_{12}. Two types of sprue can be distinguished on the basis of their causative agents. One type is celiac disease of children and nontropical sprue of adults. This type is induced by the presence of a gluten in the diet. Gluten is present in wheat, rye, oats, and barley. Gliadin, the protein in gluten that causes the disease, is composed of over 40% glutamine residues. When gliadin is removed from the diet, celiac disease and nontropical sprue are ameliorated. The second type, tropical sprue, does not respond to a gluten-free diet but can be treated with folate and B_{12}, usually taken in combination with an antibiotic such as tetracycline.

Megaloblastic anemia The megaloblastic anemia seen in sprue is often associated with deficiencies of both folate and B_{12}. The anemia is characterized by the appearance of large immature erythrocytes in which the size of both the nuclei and the cytoplasm is increased. Maturation is arrested by a slowing of DNA synthesis in the S phase. The bone marrow accumulates abnormal precursors of the polymorphonuclear leukocytes with hypersegmented nuclei, and giant platelets appear in the peripheral blood. Almost any defect in DNA synthesis will cause megaloblastic anemia, but the usual cause is deficiency in either folate or B_{12}, or both. The lack of these vitamins is also reflected in other fast-growing cells, especially those of the gastrointestinal tract and the epithelial cells of the mouth. Glossitis, cheilosis, and stomatitis are typical signs.

Absorption of vitamin B_{12} A B_{12} coenzyme is involved in the utilization of 5-methyl THF for methionine synthesis from homocysteine (see p. 563). In B_{12} deficiency, 5-methyl THF accumulates in the serum. According to the methyl-trap hypothesis, all the body's folate becomes tied up as 5-methyl THF and is thus unavailable for other THF-requiring reactions. The administration of large doses of folate will correct the hematologic symptoms of megaloblastic anemia. However, if the folate deficiency is secondary to B_{12} deficiency, the irreversible neurologic damage caused by B_{12} deficiency will proceed undetected. Megaloblastic anemia is caused by a folate deficiency. This deficiency may be related to inadequate folate intake and absorption, or it may be secondary to a B_{12} deficiency.

Vitamin B_{12} is readily absorbed when administered subcutaneously or intramuscularly. Absorption from the gastrointestinal tract is more complex. A glycoprotein secreted by the gastric mucosa, called *intrinsic factor,* binds B_{12} and enhances its absorption by the ileum. Insufficient amounts of B_{12} are transported by other mechanisms. In the absence of intrinsic factor, B_{12} cannot be absorbed unless it is given in amounts greater than those present in the ordinary diet.

1. Folate polyglutamates Folic acid absorption is also complex. The derivatives of folate in most foods are the folate polyglutamate forms of 5-methyl THF and 10-formyl THF. Very little unmodified folate occurs in food. As many as six glutamyl residues are joined in peptide linkage through the γ-carboxyl group. Folate polyglutamates are the intracellular forms of the folate coenzymes.

Folate polyglutamate

2. and 3. Absorption of folate polyglutamates; γ-L-glutamyl-carboxypeptidase

The folate polyglutamates cannot be absorbed from the intestine until all but one of the glutamate residues are hydrolyzed off. Therefore the kind of folate present in most foods is not absorbed well in patients with tropical sprue. An intestinal γ-L-glutamyl-carboxypeptidase is the enzyme that ordinarily converts folate polyglutamate to folate. In much of the medical literature this enzyme is referred to by the trivial name *conjugase*. This enzyme is located in the brush border cells of the intestinal mucosa. Diseases that cause degeneration of the mucosa (for example, intestinal cancer, tropical or nontropical sprue) cause loss of conjugase and with it a decreased ability to absorb folate or its derivatives.

4. Dietary folate Dietary folate is used so poorly in tropical sprue that orally administered folate polyglutamate in amounts as high as 3.4 μmol (1.5 mg)/day produces no effect. However, folate lacking all but one glutamyl residue may produce hematologic responses in patients when only 56 nmol (25 μg)/day are given orally. This is the usual form of commercial folic acid preparations.

Inhibition of conjugase Some drugs and metabolites inhibit conjugase activity so severely that serum folate levels will be drastically reduced. These drugs inhibit the enzyme directly so that their effects are seen even without mucosal degeneration. Sodium sulfobromophthalein, a substance used in a liver function test, drastically lowers serum folate levels.

5. Conjugase inhibitor Certain bile salts also inhibit conjugase, and this may be important in patients with tropical sprue. In Chapter 11 it was noted that two bile acids, deoxycholate and lithocholate, are produced from cholate and chenodeoxycholate, respectively, by enzymes produced by intestinal bacteria. These two bile acids inhibit conjugase more severely than the other two bile acids.

6. Speculations on pathogenesis of tropical sprue Tropical sprue is a disease that can occur in epidemic proportions, and it responds well to vigorous antibiotic therapy. This disease also responds to folate therapy. Some investigators believe that other factors or toxins, probably of microbial origin, are causative agents of tropical sprue. Speculation as to the chemical nature of these toxins suggests that they are antifolates or unsaturated fats. Although a microbial origin for the disease has not been proved conclusively, circumstantial evidence strongly favors it.

References   Halstead CH: Intestinal absorption and malabsorption of folates, Annu Rev Med 31:79, 1980.
Said HM et al: Folate transport by human intestinal brush-border membrane vesicles, Am J Physiol 252(2):G227, 1987.
Wang TT et al: Comparison of folate conjugase activities in human, pig, rat and monkey intestine, J Nutr 115:814, 1985.

Case 3 — Defect in B_{12} coenzyme synthesis*

A male infant hospitalized from age 4 wk until his death at age 7½ wk, was found to have very low serum, brain, and liver concentrations of methionine. He had methylmalonic aciduria. Serum homocystine and cystathionine were elevated, but other amino acid concentrations were normal. The total B_{12} content of the liver and kidneys was normal, but the concentrations of 5'-deoxyadenosylcobalamin were less than 10% of controls. Folic acid was administered on the fifteenth day of hospitalization; 5 days later the serum 5-methyl THF was well above normal. Postmortem examination of extracts from the liver and kidneys were abnormally low in 5-methyl THF methyltransferase.

Biochemical questions

1. What metabolic abnormality is suggested by the low methionine and high homocystine levels? What is the coenzyme?
2. What are two well-established enzymatic reactions in animals that require B_{12}-containing coenzymes?
3. Considering the laboratory values presented, what reaction would you expect to be blocked in this infant? Explain.
4. Are these findings consistent with the methyl-trap hypothesis? Explain.

Case discussion

1. Low methionine and high homocystine levels In this patient these low levels suggest a defect related in some way to the synthesis of methionine from homocysteine (see the reaction on p. 563). Homocystine, the disulfide form of homocysteine, often is present in situations where homocysteine accumulates, because the sulfhydryl group of the latter compound is very sensitive to oxidation. Because the vitamin B_{12} findings are also abnormal, it is possible that the B_{12} coenzyme–directed synthesis of methionine is disturbed. This prediction was borne out by a decreased activity of 5-methyl THF methyltransferase. The genetic defect, however, could be in either the enzyme itself or the availability of a coenzyme essential to the reaction. Consequently, to fully understand the biochemical features of this case, it is necessary to describe the structure and functions of vitamin B_{12} coenzymes.

Most of the modern literature now designates the group of cobalt-containing compounds to which vitamin B_{12} belongs as *cobalamin*. The structure of a cobalamin is given on p. 23, but it is convenient to abbreviate it as follows:

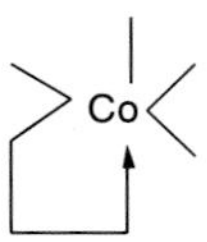

The four straight lines on the right and left sides, which are directed at the cobalt atom, represent bonds to four nitrogen atoms in the corrin ring structure. The corrin ring system has structural features in common with the porphyrin ring system but also differs in many respects. The arrow at the bottom of the cobalt atom represents a

*Case description adapted from Mudd HS et al: Biochem Biophys Res Commun 35:121, 1969.

coordinate covalent bond from a nitrogen atom of the dimethylbenzimidazole moiety. A corrinoid compound that contains dimethylbenzimidazole is called a cobalamin.

Vitamin B_{12}, as it occurs in vitamin capsules, is cyanocobalamin and is designated as:

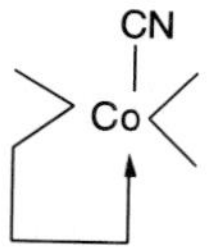

The cyanide substituent arises during the isolation of the vitamin from its natural commercial sources, usually bacteria. It is not found on cobalamin that occurs in the body and is not toxic in the usual concentrations of cobalamin taken in vitamin capsules.

The three cobalamins found in the body are designated as follows:

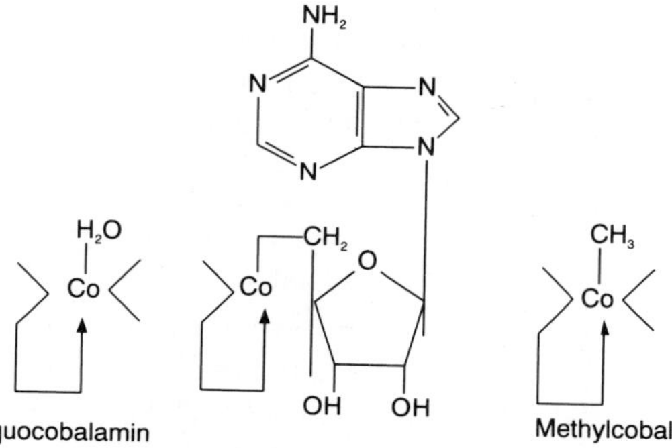

Aquocobalamin

5'-Deoxy-5'adenosylcobalamin
(Adenosylcobalamin)

Methylcobalamin

2. Coenzyme reactions Only methylcobalamin and adenosylcobalamin are coenzymes. These coenzymes are normally synthesized from dietary cyanocobalamin or aquocobalamin. Two enzymatic reactions have been conclusively shown to require cobalamin coenzymes in humans. One requires adenosylcobalamin, and the other methylcobalamin.

Adenosylcobalamin is required for the isomerization of methylmalonyl CoA to succinyl CoA. This reaction is important in using the methylmalonyl CoA synthesized by the carboxylation of propionyl CoA so that odd chain fatty acids and certain amino acids, whose catabolism yields propionate, can be metabolized. The methylmalonic aciduria seen in this case is suggestive of a defect in this 5'-deoxyadenosylcobalamin-requiring reaction.

The reaction catalyzed by 5-methyl THF-homocysteine methyltransferase contains methylcobalamin as a tightly bound prosthetic group. When isolated, the holoenzyme appears to have the cobalamin present as aquocobalamin. To activate the prosthetic group, it must be reduced with a reduced flavin and methylated with S-adenosylmethionine (S-AdoMet).

Aquocobalamin + S-AdoMet $\xrightarrow[\text{FADH}_2 \quad \text{FAD}]{}$ Methylcobalamin + S-Adenosylhomocysteine

The methylcobalamin prosthetic group can then transfer its methyl group to homocysteine and in turn be remethylated by 5-methyl THF:

1. Methylcobalamin + Homocysteine → Methionine + Reduced cobalamin (cob[I]alamin)
2. Cob[I]alamin + 5-methyl THF → Methylcobalamin + THF

Although the transferase can catalyze many cycles of methyl transfer in this way, the prosthetic group occasionally reverts to an inactive form, again necessitating further reactivation by $FADH_2$ and S-AdoMet. The chemical details of this inactivation-reactivation process are obscure, but it is probably related to the very marked nucleophilic properties of the cob[I]alamin intermediate that occasionally cause the latter to undergo a side reaction.

3. The blocked reaction Considering these facts, the data in this case indicate a defect in the conversion of cobalamin to both its coenzymatic forms. Total cobalamin content of liver and kidneys was normal, but $5'$-deoxyadenosylcobalamin was low; thus the methylmalonic aciduria resulted. Although methylcobalamin was not determined in this case, a similar more recent case showed it to be low. Thus the low activity of the 5-methyl THF methyltransferase was probably caused by the lack of methylcobalamin. The reactions involved in the conversion of aquocobalamin to cobalamin coenzymes are not well understood in the human. Probably the defect in this case is in a reaction that produces an intermediate common to the synthesis of both coenzymes. This intermediate (see previous numbered equation) is probably a reduced cobalamin, and the reducing agent is likely $FADH_2$.

4. The methyl-trap hypothesis Even though the lack of activity of 5-methyl THF-homocysteine methyltransferase is secondary to a primary defect in the synthesis of cobalamin methyl, these results are in accord with the methyl-trap hypothesis. Thus liver methionine was low and both homocystine and 5-methyl THF concentrations were high because the methylcobamide-dependent transferase reaction was blocked. It is expected that 5-methyl THF accumulated, and the concentration of other folate derivatives decreased, because all the folate was "trapped" as methyl THF. If the patient had lived, it is expected that the deficiency of folate derivatives other than 5-methyl THF would have produced megaloblastic anemia.

References

Rosenblatt DS and Cooper BA: Inherited disorders of vitamin B_{12} metabolism, Blood Revs 1:172-182, 1987.

Scriver CR et al, editors: The metabolic basis of inherited disease, ed 6, New York, 1989, McGraw-Hill Book Co.

Case 4   Adenosine deaminase in immunodeficiency

A 22-month-old girl of a consanguineous mating was found to have a severe combined immunodeficiency disease. This resulted in recurring respiratory infections. Injections of diphtheria-pertussis-tetanus (DPT) and typhoid vaccine produced only a minimal response. During a search for a suitable donor for a bone marrow transplant, lysates of the girl's erythrocytes were found to lack detectable adenosine deaminase activity. Her mother and father both showed approximately 50% of the normal red cell adenosine deaminase activity.

Biochemical questions

1. How would you characterize the pattern of inheritance for adenosine deaminase?
2. What is the reaction catalyzed by adenosine deaminase?
3. One possible explanation of these data is based on the observation that adenosine is inhibitory to DNA methylation in cultured lymphoid cells. Explain how elevated levels of adenosine might inhibit DNA methylation.
4. The toxic substance in the patient may be dATP, which accumulates because deoxyadenosine cannot be deaminated. Ribonucleoside reductase is very sensitive to dATP, and the inhibition of this enzyme might account for the seriousness of the disease. Suggest a reasonable therapy based on these assumptions.

Case discussion

Inherited immunodeficiency diseases may be caused by defects in T cells, B cells, or both. Severe combined immunodeficiency disease (SCID) affects both cell types and is the disease of this patient. Many different types of immunodeficiency are recognized, but in most cases the molecular lesion is not known. The finding of a partial or complete lack of adenosine deaminase activity in some patients was the first clue to the important relationship of purine catabolism with immune function. This relationship was further emphasized by the later discovery of a purine nucleoside phosphorylase deficiency in immunodeficient patients with T cell abnormalities only. Apparently, lymphoid tissues and cells are sensitive to disturbances in purine catabolism, because other tissues seem to be less affected by a lack of adenosine deaminase or purine nucleoside phosphorylase.

1. Inherited SCID These diseases may be either autosomal or X-linked recessive. The adenosine deaminase deficiency form makes up approximately one third of the cases and shows an autosomal inheritance. The structural gene for adenosine deaminase is located on chromosome 20. The disease is quite rare, with an incidence of 1 in 100,000 live births.

2. Adenosine deaminase This catalyzes the conversion of adenosine to inosine and deoxyadenosine to deoxyinosine. Purine nucleoside phosphorylase catalyzes a phosphorylytic cleavage of inosine and deoxyinosine to yield hypoxanthine and either ribose 1-phosphate or deoxyribose 1-phosphate.

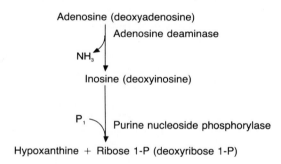

Pathogenesis Defects in these enzymes should result in the accumulation of adenosine, deoxyadenosine, and, in the case of purine nucleoside phosphorylase, inosine. This prediction is borne out by the high concentrations of deoxyadenosine in the urine of patients. Excessive adenosine can be deaminated through a different route by conversion to AMP and then catabolized by adenylate deaminase (see Figure 13.15). Excessive deoxyadenosine, however, accumulates and is either excreted or converted to dATP. High concentrations of dATP are found in lymphocytes of patients with adenosine deaminase deficiency disease. It is believed that the ribonucleotide reductase of these cells is inhibited by the abnormally high concentrations of one of its products. Recall that ribonucleotide reductase is exquisitely controlled by balanced concentrations of its products. Normal lymphocytes can be treated in cell culture with adenosine deaminase inhibitors to provide a laboratory model of the disease; when such cultured cells are assayed for deoxyribonucleoside triphosphates, the concentration of dATP is very high, as expected, and the concentrations of dGTP, dCTP, and dTTP are low. This is further evidence that the elevated concentration of dATP seen in the disease may inhibit ribonucleotide reductase.

More complex theories concerning the pathogenesis of adenosine deaminase deficiency have been proposed. An early idea considered that cAMP levels were increased in the disease and that this high level of cAMP led to an inhibition of mitogenesis. Yet the cAMP concentrations in the disease are not sufficiently increased to have deleterious effects.

3. DNA methylation Another idea suggests that the accumulated adenosine forces the formation of *S*-adenosylhomocysteine by stimulating the back reaction of *S*-adenosylhomocysteine hydrolase (see p. 385). This is made likely by K_{eq} of the hydrolytic reaction, which is only 1.4×10^{-6} mol/L. The product *S*-adenosylhomocysteine is a potent inhibitor of the methylation of RNA and DNA. Possibly lymphoid tissues are especially sensitive to undermethylated DNA.

4. Therapy The most likely explanation of the physiologic defects in the disease is an inhibition of ribonucleotide reductase, resulting in unbalanced cellular concentrations of deoxyribonucleoside triphosphates. This suggests that a possible therapy would be the administration of the deoxynucleosides found in low amounts. So far this has not been effective, but the problem may be in getting the proper amounts of each substance to the appropriate cells.

Perhaps there will be more success in using the knowledge obtained in studying this disease to treat the much larger group of people suffering from acute lymphoblastic leukemia of T cell origin. This leukemia has been quite resistant to conventional treatment but might respond to deoxycoformycin, a known inhibitor of adenosine deaminase.

References

Bonthron DT et al: Identification of a point mutation in the adenosine deaminase gene responsible for immunodeficiency, J Clin Invest 76:894, 1985.

Kantoff PW et al: Prospects for gene therapy for immunodeficiency diseases, Annu Rev Immunol 6:58-94, 1988.

Martin DW and Gelfand EW: Biochemistry of diseases of immunodevelopment, Annu Rev Biochem 50:845, 1981.

Snyder FF et al: Substrate inhibition of adenosine phosphorylation in adenosine deaminase deficiency and adenosine-mediated inhibition of PP-ribose-P dependent nucleotide synthesis in hypoxanthine phosphoribosyl transferase deficient erythrocytes, J Inherited Metab Dis 11:174, 1988.

Case 5   Orotic aciduria*

A male infant was delivered normally at full term. He was found to have severe anemia. The red blood cell count was 2.55 million/mm³, and hemoglobin was 0.9 mmol/L (6 g/dl). He was given antibiotics and transfusions. Despite antibiotics the anemia worsened. There was no response following treatment with B_{12}, folic acid or pyridoxine.

A prominent feature of this child's urine was a crystalline sediment, which was found to be orotic acid. Orotic acid in amounts as high as 9.6 mmol (1.5 g) was excreted daily (normal, 9 μmol [1.4 mg/day]).

Treatment with uracil produced no hematologic signs of remission, but the daily excretion of orotic acid dropped from 600 to 400 mg.

While partial remission was induced by prednisolone, the infant was given a yeast extract containing substantial amounts of uridylic and cytidylic acids. There was a striking reduction in the excretion of orotic acid, and the bone marrow findings were almost normal. The hemoglobin level rose to 2.1 mmol/L (14 g/dl), and the hematocrit was 0.44 (44%). The child gained weight and became active.

*Case description adapted from Huguley CM et al: Blood 14:615, 1959.

Biochemical questions

1. What might cause the high urinary levels of orotic acid? Consider the elevated excretion in terms of abnormalities in absorption, clearance, biosynthesis, and catabolism. What kind of defect is most likely involved in this case? Why?

◆ 2 How would you prove that the crystals in the urine sediment were orotic acid? Use several criteria.

3 Several cases of orotic aciduria have been described since this report. Many of these patients were successfully treated with uridine rather than with the uridylate-cytidylate mixture used in this case. Why would you expect uridine to be more effective than uracil, the nucleotide mixture, or any other ribonucleoside? What adverse side effect of the nucleotide mixture is avoided by using uridine?

4 Why are the symptoms in this patient like those in patients with deficiencies in either folate or B_{12}?

Reference

Scriver CR et al, editors: The metabolic basis of inherited disease, ed 6, New York, 1989, McGraw-Hill Book Co.

Case 6   Excessive purine synthesis in gout*

The patients were brothers, aged 30 and 35 yr, who had suffered from gouty arthritis since age 20. High concentrations of urine and serum uric acid were successfully reduced with allopurinol. Their erythrocyte hypoxanthine guanine phosphoribosyltransferase levels were normal. The content of PRPP in their erythrocytes, however, was much above normal. This was apparently the result of a mutation in the gene for PRPP synthetase that causes the production of a novel enzyme, because the synthetase from the patients had a *greater* than normal activity. The brothers' mother and father were without symptoms and responded normally to all the tests.

Biochemical questions

1. What biochemical reactions use PRPP as a substrate? What would be the consequence of abnormally large amounts of PRPP on these reactions?
2. Excessive amounts of PRPP are also produced in the Lesch-Nyhan syndrome. How do the accumulations in the Lesch-Nyhan disease differ from those in this case?
3. The increased activity of PRPP synthetase in these patients could be caused by either the presence of quantitatively more enzyme or an enzyme that was present in the same concentration but was more efficient. What would one need to do to test these alternatives?

References

Becker MA et al: Mechanisms of accelerated purine nucleotide synthesis in human fibroblasts with superactive phosphoribosylpyrophosphate synthetase, J Biol Chem 262:5596, 1987.
Becker MA et al: Inherited superactivity of phosphoribosylpyrophosphate synthetase: association of uric acid overproduction and sensorineural deafness, Am J Med 85:383, 1988.

Case 7   Lesch-Nyhan syndrome

At the age of 6 mo a boy showed signs of slow motor development. His mother had noticed orange crystals on his diapers, but she did not report this to the child's pediatrician until questioned about it several months later when she became concerned about the boy's failure to develop and his compulsive urge to bite at his fingers and

*Case description adapted from Sperling O et al: Biochem Med 6:310, 1972.

lips. Again, after questioning, the mother revealed that she had a younger brother with similar symptoms.

The Lesch-Nyhan syndrome was suspected so the output of urinary uric acid was measured, normalized to creatinine output, and found to be at least twofold increased over normal. Serum urate concentration was 0.59 mmol/L (10 mg/dl), abnormally high for a boy of this age. Further laboratory tests and examination indicated that the child also had megaloblastic anemia. Because of this unusual disease, the child was referred to the local university hospital where as part of additional testing it was discovered that his urine contained excessive amounts of 5-amino imidazole 4-carboxamide.

Biochemical questions

1. One of the earliest signs of the Lesch-Nyhan syndrome, and one often overlooked, is the appearance of orange crystals in the diapers. What are these crystals?
2. Is the family history of this patient consistent with the known X-linked recessive inheritance of this patient? Explain.
3. The enzyme defective in patients with the Lesch-Nyhan syndrome is hypoxanthine guanine phosphoribosyltransferase (HPRT). Explain how a defect in this enzyme leads to excessive urate excretion.
◆ 4. Immunologic assay using antisera prepared against human HPRT fails to show cross-reacting material in most patients with Lesch-Nyhan syndrome. Propose explanations for this observation considering (a) the specificity of the antigen-antibody reaction, (b) regulation of gene expression, and (c) lability and turnover of the mutant enzyme.
5. Offer an explanation for the megaloblastic anemia and the increased excretion of 5-aminoimidazole 4-carboxamide.
6. Adenine has been used experimentally in the treatment of patients with Lesch-Nyhan syndrome. This treatment produces a major problem, the formation of very insoluble crystals of 2,8-dioxyadenine. How is adenine converted to 2,8-dioxyadenine? How might this be prevented by administering another drug?
7. Consider the possibility of using adenosine instead of adenine in treatment. Explain the differences and difficulties that might be involved.
8. A patient with Lesch-Nyhan syndrome was found to have an unusual kind of HPRT. At first the erythrocyte enzyme seemed to have normal activity. It was then found that its K_m for guanine was 4.8×10^{-5} mol/L, whereas the normal enzyme had a K_m of 5×10^{-6} mol/L. What does this mean and what would you expect to be the consequences for the patient?
◆ 9. Erythrocytes from female carriers of the Lesch-Nyhan syndrome have normal levels of HPRT, yet their skin fibroblasts have half the normal activity. Offer an explanation for this based on a consideration of erythrocyte precursors and the Lyon hypothesis.
10. Predict the electrophoretic mobility of the following mutant forms of HPRT compared to normal HPRT.

Mutant form	Amino acid substitution
HPRT$_{Toronto}$	Arg → Gly$_{50}$
HPRT$_{Kinston}$	Asp → Asn$_{193}$
HPRT$_{London}$	Ser → Leu$_{109}$
HPRT$_{Yale}$	Gly → Arg$_{71}$

References

Fujimori S et al: Identification of a single nucleotide change in the hypoxanthine-guanine phosphoribosyltransferase gene (HPRT$_{Yale}$) responsible for Lesch-Nyhan syndrome, J Clin Invest 83:11, 1989.

Jankovic J: Orofacial and other self-mutilations, Adv Neurol 49:365, 1988.

Scriver CR et al, editors: The metabolic basis of inherited disease, ed 6, New York, 1989, McGraw-Hill Book Co.

Wilson JM et al: Hypoxanthine-guanine phosphoribosyltransferase deficiency, N Engl J Med 309:900, 1983.

Case 8	Chemotherapy of breast cancer

Several years ago a 45-year-old woman had a radical mastectomy for a potentially curable breast carcinoma. Axillary nodes were involved, which is a sign that the patient has disseminated micrometastases. After surgery prolonged cyclic combination chemotherapy was started. The patient was given a combination of cyclophosphamide (orally for 14 days), methotrexate, and fluorouracil (intravenously for 8 days). After a 2 wk rest without drugs, another cycle of the drug combination was given. Drug toxicity and dosage adjustments were analyzed with a number of hematologic tests. Twelve cycles of treatment were given, and after 27 mo the patient, as well as 94.7% of a large group of similarly treated women, was apparently free of disease. A control group of patients who had similar surgery but no chemotherapy showed a treatment failure of 24% after the same period.

Biochemical questions

1. Explain how cyclophosphamide is effective in treating cancer.
2. Methotrexate is a folate analogue. How is it effective in cancer chemotherapy?
3. Trace the route that fluorouracil takes in the blood and the interconversions it undergoes to manifest its toxicity.
♦ 4. Discuss the specificity of cancer versus normal cells for the chemotherapeutic agents. What is meant by "acceptable toxicity"?
5. Why is this particular combination of drugs more effective than the use of a single drug?
♦ 6. Cancer chemotherapy is more effective in treating micrometastases rather than large tumors. Why?

References

Antman K and Gale RP: Advanced breast cancer: high-dose chemotherapy and bone marrow autotransplants, Ann Intern Med 108:570, 1988.

Bonadonna G et al: Combination chemotherapy as an adjuvant treatment in operable breast cancer, N Engl J Med 294:405, 1976.

DeVita VT, Henney JE, and Weiss RB: Advances in multinodal primary management of cancer, Adv Intern Med 26:115, 1980.

Schabel FM Jr: Concepts for systemic treatment of micrometastases, Cancer 35:15, 1975.

Case 9	Methotrexate treatment of adenocarcinoma

A patient has an advanced stage of cancer. He was treated with a combination of methotrexate and thymidine, because conventional therapy had failed. Informed consent was obtained for these experimental studies. It is known that methotrexate toxicity in cultured bone marrow cells can be relieved with thymidine, and this was the rationale for the treatment.

Biochemical questions

1. What enzymatic reaction is inhibited by methotrexate?
2. Offer an explanation for why thymidine reverses the toxicity of methotrexate.
3. What other metabolites might reverse the methotrexate inhibition?

◆4. Why might thymidine be more useful than leucovorin (citrovorum factor) in "rescuing" patients treated with very large amounts of methotrexate?

5. What thymidine catabolite would you expect to be present in the urine of the patient?

6. Explain why the administration of both methotrexate and thymidine may be more effective in treating some tumors than methotrexate alone. Recall the use of thymidine with 5-fluorouracil therapy discussed earlier.

◆7. Explain how the inactivation of adenosyl cobalamin with nitrous acid potentiates the effectiveness of antifolates for inhibiting dTMP synthesis.

References

Ermens AA et al: Effect of cobalamin inactivation on folate metabolism of leukemic cells, Leuk Res 12:905, 1988.

Kamen BA and Winick NJ: High dose methotrexate therapy: insecure rationale? Biochem Pharmacol 37:2713, 1988.

Martin DS et al: An overview of thymidine, Cancer 45(suppl 5):1117, 1980.

Schroeder H and Fogh K: Methotrexate and its polyglutamate derivatives in erythrocytes during and after weekly low-dose oral methotrexate therapy of children with acute lymphoblastic leukemia, Cancer Chemother Pharmacol 21:145, 1988.

◆**Case 10**　　　　Folate deficiency in alcoholism

A 57-year-old man was admitted to the hospital because of weakness and shortness of breath. The patient had a 30 yr record of alcoholism and had been previously hospitalized for megaloblastic anemia and scurvy. On admission his hematocrit was 28%, reticulocytes 22%, and his leukocyte count was $4450/mm^3$. The serum folate was less than 2.3 nmol/L (1×10^{-9} g/ml); serum B_{12} was 0.085 nmol/L (116×10^{-12} g/ml).

After recovery the patient volunteered for a metabolic study. He was put on a low-folate diet (11.2 nmol [5 μg]/day), and on the tenth day the reticulocyte count had fallen to 11.8%. Muscatel wine (32 oz/day) was added to the diet. On day 16, the reticulocyte count was 0.7%. On day 23, intramuscular injections of folate (168 nmol [75 μg]/day) were begun with no change in reticulocytes, leukocytes, or platelets. The folate dose was increased on day 34. Several days later the reticulocytes went to 18.4%, the platelet count rose, and the leukocytes increased.

Biochemical questions ◆1. Is the improvement seen in alcoholics with megaloblastic anemia the result of the folate in the hospital diet or because the alcohol, which might have been acting as a hemosuppressant, has been removed from the diet? What is the minimum daily requirement for folic acid? How does this compare with the folate level when the patient was taking both folate and muscatel? What can be concluded from this observation?

2. What conclusion can be made about the relationships of alcohol and folic acid to hemopoiesis?

3. Bertino and co-workers studied the effects of alcohol on folate metabolism in liver and bone marrow cells. They found that 1.5% (w/v) ethanol inhibited the in vitro incorporation of ^{14}C-formate, but not of ^3H-thymidine, into the nucleic acids of bone marrow cells. From these data, what can you say about the metabolic step in nucleic acid biosynthesis that is inhibited by alcohol? Where in this scheme might alcohol and folate interact with the same enzyme?

4. Bertino and associates also found that in liver extracts the enzyme THF formylase was inhibited by 1.5% ethanol. The inhibition by alcohol was competitive in

respect to formate. What steps in nucleotide synthesis would be blocked as a result of this inhibition? Theoretically, how might the ethanol inhibition be overcome? Consider as many ways as you can. Would any of these be practical?

5. It has been suggested that folate be added to fortified wines, since the most life-threatening cases of folate deficiency are found in chronic alcoholics. Why might this be dangerous? Consider the methyl-trap hypothesis.

6. Megaloblastic anemia is much less common in alcoholics who drink beer rather than other forms of alcohol. Explain.

References   Bertino JR et al: Effect of ethanol on folate metabolism, J Clin Invest 44:1028, 1965.

Halsted CH: Folate deficiency in alcoholism, Am J Clin Nutr 33:2736, 1980.

Hoyumpa AM et al: Mechanisms of vitamin deficiencies in alcoholism, Alcohol Clin Exp Res 10:573, 1986.

Kaunitz JD and Lindenbaum J: The bioavailability of folic acid added to wine, Ann Intern Med 87:542, 1977.

Case 11   Purine nucleoside phosphorylase deficiency

A 7-year-old girl had severely defective T cell immunity but normal B cell function. Urate concentrations of both plasma and urine were low; however, the urine contained substantial amounts of purine nucleosides, including N-9-ribosyl uric acid, and the plasma contained inosine at 1000 times the normal concentration. Neither erythrocytes nor skin fibroblasts contained detectable amounts of purine nucleoside phosphorylase. The patient as a whole overproduced purines, but her fibroblasts made normal amounts. The patient's erythrocytes produced abnormally large amounts of dGTP.

Biochemical questions

1. What is the reaction catalyzed by purine nucleoside phosphorylase?
♦2. Explain the high serum levels of inosine and the presence of N-9-ribosyl uric acid in the urine.
3. The explanation for the overproduction of purines in this case may be similar to purine overproduction in the Lesch-Nyhan syndrome. Explain.
4. Why are the erythrocyte dGTP levels so high?
5. Adenosine deaminase deficiency (case 4) causes defects in immune function similar to those seen in this case. In adenosine deaminase deficiency the toxic agent is believed to be dATP. Propose an explanation for the defective immune function in this case.

References   Cohen A et al: dGTP as a possible toxic metabolite in the immunodeficiency associated with purine nucleoside phosphorylase deficiency, J Clin Invest 61:1405, 1978.

Williams SR et al: Molecular basis of a human purine nucleoside phosphorylase deficiency, Cold Spring Harbor Symp Quant Biol LI: 1059, 1986.

See also references to case 4.

♦Case 12   Adenine phosphoribosyl-transferase deficiency

A 4-year-old boy had passed solid material in his urine since birth. Uric acid concentrations of the plasma and urine were normal. Intelligence was normal, and there was no evidence of self-mutilation. The stones were identified as 2,8-dihydroxyadenine. A defect in adenine phosphoribosyltransferase was suspected. Erythrocyte lysates showed no detectable activity of this enzyme. Erythrocytes from the boy's older brother also lacked this enzymatic activity. His mother had only 47% of the normal amount of the enzyme, and his father had 41% of the normal activity.

Biochemical questions

1. Explain how a lack of adenine phosphoribosyltransferase can lead to the excretion of abnormally large amounts of 2,8-dihydroxyadenine.
2. Compare the solubilities of 2,8-dihydroxyadenine and uric acid.
3. Stone formation was greatly reduced by treatment with allopurinol and a low-protein diet. Explain the biochemical basis for this treatment.
4. Predict whether this disease is inherited as a recessive or dominant trait, autosomal, or X linked.

References

Broderick TP et al: Comparative anatomy of the human APRT gene and enzyme: nucleotide sequence divergence and conservation of a nonrandom CpG dinucleotide arrangement, Proc Natl Acad Sci USA 84:3349, 1987.

Van Acker KJ et al: Complete deficiency of adenine phosphoribosyltransferase, N Engl J Med 297:127, 1977.

Additional questions and problems

1. Compare and contrast DNA and RNA in terms of their chemical compositions, structures, and functions.
2. How are nucleic acids digested?
3. Describe those biosynthetic reactions that require folate coenzymes.
4. What biochemical reactions require 5-phosphoribosyl pyrophosphate, and how are these reactions affected in the Lesch-Nyhan syndrome?
5. Review the biosynthesis of phosphorylated coenzymes that arise from the B vitamins.
6. Explain why folate analogues are useful in cancer chemotherapy.
7. In cancer chemotherapy, leucovorin (5-formyl THF), a derivative that can be converted to the other coenzymatic forms of folate, is given to counteract the effects of large doses of an antifolate. Why is folic acid itself not used? Explain in terms of enzyme functions.
8. Some years ago *p*-aminobenzoic acid was thought by some to be a vitamin. If this substance were included in vitamin preparations, under what conditions would its ingestion be hazardous?
9. What tests would you perform to determine if an unknown viral nucleic acid were DNA or RNA and whether it were single or double stranded?
10. Cultured cells resistant to the growth-inhibitory analogue 8-azaguanine were found to lack activity for hypoxanthine-guanine phosphoribosyltransferase. Explain why.
11. Explain how patients with glycogen storage disease (glucose 6-phosphatase deficiency) develop gout in early adult life.
12. Busulfan and dimethyl sulfate are alkylating agents. How would a bifunctional alkylating agent such as busulfan compare with a monofunctional alkylating agent such as dimethyl sulfate in reacting with DNA?

Multiple choice problems

The patient is a middle-aged male with the classic symptoms of gouty arthritis. Problems 1 to 10 refer to this patient and others with gout.

1. Active gout is *generally* associated with:
    1. Hyperuricemia.
    2. Increased orotic acid synthesis.
    3. Urate crystals in joint fluid.
    4. Increased intestinal xanthine oxidase activity.
    The best answer is:

    A. 1, 2, and 3.
    B. 1 and 3.
    C. 2 and 4.
    D. 4 only.
    E. All are correct.

2. Lysates of the patient's erythrocytes were unable to convert guanine to guanylic acid. Possibly the defective enzyme is:
    A. Xanthine oxidase.
    B. PRPP amidotransferase.
    C. Adenine phosphoribosyl transferase.
    D. Hypoxanthine phosphoribosyl transferase (HPRT).
    E. Purine nucleoside phosphorylase.

3. The patient was treated with allopurinol, a hypoxanthine analogue. Allopurinol inhibits the conversion of:
    1. Hypoxanthine to IMP.
    2. Xanthine to uric acid.
    3. Adenine to hypoxanthine.
    4. Hypoxanthine to xanthine.
    The best answer is:
        A. 1, 2, and 3.
        B. 1 and 3.
        C. 2 and 4.
        D. 4 only.
        E. All are correct.

4. It has been proposed that a factor in the pathogenesis of gout is a deficiency of renal ammonium production leading to an acid urine, a common sign in gout. Thus the substrate used for ammonium production in the kidney would be spared, and it could be used for purine synthesis. This substrate is:
    A. 5-Phosphoribosyl amine.
    B. Glycine.
    C. Uric acid.
    D. Glutamate.
    E. Glutamine.

5. In patients with secondary gout a considerable amount of pyrimidine nucleotides are also synthesized. The pyrimidine synthetic pathway differs from the purine biosynthetic pathway in that:
    A. The first reaction in purine synthesis is inhibited by the feedback of nucleotides.
    B. PRPP is the ribose donor for only purine synthesis.
    C. The purine bases, but not the pyrimidine bases, use asparate to furnish a ring nitrogen atom.
    D. All the intermediates in the purine pathway are ribose phosphate derivatives.
    E. Folic acid coenzymes function only in pyrimidine biosynthesis.

6. Although most cases of gout are idiopathic, a few patients have a *more* active form of glutathione reductase in their erythrocytes. The function of glutathione in the red cell is highly dependent on the functional group that is part of the amino acid:
    A. Methionine.
    B. Glycine.

C. Cysteine.
D. Glutamate.
E. Glutamine.

7. Glutathione reductase requires a substrate derived from a metabolic cycle that provides precursors for the synthesis of both purine and pyrimidine nucleotides. This cycle is the:
A. Pentose phosphate pathway.
B. Krebs cycle.
C. Urea cycle.
D. Glycolytic pathway.
E. β-Oxidation of fatty acids.

8. One pK_a of uric acid is 5.4. This means that:
1. The plasma uric acid exists mostly as undissociated uric acid.
2. Plasma uric acid exists mostly in the dissociated form.
3. The crystals in the patient's urine are sodium urate.
4. The crystals in the urine are uric acid.
The best answer is:
    A. 1 and 4.
    B. 1 and 3.
    C. 2 and 4.
    D. 4 only.
    E. All are correct.

9. Although diet therapy in gout is often not very effective, one suspects that a high-protein diet might stimulate purine synthesis by providing the amino acid precursors. Amino acids that serve as precursors of the purine ring include:
1. Aspartate.
2. Glutamine.
3. Glycine.
4. Leucine.
The best answer is:
    A. 1, 2, and 3.
    B. 1 and 3.
    C. 2 and 4.
    D. 4 only.
    E. All are correct.

10. Some patients with gout have a partial defect in hypoxanthine-guanine phosphoribosyl transferase, an enzyme almost totally defective in the Lesch-Nyhan syndrome. The products of the reaction catalyzed by this enzyme are:
1. AMP.
2. IMP.
3. Inorganic phosphate.
4. Inorganic pyrophosphate.
The best answer is:
    A. 1, 2, and 3.
    B. 1 and 3.
    C. 2 and 4.
    D. 4 only.
    E. All are correct.

Chapter 14

Structure and synthesis of DNA

Objectives

1 To explain how the structure and metabolism of DNA relates to the functioning of the genetic apparatus
2 To show how DNA ultimately controls all metabolic functions
3 To describe how mutations in DNA lead to the inheritance of certain diseases
4 To describe the molecular action of drugs and antibiotics that affect DNA synthesis
5 To explain how recombinant DNA technology can be applied to human genetic diseases in respect to diagnosis, carrier detection, and therapy

The discovery of the base-paired, double-helical structure of deoxyribonucleic acid (DNA) provides the theoretic framework for determining how the information coded into DNA sequences is replicated and how these sequences direct the synthesis of ribonucleic acid (RNA) and proteins. Already clinical medicine has taken advantage of many of these discoveries, and the future promises much more. For example, the biochemistry of the nucleic acids is central to an understanding of virus-induced diseases, the immune response, the mechanism of action of drugs and antibiotics, and the spectrum of inherited diseases.

In approaching the study of the molecular mechanisms of heredity, this chapter first discusses the structural and functional roles of the genetic material, DNA. This includes an analysis of its replication and susceptibility to mutation. The health-related aspects of the use of recombinant DNA techniques are considered, and examples of their use in the analysis of several human genetic diseases are used to illustrate the biochemical side of genetics.

Functional roles of DNA

DNA as the genetic material

The nucleic acids were recognized as chemical substances more than 70 years before DNA was found to be responsible for the transmission of inherited characteristics. Later it was suspected that DNA might be the genetic material because of its high concentration in chromosomes and in some viruses. The premise was complicated, however, because the concentration of protein in these structures was also high. Furthermore, RNA but not DNA was found in some viruses. Indirect evidence pointed to a role for nucleic acids as the transmitters of biologic information; the wavelengths of light in the ultraviolet region that are the most mutagenic are the same wavelengths at which nucleic acids absorb the most light energy.

Constancy of DNA concentration One property expected of the genetic material is a constancy of amount in every cell of the body under every environmental situation. DNA, not RNA or protein, fulfills this expectation. Its content per nucleus is the same in every cell except the germ cells, which have exactly half that found in the somatic cells. Again, this is expected if progeny obtain half their characteristics from each parent. This constancy is so dependable that the measurement of the DNA concentration in a tissue can be used to calculate the number of nuclei and thus the number of cells. This works well for diploid cells such as those of the kidney, but corrections must be made for polyploid mammalian liver or cancer cells.

Transformation of cells with DNA The best evidence that exogenous DNA can produce permanent changes in cells came from the experiments of Avery et al. DNA from one strain of bacterial cells was used to transform a different strain of cells so that they came to resemble the strain from which the DNA was derived. In the original experiment, DNA was isolated from cells of a strain of *Diplococcus pneumoniae* that contained a characteristic complex polysaccharide on their surfaces. This polysaccharide made the cells pathogenic for mice and gave a glistening, smooth appearance to colonies formed by these cells on nutrient agar. When the polysaccharide was missing, as it was in some other strains of the microorganism, the colonies were rough in appearance and the cells were harmless when injected into mice. When DNA from the smooth cells was added to rough cells, the DNA entered some of the cells and became a permanent part of their genetic apparatus; subsequent generations were permanently changed to pathogenic cells that formed smooth colonies. This process is called bacterial *transformation*.

Subsequently, similar experiments were done with viral nucleic acids. The pure viral nucleic acid, when added to cells, led to the synthesis of complete virus particles; the protein coat was not required. This process is called *transfection*. More recently, DNA has been used in cell-free extracts to program the synthesis of RNA that functions as the template for the synthesis of proteins characteristic of the DNA template. Considering all this evidence, DNA undoubtedly is a carrier of genetic information.

Cellular location of DNA

Most of the DNA of animal cells is found in the nucleus, where DNA is the major constituent of the chromosomes. On the other hand, most of the RNA is located in the cytoplasm. Nuclear DNA exists as a thin, double helix only 2 nm wide. The double helix is folded and complexed with protein to form chromosomal strands approximately 100 to 200 nm in diameter. Each chromosome contains a single DNA duplex. The human chromosomes vary in size; the smallest contains approximately 4.6×10^7 base pairs of DNA, and the largest 2.4×10^8 base pairs. In contrast, the *Escherichia coli* chromosome has 4.5×10^6 base pairs. The DNA of the chromosomes is tightly packed and associated with both histone and nonhistone proteins.

The amount of genomic DNA in a particular organism is roughly proportional to the complexity of the organism. Table 14.1 shows the content of DNA in the genomes of several widely different organisms. The data are normalized to a haploid set of chromosomes, since some cells listed are haploid and others are diploid. The DNA content

Table 14.1   DNA content of some cells and viruses

Source of DNA	Haploid size of genome, base pairs
Viruses	
SV40	5×10^3
Papilloma (wart)	8×10^3
Adenoviruses	2.1×10^4
Herpesviruses	1.56×10^5
Poxviruses	2.4×10^5
Cells	
Escherichia coli	4.5×10^6
Yeast	1.3×10^7
Drosophila	1.6×10^8
Human	3.2×10^9
Animal mitochondria	1.5×10^4

of a few viruses is given for comparison. The size of DNA is often measured in base pairs, since cellular DNA is all double stranded and base paired. Thus the number of base pairs in a DNA molecule is a precise measure of the number of mononucleotides that comprise the polynucleotide chain. On the average, one base pair represents about 600 daltons; thus, to estimate the molecular weight of a DNA, the number of base pairs is multiplied by 600.

Histones and chromosome structure The chromosome structure is visible only during the mitotic portion of the cell cycle. The constituent parts of the chromosomes are nucleoprotein fibers called *chromatin*. When condensed, chromatin forms a microscopically visible chromosome-like structure. The chromosomes are composed of DNA, RNA, and proteins. The relative amounts of the three vary, but chromatin is primarily protein and DNA.

Nucleosomes The histones are the major proteins associated with the chromosomes. These small, basic proteins can be separated into five groups by polyacrylamide gel electrophoresis. All five histone groups are found in every eukaryotic cell. These groups are called H1, H2A, H2B, H3, and H4. Each histone is present in equimolar amounts except for H1, which is present in approximately half the concentration of the others. Furthermore, the H1 electrophoretic band is composed of many similar but slightly different proteins. In this respect the H1 group differs from the other histone groups, which are each single proteins. The histone groups differ in their relative content of lysine and arginine residues. Table 14.2 lists some of the properties of these chromosomal proteins. The sequences of histones H2A, H2B, H3, and H4 are greatly conserved between species, even though an organism might have several genes for the same histone. This virtual sequence identity testifies to a very similar and essential function for the four histones in all eukaryotic species. A clue to this function is the ability of these histones to associate at high ionic strength to form an octomer containing two copies of each of the four histone groups.

Histones of the nucleosome core The four histone groups that are composed of homogeneous proteins, H2A, H2B, H3, and H4, make up the nucleosome core. Each core consists of two copies of the four histones. The double-stranded DNA is wrapped twice around each core in a left-handed superhelix. A superhelix is the name given to the additional helix made by the double-stranded, helical DNA as it is wrapped around the nucleosome core. A familiar superhelix in everyday life is a twisted spiral telephone cord. The nucleosome core of histones do not recognize specific DNA structures; rather, they can bind to any stretch of DNA as long as it is not too close to a neighboring nucleosome. The order of contact of histones to the DNA is as follows:

Table 14.2 Properties of animal histones

Electrophoretic group	Mass (kilo-daltons)	Lysine (%)	Arginine (%)
H1	21	27	2
H2A	14.5	11	9
H2B	13.8	16	6
H3	15.2	10	15
H4	11.3	10	14

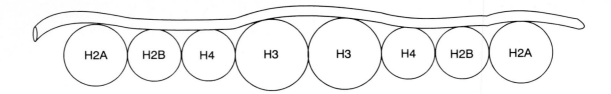

Protein-protein interactions between the histone subunits are undoubtedly important in promoting formation of a nucleosome in which 146 base pairs of DNA are coiled around the outside of the histone core. One molecule of histone H1 binds to an exterior region of each nucleosome, but histone H1 is not needed to determine nucleosome structure. The distance between nucleosomes is approximately 200 base pairs; consequently, in electron micrographs, nucleosomes resemble evenly spaced beads on a string of DNA (Figure 14.1, *A*). Neutron and x-ray diffraction data are also consistent with this structure.

The histone core protects the DNA bound to the nucleosome from digestion by pancreatic deoxyribonuclease (DNase) I or micrococcal nuclease. Nucleases, however, will cleave the linker DNA that connects the nucleosome subunits to one another.

Nucleosomes can be reconstructed in the laboratory from DNA and pure histones. Histone H1 is not necessary for the reconstruction, which further shows that H1 is an accessory protein and not a major structural part of the nucleosome subunit.

The primary function of the nucleosomes is to condense DNA. Further condensation of nucleosome DNA requires nonhistone nuclear proteins. These proteins make up a scaffoldlike structure around an additional helix consisting of coiled nucleosomes. This produces a structure that resembles a solenoid, with six nucleosome subunits per turn (Figure 14.1, *B*). The solenoid structure can form large loops that give additional structure to the incipient chromosome.

Banded chromosomes Although it is not known how the characteristic banded structure of a chromosome is related to its function, the DNA of a single chromosome probably consists of a single DNA duplex running from one end of the chromosome to the other. These bands, which can be seen microscopically after staining with fluorescent dyes such as Giemsa or quinacrine, are believed to represent regions of heterochromatin complexed with histones and nonhistone proteins.

Nonhistone proteins The nucleus contains a large number of proteins other than histones. These so-called nonhistone proteins may or may not be tightly associated with the chromosomes. For example, the nucleus contains enzymes associated with the synthesis of RNA and DNA; these are nonhistone proteins, but they are not part of the structure of chromosomes. One group of nonhistone proteins are the high mobility group (HMG) proteins, named for their rapid movement on polyacrylamide gel electrophoresis. The HMG proteins, but not histone H1, are associated with the chromatin that is most active in RNA synthesis.

Mitochondrial nucleic acid Not all the cellular DNA is in the nucleus; some is found in the mitochondria. In addition, mitochondria contain RNA as well as several enzymes used for protein synthesis. Interestingly, mitochondrial RNA and DNA bear a closer resemblance to the nucleic acid of bacterial cells than they do to animal cells. For example, the rather small DNA molecule of the mitochondrion is circular and does not form nucleosomes. Its information is contained in approximately 16,500 nucleotides that function in the synthesis of two ribosomal and 22 transfer RNAs (tRNAs). In addition, mitochondrial DNA codes for the synthesis of 13 proteins, all components of the respi-

Figure 14.1

Nucleosome structures. **A,** Double-stranded DNA is coiled in a left-handed superhelix around a histone protein core. **B,** Solenoid-like structure of coiled, nucleosome-containing DNA.

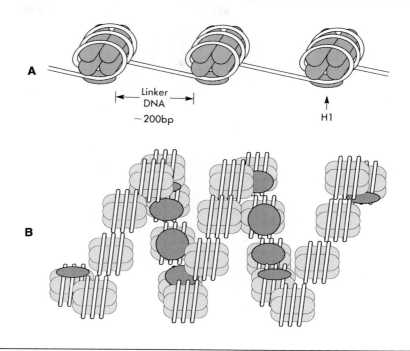

A

Linker
DNA

~200bp

H1

B

ratory chain and the oxidative phosphorylation system. Still, mitochondrial DNA does not contain sufficient information for the synthesis of all mitochondrial proteins; most are coded by nuclear genes. Most mitochondrial proteins are synthesized in the cytosol from nuclear-derived messenger RNAs (mRNAs) and then transported into the mitochondria, where they contribute to both the structural and the functional elements of this organelle. Because mitochondria are inherited cytoplasmically, an individual does not necessarily receive mitochondrial nucleic acid equally from each parent. In fact, mitochondria are inherited maternally.

Clinical comment

Leber's hereditary optic myopathy This disease is one of several myopathies caused by defects in the mitochondrial genome; thus they are called *mitochondrial myopathies*. Leber's hereditary optic neuropathy causes blindness in young males more often than females, even though the disease is transmitted maternally. The myopathy is caused by a single base mutation at position 11,778 of the mitochondrial DNA that changes an arginine codon to a histidine codon in subunit four of NADH–coenzyme Q oxidoreductase. This enzyme is part of complex I of the respiratory chain. (See also Singh, G et al: N Engl J Med 320:1300, 1989.)

Other conformations of DNA

A-DNA The Watson-Crick model of DNA is based on the x-ray diffraction patterns of B-DNA. Most DNA is B-DNA; however, DNA may take on two other conformations, A-DNA and Z-DNA. These conformations are greatly favored by the base sequence or by bound proteins. When B-DNA is slightly dehydrated in the laboratory, it takes on the A conformation. A-DNA is very similar to B-DNA except that the base pairs are not stacked perpendicular to the helix axis; rather, they are tilted because the deoxyribose moiety "puckers" differently. An A-DNA helix is wider and shorter than the B-DNA helix.

Z-DNA This conformation differs more radically. It is a left-handed helix instead of the right-handed conformation of A-DNA and B-DNA. The Z-DNA conformation exists only along a string of alternating purines and pyrimidines, especially several guanine-cytosine residues in a row. An alternating dinucleotide sequence results where the external phosphate groups zigzag, thus Z-DNA. This structure results from alternating *anti* and *syn* conformations of the glycosidic bonds. In A- and B-DNA the conformations of the glycosidic bonds are all *anti*.

In Z-DNA, guanine residues are *syn,* whereas cytosine and thymine residues are *anti*. Eukaryotic DNA contains several alternating purine-pyrimidine sequences consistent with the Z-DNA conformation; however, the biologic significance of Z-DNA is still unclear.

The "central dogma"   DNA has two broad functions: replication and expression. First, DNA must be able to replicate itself so that the information coded into its primary structure is transmitted faithfully to progeny cells. Second, this information must be expressed in some useful way. The method for this expression is through RNA intermediaries, which in turn act as templates for the synthesis of every protein in the body. The relationships of DNA to RNA and to protein are often expressed in a graphic syllogism called the "central dogma." The concept was proposed by Crick in 1958 and was revised in 1970 to accommodate the discovery of the RNA-dependent DNA polymerase. Crick's original theory suggested that the flow of information was always from RNA to protein and could not be reversed, yet it allowed for the possibility of DNA synthesis from RNA.

Genetic expression involves the transfer of information by the processes of transcription and translation. *Transcription* is the process that transfers information using the same four-letter language of the nucleic acids; that is, one strand of DNA serves as a template for the synthesis of an RNA strand, the sequence of which is analogous to one DNA

Figure 14.2   The "central dogma."

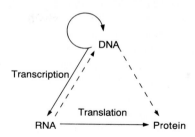

strand and complementary to the other. Transcription is "reversible" in a few cases. The dashed line in Figure 14.2 represents the synthesis of DNA from information contained in the RNA of certain tumor viruses. Information flow in the direction of RNA to protein is termed *translation,* since the four-letter language of the nucleic acids must be converted to the different 20-letter language of the amino acids that make up proteins. The process of translation is always unidirectional. Single-stranded DNA templates can be translated in the laboratory, but no evidence exists for such a function in vivo; consequently the line between DNA and protein in Figure 14.2 is dashed.

DNA synthesis   Knowing that DNA was the hereditary material gave no clues as to how the molecule might reproduce itself until Watson and Crick proposed their model for the structure of DNA. In this model the DNA strands are arranged in an antiparallel fashion and are base paired along their entire length in the form of a double helix. Recall from Chapter 13 that the molar concentration of adenine equals that of thymine, and the cytosine concentration is the same as that of guanine. Base pairing of adenine with thymine and of cytosine with guanine yields a structure in which the sequence of one strand can be automatically determined if the sequence of the other strand is known. The importance of this concept in the replication of DNA and in the synthesis of RNA strands of complementary sequences was recognized immediately, but several years were required for the enzymatic studies that gave unequivocal proof.

Semiconservative replication

Conservative replication

Double-stranded DNA

Maternal strands ———

Daughter strands --------

Strand separation   Implicit in the functioning of the Watson-Crick DNA model is the idea that the strands of a DNA molecule must separate and new daughter strands must be synthesized in response to the sequence of bases in the mother strand. This is called *semiconservative replication.* Still, *conservative replication,* in which both strands of a daughter molecule are newly synthesized, could not be ruled out by consideration of the structure of DNA alone.

The experiments of Meselson and Stahl proved replication to be semiconservative. These consisted of growing *E. coli* cells in a medium containing $^{15}NH_4Cl$, so that the nitrogen atoms of the purine and pyrimidine bases of the DNA were heavily labeled. Cells were then transferred to a medium containing the usual light $^{14}NH_4Cl$ and grown for one or more generations. The DNA was then prepared and separated by density-gradient equilibrium centrifugation in a solution of cesium chloride. After one generation the progeny DNA separated in such a way that all of it appeared at a position midway between the very heavy parental DNA and the light DNA of a control culture. Because all the DNA existed as a hybrid that contained one heavy strand and one light strand, DNA clearly was replicated by a semiconservative mechanism.

This presented a more difficult problem: How do the double-helical strands separate during DNA synthesis? In a rapidly growing cell such as *E. coli* it has been calculated that if the strands separate by untwisting, the molecule would have to rotate at 10,000 rpm, a rate that is highly improbable. The answer to this problem lies in an understanding of the mechanism of DNA replication at the enzyme level. We will return to this subject after first considering the enzymes involved in DNA synthesis.

DNA polymerases

DNA synthesis is more complex than originally thought. One reason is that DNA replication requires many different enzymes, not just DNA polymerase. For example, replication requires enzymes that coordinate the growth of cell membranes with DNA synthesis. Other enzymes and proteins initiate the synthesis of small RNA primers that bind to single-stranded DNA. Additional enzymes are needed to remove the RNA primers from the growing deoxyribonucleotide chain, fill in the small regions vacated by the RNA primers, seal the strands together, and aid in the untwisting of the DNA helix. More than one enzyme may be required for each of these functions, and this list of functions is not meant to be complete.

Bacterial DNA polymerases The mechanisms involved in DNA synthesis are most easily understood by considering the DNA polymerases. The most extensively studied are the three DNA polymerases, I, II, and III, from *E. coli*. Some ambiguity still exists about the essentiality of the specific roles played by each of the polymerases. One complicating feature is that it is difficult to distinguish the polymerases from enzymes that function exclusively to repair damaged DNA, primarily because some of the processes that occur during DNA replication are identical to events necessary for DNA repair. However, all the DNA polymerasees require a DNA template and all four of the deoxyribonucleoside triphosphates. Synthesis proceeds from the 5' to the 3' end of the growing polynucleotide, and inorganic pyrophosphate (PP_i) is a product of the reaction. These features are illustrated in Figure 14.3 for DNA polymerase I, but they apply to other polymerases as well. Polynucleotides formed using radioactive deoxyribonucleoside triphosphates have sequences identical to those of one strand of the DNA template and complementary to sequences of the other strand. Both strands are labeled in vitro.

DNA polymerase I is a nonessential enzyme, since viable *E. coli* mutants lack it (*pol* A⁻). This conclusion is complicated, however, since the enzyme catalyzes three separate chemical reactions. It polymerizes deoxyribonucleoside triphosphates, and it has two exonucleolytic activities, a 3' to 5' activity and a 5' to 3' activity. The *pol* A⁻ mutants lack only the polymerization activity. Other mutants lacking both the polymerase and the 5' to 3' exonuclease activity are lethal. Thus the exonuclease function is the more important one. This fits with the role of this enzyme in removing damaged DNA segments (DNA

Figure 14.3   Reaction catalyzed by DNA polymerase I.

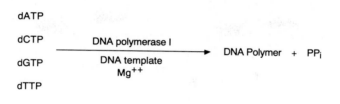

repair) and in removing covalently attached RNA from DNA chains. We will later see that small RNAs serve as primers of DNA synthesis.

DNA polymerase I has been purified to homogeneity. When the pure enzyme is treated with subtilisin, a proteolytic enzyme from *Bacillus subtilis,* the polymerase is cleaved into two pieces. The small fragment retains the 5′ to 3′ nuclease activity, whereas the larger piece, called a Klenow fragment, has both polymerase activity and the 3′ to 5′ exonuclease activity. The Klenow fragment is sold commercially for use in labeling DNA for use in detecting recombinant DNA (see later discussion).

DNA polymerase II is more likely needed for the repair synthesis of DNA. Repair synthesis requires excision of the damaged DNA, the synthesis of a fresh replacement segment complementary to the remaining single strand, and the sealing of the replacement segment to the larger polynucleotide chain. DNA polymerase II does not have 5′ to 3′ exonuclease activity. Mutants deficient in DNA polymerase II activity, as determined by in vitro assay, grow well; therefore the enzyme does not seem to have an indispensable function in the cell.

DNA polymerase III has all the enzymatic activities of DNA polymerase I. A subunit of the enzyme is the product of the *dna* E gene. Temperature-sensitive mutations of this gene testify to the importance of DNA polymerase III. A temperature-sensitive mutant is one that grows at 30° C but fails to grow at 42° C, a temperature not lethal for wild-type *E. coli.* The failure to grow at the higher temperature is caused by a mutation in the gene for DNA polymerase III so that a very heat-labile enzyme is produced. Since this appears to be the only mutation in this strain, DNA polymerase III is the only enzyme inactivated at 42° C; thus the enzyme is essential to the organism.

Template and primer At this point it is necessary to make a distinction between the meanings of template and primer. The word *template* refers to the structural sequence of the polymerized monomeric units of a macromolecule that provides the pattern for the synthesis of another macromolecule with a complementary or characteristic sequence. The word *primer,* on the other hand, refers to a polymeric molecule that contains the growing point for the further addition of monomeric units. Glycogen is an example of a primer to which glucose units are added; however, glycogen has no template activity. Under certain circumstances DNA has both primer and template activities. For example, the addition of mononucleotides is to the 3′ end of the growing DNA primer. This presents a problem with regard to how the other strand is synthesized. Biochemists have looked hard but unsuccessfully for an enzyme that can add deoxyribonucleotides onto the 5′ end of DNA primers. Such a primer should contain a triphosphate on the hydroxyl group of the 5′ end. Although a very active 5′-exonuclease, actually part of DNA polymerase I, has made the search for such an activated 5′ end extremely difficult, investigators conclude that a polymerase able to use such a primer probably does not exist. On the contrary, good evidence suggests that the synthesis of both strands is by the known DNA polymerases.

Stages of DNA synthesis

Origin of replication In *E. coli* cells, DNA replication starts at a specific site called *ori*C. The *ori*C locus contains only 245 base pairs. Similar sequences are responsible for initiating the synthesis of plasmid and bacteriophage DNA. The *ori*C nucleotide sequence binds several units of the tetrameric form of the *dna*A protein. This protein is named for the gene that encodes it. The *dna*B and dnaC proteins then bind to the complex. As a result of binding these proteins, a portion of the helical DNA is unwound. This forces the rest of the DNA into a left-handed double helix that wraps around the proteins to give a structure resembling the histone-containing nucleosome of eukaryotic cells. The exposed single-stranded DNA is stabilized by the binding of a 74 kDa, single-stranded DNA-binding protein called SSB.

RNA primers All DNA polymerases add mononucleotides to the 3′end of an existing primer. Consequently a special primer is needed for DNA to replicate in its entirety. RNA polymerases can initiate polymer synthesis without a primer; thus short RNA primers are used to initiate DNA synthesis.

The RNA oligonucleotides are complementary to a sequence on one of the strands of the DNA template and base pair with a portion of the DNA molecule. Subsequently, deoxyribonucleotides are covalently attached to the RNA primer. The synthesis of the primer itself is catalyzed by a special RNA polymerase called *primase*. Similar RNA polymerase–like enzymes are used to prime the synthesis of certain viral DNAs and eukaryotic DNA.

The *dna*B and *dna*C proteins, as well as at least a few other accessory proteins, are required for the primase to initiate RNA synthesis from the DNA template. This enzyme complex, called a *primasome,* is very large, almost as large as the DNA polymerase III holoenzyme (800 kDa), which joins the primasome and catalyzes the addition of monodeoxyribonucleotides to both growing strands. No problem exists in visualizing the addition of oligonucleotide monomers to the RNA primer at the 5′ end of the continuous strand, since this DNA polymerase is a highly accurate, processive enzyme. *Processivity* means that the polymerase can rapidly add many mononucleotides to the primer, more than 1000 per second, before dissociating from it. On the other strand, DNA synthesis must proceed away from the replicative fork; however, if the template strand is looped back toward the replicative fork, subunits of the DNA polymerase could add nucleotides to both growing strands. Addition to an RNA primer would continue until synthesis was blocked by the previously made primer and its attached oligodeoxyribonucleotide (Figure 14.4). This stalling might trigger the synthesis of a new RNA primer and the addition of deoxynucleotides to it.

The mechanism of DNA synthesis is known in considerable detail so the steps illustrated in Figure 14.4 are very much simplified.

Bidirectional synthesis DNA synthesis occurs in both directions at each of the replicating forks. Once a DNA strand has been primed, synthesis toward the replicating fork can be visualized as continuous. Growth of the opposite, lagging strand occurs in discontinuous bursts, each burst primed by a short RNA segment. DNA synthesis visualized by electron microscopy gives the appearance of an "eye" or several eyes along a DNA template and resembles this: . Eyes are thought to be regions of DNA where recent synthesis has been initiated. Synthesis from an eye is bidirectional. Thus as the eye enlarges, DNA synthesis along either new strand may be considered continuous where

Figure 14.4

A single unit of DNA polymerase III complex synthesizes both new strands of DNA, one continuously and the other in short pieces. Deoxynucleotide addition to the daughter strands is indicated by vertical lines across the strands.

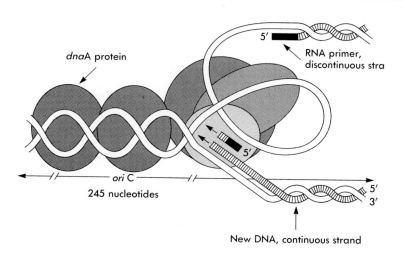

the DNA polymerase is close to and moving toward the replicating fork, and it may be considered discontinuous where the DNA polymerase is close to a replicating fork but moving away from the fork. Consequently, as an eye enlarges, DNA synthesis is more or less continuous at one end of a growing strand but discontinuous at the other end of the same strand.

Removal of the RNA primer and ligation of the DNA fragments The end result is newly synthesized DNA that is interspersed with segments of RNA and that is discontinuous but base paired with an intact parental strand. Subsequently, the 5′ exonucleolytic activity of DNA polymerase I removes the RNA segment, and either DNA polymerase I or II fills the gap vacated by the RNA. DNA ligase (sometimes called polynucleotide ligase) is required to join these short pieces into phosphodiester linkage. The ligation reaction shown in the following diagram requires that energy be supplied from ATP. This enzyme also occurs in animal cells.

$$\text{ATP} + \text{Ligase} \rightleftharpoons \text{Ligase} \sim \text{AMP} + \text{PP}_i$$

DNA ligase is not only important in DNA replication; it is also used to seal deoxyribonucleotide segments in the crossover events during gene recombination. The enzyme also functions to close breaks in segments of DNA undergoing repair and is required to join the ends of mitochondrial DNA to form their characteristic circular structure.

Topoisomerases Armed with this information, the unwinding problem mentioned earlier can be reconsidered. By the alternating action of endonucleolytic and ligase activities, the unwinding of DNA could be reduced to an untwisting of only a small part of the double helix at any given time. Both activities are part of the enzyme called topoisomerase I.

Topoisomerase I This enzyme releases the torque developed during the unwinding required for replication. This torque introduces superhelices into DNA. A superhelix can be visualized as a helix on top of the basic DNA helix. The enzyme first introduces a single-strand break in the superhelix. This is not a hydrolytic cleavage but rather a transesterification of the 5′ phosphoryl at one end of the broken strand to a tyrosine hydroxyl group, thus conserving the energy of the phosphodiester bond. The single-strand break relieves the torque as the broken strand with the enzyme still attached rotates about the unbroken strand. When the strain on the double helix is relaxed, the enzyme transfers the 5′ phosphorylated end back to the polynucleotide chain and dissociates from the DNA duplex. Actually the broken strand need only pass through the neighboring intact strand and reseal to remove one superhelical turn. To relieve the tension of several superhelical turns, as occurs during DNA replication, the topoisomerase catalyzes several "nicking-closing" reactions. Topoisomerase I recognizes either positively or negatively supercoiled DNA. Topoisomerase I activity has also been found in the nuclei of animal cells.

Topoisomerase II Another enzyme, called topoisomerase II, or DNA gyrase, also plays a role in the unwinding of replicating DNA. Although topoisomerase I can relieve the positive superhelical torsion introduced into DNA as a result of unwinding, topoisomerase II can introduce negative superhelices ahead of the replicating fork. This relieves the twisting pressure of DNA replication before it can develop. The enzymes also differ in that topoisomerase I does not require high energy in the form of adenosine 5′-triphosphate (ATP), whereas topoisomerase II does, since energy is required to make negatively supercoiled DNA. This torsional energy is conserved in the negative superhelices found in most naturally occurring DNA, such as the DNA of nucleosomes. The topoisomerases also differ in that enzyme I cleaves only one DNA strand, whereas enzyme II cleaves both. Approximately 200 base pairs of DNA coil about topoisomerase II, much as occurs with the DNA in a nucleosome. Both strands are opened, and the 5′ phosphoryl groups are linked to tyrosine hydroxyl groups on the enzyme. A DNA segment is passed through both the anchored but cleaved ends. This passage is always in the same direction, so that only a negative superhelix forms when the strands are resealed. A DNA gyrase–like activity has been isolated from animal cells.

DNA synthesis in animal cells

The replication process in animal cells is necessarily more complex than in bacteria because several chromosomes must be replicated. DNA synthesis in animal cells also differs in that several origins of replication occur within a single chromosome rather than the single site in *E. coli*. This speeds up the duplication of the animal genome, which is approximately 1000 times larger than that of bacteria. The eukaryotic origins of replication have a high affinity for the nuclear matrix, the nucleoprotein material that remains after nuclei have been washed with a high concentration of salt.

DNA polymerases from several different animal cells have been isolated and studied. The three DNA polymerases of animal cells, called α, β, and γ, can be distinguished

Table 14.3   Comparison of properties of DNA polymerases from animal tissues

Property	DNA polymerase α	DNA polymerase β	DNA polymerase γ
Molecular weight	155,000, plus three other subunits	43,000	193,000 (four oligomers)
Template specificity	Nicked DNA template, RNA primer	Nicked DNA template, DNA primer	Ribonucleotide template and DNA primer
Deoxyribonucleoside triphosphate dependence	All four required	All four work, but single nucleotide will incorporate	All four
Inhibition by sulfhydryl reagents	Sensitive	Less sensitive	Sensitive

by their molecular weights, template specificity, and sensitivity to sulfhydryl reagents. Table 14.3 compares the three in regard to these differences. DNA polymerase α is probably the most important for DNA replication. This enzyme shares many functional properties with DNA polymerase III of *E. coli*.

DNA polymerase α Even though DNA polymerase α and its associated subunits have not been purified to homogeneity, much is known about the function of this enzyme complex. The concentration of DNA polymerase α is higher than that of the other two polymerases. One of the associated subunits has primase activity capable of making short RNA primers. At first this enzyme was thought to be present in the cytoplasm of cells, but with special precautions it can be isolated from the nuclei. Unlike *E. coli* DNA polymerase I, polymerase α has no associated nuclease acitivity. It is membrane-bound, however, and fractionates with ribonucleotide reductase, dTMP synthase, and thymidylate kinase, enzymes important in the synthesis of DNA precursors. The synthesis of polymerase α increases greatly in regenerating liver and in other rapidly dividing cells.

DNA polymerase β This is a smaller, stable enzyme that has been highly purified. It is immunologically distinct from the other polymerases, indicating that it is not merely a subunit of the larger polymerases. Polymerase β is undoubtedly a repair enzyme.

DNA polymerase γ This is the enzyme responsible for the synthesis of mitochondrial DNA and the DNA of some viruses, such as adenoviruses. Polymerase γ is very large and consists of a tetramer of identical oligomers, each having a molecular weight of 47,000. Synthetic ribonucleotides are very effective templates in the laboratory, but this mitochondrial enzyme differs from reverse transcriptase in that natural RNAs are poor templates.

Reverse transcriptase This enzyme is associated with the virions of RNA tumor viruses such as the Rous sarcoma virus (RSV). The enzyme has remarkable enzymatic activity in that it can catalyze several seemingly diverse steps in the synthesis of double-stranded DNA from the single-stranded RNA viral genome. The enzyme uses a tRNA for tryptophan as a primer to make a copy of DNA that is complementary to the viral RNA. The resulting RNA-DNA hybrid is converted to a double-stranded DNA molecule by ribonuclease (RNase)H and DNA-dependent DNA polymerase activities that are intrinsic to reverse transcriptase.

Replication of linear eukaryotic chromosomes Eukaryotic chromosomes, unlike their bacterial counterparts, are linear rather than circular. Since RNA oligonucleotides prime both prokaryotic and eukaryotic DNA synthesis, the 5′ termini of the daughter strands are incomplete in that they lack the DNA sequences that correspond to the RNA primers.

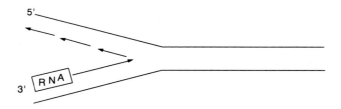

In protozoa this problem is solved by the addition of preexisting oligodeoxynucleotide blocks to the 3′ ends of DNA. These blocks are composed of tandemly repeated units of $(T_2G_4)_n$ or $(T_4G_4)_n$, where n is approximately 50. The enzyme that adds these polymers requires a primer but not a template. These oligonucleotide block polymers are called *telomers*. Another DNA polymerase,

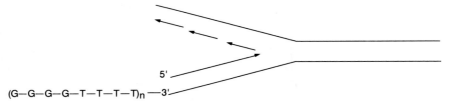

which is template dependent, copies the G_4T_4 units, synthesizing a complementary loop of C_4A_4. This looping back allows the 5′ end of the genomic daughter strand to be finished.

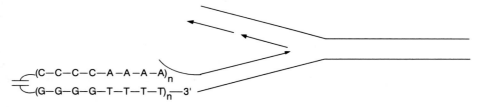

DNA ligase joins both ends of the telomere to the daughter strand, but the loop is subsequently cleaved to give flush-ended telomeres that consist of one strand of G_4T_4 and another of C_4A_4.

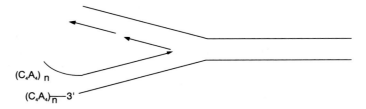

Nucleosome formation The synthesis of DNA and of histone proteins is coordinated. Duplication of a genome requires doubling the amount of histone proteins. During DNA synthesis the parental histones remain associated with only the growing strand made continuously at the replicative fork. The DNA of this new strand immediately hybridizes to one parental strand; thus the "parental" histones tend to stay associated with the DNA structure that remains essentially double stranded throughout the replication process. The other daughter strand is made in bits and pieces, and at any one time during replication it might contain considerable amounts of single-stranded DNA. As the segments on the lagging strand are finished and ligated together, the structure now binds newly synthesized "daughter" histones.

The association of histones to one or the other of the strands can be distinguished using electron microscopy of material from cells grown under conditions where protein (i.e. histone) synthesis is inhibited but DNA synthesis is not inhibited.

Information stored in eukaryotic genes Most of the *E. coli* genome codes for mRNAs that are translated into proteins, but this is not the case for the animal genome. Since animals are more complex, they probably require 50 to 100 times more protein than bacteria, but their genomes are more than 500 times larger. For example, the single chromosome of *E. coli* contains 4.5 million base pairs, whereas the 23 haploid human chromosomes contain 2400 million base pairs. The animal genome does have duplicate genes for many proteins, and the genes for ribosomal RNA and for some tRNAs are repeated many times, but the function of most animal DNA is still unknown.

Even the genes that code for proteins are more complex in vertebrates than in bacteria. Most, but not all, are expressed as long RNA molecules that are reduced in size by splicing together the coding segments. This yields a continuous template RNA that is sequentially decoded by protein-synthesizing enzymes (see Chapter 15).

DNA that has no apparent function as a template for cellular RNAs is sometimes called *nongenetic DNA*. Nongenetic DNA includes the pseudogenes. *Pseudogenes* are genes that cannot be expressed because they lack sequences necessary for RNA modification or protein synthesis initiation or because they contain protein synthesis "stop" signals in the middle of coding sequences. Nongenetic DNA also includes much repetitive DNA; about 30% of human DNA is repetitive. An example is the human *Alu* sequence. This 300–base pair sequence is repeated almost a million times at many places throughout the genome. Mouse satellite DNA is composed of a repeated sequence of similar size and number, but its sequence is tandemly repeated. The *Alu* sequence is named for the restriction endonuclease that cleaves at a single site within each repeated segment to yield many, almost identical copies of 300 base pairs each. Individual *Alu* sequences are homologous to one another by about 85%. *Alu* sequences are sometimes transcribed into RNA. A small cytoplasmic RNA, called 7SL RNA, that functions as part of a protein-secreting system is homologous with the *Alu* sequences at its 3′ and 5′ end; consequently the *Alu* sequences originally may have been derived from the 7SL RNA gene. What role, if any, they now play is unknown.

Transposable genetic elements Mobile genetic elements further complicate the organization of the chromosome. Mobile genetic elements are relatively small pieces of DNA that have characteristic sequences at either end. These pieces of DNA can move from one gene, or larger piece of DNA, to other locations, even on a different chromosome. The short sequences that flank the genetic elements are cleaved by an endonuclease to give staggered ends that base pair with complementary strands, a result of the nucleolytic cleavage of the target DNA by the same or a similar enzyme. A recombinational event (i.e., a crossing-over reaction) serves to transfer the small genetic element to its new location.

Many examples of mobile elements are found in bacteria, where they are called *transposons*. Bacterial transposons have terminal repeat sequences that both code for the enzymes catalyzing the process of transposition (transposases) and physically interact with these enzymes to bring them to the DNA target site. At this site the DNA-bound transposase presumably catalyzes the endonucleolytic cleavage of the terminal repeat sequence of the transposon and also catalyzes a similar sequence in the target DNA.

Perhaps the best examples of genetic transposition in animal cells are the integration and subsequent removal of DNA programmed by RNA retroviruses to and from any number of sites on eukaryotic chromosomal DNA. The single-stranded RNA retrovirus uses the enzyme reverse transcriptase to make a complementary DNA copy of itself. The RNA present in the DNA-RNA hybrid is rapidly degraded, leaving a single-stranded DNA, which integrates into the host genome when copied. Integration is random in respect to the host DNA. Transcription of the integrated viral genome produces many copies of viral RNA, which can be packaged into virus particles; the whole process is repeated in other cells or other organisms *(horizontal transmission)*. Sometimes the provirus is carried in germ line cells, where the sequences might be transmitted to new generations *(vertical transmission)*. In the case of the retroviruses, the enzymes needed for the movement of the proviral DNA are coded within the transposon, not in the repeated terminal sequences, as they are in bacteria.

The animal transposons, as with other transposons, can carry along pieces of the host DNA. For example, cellular oncogenes (abbreviated c-*onc*) are sometimes carried along with the proviral DNA when it is excised. These host sequences are maintained and carried, as RNA, in the retrovirus, where over time and after many passages they are extensively modified. Oncogenes isolated from retroviruses are abbreviated v-*onc*. More than 20 different v-*onc* genes have been found in the retroviruses of different experimental animals. These oncogenes code for proteins that lead to the transformation of normal cells to cancer cells. The various oncogene proteins have diverse functions and may be found in different parts of the cell. Several are tyrosine-specific protein kinases or other protein kinases, some bind guanine nucleotides and have guanosine triphosphatase (GTPase) activity, whereas others may be derivatives of normal hormone receptors or protein growth factors.

Molecular basis of mutation

On rare occasions a base may be changed or modified in the DNA sequence. As seen in Chapter 15 when protein synthesis is considered, such a change in the structural gene for a protein could lead to the insertion of the wrong amino acid. If changed at a crucial position, the resulting protein will be unable to function. If the amino acid replacement occurs at a less important position, activity may be diminished or not affected at all. Mutations are responsible for dozens of known genetic diseases and undoubtedly for many more yet to be discovered. Usually these changes are subtle so they cannot be detected cytologically at the level of the chromosome. Gross chromosomal abnormalities do occur and are very important in the health sciences, but generally they are not inherited in the classic mendelian way. Rather, most are caused by nondisjunction, that is, a failure of either the egg or the sperm to receive an exact set of haploid chromosomes or of a mitotic cell to receive an exact diploid set early in development. Many others are caused by translocations that are also difficult to predict.

Mutations are caused by both chemical and physical agents, although the action of even the physical agents, (e.g., ionizing radiation) can usually be explained by a chemical mechanism. Regardless of the agent used to produce a mutation, none is selective in the sense that it can specifically mutate one gene and not another. Because all genes are composed of only four different types of purine or pyrimidine bases, an agent that may react specifically with only one of the four could potentially cause mutations in every gene. Mutations are essentially random events. During our evolution the selective pressures of nature eliminated an astronomic number of deleterious mutations. The smaller number of beneficial mutations gave primitive life a survival advantage over competitors and allowed for the eventual emergence of intelligent beings. Consequently, in a highly evolved species such as humans, most mutations produce deleterious effects.

Mutagens

Purine and pyrimidine analogues Mutations may be produced in many ways. Bases may be deleted or new ones may be inserted; more frequently an existing base may be chemically modified so that on replication, improper base pairing will cause a different base to appear at the modified position. The latter type of mutation is called a *replacement*. When a purine is replaced by another purine or a pyrimidine by a different pyrimidine, the change is called a *transition*. A *transversion* is a change from pyrimidine to purine or purine to pyrimidine.

Many of the mutations caused by artificially produced base analogues are transitions. Mutations are produced by base analogues in one of two different ways. On entering the cell, a base analogue is converted to a nucleoside triphosphate that base pairs, perhaps incorrectly, with a DNA template and is inserted into the nucleotide chain. This is one way in which the mutation can be produced. The other requires an additional round of replication so that an improper base pair forms as a result of the previously incorporated analogue. The result in both cases is a permanently modified DNA.

As might be expected, base analogues can also inhibit DNA synthesis and cell multiplication. It is this feature that has stimulated organic chemists to create hundreds of different base analogues in the hope that some may be useful for inhibiting rapidly proliferating cancer cells. Examples of base analogues that have some usefulness in cancer chemotherapy and that are also mutagenic are 6-mercaptopurine and 2-aminopurine.

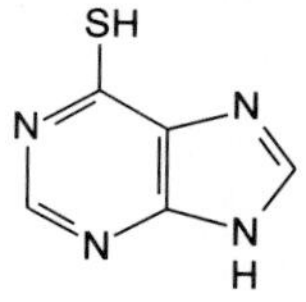

6-Mercaptopurine 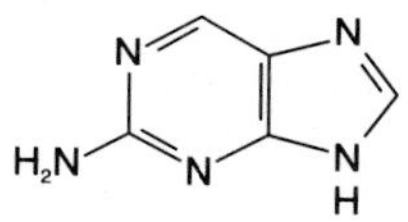   2-Aminopurine

Not all analogues become active against cancer cells through incorporation into nucleic acid. As seen in Chapter 13, some analogues block the synthesis of normal purine and pyrimidine nucleotides; for example, 8-azaguanine blocks guanosine monophosphate (GMP) synthesis and 6-mercaptopurine inhibits adenosine monophosphate (AMP) synthesis.

Alkylating agents Alkylating agents are also mutagenic substances that have been used in cancer chemotherapy. Alkylating agents such as nitrogen or sulfur mustards chiefly cause transversions. Bifunctional compounds such as those shown next produce cross-links between DNA strands or between a DNA strand and any other reactive group in the vicinity.

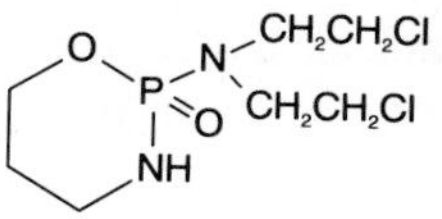

Cyclophosphamide 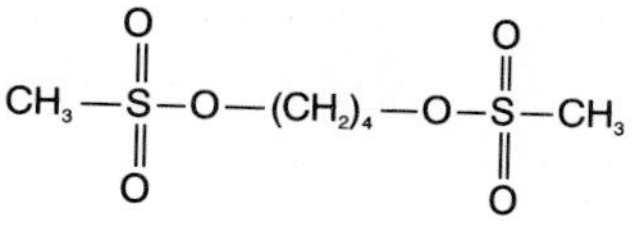   Busulfan

The mechanism of action of alkylating agents is complex. Adenine and guanine are easily alkylated. Guanine is alkylated primarily at position 7 and adenine at position 3. The reaction produces an exceedingly labile glycosidic bond. Splitting of this bond leads to depurination.

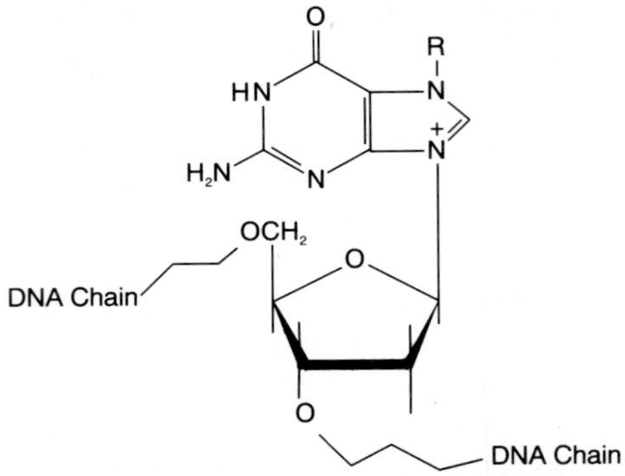

In those cases where alkylation does not lead to depurination, it is more likely that the mutation will be of the transition type. However, when depurination does occur, on replication the position opposite the gap might be filled by any one of the four bases. This accounts for the transversions often caused by these agents.

Dyes Acridine dyes such as the antimalarial agent quinacrine (Atabrine) shown next are large planar aromatic compounds that intercalate or sandwich themselves between the stacked bases of the helix.

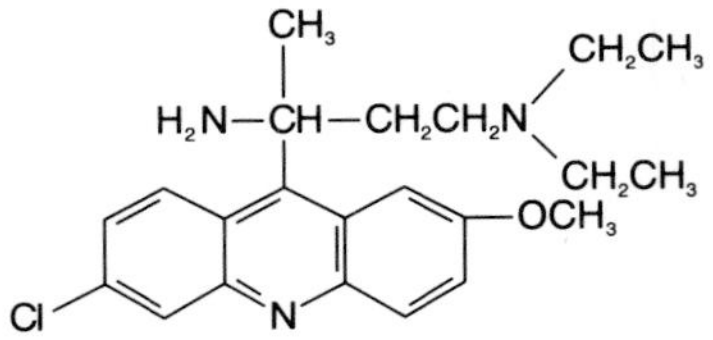

On replication, *insertion* or *deletion* of bases may occur. Chain scission and chromosome breaks are also possible. Quinacrine is useful in human cytogenetics, since it intercalates significantly into the heterochromatin of the Y chromosome, making it fluoresce and rendering it identifiable cytologically. Detection of the Y chromosome is important in prenatal sex determination.

Other dyes present in our environment are potentially mutagenic. For example, some hair dyes were shown to be mutagenic for *E. coli*.

Physical agents

Growing tissues are most sensitive to ionizing radiation. DNA synthesis is inhibited, yet the action of x-rays is indirect. They produce free radicals, which in turn react with DNA and thus produce point mutations or chromosomal breaks.

Large doses of ultraviolet light can damage DNA. In humans this damage is confined to the skin, since, unlike x-rays, ultraviolet light is easily absorbed. The chemical lesion in this case is the formation of dimers between adjacent thymine residues on the same DNA strand. Unless corrected or removed, these dimers will stop DNA synthesis.

DNA repair

Because most mutations are very damaging, even the simplest organisms have enzyme systems that repair DNA. These DNA repair systems are important because genetic defects in them can cause some human diseases.

Excision repair

The excision repair system consists of several enzymes, each involved in several steps. First, the error must be recognized. For example, an endonuclease binds regions of the DNA that contain thymine dimers and cleaves at the 5′ sides of the dimers. A DNA polymerase activity replaces that portion of the DNA strand that had contained the thymine dimer. An exonuclease then removes the piece of DNA containing the dimer, and a DNA ligase rejoins the repaired and restored DNA strand. These reactions are diagrammed in Figure 14.5. This form of nucleotide repair also acts on other types of damaged DNA, such as carcinogen-DNA adducts, and removes them by chain scission, patching, and ligation.

Some damaged bases, particularly alkylated purine bases, are removed by *N*-glycosylases. The gapped chain is cleaved by apurinic endonucleases and the defective strand patched and ligated. Single-strand breaks are repaired by analogous excision repair mechanisms. Mitomycin D and platinum complexes used in cancer therapy can cause DNA-DNA cross-links between bases on opposite strands. These cross-linked bases can be excised and repaired, first on one strand and then on the other. The repair is error free unless the drugs have cross-linked directly opposing bases.

Figure 14.5

Repair of DNA inactivated by ultraviolet light. Light causes the dimerization of adjacent thymine residues that block DNA replication. The four enzymes shown are involved in removal and replacement of a portion of the DNA that contains the dimer.

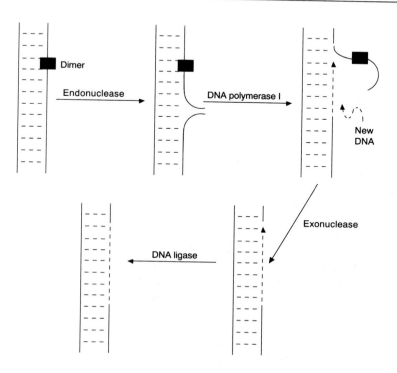

Postreplication repair   Sometimes damaged DNA is replicated before it can be repaired. When this happens, the replicating strand stops at the site of damage, skips over the damaged base, and completes synthesis of the new strand. The new daughter and old maternal strands separate, and eventually the missing base is added, postreplicatively. The complementary maternal strand still contains the damaged DNA, so this mechanism is not, strictly speaking, a repair mechanism, even though it allows synthesis of normal DNA. Eventually the damaged DNA is repaired by another mechanism.

Photoreactivation   This system acts directly on DNA damaged by ultraviolet light to restore the damaged base to its original state without actually replacing it. Because this system operates only on ultraviolet light–damaged DNA, it plays a limited role in repairing human DNA. A light-activated DNA photolyase catalyzes the conversion of thymine dimers to monomers.

DNA glycosylases   The DNA bases that contain amino groups tend to deaminate spontaneously. In particular, cytosine significantly deaminates to uracil, but adenine and guanine can also deaminate to hypoxanthine and xanthine, respectively. If not corrected, the new bases can cause serious mutations on replication. Fortunately, highly specific enzymes recognize these bases as being foreign to DNA, and they catalyze the hydrolysis of the N-glycosyl bonds that connect the bases to the DNA polymer. This produces DNA polymers with a few skipped bases. Another repair enzyme, an endonuclease, recognizes the skipped bases and cleaves the chain to leave a 3′ hydroxyl group on the 5′ adjacent nucleotide. A DNA polymerase now fills in the missing mononucleotide, taking its instructions from the intact complementary strand.

Defects have been found in these mechanisms that cause various human diseases. For example, patients with the genetic disease xeroderma pigmentosum are especially sensitive to ultraviolet light and develop skin cancer. Skin fibroblasts cultured from these patients have been shown to be defective in DNA repair.

Chemical carcinogenesis   Most human cancer is caused by substances in the environment, chemicals, viruses, or radiation. All these carcinogens affect DNA. Sometimes we can relate a substance to a specific type of cancer; for example, cigarette smoking to lung cancer. In other cases it is more difficult to draw a cause-and-effect relationship. Nevertheless, the evidence is overwhelming that carcinogens cause cancer by interacting with DNA. When the DNA modified by these agents cannot be repaired, cancer often results.

Carcinogenesis develops in three stages: initiation, promotion, and progression. Chemical substances can act at the initiation stage or at the promotion stage. Moreover, some chemicals have both initiating and promoting activities.

Initiating agents   Initiating agents alter the native molecular structure of DNA. They may cause an accumulation of somatic mutations over the lifetime of an individual. Initiating agents may be physical, biologic, or chemical—for example, ionizing radiation, tumor viruses, and cyclophosphamide. Chemical initiating agents have been extensively studied. These substances either interact directly with the DNA or are enzymatically modified to produce a metabolite that interacts with the DNA. The interaction is often covalent, although noncovalent reactions are possible. The resultant action of the initiating agent with DNA causes an irreversible change similar to a mutation; however, an observable mutational event, that is, a phenotype, is not an obligatory step in the initiation process. Furthermore, initiation does not in itself cause cancer. Instead it programs the cell so that subsequent reaction at a later time with a promoting agent starts the formation of cancerous cells.

Promoting agents

Unlike initiating agents, promoting agents do not interact directly with DNA, but rather influence the expression of the genetic information coded in DNA. Promoting agents include a variety of substances, such as hormones, protein growth factors, drugs, and plant products. Asbestos, cigarette smoke, alcohol, and phorbol esters are examples of promoting agents. These substances influence genetic expression by binding to receptors on cell surfaces or in the cytoplasm or nucleus. In contrast to initiating agents, the action of promoting agents is reversible. Some promoting agents, (e.g., estrogen and prolactin) are very specific so that they promote the formation of a tumor only in their target tissues. Other promoting agents do not act through a receptor mechanism, are nonspecific, and can promote tumor formation in a variety of tissues (e.g., iodoacetate). Table 14.4 summarizes epidemiologic data on some promoting agents.

Phorbol esters are promoters that interact with cellular receptors and activate protein kinase C. Usually protein kinase C is activated by Ca^{++} and diacylglycerol, both of which result from the hydrolysis of phosphoinositides catalyzed by phospholipase C. Phospholipase C is normally activated by several different growth factors (see Chapter 17). Thus phorbol esters bypass a tightly regulated step in the control of cell growth. Since protein kinase C phosphorylates various proteins, it is not known how this activity participates in establishing a cancerous line of cells.

Following promotion, cells go through a stage in carcinogenesis called *progression*. During this stage, there is a karyotype change from diploid to aneuploid that is associated with metastasis and morphologic changes. A summary of the carcinogenic events is illustrated in Figure 14.6.

Oncogenes

Oncogenes are genes involved in the transformation of normal cells to tumor cells. They do this by affecting cell growth and differentiation. Oncogenes were discovered as part of the genomes of RNA tumor viruses. Recall that RNA tumor viruses are propagated through DNA intermediates that are synthesized in reactions mediated by reverse transcriptase, a RNA-dependent DNA polymerase that is part of the virion. Thus RNA tumor viruses are called retroviruses. Not all retroviruses transform normal cells into tumor cells; for example, the human immunodeficiency virus (HIV) causes acquired immunodeficiency syndrome (AIDS).

The Rous sarcoma virus (RSV) is a retrovirus that contains an oncogene called v-*src*. Oncogenes that are part of the genomes of retroviruses are abbreviated v-*onc* whereas

Table 14.4	Examples of promoters with their associated neoplasms in humans	
	Agent	**Neoplasm**
	Excessive dietary fat (calories)	General increase in incidence of cancer
	Alcoholic beverages	Oral, liver, and esophageal cancer
	Cigarette smoke	Bronchogenic carcinoma, esophageal, and bladder cancer
	Synthetic estrogens	Liver adenomas
	Asbestos	Bronchogenic carcinoma and mesothelioma
	Not yet proved in humans	
	Saccharin	Bladder
	Tetradecanoylphorbol acetate (a phorbol ester)	Skin

Adapted from Pitot HC, Goldsworthy T, and Moran S: J Supramolec Struct Cell Biochem 17:133, 1981.

Figure 14.6   Karyotypic changes during initiation, promotion, and progression
stages of carcinogenesis. (Adapted from Pitot HC, Goldsworthy T,
and Moran S: J Supramolec Struct Cell Biochem 17:133, 1981.)

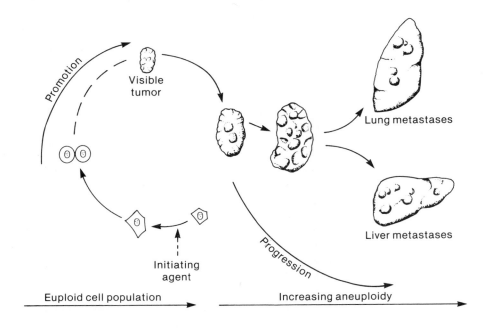

oncogenes of cellular genomes are designated c-*onc*. Cellular oncogenes are often called
proto-oncogenes to distinguish them from their viral counterparts and to emphasize that
the viral oncogenes originated at one time from cellular oncogenes.

Evidence for the first oncogene was found when certain mutants of RSV failed to
transform cells at high temperature but did transform cells at low temperature. The mutant
virus replicated well at both temperatures. This means that the virus contained a temperature-sensitive mutation in a gene that coded for a sarcoma-producing protein, that
is, for v-*src*.

Normal cells also were found to contain *src*-like genes when their DNA was hybridized
with labeled nucleic acid probes from the v-*src* gene. As with other cellular genes, the
c-*onc* genes were interrupted with introns. v-*Onc* genes lack introns. Consequently, in
the distant past c-*onc* genes must have been transferred to the retroviruses. If the transfer
had been from the virus to the cell, c-*onc* genes probably would not contain introns.

The retroviral *onc* gene products have diverse functions, depending on the particular
oncogene. A few representative oncogenes and the functions of their protein products are
shown in Table 14.5. Notice that the oncogene proteins are all involved in some aspect
of cellular growth promotion. For example, the tyrosine kinase activity of *src* is a property
shared by several receptors for hormones (e.g., insulin) and for growth factors (e.g.,
epidermal growth factor, or EGF). Some oncogenes code for subunits of growth factors;
an example is the v-*sis* protein, which is homologous to a subunit of platelet-derived
growth factor (PDGF).

The oncogene v-*erb*B, unlike c-*erb*B, codes for a shortened form of the EGF receptor
protein. Usually the binding of EGF is required to turn on the tyrosine kinase activity of

Table 14.5 Oncogenes of RNA tumor viruses

Virus	Oncogene	Oncogene protein product
Abelson murine leukemia virus	*abl*	Tyrosine kinase
Rous sarcoma virus	*src*	Tyrosine kinase
Avian erythroblastosis virus	*erb*B	Receptor for growth factor; tyrosine kinase activity
Simian sarcoma virus	sis	Growth factor, subunit of PDGF
Harvey murine sarcoma virus	Ha-*ras*	Guanine nucleotide-binding protein
Avian myelocytomatosis virus	*myc*	Nuclear protein

the EGF receptor protein, the one coded by c-*erb*B. However, the tyrosine kinase activity of the receptor derived from v-*erb*B is switched on permanently, even in the absence of EGF. The uncontrolled growth characteristic of cancer cells results.

Other oncogenes code for proteins that bind guanine nucleotides, and others code for nuclear proteins. The guanine nucleotide-binding proteins, the so-called G proteins, affect several key reactions. Some G proteins are stimulatory, whereas others are inhibitory. For example, they link hormone receptors to adenylate cyclase (see Chapter 17), they translocate molecules in protein synthesis, and they regulate cell proliferation. The functions of the nuclear oncogene proteins are not as well characterized, but they also are probably involved in key reactions of gene expression that regulate growth.

A unifying theme of the role of oncogene function in cell transformation is that the expression of the proto-oncogenes is normally tightly regulated. Anything that disrupts the regulation of oncogene expression could potentially cause cancer. For example, a chromosome break or a mutation, in either somatic or germinal cells, at or near an oncogene might lead to aberrant expression of the oncogene (see case 5). Similarly, a tumor virus could introduce an unregulated oncogene into cells and also cause cancer.

DNA sequence analysis

Some of the chemical reactions that cause mutations by modifying purine and pyrimidine bases can be used in the laboratory for sequencing DNA. One of the most commonly used methods for sequencing DNA is the chemical technique developed by Maxam and Gilbert. Since DNA molecules are extremely large, they are first broken into pieces more easily sequenced. This is done with restriction endonucleases (see Chapter 13), and the fragments are separated one from the other. A double-stranded DNA fragment is then labeled at its 5′ end using ATP (γ-^{32}P) and a polynucleotide kinase derived from bacteriophage T4-infected cells. The DNA fragment is denatured to yield single-stranded pieces, and these are separated from one another by electrophoresis on neutral gels of polyacrylamide:

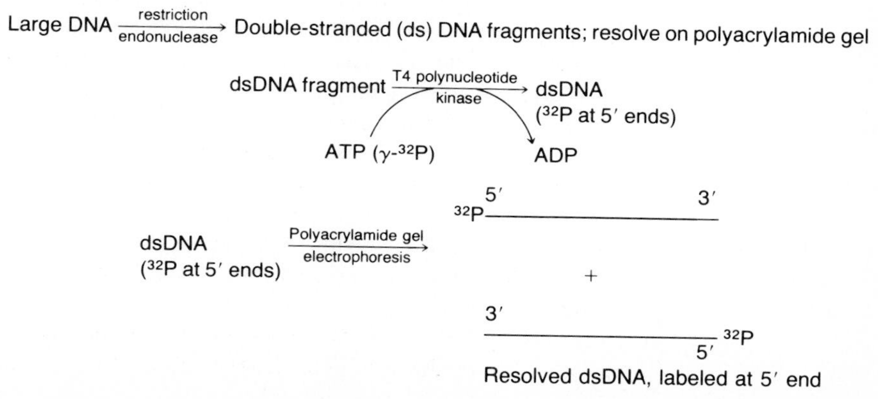

Large DNA $\xrightarrow[\text{endonuclease}]{\text{restriction}}$ Double-stranded (ds) DNA fragments; resolve on polyacrylamide gel

dsDNA fragment $\xrightarrow[\text{kinase}]{\text{T4 polynucleotide}}$ dsDNA (^{32}P at 5′ ends)

ATP (γ-^{32}P)   ADP

^{32}P————5′————————————3′————

dsDNA (^{32}P at 5′ ends) $\xrightarrow[\text{electrophoresis}]{\text{Polyacrylamide gel}}$

+

3′————————————————5′ ^{32}P

Resolved dsDNA, labeled at 5′ end

Each 5'-labeled single-stranded DNA is then sequenced. The method is based on two principles:

(1) polyacrylamide gel electrophoresis in 7 mol/L urea is capable of separating oligodeoxyribonucleotides that differ in size by just a single base.

(2) DNA can be chemically reacted so that chain cleavages occur at specific bases.

In a typical experiment (Fig. 14.7) a portion of the 5'-labeled single-stranded DNA is reacted with dimethylsulfate under conditions where many but not all of the purine bases react. This reagent alkylates the N7 position of guanine and the N3 position of adenine. Depurination of methylated guanine residues predominates when an aliquot is heated at neutral pH. Methylated adenine bases are preferentially removed by treatment with dilute acid. Both samples are then cleaved at the depurinated sites by alkaline hydrolysis. One now has two samples: one that contains DNA fragments all 3'-terminated at adenine residues and another consisting of DNA fragments all terminated at guanine residues. In practice, however, a few adenines are removed with the guanines. The deadenylation reaction is more specific. Since the ^{32}P label is at the 5' end of the DNA and the depurination reactions are not allowed to go to completion, the mixture of labeled oligonucleotides will range in size from the very largest, cleaved at the 3' purine residue most distal to the label, to the smallest, where the purine residue most proximal to the label has been removed.

An analogous series of reactions is used to produce depyrimidinated DNA fragments. Hydrazine is used in these reactions, since both cytosine and thymine react with hydrazine. The bases are cleaved to yield urea and a pyrazole ring. The deoxyribose moiety is left as a hydrazone. Piperidine, which reacts with the hydrazone, is used to cleave the nucleotide chain. Cytosines react specifically with hydrazine in 5 mol/L NaCl, but no specific reaction exists for thymines. Consequently, one aliquot yields labeled oligonucleotides 3'-terminated at cytosines, whereas a second aliquot contains nucleotides cleaved in the absence of NaCl at both cytosine and thymine residues. These reactions are summarized in Figure 14.7.

Each of four aliquots terminated at A, G, C, and C + T are denatured and loaded onto slabs of polyacrylamide gel. After electrophoresis, the radioactively labeled nucleotides are visualized by exposing the gel to an x-ray film. The resultant autoradiogram shows the position of each labeled oligonucleotide. An example of such an autoradiogram is shown in Figure 14.8. The longer polynucleotides are shown at the top of the diagram, and the shorter, faster moving polynucleotides are at the bottom. Because the chemical reactions do not go to completion, the autoradiogram displays a whole spectrum of polynucleotides, some of which have been cleaved at only a single base and others that have been cleaved at several. Each band differs from the adjacent band by one nucleotide. Consequently, the DNA sequence, 5' to 3', can be read starting from the bottom. Figure 14.8 shows only a portion of the gel.

Sequences of more than 350 bases can be obtained from some gels. The sequence of the complementary DNA strand is often determined as a check on accuracy of the first analysis. The primary structure of thousands of deoxyribonucleotides can be deduced by the analysis of overlapping DNA fragments generated by different restriction endonucleases. This technology is useful in determining the proximity of one gene to another and in studying mechanisms of gene expression.

Recombinant DNA technology in medicine

Thirty years ago the thought of sequencing all the genes in the human genome seemed as impossible as a trip to the moon. Now it is in the realm of possibility, and scientists are seriously talking about sequencing the entire human genome, all 23 chromosomes, amounting to 2,400 million base pairs (2.4×10^6 kb). Such plans are not based on the

Figure 14.7

Summary of reactions used to prepare oligonucleotides specifically cleaved at a particular base.

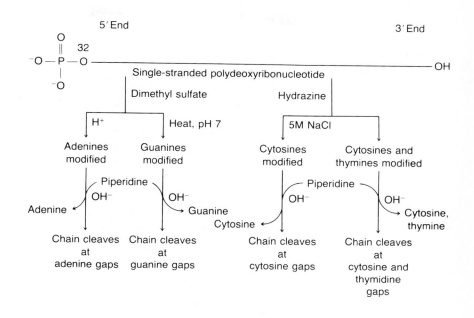

Figure 14.8

Autoradiogram of a Maxam-Gilbert sequencing gel. The large letters at the top, near the point of application, indicate which base was removed to generate the 3′ terminals of labeled nucleotides in that column. Smaller nucleotides migrate to the bottom of the gel, and the material in each labeled band is one nucleotide shorter than that in the band above it. The direction of electrophoresis is from top to bottom. The 5′ to 3′ sequence is revealed by starting at the bottom and reading up. (Courtesy of Drs. Brian Nichols and John Donelson.)

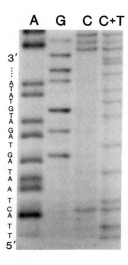

physical chemical separation of thousands of genes from human tissue, but rather on the isolation of genes or short pieces of DNA from a relatively small number of cells. These pieces of DNA can be cleanly separated from one another and then amplified to obtain sufficient material by using the powerful biologic methods of recombinant DNA technology.

Restriction endonucleases

Restriction endonucleases are at the core of recombinant DNA technology. The specificity of some of these enzymes is described in Chapter 13. Fortunately the restriction endonucleases cleave DNA at relatively few sites. The DNA fragments that result are usually long enough to be useful but short enough to analyze and even sequence. The mixture of DNA fragments following restriction nuclease digestion are resolved by electrophoresis on gels of polyacrylamide or agarose. The dye ethidium bromide is often used to detect the resolved fragments on the gel. For further analysis, individual fragments can be cut out of the gel and separated from the gel matrix and dye.

Restriction maps

Now commercially available are many restriction enzymes, named for the bacterial species from which they are isolated and possessing a wide range of specificity. Because the enzymes recognize such different nucleotide sequences, they can be used individually and in combination to develop "maps" of a particular DNA. For example, a 12 kb viral DNA might be cut into two pieces of 2 kb and 10 kb by endonuclease "A"; however, endonuclease "B" might cleave it into 5 kb and 7 kb fragments. All these can be resolved from one another by gel electrophoresis. Figure 14.9 illustrates these cleavages and shows that there are two different arrangements for the cleavages catalyzed by each nuclease. When endonuclease "A" and endonuclease "B" are mixed together during the digestion and the products separated by electrophoresis, the size of the fragments allows one to determine the relative position of the cleavage sites produced by each of the restriction enzymes. In this example the 2 kb fragment is assumed to be located at the left end of thee original DNA molecule. The ends of the fragments in Figure 14.9 are labeled L, for left, and R, for right.

Cloning of recombinant DNA

The most useful restriction enzymes cleave both strands of duplex DNA, but the nucleotide breaks are not directly opposite one another; the cleavages occur a few base pairs from a point of symmetry. These staggered ends can base pair to hold the DNA together weakly, or they can base pair with other DNA preparations that have been cut with the same type of nuclease. An example of joining DNA segments in this way is shown in Figure 14.10. In the laboratory the broken strands can be covalently joined together by treating them with DNA ligase. In this way foreign DNA can be linked to the DNA of plasmids or bacteriophages (bacterial viruses). This is illustrated in Figure 14.11, which shows some of the genes and restriction sites of the commonly used plasmid pBR322. Ap^R is a gene for ampicillin resistance, Tc^R is the gene for tetracycline resistance, and *ori* denotes the origin of replication sequences necessary for the plasmid to replicate its DNA separately from the bacterial DNA.

The bacteriophages or plasmids that have foreign DNA built into them in this manner are called *vectors*. Usually bacteriophages insert DNA very efficiently into their host bacteria. A single bacteriophage DNA can be replicated approximately 100-fold or more per cell depending on the type of phage and its host.

Plasmid vectors Plasmids are small, circular DNAs that range in size from 2 to approximately 100 kb. They also can be amplified to approximately 20 copies per cell. In addition, plasmids can be introduced into bacterial cells, although less efficiently than phage vectors. Some plasmids can be used to introduce recombinant DNA into yeast cells.

Figure 14.9   The construction of restriction maps. The relative position of cleavage sites for two hypothetic restriction enzymes is illustrated for a 12 kb piece of DNA. Endonuclease "A" cleaves at a single site to produce a 2 kb fragment and a 10 kb fragment. The 2 kb fragment is known to come from the left side of the molecule, indicated by *L*. The right end of the molecule is indicated by *R*. Endonuclease "B" cleaves the 12 kb DNA also at a single site, which yields a 5 kb fragment and 7 kb fragment, but it is not known whether the 5 kb fragment originates from the L or R end. Mixing the two nucleases before digestion will produce three fragments of DNA that can be analyzed on sizing gels. Two patterns are possible. In one case DNA fragments of 2, 3, and 7 kb would show when the gel is stained for DNA. In the other case only 2 and 5 kb fragments would appear.

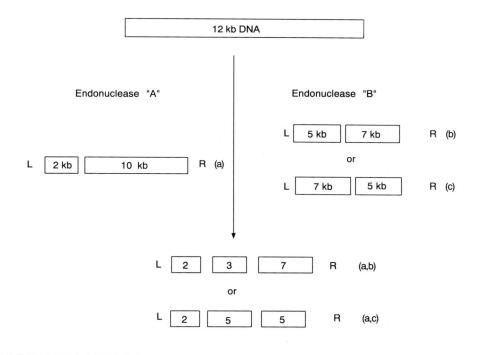

Clones One usually tries to adjust experimental conditions so that only one plasmid or phage is introduced into a single cell. When that cell replicates on a plate of nutrient agar, it will produce millions of cells until a colony becomes visible to the naked eye. If cells are sufficiently diluted before they are spread onto the plate, each colony will consist of a *clone* derived from a single ancestral cell. If each cell originally contained only one copy of the phage or plasmid vector, each colony will contain recombinant DNA that is homogeneous. That is, each colony will contain *cloned DNA*.

Clone selection using antibiotic resistance Plasmids found in nature often carry genes that code for proteins that make their hosts resistant to antibiotics. These genes are used to good advantage to select bacterial colonies that contain recombinant plasmids. For example, a restriction site may occur within a gene for tetracyline resistance. Figure 14.11 shows that a SalI site exists within the tetracycline resistance gene of plasmid

Figure 14.10   The formation of recombinant DNA using the restriction endonuclease
*Bam*HI.

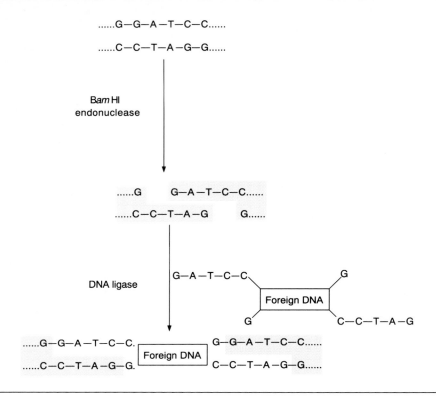

pBR322. When foreign DNA is inserted at this site, it interrupts the gene for tetracycline
resistance so that the cells are now sensitive to tetracycline.

Tetracycline sensitivity can be detected by plating the vector-containing cells onto agar
containing ampicillin but not tetracycline. All colonies on such plates will contain either
the original plasmid or plasmids containing recombinant DNA. Colonies are then replicate-
plated onto agar that contains tetracycline. Colonies that are sensitive to this antibiotic
can be determined from the differences in the pattern of colonies appearing on the two
plates. The antibiotic-sensitive colonies can be picked off and used for further experiments.
These colonies will contain plasmids carrying foreign DNA.

Phage and cosmid vectors Plasmids usually are able to carry much less foreign DNA
than a bacteriophage such as lambda. Thus plasmid vectors mainly are used to obtain
pieces of DNA (up to 3 kb in size) suitable for sequencing, to subclone larger DNAs
carried by phage vectors, or to clone single expressible genes derived from complementary
DNAs (cDNAs). On the other hand, vectors of bacteriophage lambda can accommodate
up to 23 kb of foreign DNA. The amount of DNA that can be inserted depends on the
amount of nonessential lambda genes that can be replaced and the total amount of DNA

Figure 14.11

Some genes and restriction sites on the *E. coli* plasma pBR322. Eco RI, Pst I, and Sal I are restriction endonucleases and the sites cleaved by these enzymes. (See text for abbreviations).

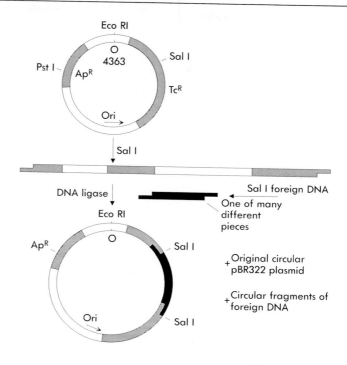

that can be packaged into a particle able to be covered with phage coat proteins. Some eukaryotic genomic genes have such long intervening sequences that their DNA may not fit into one phage head. For cloning these large genes, cosmid vectors are often used. Cosmids can accommodate very large pieces of genomic DNA. They contain only a small part of the lambda DNA, the two cohesive ends (*cos* gene) that allow any DNA to be packaged into particles in vitro. As with bacteriophages, these particles are very efficient vehicles for inserting DNA into *E. coli* cells; however, unlike phages, they cannot make infectious particles in vivo. Cosmids can carry up to 40 kb of recombinant DNA. They are engineered to contain an origin of replication, so the recombinant DNA can be amplified, and an antibiotic resistance gene, so they can be easily selected. Once in the cell, cosmids behave more as plasmids do.

cDNA is made in the laboratory from mRNAs using the enzyme reverse transcriptase. Unlike eukaryotic genomic DNA, cDNA does not contain intervening sequences. A complete cDNA may contain all the information needed to synthesize a mRNA for a specific protein. If the cDNA or its vector is modified so that the recombinant DNA has all the sequences necessary for RNA and protein synthesis, one can synthesize in bacterial cells the protein encoded in the recombinant cDNA. Such a vector is called an *expression vector*. Expression vectors can be used to obtain large quantities of a protein that might be very difficult to isolate by conventional techniques. This is possible because bacterial cells contain many copies of the recombinant gene and because RNA and protein synthesis occurs very rapidly in bacteria.

Libraries of genomic DNA

A recombinant gene library is made by digesting cDNA or genomic DNA with one or more restriction enzymes, ligating the fragments to a vector, and introducing the vector into appropriate bacterial or yeast cells so that each cell will contain a single recombinant DNA. A culture of millions of cells may contain fragments of all the cDNAs or all the fragments of a genome. These libraries can be screened to isolate a clone of interest by using a variety of techniques.

Genomic DNA is much more complex than cDNA. Since cDNAs are synthesized in the laboratory from mRNAs using reverse transcriptase, they are no more diverse than the number of mRNAs present at the time of isolation. However, only about 2% of the mammalian genome codes for the synthesis of proteins and their mRNAs. Thus genomic DNA libraries are at least 50 times more diverse than cDNA libraries. cDNA libraries are not easier to make, however, both because of the inherent instability of mRNA compared to DNA and because reverse transcriptase prematurely terminates when copying long mRNAs.

The otherwise impossible task of analyzing the 4 to 8×10^9 base pairs of human genomic DNA can be simplified. Many of the 23 human chromosomes can be physically separated from each other, their DNA isolated, and genomic libraries made from them. Also, a technique called pulsed-field electrophoresis allows the separation of DNAs as large as 10^6 base pairs. This procedure uses short pulses of current at alternating angles to the direction of movement of the DNA. The long DNA molecules take longer to reorient in the agarose gel matrix than the shorter molecules, and thus migrate more slowly.

Detection of recombinant DNA

Recombinant DNA in bacterial colonies or in bacteriophage plaques is most frequently detected using labeled hybridization probes. The hybridization probes may be composed of RNA, DNA, or oligonucleotides that have been labeled. Often a DNA restriction fragment is labeled in vitro using commercially available enzymes and radioisotopes. Smaller deoxyribonucleotides can be synthesized chemically using automated machines called DNA synthesizers. Methods are available for labeling either end of a nucleotide with ^{32}P substrates. Heavily labeled DNA probes can be made using α (^{32}P)-deoxyribonucleoside triphosphates and the Klenow fragment (see p. 607) of the *E. coli* DNA polymerase I. Small, random oligonucleotides are added as primers to the denatured DNA template so that large segments of cDNA are labeled with ^{32}P, not just the ends of the oligonucleotides.

A Petri dish containing bacterial colonies is blotted with nitrocellulose paper. This transfers a large portion of each colony to the paper, which is saturated with a solution that lyses (breaks open) the cells. The DNA of the lysed colonies is denatured with alkali. The nitrocellulose paper is neutralized, washed, and the paper either baked in an oven or treated with ultraviolet light to immobilize the denatured DNA. The DNA on the paper is hybridized with the labeled probe of interest, and the excess label is washed off. The dried paper is exposed to photographic film and the film developed. The exposed spots on the film can be matched with the colonies on the master plate and colonies picked off for further study.

A very similar protocol can be used to detect a particular DNA or RNA that has been resolved by electrophoresis. E.M. Southern developed a method for detecting individual DNAs that had been resolved by gel electrophoresis. The gel was blotted to nitrocellulose paper and the DNA on the paper hybridized with a labeled DNA probe. This procedure of analyzing DNA resolved by electrophoresis is called *Southern blotting*. Procedures were developed later for transferring by blotting RNAs that had been resolved by electrophoresis. This procedure was called *Northern blotting*. *Western blotting* is the analogous transfer of electrophoretically resolved proteins to nitrocellulose paper. In this case the proteins are usually detected immunologically.

Restriction fragment length polymorphism The Southern blotting of genomic DNA is used to detect restriction fragment length polymorphisms (RFLPs). *Polymorphism* is a genetic term indicating variation of a gene product within a population. Such variation may result in observable phenotypes; if serious, these are called mutations. In humans polymorphisms usually are not associated with an easily identifiable phenotype. Many polymorphisms occur in clinically normal people as part of their natural diversity. It has been estimated that a normal individual may be heterogeneous at 20% of his or her genes. The resulting enzymes are different but normal. Some loci, such as α-haptoglobin, are so diverse that no single allele can serve as the "normal" standard. Usually a single gene product must be represented in less than 1% of the cases examined for it to be considered polymorphic.

With the advent of electrophoresis, many polymorphisms were detected by the abnormal migration of a protein or enzyme. Many more polymorphisms now can be detected by the gel electrophoresis of DNA restriction fragments followed by Southern blotting. These DNA fragments show different mobilities on the gel, and thus they differ in length. For this reason they are called *length polymorphisms*. Length differences can result either from a mutation at an existing restriction site so that it is no longer recognized by the enzyme or from a mutation at another position to create a new restriction site. In the former case the restriction fragment would be longer; in the latter case it would be shorter.

RFLPs are often a reflection of individual genetic diversity and are not related to a clinical phenotype, but occasionally they can be diagnostic of an inherited disease. This technique is relatively new; yet, it has been applied to the prenatal detection of sickle cell anemia, thalassemia, phenylketonuria, α₁-antitrypsin deficiency, Huntington's chorea, Duchenne muscular dystrophy, hemophilia A and B, cystic fibrosis, and several other diseases.

Clinical comment

Prenatal diagnosis of sickle cell disease Sickle cell disease is caused by a mutation that results in the substitution of a valine residue for a glutamate residue in the sixth position of the hemoglobin β-chain. This results from the substitution of a T for an A in the glutamate codon. When (1) DNA from a patient with sickle cell disease is digested with the restriction enzyme *Mst*II, (2) the fragments separated by gel electrophoresis, (3) the gel blotted under denaturing conditions to nitrocellulose paper, and (4) the resolved DNA fragments hybridized with radioactive human globin DNA, one finds a new 376–base pair fragment instead of the normal 175–base pair DNA fragment. This indicates that a site usually recognized by the enzyme *Mst*II has been changed as a result of the mutation. *Mst*II requires the following sequence for its endonuclease action:

$$\ldots\ldots\text{C-C}\qquad\text{T-N-A-G-G}\ldots\ldots$$

The letter N indicates that any nucleotide at that position satisfies the specificity of *Mst*II. The normal sequence of the DNA of the β-chain of hemoglobin A near the site of the sickle cell mutation is:

$$\ldots\ldots\text{C-C}\qquad\text{T-G-A-G-G}\ldots\ldots$$

This sequence is cleaved by *Mst*II; however, the mutation to sickle cell anemia results in a change to the sequence:

$$\ldots\ldots\text{C-C-T-G-T-G-G}\ldots\ldots$$

This sequence cannot be cleaved by *Mst*II.

Restriction analysis can be used to detect sickle cell disease prenatally, since the DNA of all cells, including amniotic cells, carries the mutant DNA. It is much more difficult to obtain fetal blood for the analysis of the mutant hemoglobin A β-chain. Furthermore, fetal blood is composed mostly of fetal hemoglobin, since hemoglobin A is made later in development.

Reverse genetics

Originally this term applied to the modification in vitro of a piece of DNA of unknown function, with the subsequent identification of its function by introducing it back into cells. In human genetics the term is now used to describe the determination of the cause of an inherited disease by starting with the responsible gene and tracing it back to a defective enzyme. In the past, human geneticists first recognized the defective enzyme and then located the responsible gene.

Reverse genetics has been applied to diseases such as Duchenne muscular dystrophy and cystic fibrosis, in which the responsible enzymes are unknown and the disease results from a significant deletion. By combining RFLP analysis with cytogenetics, it has been possible to increasingly narrow the location of the defective genes to small regions on the affected chromosomes.

Polymerase chain reaction

Polymerase chain reaction (PCR) is a technique for amplifying small amounts of DNA when only small amounts of human tissue or cells are available. One needs to know the sequence of the DNA in question. Using the sequence, one chemically synthesizes two oligodeoxyribonucleotides of approximately 20 to 25 residues that are complementary to each of the two ends of the DNA of interest. These are annealed (heated, then cooled slowly) to the ends and serve as primers for the in vitro copying of each DNA strand using a heat-stable DNA polymerase from *Thermus aquaticus* (*Taq* polymerase). The chemically synthesized primers are in excess so that when the reaction mixture is heated to 63° C, the DNA strands separate and reanneal with more primer on cooling. This process of heating, cooling, and synthesis can be repeated many times, and in the process the DNA fragment of interest is greatly amplified. So much of a particular DNA can be made that it can be detected by procedures that do not require radioisotopes. The polymerase chain reaction has been applied to the detection of carriers of hemophilia A, to the prenatal diagnosis of sickle cell disease, and to fetal sex determination, among others.

Cloning and sequencing the human genome

Molecular biologists around the world have proposed the long-term goal of sequencing every gene on every human chromosome. Such a project will require international co-operation on a grand scale, as well as the development of automated machines to do most of the time-consuming work. At present such a project is theoretically possible, but the necessary machines have not yet been built. The goal is a worthy one because it will provide the framework to analyze every human genetic disease. Literally thousands of DNA probes would be available for screening patients, carriers, and others at potential risk. The sequences would also provide the information needed for genetic therapy when such procedures become available.

Bibliography

Adams RLP et al: The biochemistry of the nucleic acids, ed 10, London, 1986, Chapman and Hall Ltd.

Antonarakis SE et al: Genetic diseases: diagnosis by restriction endonuclease analysis, J Pediatr 100:845, 1982.

Butler WJ and McDonough PG: The new genetics: molecular technology and reproductive biology, Fertil Steril 51:375, 1989.

DuBridge RB and Calos MP: Molecular approaches to the study of gene mutation in human cells, Trends Genet 3:293, 1987.

Epstein CJ et al: Recent developments in the prenatal diagnosis of genetic diseases and birth defects, Annu Rev Genet 17:49, 1983.

Gilbert W: DNA sequencing and gene structure, Science 214:1305, 1981.

Itakura K et al: Synthesis and use of synthetic oligonucleotides, Annu Rev Biochem 53:323, 1984.

Kazazian HH: The nature of mutation, Hosp Pract 20:55, 1985.

Kogan SC et al: An improved method for prenatal diagnosis of genetic diseases by analysis of amplified DNA sequences, N Engl J Med 317:985, 1987.

Kornberg A: DNA replication (Minireview compendium), J Biol Chem 263:1, 1988.

McHenry CS: DNA polymerase III holoenzyme of *Escherichia coli*, Annu Rev Biochem 57:519, 1988.

McKusick VA: The new genetics and clinical medicine: a summing up, Hosp Pract 23:177, 1988.

Newport JW and Forbes DJ: The nucleus: structure, function, and dynamics, Annu Rev Biochem 56:535, 1987.

Orkin SH: Reverse genetics and human disease, Cell 47:845, 1986.

Scriver CR et al, editors: The metabolic basis of inherited disease, ed 6, New York, 1989, McGraw-Hill Book Co.

Watson JD et al: The molecular biology of the gene, ed 4, Menlo Park, Calif, 1987, Benjamin/Cummings Publishing Co, Inc.

Clinical examples

A diamond (♦) on a case or a question indicates that literature search beyond this text is necessary for full understanding.

Case 1	Xeroderma pigmentosum

A 45-year-old woman, who had lived her entire life on a farm, exhibited the skin manifestations of xeroderma pigmentosum (XP) without the fairly common neurologic symptoms. She was heavily freckled and had a lengthy history of skin cancer. In the past 5 yr she had 35 neoplasms excised from sun-exposed areas of her skin. Recently she developed an ocular involvement that required a corneal transplant.

Biochemical questions

1 What is the molecular defect in XP, and how is it related to a predisposition to cancer?

2 Describe the enzymatic systems involved in repairing ultraviolet-damaged DNA. How is DNA damaged by ultraviolet light? What is a thymine dimer?

3 Propose a biochemical explanation for the gentic heterogeneity of XP.

Case discussion

XP is an autosomal recessive human skin disease characterized by sensitivity to sunlight and by the development of multiple cutaneous neoplasms. It occurs in about one person in 250,000 of the general population, often in consanguineous matings, and is found worldwide in all races. Although the disease is detected by the effects it has on the exposed areas of the skin and eyes, it also can have systemic effects. For example, it causes several neurologic abnormalities, including progressive mental deficiency, deafness, ataxia, and retarded growth. The neurologic manifestations are characteristic of a variant form called the De Sanctis-Cacchione syndrome. Treatment involves minimizing exposure to sunlight and the removal of the tumors when they appear, but no cure exists.

1. Molecular XP defect The molecular defect in XP is associated with the excision repair of DNA damaged by exposure to ultraviolet light. The association of this defect with a strong predisposition to cancer has attracted much attention to this otherwise rather obscure disease. Exposure to sunlight is well correlated to the incidence of the two most common skin cancers, basal cell carcinoma and squamous cell carcinoma. The tumors of XP patients are no different from those of other individuals without the disease; only the incidence of tumors in XP patients is greatly increased. Patients with XP also develop malignant melanomas. Exactly how ultraviolet light induces skin cancer is not known, but it is suspected that the mutagenicity of ultraviolet light and its ability to induce cancer may both be the consequence of ultraviolet-modified DNA. In this connection the excessive freckling seen in XP patients may be a manifestation of the mutagenic effects of ultraviolet light. Freckles, even in normal persons, are collections of large melanocytes with melanosomes that are unusually large and dark. It is thought that each freckle represents a clone of melanocytes that arose from a single cell that had undergone an ultraviolet-induced mutation. The XP patient also has patches of hypopigmentation that could arise from mutations in pigment formation.

Cultured skin cells taken from XP patients are exceptionally sensitive to killing, mutagenesis, and chromosomal aberrations when exposed to ultraviolet light. These cells are also sensitive to chemicals that form large adducts with DNA. These chemicals were described earlier (see p. 618) as initiating agents of chemical carcinogenesis.

2. Repair of ultraviolet-damaged DNA Much is known about how ultraviolet light modifies bacterial DNA. The 5-6 double bond of cytosine can be hydrated, and DNA-protein cross-links may form, but the most significant lesion is the cross-linking of adjacent pyrimidine residues in DNA. Adjacent thymine residues on the same DNA strand dimerize to yield the following product:

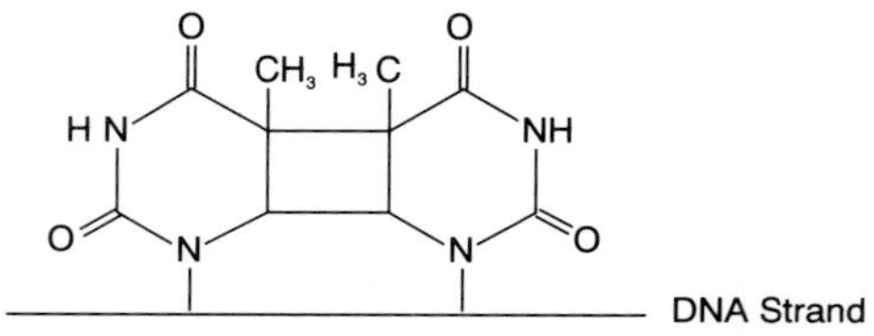

Both bacterial and animal cells have evolved enzyme systems to cope with ultraviolet-induced thymine dimers. At least three different enzyme systems neutralize the effects of thymine dimers. One, called *excision repair,* is discussed on p. 617. In excision repair an endonuclease recognizes the damaged DNA and breaks the nucleotide chain close to and on the 5′ side of the lesion. The damaged single strand peels off the DNA duplex, and a DNA polymerase catalyzes the synthesis of a polynucleotide to replace the damaged segment. A 5′-exonuclease (perhaps part of the DNA polymerase) hydrolyzes away the damaged single-stranded segment, and a DNA ligase seals the break, yielding a complete DNA duplex. A genetic defect in excision repair is responsible for XP.

Another way cells cope with damaged DNA strands is by *postreplication repair.* If the DNA is in the process of being replicated before being damaged, the DNA polymerase will skip over thymine dimers subsequently produced, leaving a slight gap in the newly synthesized strand. This gap can be later filled. The damaged DNA is not actually repaired by this mechanism, even though a correct copy of the damaged strand is made.

A third method for dealing with thymine dimers is called *photoreactivation.* The enzyme responsible for this form of repair has recently been found in human cells. The photoreacting enzyme simply converts dimers to monomers before they can do any harm. The enzyme binds to thymine dimers in DNA and catalyzes the formation of the thymine monomers when exposed to light of wavelengths 300 to 600 nm.

3. Genetic variants Many genetic variants in XP exist. At least 10 genetic complementation groups attest to the complexity of DNA excision repair in humans. It is likely that our understanding of excision repair is very superficial. The excision reaction possibly is catalyzed by an enzyme complex, or perhaps by a variety of enzyme complexes, so that a defect in any protein component would affect the catalytic activity of the complex. For example, cells from patients of complementation group E lack a nuclear protein that normally binds to damaged DNA. Consequently the defective protein in this variant functions at the stage of binding and recognition (Chus and Chang, 1988).

Several other diseases seem to be related to genetic defects in DNA repair. These are Cockayne's syndrome, which has two known complementation groups; ataxia telangiectasia, with five complementation groups; and Fanconi's syndrome, which has two variants. The molecular defects in these three diseases are unknown; however, a defect in DNA ligase I is responsible for Bloom's syndrome. Bloom's syndrome is characterized by photosensitivity and facial lesions.

None of the genes responsible for these diseases has been isolated.

References

Chu G and Chang E: Xeroderma pigmentosum group E cells lack a nuclear factor that binds to damaged DNA, Science 242:564, 1988.

Cleaver JE and Kraemer KH: In Scriver CR et al, editors: The metabolic basis of inherited disease, ed 6, New York, 1989, McGraw-Hill Book Co.

Rubin JS: The molecular genetics of the incision step in the DNA excision repair process, Int J Radiat Biol 54:309, 1988.

Timme TL and Moses RE: Diseases with DNA damage–processing defects, Am J Med Sci 295:40, 1988.

Case 2 — Chronic granulomatous disease

The patient, B.B., was a 15-year-old boy with a complex of X-linked inherited diseases. The boy was adopted at 10 days; no family history was available. Signs of Duchenne muscular dystrophy were seen at ages 1 and 2 years, retinitis pigmentosa at age 2, and a diagnosis of chronic granulomatous disease (CGD) was made at age 3. A small deletion in the X chromosome was found in somatic cell hybrids with Chinese hamster cells. This analysis limited the affected region to approximately 3000 kb of DNA. The locus was assigned to the position Xp21.1 (X, the X chromosome; p, short arm; 21.1, region 2, band 1, subband .1).

(Case adapted from Francke et al, 1985.)

Molecular genetics of X-linked chronic granulomatous disease A series of bacteriophage clones were made from seven regions of Xp21 that were absent from the patient. These clones spanned about 10% of the deleted region (i.e., 250 kb). Southern blot analysis was done using several single-copy probes made from these clones. DNA complementary to one probe was missing from both the X-chromosome–containing cell hybrid and from samples of the patient's DNA.

Radioactive cDNA probes were made from mRNA isolated from HL60 cells. These human cells had been treated to induce large amounts of the oxidase system. This mixture of cDNAs contained those associated with the oxidative response of the phagocytes, as well as cDNAs complementary to all the other noninduced mRNAs in the cells, that is, the constitutive mRNAs. The constitutive cDNAs were removed by subtractive hybridization using mRNA from an Epstein-Barr virus–transformed B-cell line that had been prepared from another patient with a similar deletion in the X chromosome. The cDNA-mRNA hybrids were removed, leaving labeled cDNAs for approximately 500 mRNAs.

These enriched and labeled cDNAs were used to probe the DNA from the Xp21 clones. One clone, pERT 379 (phenol-enhanced reassociation technique), responded positively. It was used to prepare nonrepetitive DNA fragments as hybridization probes for the Northern analysis of the HL60 mRNAs that had been used to make the enriched cDNAs. A 5 kb mRNA was found; it was named the 379 transcript.

cDNAs complementary to the 5 kb mRNA were isolated from libraries prepared from the mRNA of induced HL60 cells. These libaries were made using bacteriophage lambda expression vectors, *gt*10 and *gt*11. One bacteriophage clone was found that apparently contained the DNA for the entire 5 kb RNA. Using DNA from this clone as a probe, it was found that the 5 kb RNA was abundant in normal phagocytic cells and in the induced HL60 cells, but it was missing from several nonphagocytic cells and from three of four patients with X-linked CGD. The fourth patient made a 379 transcript, but it contained a smaller internal deletion.

The DNA sequence corresponding to the 379 transcript showed an open reading frame that could code for a protein of 468 amino acids and with a molecular weight of 54,000. This protein showed no sequence homology with any known proteins.

(These experiments were reported by Royer-Pokora et al, 1986.)

Biochemical questions

1. Explain how a chromosome deletion can cause three biochemically unrelated genetic disorders in the same patient.
◆ 2. How was it possible to isolate clones containing DNA that is deleted from the patient's X chromosome?
3. Earlier biochemical work had suggested that the defective protein in CGD was a cytochrome b-245. Subsequently the α and β subunits of this protein were separated and the sequence of the 42 N-terminal amino acids determined. The coding region for these 42 amino acids was present in a 5′ region of the 3779 transcript sequence, which had been originally described as noncoding. What might account for the description of this region as noncoding?
4. The correct assignment of the N-terminal amino acids, when matched with the DNA sequences, resulted in the addition of 101 more amino acids to give a protein with a molecular weight of 65,000. The β-subunit of cytochrome b-245 shows a diffuse molecular weight ranging from 60 to 90 kDa on polyacrylamide gels. What probably accounts for the apparently different molecular weights of this protein?

Case discussion

CGD is characterized by a severe predisposition to infection. Two forms of this genetic syndrome exist: an X-linked form representing approximately two thirds of the cases and an autosomal recessive form that comprises the rest. The biochemical defect in both forms of the disease affects a complex oxidase system that generates superoxide radicals that kill microorganisms engulfed by phagocytic cells. (See the clinical comment in Chapter 5, p. 203.) The defective genes in both forms of CGD have been cloned.

The technique of reverse genetics was used to obtain the amino acid sequence of the missing protein in the X-linked form of the disease without knowing precisely which protein was associated with the disease.

1. The chromosome deletion The constellation of clinical abnormalities suggests that this rather unusual patient has a chromosome deletion that contains the genes that are usually defective in most cases of the disease. These genes are rather close together on the X-chromosome. The diseases are biochemically distinct because they are caused by defects in the genes for different enzymes. The diseases are related only by their close association on the X chromosome.

2. Subtractive hybridization Subtractive, or competitive, hybridization can be used to hybridize an enormous amount of background DNA and thus remove it from the DNA of interest. This technique was used successfully by Kunkel and co-workers (see Royer-Pokora et al, 1986) to isolate the gene defective in Duchenne muscular dystrophy. Briefly, the DNA from normal X chromosomes was cleaved with the restriction enzyme *Mbo*II to generate many fragments with overlapping, or "sticky," ends. Potentially the overlapping ends of these fragments can base pair intramolecularly to form circles or intermolecularly to form longer fragments. Separately, the DNA from the X chromosome of a patient with a deletion in the X chromosome is also fragmented; however, this DNA is fragmented by shearing forces that produce blunt-ended (nonoverlapping) pieces. Both sets of DNA fragments are mixed together, heated to denature them and allow the strands to separate, and allowed to rehybridize. The sheared DNA with the deletion is used somewhat in excess. Potentially all normal DNA fragments can hybridize with the sheared DNA, except for those sequences that are missing because of the chromosomal deletion. The nonhybridized fragments can be built into plasmids by virtue of their sticky ends and the plasmids cloned.

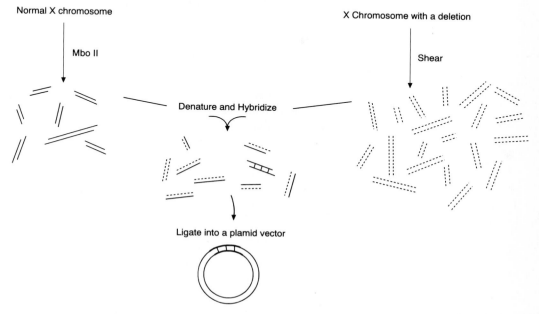

3. The 379 transcript codes for the β-subunit of cytochrome b-245 The misassignment of the protein synthesis start signal was the result of a nucleotide sequencing error. Antibody made to the fusion protein product of this gene cross-reacts with the authentic β-subunit of cytochrome b-245 and thus confirms this assignment.

4. The diffuse molecular weight of cytochrome b-245 isolated from cells The diffuse pattern of the β-subunit on polyacrylamide gels results from the covalent attachment of variable amounts of complex polysaccharides to the protein. The subunit contains approximately 21% carbohydrate, which disproportionately increases its molecular weight. Recombinant proteins are usually expressed in bacterial cells that do not have the necessary enzymes to glycosylate proteins. As a result, such recombinant proteins may have different properties than the native proteins isolated from animal cells.

References

Francke U et al: Minor Xp21 chromosome deletion in a male associated with expression of Duchenne muscular dystrophy, chronic granulomatous disease, retinitis pigmentosa, and McLeod syndrome, Am J Hum Genet 37:250, 1985.

Orkin H: Molecular genetics of chronic granulomatous disease, Annu Rev Immunol 7:227, 1989.

McKusick V: The new genetics and clinical medicine: a summing up, Hosp Pract 23:177, 1988.

Royer-Pokora B et al: Cloning the gene for an inherited human disorder—chronic granulomatous disease—on the basis of its chromosomal location, Nature 322:32, 1986.

Segal AW: The molecular and cellular pathology of chronic granulomatous disease (review), Eur J Clin Invest 18:433, 1988.

Case 3   Cystic fibrosis

The typical patient is a young white person who produces thick, mucous secretions. The disease is associated with chronic pulmonary difficulties, failure of the pancreas to secrete enzymes, and an increased loss of salt in sweat. Cystic fibrosis is the most common of the lethal inherited diseases in northern Europe and North America. The defective gene is inherited as an autosomal recessive trait. Recently, without knowledge of the defective protein, the disease was mapped to human chromosome 7, between 7q2.2 and 7q3.1, a distance of 2000 to 3000 kb. Subsequently, it has been further narrowed to a region of approximately 40 kb. Most cases of cystic fibrosis are believed to be caused by a single mutational event, so the new DNA probes are sufficiently precise to detect the carrier state in the general population.

Biochemical questions

1. At first many known RFLPs for conventional markers (e.g., for serum enzymes) were used to probe DNA from patients with cystic fibrosis. No linkage to other diseases could be found, but about 40% of the human genome was eliminated as *not* containing the defective gene. What is meant by linkage? How were these experiments done?
2. How many genes might be accommodated in a 40 kb stretch of DNA? Assume an average number and size of introns.
◆ 3. Estivill and co-workers searched for the cystic fibrosis gene among nonmethylated regions near the cystic fibrosis locus using a methylation-sensitive restriction enzyme. What is the rationale for searching in a region that lacked methylated DNA?

References

Estivill X et al: A candidate for the cystic fibrosis locus isolated by selection for methylation-free islands, Nature 326:840, 1987.
Johnson JP: Genetic counseling using linked DNA probes: cystic fibrosis as a prototype, J Pediatr 113:957, 1988.
McPherson MA: Recent advances in cystic fibrosis, J Inher Metab Dis 11suppl 1:94, 1988.
Ostrer H and Hejtmancik JF: Prenatal diagnosis and carrier detection of genetic diseases by analysis of deoxyribonucleic acid, J Pediatr 112:679, 1988.

Case 4   Acute lymphoblastic leukemia

A 5-year-old girl was admitted to the hospital suffering from loss of appetite, weakness, pain in the joints, and fever. The child's spleen, liver, and lymph nodes were found to be enlarged. A diagnosis of acute lymphoblastic leukemia was made. The patient was immediately given prednisone, vincristine, and L-asparaginase. Allopurinol was begun as a prophylactic measure. On induction of remission, central nervous system prophylaxis involved the injection of methotrexate and radiation treatment. Maintenance therapy involved the administration of 6-mercaptopurine and methotrexate, with occasional pulses of vincristine and prednisone for a period of 2½ to 3 yr.

Biochemical questions

1. What is the mechanism of action of the drugs used in this case?
2. The compounds listed have some serious side effects when high dosages are administered. Why?
3. What are the side effects of cancer therapy in a 5-year-old child?
4. Why was allopurinol administered?

5. Why must the 6-mercaptopurine level be reduced if allopurinol is used in the treatment?

References

Jacobs AD and Gale RP: Recent advances in the biology and treatment of acute lymphoblastic anemia in adults, N Engl J Med 311:1219, 1984.

Linker CA: Improved results of treatment of adult acute lymphoblastic leukemia Blood 69:1242, 1987.

Mayer RJ: Allogeneic transplantation versus intensive chemotherapy in first-remission acute leukemia: is there a "best choice"? J Clin Oncol 6:1532, 1988.

◆ Case 5 — Blastic chronic myelogenous leukemia

Patients with Philadelphia chromosome–positive blastic chronic myelogenous leukemia were studied for the effectiveness of treatment with interferon. Approximately 60% to 70% of the patients treated responded. A good predictor of responsive patients is the presence of the enzyme terminal deoxynucleotidyl transferase in peripheral blood or bone marrow blast cells and the disappearance of the Philadelphia chromosome.

Biochemical questions

1. What is the Philadelphia chromosome? How is it and other chromosomal translocations related to disease?
2. What is terminal deoxynucleotidyl transferase?
3. What is the significance of the *bcr*-c-*abl* fusion protein?
4. Explain how the tyrosine protein kinase activity of the fusion protein might contribute to leukemia.

References

Beard MEJ and Fitzgerald PH: The Philadelphia chromosome: a brief review, Aust NZ J Med 18:617, 1988.

Browett PJ et al: Chromosome 22 breakpoints in variant Philadelphia translocations and Philadelphia-negative chronic myeloid leukemia, Cancer Genet Cytogenet 37:169, 1989.

Croce CM: Chromosomal translocations, oncogenes and B-cell tumors, Hosp Pract 20:41, 1985.

Heisterkamp N et al: The *bcr* gene in Philadelphia chromosome positive acute lymphoblastic leukemia, Blood 73:1307, 1989.

Additional questions and problems

1. Compare and contrast the different DNA polymerases.
2. Distinguish templates from primers and give examples of each.
3. In a genetically determined disease such as phenylketonuria, is the critical liver enzyme missing or just inactive? (See Ledley FD et al: Science 228:77, 1985.)
4. In X-linked disease, why is the male usually affected and the female not?
◆ 5. Several carcinogenic compounds were found to be mutagenic in a bacterial test system (the Ames test), but only after activation by an extract of rat liver microsomes containing NADPH. What type of biochemical reactions are necessary to convert these compounds to mutagens?
◆ 6. Restriction endonuclease mapping has been used for the prenatal diagnosis of thalassemias caused by globin-gene deletions. Explain how this is done.
◆ 7. Discuss the advantages and disadvantages of enzyme replacement therapy using red cells into which have been incorporated enzymes that the patient cannot make.
◆ 8. Describe the molecular mechanisms by which photochemotherapy for psoriasis might induce cutaneous carcinomas.

Multiple choice
problems

The problems in this section deal with the prenatal diagnosis of cystic fibrosis (CF) using RFLPs which is done routinely in many large medical centers.

1. RFLPs are detected using labeled DNA probes. The labeled DNA must be complementary to:
    A. All the fragments of the patient's DNA cleaved by the restriction enzyme selected.
    B. The DNA sequence around a point mutation.
    C. DNA within the mutated gene.
    D. DNA within the mutated gene or close enough to the gene so that at least one of the restriction fragments differs in size from the normal.
    E. Long DNA fragments from the proband and short DNA fragments from normal controls.

2. DNA restriction fragments are usually separated one from the other by:
    A. Paper chromatography.
    B. High-speed centrifugation.
    C. Electrophoresis in agarose gels.
    D. High-performance liquid chromatography.
    E. thin-layer chromatography.

3. Before the separated restriction fragments are hybridized to the labeled DNA probe, they must be:
    A. Denatured with alkali.
    B. Carefully maintained in their native state.
    C. Stained with ethidium bromide.
    D. Treated with a nuclease specific for single-stranded DNA.
    E. Chemically cross-linked to hold the double helix together.

4. A DNA probe called KM-19 (See Feldman GL et al: Lancet ii:102, 1988) detects a polymorphism of *Pst*I-cut DNA that is very useful in the prenatal diagnosis of CF. DNA surrounding the *Pst*I cut of KM-19 has been sequenced and complementary oligonucleotides chemically synthesized to prime the polymerase chain reaction (PCR). The PCR is especially useful in prenatal diagnosis because:
    1. Small amounts of fetal DNA can be amplified to a concentration that can be analyzed.
    2. DNA generated by the PCR can always be more accurately analyzed than DNA isolated from fetal cells.
    3. The amplified DNA can be analyzed within 24 hr, whereas Southern blotting methods may take up to 10 days.
    4. The PCR will amplify DNA from the proband only and will not contaminate DNA from the mother.
    The best answer is:
    A. 1, 2, and 3.
    B. 1 and 3.
    C. 2 and 4.
    D. 4 only.
    E. All are correct.

5. The synthetic primers used for the PCR differ from cellular primers of DNA synthesis in that:
    1. Cellular primers are never RNA.
    2. The synthetic primers are removed by the action of RNase H.

3. The synthetic primers are RNA oligonucleotides.
4. The synthetic primers are DNA oligonucleotides.
   The best answer is:
   A. 1, 2, and 3.
   B. 1 and 3.
   C. 2 and 4.
   D. 4 only.
   E. All are correct.

6. Two oligonucleotide primers were used to amplify a DNA fragement that contained the informative *Pst*I site. The two primers will function best when they are complementary to the:
   A. 5′ end of one template strand and to the 3′ end of the same strand.
   B. 5′ end of one strand and the 3′ strand of the other.
   C. 5′ ends of both strands.
   D. 3′ ends of both strands.
   E. Middle or the ends of either template strand.

7. The conditions for the PCR were annealing for 2 min at 55° C, DNA extension for 5 min at 72° C, and denaturation for 1 min at 94° C. Fifteen cycles of this treatment were done, more polymerase was added, and 15 additional cycles completed. This protocol requires that:
   A. The DNA polymerase be from a heat-resistant microorganism.
   B. The template DNA be very short so that it will separate from the primer at 55° C.
   C. The reaction mixture also contain the *Pst*I restriction enzyme.
   D. The template DNA be present in molar excess of the primer.
   E. Radiolabeled deoxyribonucleoside triphosphates be used to label the amplified DNA.

8. The product of the amplification reaction is 0.95 kb. The amplified DNA from a normal homozygote does not contain a *Pst*I site. Amplified DNA from a CF homozygote contains a *Pst*I site, which yields fragments of 0.65 kb and 0.30 kb. CF is an autosomal recessive disease, so the most informative pattern for the *Pst*I-treated amplified DNA of the clinically normal parents of a child with CF is:
   A. DNA of 0.95 kb for both parents.
   B. Approximately half the DNA of both parents being 0.95 kb and the other half 0.65 kb and 0.30 kb.
   C. DNA of one parent being 0.95 kb, 0.65 kb, and 0.30 kb, and that of the other parent all 0.95 kb.
   D. DNA of both parents being 0.65 kb and 0.30 kb.
   E. DNA of one parent being 0.65 kb and 0.30 kb, and that of the other parent 0.95 kb, 0.65 kb, and 0.30 kb.

9. Prenatal screening of amplified DNA from an at-risk fetus of the parents in problem 8 was done. The *Pst*I-treated DNA showed only one band at 0.95 kb. This suggests that the fetus is:
   A. Homozygous for CF.
   B. Heterozygous for CF.
   C. Homozygous and normal for CF.
   D. Either heterozygous or homozygous for CF and that Southern blotting analysis

is necessary to distinguish between these alternatives.
E. Either heterozygous for CF or normal and that Southern blotting using more markers is necessary to distinguish these alternatives.

10. The use of the PCR with the KM-19 polymorphism is useful in the prenatal diagnosis of about 70% of the families at risk for CF. The major reason why it is not possible to detect all cases of CF is because:
A. The gene for CF has not been cloned, although it is close to KM-19.
B. Many cases of CF are the result of double mutations that introduce another *Pst*I site.
C. In certain individuals some *Pst*I is methylated.
D. Multiple CF genes exist.
E. It is still not known which protein is defective in CF.

Chapter 15

RNA and protein biosynthesis

Objectives

1 To explain how the primary structure of DNA is reflected in the synthesis of RNA and how RNAs contribute to the synthesis of proteins
2 To describe the synthesis of RNA molecules and how this synthesis is regulated
3 To show how RNA molecules are modified following their synthesis
4 To explain how drugs and antibotics affect RNA and protein synthesis
5 To describe how viruses express their genetic information and influence cellular metabolism.

The master molecule deoxyribonucleic acid (DNA) has two major functions: it must (1) replicate and (2) express itself. Chapter 14 focuses on the replication function; this chapter describes the second function, expression. The expression of genetic information requires (1) that the information contained in the nucleotide sequence of a gene be converted to an analogous sequence in ribonucleic acid (RNA) and (2) that this sequence be used to direct the synthesis of a specific protein. The first process is called *transcription* and the second process, *translation*. A *transcription unit* is the portion of a DNA sequence that codes for the synthesis of an RNA. The RNA that is made from a transcription unit is called a *transcript*.

RNA synthesis: expression of the genetic material
Heterochromatin of eukaryotic cells

Many years ago cytologists observed that some regions of the chromosomes appeared tightly packed (heterochromatin) and visible during the interphase portion of the cell cycle, whereas other regions (euchromatin) lost their visibility when the cells entered interphase. These investigators believed that the condensed heterochromatin was not expressed but that the euchromatin was expressed. This suspicion has been confirmed by the observation that the tightly packed DNA is not transcribed into RNA.

It is not known how euchromatin and heterochromatin differ chemically; their differences are not caused simply by the attachment of histone proteins, since these proteins are bound to both types of chromatin.

The heterochromatin and the DNA sequences within it should not be thought of as useless, however. In embryonic life much of what later becomes heterochromatin does appear to be expressed. One of the most striking examples of a totally heterochromatic chromosome has already been described (see Chapter 2), that of the human X chromosome. In females one of the two X chromosomes exists as a highly condensed Barr body, the DNA of which is not expressed. No evidence exists for the total inactivation of other chromosomes; however, some observations indicate that small segments of an autosomal chromosome may be inactive in one of a pair of chromosomes.

Those regions of euchromatin that are transcriptionally active are loose, open nucleosome structures that are accessible to deoxyribonuclease I (DNase l) treatment in cell-free extracts. The hypersensitivity to nucleases seems to be mostly an intrinsic property of the DNA sequence itself; nevertheless, some of the hypersensitivity probably results from changes in the nucleosome structure caused by the covalent modification of constituent histones. Such modifications may cause histones to leave the DNA. They may then be replaced by other proteins that function during transcription and that also help to maintain the open DNA structure.

Messenger RNA

RNA is an intermediary between DNA and the synthesis of proteins. This template RNA, or messenger RNA, is a direct reflection of the instructions "written" in DNA. The separate nucleus in animal cells and the fact that cells rich in RNA are very active in protein synthesis argue circumstantially for this intermediary role. Direct evidence for a role of RNA templates in protein synthesis comes from in vitro experiments showing that RNA, but not DNA, serves as a template for protein synthesis. The three types of RNA are transfer RNA (tRNA or sRNA), messenger RNA (mRNA), and ribosomal RNA (rRNA). All are involved in protein synthesis. The tRNA carries the amino acids as high-energy esters to a site on the ribosome, a particle composed of about 60% RNA and 40% protein. Here the tRNA base pairs in a specific way with the mRNA, the template RNA.

RNA polymerase

The DNA-dependent RNA polymerase was discovered in rat liver as a catalyst for the incorporation of radioactive ribonucleoside triphosphates into large polymers that had all the characteristics of RNA. The reaction is shown in Figure 15.1. Mechanistically, it resembles DNA polymerase; addition is to the 3' end of the growing chain, and a DNA template as well as all four of the ribonucleoside triphosphates are required. The RNA product is of high molecular weight, attacked by RNase and alkali, and has a base composition similar to one of the strands of the DNA template. If the RNA product is heated and slowly cooled in the presence of the DNA template, a double-stranded *hybrid* is produced; one strand is DNA and the other is RNA. The RNA product hybridizes only to the template DNA and not to other DNAs.

Inhibitors Several inhibitors have been used to study the mechanism of transcription. The cancer chemotherapeutic agent actinomycin D inhibits the elongation of the RNA chain by binding to deoxyguanosine residues on the DNA template in a manner that prevents movement of the RNA polymerase along the DNA molecule. Actinomycin D is valuable clinically as an anticancer agent and in the laboratory, where it is used to separate the process of transcription from translation. It is generally assumed that acti-nomycin D inhibits transcription but not translation, although the drug has some side effects. Proflavin, a member of the family of acridine dyes mentioned in connection with mutagenesis, also can block the elongation of RNA chains by binding to the DNA template. Proflavin is less specific than actinomycin D and binds to groups other than deoxyguanine residues.

Rifamycin and its derivatives (rifampicin) inhibit the RNA polymerases of microorganisms and mitochondria by preventing proper initiation of the synthesis of RNA chains. Rifampicin is useful in the treatment of tuberculosis, since it probably inhibits RNA synthesis in tubercle bacilli. It is also one of the few antiviral agents. Poxviruses replicate their RNA in the cytoplasm of animal cells. The enzyme involved is induced by the virus and is sensitive to rifampicin.

Figure 15.1   DNA-dependent RNA polymerase.

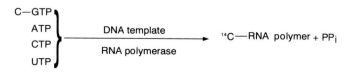

Subunits The RNA polymerase from *Escherichia coli* has been well characterized; it is composed of several different kinds of subunits. All the subunits can be physically separated one from the other. Table 15.1 lists the subunits with their molecular weights and their possible functions. The *holoenzyme* is designated $\alpha_2\beta\beta'\sigma$; it has a molecular weight of about 450,000 and promotes *asymmetric* synthesis of RNA; that is, only one strand of DNA is copied. The *core enzyme* is designated $\alpha_2\beta\beta'$. It has activity in vitro with nicked DNA templates, but since it lacks the σ-subunit, initiation is faulty and often both strands of DNA are copied, yielding a double-stranded RNA, an artifact of the incomplete in vitro conditions.

Mechanism of RNA synthesis

Synthesis starts by the attachment of the holoenzyme to characteristic sites on the DNA template under the influence of the σ-subunit. These sites are called *promoter* sites. Although only one of the DNA strands is read out, it is not always the same one.

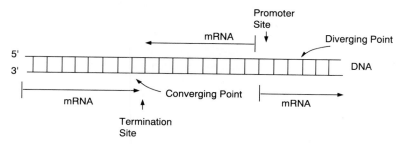

Thus synthesis on complementary strands may converge to a point or diverge from another point.

The σ-subunit itself has affinity for DNA only when it is part of the polymerase. Since the β'-subunit has a high affinity for DNA, the major forces holding the RNA polymerase to the template must be between the β'-subunit and DNA. The σ-factor must ensure that binding occurs at promoter sites and nowhere else. The β-subunit interacts with rifamycin or its derivatives. Because these compounds specifically block initiation of RNA synthesis, the β-subunit must be directly concerned with forming the first intranucleotide bond. Growth of the RNA strand is at the 3' end of the new chain. The terminal 5' nucleotide is always a guanylic or adenylic acid residue. Once this bond forms, the σ-subunit comes off and awaits the termination of synthesis of the polynucleotide chain and the release of the core enzyme. The core enzyme and the σ-factor then interact, and the holoenzyme that is formed initiates RNA synthesis all over again. The σ-factor can be thought of as cycling on and off the RNA polymerase. Elongation of the polyribonucleotide results in an RNA strand that is complementary and antiparallel to the DNA strand copied.

E. coli has a σ-factor of 70 kDa that serves to initiate most transcription; however, the expression of heat-shock genes requires another σ-factor. *Bacillus subtilis* and some of its bacteriophages code for multiple σ-factors.

Table 15.1 Properties *E. coli* RNA polymerase subunits

Subunit	Molecular weight	Property
α	36,000	?
β	150,000	Binds rifamycin
β	155,000	Binds DNA
σ	70,000	Proper initiation

The termination of RNA synthesis in *E. coli,* where it is best understood, sometimes involves another protein called rho (ρ), which is, however, not a part of the RNA polymerase. Rho is composed of six oligomers of 46 kDa each. It binds to nascent RNA that lacks secondary structure. Rho then moves along the nascent RNA, probably accompanied by the hydrolysis of adenosine triphosphate (ATP), until it engages an RNA polymerase enzyme stalled at a termination site. Release of the newly synthesized RNA with ρ-factor follows. In the absence of ρ, the RNA chain is not terminated, and abnormally long RNA molecules are produced. These may contain the information for the synthesis of several proteins. In bacterial cells, long mRNAs of this type exist naturally. They are called *polycistronic* because they contain the transcripts of several genes or cistrons. No good evidence exists as yet showing the presence of polycistronic mRNAs in animal cells. On the contrary, the mammalian RNAs that have been isolated are *monocistronic.*

Posttranscriptional modification of mRNA

Most eukaryotic mRNA is unusual because of its distinctive end groups, both of which are modified after the mRNA has been transcribed from the DNA template. An increasingly large number of virus-induced mRNAs as well as most cellular messengers have been found to have a 7-methyl guanylic acid residue at their 5′ ends. The methylated guanylic acid residue is linked through a phosphodiester bond connecting its 5′-hydroxyl group to the 5′-triphosphate end of the mRNA. These "capped" messengers are resistant to some nucleases and greatly stimulate the translation of the messenger.

A 5′ - capped mRNA

The 3′ end of most animal mRNAs contains approximately 200 adenylic acid residues. This poly A segment, as with the 5′ "cap," is added posttranscriptionally by enzymes that recognize mRNA but not rRNA or tRNA. The poly A tail probably is important in transporting mRNA from the nucleus to the cytoplasm; it is not essential for translation, since several mRNAs (e.g., histone mRNA) lack poly A tails.

RNA splicing

Perhaps the most unusual posttranslational modification is the removal of large polynucleotide pieces from the internal portion of the RNA. It is easy to visualize how an RNA matures by removal or addition of nucleotides at the ends of the polymer, but it is more difficult to imagine how pieces are taken out of the middle. Basically the process involves the removal of a stretch of apparently "nonsense" ribonucleotides from a central portion of the RNA followed by the rejoining, or splicing, of the remaining pieces of RNA. The process is called *splicing,* and the piece removed is called an *intervening sequence* or an *intron.* The coding portions of the sequence are called *exons.* The intervening sequence has few known functions and is usually degraded. It appears that most precursor forms of mRNA contain intervening sequences; histone mRNAs are again notable exceptions. A single mRNA precursor may contain several intervening sequences, for example, the

Figure 15.2

Maturation of chick ovalbumin mRNA precursor. Coding sequences are shown as heavy shaded bars connected by lines representing intervening sequences.

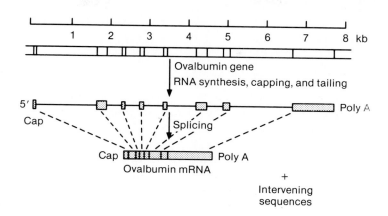

chicken ovalbumin mRNA is split eight times. The maturation of chick ovalbumin mRNA is illustrated in Figure 15.2. Intervening sequences in mRNA precursors vary in length from less than 100 to more than 1000 mononucleotides. Intervening sequences are also present in the precursors of yeast rRNA and in yeast tRNAs. In the latter the sequence removed is much shorter, about 15 nucleotides.

The enzymes responsible for splicing RNA precursors function in the nucleus. They have remarkable specificity. The intervening sequences may loop out to form a region of secondary structure. The enzyme recognizes this structure and cleaves at the intervening sequence in such a way as to preserve the reading frame of the messenger. If the enzyme were to make a mistake of even one nucleotide, a nonsense-containing protein could result and the effect might be devastating. The splicing enzyme complex has two activities: (1) to cut out the intervening sequences and (2) to reseal the "sense" strands. Possibly two associated proteins participate in this function. Most splicing involves RNA, not DNA. Obviously DNA must contain all the genetic information for the synthesis of the precursor RNAs, the part that eventually becomes mRNA as well as the intervening sequences themselves. However, evidence suggests that the genes for immunoglobulins might be rearranged at the DNA level during embryogenesis by somatic recombination, a somewhat analogous process.

Splice junction sequences The splicing of RNA requires a remarkable fidelity, so it is not surprising the nucleotide sequences at the splice junctions are highly conserved. Every mammalian intron has a GU sequence at the 5′ end and an AG sequence at the 3′ end.

$$5'—GU—Intron—AG—3'$$

In addition to these invariant nucleotides, other highly conserved sequences exist on either side of both splice junctions. The 5′ consensus sequence contains nine nucleotides; three of the nine are part of the intron and include the GU sequence. The 3′ consensus sequence

Figure 15.3   Consensus sequences at splice junctions. The circled bases are the
                             invariar t ends of the intron. *Py* indicates any pyrimidine and *N*, any
                             base.

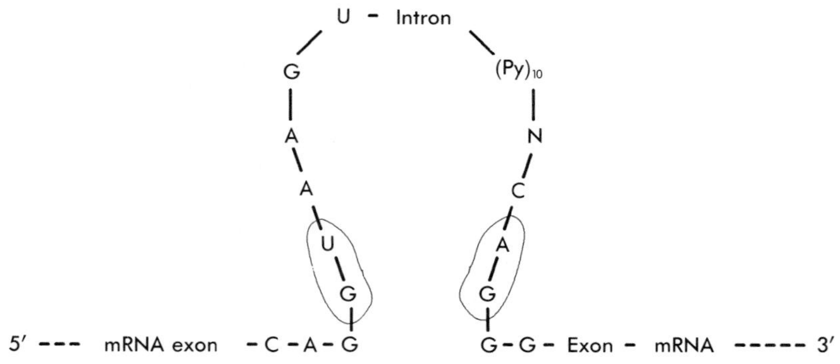

Figure: Consensus sequences at splice junctions, showing the intron loop with U at the top, and the sequences G–A–A–U–G (circled U–G) on the 5' side and (Py)$_{10}$–N–C–A–G (circled A–G) on the 3' side, connecting 5' --- mRNA exon – C – A – G to G – G – Exon – mRNA ----- 3'.

contains 16 nucleotides; 15 of the 16 are part of the intron and include the AG sequence. Typical consensus sequences near the splice sites are illustrated in Figure 15.3.

Spliceosomes Spliceosomes are the large enzyme complexes that catalyze the removal of introns and the joining of exons. In addition to the precursor mRNA, spliceosomes contain three ribonucleoprotein complexes, abbreviated snRNPs (pronounced "snurps"), for *small nuclear ribonucleoproteins*. Each snRNP contains a different small nuclear RNA; these are abbreviated U1, U2, and U5. Other small nuclear RNAs, such as U4 and U6, are part of the snRNPs that play important roles in the assembly of the spliceosome during a process that requires ATP.

The U1 RNA (165 nucleotides) contains a sequence that is complementary to the 5' splice site, whereas the U2 snRNP binds to the polypyrimidine sequence near the 3' splice site and to a so-called branch site that is only 20 or so bases "upstream." U5 snRNP also binds at the 3' splice site. Figure 15.3 shows the 5' and 3' splice sites.

The lariat intermediate The branch site of the precursor mRNA is located within the intron between the two splice sites. The branch site contains an adenylic acid residue that is important in forming a "lariat" intermediate. It is called a lariat because when the intron is cleaved at the 5' splice site, the 5'-phosphoryl group of the intron is transesterified to the 2'-hydroxyl of the adenylic acid residue in the branch site. Figure 15.4 shows this feature.

Clearly, two phosphodiester bonds must be broken to remove an intron, but also note (Figure 15.4) that two new phosphodiester bonds are formed: one between the two exons and the other between the 5' end of the intron and a 2'-hydroxyl group on the adenylic acid residue within the branch site. Consequently, in the simplest sense, RNA splicing consists of two transesterification reactions.

Catalytic RNA: self-splicing reactions Some precursor RNAs in lower organisms such as yeast, fungi, and *Tetrahymena* can be spliced in reactions catalyzed by RNA, not protein enzymes. This was the first evidence that biologic catalysts may be composed

Figure 15.4

Steps in splicing together two exons. Exons are boxed. The branch site is shaded. Y, any pyrimidine; R, any purine; N, any nucleotide.

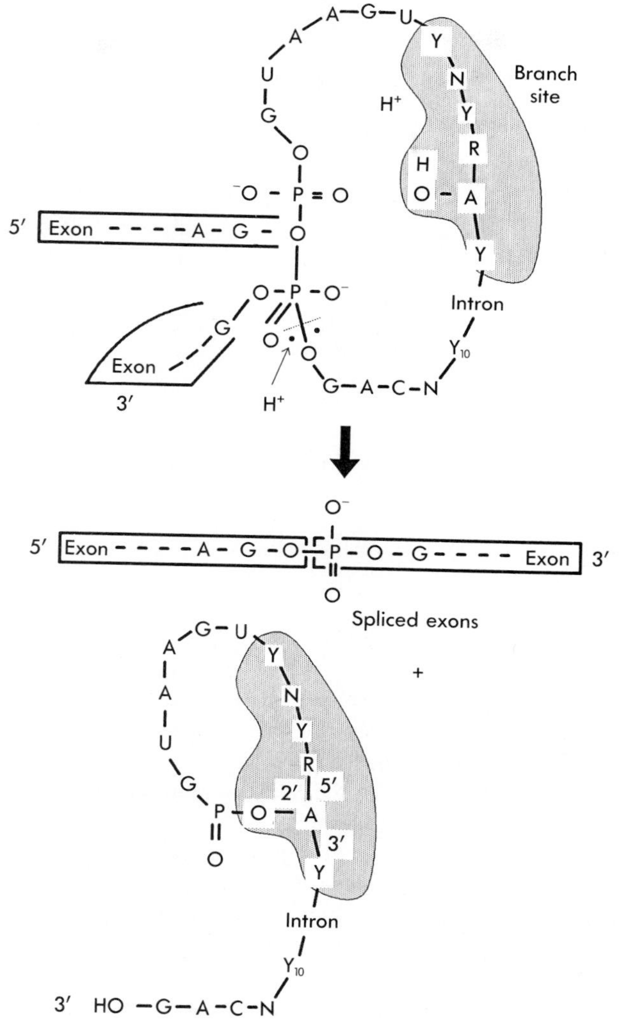

entirely of RNA rather than protein. Self-splicing precursor RNAs are of two types, called group I and group II. Group I RNAs require in addition a guanosine or guanylic acid (GMP, GDP, or GTP) cofactor. The 3'-hydroxyl of the guanosine cofactor acts as a nucleophile to initiate the transesterification reactions and actually becomes incorporated into a part of the excised intron. Group II RNAs undergo a self-splicing reaction that is virtually the same as the spliceosome catalyzed reaction; that is, a lariat forms at a 2'-hydroxyl group of an adenylic acid residue within the branch point.

Operons and control of RNA synthesis

In bacteria the synthesis of polycistronic mRNAs is regulated in a special way. The polycistronic message is coded by a continuous stretch of DNA that contains the infor-

Figure 15.5   Genes of the lactose operon. The arrangement of the genes
concerned with regulation of the lactose operon are shown as they
occur on the *E. coli* chromosome. The "i" gene codes for the
repressor and has its own promoter site. *cAMP,* Cyclic adenosine
monophosphate; *CAP,* catabolite gene—activator protein.

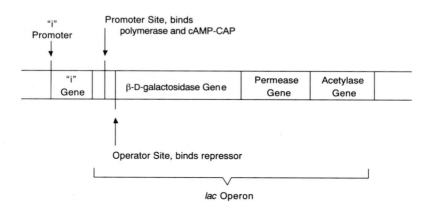

mation for the synthesis of several proteins that are required for a particular metabolic
function. Such a set of genes is called an *operon.*

The lactose, or *lac,* operon of *E. coli* can serve as a representative example. Usually
E. coli cells are grown in the laboratory in a simple medium containing glucose and a
few salts. The organism can make all its essential cellular constituents from this medium.
Most strains of *E. coli* can also grow well on lactose or other galactosides; however, this
requires that they induce three enzymes that are concerned with lactose metabolism. These
are a β-D-galactosidase needed to hydrolyze the lactose to D-glucose and D-galactose, a
D-galactoside permease to facilitate the movement of lactose into the cell, and a transacetylase whose function is not well understood. The structural genes that code for these
proteins are contiguous to an *operator* site of about 35 base pairs and a *promoter* site for
attaching the RNA polymerase. The arrangements of these genes are shown in Figure
15.5. Next to the *lac* operon is the "i" gene, which codes for the synthesis of a *repressor*
protein, a tetramer with molecular weight of about 150,000. The *lac* repressor has a high
affinity for about 25 base pairs of the *lac* operator site and for no other operators. When
it is bound to the operator, movement of the RNA polymerase along the template and
into the operon is blocked. When the repressed cells are given a diet of lactose instead
of glucose, lactose induces the synthesis of the proteins coded by the *lac* operon. In this
role, lactose is called an *inducer.* Inducers need not always be sugars, but they are always
small molecules. The inducer, either lactose or its analogues, specifically interacts with
the *lac* repressor to effect its removal from the operator. The RNA polymerase then
transcribes the derepressed cistron. The control region of the *lac* operon has been sequenced. The following diagram shows the relative arrangement of the protein binding
sites in the control region.

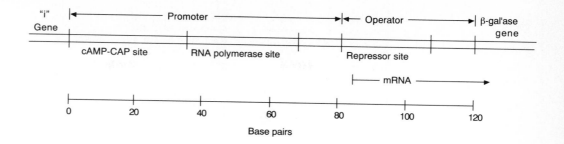

So far the type of control described is of a negative type; that is, the expression of the *lac* operon is normally prevented by a repressor protein. Actually the situation is more complicated. Control of the *lac* operon also has an element of positive control, and the positive controlling element is cyclic adenosine monophosphate (cAMP). Cells grown on glucose have low concentrations of cAMP. On induction with lactose, cAMP binds to a protein called a catabolite gene–activator protein (CAP), a dimer with molecular weight of 45,000. The cAMP-CAP complex binds to DNA, probably at the promoter site, in such a way that the bound RNA polymerase can initiate RNA synthesis. Thus expression of the *lac* operon requires both the inducer lactose to remove the repressor protein and cAMP to activate CAP so that RNA synthesis is initiated. The function of cAMP in directly stimulating the transcription of certain genes might occur only in bacteria. Cyclic AMP functions differently in animal cells.

Analogous to enzyme induction is the process of end product repression. For this kind of control the end product of a metabolic sequence, for example, an amino acid, must be synthesized in excess. The excessive amount of the amino acid interacts with a repressor protein; in so doing, it *activates* the repressor to block the functioning of an operon concerned with making the enzymes used in the synthesis of the amino acid. The small molecule end product is called a *corepressor*.

Transfer and ribosomal RNA synthesis

Thus far the synthesis of RNA in the simplest sense, that is, single-stranded, unmodified RNAs such as mRNA, has been considered. The rRNAs and tRNAs are the most abundant and the most stable of the cellular RNAs. Their synthesis and maturation differ somewhat from mRNA.

Transfer RNA structure Much is known about the structure of the relatively small (molecular weights, 25,000 to 30,000) tRNAs. Since several amino acids can be attached to multiple species of tRNA, probably more than 40 different types of tRNA exist. The reason for this will be apparent later when the genetic code is discussed. The complete nucleotide sequences of many tRNAs have been determined, and knowledge of the sequences has proved valuable.

The sequence of the yeast tRNA for alanine, as shown in Figure 15.6, was the first reported. When the sequence is arranged to give the maximum number of base pairs, the "cloverleaf" pattern illustrated in Figure 15.6 results. All the tRNAs that have since been sequenced can be arranged in this pattern. Other similarities have also been noted; every tRNA contains a CCA grouping at the 3' end. It is at this end of the molecule where the amino acid is attached through its acyl group to the 2' or 3'-hydroxyl of the terminal adenosine, forming a high-energy ester.

Figure 15.6   Nucleotide sequence of alanine tRNA. See text for discussion and abbreviations. (From Holley R et al: Science 147:1462, 1965. Copyright 1965 by the American Association for the Advancement of Science.)

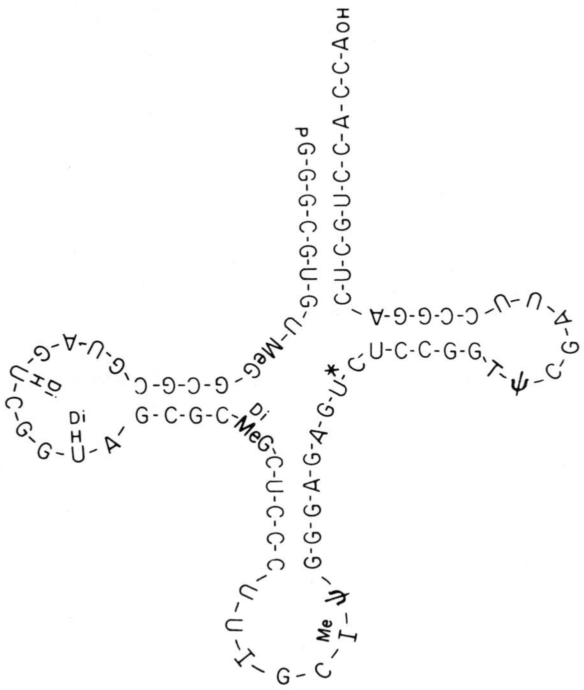

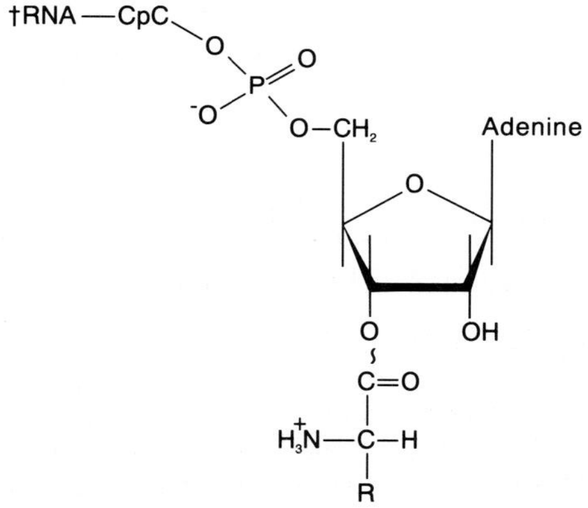

The 5' end of almost every tRNA contains a guanine residue. Several unusual bases exist in tRNA, and these proved useful in elucidating the structure. Every tRNA contains a common T-ψ-C-G tetranucleotide in the right-hand loop. T represents ribosyl thymidine, and ψ represents pseudouridine (5-ribosyl uracil). This sequence is probably part of a binding site that recognizes features of the ribosome involved in forming the peptide bond.

The *anticodon* is located midway in the molecule. It is in a single-stranded region that base pairs with the *codon* (a code word consisting of three nucleotides) of the mRNA in an antiparallel fashion. Note that the anticodon of yeast alanine-tRNA is I-G-C. The inosinic acid residue base pairs much like guanylic acid so that the codon G-C-G in an mRNA would specify alanine.

The synthesis of tRNA and rRNA are alike in that both are synthesized as larger precursor molecules and both contain modified or unusual bases. The large precursor molecules are first cleaved by nucleases to the proper size, and then other enzymes modify the appropriate bases. Cells contain several different methylating enzymes that use *S*-adenosylmethionine to methylate tRNA. Foreign nucleic acid such as that found in a virus can be overmethylated when introduced into a cell. Occasionally this acts as a defensive measure on the part of the host to eliminate foreign nucleic acid; that is, undermethylated foreign DNA is hydrolyzed by host restriction endonucleases before it can be methylated. One theory proposes that the transformation of cells by cancer viruses may occur because of a failure in performing specific methylations.

Biosynthesis and maturation of rRNA Very little of the genetic potential of cells is used to make tRNA, perhaps one or a few genes for each tRNA. Much more of the cell's DNA is used to make rRNA. As many as 100 to 1000 identical genes exist for rRNA in vertebrate oocytes. The site of rRNA synthesis is the nucleolus, where each multiple gene is transcribed sequentially but separately from a strand of DNA that has been amplified many times. Each nucleolus contains a single circular DNA that contains dozens of tandemly arranged genes for rRNA. Figure 15.7 illustrates the growth of several strands of rRNA of varying lengths. The electron micrograph shows that the redundant genes for rRNA are separated by a short DNA segment. The length of a single gene is shown by the line designated *M*. The regular head-to-tail sequence of the fernlike pattern of RNAs emerging from the central DNA strand indicates that these genes are all read out in the same direction. The longer "leaves" on the fern indicate those RNA molecules whose synthesis is almost completed. The short strands at the other end of the matrix are RNA strands just starting to grow. Enlargements of these electron micrographs indicate that each growing RNA has what appears to be a molecule of RNA polymerase attached at the junction made by the RNA strand and the DNA. The 28S and 18S RNAs found in the ribosome are made as one piece of 45S RNA.

Several degradative steps yield 28S and 18S RNAs. Ribosomal proteins made in the cytoplasm are transported into the nucleoli, where part of the ribosomal subunits are assembled. Figure 15.8 outlines the maturation of ribosomal RNA in the nucleoli of human HeLa cells. As with tRNA, rRNA is also methylated. This takes place in the nucleoli while the RNA exists as a 45S precursor. Methylation requires *S*-adenosylmethionine. On methylation, the ribosomal RNA precursor is able to function as a substrate for nucleolar nucleases that cleave the molecule into its mature pieces of 23S, 18S, and 5.8S size.

RNA polymerases of animal cells The polymerases from mammalian tissues are tightly bound to the deoxyribonucleoprotein of the nucleus and aggregate easily; however, soluble preparations can be obtained by treating the extracts with sound waves (sonication) under appropriate conditions. Three DNA-dependent RNA polymerases derived from rat liver have been isolated and separated from one another. RNA polymerase I is found in the nucleolus; it is involved in the synthesis of the 45S rRNA precursor. RNA polymerases II and III are present in the nucleoplasm and are more concerned with the synthesis of

Figure 15.7   Electron micrograph of a nuclear core with growing strands of rRNA
attached to their sequential DNA. The line designated *M* shows the
length of the single rRNA gene. (× 25,500.) (From Miller OL and
Beatty B: J Cell Physiol 74:225, 1969.)

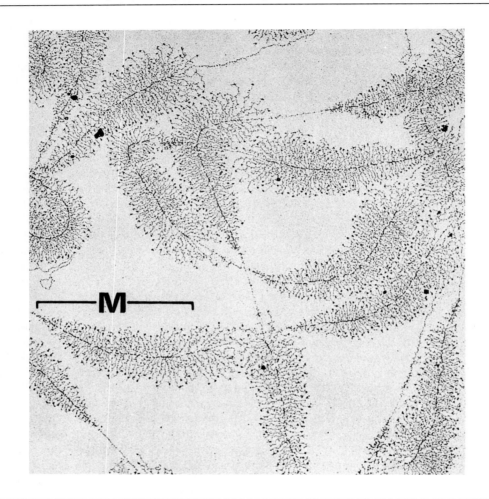

mRNA and tRNA, respectively. They are specifically inhibited by the antibiotic
α-amanitin. A fourth RNA polymerase has been found in mitochondria. All the mammalian polymerases have several properties in common; they are inhibited by actinomycin
D, DNA is used as the template, and all four ribonucleoside triphosphates are required.
However, polymerases I, II, and III also have many properties that distinguish them from
each other. These are summarized in Table 15.2.

Clinical comment   **Mushroom poisoning** Thousands of mushroom species are toxic when eaten. One of the
most dangerous species is *Amanita phalloides*. On ingestion, this mushroom causes severe
abdominal cramps, vomiting, and diarrhea within 12 to 24 hours. About 20% of the cases
are fatal. Toxicity is primarily caused by the toxin lα-amanitin, which is not inactivated

Figure 15.8 Maturation of HeLa cell ribosomal RNA.

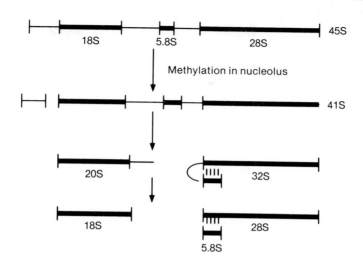

Table 15.2 Comparisons between RNA polymerases I, II, and III from rat liver

Property	Polymerase I	Polymerases II and III
Cellular location	Nucleolus	Nucleoplasm
Optimum ionic strength	Low	High
Optimum Mn^{++}/Mg^{++} activity ratio	2	5
Sensitivity to α-amantin	Insensitive	Sensitive; polymerase III less sensitive
RNA synthesized	Large rRNA precursors	Polymerase II: mRNA precursors Polymerase III: tRNA and 5S RNA

by cooking the mushrooms. This cyclic octapeptide is a potent inhibitor of RNA polymerase II. It binds very tightly to the enzyme, blocking the elongation phase of transcription. RNA polymerase III is less sensitive to the toxin, and RNA polymerase I is not inhibited at all. Toxicity undoubtedly results from inhibition of the synthesis of important mRNA precursors.

Factors regulating eukaryotic transcriptional activity

All the nucleus-associated RNA polymerases are large enzymes composed of several subunits and with molecular weights of approximately 500,000. In contrast to the bacterial enzyme, little is known about the function of the individual subunits of the eukaryotic RNA polymerases. The accessory proteins that regulate the activity of the eukaryotic RNA polymerases are the subject of intensive investigation. The regulation of eukaryotic RNA synthesis differs from the regulation of bacterial RNA synthesis.

Nucleosome structure An open nucleosome structure is required for transcription by the eukaryotic RNA polymerases. The open structure is usually much larger than the size of the transcription element itself. Two nonhistone proteins, HMG14 and HMG17, are involved; HMG stands for *high mobility group,* a designation given to several small nuclear proteins that migrate rapidly during electrophoresis on polyacrylamide gels. The modification of histones by acetylation or by ubiquinylation is also probably important at this stage. Ubiquitin is a small protein that marks the proteins for proteolytic destruction when covalently attached to cellular proteins.

Cis and trans regulation The terms *cis* and *trans* come from genetics, not chemistry. They refer to the linkage of genetic markers. In the context of gene expression, *cis*-acting regulatory sequences are segments of DNA that are important in starting transcription. *Trans*-acting factors refer to non-DNA substances, usually proteins, that interact with the *cis* sequences. Consequently, *trans*-acting factors are often detected as proteins that bind specifically to certain DNA sequences.

Eukaryotic promoters Promoter sequences for eukaryotic genes are larger and located further from transcription start sites than the promoters of bacterial genes. Eukaryotic promoters are usually located within at least 200 to 300 base pairs upstream of the 5′ site on the mRNA that later becomes capped with a methyl guanylic acid residue. Sequences of 6 to 20 nucleotides (*motifs*) within promoters may be repeated several times. These motifs most likely bind specifically to proteins to regulate either constitutive or induced transcription.

The promoters for RNA polymerase II contain two, and sometimes three, major conserved sequences. One sequence is the TATA, or ATA, box, located about −30 nucleotides upstream of the transcription start site. The ATA box has the following consensus sequence:

5′ - T A T A [A or T] A [A or T] - 3′

The other sequence is the CAAT, or CAT, box. It is located at −70 to −90 base pairs upstream of the transcription start site and has the following consensus sequence.

5′ - G G [T or C] C A A T C T - 3′

Sometimes an additional GC-rich conserved sequence (the GC box) is present upstream between −40 and −110 base pairs from the transcription start site. Changes introduced in the sequence of any one of these elements result in a greatly decreased rate of transcription in vitro.

Clinical comment

β-Thalassemia caused by a promoter mutation β-Thalassemia results from an inherited defect in the synthesis of hemoglobin. Insufficient amounts of β-globin are made. Many genetic forms of β-thalassemia produce this phenotype (see case 1). One form of the disease is caused by a mutation in the ATA box of the β-globin promoter. The normal sequence ATAAAA is mutated to ATGAAA. This mutation results in an 80% decrease in β-globin mRNA. The resulting disease is relatively mild. It is called β⁺-thalassemia to indicate that some β-globin chains are still made.

Enhancers Enhancers are *cis*-acting DNA segments that increase the activities of nearby promoters. Enhancers are not promoters themselves, but in some cases they stimulate transcription up to 100-fold. They function in either orientation, 5′ to 3′ or 3′ to 5′, and they can activate promoters that are located several hundred base pairs away, although with reduced activity. Usually enhancers are found upstream from their promoters, but a few enhancer-like sequences are found at downstream positions. Some enhancers are specific for a certain promoter or cell type, whereas others can stimulate many different promoters. Enhancers do not have strong sequence homologies with one another, but they often consist of six to eight nucleotides that are tandemly repeated.

Experimentally, enhancers are recognized as the binding sites for *trans*-acting substances, such as a hormone-receptor complex, or as a DNA segment that can bring a gene under the control of a stimulus such as heat shock or a steroid hormone. Enhancers exist for genes transcribed by RNA polymerase I as well as for genes recognized by RNA polymerase II.

The mode of action of enhancers is unclear, but it has been speculated that the binding of specific proteins to enhancers creates a complex structure to which the RNA polymerase can bind in a way that enhances its ability to engage the promoter region of the DNA template. Alternatively, the binding of specific proteins to enhancers might cause the DNA to supercoil in such a way as to make it easier for the RNA polymerase to melt out the double-stranded DNA near the promoter.

***Trans*-acting factors** RNA polymerase II cannot initiate RNA synthesis by itself. Several accessory proteins are needed. Some of the accessory proteins are generally required for the translation of any gene, whereas other auxiliary proteins are needed for the expression of just one or a few genes. These proteins often recognize and bind to sequences contained in both promoters and enhancers. These short sequences of 6 to 20 base pairs, the motifs, bind very specifically to proteins called *trans*-acting factors and in so binding stimulate transcription.

Transcription factors are *trans*-acting factors that stimulate the binding of RNA polymerase to promoters. The transcription factors are specific for the particular RNA polymerase, so some factors stimulate mRNA synthesis, whereas others stimulate the synthesis of tRNA or rRNA. Many such factors exist. Table 15.3 lists some of the transcription factors that have been isolated from human cells.

Table 15-3　　　　　　　　　　Some human transcription factors

Factor	Properties
RNA polymerase I: ribosomal RNA genes	
SLI	Not sequence specific; binds to -170 to -120 base pairs; -45 to $+20$
UBF-I	Binds to -120 to -105; interacts with SLI on the template DNA
TFID	Binds to -35 to -15 region
TFIC	Binds to the TFID-DNA complex
RNA polymerase II: mRNA genes	
Sp1	Binds to CCGCCC (GC box)
CTF	Family of factors that bind CCAAT (CAAT box)
TFIIB	Interacts with polymerase
TFIIA	Interacts with TFIID, then binds DNA
TFIID	Binds TATA box, interacts with TFIIA
TFIIE, TFIIF	Interact with polymerase and each other
SII, RAP 38	Elongation factors
RNA polymerase III: tRNA and 5S RNA genes	
TFIIIC	Binds internal control regions of all class III genes; a large complex
TFIIIB	Protein-protein interactions
TFIIIA	Binds control regions of 5S RNA genes

An example of a *trans*-acting protein, and a rather general transcription factor for RNA polymerase II, is the Sp1 protein isolated from HeLa cell nuclei. This protein can activate the transcription of several different genes. Sp1 recognizes and binds to promoters that contain the GC box, a CCGCCC sequence that may be tandemly repeated several times. Good evidence suggests that the hypomethylation of cytidylic acid residues in promoter sequences allows transcription to occur and that hypermethylation inhibits transcription. Thus it is possible that methylation of cytidine residues in the GC box are important in shutting down transcription by preventing the attachment of protein Sp1.

The meanings of some of the terms used to describe gene expression and RNA synthesis are listed next.

cis element, *cis*-acting factor	Short sequence of DNA important for regulating the expression of a nearby gene.
trans-acting factor	Factor (usually a protein) that binds to a *cis* element, or some other DNA sequence, and in so doing regulates the expression of a gene.
promoter	Short DNA sequence at which RNA polymerase binds.
TATA or ATA box	Short DNA sequence (TATA) found in almost all promoters for RNA polymerase II and is located about 30 nucleotides upstream of the mRNA start site.
CAAT box and GC box	Short DNA sequences containing CAAT and GTC bases found upstream from many promoters for RNA polymerase II: upstream location varies widely.
transcription factors	Proteins that bind to promoters and stimulate the activity of RNA polymerase.
enhancers	DNA sequences that stimulate promoter function but are not promoters themselves. They may act at a distance of several kb.

Protein biosynthesis

All the different RNAs—tRNA, mRNA, and rRNA (as part of the ribosome)—are involved in the synthesis of proteins. The process of protein biosynthesis is called *translation* because information must be transferred from the four-letter language of the nucleic acids to the 20-letter language of the amino acid constituents of the proteins.

Aminoacyl-tRNA synthases

Amino acids themselves have no special affinity for nucleic acids; the proper amino acid must therefore be combined with the proper tRNA under the influence of an aminoacyl-tRNA synthase. Once an amino acid is attached to a tRNA, all the specificity in recognizing the mRNA, the ribosome, and peptide bond–forming enzymes is a property of the tRNA, not the amino acid. Thus the aminoacyl-tRNA synthases must be highly specific both for the amino acid and for the tRNA. Only L-amino acids are recognized, not the D isomers. Also, none of the synthases will function with peptides or amino acids without free α-amino groups.

There are 20 genetically important amino acids. All other amino acids in proteins are derived from the set of 20; for example, the hydroxyproline found in collagen is formed from prolyl residues after they become part of the protein chain.

With a few exceptions, one aminoacyl-tRNA synthase exists for each amino acid; however, a synthase can recognize all the tRNA acceptors for a particular L-amino acid. For example, the methionyl-tRNA synthase recognizes only methionine, but it can aminoacylate both of the tRNAs for methionine: $tRNA^F$, which functions to initiate the synthesis of protein chains, and $tRNA^M$, which inserts methionyl residues at internal positions in the growing polypeptide chain. These functions are considered in some detail later. The synthases also have a high specificity for the ribonucleoside triphosphate; only ATP will function. The reaction catalyzed is shown in Figure 15.9.

Figure 15.9   Attachment of amino acid to tRNA.

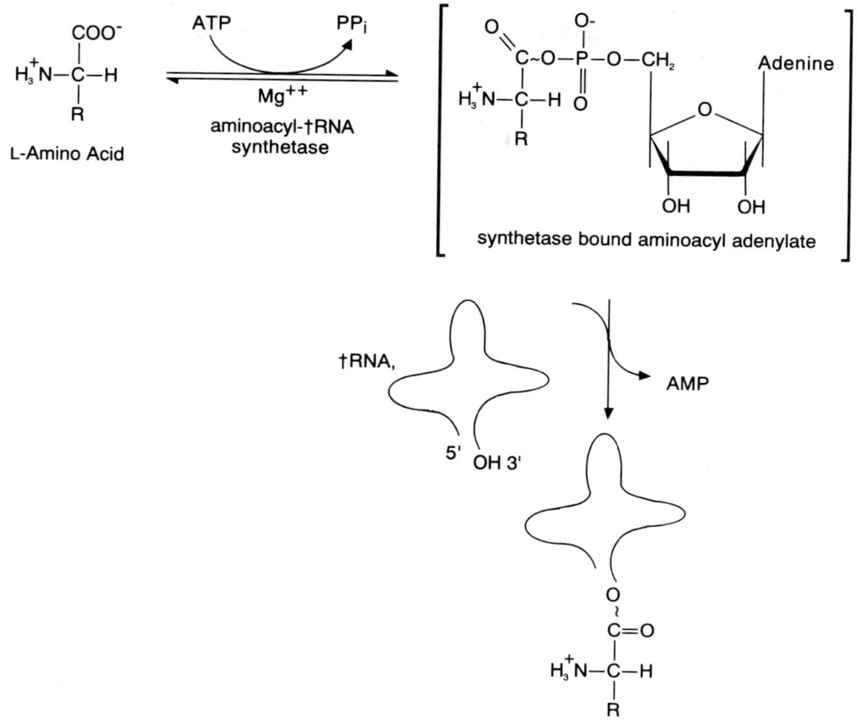

Aminoacyl-tRNA molecules have a large, negative free energy of hydrolysis that is used to drive the synthesis of the peptide bond between two α-L-aminoacyl residues. The formation of the aminoacyl-tRNAs is coupled to the hydrolysis of ATP with the intermediate formation of aminoacyl adenylate and inorganic pyrophosphate (PP$_i$). Transacylation to tRNA liberates AMP (Figure 15.9). One would predict that the reaction is reversible, and this is true in vitro. However, in the cell a high concentration of aminoacyl-tRNA is maintained as a result of extremely active pyrophosphatases that convert pyrophosphate to inorganic phosphate (P$_i$), removing a product of the forward reaction and thus preventing reversal.

Ribosome: site of protein synthesis

Aminoacyl-tRNAs are condensed into protein on ribosomal particles, not in solution. This aspect of protein synthesis adds to its complexity, since enzymatic reactions that occur in solution are much easier to study. The functional ribosome consists of two rather large particles. In animal cells these subparticles have sedimentation coefficients of 40S and 60S. In bacteria and in mitochondria the subparticles sediment at 30S and 50S. The subunits are derived from the functional particle, which is 80S in all parts of the mammalian cell except the mitochondria, where it is 70S. Even though these particles are large (about 200 Å in diameter), high-speed ultracentrifuges are required to sediment them. Figure 15.10 shows the composition of a mammalian ribosome. The particle as a

Figure 15.10   Composition of mammalian ribosome.

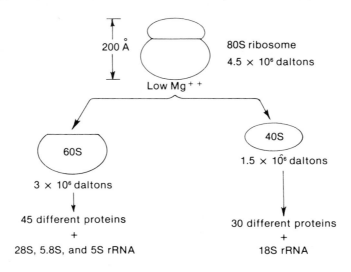

whole is about half protein and half rRNA. The 40S subparticle contains a single RNA that sediments at 18S and about 30 different proteins. The larger 60S subparticle contains three RNAs that sediment at 28S, 5.8S, and 5S. The sequences of the ribosomal RNAs are known, but few clues exist as to how they function. In addition, the larger subunit contains about 45 different proteins.

By lowering the magnesium ion concentration, the 80S ribosome dissociates into its constituent subparticles. This reaction is easily reversed and, as will be seen, occurs during the process of protein synthesis. Complete removal of magnesium ions breaks down the ribosome further, but strong dissociating solvents are needed to separate the RNAs from all of the proteins and the proteins from one another. Under special conditions the protein and RNA parts of the small and large subunits from bacteria can reassemble in vitro. Consequently the assembly of this complex subcellular organelle resembles virus assembly in that both processes occur spontaneously when all the parts are present.

A single peptide chain grows from an 80S ribosome. On completion of the chain the ribosome dissociates into subparticles and becomes available for the initiation of a new protein chain.

The ribosome is a rather nonspecific protein-synthesizing particle. Certain mRNAs are translated more efficiently than others, but by and large the particle is capable of making any protein of the species in which it is found. Animal ribosomes will not work with peptide polymerizing factors from bacteria, but liver ribosomes appear to function the same as ribosomes from other animal tissues or organs. A smaller bacteria-like ribosome is found in mitochondria. Protein synthesis in mitochondria resembles more protein synthesis in bacteria, since these ribosomes and their polypeptide-synthesizing enzymes are exchangeable and protein synthesis on the mitochondrial ribosomes is sensitive to inhibition by antibiotics that specifically block bacterial but not animal protein synthesis.

Polyribosome   Only one polypeptide chain grows from each ribosome, but a strand of mRNA can accommodate several ribosomes. An mRNA binds monomeric ribosomes in proportion to its length, and each ribosome holds a growing protein chain. The growing chain may be at one of several stages of completion. Diagrammatically such a polyribosome would appear as follows:

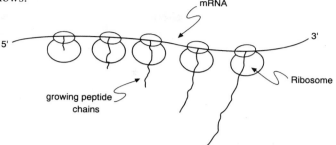

The monocistronic mRNAs of animal cells sometimes produce very long polypeptide chains. The latter must be cleaved after synthesis to become active (e.g., proinsulin, trypsinogen, and blood-clotting factors); some chains can be cleaved to yield more than one functional protein. Poliovirus makes a giant protein that is subsequently cleaved to form several functional proteins. In any case, it is the primary sequence of a protein that determines its tertiary structure and thus its biologically active conformation. It is likely that the amino acid sequence of proenzymes plays an important role in forming a highly specific conformation that, on activation, yields the functional enzyme. The active enzyme might not have been able to achieve this conformation had it not existed as a proenzyme. For example, with the A and B chains of insulin, it is difficult to cross-link their appropriate cysteine residues in the laboratory; but proinsulin can cross-link cysteines intramolecularly with great facility. Later proinsulin is cleaved to the A and B chains.

Protein synthesis starts by attaching a ribosome near the 5′ end of an mRNA. Recall that this is the end of the mRNA that is first synthesized from the DNA template. Thus in bacterial cells an mRNA molecule need not be completely synthesized before it starts to be read out. In animal cells, in which the mRNA is made in the nucleus, transcription and translation are not coupled. Much mRNA is transported to the cytoplasm in association with transport proteins.

As a ribosome directs the addition of aminoacyl groups to a growing peptidyl-tRNA, it moves along the message, decoding it from the 5′ to the 3′ end until it has moved far enough for another ribosome to be bound onto the initiating site just vacated. Soon the messenger is completely loaded with ribosomes. These polyribosomes can be seen with the electron microscope, and the number of ribosomes on them counted. The number of ribosomes on a polyribosome is proportional to the size of the protein being synthesized. Thus the β-globin chain of hemoglobin consists of about 150 amino acid residues, and its polyribosome holds five ribosome monomers. A major chain of myosin contains about 1800 amino acid residues, and its polyribosome holds more than 100 monomeric ribosomes.

Initiation of protein synthesis

Protein synthesis starts at the amino end of the peptide and progresses by the addition of amino acids at the carboxyl end.

All protein chains start with the same amino acid, methionine, at the N-terminal position. Very often the methionine residue is cleaved off after the growing polypeptide chain has been somewhat extended; consequently, proteins isolated from cells contain amino acids other than methionine at their N-terminals. Methionine has two acceptor tRNAs, tRNA^M and tRNA^F. Only Met-tRNA^F is involved in initiation.

Initiation of protein synthesis in prokaryotes In bacterial cells and in mitochondria, formylmethionyl-tRNA^F initiates the synthesis of every protein chain. In animal cells it is not necessary to formylate Met-tRNA^F to initiate protein synthesis. All chains are started with Met-tRNA^F, however. That the initiating tRNA in animal cells is really Met-tRNA^F and not Met-tRNA^M is indicated by the observation that the transformylase from *E. coli* can formylate the Met-tRNA^F of animal cells. Following is the reaction catalyzed by this enzyme:

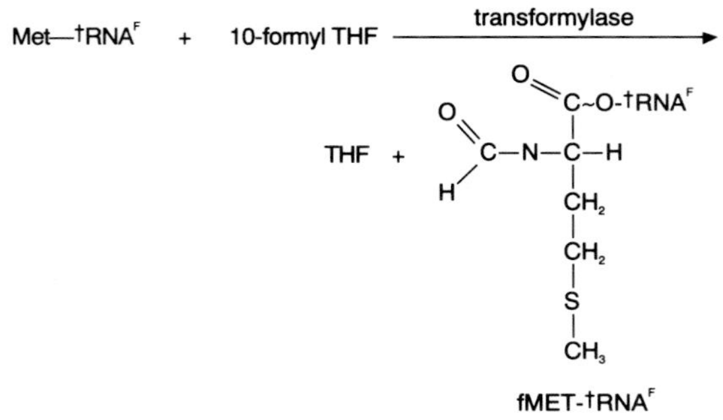

In addition to fMet-tRNA^F, the initiation of protein synthesis in bacterial cells also requires three proteins that are bound to the ribosomes but are not generally considered to be structural parts of the ribosome itself. One factor, initiation factor 3 (IF-3), is required for the recognition of mRNA. IF-1 and IF-2 are required for positioning mRNA and fMet-tRNA on the ribosomes. A diagram of the essentials of the initiation reactions is shown in Figure 15.11. To the extent that the analogous experiments have been done, the IFs from mammalian cells resemble in a functional sense those from bacteria. Because the initiation reactions in the bacterial system are simpler, Figure 15.11 describes the process in bacteria.

Despite much experimentation, the early steps in protein synthesis initiation are somewhat uncertain. Most likely the 30S ribosome first interacts with fMet-tRNA, not mRNA. In a later step the mRNA binds. The messenger is bound in a functional way in a reaction that also requires the other two initiation factors. The larger subunit is added in a reaction in which GTP is split to give GDP and P_i; IF-2 acts as a GTPase in this reaction. The hydrolysis of GTP in this process produces a conformational change in the 50S particle so that the final initiation complex is accommodated to accept the attachment of the succeeding aminoacyl-tRNA programmed by the messenger.

Initiation of protein synthesis in eukaryotes Protein synthesis initiation on animal ribosomes requires seven to nine different initiation factors, many of which have been highly purified. These factors are abbreviated eIFs for *eukaryotic initiation factors*. One of these factors, eIF-3, is very large, having several subunits and a molecular weight greater than 300,000. Among the more interesting of these factors is eIF-2. This factor binds GTP and Met-tRNA^F, carrying these substances to the ribosome. The activity of eIF-2 may be inhibited when phosphorylated by cAMP-independent protein kinases.

Figure 15.11   Initiation of protein synthesis *(E. coli).*

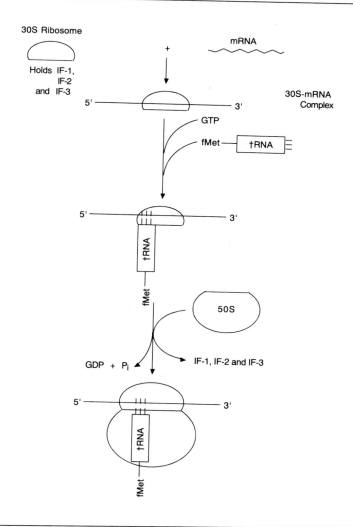

Ordinarily these protein kinases are present as latent, inactive enzymes. One of the eIF-2 protein kinases becomes activated when the hemin concentration is lowered. This leads to an inhibition of globin synthesis in reticulocytes.

Clinical comment **Globin synthesis in iron deficiency** The ratio of α- to β-globin chains was 0.74 in the peripheral erythrocytes of 11 patients with iron deficiency anemia (Benbassat I et al: Blood 44:551, 1974). The normal ratio is approximately 1.0.

Hemoglobin is actively synthesized in reticulocytes, which are precursor cells of mature erythrocytes. Equal amounts of α- and β-globin chains are made despite the presence of 40% more α-globin mRNA than β-globin mRNA. Furthermore, the polyribosomes for α-chains accommodates three monomeric ribosomes, whereas the polysomes for β-chains hold an average of five monomers. When a mixture of α- and β-mRNA is added to a reticulocyte cell-free translation system, more β-globin is made.

Iron deficiency possibly lowers the hemin concentration, which activates a protein kinase for eIF-2. This would effectively lower the concentration of nonphosphorylated eIF-2. Thus lowered eIF-2 or eIF-2B, a factor required for the catalytic recycling of eIF-2, might be directly or indirectly responsible for the relatively greater synthesis of β-globin versus α-globin in patients with iron deficiency anemia.

In interferon-treated cells a special eIF-2 kinase is made that is inactive until stimulated by double-stranded RNA. Double-stranded RNA in the cytoplasm is evidence of a virus infection. Thus, when an interferon-treated cell is infected with a virus, double-stranded RNA activates the eIF-2 kinase, and eIF-2 is phosphorylated. The phosphorylated eIF-2 fails to recycle catalytically during protein synthesis initiation, and the cell fails to make viral proteins.

Table 15.4 summarizes some of the functions of individual eIFs.

Messenger RNA binding to ribosomes

mRNA binding to prokaryotic ribosomes An mRNA is attached to the ribosome so that the initiating sequence is presented in the proper reading frame. The first amino acid inserted into peptide linkage is a methionyl residue coded by an AUG or GUG codon. Not every AUG sequence in an mRNA programs the start of a new protein chain, since AUG sequences also code for methionyl residues that occur at internal positions of the polypeptide chain. If one considers AUG and GUG sequences, it is obvious that several might occur within a cistron. For example, AUG occurs six times and GUG five times in the coat cistron of R17 bacteriophage RNA; but protein synthesis initiates faithfully, even in vitro, at the proper AUG codon. As one might guess, some RNAs cannot function at all as mRNA. For example, rRNA contains several AUG and GUG sequences, but because of other structural features, this RNA is not translated. The secondary structure of the RNA plays a major role in presenting single-stranded AUG codons to the ribosome. Thus formaldehyde-treated R17 RNA initiates at a few additional sites because formaldehyde reacts with the RNA to reduce its secondary structure sufficiently to generate

Table 15.4   Functions of eukaryotic initiation factors

Initiation factors	Molecular weight (kDa)	Functions
eIF-1	15	Stabilizes the 43S preinitiation complex
eIF-2	125	Binds GTP, Met-tRNA, then attaches to the 43S complex
eIF-2B (GEF)	280	Guanine nucleotide exchange factor (GEF); multisubunit complex that replaces GDP with GTP on eIF-2
eIF-3	700	Forms 43S complex; binding of mRNA
eIF-4B	80 (subunit)	Stimulates translation; part of cap-binding complex; binds AUG
Cap-binding proteins (CBPs)		
CBP I (eIF-4E)	24	Binds to 5′ guanine cap on mRNA; requires ATP
CBP II (eIF-4F) Subunits:	24	Same as CPB I (eIF-4D)
	50 (eIF-4A$_c$)*	Immunologically similar to eIF-4A
	220	Stimulates translation; an ATPase

*eIF-4A$_c$ is found in the cap-binding complex; eIF-4F is found free; it binds to mRNA in an ATP-dependent manner.

single-stranded regions containing AUG and GUG sequences.

In addition to secondary structure, other features of the RNA are important for translation. For example, ribosomes from *Bacillus stearothermophilus* translate only the minor coat cistron of formaldehyde-treated R17 RNA, whereas several other AUG segments are translated by *E. coli* ribosomes. This suggests that the ribosome recognizes primary sequences on the mRNA in addition to the initiating codons. These sequences must be located on the 5′ side of the initiating codon. The sequences on the 3′ side would be expected to vary considerably between different mRNAs, since it is this region that contains the code words for the protein to be synthesized. Every bacterial mRNA that has been sequenced contains a polypurine-rich tract of about five to seven bases located approximately 10 nucleotides from the 5′ side of the initiating AUG. These polypurines base-pair with a complementary sequence found near the 3′ end of the 16S RNA of the small subunit:

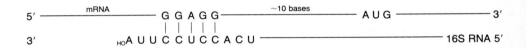

These regions are sometimes called Shine-Dalgarno sequences after the scientists who first proposed that mRNAs were recognized in this way by the ribosome.

mRNA binding to eukaryotic ribosomes Animal mRNAs do not contain these polypurine tracts, but since animal mRNAs are monocistronic, it is believed that the ribosome recognizes the 5′ cap, or a structure analogous to a cap, and initiates at the AUG sequence closest to the 5′ end. Several initiation factors or cap-binding proteins are required for this (Table 15.4). Because these factors function as large, multisubunit complexes, a few of the dissociated subunits were earlier identified as separate factors. For example, eIF-4E is a subunit of the larger eIF-4F multiprotein complex. Similarly the eIF-4F complex is the same as cap-binding protein II (CBP II), and eIF-4E is the same as CBP I.

mRNA must be single stranded for ribosomes to bind Experimental evidence indicates that CBP I, as part of the larger CBP II, binds to mRNA first. It binds to the 5′ guanine cap. This is followed by the binding of eIF-4B in an ATP-dependent reaction. This reaction displaces most of the CBP II complex, but it leaves eIF-4A$_c$ attached to the mRNA. Many additional copies of eIF-4A$_F$ bind to the 5′ end of the mRNA in ATP-dependent reactions that lead to the removal of mRNA secondary structure. This exposes the initiating AUG codon on the mRNA to which eIF-4B binds. The activated mRNA is now receptive to binding to the 43S preinitiation complex. These reactions are summarized in Figure 15.12.

43S preinitiation complex This is a complex of the 40S ribosome subunit, eIF-3, eIF-4C, and the eIF-2 ternary complex. The eIF-2 ternary complex consists of GTP and Met-tRNA bound to eIF-2. The 43S preinitiation complex attaches to the activated mRNA complex (Figure 15-12) to form a 48S initiation complex. This in turn accepts the large 60S ribosome subunit to form an 80S initiation complex, and in the process GTP is hydrolyzed and released as an eIF-2–GDP complex. The 80S initiation complex holds the growing nascent polypeptide chain.

Figure 15.12   mRNA binding to initiating eukaryotic ribosomes. See Table 15.4 for
abbreviations of the factors.

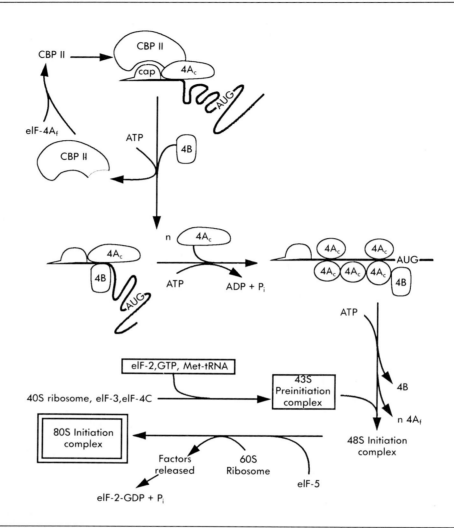

The scanning hypothesis In bacteria the Shine-Dalgarno sequences direct the ribosome to bind a downstream AUG. The analogous sequence in eukaryotic mRNAs is:

$$. . . . .CCRCC\text{-}AUG\text{-}G. . . .$$

This sequence is located near the 5′ cap. The 43S preinitiation complex, with its bound Met-tRNA, binds near the 5′ cap and moves along the mRNA "scanning" for the first appropriate AUG in the consensus sequence just shown. The anticodon of the ribosome-bound Met-tRNA base-pairs with the AUG and allows the 60S ribosome subunit to attach.

Protein chain elongation

Two sites exist for tRNA attachment on the ribosome. One usually contains the peptidyl-tRNA; it is closest to the 5′ end of the mRNA and is called the *P site*. After the initiation reactions shown in Figure 15.11, fMet-tRNA is in the P site, or Met-tRNA is in the P

Figure 15.13   Peptide chain elongation.

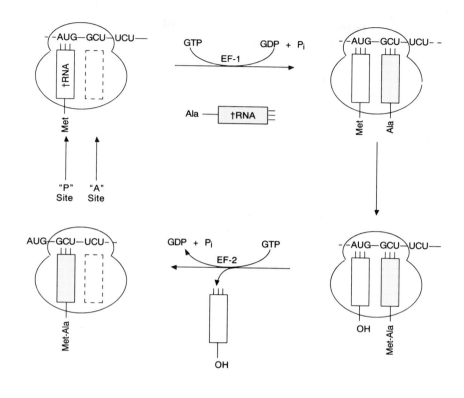

site of eukaryotic ribosomes. The other site, the aminoacyl-tRNA or *A site,* is on the 3′ side of the P site (Figure 15.13). The next aminoacyl-tRNA added to the growing chain is attached to this site under the influence of GTP and elongation factor 1 (EF-1), one of two protein elongation factors that work on the ribosome during the steps of polypeptide chain elongation. During this reaction, GTP is split to GDP and P_i.

The methionyl group is transferred to the nucleophilic amino group of the aminoacyl-tRNA in the A site. In subsequent steps the growing peptidyl group is transferred to the next aminoacyl-tRNA. This reaction is catalyzed by the ribosome itself, specifically by the larger subunit. The result is a peptidyl-tRNA, now one amino acid longer, resting in the A site, whereas the P site contains the deacylated tRNA that formerly held the peptidyl chain. This deacylated tRNA is removed in a reaction that requires EF-2 and GTP and that is concomitant with the translocation to the P site of the new peptidyl-tRNA still hydrogen-bonded to its codon on the mRNA. The cycle is now completed, and a new codon is brought into apposition with the A site. These reactions are shown diagrammatically in Figure 15.13. In bacteria EF-1 is called EF-T_u (EF-T_s) and EF-2 is called EF-G.

Termination of protein synthesis

Termination of protein synthesis in both animal and bacterial systems requires one or more protein release factors. The release factors recognize the termination codons UAA, UAG, and UGA. Release factors 1 and 2 (RF-1 and RF-2) are found in *E. coli*. Either of the factors will function to release the peptidyl group from the tRNA bound to the ribosome, probably by assisting in the catalysis of a hydrolytic reaction. RF-1 recognizes the termination codons UAA or UAG, and RF-2 recognizes UAA or UGA. It is believed that the synthesis of most proteins is terminated by the UAA signal, so that either factor might serve to stop the synthesis of most proteins.

Only a single RF has been isolated from mammalian cells. No special tRNAs are required, but termination is stimulated by GTP. The peptidyl transferase of the ribosome probably catalyzes the hydrolysis of the peptidyl ester bond.

Posttranslational processing of secretory proteins

Many of the proteins secreted from cells are believed to be synthesized with extra amino acids at their N-terminals. These N-terminal extensions contain unusually large amounts of hydrophobic amino acids that are capable of interaction with nonpolar portions of membranes. The polypeptide extension of 15 to 30 amino acid residues is thought to represent the signal for the selection and binding to membranes of those proteins destined for secretion; for this reason the N-terminal extension is called a *signal peptide*. The signal peptides were discovered by analyzing the protein products synthesized by mRNA-dependent cell-free amino acid incorporating systems that are relatively free of membranes. Signal peptides are almost impossible to find in whole cells or tissues because they are rapidly removed by membrane-associated peptidases either during the secretory process or immediately thereafter. Proteins synthesized with signal peptides include several polypeptide hormones (e.g., insulin), albumin, collagen, immunoglobulins, and certain viral coat proteins.

Antibiotics and protein synthesis

Several antibiotics recognize differences between animal and bacterial protein synthesis by specifically inhibiting one or the other. The health scientist is most interested in those antibiotics that act specifically against bacteria. Table 15.5 lists a few commonly used antibiotics and the step in protein synthesis they inhibit.

Diphtheria toxin, although not an antibiotic, is another substance that inhibits protein synthesis. It is made in bacteria; thus it does not inhibit bacterial protein synthesis. The toxin is an enzyme that is exceedingly deadly, even in small amounts. It catalyzes the rather unusual reaction shown in Figure 15.14, which leads to the inactivation of the animal (EF-2). The glycosidic bond in the oxidized form of nicotinamide adenine dinucleotide (NAD^+) that holds the ribose moiety to the pyridine ring is transferred to a modified histidine residue in EF-2 and in the process inactivates EF-2. The extraordinary lethal effects of this toxin emphasizes that protein synthesis is a very important process.

Genetic code

Codons Three nucleotide bases are required to specify the insertion of an amino acid into a polypeptide chain. These groups of three are called codons, or *code words*. Each series of three bases is read in a linear sequential manner with no overlapping of code

Table 15.5	Antibiotics that inhibit protein synthesis	
	Antibiotic	**Step inhibited**
	Chloramphenicol	Ribosomal peptidyl transferase
	Streptomycin	Initiation; causes misreading of code
	Tetracycline	Prevents aminoacyl-tRNA attachment to ribosome
	Puromycin (also inhibits animal cells)	Accepts growing peptidyl chain in place of aminoacyl-tRNA; chain terminates prematurely
	Cycloheximide (inhibits only animal cells)	Ribosomal peptidyl transferase

words. Because there are four different bases in RNA, the maximum number of three-letter code words is 4^3, or 64. Sixty-one of these words are used to specify the 20 amino acids; thus some amino acids have several code words and the code is said to be degenerate. The code is universal in that essentially the same code word dictionary is used in all species tested. Consequently the genetic code can be described as triplet, nonoverlapping, degenerate, and universal. Table 15.6 lists the codon assignments for the 20 amino acids. Notice that the first two letters of a code word are very specific but that often the same amino acid is coded regardless of the third nucleotide.

Missense and nonsense mutations The three codons without amino acid assignments function as chain terminator signals. Some mutations will cause a base to change so that a terminator codon is generated. This often represents a serious mutation, since it results in premature chain termination; these mutations have been called "nonsense" mutations because a code word has been created for which no amino acid exists. A mutation that changes a base so that a new amino acid is now specified by the code is called a "missense" mutation. Missense mutations often result in altered or reduced enzymatic activity and less frequently in a complete absense of activity. Considering the degeneracy of the code, one can predict that almost one third of all base replacements will probably cause no change at all in the protein made, since they will occur in the third nucleotide of the codon.

Codon-anticodon interactions The genetic code assumes that each codon base-pairs in antiparallel fashion with the anticodon of the tRNAs that are specific for the amino acid corresponding to the code word. It was found, however, that when purified tRNAs became available, a single tRNA could recognize several code words. For example, the tRNA for alanine, whose structure is represented in Figure 15.6, recognizes GCU, GCC, and GCA. The anticodon for this tRNA is IGC, and the base-paired structures would be:

Figure 15.14   Action of diphtheria toxin.

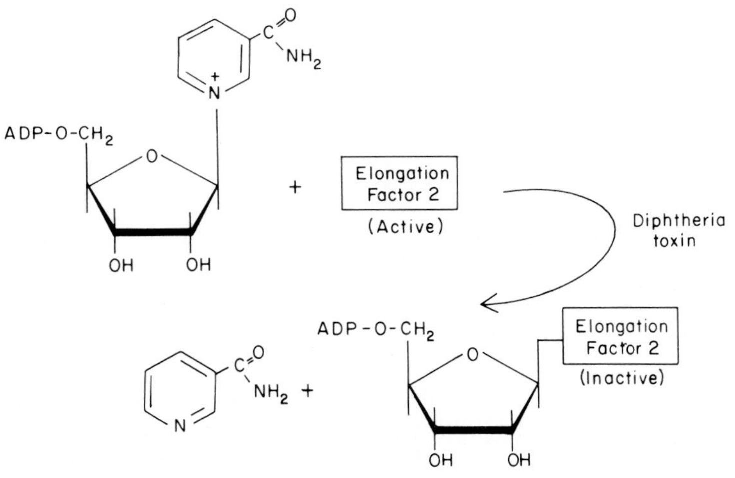

Table 15.6 Genetic code

First position (5′ end)	Second position				Third position (3′ end)
	U	**C**	**A**	**G**	
U	Phe	Ser	Tyr	Cys	U
	Phe	Ser	Tyr	Cys	C
	Leu	Ser	Term*	Term*	A
	Leu	Ser	Term*	Trp	G
C	Leu	Pro	His	Arg	U
	Leu	Pro	His	Arg	C
	Leu	Pro	Gln	Arg	A
	Leu	Pro	Gln	Arg	G
A	Ile	Thr	Asn	Ser	U
	Ile	Thr	Asn	Ser	C
	Ile	Thr	Lys	Arg	A
	Met	Thr	Lys	Arg	G
G	Val	Ala	Asp	Gly	U
	Val	Ala	Asp	Gly	C
	Val	Ala	Glu	Gly	A
	Val	Ala	Glu	Gly	G

*Chain-terminating codons.

Notice that the nonstandard base pairing is in the third position of the codon, the position that has the least effect on specifying a particular amino acid (Table 15.6). Crick has proposed a hypothesis to explain these data. It is called the *wobble hypothesis* because it predicts that a "wobble" in the base pairing in the third position of the code word might account for the lessened specificity. On the basis of the analysis of several other tRNAs, Crick proposed the following rules:

Third position of anticodon in tRNA	Third position of codon in mRNA
U or psi	A or G
C	G
A	U
G	C or U
I	C, U, or A

Assuming that the wobble hypothesis holds, it is not necessary to have 61 different types of tRNA to read all the 61 possible code words.

The genetic code for the most part is universal. The codon assignments established for *E. coli* are consistent with the known amino acid replacements in many abnormal human hemoglobins and mutant coat proteins of the tobacco mosaic virus. For most organisms, both prokaryotic and eukaryotic, the code is universal; however, minor variations occur in some organisms and in the synthesis of proteins coded by mitochondrial DNA. The relative use of multiple codons for the same amino acid varies significantly from species to species and may even vary between proteins of the same species.

Synthesis of specific proteins

Up to this point the description of protein synthesis has been general; usually only those features common to the synthesis of all proteins have been considered. Many proteins of interest to health scientists, however, have special biosynthetic features that are peculiar only to them. Antibody proteins, for example, are formed in response to antigens, and virus proteins are made from instructions written in viral nucleic acids.

Immunoglobulins

Antibodies, or immunoglobulins, make up the γ-globulin fraction of the plasma. Chapter 2 considers the structure of these proteins but says little about their functions. These are special defensive proteins that are synthesized in response to exposure to a foreign material, usually a protein or a complex carbohydrate. The foreign material is called an *antigen*. The formation of antibodies affords immunity against the antigen, and this response is usually protective. For example, a human exposed to the rubeola virus for the first time will usually develop measles. The individual's body then manufactures antibodies against this virus that help in recovery from the acute illness. Moreover, some plasma cells, which are derived from lymphocytes, retain the memory for making antibodies rapidly against rubeola virus if the individual is later challenged by the virus. In this way, immunity to measles results, and contact with the rubeola virus later will not cause the acute illness.

Immunization Humans can be protected against certain harmful diseases in a similar fashion by the administration of suitable antigens, a process known as *immunization*. Cowpox, a harmless virus, is injected to trigger the formation of antibodies against it. These same antibodies also attack the smallpox virus; therefore this immunization affords protection against smallpox. In other situations, immunization is carried out by administration of inactivated bacterial toxin (tetanus toxoid), live virus in attenuated form (Sabin polio vaccine), or fully active material in an amount too low to produce serious illness (desensitization for hay fever). In these cases, persons manufacture their own antibodies against these antigens.

In other situations it is possible to administer preformed antibodies *(passive immunity)* that are isolated either from a human who already has had the disease (mumps hyperimmune serum) or from an animal who has been injected previously with the antigen (rabies vaccine, tetanus antitoxin). This procedure is used when there is not enough time to allow the body to make its own antibodies, for example, for a person not previously immunized against tetanus who steps on a contaminated rusty nail. This procedure is often risky, since the recipient may develop antibodies to the injected serum (which usually contains foreign proteins) and become very ill (serum sickness). Finally, the immune response may actually be harmful in certain situations, since antibodies formed against a transplanted heart or kidney (foreign tissue) may lead to rejection of the transplanted organ by the new host.

Organization and rearrangement of immunoglobulin genes Portions of the DNA sequences that code for immunoglobulins are physically separated from one another and must be rearranged and brought close together by breakage-reunion mechanisms before they are expressed.

searches for lymphocytes that have the potential to synthesize antibody molecules that are specific for the particular antigen.

The commitment of a particular lymphocyte to make the specific antigen was made earlier in development. At that time several different variable (V) genes were spliced onto constant (C) genes. This created cells that had an antibody gene (complete VC gene) composed of a piece of DNA having a constant sequence (C gene) connected to a piece of DNA that might originate from one of several V genes.

In embryonic cells the V and C genes are located far apart, but in B lymphocytes, the antibody-producing cells, the genes are found much closer together.

Even more diversity of antibodies can be generated from joining (J) genes. A tandem array of several small J genes is located close to the C region. The J genes encode a small part of the region that can vary in immunoglobulin genes; that is, the variable region of light-chain genes is made up of a V gene linked to a J gene. Thus the kappa light-chain V gene is spliced to one of five different J genes before being spliced to the C gene.

The RNA made from the rearranged immunoglobulin gene still contains intervening sequences that are removed to form functional mRNA. The way in which an antigen triggers the synthesis of a specific antibody is beyond the scope of this book and is dealt with in immunology texts. For the present it is sufficient to state that an antigen directs B lymphocytes *(plasma cells)* to synthesize a specific immunoglobulin. Each clone of plasma cells makes only one species of L and H chain, and this property has allowed the isolation of monoclonal antibodies. Monoclonal antibodies promise to become increasingly important in both biology and medicine.

Immunoglobulin-associated diseases Certain diseases are characterized by an overproduction of immunoglobulins. One such disease is a plasma cell cancer, *multiple myeloma*. In this condition, L chains are overproduced by the proliferating plasma cells. These are released into the blood without combining with H chains, and they are excreted in the urine because of their relatively small size. Urinary L chain protein is called Bence Jones protein, and it has a characteristic property. Bence Jones protein is precipitated at 60° C, but on further warming, the precipitate dissolves. This simple test has been in use for many years in the diagnosis of multiple myeloma. A *monoclonal gammopathy* is a condition in which only one immunoglobulin is present in the plasma in excess. This condition results from a cancer that has developed from a single clone of plasma cells.

Virus replication   Viruses are the most frequent cause of human illness. It is estimated that 60% of all illness caused by viruses are not even detected clinically. There are 50 or more different disease syndromes caused by viruses. Although some common virus diseases are not serious, others (e.g., smallpox and influenza) have killed millions of people. The natural immunity processes are the best protection against viruses.

Viruses are small particles (1 to 300 genes); many are the size of ribosomes. They all contain nucleic acid covered by a protein or lipoprotein coat. The nucleic acid may be either double- or single-stranded DNA or RNA, but not both. Some viruses actually carry enzymes with them as part of the virus particle. Perhaps the most interesting example of this is the reverse transcriptase that can cause the host cell to make DNA from the RNA template contained in RNA tumor viruses. The protein coat is probably the most important antigenic component of the virus.

Variation of viruses Mutational events and recombination of viral nucleic acid often affect the coat proteins. Thus the amino acid sequence of the coat proteins might change considerably over time, and a person's antibodies against the ancestors of the virus may no longer be able to react with the coat proteins of the present virus and inactivate it. Furthermore, large portions of the RNA of segmented viruses may exchange by passage through an animal vector. For this reason, immunity against a particular virus may not be very long lasting. In general, in order that immunization against a virus be effective, the antigen should be prepared using precisely the same strain against which protection is sought.

Since the genes of a virus may be either RNA or DNA, viruses and the host often synthesize their nucleic acids differently. Even the smallest viruses contain the information in their nucleic acid for the synthesis of one or more coat proteins, and virtually all viruses code for the synthesis of an RNA or DNA polymerase used to replicate their nucleic acid. The synthesis of double-stranded DNA virus particles is straightforward and follows a mechanism similar to that used by host DNA. Virus particles that contain single-stranded RNA, on the other hand, present some special problems, since these viruses may be either of two types. The genomic RNA may be either a plus strand, that is, an information-carrying strand, or a minus type, the complement of the information strand. A special RNA polymerase may synthesize an RNA chain complementary to the RNA chain found in the virus. The complementary strand serves as the template for the synthesis of dozens of genomic strands. Coat proteins are made from the single-stranded plus strands. These proteins have high affinity for the genomic RNA particles and interact with them to form the virus coat, or capsid. Many new viruses result, the cell lyses, the new virus particles are released, and neighboring cells become infected.

Interferons Many viruses cause the cells that they infect to produce a type of immunity substance, a group of proteins called interferons. The double-stranded nucleic acid of a virus (or a synthetic polynucleotide such as poly I:C) somehow stimulates the expression of a host gene for interferon. The interferon mRNA is made and carried to the ribosome where interferon proteins are produced. These proteins can find their way to other cells and there prevent the replication of viruses. Interferon acts by triggering the synthesis of "antiviral" proteins. Some "antiviral" proteins inhibit virus replication at the level of translation. Interferon is more active in preventing viral infections than stopping ones that are well advanced. Virtually every group of viruses is capable of eliciting interferon production. Interferon produced in response to a specific virus is active against many different viruses, indicating that it must function by blocking reactions of fundamental importance in the replication of any virus. Human interferon genes have been cloned into *E. coli* cells, where they elicit the formation of relatively large amounts of interferon. The bacteria-produced human interferon is useful in the treatment of some viral diseases and a few forms of leukemia and other cancers.

Genetic analysis of human disease

We often think of genetically transmitted diseases as rare and of little importance. True, inherited diseases occur infrequently, but when all the genetically transmitted diseases are considered together, they no longer seem so rare.

Data on the frequency with which genetic diseases occur in humans are inaccurate for several reasons, including poor diagnosis, unusual distribution of population types, and unreported cases. Frequency data are usually reported as either the incidence or prevalence of a given disease. These terms are often confused, but they have precise meaning. The *incidence* of a disease relates to the number of cases per number of live births, whereas the *prevalence* is the number of cases in a given population at a given time.

The incidence of a disease may be difficult to determine if the disease is not physically recognizable at birth and if no test is available to determine its prospective development. An example of a disease not physically recognizable at birth is Huntington's chorea, which rarely develops before the patient is 35 years of age. Other errors in determining the incidence of a disease are related to the chemical tests used. Some phenylketonuric children escape detection because the FeCl$_3$ urine test is used instead of the Guthrie test; the latter depends on a more reliable microbiologic assay for phenylalanine metabolites. In other cases the appropriate tests are available, but the infant suffers irreversible damage before symptoms indicate that the test should have been done. An example of this problem is galactosemia.

The prevalence of a disease is even more difficult to determine because large populations are difficult to screen, especially for a rather rare disease. Furthermore, patients may refuse to be studied, or if they are cooperative, they may not be aware that they carry a mutant gene; they may have a subclinical form of a disease, such as myotonic dystrophy. As we have already seen, the incidence and prevalence of genetically transmitted diseases vary with race, geographic area, or sex. Table 15.7 lists frequency data for some of these diseases.

Classification of genetic diseases

The simplest means of classifying genetic diseases is according to whether the defective gene is on the X chromosome or an autosomal chromosome and whether the disease is dominant or recessive. Furthermore, such information is often useful in counseling the

Table 15.7 Incidence and prevalence of some genetically transmitted diseases*

Disease	Incidence (1 in *n* live births)	Prevalence (1 in *n* people)
Autosomal recessive		
Cystic fibrosis	2000	3000
Phenylketonuria	10,000	40,000
Galactosemia (transferase deficient)		70,000
Wilson's disease		4×10^6 (United States)
		1×10^5 (Romania)
Von Gierke's disease		4×10^5
Tay-Sachs disease		6000 (Ashkenazi Jews)
		5×10^5 (Gentiles)
Sickle cell anemia		70 to 300 (American blacks)
Albinism		20,000
Galactokinase deficiency		1×10^5
Mucopolysaccharidosis I (Hurler's syndrome)		40,000
Autosomal dominant		
Neurofibromatosis		3300 to 4000
Huntington's chorea		20,000 to 25,000 (United States)
		3×10^5 (Japan)
Charcot-Marie-Tooth disease		5000
Achondroplasia	10,000	
X-linked recessive		
Duchenne's muscular dystrophy		6000 to 10,000
Color blindness		16 (American men)
		200 (American women)
Hemophilia A		20,000
Mucopolysaccharidosis II (Hunter's syndrome)		50,000

*For further information see Bergsma D; Birth defects: atlas and compendium, New York, 1973, The National Foundation.

family or in detecting other family members who have the disease. This classification is used in grouping the diseases listed in Table 15.7. Although many of the conditions such as color blindness are not life threatening, they may represent situations that would lead a person to seek medical advice. It is worth noting that the most benign diseases often occur at a very high frequency.

Perhaps the best means of classifying genetic diseases is on the basis of the defective protein responsible for the condition. This has been done for more than 100 human genetic diseases, not including the hemoglobinopathies or other red cell defects. In addition, more than 80 variants of glucose 6-phosphate dehydrogenase are known. These intracistronic variants are caused by mutation at different sites in the gene responsible for programming the synthesis of the enzyme. In many genetic diseases the amount of enzymatic activity lost is in direct proportion to the severity of the disease.

Cytogenetic diseases

The cytogenetic diseases are not strictly biochemical diseases, but since they are detected as gross cytologic abnormalities of the chromosomes, they can be thought of as affecting protein and nucleic acid metabolism. Some occur quite often, and the frequency data listed in Table 15.8 emphasize the importance of these diseases and provide a basis for comparing these data with similar data on the frequency of inherited genetic diseases (Table 15.7).

Bibliography

Chambliss G et al, editors: Ribosomes: structure, function and genetics, Baltimore, 1980, University Park Press.

Geiduschek EP and Tocchini-Valentini GP: Transcription by RNA polymerase III, Annu Rev Biochem 57:873, 1988.

Gross DS and Garrard WT: Nuclease hypersensitive sites in chromatin, Annu Rev Biochem 57:159, 1988.

Kozak M: Comparison of initiation of protein synthesis in procaryotes, eucaryotes and organelles, Microbiol Rev 47:1, 1983.

Moldave K: Eukaryotic protein synthesis, Annu Rev Biochem 54:1109, 1985.

Pestka S et al: Interferons and their actions, Annu Rev Biochem 57:727, 1987.

Ross J: The turnover of messenger RNA, Sci Am 260(4):48, 1989.

Schimmel P: Aminoacyl tRNA synthetases: general scheme of structure-function relationships in the polypeptides and recognition of transfer RNAs, Annu Rev Biochem 56:125, 1987.

Sonenberg N: Cap-binding proteins of eukaryotic mRNA: functions in initiation and control of translation, Prog Nucleic Acid Res Mol Biol 35:173, 1988.

Spirin AS: Ribosome structure and protein synthesis, Menlo Park, Calif, 1986, The Benjamin/Cummings Publishing Co.

Table 15.8 Incidence and prevalence of some cytogenetic diseases

Cytogenetic disease (chromosome abnormality)	Incidence (1 in n live births)	Prevalence (1 in n people)
Trisomy 21 (Down's snydrome or mongolism)	770	2000 to 3000
XXY	800	
XYY	800	
XXX	1000	
Trisomy 18	3000	
Trisomy D	5000	
Monosomy X	5000	

Stamatoyannopoulos G et al, editors: The molecular basis of blood diseases, Philadelphia, 1987, WB Saunders Co.

Steitz JA: "Snurps," Sci Am 258(6):56, 1988.

Watson JD et al: The molecular biology of the gene, ed 4, Menlo Park, Calif, 1987, The Benjamin/Cummings Publishing Co.

Weissbach H and Pestka S editors: Molecular mechanisms of protein biosynthesis, New York, 1977, Academic Press, Inc.

Clinical examples

A diamond (♦) on a case or a question indicates that literature search beyond this text is necessary for full understanding.

Case 1	β-Thalassemia

A 5-year-old child of Italian descent developed an acute upper respiratory infection. When examined, he was severely anemic with a temperature of 40° C. The liver and spleen were enlarged. He grew rapidly worse and died on the fourth day of his illness. A twin brother and a 2-year-old brother suffered from the same type of anemia.

Biochemical questions

1. What is thalassemia? What cells are primarily affected?
2. How does α-thalassemia differ from β-thalassemia?
3. Explain the inheritance of the thalassemias. What is the difference between thalassemia minor and thalassemia major? Which form would you expect in the thalassemia heterozygotes? In the homozygotes? Does the child in this case have the minor or the major form of the disease?
4. β-Thalassemia is characterized by abnormalities in gene expression. Suggest ways in which this control might be disturbed. Consider several alternatives and place these in the framework of what you know about the mechanisms of gene expression.

Case discussion

1. **What is thalassemia?** Thalassemia is a group of related inherited disorders characterized by the reduced synthesis of one or more globin chains leading to severe anemia early in life. The thalassemias are unlike other hemoglobinopathies, such as sickle cell anemia, in which a specific amino acid is substituted by another amino acid in one of the globin chains. In contrast, the thalassemias are diseases caused by defects in the expression of globin genes or in the translation of globin mRNAs.

The cells affected in these disorders are those that synthesize hemoglobin: reticulocytes and their precursors.

2. **α- and β-Thalassemia** Most thalassemic hemoglobin seems normal in respect to the primary structure of the globin subunits; the abnormality is in the relative amounts of α- and β- or β-like chains. Recall that the normal hemoglobins all contain the same α-chain, but the non-α-chain may vary. For example, in hemoglobin A, the major hemoglobin of adults, the composition is α_2,β_2; the composition of hemoglobin A_2 is α_2,δ_2; and that of fetal hemoglobin F, α_2, γ_2. In some forms of thalassemia the synthesis of a particular globin subunit is affected. For example, disease variants affect individual α-, β-, γ-, and δ-subunits. In α-thalassemia the rate of synthesis of α-chains is lower than normal; in β-thalassemia the rate of synthesis of β-chains is low.

Globin genes The genes for the human globin subunits occur on two chromosomes. There are four α-genes per diploid genome. These genes are located as pairs 2.5 kb apart on chromosome 16. The genes for the β- and β-like—globin molecules are located close together on the short arm of chromosome 11. The gene order is 5'-ε-^{G}γ-^{A}γ-δ-β-3', where ε is the embryonic hemoglobin, γ the fetal, and δ and β adult globins. At least two different forms of the normal fetal γ-subunit exist: ^{G}γ, having a glycyl residue at position 136, and ^{A}γ, an alanyl residue at position 136. Each gene is separated from the other by DNA spacers that vary in length from approximately 3 to 5 kb, for the spacers between the γ-genes and the δ,β-genes, to approximately 14 kb between ε,γ and γ,δ. Thus a large amount of nontranscribed spacer DNA separates the embryonic ε-gene from the fetal γ-genes and a large spacer between the fetal γ-genes and the adult δ- and β-genes.

The δ- and β-chains are known to be linked, since a subunit in hemoglobin Lepore contains the N-terminal sequence of the δ-chain and the C-terminal sequence of the β-chain. Hemoglobin Lepore could arise by a crossing-over between the homologous chromosomes that carry these two linked genes. The crossover would occur between the δ-gene on one chromosome and the β-gene on the other.

3. Inheritance of thalassemia All the thalassemias are inherited in simple mendelian fashion as autosomal recessive defects. They are the most common single-gene disorders in the world. The incidence of the disease is difficult to measure accurately. The disorder occurs most often in the semitropical areas of Africa, the Mediterranean regions, and southeast Asia, where gene frequencies of 15% may occur. At these gene frequencies, approximatetly 25% of newborns are carriers of the disease.

Previously *thalassemia major* was used to describe the disease in the homozygote and *thalassemia minor* to describe the much milder symptoms of the heterozygote. We know now that milder homozygous forms of thalassemia exist. In fact, thalassemia should be thought of as a syndrome that can be caused by any of many different molecular events, all of which result in a similar phenotype. The terms thalassemia major and minor are now most often used to refer to the clinical severity of the diseases.

Thalassemia minor is usually not serious; however, individuals with thalassemia major, such as the boy in this case, do not survive to reproductive age. Despite this, thalassemia is almost as prevalent in certain Mediterranean and southeast Asian kindreds as sickle cell anemia in people of African ancestry. As with sickle cell anemia, heterozygotes may be more resistant to malaria so that prevalence of the disease is high in regions of the world where malaria is endemic.

No effective treatment exists for thalassemia, so prevention of the disease by genetic counseling is important.

4. Gene expression in thalassemia The thalassemia syndrome is fairly easy to recognize. Identification of the specific variant is more difficult. The low levels of a particular subunit are detected by electrophoresis of extracts of erythrocytes. Red cell precursors also show abnormal ratios of globin subunits, and because globin synthesis occurs in the precursor cells but not in the mature erythrocytes, investigators have focused attention on precursor cells such as reticulocytes in efforts to determine the nature of the genetic defects in this disease.

α-Thalassemia α-Thalassemia can result from defects in the expression of one or all of the four genes for the α-globin subunits (Table 15.9). When only one or two α-genes are expressed, the result is not very serious; however, if three or all four α-genes are defective, severe anemia or death in utero is the result. A defect in three α-genes causes hemoglobin H (HbH) disease. The erythrocytes of these individuals contain large amounts of HbH, which is a tetramer composed of only β-chains. In

Table 15.9   Genes affected in α-thalassemia

Disease	Clinical signs	Number of genes affected
α-Thalassemia-2	None; silent carrier	1
α-Thalassemia-1 (α-thalassemia trait)	Very mild anemia	2
HbH (β_4) disease	Chronic anemia	3
Hydrops fetalis	Death in utero	4

hydrops fetalis, where all four α-globin genes are inactive, the major hemoglobins found in the fetus are HbBart's (γ_4) and HbH (β_4). No α-chains are found.

In most forms of α-thalassemia the genes are partially or totally deleted. Deletion patterns may be quite complex in the different forms of α-thalassemia. For example, the deletion of two α-genes can be either —α/—α, where the deletions are on different homologous chromosomes, or ——/αα, where the deletions are on the same chromosome. The latter occurs more frequently in Chinese populations, where the occurrence of hydrops fetalis (——/——) is high in marriages of α-thalassemia-1 patients (——/αα). On the other hand, black populations have more *trans* (—α/—α) gene arrangements and consequently hydrops fetalis is rare.

Nondeletion forms of α-thalassemia occur less often. The nondeletion forms affect several different stages of gene expression. For example, some mutations lead to mRNA-processing defects where a new donor site within the first intron is used. The resulting mRNA is very unstable. Other examples include mutation in the initiator codon, amino acid substitution that leads to unstable α-chains, and chain-termination mutations that produce abnormally long protein chains.

β-Thalassemia The β-thalassemias are roughly divided into two broad types: β^+-thalassemia, where β-globin chains are in low concentration, and β^0-thalassemia, where the erythrocyte contains no detectable β-chains. In both types there is an increased synthesis of δ- and γ-globin to compensate partially for the lack of β-chains. Despite this compensation, an abnormally high amount of α-globin is present, which precipitates out and causes the red cell to hemolyze. In β^+-thalassemia the amount of β-globin mRNA is decreased from normal in about the same proportion as the amount of β-globin made, whereas in β^0-thalassemia mRNA levels depend on the disease variant. In some β^0-patients, β-globin mRNA is made or partial β-mRNA sequences are made, but in other patients β-mRNA cannot be found. In β-mRNA–negative patients, the absence of mRNA could be caused by defects in transcription or mRNA processing or by a small deletion in the β-globin structural gene. In at least one β-mRNA–positive patient, the defect seems to result from mutation of codon 17 to an amber termination codon. This causes premature termination of the synthesis of the β-globin chain. Other types of β^0-thalassemia may be caused by defects in translation factors (Ferrara type) or partial deletions of mRNA sequences. β-Thalassemia can be caused by mutation in any of several loci that affect the expression of β-globin. Table 15.10 summarizes some of these variants, the steps affected, and whether the variant is β^0 or β^+.

The explanation of the molecular defect in one type of β^+-thalassemia involves a single base mutation in an intron of a β-globin RNA transcript that leads to improper

Table 15.10   Molecular defects in some variants of β-thalassemia

	β-Thalassemia type
Gene deletions	
In one of these rare forms, 619 base pairs of DNA are deleted from the 3' end of the β-globin gene.	β⁰
β-Globin mRNA processing	
Splice site mutations	
At invariant sequences	
5'-GT AT or TT	
At consensus sequences	β⁰
Cryptic sites in introns: a mutation in an intron creates a favored splice site (see Figure 15.15).	β⁺
Polyadenylation signal site	β⁺ and β⁰
AAUAAA or AACAAA	
	β⁺
Transcription	
Mutations that slow transcription occur as single base substitutions in front of the cap site, near or within the CAAT or ATA boxes.	
	β⁺
Translation	
Nonsense mutations cause premature chain termination.	
Frame-shift mutations, insertions or deletions, cause change in reading frame.	β⁰
	β⁰

processing of the mRNA precursor. This is illustrated in Figure 15.15, which outlines the steps in globin gene expression. More than 70% of the globin gene does not contain coding information and must be removed. The primary RNA transcript is approximately 1.5 kb in length. The dark portions in Figure 15.15 represent the coding sequences. Notice that three coding sequences (exons) and two noncoding sequences (introns) are present. The nucleotide sequences at the junctions between the globin introns and exons are highly conserved and probably help provide the high specificity needed to cut out the introns precisely and ligate the exons (splicing). The 3'-splice junction between intron I and exon II is CCCUUAG. The mutation in the β-globin gene of this patient led to a single base change within intron I and 21 bases from the 3'-splice junction (Figure 15.15). This led to the creation of a new splice junction at a different place on the RNA. This new sequence is CUAUUAG (new base is underlined). About 90% of the mRNA in this patient was incorrectly spliced at the new splice junction. This produced a β-mRNA that was 19 nucleotides longer. The nucleotide insertion also changed the translation reading frame to create a new in-phase termination site at codon 36 of β⁺-globin mRNA. One predicts that a short β-globin peptide would be made. The abnormally spliced β⁺-globin mRNA is also unstable, probably because the new secondary structure of the RNA makes it susceptible to nucleases.

References

Honig GR and Adams JG III: Human hemoglobin genes, Wien, Austria, 1986, Springer-Verlag.

Orkin SH: Disorders of hemoglobin synthesis: the thalassemias. In Stamatoyannopoulos G et al, editors: The molecular basis of blood diseases, Philadelphia, 1987, WB Saunders Co.

Weatherall DJ et al: The hemoglobinopathies. In Scriver CR et al, editors: The metabolic basis of inherited disease, ed 6, New York, 1989, McGraw-Hill Information Services Co.

Figure 15.15   Transcription and processing of the β-globin gene in β⁺-thalassemia.
A G to A mutation within intron I creates a new splice site. Splicing
occurs either at the original site (10%) to yield normal β-globin mRNA
or at the new site (90%) to give an abnormally long mRNA.

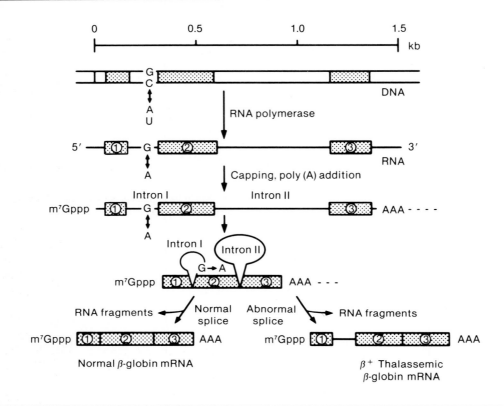

Case 2   Systemic lupus erythematosus

A 63-year-old woman with episodic swelling and pain of the hands, wrists, and knees
showed a positive test for antinuclear antibody, but tests for lupus erythematosus cells
were negative. However, a diagnosis of systemic lupus erythematosus (SLE) was made
based on a constellation of clinical and laboratory findings. Prednisone therapy was
started. Six months later the patient was admitted to the hospital with pleuritic pain
and purpura. Serum total protein was 56 g/L, and electrophoresis showed 66%
albumin and 4% γ-globulin. After an additional 6 mo she was admitted again with
shortness of breath, chest pain, and fever. A sputum culture grew *Diplococcus
pneumoniae* and coagulase-positive *Staphylococcus aureus*. Despite intensive antibiotic
and supportive therapy, the patient died 1 wk later. At autopsy, microscopic
examination revealed a lymphoma of a mixed cell type and a pathologic lesion typical
of SLE.

Biochemical questions   1. Patients with SLE produce antibodies to single-and double-stranded DNA,
especially viral nucleic acids. Some patients will also show antibodies against
RNA:DNA hybrids. Considering the relationship of lymphoma to SLE in this case,

Table 15.11 Patients with SLE who have antibodies to nucleic acids or to associated proteins

Antibodies against*	Percentage of patients	Antigens
Nuclear material	95	
DNA	70	
Sm	30	
RNP	40	snRNA-associated peptides
Ro	30	Peptides attached to U1 RNA
La	10	RNA polymerase
Histones	70	Protein complexed to RNA
		In drug-induced SLE

*Sm, Ro, and La are abbreviations for the nuclear antigens listed.

describe a possible role for the RNA-dependent DNA polymerase in producing these results.

2. Symptoms of lupus are often noticed after exposure to sunlight. Can you offer an explanation based on ultraviolet (UV) light activation of a latent virus?

3. Patients with SLE often have autoantibodies against nuclear or cytoplasmic ribonucleoprotein complexes containing small RNAs. One of these RNAs, U1RNA, is complementary at its 5′ end to the common sequence at heterogeneous nuclear RNA splice junctions (i.e., UUCAG ↓ GT). The autoantibodies have been found to inhibit splicing in cultured cells. Describe the implications of this finding.

Case discussion

SLE is an inflammatory autoimmune disease that is caused by antigen-antibody complexes being trapped in the capillaries of visceral organs. Several organs may be affected. Viruses possibly trigger the disease in susceptible individuals as they do in mice. Many patients display antibodies against viral antigens, but 95% have antibodies to cellular nuclear antigens. Table 15.11 lists the percentage of patients that make antibodies to nuclear proteins and nucleic acids. Despite extensive experimentation, little is known about the molecular pathophysiology of the disease; consequently, these questions and answers are rather speculative.

1. Viral etiology Several RNA tumor viruses carry RNA-dependent DNA polymerases as part of their virions. If the lymphoma had been triggered by an RNA tumor virus, RNA:DNA hybrids could have been produced and the patient might have made antibodies in response to them. Lymphomas are cancers of the lymphoid tissue, the sites of antibody formation.

2. Activation of a latent virus UV light could activate a latent virus, since it is known that UV light can activate latent bacterial viruses. Also, the lupus antibodies might be formed against a piece of excised DNA containing a UV-generated thymine dimer.

3. Small nuclear RNAs and splicing It is possible that small nuclear RNAs (snRNA) function in removing intervening sequences from mRNAs by forming base-paired structures that could be ligated immediately following excision of the intervening sequence. If patients with SLE make antibodies against snRNA, they may have a defect in removing intervening sequences from their mRNAs.

References

Hochberg MC et al: Systemic lupus erythematosus: a review of clinico-laboratory features and immunogenetic markers in 150 patients with emphasis on demographic subsets, Medicine 64:285, 1985.

Kalunian KC et al: Idiotypic characteristics of immunoglobulins associated with systemic lupus erythematosus: studies of antibodies deposited in glomeruli, Arthritis Rheum 32:513, 1989.

Steitz JA: "Snurps," Sci Am 258(6):56, 1988.

Case 3   Diphtheria

A 9-year-old daughter of a migrant worker was hospitalized for diphtheria during an epidemic in a southwestern city. Nausea, chills, vomiting, headache, and a sore throat brought the girl to the hospital. The examining physician noted a tenacious gray membrane near the tonsils. Her leukocyte count was elevated. The patient had never been immunized against diphtheria. Antibiotic treatment with penicillin and erythromycin was begun as well as intravenous administration of diphtheria antitoxin.

Diphtheria is caused by infection with *Corynebacterium diphtheriae* strains that are lysogenic for or infected with a bacteriophage carrying the *tox* gene. The viral gene codes for the synthesis of a protein toxin that is secreted from the bacterial cell.

Biochemical questions ◆ 1. What is the mechanism of action of erythromycin?

2. Why is it necessary to administer diphtheria antitoxin, since the antibiotics will eventually kill all the infectious bacteria?

3. Diphtheria toxin is an enzyme. What reaction does it catalyze? How does this affect protein synthesis?

◆ 4. Diphtheria toxoid is used to immunize healthy young children against diphtheria. It is made by treating diphtheria toxin with formaldehyde. What effect would you predict formaldehyde would have on the enzymatic activity of the toxin? Why? What effect does formaldehyde have on the serologic specificity or immunogenicity of the toxin?

5. Diphtheria toxin has a molecular weight of 62,000. Brief treatment with trypsin yields two fragments. The N-terminal fragment, A, has a molecular weight of 24,000 and retains enzymatic activity. The C-terminal fragment, B, has a molecular weight of 38,000 but is without enzymatic activity. The fragments, when assayed one at a time, do not inhibit cultured human cells, whereas the intact toxin molecule is inhibitory. Rats and mice are relatively resistant to the disease. Cultured mouse L cells are insensitive to both fragments as well as to the intact toxin; however, protein synthesis in cell-free extracts of L cells is sensitive to either the A fragment or the intact toxin. Describe a function for the B portion of the toxin that explains these data.

References   Pappenheimer AM: Diphtheria toxin, Annu Rev Biochem 46:69, 1977.

Pappenheimber AM: Diphtheria: molecular biology of an infectious process, Trends Biochem Sci 3:N220, 1978.

◆ Case 4   Transmission of herpes simplex virus

Two newborns developed herpes simplex virus infections while in the hospital. One child recovered and the other died. Autopsy showed a disseminated herpes infection. Epidemiologists needed to know whether the infants were infected by the same viral strain. Samples from both infants were used to culture the virus. Radioisotopically labeled DNA from the culture virus was analyzed with restriction enzymes, and the results indicated that the two infants were infected by the same strain.

Biochemical questions

1. How are restriction endonucleases used to distinguish between the many strains of herpes simplex virus?
2. Vmw65 is a virion phosphoprotein that greatly stimulates the transcription of the immediate early genes of herpes simplex virus 1 (HSV-1). Purified Vmw65 does not itself bind to viral DNA promoters, but in infected cells it is found associated with the viral promoter TAATGARAT. Explain how this is possible.

References

O'Hare P and Goding CR: Herpes simplex virus regulatory elements and the immunoglobulin octamer domain bind a common factor and are both targets for virion *trans* activation, Cell 52:435, 1988.

Roizman C and Buchman T: The molecular epidemiology of herpes simplex viruses, Hosp Pract 14:95, 1979.

Additional questions and problems

1. What steps in protein synthesis require GTP?
2. What would you expect the effect of chloramphenicol to be on (a) cytoplasmic eukaryotic ribosomes, (b) bacterial ribosomes, and (c) mitochondrial ribosomes?
3. Why are drugs relatively ineffective against viral disease?
4. What is necessary for a substance to be an antibiotic?
5. In a genetically determined disease such as phenylketonuria, is the critical liver enzyme missing or just inactive? How is this related to mental retardation? How can this situation be corrected?
6. How is a dominant or recessive trait expressed at the molecular level in terms of the protein product? In X-linked genetic diseases, why is the male usually affected and the female not?
7. Sometimes the following drugs are used in cancer chemotherapy. How do they function?

L-Asparaginase	Busulfan
Actinomycin D	Hydroxyurea
Arabinosyl cytosine	Estrogens and androgens
5-Fluoruracil	6-Thioguanine
Amethopterin	

8. Hemoglobin Constant Spring contains an extraordinarily large α-subunit. This subunit contains an additional 31 amino acid residues at the C-terminal end of the usual α-chain. Propose explanations for this abnormality. Consider mechanisms nvolving protein synthesis, chain termination, and DNA recombination.

Multiple choice problems

M.D., a 5-year-old Italian boy, was diagnosed as having thalassemia, an inherited erythroblastic anemia. Hemoglobin synthesis is defective in this disease, so only small amounts are present in the maturing red cells. In α-thalassemia, synthesis of α-hemoglobin chains is defective, whereas in β-thalassemia, the defect is in the synthesis of β-hemoglobin.

The following problems apply to this case.

1. Proof that a change in the ratio of α- to β-chains accounts for the pathologic changes seen in patients with thalassemia can be obtained by electrophoretic separation of the patient's hemoglobin. The *normal* distribution of α- and β-chains in the adult human hemoglobin (HbA) is:
   A. α_1, β_3.
   B. α_2, β_2.
   C. α_3, β_1.

D. α_4.

E. β_4.

2. Although the relative amounts of α- and β-chains change in thalassemia, the subunits themselves migrate on electrophoresis similar to normal chains. Thus one can conclude that the genetic defect in this disease results from a mutation in a gene for:
   1. Abnormal γ-chains.
   2. α-Chains in α-thalassemia and β-chains in β-thalassemia.
   3. Heme biosynthesis.
   4. A protein that controls α- or β-chain synthesis.
      The best answer is:
      A. 1, 2, and 3.
      B. 1 and 3.
      C. 2 and 4.
      D. 4 only.
      E. All are correct.

3. The ribosomes from thalassemic reticulocytes are just as active as normal ribosomes in translating the artificial mRNA polyuridylate (a homopolymer consisting of only uridylic acid residues). UUU is the code word for phenylalanine. The translation of polyuridylate with these ribosomes also requires:
   1. Phenylalanyl-tRNA.
   2. GTP.
   3. Elongation factors 1 and 2.
   4. All the aminoacyl-tRNA synthases.
      The best answer is:
      A. 1, 2, and 3.
      B. 1 and 3.
      C. 2 and 4.
      D. 4 only.
      E. All are correct.

4. It was found that the crude mRNAs from β-thalassemic reticulocytes program the synthesis of more α- than β-chains in vitro. From this finding the following interpretations are possible:
   1. Less mRNA for β-chains is synthesized than normal.
   2. More mRNA for β-chains is degraded than normal.
   3. mRNA for β-chains is translated less efficiently than normal.
   4. mRNA for β-chains has a lower affinity for ribosomes than normal.
      The best answer is:
      A. 1, 2, and 3.
      B. 1 and 3.
      C. 2 and 4.
      D. 4 only.
      E. All are correct.

5. Completed α- or β-chains are found in the reticulocytes of thalassemic patients. These chains are *not* bound to polyribosomes. This means that:
   A. Chain termination is defective.
   B. Chain initiation is defective.
   C. Chain elongation is defective.

    D. Chain termination is *not* defective.

    E. Heme synthesis is defective.

6. In the β-thalassemia of Ferrara, β-chains are completely missing. However, if poly-ribosomes from the reticulocytes of these patients are mixed in vitro with crude supernatant enzymes from normal reticulocytes, normal amounts of β-chains are made. (This enzyme fraction also contains tRNA but not globin mRNA.) Possible explanations for this finding are:

    1. A defect exists in an initiation factor that recognizes β-chain mRNA.

    2. A rare tRNA used in the readout of β-chain mRNA has mutated so it cannot function.

    3. A mutated aminoacyl-tRNA synthase has a lowered affinity for a rare tRNA used to decode β-chain mRNA.

    4. The gene for β-chains in this disease is deleted.

    The best answer is:

    A. 1, 2, and 3.

    B. 1 and 3.

    C. 2 and 4.

    D. 4 only.

    E. All are correct.

7. Although a lowered amount of globin than normal is made in thalassemia, heme synthesis is also reduced as a secondary response. The rate-limiting reaction in heme biosynthesis is catalyzed by:

    A. Ferrochelatase.

    B. δ-Aminolevulinate dehydrase.

    C. δ-Aminolevulinate synthase.

    D. Porphobilinogenase.

    E. Apoferritinase.

8. It is possible to be a heterozygote for both sickle cell trait and β-thalassemia. In such a person the abnormal β-chains caused by the sickle cell trait have a substitution of a valine where normally a glutamate residue is present. When such hemoglobin is submitted to electrophoresis at neutral pH, the *abnormal* hemoglobin chains will migrate:

    A. The same as normal adult hemoglobin (HbA).

    B. Closer to the anode (positive pole) than normal.

    C. Farther from the anode than normal.

    D. Not at all; it will aggregate at the origin.

    E. The same as fetal hemoglobin (HbF).

9. The glutamate to valine change in the sickle cell hemoglobin accounts for the sickling of the cells seen in this disease. The lowered solubility of sickle cell hemoglobin occurs because the mutant protein is:

    A. More heat sensitive than normal.

    B. More susceptible to proteolytic enzymes than normal.

    C. More hydrophilic than normal.

    D. More hydrophobic than normal.

    E. Synthesized in excessive amounts.

10. One code word for glutamate is GAG, and a code word for valine is GUG. If we assume that the mutation in sickle cell disease is in this codon, the type of mutation must be a:
    A. Transversion.
    B. Transition.
    C. Nonsense mutation.
    D. Deletion.
    E. Insertion.

Chapter 16

Lipoproteins

Objectives

1 To discuss the principles of lipid transport in the circulatory system
2 To describe the composition, structure, metabolism, and function of the main classes of plasma lipoproteins
3 To describe the clinically relevant defects in lipoprotein metabolism and their relationship to disease

A large amount of lipid must be moved from one organ to another, and the circulatory system is the transportation network through which this movement takes place. Plasma, the transport vehicle in the circulatory system, is an aqueous system. It is ideal for water-soluble substances, such as glucose and amino acids, that are transported in the circulation. Lipids, however, are not water-soluble, and the body is faced with the difficulty of having to transport large amounts of lipids through the aqueous plasma system.

To overcome this problem, mechanisms have evolved whereby lipids are transported in the plasma as macromolecular complexes that are miscible with water. The simplest system is a complex of fatty acid and albumin, the most abundant protein in the plasma. This is called the plasma free fatty acid, commonly abbreviated as FFA. The other system is composed of micelles containing a mixture of lipids and specific proteins. These are called plasma lipoproteins, and they transport cholesterol and triacylglycerol in the circulation.

Lipid transport and lipoproteins would not merit a separate chapter in a general biochemistry textbook. In the health sciences, however, this subject assumes a disproportionate importance. As many as 50% of the adult population in developed countries have lipid transport abnormalities, which are serious risk factors for atherosclerosis and its main complication, coronary heart disease. Furthermore, lipid transport abnormalities are associated with several other very common diseases, including diabetes, obesity, and alcoholism. There also are a number of serious genetic defects involving lipid transport. For these reasons, the health sciences student must acquire a detailed understanding of lipoproteins and the biochemistry of lipid transport.

Triacylglycerol is a major component involved in lipid transport, and its concentration in the plasma often is measured clinically. However, the substance is not called triacylglycerol in present clinical practice; it is commonly called triglyceride. Because much of the material to be presented is directly related to clinical situations, the term *triglyceride* will be used instead of triacylglycerol throughout the chapter.

Lipid transport

Lipids must be carried through the blood plasma from one tissue to another. An overview of this process, known as lipid transport, is presented in Figure 16.1. The transport process serves four main purposes. Dietary triglyceride must be transported from the intestine to other tissues in the body. Triglyceride formed in the liver must be secreted and subsequently deposited for storage in adipose tissue. Fatty acid stored as triglyceride in the adipose tissue must be taken to other tissues when they require a source of energy. Finally, cholesterol must be moved from one place to another in the body. The dietary cholesterol that is absorbed by the intestine has to be delivered to the liver. Cholesterol

685

Figure 16.1   Integration of lipid transport in the blood plasma.

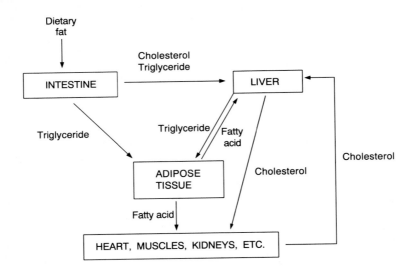

formed in the liver has to be taken to other tissues for membrane or steroid hormone synthesis. In addition, excess cholesterol contained in extrahepatic tissues or generated in the plasma as a consequence of lipoprotein catabolism must be taken to the liver, where it can be excreted in the bile.

One of the systems that mediate lipid transport is the plasma FFA. This system carries fatty acid from the adipocytes to other tissues, primarily for oxidation to provide energy. The other lipid transport system consists of a group of plasma lipoproteins that primarily carry cholesterol and triglycerides.

Free fatty acid (FFA)   From 95% to 98% of the fatty acid present in the blood plasma is contained in fatty acid esters, such as triglycerides, phospholipids, and cholesteryl esters. These fatty acid esters are present in the plasma lipoproteins. A small amount of fatty acid, about 2% to 5%, is present in unesterified form and is bound almost entirely to plasma albumin. This component is known as the plasma FFA. FFA is a very important source of fatty acid for the tissue under certain metabolic conditions.

FFA is the form in which fat stored in the adipocytes is released for use elsewhere in the body (see p 451). As the fatty acid enters the plasma, it physically binds to the protein albumin. Albumin has a very large capacity to bind fatty acids, but under most conditions an average of 0.5 to 1.5 fatty acids are bound by each albumin molecule. The amount of FFA transported is high during fasting and vigorous exercise, conditions in which the tissues require large amounts of fatty acid for energy. By contrast, the plasma FFA concentration is low after eating when there is enough glucose to satisfy the energy needs of the tissues. Though the binding between FFA and albumin is very strong, with dissociation constants in the range of 10 nM to 10 μM, a small amount of fatty acid is unbound and in equilibrium with the FFA-albumin complex.

$$\text{FFA/albumin} \rightleftharpoons \text{FA} + \text{albumin}$$

Figure 16.2 Metabolism of plasma FFA.

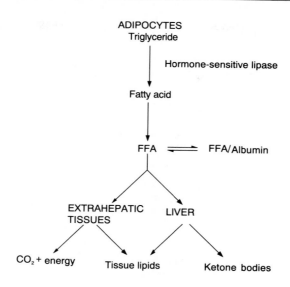

The unbound material is readily available for uptake by the tissues. As this occurs, FFA rapidly dissociates from albumin to reestablish the equilibrium and thereby also becomes available for uptake. In this way, the circulating FFA can be quickly taken up by the tissues; the half-life of plasma FFA is only 1 to 2 minutes. Much of the energy needed by the tissues in the fasting state is supplied by the plasma FFA. The FFA taken up also can be utilized for the synthesis of tissue lipids, such as phospholipids and triglycerides. In addition, FFA taken up by the liver provides most of the substrate for ketone body production (see p 454). The brain is the only tissue that does not utilize large amounts of FFA for energy. Instead, the brain oxidizes the ketone bodies that are formed by the liver from the circulating FFA. An overview of plasma FFA metabolism is presented in Figure 16.2.

Lipoproteins

The plasma lipoproteins are large complexes made up of physically combined lipid and protein. They have a micellar structure, with nonpolar lipids contained in a hydrophobic core that is surrounded by amphipathic lipids and proteins. The hydrophilic protein and lipid components serve to carry the nonpolar lipids, triglycerides and cholesteryl esters, through the aqueous environment, much as a boat can carry someone who cannot swim across a river.

Lipoprotein structure

Triglycerides and cholesteryl esters are located in the central core of the macromolecular complex. The nonpolar lipids are surrounded by a surface coat of phospholipids, unesterified (free) cholesterol, and one or more types of apolipoproteins that are commonly referred to as apoproteins. Phospholipids, free cholesterol, and the apoproteins are amphipathic. Their nonpolar groups, the fatty acyl chains of the phospholipids, the ring structure of cholesterol, and the hydrophobic aminoacyl chains of the apoproteins, interact with the triglycerides and cholesteryl esters in the central core. Conversely, the polar groups of the surface coat lipids and apoproteins interact with the water and ionic constituents of the plasma, thereby solubilizing the macromolecular complex in the aqueous environment.

A schematic representation of the structure of one type of plasma lipoprotein, human-high-density lipoprotein$_3$ (HDL$_3$), is presented in Figure 16.3. This model shows how the phospholipids (PL), free cholesterol (FC), and apoprotein groups are fitted together to form the surface coat. The polar groups of these amphipathic molecules comprise the outer layer of this coat and project outward to contact the surrounding plasma, whereas their nonpolar components form an inner layer that is in contact with the central lipid core. Likewise, the model illustrates how the triglycerides (TG) and cholesteryl esters (CE) pack together to make up the nonpolar lipid core in the center of the macromolecular complex. Although the size and precise molecular details vary, this type of micellar model represents the structure of the other classes of plasma lipoproteins as well.

Classes of plasma lipoproteins

There are five main classes of plasma lipoproteins: chylomicrons, very low–density lipoproteins (VLDL), intermediate–density lipoproteins (IDL), low–density lipoproteins (LDL), and high–density lipoproteins (HDL), as indicated in Table 16.1 These five lipoprotein classes can be separated by ultracentrifugation, as illustrated in Figure 16.4. Several other components and subfractions exist. One of these is the chylomicron remnants, which are formed from chylomicrons and have a density in the range of 0.94 to

Figure 16.3   Structural model of human HDL$_3$ with dimensions derived from space-filling atomic models. *PL*, phospholipids; *FC*, free cholesterol; *CE*, cholesteryl esters; *TG*, triglycerides. (From Edelstein C et al: J Lipid Res 20:143, 1979.)

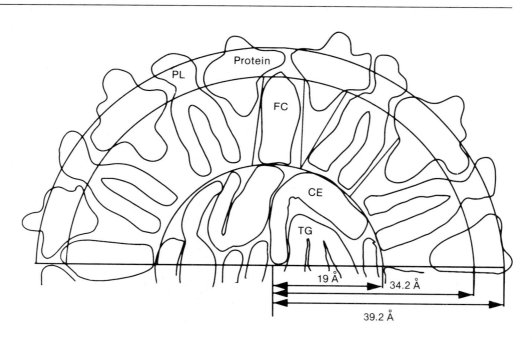

1.006. Therefore the density of the chylomicron remnants overlaps with that of the VLDL. The HDL fraction also is not homogeneous, as is evident from the biphasic peak seen in Figure 16.4. The two HDL subclasses have different functions and properties.

Electrophoretic separation The main classes of plasma lipoproteins also can be separated by electrophoresis, as is illustrated schematically in Figure 16.5. Four major bands

Table 16.1   Properties of plasma lipoproteins

Lipoprotein	Density range (g/ml)	Flotation units (SF)*	Major lipids	Electrophoretic mobility	Major apoproteins
Chylomicron	<0.94	>400	Triglyceride	0	B-48, A-I, IV
Very low density (VLDL)	0.94 to 1.006	20 to 400	Triglyceride	Pre-β	B-100, E, C-I, II, III
Intermediate density (IDL)	1.006 to 1.019	12 to 20	Triglyceride and cholesteryl esters	β	B-100, E
Low density (LDL)	1.019 to 1.063	0 to 12	Cholesteryl esters	β	B-100
High density (HDL)	1.063 to 1.21	—	Phospholipids and cholesterol	α_1	A-I, II

*Flotation at density 1.063.

Figure 16.4   Separation of the major plasma lipoprotein classes normally present in plasma by density fractionation. *VLDL,* very low–density lipoproteins; *IDL,* intermediate-density lipoproteins; *LDL,* low–density lipoproteins; *HDL,* high-density lipoproteins. Usual density ranges are chylomicrons, <0.94; VLDL and chylomicron remnants, 0.94-1.006; IDL, 1.006-1.019; LDL 1.019-1.063; HDL₂ 1.063-1.125; HDL₃, 1.125-1.21.

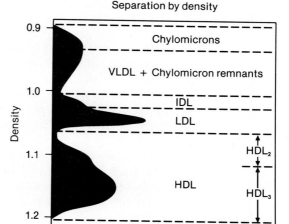

Figure 16.5   Schematic representation of lipoprotein separation by plasma
                                   electrophoresis.

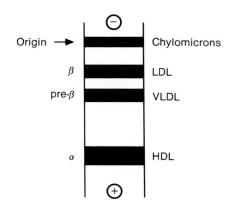

can be visualized following plasma electrophoresis. One band remains at the origin and contains the chylomicrons. Another, called β-lipoproteins, migrates with the β-globulins and contains the LDL. A third band migrates in front of the β-region and is called the pre-β-lipoproteins. This fraction contains the VLDL. The fourth band, called α-lipoproteins, migrates with the α-globulins and contains HDL. Any IDL that is present migrates with the β-lipoproteins. Chylomicron remnants spread over the β- and pre-β-regions. Plasma lipoprotein analyses are made routinely in clinical laboratories by electrophoresis on cellulose acetate, and the lipid-containing regions are made visible with a stain such as oil red O.

Chemical and physical properties

The properties of the five main classes of plasma lipoproteins are listed in Table 16.1 and the composition of four of these lipoproteins is given in Table 16.2. The compositions are average values, because each of the lipoprotein classes are made up of a spectrum of molecules that vary in size and lipid content.

Apolipoproteins Letters are used to designate the various apoproteins, for example, apo-A, apo-B, etc. Although it was originally thought that each class contained a single apoprotein component, it was found subsequently that many were heterogeneous. This has given rise to the use of a Roman numeral to distinguish the several members within a class, such as apo-A-I, apo-A-II, etc. Apo-E and apo-A-II appear to be derived from the same precursor, an apoprotein that has a molecular weight of 46,000. Table 16.3 lists the known apoproteins.

Information concerning function is available for only some of the apoproteins. Apo-A-1 is the main structural protein of HDL and is an activator of lecithin-cholesterol acyltransferase. Apo-B-100 is the main structural component of VLDL, IDL, and LDL. In addition, apo-B-100 is involved in the binding of LDL to high-affinity cell surface LDL receptors. Apo-B-100 is synthesized in the liver. A smaller form, apo-B-48, is synthesized by the intestine. It is the main B component in chylomicrons. Apo-C-II is the activator of lipoprotein lipase. Apo-C-III appears to inhibit VLDL uptake by the liver.

Table 16.2   Average lipid and protein composition of plasma lipoproteins

Component	Average composition (%)				
	Chylomicrons	**VLDL**	**IDL**	**LDL**	**HDL**
Protein	2	9	21	21	50
Triglycerides	84	54	19	11	4
Cholesterol	2	7	8	8	2
Cholesteryl esters	5	12	27	37	20
Phospholipids	7	18	25	22	24

Table 16.3   Apoproteins of the plasma lipoproteins

Apoprotein	Molecular weight	Plasma lipoprotein(s) containing an appreciable amount	Major site(s) of synthesis	Function
A-I	28,300	HDL, chylomicrons	Intestine, liver	LCAT* activation
A-II	17,400	HDL	Intestine, liver	†
A-IV	44,500	Chylomicrons,	Intestine	†
B-100	549,000	LDL, VLDL, IDL	Liver	Recognition by LDL receptor
B-48	264,000	Chylomicrons, chylomicron remnants	Liver	Chylomicron formation
C-I	6630	VLDL, HDL	Intestine	†
C-II	8840	VLDL, HDL	Liver	Lipoprotein lipase activation
C-III	8760	VLDL, HDL	Liver	Inhibition of VLDL and uptake by liver
D	22,100	HDL	†	†
E	34,100	VLDL, chylomicron remnants, IDL	Liver, macrophages	Recognition by LDL and chylomicron remnant receptor

*LCAT, lecithin cholesterol acyltransferase.
†Not known.

Apo-E is the recognition factor for the binding of chylomicron remnants to their receptors in the liver and for IDL binding to the LDL receptor. The main apoproteins contained in the different lipoproteins are listed in Table 16.4.

Amphipathic helix Segments of many of the apoproteins have a conformation known as an amphipathic helix, illustrated in Figure 16.6. An Edmundson wheel diagram shows the positioning of the amino acid residues in one of the helical domains of apo-E. The amino acid sequence is such that one surface of the helix contains only uncharged residues such as leucine and valine, forming a nonpolar face. The opposite surface contains charged residues such as arginine and glutamate, forming a polar face. The nonpolar face of the helix interacts with lipids in the interior of the lipoprotein, and the polar face interacts with hydrophilic groups contained in the surface coat or with water and ions in the surrounding plasma.

Table 16.4   Distribution of apoproteins in the lipoproteins

Lipoprotein	Main apoproteins
Chylomicrons	B-48, A-I, A-IV
Chylomicron remnants	B-48, E
VLDL	B-100, E, C-I, C-II, C-III
IDL	B-100, E
LDL	B-100
HDL	A-I, A-II

Figure 16.6

The apoprotein amphipathic helix. Edmundson wheel diagram of apo-E was obtained from 66 base pair DNA repeats contained in exon 4 of the gene. Position and charge of each amino acid is shown. A nonpolar face comprises about 30% of the circumference and is formed by residues 2, 5, 9, 13, 16, and 20, as indicated on the left of helical structure. The polar face, comprising about 70% of the circumference, is formed by residues 3, 4, 6, 7, 8, 11, 12, 14, 15, 17, and 18, as indicated on the right side. Residues 1 and 21 are variable. (Modified from Breslow JL: Genetics of the human apolipoproteins. In Scanu AM and Spector AA, editors: Biochemistry and biology of plasma lipoproteins, New York, 1986, Marcel Dekker, Inc.

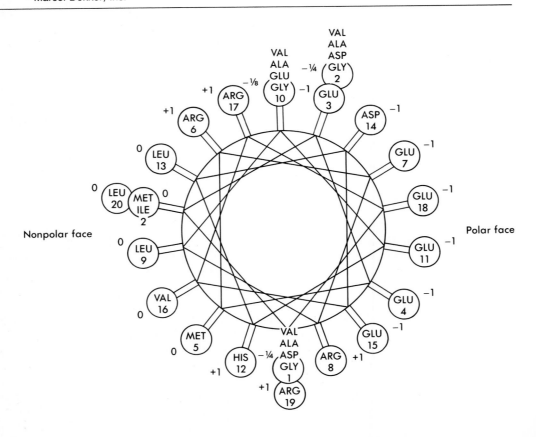

Proteolipids

A lipid-protein complex can be isolated from myelin, the membrane material of nerve axons, as well as from other tissues by extraction with organic solvents. This lipoprotein material is known as a proteolipid. Unlike the plasma lipoproteins, however, this proteolipid is insoluble in aqueous media and is soluble in lipid solvents. It contains 8% protein and 92% lipid, and the molecular weight of its protein component is 36,000. Like the plasma lipoproteins, proteolipids have a micellar structure.

Lipoprotein function

An overview of lipoprotein function is presented in Figure 16.7. Lipoproteins carry out three main functions. One is to transport the dietary fat from the intestinal mucosa, where it is absorbed, to the other tissues. Chylomicrons and chylomicron remnants perform this function. The second is to transfer triglyceride from the liver to other tissues, where the lipid can be either stored or oxidized for energy. VLDLs carry out this role. After the VLDLs deliver their triglyceride to the tissues, their remaining constitutents are returned to the liver in the form of IDL and LDL. The third system mediates reverse cholesterol transport. This system, which involves HDL and LDL, returns excess cholesterol from extrahepatic tissues to the liver.

Lipoprotein biogenesis

Chylomicrons, VDL, and the precursor form of HDL called nascent HDL are synthesized intracellularly. Chylomicrons are produced by the small intestine, VLDL by the liver, and nascent HDL by both tissues. After secretion, these lipoproteins undergo extensive remodeling in the circulation. Some lipids and apoproteins are added and others are removed, leading to the conversion of these original particles to new types of lipoproteins. Chylomicrons are converted to chylomicron remnants, VLDL to IDL and then to LDL, and nascent HDL to the mature forms of HDL, HDL_3, and HDL_2. Thus four types of lipoproteins, chylomicron remnants, IDL, LDL, and mature forms of HDL are generated in the plasma. The site of origin of each of the main lipoproteins is listed in Table 16.5.

Chylomicron synthesis The mucosal cells in the small intestine absorb the products of lipid digestion and the lipids contained in the bile. A mixture of cholesterol, fatty acid, and 2-monoglyceride enters through the villi, and triglycerides are synthesized from

Figure 16.7   Overview of lipoprotein function.

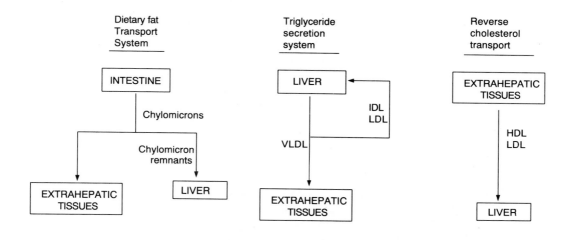

Table 16.5 Tissue of origin and function of the plasma lipoproteins.

Class	Origin	Physiologic function
Chylomicrons	Intestine	Absorption of dietary fat
Chylomicron remnants	Plasma	Delivery of dietary fat to liver
VLDL	Liver	Transport of triglyceride from liver to other tissues
IDL	Plasma	Initial product formed in VLDL catabolism
LDL	Plasma	Cholesteryl ester transport
HDL	Liver, Intestine	Removal of excess cholesterol from tissues and lipoproteins; remodeling of lipoproteins

the fatty acid and monoglyceride in the smooth endoplasmic reticulum. Cholesterol and phospholipids are added to the triglycerides in the smooth endoplasmic reticulum to form a micelle. The necessary apoproteins, apo-B-48, apo-A-I, apo-A-II, and apo-A-IV, are synthesized in the rough endoplasmic reticulum and combined with the lipid as the components move from the rough and smooth endoplasmic reticulum into the Golgi apparatus. Apparently, the availability of materials that comprise the chylomicron surface—apoproteins, phospholipid, and cholesterol— is limiting for synthesis. The size of the chylomicrons formed depends on the amount of triglyceride available in the enterocyte. When more triglyceride is available for secretion, larger chylomicrons are formed. These chylomicrons contain more triglyceride and therefore have a greater core-to-surface ratio. The chylomicrons then enter secretory vesicles and move to the basolateral plasma membrane. They are secreted into the lateral intercellular space and then pass into the lymph. Modification begins to occur immediately after secretion, so that the mature chylomicrons present in the plasma have a somewhat different lipid and apoprotein composition than the newly secreted particles.

VLDL synthesis There are two sources of the fatty acid that forms the triglyceride present in the VLDL. These are dietary carbohydrate, which is converted into fatty acid in the hepatocytes, and fatty acid taken up by the liver. The latter is derived from either the plasma FFA or the lipids contained in plasma lipoproteins. Triglycerides, phospholipids and cholesterol present in the smooth endoplasmic reticulum are combined into a micelle. The required apoproteins, apo-B-100, apo-E, apo-C-I, apo-C-II, and apo-C-III are synthesized and processed in the rough endoplasmic reticulum. Apo-B-100 is partially glycosylated in the rough endoplasmic reticulum and then added to the lipid at the junction of the rough and smooth endoplasmic reticulum. The other apoproteins are added sequentially to the lipid–Apo-B-100 complex in the Golgi apparatus, where additional glycosylation occurs and phospholipid is added. After the particle enters the secretory vesicle, it is released into space of Disse and then enters the plasma.

HDL synthesis The precursor form of HDL, called nascent HDL, is formed by the liver and small intestine; however very little is known about this process or its regulation. The liver secretes a discoid particle containing phospholipid, cholesterol, and apo-E. These discoid HDLs are remodeled into the spherical HDL$_3$ after secretion. In this process, lipids are added, cholesterol is esterified, apo-E is removed, and apo-A-I is added. Some evidence indicates that the intestine also secretes similar discoid particles, except that they contain apo-A-I instead of apo-E. These nascent HDLs also are converted to HDL_3 soon after secretion. Other data suggest that the intestine does not secrete nascent HDLs directly; rather, they come from the surface of chylomicrons that undergo remodeling immediately after secretion. More work is needed to determine which of these two mechanisms is correct.

Factors regulating lipoprotein synthesis Dietary fat increases chylomicron production by the intestine, and the production of VLDL by the liver increases when excess fatty acid is available. Unsaturated fatty acids are more effective than saturates in stimulating VLDL formation, and long-chain fatty acids produce a greater effect than medium or short-chain fatty acids. Dietary carbohydrate also increases VLDL production by increasing fatty acid synthesis and stimulating insulin release. Likewise, alcohol increases VLDL production by making more fatty acid available for triglyceride synthesis in the liver.

The effects of cholesterol on lipoprotein formation are complicated and not fully understood. In animal experiments, a high dietary cholesterol intake alters the type of particle formed. In some species, the liver produces a modified form of VLDL that is rich in cholesterol esters, called β-VLDL. In others, a large HDL particle rich in cholesterol and apo-E, called either HDL_1 or HDL_c, is produced in response to excess dietary cholesterol. By contrast, LDL, a lipoprotein that normally is present, becomes elevated in humans when the diet is high in cholesterol. The LDL particles that accumulate do not appear to have a grossly abnormal structure or composition, and it is not known whether the LDL elevation that occurs in response to a high-cholesterol diet in humans results from any change in lipoprotein formation.

Clinical comment

Abetalipoproteinemia Abetalipoproteinemia is a rare genetic abnormality involving the synthesis of lipoproteins that contain apo-B. Both chylomicrons and VLDL are affected. As a result, LDL, which is formed from VLDL in the plasma, also is deficient. Fat malabsorption occurs in abetalipoproteinemia because chylomicrons cannot be formed by the intestine. The other clinical abnormalities that occur are ataxic neuropathy, retinitis pigmentosa, and acanthocytosis. Abetalipoproteinemia is an autosomal recessive defect that is thought to result from a genetic defect in the synthesis of apo-B. Both apo-B-100 and apo-B-48 are affected because they are derived from the same gene. About half of the cases result from consanguineous marriages. Heterozygotes for abetalipoproteinemia show no clinical manifestations.

Lipoprotein metabolism

Chylomicrons Chylomicrons are synthesized in the intestinal mucosa. They contain chiefly triglycerides, and approximately 98% of their dry weight is lipid. Chylomicrons therefore have an extremely low density (<0.94). They are very large molecular complexes that refract light and give the plasma a milky appearance when they are present in excessive concentrations. They are secreted into the lymph and enter the blood plasma via the thoracic duct. Their major function is to transport dietary fat, primarily in the form of triglyceride, into the body. The other dietary lipid transported in chylomicrons is cholesterol, which is converted to cholesteryl esters before incorporation into these lipoproteins.

Chylomicron remnants Chylomicrons are catabolized in the plasma to a smaller particle that has a higher density. The catabolic product, called the chylomicron remnant, is formed when much of the triglyceride originally present in the chylomicron has been hydrolyzed by lipoprotein lipase. It contains phospholipids, cholesterol, cholesteryl esters, apo-E, and the remaining triglyceride. These remnants have a density similar to that of IDL and small VLDL and are recovered in the S_f20-400 fraction (Table 16.1) on ultracentrifugation. They are removed from the plasma by the liver through a receptor that binds apo-E but is distinct from the LDL receptor.

Chylomicron and chylomicron remnant metabolism is summarized in Figure 16.8. This illustrates that the majority of the triglycerides carried by the chylomicron are hydrolyzed by the capillary form of lipoprotein lipase, and the released fatty acids are taken up by the tissues, predominantly adipocytes and muscles. The residual particle, which is rich in cholesteryl esters and apo-E, is taken up by the liver. This is the main route whereby dietary cholesterol is delivered to the liver.

Very low–density lipoproteins VLDLs are also very large complexes. They contain about 90% lipid, 50% to 65% of which is triglyceride. VLDLs are snythesized primarily in the liver and serve to transport triglycerides from the liver to other tissues, especially to adipose tissue.

Low-density lipoproteins Most of the cholesterol contained in the blood plasma of a normal human after an overnight fast is present in LDLs. About 75% of the cholesterol in LDL is in the form of cholesteryl esters. A single apoprotein, apo-B-100, comprises 98% of the protein present in LDL. LDLs are formed in the plasma during VLDL catabolism and therefore can be thought of as a VLDL remnant.

About half of the circulating LDL is taken up by the liver, and the remainder is taken up by extrahepatic tissues. Much of the LDL uptake is mediated by the LDL receptor, which binds apo-B-100. However, some LDL uptake occurs without the intervention of the receptor; this is called non-receptor-mediated LDL uptake.

Intermediate-density lipoproteins IDLs are formed in the plasma during the conversion of VLDL to LDL. They contain amounts of triglyceride and cholesterol that are intermediate between those of VLDL and LDL. Some of the IDL formed is taken up directly by the liver, and the remainder is converted to LDL. Normally the conversion

Figure 16.8   Metabolism of chylomicrons and chylomicron remnants. *TG,* Triglycerides; *CE,* Cholesteryl esters.

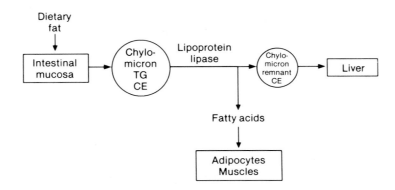

of VLDL to LDL proceeds so efficiently that appreciable quantities of IDL usually do not accumulate in the blood plasma obtained after an overnight fast. Like LDL, IDL uptake by the liver is mediated by the LDL receptor. In the case of IDL, however, apo-E mediates the binding.

Figure 16.9 summarizes VLDL, IDL, and LDL metabolism. This system of lipoproteins appears to have two major functions. One is to remove excess fatty acid from the liver. The fatty acid, which often is derived from dietary carbohydrate, is released as triglyceride in VLDL. After subsequent hydrolysis of the triglyceride by lipoprotein lipase, the fatty acid is taken up primarily by adipocytes and muscle. The other function of the system is to accept cholesteryl esters that are formed in the plasma. The cholesteryl esters are passed from HDL to VLDL. LDLs, the final product formed from VLDL, retain the cholesteryl esters and have a high cholesteryl ester content. The LDLs either supply the cholesterol to extrahepatic tissues or deliver it to the liver for excretion in the bile.

High-density lipoproteins HDLs are synthesized in the liver and intestine. They act as catalysts, facilitating the catabolism of VLDL and chylomicrons. C-apoproteins are transferred between HDL and VLDL, and lipid transfer also occurs.

Figure 16.9 VLDL metabolism. The conversion of VLDL to LDL and LDL is shown. *TG,* Triglycerides; *CE,* Cholesteryl esters.

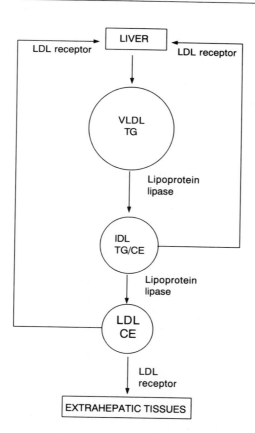

Figure 16.10

Schematic representation of the conversion of nascent HDL to spherical HDL. These reactions occur in the blood plasma during lipoprotein metabolism. Nascent HDL is derived either from the surface coat of chylomicrons or is secreted directly by the liver. Lipids and apoproteins taken up during conversion to the spherical forms are derived from the large, lower-density lipoproteins that are undergoing catabolism in the circulation. In addition, some of the cholesterol that is taken up is derived from the tissues.

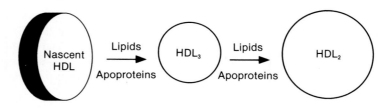

The HDLs that are released into the plasma are composed primarily of phospholipids and apoproteins, and they have a flat, discoid structure. In this form, they are called *nascent HDLs*. Some of the nascent HDLs are secreted directly into the plasma by the liver and, possibly, from the intestine. The rest are formed in the plasma from the surface coat components of chylomicrons after they are secreted by the intestinal mucosal cells. Subsequently the nascent HDLs take up lipids and apoproteins from VLDLs and chylomicrons as these large, triglyceride-rich lipoproteins are catabolized in the plasma. The HDLs also take up cholesterol released from extrahepatic tissues. The LCAT reaction takes place in HDL, and the cholesterol that is taken up is converted to cholesteryl esters that move to the center of the discoid structures. When this occurs, the HDLs assume a spherical shape (Figure 16.10).

HDL$_2$ and HDL$_3$ HDLs are separated into two fractions by ultracentrifugation: HDL$_2$ in the density range 1.060 to 1.125, and HDL$_3$ in the density range 1.125 to 1.210. HDL$_2$ is composed of 40% protein and 60% lipid, whereas HDL$_3$ contains 55% protein and 45% lipid. More than 90% of the protein present in both subclasses consists of apo-A-I and apo-A-II, with the ratio for apo-A-I to apo-A-II in both cases being 3:1. There is evidence that HDL$_3$ is converted to HDL$_2$ as a result of two processes. One is the uptake of lipids and apoproteins that are released during the catabolism of VLDL, chylomicrons, and their intermediates. The other is the uptake of free cholesterol that is released from the tissues. HDL$_3$ is a major site of the lecithin-cholesterol acyltransferase (LCAT) reaction through which cholesterol is converted to cholesteryl esters.

A schematic representation of HDL formation and metabolism is presented in Figure 16.11. According to this model, the nascent HDL are derived from either the liver or the intestine. The discord nascent HDL is converted to spherical forms, HDL$_3$ and subsequently HDL$_2$. This involves a transfer of lipids between the HDL$_3$ and VLDL that are undergoing catabolism in the plasma. Cholesteryl esters formed in the LCAT reaction are transferred to the VLDL. At the same time, triglycerides are transferred from the VLDL to the HDL. These transfers are mediated by one or more plasma lipid transfer proteins. The resulting HDL is enriched in triglyceride and is larger; it is called HDL$_2$. Two pathways appear open to the HDL$_2$. One is partial hydrolysis by hepatic lipase, resulting in the

Figure 16.11   HDL metabolism. *C*, Cholesterol; *CE*, cholesteryl ester; *TG*, triglyceride; *apos*, apoproteins.

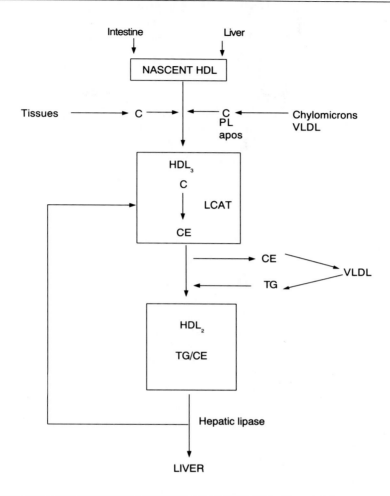

conversion of the HDL_2 back to HDL_3. The other is binding to receptors and uptake into the tissues, primarily the liver. These final steps in HDL metabolism are not fully understood and are the subjects of intensive study.

HDL heterogenity One of the factors that complicates the study of HDL metabolism is that this lipoprotein class is heterogeneous. Application of newer separation techniques, such as immunoadsorbant chromatography and gradient gel electrophoresis, indicates that many of the particles have different apoprotein compositions. For example, there are HDLs that have only apo-A-I, some with apo-A-I and apo-A-II, and still others with apo-E. Each of these may represent components with a different origin, metabolic pathway, and function. Because of this complexity, much uncertainty remains regarding HDL metabolism.

Reverse cholesterol transport An important function of HDL is the removal of excessive cholesterol from tissues and channeling the cholesterol for deposit in the liver. This is important because the steroid nucleus cannot be degraded, and the liver is the only organ that can rid the body of excess cholesterol by secreting it in the bile for excretion in the feces. The process of moving cholesterol from extrahepatic tissues to the liver is called reverse cholesterol transport. This is illustrated in Figure 16.12. In addition to HDL, LCAT, lipid transfer proteins, VLDL, and LDL play a role in reverse cholesterol transport.

Cholesterol surface transfer

The movement of cholesterol from cells to HDL occurs through surface transfer. The mechanism involves diffusion of cholesterol from the plasma membrane of the cells to the surface coat of the HDL. This actually is an exchange process, with the cholesterol being able to diffuse back and forth between the cell and lipoprotein surface. Because of the LCAT reaction, however, the cholesterol entering the HDL is rapidly converted to cholesteryl ester, after which it either moves to the core of the HDL or is transferred to

Figure 16.12

Reverse cholesterol transport. This is the process whereby excess cholesterol contained in extrahepatic tissue is taken to the liver for utilization there or excretion through the bile.

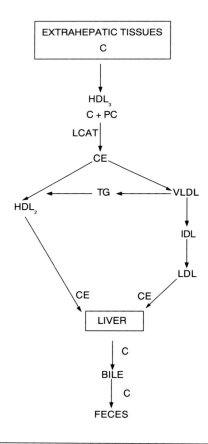

VLDL. The rapid removal of the cholesterol after it becomes associated with the HDL causes the net movement to be in the direction of cell to HDL, leading to an overall transfer of cholesterol from the tissues into the plasma.

Movement of cholesterol from one surface to another is not confined to tissues and HDL. Cholesterol also can move between the surface coats of various lipoproteins, and between the surface of blood cells such as erythrocytes and lipoproteins. Only cholesterol itself, which is amphipathic, can undergo this type of surface transfer. The movement of cholesteryl esters is much more restricted and can occur only if it is facilitated by a specific transfer protein.

Clinical comment

HDL and atherosclerosis Evidence was presented in 1951 that people who have elevated HDL levels are less prone to develop atherosclerosis. In addition, premenopausal women were shown to have higher HDL levels than men of a corresponding age. This difference, which is mediated hormonally, somehow protects the premenopausal woman against atherosclerosis. These important findings about HDL were neglected for 25 years while clinical investigators concentrated on diseases resulting from increased plasma lipoprotein concentrations, the hyperlipoproteinemias. In 1976, however, the protective effect of HDL was rediscovered. Epidemiologic studies clearly indicate that people with high HDL levels are protected against atherosclerosis. This has led to a rebirth of clinical interest in HDL, and many investigators are now trying to determine the molecular mechanism of the protective effect. The most widely held theory is that HDLs protect because they mediate reverse cholesterol transport, thereby reducing cholesterol accumulation in the arterial intima and preventing the development of atherosclerotic lesions.

Clinical comment

Tangier disease Tangier disease is a genetic abnormality involving HDL. This autosomal recessive defect was first described in a family living on Tangier Island, from which the disease takes its name. No normal HDLs are detected in the plasma of homozygotes, but apo-A-I and apo-A-II are present in the plasma and appear to be normal. Patients with Tangier disease have enlarged, orange tonsils; cholesteryl ester accumulation in the reticuloendothelial system; corneal opacities; and neuropathy. The metabolic basis for Tangier disease is unknown; it probably involves a defect in either HDL biogenesis or metabolism.

Enzymes of lipoprotein metabolism

Four enzymes mediate the metabolism and interconversion of the plasma lipoproteins. They are listed in Table 16.6. LCAT, lipoprotein lipase, and hepatic lipase act on the lipoproteins while they are in the circulation. Acid lipase is a lysosomal hydrolase and acts after the lipoprotein is taken up by receptor-mediated endocytosis.

Lecithin-cholesterol acyltransferase

LCAT (see p. 475) mediates cholesterol esterification in the plasma. The enzyme is synthesized and secreted by the liver. It acts primarily on HDL_3, where the cholesterol taken up by these HDL from the tissues or from the surface of chylomicrons and VLDL is esterified. Phosphatidylcholine contained in the surface coat of the HDL provides the fatty acid, which is directly transferred to cholesterol. The fatty acid is derived from the *sn*-2 position of the phosphatidylcholine. Because the *sn*-2 position contains mostly polyunsaturated fatty acid, the cholesteryl esters formed by the LCAT reaction are polyunsaturated and usually contain linoleic or arachidonic acid. Genetic deficiencies in LCAT occur; this disease is discussed on p. 493.

Lipoprotein lipase

The enzyme that hydrolyzes the triglycerides contained in the circulating chylomicrons and VLDL is lipoprotein lipase. It acts at the surface of the capillary endothelial cell.

Table 16.6   Enzymes mediating lipoprotein metabolism

Enzyme	Substrate(s)	Reaction	Site of action	Metabolic role
Lecithin-cholesterol acyltransferase (LCAT)	Cholesterol and phosphatidylcholine	Cholesterol esterification	Plasma lipoproteins, especially HDL	Reverse cholesterol transport
Lipoprotein lipase	Triglyceride in VLDL and chylomicrons	Hydrolysis	Capillary surface	VLDL and chylomicron degradation
Hepatic lipase	Triglyceride and phospholipids in HDL_2 and IDL	Hydrolysis	Liver sinusoids	HDL_2 and IDL catabolism
Acid lipase	Triglyceride and cholesteryl ester	Hydrolysis	Lysosomes	Catabolism of lipoproteins incorporated into tissues by receptor-mediated endocytosis

The enzyme is activated by apoprotein C-II, one of the low molecular weight proteins present in VLDL and HDL. Lipoprotein lipase is present in many tissues, including adipose tissue, mammary gland, and heart. In the adipose tissue the activity of lipoprotein lipase is increased by administration of the hormone insulin. The enzyme is formed within the tissue and then secreted. It passes through the capillary endothelium and binds firmly to the heparan carbohydrate chains of glycoproteins present at the surface of the endothelial cells. The enzyme acts on the lipoprotein triglycerides while it remains firmly bound to the endothelial cell surface. Lipoprotein lipase has specificity for fatty acids esterified at positions *sn*-1 and *sn*-3 of the triglycerides. For the most part, the enzyme hydrolyzes triglycerides to monoglycerides. The monoglycerides are transported to the liver, where they are degraded by a hepatic monoglyceride lipase.

Hepatic lipase

A lipid hydrolase with properties somewhat similar to lipoprotein lipase is present in the liver sinusoids. This enzyme, called hepatic lipase, acts on IDL and HDL_2. It hydrolyzes the triglycerides contained in these lipoproteins, as well as phosphoglycerides present in their surface coat. In this way, hepatic lipase probably is involved in the processing of IDL, either for uptake by the liver or for conversion to LDL. Likewise, hepatic lipase probably also is involved in HDL_2 processing, either for uptake by the liver or for conversion back to HDL_3.

Clinical comment

Effect of heparin When heparin is given intravenously, lipoprotein lipase activity is released into the blood plasma. This is a pharmacologic action of heparin, and heparin is not a physiologic cofactor for the enzyme. The lipolytic activity that is released is made up of two different enzymes that hydrolyze triglycerides. One of the lipases is derived from the liver and is resistant to inactivation by protamine. This is hepatic lipase. The second enzyme, which is of extrahepatic origin and is inhibited by protamine, is primarily responsible for the hydrolysis of the triglycerides in the chylomicrons and VLDLs. This is lipoprotein lipase. After heparin injection, from 45% to 95% of the lipolytic activity that appears in the blood plasma is protamine resistant, that is, it is the hepatic lipase.

Acid lipase

Acid lipase acts intracellularly after the lipoprotein is removed from the circulation by receptor-mediated endocytosis. It is contained in lysosomes and is a member of a family

Table 16.7 Lipoprotein receptors

Receptor	Recognition	Lipoprotein	Tissue	Metabolic role	Comments
LDL	Apo-B-100 Apo-E	LDL, IDL	Liver, many other tissues	Removal of LDL, IDL from circulation	Mediates delivery of cholesterol from plasma to tissues
Chylomicron remnant	Apo-E	Chylomicron remnants	Liver	Uptake of dietary fat by liver	Also called apo-E receptor
HDL	Unknown	HDL	Liver, possibly other tissues	HDL binding to cells	Existence is questionable
Scavenger	Chemically modified apo-B-100	Modified or damaged LDL	Endothelium, macrophages	Removal of chemically altered LDL	Poorly defined

of enzymes called lysosomal acid hydrolases. The enzyme hydrolyzes the triglycerides and cholesteryl esters remaining in the lipoprotein, releasing fatty acids and cholesterol for utilization within the cell.

Clinical comment

Acid lipase deficiency There are two diseases that are caused by genetic deficiency of the lysosomal acid lipase. One is Wolman's disease. It is the more severe form and manifests itself during the first month of life. Characteristic findings are anemia and massive enlargement of the liver and spleen due to accumulation of cholesteryl esters. Death usually results in a few months. The other form, which is less severe, is cholesteryl ester storage disease. In addition to liver and spleen enlargement, there is LDL elevation and accelerated cardiovascular disease. These diseases can be diagnosed by measuring the acid lipase content in the circulating leukocytes.

Lipoprotein receptors

Lipoproteins are removed from the circulation by binding to receptors located on the surface membrane of target cells. These receptors and their properties are listed in Table 16.7.

LDL receptor The best characterized lipoprotein receptor is the LDL receptor, which has been cloned and sequenced. The LDL receptor recognizes apo-B-100 and apo-E; therefore, it also is known as the apo-B/E receptor. Through binding apo-B-100, the LDL receptor mediates the clearance of LDL from the plasma. It also mediates the clearance of IDL, but in this case the recognition factor is apo-E. Even though VLDLs contain apo-B-100 and apo-E, these lipoproteins are prevented from binding to the LDL receptor, possibly because they contain apo-C-III. The LDL receptor is present in the liver as well as in a number of other tissues, including fibroblasts and monocytes. It is a 160,000 Da glycoprotein contained in indented regions of the plasma membrane known as coated pits. The under-surface of the coated pits are covered with clathrin, a protein that facilitates the formation of the endocytic vesicles. Figure 16.13 illustrates the structure of the LDL receptor. The expression of the LDL receptor is regulated by the need of the cell for cholesterol. It is down-regulated when sufficient cholesterol is available and up-regulated when the cell requires additional cholesterol. Figure 16.14 illustrates the regulation of LDL receptor expression. This regulation is controlled by either the extent to which existing receptors are recycled to the cell surface after endocytosis, or the amount of new receptors that is synthesized.

Figure 16.13

Structure of LDL receptor. The structure is divided into five domains. The first domain contains the LDL binding site. This is followed by large domain that contains homology with the epidermal growth factor (EGF) precursor. Next, is a small segment that contains a large number of O-linked carbohydrate residues. The fourth domain spans the plasma membrane, and the last domain is a short segment that projects into the cytoplasm but does not have any kinase activity. (From Brown MS and Goldstein JL: Science 232:34, ©The Nobel Foundation 1986.)

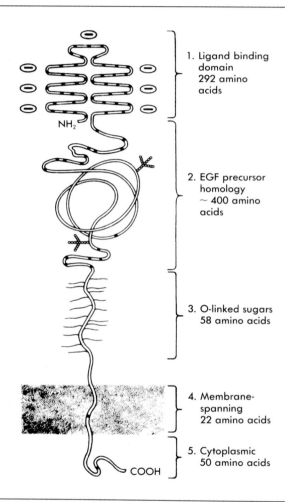

1. Ligand binding domain 292 amino acids

2. EGF precursor homology ~ 400 amino acids

3. O-linked sugars 58 amino acids

4. Membrane-spanning 22 amino acids

5. Cytoplasmic 50 amino acids

Not all of the circulating LDL is taken up through binding to the LDL receptor. Some LDLs enter the tissues by pinocytosis, without receptor binding. The exact proportion that enters by this non-receptor-mediated route depends on the concentration of circulating LDL.

LDL receptor gene There is a single copy of the LDL receptor gene in the haploid human genome. It is located on chromosome 19. The gene spans about 45 kilobase pairs and is a mosaic composed of 18 exons separated by 17 introns. Exon 1 contains the signal sequence. Exons 2 to 6 code for the LDL binding region, which consists of seven

Figure 16.14

Regulation of LDL receptor expression. The number of receptors on the cell surface is controlled by the cholesterol content of the cell. This is regulated in two ways. The receptor may be either recycled from the endosome to the plasma membrane or degraded following endocytosis. In addition, the rate of receptor synthesis in the endoplasmic reticulum is controlled by the availability of cholesterol in the cell. (From Brown, MS, and Goldstein, JL: Science 232:34, ©The Nobel Foundation 1986.)

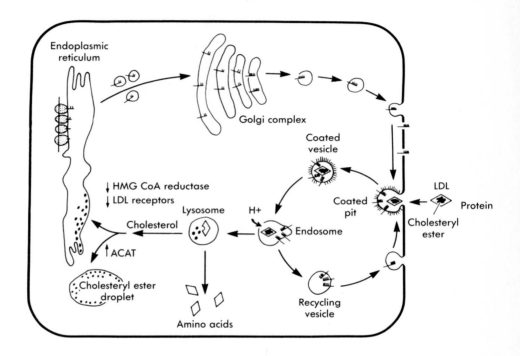

cysteine-rich, 40-amino-acid repeats. Exons 7 to 14 contain the epidermal growth factor (EFG) homology region, 15 the O-linked sugar region, and 16 the transmembrane domain, which is small and passes through the membrane only once. The cytoplasmic domain, which also is relatively small and does not contain kinase activity, is encoded in exons 17 and 18.

Clinical comment

Familial hypercholesterolemia Familial hypercholesterolemia is a genetic disease caused by a deficiency or malfunction of the LDL receptor. It leads to excessive accumulation of LDL in the plasma, producing very high levels of plasma cholesterol. This is an autosomal recessive disease that causes severe atherosclerosis. Homozygotes do not have any functioning LDL receptors and often die of coronary heart disease in childhood. Heterozygotes have half the usual number of functioning receptors and commonly develop heart disease before reaching middle age. Fortunately, only a small percentage of hypercholesterolemia is caused by the familial form.

Cloning studies indicate that this disease actually is composed of a group of LDL receptor gene defects. There are four types of defects: insertions, deletions, nonsense mutations, and missense mutations. These mutations disrupt the receptor in various ways. In some forms, the receptor is either not expressed or is not incorporated into the plasma membrane. In others, a structurally defective receptor is produced that does not properly bind to LDL. All of the forms produce the same clinical result, severe hypercholesterolemia caused by failure to remove circulating LDL effectively.

Additional lipoprotein receptors At least three other lipoprotein receptors besides the LDL receptor are thought to be involved in lipoprotein metabolism. These receptors, listed in Table 16.7, are the **chylomicron remnant, HDL,** and **scavenger** receptors. None of these additional receptors has been purified or cloned, and their role in lipoprotein metabolism is very controversial and not completely understood. The chylomicron remnant receptor is confined to the liver and mediates the uptake of chylomicron remnants. Apo-E is the recognition factor, and this receptor also is called the apo-E receptor. It differs from the LDL receptor, which also recognizes apo-E. There is considerable debate as to whether the HDL receptor, which is thought to facilitate HDL binding to cells, actually exists. The scavenger receptor mediates the uptake of LDL in which the apo-B-100 has been chemically modified, such as by acetylation.

Lipid transfer proteins

Lipid transfer proteins facilitate the movement of phospholipids, triglyceride, and cholesteryl esters between lipoproteins. Their function in lipoprotein metabolism is illustrated in Figure 16.15. At least two separate lipid transfer proteins are present in human plasma. One transfers cholesteryl esters, triglycerides, and phospholipids. The other only transfers phospholipids. Lipid transfer proteins play an important role in HDL metabolism. They transfer cholesteryl esters from HDL, after synthesis in the LCAT reaction, to VLDL. This is accompanied by transfer of triglyceride from the VLDL to HDL, possibly through the action of the same lipid transfer protein. A lipid transfer protein also transfers phospholipids from chylomicrons and VLDL to HDL as these triglyceride-rich lipoproteins are degraded in the circulation.

Abnormal lipoproteins

The most common lipoprotein diseases are associated with increases in the amount of one of the normally occurring plasma lipoproteins. These diseases are described in the next section. There also are conditions where abnormal lipoproteins accumulate in the plasma. These are β-very low–density lipoproteins (β-VLDL), high-density lipoprotein c (HDLc), lipoprotein (a) [LP(a)], and lipoprotein-X (LP-X).

β-VLDL β-VLDLs are lipoproteins that migrate on electrophoresis as β-lipoproteins but are in the VLDL density range on ultracentrifugation. Also called floating β-lipoproteins, they are derived from chylomicrons and VLDL. These lipoproteins are rich in cholesterol and are taken up primarily by macrophages. They constitute a serious risk factor for atherosclerosis.

Figure 16.15   Role of lipid transfer proteins in lipoprotein metabolism.

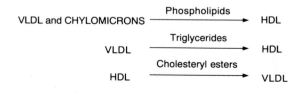

HDLc HDLc lipoproteins are a form of large HDL that are rich in cholesterol esters and contain apo-E. Produced when the diet is very high in cholesterol, these lipoproteins cause atherosclerosis. Another name for HDLc is HDL_1. HDLc occurs mainly in animals and appears to be of little importance in humans.

LP(a) LP(a) is a varient of LDL in which the apo-B contains a disulfide-linked protein called apo(a). Normal people contain very small amounts of LP(a). Elevations in LP(a) levels are associated with an increased risk of coronary heart disease. Apo(a) is related structurally to plasminogen, a proenzyme that is part of the fibrinolysis system in the plasma.

LP-X Patients who have obstructive jaundice develop a unique form of hyperlipidemia. This results from the accumulation in the plasma of an abnormal lipoprotein called LP-X. Although LP-X migrates in the β-region on electrophoresis, it can be separated from LDL by ultracentrifugation. The structure of LP-X is very different from that of the other plasma lipoproteins. LP-X is a unilamellar liposome with a diameter of 400 to 700 Å. It is composed of 94% lipid and 6% protein. The lipid is mostly phospholipid and unesterified cholesterol arranged in a bilayer. Bile acids also are present, but they account for only 2% to 3% of the lipid content. Albumin, which makes up about half of the protein content, is trapped in the aqueous core of the liposome. The remainder of the protein consists of apoproteins, particularly the C-apoproteins, that are associated with the lipid bilayer. LP-X is made by the liver and is formed within the bile canaliculi. When found in the plasma, LP-X is a diagnostic sign of obstructive liver disease.

Lipoprotein molecular biology

The genes for the apoproteins have been cloned, and their structure, properties and chromosomal locations have been determined.

Apoprotein genes

Figure 16.16 illustrates the structural organization of six of the apoprotein genes. Five of the six, apo-A-I, A-II, C-II, C-III, and E have similar structures. These genes contain four exons and three introns. The gene for apo-A-IV differs only in that exon 1 is deleted; it contains three exons and two introns corresponding to exons 2 to 4 and introns 2 and 3 of the other genes. The introns have similar locations in each of the genes. Exon 1 codes for the 5'-untranslated region. This is deleted in the apo-A-IV gene. Exon 2 contains the signal peptide, and exons 3 and 4 contain the mature peptide. The differences in the lengths of the six mRNAs are accounted for primarily by the size of the last exon. The structural similarities suggest that all of these apoprotein genes are derived from a common ancestor gene and form an apolipoprotein multigene family.

The apo-B gene has a different structural organization. It is much larger, spanning 43 kilobases, and contains 29 exons and 28 introns. The introns are concentrated in the 5' end of the gene, and a total of 19 introns are present in the first 1000 codons. Therefore the apo-B gene is not closely related to the apolipoprotein multigene family. Likewise, the apo-D gene has a different structure and also is not a member of the multigene family.

Chromosome locations The apoprotein genes are located on five chromosomes, as indicated in Table 16.8. Three of the genes are clustered on chromosome 11 on the long arm, close to 11q13. The apo A-I gene is located only 2.5 kilobase pairs upstream from the 3' end of the apo C-III gene, which is in reverse orientation and located on the opposite DNA strand. Likewise, the genes for apo-E and apo-C-II are about 2 centimorgans apart on chromosome 19. The LDL receptor gene also is located on chromosome 19.

Figure 16.16

Structure of some of the human apolipoprotein genes. Transcription occurs from left to right. The wide bars represent exons and the thin lines the 5'-flanking region, introns, and 3'-flanking region. Each exon is divided into its functional components; the solid area is the mature peptide coding region, the hatched regions code for the signal peptide, and the open areas at the 5' and 3' ends are untranslated regions. In exon 3 of the apo-A-I and apo-A-II genes, the small open area between the signal peptide and mature peptide coding regions represents the prosegment. (From Li, W-H, et al: J Lipid Res 29:245, 1988.)

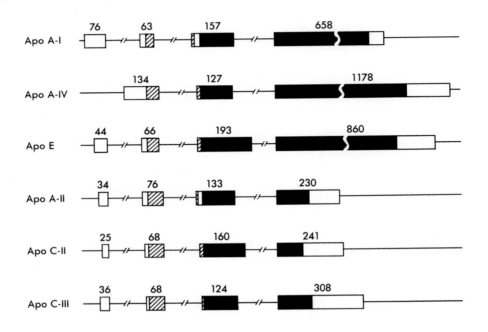

Table 16.8   Human chromosomal locations of the apoprotein genes

Chromosome	Apoprotein genes
1	A-II
2	B*
3	D
11	A-I, A-IV, C-III
19†	C-I, C-II, E

*Apo-B-100 and apo-B-48 are derived from the same gene.
†Chromosome 19 also contains the LDL receptor gene.

Figure 16.17

Schematic diagram of the RFLP involving the apo-A-I and apo-C-III genes. Normally, the two genes are located on the opposite DNA strands of chromosome 11, 2.5 kilobase pairs apart. In the mutant, a segment of the opposite strand containing the 5′ end of apo-C-III gene is inserted into the apo-A-I gene, in exon 4. The insertion gives rise to new EcoRI and PstI restriction endonuclease sites within the mutant apo-A-I gene. When the mutant gene is digested with either of these restriction enzymes and hybridized with an apo-A-I cDNA probe, the DNA fragments produced have an abnormal size. For example, the normal EcoRI fragment contains 13 kilobase pairs; the mutant fragment contain only 6.5 kilobase pairs. This difference is detected easily by autoradiography. (Modified from Breslow, JL: Genetics of the human apolipoproteins. In Scanu AM and Spector AA, editors: Biochemistry and biology of plasma lipoproteins, New York, 1986, Marcel Dekker, Inc.)

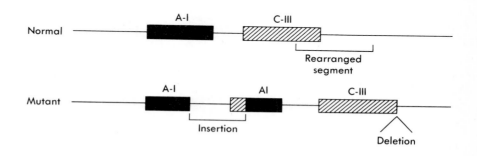

Clinical comment

Apoprotein gene restriction fragment length polymorphisms One approach to determining the cause of lipoprotein diseases is to look for polymorphisms of the apoprotein genes. This technique also is useful in screening populations for genetic defects involving lipoproteins. Leukocytes are obtained from the patient and the DNA isolated and then digested with a restriction endonuclease. The resulting DNA fragments are separated by electrophoresis on an agarose gel, transferred to nitrocellulose by blotting, hybridized with a radioactive cDNA probe, and detected by autoradiography. A restriction fragment length polymorphism (RFLP) refers to an altered pattern of gene fragments resulting from some abnormality in the structure of the genomic DNA.

The most widely studied RFLP involving the apoprotein genes involves the genes for apo-A-I and apo-C-III. This is illustrated schematically in Figure 16.17. The RFLP was detected first in patients having low plasma levels of apo-A-I, apo-C-III, and HDL. A different fragment pattern was observed when their DNA digest was hybridized with an apo-A-I cDNA probe. This results from a DNA insertion in the apo-A-I gene, so that the sizes of the fragments produced by digestion with either EcoRI or PstI are different from normal. The DNA insert in the apo-A-I gene in this RFLP is derived from the apo-C-III gene, which normally is located on the opposite DNA strand on chromosome 11, 2.5 kilobase pairs downstream.

Many other RFLPs of the genes involved in lipoprotein metabolism have been detected. These are listed in Table 16.9. Some are associated with lipoprotein abnormalities; others do not appear to cause any clinical defect.

Table 16.9   Apoprotein gene polymorphisms

Gene locus	Associated abnormalities*
Apo-A-I, C-III, A-IV	Hypertriglyceridemia Low HDL Coronary heart disease Familial combined hyperlipidemia
A-II	Hypertriglyceridemia
Apo-B	Hypercholesterolemia Coronary heart disease Hypertriglyceridemia
Apo-E, C-I, C-II	Dysbetalipoproteinemia Coronary heart disease

*In many instances, polymorphisms in these genes are not associated with detectable phenotypic lipid abnormalities or increased incidence of disease.

Diseases of lipoprotein metabolism

The most common lipoprotein abnormalities lead to the accumulation of excessive lipids in the blood plasma. These diseases collectively are known as the hyperlipidemias, or more properly, the hyperlipoproteinemias. Table 16.10 lists the phenotypic abnormalities that occur in patients with hyperlipoproteinemia. Several of these diseases are risk factors for atherosclerosis and coronary heart disease. A condition also exists in which HDL is low; this also increases the risk of coronary heart disease.

None of the phenotypic abnormalities listed in Table 16.10 are caused by only a single gene defect. However, there are single gene defects that cause specific lipoprotein abnormalities. These are listed in Table 16.11. In addition, there are a number of other rare genetic defects, such as abetalipoproteinemia (see p. 695), Tangier disease (see p. 701), and LCAT deficiency (see p. 493), that cause abnormalities in lipoprotein metabolism.

Clinical comment

Hyperlipoproteinemia and atherosclerosis The most serious consequence of an elevation in plasma lipoproteins is atherosclerosis, an arterial disease that causes coronary heart disease and stroke. Increases in LDL, IDL, and β-VLDL lead to this problem. These lipoproteins are rich in cholesterol, giving rise to the view that plasma cholesterol elevations are unhealthy. Though this is generally true, it must be remembered that HDL also contains cholesterol. High HDL levels afford protection against coronary heart disease, and elevated cholesterol in the form of HDL is beneficial, not harmful. This is the so-called "good" cholesterol, as distinct from the harmful cholesterol associated with LDL, IDL, and β-VLDL.

Integration of lipoprotein metabolism

An overview of lipoprotein metabolism is shown in Figure 16.18. This illustrates how the individual systems that have been discussed in the earlier sections operate in concert.

The pathway performs three central functions. The first, dietary fat absorption, is mediated by the chylomicron system. Chylomicron remnants, the end-product of this system, are cleared by the liver. The chylomicron remnant receptor, also called the apo-E receptor, mediates this clearance.

The second function, transfer of lipids from the liver to other tissues, is mediated by the VLDL system. IDLs are an intermediate in this system, and LDLs are the end-product. Some IDLs are cleared directly by the liver, while LDLs are removed by the liver and extrahepatic tissues. The removal of IDL and LDL is mediated by the same receptor, the

Table 16.10   Hyperlipoproteinemia phenotypes

Phenotype	Lipoprotein elevation	Major plasma lipid elevation
I	Chylomicrons	Triglycerides
IIa	LDL	Cholesterol
IIb	LDL and VLDL	Cholesterol and triglycerides
III	β-VLDL and IDL	Cholesterol and triglycerides
IV	VLDL	Triglycerides
V	VLDL and chylomicrons	Triglycerides and cholesterol

Table 16.11   Diseases of lipoprotein metabolism caused by single gene defects

Disease	Lipoprotein abnormality	Lipid abnormality	Metabolic basis	Clinical implications
Familial hypercholesterolemia	LDL elevated	Cholesterol elevated	Decrease clearance of LDL from plasma. Familial form results from genetic deficiency or abnormality in LDL receptor	Risk factor for atherosclerosis
Familial hypertriglyceridemia	VLDL elevated	Triglyceride elevated	Uncertain; VLDL overproduction or decreased catabolism	Questionable as to whether this is an independent risk factor for atherosclerosis
Familial combined hyperlipidemia	LDL and VLDL elevated	Cholesterol and triglyceride elevated	Uncertain; overproduction of apo-B-100	Risk factor for atherosclerosis
Familial dysbetalipoproteinemia	β-VLDL and IDl elevated	Cholesterol and triglyceride elevated	Decreased clearance of remnants; defective binding of apo E to LDL receptor	Risk factor for atherosclerosis
Familial lipoprotein lipase deficiency	Chylomicrons and VLDL elevated	Triglyceride elevated	Deficiency of lipoprotein lipase or apo-C-II	Acute pancreatitis
Hypoalphalipoproteinemia	HDL reduced	None	Uncertain, occasionally caused by genetic apo-A-I/apo-C-III deficiency	Risk factor for atherosclerosis

LDL receptor, which also is called the apo-B/E receptor. Apo-E is the recognition factor of IDL, whereas apo-B-100 is the component of LDL that is recognized by the receptor.

The third function is the return of excess cholesterol that accumulates in either the plasma or the extrahepatic tissues to the liver so that it can be excreted in the bile. This is mediated by HDL, working in conjunction with the VLDL system. After the excess cholesterol is taken up by HDL, it is converted to cholesteryl esters. Some HDLs are taken up directly by the liver. The remainder of the cholesteryl ester is transferred to the VLDL system through the action of a lipid transfer protein. It subsequently is taken up by the liver as a part of IDL or LDL.

Lipoprotein lipase is present on the capillary surface of extrahepatic tissues and mediates the hydrolysis of triglycerides contained in chylomicrons and VLDL. A similar enzyme, hepatic lipase, hydrolyzes triglycerides and phospholipids of lipoproteins in the hepatic microcirculation. LCAT mediates the esterification of cholesterol in the HDL.

Figure 16.18

Integration of lipoprotein metabolism. The thick arrows indicate the secretion of lipoproteins from the tissues, interconversion of lipoproteins in the plasma, and removal of the lipoproteins from the circulation by the tissues. The thin arrows signify transfer of lipids between lipoproteins, from tissues to lipoproteins, or from lipoproteins to tissues. Lipid transfer proteins mediate the transfer of lipids between lipoproteins. Lipoprotein lipase hydrolyzes triglycerides before they enter extrahepatic tissues, and hepatic lipase hydrolyzes triglycerides and phospholipids during their uptake by the liver. LDL and IDL uptake are mediated by the LDL receptor, and chylomicron remnant uptake by the liver is mediated by the chylomicron remnant receptor. *C,* Cholesterol; *CE,* cholesteryl ester; *PL,* phospholipids; *TG,* triglycerides.

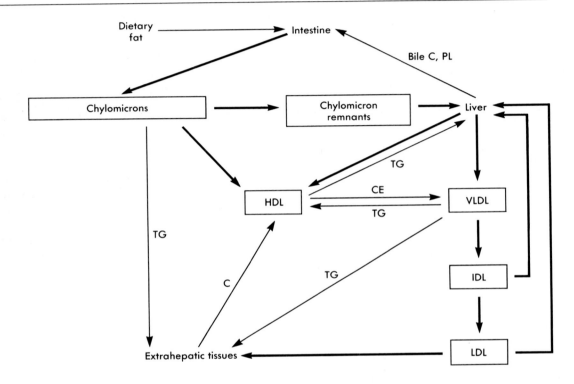

Bibliography

Breslow JL: Human apolipoprotein molecular biology and genetic variation, Annu Rev Biochem 54:699, 1985.

Brown MS and Goldstein JL: A receptor-mediated pathway for cholesterol homeostasis, Science 232:34, 1986.

Havel RJ and Kane JP: Structure and metabolism of plasma lioproteins. In Scriver CR et al, editors: The metabolic basis of inherited disease, ed 6, New York, 1989, McGraw-Hill Information Services Co.

Li W-H et al: The apoprotein multigene family: biosynthesis, structure-function relationships, and evolution, J Lipid Res 29:245, 1988.

Lusis AJ: Genetic factors affecting blood lipoproteins: the candidate gene approach, J Lipid Res 29:397, 1988.

Scanu AM and Spector AA: Biochemistry and biology of plasma lipoproteins, New York, 1986, Marcel Dekker, Inc.

Segrest JP and Albers JJ: Plasma lipoproteins. Methods in enzymology, Orlando, Fla, 1986, Academic Press, Inc.

Williams DL: Molecular biology in arteriosclerosis research, Arteriosclerosis 5:213, 1985.

Clinical examples

| Case 1 | Endogenous hypertriglyceridemia |

On routine physical examination it was noted that a 36-year-old man had a fasting plasma cholesterol concentration of 6.5 mmol/L (252 mg/dl; normal range, 3.1 to 5.7 mmol/L), and a plasma triglyceride concentration of 5.6 mmol/L (530 mg/dl; normal range, 1.1 to 2.7 mmol/L). He was otherwise asymptomatic and was judged to be healthy except for this defect. Initially his physician recommended a eucaloric diet containing 15% protein, 15% fat, 70% carbohydrate, and 300 mg of cholesterol. Although the patient rigorously followed this diet, there was no improvement in the plasma lipid concentrations; in fact, they became worse. The physician then placed the patient on a fat-free formula diet for 2 weeks. Still no improvement in plasma lipid levels occurred. Because the treatment had been unsuccessful to this point, a radical change in dietary therapy was instituted. Instead of fat being restricted, dietary carbohydrate was reduced. The new dietary composition was 20% protein, 55% fat, 25% carbohydrate, and 400 mg of cholesterol. A marked reduction in plasma lipids was observed within 48 hr; the plasma cholesterol remained in the range of 5.8 mmol/L and triglycerides in the range of 2.8 mmol/L. These values were considered to be acceptable, and no drug treatment was recommended. The patient was discharged on a low-carbohydrate diet.

Biochemical questions

1. What is hyperlipoproteinemia?
2. How would a plasma lipoprotein electrophoretic analysis have aided in the diagnosis of this problem?
3. What are the dangers of a long-term, fat-free diet?
4. Describe the type of hyperlipoproteinemia in this case and explain the improvement in plasma lipid concentrations when the patient was given a low-carbohydrate diet. Explain the persistence of elevated plasma triglyceride concentration in spite of a fat-free diet.

Case discussion

1. Definitions Hyperlipidemia is defined as any condition in which, after a fast of 12 hr, the plasma cholesterol concentration is greater than 5.7 mmol/L, the plasma triglyceride concentration is greater than 2.7 mmol/L, or both. It is presently defined in terms of plasma lipid concentrations because they can be easily measured in the clinical laboratory. However, the plasma lipids are components of plasma lipoproteins. Hyperlipidemia is a defect in lipoprotein metabolism; it actually is a hyperlipoproteinemia. The hyperlipoproteinemias are classified in two ways. The first classification is in terms of primary versus secondary etiology. A primary hyperlipoproteinemia is a disease entity of itself; it is not simply the manifestation of a well-defined illness. In contrast, a *secondary* hyperlipoproteinemia is a symptom of another disease that happens to have as one of its signs a plasma lipid abnormality. Some of the diseases that produce a secondary hyperlipoproteinemia include diabetes mellitus, hypothyroidism, and biliary obstruction. Because we are told that the patient was otherwise healthy, we must assume that his hyperlipoproteinemia is primary.

2. Plasma analysis and diagnosis The patient has a major elevation in plasma triglyceride and a minor elevation in plasma cholesterol. Therefore he may have any of the forms of hyperlipoproteinemia that are associated with an elevation in triglyceride-rich lipoproteins. *Lipoprotein electrophoresis* would distinguish between the possibilities; chylomicrons or VLDL. A normal electrophoretic pattern contains only β- and α-lipoproteins as major bands, corresponding to LDL and HDL, respectively. Although some VLDL (pre-β-lipoprotein) normally is present, this usually appears as a trace amount on electrophoresis of normal specimens.

Therefore a prominent pre-β component, signifying an elevation in VLDL, is abnormal in plasma from a fasting individual. This is illustrated in Figure 16.19. The patient described in this case has this abnormality, excessive VLDLs, which are the triglyceride-rich lipoproteins produced by the liver. Lipoprotein electrophoresis would distinguish this from a chylomicron elevation, which would have appeared as a lipoprotein band remaining at the origin of the electrophoretogram. Likewise, in a case of mixed hyperlipidemia, both chylomicron and VLDL components would be visible. Thus lipoprotein electrophoresis can help to distinguish the type of phenotypic abnormality in cases of hypertriglyceridemia.

Figure 16.19

Plasma lipoprotein electrophoretic patterns from a normal subject and patients with hypercholesterolemia and endogenous hypertriglyceridemia. The normal electrophoretic pattern contains a prominent β-lipoprotein band *(LDL)* and a lightly staining α-lipoprotein band *(HDL)*. Very little pre-β-staining material is present. By contrast, the specimen from the patient with endogenous hypertriglyceridemia contains a deeply staining pre-β-lipoprotein band due to the elevated VLDL concentration. For comparison a pattern from a patient with hypercholesterolemia is included. This contains a deeply staining β(LDL)-band.

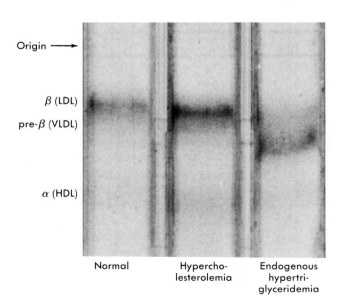

3. Dangers of fat-free diet Because of the correlation between hyperlipidemia and atherosclerosis, many people believe that dietary fat is unhealthy and should be avoided. Although this is correct for patients with hypercholesterolemia, dietary fat serves many useful purposes. Fat supplies calories in a concentrated form. Moreover, polyunsaturated fatty acids of the *linoleic* family (ω-6) are essential components in the diet; they cannot be synthesized by humans. Prostaglandins and other metabolically active eicosanoid derivatives are synthesized from these essential fatty acids, and a deficiency in dietary fat eventually could produce a deficiency of these important autocrine and paracrine mediators. The fat-soluble vitamins (A, D, E, and K) occur in the lipid portion of foods. Omission of fat from the diet may eventually produce deficiencies in fat-soluble vitamins unless vitamin replacement therapy is instituted.

4. Type of hyperlipoproteinemia Because this patient had a large elevation in plasma triglyceride concentration that was not helped by reduction in dietary fat, syndromes associated with chylomicron elevations can be excluded. Likewise, the plasma cholesterol concentration is not high enough for this to be a case of dysbetalipoproteinemias, where β-VLDLs are present. Therefore it was reasoned that the patient had endogenous hypertriglyceridemia. To support this supposition, plasma lipoprotein electrophoresis was carried out. As shown in Figure 16.19, electrophoresis established that the elevation was in VLDL. This confirmed the diagnosis of endogenous hypertriglyceridemia, an excess of VLDL in the plasma. VLDLs are the triglyceride-rich lipoproteins synthesized primarily in the liver.

Much of the fatty acid used by the liver for the production of triglycerides is synthesized from dietary carbohydrates. In this way the liver converts excess calories into fat, the major storage form of energy in the body. The VLDL then transports the newly formed triglyceride to the adipocytes for storage. It is not surprising in view of this explanation of VLDL production that the low-fat diet did not help this patient, because the source of the circulating triglyceride was primarily dietary carbohydrate, not dietary fat. For as yet unexplained reasons, the patient was sensitive to dietary carbohydrate, the result being a plasma VLDL elevation. This could result from overproduction of VLDL, a defect in the VLDL catabolic mechanism, or both.

Reference

Grundy SM and Vega GL: Hypertriglyceridemia: causes and relation to coronary heart disease, Semin Thromb Hemost 14:149, 1988.

Kane JP and Havel RJ: Disorders of the biogenesis and secretion of lipoproteins containing the β apolipoproteins. In Scriver CR et al, editors: The metabolic basis of inherited disease, ed 6, New York, 1989, McGraw-Hill Information Services Co.

Case 2   Hyperchylomicronemia

A 15-year-old boy had a long history of abdominal complaints, including bouts of abdominal pain so severe that narcotics were required for relief. These episodes were intermittent, occurring every 6 months on the average. On one occasion abdominal surgery (an exploratory laparotomy) was performed, and the patient's appendix was removed. However, this did not correct the problem. The patient had recently felt well until he suddenly developed another episode of abdominal pain. His mother stated that the illness came on 4 hr after he had eaten a meal consisting of pork chops, fried potatoes, milk, and ice cream topped with a generous serving of whipped cream. No one else in the family had been made ill by this meal. Examination of the patient indicated an acute abdominal emergency. The patient was brought to the hospital at 8:00 AM, 14 hr after his last meal. On arrival a blood specimen was drawn. Within 15 min the laboratory technician reported that valid results could not be obtained from the blood plasma because it was cloudy. The plasma was "milky," but on centrifugation for 30 min at 15,000 rpm, it cleared considerably and there was a thick band of "cream" located at the top of the specimen.

Biochemical question

1. What kind of lipid abnormality would you suspect in this patient?
2. What chemical and electrophoretic studies should be obtained on the plasma sample to aid in making the diagnosis?
3. What kind of a diet would be helpful in treating this disease?
4. Would you recommend that this boy's diet by supplemented with medium-chain triglycerides?

Case discussion

1. Lipid abnormality One would suspect the presence of hyperchylomicronemia in this case. The thick band of "cream" that floated to the top of the centrifuge tube after relatively low-speed centrifugation indicates the presence of very large, triglyceride-rich lipoproteins. Because the illness followed a meal containing a large amount of fat, it is likely that these lipoproteins are chylomicrons of intestinal origin, which are very rich in dietary fat. VLDLs are smaller particles than chylomicrons and require longer periods of centrifugation at higher forces for flotation at density 1.006. Hyperchylomicronemia is a familial disease that usually manifests itself in childhood or early adolescence. In contrast, endogenous hypertriglyceridemia and mixed hyperlipidemia, which can have a similar blood plasma picture, usually become manifest later in life and in many cases are secondary to diseases such as diabetes mellitus or alcoholism. The age of the patient, the previous history of recurrent abdominal pain, and the persistence of severe hyperlipoproteinemia after a fat-rich meal strongly suggest a diagnosis of hyperchylomicronemia.

2. Chemical, electrophoretic studies Three types of laboratory studies would support this diagnosis. First, plasma lipid determinations should be run. A large elevation in triglyceride concentrations would be expected, for chylomicrons are composed of 85% to 90% triglyceride. Lipoprotein electrophoresis should be done to demonstrate the presence of chylomicrons. These lipoproteins are cleared from the blood within 4 to 6 hr after a fatty meal in normal people and their presence in the plasma after 14 hr of fasting is grossly abnormal. If the defect were endogenous hypertriglyceridemia, an excessive pre-β-lipoprotein band would be visible and no chylomicron band would be seen on electrophoresis. Conversely, if the defect were mixed hyperlipidemia, both a chylomicron band and an excessive pre-β-lipoprotein band would be observed. Therefore lipoprotein electrophoresis should provide a definitive diagnosis in this case. To confirm the diagnosis, the patient might be given an intravenous injection of heparin to test for lipoprotein lipase activity because the metabolic defect in most cases of hyperchylomicronemia is a deficiency of lipoprotein lipase. It must be remembered that heparin injection releases both the hepatic lipase as well as lipoprotein lipase. Patients with hyperchylomicronemia are deficient only in lipoprotein lipase, which is protamine-inhibited. Therefore the lipase assay should be carried out with and without the addition of protamine.

3. Diet In hyperchylomicronemia, the problem results from an inability to clear dietary fat from the blood plasma fast enough because of a lipoprotein lipase deficiency. Therefore the treatment involves a drastic reduction in dietary fat intake. A diet rich in carbohydrate and protein and low in fat should be recommended for this patient.

4. Triglyceride supplementation Medium-chain triglycerides in the form of a dietary supplement would be useful in this situation. Medium-chain triglycerides are digested and absorbed differently from the long-chain fatty acid triglycerides contained in normal foods. After hydrolysis, the medium-chain fatty acids are not resynthesized into triglycerides by the intestinal mucosa. Instead, they are absorbed as free fatty acid in the portal blood, and they bypass the chylomicron and lipoprotein lipase steps. Because they do not circulate in the form of triglycerides, medium-chain fatty acid utilization would not be affected by the lipoprotein lipase deficiency in this disease. Therefore fat calories could be supplied to this patient in the form of medium-chain triglycerides, enabling the carbohydrate content of the diet to be reduced to a more palatable level.

Reference   Brunzell JD: Familial lipoprotein lipase deficiency and other causes of the chylomicronemia syndrome. In Scriver CR et al, editors: The metabolic basis of inherited disease, ed 6, New York, 1989, The McGraw-Hill Information Services Co.

Case 3   Familial type 3 hyperlipoproteinemia (dysbetalipoproteinemia)

A 29-year-old man was referred to a cardiologist because of a history of heart and vascular problems in his immediate family. His father had suffered from poor circulation in his legs for several years and had recently been operated on to have grafts inserted in his femoral arteries. During this hospitalization the father was noted to have hyperlipidemia. The patient had three brothers, all older than himself. One died of a myocardial infarction at age 37 years; another had lipid accumulations (xanthoma) in the palmar creases of his hands and on his elbows and buttocks; and a third brother was asymptomatic. The patient stated that he felt well and was eating the usual American diet. After an overnight fast, blood was drawn for lipid determination. The plasma total cholesterol was 8.8 mmol/L (340 mg/dl) and the plasma triglyceride concentration was 4.3 mmol/L (380 mg/dl). This test was repeated once, and similar values were obtained, suggesting the presence of dysbetalipoproteinemia.

Biochemical questions   1. What are chylomicron remnants, and how are they metabolized?
2. What function has been attributed to the apo-E contained in chylomicron remnants?
3. What are β-VLDLs?
4. What are IDLs, and how are they formed?
5. What is the mechanism of familial dysbetalipoproteinemia?

Case discussion   **1. Chylomicron remnants** These are lipoproteins that are formed from chylomicrons. They are the end-products of chylomicron degradation, formed after much of the triglyceride contained in the chylomicron core is hydrolyzed by lipoprotein lipase. The cholesteryl esters originally present in the chylomicron are retained when it is converted into the remnant. Thus this lipoprotein is rich in both triglyceride and cholesteryl esters. The chylomicron remnant acquires apo-E as it is formed. Chylomicron remnants are taken up by the liver through an endocytosis process mediated by the chylomicron remnant receptor.
    2. Apo-E function Apo-E is thought to be the recognition factor for the chylomicron remnant receptor, which also is called the apo-E receptor. Thus apo-E is required for the proper clearance of chylomicron remnants from the plasma.
    3. β-VLDLs β-VLDLs are lipoproteins that are isolated by ultracentrifugation with VLDL but migrate on electrophoresis with LDL. They are cholesterol-rich VLDLs and chylomicron remnants that are abnormal and can cause atherosclerosis. They only accumulate in the plasma in abnormal circumstances, such as cholesterol overload. β-VLDLs are taken up by macrophages. When this occurs, the macrophages fill up with cholesteryl esters and become foam cells.
    4. IDLs IDLs are intermediates in the degradation of VLDL. They are rich in

triglycerides and cholesteryl esters, and they contain apo-B-100 and apo-E. IDLs are rapidly taken up by the liver through the LDL receptor, or converted in the plasma to LDL. Therefore IDLs normally do not accumulate in the plasma of fasting individuals.

5. Disease mechanism Familial dysbetalipoproteinemia is caused by the accumulation in the plasma of remnant particles derived from the partial catabolism of VLDL, chylomicron remnants, or both. Some forms of this disease are caused by a genetic defect in apo-E that interferes with binding to the LDL receptor, reducing the removal of remnant particles that normally are cleared by apo-E-mediated endocytosis. In other cases no binding defect occurs, and the mechanism of the hyperlipoproteinemia is unknown.

Reference

Mahley RW and Rall SC, Jr: Type III hyperlipoproteinemia (dysbetalipoproteinemia): The role of apolipoprotein E in normal and abnormal lipoprotein metabolism. In Scriver CR et al, editors: The metabolic basis of inherited disease, ed 6, New York, 1989, McGraw-Hill Information Services Co.

Case 4   Familial hypercholesterolemia

A 32-year-old woman was hospitalized with an acute myocardial infarction. Subsequently, her plasma cholesterol was found to be 10.9 mMol/L (420 mg/dl), but her plasma triglyceride concentration was normal. Further analysis revealed that her LDL level was highly elevated. Coronary angiography indicated the presence of severe arteriosclerosis in all three coronary arteries. Her father and two of her five siblings also were found to have hypercholesterolemia.

Biochemical questions

1. How are LDLs formed?
2. Describe the structure and chemical composition of LDL.
3. What apoprotein does LDL contain?
4. What is the function of LDL?
5. How is LDL removed from the plasma?
6. What is the mechanism of the hypercholesterolemia in this case?

Reference

Russell DW, Esser V, and Hobbs HH: Molecular basis of familial hypercholesterolemia, Arteriosclerosis (suppl) 9:1, 1989.

Case 5   Apolipoprotein C-II deficiency

A 12-year-old girl has a history of hypertriglyceridemia. When she was seen initially, her plasma triglyceride concentration was in the range of 5.3 to 12.8 mMol/L (500 to 1200 mg/dl). A lipoprotein analysis revealed reduced levels of HDL and an elevation in chylomicrons and VLDL. Apoprotein analysis indicated a normal content of apo-C-II as measured immunochemically, but the apo-C-II had a low molecular weight and high pI on two-dimensional gel electrophoresis. Therefore, the apo-C-II was defective, and a diagnosis of apo-C-II deficiency was made. The patient was treated with a low-fat diet, and her plasma triglyceride concentration was maintained between 5.3 and 7.4 mMol/L.

Biochemical questions

1. What are apoproteins?
2. What lipoproteins ordinarily contain apo-C-II?
3. What is the function of apo-C-II?
4. Why would an apo-C-II deficiency cause hypertriglyceridemia?
5. What is the rationale for treatment with a low-fat diet in this patient?

6. Would you expect to find a restriction fragment length polymorphism in the apo-C-II gene locus in this patient?

Reference

Sprecher DL et al: Identification of an apo-C-II variant (apoC-IIBethesda) in a kindred with apo-C-II deficiency and type I hyperlipoproteinemia, J Lipid Res 29:273, 1988.

Additional questions
and problems

1. Is the FFA-albumin complex a plasma lipoprotein?
2. A patient with hypertriglyceridemia was given an intravenous injection of heparin. Within 10 minutes the plasma triglyceride concentration fell into the normal range. How did the heparin injection reduce the patient's hypertriglyceridemia?
3. Do VLDL and chylomicrons carry out the same functions?
4. What is the role of the plasma lipid transfer proteins in lipoprotein metabolism?
5. What are the similarities and differences in the metabolism of IDL and chylomicron remnants?
6. How does apo-B-100 differ from apo-B-48?
7. What is the role of lecithin-cholesterol acyltransferase (LCAT) in lipoprotein metabolism?
8. What are the main lipid and apoprotein components of VLDL? What changes occur when the VLDL is converted to LDL?
9. How is lysosomal acid lipase involved in lipoprotein metabolism?
10. Does the LDL receptor mediate the clearance of all of the lipoproteins from the circulation?

Multiple choice
problems

1. Lipoprotein triglycerides are hydrolyzed in the plasma through the action of lipoprotein lipase. This enzyme is:
   1. Present on the surface of capillary endothelial cells.
   2. Activated by apo-C-II.
   3. Released into the plasma following heparin injection.
   4. Involved in the conversion of HDL_3 to HDL_2.
      A. 1, 2, and 3.
      B. 1 and 3.
      C. 2 and 4.
      D. 4 only.
      E. All are correct.

2. The FFA present in the plasma is transported in association with:
   A. Albumin.
   B. HDL.
   C. LDL.
   D. VLDL.
   E. Chylomicrons.

3. If plasma triglycerides are formed in the intestinal mucosa from dietary fat, most of the triglyceride would be present in:
   A. VLDL.
   B. LDL.
   C. Chylomicrons.
   D. HDL.
   E. IDL.

4. What apoprotein is the recognition factor for LDL binding to the LDL receptor?
   A. Apo-E
   B. Apo-B-48
   C. Apo-C-I
   D. Apo-B-100
   E. Either apo-E or apo-B-100

5. What apoprotein is the main activator of the LCAT reaction in HDL?
   A. Apo-B-100
   B. Apo-I
   C. Apo-C-III
   D. Apo-E
   E. Apo-A-II

6. What is the main function of chylomicron remnants?
   A. Supply triglycerides for the lipoprotein lipase reaction.
   B. Take up excess cholesterol from tissues.
   C. Supply dietary cholesterol to the liver.
   D. Supply cholesterol for the lecithin-cholesterol acyltransferase (LCAT) reaction.
   E. Take up triglycerides during VLDL catabolism

7. Which of the following processes occur when VLDL is metabolized in the circulation?
   1. Some of the VLDL triglycerides are transferred to HDL
   2. Cholesteryl esters are transferred from HDL to the VLDL.
      A. 1 only.
      B. 2 only.
      C. Both are correct.
      D. Neither is correct.

8. In the structural organization of the genes comprising the apolipoprotein multigene family:
   1. Each exon codes for a segment of the mature apoprotein.
   2. None of the introns is contained within the coding sequence of the mature protein.
      A. 1 only.
      B. 2 only.
      C. Both are correct.
      D. Neither is correct.

9. The LDL receptor:
   1. Is a dimer linked by disulfide bonds.
   2. May be recycled back to the cell membrane following endocytosis.
   3. Is synthesized in increased amounts when the cell has a high content of cholesterol.
   4. Has a small transmembrane domain that passes through the membrane only once.
      A. 1, 2, and 3.
      B. 1 and 3.
      C. 2 and 4.
      D. 4 only.
      E. All are correct.

Chapter 17

Molecular endocrinology: mechanism of hormonal action

Barry H. Ginsberg

Objectives

1 To discuss the principles of feedback control and how these relate to hormonal control of metabolism
2 To describe the structure, biosynthesis, and secretion of the hormones.
3 To describe the actions of hormones at the cellular and molecular level
4 To discuss the biochemical basis of some common human endocrine diseases

Endocrinology is the branch of biologic science that deals with hormones, hormonal regulation of metabolism, and diseases associated with hormonal abnormalities. Hormones represent one of the main methods of communication, coordination, and regulation of various systems of the body. Although present only in minute concentrations, hormones often modulate metabolic processes at a variety of sites. Intelligent diagnosis and therapy of the many endocrine diseases require an understanding of normal regulation of hormone secretion and action. For example, determining the cause of a loss of menstrual cycles in a woman requires an understanding of the normal control and action of the hypothalamic-releasing hormones, prolactin, follitropin (follicle-stimulating hormone), lutropin (luteinizing hormone), estrogens, androgens, glucocorticoids, and thyroid hormones, a total of at least 15 different hormones. The purpose of this chapter is to present the basic principles of endocrine function and control at the cellular and molecular level. Specific details of the individual hormones are presented in subsequent chapters. Together these chapters provide a sound foundation to which the student can add the clinical and pathologic aspects of endocrinology.

Overview of hormonal regulation
Definitions

To discuss the characteristics of the endocrine system critically, it is first necessary to define some terms:

hormone A chemical substance, present in very low concentration in the blood, that has a regulatory effect on the metabolism of at least one specific organ or tissue at a site distant from the site of secretion. Hormones can directly alter the metabolism of cells. For example, insulin can cause cardiac cells to switch from fat to glucose as an energy source. Hormones can also alter the synthesis and secretion of other substances, such as increased secretion of thyroid-binding globulin by estrogens, or they can influence the synthesis of another hormone. When the principal action of a hormone is to increase the secretion of another hormone, the first is called a *trophic* hormone. This definition of hormones is complicated by the existence of *paracoids,* chemicals that act locally on cells in the immediate area of the secreting cell, and of *autocoids,* chemicals that act directly on the secreting cell. Neurotransmitters can be considered paracoids, and some growth factors are considered autocoids. Some substances that have classically been considered hormones, for example, cholecystokinin, have been discovered to be paracoids or neurotransmitters in the brain.

gland A group of cells specialized to secrete hormones into the bloodstream. Although it was classically thought that a single cell type could produce only a single hormone, it is now clear that some glandular cells can produce multiple hormones. Although most hormones are produced in glands that are composed of large numbers of cells, some hormones are produced by small clusters of secretory cells. For example, enteroglucagon is produced by the endocrine cells in the crypts of the villus lining of the jejunum.

receptor A protein or glycoprotein that specifically binds a hormone and produces a biologic action after such binding. These proteins are the first site of action of the hormones at the cell. They have very high specificity for the individual hormones. Receptors may be found at the plasma membrance or intracellularly, usually in the nucleus.

Cybernetics

Cybernetics is the study of human systems and the mechanical systems that copy them. Processes produce signals that can influence the process itself. This influence can increase the process, called *positive feedback,* or can reduce the process, called *negative feedback.* Virtually all the hormones are regulated by negative feedback. Many examples of both types of feedback are present in our everyday lives.

A good example of positive feedback is seen by placing the microphone of a PA system in front of the speaker. Sounds picked up by the microphone are amplified and resonate from the speaker. This signal is picked up by the microphone and is reamplified, producing a louder sound. After being picked up by the microphone, it is amplified again and again until *squelch* is produced, as the sound becomes amplified beyond the capacity of the speaker. Just as in this example, positive feedback always produces an unstable system and is rarely found in biologic systems. Ovulation is one of the few endocrine systems to utilize positive feedback (Figure 17.1, *A*).

Negative feedback is much more common and can be found in many household examples, such as a thermostat. After setting the thermostat to 72° F, the heat is produced until the temperature exceeds the setting. The thermostat then turns off the heat until the temperature falls below the setting, and then it turns on again. Negative feedback is part of most self-correcting, stable systems and of virtually all endocrine systems (Figure 17.1, *B*).

Hormones and neurotransmitters

Two parallel systems exist for regulation and integration of the body, the *endocrine system,* using hormones as signals, and the *nervous system,* using neurotransmitters. These systems have many similarities: both can regulate one or many areas of the body, both use similar and sometimes identical chemicals, and both can act on the same tissues, sometimes in identical fashion. For example, when appropriately stimulated, fat cells (adipocytes) undergo lipolysis and release triglycerides into the blood, providing an energy source for other cells. This signal can be from the sympathetic nervous system or from the release of hormones by the adrenal medulla. The end result, an increase in plasma triglycerides, is identical and occurs by similar molecular mechanisms.

Despite these similarities, substantial differences also exist between these systems (Table 17.1). The nervous systems spreads its signals electrically along defined and specific nerve pathways, using neurotransmitters only for cell-to-cell communication. The final target of the action is determined by the specific nerve pathway. The endocrine system uses the bloodstream to spread its hormonal signals (Figure 17.2). All cells with access to the blood can be affected by the hormone, but only cells with a sensitivity to the hormone are actually influenced. The speed of the nervous system is very fast; it takes less than 1 sec for a person to step on the brake of a car. In contrast, the endocrine system is slow. Even the fastest responses, such as a fright response from the adrenal medulla, takes many seonds; some responses, such as changes in some of the male hormones (androgens), take hours to occur. The duration of the signal of the nervous system is very brief, lasting seconds or less. The duration of the signals of the endocrine system are long, and many hormones, such as thyroxine, have half-lives measured in

Figure 17.1   Feedback inhibition. **A,** Positive feedback on the ovaries. At the beginning of the menstrual cycle, little estradiol is being produced. Follitropin (follicle-stimulating hormone, FSH) induces estradiol production, and the estradiol increases follitropin action, causing still greater estradiol production. **B,** Negative feedback of the thyroid hormones. Thyroid-stimulating hormone *(TSH)* causes an increase in the production of thyroxine by the thyroid gland. The thyroxine shuts off secretion of TSH, and the stimulation of thyroxine decreases until it falls below the "set point," causing an increase in TSH.

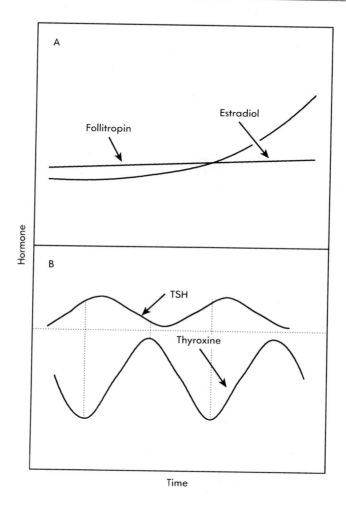

days. The target of the nervous system is usually single and specific, often a muscle. In contrast, the targets of the endocrine system are usually multiple and diffuse. Cortisol, for example, influences virtually every cell of the body. Finally, the nervous system functions mainly for motor coordination, whereas the major purpose of the endocrine system is metabolic regulation.

In many cases neurotransmitters and hormones may be similar or identical chemicals. Insulin, glucagon, and cholecystokinin, all thought to be classical hormones, may also be neurotransmitters in the brain. Similarly, dopamine, thought to be a classical neuro-

Figure 17.2 Location of the endocrine glands. The classical endocrine glands are labeled on the right, and the organs with incidental endocrine function are on the left.

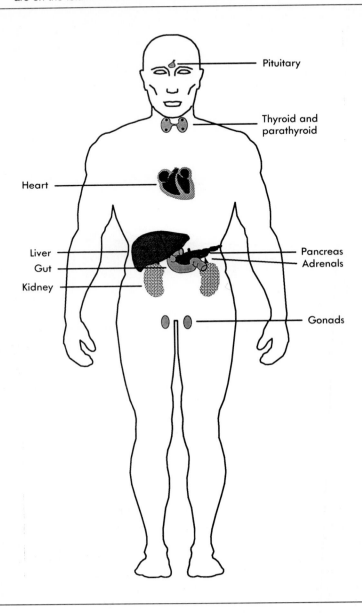

transmitter, acts as a hormone in the hypothalamic-pituitary axis. In some cases the hormones and neurotransmitters may be closely related. Norepinephrine is released by the nerve termini of the sympathetic nervous system. It is very similar in structure to epinephrine, the major hormone released by the adrenal medulla (Figure 17.3). In many tissues their actions are identical.

Figure 17.3   Structure of the catecholamines.

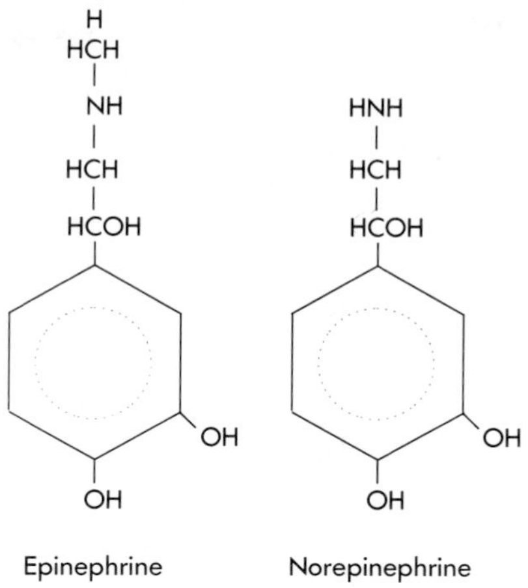

Epinephrine   Norepinephrine

Table 17.1   Properties of the regulatory systems

Property	Neuronal	Hormonal
Path	Nerves	Blood
Speed	Fast (<1 sec)	Slow (seconds to hours)
Duration	Short (<1 sec)	Long (seconds to hours)
Target	Single and specific	Multiple and general
Function	Movements	Metabolic

The chemical substances that are released in the nervous system and have a regulatory function are called neurotransmitters. Neurotransmitters function at various sites, such as the neuromuscular junction, and at synapses within the brain and autonomic nervous systems. Two criteria have been established for the identification of a substance as a neurotransmitter. First, it must be released in appropriate amounts when the nerve is stimulated. Second, it must mimic the effect of electrical stimulation of the nerve. The neurotransmitter is usually stored in vesicles contained in the nerve endings. When the electric depolarization wave reaches this region, the neurotransmitter is released by a calcium-dependent process that involves exocytosis. The released neurotransmitter combines with a specific receptor site on the membrane of the target, leading to transduction of the signal and alteration of the target cell without any electrical continuity between the two cells. The response to the neurotransmitter at the target cell probably is mediated by mechanisms that are similar or identical to those used by the endocrine system, as described in more detail later. For example, dopamine causes an increase in cyclic adenosine monophosphate (cAMP) in certain target cells, whereas acetylcholine produces an increase in cyclic guanosine monophosphate (cGMP). Although the neurotransmitter is bound tightly to its receptor on the target cell, it rapidly dissociates so that the signal can be terminated quickly. Inactivation of the neurotransmitter occurs through either reabsorption and storage in nerve endings or enzymatic degradation. The enzymes needed to synthesize the neurotransmitter are synthesized in the cell body of the neuron and sent to the axon terminals as needed. The two most common neurotransmitters are norepinephrine and acetylcholine, but many other substances are widely accepted as neurotransmitters, including catecholamines such as 3,4-dihydroxyphenethylamine, dopamine, epinephrine, amino acid derivatives such as glutamine, glycine, γ-aminobutyric acid, histamine, proline, aspartate, taurine, 5-hydroxytryptamine (serotonin), melatonin, proteins such as substance P, insulin, glucagon, cholecystokinin, and others such as octopamine, carnosine, and adenosine triphosphate (ATP).

Norepinephrine acts primarily as the chemical transmitter in the sympathetic nervous system and in the brain. Acetylcholine is the chemical transmitter at the endings of both the cholinergic and the motor nerves, as well as in the ganglia of the sympathetic nervous system. The enzyme acetylcholinesterase hydrolyzes acetylcholine at the postsynaptic terminal or motor end-plate, terminating its action. The other neurotransmitters act primarily within the central nervous system.

$$CH_3-\overset{\overset{\displaystyle O}{\|}}{C}-OCH_2-CH_2-{}^+N(CH_3)_3 \xrightarrow{\text{Acetylcholinesterase}}$$

Acetylcholine

$$CH_3-COOH \quad + \quad HO-CH_2-CH_2-{}^+N(CH_3)_3$$

Acetic acid   Choline

Hormonal action

Endocrine regulation of metabolic processes is mediated by hormones, but the metabolic alterations caused by the hormones depend on at least six factors:

    Hormones
    Glands
    Transport in blood
    Target tissue
    Feedback
    Degradation of hormone

Alterations in any of these factors can dramatically alter effectiveness of the hormone and may produce a disease state.

A variety of processes can influence the actions of the hormones (Figure 17.4). The hormones are the major carriers of information in the endocrine system, as defined earlier. They are secreted by glands, generally into the vascular system. The secretion is regulated by many factors, which may include other hormones or stimulating or inhibiting influences. Secretion may also be affected by feedback inhibition through the end effect. The hormones are distributed by the vascular system, but this is not necessarily a passive process. Some portions of the vascular system are targeted. For example, the portal circulations cause higher concentrations of hormones to reach certain organs. The hepatic portal circulation has insulin concentrations 10 times higher than the general circulation, and the liver is exposed to these higher levels. Target tissues may be single or multiple; hormones bind to receptors at the target tissues and then cause an effect. This effect may be an end action, such as alterations in glucose metabolism, or the production of another hormone. The end effect, and sometimes the second hormone, whose production is

Figure 17.4

Components of the endocrine system. The gland is influenced by both nerves and feedback by hormonal effects mediated through a sensor. The hormone is transported by the blood and goes to the target tissue. The effects of the hormone on the target tissue feed back on the gland and the sensor.

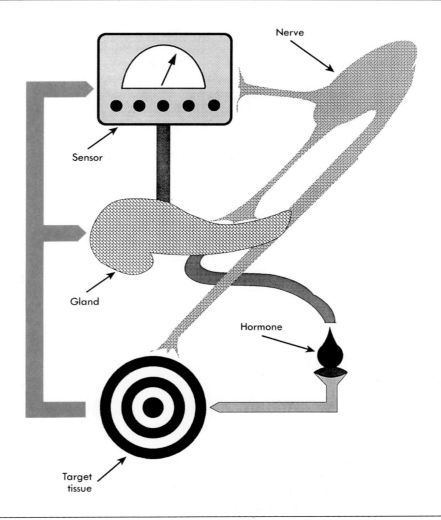

stimulated, will have negative feedback to shut off secretion of the hormone by the initial gland. Finally, it is important to realize that degradation of the hormone is an important component of the endocrine process. Signals are only significant if they have a finite life and do not continue to influence the organism indefinitely.

Hormones

Hormones can be classified into five groups according to structure, synthesis, and mechanisms of action: (1) peptides and proteins, (2) catecholamines, (3) steroids and vitamin D, (4) thyronines, and (5) others. The peptides and proteins and the catecholamines seem to have their major action at the cell surface and are covered in detail in Chapter 18. The other categories of hormones are discussed in Chapter 19. Although more than 100 different hormones exist, less than 20 are of major clinical significance, and only these are emphasized in this text.

The proteins and peptides are the largest category of hormones and consist of oligo-peptides, polypeptides, proteins, and glycoproteins (see box below). The oligopeptides, hormones with less than 20 amino acids, are cleaved from a larger protein precursor, as is the nonapeptide oxytocin. Many of these hormones are closely related to neurotransmitters and are secreted by neurons or other cells of the brain. The posterior pituitary hormones—vasopressin, oxytocin, and vasotocin—are directly secreted by neurons of the hypothalamus. The hormones that influence anterior pituitary function—thyroid-releasing hormone (TRH), gonadotropin-releasing hormone (GnRH), somatostatin, and others—are secreted by hypothalamic and other brain cells. Other oligopeptide hormones include the gastrointestinal hormones secretin and cholecystokinin.

The polypeptide hormones can be arbitrarily defined as those proteins with 21 to 150 amino acids. This is a large category that comprises most of the proteinaceous hormones made outside of the brain. It includes the hormones of calcium metabolism (parathyroid hormone and calcitonin), the hormones of glucose metabolism (insulin, glucagon, and pancreatic polypeptide), the growth factors (insulin-like growth factors I and II, epidermal growth factor, nerve growth factor, relaxin, fibroblast growth factor, and various less well-defined polypeptide growth factors), and the pituitary hormone adrenocorticotropin, or adrenocorticotropic hormone (ACTH).

The protein hormones are a fairly small category, consisting of the related proteins—growth hormone (somatotropin, GH), prolactin, and human chorionic somatomammotropin (human placental lactogen, HPL or HCS)—and the unrelated protein, renin. The remaining hormones in this category are glycoproteins: follitropin (follicle-stimulating hormone, FSH), lutropin (leutinizing hormone, LH), human chorionic gonadotropin (HCG), thyrotropin (thyroid-stimulating hormone, TSH), and erythropoietin.

The Protein Hormones

Oligopeptides
Vasopressin, oxytocin, vasotocin
Thyroid-releasing hormone
Gonadotropin-releasing hormone
Somatostatin
Cholecystokinin, secretin, vasoactive intestinal peptide (VIP), gastric inhibitory peptide (GIP)

Polypeptide hormones
Parathyroid hormone, calcitonin
Insulin, glucagon
Pancreatic polypeptide
Insulin-like growth factors I and II (IGF-I, IGF-II) epidermal growth factor (EGF), fibroblast growth factor (FGF), nerve growth factor (NEF)
ACTH

Protein hormones
Somatotropin
Prolactin
Chorionic somatomammotropin
Renin

Glycoproteins
Follitropin (FSH)
Lutropin (LH)
Chorionic gonadotropin
Thyrotropin (TSH)
Erythropoietin

Other hormones are listed in the box below.

All the polypeptides, protein hormones, and glycoproteins are synthesized by the standard protein synthesizing machinery of the cell. Many have posttranslational modifications, including amidation of the carboxyl terminus, blockage of the amino terminus, and glycosylation. All the hormones in this category bind to receptors at the plasma membrane. Many stimulate adenylate cyclase, some activate tyrosine kinases, and some have unknown mechanisms of action. The box on p. 730 lists some hormones by mechanism of action.

Glands

Hormones are synthesized in glands, which may be entire organs (e.g., the thyroid gland), parts of other organs (e.g., the cluster of secretory cells in the islets of Langerhans in the pancreas), or just a few cells (e.g., the neuroendocrine secretory cells in the crypts of the microvilli of the small intestine). Figure 17.2 shows the location of the endocrine glands. Hormones may be synthesized by specific enzymes or as proteins by the normal protein synthesizing machinery of the cell. The proteins are often posttranslationally modified to form the active hormones.

Enzymatic synthesis Most nonprotein hormones are synthesized by specific enzymatic pathways. The precursor molecules are common biochemicals, such as the amino acids tyrosine and tryptophan or cholesterol. These pathways can be very complex and require many enzymes. Cortisol synthesis, for example, requires seven enzymes that must work in a specific order. The coordinated regulation of these enzymes is not understood in humans, but it is unlikely that cells other than specialized endocrine secretory cells or their precursors could produce these hormones, even in disease states. Cholesterol is the precurosor for the steroids and vitamin D, tyrosine for the catecholamines, and thyronines and tryptophan for melatonin and serotonin. The prostaglandins are produced from highly unsaturated fatty acids.

Other Hormones

Catecholamines

Epinephrine
Norepinephrine
Dopamine

Steroids and derivatives

Cortisol, other glucocorticoids
Aldosterone, other mineralocorticoids
Testosterone, other androgens
Estradiol, other estrogens
Progesterone, other progestins
Cholecalciferol and other vitamin Ds

Thyronines

Thyroxine
Triiodothryonine

Others

Serotonin
Melatonin
Prostaglandins
Endorphins

Mechanism of Action of Some Hormones

Stimulates production of cyclic AMP

Corticotropin
Lutropin
Follitropin
Chorionic gonadotropin
Thyrotropin
Vasopressin
Parathyroid hormone
Glucagon
β-Catecholamines
Prostaglandins

Stimulates a tyrosine kinase

Insulin
Insulin-like growth factor I
Epidermal growth factor
Platelet-derived growth factor

Stimulates inositol phosphates and calcium

Thyroid-releasing hormone
Gonadotropin-releasing hormone
α-Catecholamines
Angiotensin
Acetylcholine (muscarinic)
Thrombin

Ribosomal synthesis The protein hormones are synthesized on the ribosomes of the rough endoplasmic reticulum (ER). The pathway may be very complex. Most protein hormones are synthesized as *preprohormones* and must be processed further to produce the finished hormone. *Prehormones* are produced in the ER. The *pre*-piece, also called the *leader sequence,* is the first few amino acids of the protein, often 20 to 30 amino acids in size. These amino acids are very hydrophobic and are thought to be necessary for the movement of the protein across the membrane lipid bilayer of the rough ER. The leader sequence is always rapidly cleaved from the newly synthesized protein. This process is so rapid that these early hormonal precursors are difficult to demonstrate in animal cells. The existence of a leader sequence was demonstrated when messenger ribonucleic acid (mRNA) for hormones was translated using plant ribosomes. The leader sequence has not been demonstrated to be present in the circulation in any human disease.

After the leader sequence has been cleaved, the remaining protein, called a *prohormone,* is usually considerably larger than the final hormone and often has little or no biologic activity. It is thought that the extra size is necessary to allow the protein to fold properly. Once the final disulfide linkage and quarternary structure is set, the extra portions are cleaved from the final hormone by proteolysis. This process often starts in the Golgi apparatus and continues in the secretory granule. The cleaved extra peptides remain in the secretory granule and are secreted with the hormone.

Insulin is made through this process. The first peptide synthesized, *preproinsulin,* is rapidly cleaved to a 9000-dalton peptide, *proinsulin.* After proper folding and disulfide bond formation, the extra piece in the middle of the polypeptide chain is cleaved out by a trypsinlike proteeolytic process. This produces a 5600-dalton, two-chain insulin molecule; the 2500-dalton extra peptide, called the C-peptide; and four basic amino acids. When insulin is secreted, a mixture of substances is released by the islets of Langerhans.

About 6% of proinsulin is unprocessed and is secreted intact. Equimolar ratios of insulin and C-peptide are also released. In some patients whose insulin cannot be measured, one can still measure this C-peptide to determine the amount of insulin secretion.

Many proteins and peptides may also undergo posttranslational modification. The carboxyl terminus may be amidated, as is GnRH; the amino terminus may be acetylated, as is calcitonin, or otherwise blocked; or the proteins may be glycosylated, as is TSH. No hormones are known to have covalent fatty acids attached.

Clinical comment

Production of hormones by tumors Frequently diseases are caused by excesses of hormones. They may be made by the gland that normally makes the hormone, for example, an excess of thyroid hormone produced by the thyroid gland in Grave's disease or by some other tissue, usually a cancer. If the cell making the excess hormone is normally differentiated to form this hormone, it is called *eutopic* production. Thus, excess insulin production by the islets of Langerhans or even by a tumor of the Islets (called an islet cell tumor) is eutopic. Tumors of gland tissue that produce these hormones eutopically do not need to be in the normal location. Thus, a choriocarcinoma (tumor of the placenta) producing chorionic gonadotropin would be considered eutopic, even if the tumor was metastatic to the lungs.

Occasionally, cells make hormones that are not part of their normal differentiation path. this is called *ectopic* production of hormones. These cells are almost always malignant tumors. For example, a small cell carcinoma of the lung may make chorionic gonadotropin or ACTH. Ectopic production of hormones is always of protein or peptide hormones. No well-established cases of ectopic production of steroids, catecholamines or thyronines have been reported.

Hormone secretion The protein hormones made in the rough ER with the help of the leader sequence enter vesicles that bud from the rough ER and move to the Golgi apparatus. Hormones created by enzymatic synthesis are made in the smooth ER and again appear, entering vesicles that bud and go to the Golgi apparatus. Hormones are further modified by the Golgi apparatus and packaged into the secretory vesicles. These secretory vesicles move toward the cell membrane along microtubules. Overall movement occurs at about 10 μ/minute. Secretion, at least in some cells, is prevented by microfilaments of the cell web that block fusion of the vesicles with the plasma membrane. The secretory stimulus causes clumping of the microfilaments, opening holes and allowing fusion of the vesicles and exocytosis of the hormone (Figure 17.5).

The secretory vesicles normally accumulate and may store a large amount of hormone for later release. Some glands have developed specialized methods of storing hormones. The cells of the thyroid gland form a follicle, a spherelike structure in which the cells form the surface of the sphere. The cells store the thyroid hormone as part of a complex protein called thyroglobulin, which is stored extracellularly in the center of the sphere. When the thyroid hormone is needed, the cells recover the thyroglobulin and degrade it to the thyroid hormones, which are then secreted.

Clinical comment

Visualization of glands The properties of the individual glands may be utilized clinically to visualize the glands. Such information may be necessary to localize endocrine tumors or to know the size of a specific gland. Sometimes a stored hormone can be made radioactive and visualized with special cameras that can easily view radioactivity. The thyroid gland can be visualized with radioactive iodine (^{131}I or ^{123}I) and the adrenal glands with radioactive cholesterol derivatives (Figure 17.6). Some glands have such high levels of protein synthesis that they can be visualized with radioactive amino acid derivatives. The parathyroid glands and the islets of Langerhans can be visualized with selenomethionine.

Figure 17.5   Mechanism of hormonal secretion. See text for details. *ER,* Endoplasmic reticulum.

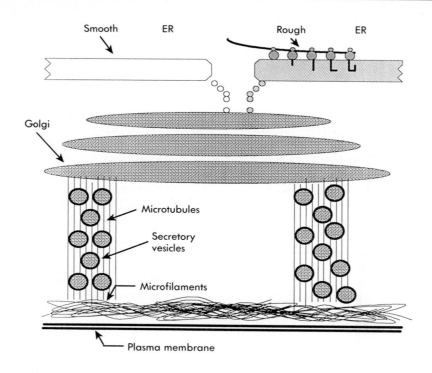

Role of the blood   The vascular system is important as the main route of dispersion of the hormones, but its role may be considerably greater. The vascular system may regulate the amount of hormone reaching the tissues, and it may have binding proteins that alter the levels of free hormone. Circulations may be specific and targeted, such as the portal circulations, which deliver very high levels of hormones to specific organs, or the general circulation, which delivers similar blood concentrations of hormones to all organs.

Portal circulation Two portal circulations exist. The hypothalamic-hypophyseal portal circulation delivers high levels of the hypothalamic-releasing factors to the pituitary gland. The levels of these factors may be 100 times higher in this blood than in the general circulation. The hepatic portal circulation brings the pancreatic and gastrointestinal hormones to the liver at levels that are about tenfold higher than in the general circulation. These special circulations may allow certain hormones to have greater effects on the pituitary gland and the liver than on other tissues.

Endothelial transport The vascular system may not be passive in the distribution of hormones. The entire vascular system is lined with a continuous layer of endothelial cells that prevents simple diffusion of the protein and other hormones. The endothelial cells specifically bind these hormones on their vascular surface and transport them across the cell, releasing them on the tissue side of the cell. Alterations in the rate at which these

Figure 17.6   Thyroid scan. A dose of radioactive iodine (^{131}I) given to a normal patient is highly concentrated by the throid gland. The gamma emissions by the iodine can be viewed by special cameras, which gives the physician information about the shape, size, and function of the gland.

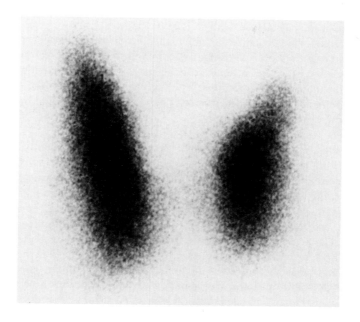

endothelial cells transport the hormones may make major differences in the levels of hormone that reach the tissues. Although this is currently an area of active research, there are no known examples of disease caused by problems with endothelial transport.

Binding proteins Certain hormones do not circulate free in the plasma but rather are tightly bound to specific binding proteins. The thyroid hormones, thyroxine and triiodothyronine, are more than 99% bound to thyroid-binding globulin and other serum proteins. Likewise, testosterone is more than 90% bound to sex hormone–binding globulin, cortisol more than 90% bound to cortisol-binding globulin, and the insulin-like growth factors more than 99% bound to specific binding proteins. The posterior pituitary hormones, vasopressin and oxytocin, are bound to the neurophysins in the hypothalamus and are transported down the pituitary stalk with them. Once released from the posterior pituitary glands, the hormones are freed from the neurophysin and are transported in blood in unbound form.

The binding of hormones to these proteins alters the properties of the hormones significantly. The plasma levels of active hormone is lowered because only the free hormone is active. The hormone often lasts longer in the circulation since the bound hormone is neither degraded nor excreted by the kidney. Finally, the binding to these proteins buffers the organism against extremes of hormone concentration.

Clinical comment   **Complications with binding proteins** The binding proteins also complicate the evaluation of the thyroid and adrenal glands and androgens. Only the free hormone is active, but most assays measure the total hormone. Since the binding proteins are usually constant, the free hormone is in direct proportion to the total hormone. However, this is not always the case. Thyroid binding is increased by oral contraceptives and is sometimes missing in patients with an X-linked recessive condition. In these cases patients with normal free hormone levels might have elevated or reduced total hormone levels.

Hormonal action at the cell

The hormones diffuse to all cells of the body. Specificity of action is conferred by specific receptors that recognize the individual hormones. *Receptors* are molecules that specifically bind a hormone or other ligand, resulting in a biologic activity. The interactions of the hormone with the receptor can occur at the plasma membrane or intracellularly, or sometimes at both locations. The protein hormones and the catecholamines bind to receptors at the plasma membrane. Steroids, vitamin D, and the thyronines bind to intracellular receptors.

Plasma membrane receptors These receptors are generally large glycoproteins embedded in the membrane. They may be composed of single polypeptide chains, such as the β-adrenergic receptor, or multiple subunits, such as the insulin receptor (Figure 17.7). Binding of the hormone to the receptor produces a second messenger that may be cAMP, protein tyrosine phosphorylation, inositol trisphosphate, or alterations in ion channels (see previous box on p. 730). Since the binding to the receptor occurs extracellularly, the signal must be passed through the membrane to the final effector, which is intracellular.

Each receptor binds a specific hormone with high affinity. Sometimes other hormones with similar structure will bind, but with lower affinity. For example, proinsulin binds to the insulin receptor with about 5% of the affinity of insulin. Occasionally a hormone may bind to an unrelated receptor with some affinity, a process called *specificity spillover*.

Adenylate cyclase Many hormones serve to increase the intracellular levels of cAMP. Adenylate cyclase, the enzyme that synthesizes cAMP from ATP, is linked to receptors by guanine nucleotide–binding proteins, called the G or N proteins. This protein is made up of three subunits; two are common to all cAMP-dependent receptor systems and one is specific for each system. After a hormone such as epinephrine binds to its specific receptor, the G protein, with guanosine diphosphate (GDP) attached, binds to the hormone receptor, forming a ternary complex. The complex binds guanosine triphosphate (GTP), liberating GDP, and the specific subunit of the G protein dissociates from the complex. The liberated, activated G-protein subunit diffuses along the membrane until it binds to an adenylate cyclase, activating the cyclase. This is an activating cycle, and the G protein is called the G_s protein because it stimulates the cyclase activity. A G_i protein works exactly the same way except that it inhibits adenylate cyclase activity. The G_s and G_i proteins work with different receptors but may bind to the same adenylate cyclase. After the adenylate cyclase is activated by the G_s or is inhibited by the G_i protein, the complex hydrolyzes the GTP to GDP, liberating phosphate, and dissociates into an inactive cyclase and the specific G-protein subunit, with GDP bound (Figure 17.8).

The activated adenylate cyclase will synthesize cAMP from ATP in the presence of magnesium. Cells that are responsive to cAMP have a cAMP-dependent protein kinase. When cAMP levels rise, this enzyme is activated and adds phosphate groups to the serine and threonine groups of other proteins, changing their activity. cAMP is normally hydrolyzed by a specific phosphodiesterase. This inactivation is inhibited by methylxanthines, such as the caffeine of coffee and the theophylline used to treat asthma.

Tyrosine kinase Some receptors produce their biologic effect by phosphorylating the tyrosine groups of proteins. These receptors are generally for anabolic peptides such as insulin and many growth factors (see previous box on page 730). These receptors may be composed of a single polypeptide chain (EGF receptor) or multiple subunits (insulin receptor). They have an extracellular domain that binds the hormone, a hydrophobic transmembrane domain of 20 to 25 amino acids, and an intracellular domain that has the tyrosine kinase activity.

After binding of the hormone, the signal is passed to the tyrosine kinase, which phosphorylates proteins and thus modifies their activity. The insulin receptor is involved in three separate types of phosphorylation:

Figure 17.7

Receptors in the cell membrane. The insulin receptor is composed four subunits, two extracellular α-subunits of 135,000 daltons that bind insulin, and two β-subunits of 90,000 daltons that cross the membrane and have the tyrosine kinase activity. The β-adrenergic receptor consists of a single polypeptide chain that crosses the membrane repeatedly.

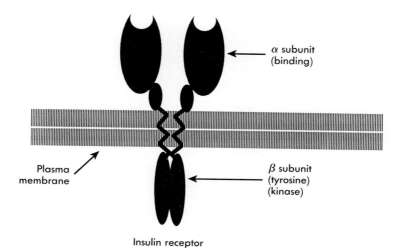

α subunit
(binding)

Plasma
membrane

β subunit
(tyrosine)
(kinase)

Insulin receptor

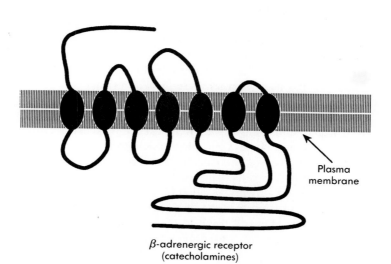

Plasma
membrane

β-adrenergic receptor
(catecholamines)

Figure 17.8　　　　　　　The adenylate cyclase system. See text for details.

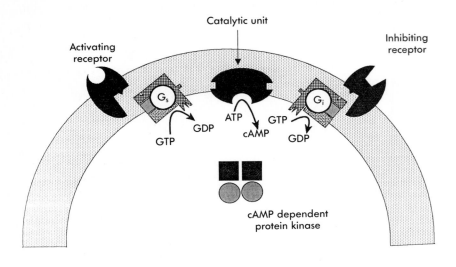

1. It can phosphorylate tyrosine groups of other proteins, altering their activity.
2. It can phosphorylate itself on specific tyrosine groups, which seems to increase its activity and may involve clustering of receptors.
3. It can be phosphorylated on serine groups by other kinases, such as protein kinase C. The serine (and possibly threonine) phosphorylation may inactivate or desensitize the receptor so that it can bind insulin but not produce a biologic effect.

Inositol phosphates The phosphatidylinositol of the plasma membrane exists in several phosphorylated forms. The most heavily phosphorylated form is phosphatidylinositol-4,5-bisphosphate (PIP_2). After the stimulation of certain hormone receptors by the binding of their hormones, a phospholipase C is activated and hydrolyzes the PIP_2 to diacylglycerol and inositol trisphosphate. The process of activation of the phospholipase C involves a protein that has many similarities to the G_s proteins.

The products of hydrolysis of PIP_2 each stimulate another pathway. Inositol trisphosphate stimulates the release of calcium from intracellular stores, raising intracellular calcium and stimulating a variety of calcium-dependent processes. The diacyglycerol stimulates protein kinase C, another protein that adds phosphate groups onto the serine and threonine groups of proteins. Increased protein kinase C activity is usually associated with rapid growth of cells, whereas increased activity of cAMP-dependent protein kinase is usually associated with cessation of cell growth (Figure 17.9).

Great similarity exists in the mechanism of action of all the hormones that act at the cell membrane: they all result in the phosphorylation of other proteins. Most act via cAMP to stimulate the serine/threonine-specific, cAMP-dependent kinase; some act to stimulate the serine/threonine-specific protein kinase C; and some act to stimulate a tyrosine kinase that is part of the receptor itself. The specificity of these kinases for both substrate proteins and sites of phosphorylation determine the final biologic outcome.

Figure 17.9   Phosphatidylinositol as a second message. Phosphatidylinositol in the plasma membrane can be sequentially phosphorylated at the 4 and 5 positions of the myoinositol. Stimulation of certain receptors, such as that caused by gonadotropin-releasing hormone *(GnRH),* working through a G protein, in turn stimulates a phospholipase C. The phosopholipase C cleaves the phosphatidylinositol-4,5-bisphosphate to inositol trisphosphate, which increases intracellular calcium and diacylgicceraol, stimulating protein kinase C.

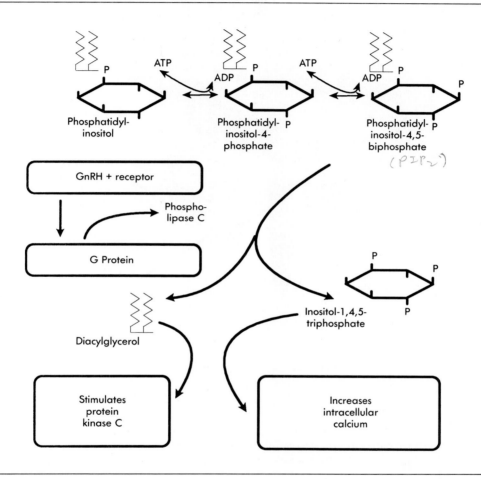

Steroid hormone receptors Steroid hormones, vitamin D, and the thyronines exert their major effects by interacting with specific intracellular receptors. The receptors for all these hormones are closely related, and belong to a "superfamily." The genes for all the receptors have been cloned and sequenced. Some of the structural properties of these receptors can be inferred from the amino acid sequence.

These receptors, in their normal, unoccupied form, are made up of multiple subunits (Figure 17.10): one or more specific hormone-binding subunits of molecular weight 60 to 80,000 and two subunits of molecular weight 90,000 that appear to be identical with the ubiquitous "heat shock protein." The binding subunits of the various steroid hormone

Figure 17.10   Steroid hormone receptor. See text for details.

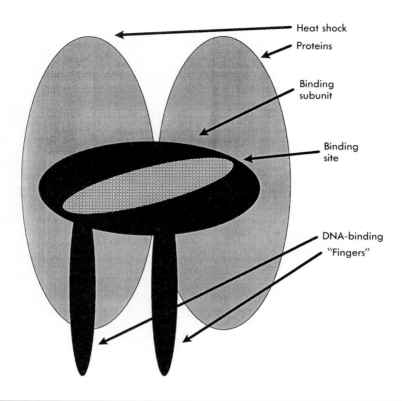

Heat shock
Proteins

Binding
subunit

Binding
site

DNA-binding
"Fingers"

receptors are very similar to each other and to the oncogenes of the *erb-A* family (see Chapter 20). They have two areas that are closely involved in the biologic activity of the receptor: (1) the binding of deoxyribonucleic acid (DNA) and (2) the binding of the specific hormone.

The hormone-binding subunit of each of the receptors has a DNA-binding region about 70 amino acids in length. This region, which is about 300 amino acids from the carboxyl terminus, is rich in cysteine, lysine, and arginine residues in a highly specific pattern. When Zn^{++} binds to this region, it produces two fingerlike structures that seem to be the specific sites of DNA binding. This region is very similar in the different steroid hormone receptors and is entirely conserved in the same receptor of different species.

The second common area is the hormone-binding region. This sequence of approximately 250 amino acids starts about 50 amino acids after the end of the DNA-binding region. It is rich in hydrophobic amino acids and is made up of α-helices and β-strands that form a hydrophobic pocket for steroid hormones and thyronines. The same hormone receptors of different species are 70% to 95% homologous. Receptors for different hormones share the same overall structure but only have 10% to 20% identical amino acids. The N-terminal sequence of the receptor is highly variable; this region probably gives the specificity for the various areas of the chromatin and for the quarternary structure of the receptor.

All the biologically important regions of the receptor are found on the binding subunit, and the function of the heat shock protein is uncertain. Some investigators believe that it serves to stabilize the receptor before hormone binding. Since this protein is abundant in cells, virtually all the newly synthesized and recycled binding subunits are quickly bound by the heat shock protein.

The receptors for the steroids, vitamin D, and thyronines are localized almost entirely in the nucleus of the cell. These hormones are generally lipid soluble and diffuse across the membranes of the cell to bind to the receptor in the nucleus. Following binding of the hormone, a transformation of the receptor occurs. If the hormone is an active form, this transformation will be an activation in which the binding subunit will dissociate from the heat shock proteins and change its conformation so that it has a higher affinity for the hormone and also has a high affinity for DNA. Most of these activated receptors will bind nonspecifically to the DNA of the nucleus, but about 100 molecules in each cell will bind to specific hormone regulatory elements that are generally located in the 5′ upstream region of the regulated gene.

The specific binding of the receptor to one or more (usually about 100) specific genes results in the increased synthesis of the specific mRNA's encoded in those genes. The binding to the hormone regulatory elements seems to alter the structure of the chromatin and allow much greater transcription of the gene. In a few cases this binding results in inhibition of mRNA for other genes. The increased mRNA that is formed migrates to the cytoplasm. Translation of these mRNA molecules on the ribosomes leads to an increase in specific proteins and an alteration in cell function.

Occasionally, specific receptors for the steroids in the plasma membrane have been reported; these receptors are similar to those for the protein hormones. Generally, these have proved to be artifacts. The exception is the progesterone receptor of the *Xenopus laevis* oocyte. This membrane-bound receptor has the correct specificity to act as a receptor for progesterone. With time other exceptions may be found.

Feedback inhibition of hormone synthesis

After a hormone has produced its effect, the effect generally serves to inhibit further synthesis of the hormone. For example, eating a meal increases blood glucose, causing an increase in insulin synthesis. The increased insulin initiates a variety of processes that serve to lower the blood glucose: increased glucose usage by the muscles, increased glycogen synthesis by the muscles and liver, and increased fat synthesis by adipocytes. The blood glucose lowered by these activities serves to inhibit the continued synthesis of insulin, completing the negative feedback. In some cases secretion of the hormone serves to inhibit directly the synthesis of that hormone. This is called *short-loop feedback*. A lack of feedback inhibition results in profound increases in hormone synthesis and usually is detrimental to the animal (see case 2 at end of the chapter).

Destruction of the hormone

For any message to be useful, it must have a finite lifetime. The ringing of your telephone is a message that informs you that someone is waiting to talk to you. When you lift the receiver to answer the phone, you destroy that message. Imagine the chaos if the phone continued to ring, even after it were answered. Since it would always be ringing, you would never know when someone was trying to call. Similarly, hormones, which are messages, must be destroyed in a timely fashion if they are to present current information to the cells. Destruction of hormones may take various forms; some are specific, some nonspecific. Certain hormones, particularly the neurotransmitters, are taken up into neurons. Many hormones are nonspecifically destroyed by the liver and kidneys via proteases or specific metabolic pathways, and other hormones are excreted into the urine. Many hormones are also specifically destroyed at the cell in which they act. This process, which is often receptor mediated, may lead to the destruction of the receptor (see following discussion on receptor's life cycle).

Regulation of hormonal action

Increased hormonal action is often the consequence of increased hormone secretion. The production of more epinephrine leads to a faster heart rate, higher blood pressure, and greater sweating. Hormonal action can also be regulated at other sites. Alterations in receptor number, affinity, or activity or changes in the hormone's destruction rate can all alter the biologic activity of the hormone. The binding reaction of a hormone (H) and receptor (R) can be written:

$$H + R \rightleftharpoons HR$$

The bioactivity of a hormone is a function of the bound hormone (HR), and the mass action expression for the binding reaction can solved for the concentration of HR:

$$\frac{[HR]}{[H][R]} = K$$

$$\text{Bioeffect} = f[HR] = fK[H][R]$$

From this expression it is clear that altering any of the factors just mentioned will alter biologic activity. Changing the hormone concentration [H] is the most common variation to alter bioeffect, but clearly, altering the receptor concentration [R], the receptor affinity K, or the ability of the receptor to pass its message into the cell f will also alter biologic activity. Alteration in each of these factors has been demonstrated to cause human disease (see case 3).

Life cycle of a receptor

The receptors for the protein hormones have a life cycle that may partly be responsible for some diseases. Figure 17.11 depicts what may be the typical life cycle of the insulin receptor. The information for the receptor is encoded in the genome and is transcribed onto a specific mRNA (arrow 1), which migrates to the rough endoplasmic reticulum, where it is translated into receptor molecules (arrow 2). These receptors are moved to the Golgi apparatus (arrow 3), where final processing and glycosylation occurs. Then they are inserted into the plasma membrane (arrow 4), in which they exist as monomers or small oligomers of receptor. Insulin binds and excites the tyrosine kinase domain of the receptor, producing an insulin action that is characteristic of the specific cell. After insulin binds to the receptor, the receptors form patches (arrow 5) that are endocytosed into specialized vesicles, called *receptorsomes* (arrow 6). The purpose of this vesicle is the destruction of insulin, but the protease is imperfect in this task, degrading 30% to 50% of the insulin receptors as well. The remaining receptors may be recycled through the Golgi apparatus back to the plasma membrane (arrow 7).

Down regulation

Since the voyage through this recycling pathway results in the net destruction of 30% to 50% of the receptors, any substance that causes the receptor to undergo recycling will reduce the number of receptors. Since the major factor that causes this is the hormone itself, elevated levels of hormones, particularly when they have been elevated for prolonged periods, typically cause reductions in the number of their own receptors, a process called *down regulation*. Most hormones that bind to plasma membrane receptors will cause down regulation to occur; prolactin is a notable exception. Steroid hormone receptors are also recycled and can undergo down regulation, but the mechanism is poorly understood.

Figure 17.11   Life cycle of a hormone receptor. See text for details.

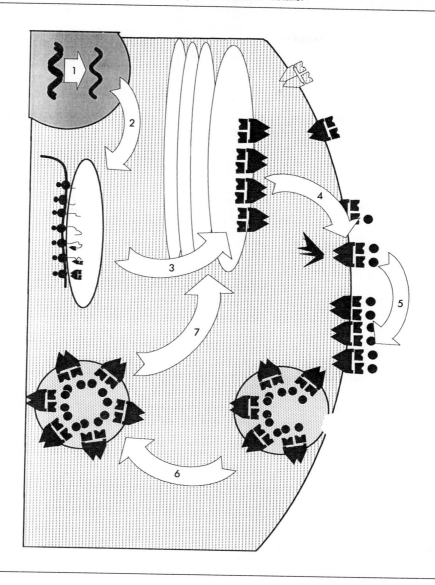

Bibliography

Baulieu EE and Mester J: Steroid hormone receptors. In DeGroot L, editor: Endocrinology, Philadelphia, 1988, WB Saunders Co.

Berridge MJ: Inositol triphosphates and diacylglycerol as second messengers, Biochem J 220:345-360, 1984.

Dernovsek K et al: Rapid transport of biologically intact insulin through cultured endothelial cells, J Clin Endocrinol Metab 58:761-765, 1984.

Kahn CR: Membrane receptors for peptide hormones. In DeGroot L, editor: Endocrinology, Philadelphia, 1988, WB Saunders Co.

Lefkowitz RJ: Clinical physiology of adrenergic receptor regulation, Am J Physiol 243:E43-R47, 1982.

Pastan I, Roth J, and Macchia V: Binding of hormone to tissue: the first step in polypeptide action, Proc Natl Acad Sci 56:1802-1809, 1966.

Ringold GM: Steroid hormone regulation of gene expression, Annu Rev Pharmacol Toxicol 25:529-566, 1985.

Schlessinger J et al: Direct visualization of binding aggregation and internalization of insulin and EFG on living fibroblasts, Proc Natl Acad Sci 75:2659-2663, 1978.

Shia MA, Rubin JB, and Pilch PF: The insulin receptor protein kinase: physiochemical requirements for activity, J Biol Chem 258:1450-1455, 1983.

Clinical examples

A diamond (♦) on a case or a question indicates that literature search beyond this text is necessary for full understanding.

Case 1	Pseudohypoparathyroidism

A 4-year-old boy was brought for evaluation of a recent seizure. He was slow in his developmental milestones and had seven seizures since birth. He urinated frequently, often 10 to 12 times per day. The family was not very close, but the father thought that some of his relatives had a similar problem. On physical examination the patient was short and overweight with a rounded face. He twitched to any touch, but twitching of the corner of the mouth was most prominent, especially on touching him in the parotid gland (Chvostek's sign). He had several bony abnormalities, including a short fourth metecarpal bone. Examination was otherwise normal. Laboratory examination showed signs of dehydration, a very low calcium concentration, high phosphate levels, and a low thyroxine level. His parathyroid hormone and thyroid-stimulating hormone (TSH) levels were extremely high.

Biochemical questions

1. What single biochemical abnormality could explain this patient's problems?
♦ 2. How could you prove that this is the problem?
3. What other problems might this biochemical defect cause?

Case discussion

1. Biochemical abnormalitiy The patient demonstrates resistance to at least two hormones. He has a high parathyroid hormone level, which should raise serum calcium concentration, but his serum calcium is low. In addition, he has a high TSH level, which should raise serum thyroxine, but his thyroxine level is low. Since he is producing these hormones, they must be inactive either because the hormones lack activity or because the cells are incapable of responding to the hormone. It would be unlikely that two distinct hormones would have lost biologic activity, and therefore the patient must have a defect in the cellular response mechanism.

The defect could be at the receptor, but each of these hormones has a distinct receptor, and it would be unlikely that the patient would have two independent receptor defects. Both the hormones work by causing the cell to produce cAMP, and therefore the most probable defect would be one of ineffective cAMP. Either he cannot produce cAMP, or the cAMP that he does produce is ineffective in the cell. Thus the defect could be in the G-stimulatory protein or the adenylate cyclase, so that cAMP cannot be produced. Alternatively, it may be in the cAMP-dependent protein kinase, so that the cAMP is ineffective.

2. Proof of the problem One could test these theories by measuring cAMP in the urine in response to a large dose of parathyroid hormone. If the cAMP rises significantly, the patient produces the cAMP and a defect must exist in the protein kinase or in a subsequent system. All the defects would also be seen in cells taken from the patient and placed into tissue culture. If a defect were present in protein kinase, the cells would not respond to a cAMP analogue, dibutryryl cAMP. One could also measure the adenylate cyclase by sodium fluoride stimulation or directly measure the G-stimulatory protein.

3. Other potential problems One might expect the response of other hormones that work through cAMP to also be influenced. Thus the patient might become hypoglycemic, and his seizures could have been caused either by hypoglycemia (low blood sugar)—from a lack of glucagon action, a lack of epinephrine action, or lack of cortisol (because of absent ACTH action)—or by the low calcium concentration. In addition, the patient might be sterile, since LH and FSH would not be active on the gonadal cells.

In reality this patient has a disease called pseudohypoparathyroidism. It is caused entirely by a defect in the G_s-stimulating protein of the adenylate cyclase system. A defect that affected the Gs protein of every endocrine system and totally blocked all the hormones just described would be lethal, and the child would not develop to the age of 4 yr. In pseudohypoparathyroidism, parathyroid hormone is always severely affected, and other hormones are involved to a milder degree. In this case one would treat the patient with 1,25-dihydroxycholecalciferol to raise his serum calcium concentration. Thyroxine also would be given to alleviate his hypothyroidism. The mental retardation that occurs in children having his disease is often mild but irreversible.

References

Albright F, Forbes AP, and Henneman PH: Pseudohypoparathyroidism, Trans Assoc Am Physicians 66:337, 1952.

Levine MA et al: Genetic deficiency of the alpha subunit of G_s as the molecular basis for Albright's hereditary osteodystrophy, Proc Natl Acad Sci 84:365, 1988.

Case 2   Lack of hormone feedback

A 13-year-old female had a history of classical symptoms of hyperthyroidism from birth. She had always been nervous, tremulous, and very thin despite an extraordinary appetite; she also had cardiac rhythm disturbances. She had noted some neck enlargement, which had worsened recently. Physical examination confirmed the symptoms; the patient was fidgeting with wet, hot skin. She had a marked stare and bulging eyes. Her thyroid gland was enlarged about twofold. Her heart rate was rapid at 140 and irregular. Laboratory examination showed a free thyroxine level of 4.7 ng/dl (normal, 0.6 to 2.1) and a TSH level of 15 μIU/ml (normal 0.5 to 5.8). A computed tomogram and magnetic resonance image of the pituitary gland were normal.

Biochemical questions

1. Describe the biochemical abnormalities that might cause clinical hyperthyroidism.
◆ 2. How would you demonstrate the abonormality?

Case discussion

1. Biochemical abnormalities Hyperactivity of a hormone is caused by either excess levels of hormone or hypersensitivity to the hormone. Excess levels of hormone can only result from abnormal feedback inhibition of hormone secretion. All other problems with excess secretion or diminished degradation would be corrected rapidly by a normal feedback inhibition system. Abnormalities of feedback inhibition may be caused by (a) production of the hormone by an abnormal cell such as a tumor cell, (b) stimulation of the hormone by a substance such as an antibody that is not subject to feedback inhibition, or (c) a defect in the secretory cell, leading to a lack of feedback inhibition. In the case of hyperthyroidism all these defects exist. Rarely, some patients have hyperthyroidism because of excess thyroid hormone secretion by thyroid tumors such as papillary thyroid carcinoma. More often, these patients' thyroid hormone is stimulated by antibodies to the TSH receptor (Graves' disease). Also rarely, patients may have a defect in feedback inhibition in an otherwise normal thyroid axis.

Hyperthyroidism, the constellation of symptoms associated with excess thyroid hormone action, usually results from excessive function of the thyroid gland. This is caused by an autoimmune process in which antibodies to the TSH receptor bind and activate it, called Graves' disease. In this condition the very high thyroid hormone level inhibits the production of TSH by feedback inhibition. The high TSH level in this patient eliminates this possibility and implies a pituitary source of the disease, most often a tumor. The computed tomogram and magnetic resonance image of the pituitary gland were normal, however, suggesting that this is not the problem. The most likely diagnosis, therefore, is a pituitary insensitivity to the feedback inhibition of the thyroid hormones. Since the thyroxine does not shut off TSH, the TSH will continue to stimulate the production of thyroxine, until a level is reached that will cause feedback inhibition.

2. Proof of abnormality One could prove this by treating the patient with additional thyroxine and demonstrating that TSH production can be inhibited. (This is a dangerous maneuver that should only be performed in the hospital.) Treating this patient with additional thyroxine gave a free thyroxine of 10 ng/dl and a TSH of 0.02 μIU/ml. Thus the pituitary gland will respond to thyroid hormone, but only at very high levels. It is resistant to the feedback inhibition of thyroxine to shut off TSH synthesis. In this case the patient was resistant to thyroid hormone at the pituitary gland but sensitive at the rest of the body, which is a very difficult condition to treat.

Reference	Gersherngorn MC and Weintraub BD: TSH-induced hyperthyroidism caused by selective resistance to thyroid hormone, J Clin Invest 56:633, 1975.

Case 3	Diabetes mellitus

A 23-year-old woman had excessive thirst, excessive urination, and a good appetite but a weight loss of 11 kg. Physical examination demonstrated that the patient was thin with dry skin and sunken eyeballs. She had a peculiar, dark, silky pigmentation over her knuckles, elbows, and the back of her neck, called *acanthosis nigricans*. The remainder of the examination was normal. A random blood sugar level was 34 mM (normal, 3.6 to 6 mM).

Biochemical questions	1. Describe the biochemical abnormalities that can lead to diabetes mellitus.
	◆ 2. How would you evaluate the possible causes of diabetes mellitus in this patient?

Case discussion	**1. Biochemical abnormalities** Diabetes mellitus is characterized by an elevated blood sugar level caused by ineffective insulin action. The ineffective insulin action may result from a lack of insulin, called type I diabetes, or insulin-dependent diabetes mellitus; or from a resistance to insulin, called type II diabetes, or non-insulin-dependent diabetes mellitus. Type I diabetes is usually caused by an autoimmune disease that destroys the cells that make insulin, and therefore patients lack insulin. A few patients have a mutation that inactivates insulin; these patients make a great amount of inactive insulin. Type II diabetes may be caused by a lack of insulin receptors or by a postreceptor defect, such as in tyrosine kinase or in the final pathway to the action of insulin (e.g., a lack of glucose transporters).

2. Evaluation of causes The first test should evaluate the amount of insulin in this patient. Most often the fasting insulin in a young diabetic patient would be immeasurable, but in this patient it is 67 μU/ml, a high level. This indicates that insulin is not lacking, but rather that an insulin is inactive or insulin resistance is present. Insulin resistance is confirmed by treating the patient with 7 units of intravenous insulin, which should drop the blood sugar by 50% in a normal person; however, it has no effect on the blood sugar level in this patient. Measurement of other hormones known to cause insulin resistance are normal, and no antibodies to insulin are present.

Since insulin resistance exists, an attempt is then made to determine if cellular defects in insulin action are present. A fat biopsy from this patient provides fat cells (adipocytes) for study. Treatment of cells with insulin demonstrates that insulin is ineffective on the cells. Many patients have abnormalities of receptor number and less frequently receptor affinity. A study of insulin receptors, however, shows them to be normal. Thus a defect exists beyond the receptor, a *postreceptor defect*. This patient has a condition described as "type C insulin resistance with acanthosis nigricans."

Reference

Truglia JA, Livingston JN, and Lockwood DH: Insulin resistance: receptor and post-binding defect in human obesity and non-insulin dependent diabetes mellitus, Am J Med 79 (suppl 2B): 13, 1985.

Case 4   Diabetic ketoacidosis in a newborn

A newborn developed respiratory distress within a few hours after birth. The baby was very small, weighing only 1.7 kg despite being full-term, and had a strange appearance, somewhat resembling a miniature leprechaun. He was breathing deeply and had fruity breath. The parents were healthy, and there was no family history of consanguinity, unusual genetic diseases or diabetes. The newborn had been wetting his diaper excessively. Physical examination revealed a rapid, weak pulse and low blood pressure, which was unmeasurable on sitting the baby up. Respiratory distress also was present, with rapid, very deep breathing, and the baby was using accessory muscles. Laboratory examination demonstrated a blood glucose level of 40 mM (normal, 3.5 to 6 mM), elevated serum ketones, and a pH of 7.0 (normal, 7.40 to 7.45), indicating diabetic ketoacidosis. The patient was given intravenous rehydration and large amounts of insulin, but he continued to deteriorate, with increasing blood sugar levels. He died after about 6 hr of therapy from cardiac arrest and had a pH of 6.5.

Biochemical questions

1. Describe possible biochemical explanations for the ineffectiveness of the insulin. Consider all possibilities and describe how you would distinguish between them.
♦ 2. After the baby had died, an insulin level, drawn before the institution of therapy, was found to be 6 mM (normal fasting <0.1 mM). How would this finding alter your discussion?

Reference

Schilling EE et al: Primary defect of insulin receptors in skin fibroblasts cultured from an infant with leprechaunism and insulin resistance, Proc Natl Acad Sci USA, 76:5877, 1979.

Case 5   Breast cancer

A 55-year-old woman sought medical attention because of a lump in her left breast. Excisional biopsy was positive for breast cancer, and one lymph node was found to have the disease. The remainder of the workup for the metastatic cancer was negative.

A biochemical evaluation of the tumor showed that it was positive for estrogen receptors. The patient received radiotherapy to the area of the tumor and was treated with tamoxifen, a steroid that binds to the estrogen receptor but does not activate it. Her tumor shrunk in size, and she made a full recovery.

Biochemical questions

1. Describe the mechanism of action of estrogens.
2. How might tamoxifen work to reduce estrogen action in the breast?
2. What other mechanisms might be used to reduce estrogen action?

Reference

Namer M et al: Increase of progesterone receptor by tamoxifen as a hormonal challenge test in breast cancer, Cancer Res 40:1760, 1980.

Case 6   Myasthenia gravis

A 34-year-old physician developed weakness in his hands and feet without any loss of sensation. He also had drooping of his eyelids. The weakness increased with use of any muscle, making exercise virtually impossible. He was diagnosed as having myasthenia gravis. A test of his serum showed antibodies to the acetylcholine receptor. His cells, even after washing off any antibodies, still demonstrated a diminished number of acetylcholine receptors. He was treated with plasmapheresis (replacing his plasma with normal plasma) and improved considerably.

Biochemical questions ◆ 1. Describe mechanisms that might account for the loss of acetylcholine receptors.
◆ 2. Contrast the various factors that might affect the action of acetylcholine, a neurotransmitter, with factors that might affect the action of a hormone.

Reference

Vincent A: Acetylcholine receptors and myasthenia gravis, Clin Endocrinol Metabol 12:56, 1983.

Case 7   Thyroid hormone resistance

A variant of the disease described in case 2 exists in which total body resistance to thyroid hormone is present. Both the pituitary gland and the cells of the body are resistant to thyroid hormone.

Biochemical questions

1. Describe the probable history and physical examination of this patient.
2. What would be the levels of thyroxine and TSH?

Additional questions and problems

1. Some diseases are caused by stimulation of the endocrine system by substances that are not subject to feedback regulation. For example, in Graves' disease, antibodies act to stimulate the TSH receptor. Describe the consequences of situations in which there is more of this antibody activity than is normally produced by TSH. What would happen if it were exactly equal? Less?
2. A scientist discovers that a peptide identical to atrial naturetic factor in the supraoptic nucleus of the hypothalamus. How would you determine if it is a neurotransmitter or hormone? How would you decide if it were made by the area or accumulated by it?
3. A baby is born with a genetic defect in the ability to alter tyrosine in any way. What types of endocrine problems might this newborn have?
4. Women taking oral contraceptives have increased amounts of binding proteins.

Which hormones will have their serum levels raised by this? Which would have their biologic actions affected?

5. What would be the consequences of a G_i protein with increased activity and decreased sensitivity to hormones?

6. During chelation therapy for removal of lead, a child with lead toxicity develops zinc deficiency. Since zinc is necessary for the structure of the DNA-binding sites of the steroid receptor, the patient will develop some clinical problems. Describe them.

7. Patients with dysautonomia have a defect in the action of epinephrine and norepinephrine. They have a slow heart rate and low blood pressure and become dizzy on standing. Describe potential sites for this dysfunction.

Multiple choice problems

1. A family has a genetic defect in the G-coupling protein of the adenylate cyclase. The action of which of the following hormones would be directly affected?
   A. Thyroxine
   B. Cortisol
   C. Insulin
   D. Epinephrine
   E. All of the above

2. Which of the following may be involved in the intracellular transmission of hormonal action?
   A. cAMP
   B. Inositol triphosphate
   C. Tyrosine kinase
   D. Protein kinase C
   E. All of the above

3. Which of the following are synthesized as glycoproteins?
   A. Lutropin
   B. Thyroxine
   C. Insulin
   D. Calcitonin
   E. All of the above

4. You have been culturing a special line of irreplaceable cells for over a month. Your laboratory technician accidentally treats them with trypsin. You have been planning receptor studies on the cells. Which of the following hormone receptors could you still study?
   A. Insulin
   B. Parathyroid hormone
   C. Cortisol
   D. Glucagon
   E. All of the above

5. Which set of hormones use the same second message?
   A. Insulin and glucagon
   B. Parathyroid hormone and vasopressin
   C. TSH and growth hormone
   D. Insulin and prolactin
   E. Growth hormone and FSH

6. Which set of hormones are all of *different* types?
   A. Insulin, glucagon, epinephrine, and estradiol
   B. FSH, cholecalciferol, thyroxine, and epinephrine
   C. TSH, testosterone, progesterone, and cholecalciferol
   D. Growth hormone, prolactin, estradiol, and cholecalciferol
   E. Triiodothyronine, thyroxine, melatonin, and insulin

7. The "second messenger" for insulin is:
   A. Cyclic AMP.
   B. Tyrosine kinase.
   C. Serine kinase.
   D. Cyclic GMP.
   E. Diacyl glycerol.

8. A new pharmaceutical firm plans to market drugs for oral administration. Which of these hormones could be used?
   1. Insulin
   2. Cortisol
   3. Growth hormone
   4. Estradiol
      A. 1, 2, and 3.
      B. 1 and 3.
      C. 2 and 4.
      D. 4 only.
      E. All are correct.

9. A 24-year-old woman comes to see you for symptoms of increased nervousness and increased appetite. A pregnancy test is positive. You are worried that she might have excess thyroid function, but you know that in pregnancy the thyroid-binding proteins are all elevated. If she were hyperthyroid you might expect to see a laboratory profile that shows:
   1. Elevated total thyroxine.
   2. Elevated free thyroxine.
   3. Elevated triiodothyronine.
   4. Diminished TSH.
      A. 1,2, and 3.
      B. 1 and 3.
      C. 2 and 4.
      D. 4 only.
      E. All are correct.

10. Diabetes mellitus could be caused by:
    1. Insufficient secretion of insulin.
    2. Loss of insulin receptors.
    3. Increased degradation of insulin.
    4. Ineffective insulin receptors.
    A. 1, 2, and 3.
    B. 1 and 3.
    C. 2 and 4.
    D. 4 only.
    E. All are correct.

Chapter 18

Molecular endocrinology: hormones active at the cell surface

Barry H. Ginsberg

Objectives

1 To explain the chemistry of the hormones acting at the cell surface
2 To describe the gene structure when it is known
3 To explain the regulation of hormone secretion
4 To understand the actions of hormones acting at the cell surface
5 To describe disease states associated with loss or overproduction of hormones

Hypothalamic and pituitary hormones

Although the hypothalamus and pituitary glands are spatially separated, they should be considered as an integral unit because they function together for optimum action. Figure 18.1 shows the anatomy of the hypothalamic-pituitary area.

Hypothalamus

The hypothalamus produces at least eight hormones: vasopressin (AVP, ADH), oxytocin, thyroid-releasing hormone (TRH), gonadotropin-releasing hormone (GnRH, LHRH), corticotropin-releasing hormone (CRH), somatropin-releasing hormone (SRH, GHRH), somatostatin (SRIH, SS), and dopamine. These are synthesized in specialized areas of the hypothalamus, as indicated in Table 18.1. The hypothalamic hormones can be separated into two categories:

1. The hypothalamic neuropeptides which are transported down the pituitary stalk in axons of cells that originate in the hypothalamus and terminate in the posterior pituitary gland where they are secreted into the blood (vasopressin and oxytocin)
2. The hypothalamic-releasing hormones, which are made in the hypothalamus, released into the hypothalamic-pituitary portal circulation, and have their effects on the secretion of hormones by the pituitary gland (TRH, GnRH, SRH, CRH, SRIH, and dopamine).

Hypothalamic neuropeptides

The hypothalamic neuropeptides, vasopressin and oxytocin, are synthesized in the cell bodies of neurons in the supraoptic and paraventricular nuclei of the hypothalamus and transported down the axons of these neurons to the posterior pituitary gland, where they are released. The two neuropeptides share many features. Both vasopressin and oxytocin are synthesized by standard methods of protein synthesis as 20,000-dalton precursor molecules. This precursor has the specific neuropeptide and neurophysin and in the case of vasopressin, an additional glycopeptide. The precursor molecule is synthesized in the neuron cell body and packaged into secretory granules. As the granules pass down the axon, the neuropeptide is cleaved from the neurophysin and glycopeptide. Each neurophysin tightly binds to its respective neuropeptide. The neurophysin cleaved from the precursor of vasopressin binds tightly to vasopressin, and the neurophysin cleaved from oxytocin binds tightly to oxytocin. This tight binding occurs only during the transport of these peptides in the secretory granule as it traverses the length of the axon. The axons terminate in the posterior pituitary gland, and the neuropeptides are stored there along with their respective neurophysin. After appropriate stimulation, the secretory granules are exocytosed, releasing both the neuropeptide and the neurophysin. Once released from

Figure 18.1

Location of some of the nuclei and areas of the hypothalamus. The hypothalamus can be divided into specific nuclei and areas that often have well-defined functions (see Table 18.1). In this diagram of a longitudinal section of the hypothalamus, the nucleii involved in hormonal secretion are labeled. *1,* Hypothalamus; *2,* arcuate nucleus; *3,* preoptic area; *4,* ventromedial nucleus; *5,* superchiasmic area; *6,* paraventricular nucleus; *7,* optic chiasm; *8,* mammillary body; *9,* posterior pituitary; *10,* anaterior pituitary.

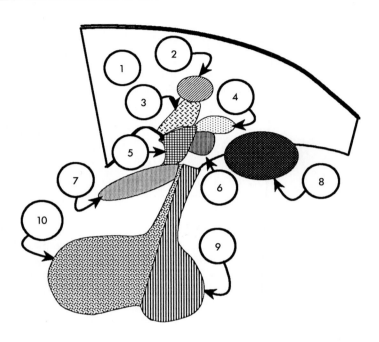

the posterior pituitary gland the neuropeptide and neurophysin are diluted and separated. The neurophysins do not act as binding proteins for oxytocin and vasopressin in blood. The total time from synthesis to secretion averages about 1.5 hr.

Vasopressin Also known as arginine vasopressin (AVP) or antidiuretic hormone (ADH), vasopressin is a cyclic nonapeptide (Figure 18.2) with a molecular weight of 1084. The gene for vasopressin contains three exons and two introns. The entire vasopressin peptide is coded in the first exon at the *N* terminus, following the signal peptide. The carboxyl-terminal glycine is blocked with an amide group, and both this amide-blocking group and the disulfide bond must be intact for biologic activity. The remainder of the gene codes for the neurophysin and additional glycopeptide.

The main biologic activity of vasopressin is antidiuresis, which causes a concentrated urine; thus it is often called antidiuretic hormone. It works at the distal tubule, increasing permeability of the tubule to water. The AVP binds to a specific receptor on the plasma membrane of the distal tubule of the nephron and activates adenylate cyclase. Through a series of phosphorylations, the first of which is performed by a cyclic AMP–dependent protein kinase, microtubules and microfilaments are rearranged. By mechanisms that are still not clear, this leads to the insertion of water-conducting particles into the luminal

Table 18.1   Classification, localization, and actions of the hypothalamic hormones

Hormone	Abbreviation	Localization	Actions*	Second messenger*
Activating releasing hormones				
Thyroid-releasing hormone	TRH	Paraventricular nucleus	Regulates TSH and prolactin	PI cycle
Gonadotropin-releasing hormone	GnRH, LHRH	Preoptic and septal areas	Regulates LH and FSH	PI cycle
Corticotropin-releasing hormone	CRH	Paraventricular nucleus, preoptic area, supraoptic nucleus	Regulates ACTH and endorphins	cAMP
Somatotropin-releasing hormone	SRH, GHRH	Arcuate nucleus	Regulates somatotropin	
Inhibiting releasing hormones				
Dopamine	—	Arcuate nucleus	Regulates prolactin (?ACTH)	cAMP
Somatostatin	SRIH, SS	Anterior periventricular area, preoptic area	Regulates prolactin and TSH	
Hypothalamic neuropeptides				
Vasopressin	AVP, ADH	Supraoptic nucleus	Regulates water balance	cAMP PI cycle
Oxytocin	—	Paraventricular nucleus	Causes uterine contractions	

*Abbreviations: *TSH*, thyroid-stimulating hormone; *LH*, luteinizing hormone; *FSH*, follicle-stimulating hormone; *ACTH*, adrenocorticotropin; *cAMP*, cyclic adenosine monophosphate; *PI*, phosphatidylinositol.

membrane. Since these portions of the nephron permeate hyperosmolar regions of the kidney, water will be withdrawn from the urine. This results in a more concentrated urine, an antidiuretic action.

Vasopressin also reacts with smooth muscle of the vascular system to cause contraction and constriction of the blood vessel. Although the sensitivity of the muscle to ADH is low and this action is not important at normal levels of ADH, some localized variations occur. The splanchnic circulation is exquisitely sensitive to ADH and can be influenced by physiologic concentrations of ADH. Clinically, patients are sometimes given ADH to specifically constrict the splanchnic circulation in a gastrointestinal hemorrhage, such as that resulting from a peptic ulcer.

The major factor regulating vasopressin secretion is the osmolality of the blood. Increases in osmolality dramatically raise the secretory rate, and decreases in osmolality suppress secretion. Decreases in blood volume or blood pressure also stimulate ADH secretion.

Clinical comment

Diseases associated with altered vasopressin secretion Alterations in ADH secretion can cause serious, even life-threatening illness. Excess ADH leads to a syndrome of inappropriate ADH (SIADH). This can be caused by head trauma, tuberculosis, tumors, and various other conditions. Too much ADH is secreted, resulting in the retention of water without the balancing effect of electrolytes such as sodium, potassium, and chloride. The net result is an expansion of body fluids and a dilution of the electrolytes, resulting in a fall in osmolality. Although normal serum sodium concentration is 140 mEq/L, with

Figure 18.2   Structure of the hypothalamic neuropeptides. The primary sequences
of arginine vasopressin; oxytocin; and d-arginine, d-asparagine
vasopressin (D,D AVP), a long-lasting agonist, are shown. Note the
similarities among them.

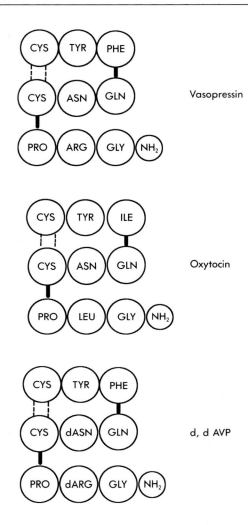

an osmolality of 290 to 295 mOsm, patients with SIADH may have much lower values. When the serum sodium concentration falls below 110 mEq/L, or the osmolality below 240 mOsm, the patient will have seizures. If it falls substantially lower than this, the patient will die.

Loss of the ability to produce ADH causes hypothalamic diabetes insipidus (see case at end of the chapter). This condition may be hereditary or may be caused by tumors of the pineal gland, other tumors or masses in the hypothalamus, neurosurgery, trauma, or infections. In each of these cases the ability of the hypothalamus to secrete ADH at the posterior pituitary gland is inhibited. In some cases, such as following pituitary surgery or trauma, the severed nerves from the hypothalamus may form new nerve endings and secrete ADH from an abnormal location. In these patients the diabetes insipidus is only temporary.

Patients with diabetes insipidus cannot concentrate their urine and secrete large amounts, up to 10 to 20 L/day. The volume of fluids may overload the urinary tract, leading to distention of the bladder and ureters, which may be permanent. In addition, since the problem often lies in the hypothalamus, these patients may have impaired thirst and may not drink enough to prevent severe dehydration. If the dehydration is severe enough, the patients may become comatose and even may die.

Oxytocin Oxytocin is very similar to vasopressin in size, gene structure, and properties. It is synthesized in the same hypothalamic nucleii as vasopressin, although each hormone appears to be made in different cells. The main biologic effects of this cyclic nonapeptide appear to be in parturition and lactation.

During the maturation of the uterus in pregnancy, the number of receptors for oxytocin increases by 100 to 200-fold. Therefore even small increases in this hormone have major physiologic significance. The release of oxytocin by the posterior pituitary gland during labor causes increased contraction of the uterus, leading to delivery. Oxytocin is often given by obstetricians to hasten progression of a stalled delivery. After birth, oxytocin is important for the appropriate delivery of milk, and suckling by the infant leads to increased oxytocin secretion. The oxytocin acts on myoepithelial cells of the breast to squeeze milk from the glands to the ducts, making it available to the infant.

No diseases are associated with an excess of oxytocin. Lack of oxytocin is poorly characterized. Although one would suspect that a lack of hormone could lead to poor progression of labor, this has not been well documented.

Hypothalamic-releasing factors

The hypothalamic regulatory factors are listed in Table 18.1. They vary in size and chemical composition. All are synthesized in the hypothalamus, usually in very specific regions, and are transported to the pituitary gland via the hypothalamic-pituitary portal circulation. All the peptide and polypeptide hormones have a blocked carboxyl terminus. The half-lives of these hormones are extremely short, often only seconds. The individual hormones are discussed in more detail along with their appropriate pituitary hormone.

Anterior pituitary hormones

The anterior pituitary gland produces six hormones (Table 18.2); all are peptides, proteins, or glycoproteins, and all are under the regulation of the hypothalamic regulatory hormones. All six are activated by the hypothalamus, and two are also inhibited by these regulatory hormones. Table 18.3 lists the stimulators and inhibitors of the pituitary hormones.

Thyrotropin Thyrotropin, also called thyroid-stimulating hormone (TSH) is one of the three glycoproteins produced by the pituitary gland. Its action increases the secretion of the thyroid hormones by the thyroid gland. Thyrotropin also increases the growth of the thyroid gland. The hormone acts by first binding to a glycolipid and then to a specific receptor on the cell surface, activating adenylate cyclase and increasing intracellular AMP.

The TSH molecule is a glycoprotein of molecular weight 30,000, with about 15% carbohydrate. It is composed of two subunits, each approximately 15,000 daltons. The subunits, designated α and β, are linked by noncovalent bonds. Both lack biologic activity alone. The α-subunit is common to all the pituitary glycoproteins and chorionic gonadotropin, and the specificity of the TSH therefore results from the β-subunit. The β-subunit is encoded in a single gene, with great homology among different animal species.

TSH secretion is stimulated by thyroid-releasing hormone (TRH) and inhibited by somatostatin and the thyroid hormones. TRH is a tripeptide, (pyro) glutamyl histidinyl prolineamide. It is blocked at both the amino terminus by the pyro ring of the glutamic

Table 18.2 Classification, localization, and actions of the pituitary hormones

Hormone	Abbreviation	Target	Actions	Properties	Second messenger
Thyrotropin	TSH	Thyroid gland	Increases thyroxine, triiodothyronine, thyroid growth	Glycoprotein Two chains 30,000 daltons	cAMP
Follitropin	FSH	Gonads	Increases spermatogenesis, follicle growth	Glycoprotein Two chains 30,000 daltons	cAMP
Lutropin	LH	Gonads	Increases estrogens, progesterone, testosterone	Glycoprotein Two chains 30,000 daltons	cAMP
Human chorionic gonadotropin	hCG	Placenta	Maintains placenta	Glycoprotein Two chains 30,000 daltons	cAMP
Somatotropin	hGH	Liver, kidney, endothelium, other	Increases somatomedin C	Protein Single chain 22,000 daltons	Unknown
Prolactin		Breast	Increases milk production	Protein Single chain 23,000 daltons	Unknown
Adrenocorticotropic hormone, adrenocorticotropin	ACTH	Adrenal cortex	Increases cortisol growth	Polypeptide Single chain 4100 daltons	cAMP

Table 18.3 Regulation of the pituitary hormones

Hormone	Stimulators	Inhibitors
Thyrotropin	TRH	SRIH, thyroxine, triiodothyronine
Follitropin	GnRH	Inhibin
Lutropin	GnRH	Estradiol
Chorionic gonadotropin	—	—
Somatotropin	SRH	SRIH Somatostatin C
Prolactin	TRH	Dopamine
ACTH	CRH	Cortisol

acid and at the carboxyl terminus by the amide. It is synthesized as a 29,000-dalton precursor molecule that contains five copies of the TRH molecule. TRH stimulates the secretion of both TSH and prolactin. TRH also seems to activate the glycosylation of TSH, and a glycosylation-poor TSH of decreased activity may be released in hypothalamic disease.

Somatostatin will block the secretion of TSH, but the most potent inhibitor of TSH release are the thyroid hormones themselves. If thyroxine or triiodothyronine are high, the pituitary gland will not secrete TSH, no matter how high the level of TRH. This is the basis for the TRH stimulation test used to detect very subtle alterations in thyroid function (Figure 18.3).

Clinical comment

Abnormalities of TSH Most thyroid problems are caused by disease of the thyroid rather than the pituitary gland. Excessive levels of TSH usually result from an inability of the thyroid gland to produce the thyroid hormones and subsequent lack of feedback inhibition by these hormones. In this conditon, called *primary hypothyroidism,* TSH levels may reach 100 ng/ml (normal, 0.5 to 5.8 ng/ml). A high TSH level may be one of the most accurate and subtle indications of thyroid disease and may precede clinical hypothyroidism by months to years. Rarely, pituitary tumors produce TSH, causing hyperthyroidism. Also, one group of patients has a resistance to thyroid hormone, as described in Chapter 17, case 2.

Figure 18.3

Thyroid-tropin-releasing hormone (TRH) test. In some patients it is necessary to learn about the responsiveness of the pituitary gland to TRH. The test for this is performed by administering 0.5 mg of TRH intravenously and measuring TSH just before the administration and every 10 to 15 min for an hour. A normal curve, labeled *euthyroid,* will demonstrate a peak level of TSH of about 10 to 25 ng/ml approximately 15 to 20 min after TRH administration. If the pituitary gland is not functioning properly, either because of pituitary disease or excess thyroid hormone (which feeds back to shut off the pituitary thyrotrophs), no stimulation will occur *(hyperthyroid).* Finally, if the pituitary gland is overactive, usually because of a lack of thyroid hormone and its feedback, an excessive production of TSH will occur, which is often late (30 to 40 min) as well *(hypothyroid).*

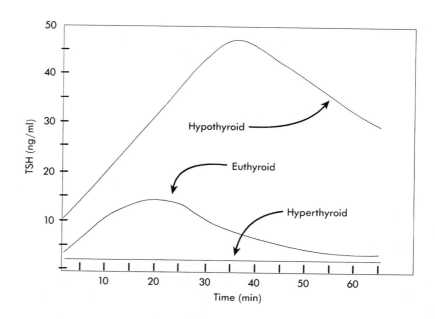

A diminished amount of TSH is found most often in patients with hyperfunction of the thyroid gland, which produces hyperthyroidism. The excess thyroid hormones feed back on the pituitary glands to shut off TSH production. Occasionally a tumor or other mass in the region of the pituitary decreases TSH secretion, resulting in hypothyroidism. This is usually caused by destruction of the hypothalamic-pituitary portal circulation, with resultant loss of TRH stimulation, or by direct pressure of the tumor on the pituitary gland.

The gonadotropins Of the three gonadotropins, two are produced by the pituitary gland and one by the placenta. The two pituitary hormones are lutropin, or luteinizing hormone (LH), and follitropin, or follicle-stimulating hormone (FSH). The placental hormone is human chorionic gonadotropin (hCG). All these hormones work in a complex fashion. In females they function primarily on the ovary to trigger the production of the ovum and to support implantation and growth of the fetus.

The structure of the gonadotropins is similar to TSH. They are glycoproteins of molecular weight about 30,000 and are composed of two subunits. The α-subunit is identical for all three hormones and is also identical to the α-subunit of TSH. The β-subunit provides the specificity of the hormone. In measuring these hormones, the better assays are specifically directed against the β-subunit. Indeed, a pregnancy test is a specific test for the β-subunit of hCG.

The gonadotropins bind to specific receptors on the ovarion and testicular cells. Different receptors exist for LH and FSH. LH acts to increase intracellular cyclic AMP. The binding of LH is of very high affinity, and some scientists believe that the hormone-receptor complex is not dissociated but rather degraded as a unit.

Biologically, FSH is a sexual "growth hormone," promoting the growth of spermatocytes in the testes and follicles in the ovary. LH acts to produce steroid hormones, promoting production of testosterone by the Leydig cells of the testes and production of progesterone by the corpus luteum of the ovary. hCG is a placental hormone that acts much as LH does. hCG has a very small amount of FSH-like activity, but this is insufficient to be useful clinically.

The main stimulus for the secretion of both LH and FSH is gonadotropin-releasing hormone (GnRH), a decapeptide with both a blocked amino and a blocked carboxyl terminus: (pyro) Glu-His-Trp-Ser-Tyr-Gly-Leu-Arg-Pro-Gly-NH$_2$. GnRH is a synthesized as the first 10 amino acids of a 10,000-dalton precursor and is specifically cleaved from the prohormone. GnRH seems to bind to a specific receptor on the pituitary gonadotrophs and stimulates the phosphatidyl-inositol cycle, increasing intracellular calcium and protein kinase C, which mediates an increase in gonadotropin secretion, particularly LH.

The regulation of the pituitary gonadotropins is still somewhat unclear, but certain peculiar patterns emerge. For the GnRH to be active, it must be given intermittently for only a few minutes of each hour. Constant stimulation by GnRH leads to desensitization of the pituitary gland and cessation of LH and FSH production (Figure 18.4). Long-lasting agonists (active compounds with similar structure) of GnRH can be used for birth control and to prevent early (precocious) puberty through decreased release of pituitary gonadotropins and decreased gonadal steroid synthesis. GnRH appears to be much more active on LH than on FSH and another as yet unidentified hypothalamic factor may stimulate FSH release.

Steroids feed back to block production of the gonadotropins. In males, testosterone will block secretion of LH. In females, both estradiol and progesterone will individually inhibit the secretion of LH, but when given together, they are much more potent than either alone. This synergy of these two hormones is the biochemical basis for the combined use of estrogens and progestins in oral contraceptives.

Figure 18.4.

Effect of gonadotropin-releasing hormone (GnRH) on lutropin (LH) secretion. GnRH is most effective when given for about 6 min of every hour. As seen in the upper panel, such administration results in a dramatic rise in LH levels. When the GnRH is given continuously, as shown in the lower panel, it actually inhibits the secretion of LH.

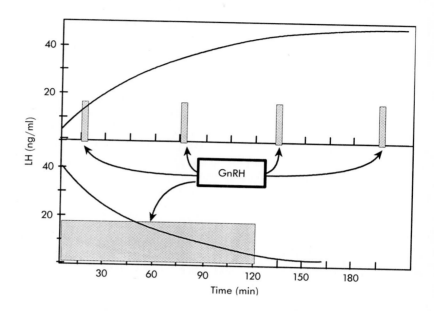

Another hormone involved in the regulation of the gonadotropins is a protein made by the rapidly growing Sertoli cells of the testis and granulosa cells of the developing ovarian follicle, called inhibin. This protein feeds back directly on the pituitary gland to shut off FSH secretion. Inhibin is a glycoprotein composed of two subunits. It exists in two forms; the form active in inhibiting FSH secretion is an α, β-dimer. The β-subunit is highly homologous with β-transforming growth factor and Müllerian inhibiting factor. Inhibin has recently been cloned, and our knowledge of its regulatory effects are likely to increase substantially in the next few years.

Clinical comment

Problems associated with gonadotropin production Overproduction of the pituitary gonadotropins is seen most frequently in postmenopausal females. The menopause is associated with ovarian failure and a lack of feedback inhibition by the ovarian factors (steroids and inhibin) on the hypothalamic-pituitary axis. Thus these woman have low estrogen and progesterone levels, a low inhibin level, and high LH and FSH levels. Testicular failure in men is also associated with a low testosterone level and no spermatogenesis but high LH and FSH levels.

Overproduction of chorionic gonadotropin is a common feature of many types of tumors, both eutopic and ectopic (see Chapter 17).

Underproduction of the pituitary gonadotropins occurs in several conditions. The most interesting of these is called Kallmann's syndrome (see case 2). In all patients with gonadotropin underproduction, prepubertal onset is associated with failure to mature sexually or to grow normally. Males do not develop an adult male habitus or muscle pattern. Phallic or testicular enlargement and spermatogenesis do not occur. Females do not develop breasts or menstruate, although a normal sexual hair pattern is present (controlled by the adrenal androgens). Postpubertal gonadotropin failure is associated with a loss of libido, impotency in males, and amenorrhea in females.

Somatotropin Somatotropin, or human growth hormone (hGH), is one of a family of closely related hormones that probably arose from a primordial gene and also includes prolactin and chorionic somatomammotropin (also known as human placental lactogen, hPL). hGH is highly specific for each species, and only primate hGH is effective in humans. The main action of hGH is to trigger growth of the organism, but it does this indirectly, raising the levels of somatomedin C (synonymous with insulin-like growth factor I, IGF-I) which in turn causes growth of the long bones and soft tissues. The overall effect of hGH is to stimulate amino acid uptake by cells and produce anabolic effects on protein metabolism. It also increases lipolysis by fat cells and causes a resistance to insulin in most of the cells in the body. hGH stimulates calcium absorption by the gut and has a small effect on breasts to stimulate milk production.

Somatotropin is a single-chain protein of molecular weight 22,000 with two intrachain disulfide bonds but no carbohydrate. There is an 83% homology between the amino acids of hGH and a placental somatotropin-like molecule, hPL. Interestingly, prolactin, the other member of this family, has only 16% homology between its amino acids and hGH. Both hGH and hPL are encoded in proximity on chromosome 17. Apparently, differential splicing of the hGH gene forms slightly varying forms of hGH (Figure 18.5) The hGH gene has been cloned, and the cDNA is used commercially to make hGH for human use.

Somatotropin synthesis and secretion is dramatically increased by the hypothalamic hormone somatotropin-releasing hormone (SRH or GHRH). This protein hormone, which was relatively elusive in being discovered, is a 44-amino-acid peptide with an amidated carboxyl terminus. It acts on the pituitary somatotrophs by binding to a highly specific receptor and activating adenylate cyclase to raise cAMP levels and stimulate the relase of hGH. Thyroid hormone and cortisol are also necessary for the synthesis and secretion of hGH. In their absence, hGH is not secreted, and the patients are clinically similar to patients who have a hGH deficiency.

The major inhibitory factor of somatotropin secretion is somatostatin. As one might imagine from its name, somatostatin was discovered based on this inhibition. It is a 14-amino-acid cyclic peptide; the intrachain disulfide bond is not necessary for biologic activity. A 28-amino-acid form also exists, which seems to be more prevalent in extra-neural tissues, whereas the 14-amino-acid form is found primarily in neural tissues. The two forms of somatostatin have different receptors. When the inhibition of secretion of hGH by somatostatin is released, a rebound hypersecretion occurs, suggesting that somatostatin does not inhibit synthesis of hGH.

Somatostatin will partially inhibit TSH release by the pituitary gland. It is also synthesized in the D cells of the islets of Langerhans in the pancreas and will inhibit insulin and glucagon secretion. Thus, in the pancreas, somatostatin is acting as a paracrine hormone. Somatostatin can also inhibit gastrin secretion by the gut and 13 other hormones at various locations in the body.

The secretion of hGH has a definite diurnal variation. Humans secrete the highest levels of hGH shortly after the onset of sleep, and secretion is greatest in sleep stages 3 and 4. The high levels of hGH cause an insulin resistance that occurs between 4 and 8 AM in most people. This is of clinical significance only in patients with diabetes mellitus and is called the *dawn phenomenon*.

Figure 18.5   Structure of the gene for human growth hormone (hGH). hGH is
made as two slighly different precursor molecules by differential gene
slicing as shown. In the smaller precursor, amino acides 31 to 46 are
spliced out of the sequence. mRNA, Messenger ribonucleic acid; K,
kilodaltons.

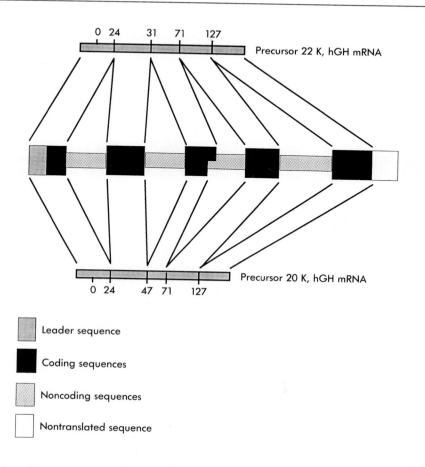

Leader sequence

Coding sequences

Noncoding sequences

Nontranslated sequence

Clinical comment

Problems associated with somatotropin production Underproduction of hGH before
puberty will cause the patient to be very short but well proportioned, a condition called
pituitary dwarfism. This condition may be caused by isolated hGH deficiency or by
panhypopituitarism (failure of all anterior pituitary hormones), which results from a tumor
or mass. These patients will respond to treatment with hGH, which must be injected a
few times each week. Since hGH is highly species specific, it initially was obtained only
from human pituitary glands but is now made by recombinant DNA technology.

Some patients closely resemble pituitary dwarfs but have very high levels of hGH.
These patients, called Laron dwarfs, have very low levels of somatomedin C and thus
appear to be resistant to the somatomedin C–generating action of hGH. Administration
of hGH will not trigger additional growth in these patients. In adults, loss of hGH secretion
usually has no effect.

Overproduction of hGH before puberty will lead to gigantism, a condition in which patients are very large but well proportioned. They have increased growth of long bones and soft tissues. After puberty, overproduction of hGH will not result in growth of the long bones because the epiphyses have closed. Instead, growth in bones that grow appositionally occurs, such as the skull. Thus these patients may have enlargement of the frontal sinuses and growth of the jaw, causing prognathism (protruding jaw) or a long, thin face. These patients also experience growth of soft tissues, which causes coarsening of facial features, and enlargement of hands and feet because of cartilage growth. Most will have glucose intolerance, and 20% will have diabetes.

Overproduction of hGH is almost always caused by large, benign tumors of the pituitary gland. Very rarely, tumors produce somatotropin-releasing hormone. Treatment is usually surgical removal of the pituitary tumor.

Prolactin Prolactin was first discovered in 1928, but only in 1971 was it separated from the chemically similar somatotropin and purified to homogeneity. In lower animals, prolactin acts to regulate water metabolism, but in humans it is responsible for the production of milk (lactation) by females. It has no known function in human males.

Prolactin is one of the members of a family of hormones with similar structure, including prolactin, somatotropin, and chorionic somatomammotropin. These three proteins probably were derived from a single primordial gene by gene duplication. This theory is further supported by the close relationship between the gene structures as well as the amino acid sequences of the proteins.

Human prolactin is a 23,000-dalton protein consisting of a single polypeptide chain. Overall, prolactin has a 16% homology to somatotropin by amino acid sequence but a much greater gene homology. The prolactin gene is found on chromosome 6 and is made up of about 10 kilobases. In contrast, the somatotropin gene is found on chromosome 17 and is made up of 2.5 kilobases. The difference in size of these two genes is caused entirely by the intervening sequences. The mRNA for both hormones is derived from five exons that have remarkable similarity.

Prolactin has a diurnal variation that is different from somatotropin. Whereas somatotropin is secreted shortly after the induction of sleep, prolactin has its greatest secretion a few hours later, with peak secretion at 5 to 7 AM. This periodicity is altered to match sleep patterns.

Prolactin is unique among anterior pituitary hormones in that the regulation of its secretion is normally inhibitory. Eliminating the hypothalamic-pituitary portal circulation, as might occur with a small tumor in the suprasellar region between the hypothalamus and pituitary, will result in the loss of secretion of all the anterior pituitary hormones except prolactin. The secretion of prolactin will increase dramatically. This is caused by the removal of the various hypothalamic factors that normally increase the secretion of most anterior pituitary hormones but block the secretion of prolactin. The major hypothalamic factor altering prolactin secretion is *prolactin-inhibiting factor,* or dopamine, a catecholamine of great prominence in the brain. Dopamine inhibits the pituitary lactotrophs by binding to a specific receptor and activating adenylate cyclase in these cells. In addition, dopamine inhibits the phosphatidylinositol cycle. Prolactin also has a short-loop negative feedback mechanism, since prolactin binds directly to cells of the median eminence and increases the secretion of dopamine, which in turn inhibits prolactin secretion.

Several factors stimulate the secretion of prolactin; the major one is TRH. This tripeptide, which has its major action on TSH release, also increases prolactin release, even at very low concentration. Conditions such as primary hypothyroidism that are associated with high TRH levels lead to increased prolactin synthesis and may even cause milk production in females. Patients with primary hyperthyroidism, which is associated with low TRH levels, have blunted prolactin secretion. TRH binds to a specific receptor on the lactotrophs and increases the phosphatidylinositol cycle. Within about 15 min after binding, a dramatic increase in the mRNA for prolactin occurs and decays within an hour. Another stimulator of prolactin is vasoactive intestinal peptide (VIP). This peptide, normally considered a gastrointestinal hormone, is found in high levels in the paraventricular nucleus of the hypothalamus and in the hypothalamic-pituitary portal circulation. Other hormones that may increase prolactin secretion include serotonin, oxytocin, substance P, endorphins, and melatonin. Currently, however, the role of all the stimulatory substances except TRH are considered speculative in humans.

Secretion of prolactin is greater in females than in males and varies with the menstrual cycle because of an increase in prolactin synthesis and secretion induced by estrogens. The estrogens even increase the number of prolactin-secreting cells. The effect of estrogens on prolactin secretion may be indirect; they may increase prolactin's sensitivity to stimulatory agents and reduce its sensitivity to inhibitory agents. Estrogens are of particular interest because they also increase prolactin receptors in many parts of the body. Therefore they not only increase prolactin secretion, but also increase sensitivity to prolactin.

Prolactin acts by binding to a specific protein receptor at the cell surface. The receptor is composed of two subunits of 41 and 88 kilodaltons, each capable of binding the hormone. Antibody studies, similar to those performed for insulin receptors (see later discussion), indicate that cross-linking of receptors may be necessary for hormonal action. Prolactin is unusual among hormones in that it does not down-regulate its own receptor. Rather, prolactin seems to up-regulate its receptor. The second messenger for prolactin remains elusive.

The main function of prolactin in humans is the induction of lactation. For this to occur, the woman must have a mature breast, with glandular tissue. Combined action of prolactin, estrogens, insulin, and cortisol is necessary for milk production. Prolactin stimulates the mRNA for the milk proteins, casein and lactalbumin, and increases the production of fatty acids for milk. Prolactin levels are maintained throughout adult life, but the role of prolactin in nonlactating humans is unknown.

Clinical comment

Disorders associated with prolactin Prolactin has a dramatic effect in blocking the pituitary gonadotrophs, and elevated prolactin levels are associated with sexual dysfunction. Hyperprolactinemia before puberty blocks sexual maturation and the pubertal growth spurt. After puberty, hyperprolactinemia is associated with a loss of libido and impotence in males and amenorrhea in females. In females it usually also causes galactorrhea (milk incontinence).

Hyperprolactinemia is usually caused by benign tumors of the pituitary gland. Small tumors, which are very common, are the major cause of the galactorrhea-amenorrhea syndrome in young women. Large tumors can press on the pituitary stalk and cause panhypopituitarism. The primary treatment for these tumors is a dopamine agonist that crosses the blood-brain barrier, called bromocriptine. It not only reduces prolactin levels, but also shrinks the size of the tumor (Figure 18.6).

Some medications, acting as dopamine antagonists, increase prolactin secretion. Although many drugs of this type exist, the most common are the antipsychotic phenothiazines, such as thorazine. They may cause clinical galactorrhea in some women, which requires discontinuation of the drug.

Adrenocorticotropin and the endorphins Adrenocorticotropin (adrenocorticotropic hormone, ACTH) was postulated to exist in the 1930s based on experiments in which removal of the pituitary gland resulted in adrenal failure in animals; this was reversed by pituitary extracts. Final purification and sequencing occurred in 1955, and subsequent sequencing of the hormone from many species demonstrated remarkable homology. Using pituitary cells in culture, the precursor molecule of ACTH was discovered to be a 32,000-

Figure 18.6   Computer tomogram of a patient with a large pituitary tumor. At the age of 18 yr this patient had no sexual development. The computed tomogram shown in **A** initially found the tumor, along with a very elevated prolactin level. She was treated with bromocriptine, which reduced the level of proactin to normal and shrunk the pituitary tumor to an insignificant mass, as shown in **B.**

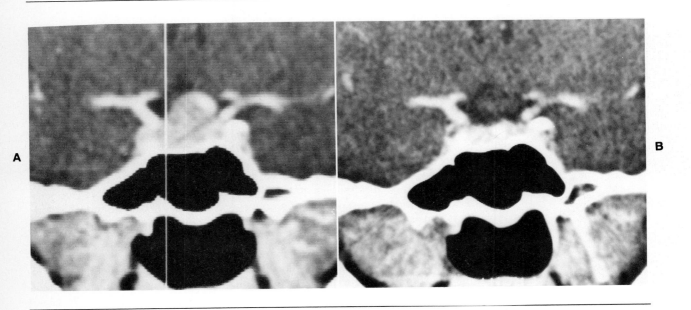

dalton glycoprotein called pro-opiomelanocortin that gave rise to several important hormones (Figure 18.7). These hormones share a common property in that they appear to be necessary for the stress response of animals. Not all these hormones are produced by all animals.

ACTH This polypeptide is a single chain of 39 amino acids. Only the amino terminal 25 amino acids are necessary for activity. ACTH binds to specific receptors on the adrenal glands and stimulates adenylate cyclase, which raises the intracellular cAMP level and thereby causes an increase in cortisol production by the adrenal cortex. High levels of ACTH have an effect resembling that of melanocyte-stimulating hormone, which causes hyperpigmentation. A very specific diurnal variation in ACTH secretion exists, with most of the daily ACTH produced just before waking in the morning. This results in a profile of cortisol which is also increased just before rising. ACTH also has a trophic effect on the adrenal cortex, causing growth of the gland.

Endorphins These peptides of 15 amino acids have endogenous morphinelike or opiatelike activity. They therefore act to increase pain tolerance during times of stress. Not surprisingly, football players have very high levels of the endorphins.

Figure 18.7

Synthesis of ACTH from pro-opiomelanocortin (POMC). POMC is synthesized as a 28.5 kilodalton (K) protein that is glycosylated to a 30 kilodalton precursor. This is proteolytically cleaved to a 20-kilodalton glycosylated pro-ACTH and B-lipotropin (B-LPH). The pro-ACTH is further cleaved to a glycosylated 16-kilodalton protein of unknown function (N-POC) and ACTH. The β-lipotropin is further cleaved to a γ-lipotropin (γ-LPH) of unknown physiological significance and the endorphins, which are hormones that serve as endogenous morphinelike compounds.

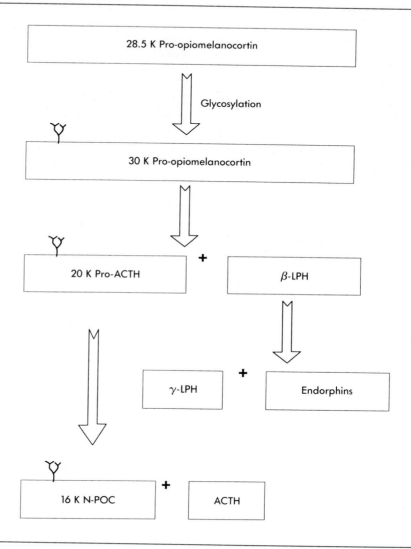

Melanocyte-stimulating hormone This hormone, MSH, is responsible for pigmentation in lower animals and comes in two forms: α-MSH, a basic form, and β-MSH, an acidic form. In lower animals these are produced in the intermediate lobe of the pituitary gland. Humans lack this portion of the pituitary, and no strong evidence suggests the existence of either peptide in humans. Regulation of pigmentation seems to be done by the ACTH molecule in humans.

These peptides all are produced by the proteolytic cleavage of pro-opiomelanocortin. This precursor is cleaved to a 20,000-dalton pro-ACTH and β-lipotropin. The latter is further cleaved to γ-lipotropin and β-endorphin. The function of the lipotropins is unclear. ACTH is produced by a series of cleavages of the pro-ACTH precursor.

The main stimulator of pro-opiomelanocortin is the hypothalamic-releasing peptide, corticotropin-releasing factor (CRF). The peptide increases both synthesis and secretion of ACTH and seems to do so very rapidly. ACTH secretion is influenced by the brain through both dopamineric and serotonergic pathways. Corticosteroids feed back directly on the pituitary gland to block the synthesis and secretion of ACTH (Figure 18.8). They do so by directly inhibiting the synthesis of the mRNA for pro-opiomelanocortin. This inhibition, if present continuously for a few weeks, can depress synthesis of ACTH for up to a year.

Humans have a single gene for pro-opiomelanocortin comprised of 7665 base pairs with three exons and two large introns, each of about 3000 bases. There are both TATA and CAT boxes slightly upstream from the initation site. Approximately 500 bases upstream from the initiation site is a 21-base region that is identical with a region found upstream to the somatotropin gene and the mouse mammary tumor virus (MMTV) gene. In the case of somatotropin and MMTV, this region is thought to bind the glucocorticoid receptor and exhibit positive control on these genes. This same region probably is used to allow negative control of the pro-opiomelanocortin gene.

Clinical comment

Disorder associated with ACTH production Overproduction of ACTH is among the more common causes of Cushing's syndrome, a condition in which too much cortisol results in a variety of clinical problems. These patients have a peculiar form of obesity, along with a general catabolic condition, poor wound healing, and immune deficiencies. The patients die from overwhelming infection. The condition is heterogeneous but commonly is associated with microadenomas of the pituitary gland. The usual problem appears to be hypothalamic insensitivity to feedback by cortisol.

High ACTH levels are also found in patients who lack adrenal cortisol production because of a lack of feedback inhibition. The resultant high levels of ACTH cause hyperpigmentation in these patients. Some patients with a depressive illness, particularly those with a family history of alcoholism, have elevated ACTH levels that do not have normal feedback inhibition by cortisol.

A lack of ACTH can be an isolated event or part of panhypopituitarism. The symptoms are similar to adrenal insufficiency but milder (see Chapter 19).

Hormones of calcium metabolism

The control of serum calcium levels is of great importance. Serum levels of ionized calcium must be kept between 1.9 and 2.6 mEq/L. Levels lower than this will lead to tetanic contractions of muscles (including the muscles of breathing), seizures, and death. Higher levels lead to depression, coma, and death. Three hormones are involved in the maintenance of calcium homeostasis: parathyroid hormone (PTH, parathormone), calcitonin, and vitamin D (Table 18.4). The last works by steroidlike mechanisms and is discussed in Chapter 19.

Parathyroid hormone

Parathyroid hormone is synthesized in the parathyroid glands, which are four small glands normally weighing about 50 mg each and found on the posterior of the thyroid gland. The parathyroid glands are derived from branchial clefts 3 and 4 and migrate to the thyroid region during early development. Sometimes one or more of these glands may migrate with the thymus gland into the chest.

Figure 18.8

Feedback axis for ACTH. Cortisol-releasing factor (CRF) is produced by the hypothalamus and causes an increase in ACTH secretion. This results in the production of more cortisol by the adrenal glands and feedback by this cortisol on both the hypothalamus and the pituitary gland to shut off ACTH production and lower cortisol secretion.

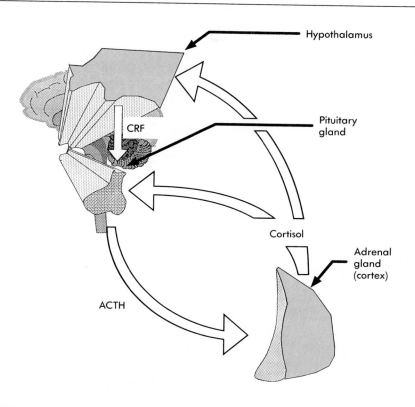

Table 18.4

Classification, localization, and actions of the hormones of calcium regulation

Hormone	Target	Actions	Properties	Second messenger
Parathyroid hormone (PTH)	Kidney, bone	Raises plasma Ca^{++} Lowers plasma PO_4^{-3}	Polypeptide Single chain 84 amino acids	cAMP
Calcitonin	Bone osteoclasts	Lowers plasma Ca^{++} (? Bone formation)	Polypeptide Single chain 32 amino acids	cAMP
Vitamin D	Gastrointestinal tract bone	Raises plasma Ca^{++} Raises plasma PO_4^{-3} (? Bone formation)	Steroidlike	nuclear receptor

PTH is secreted as an 84-amino-acid, single-chain polypeptide with a molecular weight of 9300 and no disulfide bonds. Many basic amino acids are present, giving the hormone an overall basic nature. In solution PTH appears to be mostly random coil with a small amount of α-helix and β-pleated sheet. It is synthesized as a preprohormone; PTH was the peptide on which the 25-amino-acid hydrophobic "leading" sequence was discovered. The pro-PTH has six amino acids that are cleaved off to produce PTH. PTH is most active in the intact 1 to 84 form. However, this is rapidly cleaved in peripheral tissue at position 34 to a N-terminal piece of 34 amino acids that retains 75% to 90% of biologic activity and an inactive 35 to 84 C-terminal piece that is excreted by the kidney and accumulates in kidney disease. The gene for PTH, located on chromosome 11, has two intervening sequences, both in the amino terminal half of the hormone. The mRNA for PTH in humans and other species has been cloned, and the structure of many animal PTHs have been determined from their nucleotide sequence.

PTH acts on bone and kidney to increase serum calcium levels. Thus, not surprisingly, the synthesis and secretion of PTH is stimulated by low calcium levels (hypocalcemia) and inhibited by high calcium levels (hypercalcemia). In addition, the final active form of vitamin D, 1,25-dihydroxycholecalciferol, inhibits the synthesis of PTH. The secretion of PTH is also increased by epinephrine and by low plasma magnesium levels.

Overall, PTH acts to raise serum calcium levels and lower serum phosphate levels. This action seems to be mediated through a receptor on the plasma membrane that is specific for PTH. The receptor is a glycoprotein of molecular weight about 70,000. The binding of PTH activates adenylate cyclase and raises intracellular cAMP levels. Some patients with low serum calcium levels have a resistance to PTH, which may be caused by a lack of the G_s protein of the adenylate cyclase complex (see Chapter 17, case 1).

PTH works on the kidney to produce three effects:

1. It increases the reabsorption of calcium in the renal tubules, resulting in a decrease in calcium loss in the urine and a rise in serum calcium levels.
2. It causes a phosphaturia (loss on phosphate into the urine) with a resultant decrease in serum phosphate.
3. It stimulates the 25-hydroxycholecalciferol 1-hydroxylase activity in the kidney, which adds a hydroxyl group onto vitamin D, dramatically increasing the activity of vitamin D (as 1,25-dihydroxycholecalciferol).

Thus the overall effect of PTH on the kidney results in an increased serum calcium level, decreased serum phosphate level, and increased vitamin D activity, which acts on bone and the gastrointestinal tract. PTH also acts directly on bone, where it causes both an increased resorption and an increased remodeling of bone. This raises serum levels of both calcium and phosphate.

Calcitonin

Calcitonin, initially known as thyrocalcitonin, was relatively recently discovered in 1961. Its existence as a hormone causing hypocalcemia was postulated, since perfusion of the neck with high levels of calcium resulted in a more rapid fall in serum calcium level than the removal of the parathyroid glands alone. Thus a hormone that was actively lowering calcium levels had to exist. This hormone is produced by the clear cells of the thyroid gland, which are of neural crest origin.

Calcitonin is a 32-amino-acid, single-chain polypeptide. The carboxyl-terminal proline is blocked with an amide group, which is necessary for activity. The tertiary structure of the molecule has both random coil and α-helical regions. Interestingly, calcitonin from lower animals has more random coil and is more active than human PTH in all species including humans. The molecular biosynthesis of calcitonin is complex and poorly understood. Synthesis primarily occurs from the α-calcitonin gene on chromosome 11, which is localized between the genes for catalase and PTH. The gene contains six exons, and differential splicing produces calcitonin, katacalcin, and some other related products. Katacalcin is synthesized and secreted in equimolar amounts with calcitonin, but its function remains unknown. A second β-calcitonin gene is also found on chromosome 11, which encodes a calcitonin gene-related peptide that is found primarily in the nervous system. Its function is also unknown. Calcitonin has been cloned, and recombinant human calcitonin is available as a drug for clinical use.

The regulation of calcitonin secretion is the reverse of that for PTH. Calcitonin secretion is stimulated by hypercalcemia and inhibited by hypocalcemia. In addition, gastrin and cholecystokinin stimulate calcitonin secretion, which has led to the hypothesis that calcitonin may be especially important in the handling of meal-related alterations in calcium balance.

Calcitonin produces a rapid fall in serum calcium levels in animals with high rates of bone turnover. Thus, when given to a child, calcitonin will produce hypocalcemia. In adults that do not normally have high rates of bone turnover, physiologic levels of calcitonin have no effect on calcium. The hypocalcemic effect of calcitonin seems to be mediated primarily by a dramatic inhibitory effect of this hormone on osteoclast activity, which prevents bone resorption by these cells. Overall it appears that calcitonin works to protect the skeleton during periods of high stress, such as growth and lactation. The calcitonin level decreases after menopause, a time when osteoporosis becomes more prevalent. Early studies have suggested that administration of calcitonin to postmenopausal women will diminish osteoporosis for at least short periods. Long-term studies have not been completed.

Calcitonin exerts its biologic effect on osteoclasts by binding to a specific receptor, which is a glycoprotein of molecular weight 80,000. Approximately 1 million such receptors exist per osteoclast, which may explain the exquisite sensitivity of the osteoclast to this hormone. The receptor is coupled to adenylate cyclase, and thus calcitonin will raise intracellular cAMP levels. This does not appear to be the only second message of this hormone, but others remain to be proven.

Clinical comment

Disorders associated with parathyroid hormone and calcitonin Hypercalcemia, or high blood calcium level, is a very common clinical problem usually found on routine laboratory examination. Most patients have no symptoms but, if left untreated, may develop kidney stones, kidney failure, osteoporosis, bone cysts, pancreatitis, and even coma. Hyperparathyroidism is the most common cause of hypercalcemia and is caused by a tumor or hyperplasia of the parathyroid glands. More than 90% of the tumors are benign, but some are malignant and very difficult to treat.

In patients with hyperparathyroidism the calcium level is high, with low phosphate, high chloride, and low bicarbonate levels, the latter resulting from diminished acid loss from the kidney. Twenty years ago these patients were detected much later in the disease course after symptoms appeared, and the classical symptoms of hyperparathyroidism were considered "stones, bones, moans, and groans." "Stones" suggest the pain of kidney stones, a deep cramping pain in the posterior flank that radiates into the groin. "Bones" imply deep bone pain caused by bone remodeling, cyst formation, and osteoporosis. "Moans" imply psychiatric disease, usually depression that responds to lowering the calcium level. Finally, "groans" suggest pancreatitis with deep abdominal pain radiating to the back.

Hypocalcemia can result from many conditions, but the cause is often a lack of parathyroid action. These patients have low plasma calcium and high phosphate levels. The low calcium may cause muscular irritability and tetany. If severe, it will cause seizures and even death. The patient may be unable to produce PTH or may have a resistance to it (see Chapter 17, case 1). Treatment for all these conditions is the administration of vitamin D, usually the active dihydroxyl form, and calcium.

When raised, the calcitonin level does not cause disease states, but this hormone does serve as marker for the malignant tumor, medullary thyroid carcinoma. This tumor produces calcitonin in proportion to its mass. Since the cancer sometimes affects families, it is important to screen family members for elevated calcitonin levels.

Since calcitonin blocks osteoclast activity, it is useful for treating patients with diseases in which elevated osteoclast cell activity occurs, such as Paget's disease. This painful disease, with bony deformations caused by an increased rate of bone turnover, is dramatically alleviated by injections of salmon or recombinant human calcitonin.

Pancreatic hormones

The hormones produced by the endocrine pancreas primarily regulate carbohydrate metabolism. These hormones, made in the three cell types of the islets of Langerhans, include the well-characterized hormones insulin and glucagon, somatostatin, and the poorly characterized hormone pancreatic polypeptide. Tumors of the islets of Langerhans sometimes also produce gastrin. Table 18.5 lists some actions of glucagon and insulin.

Glucagon

Glucagon is a small peptide produced by the α-cells of the islets of Langerhans. In general glucagon functions to raise blood sugar. Although a hormone with glucagon's properties was postulated early, the isolation of glucagon occurred in 1955, and the entire sequence was determined 2 years later.

Glucagon is a 29-amino-acid peptide that is identical in structure in virtually all mammals. This very highly conserved structure occurs since alterations in its primary sequence, even minor ones, destroy its function. The tertiary structure of glucagon is uncertain. In concentrated solution and in crystals, it is primarily α-helical. In dilute solution, this α-helix disappears and considerable random coil appears. Glucagon (pancreatic and gut) and the gastrointestinal hormones vasoactive intestinal peptide (VIP), gastric inhibitory peptide (GIP), and secretin have very similar amino acid sequences.

The glucagon gene consists of six exons, and the molecule is first made as a prohormone of 160 amino acids. This is rapidly processed to (1) the 10-kilodalton major proglucagon fragment that contains two glucagon-like peptides but does not contain glucagon itself and (2) another 10-kilodalton peptide, glycentin, which does contain the glucagon molecule. The glycentin is then further processed in the α-cells of the islets of Langerhans to glucagon (Figure 18.9).

Secretion of glucagon is tightly linked to the secretion of insulin by the β-cells of the islets of Langerhans. Decreases in serum glucose concentration, even when small, are potent stimuli for glucagon secretion. Such stimulation is likely to occur during fasting or during exercise-induced falls in serum glucose. This rise in glucagon is always ac-

---

Table 18.5   Actions of hormones of glucose metabolism

Tissue	Action of insulin	Action of glucagon
Liver	Glycogen synthesis Fat synthesis Albumin synthesis	Glycogenolysis Gluconeogenesis Ketogenesis
Muscle	Glycogen synthesis Glycolysis	Glycogenolysis (to lactate) Alanine cycle
Adipose tissue	Lipogenesis Antilipolysis	Lipolysis

companied by a fall in insulin secretion. Similarly, increases in serum glucose concentration inhibit glucagon secretion and stimulate insulin secretion. Ingestion of a large protein meal will lead to the secretion of both insulin and glucagon. The simultaneous increases in both hormones ensure the proper conversion of the excess glucogenic amino acids to glucose and utilization of the glucose. The hormones gastrin, cholecystokinin, and GIP may plan an important role in the simultaneous production of both hormones.

Glucagon secretion also depends on hormones and neurotransmitters. The "stress hormones" epinephrine, somatotropin, and cortisol all directly stimulate the production and secretion of glucagon. Vasopressin and the endogenous opiate β-endorphin also directly stimulate glucagon secretion. Since glucagon is secreted directly into the hepatic portal system, the level reaching the liver may be very high, up to 10 times that in the peripheral circulation.

Glucagon acts by binding to a specific receptor on the cell surface of sensitive cells. These receptors are coupled to adenylate cyclase by a G_s protein (see Chapter 17). The increased cAMP stimulates cAMP-dependent protein kinase and activates a cascade of phosphorylation, leading to alteration in many enzyme activities.

Glucagon acts in various ways on carbohydrate and lipid metabolism. In the liver increased glucagon action leads through the phosphorylation cascade to increase phosphorylation of glycogen phosphorylase and increase glycogen breakdown to glucose 1-phosphate and ultimately glucose. At the same time glycogen synthase is phosphorylated to its inactive form and glycogen synthesis ceases. Glucagon also acts on the gluconeogenic pathways. This is mediated largely through reductions in the fructose 2,6-bisphosphate concentrations, with subsequent activation of the gluconeogenic pathways and inhibition of the glycolytic pathway. Glycogenolysis raises blood sugars within 10 to 15 min, but gluconeogenesis requires 30 to 60 min. Thus the action of glucagon on the liver is to increase production of glucose, both by increasing gluconeogenesis and by increasing glycogen breakdown to glucose.

Figure 18.9   Gene structure of glucagon. See text for details.

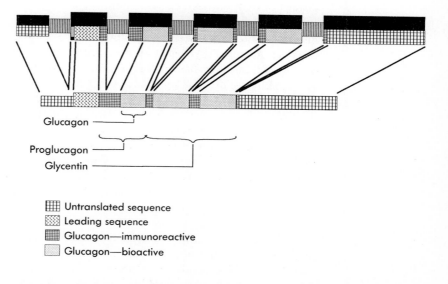

Glucagon

Proglucagon

Glycentin

▦ Untranslated sequence
▨ Leading sequence
▦ Glucagon—immunoreactive
☐ Glucagon—bioactive

Glucagon also has dramatic effects on lipogenesis (fatty acid synthesis) and ketogenesis (production of acetone, acetoacetate, and β-hydroxybutyrate), particularly in the liver. Lipogenesis is inhibited because the substrate for the production of long-chain fatty acids, acetyl coenzyme A (CoA), is diminished. Acetyl CoA production from glucose is inhibited as the glycolytic pathway is reduced. The breakdown of fatty acids, particularly to the ketones, is increased. Carnitine acyl transferase I, the rate-limiting enzyme for the transfer of fatty acids from the cytosol into the mitochondria, is glucagon sensitive and in the presence of high levels of glucagon is maximally induced. This leads to an increased metabolism of fatty acids by the mitochondria. In states of insulin deficiency, such as diabetes mellitus and starvation, the very high levels of fatty acids, coupled with the poor availability of oxaloacetic acid, lead to a flooding of the mitochondria with acetyl CoA and ketone formation, even leading to ketoacidosis (see case 6).

Insulin

Although the existence of insulin was postulated in the last century, it was finally identified in 1921 and crystalized in 1926. Insulin serves as a switch to alter metabolism from lipid to glucose utilization. Insulin has its origins in very primitive hormones, and forms of insulin have been identified in invertebrates. Evidence even exists for insulin-like molecules in prokaryotes.

In mammals insulin is produced by the β-cells of the islets of Langerhans. The embryologic origin of these cells is different from the exocrine pancreas, since the islet of Langerhans are derived from the neural crest, whereas the exocrine pancreas is derived from outpouches of the primitive gut.

Insulin is a polypeptide hormone of molecular weight 5600, with 51 amino acids in two chains. There are two interchain disulfide bonds, called the major loop, and one intrachain disulfide bond, called the minor loop (Figure 18.10). Insulin is remarkably similar from species to species. Human and chicken insulin differ by only five amino acids. Even hagfish insulin differs from human insulin by only five amino acids, (although hagfish separated from the pathway leading to humans more than 600 million years ago). Guinea pig and other histrichomorphs have insulins that are remarkably different from human insulin. Despite the similarity of the insulin itself, little similarity exists in the "pro" piece which is called the connecting peptide (C-peptide) among different species. This small peptide has less than 50% homology from species to species and is antigenically distinct in each species.

The gene for insulin resides on chromosome 11 as a single copy, although a family of insulin-like molecules exists, including the insulin-like growth factors (I and II) and relaxin. There is a TATA box 30 base pairs from the starting site for transcription and three exons. Insulin is first synthesized as preproinsulin, which contains a leader sequence of 24 amino acids. The proinsulin contains the B chain of 30 amino acids, a C- or connecting peptide, and an A chain of 21 amino acids. During the synthesis of insulin, the C-peptide is cleaved from the proinsulin but retained in the secretory granule. The proinsulin is proteolytically cleaved by an enzyme with trypsinlike specificity. The cleavage sites on either side of the C-peptide have a pair of basic amino acids that are removed from the molecule by the cleavage. When the insulin is secreted, both insulin and C-peptide are exocytosed into the blood together at equimolar concentrations. In some patients whose insulin cannot be measured because of antiinsulin antibodies in their plasma, one can still measure C-peptide (see case 4).

Because all the mammalian insulins are so similar, they are not very immunogenic in humans, and distinguishing them antigenically is difficult. Nevertheless, when injected daily, as in diabetes mellitus, patients develop antibodies to insulin. Of the medically used insulins, beef insulin is the most antigenic, since it differs from human insulin at both the minor loop (the cyclic area formed by the intrachain disulfide bond) and the B

Figure 18.10

Structure of insulin, proinsulin, and insulin-like growth factor I. These hormones share the same basic structure with almost identical A chains (light line). Insulin-like growth factor I has an additional piece at the end of the B chain.

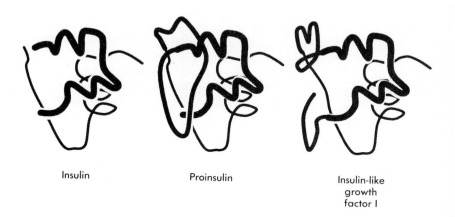

Insulin Proinsulin Insulin-like growth factor I

chain terminus. Pork insulin is less antigenic, since it differs from human insulin only in a single amino acid at the B chain terminus. Surprisingly, human insulin is also antigenic in patients with diabetes, and after years of treatment with human insulin, they develop antibodies at almost the same titers as patients using pork insulin.

The regulation of insulin secretion is complex. The most potent stimulus for secretion is elevation of the blood sugar level but even this is complicated. For example, a 50 gram glucose load given orally is a much more potent stimulator of insulin secretion than the same amount of glucose given intravenously, apparently because of an enhancement of insulin secretion by a series of hormones made by the gut in response to food. These hormones collectively are known as *candidate hormones* and include enteroglucagon, GIP and VIP.

Stimulation of insulin secretion by glucose depends on the metabolism of the sugar by the glycolytic pathway. Other fuels, such as amino acids, fatty acids, and ketone bodies, also stimulate insulin production, and this also appears to result from their metabolism in the islets of Langerhans. In addition, the β-adrenergic hormones act to increase insulin secretion. Certain drugs, including the drugs used to treat diabetes—sulfonylureas and biquanides, also act to increase insulin secretion dramatically in the normal pancreas and in patients with type II diabetes. Both cyclic AMP and calcium are necessary for the secretion of insulin.

There are no known metabolic fuels that inhibit insulin secretion, although a lack of glucose is a powerful "nonstimulus." The α-adrenergic neurotransmitters, such as norepinephrine, inhibit insulin secretion via an α_2-receptor. Somatostatin, produced by the δ-cells of the islets of Langerhans, is also a powerful inhibitor of insulin secretion.

Insulin acts by binding to specific insulin receptors on the cell surface. The receptor, discussed as a model in Chapter 17, is a glycoprotein of about 350,000 daltons. Cells that are very responsive to insulin, such as liver and fat cells, generally have 200,000 to 300,000 receptors per cell. The insulin receptor is made up of two pairs of subunits. The two α-subunits of 135,000 daltons are entirely extracellular and have the binding specificity. The two β-subunits have three regions: a small extracellular domain that combines with the α-subunits, a very small transmembrane domain of about 20 amino acids, and a large intracellular domain that is a tyrosine kinase. Insulin receptors are also found intracellularly and on the nuclear membrane. The intracellular receptors may be (1) synthetic intermediates, (2) receptors in receptorsomes, or (3) in the process of being recycled, or may have have some other, as yet unknown, function. The life cycle of the insulin receptor is discussed in Chapter 17.

Insulin acts primarily through the tyrosine kinase acitivity of its receptor, specifically phosphorylating the tyrosine groups of itself and other proteins and thereby altering their action. Glucose transporters, for example, may be recruited to the cell membrane after specific phosphorylation of the intracellular forms of the transporter, but this is not a simple process. The net result of this action is to increase glucose transport into cells. Other mediators of insulin's action have been postulated, such as calcium and an insulin mediator composed of inositol and ethanolamine (Figure 18.11). These are most likely distal mediators, whose action is initiated by a tyrosine phosphorylation.

Insulin has a variety of actions; all are anabolic and work to build up the body. All insulin actions do not occur at the same insulin concentration, and sometimes two effects on the same cell occur at different concentrations. For example, antilipolysis, the inhibition of fat breakdown, occurs in adipocytes at insulin concentrations of $5\mu U/ml$; in these same cells, however, lipogenesis, the synthesis of fats, requires about 10 times as much insulin. The major action of insulin is to shut off fat breakdown and usage and to promote glucose usage and storage (Figures 18.12 and 18.13).

The overall effect of insulin is to lower blood glucose, which it does by several pathways. In the absence of insulin, only the brain and erythrocytes use glucose for energy. In its presence, virtually all cells of the body will switch to glucose utilization for energy. Insulin increases glucose transport into cells and subsequent glycolysis. It increases glycogen synthesis by the liver and muscles and inhibits gluconeogenesis (Table 18-5).

The effect of insulin on glycolysis is complex. Insulin activates glucose transport by increasing the amount of mRNA transcribed (and therefore translated) to increase the intracellular glucose transporter pool. Insulin also increases the movement of transporters to the plasma membrane, thereby activating them. Insulin activates the glycolytic pathway by reversing the inhibition of glucagon (see previous discussion). Thus insulin activates phosphofructokinase both directly and by releasing it from the inhibition by fructose 2,6-bisphosphate. Insulin also prevents the inhibition of pyruvate kinase by glucagon and stimulates the transcription of its mRNA. Insulin acts to increase glycogen synthesis by a cascade of events initiated by the tyrosine kinase of the insulin receptors. This eventually leads to the dephosphorylation of glycogen phosphorylase, which is inactivated, and the dephosphorylation of glycogen synthase, which is activated. Insulin also blocks gluconeogenesis by inhibiting the synthesis of mRNA for phospho*enol*pyruvate carboxykinase, a key enzyme needed for the gluconeogenic pathway.

Insulin also has major effects on lipid metabolism. Even at very low concentrations, such as the levels found after an overnight fast (5 $\mu U/ml$), insulin will significantly inhibit the breakdown of fats by adipocytes. Hormone-sensitive lipase is the enzyme largely responsible for fat release by these cells, and its activity is enhanced by glucagon and epinephrine, as mediated by cAMP. Insulin is a powerful inhibitor of lipase; the mechanism of this effect is unclear. In the absence of insulin, lipase is maximally stimulated, resulting in the delivery of high concentrations of fatty acids to the liver. In the presence of high concentrations of glucagon, as frequently seen in diabetes, ketoacidosis frequently occurs (see Chapter 17, case 4, and case 6 in this chapter). Insulin also promotes lipid synthesis by the adipocytes, although this requires a higher concentration of insulin.

Insulin also has other anabolic effects. It increases synthesis of many proteins such as

Figure 18.11   The insulin mediator. Although the second messenger for insulin is the action of a specific tyrosine kinase on other proteins of the cell, there are other mediators of insulin action. Insulin's effect on pyruvate dehydrogenase, for example, appears to be mediated by the molecule shown in the shaded area. It is cleaved from an "anchor protein" by a specific phospholipase C and a protease. This mediator is a complex molecule consisting of inositol phosphate, glucosamine, a glycan, ethanolamine, and a terminal amino acid.

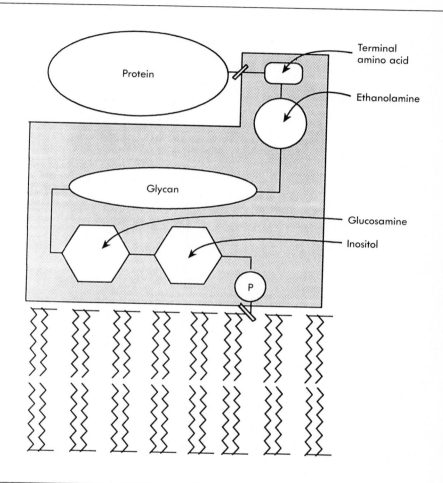

albumin by the liver and is necessary for the growth of some cells. Many of the clinical aspects of insulin are discussed in the clinical examples in Chapter 17.

Clinical comment   **Insulin preparations** Insulin is used by more than 3 million Americans each day to control their blood sugar levels, but only over the past few years have completely pure preparations become available. Commercial insulin now has less than 1 part per million (ppm) of impurities, whereas a few years ago preparations with 50,000 ppm of impurities were common. The purification of insulin makes use of the chemical stability of this small protein.

Animal insulin is harvested from beef and pork pancreas, which is frozen when the animal is slaughtered. The frozen pancreas is extracted into acidified alcohol, which

Figure 18.12   Some major actions of insulin on glucose and fat metabolism. Each of
the numbered arrows is a major control point for insulin's action. The
unnumbered arrows are often more than a single enzymatic step.
1, Insulin increases the number of glucose transporters in the
membrane; it may also increase the rate at each transporter. *2,*
Insulin alters the concentration of fructose 2,6-bisphosphate, which in
turn dramatically alters the activity of phosphofructokinase and
fructose 1,6-bisphosphatase. *3,* Insulin increases the activity of
pyruvate kinase but inhibits the activity and the synthesis of
phospho*enol*pyruvate carboxykinase. *4,* Insulin increases the activity
of pyruvate dehydrogenase, probably through the insulin mediator
(see Figure 18.11). *5,* Insulin promotes lipogenesis and forms fats for
storage from glucose through acetyl CoA. *6,* Insulin promotes the
synthesis of glycogen by increasing the activity of glycogen synthase
(independent form). *7,* Insulin inhibits the breakdown of glycogen by
inhibiting glycogen phosphorylase.

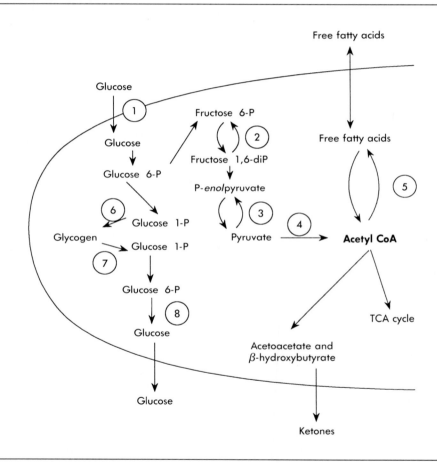

Figure 18.13

Biologic actions of insulin. Insulin initiates various actions on the cell, which occur at different insulin concentrations, even in the same cell. Antilipolysis is extremely sensitive to insulin, whereas the lipogenic effect of insulin, which causes the cell to produce fats, is far less sensitive, even in the same cell.

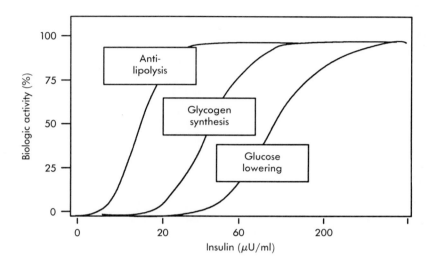

destroys most proteins but not insulin. The soluble insulin is precipitated with acetone and further purified by differential solubility in polyethylene glycol. This insulin, with about 50,000 ppm of impurities, was the commercial insulin from 1922 to 1973. Currently the insulin is further purified by sizing the insulin on gel filtration columns, followed by further separation by charge on ion exchange columns. These highly purified insulins have less than 1-ppm of impurities.

Insulin from three species are used clinically: beef, pork, and human. The first two are purified from animal pancreas. Human insulin is made either by conversion of pork insulin or by recombinant DNA techniques. Pork insulin differs from human insulin only in the carboxyl-terminal amino acid. Human insulin is generated from pork insulin by removing the incorrect amino acid, alanine, with carboxypeptidase and replacing it with the correct one, threonine, using the reverse reaction with the same enzyme. Human insulin is also produced by prokaryotes, which have the insulin gene incorporated through recombinant DNA. In *Escherichia coli,* the regions corresponding to the A and B subunits have been individually cloned, each with a trypotophan metabolism promoter. When the bacteria are exposed to tryptophan, they make either the A or the B chain, depending on the plasmid inserted. The individual peptides are purified and combined to form insulin.

Clinical preparations of insulin are designed to allow the physician considerable freedom in formulating a therapeutic plan. Currently insulins exist in two families with similar timing. Each family has a short-, intermediate-, and long-acting preparation. In the first family these preparations are crystalline zinc insulin (CZI, regular insulin), neutral protamine Hagedorn (NPH), and protamine zinc insulin (PZI, rarely used). The second family, called the lente family, consists of semilente, lente, and ultralente insulins (Figure 18.14).

Figure 18.14

Time of action of various insulin preparations. Insulins vary in the timing of their action. Short-acting insulins, as with regular insulin, peak in action at 1.5 to 2 hr; intermediate-acting insulins, as with neutral protamine Hagedorn *(NPH)* peak at 5 to 10 hr; and long-acting insulins, as with ultralente, peak at 20 to 24 hr. *CZI,* Crystalline zinc insulin.

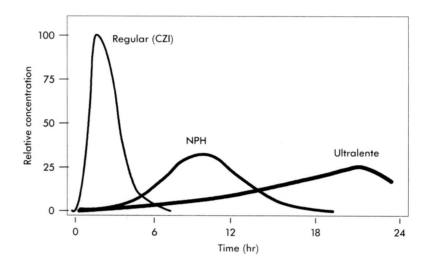

Generally these insulins are injected subcutaneously with sharp, thin needles. The injections are not painful. The short-acting insulins start to act in about 15 to 30 min, peak at 1.5 to 2 hr, and last 6 to 8 hr. Human preparations are slightly faster acting than the animal forms. Intermediate insulins start to act at 1 to 2 hr, peak at 5 to 10 hr, and last 14 to 20 hr. The long-acting preparations peak at about 24 hr and last about 48 hr.

People with diabetes are severly restricted in their ability to ingest simple carbohydrates, even though they take insulin. These sugary substances are absorbed so quickly into the bloodstream that current insulin preparations are too slow to deal with them. Three therapies currently under development may change this. First, the insulin gene can be altered so that the biosynthetic insulin cannot crystalize, but rather precipitates in an amorphous form. When this is done, the insulin is quickly absorbed from subcutaneous depots in about 15 min. By using this insulin, a diabetic person could easily ingest a candy bar. Second, insulin can be delivered to sites other than the subcutaneous tissues. Insulin delivered to the nasal muscosa is rapidly absorbed; this therapy is now in clinical trials. Finally, using implantable insulin pumps, insulin can be delivered directly into the bloodstream quickly and reproducibly. These pumps are also in clinical trials.

Catecholamines

The adrenal medulla is embryologically part of the sympathetic nervous system. Nevertheless, because it secretes epinephrine into the blood, it is also considered to be part of the endocrine system. The adrenal medulla, as with the rest of the sympathetic nervous system, is derived from neural crest and migrates to the region of the primitive adrenal gland at about 7 wk of gestation. At birth the adrenal medulla is fully functional.

Epinephrine or adrenaline, the main hormone of the adrenal medulla, and norepi-

Figure 18.15   Synthesis of the catecholamines through a series of enzymatic
reactions.

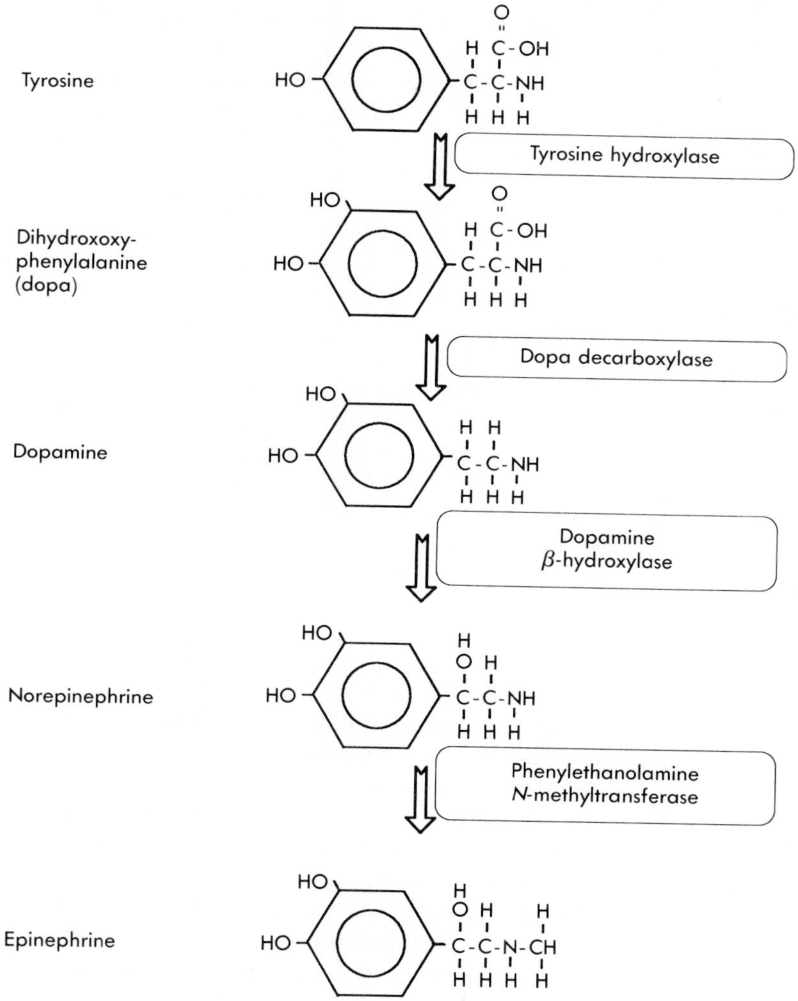

nephrine or noradrenaline, the main neurotransmitter of the sympathetic nervous system,
are involved in the response to acutely stressful situations, the "fight or flight" reaction
(see Figure 17.3). These catecholamines regulate blood pressure, cardiac output, fuel
metabolism, sweating, and pupil size and have various other important actions. The onset
of actions is within a few seconds of the stimulus, and the degradation of the substances
is equally rapid.

Epinephrine is synthesized from norepinephrine. The precursor for both is tyrosine
(Figure 18.15). The first step in synthesis is the hydroxylation of tyrosine at the ortho

position to form dihydroxyphenylalanine (L-dOPA). This is catalyzed by tyrosine hydroxylase, a cytosolic enzyme that requires tetrahydropterine as a cofactor for the oxidoreductive hydroxylation. The tyrosine hydroxylation is the rate-limiting step in catecholamine synthesis and is stimulated by cAMP. The next step in the synthesis is the decarboxylation of the side-chain to form dopamine. This is accomplished by dOPA decarboxylase, which requires pyridoxal phosphate as a cofactor. Dopamine is an active compound believed to inhibit prolactin secretion; it has a variety of physiologic effects. Dopamine is then hydroxylated on the side-chain to form norepinephrine. This is catalyzed by dopamine β-hydroxylase, which is also an oxidoreductase, requiring ascorbic acid as a cofactor. Dopamine β-hydroxylase acts within the secretory granule and is secreted with the catecholamine into the blood. Since it has a much longer half-life, plasma assays of dopamine β-hydroxylase have been used as sensitive indicators of catecholamine secretion. Finally, epinephrine is synthesized from norepinephrine by the transfer of a methyl group from methionine by the enzyme phenylethanolamine N-methyltransferase. This enzyme shows strong end-product inhibition.

The catecholamines are stored in complex secretory granules in the adrenal medulla. These granules contain the hormone at very high concentrations, as well as ATP, magnesium, calcium, enkephalins and water-soluble proteins known as chromagranins. The hormones form a stable complex in a ratio of 4:1 with magnesium ATP. In the secretory granule about 80% of the catecholamines are epinephrine and 20% norepinephrine.

The catecholamines are usually secreted in response to a strong stress stimulus, which is often transmitted through the nervous system. Unbound to any plasma proteins, they travel in the blood and reach the cells, where they interact with receptors of the cell membrane. The catecholamines interact with at least four different types of receptors, known as α_1, α_2, β_1, and β_2. The separation into α and β specificity was described many years ago based on the different actions. Norepinephrine has only α action, whereas epinephrine has both α and β. Some synthetic catecholamines, such as isoproterenol, have only β action. The subgrouping has occurred recently based on studies with inhibitors (antagonists). The β-receptor has been well characterized and is diagrammed in Figure 17.7. It is tightly linked to adenylate cyclase via the G proteins (see Figure 17.8). The second messenger for the α-receptor is less certain, but the most likely candidate appears to be the phosphatidylinositol cycle (see Figure 17.9).

The catecholamines are rapidly destroyed or inactivated by three separate mechanisms. Norepinephrine can be taken back up into nerve endings. This is a major pathway for the elimination of norepinephrine, but no comparable process exists for epinephrine. Both the catecholamines can be metabolized by either or both of two destructive enzymes. Catechol-O-methyltransferase adds a methyl group to the ortho hydroxyl group of the catecholamine, forming metanephrine (from epinephrine) or normetanephrine (from norepinephrine). These derivatives are inactive and are excreted or further metabolized by monoamine oxidase, which oxidatively removes the amino group of the side-chain, leaving a terminal-carboxyl group. Since the methyl group that distinguished epinephrine from norepinephrine is also removed, both compounds form 3,4-hydroxymandelic acid if monoamine oxidase acts directly on the catecholamine, and 3-methoxy,4-hydroxymandelic acid if both enzymes act. This last compound is usually referred to as vanillylmandelic acid (VMA.). Excess catecholamine production is often determined by the concentration of the metanephrines and VMA in the urine (Figure 18.6).

Epinephrine has dramatic effects on metabolism. It is strongly antiinsulin in its effects on the liver and adipose tissue. At the liver, epinephrine increases glycogen breakdown and increrases gluconeogenesis from lactate and amino acids. It also acts to increase lipid breakdown from fat tissue. Thus epinephrine's overall action is to increase substrate fuels, both sugars and fats. It has a very complex action on the β-cells of the islets of Langerhans.

Figure 18.16

Degradation of the catecholamines. The pathways of degradation of epinephrine and norepinephrine are shown. The two hormones of metabolism, catechol-O-methyltransferase and monoamine oxidase, work independently. Either can act on the product of the other. If both enzymes act, epinephrine and norepinephrine form a common product. The shaded areas are products that are clinically measured in the urine to demonstrate overproduction of catecholamines.

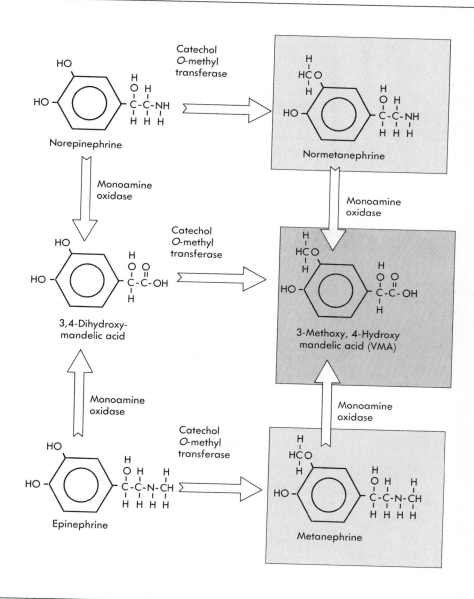

The β action of epinephrine increases insulin secretion, whereas the α action, mediated by the α_2-receptor, inhibits insulin secretion. The latter seems to predominate.

Clinical comment

Abnormal catecholamine secretion Excess catecholamine production is usually caused by a tumor of the adrenal gland, called a pheochromocytoma, which may produce either hormone or both. The most common feature of these tumors are sympathetic signs and hypertension. It is estimated that about 0.5% of all new cases of hypertension result from pheochromocytoma. The hypertension may be severe and symptomatic, and patients may have headache or visual disturbances. Patients also demonstrate other signs of catecholamine excess, such as palpitations, anxiety, tremors, and pallor. These patients often have diminished blood volume and may become dizzy on standing.

Patients may be screened for this disease by collecting urine for catecholamines, VMA, or metanephrines or by measuring plasma catecholamine concentrations. When properly performed, these are very sensitive and specific tests. The treatment of these tumors is surgical removal, but only after biochemical blockade of their activity using specific α- and β-catecholamine antagonists. Approximately 5% of these tumors are malignant and often very aggressive.

Rarely, these pheochromocytomas may be part of one of two genetic syndromes, called *multiple endocrine neoplasia* (MEN). The second form, called MEN-II, or Sipple's syndrome, is an autosomal dominant trait with variable penetrance. It consists of medullary carcinoma of the thyroid, which produces excessive calcitonin, pheochromocytoma, and hyperparathyroidism. Members of the families are usually screened by measuring calcitonin levels (see clinical comment on calcitonin).

Bibliography

Bahnsen M et al: Mechanisms of catecholamine effects on ketogenesis, Am J Physiol 247:E173, 1984.

Baly DL and Horuk R: The biology and biochemistry of the glucose transporter, Biochim Biophys Acta 947(3):571, 1988.

Bethge N, Diel F, and Usadel KH: Somatostatin—a regulatory peptide of clinical importance, J Clin Chem Clin Biochem 20(9): 603, 1982.

Blundell TL, Pitts JE, and Wood SP: The conformation and molecular biology of pancreatic hormones and homologous growth factors, CRC Crit Rev Biochem 13(2):141, 1982.

Conn PM et al: The molecular mechanism of action of gonadotropin releasing hormone (GnRH) in the pituitary, Recent Prog Horm Res 43:29, 1987.

Czech MP et al: Insulin receptor signaling: activation of multiple serine kinases, J Biol Chem 263(23):11017, 1988.

Denton RM and Brownsey RW: The role of phosphorylation in the regulation of fatty acid synthesis by insulin and other hormones, Philos Trans R Soc Lond (Biol) 302(1108):33, 1983.

Galoyan A and Srapionian R: Protein-hormonal complexes of the hypothalamus as neurochemical systems of regulation, Neurochem Res 8(12):1511, 1983.

Gershengorn MC: Thyrotropin releasing hormone: a review of the mechanisms of acute stimulation of pituitary hormone release, Mol Cell Biochem 45(3):163, 1982.

Granner DK: The molecular biology of insulin action on protein synthesis, Kidney Int (Suppl)32:S82, 1987.

Gunn I: Growth hormone deficiency, Ann Clin Biochem, 24:429, 1987.

Houslay MD: Insulin, glucagon and the receptor-mediated control of cyclic AMP concentrations in liver, Twenty-second Colworth medal lecture, Biochem Soc Trans 14(2):183, 1986.

Huffer WE: Morphology and biochemistry of bone remodeling: possible control by vitamin D, parathyroid hormone, and other substances, Lab Invest 59(4):418, 1988.

Kemper B: Molecular biology of parathyroid hormone, CRC Crit Rev Biochem 19(4):353, 1986.

Krane EJ: Diabetic ketoacidosis: biochemistry, physiology, treatment, and prevention, Pediatr Clin North Am 34(4):935, 1987.

Larner J: Insulin-signaling mechanisms: lessons from the old testament of glycogen metabolism and the new testament of molecular biology, Diabetes 37(3):262, 1988.

Leng G, Dyball RE, and Russel JA: Neurophysiology of body fluid homeostasis, Comp Biochem Physiol 90(4):781, 1988.

Ling N et al: Growth hormone releasing factors, Annu Rev Biochem 54:403, 1985.

Lips CJ et al: Evolutionary pathways of the calcitonin genes, Mol Cell Endocrinol 57(1):1, 1988.

Paladini AC, Pena C, and Poskus E: Molecular biology of growth hormone, CRC Crit Rev Biochem 15(1):25, 1983.

Richter D: Synthesis processing, and gene structure of vasopressin and oxytocin, Prog Nucleic Acid Res Mol Biol 30:245, 1983.

Sale SJ: Recent progress in our understanding of the mechanism of action of insulin, Int J Biochem, 20(9):897, 1988.

Taylor R and Agius L: The biochemistry of diabetes, Biochem J 250(3):625, 1988.

White JD et al: Biochemistry of peptide-secreting neurons, Physiol Rev 65(3):553, 1985.

Woodhead JS and Weeks I: Circulating thyrotropin as an index of thyroid function, Ann Clin Biochem 22:455, 1985.

Zarate A and Canales ES: Endocrine aspects of lactation and postpartum infertility, J Steroid Biochem 27(4):1023, 1987.

Clinical examples

A diamond (♦) on a case or question indicates that literature search beyond this text is necessary for full understanding.

Case 1 — ♦Diabetes insipidus caused by disseminated tuberculosis

A 55-year-old woman asked about her excess thirst and urination, which started 3 mo ago. She stated that she urinated at least hourly and woke up five or six times during the night to urinate. She was always thirsty and drank at least eight half-gallon bottles of diet cola each day. There was no family history of diabetes mellitus or kidney disease. On physical examination, she was a well-developed, thin, white female in mild respiratory distress with a temperature of 38.5° C (normal, 37°) but with normal blood pressure and pulse. Her skin showed erythematous, raised lesions that were not pruritic. Both her liver and her spleen were enlarged. Laboratory examination showed a serum sodium level of 151 mEq/L (normal, 138 to 145), potassium level of 4.5 mEq/L (normal, 3.5 to 5.0), chloride level of 108 mEq/L (normal 95 to 105), and bicarbonate level of 30 mEq/L (normal, 22 to 28). Her serum osmolality was 298 mOsm (normal 290 to 295) with a concomitant urine osmolality of 103 mOsm. Her blood sugar level was normal, and no glycosuria was present. There was some blood in the urine, but otherwise her kidney tests were normal. A chest roentgenogram showed tuberculosis.

Biochemical questions

1. Describe the possible abnormalities of hormones that could cause a syndrome such as this.
2. How could tuberculosis cause the syndrome?
3. How could this condition be treated?

Case discussion

1. Possible hormone abnormalities Excessive urination can be caused by (a) kidney disease; (b) excess solute load, such as the excess glucose in the urine of patients with diabetes; (c) excess water load, as in patients with psychogenic water intoxication, in which patients drink 6 to 15 L of fluid per day; or (d) diabetes insipidus. In this patient one can exclude kidney disease on the basis of the normal kidney tests (the blood in the urine is a sign of renal tuberculosis) and can exclude diabetes mellitus on the basis of normal blood and urine sugars. Thus one must distinguish between psychogenic water intoxication and diabetes insipidus. One can be reasonably certain that this patient has diabetes insipidus, since her blood osmolality is high at a time when she is not concentrating her urine. Someone with psychogenic water intoxication would have a slightly low serum osmolality with a dilute urine. One could be certain of the diagnosis by performing a water deprivation test.

The patient is hospitalized and confined to a room with no sources of water. Initially her weight is 50 kg, and her laboratory profile is as already described. After 4 hr of water deprivation, she has lost 3 kg, and her serum sodium is 155, chloride 114, potassium 4.5, and bicarbonate 31 mEq/L. Her serum osmolality has risen to 310 mOsm, but her urine osmolality has fallen to 100 mOsm. The high serum with low urine osmolality is indicative of diabetes insipidus with a loss of vasopressin action. To distinguish between an inability to produce vasopressin and a renal resistance to it, she is given 10 units of vasopressin intravenously. One hour later, her urine osmolality has risen to 485 mOsm, indicating that her kidneys are capable of responding to hormone. Thus her diabetes insipidus is caused by insufficient production of ADH.

2. Tuberculosis as the cause Tuberculosis can cause mass lesions, called tuberculomas, anywhere in the body. Since this patient has diabetes insipidus, one might expect to see such a lesion in the region of the hypothalamus and pituitary gland. A computed tomogram of the area shows a mass lesion about 1.5 cm in diameter in the region of the supraoptic nucleus of the hypothalamus. This mass has destroyed part of her hypothalamus and its ability to synthesize ADH.

3. Therapy Treatment is required for both the tuberculosis and the diabetes insipidus. The tuberculosis is treated with antibiotics, which will be required for prolonged periods. The diabetes insipidus is treated with D,D AVP (see Figure 18.2), a long-acting agonist of vasopressin that can be taken by nasal instillation. The patient will be required to take this drug twice daily, probably for the rest of her life.

References

Geheb MA: Clinical approach to the hyperosmolar patient, Crit Care Clin 3(4):797, 1987.

Germon K: Fluid and electrolyte problems associated with diabetes insipidus and syndrome of inappropriate antidiuretic hormone, Nurs Clin North Am 22(4):785, 1987.

Moses AM and Miller M: Osmotic threshold for vasopressin release as determined by hypertonic saline infusion and dehydration, Neuroendocrinology 7:219, 1971.

Moses AM and Notman DD: Diabetes insipidus and syndrome of inappropriate ADH, Adv Intern Med 27:73, 1982.

Robertson GL: Differential diagnosis of polyuria, Annu Rev Med 39:425, 1988.

Schmale H, Fehr S, and Richter D: Vasopressin biosynthesis from gene to peptide hormone, Kidney Int (Suppl)21:S8, 1987.

Case 2   ◆ Hypogonadotropic hypogonadism with midline defects (Kallmann's syndrome)

A 21-year-old woman, a star basketball player, 193 cm in height, never had a menstrual period. She was average in height, but unlike her friends, she did not stop growing at age 14. She continued to grow until about a year ago. Although she developed axillary and pubic hair at age 12, she never developed breasts. She was born with a cleft palate, which was surgically corrected, and never was able to smell anything except very strong, painful odors, such as ammonia. There was a strong family history for this condition in both males and females; her brother and three cousins had a similar syndrome. On physical examination, she was a very tall, thin, young woman, looking much younger than her stated age. She had a scar from the cleft palate repair and was unable to distinguish the smell of coffee from that of cigarettes. Her arm span was 205 cm, and her upper body length was 90 cm. She had no breast development, and her nipples were infantile. Pubic and axillary hair were normal, but on pelvic examination a very small uterus and normal ovaries were felt. Routine laboratory examination was normal. Her LH, FSH, and estradiol levels were 3.1, 4.2, and 15 ng/ml (normal for follicular phase, 3 to 19, 4 to 20, and 50 to 200 ng/ml). Prolactin, testosterone, and progesterone levels were normal for a female. A computed tomogram of the pituitary was normal, whereas one of the pelvis showed infantile female organs.

Biochemical questions

1. Described the possible abnormalities that could cause a syndrome such as this.
2. How would you distinguish among these abnormalities?
3. How would you treat this patient?

Case discussion

1. Possible abnormalities Gonadal failure is evident in this patient. She has a low level of estradiol, low gonadotropins, and no development of her sexual organs, although she does have pubic and axillary hair. She is very tall and "eunicoid," with an arm span greater than her height, very long legs, and an upper body segment smaller than her lower body segment. In this case the syndrome seems to be hereditary and accompanied by peculiar midline defects (cleft palate and disorders of smell). In evaluating possible abnormalities, one should consider secretion of GnRH by the hypothalamus, failure of secretion of LH and FSH by the pituitary gland, or failure of the ovaries. The last is not possible in this case, since ovarian failure would lead to a low estradiol level, but a lack of feedback on the hypothalamic-pituitary axis would lead to very high LH and FSH levels, which are not present. Failure of the hypothalamic-pituitary area can result from tumors or other mass lesions, hyperprolactinemia, excess androgen or progesterone production, or primary failure or delay of maturation of the gonadotropins.

In this case the computed tomogram does not demonstrate a tumor or other mass lesion. Her prolactin level is normal, as are her testosterone (and other androgens) and progesterone levels. Thus she apparently has a primary problem of the gonadotropins. The midline defects and hereditary nature suggest that she has Kallmann's syndrome, a hypothalamic disease with hypogonadotropic hypogonadism.

2. Distinguishing abnormalities GnRH cannot be measured in humans or normal animals, since it exists for only a few seconds in the pituitary portal circulation. Since one cannot measure the GnRH, one can separate hypothalamic from pituitary problems by studying the effect of adding exogenous GnRH. In this test GnRH is administered intravenously, and the effect on LH and FSH is noted. Patients with pituitary disease caused by pituitary destruction or by specific biochemical abnormalities in the phosphatidylinositol cycle or hormone synthesis/secretion will not respond to the GnRH. Patients with hypothalamic disease will still have a normal pituitary gland and will have a significant rise in LH and FSH levels. This test is positive in this patient, with a rise in both LH and FSH levels, suggesting that the problem is at the hypothalamus, and is causing a deficiency of GnRH.

3. Therapy Three treatments are available to this patient. The most physiologic therapy would be to replace the GnRH that she does not produce spontaneously. As a peptide, GnRH cannot be taken orally but must be given transmucosally (as nasal spray) or injected. It is usually given by the latter methods by special fertility pumps. These devices give small amounts of GnRH for 6 min each hour, causing the pituitary gland to cycle normally. For patients with hypothalamic failure, as in this case, this may be the best therapy, although expensive and difficult.

Patients with pituitary failure cannot respond to GnRH and must be given either estrogen replacement, such as oral contraceptives, or if they desire fertility, an LH-like hormone (hCG) and FSH, by daily injection. This patient chose estrogen/oral contraceptive therapy and after 2 yr was fully developed sexually.

References   Berezin M et al: Successful GnRH treatment in a patient with Kallmann's syndrome, who previously failed HMG/HCG treatment, Andrologia 20(4):285, 1988.

Bergstrom RW et al: Hypogonadotropic hypogonadism and anosmia (Kallmann's syndrome) associated with a marker chromosome, J Androl 8(1):55, 1987.

Byer RM: Clinical and laboratory heterogeneity in idiopathic hypogonadotrophic hypogonadism, J Clin Endocrinol Metab 43:1268, 1976.

Garg SK, Bandyopadhyay PK, and Dash RJ: Hypogonadotropic hypogonadism, Trop Geogr Med 39(3):296, 1987.

Vasquez JM and Greenblatt RB: Pituitary responsiveness to luteinizing hormone releasing hormone in different reproductive disorders: a review, J Reprod Med 30(8):591, 1985.

Case 3   ◆ Symptoms of hypocalcemia

An 18-year-old male had a problem of numbness and tingling around his mouth and in his fingertips. The problem was intermittent, usually occurring during times of great stress, such as examination periods. These episodes never occurred when he was on vacation. Aside from this problem, he had been in good health with no diarrhea. He drank about three large glasses of milk each day. His physical examination was normal, as were his laboratory values. Specifically, his total calcium concentration was 9.6 mg/dl, with an ionized calcium of 1.8 mEq/L (normal, 8.5 to 10.6 mg/dl and 1.5 to 3.5 mEq/L). After breathing rapidly for about 3 min at 30 breaths per minute, he stated that he had the same symptoms. At this time he had increased irritability of his seventh cranial nerve (Chvostek's sign) and carpal-pedal spasm on oxygen deprivation of his hand (Trousseau's sign), both indicating clinical hypocalcemia. His serum calcium concentration remained at 9.6 mg/dl, but his ionized calcium level fell to 1.1 mEq/L.

Biochemical questions   1. This patient has symptoms of hypocalcemia, but his serum calcium concentration was initially normal. Explain the mechanism.
2. How might you treat this patient?

Case discussion   **1. Mechanism for hypocalcemic symptoms** Symptoms of perioral and fingertip paresthesias are very common in patients with hypocalcemia. When seen in an intermittent pattern such as this, these symptoms are almost always caused by *hyperventilation*. Calcium balance is complicated by the binding of calcium to albumin. More than 50% of the circulating calcium is bound, and this binding to anionic sites is highly dependent on the pH of the blood. Small increases in pH can free many more anionic sites for the binding of calcium. Since this can occur in just a few seconds, parathyroid hormone is unable to compensate for the hypocalcemia, and symptoms occur. Hyperventilation can reduce the carbon dioxide of the blood and raise the pH by 0.10 to 0.15, dropping the ionized calcium by up to 30% without altering the total serum calcium concentration.

2. Therapy In this patient, as in most, the hyperventilation is related to anxiety, and the therapy should be directly at lowering the patient's anxiety about tests and other matters. Counseling will be necessary for this. Meanwhile, the patient must treat his symptoms by avoiding the diminished carbon dioxide that accompanies the hyperventilation. He can do this by breathing into a paper bag, thereby increasing the carbon dioxide content of the inspired air.

References   Edmonson JW, Brashear RF, and Li TK: Tetany: quantitative interrelationships between calcium and alkalosis, Am J Physiol 228:1082, 1975.

Fanconi A and Rose GA; The ionized, complexed and protein-bound fractions of calcium in plasma, QJ Med 17:463, 1958.

Case 4 ◆ Hypoglycemia

A 27-year-old registered nurse was admitted for hypoglycemia and was unconscious. She was accompanied by her mother, who had witnessed several of these hypoglycemic episodes. They had occurred about once a week for the past year but had become much more severe lately. Initially they were characterized by confusion and lethargy and relieved by eating, but the last one resulted in coma. The patient was brought to the emergency room, where her blood sugar level was found to be 1 mM (normal, 3.5 to 6 mM). She was treated with intravenous glucose and admitted to the hospital. The patient stated that just before the episodes occurred she felt very hungry, shaky, and sweaty and could feel her heart pounding. If she ate immediately, she could blunt the attack. The spells had no relationship to meals. She frequently skipped meals without bringing on an attack, and they never occurred in the morning before breakfast. The spells were always witnessed by her family and friends, who stated that she was absolutely normal just a few minutes before the spells. They never occurred when she was alone or working. Since admission, she had one mild spell; her blood sugar level was 2.5 mM, which was rapidly reversed with oral glucose. Until these spells started, she was in perfect health and had no other problems. She denied alcohol abuse. She did not have diabetes mellitus and denied insulin use. Physical examination revealed a thin, anxious female with normal blood pressure and pulse. No skin lesions or hepatomegaly (enlarged liver) were present. The remainder of the examination was normal. Laboratory examination, including liver and kidney tests, was normal except for the insulin and C-peptide levels drawn during the two hypoglycemic episodes, which were 28 and 25 nM (normal fasting, 0.02 to 0.1; postprandial, 0.1-0.7 nM). The C-peptide in both samples was undetectable. Computed tomograms of the chest, abdomen, and pelvis were normal.

Biochemical questions

1. Discuss the biochemical mechanisms of glucose regulation in the fasting and fed states.
2. How might the normal regulation of glucose be altered to cause hypoglycemia?
3. Which causes are the most likely ones affecting this patient? How would you disprove the others?
4. A set of duplicate insulin samples is sent to another laboratory. They call because their computed assay reported negative values, which are physiologically impossible. They are worried about their computer, but you can reassure them. Explain the strange values for insulin in the two assays.

Case discussion

1. Glucose regulation The blood glucose level is normally carefully regulated over a small range. In the fed state, most cells of the body will switch to the utilization of glucose as a metabolic fuel, as signaled to do so by increases in insulin levels. The glucose entering the bloodstream from the intestines will be used for energy by muscle and other tissues, stored as glycogen by the liver and muscles, and converted to fat and stored by the liver and adipose tissues. It will also be converted to amino acids, nucleotides, and other substances as necessary. Thus in the fed state metabolic regulation is designed to utilize or store the excess glucose. This must be done carefully to avoid overutilizing glucose and quickly to avoid excess glucose in the bloodstream, with osmotic and other consequences. In the fasted state, conservation of glucose is the primary goal. Tissues that can utilize either glucose or fat for energy switch to fat metabolism, and even tissues that normally utilize only glucose, such as the brain, alter their metabolism to obtain part of their energy from ketone bodies. This metabolic alteration is accomplished by decreases in insulin levels and increases in glucagon, somatotropin, and cortisol, which leads to increased lipolysis, decreased lipogenesis, and decreased glycogen synthesis. During the first few hours of a fast the plasma glucose level is maintained by the breakdown of liver glycogen to glucose 1-phosphate, conversion to glucose 6-phosphate, and dephosphorylation to glucose. Once the supply of liver glycogen is depleted, other sources of glucose must be utilized. Muscle glycogen is much more plentiful than liver glycogen, but muscle lacks glucose 6-phosphatase and cannot directly release glucose. Instead, the muscle metabolizes the glucose 6-phosphate to lactate, releasing the lactate into the plasma. The liver will synthesize glucose from this lactate. A small amount of glucose can also be formed from the glycerol released from the hydrolysis of triglycerols. Finally, when all glycogen from both the liver and the muscle is depleted, the only major source of glucose is the conversion of gluconeogenic amino acids to glucose. These amino acids, mostly alanine and glutamine, are derived from the catabolism of muscle proteins.

2. Alteration in glucose regulation True hypoglycemia, such as that documented in this patient by a low blood glucose level measured in the laboratory, is caused by an imbalance of glucose production and utilization. Increased glucose utilization can result from (a) excess insulin action from a tumor producing insulin, insulin injections, sulfonylurea usage, or a lack of the antiinsulin hormones (glucagon, somatotropin, cortisol, or epinephrine); (b) excess insulin-like growth factor II production by a tumor, usually a very large tumor; or (c) very large cancers that utilize glucose as their main energy source, even in the absence of insulin. Decreased glucose production can result from (a) liver or kidney failure, since these are major organs of glucose production; (b) alcoholic binges, in which the metabolism of ethanol reduces most of the cellular NAD to NADH and prevents gluconeogenesis; or (c) glycogen storage diseases, which prevent the breakdown of glycogen by either the liver or the muscle to form glucose.

3. Probable causes In view of the normal physical examination and normal chest and abdominal computer tomograms, a large tumor producing IGF-II or utilizing excess glucose is unlikely. Normal kidney and liver tests disprove serious disease of these organs, and the lack of alcoholic symptoms before the witnessed episodes disproves alcoholic binges. Glycogen storage disease is unlikely on the basis of the patient's age and lack of other symptoms and signs. Thus it appears that she has overproduction of insulin, resulting from an insulin-producing tumor (it may be too small to be detected by the tomogram), or she uses sulfonylureas or insulin. The C-peptide values are very helpful. Insulin produced by the pancreas is initially produced as proinsulin and then cleaved to insulin and C-peptide, which are secreted in equimolar amounts. If there were a tumor producing insulin or if insulin were being induced by sulfonylureas, one would expect both a high insulin level and a high C-peptide level. Commerical insulin, however, has no C-peptide. The lack of C-peptide in the face of high insulin levels implies that the patient is injecting herself with insulin.

4. Contradictory insulin values The insulin assays are initially puzzling. In one assay the values are so high that they cannot be achieved, even with large injections of insulin, and in another assay, negative values are obtained. One would suspect that the patient has been injecting herself with insulin for a year, and she probably has some antibodies to insulin. The unusual values for insulin are caused by the assumption of radioimmunoassays that the small amount of exogenous antihormone antibody added as part of the assay is the only antibody to the hormone. In this case the patient has additional antibody to insulin present in her serum, and when the serum to be tested was added to the assay, so was additional antibody. The effects of this extra antibody on the assay depend on the method of separation of the hormone bound to antibody from the free hormone. If the separation method includes hormone bound to the patient's antibody with the bound hormone, the assay will generate low or even negative values. If the antibody excludes hormone bound to the patient's antibody from the bound fraction, the assay will yield extremely high values (see Horwitz, 1989 for more details).

This patient continued to deny insulin use and was transferred to a psychiatric ward, where after 2 wk of intensive psychotherapy, she confessed to her attention-arousing behavior. With appropriate counseling, the episodes resolved.

References

Daughaday WH: Hypoglycemia in patients with non-islet cell tumors, Endocrinol Metab Clin North Am 18(1):91, 1989.

Dohm GL et at: Protein metabolism during endurance exercise, Fed Proc 44(2):48, 1985.

Forman DT: The effect of ethanol and its metabolites on carbohydrate, protein, and lipid metabolism, Ann Clin Lab Sci 18(3):18, 1988.

Horwitz DL: Factitious and artifactual hypoglycemia, Endocrinol Metab Clin North Am 18(1):203, 1989.

Samaan NA: Hypoglycemia secondary to endocrine deficiencies, Endocrinol Metab Clin North Am 18(1):145, 1989.

Service FJ: Hypoglycemic disorders: pathogenesis, diagnosis and treatment, Boston 1983, GK Hall & Co.

Wright J and Marks V: The effects of alcohol on carbohydrate metabolism, Contemp Issues Clin Biochem 1:135, 1984.

Case 5

◆ Hyperparathyroidism

A comatose 56-year-old female was transferred from the psychiatry service. The patient had a long history of depression, with three attempts at suicide by drug overdose. Over the past 20 yr she had four kidney stones that passed spontaneously. The last was analyzed and found to have a nidus of calcium phosphate. She had always complained of a variety of aches and pains, but no arthritis was found. She had a single episode of pancreatitis 2 yr ago. This was attributed to alcoholism, which she denied. Because of her depression, she was treated with electroconvulsive therapy. When the anesthesia wore off, she did not wake up. Physical examination revealed band keratopathy (calcium deposits in the cornea), diminished reflexes, and a neurologic examination consistent with a mild coma of metabolic origin. Her electrocardiogram was consistent with diffuse cardiac disease or hypercalcemia. Chest roentgenograms showed normal lungs and heart but diminished mineral content of the bones, particularly over the lateral aspects of the clavicle. Hand roentgenograms showed small cysts in the first and fourth metacarpals of the left hand and subperiosteal reabsorption of bone at multiple sites. Laboratory examination demonstrated normal sodium and potassium levels but a bicarbonate level of 17 mEq/L (normal, 20 to 28), chloride of 109 mEq/L (normal, 98 to 106), calcium of 18.3 mg/dl (normal, 8.5 to 10.5), and phosphate of 1.7 mEq/L (normal, 3.5 to 5).

Biochemical questions

◆ 1. What conditions might cause the elevated calcium level?

◆ 2. How would you confirm which of these conditions was the cause of the elevated calcium level in this patient?

◆ 3. What therapy might you use to lower the calcium concentration?

Case discussion

1. Causes of elevated calcium level The patient manifests the classical history of an elevated calcium level with kidney stones, deep bone pain, depression, and pancreatitis. One rarely sees this combination today, since most patients are diagnosed early on routine screening of laboratory values before symptoms appear. The history of this patient goes back for many years, excluding many of the more serious causes by hypercalcemia, such as cancer. The combination of high calcium and low phosphate levels suggest increased action of PTH on the kidney. This is further suggested by the high chloride and low bicarbonate levels, since PTH causes an increase in bicarbonate wasting at the distal tubule of the kidney. This excretion of bicarbonate must be accompanied by the preservation of another anion, in this case chloride, which causes hyperchloremia.

Evidence also exists for increased action of PTH on the bones. The thinning of the bones suggests diminished calcification of the bones, a condition called osteomalacia, but a normal roentgenogram is not sufficiently sensitive or specific to make the diagnosis. In this case a computed tomogram of the spine showed decreased mineralization (55% of age-matched control). The findings of thinning of the lateral aspects of the clavicle and subperiosteal reabsorption are specific for hyperparathyroidism. The latter, which appears on x-ray film as a "flea-bitten" appearance of the fingers, is caused by increased activity of the osteoclasts, which resorb the bone of the fingers.

2. Confirmation of cause To elucidate the cause of the hypercalcemia further, a PTH level (whole hormone assay) is done. The level is 52 pg/ml (normal, 0 to 55), at a time when the calcium level is 18 mg/ml. Although the value of PTH is normal, it is inappropriately high for the very elevated calcium level. Even at a calcium level of 10.5 mg/ml, the upper limit of normal, the serum calcium should have inhibited the synthesis and secretion of PTH. At a calcium level of 18 mg/ml, the PTH should be undetectable. Increased action of PTH on the kidney can be monitored by measuring cAMP in the urine. This can be corrected for excretion of cAMP elsewhere in the body, which yields the amount of cAMP produced by the kidney, called nephrogenous cAMP. In this patient the level is 15 (normal 0 to 6), which strongly suggests increased PTH action on the kidney (increased vasopressin could also raise nephrogenous cAMP, but no reason exists to suspect this). To determine the localization of the benign adenoma of the parathyroid that probably is causing this problem, a sonogram (a study of the echo patterns of sound waves bouncing off the tissues) of the neck is performed, demonstrating a 15 g tumor in the left lower parathyroid gland.

This patient became seriously hypercalcemic in this setting because of dehydration. She has been hypercalcemic for years, but by maintaining a good urine flow, she was able to keep her calcium at reasonable levels (13 to 15 mg/ml). In this case, before the electroconvulsive therapy, she was not allowed to eat or drink for 12 to 16 hr. This lack of fluid, combined with the diuretic effect of hypercalcemia, led to water loss and diminished blood flow to her kidneys. She probably decompensated under anesthesia and, with a significant increase in serum calcium, became comatose.

3. Treatment The first therapy for this patient is to rehydrate her with a physiologic salt solution called saline (0.154 M NaCl). Expanding her blood volume leads to increased kidney excretion of calcium and lowers her calcium levels. Other measures are taken to prevent the dramatic hypercalcemia before surgery, including inhibiting osteoclast activity. This could be accomplished with calcitonin at very high doses, pyrophosphate analogues called diphosphonates that coat bones, or mithromycin, a specifc osteoclast inhibitor. Surgery demonstrates the tumor in the left lower parathyroid gland. The remaining glands are normal, and after removing the tumor, the patient's serum calcium level returned to normal.

References

Broadus A: Nephrogenous cyclic AMP, Recent Prog Horm Res 36:667, 1981.

Johnson KR, Mascall GC, and Howarth AT: Differential laboratory diagnosis of hypercalcemia, CRC Crit Rev Clin Lab Sci 21(1):51, 1984.

Rao DS: Primary hyperparathyroidism: changing patterns in presentation and treatment decisions in the eighties, Henry Ford Hosp Med J 33:194, 1985.

Case 6

Diabetic ketoacidosis caused by hyperthyroidism

An 18-year-old woman arrived at the emergency room in a deep coma. Her mother said that she had been sleepy for at least the last 24 hr. The patient had diabetes mellitus for 8 yr and had been taking insulin twice daily for that time. She generally took excellent care of herself, checking her blood sugar level four times each day, and watched her diet and exercise. Her hemoglobin A_{1c} had been normal. Over the past 3 mo, however, her insulin requirements had been rising, and she was now taking 40% more insulin. Despite this, her glucose values had been in the 18 to 24 mM range (normal, 3.5 to 6), and she had been excreting large amounts of ketone bodies in her urine. She had been complaining of increasing sweating, nervousnesss, and tremors and had lost about 30 lb over the 3 mo, despite a ravenous appetite. On physical examination she was unarousable but moved all limbs to painful stimuli. She had a very fine, resting tremor and fruity breath. Her blood pressure was 120/80 with a pulse of 140, but blood pressure fell to 70/40 with a pulse of 180 on sitting her up. Her skin was hot but dry. Her eyeballs appeared sunken and her mouth was very dry. Her thyroid gland was enlarged to about 100 g (normal, 15 to 20). The remainder of the examination was unremarkable. Laboratory examination showed a blood sugar level of 32 mM (normal, 3.5 to 6), with a sodium level of 125 mEq/L (normal 137 to 145), a potassium of 3.4 mEq/L (normal, 3.5 to 5), chloride of 84 mEq/L (normal, 95 to 104), and bicarbonate of 6 mEq/L (normal, 22 to 28). The pH of her blood was 7.05 (normal, 7.40 to 7.45). Serum β-hydroxybutyrate was 15 mM (normal, <0.3). The white blood cell count was modestly elevated at 11,000, but the differential diagnosis did not indicate an infection. Chest roentgenograms and EKG were normal. Urinalysis showed a high glucose level and ketones but no infection. Thyroid tests indicated a free thyroxine of 4.7 pg/ml (normal, 0.6 to 2.1), with a TSH less than 0.1 ng/ml (normal, 0.5 to 5.8).

Biochemical questions

1. Define the syndrome occurring in this patient.
2. Explain the biochemical basis of its pathogenesis.
3. What initiated the ketoacidosis?

Case discussion

1. Definition of syndrome The patient is suffering from classical diabetic ketoacidosis. In these patients the insulin falls to levels that are so low that fat metabolism is also affected. This usually occurs in newly diagnosed diabetic persons, diabetic individuals who refuse to take their insulin, or those who have some major life stress, such as an infection, that dramatically increases their requirements for insulin.

2. Biochemical basis Two conditions are necessary for the ketoacidosis to occur: (a) insulin levels must be so low that lipolysis by the adipocytes is not inhibited, leading to high concentrations of free fatty acids in the circulation and therefore in the liver, and (b) high levels of glucagon must be present so that the enzyme carnitine acyl transferase is activated, and the free fatty acids are then transported into the mitochondria rather than being esterified to triacylglycerols.

Once the free fatty acids are in the mitochondria, they are oxidized to acetyl CoA and normally would be further oxidized by the Krebs cycle to carbon dioxide and water. Unfortunately, because of the very large amount of acetyl CoA being produced and the inhibition of glucolysis caused by the lack of insulin, insufficient oxaloacetic acid exists to metabolize all the acetyl CoA. Two of the remaining acetyl CoA molecules are joined to form an acetyl CoA, and then a third is added to form hydroxymethylglutaryl CoA. This breaks down spontaneously to acetoacetic acid and acetyl CoA. The net result of this is to join two acetyl CoA molecules into an acetoacetic acid molecule. The large amount of NADH formed by the oxidation of the fatty acids will reduce most of the acetoacetate to β-hydroxybutyrate (by a reaction parallel to pyruvate going to lactate and by the same enzyme). Some of the acetoacetate will lose carbon dioxide for form acetone.

Acetoacetic acid is a strong acid and will find its way into the bloodstream, where it will acidify the blood. The buffering systems of the body, including bicarbonate (going to carbon dioxide, phosphate, and intracellular buffering), will quickly be exhausted, and the pH of the blood will fall rapidly. Thus the patient will have a low pH and low bicarbonate concentration because of altered fat metabolism.

A second part of this syndrome results from altered glucose metabolism. The very high glucose concentration is excreted into the urine, bringing large amounts of water with it and causing serious dehydration. This explains the dry skin, dry mouth, sunken eyes, and fall in blood pressure on sitting. One could estimate that this patient has lost about 10% to 15% of her body water, or 4 to 6 L of fluid. This profound dehydration and the low pH have caused the coma. Her low sodium level is purely reactive. The very high glucose concentration has added about 40 mOsm to the osmolality of the blood, and sodium and chloride levels have fallen by a similar amount to compensate. This patient needs large amounts of fluids and moderate amounts of insulin quickly. Given proper treatment, she has a 98% chance of full recovery with no subsequent damage.

3. Initiation of ketoacidosis Typically diabetic ketoacidosis is initiated by a life stress such as an infection or the death of a close relative. In this case the patient has classic symptoms of hyperthyroidism that have been present for about 3 mo. The biochemistry and symptoms of hyperthyroidism are discussed in Chapter 19. However, hyperthyroidism increases the patient's requirements for insulin, since it causes both insulin resistance and an increased rate of insulin destruction.

References

Foster DW and McGarry JD: The metabolic derangement and treatment in diabetic ketoacidosis, N Engl J Med 309:159, 1983.

Krane EJ: Diabetic ketoacidosis: biochemistry, physiology, treatment, and prevention, Pediatr Clin North Am 34(4):935, 1987.

Mouradian M and Abourizk N: Diabetes mellitus and thyroid disease, Diabetes Care 6(5):512, 1983.

Case 7   Panhypopituitarism

A 24-year-old man sought aid for weakness, dizziness, easy burning of his skin in the sun, and inabilitity to have an erection. He was a normal, sexually active member of a motorcycle gang until his accident about 4 mos ago. He was not wearing a helmet. At admission to the hospital he was unconscious and was noted to have a basilar skull fracture. He awoke 3 days later, and his physician wanted to run further tests on him, but he signed out against medical advice. Laboratory evaluation showed a sodium level of 125 mEq/L, potassium of 4.5 mEq/L, chloride of 90 mEq/L, and bicarbonate of 25 mEq/L. Thyroid tests showed TSH was unmeasurable and thyroxine was low.

Biochemical questions

1. List all the hormones of the pituitary gland. Which might be affected by a fracture of the base of the skull?
2. How might deficiencies in these hormones explain his symptoms of dizziness, weakness, easy burning in the sun, and poor sexual function?
◆ 3. What tests would you do to confirm these diagnoses?

References

Edwards OM and Clark JD: Post-traumatic hypopituitarism: six cases and a review of the literature, Medicine (Baltimore) 65(5):281, 1986.

Keeling FP et al: Fracture of the sella turcica: a report of three cases, Clin Radiol 37(3):233, 1986.

Case 8 — Increased cyclic AMP

A 13-year-old boy was brought to the hospital by his mother because of strange behavior. He had been drinking Ultra-Jolt beverage with 10 times the caffeine as normal cola. Caffeine is a methyl-xanthine that blocks phosphodiesterase activity. It therefore increases intracellullar cAMP.

Biochemical questions

1. List the hormones that alter cAMP and their actions.
2. If the actions normally triggered by these hormones were amplified by the caffeine, what might be the consequences?

References

Cockcroft S and Stutchfield J: G-proteins, the inositol lipid signalling pathway, and secretion, Philos Trans R Soc Lond (Biol) 320(1199):247, 1988.

Meyerhoff JL et al: Regulation of pituitary cyclic AMP, plasma prolactin and POMC-derived peptide responses to stressful conditions, Adv Exp Med Biol 245:17, 1988.

Pilkis SJ, Claus TH, and el-Maghrabi MR: The role of cyclic AMP in rapid and long-term regulation of gluconeogenesis and glycolysis, Adv Second Messenger Phosphoprotein Res 22:175, 1988.

Case 9 — Acromegaly

An 18-year-old diabetic student, who is a 208 cm center on a basketball team, had diabetes for about 4 yr and had been very difficult to treat. He claimed to be following his diet and obviously exercised regularly, but he required more than 120 units of insulin each day (most insulin-dependent diabetic persons require 40 to 60) and still had high blood sugar levels. He had a large jaw and very large hands with a large amount of soft tissue. Because of his extreme insulin resistance, the physician measured antiinsulin antibodies, which were not present.

Biochemical questions

1. Discuss the hormones that might be involved in diabetes. Specificallly focus on those hormones, which when present in excess, might cause the condition.
2. A high growth hormone level is found in this patient. Discuss the regulation of growth hormone secretion.
◆ 3. What are the clinical consequences of growth hormone excess?

References

Dieguez C et al: Growth hormone and its modulation, J R Coll Physicians Lond 22(2):84, 1988.

Frohman LA and Jansson JO: Growth hormone-releasing hormone, Endocr Rev 7(3):223, 1986.

Johnston DG, Davies RR, and Prescott RW: Regulation of growth hormone secretion in man: a review, J R Soc Med 78(4):319, 1985.

Nabarro JD: Acromegaly, Clin Endocrinol (OXF) 26(4):481, 1987.

Case 10	Hypoglycemia 2

The family of a 54-year-old man sought help in evaluating his strange behavior. He had been having episodes about once per week during which he sang bawdy Italian songs and did not recognize his family. He then proceeded to drink a liter of soda and returned to normal behavior about 20 min later. In addition, he had been having spells of extreme anxiety with sweating and shaking on arising each morning. These lasted about 10 min and were relieved by eating breakfast. On one occasion, he had a similar spell when lunch was delayed for about 2 hr. These symptoms had been occurring for about 6 mo, during which he had gained 20 lb. He took no medications and had been exposed to no known toxins. He was known as a heavy binge-type drinker, often consuming more than a quart of whiskey in an evening. He tanned normally and never had liver disease. His physical examination was entirely normal. He returned the next morning, fasting, for laboratory evaluation. He was confused and agitated. His blood sugar level was 2mM. His electrolytes and renal and liver function were normal. His insulin level was 58 μU/ml (normal fasting <10); cortisol, 38 (normal for stress); somatotropin, normal; and catecholamines and glucagon, high. A C-peptide was elevated.

Biochemical questions ◆ 1. Discuss the roles of excess alcohol and insulin in this patient and describe the biochemistry of their effect.

◆ 2. How would you distinguish between the two possibilities in question 1?

References   See references for case 4.

Additional questions and problems

1. Hormones other than proteins can generally be taken orally, since they will not be digested, but proteins and polypeptides must be given by other routes. Large proteins must be injected, but polypeptides and small proteins can often be given intranasally. For each of the hormones of this chapter, discuss the possible methods of administration.

2. Which hormones are made as very large precursors that are then cleaved to oligopeptides? What is the advantage to the organism of "wasting" the synthesis of a 25,000-dalton protein to make a decapeptide?

3. Scientists are actively working on factors that increase blood vessel formation, for example, to help with coronary artery disease. Suppose such a factor were developed, but as a side effect increased the size and efficiency of the hypothalamic-pituitary portal circulation. What would be the physiologic consequences of this increase in blood flow from the hypothalamus to the pituitary gland?

4. Diabetes mellitus is associated with deficiencies of myoinositol and phosphatidylinositol. If there were also an inability to utilize the phosphatidylinositol cycle, which hormones would be affected and how?

5. ACTH and the endorphins have a common precursor molecule. Speculate on the advantages and disadvantages of such a system.

6. Although hypercalcemia is among the most common of endocrine problems, severe hypercalcemia is rare. One of the treatments for severe hypercalcemia is calcitonin. When is it likely to be most effective?

7. Certain antidepressants are monoamine oxidase inhibitors. How might they work?

Multiple choice problems

1. All the following are glycoproteins except:
   A. TSH.
   B. LH.
   C. Prolactin.
   D. FSH.
   E. hCG.

2. Which of the following hormones is made by sequential enzymatic synthesis?
   A. Insulin
   B. Vasopressin
   C. TRH
   D. Norepinephrine
   E. Somatotropin

3. Which pair of hormones work by the same second messenger?
   A. Epinephrine (β-action), norepinephrine
   B. Insulin, glucagon
   C. Prolactin, vasopressin
   D. Calcitonin, insulin
   E. Glucagon, epinephrine (β-action)

4. Which of these is the largest molecule?
   A. Corticotropin-releasing factor
   B. Gonadotropin-releasing factor?
   C. Somatotropin-releasing factor
   D. Thyrotropin-releasing factor?
   E. Dopamine

5. Somatostatin will not inhibit the secretion of:
   A. Insulin.
   B. Glucagon.
   C. Somatotropin.
   D. Prolactin.
   E. TSH.

6. A lack of glycosylation of pituitary hormones would be unlikely to cause:
   A. Impotence.
   B. Amenorrhea.
   C. Hypothyroidism.
   D. Galactorrhea.

7. On a psychiatry unit, a patient breaks into the medication cabinet and ingests large amounts of the following hormones. The physician is worried about the effects of:
   A. Insulin.
   B. Parathyroid hormone.
   C. Glucagon.
   D. Somatotropin.
   E. He is not worried.

8. Select the hormone(s) that are correctly paired with a stimulatory factor and an inhibitory factor:

Hormone	Stimulator	Inhibitor
1. ACTH	TRH	Cortisol
2. Prolactin	TRH	Dopamine
3. TSH	TRH	Dopamine
4. LH	GnRH	Estradiol

   A. 1, 2, and 3.

B. 1 and 3.
C. 2 and 4.
D. 4 only.
E. All are correct

9. A patient develops a thrombosis of the hypothalamic-pituitary portal circulation. Which hormonal axis or axes will be severely affected?
1. Sexual function
2. Salt and water metabolism
3. Thyroid function
4. Calcium metabolism
   A. 1, 2, and 3.
   B. 1 and 3.
   C. 2 and 4.
   D. 4 only.
   E. All are correct.

10. A baby is born without a certain type of hormone receptor. Loss of which of the following receptor(s) would be rapidly lethal and could not be treated by her physician?
1. Parathyroid hormone
2. TSH
3. FSH
4. Insulin
   A. 1, 2, and 3.
   B. 1 and 3
   C. 2 and 4.
   D. 4 only.
   E. All are correct.

11. A new protocol for the treatment of a rare form of cancer involves the use of the antibiotic tunicamycin, which inhibits the glycosylation of proteins. You might expect to find effects of this drug on:
1. Glucose metabolism.
2. Thyroid function.
3. Sexual function.
4. Calcium metabolism.
   A. 1, 2, and 3.
   B. 1 and 3.
   C. 2 and 4.
   D. 4 only.
   E. All are correct.

12. Glucagon mediates its role on glucose metabolism by:
1. Increasing intracellular cyclic AMP.
2. Increasing the activity of glycogen synthase.
2. Reducing the intracellular fructose 2, 6-bisphosphate.
4. Reducing the activity of glycogen phosphorylase.
   A. 1, 2, and 3.
   B. 1 and 3.
   C. 2 and 4.
   D. 4 only.
   E. All are correct.

13. Insulin affects cells by:
    1. Increasing cyclic AMP.
    2. Increasing inositol trisphosphate.
    3. Activating a serine kinase.
    4. Activating a tyrosine kinase.
        A. 1, 2, and 3.
        B. 1 and 3
        C. 2 and 4.
        D. 4 only.
        E. All are correct.

14. Which of the following hormones can be estimated by measuring a by-product or enzyme of its synthesis?
    1. Insulin
    2. Glucagon
    3. Epinephrine
    4. TSH.
        A. 1, 2, and 3.
        B. 1 and 3.
        C. 2 and 4.
        D. 4 only.
        E. All are correct.

Chapter 19

Molecular endocrinology: hormones active inside the cell

Barry H. Ginsberg

Objectives

1 To explain the chemistry of hormones that act by binding to intracellular receptors
2 To describe their synthesis
3 To discuss the regulation of their secretion
4 To explain their actions
5 To identify diseases associated with them

Thyroid hormones

The thyroid hormones are necessary for the proper functioning of almost all the systems of the human body. In most cases the action of thyroid hormone is "permissive"; that is, normal function will not occur in the absence of thyroid hormone, although no specific action can be found for the hormones. In some cases thyroid hormone works specifically to increase metabolism. The two thyroid hormones, both iodinated derivatives of tyrosine, are *thyroxine* (T_4) and *triiodothyronine* (T_3); their structures are shown in Figure 19.1. The thyroid hormones are hydrophobic molecules with a low solubility in water. Their two phenolic rings gives them a strong propensity for *pi* bonding. The hormones are not digested by humans and may be given orally to patients.

Control of secretion

The thyroid gland is stimulated by *thyroid-stimulating hormone* (TSH), which binds to both high- and low-affinity receptors on the cells of the thyroid gland. The low-affinity receptor, present in high concentration, is a ganglioside. The high-affinity receptor, present in low concentration, is a protein linked to adenylate cyclase that stimulates thyroid hormone synthesis and secretion and thyroid gland growth. It consists of two subunits of 30,000 and 50,000 daltons. One has the binding specificity for TSH, and the other is a transmembrane molecule linked to a G_s protein. The specific role of the gangliosides with low binding affinity for TSH is unclear. Some authors have speculated that they may serve to concentrate the TSH in the region of the receptor. There are no known physiologic inhibitors of thyroid gland function.

Synthesis and secretion

Thyroid hormone is synthesized in the thyroid gland by a complex mechanism. Although the precursor for both T_4 and T_3 is tyrosine, the synthesis does not occur on the free amino acid, but rather on the tyrosine residues of the large protein, *thyroglobulin*. The thyroglobulin is synthesized by the cells of the thyroid gland and stored in a follicle surrounded by glandular cells. The formation of thyroid hormone starts within this follicle, as outlined in Figure 19.2.

The first step in the synthesis of thyroid hormone is the accumulation of iodine. This is an active process requiring energy and concentrating iodine up to 10,000-fold over serum. Although the details are not complete, the iodine-concentrating mechanism apparently is linked to a Na^+/K^+-ATPase. Drugs that block the ATPase, such as ouabain, destroy the thyroid's ability to concentrate iodine. The concentration mechanism can also be inhibited by a variety of large, inorganic anions such as perchlorate and thiocyanate. Thus patients with a dietary excess of rutabaga, which contains large amounts of thio-

Figure 19.1   Structure of the thyroid hormones. The thyroid hormones are composed of iodinated thyronines. Thyroxine (T_4) is iodinated at the 3', 5' positions of both phenolic rings. Triiodothyronine (T_3) is iodinated at both positions of the inner ring but only singly iodinated on the outer ring. A thyronine singly iodinated on the inner ring but doubly iodinated on the outer ring is called reverse triiodothyronine (reverse T_3) and is inactive.

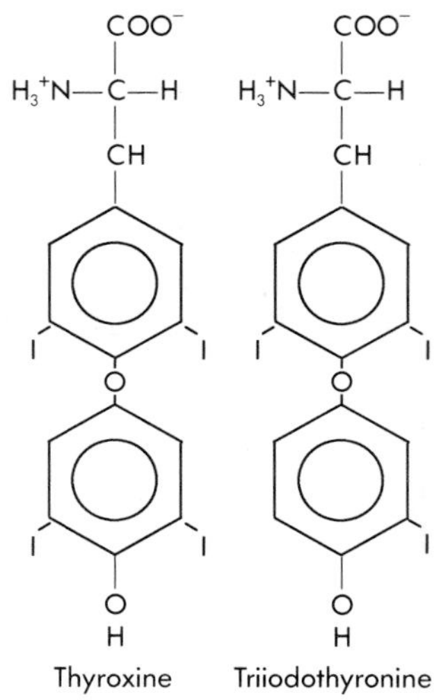

Thyroxine   Triiodothyronine

cyanate, may be hypothyroid because of a lack of intracellular iodine.

The next step in thyroid hormone synthesis involves the organification of the iodine. The site of this is not clear, but electron microscopic evidence indicates that it occurs at the apical plasma membrane or within the storage follicle itself. The iodine concentrated by the gland is in the form of iodide ions. This iodide reacts with hydrogen peroxide and tyrosine residues on thyroglobulin to form *monoiodotyrosines* (MITs) within the thyroglobulin. The activity is catalyzed by the enzyme thyroperoxidase. The hydrogen peroxide is formed by a specific enzymatic complex on the apical membranes from reduced nicotinamide adenine dinucleotide phosphate (NADPH) and oxygen. MIT undergoes a similar second reaction to form *diiodotyrosine* (DIT), which is catalyzed by the same hormone.

The last step in the formation of the thyroid hormones is the coupling of iodotyrosines to form thyronines. Coupling occurs by the translocation of the iodophenolic ring from one iodotyrosine to the phenol group of another, resulting in the formation of a thyronine and serine (at the tyrosine that has lost its phenolic group). The coupling of two DITs yields a T_4, and the coupling of a MIT to a DIT forms a T_3. MITs are usually not coupled together; this would result in an inactive compound. Under usual conditions, almost all thyroid hormone production by the thyroid gland is T_4 ($>99\%$).

The thyroglobulin, which may contain up to eight thyroxine molecules, is stored as *colloid*. It is pinocytosed by the cells and proteolytically degraded, releasing the T_4 and T_3. The thyroid cells salvage the iodine from the MIT and DIT and reuse it. An inability to salvage iodine from the iodotyrosines, although very rare, leads to iodine deficiency and hypothyroidism (see case 2 at the end of this chapter).

Figure 19.2

Synthesis of the thyroid hormones. The thyroid hormones are synthesized on thyroglobulin. Thyroperoxidase catalyzes the incorporation of iodine into the 3′ and 5′ position of tyrosine, forming primarily diiodotyrosine (DIT) and some monoiodotyrosine (MIT). Thyroperoxidase then catalyzes the movement of one iodophenolic ring onto another iodotyrosine, forming T_4 from two DITs or T_3 from a MIT and DIT. A serine is left behind on the thyroglobulin backbone. The T_4 and T_3 are released by proteolysis of the thyroglobulin.

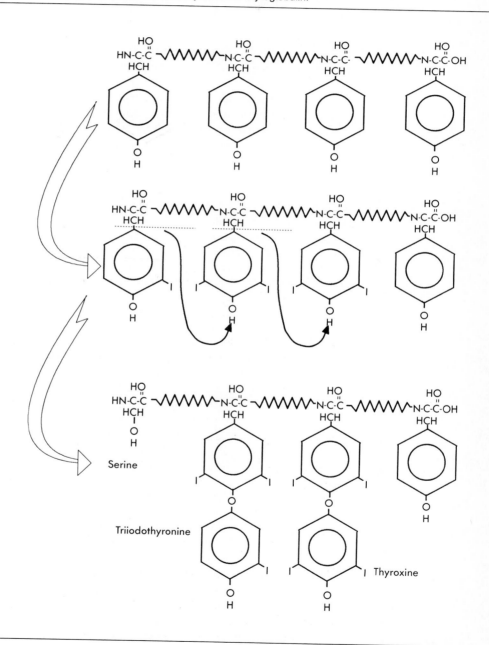

Almost all T_3 is not synthesized by the thyroid gland, but rather is formed by removal of an iodine from the outer ring of T_4 in other tissues. Many of these tissues also have the ability to remove an iodine from the inner ring to form *reverse T_3*. This is an inactive compound found in large amounts in patients with chronic disease. This unusual deiodination does not lead to thyroid disease, but it does confuse the physician because the thyroid tests may show an abnormality in the T_3 level. Patients with it are called *euthyroid sick*.

Transport

The thyroid hormones are transported in the blood almost completely bound to specific binding proteins. The bound form of the hormones are inactive, although the equilibrium reaction to release them is very rapid. The bound form of T_4 accounts for 99.97% of all T_4, and the bound form of T_3 99.8% of the circulating level of that hormone. In other words, 0.03% of T_4 and 0.2% of T_3 circulate free.

The thyroid hormones are bound to three different thyroid-binding proteins (Table 19.1). *Thyroid-binding globulin* (TBG) binds about 80% of T_4 and 55% of T_3. TBG is a glycoprotein with a molecular weight of 54,000 and a single binding site per molecule and an association constant of 1×10^{10} for thyroxine and 1×10^9 for T_3. *Thyroid-binding prealbumin* (TBPA) is a 55,000-dalton protein with two binding sites. Its association constant is two orders of magnitude lower than TBG, but because of its much greater concentration, TBPA binds approximately 15% of T_4 and 25% of T_3. Finally, albumin, with a molecular weight of 69,000, has six to eight binding sites. Although its association constant is only modest (10^5 to 10^7), albumin is present in enormous concentrations and therefore binds 5% of T_4 and 20% of T_3.

The amount of these proteins varies in some physiologic and pathologic states. When this occurs, the fraction of thyroid hormone bound changes, but the free hormone concentration remains unchanged. For example, TBG is increased by estrogens and therefore is high in women taking oral contraceptives and in pregnant women. TBG is also increased by some drugs, such as the barbiturates and diphenylhydantoin (an antiseizure drug). In these cases the total hormone concentration may be great because of higher bound hormone, but the free, active fraction is normal. The concentration of TBG is diminished by androgens and by an X-linked genetic condition.

Table 19.1   Binding proteins for steroid hormones

Protein*	Molecular weight	Hormone	Affinity	% Bound	Estrogen effect
TBG	54,000	T_4	1×10^{10}	80	Increased
		T_3	1×10^9	55	Increased
TBPA	55,000	T_4	1×10^8	15	None
		T_3	1×10^8	25	None
Albumin	69,000	T_4	1×10^6	5	None
		T_3	1×10^6	20	None
Vitamin D–binding protein	50,000	Vitamin D	—	90	None
Transcortin	50,000	Cortisol	3×10^8	75	Increased
Albumin	69,000	Cortisol	10^6	15	None
SHBG	54,000	Testosterone	10^8	75	Increased
		Estradiol	10^7		Increased

*TBG, Thryoid-binding globulin; TBPA, thyroid-binding prealbumin; SHBG, sex hormone–binding globulin.

The binding globulins serve (1) to increase the duration of the hormones, (2) to protect them from degradation and renal excretion, and (3) to buffer the organism from abrupt changes in hormone level. Since T_4 is significantly bound to proteins, it has a half-life of 4 to 7 days. T_3, with less binding to the thyroid-binding proteins, has a half-life of only 1 to 3 hours. Only the unbound form of the thyroid hormones is active; this concentration is normally kept constant, however, since this form has the feedback on the hypothalamic-pituitary axis.

Sites of action

Four different sites of action of thyroid hormone within the cell have been suggested. Binding proteins for the thyroid hormones have been found in the plasma membrane, the cytoplasm, the mitochondria, and the nucleus (Figure 19.3). Of these sites, only the nucleus has been clearly demonstrated to be a site of initial thyroid action. Thyroid has its main action by binding to specific receptors in the nucleus. At this site receptors have a specificity for various thyroid hormones that is identical to the biologic activity of these hormones. After binding of the thyroid hormone by the nuclear thyroid hormone receptor, a protein with molecular weight about 50,000, the receptor binds to specific sites on the deoxyribonucleic acid (DNA) and increases the transcription of specific genes.

The evidence for specific binding of thyroid at the other sites is less certain. The plasma membrane receptor does not appear to be functioning as a receptor, but rather as a transport system for uptake of thyroid hormone by cells. It is a high-affinity system requiring energy for the transport process. The mitochondrial receptors for thyroid hormone may be involved in the stimulation of some mitochondrial proteins, but this remains controversial. The cytoplasmic receptors are currently thought to be artifacts of cellular preparation (see Chapter 17).

Biologic effects

Thyroid hormone works on nearly every cell of the body. Nuclear receptors for thyroid hormone are universal, and thyroid hormone seems to be necessary for the appropriate functioning of almost all cells. Thus it is sometimes referred to as a *pleotropic* hormone.

Thyroid hormones stimulate the metabolic rate in most animals. This is reflected as both an increase in thermogenesis (heat production) and an increase in the rate of metabolism of other substances, such as an increased clearance rate of drugs. The increased rate of thermogenesis is thought to result from an increase in Na^+/K^+-ATPase activity. This enzyme is responsible for maintaining the normal gradient of high extracellular sodium and high intracellular potassium. The plasma membrane, however, is intrinsically "leaky"; trying to maintain an increased gradient by increasing the number of "pumps" is a futile effort that simply leads to increased hydrolysis of adenosine triphosphate (ATP). This leads to increased heat production.

Clinical comment

Thyroid disease Diseases of the thyroid gland are the second most common type of endocrine disease. They may be manifest as an excess of thyroid hormone, called *thyrotoxicosis* or *hyperthyroidism,* or as a lack of thyroid hormone, called *hypothyroidism.* In both states the problem usually lies in the thyroid gland, not in the pituitary gland. Alterations in thyroid hormone resulting from pituitary disease occur much less frequently.

Hyperthyroidism is a common disease of young females. It is usually caused by an immune alteration, leading to the production of antibodies called *thyroid-stimulating immunoglobulins* (TSIGs). These antibodies are capable of binding to the TSH receptor and stimulating adenylate cyclase. The thyroid gland is stimulated to produce thyroid hormones and to grow, but the thyroid hormone produced is not capable of shutting off the production of the TSIG. No negative feedback occurs, and the inappropriately regulated system hyperfunctions.

Such patients manifest symptoms of hypermetabolism. The most common symptoms

Figure 19.3

Sites of action of the thyroid hormones. Thyroid hormones may have four different sites of action: the plasma membrane, the mitochondria, the cytoplasm, and the nucleus. See text for details. *mRNA,* Messenger ribonucleic acid.

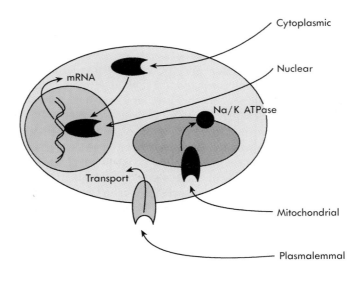

are weight loss, despite an enormous appetite; heat intolerance; an inability to adapt to even modestly warm environments; excessive sweating; and extreme nervousness. In this condition the blood levels of the thyroid hormones are very high, but the blood levels of TSH are very low because T_4 and T_3 shut off pituitary synthesis of TSH. This condition can also be caused by tumors or autonomously functioning nodules of the thyroid gland.

The levels of TSIG may be less than the equivalent of the normal TSH level. For example, suppose the TSIG stimulated the thyroid gland only about half as much as the normal TSH. Since the TSH axis is still functioning, the pituitary gland would produce about half the amount of TSH that is normally produced, and the patient would have normal thyroid hormone levels and would be considered *euthyroid*. Since the patient is producing some TSH, one could not determine that a problem existed with the axis unless one measured TSIG (not a typical clinical assay). If the TSIG were high enough, one might note that the TSH was inappropriately low for the T_4 and T_3 levels.

Hyperthyroidism caused by elevated TSIG levels is often associated with a goiter, protuberance of the eyes, and a peculiar skin condition of the anterior leg, together called *Graves' disease*. When these other features are present but the patient is not hyperthyroid, the condition is known as *euthyroid Graves' disease*.

Hypothyroidism also usually results from an autoimmune disease, in this case called *Hashimoto's thyroiditis (disease)*. The cellular immune system proceeds to destroy the cells of the thyroid gland. Early in the disease the TSH level rises as the pituitary gland senses slight falls in the levels of the thyroid hormones, with slight decreases in negative feedback. The increased TSH pushes the thyroid gland and maintains a nearly normal thyroid state, until the gland ultimately fails completely. The hallmark of early disease is a high TSH level with normal T_4 and T_3 levels.

Patients with hypothyroidism have a lowered metabolic rate. They are always cold and gain weight despite normal or diminished caloric intake. They become very susceptible to overdosage of medications because they metabolize them more slowly.

D vitamins
Synthesis of the cholecalciferols

The regulation of calcium metabolism is very complex and requires at least three hormones: parathyroid hormone, calcitonin, and vitamin D (cholecalciferol) derivatives. The first two hormones are discussed in Chapter 18. *Cholecalciferol* is a steroid derived from *7-dehydrocholesterol* (or egosterol) by a light-mediated ring cleavage. Vitamin D is only a necessary dietary factor in the absence of significant sun exposure, but in industrialized societies the amount of clothing worn and the amount of time spent indoors makes the addition of cholecalciferols to diets mandatory. Two forms of cholecalciferol exist: vitamin D_2, formed by the irradiation of the yeast steroid, ergosterol, and vitamin D_3, naturally found in animals and formed by the irradiation of 7-dehydrocholesterol. They differ only in the structure of the side-chain at the steroid 17 position (Figure 19.4).

The photolysis of the ring of 7-dehydrocholesterol is mediated by ultraviolet light with wavelengths of 290 to 315 mm. In adults it occurs primarily in the epidermis, although in neonates it also occurs in the dermis. Agents that block sunlight from reaching the skin, such as the sunscreens to protect against sunburn, also decrease the amount of photolysis of 7-dehydrocholesterol. A sunscreen with a factor of 6 to 8 will completely block intrinsic vitamin D synthesis. Deeply pigmented skin also blocks photolysis. The photolysis of 7-dehydrocholesterol produces previtamin D, which rapidly isomerizes to vitamin D_3 (Figure 19.5). This isomerization is regulated by both temperature and light; thus excessive light will produce largely inactive compounds rather than causing vitamin D intoxication.

The cholecalciferols are very soluble in fat, and large amounts of these vitamins can be stored in adipose tissues. Ingested vitamin D is absorbed in the small intestine along with fat. After ingestion the vitamin D is initially found in the chylomicron fraction of the blood plasma, but then it slowly moves to a specific vitamin D–binding protein. This protein has a higher affinity for vitamin D_3 than D_2. In states of fat malabsorption, vitamin D is also malabsorbed and a deficiency syndrome can occur. Rarely, patients using nonabsorbable oils (e.g., mineral oil) in large doses will have a deficiency of fat-soluble vitamins, including vitamin D.

Vitamin D_2 and D_3 are inactive and must be made into the active hormone by a series of hydroxylations that occur in the liver and kidney. The first hydroxylation of cholecalciferol, at the 25 position, occurs in the liver (Figure 19.6). This is catalyzed by the enzyme cholecalciferol 25-hydroxylase, an enzyme found in both the mitochondria and the microsomes of the liver, and requires molecular oxygen, NADPH, and magnesium. The enzyme is present in excess concentration, and the 25-hydroxylation does not appear to be hormonally regulated. The amount of 25-hydroxycalciferol is a constant fraction of the ingested cholecalciferol, even when enormous doses of the vitamin are given.

The kidney is the site of further activation of the cholecalciferols. The principal activation step of vitamin D metabolism occurs here, catalyzed by the enzyme 25-hydroxycholecalciferol-1-α-hydroxylase. The enzyme is found in the mitochondria and is a cytochrome P_{450} mixed-function hydroxylase. It requires both molecular oxygen and NADPH. The hormone produced by this hydroxylation, *1,25-dihydroxycholecalciferol*, is the most active form of vitamin D and is thought to account for almost all the actions of vitamin D. Not surprisingly, therefore, the enzyme that produces this hormone is carefully regulated by parathyroid hormone (PTH). In the absence of PTH, almost no activity of 1-hydroxylase occurs. This enzyme producing 1,25-dihydroxycholecalciferol is also strongly regulated by serum phosphate concentrations. High levels of serum

Figure 19.4

Structure of the cholecalciferols. Vitamin D exists in two forms: vitamin D_3, normally formed in animals by the action of sunlight on dehydrocholesterol in the skin, and vitamin D_2, formed by irradiation of the yeast sterol, ergosterol. Vitamin D_2 and D_3 are not biologically active; they must be further hydroxylated at the 1 and 25 positions by specific mixed-function oxidases to form the active hormones, 25-hydroxycholecalciferol and 1,25-dihydroxycholecalciferol.

Vitamin D2

Vitamin D3

25-Hydroxy-
cholecalciferol

1,25 Dihydroxy-
cholecalciferol

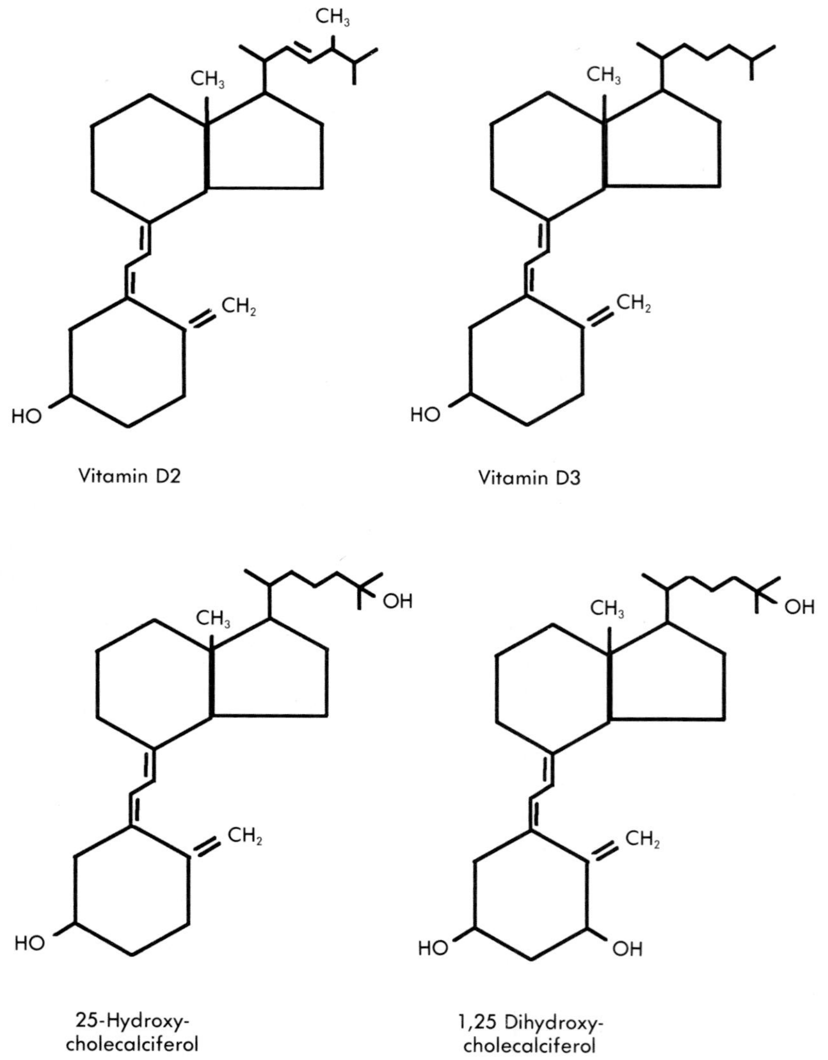

phosphate cause a reduction in the activity of the enzyme, and low serum phosphate levels stimulate the enzyme. Serum calcium concentration does not appear to directly affect this enzyme's activity but acts only through PTH.

Both 25-hydroxy- and 1,25-dihydroxycholecalciferol are further metabolized by another kidney enzyme, 24-hydroxylase, to 24,25-dihydroxy- and 1,24,25-trihydroxycholecal-

Figure 19.5

Synthesis of vitamin D_3. Vitamin D_3 can be synthesized by humans. With sufficient sunlight people can produce their needs. The starting steroid, 7-dehydrocholesterol, in the presence of light, forms previtamin D_3 by the cleavage of the 9-10 carbon-carbon bond. The previtamin D_3 then isomerizes to vitamin D_3.

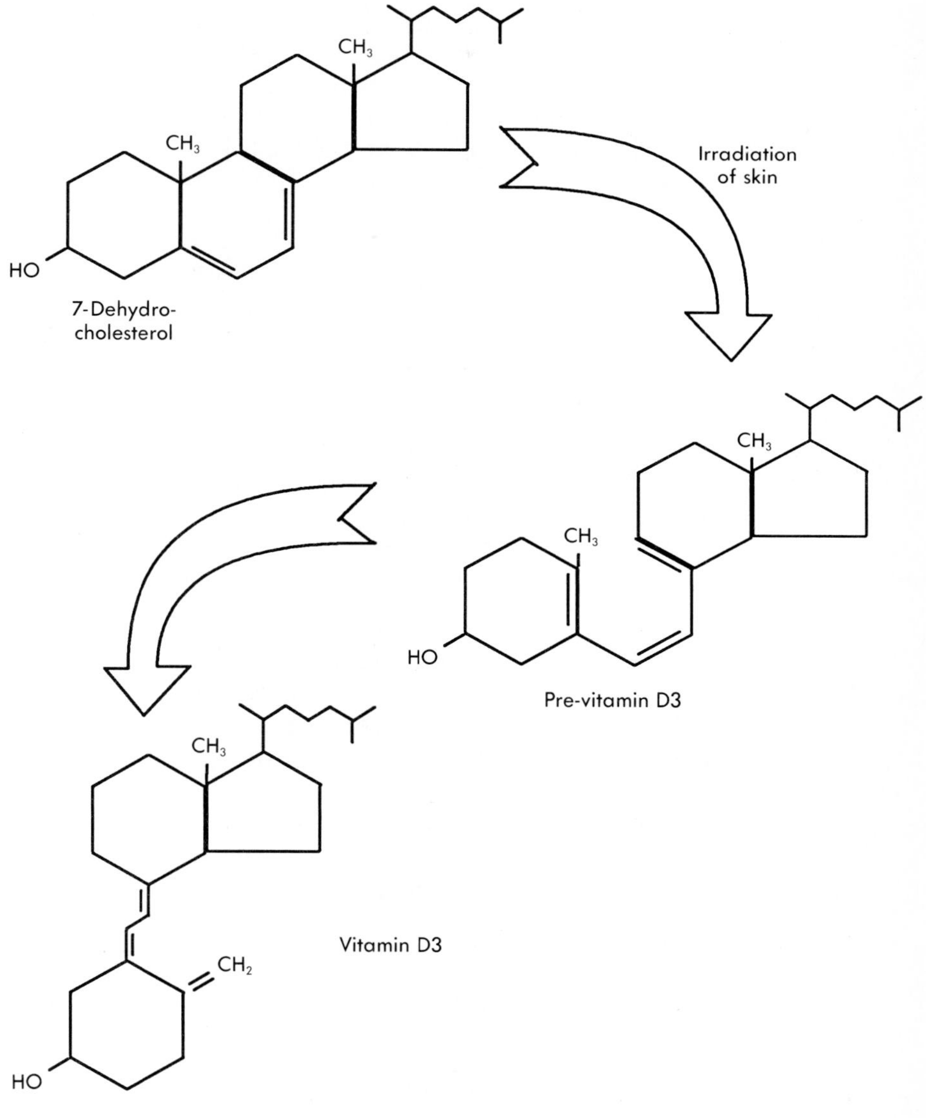

ciferol, respectively. The physiologic role of the 24-hydroxylation is presently unclear, and these metabolites are less potent than their precursors. This reaction may be the first step in the destruction of the active cholecalciferols. The side-chain of these cholecalciferols can also be metabolized at the 23- and 26-positions, and these also appear to be destructive reactions.

Transport of the cholecalciferols

Vitamin D is largely bound to a specific serum-binding protein. This protein, with a molecular weight of about 50,000, has a single site for the cholecalciferols. The binding protein binds all forms of cholecalciferol. Its main functions apparently are to protect the various forms of cholecalciferol from metabolism and renal excretion and to increase the water solubility of the unhydroxylated cholecalciferols.

Mechanism of action

A specific receptor for 1,25-dihydroxycholecalciferol exists in the nucleus of many cells. It is assumed that the hormone enters the nucleus by passive diffusion and there binds to the receptor. The receptor, with a molecular weight of about 58,000, has a much greater affinity for the dihydroxy form than the other forms of cholecalciferol. After binding the hormone, the receptor has an enhanced affinity for specific regions of the DNA, specifically activating a series of genes involved in calcium metabolism.

The overall role of vitamin D is to maintain calcium homeostasis (Figure 19.6). It does this by promoting absorption of calcium and phosphorus by the small intestine, causing resorption of calcium and phosphorus from the bone, and by increasing renal resorption of phosphate. Vitamin D does not stimulate resorption of calcium by the kidneys. In the duodenum the 1,25-dihydroxycholecalciferol stimulates the production of several proteins, including a highly specific calcium-binding protein called *calmodulin*. It also alters the viscosity of the membrane of these cells and this may influence calcium absorption. Vitamin D is not absolutely essential for the formation of bone, but together with PTH usually plays a role in bone remodeling. When given in supraphysiologic concentrations, the active forms of vitamin D induce dissolution of bone, both mineral and matrix.

Clinical comment

Vitamin D imbalances Excessive action of vitamin D is unusual but very serious. Not only is the calcium level elevated with all the problems described under hyperparathyroidism (see Chapter 18), but the phosphate level is elevated as well. Since calcium phosphate is insoluble, the increase in both ions is dangerous. Kidney stones typically occur, and if both the calcium and the phosphate levels are sufficiently elevated, calcification of the blood vessels and skin will occur. The most common cause of *hypervitaminosis D* is a massive overdose of vitamin D, either prescribed by a physician or more often illicitly obtained by the patient. Vitamin D is available without a prescription, but only at 400 units per pill. The daily requirement of vitamin D is 400 units daily, but many patients take two to five vitamin pills daily and therefore have mild overdoses of vitamin D. This is not a problem, since the feedback mechanism for calcium homeostasis autoregulates the calcium and phosphate levels. Since the 1-hydroxylation of vitamin D is a regulated step, increased intake of the vitamin, with consequent mild hypercalcemia, results in a diminished PTH level, decreased 1-hydroxylation, and a lowering of the calcium level to normal. Massive doses of vitamin D, 20,000 to 50,000 units daily, will overcome this regulation, however, and cause toxic effects. In some disease states, such as sarcoidosis, a granulomatous condition, the cells of the granuloma have an unregulated 1-hydroxylase and make large amounts of 1,25-dihydroxycholecalciferol. This results in significant hypercalcemia.

A deficiency of active vitamin D can lead to *rickets* or *osteomalacia*. Rickets develops if the deficiency occurs before completion of skeletal development; osteomalacia occurs after completion. While a child is growing, a lack of vitamin D causes diminished calcification of the bones such that the bones that form are likely to bend with chronic stress, leading to the typical bow-legged appearance of these children. Later in life a lack of vitamin D leads to a loss of mineral content of the bone, all caused by a lack of mineralization in a normal amount of matrix. The most common cause of these problems is a dietary deficiency of vitamin D along with insufficient sunlight. Problems can also

Figure 19.6

Mechanisms of calcium homeostasis. Serum calcium concentration is maintained within a very narrow range by the action of parathyroid hormone (PTH) and 1,25-dihydroxycholecalciferol (1,25-OH vitamin D). Calcium is increased by the liberation of calcium from the bones, an action of both PTH and 1,25-dihydroxycholecalciferol. Calcium is also increased as a result of the action of 1,25-dihydroxycholecalciferol on the intestinal absorption of calcium. Serum calcium is reduced by the renal loss of calcium, which is normally inhibited by PTH.

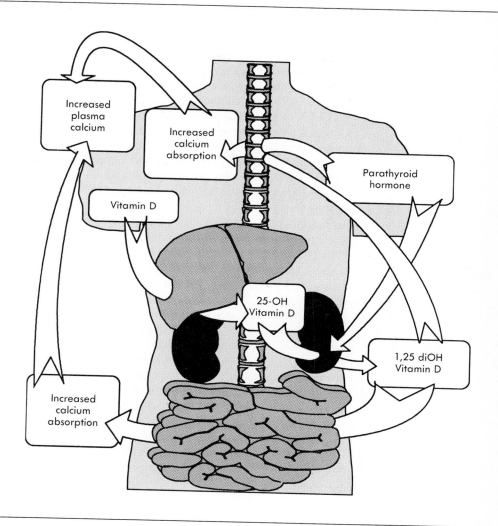

arise, however, if there is insufficient 1-hydroxylase activity, as occurs in hypoparathyroidism and severe renal disease, or if there is rapid destruction of the active 1,25-dihydroxycholecalciferol. The latter condition is known to occur with the administration of many antiseizure medications, which seem to induce microsomal hydroxylases, leading to polar, inactive metabolites of vitamin D (see case 4).

Hormones of the
adrenal cortex

The adrenal cortex synthesizes three classes of hormones with different functions (see Figure 19.7 for structures). The *glucocorticoids,* predominantly cortisol (hydrocortisone), are important in almost every aspect of biologic functioning but are most important in adaptation to stressful situations. The *mineralocorticoids,* primarily aldosterone, are important for the regulation of salt balance. The *adrenal androgens,* primarily androstenedione, are important in the development of sexual hair and libido in females.

Steroid synthesis

The nomenclature of the steroids is outlined in Figure 19.7. The basic four-ring structure is called *pregnane* if it has a two-carbon side-chain at the 17 position and *androstane* if it lacks the side-chain. Double bonds add the suffix *ene*, hydroxyl groups *ol*, and ketone groups *one*. The pathway for the synthesis of the various adrenal steroids is shown in somewhat simplified form in Figure 19.8 and in more detail in Figure 19.9. The studies of the pathways have generally been performed in bovine systems and extrapolated to humans. Most steps are catalyzed by mixed-function oxidases.

The precursor for all the adrenal steroids is *cholesterol*. The first step in steroid biosynthesis is the cleavage of the long side-chain at position 17. This is accomplished with molecular oxygen and NADPH. The oxidation probably yields a transient intermediate with hydroxyl groups at both the 20 and 22 positions. This is followed by an additional oxidative cleavage of the side-chain at the 21-22 carbon bonds and oxidation of the 21-hydroxyl group to a ketone, with formation of pregnenolone. The oxidation, performed by the cytochrome P_{450} enzyme complex, desmolase, occurs in the mitochondria.

The action of desmolase is enhanced by adrenocorticotropic hormone (ACTH), which acts by increasing intracellular cyclic adenosine monophosphate (cAMP). The next few steps of adrenal steroid biosynthesis occur in the endoplasmic reticulum (Table 19.2). Here pregnenolone is converted to 11-deoxycortisol. The process involves sequential hydroxylations at the 17-, 3-, and 21-positions, which usually occur in that order. This sequence may be altered slightly, and 3-hydroxylation can precede 17-hydroxylation.

Normally the first step in the endoplasmic reticulum is hydroxylation of pregnenolone at the 17 position to form 17-hydroxypregnenolone. This step is catalyzed by 17-α-hydroxylase, a mixed-function oxidoreductase that uses a P_{450} cofactor. This enzyme is tightly regulated by ACTH and cAMP. In the absence of stimulation by ACTH, pregnenolone is converted to corticosterone, moving down the mineralocorticoid pathway (see following discussion) by sequential 3-, 21-, and 11-hydroxylation without 17-hydroxylation. In the presence of ACTH the pregnenolone is converted to cortisol by 17-hydroxylation before the previous sequence. This explains why cortisol depends on ACTH without mineralocorticoids depending much on ACTH.

The next step in the synthesis of steroids in the endoplasmic reticulum is a hydroxylation at the 3 position. This is accompanied by an isomerization of the double bond to the 4-5 position, forming 17-hydroxyprogesterone. The enzyme is called 3-β-hydroxydehydrogenase. The 17-hydroxyprogesterone is further modified by 21-hydroxylation to form 11-deoxycortisol. This is catalyzed by 21-hydroxylase. Defects in this enzyme are the most common problems in cortisol synthesis (see case 3). The 11-deoxycortisol synthesized by these enzymes is then further modified in the mitochondria.

The last step in the synthesis of cortisol is an 11-hydroxylation, performed by 11-β-hydroxylase. Formation of other steroids requires the use of other enzymes. Corticosterone is converted to aldosterone by the action of 18-hydroxylase, followed by 18-hydroxyl reductase.

Formation of adrenal androgens is catalyzed by a 17-20 lyase, cleaving the two-carbon side-chain. This enzyme forms dihydroepiandrosterone (DHEA) from 17-hydroxypregnenolone and androstenedione from 17-hydroxyprogesterone.

Text continued on p. 815

Figure 19.7

A, Nomenclature of the steroid hormones. Although most steroid hormones have trivial names, all can also be formally named. The basic steroid nucleus has 19 or 21 carbons. The 19-carbon nucleus, with no side-chain at position 17, is referred to as *pregnane;* the 21-carbon nucleus, with a two-carbon side-chain at position 17, is called *androstane.* Double bonds add an *ene* to the name, hydroxyl groups an *ol,* and keto groups an *one.* **B,** Structures of the more important steroid hormones.

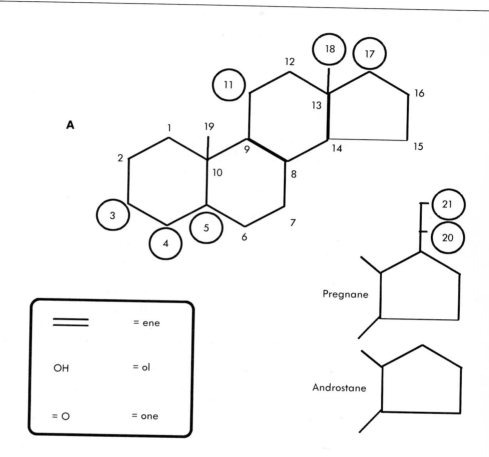

Continued on next page.

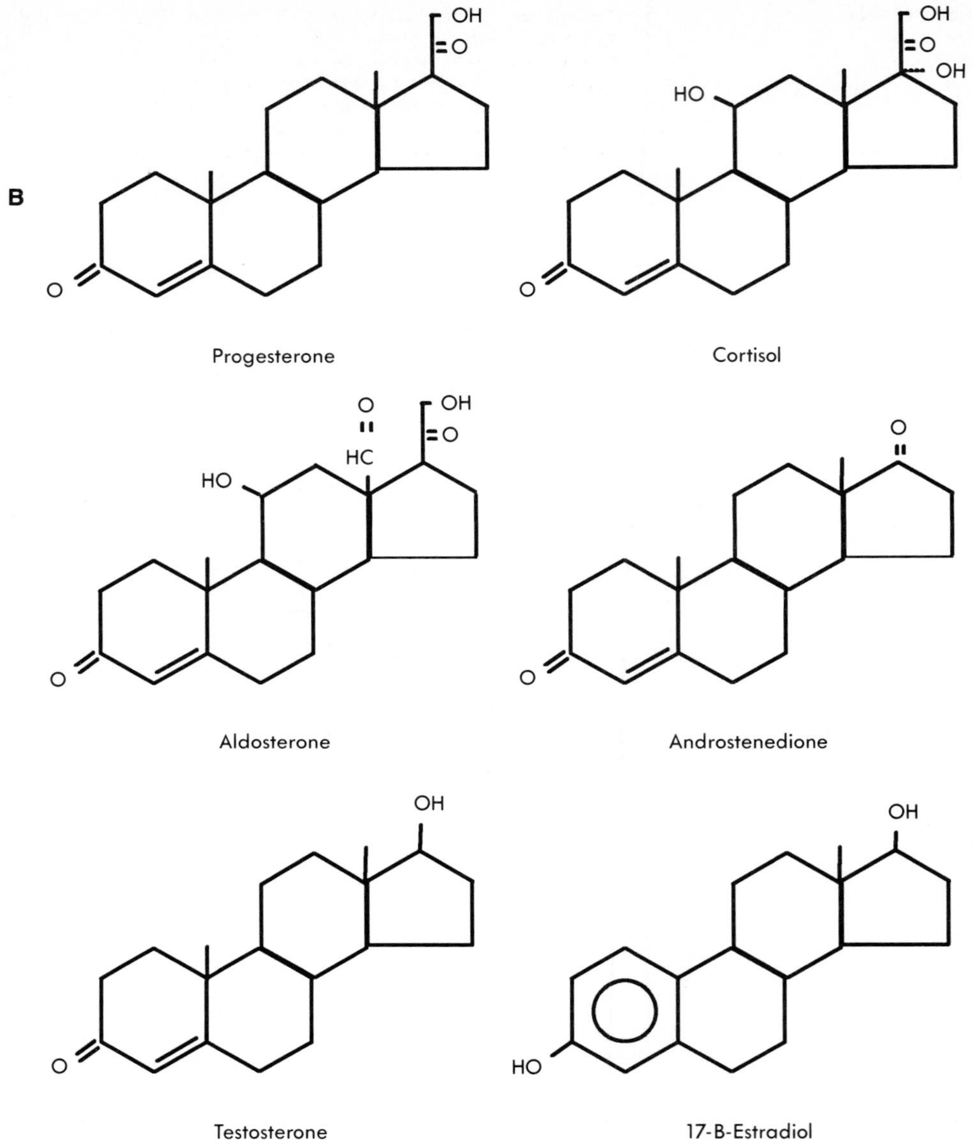

B

Progesterone

Cortisol

Aldosterone

Androstenedione

Testosterone

17-B-Estradiol

Figure 19.8

Synthesis of the adrenal steroid hormones. This simplified diagram of the pathways of adrenal steroid synthesis shows the mineralocorticoids on the left, the glucocorticoids in the middle, and the adrenal androgens on the right. The most active steroids are shown in ovals. *DHEA,* Dihydroepiandrosterone; *DHEA-S,* DHEA sulfate.

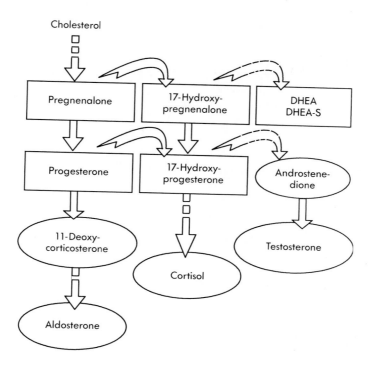

Table 19.2

Enzymes of adrenal steroid synthesis

| Enzyme | Product | Mechanism | Location | Loss of enzyme causes: | | | |
				Fatal	BP*	Male	Female
Desmolase	Pregnenolone	P_{450}†	Mito-chon-dria	Yes			
17-α-Hydroxylase	17-Hydroxypregnenolone	P_{450}	ER‡	No	High	No	Yes
3-β-Hydroxyde-hydrogenase	17-Hydroxyprogesterone	P_{450}	ER	Yes			
21-Hydroxylase	11-Deoxycortisol	P_{450}	ER	No	Low	Yes	No
11-Hydroxylase	Cortisol	P_{450}	ER	No	High	Yes	No

*Effect on blood pressure: high, hypertension; low, hypotension.
†Cytochrome P_{450}.
‡Endoplasmic reticulum.

A

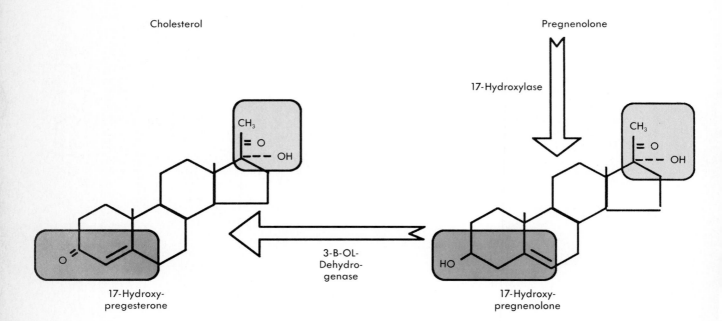

Cholesterol

Desmolase

Pregnenolone

17-Hydroxylase

3-B-OL-Dehydrogenase

17-Hydroxy-pregesterone

17-Hydroxy-pregnenolone

21-Hydroxylase

11-Deoxycortisol

11-Hydroxlase

Cortisol

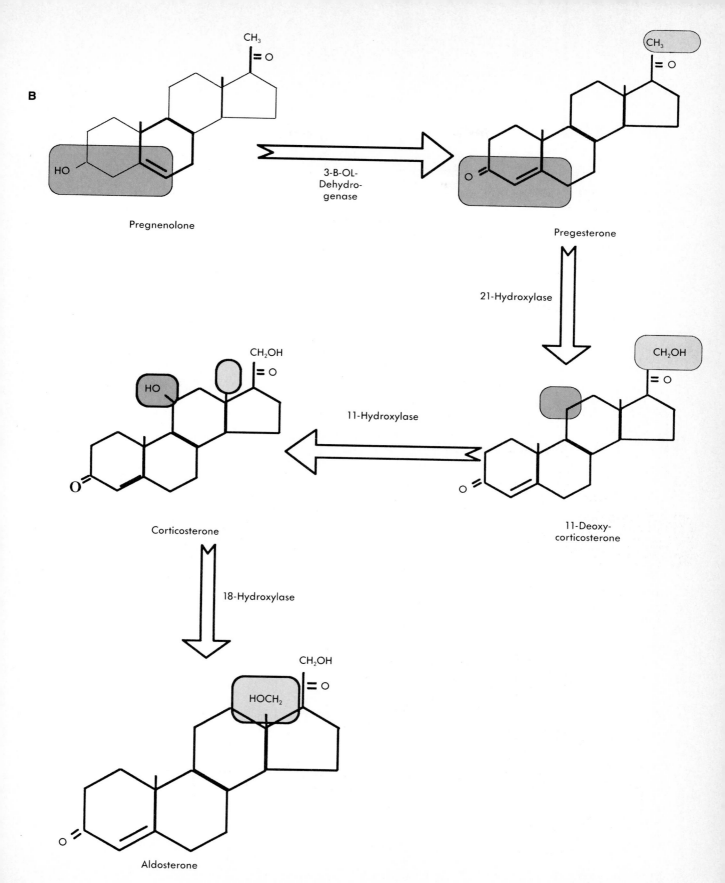

B

Pregnenolone

3-B-OL-Dehydro-genase

Pregesterone

21-Hydroxylase

11-Deoxy-corticosterone

11-Hydroxylase

Corticosterone

18-Hydroxylase

Aldosterone

Continued on next page.

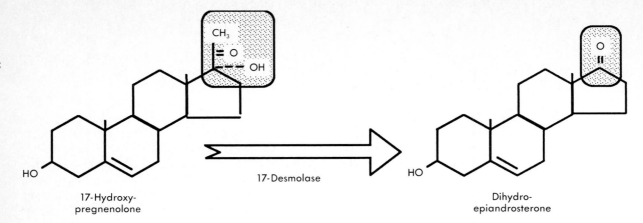

c

17-Hydroxy-
pregnenolone

17-Desmolase

Dihydro-
epiandrosterone

17-Hydroxy-
pregesterone

17-Desmolase

Androstenedione

Testosterone

Glucocorticoids

The natural glucocorticoid in humans is cortisol, or hydrocortisone. Many synthetic glucocorticoids also are used clinically. The most common are cortisone, prednisone, and dexamethasone. All work by the same mechanism.

Synthesis and secretion ACTH increases the rate of synthesis and secretion of cortisol and also causes growth of the adrenal cortex. ACTH acts by binding to a specific receptor and stimulating the production of cAMP. This increase in cAMP seems to explain all the actions of ACTH in the adrenal glands. Several inhibitors of adrenal steroid synthesis are used clinically. These include *ortho,para*-DDD; 1,1-dichloro-2-(*0*-chlorophenyl)-2-(*p*-chlorophenyl)-ethane, a compound related to the insecticide DDT; aminoglutethimide, a compound related to the hypnotic glutethimide; metyrapone; and ketoconazole, an anti-fungal agent (see case 4).

Cortisol has one of the most pronounced circadian rhythms in nature. Almost all the cortisol for the day is made in the few hours preceding awakening (Figure 19.10). The peak of ACTH is just before the peak of cortisol activity. For a person with a normal daily schedule, little cortisol remains by 4 PM. This diurnal variation is maintained relative to the sleep-and-work pattern rather than to any light-dark pattern. For example, emergency room physicians, working from midnight to 8 AM and sleeping from 3 PM to 11 PM, have their highest cortisol levels from 10 PM to 1 AM. It generally requires about 2 weeks to alter this cycle.

Transport Once secreted by the adrenal cortex, cortisol binds to a specific plasma-binding protein called *transcortin,* or *cortisol-binding globulin* (CBG), a glycosylated protein with a molecular weight of 50,000. Overall about 90% of cortisol is bound to in the plasma, 75% to transcortin, and the remainder to albumin. The transcortin has a structure homologous to thyroid-binding globulin and is similarly synthesized in the liver; its levels are increased by estrogens. The function of the transcortin is to protect the cortisol from degradation and renal excretion.

Excretion of cortisol by the kidneys occurs in proportion to the free plasma levels. This allows quantitation of the average free cortisol during the day by measuring the amount of cortisol excreted in 24 hours, a test called *urinary free cortisol*. This has become one of the most accurate measures of both excessive and deficient cortisol states. Caution must be used in collecting the sample. Since most of the cortisol is produced over just a few hours, an incomplete collection that misses some of the period of high cortisol production can be very inaccurate.

Mechanism of action The general mechanism of action of steroids is presented in Chapter 17, based on the action of estradiol. The mechanism for cortisol is more con-troversial, and some investigators believe that both cytoplasmic and nuclear receptors may exist for cortisol. The steroid enters the cell and either interacts with a cytoplasmic receptor, with activation and translocation of the receptor to the nucleus, or interacts with the cortisol receptor in the nucleus and is activated there. Activation of this protein involves the release of the heat-shock proteins from the complex and the increase in the affinity of the DNA-binding region of the receptor (the two zinc-containing "fingers") for specific *glucocorticoid response elements* (GREs). These GREs are thought to rep-resent specific glucocorticoid responsive transcriptional inducers. Studies with mouse mammary tumor virus and other systems have shown some consensus sequences that may be important in this regulation, including a hexanucleotide, T-G-T-T-C-T, and a octanucleotide, (T/A)-C-T-G-(T/A)-T-C-T. These sequences are often present in multiple copies at regions near regulated genes. The binding of the activated cortisol receptor to the GREs increases the amount of specific messenger ribonucleic acids (mRNAs) tran-

Figure 19.10   Circadian rhythm of cortisol. Maximum blood levels of ACTH occur
shortly before waking, and the cortisol levels occur later. On a normal
day virtually all the cortisol is made between 6 and 9 AM. On days
when increased stress is present, more cortisol is made for the
stress.

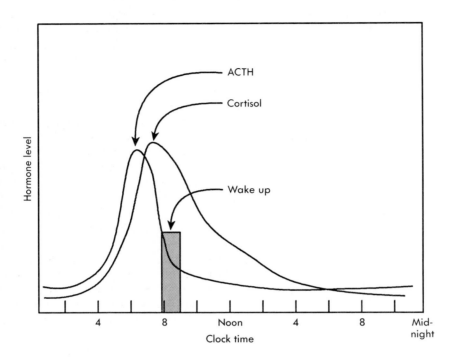

scribed. The major effect is to increase the efficiency of the initiation sites. No increase
occurs in the number of such sites, and no general increase occurs in mRNA synthesis.
No substantial data exist to suggest that the glucocorticoids can prolong the life of the
mRNA; rather, they simply increase the production rate of specific mRNA and therefore
the mRNA steady-state level. With an increase in mRNA, a subsequent increase in specific
protein synthesis and changes in cell function occur.

   Actions Glucocorticoids are also referred to as *pleotropic* because they are necessary
for the functioning of almost every cell and a variety of systems. The glucocorticoids are
necessary for the maintenance of a normal blood glucose (thus the name), since they
oppose the action of insulin and cause insulin resistance. They are also necessary for the
proper tone of blood vessels and for normal functioning of the gastrointestinal tract.
Cortisol is generally a catabolic hormone, leading to protein breakdown by cells. It is a
hormone of stress needed by the body to deal with stressful situations such as trauma,
infections, surgery, or severe emotional distress.

Clinical comment

Cortisol imbalances Alterations in cortisol production are lethal if untreated. *Hypercortisolism,* typically caused by treatment with glucocorticoids for other diseases, may also be found in Cushing's syndrome. In the latter state, cortisol levels are increased either because of a tumor (benign or malignant) that produces cortisol without the possibility of feedback inhibition or because of conditions in which the pituitary gland is resistant to feedback inhibition of ACTH by cortisol. Since cortisol has a very complex, integrated set of enzymes needed for synthesis, most tumors that produce cortisol are *eutopic,* that is, adrenal in origin (see case 4). Sometimes small cell carcinoma of the lung or a bronchial carcinoid tumor may produce an ACTH-like peptide and cause this syndrome.

The origin of the excess cortisol can often be found by measuring both cortisol and ACTH levels. Eutopic tumors producing cortisol feed back to shut off the pituitary and have low levels of ACTH. Tumors producing ACTH have high ACTH levels. Patients with pituitary resistance to feedback of cortisol also have high ACTH levels, but unlike the tumors, the ACTH is suppressible with high levels of cortisol.

Patients with excess cortisol production have a peculiar form of obesity referred to as *trunkal obesity.* There is excess deposition of fat in the abdomen, chest, and face but a paucity in the arms and legs. The insulin resistance caused by the cortisol may lead to overt diabetes. The high cortisol levels may cause poor functioning by the immune system with frequent, severe, and even life-threatening infections. In addition, the general catabolic state leads to poor wound healing.

A lack of cortisol, usually caused by autoimmune destruction of the adrenal glands, can be rapidly fatal. These patients demonstrate hypoglycemia resulting from poor gluconeogenesis, decreased blood pressure from poor vascular tone, and hyperpigmentation from high ACTH levels. Patients are unable to withstand even minor stress levels and, in response to infections, trauma, or surgery, drop their blood pressure and go into shock.

Mineralocorticoids

The adrenal steroids are necessary for the maintenance of mineral balance. *Aldosterone,* the main mineralocorticoid, causes the retention of sodium and wasting of potassium. It is produced by the sequential 21-, 11-, and 18-hydroxylation of progesterone. It does not require 17-hydroxylation (see Figure 19.9, *B*). Although ACTH will stimulate the production of aldosterone, it is not the main regulator of this hormone. The main stimulus is the volume and salt status of the individual. This is effected through the juxtaglomerular (JG) cells of the kidney. In response to a decrease in volume, the JG cells make and secrete renin, a glycoprotein enzyme. The renin proteolytically cleaves an α-2 globulin, angiotensinogen, to angiotensin I, a decapeptide. This is further cleaved by angiotensin-converting enzyme of the lung to angiotensin II, an octapeptide. The angiotensin II binds to specific receptors at the cell surface of zona glomerulosa cells of the adrenal cortex and increases the production and secretion of aldosterone. In addition, high potassium levels will increase aldosterone production.

Aldosterone binds to a specific receptor in the nucleus of affected cells. No evidence exists for cytoplasmic receptors for this hormone. The receptors increase specific mRNA synthesis and therefore protein synthesis. The main action of aldosterone is the resorption of sodium from the urine at the distal tubule and collecting ducts of the kidney. It does this by increasing the Na^+/K^+-ATPase activity at these sites. It increases the permeability of the apical membrane to sodium, increases ATP production, and also increases the activity of the ATPase (Figure 19.11).

Clinical comment

Aldosterone imbalances Alterations in mineralocorticoid activity are often associated with alterations in blood pressure. *Hyperaldosteronism,* caused by a tumor or hyperplasia of the adrenal glands, is associated with hypertension. The increase in aldosterone syn-

Figure 19.11

Regulation of aldosterone production. In the presence of diminished perfusion pressure to the kidney, the juxtaglomerular (JG) apparatus of the kidney produces renin, which proteolytically cleaves angiotensinogen to angiotensin I. The angiotensin I is further cleaved by angiotensin-converting enzyme in the lung to angiotensin II. The angiotensin II binds to specific receptors on the adrenal cells and causes increased production of aldosterone, which in turn increases salt retention and perfusion pressure to the kidneys.

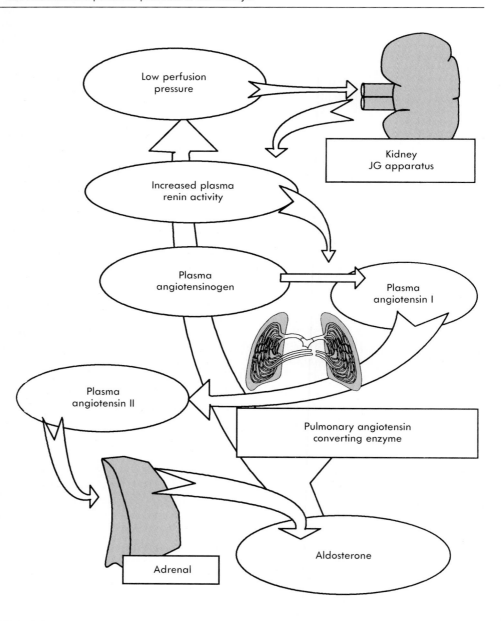

thesis leads to salt retention and expansion of the vascular volume, with subsequent high blood pressure. Curiously, this condition is usually not associated with serious edema, since the kidney is able to escape the aldosterone effect before edema occurs.

A lack of aldosterone is associated with diminished blood volume and hypotension, especially on standing *(orthostatic hypotension)*. As might be expected from the known action of the mineralocorticoids, loss of aldosterone action is associated with low plasma sodium levels, hyponatremia, and high plasma potassium levels, or hyperkalemia. Loss of mineralocorticoid alone is usually not fatal, since about 30% of the mineralocorticoid action results from a cross-reaction of cortisol with mineralocorticoid receptors. Many diseases that lead to diminished mineralocorticoid levels also lead to lowered cortisol levels and can result in medical emergencies. Examples include destruction of the adrenal glands by infection (tuberculosis), hemorrhage, or malignancy. Prompt treatment with large doses of glucocorticoid and mineralocorticoid may be life saving.

Adrenal androgens

The adrenal glands also produce some steroids with androgenic activity. These are not of great significance in males but are the major source of androgens in females, in whom they affect sexual hair production and libido. Two steroids are produced with some androgenic activity, androstenedione and DHEA (see Figure 19.8). These are formed by the cleavage of the side-chain at the 17 position. Cleavage of 17-hydroxyprogesterone yields DHEA, whereas cleavage of 11-deoxycortisol yields androstenedione.

Dihydroepiandrosterone (DHEA) is a very, very weak androgen. It is rapidly metabolized by the liver by the addition of a sulfate group to form dihydroepiandrosterone sulfate (DHEA-S). DHEA-S is very slowly excreted, with a half-life of almost 3 weeks, so that plasma levels of this compound are very high. DHEA and DHEA-S are the major components of the commonly performed test for androgens called *urinary 17-ketosteroids*. As one might imagine, this test does not measure actual androgenic potential, since the major androgen, testosterone, is a minor component of the 17-ketosteroids. It measures primarily the adrenal androgens, androstenedione and DHEA, and therefore may be of some help in separating adrenal versus other sources of hyperandrogenism. The 17-ketosteroid test has largely been replaced by specific serum assays for individual androgens.

Androstenedione is the major androgen in females. It can be aromatized to estrone, an estrogen. It is also responsible for some of the hyperandrogenic syndromes that cause excessive hair and diminished menses in some females.

Sex steroids
Androgens and male sexual function

Testosterone is the major sex steroid in males and is produced by Leydig's cells of the testis. The major stimulus for synthesis and secretion of testosterone is luteinizing hormone (LH), which binds to a specific receptor on the surface of Leydig's cells, stimulating the production of cAMP, which in turn stimulates synthesis and secretion of testosterone. Testosterone is present in very high concentrations in the testis, and such concentrations are necessary for spermatogenesis. Testosterone is also metabolized by peripheral tissues to two other active hormones, *5-α-dihydrotestosterone* (DHT) and *estradiol* (Figure 19.12). Current evidence suggests that testosterone does not directly inhibit LH secretion, but rather does so by the aromatization to estradiol in the pituitary gland, with inhibition of LH secretion by the estradiol. About 5 mg of testosterone is produced per day in males, and plasma levels are generally 3 to 7 ng/ml. Approximately 75% to 90% of the testosterone is bound to an androgen-binding protein and is inactive; only the free testosterone is active. The binding protein, a β-globulin, is produced by the Sertoli cells of the testis in response to both LH and follicle-stimulating hormone (FSH). The binding protein, called sex hormone–binding globulin or testosterone-estradiol–binding globulin (SHBG

Figure 19.12   Metabolism of testosterone. Testosterone can be metabolized in one of four ways. The most common, accounting for about 90%, is metabolism of testosterone to a variety of 17-ketosteroids (pathway 1). About 1% is metabolized by aromatase to estradiol (pathway 2). About 3% is metabolized to androstanediol, an active androgen (pathway 3). The remaining 5% is metabolized by 5-α-reductase to dihydrotestosterone (pathway 4).

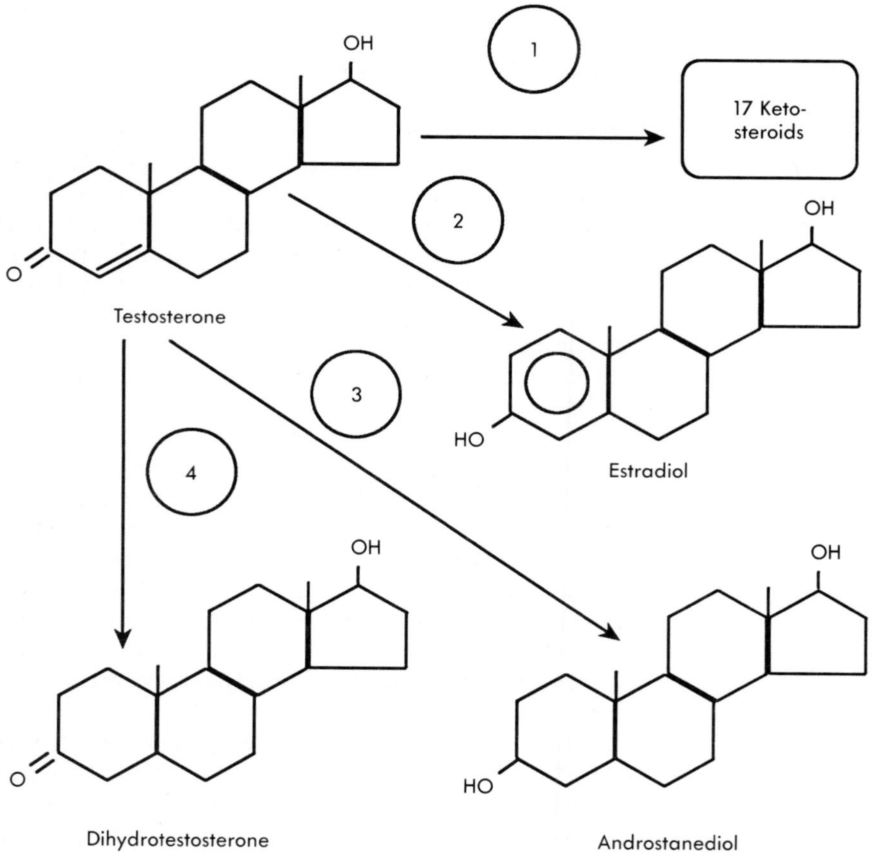

or TEBG), is unrelated to the binding globulins for the thyroid hormones and cortisol but closely related or identical to the binding protein for estradiol. It is increased by both estrogens and thyroid hormones.

Testosterone enters target cells by passive diffusion and then is transformed by the enzyme 5-α-reductase to DHT. The latter hormone is far more active on the receptor than testosterone itself and therefore appears to be the active form of the hormone. Once DHT binds to the receptor, classical schemes of activation of certain genes, increased mRNA, and increased specific protein synthesis are called into play.

Since testosterone is not the major hormone of cellular androgen activity or of feedback inhibition of LH, some have proposed that we consider it a *prohormone*, serving as substrate for estradiol or DHT production. The level of free testosterone is still important, however, since the level of this hormone determines the amount of DHT or estradiol produced.

FSH is necessary, along with very high testosterone levels, for spermatogenesis. The spermatic development and maturation is a very slow process, requiring about 70 days. The Sertoli cells of the spermatic cords are responsible for producing a hormone, *inhibin,* that is responsible for feedback inhibition of FSH. Inhibin is a recently purified hormone that exists as a heterodimer of α- and β-subunits, linked by disulfide bonds. The molecule has great homology with Müllerian inhibiting factor and with β-transforming growth factor (β-TGF).

Clinical comment

Androgen imbalances Problems with the male sexual axis can cause ambiguous genitalia or other problems. Excess androgens in adult males are not definitively associated with disease, although some association with sexual behavior problems has been reported. In children, both male and female, excess androgens cause rapid growth but also rapid maturation. These children are the tallest children in their class in elementary school because of this rapid growth. Since they also mature rapidly, however, with early closure of the epiphyses, by age 13 to 14 years they are the shortest of their peers. In females, both children and adults, excess androgens are associated with excessive facial hair *(hirsutism)* and, if significant, a male muscle pattern, male sexual hair patterns, and clitoromegaly *(virilism)*. Hirsutism occurs frequently; virilism is less common but usually associated with more serious disease.

A lack of androgen action may result from androgen deficiency or to a lack of 5-α-reductase; the latter is called *testicular feminization*. A lack of androgens in prepubertal males will result in no pubertal development. The children will not develop sexually and will continue to grow after their peers have closed their epiphyses. This results in a eunicoid appearance with greater length of legs and arms but normal growth of the spine and head. No phallic enlargement, no testicular enlargement, and no development of sexual hair occur. Testicular feminization is associated with a deficiency of androgenic action, even in utero. These genotypic males are usually born as phenotypically normal females. Often the first abnormality is the lack of development of sexual hair and primary amenorrhea (they never menstruate). Since these children are social females, they must continue life as such, and the testes are surgically removed.

Hypoandrogenic syndromes that are pituitary in origin demonstrate low plasma gonadotropins and a low testosterone level. If the problem is testicular failure, there are low testosterone levels but high plasma gonadotropins. Testicular feminization is associated with normal male testosterone levels. No clinical syndromes are described in women with low androgen levels. One would expect these women to have diminished sexual hair and diminished libido.

Estrogens and female sexual function

The estrogens are largely formed in the ovary by aromatization of androgens. Testosterone is converted to estradiol and androstenedione to estrone. The process occurs in the ovarian follicle and is cyclic in humans.

Estradiol and progesterone production and the menstrual cycle Adult human females produce a single ovum each month by a very complex and still somewhat obscure process. The process can be separated into a follicular phase, during which the ovum develops, and a luteal phase, during which implantation of the fertilized ovum occurs. At least four hormones are involved: FSH, LH, estradiol, and progesterone. Others, such as prolactin, inhibin, and androgens, may play a role. Since our understanding of the process is incomplete, other as yet unknown hormones probably are involved as well.

The follicular phase is under the control of FSH, LH, and estradiol. During this phase the ovary produces a single mature ovum and prepares it for fertilization. At the start of the cycle each month, many follicles begin to develop. Most cease development and undergo atresia. One continues to develop to a dominant follicle, inhibiting the development of all others. The process that causes the inhibition of the development of all follicles except the dominant one is not well defined. Androgens effect this process, causing the follicles to become atretic and resulting in the death of the ovum and necrosis of granulosa cells. Women with excess androgen production have low estradiol levels, do not have a menstrual cycle, and do not form ova.

The process of follicular development is primarily under the control of FSH and involves enlargement and maturation of the follicle and increased vascularization. As seen in Figure 19.13, the ovarian follicle has three types of cells: the thecal cell line outside of the follicle, the granulosa cells that cover the ovum, and the ovum itself. The first two types of cells are primarily involved in endocrine regulation in the follicle. The granulosa cells have FSH receptors, and binding of the glycoprotein hormone leads to the accumulation of cAMP, growth of the follicle, and an increase in LH receptors by these cells. The cells of the theca interna have LH but not FSH receptors; they are responsible for androgen production. The production of estradiol is performed by both the granulosa and the thecal cells and requires the presence of both LH and FSH. Plasma levels of estradiol vary during the cycle from about 50 pg/ml in early follicular phase to about 500 pg/ml in late follicular phase (Figure 19.14), with estradiol production rates of 0.1 to 1 mg/day.

After ovulation, the follicle undergoes luteinization to a corpus luteum. Under the stimulation of LH (and if pregnancy ensues, human chorionic gonadotropin, or hCG), the corpus luteum produces large amounts of progesterone and smaller amounts of estrogen. In the absence of pregnancy, at the end of 14 days, LH levels fall, the production of progesterone and estradiol diminishes, and menstruation ensues.

The production of the gonadotropins in this cycle is complex. Initially, FSH level is high and LH low. During the follicular phase, the FSH levels fall, perhaps because of increased production of inhibin by the follicle. Although LH levels remain constant, LH receptors are increasing and thus LH action is increased. Estradiol also increases, reaching a peak just before ovulation. Estradiol plays a role in increasing the sensitivity of LH to gonadotropin-releasing hormone (GnRH) and therefore is involved in positive feedback, one of the few examples of this phenomenon in nature. Immediately before ovulation, the estrogen levels fall and progesterone rises. This is accompanied by a rise in LH by a factor of 5 to 10, resulting in rupture of the follicle, release of the ovum, and subsequent formation of the corpus luteum.

Transport of female sex steroids Estradiol is largely bound to serum proteins. Most binding is to a protein closely related or identical to the β-binding globulin for testosterone, SHBG or TEBG. The estradiol is also bound to cortisol-binding globulin and to albumin, although binding to both these proteins occurs with much less affinity. The primary role of the binding appears to be protecting the estrogens from degradation or renal excretion and protecting the organism from the surges in estradiol that might otherwise accompany pulsatile production.

Mechanism of action of female sex steroids The mechanism of action of estradiol is the best studied of the actions of steroid hormones (see Chapter 17). In some tissues, such as the uterus, the concentration of receptors for estradiol may be modified by both estradiol and progesterone.

The feedback inhibition of the pituitary gland by the female sex steroids is complex,

Figure 19.13

Cells of the ovaries. The ovary has primary follicles that, under stimulation by gonadotropins, develop into primordial follicles. Most primordial follicles do not develop further but become atretic. One develops to a dominant follicle with thecal cells, a zona pellucida, granulosa cells, and a maturing ovum. After the ovum is expelled, under the influence of LH, the dominant follicle becomes a corpus luteum. See text for details of the hormones made by these various cell types.

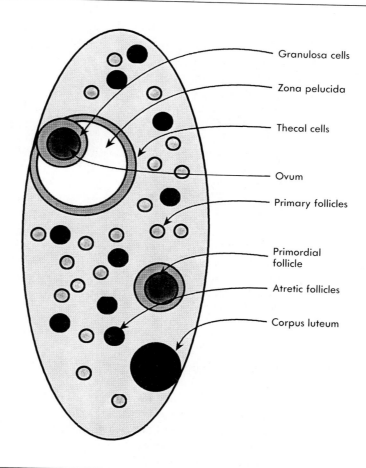

- Granulosa cells
- Zona pelucida
- Thecal cells
- Ovum
- Primary follicles
- Primordial follicle
- Atretic follicles
- Corpus luteum

since it is cyclic and altered by various factors. High levels of estradiol seem to inhibit LH secretion, although low levels sensitize LH to GnRH action. The inhibition of FSH is primarily by inhibin.

Clinical comment

Imbalances in female sex steroids Because the menstrual cycle is so complex, a variety of factors can disturb normal function. Frequently this is intentional, in the form of oral contraceptives, but alterations in pituitary, thyroid, adrenal, or androgen function can alter ovarian function.

Figure 19.14   Hormones of the menstrual cycle. See text for details.

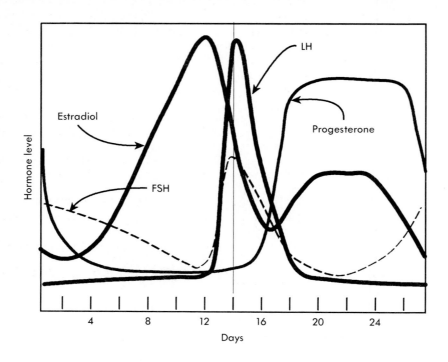

Oral contraceptives always have progestins and usually have estrogens at the lowest possible levels. The progestin and estrogens feed back to inhibit pituitary gonadotropins and stimulate endometrial development. Since FSH is inhibited, no development of ovarian follicles and no ovulation occur.

A lack of cycling can be caused by problems of the hypothalamus or pituitary gland (see Chapter 18). Hyperandrogenism leads to chronic anovulation and may even lead to polycystic ovaries.

Although the normal source of estrogens is the ovaries, estrogens can also be formed by the aromatization of androgens in adipose and other tissues or by certain rare adrenal tumors. In obese males the excess estrogen formed by the excessive adipose tissue can lead to breast formation (gynecomastia).

Carcinoma of the breast in females may respond to hormonal therapy. Many of these cancers grow much more rapidly in the presence of estrogens, and maneuvers designed to lower estrogens will increase the survival of these patients. Unfortunately, not all patients will respond to these therapies with anti-estrogenic drugs, and the therapies are toxic and expensive. In addition, if they are unlikely to succeed, valuable time may be wasted in trying them. Overall, about 40% of women with breast cancer will benefit from anti-estrogen therapy, but since the first step in estrogen action is the binding of estrogen to its receptor, tumors that lack estrogen receptors are unlikely to respond to this therapy.

Up to 75% of patients with receptors for estrogen in their breast cancer are likely to have a remission while receiving this therapy, whereas only 5% of patients without receptors will benefit. We normally measure the estrogen receptor content of most breast cancers to help us determine whether this therapy will be beneficial in an individual patient.

Bibliography

Brandon DD et al: Glucocorticoid resistance in humans and nonhuman primates, Cancer Res 49:2203s, 1989.

Chrousos GP et al: Glucocorticoids and glucocorticoid antagonists: lessons from RU 486, Kidney Int (Suppl) 26:S18, 1988.

DeLuca HF: The metabolism and functions of vitamin D, Adv Exp Med Biol 196:361, 1986.

Edwards CR et al: The specificity of the human mineralocorticoid receptor: clinical clues to a biological conundrum, J Steroid Biochem 32:213, 1989.

Franz WB: Basic review: endocrinology of the normal menstrual cycle, Prim Care 15:607, 1988.

Gravdal JA et al: Congenital hypothyroidism, J Fam Pract 29:47, 1989.

Kannan CR: Diseases of the adrenal cortex, Dis Mon 34:601, 1988.

Sheeler LR: Cushing's syndrome—1988, Cleve Clin J Med 55:329, 1988.

Silva JE: Pituitary-thyroid relationships in hypothyroidism, Baillieres Clin Endocrinol Metab 2:541, 1988.

Simpson ER and Waterman MR: Regulation of the synthesis of steroidogenic enzymes in adrenal cortical cells by ACTH, Annu Rev Physiol 50:427, 1988.

Stockigt JR and Topliss DJ: Hyperthyroidism: current drug therapy, Drugs 37:375, 1989.

Clinical examples

A diamond (♦) on a case or question indicates that literature search beyond this text is necessary for full understanding.

Case 1   Sarcoidosis with hypercalcemia

A 55-year-old woman is sent to you after passing a kidney stone. She states she has been in good health all her life, but that in evaluating possible causes for her stone, her local physician found that her serum calcium concentration was elevated. She takes no medications and, aside from some recent depression, has felt well. She has a dietary intake of calcium of about 1 g/day (normal) and does not smoke. The remainder of the history is negative. On physical examination she is a well-developed, thin, black female in no distress. Her blood pressure is 168/98, pulse 68, and respirations slightly elevated at 18/min. Examination of her head is negative except for her fundi (eye grounds), which show some granuloma. Auscultation of her lungs reveals some coarse breath sounds. The remainder of the physical examination is negative. On laboratory examination her serum electrolytes are normal; the calcium level is 13.5 (normal, 8.6 to 10.6); phosphate, 6 mg/dl (normal, 2.5 to 5); and alkaline phosphatase, 217 IU (normal <115). A 24-hour collection for urinary calcium is 450 mg (normal for this diet, <200). A chest roentgenogram shows diffuse interstitial disease, and her pulmonary function tests show a diffusion defect.

Biochemical questions ♦ 1. List the defects that could cause the hypercalcemia. Why might the phosphate and alkaline phosphatase levels be high?

♦ 2. What tests would you do to prove your diagnosis?

3. The parathyroid hormone (PTH) level is very low. What is the basis for this, and how does it influence the disease?

Case discussion

1. Defects causing hypercalcemia Hypercalcemia can be caused by disturbances of the normal hormones of calcium metabolism, by abnormal destruction of bone, or by dietary excess of calcium (only in unusual circumstances). One of the most common causes of hypercalcemia is hyperparathyroidism resulting from a tumor or hyperplasia of the parathyroid glands. In this condition (discussed in more detail in case 5 in Chapter 18) there is increased resorption of bone with increases in both calcium and phosphate, decreased excretion of calcium by the kidney, but increased excretion of phosphate. Overall this leads to an increase in calcium but a decrease in phosphate in the blood. The high rate of bone turnover causes an elevation of alkaline phosphatase, a bone enzyme. Excess vitamin D, either dietary or caused by diseases that increase hydroxylation of vitamin D to active metabolites, increases gut absorption of calcium and phosphate, slightly increases turnover of bone, and decreases renal excretion of phosphate. This leads to an elevation of calcium, phosphate, and alkaline phosphatase level in the plasma. Certain tumors may also increase serum calcium concentration, either (1) by producing PTH-like substances or (2) by direct resorption of action on bone by metastases. In both cases the calcium level is elevated; in the first the phosphate is low; in the second, high.

2. Diagnostic tests In the presence of a high serum phosphate concentration, it is unlikely that PTH or related substances are responsible. Nevertheless, it would be reasonable to measure this, and in this case the measurement is low. Measurement of vitamin D and its active derivatives shows the vitamin D level to be normal, as is the 25-hydroxycholecalciferol. The level of 1,25-dihydroxycholecalciferol is, however, five times the normal value, suggesting an increase in the activity of the 1-hydroxylase. A wash of the lung produces cells that are characteristic of sarcoidosis.

Sarcoidosis is a chronic granulomatous disease of unclear etiology. The macrophages of the granulomas, however, frequently have 1-hydroxylase, which produces the most active form of vitamin D, 1,25-dihydroxycholecalciferol. The enzyme is unregulated by the normal calcium homeostatic mechanisms, so that when significant numbers of these macrophages are present, the active vitamin D level is very high and the patient may become hypercalcemic.

3. Low PTH level The low level of PTH is absolutely appropriate. The high level of calcium suppresses the PTH and also suppresses the renal 1-hydroxylase. In this case this does not influence the disease at all, since the high levels of 1,25-dihydroxycholecalciferol caused by the unregulated 1-hydroxylase of the macrophages are responsible for the elevated calcium level.

References

Adams JS et al: Isolation and structural identification of 1,25 dihydroxycholecalciferol produced by cultured alveolar macrophages in sarcoidosis, J Clin Endocrinol Metab 60:960, 1985.
Papapoulis SE et al: 1,25 Dihydroxycholecalciferol in the pathogenesis of hypercalcemia of sarcoidosis, Lancet 1:627, 1979.

Case 2

Hypothyroid baby with a goiter

You are called by the state health authorities because a child that you delivered last week is found to be hypothyroid in the mandatory neonatal screening program for thyroid disease. The TSH level is very elevated and the T_4 level very depressed. You call the mother, who brings the child immediately. The baby is somewhat sluggish, and you note a large goiter, dry skin, a slow heart rate, and a hoarse cry. A family history reveals that an older brother and many cousins have had similar problems. The mother and father are second cousins. You find that the T_4 level is very low, but the T_3 level is only borderline low.

Biochemical questions

1. How might you evaluate the defect in this child?
2. Why is the T_3 level less affected than the T_4?

Case discussion

1. Evaluation of defect The child has a congenital thyroid abnormality, which could be caused by an absence of the thyroid gland or by an abnormality in the pathways of thyroid hormone synthesis. Since we have witnessed a goiter in this baby, we know that he has a thyroid gland and thus must have a defect in one of the steps in thyroid hormone synthesis.

The first step in the synthesis of thyroid hormone is the accumulation of iodine by the thyroid gland. This is an active process, concentrating the iodide many-fold over the level of the plasma. We can test the child's ability to concentrate iodine by giving him a small dose of radioactive iodine and checking for its accumulation by the thyroid. In this child, accumulation of radioactive iodine is greater than normal, and thus the child does not have a defect in the iodine-concentrating mechanism. The next step in the synthesis of thyroid hormones is the organification of the iodine into iodotyrosines and thyronines in thyroglobulin. If this step is abnormal, the radioactive iodine that was accumulated by the gland is still in the form of iodide and can be "flushed" from the gland with excess iodide or any of the large anionic compounds concentrated by the thyroid gland as is iodide. Sodium perchlorate, given in large doses to this patient, does not flush the radioactive iodide from the gland, and thus organification is normal. The patient's blood does have some T_4 and reasonable levels of T_3, so we know that the thyroid gland is capable of coupling the iodotyrosines to thyronines and of proteolytically cleaving the thyroglobulin. We then look at the iodinated compounds released from the thyroid gland into the patient's blood. This is easy to do, since we just gave radioactive iodine to test the activity of the thyroid gland, so the iodinated compounds all contain radioactive iodine. Chromatography of the blood shows some iodine in the position of T_4 and slightly more in the position of T_3. Surprisingly, however, large amounts of the radioactivity are found in the position of mono- and diiodotyrosines. These compounds are normally deiodinated in the thyroid gland, and failure to do so results in intrathyroidal iodine deficiency and hypothyroidism.

This deiodination defect may be limited to the thyroid gland, in which case the patient has a goiter but is euthyroid, or it may extend to all the cells of the body, in which case hypothyroidism ensues. As one might guess, this baby has a total body defect. The condition, which is rare, is usually found in families with consanguinity, and in the original family (with extensive intermarriage) was found in 40% of the children.

If untreated, these children will be severely mentally and physically retarded. They are easily treated with thyroid hormone.

2. Effect on T_3 versus T_4 With a deiodination defect the patients become iodine deficient, both in the thyroid gland and in the whole body. Under these conditions the thyroglobulin has large amounts of monoiodotyrosine. The greater the percentage of monoiodotyrosine, the larger is the proportion of T_3 produced by the thyroid gland. Since the T_3 is more potent, this serves as a stopgap measure to protect the organism from the hypothyroidism of iodine deficiency.

Reference

Ismail-Beigi F and Rahimifar M: A variant of iodotyrosine deiodinase deficiency, J Clin Endocrinol Metab 44:499-507, 1978.

Case 3   Congenital adrenal hyperplasia

A 55-year-old woman seeks your help for her "ruddy" complexion. She has had multiple medical problems in the past, many of which have not been fully evaluated. She is very short, only 147 cm tall, although she comes from a very tall family: the father is 193 cm; the mother, 172 cm; and a sister, 175 cm. She was not always short. In early elementary school she was considered tall, but as her friends grew, she did not. When she was age 13 she had problems with excessive facial hair, and at age 14 underwent a clitorectomy for an enlarged clitoris. She did not menstruate normally, and at age 16 was placed on estrogen/progesterone therapy to induce ovulation. No further workup was done at that time. She shaves daily and has always been very strong. On physical examination she is a short woman with a manlike habitus and muscle pattern. She has obvious facial and chest hair and a male balding pattern. She has a "ruddy" complexion with multiple telangiectasia (small visible blood vessels) over her face. Her vital signs are normal. Her blood pressure is 116/62. The remainder of the physical examination is normal except for the excessive hair and some continued clitoral enlargement. Pelvic examination indicates that she has a uterus. Laboratory examination shows her electrolytes and complete blood count to be normal. Her urinary free cortisol is 12 ng/g creatinine (normal, 10 to 55). Her plasma androstenedione concentration is 2000 and her DHEA-S is 9000 ng/ml (normal, <350 and <2700, respectively).

Biochemical questions   ◆ 1. Describe the biochemical abnormalities that could cause this syndrome.
                              ◆ 2. How would you test for the possible conditions?
                              ◆ 3. How would you treat it?

Case discussion   **1. Biochemical abnormalities** This patient has hyperandrogenism that was manifest from birth with rapid growth during childhood but early closure of her epiphyses and overall short stature. At puberty she demonstrated dramatic increases in androgens, which required a clitorectomy. Her blood pressure and therefore salt balance are normal.

   2. Diagnostic tests The first test one should perform is a chromosomal analysis to be sure that she is female. She does have the normal 46,XX pattern of a female. This type of syndrome is usually caused by congenital adrenal hyperplasia, in which one of the hormones involved in cortisol synthesis is deficient. This deficiency may also block synthesis of sex steroids or mineralocorticoids. In an attempt to overcome this block, excess steroids are made, which may cause salt retention or excess androgenization. In this case we know that desmolase, 17-hydroxylase, and 3-β-hydroxydehydrogenase are intact, since they are necessary for the production of androstenedione. Thus the possible enzymatic deficiencies are 21- or 11-hydroxylase. Neither of these could be completely missing. A complete 21-hydroxylase deficiency would be associated with no mineralocorticoids, severe salt wasting, and hypotension. Deficiency of the enzyme 11-hydroxylase is associated with the accumulation of the potent mineralocorticoid, 11-deoxycorticosterone, and with hypertension. However, a partial 21-hydroxylase deficiency exists, which is the most common of the enzymatic deficiencies and is associated with hyperandrogenism without salt loss. This is the most likely diagnosis.

   If there were a 21-hydroxylase deficiency, the hormones made before this step would build up. Since 17-hydroxyprogesterone is the hormone directly modified by 21-hydroxylase, it should be very high, and in this patient is 25 times normal.

   3. Treatment Since the major problem is inadequate cortisol synthesis, causing elevated ACTH and increased steroidogenesis, the appropriate therapy is to shut off the adrenal glands by giving cortisol. In practice, a synthetic form of cortisol, dexamethasone, is more effective and is generally used in these patients. Administration of dexamethasone at appropriate levels is associated with normalization of ACTH, 17-hydroxyprogesterone, and the androgens.

References

Levine LS et al: Genetic mapping of the 21-hydroxylase deficiency gene within the HLA linkage group, N Engl J Med 299:911, 1978.

Migeon CJ: Diagnosis and treatment of the adrenogenital disorders. In DeGroot L, editor: Endocrinology, Philadelphia, 1988, WB Saunders Co.

Case 4   **Osteopenia in a young man**

A 22-year-old college student is brought to the emergency room in extreme pain. He was generally feeling well, but woke up this morning, having fallen out of bed, with severe pain in the middle of his back. He denies any known trauma. His medical history reveals a seizure disorder that has been present for 10 years, for which he takes phenobarbital. Physical examination shows severe tenderness in the area of pain. A roentgenogram of the spine reveals a fracture in the first lumbar vertebrate. There is diffuse loss of calcification in all bones visible on the film. An examination by quantitative computed tomography shows the bone density of the spine to be only 55% of normal for his age. Laboratory evaluation shows a serum calcium concentration of 8.5 mg/dl (normal, 8.5 to 10.6), and a serum phosphate of 5.5 mg/dl (normal, 3.5 to 5). The serum PTH is normal, but the serum 25-hydroxycholecalciferol level is 3 ng/ml (normal, 8 to 55). A bone biopsy shows severe osteomalacia (lack of mineralization of bone with normal matrix formation).

Biochemical questions ◆ 1. List possible causes of the low 25-hydroxycholecalciferol level.
2. How does the reduction in vitamin D lead to osteomalacia?

Case discussion

1. Low 25-hydroxycholecalciferol The hormone 25-hydroxycholecalciferol is an intermediate in the formation of 1,25-dihydroxycholecalciferol from cholecalciferol. The diminished level of hormone might be caused by a decreased level of the liver hydroxylase or by increased metabolism of the 25-hydroxycholecalciferol. The patient is using phenobarbital for a seizure disorder, and this drug is known to increase the activity of liver mixed-function hydroxylases. Thus it is unlikely that the 25-hydroxylase, a mixed-function hydroxylase, would be diminished; indeed, when measured, the 25-hydroxylase is increased in activity. Studies from various laboratories have demonstrated that in patients taking phenobarbital or diphenylhydantoin (another antiseizure drug), vitamin D is more rapidly metabolized to very polar, inactive metabolites, probably by 23-hydroxylase, 26-hydroxylase, and other, poorly characterized, mixed-function hydroxylases. When measured in this patient, excess levels of very polar vitamin D metabolites are demonstrated.

2. Vitamin D and osteomalacia A reduction in active vitamin D metabolites would lead to a reduction in bone calcification through several different mechanisms. The low levels of active vitamin D metabolites would lead to a diminished absorption of calcium from the gut and increased loss of calcium into the urine. Overall this leads to a reduction in serum calcium and a diminished calcification of bone. In addition, the diminished calcium would lead to a secondary hyperparathyroidism and elevated PTH levels, with increased resorption of bone.

The clinical problem in this young man was probably caused by a seizure that occurred during the night, causing him to fall out of bed. The trauma of falling out of bed (while unconscious from the seizure) fractured his vertebrate. He will be treated with 5000 to 10,000 U of vitamin D daily to remineralize his bones to normal density.

References

Dent CE et al: Osteomalacia with long-term anticonvulsant therapy in epilepsy, Br Med J 4:69, 1970.

Hahn TJ and Halstead LR: Sequential changes in mineral metabolism and serum vitamin D metabolite concentrations produced by phenobarbital administration in the rat, Calcif Tissue Res 35:376, 1983.

Weinstein R et al: Decreased serum ionized calcium and normal vitamin D metabolite levels with anticonvulsant drug treatment, J Clin Endocrinol Metab 58:1003, 1984.

Case 5   A woman with excess hair

A 29-year-old woman complains of the onset over the past 4 years of excessive facial hair and increased darkness of the hair on her arms. During this time she has gained about 20 kg, which is particularly noticeable in her face, abdomen, and hips. She has also developed acne, a problem she never had before. Her menstrual periods have ceased. On physical examination you note that the woman is obese but has thin arms and legs. She is hypertensive. Her face is rounded ("moon" facies), and she has excessive amounts of fat at the back of her neck ("buffalo hump") and in the superclavicular region. She has bright-purple striae (stretch marks) over her abdomen. There are several poorly healed ulcers on her legs, most of which are infected. You notice that she has a male muscle pattern, a male escutcheon (diamond shape to the region of her pubic hair), and clitoromegaly. You can feel a mass about the size of a grapefruit on the right side of her abdomen. Laboratory examination shows hypernatremia with a serum sodium concentration of 151 mEq/L (normal, 137 to 145) and hypokalemia with a serum potassium of 3.0 mEq/L (normal, 3.5 to 5). Serum cortisol is 48 μg/ml at 8 AM and 18 at 4 PM (normal, 8 to 18 and 0 to 4, respectively). The testosterone level is elevated at 2 ng/ml (normal, <0.5), androstenedione is 255 ng/ml (normal, <150), and DHEA-S is 12,970 ng/ml (normal, <2700). The ACTH level is undetectable. A magnetic resonance image (MRI) shows a large mass in the area of the right adrenal gland.

Biochemical questions

1. Explain the abnormal values for the androgens and cortisol.
2. How might you treat the problem?

Case discussion

1. Abnormal androgen and cortisol values The patient has symptoms of Cushing's syndrome with central obesity, violaceous striae, poor wound healing, and infections. Physical examination shows masculinization and a large adrenal mass that is confirmed on MRI. The laboratory examination shows high values for cortisol, androstenedione, DHEA-S, and testosterone.

Cushing's syndrome has four etiologies: (1) hypothalamic resistance to cortisol, (2) ectopic production of ACTH, (3) adrenal adenomas, and (4) adrenal carcinomas. The first two are associated with a normal or elevated ACTH level, the last two with suppressed ACTH because the cortisol production is autonomous. Since the patient has suppressed ACTH and a large adrenal mass, she unfortunately has an adrenocortical carcinoma. Adrenocortical carcinomas are often poor producers of steroids, since the enzymes are not always coordinately induced, and different tumors can produce different hormones or perhaps only intermediate products. They may produce high levels of cortisol or adrenal androgens. Rarely they may even produce estrogens or mineralocorticoids.

The production of these hormones by the tumor is considered eutopic, since the adrenal gland can normally produce all these hormones. In this case the DHEA-S is extraordinarily high, whereas the other adrenal hormones are somewhat elevated. This suggests that 3-β-hydroxydehydrogenase is limiting. Thus the three hormones dependent on this enzyme, cortisol, androstenedione, and testosterone, are limited in production. The pregnenolone and 17-hydroxypregnenolone levels probably would also be elevated.

The testosterone is a strong androgen; androstenedione, a moderately potent androgen; the DHEA-S, a weak androgen. All are present in high levels and cause the excessive hair, male muscle pattern, clitoromegaly, and loss of menstruation.

2. Treatment The tumor is malignant, and its main danger lies in two problems: excess hormone production and local invasion. Patients with adrenocortical carcinoma need surgery to remove as much bulk of the tumor as possible, but because of the Cushing's syndrome they are often too sick for surgery. To reduce the effect of excess cortisol, we need to block either cortisol production or cortisol action. Although potent antagonists to estrogens and androgens exist, no such compounds exist for cortisol. We must therefore block cortisol production, and three enzyme inhibitors can be used. *Aminoglutethimide* blocks the action of desmolase, preventing the conversion of cholesterol to pregnenolone; *metyrapone* blocks the 11-hydroxylase, preventing the conversion of 11-deoxycortisol to cortisol; and *ortho,para-DDD* (a congener of DDT) blocks multiple enzymatic activities. Clinically, we usually use *o,p-DDD* until the patient has reduced action of cortisol and then surgically remove as much tumor as possible. Unfortunately, survival with these tumors is usually limited.

References

Kreiger DT: Pathophysiology of Cushing's disease, Endocr Rev 4:22, 1983.
Santen R et al: Successful medical adrenalectomy with aminoglutethimide, JAMA 230:1661, 1974.
Yamagi T et al: Serum dihydroepiandrosterone in Cushing's syndrome, J Clin Endocrinol Metab 59:1164, 1984.

Case 6   Hyperthyroidism in a small community

As an epidemiologist, you are called to a local town in northwestern Iowa. This small town of 300 people has had 35 cases of hyperthyroidism over the past 6 months. Most of the patients have been males, ages 15 to 35. You interview several patients and find various common features of hyperthyroidism, including weight loss, heat intolerance, increased sweating, and tremors. The thyroid gland in these patients is very small. This contrasts with typical patients with hyperthyroidism, who are usually female and have a large goiter. Laboratory examination shows high T_4 and T_3 levels and a suppressed TSH. There is no measurable thyroid-stimulating immunoglobulin. Statistical evaluation of a complex questionnaire that you administer to the citizens shows that the affected individuals eat more hamburgers than the unaffected controls. You individually interview each of the hyperthyroid patients, and each tells you about this special place they obtain hamburger meat that is the "best in the world." You visit the plant and find that the "best hamburger meat" is made from the neck muscles of the steer and contains about 10% thyroid gland by weight. The plant agrees to stop producing this special cut of hamburger, and the thyrotoxicosis disappears.

Biochemical questions   1. Contrast the pathogenesis of "hamburger-induced" hyperthyroidism with that of Graves' disease.

2. Do we need to worry about the inclusion of other hormones in our food? Specifically address possible problems with inclusion of the hypothalamus, pituitary, adrenal gland, and pancreas.

Case discussion

1. Hyperthyroidism and Graves' disease Hamburger-induced thyrotoxicosis is caused by the ingestion of the thyroid hormone found in the thyroid glands that were ground along with the strap muscles of the steer's neck. Thyroid hormone is stored in the thyroid gland as the altered amino acids, thyronines, within the thyroglobulin that makes up the colloid of the thyroid follicles. T_4 and T_3 are stable to digestive enzymes and can be taken orally. The thyroid gland, however, will release thyroid hormones from the thyroglobulin only after digestion. The peptic activity of the stomach and the tryptic and chymotryptic activity of the intestines are necessary to release the active hormones. Since the amount of hormone ingested depends on appetite, young men are most likely to be affected. In this condition the total T_4, free T_4, and T_3 levels would be high and the TSH level low. Measurement of the biologic activity in vivo by measuring the amount of radioactive iodine accumulated by the thyroid gland shows that the thyroid gland is not functioning, since the thyroid hormone is coming from an outside source.

In contrast to this, patients with Graves' disease are hyperthyroid because of the production of an immunoglobulin that activates the thyroid gland by binding to TSH receptors and stimulating cAMP. Many of the tests are the same as in hamburger-induced thyrotoxicosis: high total T_4, free T_4, and T_3 levels and a low TSH level. Measurement of the accumulation of iodine by the thyroid gland would demonstrate the difference—hyperfunction of the thyroid gland.

2. Other hormones in food Inclusion of glands along with ground meat is a potential problem that is discouraged by standard butchering practice and some state laws. In practice, however, most hormones are not stable to the digestive process and would be inactivated. All the proteins, glycoproteins, and polypeptide hormones would be destroyed by digestion. Therefore it is safe to include hypothalamus, pituitary, pancreas, or parathyroid glands in a chopped-meat preparation. Although cortisol is stable to digestion, it is not stored to any great extent in the adrenals and therefore does not pose a health hazard. Interestingly, the major source of extraneous hormones does not come from ingestion of glands, but rather from liver. Some chickens are treated with estrogens, which are concentrated by the liver. Ingestion of small amounts of chicken liver poses no threat, but some males who eat excessive amounts of chicken livers from chickens treated with estrogens may develop breasts.

Reference

Hay ID and Klee GG: Thyroid dysfunction, Endocrinol Metab Clin North Am 17:473, 1988.

Case 7

An infertile young woman

A 24-year-old woman complains of infertility. She and her husband have been having unprotected intercourse for 2 years. She states that she has not had any medical problems. She had a normal childhood with rapid growth at ages 11 to 13 and development of breasts at age 12. She never developed pubic or axillary hair and has never menstruated. On physical examination she appears to be a normal female but missing pubic and axillary hair. Pelvic examination shows a blind vagina with no cervix and no palpable uterus or ovaries. Laboratory examination shows an estradiol level of 30 pg/ml (normal female, 50 to 300) and a testosterone level of 3.9 ng/ml (normal female, <0.045; normal male, 3 to 7). Chromosomal analysis shows her to be 46,XY.

Biochemical questions	1. Describe the pathways of synthesis of estradiol, testosterone, and dihydrotestosterone.
	2. Is this person male or female? Defend your position.
	◆3. Describe the biochemical abnormality in this person.

References

Fuller PJ: The cloning of the steroid hormone receptors: basic and clinical implications, Aust NZ J Med 18:890, 1988.

Kaplan SA: Cell receptors, Am J Dis Child 138:P1140, 1984.

Perez-Palacios G et al: The syndromes of androgen resistance revisited, J Steroid Biochem 27:1101, 1987.

Case 8 A young woman with weakness

A 21-year-old college student comes to see you and complains of weakness, poor sleeping, weight loss, poor appetite, and crying spells for the past 2 months. She has had poor grades during the same period. Although you are reasonably sure that she is having a period of unipolar depression, you do some medical tests. All are normal except the total T_4 and T_3 levels, both of which are high. The TSH level is normal, but the thyroxine-binding globulin (TBG) level is dramatically elevated.

Biochemical questions

1. Describe the normal physiologic role of TBG.
2. Which other hormones have binding globulins? What is the effect of estrogens on each?
3. Is there any abnormality of thyroid hormone affecting this patient?

References

Musa BU et al: Excretion of corticosteroid-binding globulin and thyroxine-binding globulin and total protein in adult males with nephrosis: effect of sex hormones, J Clin Endocrinol Metab 27:768, 1967.

Oppenheimer JH: Role of plasma proteins in the binding, distribution and metabolism of thyroid hormones, N Engl J Med 278:1153, 1968.

Case 9 A male with impotence and gynecomastia

A 38-year-old male seeks your help because of an inability to have an erection. This has been happening for about a year, since he started a new job working on a chicken farm. He is taking no medications. He eats chicken livers about four times each week. On physical examination he has a moderate amount of breast tissue and small, atrophic testicles. Laboratory examination shows a testosterone level of 0.2 ng/ml (normal, 3 to 7), a low estradiol level, and few gonadotropins. The diethylstilbesterol level is very high.

Biochemical questions

1. Diethylstilbesterol is a potent estrogen found in chicken livers. Describe why this causes a decrease in testosterone and gonadotropins.
◆ 2. What caused the gynecomastia and small testes in this patient?

References

Landau RL et al: Gynecomastia and retarded sexual development resulting from a long-standing estrogen-secreting tumor, J Clin Endocrinol Metab 14:1097, 1954.

Velduis JD et al: Pathophysiology of male hypogonadism associated with endogenous hyperestrogenism, N Engl J Med 312:1371, 1985.

Additional questions and problems

1. The commonly used antithyroid medication *propylthiouracil* works by inhibiting the action of thyroperoxidase. When given to patients with hyperthyroidism, it often does not work for weeks. Explain its action in detail as well as this delay in action. Do you think the size of the goiter would correlate with the length of time of the delay? Why or why not?

2. Iodine deficiency is now rare in the United States, but both iodine deficiency and conditions associated with poor concentrating ability of the thyroid gland for iodine are associated with an increased synthesis of T_3. Explain.

3. Elderly patients with constipation often take mineral oil to increase their bowel movements. What effect does excessive use of mineral oil have on osteopenia in the elderly?

4. Prednisone, a synthetic and powerful relative of cortisol, is frequently used to treat inflammation, asthma, and a variety of chronic conditions. What is the effect of prednisone on ACTH production? What is the effect on cortisol, aldosterone, and adrenal androgen production?

5. A patient with nephrotic syndrome loses protein in his urine, including many hormone-binding globulins. What are the consequences of the loss of these binding proteins?

6. The heat-shock proteins are part of the many intracellular receptors. What might be the physiologic consequences of a deletion mutation in the genes for these proteins?

Multiple choice problems

1. There are now potent inhibitors of angiotensin-converting enzyme. These drugs would most likely be used for hypertension caused by:
   A. Pheochromocytoma with excess epinephrine.
   B. Adrenal tumor with excess aldosterone.
   C. Adrenal tumor with excess cortisol.
   D. Kidney tumor with excess renin.
   E. Hyperthyroidism.

2. A patient with elevated TSIG is most likely to have:
   A. Hypothyroidism.
   B. Hyperthyroidism.
   C. Hypocortisolism.
   D. Hypocortisolism.
   E. None of the above.

3. Which of the following are active hormones without further metabolism?
   A. Cholecalciferol
   B. Angiotensin I
   C. Cortisol
   D. 17-Hydroxyprogesterone
   E. DHEA-S

4. Which set of hormones is *mispaired* as a trophic hormone and responding hormone?
   A. ACTH and cortisol
   B. LH and testosterone
   C. PTH and 1,25-dihydroxycholecalciferol
   D. TSH and thyroxine
   E. FSH and aldosterone

5. Which of the following enzymes is (are) *correctly* linked to its direct product?
   1. 11-Hydroxylase, cortisol
   2. 18-Hydroxylase, aldosterone

3. 1-Hydroxylase, 1,25-dihydroxycholecalciferol
4. Thyroperoxidase, thyroxine
   A. 1, 2, and 3.
   B. 1 and 3.
   C. 2 and 4.
   D. 4 only.
   E. All are correct.

6. Which of the following hormones has (have) a biologic rhythm (daily, monthly, or other)?
  1. Thyroxine
  2. Cortisol
  3. Aldosterone
  4. Estradiol
   A. 1, 2, and 3.
   B. 1 and 3.
   C. 2 and 4.
   D. 4 only.
   E. All are correct.

7. Ovulation:
  1. Occurs in response to elevated LH.
  2. Depends on progesterone.
  3. Will lead to increased progesterone production.
  4. Depends primarily on inhibin.
   A. 1, 2, and 3.
   B. 1 and 3.
   C. 2 and 4.
   D. 4 only.
   E. All are correct.

8. Cyclic AMP is of major importance in the production of:
  1. Thyroxine.
  2. Cortisol.
  3. 1,25-Dihydroxycholecalciferol.
  4. Insulin.
   A. 1, 2, and 3.
   B. 1 and 3.
   C. 2 and 4.
   D. 4 only.
   E. All are correct.

9. Thyroxine functions by:
  1. Increasing the DNA polymerase of cells.
  2. Binding to a specific receptor in the plasma membrane.
  3. Increasing adenylate cyclase.
  4. Binding to a specific receptor in the nucleus.
   A. 1, 2, and 3.
   B. 1 and 3.
   C. 2 and 4.
   D. 4 only.
   E. All are correct.

10. Which of the following is (are) true?
    1. Strong sunscreens inhibit the production of vitamin D.
    2. Sodium perchlorate inhibits the production of aldosterone.
    3. Aminoglutethimide inhibits the production of cortisol.
    4. Methotrexate inhibits the production of estradiol.
        A. 1, 2, and 3.
        B. 1 and 3.
        C. 2 and 4.
        D. 4 only.
        E. All are correct.

Abbreviations

A	Adenine	CPG III	Coproporphyrinogen III
A site	Site on ribosome that holds aminoacyl tRNA; *see also* P site	CRH	Corticotropin-releasing hormone
ACAT	Acylcoenzyme A: cholesterol acyltransferase	CTP	Cytidine 5'-triphosphate
		Cys	Cysteine
ACP	Acyl carrier protein, or acid phosphatase	Cys-Gly	Cysteinylglycine
		dAMP	Deoxyadenosine 5'-monophosphate
ACTH	Adrenocorticotropin (*adrenal corticotropic hormone*)	dATP	Deoxyadenosine 5'-triphosphate
ADH	Vasopressin, the anti-diuretic hormone	dCTP	Deoxycytidine 5'-triphosphate
ADP	Adenosine 5'-diphosphate	DFP	Diisopropyl fluorophosphate
AICAR	5-Aminoimidazole-4-carboxamide ribotide	dGTP	Deoxyguanosine 5'-triphosphate
Ala	Alanine	DHF	Dihydrofolate
ALA	Aminolevulinic acid	DIT	3,5-Diiodotyrosine
ALP	Alkaline phosphatase	DNA	Deoxyribonucleic acid
ALT	Alanine aminotransferase	DNase	Deoxyribonuclease
AMP	Adenosine 5'-monophosphate	Dopa	3,4-Dihydroxyphenylalanine
Arg	Arginine	L-Dopa	L-Dihydroxyphenylalanine
Asn	Asparagine	Dopamine	3,4-Dihydroxyphenethylamine
Asp	Aspartic acid		
AST	Aspartate aminotransferase	DPG	2,3-Bisphosphoglycerate (2,3-diphosphoglycerate)
ATP	Adenosine 5'-triphosphate		
B_6	*See* Vitamin B_6	dTMP	Deoxythymidine 5'-monophosphate
B_{12}	*See* Vitamin B_{12}		
BAL	British antilewisite, 2,3-dimercaptopropanol	dTTP	Deoxythymidine 5'-triphosphate
BCCP	Biotin carboxyl-carrier protein	ECG	Electrocardiogram
		EDTA	Ethylenediaminetetraacetic acid
BMR	Basal metabolic rate		
BUN	Blood urea nitrogen	EF-1, EF-2	Elongation factors 1 and 2 in protein biosynthesis
BV	Biologic value of proteins		
C	Cytosine	ES	Enzyme-substrate complex
cAMP	Cyclic AMP, adenosine 3',5'-monophosphate	FA	Fatty acid(s)
		FAD	Flavin adenine dinucleotide
CAP	Catabolite activator protein		
CDP	Cytidine 5'-diphosphate	$FADH_2$	Reduced form of flavin adenine dinucleotide
Cer	Ceramide		
cGMP	Cyclic GMP, guanosine 3',5'-monophosphate	FAICAR	5-Formamidoimidazole-4-carboxamide ribotide
Chln	Choline	FFA	Free fatty acid(s)
CK	Creatine kinase	fGAR	Formylglycinamide ribotide
CMP-NANA	Cytidine monophosphate *N*-acetyl neuraminic acid	fMet-tRNA^F	Formylmethionyl tRNA
CoA (CoASH)	Coenzyme A	FMN	Flavin mononucleotide
COMT	Catecholamine *O*-methyltransferase	Fru	Fructose
		FSH	Follicle-stimulating hormone

G	Guanine	ITP	Inosine 5′-triphosphate
ΔG	Change in free energy of a given system	IUPAC	International Union of Pure and Applied Chemistry
GABA	γ-Aminobutyric acid	K_m	Michaelis contant of an enzyme for a given substrate
Gal	Galactose		
GalNAc	N-Acetylgalactosamine		
GAR	Glycinamide ribotide	αKG	α-Ketoglutarate
GDP	Guanosine 5′-diphosphate	lac operon	Escherichia coli operon for genes used in catabolism of lactose
GDPMan	Guanosine diphosphate mannose		
GFR	Glomerular filtration rate	LCAT	Lecithin-cholesterol acyl transferase
Glc	Glucose (if no confusion results, glucose is also indicated by G)		
		LDH	Lactate dehydrogenase
		LDL	Low-density lipoprotein
GLC	Gas-liquid chromatography	Leu	Leucine
GlcA	Glucuronic acid	LH	Luteinizing hormone
GlcNAc	N-Acetyl glucosamine	LRH/FSHRH	Leutinizing and follicle-stimulating hormone-releasing hormone
Gln	Glutamine		
Glu	Glutamate		
Gly	Glycine	LSD	Lysergic acid diethylamide
GMP	Guanosine 5′-monophosphate	Lys	Lysine
		Man	Mannose
GSH	Glutathione, γ-glutamylcysteinylglycine	MAO	Monoamine oxidase
		Mb	Myoglobin
GSSG	Gluthathione, oxidized	MD	Muscular dystrophy
GTP	Guanosine 5′-triphosphate	MDH	Malate dehydrogenase
Hb A	Hemoglobin, major adult form	MDR	Minimum daily requirement of nutrients
HBDH	Hydroxybutyrate dehydrogenase	Met	Methionine
		MIT	3-Monoiodotyrosine
Hb F	Hemoglobin, fetal form	mOsm	Milliosmole
Hb S	Hemoglobin, form found in sickle cell anemia	MRH	Melanotropine-releasing hormone
HDL	High-density lipoprotein	MRIH	Melanotropin release-inhibiting hormone
HPRT	Hypoxanthine-guanine phosphoribosyltransferase		
		mRNA	Messeinger ribonucleic acid
His	Histidine		
HMGCoA	3-Hydroxyl-3-methylglutaryl coenzyme A	MSH	Melanotropin
		NAD+	Nicotinamide adenine dinucleotide
HTC cell	Hepatoma tissue culture cells		
		NADH	Reduced form of nicotinamide adenine dinucleotide
Hyp	Hydroxyproline		
"i" gene	Escherichia coli gene for lac operon repressor		
		NADP+	Nicotinamide adenine dinucleotide phosphate
ICDH	Isocitrate dehydrogenase		
IDL	Intermediate-density lipoprotein	NADPH	Reduced form of nicotinamide adenine dinucleotide phosphate
IdoA	Iduronic acid		
IF-1, IF-2, IF-3	Protein synthesis initiation factors 1, 2, and 3	Na+,K+-ATPase	Sodium, potassium—activated adenosinetriphosphatase
IgA, IgG, IgM	Classes of immunoglobulins		
Ile	Isoleucine	NHI	Nonheme iron
IMP	Inosine 5′-monophosphate	NSILA	Nonsuppressible insulin-like activity
IPU	2-Isopropyl-4-pentenoyl urea		
IQ	Intelligence quotient	OAA	Oxaloacetate

Orn	Ornithine	SDA	Specific dynamic action of foods
P site	Site on ribosome that holds peptidyl tRNA	SDH	Succinate dehydrogenase
Ψ	Pseudouridine, 5-ribosyl uracil	Ser	Serine
		SH	Somatotropin
PBG	Porphobilinogen	SHRH	Somatotropin-releasing hormone
pCO_2	Partial pressure of carbon dioxide	SHRIH	Somatostatin, or somato-tropin release–inhibiting hormone
PDH	Pyrivate dehydrogenase		
PEP	Phosphoenolypyruvate		
PFK	Phosphofructokinase	SLE	Systemic lupus erythema-tosus
PG	Prostaglandin		
PGI_2	Prostacyclin	T_3	Triiodothyronine
Phe	Phenylalanine	T_4	Thyroxine
P_i	Inorganic phosphate ion	TBG	Thyroxine-binding globu-lin
P/O	Phosphorylation efficiency of oxidative phosphoryla-tion	THF	Tetrahydrofolate
		Thr	Threonine
pO_2	Partial pressure of oxygen	TLC	Thin-layer chromatography
polA	*Escherichia coli* DNA poly-merase I	TPP	Thiamine pyrophosphate
		TRH	Thyrotropin-releasing hor-mone
PP_i	Inorganic pyrophosphate ion		
PRH	Prolactin-release-inhibiting hormone	tRNA	Transfer ribonucleic acid
		Trp	Tryptophan
PRIH	Prolactin-release-inhibiting hormone	TSH	Thyroid-stimulating hor-mone
Pro	Proline	TXA_2	Thromboxane A_2
PRPP	Phosphoribosylpyrophos-phate	TXB_2	Thromboxane B_2
		Tyr	Tyrosine
PTA	Plasma thromboplastin ante-cedent	U	Uridine
		UDPG	Uridine diphosphate glu-cose
PTC	Plasma thromboplastin com-ponent		
		UPG III	Uroporphyrinogen III
R17	An RNA bacteriophage of *Escherichia coli*	UTP	Uridine 5′-triphosphate
RBC	Erythrocytes, red blood cells	UV	Ultraviolet
RDA	Recommended daily dietary allowance of nutrients	V_{max}	Maximum initial velocity of an enzyme reaction
RNA	Ribonucleic acid	Val	Valine
RNase	Ribonuclease	Vitamin B_1	Thiamine
rRNA	Ribosomal ribonucleic acid	Vitamin B_2	Riboflavin
30S	Small subunit of bacterial ri-bosome that sediments at force of 30 Svedberg units	Vitamin B_6	Pyridoxine
		Vitamin B_{12}	Cobalamin
		VLDL	Very low-density lipopro-tein
50S	Large subunit of bacterial ri-bosomes that sediments at force of 50 Svedberg units	VMA	4-Hydroxy-3-methoxy-mandelic acid (Vanillyl-mandelic acid)
		WBC	Leukocytes, white blood cells
SAH	*S*-Adenosylhomocysteine		
SAM	*S*-Adenosylmethionine		

Appendixes

Appendix A Recommended dietary allowances designed for the maintenance of good nutrition of practically all healthy people in the United States[a]

Category	Age (years) or Condition	Weight[b] (kg)	Weight[b] (lb)	Height[b] (cm)	Height[b] (in)	Protein (g)	Fat-Soluble Vitamins Vitamin A (μg RE)[c]	Vitamin D (μg)[d]	Vitamin E (mg α-TE)[e]	Vitamin K (μg)
Infants	0.0-0.5	6	13	60	24	13	375	7.5	3	5
	0.5-1.0	9	20	71	28	14	375	10	4	10
Children	1-3	13	29	90	35	16	400	10	6	15
	4-6	20	44	112	44	24	500	10	7	20
	7-10	28	62	132	52	28	700	10	7	30
Males	11-14	45	99	157	62	45	1000	10	10	45
	15-18	66	145	176	69	59	1000	10	10	65
	19-24	72	160	177	70	58	1000	10	10	70
	25-50	79	174	176	70	63	1000	5	10	80
	51 +	77	170	173	68	63	1000	5	10	80
Females	11-14	46	101	157	62	46	800	10	8	45
	15-18	55	120	163	64	44	800	10	8	55
	19-24	58	128	164	65	46	800	10	8	60
	25-50	63	138	163	64	50	800	5	8	65
	51 +	65	143	160	63	50	800	5	8	65
Pregnant						60	800	10	10	65
Lactating	1st 6 months					65	1300	10	12	65
	2nd 6 months					62	1200	10	11	65

From Food and Nutrition Board, National Academy of Sciences—National Research Council: Recommended dietary allowances, revised 1989, ed 9, Pub no 1694, Washington, DC, 1989.

[a]The allowances, expressed as average daily intakes over time, are intended to provide for individual variations among most normal persons as they live in the United States under usual environmental stresses. Diets should be based on a variety of common foods in order to provide other nutrients for which human requirements have been less well defined. See text for detailed discussion of allowances and of nutrients not tabulated.

[b]Weights and heights of Reference Adults are actual medians for the U.S. population of the designated age, as reported by NHANES II. The median weights and heights of those under 19 years of age were taken from Hamill et al, (1979) (see pages 16-17). The use of these figures does not imply that the height-to-weight ratios are ideal.

Water-Soluble Vitamins							Minerals						
Vitamin C (mg)	Thiamin (mg)	Riboflavin (mg)	Niacin (mg NE)[f]	Vitamin B6 (mg)	Folate (μg)	Vitamin B12 (μg)	Calcium (mg)	Phosphorus (mg)	Magnesium (mg)	Iron (mg)	Zinc (mg)	Iodine (μg)	Selenium (μg)
30	0.3	0.4	5	0.3	25	0.3	400	300	40	6	5	40	10
35	0.4	0.5	6	0.6	35	0.5	600	500	60	10	5	50	15
40	0.7	0.8	9	1.0	50	0.7	800	800	80	10	10	70	20
45	0.9	1.1	12	1.1	75	1.0	800	800	120	10	10	90	20
45	1.0	1.2	13	1.4	100	1.4	800	800	170	10	10	120	30
50	1.3	1.5	17	1.7	150	2.0	1200	1200	270	12	15	150	40
60	1.5	1.8	20	2.0	200	2.0	1200	1200	400	12	15	150	50
60	1.5	1.7	19	2.0	200	2.0	1200	1200	350	10	15	150	70
60	1.5	1.7	19	2.0	200	2.0	800	800	350	10	15	150	70
60	1.2	1.4	15	2.0	200	2.0	800	800	350	10	15	150	70
50	1.1	1.3	15	1.4	150	2.0	1200	1200	280	15	12	150	45
60	1.1	1.3	15	1.5	180	2.0	1200	1200	300	15	12	150	50
60	1.1	1.3	15	1.6	180	2.0	1200	1200	280	15	12	150	55
60	1.1	1.3	15	1.6	180	2.0	800	800	280	15	12	150	55
60	1.0	1.2	13	1.6	180	2.0	800	800	280	10	12	150	55
70	1.5	1.6	17	2.2	400	2.2	1200	1200	320	30	15	175	65
95	1.6	1.8	20	2.1	280	2.6	1200	1200	355	15	19	200	75
90	1.6	1.7	20	2.1	260	2.6	1200	1200	340	15	16	200	75

[c]Retinol equivalents. 1 retinol equivalent = 1 μg retinol or 6 μg β-carotene. See text for calculation of vitamin A activity of diets as retinol equivalents.

[d]As cholecalciferol. 10 μg cholecalciferol = 400 IU of vitamin D.

[e]α-Tocopherol equivalents. 1 mg d-α tocopherol = 1 α-TE. See text for variation in allowances and calculation of vitamin E activity of the diet as α-tocopherol equivalents.

[f]1 NE (niacin equivalent) is equal to 1 mg of niacin or 60 mg of dietary tryptophan.

Appendix B   International classification of enzymes

Main class and subclass	Prosthetic group or coenzyme	Example
1 Oxidoreductases		
1.1 Acting on $=$ CHOH donors		
1.1.1 With NAD or NADP as acceptors	NAD, NADP	Lactate or malate dehydrogenases
1.1.3 With O_2 as acceptor	FAD	Glucose oxidase
1.2 Acting on $=$ C $=$ O donors		
1.2.1 With NAD or NADP as acceptors	NAD, NADP	Glyceraldehyde 3-phosphate dehydrogenase
1.2.3 With O_2 as acceptor	FAD	Xanthine oxidase
1.3 Acting on $=$ CH—CH $=$ donors		
1.3.1 With NAD or NADP as acceptors	NAD, NADP	Dihydrouracil dehydrogenase
1.3.2 With FAD as acceptor	FAD	Acyl CoA dehydrogenase
1.4 Acting on $=$ CH—NH_2 donors		
1.4.3 With O_2 as acceptor	FAD, pyridoxal phosphate	Amino acid oxidases
2 Transferases		
2.1 Transferring 1—C groups		
2.1.1 Methyltransferases	Tetrahydrofolate (THF)	5-Methyl THF methyl transferase
2.1.2 Hydromethyl and formyl transferases	THF	Serine hydroxymethyl transferase
2.1.3 Carboxyl or carbamoyl transferases		Ornithine transcarbamoylase
2.3 Acyl transferases		Choline acetyl transferase, palmitoyl CoA-Carnitine transferase
2.4 Glycosyl transferases	UDP	Galactosyl transferase
2.6 Transferring—NH_2 groups		
2.6.1 Aminotransferases	Pyridoxal phosphate	Transaminases
2.7 Transferring phosphorus-containing groups		
2.7.1.2 ATP-glucose 6-phosphotransferase		Glucokinase
3 Hydrolases		
3.1 Cleaving ester linkages		
3.1.1 Carboxylic ester hydrolases		Lipases
3.1.3 Phosphomonoester hydrolases		Phosphatases
3.1.4 Phosphodiester hydrolases		Phosphodiesterase
3.2 Cleaving glycosides		
3.2.1 Glycosides		Amylases
3.2.2 Glycosylamine hydrolases		Nucleosidases

Appendix B   International classification of enzymes—cont'd

Main class and subclass	Prosthetic group or coenzyme	Example
3.4 Cleaving peptide bonds		
3.4.1 α-Aminopeptide amino acid hydrolases		Leucine aminopeptidase
3.4.2 α-Carboxypeptide amino acid hydrolases		Carboxypeptidases
3.4.4 Peptidopeptide hydrolases		Pepsin, trypsin
4 Lyases		
4.1 —C—C—lyases		
4.1.1 Carboxy lyases	Thiamine pyrophosphate (TPP), NAD, CoASH, lipoic acid	Pyruvate decarboxylase, α-ketoglutarate dehydrogenase complex
4.1.2 Aldehyde lyases		Aldolase
4.2—C—O—lyases		
4.2.1 Hydrolyases		Fumarase
5 Isomerases		
5.1 Racemases and epimerases		
5.1.3 Acting on carbohydrates		Ribulose 5-phosphate epimerase
5.2 *Cis-trans* isomerases		Maleylacetoacetate isomerase
5.3 Intramolecular oxidoreductases		
5.3.1 Interconverting aldoses and ketoses		Glucose phosphate isomerase
5.4 Intramolecular transferases	Methyl cobalamin	Methylmalonyl-CoA mutase
6 Ligases		
6.1 Forming C—O bonds		
6.1.1 Amino acid-tRNA ligases		Aminoacyl-tRNA synthases
6.2 Forming C—S bonds		
6.2.1 Acyl CoA ligases	CoASH	Thiokinases
6.3 Forming C—N bonds		
6.3.4.1 Xanthosine 5'-phosphate:ammonia ligase		GMP synthase
6.3.4.2 UTP:ammonia ligase (ADP)		CTP synthase
6.3.4.3 Formate:tetrahydrofolate ligase (ADP)		Formyltetrahydrofolate synthase
6.4 Forming C—C bonds		
6.4.1.1 Pyruvate:CO_2 ligase (ADP)	Biotin	Pyruvate carboxylase
6.4.1.2 Acetyl CoA:CO_2 ligase (ADP)	Biotin	Acetyl CoA carboxylase
6.4.1.3 Propionyl CoA:CO_2 ligase (ADP)	Biotin	Propionyl CoA carboxylase

Appendix C   Four-place logarithms

N	0	1	2	3	4	5	6	7	8	9	Proportional parts								
											1	2	3	4	5	6	7	8	9
10	0000	0043	0086	0128	0170	0212	0253	0294	0334	0374	*4	8	12	17	21	25	29	33	37
11	0414	0453	0492	0531	0569	0607	0645	0682	0719	0755	4	8	11	15	19	23	26	30	34
12	0792	0828	0864	0899	0934	0969	1004	1038	1072	1106	3	7	10	14	17	21	24	28	31
13	1139	1173	1206	1239	1271	1303	1335	1367	1399	1430	3	6	10	13	16	19	23	26	29
14	1461	1492	1523	1553	1584	1614	1644	1673	1703	1732	3	6	9	12	15	18	21	24	27
15	1761	1790	1818	1847	1875	1903	1931	1959	1987	2014	*3	6	8	11	14	17	20	22	25
16	2041	2068	2095	2122	2148	2175	2201	2227	2253	2279	3	5	8	11	13	16	18	21	24
17	2304	2330	2355	2380	2405	2430	2455	2480	2504	2529	2	5	7	10	12	15	17	20	22
18	2553	2577	2601	2625	2648	2672	2695	2718	2742	2765	2	5	7	9	12	14	16	19	21
19	2788	2810	2833	2856	2878	2900	2923	2945	2967	2989	2	4	7	9	11	13	16	18	20
20	3010	3032	3054	3075	3096	3118	3139	3160	3181	3201	2	4	6	8	11	13	15	17	19
21	3222	3243	3263	3284	3304	3324	3345	3365	3385	3404	2	4	6	8	10	12	14	16	18
22	3424	3444	3464	3483	3502	3522	3541	3560	3579	3598	2	4	6	8	10	12	14	15	17
23	3617	3636	3655	3674	3692	3711	3729	3747	3766	3784	2	4	6	7	9	11	13	15	17
24	3802	3820	3838	3856	3874	3892	3909	3927	3945	3962	2	4	5	7	9	11	12	14	16
25	3979	3997	4014	4031	4048	4065	4082	4099	4116	4133	2	3	5	7	9	10	12	14	15
26	4150	4166	4183	4200	4216	4232	4249	4265	4281	4298	2	3	5	7	8	10	11	13	15
27	4314	4330	4346	4362	4378	4393	4409	4425	4440	4456	2	3	5	6	8	9	11	13	14
28	4472	4487	4502	4518	4533	4548	4564	4579	4594	4609	2	3	5	6	8	9	11	12	14
29	4624	4639	4654	4669	4683	4698	4713	4728	4742	4757	1	3	4	6	7	9	10	12	13
30	4771	4786	4800	4814	4829	4843	4857	4871	4886	4900	1	3	4	6	7	9	10	11	13
31	4914	4928	4942	4955	4969	4983	4997	5011	5024	5038	1	3	4	6	7	8	10	11	12
32	5051	5065	5079	5092	5105	5119	5132	5145	5159	5172	1	3	4	5	7	8	9	11	12
33	5185	5198	5211	5224	5237	5250	5263	5276	5289	5302	1	3	4	5	6	8	9	10	12
34	5315	5328	5340	5353	5366	5378	5391	5403	5416	5428	1	3	4	5	6	8	9	10	11
35	5441	5453	5465	5478	5490	5502	5514	5527	5539	5551	1	2	4	5	6	7	9	10	11
36	5563	5575	5587	5599	5611	5623	5635	5647	5658	5670	1	2	4	5	6	7	8	10	11
37	5682	5694	5705	5717	5729	5740	5752	5763	5775	5786	1	2	3	5	6	7	8	9	10
38	5798	5809	5821	5832	5843	5855	5866	5877	5888	5899	1	2	3	5	6	7	8	9	10
39	5911	5922	5933	5944	5955	5966	5977	5988	5999	6010	1	2	3	4	5	7	8	9	10
40	6021	6031	6042	6053	6064	6075	6085	6096	6107	6117	1	2	3	4	5	6	8	9	10
41	6128	6138	6149	6160	6170	6180	6191	6201	6212	6222	1	2	3	4	5	6	7	8	9
42	6232	6243	6253	6263	6274	6284	6294	6304	6314	6325	1	2	3	4	5	6	7	8	9
43	6335	6345	6355	6365	6375	6385	6395	6405	6415	6425	1	2	3	4	5	6	7	8	9
44	6435	6444	6454	6464	6474	6484	6493	6503	6513	6522	1	2	3	4	5	6	7	8	9
45	6532	6542	6551	6561	6571	6580	6590	6599	6609	6618	1	2	3	4	5	6	7	8	9
46	6628	6637	6646	6656	6665	6675	6684	6693	6702	6712	1	2	3	4	5	6	7	7	8
47	6721	6730	6739	6749	6758	6767	6776	6785	6794	6803	1	2	3	4	5	5	6	7	8
48	6812	6821	6830	6839	6848	6857	6866	6875	6884	6893	1	2	3	4	4	5	6	7	8
49	6902	6911	6920	6928	6937	6946	6955	6964	6972	6981	1	2	3	4	4	5	6	7	8
50	6990	6998	7007	7016	7024	7033	7042	7050	7059	7067	1	2	3	3	4	5	6	7	8
51	7076	7084	7093	7101	7110	7118	7126	7135	7143	7152	1	2	3	3	4	5	6	7	8
52	7160	7168	7177	7185	7193	7202	7210	7218	7226	7235	1	2	2	3	4	5	6	7	7
53	7243	7251	7259	7267	7275	7284	7292	7300	7308	7316	1	2	2	3	4	5	6	6	7
54	7324	7332	7340	7348	7356	7364	7372	7380	7388	7396	1	2	2	3	4	5	6	6	7
N	0	1	2	3	4	5	6	7	8	9	1	2	3	4	5	6	7	8	9

*Interpolation in this section of the table is inaccurate.

Appendix C   Four-place logarithms—cont'd

N	0	1	2	3	4	5	6	7	8	9	Proportional parts								
											1	2	3	4	5	6	7	8	9
55	7404	7412	7419	7427	7435	7443	7451	7459	7466	7474	1	2	2	3	4	5	5	6	7
56	7482	7490	7497	7505	7513	7520	7528	7536	7543	7551	1	2	2	3	4	5	5	6	7
57	7559	7566	7574	7582	7589	7597	7604	7612	7619	7627	1	2	2	3	4	5	5	6	7
58	7634	7642	7649	7657	7664	7672	7679	7686	7694	7701	1	1	2	3	4	4	5	6	7
59	7709	7716	7723	7731	7738	7745	7752	7760	7767	7774	1	1	2	3	4	4	5	6	7
60	7782	7789	7796	7803	7810	7818	7825	7832	7839	7846	1	1	2	3	4	4	5	6	6
61	7853	7860	7868	7875	7882	7889	7896	7903	7910	7917	1	1	2	3	4	4	5	6	6
62	7924	7931	7938	7945	7952	7959	7966	7973	7980	7987	1	1	2	3	3	4	5	6	7
63	7993	8000	8007	8014	8021	8028	8035	8041	8048	8055	1	1	2	3	3	4	5	5	6
64	8062	8069	8075	8082	8089	8096	8102	8109	8116	8122	1	1	2	3	3	4	5	5	6
65	8129	8136	8142	8149	8156	8162	8169	8176	8182	8189	1	1	2	3	3	4	5	5	6
66	8195	8202	8209	8215	8222	8228	8235	8241	8248	8254	1	1	2	3	3	4	5	5	6
67	8261	8267	8272	8280	8287	8293	8299	8306	8312	8319	1	1	2	3	3	4	5	5	6
68	8325	8331	8338	8344	8351	8357	8363	8370	8376	8382	1	1	2	3	3	4	4	5	6
69	8388	8395	8401	8407	8414	8420	8426	8432	8439	8445	1	1	2	2	3	4	4	5	6
70	8451	8457	8463	8470	8476	8482	8488	8494	8500	8506	1	1	2	2	3	4	4	5	6
71	8513	8519	8525	8531	8537	8543	8549	8555	8561	8567	1	1	2	2	3	4	4	5	5
72	8573	8579	8585	8591	8597	8603	8609	8615	8621	8627	1	1	2	2	3	4	4	5	5
73	8633	8639	8645	8651	8657	8663	8669	8675	8681	8686	1	1	2	2	3	4	4	5	5
74	8692	8698	8704	8710	8716	8722	8727	8733	8739	8745	1	1	2	2	3	4	4	5	5
75	8751	8756	8762	8768	8774	8779	8785	8791	8797	8802	1	1	2	2	3	3	4	5	5
76	8808	8814	8820	8825	8831	8837	8842	8848	8854	8859	1	1	2	2	3	3	4	5	5
77	8865	8871	8876	8882	8887	8893	8899	8904	8910	8915	1	1	2	2	3	3	4	4	5
78	8921	8927	8932	8938	8943	8949	8954	8960	8965	8971	1	1	2	2	3	3	4	4	5
79	8976	8982	8987	8993	8998	9004	9009	9015	9020	9025	1	1	2	2	3	3	4	4	5
80	9031	9036	9042	9047	9053	9058	9063	9069	9074	9079	1	1	2	2	3	3	4	4	5
81	9085	9090	9096	9101	9106	9112	9117	9122	9128	9133	1	1	2	2	3	3	4	4	5
82	9138	9143	9149	9154	9159	9165	9170	9175	9180	9186	1	1	2	2	3	3	4	4	5
83	9191	9196	9201	9206	9212	9217	9222	9227	9232	9238	1	1	2	2	3	3	4	4	5
84	9243	9248	9253	9258	9263	9269	9274	9279	9284	9289	1	1	2	2	3	3	4	4	5
85	9294	9299	9304	9309	9315	9320	9325	9330	9335	9340	1	1	2	2	3	3	4	4	5
86	9345	9350	9355	9360	9365	9370	9375	9380	9385	9390	1	1	2	2	3	3	4	4	5
87	9395	9400	9405	9410	9415	9420	9425	9430	9435	9440	0	1	1	2	2	3	3	4	4
88	9445	9450	9455	9460	9465	9469	9474	9479	9484	9489	0	1	1	2	2	3	3	4	4
89	9494	9499	9504	9509	9513	9518	9523	9528	9533	9538	0	1	1	2	2	3	3	4	4
90	9542	9547	9552	9557	9562	9566	9571	9576	9581	9586	0	1	1	2	2	3	3	4	4
91	9590	9595	9600	9605	9609	9614	9619	9624	9628	9633	0	1	1	2	2	3	3	4	4
91	9638	9643	9647	9652	9657	9661	9666	9671	9675	9680	0	1	1	2	2	3	3	4	4
93	9685	9689	9694	9699	9703	9708	9713	9717	9722	9727	0	1	1	2	2	3	3	4	4
94	9731	9736	9741	9745	9750	9754	9759	9763	9768	9773	0	1	1	2	2	3	3	4	4
95	9777	9782	9786	9791	9795	9800	9805	9809	9814	9818	0	1	1	2	2	3	3	4	4
96	9823	9827	9832	9836	9841	9845	9850	9854	9859	5963	0	1	1	2	2	3	3	4	4
97	9868	9872	9877	9881	9886	9890	9894	9899	9903	9908	0	1	1	2	2	3	3	4	4
98	9912	9917	9921	9926	9930	9934	9939	9943	9948	9952	0	1	1	2	2	3	3	4	4
99	9956	9961	9965	9969	9974	9978	9983	9987	9991	9996	0	1	1	2	2	3	3	3	4
N	0	1	2	3	4	5	6	7	8	9	1	2	3	4	5	6	7	8	9

What is a normal laboratory value?

The normal laboratory value for a substance is that concentration which can be measured in tissue or body fluids from apparently healthy human beings. A "healthy human being" is not so easily defined. In the first place, if a series of analyses is done on a large number of individuals, the values obtained will cover some range of numbers. Further scrutiny will often reveal that the set can be divided into subsets with respect to age and sex. Sometimes there are additional influences that affect the distribution of values, for example, racial extraction, previous dietary history, and patterns of physical activity. Less frequently it is possible to show rhythmic changes produced by the menstrual cycle (certain blood electrolytes and hormones), by diurnal variations, and even by seasonal changes. It must also be accepted that humans exhibit a degree of individual variability in their biochemical makeup, just as they do in such obvious characteristics as weight, height, and personality. Although many of these factors are poorly understood, they are nevertheless real.

For these and other reasons, it is customary to present normal ranges for body constituents, with the range chosen according to simple statistical principles. These are usually applied so that the cited range of values will include 95% of those values that can be predicted to occur in the entire population. It is sometimes necessary to establish several ranges. The concentration of hemoglobin in the blood, for example, depends on both age and sex. The concentration of creatine kinase, a blood serum enzyme, appears to be affected significantly by sex but not by age. Some believe that the dependence could more properly be based on muscle mass, but the evidence is not conclusive. Still other constituents such as blood glucose seem to be independent of age and sex. Of the normal values cited on the inside of the covers, those most obviously affected by age or sex have been indicated; where no such indication appears it may be assumed that the values apply generally.

What is an abnormal laboratory value?

An abnormal value may be strictly defined as any measurement of a body constituent that falls outside the normal range in cases where a disease has been diagnosed by other means. Although correct, this is a very stringent definition. A more useful criterion of abnormality is that if any value falls more than two standard deviations from the mean of the normal range, it may be abnormal. Alternatively, such a value may be the result of a laboratory error or of individual variation.

By how much must a test result depart from the "normal" before it may be considered "abnormal"? From what has been said, it should be clear that this question has no unique answer. It depends on the test in question and, to some extent, on the patient in question. Remember always that tests of some 5% of all perfectly normal individuals would not give "normal" results. This does not mean they are sick, only that they are different.

Compilations of laboratory values A number of sources include compilations of normal and abnormal laboratory test values as well as information on the interpretation of diagnostic test results. Some of these sources are included in the references at the end of this appendix.

Conversion to and from SI units The system of SI units is gradually coming into wider use. Nevertheless, most current texts and some current journals still refer to laboratory values in older, proprietary units. Students are referred to several inexpensive and recent sources, cited in the appended reference list, that provide insight into the process of conversions from one unit system to another.

References

Anonymous: Normal reference laboratory values, N Engl J Med 302:37, 1980.

Bold AM and Wilding P: Clinical chemistry companion, St Louis, 1978, The CV Mosby Co.

Gornall AG, editor: Applied biochemistry of clinical disorders, ed 2, Philadelphia, 1986, JB Lippincott Co.

Lippert H and Lehmann P: SI units in medicine, an introduction to the international system of units with conversion tables and normal ranges, Baltimore, 1978, Urban & Schwarzenberg, Inc.

Lundberg GD: SI unit implementation—the next step, JAMA 260:73,1988.

Wallach J: Interpretations of diagnostic tests, Boston, 1986, Little, Brown & Co.

Wills ED: Wills' biochemical basis of medicine, ed 2, London: J Hywel Thomas, Brian Gillham; Boston: Wright, 1989.

Young DS: Implementation of SI units for clinical laboratory data, Ann Intern Med 106:114, 1987.

The SI, or Système International d'Unites, is a new extension of the metric system adopted in principle by the General Conference on Weights and Measures convened in 1960. The SI was proposed to provide international standardization of computation and expression of clinical chemical values, to promote uniformity of published data on a readily understood basis, and to facilitate mechanization of laboratory data transmission from points of origin to users. In addition, it replaces archaic and proprietary units and values long used by other chemists. Thus SGOT, or serum glutamate-oxaloacetate transaminase, is now replaced by AST, which stands for aspartate aminotransferase. The latter name is obviously more in accord with IUPAC rules for enzyme nomenclature. The SI was therefore adopted and recommended by the International Federation of Clinical Chemistry and the World Health Organization.

Basic units of the SI

There are seven basic units in the SI:

Parameter	Unit	Symbol
Length	Meter	m
Mass	Kilogram	kg
Time	Second	s
Electric current	Ampere	A
Temperature	Kelvin	K
Luminous intensity	Candela	cd
Amount of substance	Mole	mol

All of these had already been accepted by chemists and physicists, and international standards or definitions had been established.

Derived SI units

To meet the practical needs of clinical chemistry, additional units were derived from the basic units just defined. These derived units cover the vast majority of situations that arise in relating biochemistry to health-related problems. Following are the permitted derived units:

Parameter	Unit	Symbol	Dimensions
Volume	Liter	L	$10^{-3} m^3$
Mass	Gram	g	$10^{-3} kg$
Temperature (T)	Degrees centigrade	°C	$T-273 K$
Time	Minute, hour, day, year	min, h, d, a	s, m, h, d
Force	Newton	N	$kg \cdot m \cdot s^{-2}$
Pressure	Pascal	Pa	$N \cdot m^{-2}$
Work, energy, heat	Joule	J	$N \cdot m$
Power	Watt	W	$J \cdot s^{-1}$
Enzyme activity	Katal	kat	$mol \cdot s^{-1}$
Substance concentration	Moles/liter	mol/L	$Mol \cdot 10^3 \cdot m^{-3}$

Note carefully that in the SI such units as mEq/L as an expression of concentration and mm Hg as an expression of pressure have disappeared. Similarly, the calorie is no longer acceptable as a unit of energy.

To facilitate conversions between units, the following factors may be used, rounded off as appropriate:

1 Calorie = 4.185 J
1 mm Hg = 7.501 kPa
1 kPa = 0.133 mm Hg

Acceptable prefixes for multiples or submultiples of SI units are defined as follows:

Multiples		Submultiples	
10^6	mega (M)	10^{-18}	atto (a)
10^3	kilo (k)	10^{-15}	femto (f)
10^2	hecto (h)	10^{-12}	pico (p)
10^1	deca (da)	10^{-9}	nano (n)
		10^{-6}	micro (μ)
		10^{-3}	milli (m)
		10^{-2}	centi (c)
		10^{-1}	deci (d)

All the definitions given have been accepted more or less readily by major institutions and publications. There is, as yet, less general acceptance of another unit designed to express enzyme activity. This unit is known as the katal, which is expressed as mol/sec, so that the international unit, SI, would become 16.7 nkat · L^{-1}.

Descriptive abbreviations

Certain other internationally accepted conventions are as follows, in terms of uniform abbreviations:

Descriptor		Sample nature	
Arterial	a	Blood	B
Venous	v	Plasma	P
Capillary	c	Spinal fluid	Sp
Fasting	f	Urine	U
24-hour	d	Feces	F
Serum	s	Patient	Pt

Thus a fasting blood glucose concentration could be indicated as fB-glucose = 4.4 mmol/L. A normal serum lactate dehydrogenase concentration would be indicated as sLDH = 1500 to 3340 nkat/L.

Exceptional cases

Where it is not possible to employ rational units, the SI permits expressions such as grams (or milligrams) per liter. An example of this is the case of urinary protein excretion, where it is not possible to determine the exact nature of excreted mixtures, even though the molecular weights of individual substances may be known.

One may argue the rationality of forbidding the expression of electrolytes as mEq/L, replacing it by the permitted form of mmol/L. Nevertheless, this rule has been adopted through declaration of the mole and the liter as fundamental units of the SI. For Na^+, K^+, HCO_3^- and Cl^-, no numerical changes are involved. For Ca^{++}, values expressed as mEq/L must be divided by 2. Perhaps the best explanation in favor of the new rule is the case of inorganic phosphate, where the valence depends on the exact pH of the fluid measured. It is more precise to state the molar concentration of the phosphate ion with the pH (where known) than to guess at the exact valence of the ion. A similar rule obviously applies to the case of many organic acid metabolites, and it was on this basis that the new rule was formulated.

Conversion formulas

Conversion of older units to SI units is readily accomplished by the general equation:

$$\text{mmol/L (SI units)} = \text{mg/dl} \times 10/\text{mol wt}$$

and the inverse converting SI units to more familiar older units, is given by the general equation:

$$\text{mg/dl} = \text{mmol/L (SI units)} \times \text{mol wt}/10$$

Conversion for enzyme activity has already been discussed.

Appendix F

Normal plasma or serum values for amino acids as a function of age*
(μmol/L)

Compound	Neonates†	Infants‡	Children§	Adults‖
Alanine	329	292	234	360
β-Alanine	15	—	—	—
Arginine	54	63	53	82
Asparagine	8	19	10	Trace
½-Cystine	62	42	60	49
Glutamic acid	52	—	110	24
Glutamine	—	—	—	640
Glycine	343	213	166	284
Histidine	77	78	55	88
Hydroxyproline	32	—	25	—
Isoleucine	39	39	43	60
Leucine	72	77	85	115
Lysine	200	135	111	186
Methionine	29	18	14	21
Ornithine	91	50	33	58
Phenylalanine	78	55	42	48
Proline	183	193	106	185
Serine	163	131	94	99
Taurine	141	—	80	59
Threonine	217	177	76	138
Tryptophan	32	—	—	31
Tyrosine	69	54	43	54
Valine	136	161	162	225

*See Scriver CR and Rosenberg LE: Amino acid metabolism and its disorders, Philadelphia, 1973, WB Saunders Co.

†Data based on 25 subjects studied before first feeding.

‡Data based on 12 infants, up to 4 mo old, after a 6 to 8 hr fast.

§Data based on 9 children, 3 to 10 yr old, after an overnight fast.

‖Data based on 20 subjects ranging from 33 to 56 yr old.

Index

Boldface page numbers indicate diagrams
of chemical structure; *C* following a page
number indicates discussion in a clinical
case; *italicized* page numbers indicate
illustrations; *t* following a page number
indicates a table.

Table of normal clinical laboratory values (SI units in bold type)

	SI units	Conventional units
Alanine, serum	**0.3 to 0.6 mmol/L**	(2.6 to 5.3 mg/dl)
Albumin, serum	**30 to 50 g/L**	(3.0 to 5.0 g/dl)
	0.45 to 0.75 mmol/L	—*
Aldolase, serum (ALS)	**60 to 200 nkat/L**	(1.5 to 2.0 U/L)
Alanine aminotransferase (ALT)	**65 to 325 nkat/L**	(6.0 to 21 U/L)
Ammonia, blood	**11 to 50 μmol/L**	—
Amylase, serum	**1.5 to 4.6 μkat/L**	(96 to 290 U/L)
Amylase, urine	**2.5 to 30 μkat/L**	(160 to 2000 U/L)
Aspartate aminotransferase (AST)	**115 to 420 nkat/L**	(7.0 to 20 U/L)
Bilirubin, serum		
Direct	**0 to 7.0 μmol/L**	(0 to 0.2 mg/dl)
Total	**1.7 to 18.8 μmol/L**	(0.2 to 1.0 mg/dl)
Blood pressure (systolic/diastolic)	**16/10.6 kPa**	(120/80 mm Hg)
Calcium, total serum	**2.1 to 2.65 mmol/L**	(8.5 to 10.4 mg/dl)
Chloride, serum	**100 to 106 mmol/L**	—
Cholesterol, serum	**3.9 to 6.2 mmol/L**	(150 to 240 mg/dl)
Citrate, serum		
Adults	**80 to 160 μmol/L**	(1.5 to 3.0 mg/dl)
Children	**90 to 190 μmol/L**	(1.7 to 3.6 mg/dl)
Newborns	**150 to 310 μmol/L**	(2.8 to 5.9 mg/dl)
^{60}Co-labeled vitamin B_{12}, urinary excretion	**15% to 40% of dose administered**	—
CO_2 content, total serum	**21 to 30 mmol/L**	—
Copper, serum	**16 to 30 μmol/L**	(70 to 190 μg/dl)
Creatine, serum		
Males	**15 to 40 μmol/L**	(0.2 to 0.5 mg/dl)
Creatine kinase (CK)		
Males	**200 to 850 nkat/L**	(12 to 65 U/L)
Females	**165 to 830 nkat/L**	(10 to 50 U/L)
Creatinine, serum	**88 to 130 μmol/L**	(0.9 to 145 μg/dl)
Creatinine, urine		
Males	**8.8 to 17.0 mmol/d**	(1.0 to 2.0 g/d)
Females	**7.0 to 16.0 mmol/d**	(0.8 to 1.8 g/d)
Cysteine *plus* cystine, expressed as cysteine	**12.0 to 120 μmol/d**	(1.5 to 14.5 mg/d)
Erythrocytes (RBC)	**4 to 6 × 10^8/mm³**	—
Free fatty acids, plasma	**200 to 800 μmol/L**	—
Folic acid, serum	**14 to 34 nmol/L**	—
Fumarate	**1-17 μmol/L**	(10-200 μg/dl)
Globulins, total serum	**15 to 30 g/L**	—
Glucose, serum	**4.4 to 6.1 mmol/L**	(30 to 110 mg/dl)
Glucose, urine	**<1.38 mmol/d**	(<250 mg/d)
Hematocrit, blood		
Males	**40% to 54%**	—
Females	**37% to 47%**	—
Children	**35% to 49%**	—
Hemoglobin, blood†		
Males	**2.09 to 2.69 mmol/L**	(14 to 17 g/dl)
Females	**1.79 to 2.39 mmol/L**	(12 to 15 g/dl)
Children	**1.64 to 2.54 mmol/L**	(11 to 17 g/dl)
Immunoglobulins, serum		
IgA	**9.0 to 33 g/L**	—
IgD	**0 to 0.4 g/L**	—
IgE	**100 to 200 μg/L**	—
IgG	**7.2 to 15.0 g/L**	—
IgM	**0.5 to 2.5 g/L**	—
Insulin, serum	**6 to 23 mU/L**	—
Iron, serum	**9.8 to 27.0 μmol/L**	(55 to 150 μg/dl)

*Dash indicates the value is identical in form to the SI value.
†Based on the molecular weight of the monomer.